Grob's Basic Electronics: Fundamentals of DC and AC Circuits

Mitchel E. Schultz

Western Wisconsin Technical College

Higher Education

Boston   Burr Ridge, IL   Dubuque, IA   New York   San Francisco   St. Louis
Bangkok   Bogotá   Caracas   Kuala Lumpur   Lisbon   London   Madrid   Mexico City
Milan   Montreal   New Delhi   Santiago   Seoul   Singapore   Sydney   Taipei   Toronto

Mc Graw Hill **Higher Education**

GROB'S BASIC ELECTRONICS: FUNDAMENTALS OF DC AND AC CIRCUITS

Published by McGraw-Hill, a business unit of The McGraw-Hill Companies, Inc., 1221 Avenue of the Americas, New York, NY 10020. Copyright © 2007 by The McGraw-Hill Companies, Inc. All rights reserved. No part of this publication may be reproduced or distributed in any form or by any means, or stored in a database or retrieval system, without the prior written consent of The McGraw-Hill Companies, Inc., including, but not limited to, in any network or other electronic storage or transmission, or broadcast for distance learning.

Some ancillaries, including electronic and print components, may not be available to customers outside the United States.

This book is printed on acid-free paper.

1 2 3 4 5 6 7 8 9 0 DOW/DOW 0 9 8 7 6

ISBN-13   978-0-07-110745-7
ISBN-10   0-07-110745-2

The credits section for this book begins on page K-1 and is considered an extension of the copyright page.

www.mhhe.com

Brief Contents

Contents

Chapter 17   Capacitive Reactance   526

Chapter 18   Capacitive Circuits   548

Cumulative Review Summary

Chapter 19   Inductance   576

Chapter 20   Inductive Reactance   622

Preface

Grob's Basic Electronics: Fundamentals of DC and AC Circuits is identical to *Grob's Basic Electronics,* Tenth Edition, except that the last seven chapters covering electronic devices are not included. This edition of *Grob's Basic Electronics: Fundamentals of DC and AC Circuits* continues its rich tradition as a textbook that provides complete and comprehensive coverage of the fundamentals of electricity. The book is written for the beginning student who has little or no knowledge about the fundamental concepts of electricity. It is helpful, however, if the student has some basic knowledge of algebra and trigonometry.

Many major changes have been made in this edition of *Grob's Basic Electronics: Fundamentals of DC and AC Circuits.* A new chapter on Powers of 10 Notation has been added, many chapters have been significantly revised, some chapters have been combined to provide more streamlined coverage of a topic, and the end of chapter problems are now organized by chapter section. Also, *MultiSim 8* has been more effectively integrated into the book. All in all, this edition of *Grob's Basic Electronics: Fundamentals of DC and AC Circuits* is better than ever. Let's take a closer look at what's new.

What's New in *Grob's Basic Electronics: Fundamentals of DC and AC Circuits?*

- A new chapter entitled **"Introduction to Powers of 10"** has been added at the beginning of the book. This chapter prepares students to work with numbers expressed in powers of 10 notation. More specifically, this chapter introduces the student to scientific and engineering notation as well as the metric prefixes. Students will learn how to add, subtract, multiply and divide numbers which are expressed in powers of 10 notation.

- Every chapter begins with a list of **Important Terms** appearing within the chapter. The definition for each important term is provided at the end of the chapter. This feature allows students to quickly pin-point key terms and their corresponding definitions.

- Special **Knowledge Check Problems** have been added throughout each chapter to allow students to check their understanding of the material presented. The Knowledge-Check Problems can also be used by instructors as a review during classroom lectures. The answers to all Knowledge-Check Problems are provided at the end of each chapter.

- All new multiple-choice **Self-Test Questions** have been added to the end of each chapter. The new format of the Self-Test Questions allows the student to quickly assess their understanding of the material presented within the chapter. The answers to all Self-Test Questions appear at the back of the book.

- All **Chapter Problems** have been completely revised in this edition! All problems are now organized by chapter section, which allows the student to breakdown the topics into smaller bite size pieces. Furthermore, it allows the instructor to emphasize the areas within a chapter that he or she deems most important. The answers to all odd-numbered problems appear at the back of the book.

- The chapter previously called "Direct-Current Meters" has been renamed **Analog and Digital Multimeters.** The new chapter title reflects the changes made in this chapter. The coverage of analog meters has been condensed whereas the coverage of digital multimeters has been expanded.

- The **chapter on batteries** has been significantly revised to provide more streamlined coverage and to also reflect the changes in battery technology. The chapter now provides coverage of nickel metal hydride (NiMH) cells and fuel cells as examples.

- The chapters previously called "Magnetic Units" and "Electromagnetic Induction" have now been combined into a single chapter entitled **Electromagnetic Induction.** The content in these two chapters has been condensed based on reviewer comments. The coverage of electromagnetism is now more straightforward and streamlined than in previous editions.

- The **Appendixes** have also been completely revised in this edition. Appendix A has been revised to reflect the need for improved coverage of Electrical Symbols and Abbreviations. Appendix B provides comprehensive coverage of Solder and the Soldering Process. Appendix C: Component Values and Specifications has been added and Appendix D, Schematic Symbols has been totally revised to reflect the most current schematic symbols encountered in electronics. Appendix E, Using the Oscilloscope, has been completely rewritten and now includes coverage of digital storage oscilloscopes. And finally, Appendix F provides the student with an introduction to *MultiSim* (version 8). *MultiSim* is an interactive circuit simulation software package that allows students to create and test circuits on the computer. Appendix F introduces the student to the main features of *MultiSim* that directly relate to the study of dc/ac and semiconductor electronics.

Guided Tour for *Grob's Basic Electronics: Fundamentals of DC and AC Circuits*

CHAPTER INTRODUCTION

Each chapter begins with a short Introduction briefly outlining the main chapter topics and concepts.

I Introduction to Powers of 10

The electrical quantities you will encounter while working in the field of electronics are often extremely small or extremely large. For example, it is not at all uncommon to work with extremely small decimal numbers such as 0.000000000056 or extremely large numbers such as 1,296,000,000. To enable us to work conveniently with both very small and very large numbers, powers of 10 notation is used. With powers of 10 notation, any number, no matter how small or large, can be expressed as a decimal number multiplied by a

CHAPTER OUTLINE

The chapter Outline provides a breakdown of the individual topics covered.

Outline

- I–1 Scientific Notation
- I–2 Engineering Notation and Metric Prefixes
- I–3 Converting Between Metric Prefixes
- I–4 Addition and Subtraction Involving Powers of 10 Notation
- I–5 Multiplication and Division Involving Powers of 10 Notation
- I–6 Reciprocals with Powers of 10
- I–7 Squaring Numbers Expressed in Powers of 10 Notation
- I–8 Square Roots of Numbers Expressed in Powers of 10 Notation
- I–9 The Scientific Calculator

Objectives

After studying this chapter you should be able to

- Express any number in scientific or engineering notation.
- List the metric prefixes and their corresponding powers of 10.
- Change a power of 10 in engineering notation to its corresponding metric prefix.
- Convert between metric prefixes.
- Add and subtract numbers expressed in powers of 10 notation.
- Multiply and divide numbers expressed in powers of 10 notation.
- Determine the reciprocal of a power of 10.
- Find the square of a number expressed in powers of 10 notation.
- Find the square root of a number expressed in powers of 10 notation.
- Enter numbers written in scientific and engineering notation into your calculator.

CHAPTER OBJECTIVES

The Objectives identify expected learning outcomes for each chapter.

Objectives

After studying this chapter you should be able to

- Express any number in scientific or engineering notation.
- List the metric prefixes and their corresponding powers of 10.
- Change a power of 10 in engineering notation to its corresponding metric prefix.
- Convert between metric prefixes.
- Add and subtract numbers expressed in powers of 10 notation.
- Multiply and divide numbers expressed in powers of 10 notation.
- Determine the reciprocal of a power of 10.
- Find the square of a number expressed in powers of 10 notation.
- Find the square root of a number expressed in powers of 10 notation.
- Enter numbers written in scientific and engineering notation into your calculator.

IMPORTANT TERMS

A list of Important Terms helps students identify key terms as they read through the chapter.

Outline

I–1	Scientific Notation
I–2	Engineering Notation and Metric Prefixes
I–3	Converting Between Metric Prefixes
I–4	Addition and Subtraction Involving Powers of 10 Notation
I–5	Multiplication and Division Involving Powers of 10 Notation
I–6	Reciprocals with Powers of 10
I–7	Squaring Numbers Expressed in Powers of 10 Notation
I–8	Square Roots of Numbers Expressed in Powers of 10 Notation
I–9	The Scientific Calculator

notation to its corresponding metric prefix.
- Convert between metric prefixes.
- Add and subtract numbers expressed in powers of 10 notation.
- Multiply and divide numbers expressed in powers of 10 notation.
- Determine the reciprocal of a power of 10.
- Find the square of a number expressed in powers of 10 notation.
- Find the square root of a number expressed in powers of 10 notation.
- Enter numbers written in scientific and engineering notation into your calculator.

Important Terms

decimal notation	powers of 10
engineering notation	scientific notation
metric prefixes	

EXAMPLES

Numerous worked-out examples are provided throughout the text.

Example 1-1

Express the following numbers in scientific notation: (a) 3900 (b) 0.0000056.

ANSWER: (a) To express 3900 in scientific notation, write the number as a number between 1 and 10, which is 3.9 in this case, times a power of 10. To do this, the decimal point must be shifted three places to the left. The number of places by which the decimal point is shifted to the left indicates the positive power of 10. Therefore, $3900 = 3.9 \times 10^3$ in scientific notation.

(b) To express 0.0000056 in scientific notation, write the number as a number between 1 and 10, which is 5.6 in this case, times a power of 10. To do this, the decimal point must be shifted six places to the right. The number of places by which the decimal point is shifted to the right indicates the negative power of 10. Therefore, $0.0000056 = 5.6 \times 10^{-6}$ in scientific notation.

When expressing a number in scientific notation, remember the following rules.

Rule 1: Express the number as a number between 1 and 10 times a power of 10.

Rule 2: If the decimal point is moved to the left in the original number, make the power of 10 positive. If the decimal point is moved to the right in the original number, make the power of 10 negative.

Rule 3: The power of 10 always equals the number of places by which the decimal point has been shifted to the left or right in the original number.

Let's try another example.

Example 4-3

In Fig. 4-6, solve for R_T, I and the individual resistor voltage drops.

Figure 4-6 Circuit for Example 4-3.

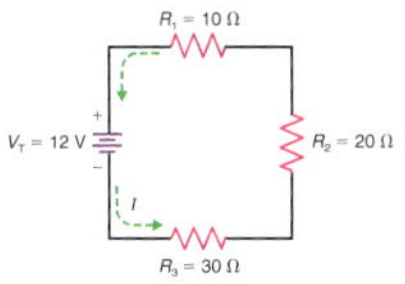

$R_1 = 10\ \Omega$
$V_T = 12\ V$
$R_2 = 20\ \Omega$
$R_3 = 30\ \Omega$

ANSWER First, find R_T by adding the individual resistance values.

$$R_T = R_1 + R_2 + R_3$$
$$= 10\ \Omega + 20\ \Omega + 30\ \Omega$$
$$= 60\ \Omega$$

Next, solve for the current, I.

$$I = \frac{V_T}{R_T}$$
$$= \frac{12\ V}{60\ \Omega}$$
$$= 200\ mA$$

Now we can solve for the individual resistor voltage drops.

$$V_1 = I \times R_1$$
$$= 200\ mA \times 10\ \Omega$$
$$= 2V$$
$$V_2 = I \times R_2$$
$$= 200\ mA \times 20\ \Omega$$

4-4 Knowledge Check

Answer at end of chapter.

A 36-V potential difference is applied across resistors R_1 and R_2 in series. How much voltage is across R_2 if R_1 has a voltage drop of 24 V?

4-4 Self-Review

Answers at end of chapter.

a. A series circuit has IR drops of 10, 20, and 30 V. How much is the applied voltage V_T of the source?
b. A 100-V source is applied across R_1 and R_2 in series. If V_1 is 25 V, how much is V_2?
c. A 120-V source is applied across three equal resistances in series. How much is each IR drop?

4–5 Polarity of *IR* Voltage Drops

When a voltage drop exists across a resistance, one end must be either more positive or more negative than the other end. Otherwise, without a potential difference no current could flow through the resistance to produce the voltage drop. The *polarity* of this *IR* voltage drop can be associated with the direction of I through R. In brief, electrons flow into the negative side of the *IR* voltage and out the positive side (see Fig. 4–8a).

KNOWLEDGE CHECK PROBLEMS

Knowledge Check problems allow students to test their understanding of important concepts. Instructors can also use the Knowledge Check problems for classroom discussion and review.

SECTION SELF-REVIEWS

Self-Review questions appear at the end of each section within a chapter, allowing students to check their understanding of the material just presented.

4-4 Self-Review

Answers at end of chapter.

a. A series circuit has *IR* drops of 10, 20, and 30 V. How much is the applied voltage V_T of the source?
b. A 100-V source is applied across R_1 and R_2 in series. If V_1 is 25 V, how much is V_2?
c. A 120-V source is applied across three equal resistances in series. How much is each *IR* drop?

4–5 Polarity of *IR* Voltage Drops

When a voltage drop exists across a resistance, one end must be either more positive or more negative than the other end. Otherwise, without a potential difference no current could flow through the resistance to produce the voltage drop. The *polarity* of this *IR* voltage drop can be associated with the direction of *I* through *R*. In brief, electrons flow into the negative side of the *IR* voltage and out the positive side (see Fig. 4–8*a*).

GOOD TO KNOW

Additional information about a topic is provided in the margins of the text.

GOOD TO KNOW

A battery is a device that converts chemical energy into electrical energy.

Figure 1–1 Positive and negative polarities for the voltage output of a typical battery.

Negativo −   Positive +

GOOD TO KNOW

Electricity is a form of energy, where energy refers to the ability to do work. More specifically, electrical energy refers to the energy associated with electric charges.

1–1 Negative and Positive Polarities

We see the effects of electricity in a battery, static charge, lightning, radio, television, and many other applications. What do they all have in common that is electrical in nature? The answer is basic particles of electric charge with opposite *polarities*. All the materials we know, including solids, liquids, and gases, contain two basic particles of electric charge: the *electron* and the *proton*. An electron is the smallest amount of electric charge having the characteristic called *negative polarity*. The proton is a basic particle with *positive polarity*.

The negative and positive polarities indicate two opposite characteristics that seem to be fundamental in all physical applications. Just as magnets have north and south poles, electric charges have the opposite polarities labeled negative and positive. The opposing characteristics provide a method of balancing one against the other to explain different physical effects.

It is the arrangement of electrons and protons as basic particles of electricity that determines the electrical characteristics of all substances. As an example, this paper has electrons and protons in it. There is no evidence of electricity, though, because the number of electrons equals the number of protons. In that case, the opposite electrical forces cancel, making the paper electrically neutral. The neutral condition means that opposing forces are exactly balanced, without any net effect either way.

When we want to use the electrical forces associated with the negative and positive charges in all matter, work must be done to separate the electrons and protons. Changing the balance of forces produces evidence of electricity. A battery, for instance, can do electrical work because its chemical energy separates electric charges to produce an excess of electrons at its negative terminal and an excess of protons at its positive terminal. With separate and opposite charges at the two terminals, electric energy can be supplied to a circuit connected to the battery. Fig. 1–1 shows a battery with its negative (−) and positive (+) terminals marked to emphasize the two opposite polarities.

■ *1–1 Self-Review*

Answers at end of chapter.

a. Is the charge of an electron positive or negative?
b. Is the charge of a proton positive or negative?
c. Is it true or false that the neutral condition means equal positive and negative charges?

1–2 Electrons and Protons in the Atom

Although there are any number of possible methods by which electrons and protons might be grouped, they assemble in specific atomic combinations for a stable arrangement. (An atom is the smallest particle of the basic elements which

PIONEERS IN ELECTRONICS

A photograph and brief statement is made about scientists and engineers whose discoveries and theories were instrumental in the development of electronics.

PIONEERS IN ELECTRONICS

In 1796 Italian Physicist *Alessandro Volta (1745–1827)* developed the first Chemical battery, which provided the first practical source of electricity.

A charge is the result of work done in separating electrons and protons. Because of the separation, stress and strain are associated with opposite charges, since normally they would be balancing each other to produce a neutral condition. We could consider that the accumulated electrons are drawn tight and are straining themselves to be attracted toward protons to return to the neutral condition. Similarly, the work of producing the charge causes a condition of stress in the protons, which are trying to attract electrons and return to the neutral condition. Because of these forces, the charge of electrons or protons has potential because it is ready to give back the work put into producing the charge. The force between charges is in the electric field.

Potential between Different Charges

When one charge is different from the other, there must be a difference of potential between them. For instance, consider a positive charge of 3 C, shown at the right in Fig. 1–7a. The charge has a certain amount of potential, corresponding to the amount of work this charge can do. The work to be done is moving some electrons, as illustrated.

Assume that a charge of 1 C can move 3 electrons. Then the charge of +3 C can attract nine electrons toward the right. However, the charge of +1 C at the opposite side can attract three electrons toward the left. The net result, then, is that six electrons can be moved toward the right to the more positive charge.

In Fig. 1–7b, one charge is 2 C, and the other charge is neutral with 0 C. For the difference of 2 C, again 2 × 3 or 6 electrons can be attracted to the positive side.

Refer to the graph in Fig. 3–5c.
a. Are the values of I on the y or x axis?
b. Is this R linear or nonlinear?
c. If the voltage across a 5-Ω resistor increases from 10 V to 20 V, what happens to I?
d. The voltage across a 5-Ω resistor is 10 V. If R is doubled to 10 Ω, what happens to I?

3–7 Electric Power

The unit of electric *power* is the *watt* (W), named after James Watt (1736–1819). One watt of power equals the work done in one second by one volt of potential difference in moving one coulomb of charge.

Remember that one coulomb per second is an ampere. Therefore power in watts equals the product of volts times amperes.

Power in watts = volts × amperes
$$P = V \times I \tag{3–4}$$

When a 6-V battery produces 2 A in a circuit, for example, the battery is generating 12 W of power.

The power formula can be used in three ways:

$$P = V \times I$$
$$I = P \div V \quad \text{or} \quad \frac{P}{V}$$
$$V = P \div I \quad \text{or} \quad \frac{P}{I}$$

PIONEERS OF ELECTRONICS

The unit of electrical power, the watt, is named for Scottish inventor and engineer *James Watt (1736–1819)*. One watt equals one joule of energy transferred in one second.

MULTISIM ICONS

The *MultiSim* icons identify those circuits for which there is a *MultiSim* activity. A CD containing *MultiSim* files is included with the textbook.

Example 5–1

||| MultiSim **Figure 5–4** Circuit for Example 5–1.

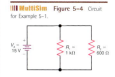

In Fig. 5–4, solve for the branch currents I_1 and I_2.

ANSWER The applied voltage, V_A, of 15 V is across both resistors R_1 and R_2. Therefore, the branch currents are calculated as $\frac{V_A}{R}$, where V_A is the applied voltage and R is the individual branch resistance.

$$I_1 = \frac{V_A}{R_1}$$
$$= \frac{15 \text{ V}}{1 \text{ } k\Omega}$$
$$= 15 \text{ mA}$$
$$I_2 = \frac{V_A}{R_2}$$
$$= \frac{15 \text{ V}}{600 \text{ } \Omega}$$
$$= 25 \text{ mA}$$

■ *5–2 Knowledge Check*

Answer at end of chapter.

If a 2-Ω resistor, R_3 is connected across points B and G in Fig. 5–3a, how much is its branch current?

■ *5–2 Self-Review*

Answers at end of chapter.

Refer to Fig. 5–3.
a. How much is the voltage across R_1?
b. How much is I_1 through R_1?
c. How much is the voltage across R_2?
d. How much is I_2 through R_2?

SUMMARY

Each chapter concludes with a comprehensive Summary of chapter topics.

Summary

- There is only one voltage V_A across all components in parallel.
- The current in each branch I_b equals the voltage V_A across the branch divided by the branch resistance R_b, or $I_b = V_A / R_b$.
- Kirchhoff's current law states that the total current I_T in a parallel circuit equals the sum of the individual branch currents. Expressed as an equation, Kirchhoff's current law is $I_T = I_1 + I_2 + I_3 + \cdots + $ etc.
- The equivalent resistance R_{EQ} of parallel branches is less than the smallest branch resistance, since all the branches must take more current from the source than any one branch.
- For only two parallel resistances of any value, $R_{EQ} = R_1 R_2 / (R_1 + R_2)$.

- For any number of *equal* parallel resistances, R_{EQ} is the value of one resistance divided by the number of resistances.
- For the general case of any number of branches, calculate R_{EQ} as V_A / I_T or use the reciprocal resistance formula:
$$R_{EQ} = \frac{1}{1/R_1 + 1/R_2 + 1/R_3 + \cdots + \text{etc.}}$$
- For any number of conductances in parallel, their values are added for G_T, in the same way as parallel branch currents are added.
- The sum of the individual values of power dissipated in parallel resistances equals the total power produced by the source.
- An open circuit in one branch results in no current through that branch, but the other branches can have their normal current. However, an

open circuit in the main line results in no current for any of the branches.
- A short circuit has zero resistance, resulting in excessive current. When one branch is short-circuited, all parallel paths are also short-circuited. The entire current is in the short circuit and no current is in the short-circuited branches.
- The voltage across a good fuse and the voltage across a closed switch are approximately 0 V. When the fuse in the main line of a parallel circuit opens, the voltage across the fuse equals the full applied voltage. Likewise, when the switch in the main line of a parallel circuit opens, the voltage across the open switch equals the full applied voltage.
- Table 5–1 compares Series and Parallel Circuits.

Table 5–1	Comparison of Series and Parallel Circuits
Series Circuit	**Parallel Circuit**
Current the same in all components	Voltage the same across all branches
V across each series R is $I \times R$	I in each branch R is V/R
$V_T = V_1 + V_2 + V_3 + \cdots + $ etc.	$I_T = I_1 + I_2 + I_3 + \cdots + $ etc.
$R_T = R_1 + R_2 + R_3 + \cdots + $ etc.	$G_T = G_1 + G_2 + G_3 + \cdots + $ etc.
R_T must be more than the largest individual R	R_{EQ} must be less than the smallest branch R
$P_T = P_1 + P_2 + P_3 + \cdots + $ etc.	$P_T = P_1 + P_2 + P_3 + \cdots + $ etc.
Applied voltage is divided into IR voltage drops	Main-line current is divided into branch currents
The largest IR drop is across the largest series R	The largest branch I is in the smallest parallel R
Open in one component causes entire circuit to be open	Open in one branch does not prevent I in other branches

Important Terms

- Equivalent Resistance, R_{EQ} - in a parallel circuit, this refers to a single resistance that would draw the same amount of current as all of the parallel connected branches.
- Kirchhoff's Current Law (KCL) - a law which states that the sum of the individual branch currents in a parallel circuit must equal the total current, I_T.

- Main Line - the pair of leads connecting all individual branches in a parallel circuit to the terminals of the applied voltage, V_A. The main line carries the total current, I_T, flowing to and from the terminals of the voltage source.
- Parallel Bank - a combination of parallel connected branches.

- Reciprocal Resistance Formula - a formula which states that the equivalent resistance, R_{EQ}, of a parallel circuit equals the reciprocal of the sum of the reciprocals of the individual branch resistances.

IMPORTANT TERMS

Definitions of Important Terms are provided at the end of each chapter.

Related Formulas

$I_1 = \dfrac{V_A}{R_1}, I_2 = \dfrac{V_A}{R_2}, I_3 = \dfrac{V_A}{R_3}$

$I_T = I_1 + I_2 + I_3 + \cdots + $ etc.

$R_{EQ} = \dfrac{V_A}{I_T}$

$R_{EQ} = \dfrac{1}{1/R_1 + 1/R_2 + 1/R_3 + \cdots + \text{etc.}}$

$R_{EQ} = \dfrac{R}{N}$ (R_{EQ} for equal branch resistances)

$R_{EQ} = \dfrac{R_1 \times R_2}{R_1 + R_2}$ (R_{EQ} for only two branch resistances)

$R_X = \dfrac{R \times R_{EQ}}{R - R_{EQ}}$

$G_T = G_1 + G_2 + G_3 + \cdots + $ etc.

$P_T = P_1 + P_2 + P_3 + \cdots + $ etc.

Self-Test

Answers at back of book.

1. A 120-kΩ resistor, R_1, and a 180-kΩ resistor, R_2, are in parallel. How much is the equivalent resistance, R_{EQ}?
   a. 72 kΩ
   b. 300 kΩ
   c. 360 kΩ
   d. 90 kΩ

   c. 9 Ω
   d. none of the above

4. Which of the following statements about parallel circuits is false?
   a. The voltage is the same across all branches in a parallel circuit.
   b. The equivalent resistance, R_{EQ}, of a parallel circuit is always smaller than the smallest branch resistance.

6. How much resistance must be connected in parallel with a 360-Ω resistor to obtain an equivalent resistance, R_{EQ}, of 120 Ω?
   a. 360 Ω
   b. 480 Ω
   c. 1.8 kΩ
   d. 180 Ω

7. If one branch of a parallel circuit

RELATED FORMULAS

Important chapter formulas are listed together at the end of the chapter.

Related Formulas

$$I_1 = \frac{V_A}{R_1}, \ I_2 = \frac{V_A}{R_2}, \ I_3 = \frac{V_A}{R_3}$$

$$I_T = I_1 + I_2 + I_3 + \cdots + \text{etc.}$$

$$R_{EQ} = \frac{V_A}{I_T}$$

$$R_{EQ} = \frac{1}{{}^1/{}_{R_1} + {}^1/{}_{R_2} + {}^1/{}_{R_3} + \cdots + \text{etc.}}$$

$$R_{EQ} = \frac{R}{N} \ (R_{EQ} \text{ for equal branch resistances})$$

$$R_{EQ} = \frac{R_1 \times R_2}{R_1 + R_2} \ (R_{EQ} \text{ for only two branch resistances})$$

$$R_x = \frac{R \times R_{EQ}}{R - R_{EQ}}$$

$$G_T = G_1 + G_2 + G_3 + \cdots + \text{etc.}$$

$$P_T = P_1 + P_2 + P_3 + \cdots + \text{etc.}$$

Self-Test

Answers at back of book.

1. A 120-kΩ resistor, R_1, and a 180-kΩ resistor, R_2, are in parallel. How much is the equivalent resistance, R_{EQ}?
   a. 72 kΩ
   b. 300 kΩ
   c. 360 kΩ
   d. 90 kΩ

2. A 100-Ω resistor, R_1, and a 300-Ω resistor, R_2, are in parallel across a dc voltage source. Which resistor dissipates more power?
   a. the 300-Ω resistor
   b. Both resistors dissipate the same amount of power.
   c. the 100-Ω resistor
   d. This is impossible to determine.

3. Three 18-Ω resistors are in parallel. How much is the equivalent resistance, R_{EQ}?
   a. 54 Ω

   c. 9 Ω
   d. none of the above

4. Which of the following statements about parallel circuits is false?
   a. The voltage is the same across all branches in a parallel circuit.
   b. The equivalent resistance, R_{EQ}, of a parallel circuit is always smaller than the smallest branch resistance.
   c. In a parallel circuit the total current, I_T, in the main line equals the sum of the individual branch currents.
   d. The equivalent resistance, R_{EQ}, of a parallel circuit decreases when one or more parallel branches are removed from the circuit.

5. Two resistors, R_1 and R_2, are in parallel with each other and a dc voltage source. If the total current, I_T, in the main line equals 6A and I_2 through R_2 is 4A, how much is I_1 through R_1?
   a. 6 A

6. How much resistance must be connected in parallel with a 360-Ω resistor to obtain an equivalent resistance, R_{EQ}, of 120 Ω?
   a. 360 Ω
   b. 480 Ω
   c. 1.8 kΩ
   d. 180 Ω

7. If one branch of a parallel circuit becomes open,
   a. all remaining branch currents increase.
   b. the voltage across the open branch will be 0 V.
   c. the remaining branch currents do not change in value.
   d. the equivalent resistance of the circuit decreases.

8. If a 10-Ω R_1, 40-Ω R_2, and 8-Ω R_3 are in parallel, calculate the total conductance, G_T, of the circuit.
   a. 250 mS

SELF-TESTS

Every chapter has a multiple-choice Self-Test, allowing quick learning assessment.

Self-Test

Answers at back of book.

1. Three resistors in series have individual values of 120 Ω, 680 Ω and 1.2 kΩ. How much is the total resistance, R_T?
   a. 1.8 kΩ
   b. 20 kΩ
   c. 2 kΩ
   d. none of the above

2. In a series circuit, the current, I, is
   a. different in each resistor.
   b. the same everywhere.
   c. the highest near the positive and negative terminals of the voltage source.
   d. different at all points along the circuit.

3. In a series circuit, the largest resistance has
   a. the largest voltage drop.
   b. the smallest voltage drop.
   c. more current than the other resistors.
   d. both a and c

4. The polarity of a resistor's voltage drop is determined by
   a. the direction of current through the resistor.
   b. how large the resistance is.
   c. how close the resistor is to the voltage source.
   d. how far away the resistor is from the voltage source.

5. A 10-Ω and 15-Ω resistor are in series across a dc voltage source. If the 10-Ω resistor has a voltage drop of 12 V, how much is the applied voltage?
   a. 18 V

   c. 0 V
   d. It cannot be determined.

7. How much is the voltage across an open component in a series circuit?
   a. The full applied voltage, V_T
   b. The voltage is slightly lower than normal.
   c. 0 V
   d. It cannot be determined.

8. A voltage of 120 V is applied across two resistors, R_1 and R_2, in series. If the voltage across R_2 equals 90 V, how much is the voltage across R_1?
   a. 90 V
   b. 30 V
   c. 120 V
   d. Cannot be determined.

9. If two series opposing voltages each have a voltage of 9 V, the net or total voltage is
   a. 0 V.
   b. 18 V.
   c. 9 V.
   d. none of these

10. On a schematic diagram, what does the chassis ground symbol represent?
    a. hot spots on the chassis.
    b. the locations in the circuit where electrons accumulate.
    c. a common return path for current in one or more circuits.
    d. none of the above

11. The notation, V_{BG}, means
    a. the voltage at Point G with respect to B.

13. A 6-V and 9-V source are connected in a series-aiding configuration. How much is the net or total voltage?
    a. -3 V
    b. $+3$ V
    c. 0 V
    d. 15 V

14. A 56-Ω and 82-Ω resistor are in series with an unknown resistor. If the total resistance of the series combination is 200 Ω, what is the value of the unknown resistor?
    a. 138 Ω
    b. 62 Ω
    c. 26 Ω
    d. It cannot be determined.

15. How much is the total resistance, R_T, of a series circuit if one of the resistors is open?
    a. infinite (∞) Ω.
    b. 0 Ω.
    c. R_T is much lower than normal.
    d. none of the above

16. If a resistor in a series circuit becomes open, how much is the voltage across each of the remaining resistors that are still good?
    a. Each good resistor has the full value of applied voltage.
    b. The applied voltage is split evenly amongst the good resistors.
    c. 0 V
    d. This is impossible to determine.

17. A 5-Ω and 10-Ω resistor are connected in series across a dc voltage source. Which resistor will dissipate more power?
    a. the 5-Ω resistor

QUESTIONS

Questions at the end of each chapter can be used for classroom discussion or homework assignment.

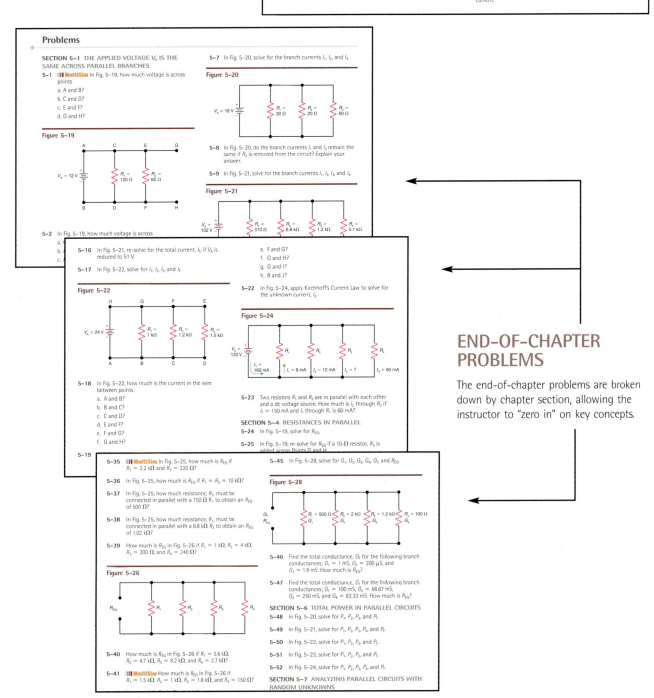

Questions

1. Name two good conductors, two good insulators, and two semiconductors.

2. In a metal conductor, what is a free electron?

3. What is the smallest unit of a compound with the same chemical characteristics?

4. Define the term ion.

5. How does the resistance of a conductor compare to that of an insulator?

6. Explain why potential difference is necessary to produce current in a circuit.

7. List three important characteristics of an electric circuit.

8. Describe the difference between an open circuit and a short circuit.

9. Is the power line voltage available in our homes a dc or an ac voltage?

10. What is the mathematical relationship between resistance and conductance?

11. Briefly describe the electric field of a static charge.

12. List at least two examples that show how static electricity can be generated.

13. What is another name for an insulator?

14. List the particles in the nucleus of an atom.

15. Explain the difference between electron flow and conventional current.

16. Define -3 C of charge and compare it to a charge of $+3$ C.

17. Why is it that protons are not considered a source of moving charges for current flow?

18. Write the formulas for each of the following statements: (a) current is the time rate of change of charge (b) charge is current accumulated over a period of time.

19. Briefly define each of the following: (a) 1 coulomb (b) 1 volt (c) 1 ampere (d) 1 ohm.

20. Describe the difference between direct and alternating current.

Problems

SECTION 5–1 THE APPLIED VOLTAGE V_A IS THE SAME ACROSS PARALLEL BRANCHES

5–1 **MultiSim** In Fig. 5–19, how much voltage is across points
   a. A and B?
   b. C and D?
   c. E and F?
   d. G and H?

Figure 5–19

5–2 In Fig. 5–19, how much voltage is across
   a.
   b.
   c.

5–7 In Fig. 5–20, solve for the branch currents I_1, I_2, and I_3.

Figure 5–20

$V_A = 18$ V
$R_1 = 30\ \Omega$
$R_2 = 20\ \Omega$
$R_3 = 60\ \Omega$

5–8 In Fig. 5–20, do the branch currents I_1 and I_3 remain the same if R_2 is removed from the circuit? Explain your answer.

5–9 In Fig. 5–21, solve for the branch currents I_1, I_2, I_3, and I_4.

Figure 5–21

$V_A = 102$ V
$R_1 = 510\ \Omega$
$R_2 = 6.8\ k\Omega$
$R_3 = 1.2\ k\Omega$
$R_4 = 5.1\ k\Omega$

5–16 In Fig. 5–21, re-solve for the total current, I_T, if V_A is reduced to 51 V.

5–17 In Fig. 5–22, solve for I_1, I_2, I_3, and I_T.

Figure 5–22

$V_A = 24$ V
$R_1 = 1\ k\Omega$
$R_2 = 1.2\ k\Omega$
$R_3 = 1.5\ k\Omega$

5–18 In Fig. 5–22, how much is the current in the wire between points
   a. A and B?
   b. B and C?
   c. C and D?
   d. E and F?
   e. F and G?
   f. G and H?

5–19

   e. F and G?
   f. G and H?
   g. G and I?
   h. B and J?

5–22 In Fig. 5–24, apply Kirchhoff's Current Law to solve for the unknown current, I_3.

Figure 5–24

$V_A = 120$ V
R_1 R_2 R_3 R_4
$I_T = 160$ mA
$I_1 = 8$ mA $I_2 = 12$ mA $I_3 = ?$ $I_4 = 60$ mA

5–23 Two resistors R_1 and R_2 are in parallel with each other and a dc voltage source. How much is I_2 through R_2 if $I_T = 150$ mA and I_1 through R_1 is 60 mA?

SECTION 5–4 RESISTANCES IN PARALLEL

5–24 In Fig. 5–19, solve for R_{EQ}.

5–25 In Fig. 5–19, re-solve for R_{EQ} if a 10-Ω resistor, R_3 is added across Points G and H.

5–35 **MultiSim** In Fig. 5–25, how much is R_{EQ} if $R_1 = 2.2\ k\Omega$ and $R_2 = 220\ \Omega$?

5–36 In Fig. 5–25, how much is R_{EQ} if $R_1 = R_2 = 10\ k\Omega$?

5–37 In Fig. 5–25, how much resistance, R_2, must be connected in parallel with a 750 Ω R_1 to obtain an R_{EQ} of 500 Ω?

5–38 In Fig. 5–25, how much resistance, R_1, must be connected in parallel with a 6.8 kΩ R_2 to obtain an R_{EQ} of 1.02 kΩ?

5–39 How much is R_{EQ} in Fig. 5–26 if $R_1 = 1\ k\Omega$, $R_2 = 4\ k\Omega$, $R_3 = 200\ \Omega$, and $R_4 = 240\ \Omega$?

Figure 5–26

R_{EQ} R_1 R_2 R_3 R_4

5–40 How much is R_{EQ} in Fig. 5–26 if $R_1 = 5.6\ k\Omega$, $R_2 = 4.7\ k\Omega$, $R_3 = 8.2\ k\Omega$, and $R_4 = 2.7\ k\Omega$?

5–41 **MultiSim** How much is R_{EQ} in Fig. 5–26 if $R_1 = 1.5\ k\Omega$, $R_2 = 1\ k\Omega$, $R_3 = 1.8\ k\Omega$, and $R_4 = 150\ \Omega$?

5–45 In Fig. 5–28, solve for G_1, G_2, G_3, G_4, G_T, and R_{EQ}.

Figure 5–28

G_T
R_{EQ}
$R_1 = 500\ \Omega$ $R_2 = 2\ k\Omega$ $R_3 = 1.2\ k\Omega$ $R_4 = 100\ \Omega$
G_1 G_2 G_3 G_4

5–46 Find the total conductance, G_T for the following branch conductances; $G_1 = 1$ mS, $G_2 = 200\ \mu$S, and $G_3 = 1.8$ mS. How much is R_{EQ}?

5–47 Find the total conductance, G_T for the following branch conductances; $G_1 = 100$ mS, $G_2 = 66.67$ mS, $G_3 = 250$ mS, and $G_4 = 83.33$ mS. How much is R_{EQ}?

SECTION 5–6 TOTAL POWER IN PARALLEL CIRCUITS

5–48 In Fig. 5–20, solve for P_1, P_2, P_3, and P_T.

5–49 In Fig. 5–21, solve for P_1, P_2, P_3, P_4, and P_T.

5–50 In Fig. 5–22, solve for P_1, P_2, P_3, and P_T.

5–51 In Fig. 5–23, solve for P_1, P_2, P_3, and P_T.

5–52 In Fig. 5–24, solve for P_1, P_2, P_3, P_4, and P_T.

SECTION 5–7 ANALYZING PARALLEL CIRCUITS WITH RANDOM UNKNOWNS

END-OF-CHAPTER PROBLEMS

The end-of-chapter problems are broken down by chapter section, allowing the instructor to "zero in" on key concepts.

CRITICAL THINKING PROBLEMS

Critical Thinking sections for each chapter provide students with more challenging problems.

4–48 In Fig. 4–46, solve for V_{AG}, V_{BG}, and V_{CG}.

Figure 4–46

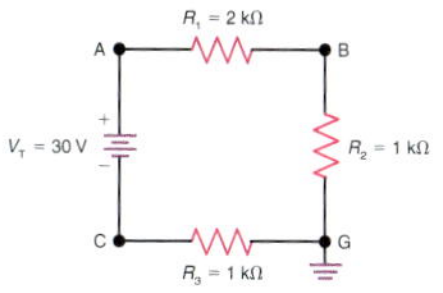

4–51 In Fig. 4–47, assume R_3 shorts. How much is
a. the total resistance, R_T?
b. the series current, I?
c. the voltage across each resistor, R_1, R_2, and R_3?

4–52 In Fig. 4–47, assume that the value of R_2 has increased but is not open. What happens to
a. the total resistance, R_T?
b. the series current, I?
c. the voltage drop across R_2?
d. the voltage drops across R_1 and R_3?

Critical Thinking

4–53 Three resistors in series have a total resistance R_T of 2.7 kΩ. If R_2 is twice the value of R_1 and R_3 is three times the value of R_2, what are the values of R_1, R_2, and R_3?

4–54 Three resistors in series have an R_T of 7 kΩ. If R_3 is 2.2 times larger than R_1 and 1.5 times larger than R_2, what are the values of R_1, R_2, and R_3?

4–55 A 100-Ω, $^1/_8$-W resistor is in series with a 330-Ω, $^1/_2$-W resistor. What is the maximum series current this circuit can handle without exceeding the wattage rating of either resistor?

4–56 A 1.5-kΩ, $^1/_2$-W resistor is in series with a 470-Ω, $^1/_4$-W resistor. What is the maximum voltage that can be applied to this series circuit without exceeding the wattage rating of either resistor?

4–57 Refer to Fig. 4–48. Select values for R_1 and V_T so that when R_2 varies from 1 kΩ to 0 Ω, the series current varies from 1 to 5 mA. V_T and R_1 are to have fixed or constant values.

Figure 4–48 Circuit diagram for Critical Thinking Prob. 4–57.

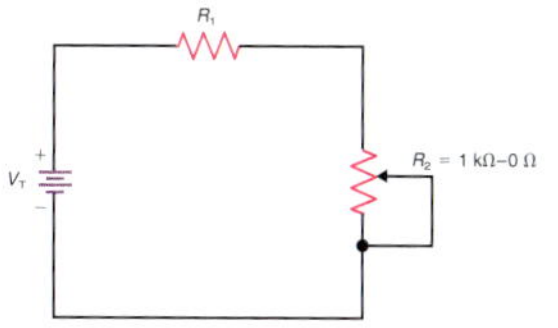

ANSWERS TO KNOWLEDGE CHECK PROBLEMS

Answers to Knowledge Check problems are provided at the end of each chapter.

Answers to Knowledge Check Problems

6–1 No, because the current through R_2 is the sum of the currents I_3 and I_4. Therefore, because I_2 is not equal to I_3 alone, R_2 and R_3 cannot be in series.

6–2 $I_1 = 800$ mA
$I_2 = 2.4$ A
$I_T = 3.2$ A

6–3 $I_2 = 480$ mA
$I_3 = 120$ mA
$I_T = 600$ mA

6–4 $R_T = 53$ Ω

6–5 $I_T = 3.64$ A

6–6 V_{CD} is negative.

6–7 V_{AB} increases.

Answers to Self-Reviews

6–1 a. $R = 1$ kΩ
b. $R = 0.5$ kΩ
c. $R_T = 1.5$ kΩ

6–2 a. 12 V
b. $I = 6$ A
c. $V_3 = 18$ V

6–3 a. $V_3 = 40$ V
b. $I = 8$ A
c. $V_1 = 4$ V

6–4 a. R_1
b. R_4
c. R_6

6–5 a. R_3
b. R_1
c. $I_2 = 4$ A
d. $V_3 = 60$ V

6–6 a. A and B are input; C and D are output.
b. Zero
c. $R_x = 71.35$ Ω
d. 99.99 Ω
e. 500 Ω

6–7 a. $I = 10$ A.
b. $I = 1.1$ A.
c. $V_{AB} = 12$ V
d. 0 V

Cumulative Summary (Chapters 1–6)

■ The electron is the most basic particle of negative electricity; the proton is the most basic particle of positive electricity. Both have the same charge, but they have opposite polarities.

■ A quantity of electrons is a negative charge; a deficiency of electrons is a

power, such as 5 W or more. Carbon-film and metal-film resistors are better than the older carbon-composition type because they have tighter tolerances, are less sensitive to temperature changes and aging, and generate less noise internally.

■ Resistors having a power rating less

two terminals. It is used to vary the current in a circuit.

■ A thermistor is a resistor whose resistance changes with changes in operating temperature. Thermistors are available with either a positive or a negative temperature coefficient (NTC).

ANSWERS TO SELF-REVIEWS

Answers to Self-Review questions are provided at the end of each chapter.

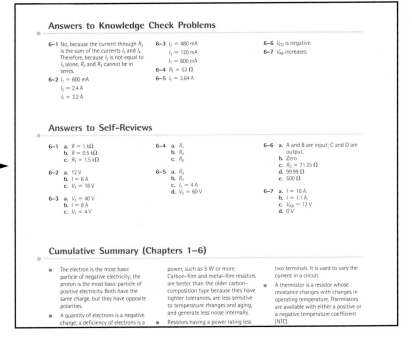

Answers to Knowledge Check Problems

6–1 No, because the current through R_2 is the sum of the currents I_3 and I_4. Therefore, because I_2 is not equal to I_3 alone, R_2 and R_3 cannot be in series.

6–2 $I_1 = 800$ mA
$I_2 = 2.4$ A
$I_t = 3.2$ A

6–3 $I_2 = 480$ mA
$I_3 = 120$ mA
$I_1 = 600$ mA

6–4 $R_T = 53$ Ω

6–5 $I_T = 3.64$ A

6–6 V_{CD} is negative.

6–7 V_{AB} increases.

Answers to Self-Reviews

6–1 a. $R = 1$ kΩ
b. $R = 0.5$ kΩ
c. $R_T = 1.5$ kΩ

6–2 a. 12 V
b. $I = 6$ A
c. $V_3 = 18$ V

6–3 a. $V_3 = 40$ V
b. $I = 8$ A
c. $V_1 = 4$ V

6–4 a. R_1
b. R_4
c. R_6

6–5 a. R_3
b. R_1
c. $I_2 = 4$ A
d. $V_3 = 60$ V

6–6 a. A and B are input; C and D are output.
b. Zero
c. $R_X = 71.35$ Ω
d. 99.99 Ω
e. 500 Ω

6–7 a. $I = 10$ A.
b. $I = 1.1$ A.
c. $V_{AB} = 12$ V
d. 0 V

Cumulative Summary (Chapters 1–6)

■ The electron is the most basic particle of negative electricity; the proton is the most basic particle of positive electricity. Both have the same charge, but they have opposite polarities.

■ A quantity of electrons is a negative charge; a deficiency of electrons is a

power, such as 5 W or more. Carbon-film and metal-film resistors are better than the older carbon-composition type because they have tighter tolerances, are less sensitive to temperature changes and aging, and generate less noise internally.

■ Resistors having a power rating less

two terminals. It is used to vary the current in a circuit.

■ A thermistor is a resistor whose resistance changes with changes in operating temperature. Thermistors are available with either a positive or a negative temperature coefficient (NTC).

Troubleshooting Challenge

Table 4–1 shows voltage measurements taken in Fig. 4–49. The first row shows the normal values that exist when the circuit is operating properly. Rows 2 to 15 are voltage measurements taken when one component in the circuit has failed. For each row, identify which component is defective and determine the type of defect that has occurred in the component.

Figure 4–49 Circuit diagram for Troubleshooting Challenge. Normal values for V_1, V_2, V_3, V_4, and V_5 are shown on schematic.

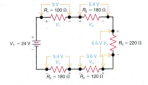

TROUBLESHOOTING CHALLENGE

Troubleshooting Challenge exercises appear in selected chapters to help students get the feel for troubleshooting real circuits.

Table 4–1	Table for Troubleshooting Challenge					
	V_1	V_2	V_3	V_4	V_5	Defective Component
	VOLTS					
1 Normal values	3	5.4	6.6	3.6	5.4	None
2 Trouble 1	0	24	0	0	0	
3 Trouble 2	4.14	7.45	0	4.96	7.45	
4 Trouble 3	3.53	6.35	7.76	0	6.35	
5 Trouble 4	24	0	0	0	0	
6 Trouble 5	0	6.17	7.54	4.11	6.17	

Ancillary Package

The following supplements are available to support *Grob's Basic Electronics: Fundamentals of DC and AC Circuits,* First Edition.

Problems Manual to Accompany Grob's Basic Electronics

This book, written by Mitchel E. Schultz, provides students and instructors with hundreds of practice problems for self-study, homework assignments, tests and review. The book is organized to correlate chapter by chapter with the textbook. Each chapter contains a number of solved illustrative problems demonstrating step-by-step how representative problems on a particular topic are solved. Following the solved problems are sets of problems for the students to solve.

Experiments in Grob's Basic Electronics: Fundamentals of DC and AC Circuits

This lab book, written by Frank Pugh and Wes Ponick, provides students and instructors with easy to follow laboratory experiments. The experiments range from an introduction to laboratory equipment to experiments dealing with filter applications. All experiments have been student tested to ensure their effectiveness. The lab book is organized to correlate with topics covered in the text, chapter by chapter.

All experiments have a *MultiSim* activity that is to be done prior to the actual physical lab activity. *MultiSim* files (version 8) are included on a bound-in CD ROM. This prepares students to work with circuit simulation software, and also to do "pre-lab" preparation before doing a physical lab exercise. *MultiSim* coverage also reflects the widespread use of circuit simulation software in today's electronics industries.

Instructor's Manual and Instructor Productivity Center CD ROM

The Instructor's Manual provides the instructor with answers to all the problems in the textbook, problems manual, and experiments manual.

The Instructor Productivity Center (IPC) contains the Instructor's Manual in electronic form; PowerPoint presentations; and electronic test banks.

Online Learning Center Website

The Online Learning Center (OLC) website contains resources for both students and instructors. Instructor resources are accessible only through a username and password system.

Students can access information about the book and its supplements; links to key websites; glossary terms; and learning activities for each chapter in the textbook.

Instructors will find a full, instructional PowerPoint presentation, arranged by chapter, which covers all of the important concepts in the book. Instructors will also find a 25-question multiple-choice test for each chapter in the book.

Acknowledgements

I would like to say thanks to everyone who took the time to respond to the extensive survey that was distributed prior to the development of this edition of *Grob's Basic Electronics: Fundamentals of DC and AC Circuits*. I realize, and greatly appreciate, the many long hours you spent poring over the manuscript. Because of your sincere commitment, this latest edition of *Grob's Basic Electronics: Fundamentals of DC and AC Circuits* is better than ever before. As you will see, this book is highly accurate, up-to-date and well organized. The following is a list of reviewers who helped make this edition of *Grob's Basic Electronics: Fundamentals of DC and AC Circuits:* the most comprehensive DC/AC textbook on the market.

Current Edition Reviewers

Bruce Clemens
  Ozark Technical College, MO

William R. Cupples
  Western Illinois University, IL

Bob Entrekin
  Grayson County College, TX

Joe Gryniuk
  Lake Washington Technical College, WA

Robert Hockman
  Texas State Technical College, TX

Patrick Hoppe
  Gateway Technical College, WI

Jerry Huff
  Owensboro Community and Technical College, KY

Thomas McGrath
  Gateway Community College, CT

Marcus Rasco
  DeVry University, TX

William Salice
  ECPI College of Technology, VA

Deborah Sharer
  University of North Carolina at Charlotte, NC

Peter So
  Centennial College, Ontario

Gilbert Ulibarri, Jr.
  Salt Lake Community College, UT

Jack Williams
  Remington College, FL

Survey Respondents

Dennis Bladine
  Clayton College and State University, GA

Roger Brown
  Chippewa Valley Technical College, WI

James Charbonneau
  Ivy Tech State College, IN

Donald Craig
  Minnesota West Community and Technical College, MN

Charles C. Curtis
  Middle Georgia Technical College, GA

James Davis
  Itawamba Community College, MS

Hector De La Rosa
  Texas State Technical College, TX

Edward Dyvig
  Iowa Central Community College, IA

Charles Fleddermann
  University of New Mexico, NM

Megan Graham
  Casper College, WY

Mohamad S. Haj-Mohamadi
  North Carolina A&T State University, NC

Linda Hollstegge
  Cincinnati State Technical and Community College, OH

Garrett D. Hunter
  Western Illinois University, IL

Eberhard Kropp
  State Fair Community College, MO

Angela Lemons
  North Carolina A&T State University, NC

David Longobardi
  Antelope Valley College, CA

Jack Malone
  Jackson Community College, MI

Daryl McCall
  Kirkwood Community College, IA

Douglas McCartney
  Central Community College, NE

Richard McKinney
  Nashville State Technical Community College, TN

Claude Mott
  Gulf Coast Community College, FL

Frank Nagel
  Tidewater Technical College, VA

Jack Nudelman
  Orange Coast College, CA

Michael O'Connor
  Lower Columbia College, WA

Terry O'Laughlin
  Madison Area Technical College, WI

Fred Payne
  Danville Area Community College, IL

Joel W. Phillips
  Bishop State Community College, AL

Michael Prichard
  Los Angeles City College, CA

Randal R. Reusser
  Gateway Technical College, WI

Frederick E. Roeser
  Central Community College, NE

Jeffrey Schwartz
  DeVry University, NY

Edward Troyan
  Lehigh Carbon Community College, PA

Ross Weber
  Minnesota State Community and Technical College, MN

Kevin Westerhaus
  Pratt Community College, KS

Mark Winans
  Central Texas College, TX

Manuscript Reviewers

Don Barrett
  DeVry University, TX

Georges C. Livanos
  Humber College, Canada

Bryan Martin
  Pittsburgh Technical College, PA

Robert Magoon
  Erie Institute of Technology, PA

Rod D. Moore
  Wichita Technical Institute, KS

Nick Smith
  Remington College, TX

Anthony Webb
  Missouri Tech, MO

In addition to the reviewers listed, I would also like to thank the highly professional staff of the McGraw-Hill Higher Education Division, especially Jonathan Plant, Tom Casson, Lindsay Roth, Jodi Rhomberg, Joyce Berendes, and Laurie Janssen. You have truly been a joy to work with throughout this whole process. Your patience, kindness and professionalism made what can be a difficult and painful process, a joyful and rewarding experience. And finally, I would like to thank John Hoeft and Pat Hoppe for their role in making this edition of *Grob's Basic Electronics: Fundamentals of DC and AC Circuits* the best ever. Thank you, Pat, for the extensive work you did in integrating *MultiSim* so thoroughly throughout the textbook, and for your ingenuity and clarity in creating the new *MultiSim* Appendix. And John, thank you for your many helpful suggestions and comments throughout the revision process. Your time and effort in proofreading is also greatly appreciated. Thanks again to all of you!

Mitchel E. Schultz

About the Author

Mitchel E. Schultz is an instructor at Western Wisconsin Technical College in La Crosse, Wisconsin where he has taught electronics for the past 17 years. Prior to teaching at Western Wisconsin Technical College, he taught electronics for 8 years at Riverland Community College in Austin, Minnesota. He has also been active in providing training for a variety of different electronic industries over the past 25 years.

Before he began teaching, Mitchel worked for several years as an electronic technician. His primary work experience was in the field of electronic communication, which included designing, testing and troubleshooting RF communications systems. Mitchel graduated in 1978 from Minnesota State, Southeast Technical College, where he earned an Associate's Degree in Electronics Technology. He also attended Winona State University, Mankato State University and the University of Minnesota. He is an ISCET Certified Electronics Technician and also holds his Extra Class Amateur Radio License.

Mitchel has authored and/or co-authored several other electronic textbooks which include: Problems in Basic Electronics, Electric Circuits: A Text and Software Problems Manual, Electronic Devices: A Text and Software Problems Manual, Basic Mathematics for Electricity and Electronics, *and* Shaum's Outline of Theory and Problems of Electronic Communication.

Dedication

This book is dedicated to Bernard Grob, the original author. Bernard, or Bernie as he was known, was a true pioneer in the field of electronics education. Bernie's trademark was his clear, easy-to-read, writing style. I am grateful to have had the opportunity to work with Bernie on the previous editions of this textbook.

Grob's Basic Electronics: Fundamentals of DC and AC Circuits

I Introduction to Powers of 10

The electrical quantities you will encounter while working in the field of electronics are often extremely small or extremely large. For example, it is not at all uncommon to work with extremely small decimal numbers such as 0.000000000056 or extremely large numbers such as 1,296,000,000. To enable us to work conveniently with both very small and very large numbers, powers of 10 notation is used. With powers of 10 notation, any number, no matter how small or large, can be expressed as a decimal number multiplied by a power of 10. A power of 10 is an exponent written above and to the right of 10, which is called the base. The power of 10 indicates how many times the base is to be multiplied by itself. For example, 10^3 means $10 \times 10 \times 10$ and 10^6 means $10 \times 10 \times 10 \times 10 \times 10 \times 10$. In electronics, the base 10 is common because multiples of 10 are used in the metric system of units.

Scientific and engineering notation are two common forms of powers of 10 notation. In electronics, engineering notation is generally more common than scientific notation because it ties in directly with the metric prefixes so often used. When a number is written in standard form without using any form of powers of 10 notation, it is said to be written in decimal notation (sometimes referred to as floating decimal notation). When selecting a calculator for solving problems in electronics, be sure to choose one that can display the answers in decimal, scientific, and engineering notation.

Objectives

After studying this chapter you should be able to

- Express any number in scientific or engineering notation.
- List the metric prefixes and their corresponding powers of 10.
- Change a power of 10 in engineering notation to its corresponding metric prefix.
- Convert between metric prefixes.
- Add and subtract numbers expressed in powers of 10 notation.
- Multiply and divide numbers expressed in powers of 10 notation.
- Determine the reciprocal of a power of 10.
- Find the square of a number expressed in powers of 10 notation.
- Find the square root of a number expressed in powers of 10 notation.
- Enter numbers written in scientific and engineering notation into your calculator.

Outline

Important Terms

decimal notation

engineering notation

metric prefixes

powers of 10

scientific notation

I–1 Scientific Notation

Before jumping directly into scientific notation, let's take a closer look at powers of 10. A power of 10 is an exponent of the base 10 and can be either positive or negative.

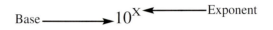

Base $\longrightarrow 10^{X} \longleftarrow$ Exponent

Positive powers of 10 are used to indicate numbers greater than 1, whereas negative powers of 10 are used to indicate numbers less than 1. Table I–1 shows the powers of 10 ranging from 10^{-12} to 10^{9} and their equivalent decimal values. In electronics, you will seldom work with powers of 10 outside this range. From Table I–1, notice that $10^{0} = 1$ and that $10^{1} = 10$. In the case of $10^{0} = 1$, it is important to realize that any number raised to the zero power equals 1. In the case of $10^{1} = 10$, it is important to note that any number written without a power is assumed to have a power of 1.

Expressing a Number in Scientific Notation

The procedure for using any form of powers of 10 notation is to write the original number as two separate factors. Scientific notation is a form of powers of 10 notation in which a number is expressed as a number between 1 and 10 times a power of 10. The power of 10 is used to place the decimal point correctly. The power of 10 indicates the number of places by which the decimal point has been moved to the left or right in the original number. If the decimal point is moved to the left in the original number, then the power of 10 will increase or become more positive. Conversely, if the decimal point is moved to the right in the original number then the power of 10 will decrease or become more negative. Let's take a look at an example.

Table I–1	Powers of 10		
$1,000,000,000 = 10^{9}$	$10 = 10^{1}$	$0.000001 = 10^{-6}$
$100,000,000 = 10^{8}$	$1 = 10^{0}$	$0.0000001 = 10^{-7}$
$10,000,000 = 10^{7}$	$0.1 = 10^{-1}$	$0.00000001 = 10^{-8}$
$1,000,000 = 10^{6}$	$0.01 = 10^{-2}$	$0.000000001 = 10^{-9}$
$100,000 = 10^{5}$	$0.001 = 10^{-3}$	$0.0000000001 = 10^{-10}$
$10,000 = 10^{4}$	$0.0001 = 10^{-4}$	$0.00000000001 = 10^{-11}$
$1,000 = 10^{3}$	$0.00001 = 10^{-5}$	$0.000000000001 = 10^{-12}$
$100 = 10^{2}$			

Example I-1

Express the following numbers in scientific notation: (a) 3900 (b) 0.0000056.

ANSWER: (a) To express 3900 in scientific notation, write the number as a number between 1 and 10, which is 3.9 in this case, times a power of 10. To do this, the decimal point must be shifted three places to the left. The number of places by which the decimal point is shifted to the left indicates the positive power of 10. Therefore, $3900 = 3.9 \times 10^3$ in scientific notation.

(b) To express 0.0000056 in scientific notation, write the number as a number between 1 and 10, which is 5.6 in this case, times a power of 10. To do this, the decimal point must be shifted six places to the right. The number of places by which the decimal point is shifted to the right indicates the negative power of 10. Therefore, $0.0000056 = 5.6 \times 10^{-6}$ in scientific notation.

When expressing a number in scientific notation, remember the following rules.

Rule 1: Express the number as a number between 1 and 10 times a power of 10.

Rule 2: If the decimal point is moved to the left in the original number, make the power of 10 positive. If the decimal point is moved to the right in the original number, make the power of 10 negative.

Rule 3: The power of 10 always equals the number of places by which the decimal point has been shifted to the left or right in the original number.

Let's try another example.

Example I-2

Express the following numbers in scientific notation: (a) 235,000 (b) 364,000,000 (c) 0.000756 (d) 0.00000000000016.

ANSWER: (a) To express the number 235,000 in scientific notation, move the decimal point 5 places to the left, which gives us a number of 2.35. Next, multiply this number by 10^5. Notice that the power of 10 is a positive 5 because the decimal point was shifted five places to the left in the original number. Therefore, $235,000 = 2.35 \times 10^5$ in scientific notation.

(b) To express 364,000,000 in scientific notation, move the decimal point eight places to the left, which gives us a number of 3.64. Next, multiply this number by 10^8. Notice that the power of 10 is a positive 8 because the decimal point was shifted eight places to the left in the original number. Therefore, $364,000,000 = 3.64 \times 10^8$ in scientific notation.

(c) To express 0.000756 in scientific notation, move the decimal point 4 places to the right, which gives us a number of 7.56. Next, multiply this number by 10^{-4}. Notice that the power of 10 is a negative 4 because the decimal point was shifted four places to the right in the original number. Therefore, $0.000756 = 7.56 \times 10^{-4}$.

(d) To express 0.00000000000016 in scientific notation, move the decimal point 13 places to the right, which gives us a number of 1.6. Next, multiply this number by 10^{-13}. Notice that the power of 10 is a negative 13 because the decimal point was shifted thirteen places to the right in the original number. Therefore, $0.00000000000016 = 1.6 \times 10^{-13}$ in scientific notation.

I–1A Knowledge Check

Answers at end of chapter.

Express the following numbers in scientific notation: (a) 76,300,000 (b) 0.0000342.

Decimal Notation

Numbers written in standard form without using any form of powers of 10 notation are said to be written in decimal notation, sometimes called floating decimal notation. In some cases, it may be necessary to change a number written in scientific notation into decimal notation. When converting from scientific to decimal notation, observe the following rules.

Rule 4: If the exponent or power of 10 is positive, move the decimal point to the right, the same number of places as the exponent.

Rule 5: If the exponent or power of 10 is negative, move the decimal point to the left, the same number of places as the exponent.

Example I–3

Convert the following numbers written in scientific notation into decimal notation: (a) 4.75×10^2 (b) 6.8×10^{-5}.

ANSWER: (a) To convert 4.75×10^2 into decimal notation, the decimal point must be shifted 2 places to the right. The decimal point is shifted to the right because the power of 10, which is 2 in this case, is positive. Therefore; $4.75 \times 10^2 = 475$ in decimal notation.

(b) To convert 6.8×10^{-5} into decimal notation, the decimal point must be shifted 5 places to the left. The decimal point is shifted to the left because the power of 10, which is -5 in this case, is negative. Therefore, $6.8 \times 10^{-5} = 0.000068$ in decimal notation.

■ *I–1B Knowledge Check*

Answers at end of chapter.

Convert the following numbers into decimal notation: (a) 2.75×10^{-5} (b) 8.41×10^4.

■ *I–1 Self-Review*

Answers at end of chapter.

a. Are positive or negative powers of 10 used to indicate numbers less than 1?
b. Are positive or negative powers of 10 used to indicate numbers greater than 1?
c. $10^0 = 1$ (True/False)
d. Express the following numbers in scientific notation: (a) 13,500 (b) 0.00825 (c) 95,600,000 (d) 0.104.
e. Convert the following numbers written in scientific notation into decimal notation: (a) 4.6×10^{-7} (b) 3.33×10^3 (c) 5.4×10^8 (d) 2.54×10^{-2}.

I–2 Engineering Notation and Metric Prefixes

Engineering notation is another form of powers of 10 notation. Engineering notation is similar to scientific notation except that in engineering notation, the powers of 10 are always multiples of 3 such as 10^{-12}, 10^{-9}, 10^{-6}, 10^{-3}, 10^{3}, 10^{6}, 10^{9}, 10^{12}, etc. More specifically, a number expressed in engineering notation is always expressed as a number between 1 and 1000 times a power of 10 which is a multiple of 3.

Example I-4

Express the following numbers in engineering notation: (a) 27,000 (b) 0.00047.

ANSWER: (a) To express the number 27,000 in engineering notation, it must be written as a number between 1 and 1000 times a power of 10 which is a multiple of 3. It is often helpful to begin by expressing the number in scientific notation: $27,000 = 2.7 \times 10^{4}$. Next, examine the power of 10 to see if it should be increased to 10^{6} or decreased to 10^{3}. If the power of 10 is increased to 10^{6}, then the decimal point in the number 2.7 would have to be shifted two places to the left. Because 0.027 is not a number between 1 and 1000, the answer of 0.027×10^{6} is not representative of engineering notation. If the power of 10 were decreased to 10^{3}, however, then the decimal point in the number 2.7 would have to be shifted one place to the right and the answer would be 27×10^{3}, which is representative of engineering notation. In summary, $27,000 = 2.7 \times 10^{4} = 27 \times 10^{3}$ in engineering notation.

    (b) To express the number 0.00047 in engineering notation, it must be written as a number between 1 and 1000 times a power of 10 which is a multiple of 3. Begin by expressing the number in scientific notation: $0.00047 = 4.7 \times 10^{-4}$. Next, examine the power of 10 to see if it should be increased to 10^{-3} or decreased to 10^{-6}. If the power of 10 were increased to 10^{-3}, then the decimal point in the number 4.7 would have to be shifted one place to the left. Because 0.47 is not a number between 1 and 1000, the answer 0.47×10^{-3} is not representative of engineering notation. If the power of 10 were decreased to 10^{-6}, however, then the decimal point in the number 4.7 would have to be shifted two places to the right and the answer would be 470×10^{-6} which is representative of engineering notation. In summary: $0.00047 = 4.7 \times 10^{-4} = 470 \times 10^{-6}$ in engineering notation.

When expressing a number in engineering notation, remember the following rules:

Rule 6: Express the original number in scientific notation first. If the power of 10 is a multiple of 3, the number appears the same in both scientific and engineering notation.

Rule 7: If the original number expressed in scientific notation does not use a power of 10 which is a multiple of 3, the power of 10 must either be increased or decreased until it is a multiple of 3. The decimal point in the numerical part of the expression must be adjusted accordingly to compensate for the change in the power of 10.

Rule 8: Each time the power of 10 is increased by 1, the decimal point in the numerical part of the expression must be moved one place to the left. Each time the power of 10 is decreased by 1, the decimal point in the numerical part of the expression must be moved one place to the right.

You know that a quantity is expressed in engineering notation when the original number is written as a number between 1 and 1000 times a power of 10 which is a multiple of 3.

Express the following numbers in engineering notation: (a) 440,000 (b) 0.068.

Metric Prefixes

The metric prefixes represent those powers of 10 that are multiples of 3. In the field of electronics, engineering notation is much more common than scientific notation because most values of voltage, current, resistance, power, etc. are specified in terms of the metric prefixes. Once a number is expressed in engineering notation, its power of 10 can be replaced directly with its corresponding metric prefix. Table I–2 lists the most common metric prefixes and their corresponding powers of 10. Notice that uppercase letters are used for the abbreviations of the prefixes involving positive powers of 10, whereas lowercase letters are used for negative powers of 10. There is one exception to the rule however; the lowercase letter "k" is used for kilo corresponding to 10^3. Because the metric prefixes are used so often in electronics, it is common practice to express the value of a given quantity in engineering notation first so that the power of 10, which is a multiple of 3, can be replaced directly with its corresponding metric prefix. For example, a resistor whose value is 33,000 Ω can be expressed in engineering notation as 33×10^3 Ω. In Table I–2 we see that the metric prefix kilo (k) corresponds to 10^3. Therefore, 33,000 Ω or 33×10^3 Ω can be expressed as 33 kΩ. (Note that the unit of resistance is the ohm with the symbol Ω). As another example, a current of 0.0000075 A can be expressed in engineering notation as 7.5×10^{-6}A. In Table I–2, we see that the metric prefix micro (μ) corresponds to 10^{-6}. Therefore, 0.0000075 A or 7.5×10^{-6} A can be expressed as 7.5 μA. (The unit of current is the ampere, abbreviated A.)

In general, when using metric prefixes to express the value of a given quantity, write the original number in engineering notation first and then substitute the appropriate metric prefix corresponding to the power of 10 involved. As this technique shows, metric prefixes are direct substitutes for the powers of 10 used in engineering notation.

Table I–3 lists many of the electrical quantities that you will encounter in your study of electronics. For each electrical quantity listed in Table I–3, take

Table I–2	Metric Prefixes	
Power of 10	**Prefix**	**Abbreviation**
10^{12}	tera	T
10^9	giga	G
10^6	mega	M
10^3	kilo	k
10^{-3}	milli	m
10^{-6}	micro	μ
10^{-9}	nano	n
10^{-12}	pico	p

Table I–3	Electrical Quantities with Their Units and Symbols	
Quantity	Unit	Symbol
Current	Ampere (A)	*I*
Voltage	Volt (V)	*V*
Resistance	Ohm (Ω)	*R*
Frequency	Hertz (Hz)	*f*
Capacitance	Farad (F)	*C*
Inductance	Henry (H)	*L*
Power	Watt (W)	*P*

special note of the unit and symbol shown. In the examples and problems that follow, we will use several numerical values with the various symbols and units from this table. Let's take a look at a few examples.

Example I-5

Express the resistance of 1,000,000 Ω using the appropriate metric prefix from Table I–2.

ANSWER: First, express 1,000,000 Ω in engineering notation: 1,000,000 Ω = 1.0×10^6 Ω. Next, replace 10^6 with its corresponding metric prefix. Because the metric prefix mega (M) corresponds to 10^6, the value of 1,000,000 Ω can be expressed as 1 MΩ. In summary, 1,000,000 Ω = 1.0×10^6 Ω = 1 MΩ.

Example I-6

Express the voltage value of 0.015 V using the appropriate metric prefix from Table I–2.

ANSWER: First, express 0.015 V in engineering notation: 0.015 V = 15×10^{-3} V. Next, replace 10^{-3} with its corresponding metric prefix. Because the metric prefix milli (m) corresponds to 10^{-3}, the value 0.015 V can be expressed as 15 mV. In summary, 0.015 V = 15×10^{-3} V = 15 mV.

Example I-7

Express the power value of 250 W using the appropriate metric prefix from Table I–2.

ANSWER: In this case, it is not necessary or desirable to use any of the metric prefixes listed in Table I–2. The reason is that 250 W cannot be expressed as a number between 1 and 1000 times a power of 10 which is a multiple of 3. In other words, 250 W cannot be expressed in engineering notation. The closest we can come is 0.25×10^3 W, which is not representative of engineering notation. Although 10^3 can be replaced with the metric prefix kilo (k), it is usually preferable to express the power as 250 W and not as 0.25 kW.

   In summary, whenever the value of a quantity lies between 1 and 1000, only the basic unit of measure should be used for the answer. As another example, 75 V should be expressed as 75 V and not as 0.075 kV or 75,000 mV, etc.

■ *I–2B Knowledge Check*

   Answers at end of chapter.

   Express the following values using the appropriate metric prefixes:
   (a) 0.050 V (b) 18,000 Ω (c) 0.000000000047 F.

■ *I–2 Self-Review*

   Answers at end of chapter.

a.   **Express the following numbers in engineering notation:**
   (a) 36,000,000 (b) 0.085 (c) 39,300 (d) 0.000093.
b.   **List the metric prefixes for each of the powers of 10 listed:**
   (a) 10^{-9} (b) 10^6 (c) 10^{-12} (d) 10^3 (e) 10^4.
c.   **Express the following values using the appropriate metric prefixes:**
   (a) 0.000010 A (b) 2,200,000 Ω (c) 0.000000045 V (d) 5600 Ω
   (e) 18 W.

I–3 Converting Between Metric Prefixes

As you have seen in the previous section, metric prefixes can be substituted for powers of 10 that are multiples of 3. This is true even when the value of the original quantity is not expressed in proper engineering notation. For example, a capacitance value of 0.047×10^{-6} F could be expressed as 0.047 μF. Also, a frequency of 1510×10^3 Hz could be expressed as 1510 kHz. Furthermore, the values of like quantities in a given circuit may be specified using different metric prefixes such as 22 kΩ and 1.5 MΩ or 0.001 μF and 3300 pF, as examples. In some cases, therefore, it may be necessary or desirable to convert from one metric prefix to another when combining values. Converting from one metric prefix to another is actually a change in the power of 10. When the power of 10 is changed, however, care must be taken to make sure that the numerical part of the expression is also changed so that the value of the original number remains the same. When converting from one metric prefix to another observe the following rule:

Rule 9: When converting from a larger metric prefix to a smaller one, increase the numerical part of the expression by the same factor by which the metric prefix has been decreased. Conversely, when converting from a smaller metric prefix to a larger one, decrease the numerical part of the expression by the same factor by which the metric prefix has been increased.

Example 1-8

Make the following conversions: (a) convert 25 mA to μA (b) convert 2700 kΩ to MΩ.

ANSWER: (a) To convert 25 mA to μA, recall that the metric prefix milli (m) corresponds to 10^{-3} and that metric prefix micro (μ) corresponds to 10^{-6}. Since 10^{-6} is less than 10^{-3} by a factor of 1000 (10^{3}), the numerical part of the expression must be increased by a factor of 1000 (10^{3}). Therefore, 25 mA = 25×10^{-3} A = $25{,}000 \times 10^{-6}$ A = 25,000 μA.

(b) To convert 2700 kΩ to MΩ, recall that the metric prefix kilo (k) corresponds to 10^{3} and that the metric prefix mega (M) corresponds to 10^{6}. Since 10^{6} is larger than 10^{3} by a factor of 1000 (10^{3}), the numerical part of the expression must be decreased by a factor of 1000 (10^{3}). Therefore, 2700 kΩ = 2700×10^{3} Ω = 2.7×10^{6} Ω = 2.7 MΩ.

■ *1–3 Knowledge Check*

Answers at end of chapter.

Make the following conversions: (a) 0.022 μF to nF (b) 1410 kHz to MHz.

■ *1–3 Self-Review*

Answers at end of chapter.

a. **Converting from one metric prefix to another is actually a change in the power of 10. (True/False)**
b. **Make the following conversions: (a) convert 2.2 MΩ to kΩ (b) convert 47,000 pF to nF (c) convert 2500 μA to mA (d) convert 6.25 mW to μW.**

1–4 Addition and Subtraction Involving Powers of 10 Notation

When adding or subtracting numbers expressed in powers of 10 notation, observe the following rule:

Rule 10: Before numbers expressed in powers of 10 notation can be added or subtracted, both terms must be expressed using the same power of 10. When both terms have the same power of 10, just add or subtract the numerical parts of each term and multiply the sum or difference by the power of 10 common to both terms. Express the final answer in the desired form of powers of 10 notation.

Let's take a look at a couple of examples.

Example I-9

Add 170×10^3 and 23×10^4. Express the final answer in scientific notation.

ANSWER: First, express both terms using either 10^3 or 10^4 as the common power of 10. Either one can be used. In this example we will use 10^3 as the common power of 10 for both terms. Rewriting 23×10^4 using 10^3 as the power of 10 gives us 230×10^3. Notice that because the power of 10 was decreased by a factor of 10, the numerical part of the expression was increased by a factor of 10. Next, add the numerical parts of each term and multiply the sum by 10^3 which is the power of 10 common to both terms. This gives us $(170 + 230) \times 10^3$ or 400×10^3. Expressing the final answer in scientific notation gives us 4.0×10^5. In summary $(170 \times 10^3) + (23 \times 10^4) = (170 \times 10^3) + (230 \times 10^3) = (170 + 230) \times 10^3 = 400 \times 10^3 = 4.0 \times 10^5$.

Example I-10

Subtract 250×10^3 from 1.5×10^6. Express the final answer in scientific notation.

ANSWER: First, express both terms using either 10^3 or 10^6 as the common power of 10. Again, either one can be used. In this example, we will use 10^6 as the common power of 10 for both terms. Rewriting 250×10^3 using 10^6 as the power of 10 gives us 0.25×10^6. Notice that because the power of 10 was increased by a factor 1000 (10^3), the numerical part of the expression was decreased by a factor of 1000 (10^3). Next, subtract 0.25 from 1.5 and multiply the difference by 10^6, which is the power of 10 common to both terms. This gives us $(1.5 - 0.25) \times 10^6$ or 1.25×10^6. Notice that the final answer is already in scientific notation. In summary, $(1.5 \times 10^6) - (250 \times 10^3) = (1.5 \times 10^6) - (0.25 \times 10^6) = (1.5 - 0.25) \times 10^6 = 1.25 \times 10^6$.

■ I-4 Knowledge Check

Answers at end of chapter.

Perform the following mathematical operations and express your answers in scientific notation: (a) $(15 \times 10^2) + (6.0 \times 10^4)$ (b) $(550 \times 10^{-3}) - (250 \times 10^{-4})$.

■ I-4 Self-Review

Answers at end of chapter.

a. Add the following terms expressed in powers of 10 notation. Express the answers in scientific notation. (a) $(470 \times 10^4) + (55 \times 10^6)$ (b) $(3.5 \times 10^{-2}) + (1500 \times 10^{-5})$.

b. Subtract the following terms expressed in powers of 10 notation. Express the answers in scientific notation. (a) $(65 \times 10^4) - (200 \times 10^3)$ (b) $(850 \times 10^{-3}) - (3500 \times 10^{-4})$.

I-5 Multiplication and Division Involving Powers of 10 Notation

When multiplying or dividing numbers expressed in powers of 10 notation, observe the following rules.

Rule 11: When multiplying numbers expressed in powers of 10 notation, multiply the numerical parts and powers of 10 separately. When multiplying powers of 10, simply add the exponents to obtain the new power of 10. Express the final answer in the desired form of powers of 10 notation.

Rule 12: When dividing numbers expressed in powers of 10 notation, divide the numerical parts and powers of 10 separately. When dividing powers of 10, subtract the power of 10 in the denominator from the power of 10 in the numerator. Express the final answer in the desired form of powers of 10 notation.

Let's take a look at a few examples.

Example I-11

Multiply (3×10^6) by (150×10^2). Express the final answer in scientific notation.

ANSWER: First multiply 3×150 to obtain 450. Next, multiply 10^6 by 10^2 to obtain $10^6 \times 10^2 = 10^{6+2} = 10^8$. To review, $(3 \times 10^6) \times (150 \times 10^2) = (3 \times 150) \times (10^6 \times 10^2) = 450 \times 10^{6+2} = 450 \times 10^8$. The final answer expressed in scientific notation is 4.5×10^{10}.

Example I-12

Divide 5.0×10^7 by 2.0×10^4. Express the final answer in scientific notation.

ANSWER: First divide 5 by 2 to obtain 2.5. Next divide 10^7 by 10^4 to obtain $10^{7-4} = 10^3$. To review, $\dfrac{5.0 \times 10^7}{2.0 \times 10^4} = \dfrac{5}{2} \times \dfrac{10^7}{10^4} = 2.5 \times 10^3$.

Notice that the final answer is already in scientific notation.

■ *I-5 Knowledge Check*

Answers at end of chapter.

Perform the following mathematical operations and express your answers in scientific notation: (a) $(5.0 \times 10^{-3}) \times (4.0 \times 10^{-4})$ (b) $(500 \times 10^6) \div (40 \times 10^3)$.

■ I-5 Self-Review

Answers at end of chapter.

a. Multiply the following numbers expressed in powers of 10 notation. Express your answers in scientific notation. (a) $(3.3 \times 10^{-2}) \times (4.0 \times 10^{-3})$ (b) $(2.7 \times 10^{2}) \times (3 \times 10^{-5})$.

b. Divide the following numbers expressed in powers of 10 notation. Express your answers in scientific notation. (a) $(7.5 \times 10^{8}) \div (3.0 \times 10^{4})$ (b) $(15 \times 10^{-6}) \div (5 \times 10^{-3})$.

I-6 Reciprocals with Powers of 10

Taking the reciprocal of a power of 10 is really just a special case of division using powers of 10 because 1 in the numerator can be written as 10^{0} since $10^{0} = 1$. With zero as the power of 10 in the numerator, taking the reciprocal results in a sign change for the power of 10 in the denominator. Let's take a look at an example to clarify this point.

Example I-13

Find the reciprocals for the following powers of 10: (a) 10^{5} (b) 10^{-3}.

ANSWER: (a) $\dfrac{1}{10^{5}} = \dfrac{10^{0}}{10^{5}} = 10^{0-5} = 10^{-5}$; therefore, $\dfrac{1}{10^{5}} = 10^{-5}$.

(b) $\dfrac{1}{10^{-3}} = \dfrac{10^{0}}{10^{-3}} = 10^{0-(-3)} = 10^{3}$; therefore, $\dfrac{1}{10^{-3}} = 10^{3}$.

Notice that in both (a) and (b), the power of 10 in the denominator is subtracted from zero which is the power of 10 in the numerator.

Here's a simple rule for reciprocals of powers of 10.

Rule 13: When taking the reciprocal of a power of 10, simply change the sign of the exponent or power of 10.

Negative Powers of 10

Recall that a power of 10 indicates how many times the base, 10, is to be multiplied by itself. For example, $10^{4} = 10 \times 10 \times 10 \times 10$. But you might ask how this definition fits with negative powers of 10. The answer is that negative powers of 10 are just reciprocals of positive powers of 10. For example,

$$10^{-4} = \frac{1}{10^{4}} = \frac{1}{10 \times 10 \times 10 \times 10}.$$

■ I-6 Knowledge Check

Answers at end of chapter.

Take the reciprocal of the following numbers: (a) 10^{-18} (b) 10^{1}.

■ *I–6 Self-Review*

Answers at end of chapter.

   a. Take the reciprocals of each of the powers of 10 listed. (a) 10^{-4} (b) 10^9 (c) 10^{-18} (d) 10^0.

I–7 Squaring Numbers Expressed in Powers of 10 Notation

When squaring a number expressed in powers of 10 notation, observe the following rule.

Rule 14: To square a number expressed in powers of 10 notation, square the numerical part of the expression and double the power of 10. Express the answer in the desired form of powers of 10 notation.

Example I–14

Square 3.0×10^4. Express the answer in scientific notation.

ANSWER: First, square 3.0 to obtain 9.0. Next, square 10^4 to obtain $(10^4)^2 = 10^8$. Therefore, $(3.0 \times 10^4)^2 = 9.0 \times 10^8$.

■ *I–7 Knowledge Check*

Answer at end of chapter.

Square the number 250×10^{-6} and express your answer in scientific notation.

■ *I–7 Self-Review*

Answers at end of chapter.

   a. Obtain the following answers and express them in scientific notation. (a) $(4.0 \times 10^{-2})^2$ (b) $(6.0 \times 10^5)^2$ (c) $(2.0 \times 10^{-3})^2$.

I–8 Square Roots of Numbers Expressed in Powers of 10 Notation

When taking the square root of a number expressed in powers of 10 notation, observe the following rule.

Rule 15: To find the square root of a number expressed in powers of 10 notation, take the square root of the numerical part of the expression and divide the power of 10 by 2. Express the answer in the desired form of powers of 10 notation.

Example I-15

Find the square root of 4×10^6. Express the answer in scientific notation.

ANSWER: $\sqrt{4 \times 10^6} = \sqrt{4} \times \sqrt{10^6} = 2 \times 10^3$

Notice that the answer is already in scientific notation.

Example I-16

Find the square root of 90×10^5. Express the answer in scientific notation.

ANSWER: The problem can be simplified if we increase the power of 10 from 10^5 to 10^6 and decrease the numerical part of the expression from 90 to 9. This gives us $\sqrt{90 \times 10^5} = \sqrt{9 \times 10^6} = \sqrt{9} \times \sqrt{10^6} = 3.0 \times 10^3$. Again, the answer is already in scientific notation.

■ **I-8 Knowledge Check**

Answer at end of chapter.

Take the square root of 490×10^{-5} and express your answer in scientific notation.

■ **I-8 Self-Review**

Answers at end of chapter.

a. Obtain the following answers and express them in scientific notation. (a) $\sqrt{36 \times 10^4}$ (b) $\sqrt{160 \times 10^{-5}}$ (c) $\sqrt{25 \times 10^{-8}}$.

I-9 The Scientific Calculator

Throughout your study of electronics, you will make calculations using numerical values that are expressed in decimal, scientific, or engineering notation. In most cases, you will want to use a scientific calculator to aid you in your calculations. Be sure to select a calculator that can perform all of the mathematical functions and operations that you will encounter in your study of electronics. Also, make sure the calculator you select can store and retrieve mathematical results from one or more memory locations. If the school or industry responsible for your training does not recommend or mandate a specific calculator, be sure to ask your instructor or supervisor for his or her recommendation on which calculator to buy. And finally, once you have purchased your calculator, carefully read the instructions that are included with it. At first, you may not understand many of your calculators functions and features, but as you progress in your studies, you will become more familiar with them.

Entering and Displaying Values

Figure I–1 shows a typical scientific calculator that would be suitable for use in a course involving the study of dc and ac electronics. Although it may not be apparent at first, almost all of the keys on the calculator are used for more than one function. For example, to turn the calculator on, press the $_{OFF}$ [ON/C] key located below and to the right of the LCD display. To turn the calculator off, press the [2ndF] key, located below and to the left of the LCD display, followed by pressing the $_{OFF}$ [ON/C] key. The $_{OFF}$ [ON/C] key is also pressed when you want to clear the LCD display of its current numerical value. The functions labeled in orange, located above or beside a key, can be accessed only by pressing the [2ndF] key first.

The calculator in Figure I–1 has four display notation systems for displaying calculation results: floating decimal notation, fixed decimal notation (FIX), scientific notation (SCI), and engineering notation (ENG). (The LCD display always shows the current notation being used.) When the FIX, SCI, or ENG symbol is displayed, the number of digits displayed to the right of the decimal point can be set to any value from 0–9. With floating decimal notation, however, there is no set number of digits displayed for any given answer. On this particular calculator, the number of digits displayed to the right of the decimal point can be set by pressing the [2ndF] key followed by the ^{TAB}[+/−] key and then the desired digit (0–9). To change between the different display notations, press the [2ndF] key followed by the ^{FSE}[.] key located in the bottom row of keys. Pressing these keys in succession allows the user to switch between floating decimal notation, fixed decimal notation (FIX), scientific notation (SCI), and engineering notation (ENG).

To enter a number expressed in either scientific or engineering notation into the calculator in Figure I–1, proceed as follows. Enter the numerical part of the expression first. Next, press the [EXP] key followed by the exponent or power of 10. If the power of 10 is negative, press the [+/−] key before entering the exponent value. That's how simple it is.

Although the calculator you may be using for your coursework is not the same as that shown in Figure I–1, the procedure for entering and displaying values will be similar. As mentioned earlier, take the time to read the instruction manual for the calculator you have chosen, and keep it with your calculator for future reference.

Figure I–1   Scientific calculator (Sharp EL-531 V)

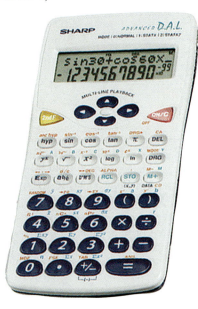

Example I-17

Show the steps for multiplying 2.5×10^4 by 3.0×10^{-3} using the calculator in Figure I–1. Display the answer in scientific (SCI) notation.

ANSWER: Set the display notation to scientific (SCI) by pressing the [2ndF] and $_{FSE}$ [.] keys in succession until SCI appears on the LCD display. Next, enter the problem using the following keying sequence: [2] [.] [5] [EXP] [4] [×] [3] [.] [0] [EXP] [±] [3] [=]. The answer is displayed in scientific notation as 7.500×10^{01}.

(The number of digits displayed to the right of the decimal point can be set to any value from 0 to 9.)

■ *I-9 Knowledge Check*

Answer at end of chapter.

Show the steps for dividing 24 by 1×10^3 using the calculator in Figure I–1. Display the answer in engineering (ENG) notation.

a. When selecting a calculator for solving problems in electronics, it is important to select one that allows you to enter and display values in decimal, scientific, or engineering notation. (True/False)

b. Show the steps for multiplying 3.3×10^{-3} by 10×10^3 using the calculator in Figure I–1. Display the answer in engineering (ENG) notation.

Summary

- A power of 10 is an exponent that is written above and to the right of 10, which is called the base.

- A power of 10 indicates how many times the base, 10, is to be multiplied by itself.

- Positive powers of 10 indicate numbers greater than 1 and negative powers of 10 indicate numbers less than 1. Also, $10^0 = 1$ and $10^1 = 10$.

- Powers of 10 notation is a convenient method for expressing very small or very large numbers as a decimal number multiplied by a power of 10.

- Scientific and engineering notation are two forms of powers of 10 notation.

- A number expressed in scientific notation is always expressed as a number between 1 and 10 times a power of 10.

- A number expressed in engineering notation is always expressed as a

number between 1 and 1000 times a power of 10 which is a multiple of 3.

- Decimal notation refers to those numbers that are written in standard form without any form of powers of 10 notation.

- Metric prefixes are letter symbols used to replace the powers of 10 that are multiples of 3. Refer to Table I–2 for a complete listing of the metric prefixes and their corresponding powers of 10.

- Converting from one metric prefix to another is a change in the power of 10 used to express a given quantity.

- Before numbers expressed in powers of 10 notation can be added or subtracted, both terms must have the same power of 10. When both terms have the same power of 10, just add or subtract the numerical parts of the expression and multiply the sum or difference by the power of 10 common to both terms.

- When multiplying numbers expressed in powers of 10 notation, multiply the numerical parts and powers of 10 separately. When multiplying powers of 10, simply add the exponents.

- When dividing numbers expressed in powers of 10 notation, divide the numerical parts and powers of 10 separately. When dividing powers of 10, simply subtract the power of 10 in the denominator from the power of 10 in the numerator.

- Taking the reciprocal of a power of 10 is the same as changing the sign of the exponent.

- To square a number expressed in powers of 10 notation, square the numerical part of the expression and double the power of 10.

- To take the square root of a number expressed in powers of 10 notation, take the square root of the numerical part of the expression and divide the power of 10 by 2.

Important Terms

- Decimal Notation – Numbers that are written in standard form without using powers of 10 notation.

- Engineering Notation – A form of powers of 10 notation in which a number is expressed as a number between 1 and 1000 times a power of 10 that is a multiple of 3.

- Metric Prefixes – Letter symbols used to replace the powers of 10 that are multiples of 3.

- Powers of 10 – A numerical representation consisting of a base of 10 and an exponent; the base 10 raised to a power.

- Scientific Notation – A form of powers of 10 notation in which a number is expressed as a number between 1 and 10 times a power of 10.

Self-Test

1. 10^4 means the same thing as
   a. 10,000.
   b. 10×4.
   c. $10 \times 10 \times 10 \times 10$.
   d. both a and c.

2. Negative powers of 10
   a. indicate numbers less than 1.
   b. are not used with engineering notation.
   c. indicate numbers greater than 1.
   d. are used only with scientific notation.

3. A number expressed in scientific notation is always expressed as a number between
   a. 1 and 1000 times a power of 10 which is a multiple of 3.
   b. 1 and 10 times a power of 10.
   c. 1 and 100 times a power of 10.
   d. 0 and 1 times a power of 10.

4. A number expressed in engineering notation is always expressed as a number between
   a. 1 and 10 times a power of 10 that is a multiple of 3.
   b. 1 and 10 times a power of 10.
   c. 1 and 1000 times a power of 10 that is a multiple of 3.
   d. 0 and 1 times a power of 10 that is a multiple of 3.

5. 10^0 equals
   a. 0.
   b. 10.
   c. 1.
   d. none of the above.

6. Metric prefixes are used only with those powers of 10 that are
   a. multiples of 3.
   b. negative.
   c. associated with scientific notation.
   d. both a and b.

7. 40×10^{-3} A is the same as
   a. 40 mA.
   b. 40 μA.
   c. 40 kA.
   d. 40 MA.

8. 3.9 MΩ is the same as
   a. 3.9×10^3 Ω.
   b. 3.9×10^6 Ω.
   c. 3,900 kΩ.
   d. both b and c.

9. A number written in standard form without any form of powers of 10 notation is said to be written in
   a. scientific notation.
   b. decimal notation.
   c. engineering notation.
   d. metric prefix notation.

10. The metric prefix pico (p) corresponds to
    a. 10^{12}.
    b. 10^{-9}.
    c. 10^{-12}.
    d. 10^{-6}.

11. Positive powers of 10
    a. indicate numbers less than 1.
    b. are not used with engineering notation.
    c. indicate numbers greater than 1.
    d. are used only with scientific notation.

12. 10^1 equals
    a. 0.
    b. 10.
    c. 1.
    d. none of the above.

13. In engineering notation, the number 0.000452 is expressed as
    a. 452×10^{-6}.
    b. 4.52×10^{-4}.
    c. 4.52×10^{-6}.
    d. 0.452×10^{-3}.

14. $(40 \times 10^2) + (5.0 \times 10^3)$ equals
    a. 90×10^3.
    b. 9.0×10^2.
    c. 20×10^5.
    d. 9.0×10^3.

15. When dividing powers of 10
    a. subtract the power of 10 in the numerator from the power of 10 in the denominator.
    b. change the sign of the power of 10 in the numerator.
    c. subtract the power of 10 in the denominator from the power of 10 in the numerator.
    d. add the exponents.

16. When multiplying powers of 10
    a. subtract the exponents.
    b. add the exponents.
    c. multiply the exponents.
    d. none of the above.

17. 10,000 μV is the same as
    a. 0.01 mV.
    b. 10 kV.
    c. 10 mV.
    d. 0.0001 V.

18. $\sqrt{81 \times 10^6}$ equals
    a. 9×10^3.
    b. 9×10^6.
    c. 9×10^2.
    d. 81×10^3.

19. $(4.0 \times 10^3)^2$ equals
    a. 16×10^5.
    b. 1.6×10^7.
    c. 4.0×10^5.
    d. 16×10^1.

20. The number 220×10^3 is the same as
    a. 2.2×10^5.
    b. 220,000.
    c. 2200.
    d. both a and b.

Questions

1. For 10^7, which is the base and which is the exponent?

2. Define: (a) scientific notation (b) engineering notation (c) decimal notation.

3. In electronics, why is engineering notation more common than scientific notation?

4. List the metric prefixes for each of the following powers of 10: (a) 10^{-3} (b) 10^3 (c) 10^{-6} (d) 10^6 (e) 10^{-9} (f) 10^9 (g) 10^{-12} (h) 10^{12}.

5. List the units and symbols for each of the following quantities: (a) frequency (b) voltage (c) power (d) resistance (e) capacitance (f) inductance (g) current.

Problems

SECTION I–1 SCIENTIFIC NOTATION

Express each of the following numbers in scientific notation:

I–1 3,500,000

I–2 678

I–3 160,000,000

I–4 0.00055

I–5 0.150

I–6 0.00000000000942

I–7 2270

I–8 42,100

I–9 0.033

I–10 0.000006

I–11 77,700,000

I–12 100

I–13 87

I–14 0.0018

I–15 0.000000095

I–16 18,200

I–17 640,000

I–18 0.011

I–19 0.00000000175

I–20 3,200,000,000,000

Convert each of the following numbers expressed in scientific notation into decimal notation.

I–21 1.65×10^{-4}

I–22 5.6×10^5

I–23 8.63×10^2

I–24 3.15×10^{-3}

I–25 1.7×10^{-9}

I–26 4.65×10^6

I–27 1.66×10^3

I–28 2.5×10^{-2}

I–29 3.3×10^{-12}

I–30 9.21×10^4

SECTION I–2 ENGINEERING NOTATION AND METRIC PREFIXES

Express each of the following numbers in engineering notation:

I–31 5500

I–32 0.0055

I–33 6,200,000

I–34 150,000

I–35 99,000

I–36 0.01

I–37 0.00075

I–38 0.55

I–39 10,000,000

I–40 0.0000000032

I–41 0.000068

I–41 92,000,000,000

I–43 270,000

I–44 0.000000000018

I–45 0.000000450

I–46 0.00010

I–47 2,570,000,000,000

I-48 20,000

I-49 0.000070

I-50 2500

Express the following values using the metric prefixes from Table I-2: (Note: The metric prefix associated with each answer must coincide with engineering notation)

I-51 1000 W

I-52 10,000 Ω

I-53 0.035 V

I-54 0.000050 A

I-55 0.000001 F

I-56 1,570,000 Hz

I-57 2,200,000 Ω

I-58 162,000 V

I-59 1,250,000,000 Hz

I-60 0.00000000033 F

I-61 0.00025 A

I-62 0.000000000061 F

I-63 0.5 W

I-64 2200 Ω

I-65 180,000 Ω

I-66 240 V

I-67 4.7 Ω

I-68 0.001 H

I-69 0.00005 W

I-70 0.0000000001 A

SECTION I-3 CONVERTING BETWEEN METRIC PREFIXES

Make the following conversions:

I-71 55,000 μA = _____ mA

I-72 10 nF = _____ pF

I-73 6800 pF = _____ μF

I-74 1.49 MHz = _____ kHz

I-75 22,000 nF = _____ μF

I-76 1500 μH = _____ mH

I-77 1.5 MΩ = _____ kΩ

I-78 2.2 GHz = _____ MHz

I-79 0.039 MΩ = _____ kΩ

I-80 5600 kΩ = _____ MΩ

I-81 7500 μA = _____ mA

I-82 1 mA = _____ μA

I-83 100 kW = _____ W

I-84 50 MW = _____ kW

I-85 4700 pF = _____ nF

I-86 560 nF = _____ μF

I-87 1296 MHz = _____ GHz

I-88 50 mH = _____ μH

I-89 7.5 μF = _____ pF

I-90 220,000 MΩ = _____ GΩ

SECTION I-4 ADDITION AND SUBTRACTION INVOLVING POWERS OF 10 NOTATION

Add the following numbers and express your answers in scientific notation:

I-91 $(25 \times 10^3) + (5.0 \times 10^4)$

I-92 $(4500 \times 10^3) + (5.0 \times 10^6)$

I-93 $(90 \times 10^{-12}) + (0.5 \times 10^{-9})$

I-94 $(15 \times 10^{-3}) + (100 \times 10^{-4})$

I-95 $(150 \times 10^{-6}) + (2.0 \times 10^{-3})$

I-96 $(150 \times 10^0) + (0.05 \times 10^3)$

Subtract the following numbers and express your answers in scientific notation:

I-97 $(100 \times 10^6) - (0.5 \times 10^8)$

I-98 $(20 \times 10^{-3}) - (5000 \times 10^{-6})$

I-99 $(180 \times 10^{-4}) - (3.5 \times 10^{-3})$

I-100 $(7.5 \times 10^2) - (0.25 \times 10^3)$

I-101 $(5.0 \times 10^4) - (240 \times 10^2)$

I-102 $(475 \times 10^{-5}) - (1500 \times 10^{-7})$

SECTION I-5 MULTIPLICATION AND DIVISION INVOLVING POWERS OF 10 NOTATION

Multiply the following numbers and express your answers in scientific notation:

I-103 $(6.0 \times 10^3) \times (3.0 \times 10^2)$

I-104 $(4.0 \times 10^{-9}) \times (2.5 \times 10^6)$

I-105 $(50 \times 10^4) \times (6.0 \times 10^3)$

I-106 $(2.2 \times 10^{-2}) \times (6.5 \times 10^0)$

I-107 $(5.0 \times 10^{-5}) \times (2.0 \times 10^{-1})$

I-108 $(100 \times 10^{-3}) \times (50 \times 10^{-6})$

Divide the following numbers and express your answers in scientific notation:

I-109 $(100 \times 10^5) \div (4.0 \times 10^2)$

I-110 $(90 \times 10^{-9}) \div (3.0 \times 10^{-5})$

I-111 $(5.0 \times 10^6) \div (40 \times 10^3)$

I-112 $(750 \times 10^{-7}) \div (3.0 \times 10^{-4})$

I-113 $(55 \times 10^9) \div (11 \times 10^2)$

I-114 $(220 \times 10^3) \div (2.0 \times 10^7)$

SECTION I-6 RECIPROCALS WITH POWERS OF 10

Find the reciprocal for each power of 10 listed.

I-115 10^4

I-116 10^{-4}

I-117 10^1

I-118 10^{-8}

I-119 10^{-7}

I-120 10^{-13}

I-121 10^{15}

I-122 10^{18}

SECTION I-7 SQUARING NUMBERS EXPRESSED IN POWERS OF 10 NOTATION

Express the following answers in scientific notation:

I-123 $(5.0 \times 10^3)^2$

I-124 $(2.5 \times 10^{-7})^2$

I-125 $(90 \times 10^4)^2$

I-126 $(7.0 \times 10^5)^2$

I-127 $(12 \times 10^{-9})^2$

I-128 $(800 \times 10^{-12})^2$

SECTION I-8 SQUARE ROOTS OF NUMBERS EXPRESSED IN POWERS OF 10 NOTATION

Express the following answers in scientific notation:

I-129 $\sqrt{40 \times 10^{-5}}$

I-130 $\sqrt{50 \times 10^4}$

I-131 $\sqrt{36 \times 10^{-12}}$

I-132 $\sqrt{49 \times 10^{-3}}$

I-133 $\sqrt{150 \times 10^{-5}}$

I-134 $\sqrt{35 \times 10^{-6}}$

SECTION I-9 THE SCIENTIFIC CALCULATOR

Show the steps for entering the following math problems using the calculator in fig. I-1. Display all answers in engineering notation.

I-135 $(15 \times 10^{-3}) \times (1.2 \times 10^3)$

I-136 $60 \div (1.5 \times 10^3)$

I-137 $12 \div (10 \times 10^3)$

I-138 $(5 \times 10^{-3}) \times (120 \times 10^3)$

I-139 $(6.5 \times 10^4) + (25 \times 10^3)$

I-140 $(2.5 \times 10^{-4}) - (50 \times 10^{-6})$

Answers to Knowledge Check Problems

I-1A a. 7.63×10^7
      b. 3.42×10^{-5}

I-1B a. 0.0000275
      b. 84,100

I-2A a. 440×10^3
      b. 68×10^{-3}

I-2B a. 50 mV
      b. 18 kΩ
      c. 47 pF

I-3   a. 0.022 μF = 22 nF
      b. 1410 kHz = 1.41 MHz

I-4   a. 6.15×10^4
      b. 5.25×10^{-1}

I-5   a. 2.0×10^{-6}
      b. 1.25×10^4

I-6   a. 10^{18}
      b. 10^{-1}

I-7   a. 6.25×10^{-8}

I-8   a. 7.0×10^{-2}

I-9   Set the display notation to engineering (ENG) by pressing the $\boxed{\text{2ndF}}$ and $^{\text{FSE}}\boxed{\cdot}$ keys in succession until ENG appears on the LCD display. Next, enter the problem using the following keying sequence: $\boxed{2}\boxed{4}\boxed{\div}\boxed{1}\boxed{\text{EXP}}\boxed{3}\boxed{=}$. The answer will be displayed as 24.000×10^{-03}.

Answers to Self-Reviews

I-1   **a.** Negative Powers of 10
    b. Positive Powers of 10
    c. True
    d. (a) 1.35×10^4
      (b) 8.25×10^{-3}
      (c) 9.56×10^7
      (d) 1.04×10^{-1}
    e. (a) 0.00000046   (b) 3330
      (c) 540,000,000   (d) 0.0254

I-2   **a.** (a) 36×10^6 (b) 85×10^{-3}
      (c) 39.3×10^3 (d) 93×10^{-6}
    b. (a) nano (n)   (b) mega (M)
      (c) pico (p)   (d) kilo (k)
      (e) none
    c. (a) $10\ \mu A$ (b) $2.2\ M\Omega$
      (c) 45 nV   (d) $5.6\ k\Omega$
      (e) 18 W

I-3   **a.** True
    b. (a) $2.2\ M\Omega = 2200\ k\Omega$
      (b) 47,000 pF = 47 nF
      (c) $2500\ \mu A = 2.5\ mA$
      (d) $6.25\ mW = 6250\ \mu W$

I-4   **a.** (a) 5.97×10^7 (b) 5.0×10^{-2}
    b. (a) 4.5×10^5 (b) 5.0×10^{-1}

I-5   **a.** (a) 1.32×10^{-4} (b) 8.1×10^{-3}
    b. (a) 2.5×10^4 (b) 3.0×10^{-3}

I-6   **a.** (a) 10^4 (b) 10^{-9}
      (c) 10^{18} (d) 10^0

I-7   **a.** (a) 1.6×10^{-3} (b) 3.6×10^{11}
      (c) 4.0×10^{-6}

I-8   **a.** (a) 6.0×10^2 (b) 4.0×10^{-2}
      (c) 5.0×10^{-4}

I-9   **a.** True
    b. Set the display notation to engineering (ENG) by pressing the 2ndF and ^{FSE} . keys in succession until ENG appears on the LCD display. Next, enter the problem using the following keying sequence: 3 . 3 EXP ± 3 × 1 0 EXP 3 =. The calculator will display the answer as 33.000×10^{00}.

Electricity

We see applications of electricity all around us, especially in the electronic products we own and operate every day. For example, we depend on electricity for lighting, heating, air conditioning and for the operation of our vehicles, cell phones, appliances, computers, and home entertainment systems to name a few. The applications of electricity are extensive and almost limitless to the imagination.

Although there are many applications of electricity, electricity itself can be explained in terms of electric charge, voltage, and current. In this chapter, you will be introduced to the basic concepts of electricity which include a discussion of the following topics: basic atomic structure, the coulomb unit of electric charge, the volt unit of potential difference, the ampere unit of current, and the ohm unit of resistance. You will also be introduced to conductors, semiconductors, insulators, and the basic characteristics of an electric circuit.

Outline

Objectives

After studying this chapter you should be able to:

- *List* the two basic particles of electric charge.
- *Describe* the basic structure of the atom.
- *Define* the terms *conductor, insulator,* and *semiconductor* and give examples of each.
- *Define* the coulomb unit of electric charge.
- *Define* potential difference and voltage and list the unit of each.
- *Define* current and list its unit of measure.
- *Describe* the difference between voltage and current.
- *Define* resistance and conductance and list the unit of each.
- *List* three important characteristics of an electric circuit.
- *Define* the difference between electron flow and conventional current.
- *Describe* the difference between direct and alternating current.

Important Terms

alternating current (ac)	conventional current	element	potential difference
ampere	coulomb	free electron	proton
atom	current	insulator	resistance
atomic number	dielectric	ion	semiconductor
circuit	direct current (dc)	molecules	siemens
compound	electron	neutron	static electricity
conductance	electron flow	nucleus	volt
conductor	electron valence	ohm	

Figure 1–1 Positive and negative polarities for the voltage output of a typical battery.

Negative –   Positive +

Figure 1–2 Electron and proton in hydrogen (H) atom.

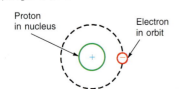

Proton in nucleus   Electron in orbit

1–1 Negative and Positive Polarities

We see the effects of electricity in a battery, static charge, lightning, radio, television, and many other applications. What do they all have in common that is electrical in nature? The answer is basic particles of electric charge with opposite *polarities*. All the materials we know, including solids, liquids, and gases, contain two basic particles of electric charge: the *electron* and the *proton*. An electron is the smallest amount of electric charge having the characteristic called *negative polarity*. The proton is a basic particle with *positive polarity*.

The negative and positive polarities indicate two opposite characteristics that seem to be fundamental in all physical applications. Just as magnets have north and south poles, electric charges have the opposite polarities labeled negative and positive. The opposing characteristics provide a method of balancing one against the other to explain different physical effects.

It is the arrangement of electrons and protons as basic particles of electricity that determines the electrical characteristics of all substances. As an example, this paper has electrons and protons in it. There is no evidence of electricity, though, because the number of electrons equals the number of protons. In that case, the opposite electrical forces cancel, making the paper electrically neutral. The neutral condition means that opposing forces are exactly balanced, without any net effect either way.

When we want to use the electrical forces associated with the negative and positive charges in all matter, work must be done to separate the electrons and protons. Changing the balance of forces produces evidence of electricity. A battery, for instance, can do electrical work because its chemical energy separates electric charges to produce an excess of electrons at its negative terminal and an excess of protons at its positive terminal. With separate and opposite charges at the two terminals, electric energy can be supplied to a circuit connected to the battery. Fig. 1–1 shows a battery with its negative (−) and positive (+) terminals marked to emphasize the two opposite polarities.

■ *1–1 Self-Review*

Answers at end of chapter.

a. Is the charge of an electron positive or negative?
b. Is the charge of a proton positive or negative?
c. Is it true or false that the neutral condition means equal positive and negative charges?

1–2 Electrons and Protons in the Atom

Although there are any number of possible methods by which electrons and protons might be grouped, they assemble in specific atomic combinations for a stable arrangement. (An atom is the smallest particle of the basic elements which forms the physical substances we know as solids, liquids, and gases.) Each stable combination of electrons and protons makes one particular type of atom. For example, Fig. 1–2 illustrates the electron and proton structure of one atom of the gas, hydrogen. This atom consists of a central mass called the *nucleus* and one electron outside. The proton in the nucleus makes it the massive and stable part of the atom because a proton is 1840 times heavier than an electron.

In Fig. 1–2, the one electron in the hydrogen atom is shown in an orbital ring around the nucleus. To account for the electrical stability of the atom, we can consider the electron as spinning around the nucleus, as planets revolve

Figure 1–3 Atomic structure showing the nucleus and its orbital rings of electrons. (*a*) Carbon (C) atom has six orbital electrons to balance six protons in nucleus. (*b*) Copper (Cu) atom has 29 protons in nucleus and 29 orbital electrons.

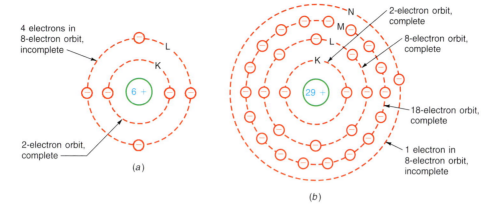

around the sun. Then the electrical force attracting the electrons in toward the nucleus is balanced by the mechanical force outward on the rotating electron. As a result, the electron stays in its orbit around the nucleus.

In an atom that has more electrons and protons than hydrogen, all protons are in the nucleus, and all the electrons are in one or more outside rings. For example, the carbon atom illustrated in Fig. 1–3*a* has six protons in the nucleus and six electrons in two outside rings. The total number of electrons in the outside rings must equal the number of protons in the nucleus in a neutral atom.

The distribution of electrons in the orbital rings determines the atom's electrical stability. Especially important is the number of electrons in the ring farthest from the nucleus. This outermost ring requires eight electrons for stability, except when there is only one ring, which has a maximum of two electrons.

In the carbon atom in Fig. 1–3*a*, with six electrons, there are just two electrons in the first ring because two is its maximum number. The remaining four electrons are in the second ring, which can have a maximum of eight electrons.

As another example, the copper atom in Fig. 1–3*b* has only one electron in the last ring, which can include eight electrons. Therefore, the outside ring of the copper atom is less stable than the outside ring of the carbon atom.

When many atoms are close together in a copper wire, the outermost orbital electrons are not sure to which atoms they belong. They can migrate easily from one atom to another at random. Such electrons that can move freely from one atom to the next are often called *free electrons*. This freedom accounts for the ability of copper to conduct electricity very easily. It is the movement of free electrons that provides electric current in a metal conductor.

The net effect in the wire itself without any applied voltage, however, is zero because of the random motion of the free electrons. When voltage is applied, it forces all the free electrons to move in the same direction to produce electron flow, which is an electric current.

Conductors, Insulators, and Semiconductors

When electrons can move easily from atom to atom in a material, the material is a *conductor*. In general, all metals are good conductors, with silver the best and copper second. Their atomic structure allows free movement of the outermost orbital electrons. Copper wire is generally used for practical conductors because it costs much less than silver. The purpose of using conductors is to allow electric current to flow with minimum opposition.

GOOD TO KNOW

Conductors have many free electrons whereas insulators have very few or none at all.

The wire conductor is used only to deliver current produced by the voltage source to a device that needs the current to function. As an example, a bulb lights only when current flows through the filament.

A material with atoms in which the electrons tend to stay in their own orbits is an *insulator* because it cannot conduct electricity very easily. However, insulators can hold or store electricity better than conductors. An insulating material, such as glass, plastic, rubber, paper, air, or mica, is also called a *dielectric,* meaning it can store electric charge.

Insulators can be useful when it is necessary to prevent current flow. In addition, for applications requiring the storage of electric charge, as in capacitors, a dielectric material must be used because a good conductor cannot store any charge.

Carbon can be considered a semiconductor, conducting less than metal conductors but more than insulators. In the same group are germanium and silicon, which are commonly used for transistors and other semiconductor components. Practically all transistors are made of silicon.

Elements

The combinations of electrons and protons forming stable atomic structures result in different kinds of elementary substances having specific characteristics. A few familiar examples are the elements hydrogen, oxygen, carbon, copper, and iron. An *element* is defined as a substance that cannot be decomposed any further by chemical action. The atom is the smallest particle of an element that still has the same characteristics as the element. *Atom* is a Greek word meaning a "particle too small to be subdivided." As an example of the fact that atoms are too small to be visible, a particle of carbon the size of a pinpoint contains many billions of atoms. The electrons and protons within the atom are even smaller.

Table 1–1 lists some more examples of elements. These are just a few out of a total of 112. Notice how the elements are grouped. The metals listed across the top row are all good conductors of electricity. Each has an atomic structure with an unstable outside ring that allows many free electrons.

Table 1–1	Examples of the Chemical Elements			
Group	**Element**	**Symbol**	**Atomic Number**	**Electron Valence**
Metal conductors, in order of conductance	Silver	Ag	47	+1
	Copper	Cu	29	+1*
	Gold	Au	79	+1*
	Aluminum	Al	13	+3
	Iron	Fe	26	+2*
Semiconductors	Carbon	C	6	±4
	Silicon	Si	14	±4
	Germanium	Ge	32	±4
Active gases	Hydrogen	H	1	±1
	Oxygen	O	8	−2
Inert gases	Helium	He	2	0
	Neon	Ne	10	0

* Some metals have more than one valence number in forming chemical compounds. Examples are cuprous or cupric copper, ferrous or ferric iron, and aurous or auric gold.

Semiconductors have four electrons in the outermost ring. This means that they neither gain nor lose electrons but share them with similar atoms. The reason is that four is exactly halfway to the stable condition of eight electrons in the outside ring.

The inert gas neon has a complete outside ring of eight electrons, which makes it chemically inactive. Remember that eight electrons in the outside ring is a stable structure.

Molecules and Compounds

A group of two or more atoms forms a molecule. For instance, two atoms of hydrogen (H) form a hydrogen molecule (H_2). When hydrogen unites chemically with oxygen, the result is water (H_2O), which is a compound. A compound, then, consists of two or more elements. The molecule is the smallest unit of a compound with the same chemical characteristics. We can have molecules for either elements or compounds. However, atoms exist only for elements.

1–2 Self-Review
Answers at end of chapter.
a. Which have more free electrons: conductors or insulators?
b. Which is the best conductor: silver, carbon, or iron?
c. Which is a semiconductor: copper, silicon, or neon?

1–3 Structure of the Atom

Our present planetary model of the atom was proposed by Niels Bohr in 1913. His contribution was joining the new ideas of a nuclear atom developed by Lord Rutherford (1871–1937) with the quantum theory of radiation developed by Max Planck (1858–1947) and Albert Einstein (1879–1955).

As illustrated in Figs. 1–2 and 1–3, the nucleus contains protons for all the positive charge in the atom. The number of protons in the nucleus is equal to the number of planetary electrons. Then the positive and negative charges are balanced because the proton and electron have equal and opposite charges. The orbits for the planetary electrons are also called *shells* or *energy levels*.

Atomic Number

This gives the number of protons or electrons required in the atom for each element. For the hydrogen atom in Fig. 1–2, the atomic number is one, which means that the nucleus has one proton balanced by one orbital electron. Similarly, the carbon atom in Fig. 1–3 with atomic number six has six protons in the nucleus and six orbital electrons. The copper atom has 29 protons and 29 electrons because its atomic number is 29. The atomic number listed for each of the elements in Table 1–1 indicates the atomic structure.

Orbital Rings

The planetary electrons are in successive shells called K, L, M, N, O, P, and Q at increasing distances outward from the nucleus. Each shell has a maximum number of electrons for stability. As indicated in Table 1–2, these stable shells correspond to inert gases, such as helium and neon.

The K shell, closest to the nucleus, is stable with two electrons, corresponding to the atomic structure for the inert gas, helium. Once the stable number of electrons has filled a shell, it cannot take any more electrons. The

Table 1–2	Shells of Orbital Electrons in the Atom	
Shell	**Maximum Electrons**	**Inert Gas**
K	2	Helium
L	8	Neon
M	8 (up to calcium) or 18	Argon
N	8, 18, or 32	Krypton
O	8 or 18	Xenon
P	8 or 18	Radon
Q	8	—

atomic structure with all its shells filled to the maximum number for stability corresponds to an inert gas.

Elements with a higher atomic number have more planetary electrons. These are in successive shells, tending to form the structure of the next inert gas in the periodic table. (The periodic table is a very useful grouping of all elements according to their chemical properties.) After the K shell has been filled with two electrons, the L shell can take up to eight electrons. Ten electrons filling the K and L shells is the atomic structure for the inert gas, neon.

The maximum number of electrons in the remaining shells can be 8, 18, or 32 for different elements, depending on their place in the periodic table. The maximum for an outermost shell, though, is always eight.

To illustrate these rules, we can use the copper atom in Fig. 1–3b as an example. There are 29 protons in the nucleus balanced by 29 planetary electrons. This number of electrons fills the K shell with two electrons, corresponding to the helium atom, and the L shell with eight electrons. The 10 electrons in these two shells correspond to the neon atom, which has an atomic number of 10. The remaining 19 electrons for the copper atom then fill the M shell with 18 electrons and one electron in the outermost N shell. These values can be summarized as follows:

$$
\begin{aligned}
\text{K shell} &= 2 \text{ electrons} \\
\text{L shell} &= 8 \text{ electrons} \\
\text{M shell} &= 18 \text{ electrons} \\
\text{N shell} &= 1 \text{ electron} \\
\text{Total} &= 29 \text{ electrons}
\end{aligned}
$$

For most elements, we can use the rule that the maximum number of electrons in a filled inner shell equals $2n^2$, where n is the shell number in sequential order outward from the nucleus. Then the maximum number of electrons in the first shell is $2 \times 1 = 2$; for the second shell $2 \times 2^2 = 8$, for the third shell $2 \times 3^2 = 18$, and for the fourth shell $2 \times 4^2 = 32$. These values apply only to an inner shell that is filled with its maximum number of electrons.

GOOD TO KNOW

Each orbital ring of electrons corresponds to a different energy level; the larger the orbit, the higher the energy level of the orbiting electrons.

Electron Valence

This value is the number of electrons in an incomplete outermost shell. A completed outer shell has a valence of zero. Copper, for instance, has a valence of one, as there is one electron in the last shell, after the inner shells have been completed with their stable number. Similarly, hydrogen has a valence of one, and carbon has a valence of four. The number of outer electrons is considered positive valence because these electrons are in addition to the stable shells.

Except for H and He, the goal of valence is eight for all atoms, as each tends to form the stable structure of eight electrons in the outside ring. For this reason, valence can also be considered the number of electrons in the outside ring needed to make eight. This value is the negative valence. As examples, the valence of copper can be considered $+1$ or -7; carbon has the valence of ± 4. The inert gases have zero valence because they all have complete outer shells.

The valence indicates how easily the atom can gain or lose electrons. For instance, atoms with a valence of $+1$ can lose this one outside electron, especially to atoms with a valence of $+7$ or -1, which need one electron to complete the outside shell with eight electrons.

Subshells

Although not shown in the illustrations, all shells except K are divided into subshells. This subdivision accounts for different types of orbits in the same shell. For instance, electrons in one subshell may have elliptical orbits, and other electrons in the same main shell have circular orbits. The subshells indicate magnetic properties of the atom.

Particles in the Nucleus

A stable nucleus (that is, one that is not radioactive) contains protons and neutrons. The neutron is electrically neutral (it has no net charge). Its mass is almost the same as that of a proton.

A proton has the positive charge of a hydrogen nucleus. The charge is the same as that of an orbital electron but of opposite polarity. There are no electrons in the nucleus. Table 1–3 lists the charge and mass for these three basic particles in all atoms. The C in the charge column is for coulombs.

■ *1–3 Self-Review*

Answers at end of chapter.

a. An element with 14 protons and 14 electrons has what atomic number?
b. What is the electron valence of an element with an atomic number of 3?
c. Except for H and He, what is the goal of valence for all atoms?

Table 1–3	Stable Particles in the Atom	
Particle	**Charge**	**Mass**
Electron, in orbital shells	0.16×10^{-18} C, negative	9.108×10^{-28} g
Proton, in nucleus	0.16×10^{-18} C, positive	1.672×10^{-24} g
Neutron, in nucleus	None	1.675×10^{-24} g

Figure 1–4 The coulomb (C) unit of electric charge. (*a*) Quantity of 6.25×10^{18} excess electrons for a negative charge of 1C. (*b*) Same amount of protons for a positive charge of 1C, caused by removing electrons from neutral atoms.

1 C of excess electrons in dielectric

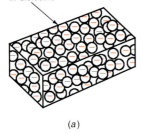

(*a*)

1 C of excess protons in dielectric

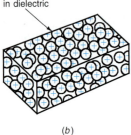

(*b*)

1–4 The Coulomb Unit of Electric Charge

If you rub a hard rubber pen or comb on a sheet of paper, the rubber will attract a corner of the paper if it is free to move easily. The paper and rubber then give evidence of a static electric charge. The work of rubbing resulted in separating electrons and protons to produce a charge of excess electrons on the surface of the rubber and a charge of excess protons on the paper.

Because paper and rubber are dielectric materials, they hold their extra electrons or protons. As a result, the paper and rubber are no longer neutral, but each has an electric charge. The resultant electric charges provide the force of attraction between the rubber and the paper. This mechanical force of attraction or repulsion between charges is the fundamental method by which electricity makes itself evident.

Any charge is an example of *static electricity* because the electrons or protons are not in motion. There are many examples. When you walk across a wool rug, your body becomes charged with an excess of electrons. Similarly, silk, fur, and glass can be rubbed to produce a static charge. This effect is more evident in dry weather, because a moist dielectric does not hold its charge so well. Also, plastic materials can be charged easily, which is why thin, light-weight plastics seem to stick to everything.

The charge of many billions of electrons or protons is necessary for common applications of electricity. Therefore, it is convenient to define a practical unit called the *coulomb* (C) as equal to the charge of 6.25×10^{18} electrons or protons stored in a dielectric (see Fig. 1–4). The analysis of static charges and their forces is called *electrostatics*.

The symbol for electric charge is Q or q, standing for quantity. For instance, a charge of 6.25×10^{18} electrons is stated as $Q = 1C$. This unit is named after Charles A. Coulomb (1736–1806), a French physicist, who measured the force between charges.

Negative and Positive Polarities

Historically, negative polarity has been assigned to the static charge produced on rubber, amber, and resinous materials in general. Positive polarity refers to the static charge produced on glass and other vitreous materials. On this basis, the electrons in all atoms are basic particles of negative charge because their polarity is the same as the charge on rubber. Protons have positive charge because the polarity is the same as the charge on glass.

Charges of Opposite Polarity Attract

If two small charged bodies of light weight are mounted so that they are free to move easily and are placed close to each other, one can be attracted to the other when the two charges have opposite polarity (Fig. 1–5*a*). In terms of electrons and protons, they tend to be attracted to each other by the force of attraction between opposite charges. Furthermore, the weight of an electron is only about $\frac{1}{1840}$ the weight of a proton. As a result, the force of attraction tends to make electrons move to protons.

Figure 1–5 Physical force between electric charges. (*a*) Opposite charges attract. (*b*) Two negative charges repel each other. (*c*) Two positive charges repel.

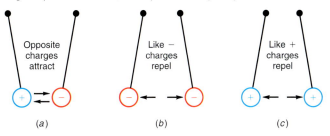

Charges of the Same Polarity Repel

In Fig. 1–5*b* and *c*, it is shown that when the two bodies have an equal amount of charge with the same polarity, they repel each other. The two negative charges repel in Fig. 1–5*b*, and two positive charges of the same value repel each other in Fig. 1–5*c*.

Polarity of a Charge

An electric charge must have either negative or positive polarity, labeled $-Q$ or $+Q$, with an excess of either electrons or protons. A neutral condition is considered zero charge. On this basis, consider the following examples, remembering that the electron is the basic particle of charge and the proton has exactly the same amount, although of opposite polarity.

Example 1–1

A neutral dielectric has added to it 12.5×10^{18} electrons. What is its charge in coulombs?

ANSWER This number of electrons is double the charge of 1 C. Therefore, $-Q = 2$ C.

Example 1–2

A dielectric has a positive charge of 12.5×10^{18} protons. What is its charge in coulombs?

ANSWER This is the same amount of charge as in Example 1 but positive. Therefore, $+Q = 2$ C.

Example 1-3

A dielectric with $+Q$ of 2 C has 12.5×10^{18} electrons added. What is its charge then?

ANSWER The 2 C of negative charge added by the electrons cancels the 2 C of positive charge, making the dielectric neutral, for $Q = 0$.

Example 1-4

A neutral dielectric has 12.5×10^{18} electrons removed. What is its charge?

ANSWER The 2 C of electron charge removed allows an excess of 12.5×10^{18} protons. Since the proton and electron have exactly the same amount of charge, now the dielectric has a positive charge of $+Q = 2$ C.

■ 1-4 Knowledge Check

Answer at end of chapter.

If 25×10^{18} electrons are added to a neutral dielectric, what is the charge stored in coulombs?

Note that we generally consider that the electrons move, rather than heavier protons. However, a loss of a given number of electrons is equivalent to a gain of the same number of protons.

Charge of an Electron

The charge of a single electron, designated Q_e, is 0.16×10^{-18} C. This value is the reciprocal of 6.25×10^{18} electrons which is the number of electrons in 1 coulomb of charge. Expressed mathematically;

$$-Q_e = 0.16 \times 10^{-18} \text{ C}$$

($-Q_e$ denotes that the charge of the electron is negative.)

It is important to note that the charge of a single proton, designated Q_P, is also equal to 0.16×10^{-18} C. However, its polarity is positive instead of negative.

In some cases, the charge of a single electron or proton will be expressed in scientific notation. In this case, $-Q_e = 1.6 \times 10^{-19}$ C. It is for convenience only that Q_e or Q_P is sometimes expressed as 0.16×10^{-18} C instead of 1.6×10^{-19} C. The convenience lies in the fact that 0.16 is the reciprocal of 6.25 and 10^{-18} is the reciprocal of 10^{18}.

Figure 1–6 Arrows to indicate electric field around a stationary charge Q.

Electric lines
of force

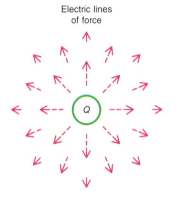

The Electric Field of a Static Charge

The ability of an electric charge to attract or repel another charge is a physical force. To help visualize this effect, lines of force are used, as shown in Fig. 1–6. All the lines form the electric field. The lines and the field are imaginary, since they cannot be seen. Just as the field of the force of gravity is not visible, however, the resulting physical effects prove that the field is there.

Each line of force in Fig. 1–6 is directed outward to indicate repulsion of another charge in the field with the same polarity as Q, either positive or negative. The lines are shorter farther away from Q to indicate that the force decreases inversely as the square of the distance. The larger the charge, the greater the force. These relations describe Coulomb's law of electrostatics.

■ 1–4 Self-Review

Answers at end of chapter.

a. How many electron charges are there in the practical unit of 1 coulomb?
b. How much is the charge in coulombs for a surplus of 18.75×10^{18} electrons?
c. Do opposite electric charges attract or repel each other?

1–5 The Volt Unit of Potential Difference

Potential refers to the possibility of doing work. Any charge has the potential to do the work of moving another charge by either attraction or repulsion. When we consider two unlike charges, they have a *difference of potential*.

A charge is the result of work done in separating electrons and protons. Because of the separation, stress and strain are associated with opposite charges, since normally they would be balancing each other to produce a neutral condition. We could consider that the accumulated electrons are drawn tight and are straining themselves to be attracted toward protons to return to the neutral condition. Similarly, the work of producing the charge causes a condition of stress in the protons, which are trying to attract electrons and return to the neutral condition. Because of these forces, the charge of electrons or protons has potential because it is ready to give back the work put into producing the charge. The force between charges is in the electric field.

Potential between Different Charges

When one charge is different from the other, there must be a difference of potential between them. For instance, consider a positive charge of 3 C, shown at the right in Fig. 1–7a. The charge has a certain amount of potential, corresponding to the amount of work this charge can do. The work to be done is moving some electrons, as illustrated.

Assume that a charge of 1 C can move 3 electrons. Then the charge of +3 C can attract nine electrons toward the right. However, the charge of +1 C at the opposite side can attract three electrons toward the left. The net result, then, is that six electrons can be moved toward the right to the more positive charge.

In Fig. 1–7b, one charge is 2 C, and the other charge is neutral with 0 C. For the difference of 2 C, again 2 × 3 or 6 electrons can be attracted to the positive side.

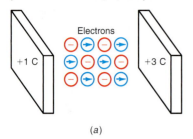

(a)

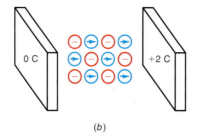

(b)

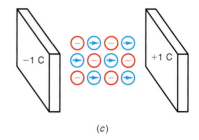

(c)

In Fig. 1–7c, the difference between the charges is still 2 C. The +1 C attracts three electrons to the right side. The −1 C repels three electrons to the right side also. This effect is really the same as attracting six electrons.

Therefore, the net number of electrons moved in the direction of the more positive charge depends on the difference of potential between the two charges. This potential difference is the same for all three cases in Fig. 1–7. Potential difference is often abbreviated PD.

The only case without any potential difference between charges occurs when they both have the same polarity and are equal in amount. Then the repelling and attracting forces cancel, and no work can be done in moving electrons between the two identical charges.

The Volt Unit

The *volt unit* of potential difference is named after Alessandro Volta (1745–1827). Fundamentally, the volt is a measure of the amount of work or energy needed to move an electric charge. By definition, when 0.7376 foot-pound (ft · lb) of work is required to move 6.25×10^{18} electrons between two points, the potential difference between those two points is one volt. (Note that 6.25×10^{18} electrons make up one coulomb of charge.) The metric unit of work or energy is the joule (J). One joule is the same amount of work or energy as 0.7376 ft · lb. Therefore, we can say that the potential difference between two points is one volt when one joule of energy is expended in moving one coulomb of charge between those two points. Expressed as a formula, $1 \text{ V} = \dfrac{1 \text{ J}}{1 \text{ C}}$.

In electronics, potential difference is commonly referred to as voltage, with the symbol V. Remember, though, that voltage is the potential difference between two points and that two terminals are necessary for a potential difference to exist. A potential difference cannot exist at only one point!

Consider the 2.2-V lead-acid cell in Fig. 1–8a. Its output of 2.2 V means that this is the amount of potential difference between the two terminals. The lead-acid cell, then, is a voltage source, or a source of electromotive force (emf). The schematic symbol for a battery or dc voltage source is shown in Fig. 1–8b.

Sometimes the symbol E is used for emf, but the standard symbol V represents any potential difference. This applies either to the voltage generated by a source or to the voltage drop across a passive component such as a resistor.

It may be helpful to think of voltage as an electrical pressure or force. The higher the voltage, the more electrical pressure or force. The electrical pressure of voltage is in the form of the attraction and repulsion of an electric charge such as an electron.

MultiSim Figure 1–8
Chemical cell as a voltage source.
(a) Voltage output is the potential difference between the two terminals.
(b) Schematic symbol of any dc voltage source with constant polarity. Longer line indicates positive side.

(a)

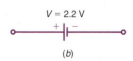

V = 2.2 V

(b)

The general equation for any voltage can be stated as

$$V = \frac{J}{C} \qquad (1\text{--}1)$$

Let's take a look at an example.

Example 1-5

What is the output voltage of a battery that expends 3.6 J of energy in moving 0.5 C of charge?

ANSWER Use Equation 1-1.

$$V = \frac{J}{C}$$
$$= \frac{3.6\ J}{0.5\ C}$$
$$= 7.2\ V$$

■ *1–5 Knowledge Check*

> *Answer at end of chapter.*
>
> **What is the output voltage of a battery that expends 15 J of energy in moving 5 C of charge?**

■ *1–5 Self-Review*

> *Answers at end of chapter.*
>
> a. **How much potential difference is there between two identical charges?**
> b. **If 27 J of energy is expended in moving 3 C of charge between two points, how much voltage is there between those two points?**

1–6 Charge in Motion Is Current

When the potential difference between two charges forces a third charge to move, the charge in motion is an *electric current*. To produce current, therefore, charge must be moved by a potential difference.

In solid materials, such as copper wire, free electrons are charges that can be forced to move with relative ease by a potential difference, since they require relatively little work to be moved. As illustrated in Fig. 1–9, if a potential difference is connected across two ends of a copper wire, the applied voltage forces the free electrons to move. This current is a drift of electrons, from the point of negative charge at one end, moving through the wire, and returning to the positive charge at the other end.

To illustrate the drift of free electrons through the wire shown in Fig. 1–9, each electron in the middle row is numbered, corresponding to a

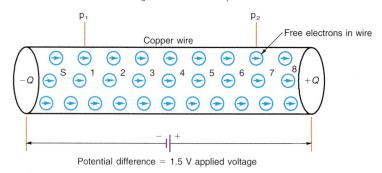

Figure 1–9 Potential difference across two ends of wire conductor causes drift of free electrons throughout the wire to produce electric current.

Potential difference = 1.5 V applied voltage

copper atom to which the free electron belongs. The electron at the left is labeled S to indicate that it comes from the negative charge of the source of potential difference. This one electron S is repelled from the negative charge $-Q$ at the left and is attracted by the positive charge $+Q$ at the right. Therefore, the potential difference of the voltage source can make electron S move toward atom 1. Now atom 1 has an extra electron. As a result, the free electron of atom 1 can then move to atom 2. In this way, there is a drift of free electrons from atom to atom. The final result is that the one free electron labeled 8 at the extreme right in Fig. 1–9 moves out from the wire to return to the positive charge of the voltage source.

Considering this case of just one electron moving, note that the electron returning to the positive side of the voltage source is not the electron labeled S that left the negative side. All electrons are the same, however, and have the same charge. Therefore, the drift of free electrons resulted in the charge of one electron moving through the wire. This charge in motion is the current. With more electrons drifting through the wire, the charge of many electrons moves, resulting in more current.

The current is a continuous flow of electrons. Only the electrons move, not the potential difference. For ordinary applications, where the wires are not long lines, the potential difference produces current instantaneously through the entire length of wire. Furthermore, the current must be the same at all points of the wire at any time.

Potential Difference Is Necessary to Produce Current

The number of free electrons that can be forced to drift through a wire to produce a moving charge depends upon the amount of potential difference across the wire. With more applied voltage, the forces of attraction and repulsion can make more free electrons drift, producing more charge in motion. A larger amount of charge moving during a given period of time means a higher value of current. Less applied voltage across the same wire results in a smaller amount of charge in motion, which is a smaller value of current. With zero potential difference across the wire, there is no current.

Two cases of zero potential difference and no current can be considered to emphasize that potential difference is needed to produce current. Assume that the copper wire is by itself, not connected to any voltage source, so that there is no potential difference across the wire. The free electrons in the wire can move from atom to atom, but this motion is random, without any organized

drift through the wire. If the wire is considered as a whole, from one end to the other, the current is zero.

As another example, suppose that the two ends of the wire have the same potential. Then free electrons cannot move to either end because both ends have the same force and there is no current through the wire. A practical example of this case of zero potential difference would be to connect both ends of the wire to just one terminal of a battery. Each end of the wire would have the same potential, and there would be no current. The conclusion, therefore, is that two connections to two points at different potentials are needed to produce a current.

The Ampere of Current

Since current is the movement of charge, the unit for stating the amount of current is defined in rate of flow of charge. When the charge moves at the rate of 6.25×10^{18} electrons flowing past a given point per second, the value of the current is one *ampere* (A). This is the same as one coulomb of charge per second. The *ampere unit* of current is named after André M. Ampère (1775–1836).

Referring back to Fig. 1–9, note that if 6.25×10^{18} free electrons move past p_1 in 1 second (s), the current is 1 A. Similarly, the current is 1 A at p_2 because the electron drift is the same throughout the wire. If twice as many electrons moved past either point in 1 s, the current would be 2 A.

The symbol for current is I or i for intensity, since the current is a measure of how intense or concentrated the electron flow is. Two amperes of current in a copper wire is a higher intensity than one ampere; a greater concentration of moving electrons results because of more electrons in motion. Sometimes current is called *amperage*. However, the current in electronic circuits is usually in smaller units, milliamperes and microamperes.

How Current Differs from Charge

Charge is a quantity of electricity accumulated in a dielectric, which is an insulator. The charge is static electricity, at rest, without any motion. When the charge moves, usually in a conductor, the current I indicates the intensity of the electricity in motion. This characteristic is a fundamental definition of current:

$$I = \frac{Q}{T} \tag{1–2}$$

where I is the current in amperes, Q is in coulombs, and time T is in seconds. It does not matter whether the moving charge is positive or negative. The only question is how much charge moves and what its rate of motion is.

In terms of practical units,

$$1\ A = \frac{1\ C}{1\ s} \tag{1–3}$$

One ampere of current results when one coulomb of charge moves past a given point in 1 s. In summary, Q represents a specific amount or quantity of electric charge, whereas the current I represents the rate at which the electric charge, such as electrons, is moving. The difference between electric charge and current is similar to the difference between miles and miles per hour.

Example 1-6

The charge of 12 C moves past a given point every second. How much is the intensity of charge flow?

ANSWER

$$I = \frac{Q}{T} = \frac{12\ C}{1\ s}$$
$$I = 12\ A$$

Example 1-7

The charge of 5 C moves past a given point in 1 s. How much is the current?

ANSWER

$$I = \frac{Q}{T} = \frac{5\ C}{1\ s}$$
$$I = 5\ A$$

■ *1-6 Knowledge Check*

Answer at end of chapter.

A charge of 0.1 C moves past a given point every 0.05 seconds. How much is the current?

The fundamental definition of current can also be used to consider the charge as equal to the product of the current multiplied by the time. Or

$$Q = I \times T \qquad\qquad (1\text{–}4)$$

In terms of practical units,

$$1\ C = 1\ A \times 1\ s \qquad\qquad (1\text{–}5)$$

One coulomb of charge results when one ampere of current accumulates charge during one second. The charge is generally accumulated in the dielectric of a capacitor or at the electrodes of a battery.

For instance, we can have a dielectric connected to conductors with a current of 0.4 A. If the current can deposit electrons for 0.2 s, the accumulated charge in the dielectric will be

$$Q = I \times T = 0.4\ A \times 0.2\ s$$
$$Q = 0.08\ C$$

The formulas $Q = IT$ for charge and $I = Q/T$ for current illustrate the fundamental nature of Q as an accumulation of static charge in an insulator, whereas I measures the intensity of moving charges in a conductor. Furthermore, current I is different from voltage V. You can have V without I, but you cannot have current without an applied voltage.

The General Nature of Current

The moving charges that provide current in metal conductors such as copper wire are the free electrons of the copper atoms. In this case, the moving charges have negative polarity. The direction of motion between two terminals for this *electron current,* therefore, is toward the more positive end. It is important to note, however, that there are examples of positive charges in motion. Common applications include current in liquids, gases, and semiconductors. For the current resulting from the motion of positive charges, its direction is opposite from the direction of electron flow. Whether negative or positive charges move, though, the current is still defined fundamentally as Q/T. Note also that the current is provided by free charges, which are easily moved by an applied voltage.

■ *1-6 Self-Review*
Answers at end of chapter.

a. The flow of 2 C/s of electron charges is how many amperes of current?
b. The symbol for current is I for intensity. (True/False)?
c. How much is the current with zero potential difference?

1–7 Resistance Is Opposition to Current

The fact that a wire conducting current can become hot is evidence that the work done by the applied voltage in producing current must be accomplished against some form of opposition. This opposition, which limits the amount of current that can be produced by the applied voltage, is called *resistance.* Conductors have very little resistance; insulators have a large amount of resistance.

The atoms of a copper wire have a large number of free electrons, which can be moved easily by a potential difference. Therefore, the copper wire has little opposition to the flow of free electrons when voltage is applied, corresponding to low resistance.

Carbon, however, has fewer free electrons than copper. When the same amount of voltage is applied to carbon as to copper, fewer electrons will flow. Just as much current can be produced in carbon by applying more voltage. For the same current, though, the higher applied voltage means that more work is necessary, causing more heat. Carbon opposes the current more than copper, therefore, and has higher resistance.

The Ohm

The practical unit of resistance is the *ohm.* A resistance that develops 0.24 calorie of heat when one ampere of current flows through it for one second has one ohm of opposition. As an example of a low resistance, a good conductor such as copper wire can have a resistance of 0.01 Ω for a 1-ft length. The resistance-wire heating element in a 600-W 120-V toaster has a resistance of 24 Ω, and the tungsten filament in a 100-W 120-V light bulb has a resistance of 144 Ω. The ohm unit is named after Georg Simon Ohm (1787–1854), a German physicist.

Figure 1–10*a* shows a wire-wound resistor. Resistors are also made with powdered carbon. They can be manufactured with values from a few ohms to millions of ohms.

PIONEERS
IN ELECTRONICS

The unit of measure for resistance, the ohm, was named for German physicist *Georg Simon Ohm (1787–1854).* Ohm is also known for his development of Ohm's law: $I = \dfrac{V}{R}$.

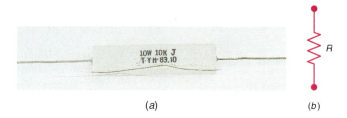

(*a*)   (*b*)

The symbol for resistance is *R*. The abbreviation used for the ohm unit is the Greek letter *omega,* written as Ω. In diagrams, resistance is indicated by a zigzag line, as shown by *R* in Fig. 1–10*b*.

Conductance

The opposite of resistance is *conductance*. The lower the resistance, the higher the conductance. Its symbol is *G*, and the unit is the *siemens* (S), named after Ernst von Siemens (1816–1892), a German inventor. (The old unit name for conductance is *mho,* which is *ohm* spelled backward.)

Specifically, *G* is the reciprocal of *R*, or $G = \dfrac{1}{R}$. Also, $R = \dfrac{1}{G}$.

Example 1-8

Calculate the resistance for the following conductance values: (a) 0.05 S (b) 0.1 S

ANSWER

$$\text{(a)} \quad R = \frac{1}{G}$$

$$= \frac{1}{0.05\ S}$$

$$= 20\ \Omega$$

$$\text{(b)} \quad R = \frac{1}{G}$$

$$= \frac{1}{0.1\ S}$$

$$= 10\ \Omega$$

Notice that a higher value of conductance corresponds to a lower value of resistance.

Example 1-9

Calculate the conductance for the following resistance values: (a) 1 kΩ (b) 5 kΩ.

ANSWER

(a) $G = \dfrac{1}{R}$

$= \dfrac{1}{1000 \ \Omega}$

$= 0.001 \text{ S or } 1 \text{ mS}$

(b) $G = \dfrac{1}{R}$

$= \dfrac{1}{5000 \ \Omega}$

$= 0.0002 \text{ S or } 200 \ \mu\text{S}$

Notice that a higher value of resistance corresponds to a lower value of conductance.

Figure 1-11 Example of an electric circuit with a battery as a voltage source connected to a light bulb as a resistance. (*a*) Wiring diagram of the closed path for current. (*b*) Schematic diagram of the circuit.

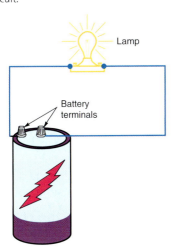

Lamp

Battery terminals

(*a*)

I

Voltage Source

+

−

$V = 1.5$ V

Resistance load
$R = 300 \ \Omega$

I

(*b*)

■ 1-7 Knowledge Check

Answers at end of chapter.

a. Calculate the resistance corresponding to a conductance of 0.004 S.
b. Calculate the conductance corresponding to a resistance of 30 Ω.

■ 1-7 Self-Review

Answers at end of chapter.

a. Which has more resistance, carbon or copper?
b. With the same voltage applied, which resistance will allow more current, 4.7 Ω or 5000 Ω?
c. What is the conductance value in siemens units for a 10-Ω R?

1-8 The Closed Circuit

In applications requiring current, the components are arranged in the form of a *circuit,* as shown in Fig. 1–11. A circuit can be defined as a path for current flow. The purpose of this circuit is to light the incandescent bulb. The bulb lights when the tungsten-filament wire inside is white hot, producing an incandescent glow.

The tungsten filament cannot produce current by itself. A source of potential difference is necessary. Since the battery produces a potential difference of 1.5 V across its two output terminals, this voltage is connected across the filament of the bulb by the two wires so that the applied voltage can produce current through the filament.

In Fig. 1–11*b,* the schematic diagram of the circuit is shown. Here the components are represented by shorthand symbols. Note the symbols for the battery and resistance. The connecting wires are shown simply as straight lines because their resistance is small enough to be neglected. A resistance of less than 0.01 Ω for the wire is practically zero compared with the 300-Ω resistance

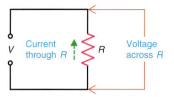

Figure 1–12 Comparison of voltage (*V*) across a resistance and the current (*I*) through *R*.

of the bulb. If the resistance of the wire must be considered, the schematic diagram includes it as additional resistance in the same current path.

Note that the schematic diagram does not look like the physical layout of the circuit. The schematic shows only the symbols for the components and their electrical connections.

Any electric circuit has three important characteristics:

1. There must be a source of potential difference. Without the applied voltage, current cannot flow.
2. There must be a complete path for current flow, from one side of the applied voltage source, through the external circuit, and returning to the other side of the voltage source.
3. The current path normally has resistance. The resistance is in the circuit either to generate heat or limit the amount of current.

How the Voltage Is Different from the Current

It is the current that moves through the circuit. The potential difference does not move.

In Fig. 1–11, the voltage across the filament resistance makes electrons flow from one side to the other. While the current is flowing around the circuit, however, the potential difference remains across the filament to do the work of moving electrons through the resistance of the filament.

The circuit is redrawn in Fig. 1–12 to emphasize the comparison between *V* and *I*. The voltage is the potential difference across the two ends of the resistance. If you want to measure the PD, just connect the two leads of a voltmeter across the resistor. However, the current is the intensity of the electron flow past any one point in the circuit. Measuring the current is not as easy. You would have to break open the path at any point and then insert the current meter to complete the circuit.

The word *across* is used with voltage because it is the potential difference between two points. There cannot be a PD at one point. However, current can be considered at one point, as the motion of charges through that point.

To illustrate the difference between *V* and *I* in another way, suppose that the circuit in Fig. 1–11 is opened by disconnecting the bulb. Now no current can flow because there is no closed path. Still, the battery has its potential difference. If you measure across the two terminals, the voltmeter will read 1.5 V even though the current is zero. This is like a battery sitting on a store shelf. Even though the battery is not producing current in a circuit, it still has a voltage output between its two terminals. This brings us to a very important conclusion: Voltage can exist without current, but current cannot exist without voltage.

The Voltage Source Maintains the Current

As current flows in a circuit, electrons leave the negative terminal of the cell or battery in Fig. 1–11, and the same number of free electrons in the conductor are returned to the positive terminal. As electrons are lost from the negative charge and gained by the positive charge, the two charges tend to neutralize each other. The chemical action inside the battery, however, continuously separates electrons and protons to maintain the negative and positive charges on the outside terminals that provide the potential difference. Otherwise, the current would neutralize the charges, resulting in no potential difference, and the current would stop. Therefore, the battery keeps the current flowing by maintaining the potential difference across the circuit. The battery is the voltage source for the circuit.

The Circuit Is a Load on the Voltage Source

We can consider the circuit as a means whereby the energy of the voltage source is carried by the current through the filament of the bulb, where the electric energy is used in producing heat energy. On this basis, the battery is the *source* in the circuit, since its voltage output represents the potential energy to be used. The part of the circuit connected to the voltage source is the *load resistance,* since it determines how much work the source will supply. In this case, the bulb's filament is the load resistance on the battery.

The current that flows through the load resistance is the *load current.* Note that a lower value of ohms for the load resistance corresponds to a higher load current. Unless noted otherwise, the term *load* by itself can be assumed generally to mean the load current. Therefore, a heavy or big load electrically means a high current load, corresponding to a large amount of work supplied by the source.

In summary, we can say that the closed circuit, normal circuit, or just a circuit is a closed path that has V to produce I with R to limit the amount of current. The circuit provides a means of using the energy of the battery as a voltage source. The battery has its potential difference V with or without the circuit. However, the battery alone is not doing any work in producing load current. The bulb alone has resistance, but without current, the bulb does not light. With the circuit, the voltage source is used to produce current to light the bulb.

Open Circuit

When any part of the path is open or broken, the circuit is incomplete because there is no conducting path. The *open circuit* can be in the connecting wires or in the bulb's filament as the load resistance. The resistance of an open circuit is infinitely high. The result is no current in an open circuit.

Short Circuit

In this case, the voltage source has a closed path across its terminals, but the resistance is practically zero. The result is too much current in a *short circuit.* Usually, the short circuit is a bypass around the load resistance. For instance, a short across the tungsten filament of a bulb produces too much current in the connecting wires but no current through the bulb. Then the bulb is shorted out. The bulb is not damaged, but the connecting wires can become hot enough to burn unless the line has a fuse as a safety precaution against too much current.

■ *1–8 Self-Review*

Answers at end of chapter.

Answer True or False for the circuit in Fig. 1–11.
a. The bulb has a PD of 1.5 V across its filament only when connected to the voltage source.
b. The battery has a PD of 1.5 V across its terminals only when connected to the bulb.
c. The battery by itself, without the wires and the bulb, has a PD of 1.5 V.

1–9 The Direction of Current

Just as a voltage source has polarity, current has a direction. The reference is with respect to the positive and negative terminals of the voltage source. The direction of the current depends on whether we consider the flow of negative electrons or the motion of positive charges in the opposite direction.

Figure 1–13 Direction of *I* in a closed circuit, shown for electron flow and conventional current. The circuit works the same way no matter which direction you consider. (*a*) Electron flow indicated with dashed arrow in diagram. (*b*) Conventional current indicated with solid arrow. (*c*) Electron flow as in (*a*) but with reversed polarity of voltage source. (*d*) Conventional *I* as in (*b*) but reversed polarity for *V.*

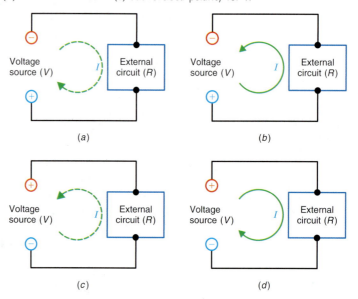

Electron Flow

As shown in Fig. 1–13*a*, the direction of electron drift for the current *I* is out from the negative side of the voltage source. The *I* flows through the external circuit with *R* and returns to the positive side of *V.* Note that this direction from the negative terminal applies to the external circuit connected to the output terminals of the voltage source. *Electron flow* is also shown in Fig. 1–13*c* with reversed polarity for *V.*

Inside the battery, the electrons move to the negative terminal because this is how the voltage source produces its potential difference. The battery is doing the work of separating charges, accumulating electrons at the negative terminal and protons at the positive terminal. Then the potential difference across the two output terminals can do the work of moving electrons around the external circuit. For the circuit outside the voltage source, however, the direction of the electron flow is from a point of negative potential to a point of positive potential.

Conventional Current

A motion of positive charges, in the opposite direction from electron flow, is considered *conventional current.* This direction is generally used for analyzing circuits in electrical engineering. The reason is based on some traditional definitions in the science of physics. By the definitions of force and work with positive values, a positive potential is considered above a negative potential. Then conventional current corresponds to a motion of positive charges "falling downhill" from a positive to a negative potential. The conventional current, therefore, is in the direction of positive charges in motion. An example is shown in Fig. 1–13*b*. The conventional *I* is out from the positive side of the voltage source, flows through the external circuit, and returns to the negative side of *V.* Conventional current is also shown in Fig. 1–13*d*, with the voltage source in reverse polarity.

Electrical engineers usually
analyze electronic circuits using
conventional current flow,
whereas electronic technicians
usually use electron flow. Both
directions of current flow produce
the same results. Which one to
use is mainly a matter of personal
preference.

Examples of Mobile Positive Charges

An ion is an atom that has either lost or gained one or more valence electrons to become electrically charged. For example, a positive ion is created when a neutral atom loses one or more valence electrons and thus becomes positively charged. Similarly, a negative ion is created when a neutral atom gains one or more valence electrons and thus becomes negatively charged. Depending on the number of valence electrons that have been added or removed, the charge of an ion may equal the charge of one electron (Q_e), two electrons ($2\,Q_e$), three electrons ($3\,Q_e$), etc. Ions can be produced by applying voltage to liquids and gases to ionize the atoms. These ions are mobile charges that can provide an electric current. Positive or negative ions are much less mobile than electrons, however, because an ion includes a complex atom with its nucleus.

An example of positive charges in motion for conventional current, therefore, is the current of positive ions in either liquids or gases. This type of current is referred to as ionization current. The positive ions in a liquid or gas flow in the direction of conventional current because they are repelled by the positive terminal of the voltage source and attracted to the negative terminal. Therefore, the mobile positive ions flow from the positive side of the voltage source to the negative side.

Another example of a mobile positive charge is the hole. Holes exist in semiconductor materials such as silicon and germanium. A hole possesses the same amount of charge as an electron but instead has positive polarity. Although the details of the hole charge are beyond the scope of this discussion, you should be aware that in semiconductors, the movement of hole charges are in the direction of conventional current.

It is important to note that protons themselves are not mobile positive charges because they are tightly bound in the nucleus of the atom and cannot be released except by nuclear forces. Therefore, a current of positive charges is a flow of either positive ions in liquids and gases or positive holes in semiconductors. Table 1–4 summarizes the different types of electric charge that can provide current in a circuit.

In this book, the current is considered as electron flow in the applications where electrons are the moving charges. A dotted or dashed arrow, as in Fig. 1–13a and c, is used to indicate the direction of electron flow for I. In Fig. 1–13b and d, the solid arrow means the direction of conventional current. These arrows are used for the unidirectional current in dc circuits. For ac circuits, the direction of current can be considered either way because I reverses direction every half-cycle with the reversals in polarity for V.

Table 1–4	Types of Electric Charges for Current			
Type of Charge	Amount of Charge	Polarity	Type of Current	Applications
Electron	$Q_e = 0.16 \times 10^{-18}$ C	Negative	Electron flow	In wire conductors
Ion	Q_e or multiples of Q_e	Positive or negative	Ion current	In liquids and gases
Hole	$Q_e = 0.16 \times 10^{-18}$ C	Positive	Hole current	In p-type semiconductors

■ *1–9 Self-Review*

Answers at end of chapter.

a. Is electron flow out from the positive or negative terminal of the voltage source?
b. Does conventional current return to the positive or negative terminal of the voltage source?
c. Is it true or false that electron flow and conventional current are in opposite directions?

1–10 Direct Current (DC) and Alternating Current (AC)

The electron flow illustrated for the circuit with a bulb in Fig. 1–11 is *direct current* because it has just one direction. The reason for the unidirectional current is that the battery maintains the same polarity of output voltage across its two terminals.

The flow of charges in one direction and the fixed polarity of applied voltage are the characteristics of a dc circuit. The current can be a flow of positive charges, rather than electrons, but the conventional direction of current does not change the fact that the charges are moving only one way.

Furthermore, the dc voltage source can change the amount of its output voltage but, with the same polarity, direct current still flows only in one direction. This type of source provides a fluctuating or pulsating dc voltage. A battery is a steady dc voltage source because it has fixed polarity and its output voltage is a steady value.

An alternating voltage source periodically reverses or alternates in polarity. The resulting *alternating current*, therefore, periodically reverses in direction. In terms of electron flow, the current always flows from the negative terminal of the voltage source, through the circuit, and back to the positive terminal, but when the generator alternates in polarity, the current must reverse its direction. The 60-cycle ac power line used in most homes is a common example. This frequency means that the voltage polarity and current direction go through 60 cycles of reversal per second.

The unit for 1 cycle per second is 1 hertz (Hz). Therefore 60 cycles per second is a frequency of 60 Hz.

The details of ac circuits are explained in Chap. 15. Direct-current circuits are analyzed first because they usually are simpler. However, the principles of dc circuits also apply to ac circuits. Both types are important because most electronic circuits include ac voltages and dc voltages. A comparison of dc and ac voltages and their waveforms is illustrated in Figs. 1–14 and 1–15. Their uses are compared in Table 1–5.

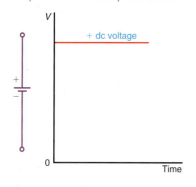

Figure 1–14 Steady dc voltage of fixed polarity, such as the output of a battery. Note schematic symbol at left.

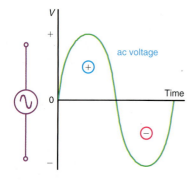

Figure 1–15 Sine wave ac voltage with alternating polarity, such as from an ac generator. Note schematic symbol at left. The ac line voltage in your home has this waveform.

■ *1–10 Self-Review*

Answers at end of chapter.

a. When the polarity of the applied voltage reverses, the direction of current flow also reverses. (True or False)
b. A battery is a dc voltage source because it cannot reverse the polarity across its output terminals. (True or False)

Table 1–5	Comparison of DC Voltage and AC Voltage	
DC Voltage	**AC Voltage**	
Fixed polarity	Reverses in polarity	
Can be steady or vary in magnitude	Varies between reversals in polarity	
Steady value cannot be stepped up or down by a transformer	Can be stepped up or down for electric power distribution	
Terminal voltages for transistor amplifiers	Signal input and output for amplifiers	
Easier to measure	Easier to amplify	
Heating effect is the same for direct or alternating current		

1–11 Sources of Electricity

There are electrons and protons in the atoms of all materials, but to do useful work, the charges must be separated to produce a *potential difference* that can make current flow. Some of the more common methods of providing electrical effects are listed here.

Static Electricity by Friction

In this method, electrons in an insulator can be separated by the work of rubbing to produce opposite charges that remain in the dielectric. Examples of how *static electricity* can be generated include combing your hair, walking across a carpeted room, or sliding two pieces of plastic across each other. An *electrostatic discharge (ESD)* occurs when one of the charged objects comes into contact with another dissimilarly charged object. The electrostatic discharge is in the form of a spark. The current from the discharge lasts for only a very short time but can be very large.

Conversion of Chemical Energy

Wet or dry cells and batteries are the applications. Here a chemical reaction produces opposite charges on two dissimilar metals, which serve as the negative and positive terminals.

Electromagnetism

Electricity and magnetism are closely related. Any moving charge has an associated *magnetic field;* also, any changing magnetic field can produce current. A motor is an example showing how current can react with a magnetic field to produce motion; a generator produces voltage by means of a conductor rotating in a magnetic field.

Photoelectricity

Some materials are photoelectric, that is, they can emit electrons when light strikes the surface. The element cesium is often used as a source of *photoelectrons*. Also, photovoltaic cells or solar cells use silicon to generate output voltage from the light input. In another effect, the resistance of the element selenium changes with light. When this is combined with a fixed voltage source, wide variations between *dark current* and *light current* can be produced. Such characteristics are the basis of many photoelectric devices, including television camera tubes, photoelectric cells, and phototransistors.

■ 1–11 Self-Review

Answers at end of chapter.

a. The excess charges at the negative terminal of a battery are _____.

b. Any moving charge has an associated _____.

c. An electrostatic discharge (ESD) is in the form of a _____.

1–12 The Digital Multimeter

As an electronics technician, you can expect to encounter many situations where it will be necessary to measure the voltage, current, or resistance in a circuit. When this is the case, a technician will most likely use a digital multimeter (DMM) to make these measurements. A DMM may be either a handheld or benchtop unit. Both types are shown in Fig. 1–16. All digital meters have numerical readouts that display the value of voltage, current, or resistance being measured.

Measuring Voltage

Fig. 1–17*a* shows a typical DMM measuring the voltage across the terminals of a battery. To measure any voltage, the meter leads are connected directly across the two points where the potential difference or voltage exists. For dc voltages, the red lead of the meter is normally connected to the positive (+) side of the potential difference, whereas the black lead is normally connected to the negative (−) side. When measuring an alternating (ac) voltage, the orientation of the meter leads does not matter since the voltage periodically reverses polarity anyway.

Figure 1–16 Typical digital multimeters (DMMs) (*a*) Handheld DMM (*b*) Benchtop DMM

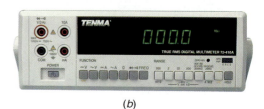

(*b*)

(*a*)

Figure 1–17   DMM measurements (a) Measuring voltage (b) Measuring current (c) Measuring resistance

(a)

(b)

(c)

Measuring Current

Figure 1–17b shows the DMM measuring the current in a simple dc circuit consisting of a battery and a resistor. Notice that the meter is connected between the positive terminal of the battery and the right lead of the resistor. Unlike voltage measurements, current measurements must be made by placing the meter in the path of the moving charges. To do this, the circuit must be broken open at some point, and then the leads of the meter must be connected across the open points to recomplete the circuit. When measuring the current in a dc circuit, the black lead of the meter should be connected to the point which traces directly back to the negative side of the potential difference. Likewise, the red lead of the meter should be connected to the point which traces directly back to the positive side of the potential difference. When measuring ac currents, the orientation of the meter leads is unimportant.

Measuring Resistance

Figure l–17c shows the DMM measuring the ohmic value of a single resistor. Note that the orientation of the meter leads is unimportant when measuring resistance. What is important is that no voltage is present across the resistance being measured, otherwise the meter could be damaged. Also, make sure that no other components are connected across the resistance being measured. If there are, the measurement will probably be both inaccurate and misleading.

■ *1–12 Self-Review*

Answers at end of chapter.

a.   When using a DMM to measure voltage, place the meter leads directly across the two points of potential difference. (True or False)

b.   When using a DMM to measure current, break open the circuit first and then insert the meter across the open points. (True or False)

c.   When using a DMM to measure the value of a single resistor, the orientation of the meter leads is extremely important. (True or False)

GOOD TO KNOW

Most DMM's have a built-in fuse that serves to protect the meter from becoming damaged when the measured current is excessively high. Excessive current could result from improperly connecting the meter into the circuit when measuring current.

Summary

- Electricity is present in all matter in the form of electrons and protons.

- The electron is the basic particle of negative charge, and the proton is the basic particle of positive charge.

- A conductor is a material in which electrons can move easily from one atom to the next.

- An insulator is a material in which electrons tend to stay in their own orbit. Another name for insulator is dielectric.

- The atomic number of an element gives the number of protons in the nucleus of the atom, balanced by an equal number of orbital electrons.

- Electron valence refers to the number of electrons in the outermost shell of an atom. Except for H and He, the goal of valence is eight for all atoms.

- Charges of opposite polarity attract, and charges of like polarity repel.

- One coulomb (C) of charge is a quantity of electricity corresponding to 6.25×10^{18} electrons or protons. The symbol for charge is Q.

- Potential difference or voltage is an electrical pressure or force that exists between two points. The unit of potential difference is the volt (V).
  $$1\ V = \frac{1\ J}{1\ C}\ \text{In general, } V = J/C$$

- Current is the rate of movement of electric charge. The symbol for current is I, and the basic unit of measure is the ampere (A).
  $$1\ A = \frac{1\ C}{1\ s}$$

- Resistance is the opposition to the flow of current. The symbol for resistance is R, and the basic unit of measure is the ohm (Ω).

- Conductance is the reciprocal of resistance. The symbol for conductance is G, and the basic unit of measure is the siemen (S).

- $R = 1/G$ and $G = 1/R$

- An electric circuit is a closed path for current flow. A voltage must be connected across a circuit to produce current flow. In the external circuit outside the voltage source, electrons flow from the negative terminal toward the positive terminal.

- A motion of positive charges, in the opposite direction of electron flow, is considered conventional current.

- Voltage can exist without current, but current cannot exist without voltage.

- Direct current has just one direction because a dc voltage source has fixed polarity. Alternating current periodically reverses in direction as the ac voltage source periodically reverses in polarity.

- Table 1–6 summarizes the main features of electric circuits.

- A digital multimeter is used to measure the voltage, current, or resistance in a circuit.

Table 1–6	Electrical Characteristics		
Characteristic	**Symbol**	**Unit**	**Description**
Charge	Q or q^{*}	Coulomb (C)	Quantity of electrons or protons; $Q = I \times T$
Current	I or i^{*}	Ampere (A)	Charge in motion; $I = Q/T$
Voltage	V or $v^{*,\dagger}$	Volt (V)	Potential difference between two unlike charges; makes charge move to produce I
Resistance	R or $r^{\ddagger}$	Ohm (Ω)	Opposition that reduces amount of current; $R = 1/G$
Conductance	G or $g^{\ddagger}$	Siemens (S)	Reciprocal of R, or $G = 1/R$

* Small letter q, i, or v is used for an instantaneous value of a varying charge, current, or voltage.

† E or e is sometimes used for a generated emf, but the standard symbol for any potential difference is V or v in the international system of units (SI).

‡ Small letter r or g is used for internal resistance or conductance of transistors.

Important Terms

- Alternating Current (ac) - a current that periodically reverses in direction as the alternating voltage periodically reverses in polarity.

- Ampere - the basic unit of current.
  $$1\,A = \frac{1\,C}{1\,s}$$

- Atom - the smallest particle of an element that still has the same characteristics as the element.

- Atomic Number - the number of protons, balanced by an equal number of electrons, in an atom.

- Circuit - a path for current flow.

- Compound - a combination of two or more elements.

- Conductance - the reciprocal of resistance.

- Conductor - any material that allows the free movement of electric charges, such as electrons, to provide an electric current.

- Conventional Current - the direction of current flow associated with positive charges in motion. The current flow direction is from a positive to a negative potential, which is in the opposite direction of electron flow.

- Coulomb - the basic unit of electric charge. $1\,C = 6.25 \times 10^{18}$ electrons or protons.

- Current - a movement of electric charges around a closed path or circuit.

- Dielectric - another name for insulator.

- Direct Current (dc) - a current flow that has just one direction.

- Electron - the most basic particle of negative charge.

- Electron Flow - the movement of electrons that provides current in a circuit. The current flow direction is from a negative to a positive potential, which is in the opposite direction of conventional current.

- Electron Valence - the number of electrons in an incomplete outermost shell of an atom.

- Element - a substance that cannot be decomposed any further by chemical action.

- Free Electron - an electron that can move freely from one atom to the next.

- Insulator - a material with atoms in which the electrons tend to stay in their own orbits.

- Ion - an atom that has either gained or lost one or more valence electrons to become electrically charged.

- Molecules - the smallest unit of a compound with the same chemical characteristics.

- Neutron - a particle contained in the nucleus of an atom that is electrically neutral.

- Nucleus - the massive, stable part of the atom that contains both protons and neutrons.

- Ohm - the unit of resistance.

- Potential Difference - a property associated with two unlike charges in close proximity to each other.

- Proton - the most basic particle of positive charge.

- Resistance - the opposition to the flow of current in an electric circuit.

- Semiconductor - a material which is neither a good conductor nor a good insulator.

- Siemens - the unit of conductance.

- Static Electricity - any charge, positive or negative that is stationary or not in motion.

- Volt - the unit of potential difference or voltage. $V = \dfrac{J}{C}$

Related Formulas

$1\,C = 6.25 \times 10^{18}$ electrons

$V = \dfrac{J}{C}$

$I = Q/T$

$Q = I \times T$

$R = 1/G$

$G = 1/R$

Self-Test

Answers at back of book.

1. **The most basic particle of negative charge is the**
   a. coulomb.
   b. electron.
   c. proton.
   d. neutron.

2. **The coulomb is a unit of**
   a. electric charge.
   b. potential difference.
   c. current.
   d. voltage.

3. **Which of the following is not a good conductor?**
   a. copper.
   b. silver.
   c. glass.
   d. gold.

4. The electron valence of a copper atom is
   a. +1.
   b. 0.
   c. ±4.
   d. −1.

5. The unit of potential difference is the
   a. volt.
   b. ampere.
   c. siemens.
   d. coulomb.

6. Which of the following statements is true?
   a. Unlike charges repel each other.
   b. Like charges repel each other.
   c. Unlike charges attract each other.
   d. both b and c.

7. In a metal conductor, such as a copper wire,
   a. positive ions are the moving charges that provide current.
   b. free electrons are the moving charges that provide current.
   c. there are no free electrons.
   d. none of the above.

8. A 100-Ω resistor has a conductance, G, of
   a. 0.01 S.
   b. 0.1 S.
   c. 0.001 S.
   d. 1 S.

9. The most basic particle of positive charge is the
   a. coulomb.
   b. electron.
   c. proton.
   d. neutron.

10. If a neutral atom loses one of its valence electrons, it becomes a(n)
    a. negative ion.
    b. electrically charged atom.
    c. positive ion.
    d. both b and c.

11. The unit of electric current is the
    a. volt.
    b. ampere.
    c. coulomb.
    d. siemens.

12. A semiconductor, such as silicon, has an electron valence of
    a. ±4.
    b. +1.
    c. −7.
    d. 0.

13. Which of the following statements is true?
    a. Current can exist without voltage.
    b. Voltage can exist without current.
    c. Current can flow through an open circuit.
    d. Both b and c.

14. The unit of resistance is the
    a. volt.
    b. coulomb.
    c. siemens.
    d. ohm.

15. Except for hydrogen (H) and helium (He) the goal of valence for an atom is
    a. 6.
    b. 1.
    c. 8.
    d. 4.

16. One ampere of current corresponds to
    a. $\dfrac{1 \text{ C}}{1 \text{ s}}$.
    b. $\dfrac{1 \text{ J}}{1 \text{ C}}$.
    c. 6.25×10^{18} electrons.
    d. 0.16×10^{-18} C/s.

17. Conventional current is considered
    a. the motion of negative charges in the opposite direction of electron flow.
    b. the motion of positive charges in the same direction as electron flow.
    c. the motion of positive charges in the opposite direction of electron flow.
    d. none of the above.

18. When using a DMM to measure the value of a resistor
    a. make sure that the resistor is in a circuit where voltage is present.
    b. make sure there is no voltage present across the resistor.
    c. make sure there is no other component connected across the leads of the resistor.
    d. both b and c.

19. In a circuit, the opposition to the flow of current is called
    a. conductance.
    b. resistance.
    c. voltage.
    d. current.

20. Aluminum, with an atomic number of 13, has
    a. 13 valence electrons.
    b. 3 valence electrons.
    c. 13 protons in its nucleus.
    d. both b and c.

21. The nucleus of an atom is made up of
    a. electrons and neutrons.
    b. ions.
    c. neutrons and protons.
    d. electrons only.

22. How much charge is accumulated in a dielectric that is charged by a 4-A current for 5 seconds?
    a. 16 C.
    b. 20 C.
    c. 1.25 C.
    d. 0.8 C.

23. A charge of 6 C moves past a given point every 0.25 seconds. How much is the current flow in amperes?
    a. 24 A.
    b. 2.4 A.
    c. 1.5 A.
    d. 12 A.

24. What is the output voltage of a battery that expends 12 J of energy in moving 1.5 C of charge?
    a. 18 V.
    b. 6 V.
    c. 125 mV.
    d. 8 V.

25. Which of the following statements is false?
    a. The resistance of an open circuit is practically zero.
    b. The resistance of a short circuit is practically zero.
    c. The resistance of an open circuit is infinitely high.
    d. There is no current in an open circuit.

Questions

1. Name two good conductors, two good insulators, and two semiconductors.

2. In a metal conductor, what is a free electron?

3. What is the smallest unit of a compound with the same chemical characteristics?

4. Define the term ion.

5. How does the resistance of a conductor compare to that of an insulator?

6. Explain why potential difference is necessary to produce current in a circuit.

7. List three important characteristics of an electric circuit.

8. Describe the difference between an open circuit and a short circuit.

9. Is the power line voltage available in our homes a dc or an ac voltage?

10. What is the mathematical relationship between resistance and conductance?

11. Briefly describe the electric field of a static charge.

12. List at least two examples that show how static electricity can be generated.

13. What is another name for an insulator?

14. List the particles in the nucleus of an atom.

15. Explain the difference between electron flow and conventional current.

16. Define -3 C of charge and compare it to a charge of $+3$ C.

17. Why is it that protons are not considered a source of moving charges for current flow?

18. Write the formulas for each of the following statements: (a) current is the time rate of change of charge (b) charge is current accumulated over a period of time.

19. Briefly define each of the following: (a) 1 coulomb (b) 1 volt (c) 1 ampere (d) 1 ohm.

20. Describe the difference between direct and alternating current.

Problems

SECTION 1–4 THE COULOMB UNIT OF ELECTRIC CHARGE

1–1 If 31.25×10^{18} electrons are removed from a neutral dielectric, how much charge is stored in coulombs?

1–2 If 18.75×10^{18} electrons are added to a neutral dielectric, how much charge is stored in coulombs?

1–3 A dielectric with a positive charge of $+5$ C has 18.75×10^{18} electrons added to it. What is the net charge of the dielectric in coulombs?

1–4 If 93.75×10^{18} electrons are removed from a neutral dielectric, how much charge is stored in coulombs?

1–5 If 37.5×10^{18} electrons are added to a neutral dielectric, how much charge is stored in coulombs?

SECTION 1–5 THE VOLT UNIT OF POTENTIAL DIFFERENCE

1–6 What is the output voltage of a battery if 10 J of energy is expended in moving 1.25 C of charge?

1–7 What is the output voltage of a battery if 6 J of energy is expended in moving 1 C of charge?

1–8 What is the output voltage of a battery if 12 J of energy is expended in moving 1 C of charge?

1–9 How much is the potential difference between two points if 0.5 J of energy is required to move 0.4 C of charge between the two points?

1–10 How much energy is expended, in joules, if a voltage of 12 V moves 1.25 C of charge between two points?

SECTION 1–6 CHARGE IN MOTION IS CURRENT

1–11 A charge of 2 C moves past a given point every 0.5 s. How much is the current?

1–12 A charge of 1 C moves past a given point every 0.1 s. How much is the current?

1–13 A charge of 0.05 C moves past a given point every 0.1 s. How much is the current?

1–14 A charge of 6 C moves past a given point every 0.3 s. How much is the current?

1–15 A charge of 0.1 C moves past a given point every 0.01 s. How much is the current?

1–16 If a current of 1.5 A charges a dielectric for 5 s, how much charge is stored in the dielectric?

1–17 If a current of 500 mA charges a dielectric for 2 s, how much charge is stored in the dielectric?

1–18 If a current of 200 μA charges a dielectric for 20 s, how much charge is stored in the dielectric?

SECTION 1–7 RESISTANCE IS OPPOSITION TO CURRENT

1–19 Calculate the resistance value in ohms for the following conductance values: (a) 0.001 S (b) 0.01 S (c) 0.1 S (d) 1 S.

1-20 Calculate the resistance value in ohms for the following conductance values: (a) 0.002 S (b) 0.004 S (c) 0.00833 S (d) 0.25 S.

1-21 Calculate the conductance value in siemens for each of the following resistance values: (a) 200 Ω (b) 100 Ω (c) 50 Ω (d) 25 Ω.

1-22 Calculate the conductance value in siemens for each of the following resistance values: (a) 1 Ω (b) 10 kΩ (c) 40 Ω (d) 0.5 Ω.

Critical Thinking

1-23 Suppose that 1000 electrons are removed from a neutral dielectric. How much charge, in coulombs, is stored in the dielectric?

1-24 How long will it take an insulator that has a charge of +5 C to charge to +30 C if the charging current is 2 A?

1-25 Assume that 6.25×10^{15} electrons flow past a given point in a conductor every 10 s. Calculate the current I in amperes.

1-26 The conductance of a wire at 100°C is one-tenth its value at 25°C. If the wire resistance equals 10 Ω at 25°C calculate the resistance of the wire at 100°C.

Answers to Knowledge Check Problems

1-4 $-Q = 4$ C

1-5 $V = 3$ V

1-6 $I = 2$ A

1-7 a. $R = 250$ Ω
   b. $G = 33.3$ mS

Answers to Self-Reviews

1-1 a. negative
   b. positive
   c. true

1-2 a. conductors
   b. silver
   c. silicon

1-3 a. 14
   b. 1
   c. 8

1-4 a. 6.25×10^{18}
   b. $-Q = 3$ C
   c. attract

1-5 a. zero
   b. 9 V

1-6 a. 2 A
   b. T
   c. zero

1-7 a. carbon
   b. 4.7 Ω
   c. $^1/_{10}$ S or 0.1 S

1-8 a. T
   b. F
   c. T

1-9 a. negative
   b. negative
   c. true

1-10 a. T
   b. T

1-11 a. electrons
   b. Magnetic field
   c. spark

1-12 a. T
   b. T
   c. F

2 Resistors

Resistors are used in a wide variety of applications in all types of electronic circuits. Their main function in any circuit, however, is to limit the amount of current or to produce a desired drop in voltage. Resistors are manufactured in a variety of shapes and sizes and have ohmic values ranging from a fraction of an ohm to several megohms. The power or wattage rating of a resistor is determined by its physical size. There is, however, no direct correlation between the physical size of a resistor and its resistance value.

In this chapter, you will be presented with an in-depth discussion of the following resistor topics; resistor types, resistor color coding, potentiometers and rheostats, power ratings, and resistor troubles.

Objectives

After studying this chapter you should be able to

- *List* several different types of resistors and describe the characteristics of each type.
- *Interpret* the resistor color code to determine the resistance and tolerance of a resistor.
- *Explain* the difference between a potentiometer and a rheostat.
- *Explain* the significance of a resistor's power rating.
- *List* the most common troubles with resistors.
- *Explain* the precautions that must be observed when measuring a resistor with an ohmmeter.

Outline

Important Terms

carbon-composition resistor
carbon-film resistor
color coding
decade resistance box
derating curve
metal-film resistor

negative temperature coefficient (NTC)
positive temperature coefficient (PTC)
potentiometer
rheostat
surface-mount resistor

taper
thermistor
tolerance
wire-wound resistor
zero-ohm resistor
zero-power resistance

Figure 2–1 Carbon-composition resistor. (*a*) Internal construction. Length is about ¼ in. without leads for ¼-W power rating. Color stripes give *R* in ohms. Tinned leads have coating of solder. (*b*) Resistors mounted on printed-circuit (PC) board.

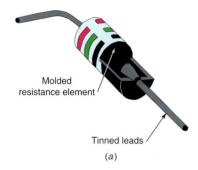

Molded resistance element

Tinned leads

(*a*)

(*b*)

2–1 Types of Resistors

The two main characteristics of a resistor are its resistance *R* in ohms and its power rating *W* in watts. Resistors are available in a very wide range of *R* values, from a fraction of an ohm to many kilohms (kΩ) and megohms (MΩ). One kilohm is 1000 Ω, and one megohm is 1,000,000 Ω. The power rating for resistors may be as high as several hundred watts or as low as ¹⁄₁₀ W.

The *R* is the resistance value required to provide the desired current or voltage. Also important is the wattage rating because it specifies the maximum power the resistor can dissipate without excessive heat. *Dissipation* means that the power is wasted, since the resultant heat is not used. Too much heat can make the resistor burn. The wattage rating of the resistor is generally more than the actual power dissipation, as a safety factor.

Most common in electronic equipment are carbon resistors with a power rating of 1 W or less. The construction is illustrated in Fig. 2–1*a*. The leads extending out from the resistor body can be inserted through the holes on a printed-circuit (PC) board for mounting as shown in Fig. 2–1*b*. The resistors on a PC board are often inserted automatically by machine. Note that resistors are not polarity-sensitive devices. This means that it does not matter which way the leads of a resistor are connected in a circuit.

Resistors with higher *R* values usually have lower wattage ratings because they have less current. As an example, a common value is 1 MΩ at ¼ W, for a resistor only ¼ in. long. The lower the power rating, the smaller the actual size of the resistor. However, the resistance value is not related to physical size. Figure 2–2 shows several carbon resistors with the same physical size but different resistance values. The different color bands on each resistor indicate a different ohmic value. The carbon resistors in Fig. 2–2 each have a power rating of ½ W, which is based on their physical size.

Wire-Wound Resistors

In this construction, a special type of wire called *resistance wire* is wrapped around an insulating core. The length of wire and its specific resistivity deter-

Figure 2–2 Carbon resistors with same physical size but different resistance values. The physical size indicates a power rating of ½ W.

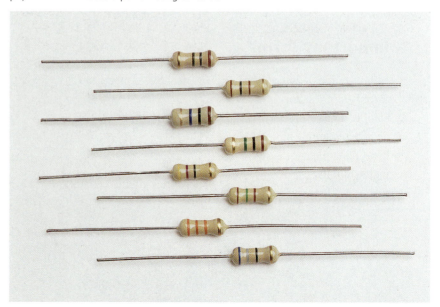

(a)

(b)

mine the *R* of the unit. Types of resistance wire include tungsten and manganin, as explained in Chap. 11, "Conductors and Insulators." The insulated core is commonly porcelain, cement, or just plain pressed paper. Bare wire is used, but the entire unit is generally encased in an insulating material. Typical fixed and variable wire-wound resistors are shown in Fig. 2–3.

Since they are generally used for high-current applications with low resistance and appreciable power, *wire-wound resistors* are available in wattage ratings from 1 W up to 100 W or more. The resistance can be less than 1 Ω up to several thousand ohms. For 2 W or less, carbon resistors are preferable because they are generally smaller and cost less.

In addition, wire-wound resistors are used where accurate, stable resistance values are necessary. Examples are precision resistors for the function of an ammeter shunt or a precision potentiometer to adjust for an exact amount of *R*.

Carbon–Composition Resistors

These resistors are made of finely divided carbon or graphite mixed with a powdered insulating material as a binder in the proportions needed for the desired *R* value. As shown in Fig. 2–1*a*, the resistor element is enclosed in a plastic case for insulation and mechanical strength. Joined to the two ends of the carbon resistance element are metal caps with leads of tinned copper wire for soldering the connections into a circuit. These are called *axial leads* because they come straight out from the ends. Carbon-composition resistors normally have a brown body and are cylindrical.

Carbon-composition resistors are commonly available in *R* values of 1 Ω to 20 MΩ. Examples are 10 Ω, 220 Ω, 4.7 kΩ, and 68 kΩ. The power rating is generally $\frac{1}{10}$, $\frac{1}{8}$, $\frac{1}{4}$, $\frac{1}{2}$, 1, or 2 W.

Film–Type Resistors

There are two kinds of film-type resistors: *carbon-film* and *metal-film resistors*. The carbon-film resistor, whose construction is shown in Fig. 2–4, is made by depositing a thin layer of carbon on an insulated substrate. The carbon film is then cut in the form of a spiral to form the resistive element. The resistance value is controlled by varying the proportion of carbon to insulator. Compared to carbon-composition resistors, carbon-film resistors have the following advantages: tighter tolerances, less sensitivity to temperature changes and aging, and they generate less noise internally.

Metal-film resistors are constructed in a manner similar to the carbon-film type. However, in a metal-film resistor, a thin film of metal is sprayed onto a ceramic substrate and then cut in the form of a spiral. The construction of a metal-film resistor is shown in Fig. 2–5. The length, thickness, and width of the

Figure 2–4 Construction of carbon-film resistor.

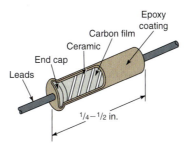

Figure 2–5 Construction of metal-film resistor.

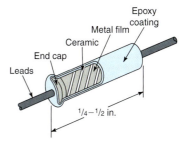

Figure 2–6 Typical chip resistors.

Figure 2–7 (a) Thermistor schematic symbol. (b) Typical thermistor shapes and sizes.

(a)

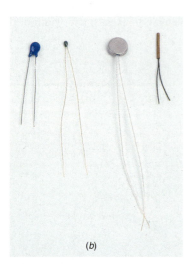

(b)

metal spiral determine the exact resistance value. Metal-film resistors offer more precise R values than carbon-film resistors. Like carbon-film resistors, metal-film resistors are affected very little by temperature changes and aging. They also generate very little noise internally. In overall performance, metal-film resistors are the best, carbon-film resistors are next, and carbon-composition resistors are last. Both carbon- and metal-film resistors can be distinguished from carbon-composition resistors by the fact that the diameter of the ends is a little larger than that of the body. Furthermore, metal-film resistors are almost always coated with a blue, light green, or red lacquer which provides electrical, mechanical, and climate protection. The body color of carbon-film resistors is usually tan.

Surface–Mount Resistors

Surface-mount resistors, also called *chip resistors,* are constructed by depositing a thick carbon film on a ceramic base. The exact resistance value is determined by the composition of the carbon itself, as well as by the amount of trimming done to the carbon deposit. The resistance can vary from a fraction of an ohm to well over a million ohms. Power dissipation ratings are typically ⅛ to ¼ W. Figure 2–6 shows typical chip resistors. Electrical connection to the resistive element is made via two leadless solder end electrodes (terminals). The end electrodes are C-shaped. The physical dimensions of a ⅛-W chip resistor are 0.125 in. long by 0.063 in. wide and approximately 0.028 in. thick. This is many times smaller than a conventional resistor having axial leads. Chip resistors are very temperature-stable and also very rugged. The end electrodes are soldered directly to the copper traces of a circuit board, hence the name *surface-mount.*

Fusible Resistors

This type is a wire-wound resistor made to burn open easily when the power rating is exceeded. It then serves the dual functions of a fuse and a resistor to limit the current.

Thermistors

A *thermistor* is a thermally sensitive resistor whose resistance value changes with changes in operating temperature. Because of the self-heating effect of current in a thermistor, the device changes resistance with changes in current. Thermistors, which are essentially semiconductors, exhibit either a *positive temperature coefficient (PTC)* or a *negative temperature coefficient (NTC).* If a thermistor has a PTC, its resistance increases as the operating temperature increases. Conversely, if a thermistor has an NTC, its resistance decreases as its operating temperature increases. How much the resistance changes with changes in operating temperature depends on the size and construction of the thermistor. Note that the resistance does not undergo instantaneous changes with changes in the operating temperature. A certain time interval, determined by the thermal mass (size) of the thermistor, is required for the resistance change. A thermistor with a small mass will change more rapidly than one with a large mass. Carbon- and metal-film resistors are different: their resistance does not change appreciably with changes in operating temperature.

Figure 2–7a shows the standard schematic symbol for a thermistor. Notice the arrow through the resistor symbol and the letter T within the circle. The arrow indicates that the resistance is variable as the temperature T changes. As shown in Fig. 2–7b, thermistors are manufactured in a wide variety of shapes and sizes. The shapes include beads, rods, disks, and washers.

Thermistors are frequently used in electronic circuits in which it is desired to provide temperature measurement, temperature control, and temperature compensation.

■ 2–1 Self-Review

Answers at end of chapter.

a. An R of 10 Ω with a 25-W rating would most likely be a wire-wound resistor. (True or False)
b. A resistance of 10,000 Ω is the same as a resistance of 10 kΩ. (True or False)
c. Which is more temperature stable, a carbon-composition or a metal-film resistor?
d. Which is larger, a 1000-Ω, ½-W or a 1000-Ω, 1-W carbon-film resistor?
e. What happens to the resistance of an NTC thermistor when its operating temperature increases?

2–2 Resistor Color Coding

Because carbon resistors are small, they are *color-coded* to mark their R value in ohms. The basis of this system is the use of colors for numerical values, as listed in Table 2–1. In memorizing the colors, note that the darkest colors, black and brown, are for the lowest numbers, zero and one, whereas white is for nine. The color coding is standardized by the Electronic Industries Alliance (EIA).

Table 2–1	Color Code	
Color		**Numerical Value**
Black		0
Brown		1
Red		2
Orange		3
Yellow		4
Green		5
Blue		6
Violet		7
Gray		8
White		9

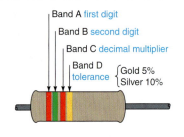

Figure 2–8 How to read color stripes on carbon resistors for *R* in ohms.

Band A first digit
Band B second digit
Band C decimal multiplier
Band D tolerance { Gold 5% / Silver 10%

▌▌▌ **MultiSim** **Figure 2–9** Examples of color-coded *R* values with percent tolerance.

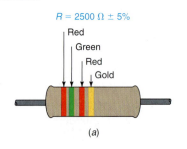

R = 2500 Ω ± 5%

Red
Green
Red
Gold

(a)

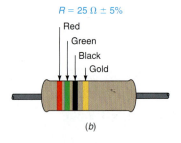

R = 25 Ω ± 5%

Red
Green
Black
Gold

(b)

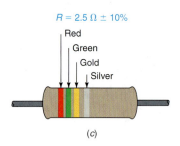

R = 2.5 Ω ± 10%

Red
Green
Gold
Silver

(c)

Resistance Color Stripes

The use of colored bands or stripes is the most common system for color-coding resistors, as shown in Fig. 2–8. The colored bands or stripes completely encircle the body of the resistor and are usually crowded toward one end. Reading from left to right, the first band closest to the edge gives the first digit in the numerical value of *R*. The next band indicates the second digit. The third band is the decimal multiplier, which tells us how many zeros to add after the first two digits.

In Fig. 2–9a, the first stripe is red for 2 and the next stripe is green for 5. The red multiplier in the third stripe means add two zeros to 25, or "this multiplier is 10^2." The result can be illustrated as follows:

Red	Green		Red	
↓	↓		↓	
2	5	×	100	= 2500

Therefore, this *R* value is 2500 Ω or 2.5 kΩ.

The example in Fig. 2–9b illustrates that black for the third stripe just means "do not add any zeros to the first two digits." Since this resistor has red, green, and black stripes, the *R* value is 25 Ω.

Resistors Under 10 Ω

For these values, the third stripe is either gold or silver, indicating a fractional decimal multiplier. When the third stripe is gold, multiply the first two digits by 0.1. In Fig. 2–9c, the *R* value is

$$25 \times 0.1 = 2.5 \ \Omega.$$

Silver means a multiplier of 0.01. If the third band in Fig. 2–9c were silver, the *R* value would be

$$25 \times 0.01 = 0.25 \ \Omega.$$

It is important to realize that the gold and silver colors represent fractional decimal multipliers only when they appear in the third stripe. Gold and silver are used most often however as a fourth stripe to indicate how accurate the *R* value is. The colors gold and silver will never appear in the first two color stripes.

Resistor Tolerance

The amount by which the actual *R* can differ from the color-coded value is the *tolerance*, usually given in percent. For instance, a 2000-Ω resistor with ±10% tolerance can have resistance 10% above or below the coded value. This *R*, therefore, is between 1800 and 2200 Ω. The calculations are as follows:

$$10\% \text{ of } 2000 \text{ is } 0.1 \times 2000 = 200.$$

For +10%, the value is

$$2000 + 200 = 2200 \ \Omega.$$

For −10%, the value is

$$2000 - 200 = 1800 \ \Omega.$$

As illustrated in Fig. 2–8, silver in the fourth band indicates a tolerance of ±10%; gold indicates ±5%. If there is no color band for tolerance, it is ±20%. The inexact value of carbon-composition resistors is a disadvantage of their economical construction. They usually cost only a few cents each, or less in larger quantities. In most circuits, though, a small difference in resistance can be tolerated.

Five-Band Color Code

Precision resistors (typically metal-film resistors) often use a five-band color code rather than the four-band code shown in Fig. 2–8. The purpose is to obtain more precise R values. With the five-band code, the first three color stripes indicate the first three digits, followed by the decimal multiplier in the fourth stripe and the tolerance in the fifth stripe. In the fifth stripe, the colors brown, red, green, blue, and violet represent the following tolerances:

Brown	±1%
Red	± 2%
Green	± 0.5%
Blue	± 0.25%
Violet	± 0.1%

Example 2-1

|||MultiSim **Figure 2–10** Five-band code.

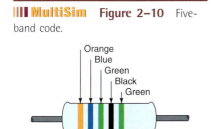

What is the resistance indicated by the five-band color code in Fig. 2–10? Also, what ohmic range is permissible for the specified tolerance?

ANSWER   The first stripe is orange for the number 3, the second stripe is blue for the number 6, and the third stripe is green for the number 5. Therefore, the first three digits of the resistance are 3, 6, and 5, respectively. The fourth stripe, which is the multiplier, is black, which means add no zeros. The fifth stripe, which indicates the resistor tolerance, is green for ±0.5%. Therefore $R = 365\ \Omega \pm 0.5\%$. The permissible ohmic range is calculated as $365 \times 0.005 = \pm 1.825\ \Omega$, or 363.175 to 366.825 Ω.

■ *2–2A Knowledge Check*

Answers at end of chapter.

Determine the resistance and tolerance for each of the following color codes:
a. **Four-band code: orange, orange, red, and gold.**
b. **Five-band code: yellow, yellow, red, red, and brown.**

Wire-Wound-Resistor Marking

Usually, wire-wound resistors are big enough to have the R value printed on the insulating case. The tolerance is generally ±5%, except for precision resistors, which have a tolerance of ± 1% or less.

Some small wire-wound resistors may be color-coded with stripes, however, like carbon resistors. In this case, the first stripe is double the width of the others to indicate a wire-wound resistor. Wirewound resistors which are color coded generally have a power rating of 4 W or less.

Preferred Resistance Values

To minimize the problem of manufacturing different R values for an almost unlimited variety of circuits, specific values are made in large quantities so that they are cheaper and more easily available than unusual sizes. For resistors of ±10%, the *preferred values* are 10, 12, 15, 18, 22, 27, 33, 39, 47, 56, 68, and

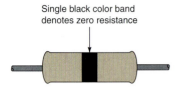

Single black color band
denotes zero resistance

GOOD TO KNOW

Even though chip resistors use a three- or four-digit code to indicate their resistance value in ohms, the digits may be too small to read with the naked eye. In other words, it may be necessary to use a magnifying device to read the value on the chip resistor.

82 with their decimal multiples. As examples, 47, 470, 4700, and 47,000 are preferred values. In this way, there is a preferred value available within 10% of any R value needed in a circuit. See Appendix C for a listing of preferred resistance values for tolerances of $\pm 20\%$, $\pm 10\%$, and $\pm 5\%$.

Zero-Ohm Resistors

Believe it or not, there is such a thing as a *zero-ohm resistor*. In fact, zero-ohm resistors are quite common. The zero-ohm value is denoted by the use of a single black band around the center of the resistor body, as shown in Fig. 2–11. Zero-ohm resistors are available in $\frac{1}{8}$- or $\frac{1}{4}$-W sizes. The actual resistance of a so-called $\frac{1}{8}$-W zero-ohm resistor is about 0.004 Ω, whereas a $\frac{1}{4}$-W zero-ohm resistor has a resistance of approximately 0.003 Ω.

But why are zero-ohm resistors used in the first place? The reason is that for most printed-circuit boards, the components are inserted by automatic insertion machines (robots) rather than by human hands. In some instances, it may be necessary to short two points on the printed-circuit board, in which case a piece of wire has to be placed between the two points. Because the robot can handle only components such as resistors, and not wires, zero-ohm resistors are used. Before zero-ohm resistors were developed, jumpers had to be installed by hand, which was time-consuming and expensive. Zero-ohm resistors may be needed as a result of an after-the-fact design change which requires new point to point connections in a circuit.

Chip Resistor Coding System

The chip resistor, shown in Fig. 2–12a, has the following identifiable features:

> Body color: white or off-white
> Dark film on one side only (usually black, but may also be dark gray or green)
> End electrodes (terminals) are C-shaped
> Three- or four-digit marking on either the film or the body side (usually the film)

The resistance value of a chip resistor is determined from the three-digit number printed on the film or body side of the component. The three digits provide the same information as the first three color stripes on a four-band resistor. This is shown in Fig. 2–l2b. The first two digits indicate the first two numbers in the numerical value of the resistance; the third digit indicates the multiplier. If a four-digit number is used, the first three digits indicate the first three numbers in the numerical value of the resistance, and the fourth digit indicates the multiplier. The letter R is used to signify a decimal point for values between 1 and 10 ohms as in $2R7 = 2.7\ \Omega$. Figure 2–12c shows the symbol used

Figure 2–12 Typical chip resistor coding system.

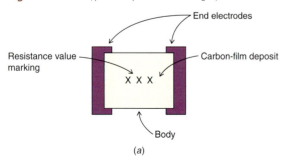

Resistance value marking
End electrodes
Carbon-film deposit
X X X
Body

(a)

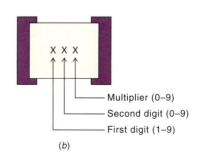

X X X
Multiplier (0–9)
Second digit (0–9)
First digit (1–9)

(b)

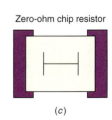

Zero-ohm chip resistor

(c)

to denote a zero-ohm chip resistor. Chip resistors are typically available in tolerances of ±1% and ±5%. It is important to note, however, that the tolerance of a chip resistor is not indicated by the three- or four-digit code.

Figure 2–13 Chip resistor with number coding.

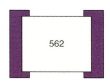

Example 2-2

Determine the resistance of the chip resistor in Fig. 2–13.

ANSWER The first two digits are 5 and 6, giving 56 as the first two numbers in the resistance value. The third digit, 2, is the multiplier, which means add 2 zeros to 56 for a resistance of 5600 Ω or 5.6 kΩ.

■ 2–2B Knowledge Check

Answers at end of chapter.

Determine the resistance of chip resistors with the following coded values:

(a) 104
(b) 2202
(c) 1R8

Thermistor Values

Thermistors are normally rated by the value of their resistance at a reference temperature T of 25°C. The value of R at 25°C is most often referred to as the zero-power resistance and is designated R_0. The term *zero-power resistance* refers to the resistance of the thermistor with zero-power dissipation. Thermistors normally do not have a code or marking system to indicate their resistance value in ohms. In rare cases, however, a three-dot code is used to indicate the value of R_0. In this case, the first and second dots indicate the first two significant digits and the third dot is the multiplier. The colors used are the same as those for carbon resistors.

■ 2–2 Self-Review

Answers at end of chapter.

a. Give the color for the number 4.
b. What tolerance does a silver stripe represent?
c. Give the multiplier for red in the third stripe.
d. Give R and the tolerance for a resistor coded with yellow, violet, brown, and gold stripes.
e. Assume that the chip resistor in Fig. 2–13 is marked 333. What is its resistance value in ohms?

Figure 2–14 Construction of variable carbon resistance control. Diameter is ³/₄ in. (*a*) External view. (*b*) Internal view of circular resistance element.

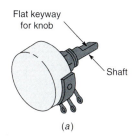

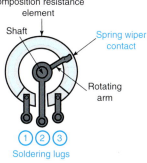

2–3 Variable Resistors

Variable resistors can be wire-wound, as in Fig. 2–3*b*, or carbon type, illustrated in Fig. 2–14. Inside the metal case of Fig. 2–14*a*, the control has a circular disk, shown in Fig. 2–14*b*, that is the carbon-composition resistance element. It can be a thin coating on pressed paper or a molded carbon disk. Joined to the two ends are the external soldering-lug terminals 1 and 3. The middle terminal is connected

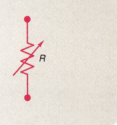
Figure 2–15 Slide control for variable R. Length is 2 in.

Figure 2–16 Decade resistance box for a wide range of R values.

to the variable arm that contacts the resistor element by a metal spring wiper. As the shaft of the control is turned, the variable arm moves the wiper to make contact at different points on the resistor element. The same idea applies to the slide control in Fig. 2–15, except that the resistor element is straight instead of circular.

When the contact moves closer to one end, the R decreases between this terminal and the variable arm. Between the two ends, however, R is not variable but always has the maximum resistance of the control.

Carbon controls are available with a total R from 1000 Ω to 5 MΩ, approximately. Their power rating is usually ½ to 2 W.

Tapered Controls

The way R varies with shaft rotation is called the *taper* of the control. With a linear taper, a one-half rotation changes R by one-half the maximum value. Similarly, all values of R change in direct proportion to rotation. For a nonlinear taper, though, R can change more gradually at one end with bigger changes at the opposite end. This effect is accomplished by different densities of carbon in the resistance element. For a volume control, its audio taper allows smaller changes in R at low settings. Then it is easier to make changes without having the volume too loud or too low.

Decade Resistance Box

As shown in Fig. 2–16, the *decade resistance box* is a convenient unit for providing any one R within a wide range of values. It can be considered test equipment for trying different R values in a circuit. Inside the box are six series strings of resistors, with one string for each dial switch.

The first dial connects in an R of 0 to 9 Ω. It is the *units* or $R \times 1$ dial.
The second dial has units of 10 from 0 to 90 Ω. It is the *tens* or $R \times 10$ dial.
The hundreds or $R \times 100$ dial has an R of 0 to 900 Ω.
The thousands or $R \times 1$ k dial has an R of 0 to 9000 Ω.
The ten-thousands or $R \times 10$ k dial provides R values of 0 to 90,000 Ω.
The one-hundred-thousands or $R \times 100$ k dial provides R values of 0 to 900,000 Ω.

The six dial sections are connected internally so that their values add to one another. Then any value from 0 to 999,999 Ω can be obtained. Note the exact values that are possible. As an example, when all six dials are on 2, the total R equals $2 + 20 + 200 + 2000 + 20,000 + 200,000 = 222,222$ Ω.

■ *2–3 Self-Review*

    Answers at end of chapter.

  a. In Fig. 2–14, which terminal provides variable R?
  b. Is an audio taper linear or nonlinear?
  c. In Fig. 2–16, how much is the total R if the $R \times 100$k and $R \times 10$k dials are set to 4 and 7, respectively, and all other dials are set to zero?

2–4 Rheostats and Potentiometers

Rheostats and potentiometers are variable resistances, either carbon or wirewound, used to vary the amount of current or voltage in a circuit. The controls can be used in either dc or ac applications.

Table 2–2	Potentiometers and Rheostats
Rheostat	**Potentiometer**
Two terminals	Three terminals
In series with load and V source	Ends are connected across V source
Varies the I	Taps off part of V

A *rheostat* is a variable R with two terminals connected in series with a load. The purpose is to vary the amount of current.

A *potentiometer*, generally called a *pot* for short, has three terminals. The fixed maximum R across the two ends is connected across a voltage source. Then the variable arm is used to vary the voltage division between the center terminal and the ends. This function of a potentiometer is compared with that of a rheostat in Table 2–2.

Rheostat Circuit

The function of the rheostat R_2 in Fig. 2–17 is to vary the amount of current through R_1. For instance, R_1 can be a small lightbulb that requires a specified I. Therefore, the two terminals of the rheostat R_2 are connected in series with R_1 and the source V to vary the total resistance R_T in the circuit. When R_T changes, I changes, as read by the meter.

In Fig. 2–17b, R_1 is 5 Ω and the rheostat R_2 varies from 0 to 5 Ω. With R_2 at its maximum of 5 Ω, then R_T equals $5 + 5 = 10$ Ω. I equals 0.15 A or 150 mA. (The method for calculating I given R and V is covered in Chap. 3, "Ohm's Law.")

When R_2 is at its minimum value of 0 Ω, R_T equals 5 Ω. Then I is 0.3 A or 300 mA for the maximum current. As a result, varying the rheostat changes the circuit resistance to vary the current through R_1. I increases as R decreases.

It is important that the rheostat have a wattage rating high enough for maximum I when R is minimum. Rheostats are often wire-wound variable resistors used to control relatively large values of current in low-resistance circuits for ac power applications.

III MultiSim **Figure 2–17** Rheostat connected in series circuit to vary the current *I*. Symbol for current meter is A for amperes. (*a*) Wiring diagram with digital meter for *I*. (*b*) Schematic diagram.

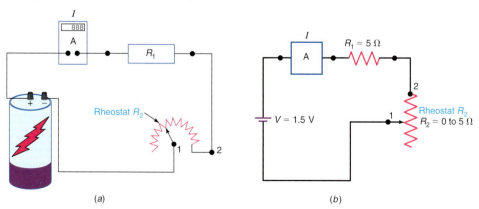

(a)   (b)

Figure 2–18 Potentiometer connected across voltage source to function as a voltage divider. (a) Wiring diagram. (b) Schematic diagram.

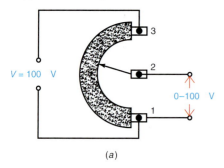

$V = 100$ V

3

2

0–100 V

1

(a)

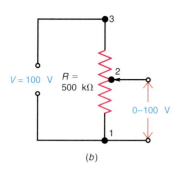

3

$V = 100$ V

$R = 500$ kΩ

2

0–100 V

1

(b)

Potentiometer Circuit

The purpose of the circuit in Fig. 2–18 is to tap off a variable part of the 100 V from the source. Consider this circuit in two parts:

1. The applied *V* is input to the two end terminals of the potentiometer.
2. The variable *V* is output between the variable arm and an end terminal.

Two pairs of connections to the three terminals are necessary, with one terminal common to the input and output. One pair connects the source *V* to the end terminals 1 and 3. The other pair of connections is between the variable arm at the center terminal and one end. This end has double connections for input and output. The other end has only an input connection.

When the variable arm is at the middle value of the 500-kΩ *R* in Fig. 2–18, the 50 V is tapped off between terminals 2 and 1 as one-half the 100-V input. The other 50 V is between terminals 2 and 3. However, this voltage is not used for output.

As the control is turned up to move the variable arm closer to terminal 3, more of the input voltage is available between 2 and 1. With the control at its maximum *R*, the voltage between 2 and 1 is the entire 100 V. Actually, terminal 2 is then the same as 3.

When the variable arm is at minimum *R*, rotated to terminal 1, the output between 2 and 1 is zero. Now all the applied voltage is across 2 and 3 with no output for the variable arm. It is important to note that the source voltage is not short-circuited. The reason is that the maximum *R* of the potentiometer is always across the applied *V*, regardless of where the variable arm is set. Typical examples of small potentiometers used in electronic circuits are shown in Fig. 2–19.

Potentiometer Used as a Rheostat

Commercial rheostats are generally wire-wound, high-wattage resistors for power applications. However, a small, low-wattage rheostat is often needed in electronic circuits. One example is a continuous tone control in a receiver. The control requires the variable series resistance of a rheostat but dissipates very little power.

A method of wiring a potentiometer as a rheostat is to connect just one end of the control and the variable arm, using only two terminals. The third terminal is open, or floating, not connected to anything.

Another method is to wire the unused terminal to the center terminal. When the variable arm is rotated, different amounts of resistance are short-circuited. This method is preferable because there is no floating resistance.

Either end of the potentiometer can be used for the rheostat. The direction of increasing *R* with shaft rotation reverses, though, for connections at opposite ends. Also, the taper is reversed on a nonlinear control.

The resistance of a potentiometer is sometimes marked on the enclosure that houses the resistance element. The marked value indicates the resistance between the outside terminals.

Figure 2–19 Small potentiometers and trimmers often used for variable controls in electronic circuits. Terminal leads are formed for insertion into a PC board.

■ *2–4 Self-Review*

Answers at end of chapter.

a. How many circuit connections to a potentiometer are needed?
b. How many circuit connections to a rheostat are needed?
c. In Fig. 2–18, with a 500-kΩ linear potentiometer, how much is the output voltage with 400 kΩ between terminals 1 and 2?

2–5 Power Rating of Resistors

In addition to having the required ohms value, a resistor should have a wattage rating high enough to dissipate the power produced by the current flowing through the resistance without becoming too hot. Carbon resistors in normal operation often become warm, but they should not get so hot that they "sweat" beads of liquid on the insulating case. Wire-wound resistors operate at very high temperatures; a typical value is 300°C for the maximum temperature. If a resistor becomes too hot because of excessive power dissipation, it can change appreciably in resistance value or burn open.

The power rating is a physical property that depends on the resistor construction, especially physical size. Note the following:

1. A larger physical size indicates a higher power rating.
2. Higher wattage resistors can operate at higher temperatures.
3. Wire-wound resistors are larger and have higher wattage ratings than carbon resistors.

For approximate sizes, a 2-W carbon resistor is about 1 in. long with a ¼ in. diameter; a ¼-W resistor is about 0.25 in. long with a diameter of 0.1 in.

For both types, a higher power rating allows a higher voltage rating. This rating gives the highest voltage that may be applied across the resistor without internal arcing. As examples for carbon resistors, the maximum voltage is 500 V for a 1-W rating, 350 V for ½-W, 250 V for ¼-W, and 150 V for ⅛-W. In wire-wound resistors, excessive voltage can produce an arc between turns; in carbon-composition resistors, the arc is between carbon granules.

Power Derating Curve

When a carbon resistor is mounted on a PC board close to other resistors and components, all of which are producing heat and enclosed in a confined space, the ambient temperature can rise appreciably above 25°C. When carbon resistors are operated at ambient temperatures of 70°C or less, the commercial power rating, indicated by the physical size, remains valid. However, for ambient temperatures greater than 70°C, the power rating must be reduced or derated. This is shown in Fig. 2–20. Notice that for ambient temperatures up to 70°C, the commercial power rating is the same (100%) as that determined by the resistor's

Figure 2–20 Resistor power derating curve.

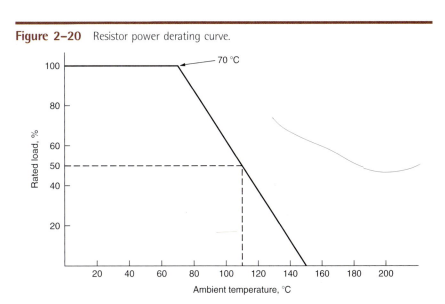

physical size. Note, however, that above 70°C, the power rating decreases linearly.

For example, at an ambient temperature of 110°C, the power rating must be reduced to 50% of its rated value. This means that a 1-kΩ, ½-W resistor operating at 110°C can safely dissipate only ¼ W of power. Therefore, the physical size of the resistor must be increased if it is to safely dissipate ½ W at 110°C. In this case, a 1-kΩ, 1-W resistor would be necessary.

The curve in Fig. 2–20, called a *power derating curve*, is supplied by the resistor manufacturer. For a ½-W carbon resistor, the power derating curve corresponds to a 6.25 mW reduction in the power rating for each degree Celsius rise in temperature above 70°C. This corresponds to a derate factor of 6.25 mW/°C.

Shelf Life

Resistors keep their characteristics almost indefinitely when not used. Without any current in a circuit to heat the resistor, it has practically no change with age. The shelf life of resistors is therefore usually no problem.

■ *2–5 Self-Review*
    Answers at end of chapter.

a. **The power rating of a resistor is mainly determined by its physical size. (True or False)**

b. **The power rating of a carbon resistor is not affected by the ambient temperature in which it operates. (True or False)**

2–6 Resistor Troubles

The most common trouble in resistors is an open. When the open resistor is a series component, there is no current in the entire series path.

Noisy Controls

In applications such as volume and tone controls, carbon controls are preferred because the smoother change in resistance results in less noise when the variable arm is rotated. With use, however, the resistance element becomes worn by the wiper contact, making the *control noisy*. When a volume or tone control makes a scratchy noise as the shaft is rotated, it indicates either a dirty or worn-out resistance element. If the control is just dirty, it can be cleaned by spraying the resistance element with a special contact cleaner. If the resistance element is worn out, the control must be replaced.

Checking Resistors with an Ohmmeter

Resistance is measured with an ohmmeter. The ohmmeter has its own voltage source so that it is always used without any external power applied to the resistance being measured. Separate the resistance from its circuit by disconnecting one lead of the resistor. Then connect the ohmmeter leads across the resistance to be measured.

An open resistor reads infinitely high ohms. For some reason, infinite ohms is often confused with zero ohms. Remember, though, that infinite ohms means an open circuit. The current is zero, but the resistance is infinitely high. Furthermore, it is practically impossible for a resistor to become short-circuited in itself. The resistor may be short-circuited by some other part of the circuit. However, the construction of resistors is such that the trouble they develop is an open circuit with infinitely high ohms.

GOOD TO KNOW

When measuring the value of a resistor in an electronic circuit, make absolutely sure that the power is off in the circuit being tested. Failure to do so could result in damage to the meter!

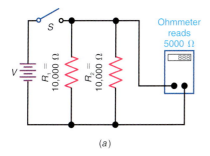

Figure 2–21 Parallel R_1 can lower the ohmmeter reading for testing R_2. (a) The two resistances R_1 and R_2 are in parallel. (b) R_2 is isolated by disconnecting one end of R_1.

Ohmmeter reads 5000 Ω

(a)

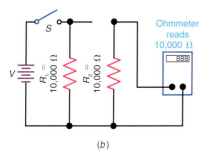

Ohmmeter reads 10,000 Ω

(b)

The ohmmeter must have an ohms scale capable of reading the resistance value, or the resistor cannot be checked. In checking a 10-MΩ resistor, for instance, if the highest R the ohmmeter can read is 1 MΩ, it will indicate infinite resistance, even if the resistor has its normal value of 10 MΩ. An ohms scale of 100 MΩ or more should be used for checking such high resistances.

To check resistors of less than 10 Ω, a low-ohms scale of about 100 Ω or less is necessary. Center scale should be 6 Ω or less. Otherwise, the ohmmeter will read a normally low resistance value as zero ohms.

When checking resistance in a circuit, it is important to be sure there are no parallel resistance paths. Otherwise, the measured resistance can be much lower than the actual resistor value, as illustrated in Fig. 2–21a. Here, the ohmmeter reads the resistance of R_2 in parallel with R_1. To check across R_2 alone, one end is disconnected, as in Fig. 2–21b.

For very high resistances, it is important not to touch the ohmmeter leads. There is no danger of shock, but the body resistance of about 50,000 Ω as a parallel path will lower the ohmmeter reading.

Changed Value of R

In many cases, the value of a carbon-composition resistor can exceed its allowed tolerance; this is caused by normal resistor heating over a long period of time. In most instances, the value change is seen as an increase in R. This is known as *aging*. As you know, carbon-film and metal-film resistors age very little. A surface-mount resistor should never be rubbed or scraped because this will remove some of the carbon deposit and change its resistance.

■ *2–6 Self-Review*

Answers at end of chapter.

a. What is the ohmmeter reading for a short circuit?
b. What is the ohmmeter reading for an open resistor?
c. Which has a higher R, an open or a short circuit?
d. Which is more likely to change in R value after many years of use, a metal-film or a carbon-composition resistor?

Summary

- The most common types of resistors include carbon-composition, carbon-film, metal-film, wire-wound, and surface-mount or chip resistors. Carbon-film and metal-film resistors are better than carbon-composition resistors because they have tighter tolerances, are less affected by temperature and aging, and generate less noise internally.

- A thermistor is a thermally sensitive resistor whose resistance value changes with temperature. If the resistance of a thermistor increases with temperature, it is said to have a positive temperature coefficient (PTC). If the resistance of a thermistor decreases with temperature, it is said to have a negative temperature coefficient (NTC).

- Wire-wound resistors are typically used in high-current applications. Wire-wound resistors are available with wattage ratings of about 1 to 100 W.

- Resistors are usually color-coded to indicate their resistance value in ohms. Either a four-band or a five-band code is used. The five-band code is used for more precise R values. Chip resistors use a three- or four-digit code to indicate their resistance value.

- Zero-ohm resistors are used with automatic insertion machines when it is desired to short two points on a printed-circuit board. Zero-ohm resistors are available in $\frac{1}{8}$- or $\frac{1}{4}$-W ratings.

- A potentiometer is a variable resistor with three terminals. It is used to vary the voltage in a circuit. A rheostat is a variable resistor with two terminals. It is used to vary the current in a circuit.

- The physical size of a resistor determines its wattage rating: the larger the physical size, the larger the wattage rating. There is no correlation between a resistor's physical size and its resistance value.

- The most common trouble in resistors is an open. An ohmmeter across the leads of an open resistor will read infinite, assuming there is no other parallel path across the resistor.

Important Terms

- **Carbon-Composition Resistor** - a type of resistor made of finely divided carbon mixed with a powdered insulating material in the correct proportion to obtain the desired resistance value.

- **Carbon-Film Resistor** - a type of resistor whose construction consists of a thin spiral layer of carbon on an insulated substrate.

- **Color Coding** - a scheme using colored bands or stripes around the body of a resistor to indicate the ohmic value and tolerance of a resistor.

- **Decade Resistance Box** - a variable resistance box whose resistance value can be varied in 1-Ω, 10-Ω, 100-Ω, 1000-Ω, 10,000-Ω, or 100,000-Ω steps.

- **Derating Curve** - a graph showing how the power rating of a resistor decreases as its operating temperature increases.

- **Metal-Film Resistor** - a type of resistor whose construction consists of a thin spiral film of metal on a ceramic substrate.

- **Negative Temperature Coefficient (NTC)** - a characteristic of a thermistor indicating that its resistance decreases with an increase in operating temperature.

- **Positive Temperature Coefficient (PTC)** - a characteristic of a thermistor indicating that its resistance increases with an increase in operating temperature.

- **Potentiometer** - a three-terminal variable resistor used to vary the voltage between the center terminal and one of the outside terminals.

- **Rheostat** - a two-terminal variable resistor used to vary the amount of current in a circuit.

- **Surface-Mount Resistor** - a type of resistor constructed by depositing a thick carbon film on a ceramic base. (A surface-mount resistor is many times smaller than a conventional resistor and has no leads that extend out from the body itself.)

- **Taper** - a word describing the way the resistance of a potentiometer or rheostat varies with the rotation of its shaft.

- **Thermistor** - a resistor whose resistance value changes with changes in its operating temperature.

- **Tolerance** - the maximum allowable percent difference between the measured and coded values of resistance.

- **Wire-Wound Resistor** - a type of resistor whose construction consists of resistance wire wrapped on an insulating core.

- **Zero-Ohm Resistor** - a resistor whose ohmic value is approximately zero ohms.

- **Zero-Power Resistance** - the resistance of a thermistor with zero-power dissipation, designated R_0.

Self-Test

Answers at back of book.

1. **A carbon composition resistor having only three color stripes has a tolerance of**
   a. ± 5%.
   b. ± 20%.
   c. ± 10%.
   d. ± 100%.

2. **A resistor with a power rating of 25 W is most likely a**
   a. carbon-composition resistor.
   b. metal-film resistor.
   c. surface-mount resistor.
   d. wire-wound resistor.

3. **When checked with an ohmmeter, an open resistor measures**
   a. infinite resistance.
   b. its color-coded value.
   c. zero resistance.
   d. less than its color-coded value.

4. **One precaution to observe when checking resistors with an ohmmeter is to**
   a. check high resistances on the lowest ohms range.
   b. check low resistances on the highest ohms range.
   c. disconnect all parallel paths.
   d. make sure your fingers are touching each test lead.

5. **A chip resistor is marked 394. Its resistance value is**
   a. 39.4 Ω.
   b. 394 Ω.
   c. 390,000 Ω.
   d. 39,000 Ω.

6. **A carbon-film resistor is color-coded with red, violet, black, and gold stripes. What are its resistance and tolerance?**
   a. 27 Ω ± 5%.
   b. 270 Ω ± 5%.
   c. 270 Ω ± 10%.
   d. 27 Ω ± 10%.

7. **A potentiometer is a**
   a. three-terminal device used to vary the voltage in a circuit.
   b. two-terminal device used to vary the current in a circuit.
   c. fixed resistor.
   d. two-terminal device used to vary the voltage in a circuit.

8. **A metal-film resistor is color-coded with brown, green, red, brown, and blue stripes. What are its resistance and tolerance?**
   a. 1500 Ω ± 1.25%.
   b. 152 Ω ± 1%.
   c. 1521 Ω ± 0.5%.
   d. 1520 Ω ± 0.25%.

9. **Which of the following resistors has the smallest physical size?**
   a. wire-wound resistors.
   b. carbon-composition resistors.
   c. surface-mount resistors.
   d. potentiometers.

10. **Which of the following statements is true?**
    a. Resistors always have axial leads.
    b. Resistors are always made from carbon.
    c. There is no correlation between the physical size of a resistor and its resistance value.
    d. The shelf life of a resistor is about 1 year.

11. **If a thermistor has a negative temperature coefficient (NTC), its resistance**
    a. increases with an increase in operating temperature.
    b. decreases with a decrease in operating temperature.
    c. decreases with an increase in operating temperature.
    d. is unaffected by its operating temperature.

12. **With the four-band resistor color code, gold in the third stripe corresponds to a**
    a. fractional multiplier of 0.01.
    b. fractional multiplier of 0.1.
    c. decimal multiplier of 10.
    d. resistor tolerance of ± 10%.

13. **Which of the following axial-lead resistor types usually has a blue, light green, or red body?**
    a. wire-wound resistors.
    b. carbon-composition resistors.
    c. carbon-film resistors.
    d. metal-film resistors.

14. **A surface-mount resistor has a coded value of 4R7. This indicates a resistance of**
    a. 4.7 Ω.
    b. 4.7 kΩ.
    c. 4.7 MΩ.
    d. none of the above.

15. **Reading from left to right, the colored bands on a resistor are yellow, violet, brown and gold. If the resistor measures 513Ω with an ohmmeter, it is**
    a. well within tolerance.
    b. out of tolerance.
    c. right on the money.
    d. close enough to be considered within tolerance.

Questions

1. List five different types of fixed resistors.

2. List the advantages of using a metal-film resistor versus a carbon-composition resistor.

3. Draw the schematic symbols for a (a) fixed resistor (b) potentiometer (c) rheostat (d) thermistor.

4. How can a technician identify a wire-wound resistor that is color-coded?

5. Explain an application using a decade resistance box.

6. List the differences between a potentiometer and a rheostat.

7. For resistors using the four-band code, what are the values for gold and silver as fractional decimal multipliers in the third band?

8. Briefly describe how you would check to see whether a 1-MΩ resistor is open or not. Give two precautions to make sure the test is not misleading.

9. Define the term "zero-power resistance" as it relates to thermistors.

10. Explain how the ambient temperature affects the power rating of a carbon resistor.

Problems

Answers to odd-numbered problems at back of book.

SECTION 2–2 RESISTOR COLOR CODING

2–1 Indicate the resistance and tolerance for each resistor shown in Fig. 2–22.

2–2 Indicate the resistance and tolerance for each resistor shown in Fig. 2–23.

2–3 Indicate the resistance for each chip resistor shown in Fig. 2–24.

2–4 Calculate the permissible ohmic range of a resistor whose resistance value and tolerance are (a) 3.9 kΩ ± 5% (b) 100 Ω ± 10% (c) 120 kΩ ± 2% (d) 2.2 Ω ± 5% (e) 75 Ω ± 1%.

2–5 Using the four-band code, indicate the colors of the bands for each of the following resistors: (a) 10 kΩ ± 5% (b) 2.7 Ω ± 5% (c) 5.6 kΩ ± 10% (d) 1.5 MΩ ± 5% (e) 0.22 Ω ± 5%.

2–6 Using the five-band code, indicate the colors of the bands for each of the following resistors: (a) 110 Ω ± 1% (b) 34 kΩ ± 0.5% (c) 82.5 kΩ ± 2% (d) 62.6 Ω ± 1% (e) 105 kΩ ± 0.1%.

SECTION 2–3 VARIABLE RESISTORS.

2–7 Refer to Fig. 2–25 on page 78. Indicate the total resistance R_T for each of the different dial settings in Table 2–3, p. 78.

SECTION 2–4 RHEOSTATS AND POTENTIOMETERS

2–8 Show two different ways to wire a potentiometer so that it will work as a rheostat.

Figure 2–22 Resistors for Prob. 2–1.

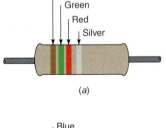

(a)

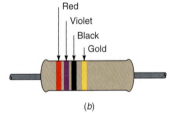

(b)

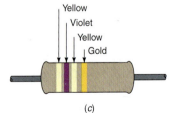

(c)

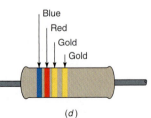

(d)

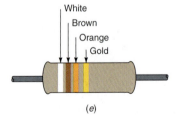

(e)

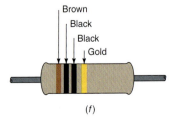

(f)

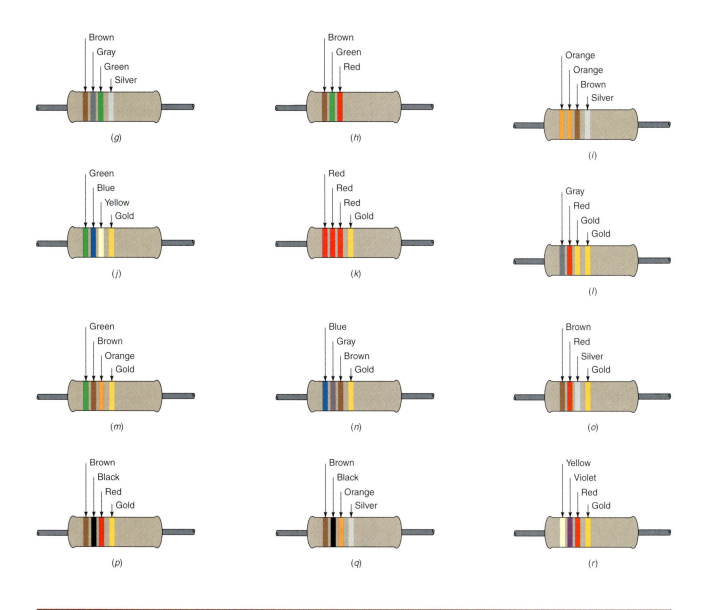

(g) Brown Gray Green Silver

(h) Brown Green Red

(i) Orange Orange Brown Silver

(j) Green Blue Yellow Gold

(k) Red Red Red Gold

(l) Gray Red Gold Gold

(m) Green Brown Orange Gold

(n) Blue Gray Brown Gold

(o) Brown Red Silver Gold

(p) Brown Black Red Gold

(q) Brown Black Orange Silver

(r) Yellow Violet Red Gold

Figure 2–23 Resistors for Prob. 2–2.

(a) Red Red Blue Brown Green

(b) Brown Black Green Gold Violet

(c) Blue Red Blue Orange Brown

(d) Green Red Orange Silver Blue

(e) Brown Black Black Brown Brown

(f) Orange Violet Yellow Red Brown

Figure 2–23 (*Continued*)

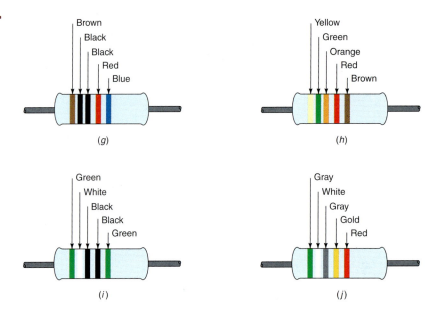

(g)

(h)

(i)

(j)

Figure 2–24 Chip resistors for Prob. 2–3.

474	122
(a)	(b)
331	1002
(c)	(d)

Figure 2–25 Decade resistance box.

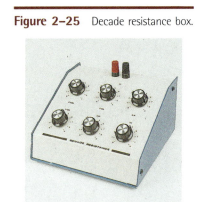

Table 2–3	Decade Resistance Box Dial Settings for Problem 2–7				
	$R \times 100$ k	$R \times 10$ k	$R \times 1$ k	$R \times 100$	$R \times 10$	$R \times 1$
(a)	6	8	0	2	2	5
(b)	0	0	8	2	5	0
(c)	0	1	8	5	0	3
(d)	2	7	5	0	6	0
(e)	0	6	2	9	8	4

Critical Thinking

2–9 A manufacturer of carbon-film resistors specifies a maximum working voltage of 250 V for all its $1/4$-W resistors. Exceeding 250 V causes internal arcing within the resistor. Above what minimum resistance will the maximum working voltage be exceeded before its $1/4$-W power dissipation rating is exceeded? **Hint:** The maximum voltage which produces the rated power dissipation can be calculated as $V_{max} = \sqrt{P \times R}$.

2–10 What is the power rating of a $1/2$-W carbon resistor if it is used at an ambient temperature of 120°C.

Answers to Knowledge Check Problems

2–2A a. 3.3 kΩ ±5%
   b. 44.2 kΩ ±1%

2–2B a. 100 kΩ
   b. 22 kΩ
   c. 1.8 Ω

Answers to Self-Reviews

2–1 a. T
   b. T
   c. metal-film
   d. 1000-Ω, 1-W
   e. R decreases

2–2 a. yellow
   b. ±10%
   c. 100
   d. 470 Ω ±5%
   e. 33,000 Ω or 33 kΩ

2–3 a. terminal 2
   b. nonlinear
   c. 470,000 Ω or 470 kΩ

2–4 a. four connections to three terminals
   b. two
   c. 80 V

2–5 a. T
   b. F

2–6 a. 0 Ω
   b. infinite ohms
   c. open circuit
   d. carbon-composition resistor

3 Ohm's Law

The mathematical relationship between voltage, current, and resistance was discovered in 1826 by Georg Simon Ohm. The relationship, known as Ohm's law, is the basic foundation for all circuit analysis in electronics. Ohm's law, which is the basis of this chapter, states that the amount of current, I, is directly proportional to the voltage, V, and inversely proportional to the resistance, R. Expressed mathematically, Ohm's law is stated as

$$I = \frac{V}{R}.$$

Besides the coverage of Ohm's law, this chapter also introduces you to the concept of power. Power can be defined as the time rate of doing work. The symbol for power is P and the unit is the watt. All of the mathematical relationships that exist between V, I, R, and P are covered in this chapter.

In addition to Ohm's law and power, this chapter also discusses electric shock and open- and short-circuit troubles.

Objectives

After studying this chapter you should be able to

- *List* the three forms of Ohm's law.
- *Use* Ohm's law to calculate the current, voltage, or resistance in a circuit.
- *List* the multiple and submultiple units of voltage, current, and resistance.
- *Explain* the difference between a linear and a nonlinear resistance.
- *Explain* the difference between work and power and list the units of each.
- *Calculate* the power in a circuit when the voltage and current, current and resistance, or voltage and resistance are known.
- *Determine* the required resistance and appropriate wattage rating of a resistor.
- *Identify* the shock hazards associated with working with electricity.
- *Explain* the difference between an open circuit and short circuit.

Outline

Important Terms

ampere

electron volt (eV)

horsepower (hp)

inverse relation

joule

kilowatt-hour (kWh)

linear proportion

linear resistance

maximum working voltage rating

nonlinear resistance

ohm

open circuit

power

short circuit

volt

volt-ampere characteristic

watt

3–1 The Current $I = V/R$

If we keep the same resistance in a circuit but vary the voltage, the current will vary. The circuit in Fig. 3–1 demonstrates this idea. The applied voltage V can be varied from 0 to 12 V, as an example. The bulb has a 12-V filament, which requires this much voltage for its normal current to light with normal intensity. The meter I indicates the amount of current in the circuit for the bulb.

With 12 V applied, the bulb lights, indicating normal current. When V is reduced to 10 V, there is less light because of less I. As V decreases, the bulb becomes dimmer. For zero volts applied, there is no current and the bulb cannot light. In summary, the changing brilliance of the bulb shows that the current varies with the changes in applied voltage.

For the general case of any V and R, Ohm's law is

$$I = \frac{V}{R} \qquad\qquad\qquad (3\text{--}1)$$

where I is the amount of current through the resistance R connected across the source of potential difference V. With volts as the practical unit for V and ohms for R, the amount of current I is in amperes. Therefore,

$$\text{Amperes} = \frac{\text{volts}}{\text{ohms}}$$

This formula says simply to divide the voltage across R by the ohms of resistance between the two points of potential difference to calculate the amperes of current through R. In Fig. 3–2, for instance, with 6 V applied across a 3-Ω resistance, by Ohm's law, the amount of current I equals $\frac{6}{3}$ or 2 A.

High Voltage but Low Current

It is important to realize that with high voltage, the current can have a low value when there is a very high resistance in the circuit. For example, 1000 V applied across 1,000,000 Ω results in a current of only $\frac{1}{1000}$ A. By Ohm's law,

$$\begin{aligned} I &= \frac{V}{R} \\ &= \frac{1000 \text{ V}}{1{,}000{,}000 \text{ }\Omega} = \frac{1}{1000} \\ I &= 0.001 \text{ A} \end{aligned}$$

The practical fact is that high-voltage circuits usually do have small values of current in electronic equipment. Otherwise, tremendous amounts of power would be necessary.

Figure 3–1 Increasing the applied voltage V produces more current I to light the bulb with more intensity.

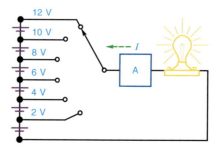

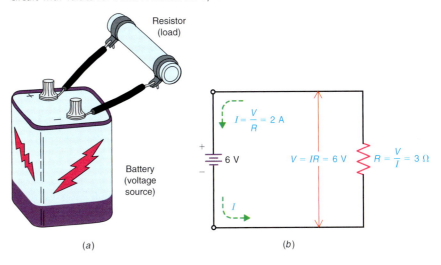

MultiSim **Figure 3–2** Example of using Ohm's law. (*a*) Wiring diagram of a circuit with a 6-V battery for *V* applied across a load *R*. (*b*) Schematic diagram of the circuit with values for *I* and *R* calculated by Ohm's law.

CALCULATOR

To do a division problem like *V/R* in Example 3–1 on the calculator, punch in the number 120 for the numerator, then press the key ⊕ for division before punching in 8 for the denominator. Finally, press the ⊜ key for the answer of 15 on the display. The numerator must be punched in first.

Low Voltage but High Current

At the opposite extreme, a low value of voltage in a very low resistance circuit can produce a very high current. A 6-V battery connected across a resistance of 0.01 Ω produces 600 A of current:

$$I = \frac{V}{R}$$
$$= \frac{6 \text{ V}}{0.01 \text{ Ω}}$$
$$I = 600 \text{ A}$$

Less *I* with More *R*

Note the values of *I* in the following two examples also:

Example 3–1

A heater with a resistance of 8 Ω is connected across the 120-V power line. How much is current *I*?

ANSWER

$$I = \frac{V}{R} = \frac{120 \text{ V}}{8 \text{ Ω}}$$
$$I = 15 \text{ A}$$

Example 3-2

A small lightbulb with a resistance of 2400 Ω is connected across the same 120-V power line. How much is current I?

ANSWER

$$I = \frac{V}{R} = \frac{120 \text{ V}}{2400 \text{ }\Omega}$$
$$I = 0.05 \text{ A}$$

Although both cases have the same 120 V applied, the current is much less in Example 2 because of the higher resistance.

■ 3-1 Knowledge Check

Answer at end of chapter.

How much is the current, I, in a 150-Ω resistor if the voltage, V, is 24 V?

Typical *V* and *I*

Transistors and integrated circuits generally operate with a dc supply of 5, 6, 9, 12, 15, 24, or 50 V. The current is usually in millionths or thousandths of one ampere up to about 5 A.

■ 3-1 Self-Review

Answers at end of chapter.

a. Calculate I for 24 V applied across 8 Ω.
b. Calculate I for 12 V applied across 8 Ω.
c. Calculate I for 24 V applied across 12 Ω.
d. Calculate I for 6 V applied across 1 Ω.

3-2 The Voltage $V = IR$

Referring back to Fig. 3–2, the voltage across R must be the same as the source V because the resistance is connected directly across the battery. The numerical value of this V is equal to the product $I \times R$. For instance, the IR voltage in Fig. 3–2 is 2 A × 3 Ω, which equals the 6 V of the applied voltage. The formula is

$$V = IR \tag{3-2}$$

With I in ampere units and R in ohms, their product V is in volts. Actually, this must be so because the I value equal to V/R is the amount that allows the IR product to be the same as the voltage across R.

Beside the numerical calculations possible with the IR formula, it is useful to consider that the IR product means voltage. Whenever there is current through a resistance, it must have a potential difference across its two ends equal to the IR product. If there were no potential difference, no electrons could flow to produce the current.

Example 3-3

If a 12-Ω resistor is carrying a current of 2.5 A, how much is its voltage?

ANSWER

$$V = IR$$
$$= 2.5 \text{ A} \times 12 \text{ } \Omega$$
$$= 30 \text{ V}$$

CALCULATOR

To do a multiplication problem like $I \times R$ in example 3-3 on the calculator, punch in the factor 2.5, then press the ⊗ key for multiplication before punching in 12 for the other factor. Finally, press the ⊜ key for the answer of 30 on the display. The factors can be multiplied in any order.

■ *3-2 Knowledge Check*

> *Answer at end of chapter.*
>
> **How much voltage is across a 470-Ω resistor if its current is 0.05 A?**

■ *3-2 Self-Review*

> *Answers at end of chapter.*
>
> a. Calculate V for 0.002 A through 1000 Ω.
> b. Calculate V for 0.004 A through 1000 Ω.
> c. Calculate V for 0.002 A through 2000 Ω.

3-3 The Resistance $R = V/I$

As the third and final version of Ohm's law, the three factors V, I, and R are related by the formula

$$R = \frac{V}{I} \tag{3-3}$$

In Fig. 3–2, R is 3 Ω because 6 V applied across the resistance produces 2 A through it. Whenever V and I are known, the resistance can be calculated as the voltage across R divided by the current through it.

Physically, a resistance can be considered some material whose elements have an atomic structure that allows free electrons to drift through it with more or less force applied. Electrically, though, a more practical way of considering resistance is simply as a V/I ratio. Anything that allows 1 A of current with 10 V applied has a resistance of 10 Ω. This V/I ratio of 10 Ω is its characteristic. If the voltage is doubled to 20 V, the current will also double to 2 A, providing the same V/I ratio of a 10-Ω resistance.

Furthermore, we do not need to know the physical construction of a resistance to analyze its effect in a circuit, so long as we know its V/I ratio. This idea is illustrated in Fig. 3–3. Here, a box with some unknown material in it is connected in a circuit where we can measure the 12 V applied across the box and the 3 A of current through it. The resistance is 12V/3A, or 4 Ω. There may be liquid, gas, metal, powder, or any other material in the box; but electrically the box is just a 4-Ω resistance because its V/I ratio is 4.

GOOD TO KNOW

Since R and G are reciprocals of each other, the conductance, G, of a circuit can be calculated as

$$G = \frac{I}{V}.$$

Figure 3–3 The resistance R of any component is its V/I ratio.

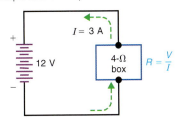

Example 3-4

How much is the resistance of a lightbulb if it draws 0.16 A from a 12-V battery?

ANSWER

$$R = \frac{V}{I}$$
$$= \frac{12 \text{ V}}{0.16 \text{ A}}$$
$$= 75 \text{ }\Omega$$

■ *3–3 Knowledge Check*

Answer at end of chapter.

A heater draws 8 A of current from the 120-V power line. How much is its resistance?

■ *3–3 Self-Review*

Answers at end of chapter.

a. Calculate *R* for 12 V with 0.003 A.
b. Calculate *R* for 12 V with 0.006 A.
c. Calculate *R* for 12 V with 0.001 A.

3–4 Practical Units

The three forms of Ohm's law can be used to define the practical units of current, potential difference, and resistance as follows:

$$1 \text{ ampere} = \frac{1 \text{ volt}}{1 \text{ ohm}}$$
$$1 \text{ volt} = 1 \text{ ampere} \times 1 \text{ ohm}$$
$$1 \text{ ohm} = \frac{1 \text{ volt}}{1 \text{ ampere}}$$

One **ampere** is the amount of current through a one-ohm resistance that has one volt of potential difference applied across it.

One **volt** is the potential difference across a one-ohm resistance that has one ampere of current through it.

One **ohm** is the amount of opposition in a resistance that has a *V/I* ratio of 1, allowing one ampere of current with one volt applied.

In summary, the circle diagram in Fig. 3–4 for *V = IR* can be helpful in using Ohm's law. Put your finger on the unknown quantity and the desired formula remains. The three possibilities are

Cover *V* and you have *IR*.
Cover *I* and you have *V/R*.
Cover *R* and you have *V/I*.

Figure 3–4 A circle diagram to help in memorizing the Ohm's law formulas *V = IR*, *I = V/R*, and *R = V/I*. The *V* is always at the top.

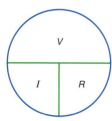

■ 3–4 Knowledge Check

■ 3–4 Knowledge Check

Answers at end of chapter.

Solve for the unknown quantity in each of the following problems.
(a) $V = 100$ V, $R = 1,000\ \Omega$, $I = ?$
(b) $I = 0.004$ A, $R = 15,000\ \Omega$, $V = ?$
(c) $V = 75$ V, $I = 0.2$ A, $R = ?$

■ 3–4 Self-Review

Answers at end of chapter.

a. Calculate V for 0.007 A through 5000 Ω.
b. Calculate the amount of I for 12,000 V across 6,000,000 Ω.
c. Calculate R for 8 V with 0.004 A.

3–5 Multiple and Submultiple Units

The basic units—ampere, volt, and ohm—are practical values in most electric power circuits, but in many electronics applications, these units are either too small or too big. As examples, resistances can be a few million ohms, the output of a high-voltage supply in a computer monitor is about 20,000 V, and the current in transistors is generally thousandths or millionths of an ampere.

In such cases, it is often helpful to use multiples and submultiples of the basic units. These multiple and submultiple values are based on the metric system of units discussed earlier. The common conversions for V, I, and R are summarized at the end of this chapter, but a complete listing of all metric prefixes is in the Introduction to Powers of 10 chapter.

Example 3-5   ‖‖ MultiSim

The I of 8 mA flows through a 5-kΩ R. How much is the IR voltage?

ANSWER

$$V = IR = 8 \times 10^{-3} \times 5 \times 10^{3} = 8 \times 5$$
$$V = 40\ \text{V}$$

In general, milliamperes multiplied by kilohms results in volts for the answer, as 10^{-3} and 10^{3} cancel.

Example 3-6

How much current is produced by 60 V across 12 kΩ?

ANSWER

$$I = \frac{V}{R} = \frac{60}{12 \times 10^3}$$
$$= 5 \times 10^{-3} = 5 \text{ mA}$$

Note that volts across kilohms produces milliamperes of current. Similarly, volts across megohms produces microamperes.

In summary, common combinations to calculate the current I are

$$\frac{V}{k\Omega} = mA \quad \text{and} \quad \frac{V}{M\Omega} = \mu A$$

Also, common combinations to calculate IR voltage are

$$mA \times k\Omega = V$$
$$\mu A \times M\Omega = V$$

These relationships occur often in electronic circuits because the current is generally in units of milliamperes or microamperes. A useful relationship to remember is that 1 mA is equal to 1000 μA.

■ 3–5 Knowledge Check

Answer at end of chapter.

How much is the current in a 1.2-kΩ resistor if it has a voltage drop of 60 mV?

■ 3–5 Self-Review

Answers at end of chapter.

a. Change the following to basic units with powers of 10 instead of metric prefixes: 6 mA, 5 kΩ, and 3 μA.
b. Change the following powers of 10 to units with metric prefixes: 6×10^{-3} A, 5×10^3 Ω, and 3×10^{-6} A.
c. Which is larger, 2 mA or 20 μA?
d. How much current flows in a 560-kΩ resistor if the voltage is 70 V?

3–6 The Linear Proportion between V and I

The Ohm's law formula $I = V/R$ states that V and I are directly proportional for any one value of R. This relation between V and I can be analyzed by using a fixed resistance of 2 Ω for R_L, as in Fig. 3–5. Then when V is varied, the meter shows I values directly proportional to V. For instance, with 12 V, I equals 6 A; for 10 V, the current is 5 A; an 8-V potential difference produces 4 A.

All the values of V and I are listed in the table in Fig. 3–5b and plotted in the graph in Fig. 3–5c. The I values are one-half the V values because R is 2 Ω. However, I is zero with zero volts applied.

|||| MultiSim **Figure 3–5** Experiment to show that *I* increases in direct proportion to *V* with the same *R*. (*a*) Circuit with variable *V* but constant *R*. (*b*) Table of increasing *I* for higher *V*. (*c*) Graph of *V* and *I* values. This is a linear volt-ampere characteristic. It shows a direct proportion between *V* and *I*.

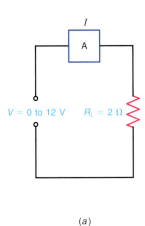

(*a*)

Volts V	Ohms Ω	Amperes A
0	2	0
2	2	1
4	2	2
6	2	3
8	2	4
10	2	5
12	2	6

(*b*)

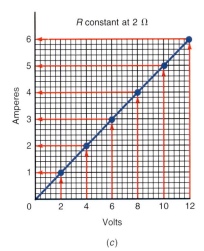

(*c*)

Plotting the Graph

The voltage values for *V* are marked on the horizontal axis, called the *x axis* or *abscissa*. The current values *I* are on the vertical axis, called the *y axis* or *ordinate*.

Because the values for *V* and *I* depend on each other, they are variable factors. The independent variable here is *V* because we assign values of voltage and note the resulting current. Generally, the independent variable is plotted on the *x* axis, which is why the *V* values are shown here horizontally and the *I* values are on the ordinate.

The two scales need not be the same. The only requirement is that equal distances on each scale represent equal changes in magnitude. On the *x* axis here, 2-V steps are chosen, whereas the *y* axis has 1-A scale divisions. The zero point at the origin is the reference.

The plotted points in the graph show the values in the table. For instance, the lowest point is 2 V horizontally from the origin, and 1 A up. Similarly, the next point is at the intersection of the 4-V mark and the 2-A mark.

A line joining these plotted points includes all values of *I*, for any value of *V*, with *R* constant at 2 Ω. This also applies to values not listed in the table. For instance, if we take the value of 7 V up to the straight line and over to the *I* axis, the graph shows 3.5 A for *I*.

Volt-Ampere Characteristic

The graph in Fig. 3–5c is called the *volt-ampere characteristic* of *R*. It shows how much current the resistor allows for different voltages. Multiple and submultiple units of *V* and *I* can be used, though. For transistors, the units of *I* are often milliamperes or microamperes.

Linear Resistance

The *straight-line (linear) graph* in Fig. 3–5 shows that *R* is a linear resistor. A linear resistance has a constant value of ohms. Its *R* does not change with the applied voltage. Then *V* and *I* are directly proportional. Doubling the value of *V* from 4 to 8 V results in twice the current, from 2 to 4 A. Similarly, three or four times the value of *V* will produce three or four times *I*, for a proportional increase in current.

GOOD TO KNOW

In Fig. 3–5c, the slope of the straight line increases as *R* decreases. Conversely, the slope decreases as *R* increases. For any value of *R*, the slope of the straight line can be calculated as $\Delta I/\Delta V$ or $\frac{1}{R}$.

Nonlinear Resistance

This type has a nonlinear volt-ampere characteristic. As an example, the resistance of the tungsten filament in a light bulb is nonlinear. The reason is that R increases with more current as the filament becomes hotter. Increasing the applied voltage does produce more current, but I does not increase in the same proportion as the increase in V. Another example of a nonlinear resistor is a thermistor.

Inverse Relation between *I* and *R*

Whether R is linear or not, the current I is less for more R, with the applied voltage constant. This is an inverse relation, that is, I goes down as R goes up. Remember that in the formula $I = V/R$, the resistance is in the denominator. A higher value of R actually lowers the value of the complete fraction.

As an example, let V be constant at 1 V. Then I is equal to the fraction $1/R$. As R increases, the values of I decrease. For R of 2 Ω, I is ½ or 0.5 A. For a higher R of 10 Ω, I will be lower at ¹⁄₁₀ or 0.1 A.

■ *3-6 Knowledge Check*

> *Answer at end of chapter.*
>
> In Fig. 3–5, how much is the current, I, if V = 7 V? How much is the current if V is doubled to 14 V?

■ *3-6 Self-Review*

> *Answers at end of chapter.*
>
> Refer to the graph in Fig. 3–5c.
> a. Are the values of I on the y or x axis?
> b. Is this R linear or nonlinear?
> c. If the voltage across a 5-Ω resistor increases from 10 V to 20 V, what happens to I?
> d. The voltage across a 5-Ω resistor is 10 V. If R is doubled to 10 Ω, what happens to I?

3–7 Electric Power

The unit of electric *power* is the *watt* (W), named after James Watt (1736–1819). One watt of power equals the work done in one second by one volt of potential difference in moving one coulomb of charge.

Remember that one coulomb per second is an ampere. Therefore power in watts equals the product of volts times amperes.

$$\text{Power in watts} = \text{volts} \times \text{amperes}$$
$$P = V \times I \qquad (3\text{--}4)$$

When a 6-V battery produces 2 A in a circuit, for example, the battery is generating 12 W of power.

The power formula can be used in three ways:

$$P = V \times I$$
$$I = P \div V \quad \text{or} \quad \frac{P}{V}$$
$$V = P \div I \quad \text{or} \quad \frac{P}{I}$$

Which formula to use depends on whether you want to calculate P, I, or V. Note the following examples:

Example 3-7

A toaster takes 10 A from the 120-V power line. How much power is used?

ANSWER

$$P = V \times I = 120 \text{ V} \times 10 \text{ A}$$
$$P = 1200 \text{ W}$$

Example 3-8

How much current flows in the filament of a 300-W bulb connected to the 120-V power line?

ANSWER

$$I = \frac{P}{V} = \frac{300 \text{ W}}{120 \text{ V}}$$
$$I = 2.5 \text{ A}$$

Example 3-9

How much current flows in the filament of a 60-W bulb connected to the 120-V power line?

ANSWER

$$I = \frac{P}{V} = \frac{60 \text{ W}}{120 \text{ V}}$$
$$I = 0.5 \text{ A}$$

Note that the lower wattage bulb uses less current.

Work and Power

Work and energy are essentially the same with identical units. Power is different, however, because it is the time rate of doing work.

As an example of work, if you move 100 lb a distance of 10 ft, the work is 100 lb × 10 ft or 1000 ft · lb, regardless of how fast or how slowly the work is done. Note that the unit of work is foot-pounds, without any reference to time.

However, power equals the work divided by the time it takes to do the work. If it takes 1 s, the power in this example is 1000 ft · lb/s; if the work takes 2 s, the power is 1000 ft · lb in 2 s, or 500 ft · lb/s.

Similarly, electric power is the rate at which charge is forced to move by voltage. This is why power in watts is the product of volts and amperes. The voltage states the amount of work per unit of charge; the current value includes the rate at which the charge is moved.

Watts and Horsepower Units

A further example of how electric power corresponds to mechanical power is the fact that

$$746 \text{ W} = 1 \text{ hp} = 550 \text{ ft} \cdot \text{lb/s}$$

This relation can be remembered more easily as 1 hp equals approximately ¾ kilowatt (kW). One kilowatt = 1000 W.

Practical Units of Power and Work

Starting with the watt, we can develop several other important units. The fundamental principle to remember is that power is the time rate of doing work, whereas work is power used during a period of time. The formulas are

$$\text{Power} = \frac{\text{work}}{\text{time}} \tag{3–5}$$

and

$$\text{Work} = \text{power} \times \text{time} \tag{3–6}$$

With the watt unit for power, one watt used during one second equals the work of one joule. Or one watt is one joule per second. Therefore, 1 W = 1 J/s. The **joule** is a basic practical unit of work or energy.

To summarize these practical definitions,

1 joule = 1 watt · second
1 watt = 1 joule/second

In terms of charge and current,

1 joule = 1 volt · coulomb
1 watt = 1 volt · ampere

Remember that the ampere unit includes time in the denominator, since the formula is 1 ampere = 1 coulomb/second.

Electron Volt (eV)

This unit of work can be used for an individual electron, rather than the large quantity of electrons in a coulomb. An electron is charge, and the volt is potential difference. Therefore, 1 eV is the amount of work required to move an electron between two points that have a potential difference of one volt.

The number of electrons in one coulomb for the joule unit equals 6.25×10^{18}. Also, the work of one joule is a volt-coulomb. Therefore, the number of electron volts equal to one joule must be 6.25×10^{18}. As a formula,

$$1 \text{ J} = 6.25 \times 10^{18} \text{ eV}$$

PIONEERS
OF ELECTRONICS

The SI unit of measure for electrical energy is the joule. Named for English physicist, *James Prescott Joule (1818–1889)*, one joule (J) is equal to one volt-coulomb.

Either the electron volt or the joule unit of work is the product of charge times voltage, but the watt unit of power is the product of voltage times current. The division by time to convert work to power corresponds to the division by time that converts charge to current.

Kilowatt-Hours

This is a unit commonly used for large amounts of electrical work or energy. The amount is calculated simply as the product of the power in kilowatts multiplied by the time in hours during which the power is used. As an example, if a lightbulb uses 300 W or 0.3 kW for 4 hours (h), the amount of energy is 0.3×4, which equals 1.2 kWh.

We pay for electricity in kilowatt-hours of energy. The power-line voltage is constant at 120 V. However, more appliances and light bulbs require more current because they all add in the main line to increase the power.

Suppose that the total load current in the main line equals 20 A. Then the power in watts from the 120-V line is

$$P = 120 \text{ V} \times 20 \text{ A}$$
$$P = 2400 \text{ W or } 2.4 \text{ kW}$$

If this power is used for 5h, then the energy or work supplied equals $2.4 \times 5 = 12$ kWh. If the cost of electricity is 6¢/kWh, then 12 kWh of electricity will cost $0.06 \times 12 = 0.72$ or 72¢. This charge is for a 20-A load current from the 120-V line during the time of 5 h.

Example 3-10

Assuming that the cost of electricity is 6 ¢/kWh, how much will it cost to light a 100-W lightbulb for 30 days?

ANSWER   The first step in solving this problem is to express 100 W as 0.1 kW. The next step is to find the total number of hours in 30 days. Since there are 24 hours in a day, the total number of hours for which the light is on is calculated as

$$\text{Total Hours} = \frac{24 \text{ h}}{\text{day}} \times 30 \text{ days} = 720 \text{ h}$$

Next, calculate the number of kWh as

$$\begin{aligned} \text{kWh} &= \text{kW} \times \text{h} \\ &= 0.1 \text{ kW} \times 720 \text{ h} \\ &= 72 \text{ kWh} \end{aligned}$$

And finally, determine the cost. (Note that 6¢ = $0.06.)

$$\begin{aligned} \text{Cost} &= \text{kWh} \times \frac{\text{cost}}{\text{kWh}} \\ &= 72 \text{ kWh} \times \frac{0.06}{\text{kWh}} \\ &= \$4.32 \end{aligned}$$

Answer at end of chapter.

How much will it cost to operate a 1500-W heater for 1 day? The cost of electricity is 7¢/kWh.

■ *3–7B Knowledge Check*

Answer at end of chapter.

What is the power rating of a soldering iron if it draws 250 mA of current from the 120-V power-line?

■ *3–7 Self-Review*

Answers at end of chapter.

a. An electric heater takes 15 A from the 120-V power line. Calculate the amount of power used.
b. How much is the load current for a 100-W bulb connected to the 120-V power line?
c. How many watts is the power of 200 J/s equal to?
d. How much will it cost to operate a 300-W light bulb for 48 h if the cost of electricity is 7¢/kWh?

GOOD TO KNOW

The power dissipated by a resistance is proportional to I^2. In other words, if the current, I, carried by a resistor is doubled, the power dissipation in the resistance increases by a factor of 4.

GOOD TO KNOW

The power dissipated by a resistance is proportional to V^2. In other words, if the voltage, V, across a resistor is doubled, the power dissipation in the resistance increases by a factor of 4.

3–8 Power Dissipation in Resistance

When current flows in a resistance, heat is produced because friction between the moving free electrons and the atoms obstructs the path of electron flow. The heat is evidence that power is used in producing current. This is how a fuse opens, as heat resulting from excessive current melts the metal link in the fuse.

The power is generated by the source of applied voltage and consumed in the resistance as heat. As much power as the resistance dissipates in heat must be supplied by the voltage source; otherwise, it cannot maintain the potential difference required to produce the current.

The correspondence between electric power and heat is indicated by the fact that 1W used during 1 s is equivalent to 0.24 calorie of heat energy. The electric energy converted to heat is considered dissipated or used up because the calories of heat cannot be returned to the circuit as electric energy.

Since power is dissipated in the resistance of a circuit, it is convenient to express the power in terms of the resistance R. The formula $P = V \times I$ can be rearranged as follows:

Substituting IR for V,

$$P = V \times I = IR \times I$$
$$P = I^2R \tag{3–7}$$

This is a common form of the power formula because of the heat produced by current in a resistance.

For another form, substitute V/R for I. Then

$$P = V \times I = V \times \frac{V}{R}$$
$$P = \frac{V^2}{R} \tag{3–8}$$

In all the formulas, V is the voltage across R in ohms, producing the current I in amperes, for power in watts.

║║ MultiSim **Figure 3–6** Calculating the electric power in a circuit as $P = V \times I$, $P = I^2R$, or $P = V^2/R$.

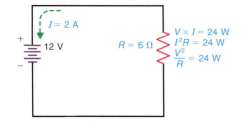

$I = 2$ A

12 V

$R = 6\ \Omega$

$V \times I = 24$ W
$I^2R = 24$ W
$\dfrac{V^2}{R} = 24$ W

Any one of the three formulas (3–4), (3–7), and (3–8) can be used to calculate the power dissipated in a resistance. The one to be used is a matter of convenience, depending on which factors are known.

In Fig. 3–6, for example, the power dissipated with 2 A through the resistance and 12 V across it is $2 \times 12 = 24$ W.

Or, calculating in terms of just the current and resistance, the power is the product of 2 squared, or 4, times 6, which equals 24 W.

Using the voltage and resistance, the power can be calculated as 12 squared, or 144, divided by 6, which also equals 24 W.

No matter which formula is used, 24 W of power is dissipated as heat. This amount of power must be generated continuously by the battery to maintain the potential difference of 12 V that produces the 2-A current against the opposition of 6 Ω.

In some applications, electric power dissipation is desirable because the component must produce heat to do its job. For instance, a 600-W toaster must dissipate this amount of power to produce the necessary amount of heat. Similarly, a 300-W lightbulb must dissipate this power to make the filament white-hot so that it will have the incandescent glow that furnishes the light. In other applications,

Example 3-11 |||| MultiSim

Calculate the power in a circuit where the source of 100 V produces 2 A in a 50-Ω R.

ANSWER

$$P = I^2R = 2 \times 2 \times 50 = 4 \times 50$$
$$P = 200 \text{ W}$$

This means that the source delivers 200 W of power to the resistance and the resistance dissipates 200 W as heat.

Example 3-12 |||| MultiSim

Calculate the power in a circuit in which the same source of 100 V produces 4 A in a 25-Ω R.

ANSWER

$$P = I^2R = 4^2 \times 25 = 16 \times 25$$
$$P = 400 \text{ W}$$

Note the higher power in Example 3–12 because of more I, even though R is less than that in Example 11.

however, the heat may be just an undesirable by-product of the need to provide current through the resistance in a circuit. In any case, though, whenever there is current I in a resistance R, it dissipates the amount of power P equal to I^2R.

Components that use the power dissipated in their resistance, such as lightbulbs and toasters, are generally rated in terms of power. The power rating is given at normal applied voltage, which is usually the 120 V of the power line. For instance, a 600-W, 120-V toaster has this rating because it dissipates 600 W in the resistance of the heating element when connected across 120 V.

Note this interesting point about the power relations. The lower the source voltage, the higher the current required for the same power. The reason is that $P = V \times I$. For instance, an electric heater rated at 240 W from a 120-V power line takes 240 W/120 V = 2 A of current from the source. However, the same 240 W from a 12-V source, as in a car or boat, requires 240 W/12 V = 20 A. More current must be supplied by a source with lower voltage, to provide a specified amount of power.

■ 3–8 Knowledge Check

Answer at end of chapter.

How much power is dissipated by a 1-kΩ resistance if its current is 20 mA?

■ 3–8 Self-Review

Answers at end of chapter.

a. Current I is 2 A in a 5-Ω R. Calculate P.
b. Voltage V is 10 V across a 5-Ω R. Calculate P.
c. Resistance R has 10 V with 2 A. Calculate the values for P and R.

3–9 Power Formulas

To calculate I or R for components rated in terms of power at a specified voltage, it may be convenient to use the power formulas in different forms. There are three basic power formulas, but each can be in three forms for nine combinations.

$$P = VI \qquad\qquad P = I^2R \qquad\qquad P = \frac{V^2}{R}$$

$$\text{or} \quad I = \frac{P}{V} \quad \text{or} \quad R = \frac{P}{I^2} \quad \text{or} \quad R = \frac{V^2}{P}$$

$$\text{or} \quad V = \frac{P}{I} \quad \text{or} \quad I = \sqrt{\frac{P}{R}} \quad \text{or} \quad V = \sqrt{PR}$$

Example 3–13

How much current is needed for a 600-W, 120-V toaster?

ANSWER

$$I = \frac{P}{V} = \frac{600}{120}$$
$$I = 5 \text{ A}$$

Example 3-14

How much is the resistance of a 600-W, 120-V toaster?

ANSWER

$$R = \frac{V^2}{P} = \frac{(120)^2}{600} = \frac{14{,}400}{600}$$
$$R = 24 \ \Omega$$

Example 3-15

How much current is needed for a 24-Ω R that dissipates 600 W?

ANSWER

$$I = \sqrt{\frac{P}{R}} = \sqrt{\frac{600 \text{ W}}{24 \ \Omega}} = \sqrt{25}$$
$$I = 5 \text{ A}$$

■ *3–9 Knowledge Check*

Answer at end of chapter.

If a 1-kΩ resistor is dissipating 2.5 W of power, how much is its current?

Note that all these formulas are based on Ohm's law $V = IR$ and the power formula $P = VI$. The following example with a 300-W bulb also illustrates this idea. Refer to Fig. 3–7. The bulb is connected across the 120-V line. Its 300-W filament requires a current of 2.5 A, equal to P/V. These calculations are

$$I = \frac{P}{V} = \frac{300 \text{ W}}{120 \text{ V}} = 2.5 \text{ A}$$

The proof is that the VI product is 120×2.5, which equals 300 W.

CALCULATOR

To use the calculator for a problem like Example 3-14 that involves a square and division for V^2/R, use the following procedure:

■ Punch in the V value of 120.

■ Press the key marked $\widehat{x^2}$ for the square of 120, equal to 14,400 on the display.

■ Next, press the division $\oplus$ key.

■ Punch in the value of 600 for R.

■ Finally, press the $\ominus$ key for the answer of 24 on the display. Be sure to square only the numerator before dividing.

Figure 3–7 All formulas are based on Ohm's law.

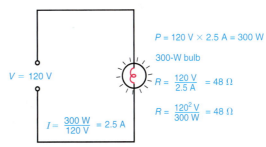

$P = 120 \text{ V} \times 2.5 \text{ A} = 300 \text{ W}$

300-W bulb

$V = 120 \text{ V}$

$R = \dfrac{120 \text{ V}}{2.5 \text{ A}} = 48 \ \Omega$

$R = \dfrac{120^2 \text{ V}}{300 \text{ W}} = 48 \ \Omega$

$I = \dfrac{300 \text{ W}}{120 \text{ V}} = 2.5 \text{ A}$

Furthermore, the resistance of the filament, equal to V/I, is 48 Ω. These calculations are

$$R = \frac{V}{I} = \frac{120 \text{ V}}{2.5 \text{ A}} = 48 \text{ }\Omega$$

If we use the power formula $R = V^2/P$, the answer is the same 48 Ω. These calculations are

$$R = \frac{V^2}{P} = \frac{120^2}{300}$$
$$R = \frac{14,400}{300} = 48 \text{ }\Omega$$

In any case, when this bulb is connected across 120 V so that it can dissipate its rated power, the bulb draws 2.5 A from the power line and the resistance of the white-hot filament is 48 Ω.

■ *3–9 Self-Review*
   Answers at end of chapter.

a. How much is the *R* of a 100-W, 120-V light bulb?
b. How much power is dissipated by a 2-Ω *R* with 10 V across it?
c. Calculate *P* for 2 A of *I* through a 2-Ω resistor.

3–10 Choosing a Resistor for a Circuit

When choosing a resistor for a circuit, first determine the required resistance value as $R = \dfrac{V}{I}$. Next, calculate the amount of power dissipated by the resistor using any one of the power formulas. Then, select a wattage rating for the resistor which will provide a reasonable amount of cushion between the actual power dissipation and the power rating of the resistor. Ideally, the power dissipation in a resistor should never be more than 50% of its power rating, which is a safety factor of 2. A safety factor of 2 allows the resistor to operate at a cooler temperature and thus last longer without breaking down from excessive heat. In practice, however, as long as the safety factor is reasonably close to 2, the resistor will not overheat.

Example 3-16

Determine the required resistance and appropriate wattage rating of a resistor to meet the following requirements: The resistor must have a 30-V IR drop when its current is 20 mA. The resistors available have the following wattage ratings: 1/8, 1/4, 1/2, 1, and 2 W.

ANSWER   First, calculate the required resistance.

$$R = \frac{V}{I}$$
$$= \frac{30 \text{ V}}{20 \text{ mA}}$$
$$= 1.5 \text{ k}\Omega$$

Next, calculate the power dissipated by the resistor using the formula $P = I^2R$.

$$P = I^2R$$
$$= (20 \text{ mA})^2 \times 1.5 \text{ k}\Omega$$
$$= 0.6 \text{ W or } 600 \text{ mW}$$

Now, select a suitable wattage rating for the resistor. In this example, a 1-W rating provides a safety factor which is reasonably close to 2. A resistor with a higher wattage rating could be used if there is space available for it to be mounted. In summary, a 1.5-kΩ, 1-W resistor will safely dissipate 600 mW of power while providing an *IR* voltage of 30 V when the current is 20 mA.

Maximum Working Voltage Rating

The maximum working voltage rating of a resistor is the maximum allowable voltage that the resistor can safely withstand without internal arcing. The higher the wattage rating of the resistor, the higher the maximum working voltage rating. For carbon-film resistors, the following voltage ratings are typical:

$$1/8 \text{ W} - 150 \text{ V}$$
$$1/4 \text{ W} - 250 \text{ V}$$
$$1/2 \text{ W} - 350 \text{ V}$$
$$1 \text{ W} - 500 \text{ V}$$

It is interesting to note that with very large resistance values, the maximum working voltage rating may actually be exceeded before the power rating is exceeded. For example, a 1 MΩ, ¼ W carbon-film resistor with a maximum working voltage rating of 250 V, does not dissipate ¼ W of power until its voltage equals 500 V. Since 500 V exceeds its 250 V rating, internal arcing will occur within the resistor. Therefore, 250 V rather than 500 V is the maximum voltage that can safely be applied across this resistor. With 250 V across the 1-MΩ resistor, the actual power dissipation is 1/16 W which is only one-fourth its power rating.

For any resistor, the maximum voltage which produces the rated power dissipation is calculated as

$$V_{\text{max}} = \sqrt{P_{\text{rating}} \times R}$$

Exceeding V_{max} causes the resistor's power dissipation to exceed its power rating. Except for very large resistance values, the maximum working voltage rating is usually much larger than the maximum voltage that produces the rated power dissipation.

Example 3-17

Determine the required resistance and appropriate wattage rating of a carbon-film resistor to meet the following requirements: The resistor must have a 225 V *IR* drop when its current is 150 μA. The resistors available have the following wattage ratings: 1/8, 1/4, 1/2, 1, and 2 W.

ANSWER First, calculate the required resistance.

$$R = \frac{V}{I}$$
$$= \frac{225 \text{ V}}{150 \text{ }\mu\text{A}}$$
$$= 1.5 \text{ M}\Omega$$

Next, calculate the power dissipated by the resistor using the formula $P = I^2R$.

$$P = I^2R$$
$$= (150 \text{ }\mu\text{A})^2 \times 1.5 \text{ M}\Omega$$
$$= 33.75 \text{ mW}$$

Now, select a suitable wattage rating for the resistor.

In this application a 1/8-W (125 mW) resistor could be considered because it will provide a safety factor of nearly 4. However, a 1/8-W resistor could not be used because its maximum working voltage rating is only 150 V and the resistor must be able to withstand a voltage of 225 V. Therefore, a higher wattage rating must be chosen just because it will have a higher maximum working voltage rating. In this application, a 1/2-W resistor would be a reasonable choice because it has a 350-V rating. A 1/4-W resistor provides a 250-V rating which is only 25 V more than the actual voltage across the resistor. It's a good idea to play it safe and go with the higher voltage rating offered by the 1/2-W resistor. In summary, a 1.5 MΩ, 1/2-W resistor will safely dissipate 33.75 mW of power as well as withstand a voltage of 225 V.

■ *3–10 Knowledge Check*

Answer at end of chapter.

A certain resistor is required to provide a 20 V *IR* drop when its current is 10 mA. Determine the required resistance and appropriate wattage rating of the resistor. The resistors available have the following wattage ratings: 1/8, 1/4, 1/2, 1, and 2 W.

■ *3–10 Self-Review*

Answers at end of chapter.

a. What is the maximum voltage that a 10-kΩ, ¼-W resistor can safely handle without exceeding its power rating? If the resistor has a 250-V maximum working voltage rating, is this rating being exceeded?

b. Determine the required resistance and appropriate wattage rating of a carbon-film resistor for the following conditions: the *IR* voltage must equal 100 V when the current is 100 μA. The available wattage ratings for the resistor are 1/8, 1/4, 1/2, 1, and 2 W.

Figure 3–8 Physiological effects of electric current.

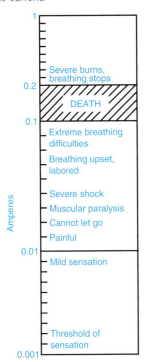

3–11 Electric Shock

While you are working on electric circuits, there is often the possibility of receiving an electric shock by touching the "live" conductors when the power is on. The shock is a sudden involuntary contraction of the muscles, with a feeling of pain, caused by current through the body. If severe enough, the shock can be fatal. Safety first, therefore, should always be the rule.

The greatest shock hazard is from high-voltage circuits that can supply appreciable amounts of power. The resistance of the human body is also an important factor. If you hold a conducting wire in each hand, the resistance of the body across the conductors is about 10,000 to 50,000 Ω. Holding the conductors tighter lowers the resistance. If you hold only one conductor, your resistance is much higher. It follows that the higher the body resistance, the smaller the current that can flow through you.

A safety tip, therefore, is to work with only one of your hands if the power is on. Place the other hand behind your back or in your pocket. Therefore, if a live circuit is touched with only one hand, the current will normally not flow directly through the heart. Also, keep yourself insulated from earth ground when working on power-line circuits, since one side of the power line is connected to earth ground. The final and best safety rule is to work on circuits with the power disconnected if at all possible and make resistance tests.

Note that it is current through the body, not through the circuit, which causes the electric shock. This is why high-voltage circuits are most important, since sufficient potential difference can produce a dangerous amount of current through the relatively high resistance of the body. For instance, 500 V across a body resistance of 25,000 Ω produces 0.02 A, or 20 mA, which can be fatal. As little as 1 mA through the body can cause an electric shock. The chart shown in Fig. 3–8 is a visual representation of the physiological effects of an electric current on the human body. As the chart shows, the threshold of sensation occurs when the current through the body is only slightly above 0.001 A or 1 mA. Slightly above 10mA, the sensation of current through the body becomes painful and the person can no longer let go or free him or herself from the circuit. When the current through the body exceeds approximately 100 mA, the result is usually death.

In addition to high voltage, the other important consideration in how dangerous the shock can be is the amount of power the source can supply. A current of 0.02 A through 25,000 Ω means that the body resistance dissipates 10 W. If the source cannot supply 10 W, its output voltage drops with the excessive current load. Then the current is reduced to the amount corresponding to the amount of power the source can produce.

In summary, then, the greatest danger is from a source having an output of more than about 30 V with enough power to maintain the load current through the body when it is connected across the applied voltage. In general, components that can supply high power are physically big because of the need for dissipating heat.

■ *3–11 Self-Review*

Answers at end of chapter.

a. The potential difference of 120 V is more dangerous than 12 V for electric shock. (True/False)
b. Resistance in a circuit should be measured with its power off. (True/False)

Figure 3–9 Effect of an open circuit. (*a*) Normal circuit with current of 2 A for 10 V across 5 Ω. (*b*) Open circuit with no current and infinitely high resistance.

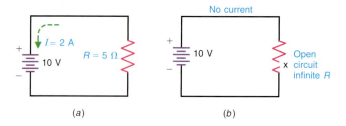

(*a*) (*b*)

3–12 Open-Circuit and Short-Circuit Troubles

Ohm's law is useful for calculating *I*, *V*, and *R* in a closed circuit with normal values. However, an open circuit or a short circuit causes trouble that can be summarized as follows: An open circuit (Fig. 3–9) has zero *I* because *R* is infinitely high. It does not matter how much the *V* is. A short circuit has zero *R*, which causes excessively high *I* in the short-circuit path because of no resistance (Fig. 3–10).

In Fig. 3–9*a*, the circuit is normal with *I* of 2 A produced by 10 V applied across *R* of 5 Ω. However, the resistor is shown open in Fig. 3–9*b*. Then the path for current has infinitely high resistance and there is no current in any part of the circuit. The trouble can be caused by an internal open in the resistor or a break in the wire conductors.

In Fig. 3–10*a*, the same normal circuit is shown with *I* of 2 A. In Fig. 3–10*b*, however, there is a short-circuit path across *R* with zero resistance. The result is excessively high current in the short-circuit path, including the wire conductors. It may be surprising, but there is no current in the resistor itself because all the current is in the zero-resistance path around it.

Theoretically, the amount of current could be infinitely high with no *R*, but the voltage source can supply only a limited amount of *I* before it loses its ability to provide voltage output. The wire conductors may become hot enough to burn open, which would open the circuit. Also, if there is any fuse in the circuit, it will open because of the excessive current produced by the short circuit.

Note that the resistor itself is not likely to develop a short circuit because of the nature of its construction. However, the wire conductors may touch, or some other component in a circuit connected across the resistor may become short-circuited.

Figure 3–10 Effect of a short circuit. (*a*) Normal circuit with current of 2 A for 10 V across 5 Ω. (*b*) Short circuit with zero resistance and excessively high current.

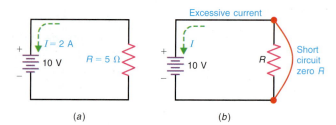

(*a*) (*b*)

■ 3-12 Self-Review

Answers at end of chapter.

a. An open circuit has zero current. (True/False)
b. A short circuit has excessive current. (True/False)
c. An open circuit and a short circuit have opposite effects on resistance and current. (True/False)

Summary

- The three forms of Ohm's law are $I = V/R$, $V = IR$, and $R = V/I$.

- One ampere is the amount of current produced by one volt of potential difference across one ohm of resistance. This current of 1 A is the same as 1 C/s.

- With R constant, the amount of I increases in direct proportion as V increases. This linear relation between V and I is shown by the graph in Fig. 3–5.

- With V constant, the current I decreases as R increases. This is an inverse relation.

- Power is the time rate of doing work or using energy. The unit is the watt. One watt equals 1 V × 1 A. Also, watts = joules per second.

- The unit of work or energy is the joule. One joule equals 1 W × 1 s.

- The most common multiples and submultiples of the practical units are listed in Table 3–1.

- Voltage applied across your body can produce a dangerous electric shock. Whenever possible, shut off the power and make resistance tests. If the power must be on, use only one hand when making measurements. Place your other hand behind your back or in your pocket.

- Table 3–2 summarizes the practical units of electricity.

- An open circuit has no current and infinitely high R. A short circuit has zero resistance and excessively high current.

Table 3–1	Summary of Conversion Factors		
Prefix	Symbol	Relation to Basic Unit	Examples
mega	M	1,000,000 or 1×10^6	5 MΩ (megohms) = 5,000,000 ohms = 5×10^6 ohms
kilo	k	1000 or 1×10^3	18 kV (kilovolts) = 18,000 volts = 18×10^3 volts
milli	m	0.001 or 1×10^{-3}	48 mA (milliamperes) = 48×10^{-3} ampere = 0.048 ampere
micro	μ	0.000 001 or 1×10^{-6}	15 μV (microvolts) = 15×10^{-6} volt = 0.000 015 volt

Table 3–2	Summary of Practical Units of Electricity				
Coulomb	Ampere	Volt	Watt	Ohm	Siemens
6.25×10^{18} electrons	$\dfrac{\text{coulomb}}{\text{second}}$	$\dfrac{\text{joule}}{\text{coulomb}}$	$\dfrac{\text{joule}}{\text{second}}$	$\dfrac{\text{volt}}{\text{ampere}}$	$\dfrac{\text{ampere}}{\text{volt}}$

Important Terms

- Ampere - the basic unit of current.
  $1\ A = \dfrac{1\ V}{1\ \Omega}$.

- Electron Volt (eV) - a small unit of work or energy that represents the amount of work required to move a single electron between two points having a potential difference of 1 volt.

- Horsepower (hp) - a unit of mechanical power corresponding to 550 ft. · lbs/s. In terms of electric power, 1 hp = 746 W.

- Inverse Relation - a relation in which the quotient of a fraction decreases as the value in the denominator increases with the numerator constant. In the equation $I = \dfrac{V}{R}$, I and R are inversely related because I decreases as R increases with V constant.

- Joule - a practical unit of work or energy. 1 J = 1 W · 1 s.

- Kilowatt-Hour (kWh) - a large unit of electrical energy corresponding to 1 kW · 1 h.

- Linear Proportion - a relation between two quantities which shows how equal changes in one quantity produce equal changes in the other. In the equation $I = \dfrac{V}{R}$, I and V are directly proportional because equal changes in V produce equal changes in I with R constant.

- Linear Resistance - a resistance with a constant value of ohms.

- Maximum Working Voltage Rating - the maximum allowable voltage that a resistor can safely withstand without internal arcing.

- Nonlinear Resistance - a resistance whose value changes as a result of current producing power dissipation and heat in the resistance.

- Ohm - the basic unit of resistance.
  $1\ \Omega = \dfrac{1\ V}{1\ A}$.

- Open Circuit - a broken or incomplete current path with infinitely high resistance.

- Power - the time rate of doing work.
  $Power = \dfrac{Work}{Time}$.

- Short Circuit - a very low resistance path around or across a component such as a resistor. A short circuit with very low R can have excessively high current.

- Volt - the basic unit of potential difference or voltage. 1 V = 1 A · 1 Ω.

- Volt-Ampere Characteristic - a graph showing how much current a resistor allows for different voltages.

- Watt - the basic unit of electric power. $1\ W = \dfrac{1\ J}{s}$.

Related Formulas

$I = \dfrac{V}{R}$

$V = I \times R$

$R = \dfrac{V}{I}$

$P = V \times I$

$I = \dfrac{P}{V}$

$V = \dfrac{P}{I}$

1 hp = 746 W

1 J = 1 W × 1 s

$1\ W = \dfrac{1\ J}{1\ s}$

1 J = 1 V × 1 C

$1\ J = 6.25 \times 10^{18}\ eV$

$P = I^2 R$

$I = \sqrt{\dfrac{P}{R}}$

$R = \dfrac{P}{I^2}$

$P = \dfrac{V^2}{R}$

$V = \sqrt{PR}$

$R = \dfrac{V^2}{P}$

Self-Test

Answers at back of book.

1. With 24 V across a 1-kΩ resistor, the current, I, equals
   a. 0.24 A.
   b. 2.4 mA.
   c. 24 mA.
   d. 24 μA.

2. With 30 μA of current in a 120-kΩ resistor, the voltage, V, equals
   a. 360 mV.
   b. 3.6 kV.
   c. 0.036 V.
   d. 3.6 V.

3. How much is the resistance in a circuit if 15 V of potential difference produces 500 μA of current?
   a. 30 kΩ.
   b. 3 MΩ.
   c. 300 kΩ.
   d. 3 kΩ.

4. A current of 1000 μA equals
   a. 1 A.
   b. 1 mA.
   c. 0.01 A.
   d. none of the above.

5. One horsepower equals
   a. 746 W.
   b. 550 ft · lb/s.
   c. approximately 3/4 kW.
   d. all of the above.

6. With R constant
   a. I and P are inversely related.
   b. V and I are directly proportional.
   c. V and I are inversely proportional.
   d. none of the above.

7. One watt of power equals
   a. $1 V \times 1 A$.
   b. $\dfrac{1 J}{s}$.
   c. $\dfrac{1 C}{s}$.
   d. both a and b.

8. A 10-Ω resistor dissipates 1 W of power when connected to a dc voltage source. If the value of dc voltage is doubled, the resistor will dissipate
   a. 1 W.
   b. 2 W.
   c. 4 W.
   d. 10 W.

9. If the voltage across a variable resistance is held constant, the current, I, is
   a. inversely proportional to resistance.
   b. directly proportional to resistance.
   c. the same for all values of resistance.
   d. both a and b.

10. A resistor must provide a voltage drop of 27 V when the current is 10 mA. Which of the following resistors will provide the required resistance and appropriate wattage rating?
    a. 2.7 kΩ, 1/8 W.
    b. 270 Ω, 1/2 W.
    c. 2.7 kΩ, 1/2 W.
    d. 2.7 kΩ, 1/4 W.

11. The resistance of an open circuit is
    a. approximately 0 Ω.
    b. infinitely high.
    c. very low.
    d. none of the above.

12. The current in an open circuit is
    a. normally very high because the resistance of an open circuit is 0 Ω.
    b. usually high enough to blow the circuit fuse.
    c. zero.
    d. slightly below normal.

13. Which of the following safety rules should be observed while working on a live electric circuit?
    a. keep yourself well insulated from earth ground.
    b. when making measurements in a live circuit place one hand behind your back or in your pocket.
    c. only make resistance measurements in a live circuit.
    d. both a and b.

14. How much current does a 75-W lightbulb draw from the 120 V power line?
    a. 625 mA.
    b. 1.6 A.
    c. 160 mA.
    d. 62.5 mA.

15. The resistance of a short circuit is
    a. infinitely high.
    b. very high.
    c. usually above 1 kΩ.
    d. approximately zero.

16. Which of the following is considered a linear resistance?
    a. lightbulb.
    b. thermistor.
    c. 1-kΩ, $^1/_2$-W carbon-film resistor.
    d. both a and b above.

17. How much will it cost to operate a 4-kW air-conditioner for 12 hours if the cost of electricity is 7¢/kwh?
    a. $3.36.
    b. 33¢.
    c. $8.24.
    d. $4.80.

18. What is the maximum voltage a 150-Ω, 1/8-W resistor can safely handle without exceeding its power rating? (Assume no power rating safety factor.)
    a. 18.75 V.
    b. 4.33 V.
    c. 6.1 V.
    d. 150 V.

19. Which of the following voltages provides the greatest danger in terms of electric shock?
    a. 12 V.
    b. 10,000 mV.
    c. 120 V.
    d. 9 V.

20. If a short circuit is placed across the leads of a resistor, the current in the resistor itself would be
    a. zero.
    b. much higher than normal.
    c. the same as normal.
    d. excessively high.

Questions

1. State the three forms of Ohm's law relating V, I, and R.

2. (a) Why does higher applied voltage with the same resistance result in more current? (b) Why does more resistance with the same applied voltage result in less current?

3. Calculate the resistance of a 300-W bulb connected across the 120-V power line, using two different methods to arrive at the same answer.

4. State which unit in each of the following pairs is larger: (a) volt or kilovolt; (b) ampere or milliampere; (c) ohm or megohm; (d) volt or microvolt; (e) siemens or microsiemens; (f) electron volt or joule; (g) watt or kilowatt; (h) kilowatt-hour or joule; (i) volt or millivolt; (j) megohm or kilohm.

5. State two safety precautions to follow when working on electric circuits.

6. Referring back to the resistor shown in Fig. 1–10 in Chap. 1, suppose that it is not marked. How could you determine its resistance by Ohm's law? Show your calculations that result in the V/I ratio of 10 kΩ. However, do not exceed the power rating of 10 W.

7. Give three formulas for electric power.

8. What is the difference between work and power? Give two units for each.

9. Prove that 1 kWh is equal to 3.6×10^6J.

10. Give the metric prefixes for 10^{-6}, 10^{-3}, 10^3, and 10^6.

11. Which two units in Table 3–2 are reciprocals of each other?

12. A circuit has a constant R of 5000 Ω, and V is varied from 0 to 50 V in 10-V steps. Make a table listing the values of I for each value of V. Then draw a graph plotting these values of milliamperes vs. volts. (This graph should be like Fig. 3–5c.)

13. Give the voltage and power rating for at least two types of electrical equipment.

14. Which uses more current from the 120-V power line, a 600-W toaster or a 200-W television receiver?

15. Give a definition for a short circuit and for an open circuit.

16. Compare the R of zero ohms and infinite ohms.

17. Derive the formula $P = I^2R$ from $P = IV$ by using an Ohm's-law formula.

18. Explain why a thermistor is a nonlinear resistance.

19. What is meant by the maximum working voltage rating of a resistor?

20. Why do resistors often have a safety factor of 2 in regard to their power rating?

Problems

SECTION 3–1 THE CURRENT $I = V/R$

In Probs. 3–1 – 3–5, solve for the current, I, when V and R are known. As a visual aid, it may be helpful to insert the values of V and R into Fig. 3–11 when solving for I.

3–1 **MultiSim** a. $V = 10$ V, $R = 5\ \Omega$, $I = ?$
　　　b. $V = 9$ V, $R = 3\ \Omega$, $I = ?$
　　　c. $V = 24$ V, $R = 3\ \Omega$, $I = ?$
　　　d. $V = 36$ V, $R = 9\ \Omega$, $I = ?$

3–2. **MultiSim** a. $V = 18$ V, $R = 3\ \Omega$, $I = ?$
　　　b. $V = 16$ V, $R = 16\ \Omega$, $I = ?$
　　　c. $V = 90$ V, $R = 450\ \Omega$, $I = ?$
　　　d. $V = 12$ V, $R = 30\ \Omega$, $I = ?$

3–3 **MultiSim** a. $V = 15$ V, $R = 3{,}000\ \Omega$, $I = ?$
　　　b. $V = 120$ V, $R = 6{,}000\ \Omega$, $I = ?$
　　　c. $V = 27$ V, $R = 9{,}000\ \Omega$, $I = ?$
　　　d. $V = 150$ V, $R = 10{,}000\ \Omega$, $I = ?$

3–4 If a 100-Ω resistor is connected across the terminals of a 12-V battery, how much is the current, I?

3–5 If one branch of a 120-V power line is protected by a 20-A fuse, will the fuse carry an 8-Ω load?

SECTION 3–2 THE VOLTAGE $V = IR$

In Probs. 3–6 – 3–10, solve for the voltage, V, when I and R are known. As a visual aid, it may be helpful to insert the values of I and R into Fig. 3–12 when solving for V.

3–6 **MultiSim** a. $I = 2$ A, $R = 5\ \Omega$, $V = ?$
　　　b. $I = 6$ A, $R = 8\ \Omega$, $V = ?$
　　　c. $I = 9$ A, $R = 20\ \Omega$, $V = ?$
　　　d. $I = 4$ A, $R = 15\ \Omega$, $V = ?$

3–7 **MultiSim** a. $I = 5$ A, $R = 10\ \Omega$, $V = ?$
　　　b. $I = 10$ A, $R = 3\ \Omega$, $V = ?$
　　　c. $I = 4$ A, $R = 2.5\ \Omega$, $V = ?$
　　　d. $I = 1.5$ A, $R = 5\ \Omega$, $V = ?$

Figure 3–11　Figure for Probs. 3–1 – 3–5.

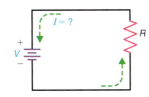

Figure 3–12　Figure for Probs. 3–6 – 3–10.

Figure 3–13   Figure for Probs. 3–11 – 3–15.

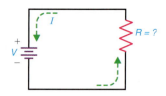

Figure 3–14   Circuit diagram for Prob. 3–21.

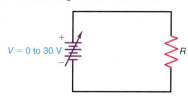

3–8 |||| **MultiSim** a. $I = 0.05$ A, $R = 1200$ Ω, $V = ?$
     b. $I = 0.2$ A, $R = 470$ Ω, $V = ?$
     c. $I = 0.01$ A, $R = 15,000$ Ω, $V = ?$
     d. $I = 0.006$ A, $R = 2200$ Ω, $V = ?$

3–9   How much voltage is developed across a 1000-Ω resistor if it has a current of 0.01 A?

3–10   A lightbulb drawing 1.25 A of current has a resistance of 96 Ω. How much is the voltage across the lightbulb?

SECTION 3–3 THE RESISTANCE $R = V/I$

In Probs. 3–11 – 3–15, solve for the resistance, R, when V and I are known. As a visual aid, it may be helpful to insert the values of V and I into Fig. 3–13 when solving for R.

3–11   a. $V = 14$ V, $I = 2$ A, $R = ?$
   b. $V = 25$ V, $I = 5$ A, $R = ?$
   c. $V = 6$ V, $I = 1.5$ A, $R = ?$
   d. $V = 24$ V, $I = 4$ A, $R = ?$

3–12   a. $V = 36$ V, $I = 9$ A, $R = ?$
   b. $V = 45$ V, $I = 5$ A, $R = ?$
   c. $V = 100$ V, $I = 2$ A, $R = ?$
   d. $V = 240$ V, $I = 20$ A, $R = ?$

3–13   a. $V = 12$ V, $I = 0.002$ A, $R = ?$
   b. $V = 16$ V, $I = 0.08$ A, $R = ?$
   c. $V = 50$ V, $I = 0.02$ A, $R = ?$
   d. $V = 45$ V, $I = 0.009$ A, $R = ?$

3–14   How much is the resistance of a motor if it draws 2 A of current from the 120-V power line?

3–15   If a CD player draws 1.6 A of current from a 13.6-V dc source, how much is its resistance?

SECTION 3–5 MULTIPLE AND SUBMULTIPLE UNITS

In Probs. 3–16 – 3–20, solve for the unknowns listed. As a visual aid, it may be helpful to insert the known values of I, V, or R into Figs. 3–11, 3–12, or 3–13 when solving for the unknown quantity.

3–16   a. $V = 10$ V, $R = 100$ kΩ, $I = ?$
   b. $V = 15$ V, $R = 2$ kΩ, $I = ?$
   c. $I = 200$ μA, $R = 3.3$ MΩ, $V = ?$
   d. $V = 5.4$ V, $I = 2$ mA, $R = ?$

3–17   a. $V = 120$ V, $R = 1.5$ kΩ, $I = ?$
   b. $I = 50$ μA, $R = 390$ kΩ, $V = ?$
   c. $I = 2.5$ mA, $R = 1.2$ kΩ, $V = ?$
   d. $V = 99$ V, $I = 3$ mA, $R = ?$

3–18   a. $V = 24$ V, $I = 800$ μA, $R = ?$
   b. $V = 160$ mV, $I = 8$ μA, $R = ?$
   c. $V = 13.5$ V, $R = 300$ Ω, $I = ?$
   d. $I = 30$ mA, $R = 1.8$ kΩ, $V = ?$

3–19   How much is the current, I, in a 470-kΩ resistor if its voltage is 23.5 V?

3–20   How much voltage will be dropped across a 40-kΩ resistance whose current is 250 μA?

SECTION 3–6 THE LINEAR PROPORTION BETWEEN V AND I

3–21   Refer to Fig. 3–14. Draw a graph of the I and V values if (a) $R = 2.5$ Ω; (b) $R = 5$ Ω; (c) $R = 10$ Ω. In each case, the voltage source is to be varied in 5-V steps from 0 to 30 V.

3–22   Refer to Fig. 3–15. Draw a graph of the I and R values when R is varied in 2-Ω steps from 2 to 12 Ω. (V is constant at 12 V.)

SECTION 3–7 ELECTRIC POWER

In Probs. 3–23 – 3–31, solve for the unknowns listed.

3–23   a. $V = 120$ V, $I = 12.5$ A, $P = ?$
   b. $V = 120$ V, $I = 625$ mA, $P = ?$
   c. $P = 1.2$ kW, $V = 120$ V, $I = ?$
   d. $P = 100$ W, $I = 8.33$ A, $V = ?$

3–24   a. $V = 24$ V, $I = 25$ mA, $P = ?$
   b. $P = 6$ W, $V = 12$ V, $I = ?$
   c. $P = 10$ W, $I = 100$ mA, $V = ?$
   d. $P = 50$ W, $V = 9$ V, $I = ?$

Figure 3–15   Circuit diagram for Prob. 3–22.

3–25 a. $V = 15.81$ V, $P = 500$ mW, $I = ?$

     b. $P = 100$ mW, $V = 50$ V, $I = ?$

     c. $V = 75$ mV, $I = 2$ mA, $P = ?$

     d. $P = 20$ mW, $I = 100$ μA, $V = ?$

3–26 How much current do each of the following lightbulbs draw from the 120-V power line?

     a. 60 W bulb

     b. 75 W bulb

     c. 100 W bulb

     d. 300 W bulb

3–27 How much is the output voltage of a power supply if it supplies 75 W of power while delivering a current of 5 A?

3–28 How much power is consumed by a 12-V incandescent lamp if it draws 150 mA of current when lit?

3–29 How much will it cost to operate a 1,500 W quartz heater for 48 h if the cost of electricity is 7¢/kWh?

3–30 How much does it cost to light a 300-W lightbulb for 30 days if the cost of electricity is 7¢/kWh?

3–31 How much will it cost to run an electric motor for 10 days if the motor draws 15 A of current from the 240 V power line? The cost of electricity is 7.5¢/kWh.

SECTION 3–8 POWER DISSIPATION IN RESISTANCE

In Probs. 3–32 – 3–38, solve for the power, P, dissipated by the resistance, R.

3–32 a. $I = 1$ A, $R = 100$ Ω, $P = ?$

     b. $I = 20$ mA, $R = 1$ kΩ, $P = ?$

     c. $V = 5$ V, $R = 150$ Ω, $P = ?$

     d. $V = 22.36$ V, $R = 1$ kΩ, $P = ?$

3–33 a. $I = 300$ μA, $R = 22$ kΩ, $P = ?$

     b. $I = 50$ mA, $R = 270$ Ω, $P = ?$

     c. $V = 70$ V, $R = 200$ kΩ, $P = ?$

     d. $V = 8$ V, $R = 50$ Ω, $P = ?$

3–34 a. $I = 40$ mA, $R = 10$ kΩ, $P = ?$

     b. $I = 3.33$ A, $R = 20$ Ω, $P = ?$

     c. $V = 100$ mV, $R = 10$ Ω, $P = ?$

     d. $V = 1$ kV, $R = 10$ MΩ, $P = ?$

3–35 How much power is dissipated by a 5.6-kΩ resistor whose current is 9.45 mA?

3–36 How much power is dissipated by a 50-Ω load if the voltage across the load is 100 V?

3–37 How much power is dissipated by a 600-Ω load if the voltage across the load is 36 V?

3–38 How much power is dissipated by an 8-Ω load if the current in the load is 200 mA?

SECTION 3–9 POWER FORMULAS

In Probs. 3–39 – 3–51, solve for the unknowns listed.

3–39 a. $P = 250$ mW, $R = 10$ kΩ, $I = ?$

     b. $P = 100$ W, $V = 120$ V, $R = ?$

     c. $P = 125$ mW, $I = 20$ mA, $R = ?$

     d. $P = 1$ kW, $R = 50$ Ω, $V = ?$

3–40 a. $P = 500$ μW, $V = 10$ V, $R = ?$

     b. $P = 150$ mW, $I = 25$ mA, $R = ?$

     c. $P = 300$ W, $R = 100$ Ω, $V = ?$

     d. $P = 500$ mW, $R = 3.3$ kΩ, $I = ?$

3–41 a. $P = 50$ W, $R = 40$ Ω, $V = ?$

     b. $P = 2$ W, $R = 2$ kΩ, $V = ?$

     c. $P = 50$ mW, $V = 500$ V, $I = ?$

     d. $P = 50$ mW, $R = 312.5$ kΩ, $I = ?$

3–42 Calculate the maximum current that a 1-kΩ, 1 W carbon resistor can safely handle without exceeding its power rating.

3–43 Calculate the maximum current that a 22-kΩ, 1/8 W resistor can safely handle without exceeding its power rating.

3–44 What is the hot resistance of a 60 W, 120 V lightbulb?

3–45 A 50-Ω load dissipates 200 W of power. How much voltage is across the load?

3–46 Calculate the maximum voltage that a 390-Ω, 1/2 W resistor can safely handle without exceeding its power rating.

3–47 What is the resistance of a device that dissipates 1.2 kW of power when its current is 10 A?

3–48 How much current does a 960 W coffeemaker draw from the 120 V power line?

3–49 How much voltage is across a resistor if it dissipates 2 W of power when the current is 40 mA?

3–50 If a 4-Ω speaker dissipates 15 W of power, how much voltage is across the speaker?

3–51 What is the resistance of a 20 W, 12 V halogen lamp?

SECTION 3–10 CHOOSING A RESISTOR FOR A CIRCUIT

In Probs. 3–52 – 3–60, determine the required resistance and appropriate wattage rating of a carbon-film resistor for the specific requirements listed. For all problems, assume that the following wattage ratings are available: 1/8 W, 1/4 W, 1/2 W, 1 W, and 2 W. (Assume the maximum working voltage ratings listed on page 99.)

3–52 Required values of V and I are 54 V and 2 mA.

3–53 Required values of V and I are 12 V and 10 mA.

3–54 Required values of V and I are 390 V and 1 mA.

3–55 Required values of V and I are 36 V and 18 mA.

3–56 Required values of V and I are 340 V and 500 μA.

3-57 Required values of V and I are 3 V and 20 mA.

3-58 Required values of V and I are 33 V and 18.33 mA.

3-59 Required values of V and I are 264 V and 120 μA.

3-60 Required values of V and I are 9.8 V and 1.75 mA.

Critical Thinking

3-61 The percent efficiency of a motor can be calculated as

$$\% \text{ efficiency} = \frac{\text{power out}}{\text{power in}} \times 100$$

where power out represents horsepower (hp). Calculate the current drawn by a 5-hp, 240-V motor that is 72% efficient.

3-62 A $^1/_2$-hp, 120-V motor draws 4.67 A when it is running. Calculate the motor's efficiency.

3-63 A $^3/_4$-hp motor with an efficiency of 75% runs 20% of the time during a 30-day period. If the cost of electricity is 7¢/kWh, how much will it cost the user?

3-64 An appliance uses 14.4×10^6 J of energy for 1 day. How much will this cost the user if the cost of electricity is 6.5¢/kWh?

3-65 A certain 1-kΩ resistor has a power rating of $^1/_2$ W for temperatures up to 70°C. Above 70°C, however, the power rating must be reduced by a factor of 6.25 mW/°C. Calculate the maximum current that the resistor can allow at 120°C without exceeding its power dissipation rating at this temperature.

Answers to Knowledge Check Problems

3–1 $I = 0.16$ A

3–2 $V = 23.5$ V

3–3 $R = 15\ \Omega$

3–4 **a.** $I = 0.1$ A

    b. $V = 60$ V

    c. $R = 375\ \Omega$

3–5 $I = 50\ \mu$A

3–6 $I = 3.5$ A, 7 A

3–7A $2.52

3–7B $P = 30$ W

3–8 $P = 400$ mW

3–9 $I = 50$ mA

3–10 $R = 2\ k\Omega$. Best choice for power rating is $^1/_2$ W.

Answers to Self-Reviews

3–1 **a.** 3 A
    b. 1.5 A
    c. 2 A
    d. 6 A

3–2 **a.** 2 V
    b. 4 V
    c. 4 V

3–3 **a.** 4000 Ω
    b. 2000 Ω
    c. 12,000 Ω

3–4 **a.** 35 V
    b. 0.002 A
    c. 2000 Ω

3–5 **a.** See Prob. **b**
    b. See Prob. **a**
    c. 2 mA
    d. 125 μA

3–6 **a.** y axis
    b. linear
    c. I doubles from 2A to 4A
    d. I is halved from 2A to 1A

3–7 **a.** 1.8 kW
    b. 0.83 A
    c. 200 W
    d. $1.01 (approx.)

3–8 **a.** 20 W
    b. 20 W
    c. 20 W and 5 Ω

3–9 **a.** 144 Ω
    b. 50 W
    c. 8 W

3–10 **a.** 50 V, No
    b. $R = 1\ M\Omega$, $P_{Rating} = 1/8$ W

3–11 **a.** T
    b. T

3–12 **a.** T
    b. T
    c. T

4

Series Circuits

A series circuit is any circuit that provides only one path for current flow. An example of a series circuit is shown in Fig. 4–1. Here two resistors are connected end to end with their opposite ends connected across the terminals of a voltage source. Figure 4–1*a* shows the pictorial wiring diagram, and Fig. 4–1*b* shows the schematic diagram. The small dots in Fig. 4–1*b* represent free electrons. Notice that the free electrons have only one path to follow as they leave the negative terminal of the voltage source, flow through resistors R_2 and R_1, and return to the positive terminal. Since there is only one path for electrons to follow, the current, I, must be the same in all parts of a series circuit. To solve for the values of voltage, current, or resistance in a series circuit, we can apply Ohm's law. This chapter covers all of the characteristics of series circuits including important information about how to troubleshoot a series circuit containing a defective component.

Objectives

After studying this chapter you should be able to

- *Explain* why the current is the same in all parts of a series circuit.

- *Calculate* the total resistance of a series circuit.

- *Calculate* the current in a series circuit.

- *Determine* the individual resistor voltage drops in a series circuit.

- *Apply* Kirchhoff's voltage law to series circuits.

- *Determine* the polarity of a resistor's *IR* voltage drop.

- *Calculate* the total power dissipated in a series circuit.

- *Determine* the net voltage of series-aiding and series-opposing voltage sources.

- *Solve* for the voltage, current, resistance, and power in a series circuit having random unknowns.

- *Define* the terms earth ground and chassis ground.

- *Calculate* the voltage at a given point with respect to ground in a series circuit.

- *Describe* the effect of an open in a series circuit.

- *Describe* the effect of a short in a series circuit.

- *Troubleshoot* series circuits containing opens and shorts.

Outline

Important Terms

chassis ground

double subscript notation

earth ground

Kirchhoff's voltage law (KVL)

series-aiding voltages

series components

series-opposing voltages

series string

troubleshooting

voltage drop

voltage polarity

4–1 Why *I* Is the Same in All Parts of a Series Circuit

An electric current is a movement of charges between two points, produced by the applied voltage. When components are connected in successive order, as in Fig. 4–1, they form a series circuit. The resistors R_1 and R_2 are in series with each other and the battery.

In Fig. 4–2a, the battery supplies the potential difference that forces free electrons to drift from the negative terminal at A, toward B, through the connecting wires and resistances R_3, R_2, and R_1, back to the positive battery terminal at J. At the negative battery terminal, its negative charge repels electrons. Therefore, free electrons in the atoms of the wire at this terminal are repelled from A toward B. Similarly, free electrons at point B can then repel adjacent electrons, producing an electron drift toward C and away from the negative battery terminal.

At the same time, the positive charge of the positive battery terminal attracts free electrons, causing electrons to drift toward I and J. As a result, the free electrons in R_1, R_2, and R_3 are forced to drift toward the positive terminal.

The positive terminal of the battery attracts electrons just as much as the negative side of the battery repels electrons. Therefore, the motion of free electrons in the circuit starts at the same time and at the same speed in all parts of the circuit.

The electrons returning to the positive battery terminal are not the same electrons as those leaving the negative terminal. Free electrons in the wire are forced to move to the positive terminal because of the potential difference of the battery.

The free electrons moving away from one point are continuously replaced by free electrons flowing from an adjacent point in the series circuit. All electrons have the same speed as those leaving the battery. In all parts of the circuit, therefore, the electron drift is the same. An equal number of electrons move at one time with the same speed. That is why the current is the same in all parts of the series circuit.

In Fig. 4–2b, when the current is 2 A, for example, this is the value of the current through R_1, R_2, R_3, and the battery at the same instant. Not only is the amount of current the same throughout, but the current in all parts of a series circuit cannot differ in any way because there is just one current path for the entire circuit. Figure 4–2c shows how to assemble axial-lead resistors on a lab prototype board to form a series circuit.

Figure 4–1   A series circuit. (*a*) Pictorial wiring diagram. (*b*) Schematic diagram.

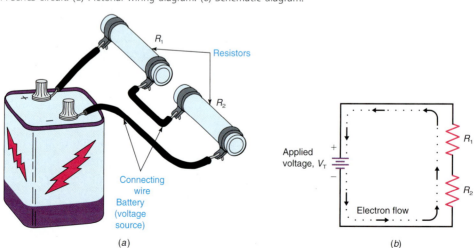

(*a*)   (*b*)

There is only one current through R_1, R_2, and R_3 in series. (*a*) Electron drift is the same in all parts of a series circuit. (*b*) Current *I* is the same at all points in a series circuit. (*c*) A series circuit assembled on a lab prototype board, using axial-lead resistors.

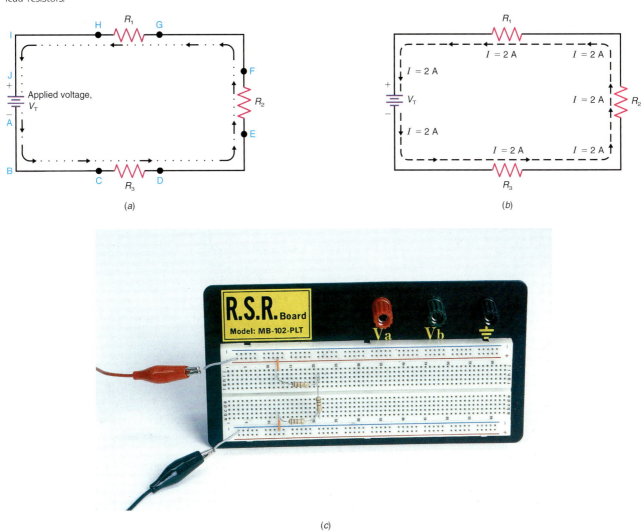

(c)

The order in which components are connected in series does not affect the current. In Fig. 4–3*b*, resistances R_1 and R_2 are connected in reverse order compared with Fig. 4–3*a*, but in both cases they are in series. The current through each is the same because there is only one path for the electron flow. Similarly, R_3, R_4, and R_5 are in series and have the same current for the connections shown in Fig. 4–3*c*, *d*, and *e*. Furthermore, the resistances need not be equal.

Figure 4–3 Examples of series connections: R_1 and R_2 are in series in both (*a*) and (*b*); also, R_3, R_4, and R_5 are in series in (*c*), (*d*), and (*e*).

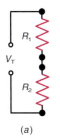
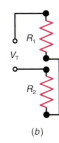
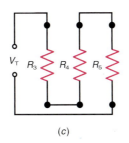
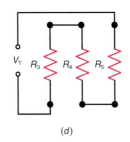
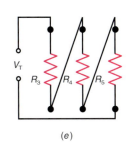

(a)   (b)   (c)   (d)   (e)

The question of whether a component is first, second, or last in a series circuit has no meaning in terms of current. The reason is that I is the same amount at the same time in all series components.

In fact, **series components** can be defined as those in the same current path. The path is from one side of the voltage source, through the series components, and back to the other side of the applied voltage. However, the series path must not have any point at which the current can branch off to another path in parallel.

■ 4–1 Knowledge Check

Answer at end of chapter.

Why is the current the same in all parts of a series circuit?

■ 4–1 Self-Review

Answers at end of chapter.

a. In Fig. 4–2b, name five parts that have the I of 2 A.
b. In Fig. 4–3e, when I in R_5 is 5 A, then I in R_3 is _____ A.
c. In Fig. 4–4b below, how much is the I in R_2?

4–2 Total R Equals the Sum of All Series Resistances

When a series circuit is connected across a voltage source, as shown in Fig. 4–3, the free electrons forming the current must drift through all the series resistances. This path is the only way the electrons can return to the battery. With two or more resistances in the same current path, therefore, the total resistance across the voltage source is the opposition of all the resistances.

Specifically, the total resistance R_T of a series string is equal to the sum of the individual resistances. This rule is illustrated in Fig. 4–4. In Fig. 4–4b, 2 Ω is added in series with the 3 Ω of Fig. 4–4a, producing the total resistance of 5 Ω. The total opposition of R_1 and R_2 limiting the amount of current is the same as though a 5-Ω resistance were used, as shown in the equivalent circuit in Fig. 4–4c.

Series String

A combination of series resistances is often called a **string.** The string resistance equals the sum of the individual resistances. For instance, R_1 and R_2 in Fig. 4–4b form a series string having an R_T of 5 Ω. A string can have two or more resistors.

IIII **MultiSim** **Figure 4–4** Series resistances are added for the total R_T. (a) R_1 alone is 3 Ω. (b) R_1 and R_2 in series total 5 Ω. (c) The R_T of 5 Ω is the same as one resistance of 5 Ω between points A and B.

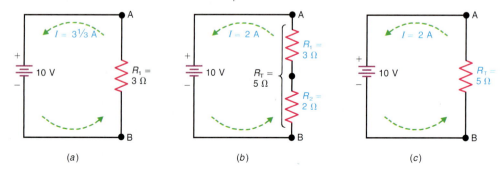

(a)   (b)   (c)

By Ohm's law, the amount of current between two points in a circuit equals the potential difference divided by the resistance between these points. Because the entire string is connected across the voltage source, the current equals the voltage applied across the entire string divided by the total series resistance of the string. Between points A and B in Fig. 4–4, for example, 10 V is applied across 5 Ω in Fig. 4–4b and c to produce 2 A. This current flows through R_1 and R_2 in one series path.

Series Resistance Formula

In summary, the *total resistance* of a series string equals the sum of the individual resistances. The formula is

$$R_T = R_1 + R_2 + R_3 + \cdots + \text{etc}. \tag{4–1}$$

where R_T is the total resistance and R_1, R_2, and R_3 are individual series resistances.

This formula applies to any number of resistances, whether equal or not, as long as they are in the same series string. Note that R_T is the resistance to use in calculating the current in a series string. Then Ohm's law is

$$I = \frac{V_T}{R_T} \tag{4–2}$$

where R_T is the sum of all the resistances, V_T is the voltage applied across the total resistance, and I is the current in all parts of the string.

Note that adding series resistance reduces the current. In Fig. 4–4a the 3-Ω R_1 allows 10 V to produce $3\frac{1}{3}$ A. However, I is reduced to 2 A when the 2-Ω R_2 is added for a total series resistance of 5 Ω opposing the 10-V source.

Example 4–1

IIII MultiSim

Two resistances R_1 and R_2 of 5 Ω each and R_3 of 10 Ω are in series. How much is R_T?

ANSWER

$$R_T = R_1 + R_2 + R_3 = 5 + 5 + 10$$
$$R_T = 20 \ \Omega$$

Example 4–2

IIII MultiSim

With 80 V applied across the series string of Example 4–1, how much is the current in R_3?

ANSWER

$$I = \frac{V_T}{R_T} = \frac{80 \text{ V}}{20 \ \Omega}$$
$$I = 4 \text{ A}$$

This 4-A current is the same in R_3, R_2, R_1, or any part of the series circuit.

■ *4–2 Knowledge Check*

> **Answer at end of chapter.**
>
> Three 1-kΩ resistors are connected in series across an applied voltage of 12 V. Calculate R_T and I.

■ *4–2 Self-Review*

> **Answers at end of chapter.**
>
> a. An applied voltage of 10 V is across a 5-kΩ resistor, R_1. How much is the current?
> b. A 2-kΩ R_2 and 3-kΩ R_3 are added in series with R_1 in part a. Calculate R_T.
> c. Calculate I in R_1, R_2, and R_3.

4–3 Series *IR* Voltage Drops

With current I through a resistance, by Ohm's law, the voltage across R is equal to $I \times R$. This rule is illustrated in Fig. 4–5 for a string of two resistors. In this circuit, I is 1 A because the applied V_T of 10 V is across the total R_T of 10 Ω, equal to the 4-Ω R_1 plus the 6-Ω R_2. Then I is 10 V/10 Ω = 1 A.

For each *IR* voltage in Fig. 4–5, multiply each R by the 1 A of current in the series circuit. Then

$$V_1 = IR_1 = 1 \text{ A} \times 4 \, \Omega = 4 \text{ V}$$
$$V_2 = IR_2 = 1 \text{ A} \times 6 \, \Omega = 6 \text{ V}$$

The V_1 of 4 V is across the 4 Ω of R_1. Also, the V_2 of 6 V is across the 6 Ω of R_2. The two voltages V_1 and V_2 are in series.

The *IR* voltage across each resistance is called an *IR drop*, or a *voltage drop*, because it reduces the potential difference available for the remaining resistances in the series circuit. Note that the symbols V_1 and V_2 are used for the voltage drops across each resistor to distinguish them from the source V_T applied across both resistors.

In Fig. 4–5, the V_T of 10 V is applied across the total series resistance of R_1 and R_2. However, because of the *IR* voltage drop of 4 V across R_1, the potential difference across R_2 is only 6 V. The positive potential drops from 10 V at point A, with respect to the common reference point at C, down to 6 V at point B with reference to point C. The potential difference of 6 V between B and the reference at C is the voltage across R_2.

‖‖ **MultiSim** **Figure 4–5** An example of *IR* voltage drops V_1 and V_2 in a series circuit.

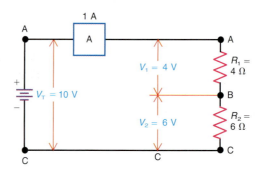

Similarly, there is an *IR* voltage drop of 6 V across R_2. The positive potential drops from 6 V at point B with respect to point C, down to 0 V at point C with respect to itself. The potential difference between any two points on the return line to the battery must be zero because the wire has practically zero resistance and therefore no *IR* drop.

Note that voltage must be applied by a source of *potential difference* such as the battery to produce current and have an *IR* voltage drop across the resistance. With no current through a resistor, the resistor has only resistance. There is no potential difference across the two ends of the resistor.

The *IR* drop of 4 V across R_1 in Fig. 4–5 represents that part of the applied voltage used to produce the current of 1 A through the 4-Ω resistance. Also, the *IR* drop across R_2 is 6 V because this much voltage allows 1 A in the 6-Ω resistance. The *IR* drop is more in R_2 because more potential difference is necessary to produce the same amount of current in the higher resistance. For series circuits, in general, the highest R has the largest *IR* voltage drop across it.

Example 4-3

In Fig. 4–6, solve for R_T, I, and the individual resistor voltage drops.

Figure 4–6 Circuit for Example 4-3.

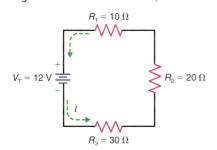

$R_1 = 10\ \Omega$
$V_T = 12$ V
$R_2 = 20\ \Omega$
I
$R_3 = 30\ \Omega$

ANSWER First, find R_T by adding the individual resistance values.

$$R_T = R_1 + R_2 + R_3$$
$$= 10\ \Omega + 20\ \Omega + 30\ \Omega$$
$$= 60\ \Omega$$

Next, solve for the current, I.

$$I = \frac{V_T}{R_T}$$
$$= \frac{12\ \text{V}}{60\ \Omega}$$
$$= 200\ \text{mA}$$

Now we can solve for the individual resistor voltage drops.

$$V_1 = I \times R_1$$
$$= 200\ \text{mA} \times 10\ \Omega$$
$$= 2\text{V}$$
$$V_2 = I \times R_2$$
$$= 200\ \text{mA} \times 20\ \Omega$$
$$= 4\text{V}$$
$$V_3 = I \times R_3$$
$$= 200\ \text{mA} \times 30\ \Omega$$
$$= 6\ \text{V}$$

Notice that the individual voltage drops are proportional to the series resistance values. For example, because R_3 is 30 Ω, and R_1 is 10 Ω, V_3 will be three times larger than V_1. With the same current through all the resistors, the largest resistance must have the largest voltage drop.

■ *4–3 Knowledge Check*

    Answer at end of chapter.

In Fig. 4–5, suppose that the applied voltage is increased to 25 V. What are the new values for I, V_1, and V_2?

■ *4–3 Self-Review*

    Answers at end of chapter.

Refer to Fig. 4–5.
a. How much is the sum of V_1 and V_2?
b. Calculate I as V_T/R_T.
c. How much is I through R_1?
d. How much is I through R_2?

4–4 Kirchhoff's Voltage Law (KVL)

Kirchhoff's voltage law states that the sum of all resistor voltage drops in a series circuit equals the **applied voltage.** Expressed as an equation, Kirchhoff's voltage law is

$$V_T = V_1 + V_2 + V_3 + \cdots + \text{etc.} \qquad (4\text{–}3)$$

where V_T is the applied voltage and V_1, V_2, V_3 ... are the individual IR voltage drops.

Example 4–4

A voltage source produces an IR drop of 40 V across a 20-Ω R_1, 60 V across a 30-Ω R_2, and 180 V across a 90-Ω R_3, all in series. According to Kirchhoff's voltage law, how much is the applied voltage V_T?

ANSWER

$$V_T = 40 \text{ V} + 60 \text{ V} + 180 \text{ V}$$
$$V_T = 280 \text{ V}$$

Note that the IR drop across each R results from the same current of 2 A, produced by 280 V across the total R_T of 140 Ω.

Example 4–5

An applied V_T of 120 V produces IR drops across two series resistors R_1 and R_2. If the voltage drop across R_1 is 40 V, how much is the voltage drop across R_2?

ANSWER Since V_1 and V_2 must total 120 V and V_1 is 40 V, the voltage drop across R_2 must be the difference between 120 V and 40 V, or

$$V_2 = V_T - V_1 = 120 \text{ V} - 40 \text{ V}$$
$$V_2 = 80 \text{ V}$$

Figure 4-7 Series string of two 120-V light bulbs operating from a 240-V line. (*a*) Wiring diagram. (*b*) Schematic diagram.

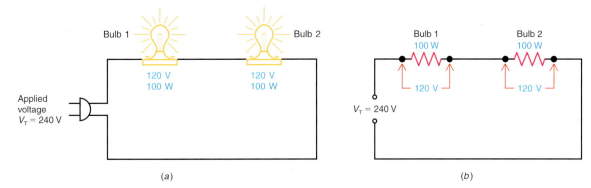

(*a*) (*b*)

It is logical that V_T is the sum of the series IR drops. The current I is the same in all series components. For this reason, the total of all series voltages V_T is needed to produce the same I in the total of all series resistances R_T as the I that each resistor voltage produces in its R.

A practical application of voltages in a series circuit is illustrated in Fig. 4–7. In this circuit, two 120-V lightbulbs are operated from a 240-V line. If one bulb were connected to 240 V, the filament would burn out. With the two bulbs in series, however, each has 120 V for proper operation. The two 120-V drops across the bulbs in series add to equal the applied voltage of 240 V.

Note: A more detailed explanation of Kirchhoff's voltage law is provided in Chap. 9 (Sec. 9–2).

■ *4–4 Knowledge Check*

Answer at end of chapter.

A 36-V potential difference is applied across resistors R_1 and R_2 in series. How much voltage is across R_2 if R_1 has a voltage drop of 24 V?

■ *4–4 Self-Review*

Answers at end of chapter.

a. A series circuit has IR drops of 10, 20, and 30 V. How much is the applied voltage V_T of the source?
b. A 100-V source is applied across R_1 and R_2 in series. If V_1 is 25 V, how much is V_2?
c. A 120-V source is applied across three equal resistances in series. How much is each IR drop?

4–5 Polarity of *IR* Voltage Drops

When a voltage drop exists across a resistance, one end must be either more positive or more negative than the other end. Otherwise, without a potential difference no current could flow through the resistance to produce the voltage drop. The *polarity* of this *IR* voltage drop can be associated with the direction of *I* through *R*. In brief, electrons flow into the negative side of the *IR* voltage and out the positive side (see Fig. 4–8*a*).

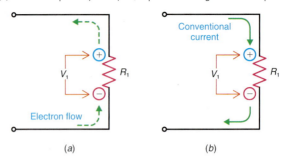

(*a*)   (*b*)

If we want to consider conventional current, with positive charges moving in the opposite direction from electron flow, the rule is reversed for the positive charges. See Fig. 4–8*b*. Here the positive charges for *I* are moving into the positive side of the *IR* voltage.

However, for either electron flow or conventional current, the actual polarity of the *IR* drop is the same. In both *a* and *b* of Fig. 4–8, the top end of *R* in the diagrams is positive since this is the positive terminal of the source producing the current. After all, the resistor does not know which direction of current we are thinking of.

A series circuit with two *IR* voltage drops is shown in Fig. 4–9. We can analyze these polarities in terms of electron flow. The electrons move from the negative terminal of the source V_T through R_2 from point C to D. Electrons move into C and out from D. Therefore C is the negative side of the voltage drop across R_2. Similarly, for the *IR* voltage drop across R_1, point E is the negative side, compared with point F.

A more fundamental way to consider the polarity of *IR* voltage drops in a circuit is the fact that between any two points the one nearer to the positive terminal of the voltage source is more positive; also, the point nearer to the negative terminal of the applied voltage is more negative. A point nearer the terminal means that there is less resistance in its path.

In Fig. 4–9 point C is nearer to the negative battery terminal than point D. The reason is that C has no resistance to B, whereas the path from D to B includes the resistance of R_2. Similarly, point F is nearer to the positive battery terminal than point E, which makes F more positive than E.

Notice that points D and E in Fig. 4–9 are marked with both plus and minus polarities. The plus polarity at D indicates that it is more positive than C. This polarity, however, is shown just for the voltage across R_2. Point D cannot be more positive than points F and A. The positive terminal of the applied voltage must be the most positive point because the battery is generating the *positive potential* for the entire circuit.

Similarly, points B and C must have the most negative potential in the entire string, since point B is the negative terminal of the applied voltage. Actually, the plus polarity marked at D means only that this end of R_2 is less negative than C by the amount of voltage drop across R_2.

Consider the potential difference between E and D in Fig. 4–9, which is only a piece of wire. This voltage is zero because there is no resistance between these two points. Without any resistance here, the current cannot produce the *IR* drop necessary for a difference in potential. Points E and D are, therefore, the same electrically since they have the same potential.

Figure 4–9 Example of two *IR* voltage drops in series. Electron flow shown for direction of *I*.

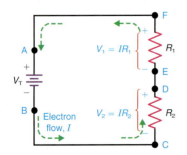

When we go around the external circuit from the negative terminal of V_T, with electron flow, the voltage drops are drops in *negative potential*. For the opposite direction, starting from the positive terminal of V_T, the voltage drops are drops in positive potential. Either way, the voltage drop of each series R is its proportional part of the V_T needed for the one value of current in all resistances.

■ *4–5 Knowledge Check*

Answer at end of chapter.

Is the polarity of an *IR* voltage drop the same for both electron flow and conventional current?

■ *4–5 Self-Review*

Answers at end of chapter.

Refer to Fig. 4–9.
a. Which point in the circuit is the most negative?
b. Which point in the circuit is the most positive?
c. Which is more negative, point D or F?

4–6 Total Power in a Series Circuit

The power needed to produce current in each series resistor is used up in the form of heat. Therefore, the *total power* used is the sum of the individual values of power dissipated in each part of the circuit. As a formula,

$$P_T = P_1 + P_2 + P_3 + \cdots + \text{etc.} \tag{4-4}$$

As an example, in Fig. 4–10, R_1 dissipates 40 W for P_1, equal to 20 V × 2 A for the *VI* product. Or, the P_1 calculated as I^2R is $(2 \times 2) \times 10 = 40$ W. Also, P_1 is V^2/R, or $(20 \times 20)/10 = 40$ W.

Similarly, P_2 for R_2 is 80 W. This value is 40 × 2 for *VI*, $(2 \times 2) \times 20$ for I^2R, or $(40 \times 40)/20$ for V^2/R. P_2 must be more than P_1 because R_2 is more than R_1 with the same current.

The total power dissipated by R_1 and R_2, then, is $40 + 80 = 120$ W. This power is generated by the source of applied voltage.

The total power can also be calculated as $V_T \times I$. The reason is that V_T is the sum of all series voltages and I is the same in all series components. In this case, then, $P_T = V_T \times I = 60 \times 2 = 120$ W.

The total power here is 120 W, calculated either from the total voltage or from the sum of P_1 and P_2. This is the amount of power produced by the battery. The voltage source produces this power, equal to the amount used by the resistors.

Figure 4–10 The sum of the individual powers P_1 and P_2 used in each resistance equals the total power P_T produced by the source.

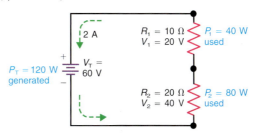

An applied voltage of 15 V is across a 100-Ω R_1 and 400-Ω R_2 in series. Calculate P_1, P_2, and P_T.

a. Each of three equal resistances dissipates 2 W. How much P_T is supplied by the source?
b. A 1-kΩ R_1 and 40-kΩ R_2 are in series with a 50-V source. Which R dissipates more power?

4–7 Series–Aiding and Series–Opposing Voltages

Series-aiding voltages are connected with polarities that allow current in the same direction. In Fig. 4–11a, the 6 V of V_1 alone could produce a 3-A electron flow from the negative terminal, with the 2-Ω R. Also, the 8 V of V_2 could produce 4 A in the same direction. The total I then is 7 A.

Instead of adding the currents, however, the voltages V_1 and V_2 can be added, for a V_T of $6 + 8 = 14$ V. This 14 V produces 7 A in all parts of the series circuit with a resistance of 2 Ω. Then I is $14/2 = 7$ A.

Voltages are connected series-aiding when the plus terminal of one is connected to the negative terminal of the next. They can be added for a total equivalent voltage. This idea applies in the same way to voltage sources, such as batteries, and to voltage drops across resistances. Any number of voltages can be added, as long as they are connected with series-aiding polarities.

Series-opposing voltages are subtracted, as shown in Fig. 4–11b. Notice here that the positive terminals of V_1 and V_2 are connected. Subtract the smaller from the larger value, and give the net V the polarity of the larger voltage. In this example, V_T is $8 - 6 = 2$ V. The polarity of V_T is the same as V_2 because its voltage is higher than V_1.

If two series-opposing voltages are equal, the net voltage will be zero. In effect, one voltage balances out the other. The current I also is zero, without any net potential difference.

Figure 4–11 Example of voltage sources V_1 and V_2 in series. (a) Note the connections for series-aiding polarities. Here 8 V + 6 V = 14 V for the total V_T. (b) Connections for series-opposing polarities. Now 8 V − 6 V = 2 V for V_T.

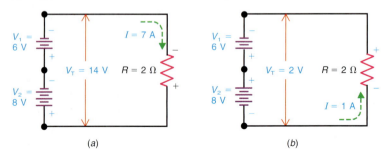

(a) (b)

■ *4–7 Knowledge Check*

Answer at end of chapter.

A 4-V V_1 and 18-V V_2 are connected in a series-aiding configuration. How much is V_T?

■ *4–7 Self-Review*

Answers at end of chapter.

a. Voltage V_1 of 40 V is series-aiding with V_2 of 60 V. How much is V_T?
b. The same V_1 and V_2 are connected series-opposing. How much is V_T?

4–8 Analyzing Series Circuits with Random Unknowns

Figure 4–12 Analyzing a series circuit to find I, V_1, V_2, P_1, and P_2. See text for solution.

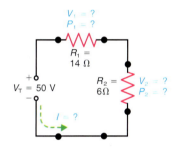

Refer to Fig. 4–12. Suppose that the source V_T of 50 V is known, with a 14-Ω R_1 and 6-Ω R_2. The problem is to find R_T, I, the individual voltage drops V_1 and V_2 across each resistor, and the power dissipated.

We must know the total resistance R_T to calculate I because the total applied voltage V_T is given. This V_T is applied across the total resistance R_T. In this example, R_T is $14 + 6 = 20\ \Omega$.

Now I can be calculated as V_T/R_T, or 50/20, which equals 2.5 A. This 2.5-A I flows through R_1 and R_2.

The individual voltage drops are

$$V_1 = IR_1 = 2.5 \times 14 = 35\ \text{V}$$
$$V_2 = IR_2 = 2.5 \times 6 = 15\ \text{V}$$

Note that V_1 and V_2 total 50 V, equal to the applied V_T.

The calculations to find the power dissipated in each resistor are as follows:

$$P_1 = V_1 \times I = 35 \times 2.5 = 87.5\ \text{W}$$
$$P_2 = V_2 \times I = 15 \times 2.5 = 37.5\ \text{W}$$

These two values of dissipated power total 125 W. The power generated by the source equals $V_T \times I$ or 50×2.5, which is also 125 W.

General Methods for Series Circuits

For other types of problems with series circuits, it is useful to remember the following:

1. When you know the I for one component, use this value for I in all components, for the current is the same in all parts of a series circuit.
2. To calculate I, the total V_T can be divided by the total R_T, or an individual IR drop can be divided by its R. For instance, the current in Fig. 4–12 could be calculated as V_2/R_2 or 15/6, which equals the same 2.5 A for I. However, do not mix a total value for the entire circuit with an individual value for only part of the circuit.
3. When you know the individual voltage drops around the circuit, these can be added to equal the applied V_T. This also means that a known voltage drop can be subtracted from the total V_T to find the remaining voltage drop.

GOOD TO KNOW

Solving a series circuit with random unknowns is similar to solving a crossword puzzle. Random clues are given for solving some of the values in the circuit and then all of the clues are pieced together for the entire solution to the problem.

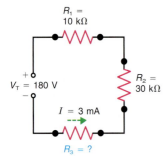

Figure 4–13 Find the resistance of R_3. See text for the analysis of this series circuit.

These principles are illustrated by the problem in Fig. 4–13. In this circuit, R_1 and R_2 are known but not R_3. However, the current through R_3 is given as 3 mA.

With just this information, all values in this circuit can be calculated. The I of 3 mA is the same in all three series resistances. Therefore,

$$V_1 = 3 \text{ mA} \times 10 \text{ k}\Omega = 30 \text{ V}$$
$$V_2 = 3 \text{ mA} \times 30 \text{ k}\Omega = 90 \text{ V}$$

The sum of V_1 and V_2 is $30 + 90 = 120$ V. This 120 V plus V_3 must total 180 V. Therefore, V_3 is $180 - 120 = 60$ V.

With 60 V for V_3, equal to IR_3, then R_3 must be 60/0.003, equal to 20,000 Ω or 20 kΩ. The total circuit resistance is 60 kΩ, which results in the current of 3 mA with 180 V applied, as specified in the circuit.

Another way of doing this problem is to find R_T first. The equation $I = V_T/R_T$ can be inverted to calculate R_T.

$$R_T = \frac{V_T}{I}$$

With a 3-mA I and 180 V for V_T, the value of R_T must be 180 V/3 mA = 60 kΩ. Then R_3 is 60 kΩ − 40kΩ = 20 kΩ.

The power dissipated in each resistance is 90 mW in R_1, 270 mW in R_2, and 180 mW in R_3. The total power is $90 + 270 + 180 = 540$ mW.

Series Voltage–Dropping Resistors

A common application of series circuits is to use a resistance to drop the voltage from the source V_T to a lower value, as in Fig. 4–14. The load R_L here represents a radio that operates normally with a 9-V battery. When the radio is on, the dc load current with 9 V applied is 18 mA. Therefore, the requirements are 9 V at 18 mA as the load.

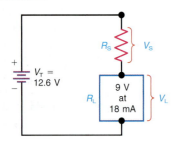

Figure 4–14 Example of a series voltage-dropping resistor R_S used to drop V_T of 12.6 V to 9 V for R_L. See text for calculations.

To operate this radio from 12.6 V, the voltage-dropping resistor R_S is inserted in series to provide a voltage drop V_S that will make V_L equal to 9 V. The required voltage drop for V_S is the difference between V_L and the higher V_T. As a formula,

$$V_S = V_T - V_L = 12.6 - 9 = 3.6 \text{ V}$$

Furthermore, this voltage drop of 3.6 V must be provided with a current of 18 mA, for the current is the same through R_S and R_L. To calculate R_S,

$$R_S = \frac{3.6 \text{ V}}{18 \text{ mA}} = 0.2 \text{ k}\Omega = 200 \text{ }\Omega$$

Circuit with Voltage Sources in Series

See Fig. 4–15. Note that V_1 and V_2 are series-opposing, with + to + through R_1. Their net effect, then, is 0 V. Therefore, V_T consists only of V_3, equal to 4.5 V. The total R is $2 + 1 + 2 = 5$ kΩ for R_T. Finally, I is V_T/R_T or 4.5 V/5 kΩ, which is equal to 0.9 mA, or 900 μA.

■ 4–8 Knowledge Check

Answer at end of chapter.

How much resistance, R_1, must be added in series with a 150-Ω R_2 to limit the current from a 24-V source to 96 mA?

Figure 4–15 Finding the I for this series circuit with three voltage sources. See text for solution.

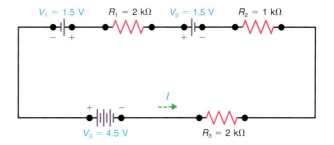

4-8 Self-Review

Answers at end of chapter.

Refer to Fig. 4–13.
a. Calculate V_1 across R_1.
b. Calculate V_2 across R_2.
c. How much is V_3?

4–9 Ground Connections in Electrical and Electronic Systems

In most electrical and electronic systems, one side of the voltage source is connected to ground. For example, one side of the voltage source of the 120-Vac power line in residential wiring is connected directly to *earth ground*. The reason for doing this is to reduce the possibility of electric shock. The connection to earth ground is usually made by driving copper rods into the ground and connecting the ground wire of the electrical system to these rods. The schematic symbol used for earth ground is shown in Fig. 4–16. In electronic circuits, however, not all ground connections are necessarily earth ground connections. The pitchfork-like symbol shown in Fig. 4–16 is considered by many people to be the most appropriate symbol for a metal chassis or copper foil ground on printed-circuit boards. This *chassis ground symbol* represents a common return path for current and may or may not be connected to an actual earth ground. Another ground symbol, common ground, is shown in Fig. 4–16. This is just another symbol used to represent a common return path for current in a circuit. In all cases, ground is assumed to be at a potential of 0 V, regardless of which symbol is shown. Some schematic diagrams may use two or all three of the ground symbols shown in Fig. 4–16. In this type of circuit, each ground represents a common return path for only those circuits using the same ground symbol. When more than one type of ground symbol is shown on a schematic diagram, it is important to realize that each one is electrically isolated from the other. The term *electrically isolated* means that the resistance between each ground or common point is infinite ohms.

Although standards defining the use of each ground symbol in Fig. 4–16 have been set, the use of these symbols in the electronics industry seems to be inconsistent with their definitions. In other words, a schematic may show the earth ground symbol, even though it is a chassis ground connection. Regardless of the symbol used, the main thing to remember is that the symbol represents a common return path for current in a given circuit. In this text, the earth ground symbol shown in Fig. 4–16 has been arbitrarily chosen as the symbol representing a common return path for current.

Figure 4–16 Ground symbols.

earth ground   chassis ground   common ground

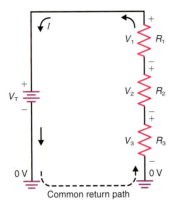

Common return path

Figure 4–17 shows a series circuit employing the earth ground symbol. Since each ground symbol represents the same electrical potential of 0 V, the negative terminal of V_T and the bottom end of R_3 are actually connected to the same point electrically. Electrons leaving the bottom of V_T flow through the common return path represented by the ground symbol and return to the bottom or R_3, as shown in the figure. One of the main reasons for using ground connections in electronic circuits is to simplify the wiring.

Voltages Measured with Respect to Ground

When a circuit has a ground as a common return, we generally measure the voltages with respect to this ground. The circuit in Fig. 4–18a, is called a *voltage divider*. Let us consider this circuit without any ground, and then analyze the effect of grounding different points on the divider. It is important to realize that this circuit operates the same way with or without the ground. The only factor that changes is the reference point for measuring the voltages.

In Fig. 4–18a, the three 10-Ω resistances R_1, R_2, and R_3 divide the 30-V source equally. Then each voltage drop is 30/3 = 10 V for V_1, V_2, and V_3. The polarity of each resistor voltage drop is positive at the top and negative at the bottom, the same as V_T. As you recall, the polarity of a resistor's voltage drop is determined by the direction of current flow.

If we want to consider the current, I is 30/30 = 1 A. Each IR drop is $1 \times 10 = 10$ V for V_1, V_2, and V_3.

Positive Voltages to Negative Ground

In Fig. 4–18b, the negative side of V_T is grounded and the bottom end of R_1 is also grounded to complete the circuit. The ground is at point A. Note that the individual voltages V_1, V_2, and V_3 are still 10 V each. Also, the current is still 1 A. The direction of current is also the same, from the negative side of V_T, through the common ground, to the bottom end of R_1. The only effect of the ground here is to provide a conducting path from one side of the source to one side of the load.

With the ground in Fig. 4–18b, though, it is useful to consider the voltages with respect to ground. In other words, the ground at point A will now be the reference for all voltages. When a voltage is indicated for only one point in a circuit, generally the other point is assumed to be ground. We must have two points for a potential difference.

Let us consider the voltages at points B, C, and D. The voltage at B to ground is V_{BA}. This *double subscript notation* indicates that we measure at B with respect to A. In general, the first letter indicates the point of measurement and the second letter is the reference point.

Then V_{BA} is +10 V. The positive sign is used here to emphasize the polarity. The value of 10 V for V_{BA} is the same as V_1 across R_1 because points B and A are across R_1. However, V_1 as the voltage across R_1 cannot be given any polarity without a reference point.

When we consider the voltage at C, then, V_{CA} is +20 V. This voltage equals $V_1 + V_2$. Also, for point D at the top, V_{DA} is +30 V for $V_1 + V_2 + V_3$.

Positive and Negative Voltages to a Grounded Tap

In Fig. 4–18c point B in the divider is grounded. The purpose is to have the divider supply negative and positive voltages with respect to ground. The negative voltage here is V_{AB}, which equals −10 V. This value is the same 10 V as V_1, but V_{AB} is the voltage at the negative end A with respect to the positive end B. The other voltages in the divider are $V_{CB} = +10$ V and $V_{DB} = +20$ V.

Figure 4–18 An example of calculating dc voltages measured with respect to ground. (*a*) Series circuit with no ground connection. (*b*) Negative side of V_T grounded to make all voltages positive with respect to ground. (*c*) Positive and negative voltages with respect to ground at point B. (*d*) Positive side of V_T grounded; all voltages are negative to ground.

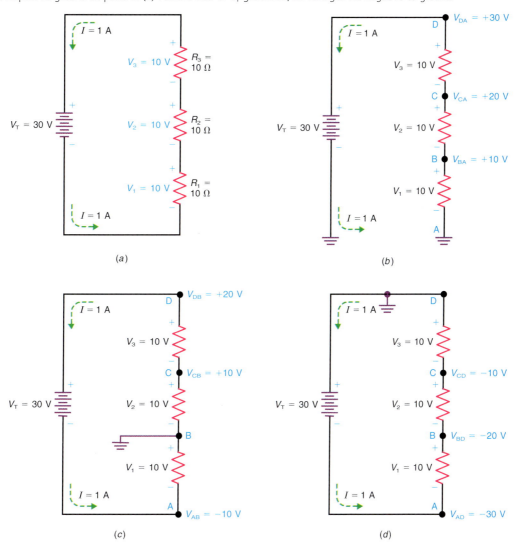

We can consider the ground at B as a dividing point for positive and negative voltages. For all points toward the positive side of V_T, any voltage is positive to ground. Going the other way, at all points toward the negative side of V_T, any voltage is negative to ground.

Negative Voltages to Positive Ground

In Fig. 4–18*d*, point D at the top of the divider is grounded, which is the same as grounding the positive side of the source V_T. The voltage source here is *inverted,* compared with Fig. 4–18*b*, as the opposite side is grounded. In Fig. 4–18*d*, all voltages on the divider are negative to ground. Here, $V_{CD} = -10$ V, $V_{BD} = -20$ V, and $V_{AD} = -30$ V. Any point in the circuit must be more negative than the positive terminal of the source, even when this terminal is grounded.

GOOD TO KNOW

The ground connections in Fig. 4–18*b*, 4–18*c* and 4–18*d* will not change the resistance, current or voltage values from those shown in the original circuit of Fig. 4–18*a*. The only thing that changes is the reference point for making voltage measurements.

■ *4–9 Knowledge Check*

Answer at end of chapter.

In Fig. 4–18*c* assume the ground connection is moved from point B to C. How much are V_{DC}, V_{BC}, and V_{AC}?

■ *4–9 Self-Review*

Answers at end of chapter.

Refer to Fig. 4–18*c* and give the voltage and polarity for

a. A to ground.
b. B to ground.
c. D to ground.
d. V_{DA} across V_T.

GOOD TO KNOW

Everyone who works with electronic hardware must be able to troubleshoot electronic equipment. The service technician, the production tester, the custom engineer, and the engineering designer all do some troubleshooting as part of their jobs.

4–10 Troubleshooting: Opens and Shorts in Series Circuits

In many cases, electronic technicians are required to repair a piece of equipment that is no longer operating properly. The technician is expected to troubleshoot the equipment and restore it to its original operating condition. To *troubleshoot* means "to diagnose or analyze." For example, a technician may diagnose a failed electronic circuit by using a digital multimeter (DMM) to make voltage, current, and resistance measurements. Once the defective component has been located, it is removed and replaced with a good one. But here is one very important point that needs to be made about troubleshooting: To troubleshoot a defective circuit, you must understand how the circuit is supposed to work in the first place. Without this knowledge, your troubleshooting efforts could be nothing more than guesswork. What we will do next is analyze the effects of both opens and shorts in series circuits.

The Effect of an Open in a Series Circuit

An *open circuit* is a break in the current path. The resistance of an open circuit is extremely high because the air between the open points is a very good insulator. Air can have billions of ohms of resistance. For a series circuit, a break in the current path means zero current in all components.

Figure 4–19*a*, shows a series circuit that is operating normally. With 40 V of applied voltage and 40 Ω of total resistance, the series current is 40 V/40 Ω = 1 A. This produces the following *IR* voltage drops across

Figure 4–19 Effect of an open in a series circuit. (*a*) Normal circuit with 1-A series current. (*b*) Open path between points P_1 and P_2 results in zero current in all parts of the circuit.

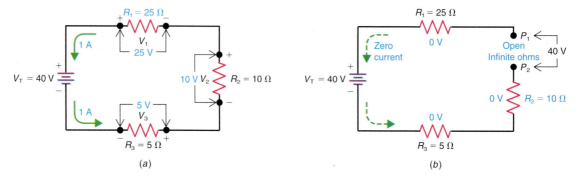

(a)   (b)

R_1, R_2, and R_3: $V_1 = 1 \text{ A} \times 25 \, \Omega = 25$ V, $V_2 = 1 \text{ A} \times 10 \, \Omega = 10$ V, and $V_3 = 1 \text{ A} \times 5 \, \Omega = 5$ V.

Now consider the effect of an open circuit between points P_1 and P_2 in Fig. 4–19b. Because there is practically infinite resistance between the open points, the current in the entire series circuit is zero. With zero current throughout the series circuit, each resistor's IR voltage will be 0 V even though the applied voltage is still 40 V. To calculate V_1, V_2, and V_3 in Fig. 4–19b, simply use 0 A for I. Then, $V_1 = 0 \text{ A} \times 25 \, \Omega = 0$ V, $V_2 = 0 \text{ A} \times 10 \, \Omega = 0$ V, and $V_3 = 0 \text{ A} \times 5 \, \Omega = 0$ V. But how much voltage is across points P_1 and P_2? The answer is 40 V. This might surprise you, but here's the proof: Let's assume that the resistance between P_1 and P_2 is $40 \times 10^9 \, \Omega$, which is 40 GΩ (40 gigohms). Since the total resistance of a series circuit equals the sum of the series resistances, R_T is the sum of 25 Ω, 15 Ω, 10 Ω, and 40 GΩ. Since the 40 GΩ of resistance between P_1 and P_2 is so much larger than the other resistances, it is essentially the total resistance of the series circuit. Then the series current I is calculated as 40 V/40 GΩ $= 1 \times 10^{-9}$ A $= 1$ nA. For all practical purposes, the current I is zero. This is the value of current in the entire series circuit. This small current produces about 0 V across R_1, R_2, and R_3, but across the open points P_1 and P_2, where the resistance is high, the voltage is calculated as $V_{\text{open}} = 1 \times 10^{-9}$ A $\times 40 \times 10^9 \, \Omega = 40$ V.

In summary, here is the effect of an open in a series circuit:

1. The current I is zero in all components.
2. The voltage drop across each good component is 0 V.
3. The voltage across the open points equals the applied voltage.

The Applied Voltage V_T is Still Present with Zero Current

The open circuit in Fig. 4–19b is another example of how voltage and current are different. There is no current with the open circuit because there is no complete path for current flow between the two battery terminals. However, the battery still has its potential difference of 40 V across the positive and negative terminals. In other words, the applied voltage V_T is still present with or without current in the external circuit. If you measure V_T with a voltmeter, it will measure 40 V regardless of whether the circuit is closed, as in Fig. 4–19a, or open, as in Fig. 4–19b.

The same idea applies to the 120-Vac voltage from the power line in our homes. The 120 V potential difference is available from the terminals of the wall outlet. If you connect a lamp or appliance to the outlet, current will flow in those circuits. When there is nothing connected, though, the 120 V potential is still present at the outlet. If you accidentally touch the metal terminals of the outlet when nothing else is connected, you will get an electric shock. The power company is maintaining the 120 V at the outlets as a source to produce current in any circuit that is plugged into the outlet.

Example 4–6

Assume that the series circuit in Fig. 4–20 has failed. A technician troubleshooting the circuit used a voltmeter to record the following resistor voltage drops.

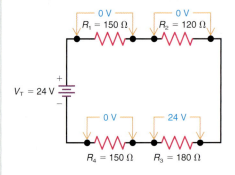

Figure 4–20 Series circuit for Example 4–6.

$V_1 = 0$ V
$V_2 = 0$ V
$V_3 = 24$ V
$V_4 = 0$ V

Based on these voltmeter readings, which component is defective and what type of defect is it? (Assume that only one component is defective.)

ANSWER To help understand which component is defective, let's calculate what the values of V_1, V_2, V_3, and V_4 are supposed to be. Begin by calculating R_T and I.

$$R_T = R_1 + R_2 + R_3 + R_4$$
$$= 150\ \Omega + 120\ \Omega + 180\ \Omega + 150\ \Omega$$
$$R_T = 600\ \Omega$$
$$I = \frac{V_T}{R_T}$$
$$= \frac{24\ \text{V}}{600\ \Omega}$$
$$I = 40\ \text{mA}$$

Next,

$$V_1 = I \times R_1$$
$$= 40\ \text{mA} \times 150\ \Omega$$
$$V_1 = 6\ \text{V}$$
$$V_2 = I \times R_2$$
$$= 40\ \text{mA} \times 120\ \Omega$$
$$V_2 = 4.8\ \text{V}$$
$$V_3 = I \times R_3$$
$$= 40\ \text{mA} \times 180\ \Omega$$
$$V_3 = 7.2\ \text{V}$$
$$V_4 = I \times R_4$$
$$= 40\ \text{mA} \times 150\ \Omega$$
$$V_4 = 6\ \text{V}$$

Next, compare the calculated values with those measured in Fig. 4–20. When the circuit is operating normally, V_1, V_2, and V_4 should measure 6 V, 4.8 V, and 6 V, respectively. Instead, the measurements made in Fig. 4–20 show that each of these voltages is 0 V. This indicates that the current I in the circuit must be zero, caused by an open somewhere in the circuit. The reason that V_1, V_2, and V_4 are 0 V is simple: $V = I \times R$. If $I = 0$ A, then each good resistor must have a voltage drop of 0 V. The measured value of V_3 is 24 V, which is considerably higher than its calculated value of 7.2 V. Because V_3 is dropping the full value of the applied voltage, it must be open. The reason the open R_3 will drop the full 24 V is that it has billions of ohms of resistance and, in a series circuit, the largest resistance drops the most voltage. Since the open resistance of R_3 is so much higher than the values of R_1, R_2, and R_4, it will drop the full 24 V of applied voltage.

The Effect of a Short in a Series Circuit

A **short circuit** is an extremely low resistance path for current flow. The resistance of a short is assumed to be 0 Ω. This is in contrast to an open, which is

Figure 4–21 Series circuit of Fig. 4–19 with R_2 shorted.

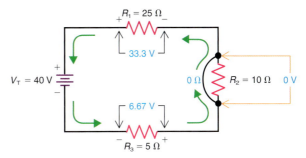

assumed to have a resistance of infinite ohms. Let's reconsider the circuit in Fig. 4–19 with R_2 shorted. The circuit is redrawn for your convenience in Fig. 4–21. Recall from Fig. 4–19a that the normal values of V_1, V_2, and V_3 are 25 V, 10 V, and 5 V, respectively. With the 10-Ω R_2 shorted, the total resistance R_T will decrease from 40 Ω to 30 Ω. This will cause the series current to increase from 1 A to 1.33 A. This is calculated as 40 V/30 Ω = 1.33 A. The increase in current will cause the voltage drop across resistors R_1 and R_3 to increase from their normal values. The new voltage drops across R_1 and R_3 with R_2 shorted are calculated as follows:

$$V_1 = I \times R_1 = 1.33\ A \times 25\ \Omega \qquad V_3 = I \times R_3 = 1.33\ A \times 5\ \Omega$$
$$V_1 = 33.3\ V \qquad\qquad\qquad\qquad V_3 = 6.67\ V$$

The voltage drop across the shorted R_2 is 0 V because the short across R_2 effectively makes its resistance value 0 Ω. Then,

$$V_2 = I \times R_2 = 1.33\ A \times 0\ \Omega$$
$$V_2 = 0\ V$$

In summary, here is the effect of a short in a series circuit:

1. The current I increases above its normal value.
2. The voltage drop across each good component increases.
3. The voltage drop across the shorted component drops to 0 V.

Example 4-7

Assume that the series circuit in Fig. 4–22 has failed. A technician troubleshooting the circuit used a voltmeter to record the following resistor voltage drops:

$$V_1 = 8\ V$$
$$V_2 = 6.4\ V$$
$$V_3 = 9.6\ V$$
$$V_4 = 0\ V$$

Based on the voltmeter readings, which component is defective and what type of defect is it? (Assume that only one component is defective.)

ANSWER This is the same circuit used in Example 4–6. Therefore, the normal values for V_1, V_2, V_3, and V_4 are 6 V, 4.8 V, 7.2 V, and 6 V, respectively. Comparing the calculated values with those measured in Fig. 4–22 reveals that V_1, V_2, and V_3 have increased from their normal values. This indicates that the current has increased, which is why we have a larger voltage drop across these

Figure 4-22 Series circuit for Example 4-7.

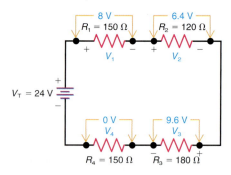

resistors. The measured value of 0 V for V_4 shows a significant drop from its normal value of 6 V. The only way this resistor can have 0 V, when all other resistors show an increase in voltage, is if R_4 is shorted. Then $V_4 = I \times R_4 = I \times 0\,\Omega = 0\,\text{V}$.

General Rules for Troubleshooting Series Circuits

When troubleshooting a series circuit containing three or more resistors, remember this important rule: The defective component will have a voltage drop that will change in the opposite direction as compared to the good components. In other words, in a series circuit containing an open, all the good components will have a voltage decrease from their normal value to 0 V. The defective component will have a voltage increase from its normal value to the full applied voltage. Likewise, in a series circuit containing a short, all good components will have a voltage increase from their normal values and the defective component's voltage drop will decrease from its normal value to 0 V. The point to be made here is simple: The component whose voltage changes in the opposite direction of the other components is the defective component. In the case of an open resistor, the voltage drop increases to the value of the applied voltage and all other resistor voltages decrease to 0 V. In the case of a short, all good components show their voltage drops increasing, whereas the shorted component shows a voltage decrease to 0 V. This same general rule applies to a series circuit that has components whose resistances have increased or decreased from their normal values but are neither open or shorted.

■ 4-10 Knowledge Check

Answer at end of chapter.

If R_3 shorts in Fig. 4–19a, how much voltage would be measured across each resistor?

■ 4-10 Self-Review

Answers at end of chapter.

a. In Fig. 4–20, how much voltage is across R_1 if it is open?
b. In Fig. 4–20, how much voltage is across R_2 if it is shorted?
c. In Fig. 4–20, does the voltage across R_3 increase, decrease, or stay the same if the value of R_1 increases?

Summary

- There is only one current I in a series circuit: $I = V_T/R_T$, where V_T is the voltage applied across the total series resistance R_T. This I is the same in all the series components.

- The total resistance R_T of a series string is the sum of the individual resistances.

- Kirchhoff's voltage law states that the applied voltage V_T equals the sum of the IR voltage drops in a series circuit.

- The negative side of an IR voltage drop is where electrons flow in, attracted to the positive side at the opposite end.

- The sum of the individual values of power used in the individual resistances equals the total power supplied by the source.

- Series-aiding voltages are added; series-opposing voltages are subtracted.

- An open circuit results in no current in all parts of the series circuit.

- For an open in a series circuit, the voltage across the two open terminals is equal to the applied voltage, and the voltage across the remaining components is 0 V.

- A short in a series circuit causes the current to increase above its normal value. The voltage drop across the shorted component decreases to 0 V, and the voltage drop across the remaining components increases.

Important Terms

- Chassis Ground - a common return path for current in a circuit. The common return path is often a direct connection to a metal chassis or frame or perhaps a copper foil trace on a printed circuit board. The symbol for chassis ground is ⏚.

- Double Subscript Notation - a notational system that identifies the points in the circuit where a voltage measurement is to be taken, i.e., V_{AG}. The first letter in the subscript indicates the point in the circuit where the measurement is to be taken, and the second letter indicates the point of reference.

- Earth Ground - a direct connection to the earth usually made by driving copper rods into the earth and then connecting the ground wire of an electrical system to this point. The earth ground connection can serve as a common return path for the current in a circuit. The symbol for earth ground is ⏚.

- Kirchhoff's Voltage Law - a law stating that the sum of the voltage drops in a series circuit must equal the applied voltage.

- Series-Aiding Voltages - voltage sources that are connected so that the polarities of the individual sources aid each other in producing current in the same direction in the circuit.

- Series Components - components that are connected in the same current path.

- Series-Opposing Voltages - voltage sources that are connected so that the polarities of the individual sources will oppose each other in producing current flow in the circuit.

- Series String - a combination of series resistances.

- Troubleshooting - a term that refers to diagnosing or analyzing a faulty electronic circuit.

- Voltage Drop - a voltage across a resistor equal to the product of the current, I, and the resistance, R.

- Voltage Polarity - a term to describe the positive and negative ends of a potential difference across a component such as a resistor.

Related Formulas

$R_T = R_1 + R_2 + R_3 + \cdots + \text{etc.}$

$R_T = \dfrac{V_T}{I}$

$I = \dfrac{V_T}{R_T}$

$V_R = I \times R$

$V_T = V_1 + V_2 + V_3 + \cdots + \text{etc.}$

$P_T = P_1 + P_2 + P_3 + \cdots + \text{etc.}$

Self-Test

Answers at back of book.

1. Three resistors in series have individual values of 120 Ω, 680 Ω and 1.2 kΩ. How much is the total resistance, R_T?
   a. 1.8 kΩ.
   b. 20 kΩ.
   c. 2 kΩ.
   d. none of the above.

2. In a series circuit, the current, I, is
   a. different in each resistor.
   b. the same everywhere.
   c. the highest near the positive and negative terminals of the voltage source.
   d. different at all points along the circuit.

3. In a series circuit, the largest resistance has
   a. the largest voltage drop.
   b. the smallest voltage drop.
   c. more current than the other resistors.
   d. both a and c.

4. The polarity of a resistor's voltage drop is determined by
   a. the direction of current through the resistor.
   b. how large the resistance is.
   c. how close the resistor is to the voltage source.
   d. how far away the resistor is from the voltage source.

5. A 10-Ω and 15-Ω resistor are in series across a dc voltage source. If the 10-Ω resistor has a voltage drop of 12 V, how much is the applied voltage?
   a. 18 V.
   b. 12 V.
   c. 30 V.
   d. The value of applied voltage cannot be determined.

6. How much is the voltage across a shorted component in a series circuit?
   a. The full applied voltage, V_T.
   b. The voltage is slightly higher than normal.
   c. 0 V.
   d. It cannot be determined.

7. How much is the voltage across an open component in a series circuit?
   a. The full applied voltage, V_T.
   b. The voltage is slightly lower than normal.
   c. 0 V.
   d. It cannot be determined.

8. A voltage of 120 V is applied across two resistors, R_1 and R_2, in series. If the voltage across R_2 equals 90 V, how much is the voltage across R_1?
   a. 90 V.
   b. 30 V.
   c. 120 V.
   d. Cannot be determined.

9. If two series opposing voltages each have a voltage of 9 V, the net or total voltage is
   a. 0 V.
   b. 18 V.
   c. 9 V.
   d. none of these.

10. On a schematic diagram, what does the chassis ground symbol represent?
    a. hot spots on the chassis.
    b. the locations in the circuit where electrons accumulate.
    c. a common return path for current in one or more circuits.
    d. none of the above.

11. The notation, V_{BG}, means
    a. the voltage at Point G with respect to B.
    b. the voltage at Point B with respect to G.
    c. the battery (b) or generator (G) voltage.
    d. none of the above.

12. If a resistor in a series circuit is shorted, the series current, I
    a. decreases.
    b. stays the same.
    c. increases.
    d. drops to zero.

13. A 6-V and 9-V source are connected in a series-aiding configuration. How much is the net or total voltage?
    a. −3 V.
    b. +3 V.
    c. 0 V.
    d. 15 V.

14. A 56-Ω and 82-Ω resistor are in series with an unknown resistor. If the total resistance of the series combination is 200 Ω, what is the value of the unknown resistor?
    a. 138 Ω.
    b. 62 Ω.
    c. 26 Ω.
    d. It cannot be determined.

15. How much is the total resistance, R_T, of a series circuit if one of the resistors is open?
    a. infinite (∞) Ω.
    b. 0 Ω.
    c. R_T is much lower than normal.
    d. none of the above.

16. If a resistor in a series circuit becomes open, how much is the voltage across each of the remaining resistors that are still good?
    a. Each good resistor has the full value of applied voltage.
    b. The applied voltage is split evenly amongst the good resistors.
    c. 0 V.
    d. This is impossible to determine.

17. A 5-Ω and 10-Ω resistor are connected in series across a dc voltage source. Which resistor will dissipate more power?
    a. the 5-Ω resistor.
    b. the 10-Ω resistor.
    c. It depends on how much the current is.
    d. They will both dissipate the same amount of power.

18. Which of the following equations can be used to determine the total power in a series circuit?
    a. $P_T = P_1 + P_2 + P_3 + \cdots + $ etc.
    b. $P_T = V_T \times I$.
    c. $P_T = I^2 R_T$.
    d. all of the above.

19. Using electron flow, the polarity of a resistor's voltage drop is

   a. positive on the side where electrons enter and negative on the side where they leave.

   b. negative on the side were electrons enter and positive on the side where they leave.

   c. opposite to that obtained with conventional current flow.

   d. both b and c.

20. The schematic symbol for earth ground is

   a.

   b.

   c.

   d.

Questions

1. Show how to connect two resistances in series across a dc voltage source.

2. State three rules for the current, voltage, and resistance in a series circuit.

3. For a given amount of current, why does a higher resistance have a larger voltage drop across it?

4. Two 300-W, 120-V light bulbs are connected in series across a 240-V line. If the filament of one bulb burns open, will the other bulb light? Why? With the open circuit, how much is the voltage across the source and across each bulb?

5. Prove that if $V_T = V_1 + V_2 + V_3$, then $R_T = R_1 + R_2 + R_3$.

6. State briefly a rule for determining the polarity of the voltage drop across each resistor in a series circuit.

7. State briefly a rule to determine when voltages are series-aiding.

8. In a series string, why does the largest R dissipate the most power?

9. Give one application of series circuits.

10. In Fig. 4–18, explain why R_T, I, V_1, V_2, and V_3 are not affected by the placement of the ground at different points in the circuit.

Problems

SECTION 4–1 WHY *I* IS THE SAME IN ALL PARTS OF A SERIES CIRCUIT

4–1 ‖‖ **MultiSim** In Fig. 4–23, how much is the current, *I*, at each of the following points?

   a. Point A

   b. Point B

   c. Point C

   d. Point D

   e. Point E

   f. Point F

4–2 In Fig. 4–23, how much is the current, *I*, through each of the following resistors?

   a. R_1

   b. R_2

   c. R_3

4–3 If R_1 and R_3 are interchanged in Fig. 4–23, how much is the current, *I*, in the circuit?

SECTION 4–2 TOTAL *R* EQUALS THE SUM OF ALL SERIES RESISTANCES

4–4 ‖‖ **MultiSim** In Fig. 4–24, solve for R_T and *I*.

4–5 ‖‖ **MultiSim** Recalculate the values for R_T and *I* in Fig. 4–24 if $R_1 = 220\ \Omega$ and $R_2 = 680\ \Omega$.

Figure 4–23

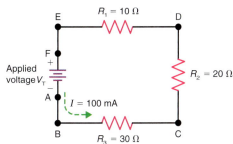

Figure 4–24

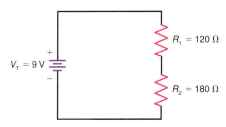

Figure 4–25

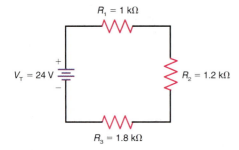

$R_1 = 1$ kΩ

$V_T = 24$ V

$R_2 = 1.2$ kΩ

$R_3 = 1.8$ kΩ

4–6 **MultiSim** In Fig. 4–25, solve for R_T and I.

4–7 **MultiSim** What are the new values for R_T and I in Fig. 4–25 if a 2-kΩ resistor, R_4, is added to the series circuit?

4–8 In Fig. 4–26, solve for R_T and I.

Figure 4–26

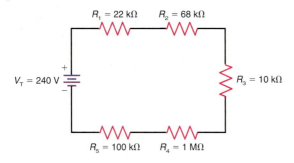

$R_1 = 22$ kΩ $R_2 = 68$ kΩ

$V_T = 240$ V

$R_3 = 10$ kΩ

$R_5 = 100$ kΩ $R_4 = 1$ MΩ

4–9 Recalculate the values for R_T and I in Fig. 4–26 if R_4 is changed to 100 kΩ.

SECTION 4–3 SERIES *IR* VOLTAGE DROPS

4–10 **MultiSim** In Fig. 4–24, find the voltage drops across R_1 and R_2.

4–11 **MultiSim** In Fig. 4–25, find the voltage drops across R_1, R_2, and R_3.

4–12 In Fig. 4–26, find the voltage drops across R_1, R_2, R_3, R_4, and R_5.

4–13 In Fig. 4–27, solve for R_T, I, V_1, V_2, and V_3.

Figure 4–27

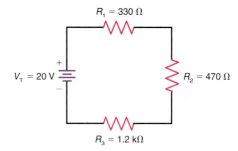

$R_1 = 330$ Ω

$V_T = 20$ V

$R_2 = 470$ Ω

$R_3 = 1.2$ kΩ

4–14 In Fig. 4–27, recalculate the values for R_T, I, V_1, V_2, and V_3 if V_T is increased to 60 V.

4–15 In Fig. 4–28, solve for R_T, I, V_1, V_2, V_3, and V_4.

Figure 4–28

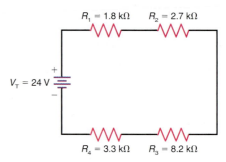

$R_1 = 1.8$ kΩ $R_2 = 2.7$ kΩ

$V_T = 24$ V

$R_4 = 3.3$ kΩ $R_3 = 8.2$ kΩ

SECTION 4–4 KIRCHHOFF'S VOLTAGE LAW (KVL)

4–16 Using Kirchhoff's Voltage Law, determine the value of the applied voltage, V_T, in Fig. 4–29.

Figure 4–29

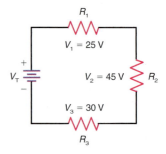

R_1

$V_1 = 25$ V

V_T

$V_2 = 45$ V R_2

$V_3 = 30$ V

R_3

4–17 If $V_1 = 2$ V, $V_2 = 6$ V, and $V_3 = 7$ V in Fig. 4–29, how much is V_T?

4–18 Determine the voltage, V_2, in Fig. 4–30.

Figure 4–30

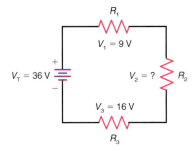

4–19 |||| **MultiSim** In Fig. 4–31, solve for the individual resistor voltage drops. Then, using Kirchhoff's Voltage Law, find V_T.

Figure 4–31

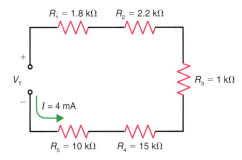

4–20 An applied voltage of 15 V is connected across resistors R_1 and R_2 in series. If $V_2 = 3$ V, how much is V_1?

SECTION 4–5 POLARITY OF *IR* VOLTAGE DROPS

4–21 In Fig. 4–32,
   a. Solve for R_T, I, V_1, V_2, and V_3.
   b. Indicate the direction of electron flow through R_1, R_2, and R_3.
   c. Write the values of V_1, V_2, and V_3 next to resistors R_1, R_2, and R_3.
   d. Indicate the polarity of each resistor voltage drop.

Figure 4–32

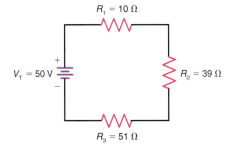

4–22 In Fig. 4–32, indicate the polarity for each resistor voltage drop using conventional current flow. Are the polarities opposite to those obtained with electron flow or are they the same?

4–23 If the polarity of V_T is reversed in Fig. 4–32, what happens to the polarity of the resistor voltage drops and why?

SECTION 4–6 TOTAL POWER IN A SERIES CIRCUIT

4–24 In Fig. 4–24, calculate P_1, P_2, and P_T.

4–25 In Fig. 4–25, calculate P_1, P_2, P_3, and P_T.

4–26 In Fig. 4–26, calculate P_1, P_2, P_3, P_4, P_5, and P_T.

4–27 In Fig. 4–27, calculate P_1, P_2, P_3, and P_T.

4–28 In Fig. 4–28, calculate P_1, P_2, P_3, P_4, and P_T.

SECTION 4–7 SERIES–AIDING AND SERIES–OPPOSING VOLTAGES

4–29 |||| **MultiSim** In Fig. 4–33,
   a. How much is the net or total voltage, V_T across R_1?
   b. How much is the current, I, in the circuit?
   c. What is the direction of electron flow through R_1?

Figure 4–33

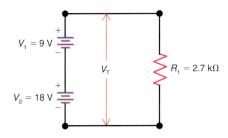

4–30 |||| **MultiSim** In Fig. 4–34,
   a. How much is the net or total voltage, V_T, across R_1?
   b. How much is the current, I, in the circuit?
   c. What is the direction of electron flow through R_1?

Figure 4–34

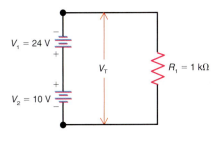

4-31 In Fig. 4-34, assume that V_2 is increased to 30 V. What is

   a. The net or total voltage, V_T, across R_1?

   b. The current, I, in the circuit?

   c. The direction of electron flow through R_1?

4-32 In Fig. 4-35,

   a. How much is the net or total voltage, V_T, across R_1?

   b. How much is the current, I, in the circuit?

   c. What is the direction of electron flow through R_1?

Figure 4-35

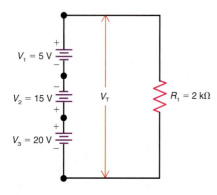

4-33 In Fig. 4-36,

   a. How much is the net or total voltage, V_T, across R_1 and R_2 in series?

   b. How much is the current, I, in the circuit?

   c. What is the direction of electron flow through R_1 and R_2?

   d. Calculate the voltage drops across R_1 and R_2.

Figure 4-36

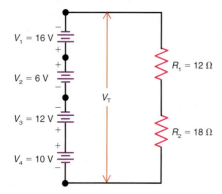

SECTION 4-8 ANALYZING SERIES CIRCUITS WITH RANDOM UNKNOWNS

4-34 In Fig. 4-37, calculate the value for the series resistor, R_s that will allow a 12-V, 150-mA CD player to be operated from a 30-V supply.

Figure 4-37

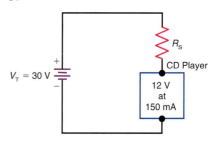

4-35 In Fig. 4-38, solve for I, V_1, V_2, V_3, V_T, R_3, P_T, P_2, and P_3.

Figure 4-38

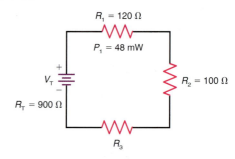

4-36 In Fig. 4-39, solve for R_T, V_1, V_3, V_4, R_2, R_3, P_T, P_1, P_2, P_3, and P_4.

Figure 4-39

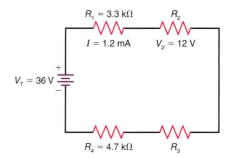

4-37 In Fig. 4–40, solve for I, R_T, V_T, V_2, V_3, V_4, R_4, P_1, P_2, P_3, and P_4.

Figure 4–40

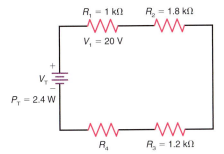

4-38 In Fig. 4–41, solve for I, R_T, R_2, V_2, V_3, P_1, P_2, P_3, and P_T.

Figure 4–41

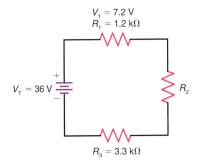

4-39 In Fig. 4–42, solve for R_3, I, V_T, V_1, V_2, V_3, V_4, P_1, P_3, P_4, and P_T.

Figure 4–42

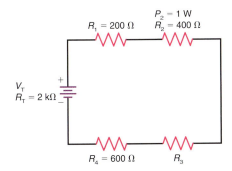

4-40 A 120-Ω resistor is in series with an unknown resistor. The voltage drop across the unknown resistor is 12 V and the power dissipated by the 120-Ω resistor is 4.8 W. Calculate the value of the unknown resistor.

4-41 A 1.5-kΩ resistor is in series with an unknown resistance. The applied voltage, V_T, equals 36 V and the series current is 14.4 mA. Calculate the value of the unknown resistor.

4-42 How much resistance must be added in series with a 6.3-V, 150-mA lightbulb if the bulb is to be operated from a 12-V source?

4-43 A 1-kΩ and 1.5-kΩ resistor are in series. If the total power dissipated by the resistors is 250 mW, how much is the applied voltage, V_T?

4-44 A 22-Ω resistor is in series with a 12-V motor that is drawing 150 mA of current. How much is the applied voltage, V_T?

SECTION 4–9 GROUND CONNECTIONS IN ELECTRICAL AND ELECTRONIC SYSTEMS

4-45 In Fig. 4–43, solve for V_{AG}, V_{BG}, and V_{CG}.

Figure 4–43

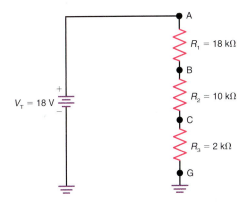

4-46 In Fig. 4–44, solve for V_{AG}, V_{BG}, and V_{CG}.

Figure 4–44

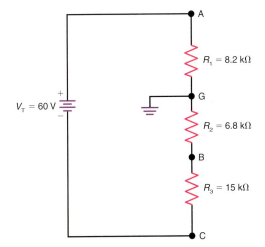

4-47 In Fig. 4-45, solve for V_{AG}, V_{BG}, V_{CG}, and V_{DG}.

Figure 4-45

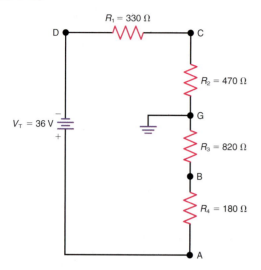

4-48 In Fig. 4-46, solve for V_{AG}, V_{BG}, and V_{CG}.

Figure 4-46

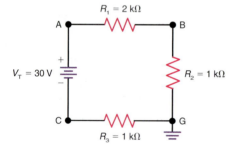

SECTION 4-10 TROUBLESHOOTING: OPENS AND SHORTS IN SERIES CIRCUITS

4-49 In Fig. 4-47, solve for R_T, I, V_1, V_2, and V_3.

Figure 4-47

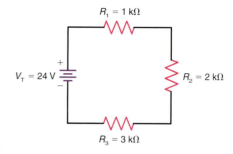

4-50 In Fig. 4-47, assume R_1 becomes open. How much is
a. the total resistance, R_T?
b. the series current, I?
c. the voltage across each resistor, R_1, R_2, and R_3?

4-51 In Fig. 4-47, assume R_3 shorts. How much is
a. the total resistance, R_T?
b. the series current, I?
c. the voltage across each resistor, R_1, R_2, and R_3?

4-52 In Fig. 4-47, assume that the value of R_2 has increased but is not open. What happens to
a. the total resistance, R_T?
b. the series current, I?
c. the voltage drop across R_2?
d. the voltage drops across R_1 and R_3?

Critical Thinking

4-53 Three resistors in series have a total resistance R_T of 2.7 kΩ. If R_2 is twice the value of R_1 and R_3 is three times the value of R_2, what are the values of R_1, R_2, and R_3?

4-54 Three resistors in series have an R_T of 7 kΩ. If R_3 is 2.2 times larger than R_1 and 1.5 times larger than R_2, what are the values of R_1, R_2, and R_3?

4-55 A 100-Ω, $^1/_8$-W resistor is in series with a 330-Ω, $^1/_2$-W resistor. What is the maximum series current this circuit can handle without exceeding the wattage rating of either resistor?

4-56 A 1.5-kΩ, $^1/_2$-W resistor is in series with a 470-Ω, $^1/_4$-W resistor. What is the maximum voltage that can be applied to this series circuit without exceeding the wattage rating of either resistor?

4-57 Refer to Fig. 4-48. Select values for R_1 and V_T so that when R_2 varies from 1 kΩ to 0 Ω, the series current varies from 1 to 5 mA. V_T and R_1 are to have fixed or constant values.

Figure 4-48 Circuit diagram for Critical Thinking Prob. 4-57.

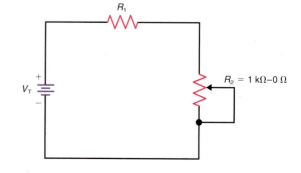

Troubleshooting Challenge

Table 4–1 shows voltage measurements taken in Fig. 4–49. The first row shows the normal values that exist when the circuit is operating properly. Rows 2 to 15 are voltage measurements taken when one component in the circuit has failed. For each row, identify which component is defective and determine the type of defect that has occurred in the component.

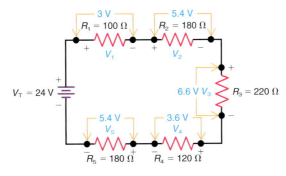

Figure 4–49   Circuit diagram for Troubleshooting Challenge. Normal values for V_1, V_2, V_3, V_4, and V_5 are shown on schematic.

Table 4–1	Table for Troubleshooting Challenge					
	V_1	V_2	V_3	V_4	V_5	Defective Component
			VOLTS			
1 Normal values	3	5.4	6.6	3.6	5.4	None
2 Trouble 1	0	24	0	0	0	
3 Trouble 2	4.14	7.45	0	4.96	7.45	
4 Trouble 3	3.53	6.35	7.76	0	6.35	
5 Trouble 4	24	0	0	0	0	
6 Trouble 5	0	6.17	7.54	4.11	6.17	
7 Trouble 6	0	0	0	24	0	
8 Trouble 7	3.87	0	8.52	4.64	6.97	
9 Trouble 8	0	0	24	0	0	
10 Trouble 9	0	0	0	0	24	
11 Trouble 10	2.4	4.32	5.28	2.88	9.12	
12 Trouble 11	4	7.2	0.8	4.8	7.2	
13 Trouble 12	3.87	6.97	8.52	4.64	0	
14 Trouble 13	15.6	2.16	2.64	1.44	2.16	
15 Trouble 14	3.43	6.17	7.55	0.68	6.17	

Answers to Knowledge Check Problems

4–1 Because there is only one path for current flow.

4–2 $R_T = 3$ kΩ, $I = 4$ mA

4–3 $I = 2.5$ A; $V_1 = 10$ V; $V_2 = 15$ V

4–4 $V_2 = 12$ V

4–5 Yes

4–6 $P_1 = 90$ mW; $P_2 = 360$ mW; $P_T = 450$ mW

4–7 $V_T = 22$ V

4–8 $R_1 = 100 \ \Omega$

4–9 $V_{DC} = +10$ V; $V_{BC} = -10$ V; $V_{AC} = -20$ V

4–10 $V_1 = 28.57$ V; $V_2 = 11.43$ V; $V_3 = 0$ V

Answers to Self-Reviews

4–1
a. R_1, R_2, R_3, V_T and the wires
b. $I = 5$ A
c. 2 A

4–2
a. $I = 2$ mA
b. $R_T = 10$ kΩ
c. $I = 1$ mA

4–3
a. 10 V.
b. $I = 1$ A
c. $I = 1$ A
d. $I = 1$ A

4–4
a. 60 V
b. 75 V
c. 40 V

4–5
a. point B or C
b. point A or F
c. point D

4–6
a. $P_T = 6$ W
b. R_2

4–7
a. 100 V
b. 20 V

4–8
a. $V_1 = 30$ V
b. $V_2 = 90$ V
c. $V_3 = 60$ V

4–9
a. -10 V
b. 0 V
c. $+20$ V
d. $+30$ V

4–10
a. 24 V
b. 0 V
c. decreases

5

Parallel Circuits

A parallel circuit is any circuit that provides one common voltage across all components. Each component across the voltage source provides a separate path or branch for current flow. The individual branch currents are calculated as $\dfrac{V_A}{R}$ where V_A is the applied voltage and R is the individual branch resistance. The total current, I_T, supplied by the applied voltage, must equal the sum of all individual branch currents.

The equivalent resistance of a parallel circuit equals the applied voltage, V_A, divided by the total current, I_T. The term equivalent resistance refers to a single resistance that would draw the same amount of current as all of the parallel connected branches. The equivalent resistance of a parallel circuit is designated R_{EQ}.

This chapter covers all of the characteristics of parallel circuits, including important information about how to troubleshoot a parallel circuit containing a defective component.

Objectives

After studying this chapter you should be able to

- *Explain* why voltage is the same across all branches in a parallel circuit.

- *Calculate* the individual branch currents in a parallel circuit.

- *Calculate* the total current in a parallel circuit using Kirchhoff's current law.

- *Calculate* the equivalent resistance of two or more resistors in parallel.

- *Explain* why the equivalent resistance of a parallel circuit is always less than the smallest branch resistance.

- *Calculate* the total conductance of a parallel circuit.

- *Calculate* the total power in a parallel circuit.

- *Solve* for the voltage, current, power, and resistance in a parallel circuit having random unknowns.

- *Describe* the effects of an open and short in a parallel circuit.

- *Troubleshoot* parallel circuits containing opens and shorts.

Outline

Important Terms

equivalent resistance, R_{EQ}

Kirchhoff's current law (KCL)

main line

parallel bank

reciprocal resistance formula

5–1 The Applied Voltage V_A Is the Same across Parallel Branches

A parallel circuit is formed when two or more components are connected across a voltage source, as shown in Fig. 5–1. In this figure, R_1 and R_2 are in parallel with each other and a 1.5-V battery. In Fig. 5–1b, the points A, B, C, and E are equivalent to a direct connection at the positive terminal of the battery because the connecting wires have practically no resistance. Similarly, points H, G, D, and F are the same as a direct connection at the negative battery terminal. Since R_1 and R_2 are directly connected across the two terminals of the battery, both resistances must have the same potential difference as the battery. It follows that the voltage is the same across components connected in parallel. The parallel circuit arrangement is used, therefore, to connect components that require the same voltage.

A common application of parallel circuits is typical house wiring to the power line, with many lights and appliances connected across the 120 V source (Fig. 5–2). The wall receptacle has a potential difference of 120 V across each pair of terminals. Therefore, any resistance connected to an outlet has an applied voltage of 120 V. The lightbulb is connected to one outlet and the toaster to another outlet, but both have the same applied voltage of 120 V. Therefore, each operates independently of any other appliance, with all the individual branch circuits connected across the 120 V line.

||| MultiSim **Figure 5–1** Example of a parallel circuit with two resistors. (*a*) Wiring diagram. (*b*) Schematic diagram.

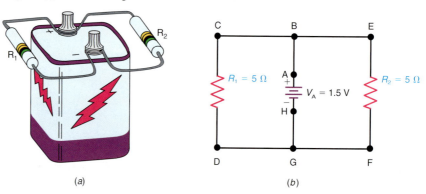

(a)   (b)

Figure 5–2 Lightbulb and toaster connected in parallel with the 120-V line. (*a*) Wiring diagram. (*b*) Schematic diagram.

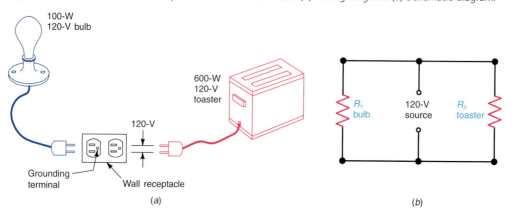

(a)   (b)

■ *5–1 Knowledge Check*

Answer at end of chapter.

If another 5-Ω resistor, R_3, is connected across points E and F in Fig. 5–1b, how much is its voltage?

■ *5–1 Self-Review*

Answers at end of chapter.

a. In Fig. 5–1, how much is the common voltage across R_1 and R_2?
b. In Fig. 5–2, how much is the common voltage across the bulb and the toaster?
c. How many parallel branch circuits are connected across the voltage source in Figs. 5–1 and 5–2?

5–2 Each Branch *I* Equals V_A/R

In applying Ohm's law, it is important to note that the current equals the voltage applied across the circuit divided by the resistance between the two points where that voltage is applied. In Fig. 5–3a, 10 V is applied across the 5 Ω of R_2, resulting in the current of 2 A between points E and F through R_2. The battery voltage is also applied across the parallel resistance of R_1, applying 10 V across 10 Ω. Through R_1, therefore, the current is 1 A between points C and D. The current has a different value through R_1, with the same applied voltage, because the resistance is different. These values are calculated as follows:

$$I_1 = \frac{V_A}{R_1} = \frac{10}{10} = 1 \text{ A}$$

$$I_2 = \frac{V_A}{R_2} = \frac{10}{5} = 2 \text{ A}$$

Figure 5–3b shows how to assemble axial-lead resistors on a lab prototype board to form a parallel circuit.

Just as in a circuit with one resistance, any branch that has less *R* allows more *I*. If R_1 and R_2 were equal, however, the two branch currents would have the same value. For instance, in Fig. 5–1b each branch has its own current equal to 1.5 V/5 Ω = 0.3 A.

The *I* can be different in parallel circuits that have different *R* because *V* is the same across all the branches. Any voltage source generates a potential difference across its two terminals. This voltage does not move. Only *I* flows around the circuit. The source voltage is available to make electrons move

GOOD TO KNOW

In a parallel circuit, the branch with the lowest resistance always has the most current. This must be true since each branch current is calculated as $\dfrac{V_A}{R}$ where V_A is the same across all branches.

Figure 5–3 Parallel circuit. (*a*) The current in each parallel branch equals the applied voltage V_A divided by each branch resistance *R*. (*b*) Axial-lead resistors assembled on a lab prototype board, forming a parallel circuit.

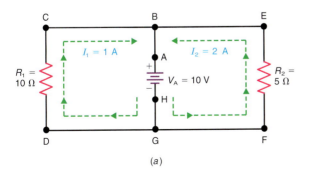

(*a*)

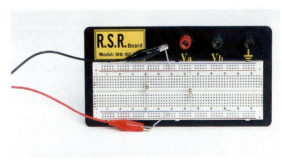

(*b*)

around any closed path connected to the terminals of the source. The amount of I in each separate path depends on the amount of R in each branch.

Example 5-1

MultiSim **Figure 5-4** Circuit for Example 5-1.

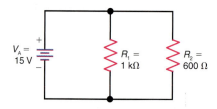

In Fig. 5–4, solve for the branch currents I_1 and I_2.

ANSWER The applied voltage, V_A, of 15 V is across both resistors R_1 and R_2. Therefore, the branch currents are calculated as $\dfrac{V_A}{R}$, where V_A is the applied voltage and R is the individual branch resistance.

$$
\begin{aligned}
I_1 &= \frac{V_A}{R_1} \\
&= \frac{15\ \text{V}}{1\ k\Omega} \\
&= 15\ \text{mA} \\
I_2 &= \frac{V_A}{R_2} \\
&= \frac{15\ \text{V}}{600\ \Omega} \\
&= 25\ \text{mA}
\end{aligned}
$$

■ *5-2 Knowledge Check*

Answer at end of chapter.

If a 2-Ω resistor, R_3 is connected across points B and G in Fig. 5–3a, how much is its branch current?

■ *5-2 Self-Review*

Answers at end of chapter.

Refer to Fig. 5–3.
a. How much is the voltage across R_1?
b. How much is I_1 through R_1?
c. How much is the voltage across R_2?
d. How much is I_2 through R_2?

5-3 Kirchhoff's Current Law (KCL)

Components to be connected in parallel are usually wired directly across each other, with the entire parallel combination connected to the voltage source, as illustrated in Fig. 5–5. This circuit is equivalent to wiring each parallel branch directly to the voltage source, as shown in Fig. 5–1, when the connecting wires have essentially zero resistance.

Figure 5–5 The current in the main line equals the sum of the branch currents. Note that from G to A at the bottom of this diagram is the negative side of the main line, and from B to F at the top is the positive side. (*a*) Wiring diagram. Arrows inside the lines indicate current in the main line for R_1; arrows outside indicate current for R_2. (*b*) Schematic diagram. I_T is the total line current for both R_1 and R_2.

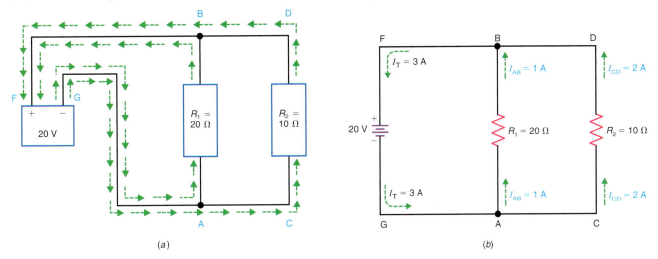

(a)   (b)

GOOD TO KNOW

As more branches are added to a parallel circuit the total current, I_T, increases.

The advantage of having only one pair of connecting leads to the source for all the parallel branches is that usually less wire is necessary. The pair of leads connecting all the branches to the terminals of the voltage source is the **main line.** In Fig. 5–5, the wires from G to A on the negative side and from B to F in the return path form the main line.

In Fig. 5–5*b*, with 20 Ω of resistance for R_1 connected across the 20-V battery, the current through R_1 must be 20 V/20 Ω = 1 A. This current is electron flow from the negative terminal of the source, through R_1, and back to the positive battery terminal. Similarly, the R_2 branch of 10 Ω across the battery has its own branch current of 20 V/10 Ω = 2 A. This current flows from the negative terminal of the source, through R_2, and back to the positive terminal, since it is a separate path for electron flow.

All current in the circuit, however, must come from one side of the voltage source and return to the opposite side for a complete path. In the main line, therefore, the amount of current is equal to the total of the branch currents.

For example, in Fig. 5–5*b*, the total current in the line from point G to point A is 3 A. The total current at branch point A subdivides into its component branch currents for each of the branch resistances. Through the path of R_1 from A to B the current is 1 A. The other branch path ACDB through R_2 has a current of 2 A. At the branch point B, the electron flow from both parallel branches combines, so that the current in the main-line return path from B to F has the same value of 3 A as in the other side of the main line.

Kirchhoff's current law (KCL) states that the total current I_T in the main line in a parallel circuit equals the sum of the individual branch currents. Expressed as an equation, Kirchhoff's current law is:

$$I_T = I_1 + I_2 + I_3 + \cdots + \text{ etc.} \qquad (5\text{--}1)$$

where I_T is the total current and I_1, I_2, I_3 . . . are the individual branch currents. Kirchhoff's current law applies to any number of **parallel branches,** whether the resistances in the branches are equal or unequal.

Example 5-2

An R_1 of 20 Ω, an R_2 of 40 Ω, and an R_3 of 60 Ω are connected in parallel across the 120-V power line. Using Kirchhoff's current law, determine the total current I_T.

ANSWER   Current I_1 for the R_1 branch is 120/20 or 6 A. Similarly, I_2 is 120/40 or 3 A, and I_3 is 120/60 or 2 A. The total current in the main line is

$$I_T = I_1 + I_2 + I_3 = 6 + 3 + 2$$
$$I_T = 11 \text{ A}$$

Example 5-3

Two branches R_1 and R_2 across the 120-V power line draw a total line current I_T of 15 A. The R_1 branch takes 10 A. How much is the current I_2 in the R_2 branch?

ANSWER $I_2 = I_T - I_1 = 15 - 10$
$$I_2 = 5 \text{ A}$$

With two branch currents, one must equal the difference between I_T and the other branch current.

Example 5-4

Three parallel branch currents are 0.1 A, 500 mA, and 800 μA. Using Kirchhoff's current law, calculate I_T.

ANSWER   All values must be in the same units to be added. In this case, all units will be converted to milliamperes: 0.1 A = 100 mA and 800 μA = 0.8 mA. Applying Kirchhoff's current law

$$I_T = 100 + 500 + 0.8$$
$$I_T = 600.8 \text{ mA}$$

You can convert the currents to A, mA, or μA units, as long as the same unit is used for adding all currents.

■ 5–3 Knowledge Check

Answer at end of chapter.

A parallel circuit has the following branch currents: $I_1 = 1.5$ A, $I_2 = 350$ mA, $I_3 = 100$ mA, and $I_4 = 50$ mA. How much is I_T?

■ 5–3 Self-Review

Answers at end of chapter.

a. Branch currents in a parallel circuit are 1 A for I_1, 2 A for I_2, and 3 A for I_3, How much is I_T?
b. Assume $I_T = 6$ A for three branch currents; I_1 is 1 A, and I_2 is 2 A. How much is I_3?
c. Branch currents in a parallel circuit are 1 A for I_1 and 200 mA for I_2. How much is I_T?

5–4 Resistances in Parallel

The combined equivalent resistance across the main line in a parallel circuit can be found by Ohm's law: *Divide the common voltage across the parallel resistances by the total current of all the branches.* Referring to Fig. 5–6a, note that the parallel resistance of R_1 with R_2, indicated by the equivalent resistance R_{EQ}, is the opposition to the total current in the main line. In this example, V_A/I_T is 60 V/3 A = 20 Ω for R_{EQ}.

The total load connected to the source voltage is the same as though one equivalent resistance of 20 Ω were connected across the main line. This is illustrated by the equivalent circuit in Fig. 5–6b. For any number of parallel resistances of any value, use the following equation,

$$R_{EQ} = \frac{V_A}{I_T} \tag{5–2}$$

where I_T is the sum of all the branch currents and R_{EQ} is the equivalent resistance of all parallel branches across the applied voltage source V_A.

The first step in solving for R_{EQ} is to add all the parallel branch currents to find the I_T being delivered by the voltage source. The voltage source thinks that it is connected to a single resistance whose value allows I_T to flow in the circuit according to Ohm's law. This single resistance is R_{EQ}. An illustrative example of a circuit with two parallel branches will be used to show how R_{EQ} is calculated.

GOOD TO KNOW

The statement "current always takes the path of least resistance" is not always true. If it were, all the current in a parallel circuit would flow in the lowest branch resistance only.

|||| **MultiSim**   **Figure 5–6**   Resistances in parallel. (*a*) Combination of R_1 and R_2 is the total R_{EQ} for the main line. (*b*) Equivalent circuit showing R_{EQ} drawing the same 3-A I_T as the parallel combination of R_1 and R_2 in (*a*).

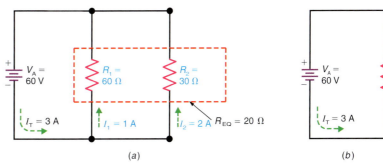

(a)   (b)

Example 5-5

Two branches, each with a 5-A current, are connected across a 90-V source. How much is the equivalent resistance R_{EQ}?

ANSWER   The total line current I_T is $5 + 5 = 10$ A. Then,

$$R_{EQ} = \frac{V_A}{I_T} = \frac{90}{10}$$
$$R_{EQ} = 9\ \Omega$$

Parallel Bank

A combination of parallel branches is often called a **bank.** In Fig. 5–6, the bank consists of the 60-Ω R_1 and 30-Ω R_2 in parallel. Their combined parallel resistance R_{EQ} is the bank resistance, equal to 20 Ω in this example. A bank can have two or more parallel resistors.

When a circuit has more current with the same applied voltage, this greater value of I corresponds to less R because of their inverse relation. Therefore, the combination of parallel resistances R_{EQ} for the bank is always less than the smallest individual branch resistance. The reason is that I_T must be more than any one branch current.

Why R_{EQ} Is Less than Any Branch R

It may seem unusual at first that putting more resistance into a circuit lowers the equivalent resistance. This feature of parallel circuits is illustrated in Fig. 5–7. Note that equal resistances of 30 Ω each are added across the source voltage, one branch at a time. The circuit in Fig. 5–7a has just R_1, which allows 2 A with 60 V applied. In Fig. 5–7b, the R_2 branch is added across the same V_A.

Figure 5–7   How adding parallel branches of resistors increases I_T but decreases R_{EQ}. (a) One resistor. (b) Two branches. (c) Three branches. (d) Equivalent circuit of the three branches in (c).

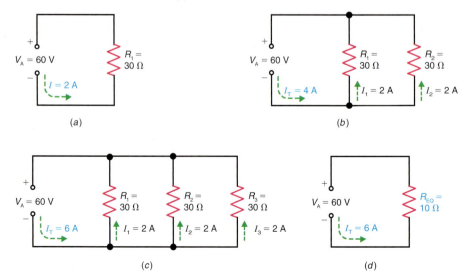

This branch also has 2 A. Now the parallel circuit has a 4-A total line current because of $I_1 + I_2$. Then the third branch, which also takes 2 A for I_3, is added in Fig. 5–7c. The combined circuit with three branches, therefore, requires a total load current of 6 A, which is supplied by the voltage source.

The combined resistance across the source, then, is V_A/I_T, which is 60/6, or 10 Ω. This equivalent resistance R_{EQ}, representing the entire load on the voltage source, is shown in Fig. 5–7d. More resistance branches reduce the combined resistance of the parallel circuit because more current is required from the same voltage source.

Reciprocal Resistance Formula

We can derive the **reciprocal resistance formula** from the fact that I_T is the sum of all the branch currents, or,

$$I_T = I_1 + I_2 + I_3 + \cdots + \text{ etc.}$$

However, $I_T = V/R_{EQ}$. Also, each $I = V/R$. Substituting V/R_{EQ} for I_T on the left side of the formula and V/R for each branch I on the right side, the result is

$$\frac{V}{R_{EQ}} = \frac{V}{R_1} + \frac{V}{R_2} + \frac{V}{R_3} + \cdots + \text{ etc.}$$

Dividing by V because the voltage is the same across all the resistances gives us:

$$\frac{1}{R_{EQ}} = \frac{1}{R_1} + \frac{1}{R_2} + \frac{1}{R_3} + \cdots + \text{ etc.}$$

Next, solve for R_{EQ}.

$$R_{EQ} = \frac{1}{\frac{1}{R_1} + \frac{1}{R_2} + \frac{1}{R_3} + \cdots + \text{ etc.}} \tag{5–3}$$

This reciprocal formula applies to any number of parallel resistances of any value. Using the values in Fig. 5–8a as an example,

$$R_{EQ} = \frac{1}{\frac{1}{20} + \frac{1}{10} + \frac{1}{10}} = 4\ \Omega$$

IIII MultiSim **Figure 5–8** Two methods of combining parallel resistances to find R_{EQ}. (a) Using the reciprocal resistance formula to calculate R_{EQ} as 4 Ω. (b) Using the total line current method with an assumed line voltage of 20 V gives the same 4 Ω for R_{EQ}.

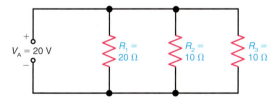

$$R_{EQ} = \frac{1}{\frac{1}{R_1} + \frac{1}{R_2} + \frac{1}{R_3}}$$

$$R_{EQ} = 4\ \Omega$$

(a)

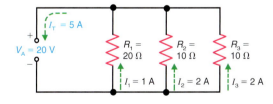

$$R_{EQ} = \frac{V_A}{I_T} = \frac{20\ V}{5\ A}$$

$$R_{EQ} = 4\ \Omega$$

(b)

Figure 5–9 For the special case of all branches having the same resistance, just divide R by the number of branches to find R_{EQ}. Here, $R_{EQ} = 60\ k\Omega/3 = 20\ k\Omega$.

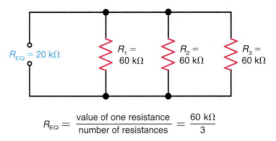

$$R_{EQ} = \frac{\text{value of one resistance}}{\text{number of resistances}} = \frac{60\ k\Omega}{3}$$

Total–Current Method

It may be easier to work without fractions. Figure 5–8*b* shows how this same problem can be calculated in terms of total current instead of by the reciprocal formula. Although the applied voltage is not always known, any convenient value can be assumed because it cancels in the calculations. It is usually simplest to assume an applied voltage of the same numerical value as the highest resistance. Then one assumed branch current will automatically be 1 A and the other branch currents will be more, eliminating fractions less than 1 in the calculations.

In Fig. 5–8*b*, the highest branch R is 20 Ω. Therefore, assume 20 V for the applied voltage. Then the branch currents are 1 A in R_1, 2 A in R_2, and 2 A in R_3. Their sum is $1 + 2 + 2 = 5$ A for I_T. The combined resistance R_{EQ} across the main line is V_A/I_T, or 20 V/5 A = 4 Ω. This is the same value calculated with the reciprocal resistance formula.

Special Case of Equal R in All Branches

If R is equal in all branches, the combined R_{EQ} equals the value of one branch resistance divided by the number of branches.

$$R_{EQ} = \frac{R}{n}$$

where R is the resistance in one branch and n is the number of branches.

This rule is illustrated in Fig. 5–9, where three 60-kΩ resistances in parallel equal 20 kΩ.

The rule applies to any number of parallel resistances, but they must all be equal. As another example, five 60-Ω resistances in parallel have the combined resistance of 60/5, or 12 Ω. A common application is two equal resistors wired in a parallel bank for R_{EQ} equal to one-half R.

Special Case of Only Two Branches

When there are two parallel resistances and they are not equal, it is usually quicker to calculate the combined resistance by the method shown in Fig. 5–10. This rule says that the combination of two parallel resistances is their product divided by their sum.

$$R_{EQ} = \frac{R_1 \times R_2}{R_1 + R_2} \qquad (5\text{–}4)$$

where R_{EQ} is in the same units as all the individual resistances. For the example in Fig. 5–10,

Figure 5–10 For the special case of only two branch resistances, of any values, R_{EQ} equals their product divided by the sum. Here, $R_{EQ} = 2400/100 = 24\Omega$.

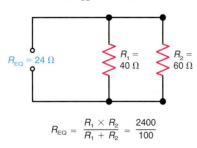

$$R_{EQ} = \frac{R_1 \times R_2}{R_1 + R_2} = \frac{2400}{100}$$

Figure 5–11 An example of parallel resistance calculations with four branches. (a) Original circuit. (b) Resistors combined into two branches. (c) Equivalent circuit reduces to one R_{EQ} for all the branches.

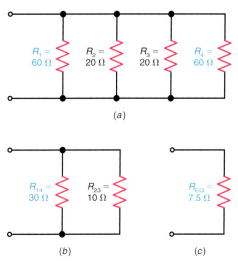

$$R_{EQ} = \frac{R_1 \times R_2}{R_1 + R_2} = \frac{40 \times 60}{40 + 60} = \frac{2400}{100}$$

$$R_{EQ} = 24 \; \Omega$$

Each R can have any value, but there must be only two resistances.

Short-Cut Calculations

Figure 5–11 shows how these special rules can help in reducing parallel branches to a simpler equivalent circuit. In Fig. 5–11a, the 60-Ω R_1 and R_4 are equal and in parallel. Therefore, they are equivalent to the 30-Ω R_{14} in Fig. 5–11b. Similarly, the 20-Ω R_2 and R_3 are equivalent to the 10 Ω of R_{23}. The circuit in Fig. 5–11a is equivalent to the simpler circuit in Fig. 5–11b with just the two parallel resistances of 30 and 10 Ω.

Finally, the combined resistance for these two equals their product divided by their sum, which is 300/40 or 7.5 Ω, as shown in Fig. 5–11c. This value of R_{EQ} in Fig. 5–11c is equivalent to the combination of the four branches in Fig. 5–11a. If you connect a voltage source across either circuit, the current in the main line will be the same for both cases.

The order of connections for parallel resistances does not matter in determining R_{EQ}. There is no question as to which is first or last because they are all across the same voltage source and receive their current at the same time.

Finding an Unknown Branch Resistance

In some cases with two parallel resistors, it is useful to be able to determine what size R_X to connect in parallel with a known R to obtain a required value of R_{EQ}. Then the factors can be transposed as follows:

$$R_X = \frac{R \times R_{EQ}}{R - R_{EQ}} \tag{5–5}$$

This formula is just another way of writing Formula (5–4).

Example 5-6

What R_X in parallel with 40 Ω will provide an R_{EQ} of 24 Ω?

ANSWER
$$R_X = \frac{R \times R_{EQ}}{R - R_{EQ}} = \frac{40 \times 24}{40 - 24} = \frac{960}{16}$$
$$R_X = 60 \ \Omega$$

This problem corresponds to the circuit shown before in Fig. 5–10.

Note that Formula (5–5) for R_X has a product over a difference. The R_{EQ} is subtracted because it is the smallest R. Remember that both Formulas (5–4) and (5-5) can be used with only two parallel branches.

Example 5-7

What R in parallel with 50 kΩ will provide an R_{EQ} of 25 kΩ?

ANSWER $R = 50 \ \text{k}\Omega$

Two equal resistances in parallel have R_{EQ} equal to one-half R.

■ **5–4A Knowledge Check**

Answer at end of chapter.

A 1.2-kΩ resistor, R_1, is in parallel with a 6.8-kΩ R_2. How much is R_{EQ}?

■ **5–4B Knowledge Check**

Answer at end of chapter.

The following resistors are in parallel. $R_1 = 150 \ \Omega$, $R_2 = 60 \ \Omega$, $R_3 = 100 \ \Omega$ and $R_4 = 120 \ \Omega$. How much is R_{EQ}?

■ **5–4C Knowledge Check**

Answer at end of chapter.

How much resistance, R_X, must be connected in parallel with a 3.3-kΩ resistor to obtain an equivalent resistance of 1.32 kΩ?

■ **5–4 Self-Review**

Answers at end of chapter.

a. Find R_{EQ} for three 4.7-MΩ resistances in parallel.
b. Find R_{EQ} for 3 MΩ in parallel with 2 MΩ.
c. Find R_{EQ} for two parallel 20-Ω resistances in parallel with 10 Ω.

5–5 Conductances in Parallel

Since conductance G is equal to $1/R$, the reciprocal resistance Formula (5–3) can be stated for conductance as $R_{EQ} = \dfrac{1}{G_T}$ where G_T is calculated as

$$G_T = G_1 + G_2 + G_3 + \cdots + \text{etc.} \qquad (5\text{--}6)$$

With R in ohms, G is in siemens. For the example in Fig. 5–12, G_1 is $1/20 = 0.05$, G_2 is $1/5 = 0.2$, and G_3 is $1/2 = 0.5$. Then

$$G_T = 0.05 + 0.2 + 0.5 = 0.75 \text{ S}$$

Notice that adding the conductances does not require reciprocals. Each value of G is the reciprocal of R.

The reason why *parallel conductances* are added directly can be illustrated by assuming a 1-V source across all branches. Then calculating the values of $1/R$ for the conductances gives the same values as calculating the branch currents. These values are added for the total I_T or G_T.

Working with G may be more convenient than working with R in parallel circuits, since it avoids the use of the reciprocal formula for R_{EQ}. Each branch current is directly proportional to its conductance. This idea corresponds to the fact that each voltage drop in series circuits is directly proportional to each of the series resistances. An example of the currents for parallel conductances is shown in Fig. 5–13. Note that the branch with G of 4 S has twice as much current as the 2-S branches because the branch conductance is doubled.

■ 5–5A Knowledge Check

Answer at end of chapter.

How much is the equivalent resistance, R_{EQ}, in Fig. 5–12?

■ 5–5B Knowledge Check

Answer at end of chapter.

If a 4-Ω resistor, R_4, is added in parallel with R_1, R_2, and R_3 in Fig. 5–12, how much is G_T? What about R_{EQ}?

Figure 5–12 Conductances G_1, G_2, and G_3 in parallel are added for the total G_T.

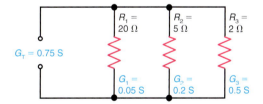

Figure 5–13 Example of how parallel branch currents are directly proportional to each branch conductance G.

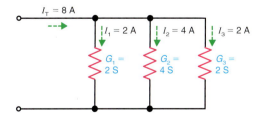

a. If G_1 is 2 S and G_2 in parallel is 4 S, calculate G_T.
b. If G_1 is 0.05 μS, G_2 is 0.2 μS, and G_3 is 0.5 μS, all in parallel, find G_T and its equivalent R_{EQ}.
c. If G_T is 4 μS for a parallel circuit, how much is R_{EQ}?

5–6 Total Power in Parallel Circuits

Since the power dissipated in the branch resistances must come from the voltage source, the **total power** equals the sum of the individual values of power in each branch. This rule is illustrated in Fig. 5–14. We can also use this circuit as an example of applying the rules of current, voltage, and resistance for a parallel circuit.

The applied 10 V is across the 10-Ω R_1 and 5-Ω R_2 in Fig. 5–14. The branch current I_1 then is V_A/R_1 or 10/10, which equals 1 A. Similarly, I_2 is 10/5, or 2 A. The total I_T is 1 + 2 = 3 A. If we want to find R_{EQ}, it equals V_A/I_T or 10/3, which is 3⅓ Ω.

The power dissipated in each branch R is $V_A \times I$. In the R_1 branch, I_1 is 10/10 = 1 A. Then P_1 is $V_A \times I_1$ or 10 × 1 = 10 W.

For the R_2 branch, I_2 is 10/5 = 2 A. Then P_2 is $V_A \times I_2$ or 10 × 2 = 20 W.

Adding P_1 and P_2, the answer is 10 + 20 = 30 W. This P_T is the total power dissipated in both branches.

This value of 30 W for P_T is also the total power supplied by the voltage source by means of its total line current I_T. With this method, the total power is $V_A \times I_T$ or 10 × 3 = 30 W for P_T. The 30 W of power supplied by the voltage source is dissipated or used up in the branch resistances.

It is interesting to note that in a parallel circuit, the smallest branch resistance will always dissipate the most power. Since $P = \dfrac{V^2}{R}$ and V is the same across all parallel branches, a smaller value of R in the denominator will result in a larger amount of power dissipation.

Note also that in both parallel and series circuits, the sum of the individual values of power dissipated in the circuit equals the total power generated by the source. This can be stated as a formula

$$P_T = P_1 + P_2 + P_3 + \cdots + \text{etc.} \tag{5–7}$$

The series or parallel connections can alter the distribution of voltage or current, but power is the rate at which energy is supplied. The circuit arrangement cannot change the fact that all the energy in the circuit comes from the source.

Figure 5–14 The sum of the power values P_1 and P_2 used in each branch equals the total power P_T produced by the source.

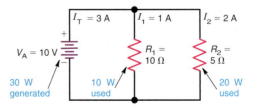

5-6 Knowledge Check

Answer at end of chapter.

In Fig. 5–14 assume V_A is increased to 20 V. What are the new values for P_1, P_2, and P_T?

5-6 Self-Review

Answers at end of chapter.

a. Two parallel branches each have 2 A at 120 V. How much is P_T?
b. Three parallel branches of 10, 20, and 30 Ω have 60 V applied. How much is P_T?
c. Two parallel branches dissipate a power of 15 W each. How much is P_T?

5-7 Analyzing Parallel Circuits with Random Unknowns

For many types of problems with parallel circuits, it is useful to remember the following points.

1. When you know the voltage across one branch, this voltage is across all the branches. There can be only one voltage across branch points with the same potential difference.
2. If you know I_T and one of the branch currents I_1, you can find I_2 by subtracting I_1 from I_T.

The circuit in Fig. 5–15 illustrates these points. The problem is to find the applied voltage V_A and the value of R_3. Of the three branch resistances, only R_1 and R_2 are known. However, since I_2 is given as 2 A, the I_2R_2 voltage must be $2 \times 60 = 120$ V.

Although the applied voltage is not given, this must also be 120 V. The voltage across all the parallel branches is the same 120 V that is across the R_2 branch.

Now I_1 can be calculated as V_A/R_1. This is $120/30 = 4$ A for I_1.

Current I_T is given as 7 A. The two branches take $2 + 4 = 6$ A. The third branch current through R_3 must be $7 - 6 = 1$ A for I_3.

Now R_3 can be calculated as V_A/I_3. This is $120/1 = 120$ Ω for R_3.

5-7 Knowledge Check

Answer at end of chapter.

In Fig. 5–15, assume $I_2 = 1.5$ A instead of 2 A. I_T, R_1, and R_2 remain the same. Recalculate V_A, I_1, I_3, and R_3.

Figure 5–15 Analyzing a parallel circuit. What are the values for V_A and R_3? See solution in text.

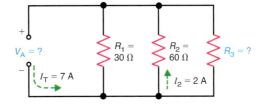

■ 5-7 Self-Review

Answers at end of chapter.

Refer to Fig. 5-15.
a. How much is V_2 across R_2?
b. How much is I_1 through R_1?
c. How much is I_T?

5-8 Troubleshooting: Opens and Shorts in Parallel Circuits

In a parallel circuit, the effect of an open or a short is much different from that in a series circuit. For example, if one branch of a parallel circuit opens, the other branch currents remain the same. The reason is that the other branches still have the same applied voltage even though one branch has effectively been removed from the circuit. Also, if one branch of a parallel circuit becomes shorted, all branches are effectively shorted. The result is excessive current in the shorted branch and zero current in all other branches. In most cases, a fuse will be placed in the main line that will burn open (blow) when its current rating is exceeded. When the fuse blows, the applied voltage is removed from each of the parallel-connected branches. The effects of opens and shorts are examined in more detail in the following paragraphs.

The Effect of an Open in a Parallel Circuit

An open in any circuit is an infinite resistance that results in no current. However, in parallel circuits there is a difference between an open circuit in the main line and an open circuit in a parallel branch. These two cases are illustrated in Fig. 5-16. In Fig. 5-16a the open circuit in the main line prevents any electron flow in the line to all the branches. The current is zero in every branch, therefore, and none of the bulbs can light.

However, in Fig. 5-16b the open is in the branch circuit for bulb 1. The **open branch** circuit has no current, then, and this bulb cannot light. The current in all the other parallel branches is normal, though, because each is connected to the voltage source. Therefore, the other bulbs light.

These circuits show the advantage of wiring components in parallel. An open in one component opens only one branch, whereas the other parallel branches have their normal voltage and current.

The Effect of a Short in a Parallel Circuit

A **short circuit** has practically zero resistance. Its effect, therefore, is to allow excessive current in the shorted circuit. Consider the example in Fig. 5-17. In

Figure 5-16 Effect of an open in a parallel circuit. (*a*) Open path in the main line—no current and no light for all bulbs. (*b*) Open path in any branch—bulb for that branch does not light, but the other two bulbs operate normally.

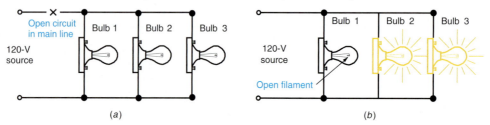

Figure 5-17 Effect of a short circuit across parallel branches. (*a*) Normal circuit. (*b*) Short circuit across points G and H shorts out all the branches.

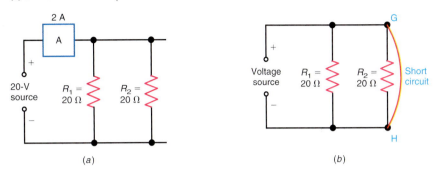

(*a*)   (*b*)

Fig. 5–17*a*, the circuit is normal, with 1 A in each branch and 2 A for the total line current. However, suppose that the conducting wire at point G accidentally makes contact with the wire at point H, as shown in Fig. 5–17*b*. Since the wire is an excellent conductor, the short circuit results in practically zero resistance between points G and H. These two points are connected directly across the voltage source. Since the short circuit provides practically no opposition to current, the applied voltage could produce an infinitely high value of current through this current path.

The Short-Circuit Current

Practically, the amount of current is limited by the small resistance of the wire. Also, the source usually cannot maintain its output voltage while supplying much more than its rated load current. Still, the amount of current can be dangerously high. For instance, the short-circuit current might be more than 100 A instead of the normal line current of 2 A in Fig. 5–17*a*. Because of the short circuit, excessive current flows in the voltage source, in the line to the short circuit at point H, through the short circuit, and in the line returning to the source from G. Because of the large amount of current, the wires can become hot enough to ignite and burn the insulation covering the wire. There should be a fuse that would open if there is too much current in the main line because of a short circuit across any of the branches.

The Short-Circuited Components Have No Current

For the short circuit in Fig. 5–17*b*, the *I* is 0 A in the parallel resistors R_1 and R_2. The reason is that the short circuit is a parallel path with practically zero resistance. Then all the current flows in this path, bypassing the resistors R_1 and R_2. Therefore R_1 and R_2 are short-circuited or *shorted out* of the circuit. They cannot function without their normal current. If they were filament resistances of light bulbs or heaters, they would not light without any current.

The short-circuited components are not damaged, however. They do not even have any current passing through them. Assuming that the short circuit has not damaged the voltage source and the wiring for the circuit, the components can operate again when the circuit is restored to normal by removing the short circuit.

All Parallel Branches Are Short-Circuited

If there were only one *R* in Fig. 5–17 or any number of parallel components, they would all be shorted out by the short circuit across points G and H.

Therefore, a short circuit across one branch in a parallel circuit shorts out all parallel branches.

This idea also applies to a short circuit across the voltage source in any type of circuit. Then the entire circuit is shorted out.

Troubleshooting Procedures for Parallel Circuits

When a component fails in a parallel circuit, voltage, current, and resistance measurements can be made to locate the defective component. To begin our analysis, let's refer to the parallel circuit in Fig. 5–18a, which is normal. The individual branch currents I_1, I_2, I_3, and I_4 are calculated as follows:

$$I_1 = \frac{120 \text{ V}}{20 \text{ }\Omega} = 6 \text{ A}$$

$$I_2 = \frac{120 \text{ V}}{15 \text{ }\Omega} = 8 \text{ A}$$

$$I_3 = \frac{120 \text{ V}}{30 \text{ }\Omega} = 4 \text{ A}$$

$$I_4 = \frac{120 \text{ V}}{60 \text{ }\Omega} = 2 \text{ A}$$

By Kirchhoff's current law, the total current I_T equals 6 A + 8 A + 4 A + 2 A = 20 A. The total current I_T of 20 A is indicated by the ammeter M_1, which is placed in the main line between points J and K. The fuse F_1 between points A and B in the main line can safely carry 20 A, since its maximum rated current is 25 A, as shown.

Now consider the effect of an open branch between points D and I in Fig. 5–18b. With R_2 open, the branch current I_2 is 0 A. Also, the ammeter M_1 shows a total current I_T of 12 A, which is 8 A less than its normal value. This makes sense because I_2 is normally 8 A. Notice that with R_2 open, all other branch currents remain the same. This is because each branch is still connected to the applied voltage of 120 V. It is important to realize that voltage measurements across the individual branches would not help determine which branch is open because even the open branch between points D and I will measure 120 V.

In most cases, the components in a parallel circuit provide a visual indication of failure. If a lamp burns open, it doesn't light. If a motor opens, it stops running. In these cases, the defective component is easy to spot.

In summary, here is the effect of an open branch in a parallel circuit.

1. The current in the open branch drops to 0 A.
2. The total current I_T decreases by an amount equal to the value normally drawn by the now open branch.
3. The current in all remaining branches remains the same.
4. The applied voltage remains present across all branches whether they are open or not.

Next, let's consider the effect of an open between two branch points such as points D and E in Fig. 5–18c. With an open between these two points, the current through branch resistors R_3 and R_4 will be 0 A. Since $I_3 = 4$ A and $I_4 = 2$ A normally, the total current indicated by M_1 will drop from 20 A to 14 A as shown. The reason that I_3 and I_4 are now 0 A is that the applied voltage has effectively been removed from these two branches. If a voltmeter were placed across either points E and H or F and G, it would read 0V. A voltmeter placed across points D and E would measure 120 V, however. This is indicated by the voltmeter M_2 as shown. The reason M_2 measures 120 V between points D and E is explained as follows: Notice that the positive (red) lead of M_2 is connected through S_1 and F_1 to the positive side of the applied voltage. Also,

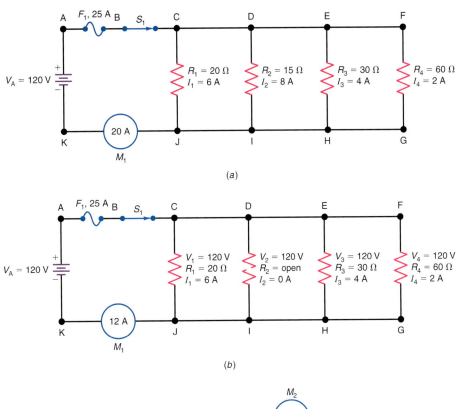

Figure 5–18 Parallel circuit for troubleshooting analysis. (*a*) Normal circuit values; (*b*) circuit values with branch R_2 open; (*c*) circuit values with an open between points D and E; (*d*) circuit showing the effects of a shorted branch.

(*a*)

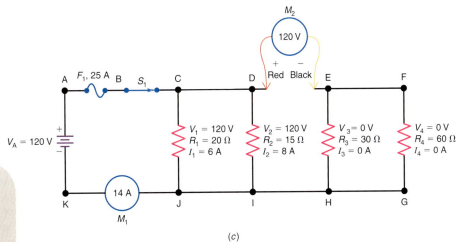

(*b*)

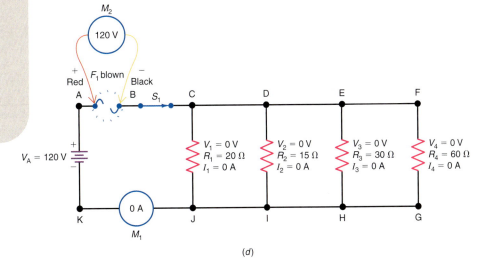

(*c*)

GOOD TO KNOW

A fuse is a safety device which serves to protect the circuit components and wiring in the event of a short circuit. Excessive current melts the fuse element which blows the fuse. With the fuse blown, there is no voltage across any of the parallel connected branches.

(*d*)

the negative (black) lead of M_2 is connected to the top of resistors R_3 and R_4. Since the voltage across R_3 and R_4 is 0 V, the negative lead of M_2 is in effect connected to the negative side of the applied voltage. In other words, M_2 is effectively connected directly across the 120-V source.

Example 5-8

In Fig. 5–18a, suppose that the ammeter M_1 reads 16 A instead of 20 A as it should. What could be wrong with the circuit?

ANSWER Notice that the current I_3 is supposed to be 4 A. If R_3 is open, this explains why M_1 reads a current that is 4 A less than its normal value. To confirm that R_3 is open; open S_1 and disconnect the top lead of R_3 from point E. Next place an ammeter between the top of R_3 and point E. Now, close S_1. If I_3 measures 0 A, you know that R_3 is open. If I_3 measures 4 A, you know that one of the other branches is drawing less current than it should. In this case, the next step would be to measure each of the remaining branch currents to find the defective component.

Consider the circuit in Fig. 5–18d. Notice that the fuse F_1 is blown and the ammeter M_1 reads 0 A. Notice also that the voltage across each branch measures 0 V and the voltage across the blown fuse measures 120 V as indicated by the voltmeter M_2. What could cause this? The most likely answer is that one of the parallel-connected branches has become short-circuited. This would cause the total current to rise well above the 25-A current rating of the fuse, thus causing it to blow. But how do we go about finding out which branch is shorted? There are at least three different approaches. Here's the first one: Start by opening switch S_1 and replacing the bad fuse. Next, with S_1 still open, disconnect all but one of the four parallel branches. For example, disconnect branch resistors R_1, R_2, and R_3 along the top (at points C, D, and E). With R_4 still connected, close S_1. If the fuse blows, you know R_4 is shorted! If the fuse does not blow, with only R_4 connected, open S_1 and reconnect R_3 to point E. Then, close S_1 and see if the fuse blows.

Repeat this procedure with branch resistors R_1 and R_2 until the shorted branch is identified. The shorted branch will blow the fuse when it is reconnected at the top (along points C, D, E, and F) with S_1 closed. Although this troubleshooting procedure is effective in locating the shorted branch, another fuse has been blown and this will cost you or the customer money.

Here's another approach to finding the shorted branch. Open S_1 and replace the bad fuse. Next, measure the resistance of each branch separately. It is important to remember that when you make resistance measurements in a parallel circuit, one end of each branch must be disconnected from the circuit so that the rest of the circuit does not affect the individual branch measurement. The branch that measures 0 Ω is obviously the shorted branch. With this approach, another fuse will not get blown.

Here is yet another approach that could be used to locate the shorted branch in Fig. 5–18d. With S_1 open, place an ohmmeter across points C and J.

With a shorted branch, the ohmmeter will measure 0 Ω. To determine which branch is shorted, remove one branch at a time until the ohmmeter shows a value other than 0 Ω. The shorted component is located when removal of a given branch causes the ohmmeter to show a normal resistance.

In summary, here is the effect of a shorted branch in a parallel circuit:

1. The fuse in the main line will blow, resulting in zero current in the main line as well as in each parallel-connected branch.
2. The voltage across each branch will equal 0 V, and the voltage across the blown fuse will equal the applied voltage.
3. With power removed from the circuit, an ohmmeter will measure 0 Ω across all the branches.

Before leaving the topic of troubleshooting parallel circuits, one more point should be made about the fuse F_1 and the switch S_1 in Fig. 5–18a: The resistance of a good fuse and the resistance across the closed contacts of a switch are practically 0 Ω. Therefore, the voltage drop across a good fuse or a closed switch is approximately 0 V. This can be proven with Ohm's law, since $V = I \times R$. If $R = 0\ Ω$, then $V = I \times 0\ Ω = 0\ V$. When a fuse blows or a switch opens, the resistance increases to such a high value that it is considered infinite. When used in the main line of a parallel circuit, the voltage across an open switch or a blown fuse is the same as the applied voltage. One way to reason this out logically is to treat all parallel branches as a single equivalent resistance R_{EQ} in series with the switch and fuse. The result is a simple series circuit. Then, if either the fuse or the switch opens, apply the rules of an open to a series circuit. As you recall from your study of series circuits, the voltage across an open equals the applied voltage.

■ 5–8A Knowledge Check

Answer at end of chapter.

How much voltage will exist across R_4 in Fig. 5–18a if it is open?

■ 5–8B Knowledge Check

Answer at end of chapter.

How much voltage exists across the fuse, F_1 in Fig. 5–18b?

■ 5–8 Self-Review

Answers at end of chapter.

a. In Fig. 5–16b, how much voltage is across bulb 1?
b. In Fig. 5–17b, how much is the resistance across points G and H?
c. In Fig. 5–18a, how much current will M_1 show if the wire between points C and D is removed?
d. With reference to Question c, how much voltage would be measured across R_4? Across points C and D?
e. In Fig. 5–18a, how much voltage will be measured across points A and B, assuming the fuse is blown?

Summary

- There is only one voltage V_A across all components in parallel.

- The current in each branch I_b equals the voltage V_A across the branch divided by the branch resistance R_b, or $I_b = V_A/R_b$.

- Kirchhoff's current law states that the total current I_T in a parallel circuit equals the sum of the individual branch currents. Expressed as an equation, Kirchhoff's current law is $I_T = I_1 + I_2 + I_3 + \cdots + $ etc.

- The equivalent resistance R_{EQ} of parallel branches is less than the smallest branch resistance, since all the branches must take more current from the source than any one branch.

- For only *two* parallel resistances of any value, $R_{EQ} = R_1R_2/(R_1 + R_2)$.

- For any number of *equal* parallel resistances, R_{EQ} is the value of one resistance divided by the number of resistances.

- For the general case of any number of branches, calculate R_{EQ} as V_A/I_T or use the reciprocal resistance formula:

$$R_{EQ} = \cfrac{1}{{}^1/_{R_1} + {}^1/_{R_2} + {}^1/_{R_3} + \cdots + \text{etc.}}$$

- For any number of conductances in parallel, their values are added for G_T, in the same way as parallel branch currents are added.

- The sum of the individual values of power dissipated in parallel resistances equals the total power produced by the source.

- An open circuit in one branch results in no current through that branch, but the other branches can have their normal current. However, an open circuit in the main line results in no current for any of the branches.

- A short circuit has zero resistance, resulting in excessive current. When one branch is short-circuited, all parallel paths are also short-circuited. The entire current is in the short circuit and no current is in the short-circuited branches.

- The voltage across a good fuse and the voltage across a closed switch are approximately 0 V. When the fuse in the main line of a parallel circuit opens, the voltage across the fuse equals the full applied voltage. Likewise, when the switch in the main line of a parallel circuit opens, the voltage across the open switch equals the full applied voltage.

- Table 5–1 compares Series and Parallel Circuits.

Table 5–1	Comparison of Series and Parallel Circuits
Series Circuit	**Parallel Circuit**
Current the same in all components	Voltage the same across all branches
V across each series R is $I \times R$	I in each branch R is V/R
$V_T = V_1 + V_2 + V_3 + \cdots + $ etc.	$I_T = I_1 + I_2 + I_3 + \cdots + $ etc.
$R_T = R_1 + R_2 + R_3 + \cdots + $ etc.	$G_T = G_1 + G_2 + G_3 + \cdots + $ etc.
R_T must be more than the largest individual R	R_{EQ} must be less than the smallest branch R
$P_T = P_1 + P_2 + P_3 + \cdots + $ etc.	$P_T = P_1 + P_2 + P_3 + \cdots + $ etc.
Applied voltage is divided into IR voltage drops	Main-line current is divided into branch currents
The largest IR drop is across the largest series R	The largest branch I is in the smallest parallel R
Open in one component causes entire circuit to be open	Open in one branch does not prevent I in other branches

Important Terms

- Equivalent Resistance, R_{EQ} - in a parallel circuit, this refers to a single resistance that would draw the same amount of current as all of the parallel connected branches.

- Kirchhoff's Current Law (KCL) - a law which states that the sum of the individual branch currents in a parallel circuit must equal the total current, I_T.

- Main Line - the pair of leads connecting all individual branches in a parallel circuit to the terminals of the applied voltage, V_A. The main line carries the total current, I_T, flowing to and from the terminals of the voltage source.

- Parallel Bank - a combination of parallel connected branches.

- Reciprocal Resistance Formula - a formula which states that the equivalent resistance, R_{EQ}, of a parallel circuit equals the reciprocal of the sum of the reciprocals of the individual branch resistances.

Related Formulas

$$I_1 = \frac{V_A}{R_1}, \; I_2 = \frac{V_A}{R_2}, \; I_3 = \frac{V_A}{R_3}$$

$$I_T = I_1 + I_2 + I_3 + \cdots + \text{etc.}$$

$$R_{EQ} = \frac{V_A}{I_T}$$

$$R_{EQ} = \frac{1}{{}^1/_{R_1} + {}^1/_{R_2} + {}^1/_{R_3} + \cdots + \text{etc.}}$$

$$R_{EQ} = \frac{R}{N} \; (R_{EQ} \text{ for equal branch resistances})$$

$$R_{EQ} = \frac{R_1 \times R_2}{R_1 + R_2} \; (R_{EQ} \text{ for only two branch resistances})$$

$$R_X = \frac{R \times R_{EQ}}{R - R_{EQ}}$$

$$G_T = G_1 + G_2 + G_3 + \cdots + \text{etc.}$$

$$P_T = P_1 + P_2 + P_3 + \cdots + \text{etc.}$$

Self-Test

Answers at back of book.

1. A 120-kΩ resistor, R_1, and a 180-kΩ resistor, R_2, are in parallel. How much is the equivalent resistance, R_{EQ}?
   a. 72 kΩ.
   b. 300 kΩ.
   c. 360 kΩ.
   d. 90 kΩ.

2. A 100-Ω resistor, R_1, and a 300-Ω resistor, R_2, are in parallel across a dc voltage source. Which resistor dissipates more power?
   a. the 300-Ω resistor.
   b. Both resistors dissipate the same amount of power.
   c. the 100-Ω resistor.
   d. This is impossible to determine.

3. Three 18-Ω resistors are in parallel. How much is the equivalent resistance, R_{EQ}?
   a. 54 Ω.
   b. 6 Ω.

   c. 9 Ω.
   d. none of the above.

4. Which of the following statements about parallel circuits is false?
   a. The voltage is the same across all branches in a parallel circuit.
   b. The equivalent resistance, R_{EQ}, of a parallel circuit is always smaller than the smallest branch resistance.
   c. In a parallel circuit the total current, I_T, in the main line equals the sum of the individual branch currents.
   d. The equivalent resistance, R_{EQ}, of a parallel circuit decreases when one or more parallel branches are removed from the circuit.

5. Two resistors, R_1 and R_2, are in parallel with each other and a dc voltage source. If the total current, I_T, in the main line equals 6A and I_2 through R_2 is 4A, how much is I_1 through R_1?
   a. 6 A.
   b. 2 A.
   c. 4 A.
   d. I_1 cannot be determined.

6. How much resistance must be connected in parallel with a 360-Ω resistor to obtain an equivalent resistance, R_{EQ}, of 120 Ω?
   a. 360 Ω.
   b. 480 Ω.
   c. 1.8 kΩ.
   d. 180 Ω.

7. If one branch of a parallel circuit becomes open,
   a. all remaining branch currents increase.
   b. the voltage across the open branch will be 0 V.
   c. the remaining branch currents do not change in value.
   d. the equivalent resistance of the circuit decreases.

8. If a 10-Ω R_1, 40-Ω R_2, and 8-Ω R_3 are in parallel, calculate the total conductance, G_T, of the circuit.
   a. 250 mS.
   b. 58 S.
   c. 4 Ω.
   d. 0.25 μS.

9. Which of the following formulas can be used to determine the total power, P_T, dissipated by a parallel circuit.
   a. $P_T = V_A \times I_T$.
   b. $P_T = P_1 + P_2 + P_3 + \cdots + \text{etc}.$
   c. $P_T = \dfrac{V_A^2}{R_{EQ}}$.
   d. all of the above.

10. A 20-Ω R_1, 50-Ω R_2, and 100-Ω R_3 are connected in parallel. If R_2 is short-circuited, what is the equivalent resistance, R_{EQ}, of the circuit?
    a. approximately 0 Ω.
    b. infinite (∞) Ω.
    c. 12.5 Ω.
    d. R_{EQ} cannot be determined.

11. If the fuse in the main line of a parallel circuit opens,
    a. the voltage across each branch will be 0 V.
    b. the current in each branch will be zero.
    c. the current in each branch will increase to offset the decrease in total current.
    d. both a and b above.

12. A 100-Ω R_1 and a 150-Ω R_2 are in parallel. If the current, I_1, through R_1 is 24 mA, how much is the total current, I_T?
    a. 16 mA.
    b. 40 mA.
    c. 9.6 mA.
    d. I_T cannot be determined.

13. A 2.2-kΩ R_1 is in parallel with a 3.3-kΩ R_2. If these two resistors carry a total current of 7.5 mA, how much is the applied voltage, V_A?
    a. 16.5 V.
    b. 24.75 V.
    c. 9.9 V.
    d. 41.25 V.

14. How many 120-Ω resistors must be connected in parallel to obtain an equivalent resistance, R_{EQ}, of 15 Ω?
    a. 15.
    b. 8.
    c. 12.
    d. 6.

15. A 220-Ω R_1, 2.2-kΩ R_2, and 200-Ω R_3 are connected across 15 V of applied voltage. What happens to R_{EQ} if the applied voltage is doubled to 30 V?
    a. R_{EQ} doubles.
    b. R_{EQ} cuts in half.
    c. R_{EQ} does not change.
    d. R_{EQ} increases but is not double its original value.

16. If one branch of a parallel circuit opens, the total current, I_T,
    a. doesn't change.
    b. decreases.
    c. increases.
    d. goes to zero.

17. In a normally operating parallel circuit, the individual branch currents are
    a. independent of each other.
    b. not affected by the value of the applied voltage.
    c. larger than the total current, I_T.
    d. none of the above.

18. If the total conductance, G_T, of a parallel circuit is 200 μS, how much is R_{EQ}?
    a. 500 Ω.
    b. 200 kΩ.
    c. 5 kΩ.
    d. 500 kΩ.

19. If one branch of a parallel circuit is short-circuited,
    a. the fuse in the main line will blow.
    b. the voltage across the short-circuited branch will measure the full value of applied voltage.
    c. all the remaining branches are effectively short-circuited as well.
    d. both a and c.

20. Two lightbulbs in parallel with the 120-V power line are rated at 60 W and 100 W, respectively. What is the equivalent resistance, R_{EQ}, of the bulbs when they are lit?
    a. 144 Ω.
    b. 90 Ω.
    c. 213.3 Ω.
    d. There is not enough information to calculate R_{EQ}.

Questions

1. Draw a wiring diagram showing three resistances connected in parallel across a battery. Indicate each branch and the main line.

2. State two rules for the voltage and current values in a parallel circuit.

3. Explain briefly why the current is the same in both sides of the main line that connects the voltage source to the parallel branches.

4. (a) Show how to connect three equal resistances for a combined equivalent resistance one-third the value of one resistance. (b) Show how to connect three equal resistances for a combined equivalent resistance three times the value of one resistance.

5. Why can the current in parallel branches be different when they all have the same applied voltage?

6. Why does the current increase in the voltage source as more parallel branches are added to the circuit?

7. Show how the formula

$$R_{EQ} = R_1 R_2 / (R_1 + R_2)$$

is derived from the reciprocal formula

$$\frac{1}{R_{EQ}} = \frac{1}{R_1} + \frac{1}{R_2}$$

8. Redraw Fig. 5–17 with five parallel resistors R_1 to R_5 and explain why they all would be shorted out with a short circuit across R_3.

9. State briefly why the total power equals the sum of the individual values of power, whether a series circuit or a parallel circuit is used.

10. Explain why an open in the main line disables all the branches, but an open in one branch affects only that branch current.

11. Give two differences between an open circuit and a short circuit.

12. List as many differences as you can in comparing series circuits with parallel circuits.

13. Why are household appliances connected to the 120-V power line in parallel instead of in series?

14. Give one advantage and one disadvantage of parallel connections.

15. A 5-Ω and a 10-Ω resistor are in parallel across a dc voltage source. Which resistor will dissipate more power? Provide proof with your answer.

Problems

SECTION 5–1 THE APPLIED VOLTAGE V_A IS THE SAME ACROSS PARALLEL BRANCHES

5–1 ||| **MultiSim** In Fig. 5–19, how much voltage is across points
   a. A and B?
   b. C and D?
   c. E and F?
   d. G and H?

Figure 5–19

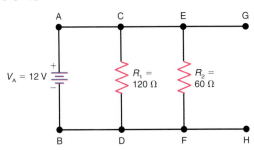

5–2 In Fig. 5–19, how much voltage is across
   a. the terminals of the voltage source?
   b. R_1?
   c. R_2?

5–3 In Fig. 5–19, how much voltage will be measured across Points C and D if R_1 is removed from the circuit?

SECTION 5–2 EACH BRANCH I EQUALS $\dfrac{V_A}{R}$

5–4 In Fig. 5–19, solve for the branch currents, I_1 and I_2.

5–5 In Fig. 5–19, explain why I_2 is double the value of I_1.

5–6 In Fig. 5–19, assume a 10-Ω resistor, R_3, is added across Points G and H.
   a. Calculate the branch current, I_3.
   b. Explain how the branch currents, I_1 and I_2 are affected by the addition of R_3.

5–7 In Fig. 5–20, solve for the branch currents I_1, I_2, and I_3.

Figure 5–20

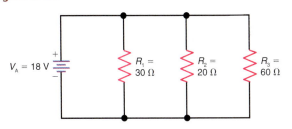

5–8 In Fig. 5–20, do the branch currents I_1 and I_3 remain the same if R_2 is removed from the circuit? Explain your answer.

5–9 In Fig. 5–21, solve for the branch currents I_1, I_2, I_3, and I_4.

Figure 5–21

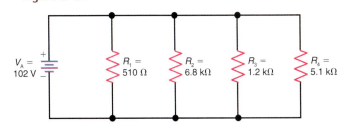

5–10 Recalculate the values for I_1, I_2, I_3, and I_4 in Fig. 5–21 if the applied voltage, V_A, is reduced to 51V.

SECTION 5–3 KIRCHHOFF'S CURRENT LAW (KCL)

5–11 ||| **MultiSim** In Fig. 5–19, solve for the total current, I_T.

5–12 ||| **MultiSim** In Fig. 5–19 re-solve for the total current, I_T, if a 10-Ω resistor, R_3, is added across Points G and H.

5–13 In Fig. 5–20, solve for the total current, I_T.

5–14 In Fig. 5–20, re-solve for the total current, I_T, if R_2 is removed from the circuit.

5–15 In Fig. 5–21, solve for the total current, I_T.

5–16 In Fig. 5–21, re-solve for the total current, I_T, if V_A is reduced to 51 V.

5–17 In Fig. 5–22, solve for I_1, I_2, I_3, and I_T.

Figure 5–22

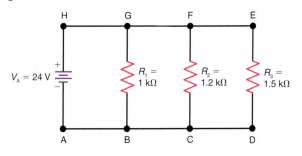

5–18 In Fig. 5–22, how much is the current in the wire between points
  a. A and B?
  b. B and C?
  c. C and D?
  d. E and F?
  e. F and G?
  f. G and H?

5–19 In Fig. 5–22 assume that a 100-Ω resistor, R_4, is added to the right of resistor, R_3. How much is the current in the wire between points
  a. A and B?
  b. B and C?
  c. C and D?
  d. E and F?
  e. F and G?
  f. G and H?

5–20 In Fig. 5–23, solve for I_1, I_2, I_3, and I_T.

Figure 5–23

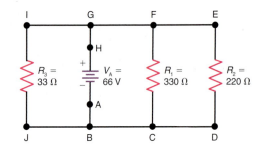

5–21 In Fig. 5–23, how much is the current in the wire between points
  a. A and B?
  b. B and C?
  c. C and D?
  d. E and F?

  e. F and G?
  f. G and H?
  g. G and I?
  h. B and J?

5–22 In Fig. 5–24, apply Kirchhoff's Current Law to solve for the unknown current, I_3.

Figure 5–24

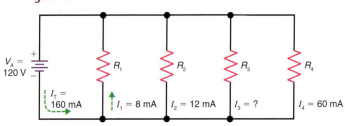

5–23 Two resistors R_1 and R_2 are in parallel with each other and a dc voltage source. How much is I_2 through R_2 if $I_T = 150$ mA and I_1 through R_1 is 60 mA?

SECTION 5–4 RESISTANCES IN PARALLEL

5–24 In Fig. 5–19, solve for R_{EQ}.

5–25 In Fig. 5–19, re-solve for R_{EQ} if a 10-Ω resistor, R_3 is added across Points G and H.

5–26 In Fig. 5–20, solve for R_{EQ}.

5–27 In Fig. 5–20, re-solve for R_{EQ} if R_2 is removed from the circuit.

5–28 In Fig. 5–21, solve for R_{EQ}.

5–29 In Fig. 5–21, re-solve for R_{EQ} if V_A is reduced to 51 V.

5–30 In Fig. 5–22, solve for R_{EQ}.

5–31 In Fig. 5–23, solve for R_{EQ}.

5–32 In Fig. 5–24, solve for R_{EQ}.

5–33 **|||| MultiSim** In Fig. 5–25, how much is R_{EQ} if $R_1 = 100$ Ω and $R_2 = 25$ Ω?

Figure 5–25

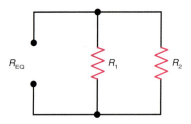

5–34 **|||| MultiSim** In Fig. 5–25, how much is R_{EQ} if $R_1 = 1.5$ MΩ and $R_2 = 1$ MΩ?

5–35 ∥∥ **MultiSim** In Fig. 5–25, how much is R_{EQ} if $R_1 = 2.2$ kΩ and $R_2 = 220$ Ω?

5–36 In Fig. 5–25, how much is R_{EQ} if $R_1 = R_2 = 10$ kΩ?

5–37 In Fig. 5–25, how much resistance, R_2, must be connected in parallel with a 750 Ω R_1 to obtain an R_{EQ} of 500 Ω?

5–38 In Fig. 5–25, how much resistance, R_1, must be connected in parallel with a 6.8 kΩ R_2 to obtain an R_{EQ} of 1.02 kΩ?

5–39 How much is R_{EQ} in Fig. 5–26 if $R_1 = 1$ kΩ, $R_2 = 4$ kΩ, $R_3 = 200$ Ω, and $R_4 = 240$ Ω?

Figure 5–26

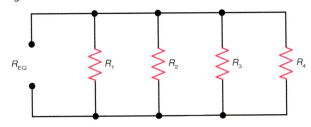

5–40 How much is R_{EQ} in Fig. 5–26 if $R_1 = 5.6$ kΩ, $R_2 = 4.7$ kΩ, $R_3 = 8.2$ kΩ, and $R_4 = 2.7$ kΩ?

5–41 ∥∥ **MultiSim** How much is R_{EQ} in Fig. 5–26 if $R_1 = 1.5$ kΩ, $R_2 = 1$ kΩ, $R_3 = 1.8$ kΩ, and $R_4 = 150$ Ω?

5–42 How much is R_{EQ} in Fig. 5–26 if $R_1 = R_2 = R_3 = R_4 = 2.2$ kΩ?

5–43 A technician is using an ohmmeter to measure a variety of different resistor values. Assume the technician has a body resistance of 750 kΩ. How much resistance will the ohmmeter read if the fingers of the technician touch the leads of the ohmmeter when measuring the following resistors:
   a. 270 Ω.
   b. 390 kΩ.
   c. 2.2 MΩ.
   d. 1.5 kΩ.
   e. 10 kΩ.

SECTION 5–5 CONDUCTANCES IN PARALLEL

5–44 In Fig. 5–27, solve for G_1, G_2, G_3, G_T, and R_{EQ}.

Figure 5–27

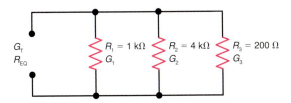

5–45 In Fig. 5–28, solve for G_1, G_2, G_3, G_4, G_T, and R_{EQ}.

Figure 5–28

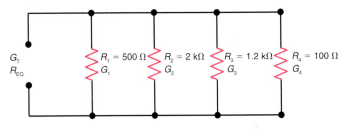

5–46 Find the total conductance, G_T for the following branch conductances; $G_1 = 1$ mS, $G_2 = 200$ μS, and $G_3 = 1.8$ mS. How much is R_{EQ}?

5–47 Find the total conductance, G_T for the following branch conductances; $G_1 = 100$ mS, $G_2 = 66.67$ mS, $G_3 = 250$ mS, and $G_4 = 83.33$ mS. How much is R_{EQ}?

SECTION 5–6 TOTAL POWER IN PARALLEL CIRCUITS

5–48 In Fig. 5–20, solve for P_1, P_2, P_3, and P_T.

5–49 In Fig. 5–21, solve for P_1, P_2, P_3, P_4, and P_T.

5–50 In Fig. 5–22, solve for P_1, P_2, P_3, and P_T.

5–51 In Fig. 5–23, solve for P_1, P_2, P_3, and P_T.

5–52 In Fig. 5–24, solve for P_1, P_2, P_3, P_4, and P_T.

SECTION 5–7 ANALYZING PARALLEL CIRCUITS WITH RANDOM UNKNOWNS

5–53 In Fig. 5–29, solve for V_A, R_1, I_2, R_{EQ}, P_1, P_2, and P_T.

Figure 5–29

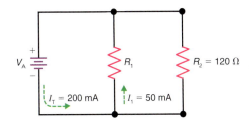

5–54 In Fig. 5–30, solve for V_A, I_1, I_2, R_2, I_T, P_2, and P_T.

Figure 5–30

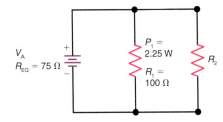

5–55 In Fig. 5–31, solve for R_3, V_A, I_1, I_2, I_T, P_1, P_2, P_3, and P_T.

5–59 In Fig. 5–35, solve for V_A, I_1, I_2, I_4, R_1, R_3, and R_{EQ}.

Figure 5–31

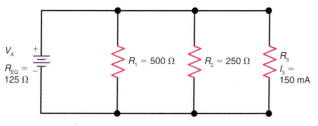

Figure 5–35

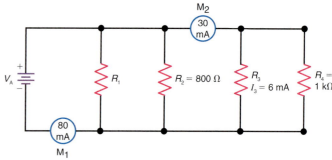

5–56 In Fig. 5–32, solve for I_T, I_1, I_2, R_1, R_2, R_3, P_2, P_3, and P_T.

Figure 5–32

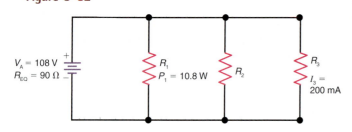

SECTION 5–8 TROUBLESHOOTING: OPENS AND SHORTS IN PARALLEL CIRCUITS

5–60 Figure 5–36 shows a parallel circuit with its normal operating voltages and currents. Notice that the fuse in the main line has a 25 A rating. What happens to the circuit components and their voltages and currents if

a. the appliance in Branch 3 shorts?

b. the motor in Branch 2 burns out and becomes an open?

c. the wire between Points C and E develops an open?

d. the motor in Branch 2 develops a problem and begins drawing 16 A of current?

5–57 In Fig. 5–33, solve for I_T, I_1, I_2, I_4, R_3, R_4, P_1, P_2, P_3, P_4, and P_T.

Figure 5–33

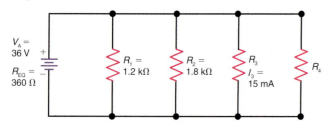

5–58 In Fig. 5–34, solve for V_A, I_1, I_2, R_2, R_3, I_4, and R_{EQ}.

Figure 5–34

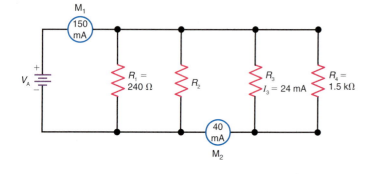

Figure 5–36

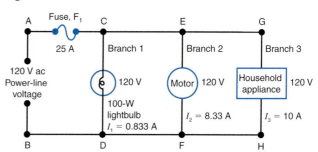

Critical Thinking

5-61 A 180-Ω, 1/4-W resistor is in parallel with 1-kΩ, 1/2-W and 12-kΩ, 2-W resistors. What is the maximum total current, I_T, that this parallel combination can have before the wattage rating of any resistor is exceeded?

5-62 A 470-Ω, 1/8-W resistor is in parallel with 1-kΩ 1/4-W and 1.5-kΩ, 1/2-W resistors. What is the maximum voltage, V, that can be applied to this circuit without exceeding the wattage rating of any resistor?

5-63 Three resistors in parallel have a combined equivalent resistance R_{EQ} of 1 kΩ. If R_2 is twice the value of R_3 and three times the value of R_1, what are the values for R_1, R_2, and R_3?

5-64 Three resistors in parallel have a combined equivalent resistance R_{EQ} of 4Ω. If the conductance, G_1, is one-fourth that of G_2 and one-fifth that of G_3, what are the values of R_1, R_2, and R_3?

5-65 A voltage source is connected in parallel across four resistors R_1, R_2, R_3, and R_4. The currents are labeled I_1, I_2, I_3, and I_4, respectively. If $I_2 = 2I_1$, $I_3 = 2I_2$, and $I_4 = 2I_3$, calculate the values for R_1, R_2, R_3, and R_4 if $R_{EQ} = 1$ kΩ.

Troubleshooting Challenge

Figure 5-37 shows a parallel circuit with its normal operating voltages and currents. Notice the placement of the meters M_1, M_2, and M_3 in the circuit. M_1 measures the total current I_T, M_2 measures the applied voltage V_A, and M_3 measures the current between points C and D. The following problems deal with troubleshooting the parallel circuit in Fig. 5-37.

Figure 5-37 Circuit diagram for troubleshooting challenge. Normal values for I_1, I_2, I_3, and I_4 are shown on schematic.

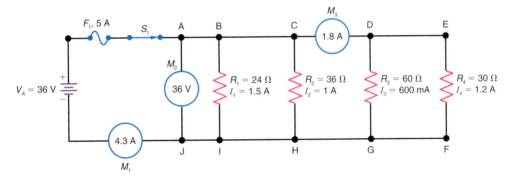

5-66 If M_1 measures 2.8 A, M_2 measures 36 V, and M_3 measures 1.8 A, which component has most likely failed? How is the component defective?

5-67 If M_1 measures 2.5 A, M_2 measures 36 V, and M_3 measures 0 A, what is most likely wrong? How could you isolate the trouble by making voltage measurements?

5-68 If M_1 measures 3.3 A, M_2 measures 36 V, and M_3 measures 1.8 A, which component has most likely failed? How is the component defective?

5-69 If the fuse F_1 is blown, **(a)** How much current will be measured by M_1 and M_3? **(b)** How much voltage will be measured by M_2? **(c)** How much voltage will be measured across the blown fuse? **(d)** What is most likely to have caused the blown fuse? **(e)** Using resistance measurements, outline a procedure for finding the defective component.

5-70 If M_1 and M_3 measure 0 A but M_2 measures 36 V, what is most likely wrong? How could you isolate the trouble by making voltage measurements?

5-71 If the fuse F_1 has blown because of a shorted branch, how much resistance would be measured across points B and I? Without using resistance measurements, how could the shorted branch be identified?

5-72 If the wire connecting points F and G opens, **(a)** How much current will M_3 show? **(b)** How much voltage would be measured across R_4? **(c)** How much voltage would be measured across points D and E? **(d)** How much voltage would be measured across points F and G?

5-73 Assuming that the circuit is operating normally, how much voltage would be measured across, **(a)** the fuse F_1; **(b)** the switch S_1?

5-74 If the branch resistor R_3 opens, **(a)** How much voltage would be measured across R_3? **(b)** How much current would be indicated by M_1 and M_3?

5-75 If the wire between points B and C breaks open, **(a)** How much current will be measured by M_1 and M_3? **(b)** How much voltage would be measured across points B and C? **(c)** How much voltage will be measured across points C and H?

Answers to Knowledge Check Problems

5–1 1.5 V

5–2 5 A

5–3 $I_T = 2$ A

5–4A $R_{EQ} = 1.02$ kΩ

5–4B $R_{EQ} = 24$ Ω

5–4C $R_X = 2.2$ kΩ

5–5A $R_{EQ} = 1.33$ Ω

5–5B $G_T = 1$ S, $R_{EQ} = 1$ Ω

5–6 $P_1 = 40$ W
$P_2 = 80$ W
$P_T = 120$ W

5–7 $V_A = 90$ V
$I_1 = 3$ A
$I_3 = 2.5$ A
$R_3 = 36$ Ω

5–8A $V_4 = 120$ V

5–8B 0 V

Answers to Self-Reviews

5–1
a. 1.5 V
b. 120 V
c. Two each

5–2
a. 10 V
b. 1 A
c. 10 V
d. 2 A

5–3
a. $I_T = 6$ A
b. $I_3 = 3$ A
c. $I_T = 1.2$ A

5–4
a. $R_{EQ} = 1.57$ MΩ
b. $R_{EQ} = 1.2$ MΩ
c. $R_{EQ} = 5$ Ω

5–5
a. $G_T = 6$ S
b. $G_T = 0.75$ μS
$R_{EQ} = 1.33$ MΩ
c. $R_{EQ} = 0.25$ MΩ

5–6
a. 480 W
b. 660 W
c. 30 W

5–7
a. 120 V
b. $I_1 = 4$ A
c. 7 A

5–8
a. 120 V
b. 0 Ω
c. 6 A
d. 0 V, 120 V
e. 120 V

chapter 6

Series-Parallel Circuits

A series-parallel circuit, also called a combination circuit, is any circuit that combines both series and parallel connections. Although many applications exist for series or parallel circuits alone, most electronic circuits are actually a combination of the two. In general, series-parallel or combination circuits are used when it is necessary to obtain different voltage and current values from a single supply voltage, V_T. When analyzing combination circuits, the individual laws of series and parallel circuits can be applied to produce a much simpler overall circuit.

In this chapter you will be presented with several different series-parallel combinations. For each type of combination circuit shown, you will learn how to solve for the unknown values of voltage, current, and resistance. You will also learn about a special circuit called the Wheatstone bridge. As you will see, this circuit has several very interesting applications in electronics. And finally, you will learn how to troubleshoot a series-parallel circuit containing both open and shorted components.

Objectives

After studying this chapter you should be able to

- *Determine* the total resistance of a series-parallel circuit.

- *Calculate* the voltage, current, resistance, and power in a series-parallel circuit.

- *Calculate* the voltage, current, resistance, and power in a series-parallel circuit having random unknowns.

- *Explain* how a Wheatstone bridge can be used to determine the value of an unknown resistor.

- *List* other applications of balanced bridge circuits.

- *Describe* the effects of opens and shorts in series-parallel circuits.

- *Troubleshoot* series-parallel circuits containing opens and shorts.

Outline

Important Terms

balanced bridge	ratio arm	strings in parallel
banks in series	standard resistor	Wheatstone bridge

6–1 Finding R_T for Series-Parallel Resistances

In Fig. 6–1, R_1 is in series with R_2. Also, R_3 is in parallel with R_4. However, R_2 is *not* in series with either R_3 or R_4. The reason is that the current through R_2 is equal to the sum of the branch currents I_3 and I_4 flowing into and away from point A. (See Fig. 6–1b). As a result, the current through R_3 must be less than the current through R_2. Therefore, R_2 and R_3 cannot be in series because they do not have the same current. For the same reason, R_4 also cannot be in series with R_2. However, because the current in R_1 and R_2 is the same as the current flowing to and from the terminals of the voltage source, R_1, R_2, and V_T are in series.

The wiring is shown in Fig. 6–1a and the schematic diagram in Fig. 6–1b. To find R_T, we add the series resistances and combine the parallel resistances.

In Fig. 6–1c, the 0.5-kΩ R_1 and 0.5-kΩ R_2 in series total 1 kΩ for R_{1-2}. The calculations are

$$0.5 \text{ k}\Omega + 0.5 \text{ k}\Omega = 1 \text{ k}\Omega$$

Also, the 1-kΩ R_3 in parallel with the 1-kΩ R_4 can be combined, for an equivalent resistance of 0.5 kΩ for R_{3-4}, as in Fig. 6–1d. The calculations are

$$\frac{1 \text{ k}\Omega}{2} = 0.5 \text{ k}\Omega$$

||| MultiSim   **Figure 6–1**   Example of a series-parallel circuit. (*a*) Wiring of a series-parallel circuit. (*b*) Schematic diagram of a series-parallel circuit. (*c*) Schematic with R_1 and R_2 in series added for R_{1-2}. (*d*) Schematic with R_3 and R_4 in parallel combined for R_{3-4}. (*e*) Axial-lead resistors assembled on a lab prototype board to form the series-parallel circuit shown in part c.

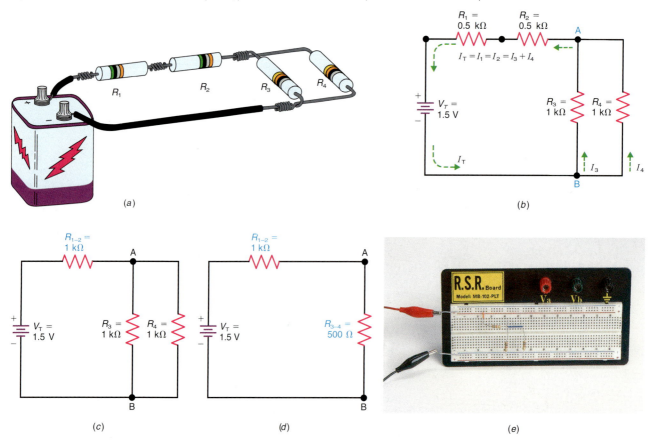

(a)

(b)

(c)

(d)

(e)

This parallel $R_{3\text{-}4}$ combination of 0.5 kΩ is then added to the series $R_{1\text{-}2}$ combination for the final R_T value of 1.5 kΩ. The calculations are

$$0.5 \text{ k}\Omega + 1 \text{ k}\Omega = 1.5 \text{ k}\Omega$$

The 1.5 kΩ is the R_T of the entire circuit connected across the V_T of 1.5 V.

With R_T known to be 1.5 kΩ, we can find I_T in the main line produced by 1.5 V. Then

$$I_T = \frac{V_T}{R_T} = \frac{1.5 \text{ V}}{1.5 \text{ k}\Omega} = 1 \text{ mA}$$

This 1-mA I_T is the current through resistors R_1 and R_2 in Fig. 6–1a and b or $R_{1\text{-}2}$ in Fig. 6–1c.

At branch point B, at the bottom of the diagram in Fig. 6–1b, the 1 mA of electron flow for I_T divides into two branch currents for R_3 and R_4. Since these two branch resistances are equal, I_T divides into two equal parts of 0.5 mA each. At branch point A at the top of the diagram, the two 0.5-mA branch currents combine to equal the 1-mA I_T in the main line, returning to the source V_T.

Figure 6–1e shows axial-lead resistors assembled on a lab prototype board to form the series-parallel circuit shown in part c.

■ **6–1 Knowledge Check**

Answer at end of chapter.

In Fig. 6–1b is R_2 in series with R_3? Why?

■ **6–1 Self-Review**

Answers at end of chapter.

Refer to Fig. 6–1b.
  a. Calculate the series R of R_1 and R_2.
  b. Calculate the parallel R of R_3 and R_4.
  c. Calculate R_T across the source V_T.

6–2 Resistance Strings in Parallel

More details about the voltages and currents in a series-parallel circuit are illustrated in Fig. 6–2, which shows two identical series **strings in parallel.** Suppose that four 120-V, 100-W light bulbs are to be wired with a voltage source that produces 240 V. Each bulb needs 120 V for normal brilliance. If the bulbs were connected directly across the source, each would have the applied voltage of 240 V. This would cause excessive current in all the bulbs that could result in burned-out filaments.

Figure 6–2 Two identical series strings in parallel. All bulbs have a 120-V, 100-W rating. (a) Wiring diagram. (b) Schematic diagram.

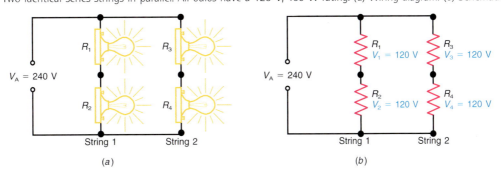

(a)   (b)

Figure 6–3 Series string in parallel with another branch. (*a*) Schematic diagram. (*b*) Equivalent circuit.

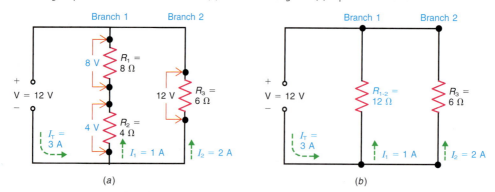

(*a*)   (*b*)

If the four bulbs were connected in series, each would have a potential difference of 60 V, or one-fourth the applied voltage. With too low a voltage, there would be insufficient current for normal operation, and the bulbs would not operate at normal brilliance.

However, two bulbs in series across the 240-V line provide 120 V for each filament, which is the normal operating voltage. Therefore, the four bulbs are wired in strings of two in series, with the two strings in parallel across the 240-V source. Both strings have 240 V applied. In each string, two series bulbs divide the 240 V equally to provide the required 120 V for normal operation.

Another example is illustrated in Fig. 6–3. This circuit has just two parallel branches. One branch includes R_1 in series with R_2. The other branch has just the one resistance R_3. Ohm's law can be applied to each branch.

Branch Currents I_1 and I_2

Each branch current equals the voltage applied across the branch divided by the total resistance in the branch. In branch 1, R_1 and R_2 total $8 + 4 = 12\ \Omega$. With 12 V applied, this branch current I_1 is $12/12 = 1$ A. Branch 2 has only the 6-Ω R_3. Then I_2 in this branch is $12/6 = 2$ A.

Series Voltage Drops in a Branch

For any one resistance in a string, the current in the string multiplied by the resistance equals the IR voltage drop across that particular resistance. Also, the sum of the series IR drops in the string equals the voltage across the entire string.

Branch 1 is a string with R_1 and R_2 in series. The I_1R_1 drop equals 8 V, whereas the I_1R_2 drop is 4 V. These drops of 8 and 4 V add to equal the 12 V applied. The voltage across the R_3 branch is also the same 12 V.

Calculating I_T

The total line current equals the sum of the branch currents for all parallel strings. Here I_T is 3 A, equal to the sum of 1 A in branch 1 and 2 A in branch 2.

Calculating R_T

The resistance of the total series-parallel circuit across the voltage source equals the applied voltage divided by the total line current. In Fig. 6–3, $R_T = 12$ V/3 A, or 4 Ω. This resistance can also be calculated as 12 Ω in parallel with 6 Ω. Using the product divided by the sum formula, $72/18 = 4\ \Omega$ for the equivalent combined R_T.

Applying Ohm's Law

There can be any number of parallel strings and more than two series resistances in a string. Still, Ohm's law can be used in the same way for the series and parallel parts of the circuit. The series parts have the same current. The parallel parts have the same voltage. Remember that for V/R the R must include all the resistance across the two terminals of V.

■ 6–2 Knowledge Check

Answer at end of chapter.

In Fig. 6–3A recalculate I_1, I_2, and I_T if $R_1 = 10\ \Omega$, $R_2 = 20\ \Omega$, $R_3 = 10\ \Omega$, and $V_T = 24$ V.

■ 6–2 Self-Review

Answers at end of chapter.

Refer to Fig. 6–3a.
a. How much is the voltage across R_3?
b. If I in R_2 were 6 A, what would I in R_1 be?
c. If the source voltage were 18 V, what would V_3 be across R_3?

6–3 Resistance Banks in Series

In Fig. 6–4a, the group of parallel resistances R_2 and R_3 is a bank. This is in series with R_1 because the total current of the bank must go through R_1.

The circuit here has R_2 and R_3 in parallel in one bank so that these two resistances will have the same potential difference of 20 V across them. The source applies 24 V, but there is a 4-V drop across R_1.

The two series voltage drops of 4 V across R_1 and 20 V across the bank add to equal the applied voltage of 24 V. The purpose of a circuit like this is to provide the same voltage for two or more resistances in a bank, where the bank voltage must be less than the applied voltage by the amount of the IR drop across any series resistance.

To find the resistance of the entire circuit, combine the parallel resistances in each bank and add the series resistance. As shown in Fig. 6–4b, the two 10-Ω resistances, R_2 and R_3 in parallel, are equivalent to 5 Ω. Since the bank resistance of 5 Ω is in series with 1 Ω for R_1, the total resistance is 6 Ω across the 24-V source. Therefore, the main-line current is 24 V/6 Ω, which equals 4 A.

GOOD TO KNOW

When a parallel bank exists in a series path, both resistors have the same voltage but the individual branch currents are less than the series current. The branch currents add, however, to equal the series current entering and leaving the parallel bank.

Figure 6–4 Parallel bank of R_2 and R_3 in series with R_1. (a) Original circuit. (b) Equivalent circuit.

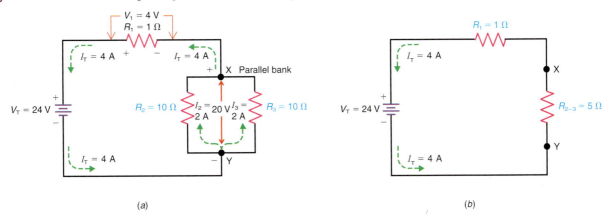

(a)

(b)

The total line current of 4 A divides into two parts of 2 A each in the parallel resistances R_2 and R_3. Note that each branch current equals the bank voltage divided by the branch resistance. For this bank, $20/10 = 2$ A for each branch.

The branch currents, I_2 and I_3, are combined in the main line to provide the total 4 A in R_1. This is the same total current flowing in the main line, in the source, into the bank, and out of the bank.

There can be more than two parallel resistances in a bank and any number of **banks in series.** Still, Ohm's law can be applied in the same way to the series and parallel parts of the circuit. The general procedure for circuits of this type is to find the equivalent resistance of each bank and then add all series resistances.

■ *6–3 Knowledge Check*

Answer at end of chapter.

In Fig. 6–4A, recalculate I_2, I_3 and I_T if $R_1 = 12\ \Omega$, $R_2 = 10\ \Omega$, $R_3 = 40\ \Omega$, and $V_T = 12$ V.

■ *6–3 Self-Review*

Answers at end of chapter.

Refer to Fig. 6–4a.
a. If V_2 across R_2 were 40 V, what would V_3 across R_3 be?
b. If I in R_2 were 4 A, with 4 A in R_3, what would I in R_1 be?
c. How much is V_1 across R_1 in Fig. 6–4b?

6–4 Resistance Banks and Strings in Series–Parallel

In the solution of such circuits, the most important fact to know is which components are in series with each other and which parts of the circuit are parallel branches. The series components must be in one current path without any branch points. A branch point such as point A or B in Fig. 6–5 is common to two or more current paths. For instance, R_1 and R_6 are *not* in series with each other. They do not have the same current because the current through R_1 equals the sum of the branch currents, I_5 and I_6, flowing into and away from point A. Similarly, R_5 is not in series with R_2 because of the branch point B.

To find the currents and voltages in Fig. 6–5, first find R_T to calculate the main-line current I_T as V_T/R_T. In calculating R_T, start reducing the branch farthest from the source and work toward the applied voltage. The reason for following this order is that you cannot tell how much resistance is in series with R_1 and R_2 until the parallel branches are reduced to their equivalent resistance. If no source voltage is shown, R_T can still be calculated from the outside in toward the open terminals where a source would be connected.

To calculate R_T in Fig. 6–5, the steps are as follows:

1. The bank of the 12-Ω R_3 and 12-Ω R_4 in parallel in Fig. 6–5a is equal to the 6-Ω R_7 in Fig. 6–5b.
2. The 6-Ω R_7 and 4-Ω R_6 in series in the same current path total 10 Ω for R_{13} in Fig. 6–5c.
3. The 10-Ω R_{13} is in parallel with the 10-Ω R_5, across the branch points A and B. Their equivalent resistance, then, is the 5-Ω R_{18} in Fig. 6–5d.
4. Now the circuit in Fig. 6–5d has just the 15-Ω R_1, 5-Ω R_{18}, and 30-Ω R_2 in series. These resistances total 50 Ω for R_T, as shown in Fig. 6–5e.
5. With a 50-Ω R_T across the 100-V source, the line current I_T is equal to $100/50 = 2$ A.

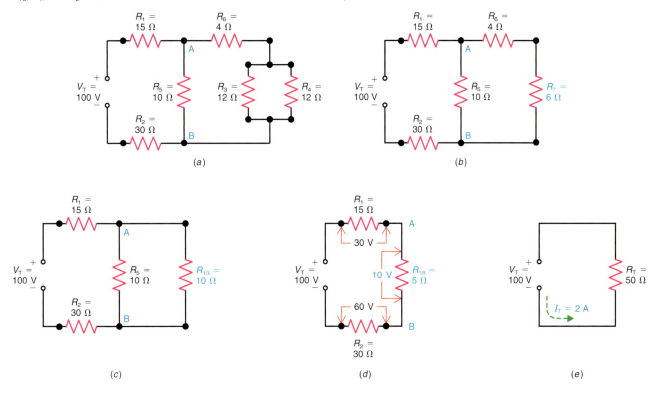

⦀ MultiSim **Figure 6–5** Reducing a series-parallel circuit to an equivalent series circuit to find the R_T. (*a*) Actual circuit. (*b*) R_3 and R_4 in parallel combined for the equivalent R_7. (*c*) R_7 and R_6 in series added for R_{13}. (*d*) R_{13} and R_5 in parallel combined for R_{18}. (*e*) The R_{18}, R_1, and R_2 in series are added for the total resistance of 50 Ω for R_T.

To see the individual currents and voltages, we can use the I_T of 2 A for the equivalent circuit in Fig. 6–5*d*. Now we work from the source V out toward the branches. The reason is that I_T can be used to find the voltage drops in the main line. The IR voltage drops here are

$$V_1 = I_T R_1 = 2 \times 15 = 30 \text{ V}$$
$$V_{18} = I_T R_{18} = 2 \times 5 = 10 \text{ V}$$
$$V_2 = I_T R_2 = 2 \times 30 = 60 \text{ V}$$

The 10-V drop across R_{18} is actually the potential difference between branch points A and B. This means 10 V across R_5 and R_{13} in Fig. 6–5*c*. The 10 V produces 1 A in the 10-Ω R_5 branch. The same 10 V is also across the R_{13} branch.

Remember that the R_{13} branch is actually the string of R_6 in series with the R_3–R_4 bank. Since this branch resistance is 10 Ω with 10 V across it, the branch current here is 1 A. The 1 A through the 4 Ω of R_6 produces a voltage drop of 4 V. The remaining 6-V IR drop is across the R_3–R_4 bank. With 6 V across the 12-Ω R_3, its current is ½ A; the current is also ½ A in R_4.

Tracing all the current paths from the voltage source, the main-line current, I_T, through R_1 and R_2 is 2 A. The 2 A I_T flowing into point B subdivides into two separate branch currents: 1 A of the 2 A I_T flows up through resistor, R_5. The other 1 A flows into the branch containing resistors R_3, R_4, and R_6. Because resistors R_3 and R_4 are in parallel, the 1-A branch current subdivides further into ½ A for I_3 and ½ A for I_4. The currents I_3 and I_4 recombine to flow up through resistor R_6. At the branch point A, I_5 and I_6 combine resulting in the 2-A total current, I_T, flowing through R_1 back to the positive terminal of the voltage source.

Answer at end of chapter.

In Fig 6–5a, how much is R_T if R_5 is changed to 40 Ω?

■ 6–4 *Self-Review*

Answers at end of chapter.

Refer to Fig. 6–5a.
a. Which R is in series with R_2?
b. Which R is in parallel with R_3?
c. Which R is in series with the $R_3 R_4$ bank?

6–5 Analyzing Series-Parallel Circuits with Random Unknowns

The circuits in Figs. 6–6 to 6–9 will be solved now. The following principles are illustrated:

1. With parallel strings across the main line, the branch currents and I_T can be found without R_T (see Figs. 6–6 and 6–7).
2. When parallel strings have series resistance in the main line, R_T must be calculated to find I_T, assuming no branch currents are known (see Fig. 6–9).
3. The source voltage is applied across the R_T of the entire circuit, producing an I_T that flows only in the main line.
4. Any individual series R has its own IR drop that must be less than the total V_T. In addition, any individual branch current must be less than I_T.

Solution for Figure 6–6

The problem here is to calculate the branch currents I_1 and I_{2-3}, total line current I_T, and the voltage drops V_1, V_2, and V_3. This order will be used for the calculations because we can find the branch currents from the 90 V across the known branch resistances.

In the 30-Ω branch of R_1, the branch current is 90/30 = 3 A for I_1. The other branch resistance, with a 20-Ω R_2 and a 25-Ω R_3, totals 45 Ω. This branch current then is 90/45 = 2 A for I_{2-3}. In the main line, I_T is 3 A + 2 A, which is equal to 5 A.

For the branch voltages, V_1 must be the same as V_A, equal to 90 V, or $V_1 = I_1 R_1$, which is 3 × 30 = 90 V.

Figure 6–6 Finding all currents and voltages by calculating the branch currents first. See text for solution.

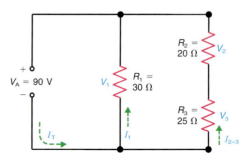

Figure 6-7 Finding the applied voltage V_A and then V_4 and R_4 from I_2 and the branch voltages. See text for the calculations.

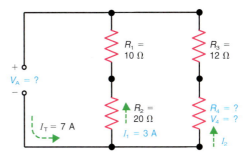

In the other branch, the 2-A I_{2-3} flows through the 20-Ω R_2 and the 25-Ω R_3. Therefore, V_2 is $2 \times 20 = 40$ V. Also, V_3 is $2 \times 25 = 50$ V. Note that these 40-V and 50-V series *IR* drops in one branch add to equal the 90-V source.

If we want to know R_T, it can be calculated as V_A/I_T. Then 90 V/5 A equals 18 Ω. Or R_T can be calculated by combining the branch resistances of 30 Ω in parallel with 45 Ω. Then, using the product-divided-by-sum formula, R_T is $(30 \times 45)/(30 + 45)$ or 1350/75, which equals the same value of 18 Ω for R_T.

Solution for Figure 6-7

To find the applied voltage first, the I_1 branch current is given. This 3-A current through the 10-Ω R_1 produces a 30-V drop V_1 across R_1. The same 3-A current through the 20-Ω R_2 produces 60 V for V_2 across R_2. The 30-V and 60-V drops are in series with each other across the applied voltage. Therefore, V_A equals the sum of $30 + 60$, or 90 V. This 90 V is also across the other branch combining R_3 and R_4 in series.

The other branch current I_2 in Fig. 6-7 must be 4 A, equal to the 7-A I_T minus the 3-A I_1. With 4 A for I_2, the voltage drop across the 12-Ω R_3 equals 48 V for V_3. Then the voltage across R_4 is $90 - 48$, or 42 V for V_4, as the sum of V_3 and V_4 must equal the applied 90 V.

Finally, with 42 V across R_4 and 4 A through it, this resistance equals 42/4, or 10.5 Ω. Note that 10.5 Ω for R_4 added to the 12 Ω of R_3 equals 22.5 Ω, which allows 90/22.5 or a 4-A branch current for I_2.

Solution for Figure 6-8

The division of branch currents also applies to Fig. 6-8, but the main principle here is that the voltage must be the same across R_1 and R_2 in parallel. For the branch currents, I_2 is 2 A, equal to the 6-A I_T minus the 4-A I_1. The voltage across the 10-Ω R_1 is 4×10, or 40 V. This same voltage is also across R_2. With 40 V across R_2 and 2 A through it, R_2 equals 40/2 or 20 Ω.

We can also find V_T in Fig. 6-8 from R_1, R_2, and R_3. The 6-A I_T through R_3 produces a voltage drop of 60 V for V_3. Also, the voltage across the parallel bank with R_1 and R_2 has been calculated as 40 V. This 40 V across the bank in series with 60 V across R_3 totals 100 V for the applied voltage.

Figure 6-8 Finding R_2 in the parallel bank and its I_2. See text for solution.

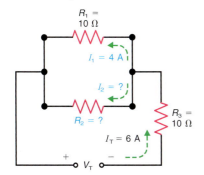

Solution for Figure 6-9

To find all currents and voltage drops, we need R_T to calculate I_T through R_6 in the main line. Combining resistances for R_T, we start with R_1 and R_2 and work in toward the source. Add the 8-Ω R_1 and 8-Ω R_2 in series with each other for 16 Ω. This 16 Ω combined with the 16-Ω R_3 in parallel equals 8 Ω between

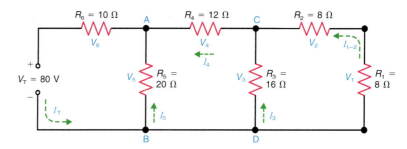

Figure 6–9 Finding all currents and voltages by calculating R_T and then I_T to find V_6 across R_6 in the main line.

points C and D. Add this 8 Ω to the series 12-Ω R_4 for 20 Ω. This 20 Ω with the parallel 20-Ω R_5 equals 10 Ω between points A and B. Add this 10 Ω in series with the 10-Ω R_6, to make R_T of 20 Ω for the entire series-parallel circuit.

Current I_T in the main line is V_T/R_T, or 80/20, which equals 4 A. This 4-A I_T flows through the 10-Ω R_6, producing a 40-V IR drop for V_6.

Now that we know I_T and V_6 in the main line, we use these values to calculate all other voltages and currents. Start from the main line, where we know the current, and work outward from the source. To find V_5, the IR drop of 40 V for V_6 in the main line is subtracted from the source voltage. The reason is that V_5 and V_6 must add to equal the 80 V of V_T. Then V_5 is $80 - 40 = 40$ V.

Voltages V_5 and V_6 happen to be equal at 40 V each. They split the 80 V in half because the 10-Ω R_6 equals the combined resistance of 10 Ω between branch points A and B.

With V_5 known to be 40 V, then I_5 through the 20-Ω R_5 is $40/20 = 2$ A. Since I_5 is 2 A and I_T is 4 A, I_4 must be 2 A also, equal to the difference between I_T and I_5. The current flowing into point A equals the sum of the branch currents I_4 and I_5.

The 2-A I_4 through the 12-Ω R_4 produces an IR drop equal to $2 \times 12 = 24$ V for V_4. Note now that V_4 and V_3 must add to equal V_5. The reason is that both V_5 and the path with V_4 and V_3 are across the same two points AB or AD. Since the potential difference across any two points is the same regardless of the path, $V_5 = V_4 + V_3$. To find V_3 now, we can subtract the 24 V of V_4 from the 40 V of V_5. Then $40 - 24 = 16$ V for V_3.

With 16 V for V_3 across the 16-Ω R_3, its current I_3 is 1 A. Also, I_{1-2} in the branch with R_1 and R_2 is equal to 1 A. The 2-A I_4 consists of the sum of the branch currents, I_3 and I_{1-2}, flowing into point C.

Finally, with 1A through the 8-Ω R_2 and 8-Ω R_1, their voltage drops are $V_2 = 8$ V and $V_1 = 8$ V. Note that the 8 V of V_1 in series with the 8 V of V_2 add to equal the 16-V potential difference V_3 between points C and D.

All answers for the solution of Fig. 6–9 are summarized below:

$R_T = 20\ \Omega$	$I_T = 4$ A	$V_6 = 40$ V
$V_5 = 40$ V	$I_5 = 2$ A	$I_4 = 2$ A
$V_4 = 24$ V	$V_3 = 16$ V	$I_3 = 1$ A
$I_{1-2} = 1$ A	$V_2 = 8$ V	$V_1 = 8$ V

■ 6–5 Knowledge Check

Answer at end of chapter.

In Fig. 6–9, how much is I_T if R_4 is changed to 22 Ω?

a. In Fig. 6–6, which R is in series with R_2?
b. In Fig. 6–6, which R is across V_A?
c. In Fig. 6–7, how much is I_2?
d. In Fig. 6–8, how much is V_3?

6–6 The Wheatstone Bridge

A **Wheatstone* bridge** is a circuit that is used to determine the value of an unknown resistance. A typical Wheatstone bridge is shown in Fig. 6–10. Notice that four resistors are configured in a diamond-like arrangement, which is typically how the Wheatstone bridge is drawn. In Fig. 6–10, the applied voltage V_T is connected to terminals A and B, which are considered the input terminals to the Wheatstone bridge. A very sensitive zero-centered current meter M_1, called a *galvanometer,* is connected between terminals C and D, which are considered the output terminals.

As shown in Fig. 6–10, the unknown resistor R_X is placed in the same branch as a variable **standard resistor** R_S. It is important to note that the standard resistor R_S is a precision resistance variable from 0–9999 Ω in 1-Ω steps. In the other branch, resistors R_1 and R_2 make up what is known as the **ratio arm.** Resistors R_1 and R_2 are also precision resistors having very tight resistance tolerances. To determine the value of an unknown resistance R_X, adjust the standard resistor R_S until the current in M_1 reads exactly 0 μA. With zero current in M_1 the Wheatstone bridge is said to be balanced. But how does the balanced condition provide the value of the unknown resistance R_X? Good question. With zero current in M_1, the voltage division among resistors R_X and R_S is identical to the voltage division among the ratio arm resistors R_1 and R_2. When the voltage division in the R_X–R_S branch is identical to the voltage division in the R_1–R_2 branch, the potential difference between points C and D will equal 0 V. With a potential difference of 0 V across points C and D, the current in M_1 will read 0 μA, which is the balanced condition. At balance, the equal voltage ratios can be stated as

$$\frac{I_1 R_X}{I_1 R_S} = \frac{I_2 R_1}{I_2 R_2}$$

Figure 6–10 Wheatstone bridge.

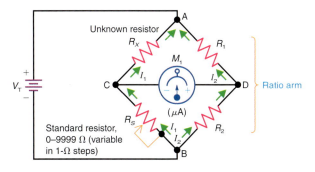

---

* Sir Charles Wheatstone (1802–1875), English physicist and inventor.

Since I_1 and I_2 cancel in the equation, this yields

$$\frac{R_X}{R_S} = \frac{R_1}{R_2}$$

Solving for R_X gives us

$$R_X = R_S \times \frac{R_1}{R_2} \qquad (6\text{--}1)$$

The ratio arm R_1/R_2 can be varied in most cases, typically in multiples of 10, such as 100/1, 10/1, 1/1, 1/10, and 1/100. However, the bridge is still balanced by varying the standard resistor R_S. The placement accuracy of the measurement of R_X is determined by the R_1/R_2 ratio. For example, if $R_1/R_2 = 1/10$, the value of R_X is accurate to within $\pm 0.1\ \Omega$. Likewise, if $R_1/R_2 = 1/100$, the value of R_X will be accurate to within $\pm 0.01\ \Omega$. The R_1/R_2 ratio also determines the maximum unknown resistance that can be measured. Expressed as an equation.

$$R_{X(\text{max})} = R_{S(\text{max})} \times \frac{R_1}{R_2} \qquad (6\text{--}2)$$

Example 6–1

In Fig. 6–11, the current in M_1 reads $0\ \mu A$ with the standard resistor R_S adjusted to $5642\ \Omega$. What is the value of the unknown resistor R_X?

ANSWER Using Formula (6-1), R_X is calculated as follows:

$$R_X = R_S \times \frac{R_1}{R_2}$$

$$= 5642\ \Omega \times \frac{1\ k\Omega}{10\ k\Omega}$$

$$R_X = 564.2\ \Omega$$

||| **MultiSim**   **Figure 6–11**   Wheatstone bridge. See Examples 6–1 and 6–2.

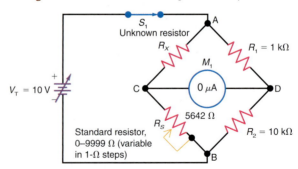

Example 6-2

In Fig. 6-11, what is the maximum unknown resistance R_X that can be measured for the ratio arm values shown?

ANSWER $R_{X(max)} = R_{S(max)} \times \dfrac{R_1}{R_2}$

$$= 9999\ \Omega \times \dfrac{1\ k\Omega}{10\ k\Omega}$$

$$R_{X(max)} = 999.9\ \Omega$$

If R_X is larger than 999.9 Ω, the bridge cannot be balanced because the voltage division will be greater than 1/10 in this branch. In other words, the current in M_1 cannot be adjusted to 0 μA. To measure an unknown resistance whose value is greater than 999.9 Ω, you would need to change the ratio arm fraction to 1/1, 10/1, or something higher.

Note that when the Wheatstone bridge is balanced, it can be analyzed simply as two series strings in parallel. The reason is that when the current through M_1 is zero, the path between points C and D is effectively open. When current flows through M_1, however, the bridge circuit must be analyzed by other methods described in Chaps. 9 and 10.

Other Balanced Bridge Applications

There are many other applications in electronics for **balanced bridge** circuits. For example, a variety of sensors are used in bridge circuits for detecting changes in pressure, flow, light, temperature, etc. These sensors are used as one of the resistors in a bridge circuit. Furthermore, the bridge can be balanced or zeroed at some desired reference level of pressure, flow, light, or temperature. Then, when the condition being sensed changes, the bridge becomes unbalanced and causes a voltage to appear at the output terminals (C and D). This output voltage is then fed to the input of an amplifier or other device that modifies the condition being monitored, thus bringing the system back to its original preset level.

Consider the temperature control circuit in Fig. 6-12. In this circuit, a variable resistor R_3 is in the same branch as a negative temperature coefficient (NTC) thermistor whose resistance value at 25°C (R_0) equals 5 kΩ as shown. Assume that R_3 is adjusted to provide balance when the ambient (surrounding) temperature T_A equals 25°C. Remember, when the bridge is balanced, the output voltage across terminals C and D is 0 V. This voltage is fed to the input of an amplifier as shown. With 0 V into the amplifier, 0 V comes out of the amplifier.

Now let's consider what happens when the ambient temperature T_A increases above 25°C, say to 30°C. The increase in temperature causes the resistance of the thermistor to decrease, since it has an NTC. With a decrease in the thermistor's resistance, the voltage at point C decreases. However, the voltage at point D does not change because R_1 and R_2 are ordinary resistors. The result is that the output voltage V_{CD} goes negative. This negative voltage is fed into the amplifier, which in turn produces a positive output voltage. The positive output voltage from the amplifier turns on a cooling fan or air-conditioning unit. The air-conditioning unit remains on until the ambient temperature decreases to its original value of 25°C. As the temperature drops back to 25°C, the resistance of the thermistor increases to its original value, thus causing the voltage V_{CD} to return to 0 V. This shuts off the air conditioner.

Figure 6–12 Temperature control circuit using the balanced bridge concept.

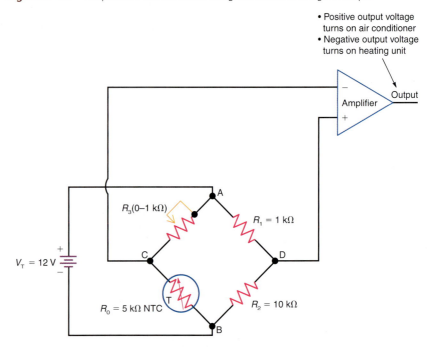

- Positive output voltage turns on air conditioner
- Negative output voltage turns on heating unit

Next, let's consider what happens when the ambient temperature T_A decreases below 25°C, say to 20°C. The decrease in temperature causes the resistance of the thermistor to increase, thus making the voltage at point C more positive. The result is that V_{CD} goes positive. This positive voltage is fed into the amplifier, which in turn produces a negative output voltage. The negative output voltage from the amplifier turns on a heating unit, which remains on until the ambient temperature returns to its original value of 25°C. Although the details of the temperature-control circuit in Fig. 6–12 are rather vague, you should get the idea of how a balanced bridge circuit containing a thermistor could be used to control the temperature in a room. There are almost unlimited applications for balanced bridge circuits in electronics.

■ 6-6 Knowledge Check

Answer at end of chapter.

In Fig. 6–12, what is the polarity of the voltage, V_{CD}, if the temperature is 50°C?

■ 6-6 Self-Review

Answers at end of chapter.

a. In Fig. 6–10, which terminals are the input terminals? Which terminals are the output terminals?
b. In reference to Fig. 6–10, how much current flows in M_1 when the bridge is balanced?
c. In Fig. 6–11, assume $R_1 = 100\ \Omega$ and $R_2 = 10\ k\Omega$. If the bridge is balanced by adjusting R_S to 7135 Ω, what is the value of R_X?
d. In reference to Question c, what is the maximum unknown resistance that can be measured for the circuit values given?
e. In reference to Fig. 6–12, to what value must the resistor R_3 be adjusted to provide 0 V output at 25°C?

6–7 Troubleshooting: Opens and Shorts in Series-Parallel Circuits

A short circuit has practically zero resistance. Its effect, therefore, is to allow excessive current. An open circuit has the opposite effect because an open circuit has infinitely high resistance with practically zero current. Furthermore, in series-parallel circuits, an open or short circuit in one path changes the circuit for the other resistances. For example, in Fig. 6–13, the series-parallel circuit in Fig. 6–13a becomes a series circuit with only R_1 when there is a short circuit between terminals A and B. As an example of an open circuit, the series-parallel circuit in Fig. 6–14a becomes a series circuit with just R_1 and R_2 when there is an open circuit between terminals C and D.

Effect of a Short Circuit

We can solve the series-parallel circuit in Fig. 6–13a to see the effect of the short circuit. For the normal circuit with S_1 open, R_2 and R_3 are in parallel. Although R_3 is drawn horizontally, both ends are across R_2. The switch S_1 has no effect as a parallel branch here because it is open.

Figure 6–13 Effect of a short circuit with series-parallel connections. (a) Normal circuit with S_1 open. (b) Circuit with short between points A and B when S_1 is closed; now R_2 and R_3 are short-circuited.

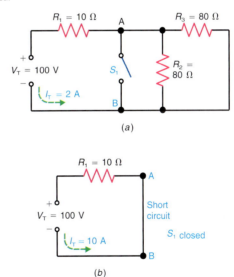

(a)

(b)

Figure 6–14 Effect of an open path in a series-parallel circuit. (a) Normal circuit with S_2 closed. (b) Series circuit with R_1 and R_2 when S_2 is open. Now R_3 in the open path has no current and zero IR voltage drop.

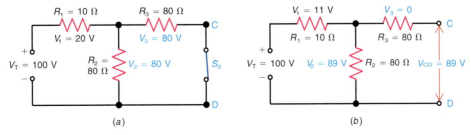

(a)   (b)

The combined resistance of the 80-Ω R_2 in parallel with the 80-Ω R_3 is equivalent to 40 Ω. This 40 Ω for the bank resistance is in series with the 10-Ω R_1. Then R_T is $40 + 10 = 50$ Ω.

In the main line, I_T is $100/50 = 2$ A. Then V_1 across the 10-Ω R_1 in the main line is $2 \times 10 = 20$ V. The remaining 80 V is across R_2 and R_3 as a parallel bank. As a result, $V_2 = 80$ V and $V_3 = 80$ V.

Now consider the effect of closing switch S_1. A closed switch has zero resistance. Not only is R_2 short-circuited, but R_3 in the bank with R_2 is also short-circuited. The closed switch short-circuits everything connected between terminals A and B. The result is the series circuit shown in Fig. 6–13b.

Now the 10-Ω R_1 is the only opposition to current. I equals V/R_1, which is $100/10 = 10$ A. This 10 A flows through the closed switch, through R_1, and back to the positive terminal of the voltage source. With 10 A through R_1, instead of its normal 2 A, the excessive current can cause excessive heat in R_1. There is no current through R_2 and R_3, as they are short-circuited out of the path for current.

Effect of an Open Circuit

Figure 6–14a shows the same series-parallel circuit as Fig. 6–13a, except that switch S_2 is used now to connect R_3 in parallel with R_2. With S_2 closed for normal operation, all currents and voltages have the values calculated for the series-parallel circuit. However, let us consider the effect of opening S_2, as shown in Fig. 6–14b. An open switch has infinitely high resistance. Now there is an open circuit between terminals C and D. Furthermore, because R_3 is in the open path, its 80 Ω cannot be considered in parallel with R_2.

The circuit with S_2 open in Fig. 6–14b is really the same as having only R_1 and R_2 in series with the 100-V source. The open path with R_3 has no effect as a parallel branch because no current flows through R_3.

We can consider R_1 and R_2 in series as a voltage divider, where each IR drop is proportional to its resistance. The total series R is $80 + 10 = 90$ Ω. The 10-Ω R_1 is $10/90$ or ⅑ of the total R and the applied V_T. Then V_1 is ⅑ $\times$ 100 V = 11 V and V_2 is ⅚ $\times$ 100 V = 89 V, approximately. The 11-V drop for V_1 and 89-V drop for V_2 add to equal the 100 V of the applied voltage.

Note that V_3 is zero. Without any current through R_3, it cannot have any voltage drop.

Furthermore, the voltage across the open terminals C and D is the same 89 V as the potential difference V_2 across R_2. Since there is no voltage drop across R_3, terminal C has the same potential as the top terminal of R_2. Terminal D is directly connected to the bottom end of resistor R_2. Therefore, the potential difference from terminal C to terminal D is the same 89 V that appears across resistor R_2.

Troubleshooting Procedures for Series–Parallel Circuits

The procedure for troubleshooting series-parallel circuits containing opens and shorts is a combination of the procedures used to troubleshoot individual series and parallel circuits. Figure 6–15a shows a series-parallel circuit with its normal operating voltages and currents. Across points A and B, the equivalent resistance R_{EQ} of R_2 and R_3 in parallel is calculated as

$$R_{EQ} = \frac{R_2 \times R_3}{R_2 + R_3}$$
$$= \frac{100\ \Omega \times 150\ \Omega}{100\ \Omega + 150\ \Omega}$$
$$R_{EQ} = 60\ \Omega$$

Figure 6–15 Series-parallel circuit for troubleshooting analysis. (*a*) Normal circuit voltages and currents; (*b*) circuit voltages with R_3 open between points A and B; (*c*) circuit voltages with R_2 or R_3 shorted between points A and B.

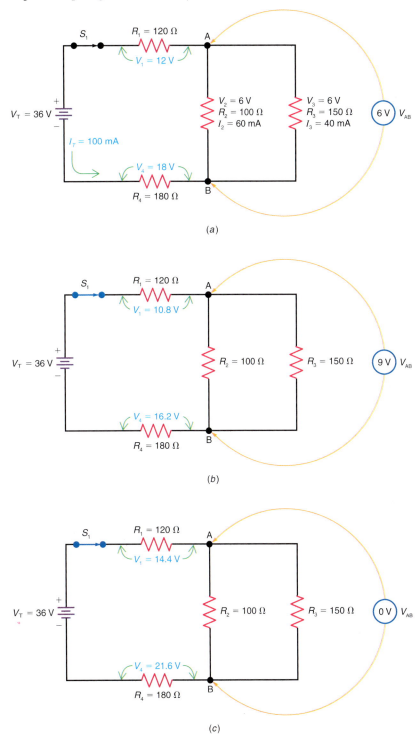

(a)

(b)

(c)

Since R_2 and R_3 are in parallel across points A and B, this equivalent resistance is designated R_{AB}. Therefore, $R_{AB} = 60\ \Omega$. The total resistance, R_T, is:

$$R_T = R_1 + R_{AB} + R_4$$
$$= 120\ \Omega + 60\ \Omega + 180\ \Omega$$
$$R_T = 360\ \Omega$$

The total current, I_T, is:

$$I_T = \frac{V_T}{R_T}$$
$$= \frac{36\ \text{V}}{360\ \Omega}$$
$$I_T = 100\ \text{mA}$$

The voltage drops across the individual resistors are calculated as

$$V_1 = I_T \times R_1$$
$$= 100\ \text{mA} \times 120\ \Omega$$
$$V_1 = 12\ \text{V}$$
$$V_2 = V_3 = V_{AB} = I_T \times R_{AB}$$
$$= 100\ \text{mA} \times 60\ \Omega$$
$$= 6\ \text{V}$$
$$V_4 = I_T \times R_4$$
$$= 100\ \text{mA} \times 180\ \Omega$$
$$= 18\ \text{V}$$

The current in resistors R_2 and R_3 across points A and B can be found as follows:

$$I_2 = \frac{V_{AB}}{R_2}$$
$$= \frac{6\ \text{V}}{100\ \Omega}$$
$$I_2 = 60\ \text{mA}$$
$$I_3 = \frac{V_{AB}}{R_3}$$
$$= \frac{6\ \text{V}}{150\ \Omega}$$
$$I_3 = 40\ \text{mA}$$

Example 6-3

Assume that the series-parallel circuit in Fig. 6–15a has failed. A technician troubleshooting the circuit has measured the following voltages:

$$V_1 = 10.8 \text{ V}$$
$$V_{AB} = 9 \text{ V}$$
$$V_4 = 16.2 \text{ V}$$

These voltage readings are shown in Fig. 6–15b. Based on the voltmeter readings shown, which component is defective and what type of defect does it have?

ANSWER If we consider the resistance between points A and B as a single resistance, the circuit can be analyzed as if it were a simple series circuit. Notice that V_1 and V_4 have decreased from their normal values of 12 V and 18 V respectively, whereas the voltage V_{AB} across R_2 and R_3 has increased from 6 V to 9 V.

Recall that in a series circuit containing three or more components, the voltage across the defective component changes in a direction that is opposite to the direction of the change in voltage across the good components. Since the voltages V_1 and V_4 have decreased and the voltage V_{AB} has increased, the defective component must be either R_2 or R_3 across points A and B.

The increase in voltage across points A and B tells us that the resistance between points A and B must have increased. The increase in the resistance R_{AB} could be the result of an open in either R_2 or R_3.

But how do we know which resistor is open? At least three approaches may be used to find this out. One approach would be to calculate the resistance across points A and B. To do this, find the total current in either R_1 or R_4. Let's find I_T in R_1.

$$I_T = \frac{V_1}{R_1}$$

$$= \frac{10.8 \text{ V}}{120 \text{ }\Omega}$$

$$I_T = 90 \text{ mA}$$

Next, divide the measured voltage V_{AB} by I_T to find R_{AB}.

$$R_{AB} = \frac{V_{AB}}{I_T}$$

$$= \frac{9 \text{ V}}{90 \text{ mA}}$$

$$R_{AB} = 100 \text{ }\Omega$$

Notice that the value of R_{AB} is the same as that of R_2. This means, of course, that R_3 must be open.

Another approach to finding which resistor is open would be to open the switch S_1 and measure the resistance across points A and B. This measurement would show that the resistance R_{AB} equals 100 Ω, again indicating that the resistor R_3 must be open.

The only other approach to determine which resistor is open would be to measure the currents I_2 and I_3 with the switch S_1 closed. In Fig. 6–15b, the current I_2 would measure 90 mA, whereas the current I_3 would measure 0 mA. With $I_3 = 0$ mA, R_3 must be open.

Example 6-4

Assume that the series-parallel circuit in Fig. 6–15a has failed. A technician troubleshooting the circuit has measured the following voltages:

$$V_1 = 14.4 \text{ V}$$
$$V_{AB} = 0 \text{ V}$$
$$V_4 = 21.6 \text{ V}$$

These voltage readings are shown in Fig. 6–15c. Based on the voltmeter readings shown, which component is defective and what type of defect does it have?

ANSWER Since the voltages V_1 and V_4 have both increased, and the voltage V_{AB} has decreased, the defective component must be either R_2 or R_3 across points A and B. Because the voltage V_{AB} is 0 V, either R_2 or R_3 must be shorted.

But how can we find out which resistor is shorted? One way would be to measure the currents I_2 and I_3. The shorted component is the one with all the current.

Another way to find out which resistor is shorted would be to open the switch S_1 and measure the resistance across points A and B. Disconnect one lead of either R_2 or R_3 from point A while observing the ohmmeter. If removing the top lead of R_3 from point A still shows a reading of 0 Ω, then you know that R_2 must be shorted. Similarly, if removing the top lead of R_2 from point A (with R_3 still connected at point A) still produces a reading of 0 Ω, then you know that R_3 is shorted.

■ *6-7 Knowledge Check*

> *Answer at end of chapter.*

> In Fig. 6–15a what happens to the voltage, V_{AB}, if R_2 increases in value?

■ *6-7 Self-Review*

> *Answers at end of chapter.*

> a. In Fig. 6–13, the short circuit increases I_T from 2 A to what value?
> b. In Fig. 6–14, the open branch reduces I_T from 2 A to what value?
> c. In Fig. 6–15a, what is the voltage across points A and B if R_4 shorts?
> d. In Fig. 6–15a, what is the voltage V_{AB} if R_1 opens?

Summary

- In circuits combining series and parallel connections, the components in one current path without any branch points are in series; the parts of the circuit connected across the same two branch points are in parallel.

- To calculate R_T in a series-parallel circuit with R in the main line, combine resistances from the outside back toward the source.

- When the potential is the same at the two ends of a resistance, its voltage is zero. If no current flows through a resistance, it cannot have any IR voltage drop.

- A Wheatstone bridge circuit has two input terminals and two output terminals. When balanced, the Wheatstone bridge can be analyzed simply as two series strings in

parallel. The Wheatstone bridge finds many uses in applications where comparison measurements are needed.

- The procedure for troubleshooting series-parallel circuits is a combination of the procedures used to troubleshoot series and parallel circuits.

Important Terms

- **Balanced Bridge** - a circuit consisting of two series strings in parallel. The balanced condition occurs when the voltage ratio in each series string is identical. The output from the bridge is taken between the centers of each series string. When the voltage ratios in each series string are identical, the output voltage is zero, and the bridge circuit is said to be balanced.

- **Banks in Series** - parallel resistor banks that are connected in series with each other.

- **Ratio Arm** - accurate, stable resistors in one leg of a Wheatstone bridge or bridge circuit in general. The ratio arm fraction, R_1/R_2, can be varied in most cases, typically in multiples of 10. The ratio arm fraction in a Wheatstone bridge determines two things: the placement accuracy of the measurement of an unknown resistor, R_X, and the maximum unknown resistance, $R_{X(max)}$, that can be measured.

- **Standard Resistor** - a variable resistor in one leg of a Wheatstone bridge that is varied to provide equal voltage ratios in both series strings of the bridge. With equal voltage ratios in each series string, the bridge is said to be balanced.

- **Strings in Parallel** - series resistor strings that are connected in parallel with each other.

- **Wheatstone Bridge** - a balanced bridge circuit that can be used to find the value of an unknown resistor.

Related Formulas

$$R_X = R_S \times \frac{R_1}{R_2}$$

$$R_{X(max)} = R_{S(max)} \times \frac{R_1}{R_2}$$

Self-Test

Answers at back of book.

QUESTIONS 1–12 REFER TO FIGURE 6–16.

1. In Fig. 6–16,
   a. R_1 and R_2 are in series.
   b. R_3 and R_4 are in series.
   c. R_1 and R_4 are in series.
   d. R_2 and R_4 are in series.

2. In Fig. 6–16,
   a. R_2, R_3, and V_T are in parallel.
   b. R_2 and R_3 are in parallel.
   c. R_2 and R_3 are in series.
   d. R_1 and R_4 are in parallel.

3. In Fig. 6–16 the total resistance, R_T, equals
   a. 1.6 kΩ.
   b. 3.88 kΩ.
   c. 10 kΩ.
   d. none of the above.

4. In Fig. 6–16, the total current, I_T, equals
   a. 6.19 mA.
   b. 150 mA.
   c. 15 mA.
   d. 25 mA.

5. In Fig. 6–16, how much voltage is across points A and B?
   a. 12 V.
   b. 18 V.
   c. 13.8 V.
   d. 10.8 V.

Figure 6–16

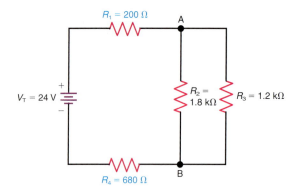

6. In Fig. 6–16, how much is I_2 through R_2?

   a. 9 mA.

   b. 15 mA.

   c. 6 mA.

   d. 10.8 mA.

7. In Fig. 6–16, how much is I_3 through R_3?

   a. 9 mA.

   b. 15 mA.

   c. 6 mA.

   d. 45 mA.

8. If R_4 shorts in Fig. 6–16, the voltage, V_{AB}

   a. increases.

   b. decreases.

   c. stays the same.

   d. increases to 24 V.

9. If R_2 becomes open in Fig. 6–16,

   a. the voltage across points A and B will decrease.

   b. the resistors R_1, R_3, and R_4 will be in series.

   c. the total resistance, R_T, will decrease.

   d. the voltage across points A and B will measure 24 V.

10. If R_1 opens in Fig. 6–16,

    a. the voltage across R_1 will measure 0 V.

    b. the voltage across R_4 will measure 0 V.

    c. the voltage across points A and B will measure 0 V.

    d. both b and c.

11. If R_3 becomes open in Fig. 6–16, what happens to the voltage across points A and B?

    a. It decreases.

    b. It increases.

    c. It stays the same.

    d. none of the above.

12. If R_2 shorts in Fig. 6–16,

    a. the voltage, V_{AB}, decreases to 0 V.

    b. the total current, I_T, flows through R_3.

    c. the current, I_3, in R_3 is zero.

    d. both a and c.

QUESTIONS 13–20 REFER TO FIGURE 6–17.

13. In Fig. 6–17, how much voltage exists between terminals C and D when the bridge is balanced?

    a. 0 V.

    b. 10.9 V.

    c. 2.18 V.

    d. 12 V.

14. In Fig. 6–17, assume that the current in M_1 is zero when R_S is adjusted to 55,943 Ω. What is the value of the unknown resistor, R_X?

    a. 55,943 Ω.

    b. 559.43 Ω.

    c. 5,594.3 Ω.

    d. 10k Ω.

Figure 6–17

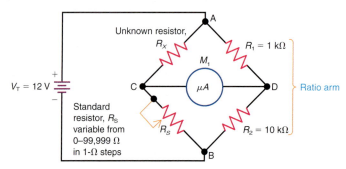

15. In Fig. 6–17, assume that the bridge is balanced when R_S is adjusted to 15,000 Ω. How much is the total current, I_T, flowing to and from the terminals of the voltage source, V_T?

   a. zero.

   b. approximately 727.27 μA.

   c. approximately 1.09 mA.

   d. approximately 1.82 mA.

16. In Fig. 6–17, what is the maximum unknown resistor, $R_{X(max)}$, that can be measured for the resistor values shown in the ratio arm?

   a. 99.99 Ω.

   b. 9,999.9 Ω.

   c. 99,999 Ω.

   d. 999,999 Ω.

17. In Fig. 6–17, the ratio R_1/R_2 determines

   a. the placement accuracy of the measurement of R_X.

   b. the maximum unknown resistor, $R_{X(max)}$, that can be measured.

   c. the amount of voltage available across terminals A and B.

   d. both a and b.

18. In Fig. 6–17, assume that the standard resistor, R_S, has been adjusted so that the current in M_1 is exactly 0 μA. How much voltage exists at terminal C with respect to terminal B?

   a. 1.1 V.

   b. 0 V.

   c. 10.9 V.

   d. none of the above.

19. In Fig. 6–17, assume that the ratio arm resistors, R_1 and R_2, are interchanged. What is the value of the unknown resistor, R_X, if R_S equals 33,950 Ω when the bridge is balanced?

   a. 339.5 kΩ.

   b. 3.395 kΩ.

   c. 33,950 Ω.

   d. none of the above.

20. In Fig. 6–17, assume that the standard resistor, R_S, cannot be adjusted high enough to provide a balanced condition. What modification must be made to the circuit?

   a. Change the ratio arm fraction R_1/R_2, from 1/10 to 1/100 or something less.

   b. Change the ratio arm fraction, R_1/R_2 from 1/10 to 1/1, 10/1 or something greater.

   c. Reverse the polarity of the applied voltage, V_T.

   d. none of the above.

Questions

1. In a series-parallel circuit, how can you tell which resistances are in series and which are in parallel?

2. Draw a schematic diagram showing two resistances in a bank that is in series with one resistance.

3. Draw a diagram showing how to connect three resistances of equal value so that the combined resistance will be 1½ times the resistance of one unit.

4. Draw a diagram showing two strings in parallel across a voltage source, where each string has three series resistances.

5. Explain why components are connected in series-parallel, showing a circuit as an example of your explanation.

6. Give two differences between a short circuit and an open circuit.

7. Explain the difference between voltage division and current division.

8. In Fig. 6–12, assume that the thermistor has a positive temperature coefficient (PTC). Explain what happens to the voltage V_{CD} (a) if the ambient temperature decreases; (b) if the ambient temperature increases.

9. Draw a circuit with nine 40-V, 100-W bulbs connected to a 120-V source.

10. (a) Two 10-Ω resistors are in series with a 100-V source. If a third 10-Ω R is added in series, explain why I will decrease. (b) The same two 10-Ω resistors are in parallel with the 100-V source. If a third 10-Ω R is added in parallel, explain why I_T will increase.

Problems

SECTION 6–1 FINDING R_T FOR SERIES-PARALLEL RESISTANCES

6–1 In Fig. 6–18, identify which components are in series and which ones are in parallel.

Figure 6–18

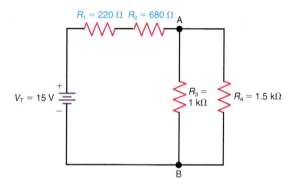

6–2 In Fig. 6–18,
- a. how much is the total resistance of just R_1 and R_2?
- b. what is the equivalent resistance of R_3 and R_4 across points A and B?
- c. how much is the total resistance, R_T, of the entire circuit?
- d. how much is the total current, I_T, in the circuit?
- e. how much current flows into point B?
- f. how much current flows away from point A?

6–3 ‖‖ **MultiSim** In Fig. 6–18, solve for the following: I_1, I_2, V_1, V_2, V_3, V_4, I_3, and I_4.

6–4 In Fig. 6–19, identify which components are in series and which ones are in parallel.

Figure 6–19

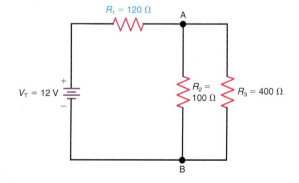

6–5 In Fig. 6–19,
- a. what is the equivalent resistance of R_2 and R_3 across points A and B?
- b. how much is the total resistance, R_T of the entire circuit?
- c. how much is the total current, I_T, in the circuit?
- d. how much current flows into point B and away from point A?

6–6 ‖‖ **MultiSim** In Fig. 6–19, solve for I_1, V_1, V_2, V_3, I_2, and I_3.

6–7 In Fig. 6–19, solve for P_1, P_2, P_3, and P_T.

6–8 In Fig. 6–20, identify which components are in series and which ones are in parallel.

Figure 6–20

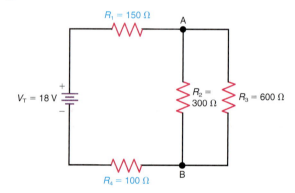

6–9 In Fig. 6–20,
- a. how much is the total resistance of just R_1 and R_4?
- b. what is the equivalent resistance of R_2 and R_3 across points A and B?
- c. how much is the total resistance, R_T, of the entire circuit?
- d. how much is the total current, I_T, in the circuit?
- e. how much current flows into point B and away from point A?

6–10 In Fig. 6–20, solve for the following: I_1, V_1, V_2, V_3, I_2, I_3, I_4, V_4, P_1, P_2, P_3, P_4, and P_T.

SECTION 6–2 RESISTANCE STRINGS IN PARALLEL

6–11 In Fig. 6–21,

    a. what is the total resistance of branch 1?

    b. what is the resistance of branch 2?

    c. how much are the branch currents I_1 and I_2?

    d. how much is the total current, I_T, in the circuit?

    e. how much is the total resistance, R_T, of the entire circuit?

    f. what are the values of V_1, V_2, and V_3?

Figure 6–21

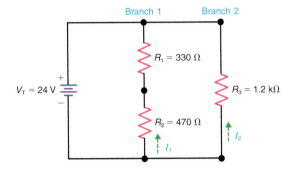

6–12 In Fig. 6–22,

    a. what is the total resistance of branch 1?

    b. what is the total resistance of branch 2?

    c. how much are the branch currents I_1 and I_2?

    d. how much is the total current, I_T, in the circuit?

    e. how much is the total resistance, R_T, of the entire circuit?

    f. what are the values of V_1, V_2, V_3, and V_4?

Figure 6–22

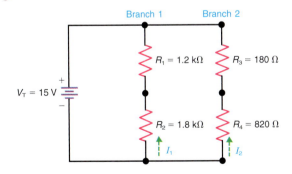

6–13 In Fig. 6–23,

    a. what is the total resistance of branch 1?

    b. what is the total resistance of branch 2?

    c. how much are the branch currents I_1 and I_2?

    d. how much is the total current, I_T, in the circuit?

    e. how much is the total resistance, R_T, of the entire circuit?

    f. what are the values of V_1, V_2, V_3, and V_4?

Figure 6–23

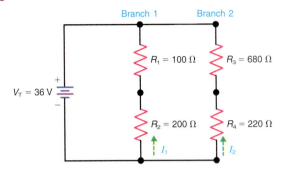

6–14 In Fig. 6–24, solve for the following:

    a. branch currents I_1, I_2 and the total current, I_T

    b. R_T

    c. V_1, V_2, V_3, and V_4

    d. P_1, P_2, P_3, P_4, and P_T

Figure 6–24

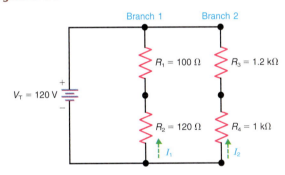

6–15 In Fig. 6–25, solve for the following:

    a. branch currents I_1, I_2, I_3 and the total current, I_T

    b. R_T

    c. V_1, V_2, V_3, V_4, and V_5

Figure 6–25

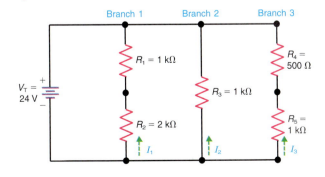

SECTION 6–3 RESISTANCE BANKS IN SERIES

6–16 In Fig. 6–26,

    a. what is the equivalent resistance of R_1 and R_2 in parallel across points A and B?

    b. what is the total resistance, R_T, of the circuit?

    c. what is the total current, I_T, in the circuit?

    d. how much voltage exists across points A and B?

    e. how much voltage is dropped across R_3?

    f. solve for I_1 and I_2.

    g. how much current flows into point B and away from point A?

Figure 6–26

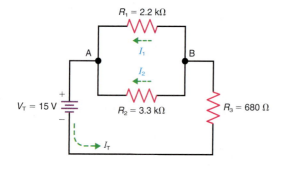

6–17 In Fig. 6–27,

    a. what is the equivalent resistance of R_2 and R_3 in parallel across points A and B?

    b. what is the total resistance, R_T, of the circuit?

    c. what is the total current, I_T, in the circuit?

    d. how much voltage exists across points A and B?

    e. how much voltage is dropped across R_1?

    f. solve for I_2 and I_3.

    g. how much current flows into point B and away from point A?

Figure 6–27

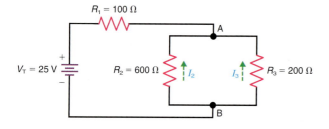

6–18 In Fig. 6–28, solve for

    a. R_T.

    b. I_T.

    c. V_1, V_2, and V_3.

    d. I_1 and I_2.

Figure 6–28

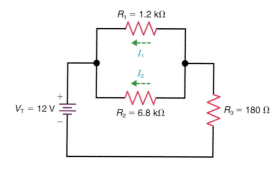

6–19 In Fig. 6–29, solve for,

    a. R_T.

    b. I_T.

    c. V_1, V_2, and V_3.

    d. I_2 and I_3.

Figure 6–29

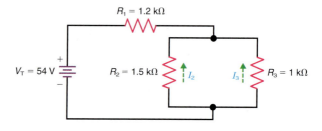

SECTION 6–4 RESISTANCE BANKS AND STRINGS IN SERIES-PARALLEL

6–20 In Fig. 6–30, solve for R_T, I_T, V_1, V_2, V_3, V_4, I_1, I_2, I_3, and I_4.

Figure 6–30

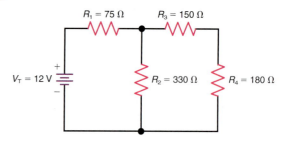

6–21 In Fig. 6–31, solve for R_T, I_T, V_1, V_2, V_3, V_4, V_5, I_1, I_2, I_3, I_4, and I_5.

6–24 In Fig. 6–34, solve for R_T, I_T, V_1, V_2, V_3, V_4, V_5, V_6, I_1, I_2, I_3, I_4, I_5, and I_6.

Figure 6–31

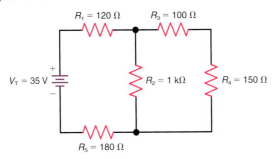

Figure 6–34

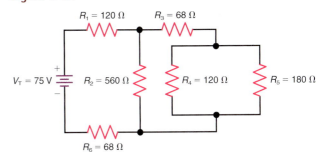

6–22 In Fig. 6–32, solve for R_T, I_T, V_1, V_2, V_3, V_4, V_5, V_6, I_1, I_2, I_3, I_4, I_5, and I_6.

6–25 In Fig. 6–35, solve for R_T, I_T, V_1, V_2, V_3, V_4, V_5, V_6, V_7, I_1, I_2, I_3, I_4, I_5, I_6, and I_7.

Figure 6–32

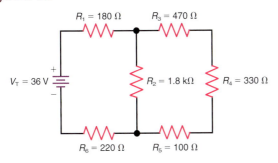

Figure 6–35

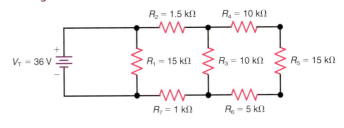

6–23 In Fig. 6–33, solve for R_T, I_T, V_1, V_2, V_3, V_4, V_5, I_1, I_2, I_3, I_4, and I_5.

6–26 In Fig. 6–36, solve for R_T, I_1, I_2, I_3, V_1, V_2, V_3, and the voltage, V_{AB}.

Figure 6–36

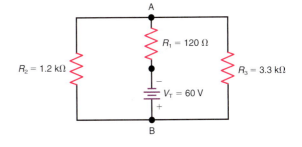

Figure 6–33

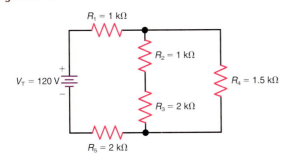

6–27 In Fig. 6–37, solve for R_T, I_T, V_1, V_2, V_3, V_4, V_5, V_6, I_1, I_2, I_3, I_4, I_5, and I_6.

6–30 In Fig. 6–40, solve for R_2, V_1, V_2, V_3, I_3, I_T, and V_T.

Figure 6–37

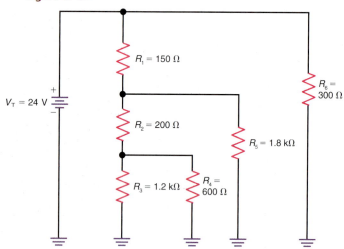

Figure 6–40

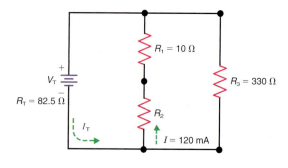

6–31 In Fig. 6–41, solve for R_T, I_T, V_T, V_1, V_2, V_3, V_4, V_5, V_6, I_2, I_3, I_4, and I_5.

Figure 6–41

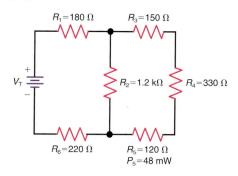

SECTION 6–5 ANALYZING SERIES–PARALLEL CIRCUITS WITH RANDOM UNKNOWNS

6–28 In Fig. 6–38 solve for V_1, V_2, V_3, I_2, R_3, R_T, and V_T.

6–32 In Fig. 6–42, solve for R_T, I_T, V_1, V_2, V_3, V_4, V_5, V_6, I_1, I_2, I_3, I_4, I_5, and I_6.

Figure 6–38

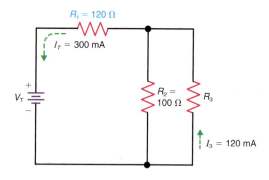

Figure 6–42

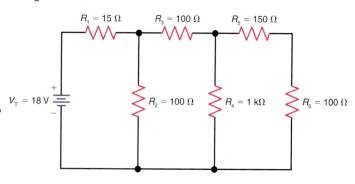

6–29 In Fig. 6–39, solve for R_T, I_T, V_T, V_1, V_2, V_4, I_2, and I_3.

Figure 6–39

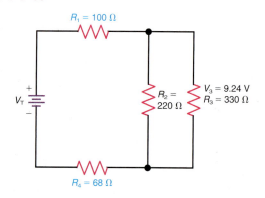

6-33 In Fig. 6-43, solve for I_T, R_T, R_2, V_2, V_3, V_4, V_5, V_6, I_2, I_3, I_4, I_5, and I_6.

Figure 6-43

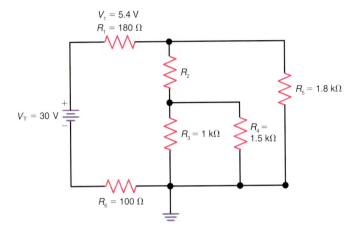

$V_1 = 5.4$ V
$R_1 = 180$ Ω
$R_5 = 1.8$ kΩ
$V_T = 30$ V
$R_3 = 1$ kΩ
$R_4 = 1.5$ kΩ
$R_6 = 100$ Ω

SECTION 6-6 THE WHEATSTONE BRIDGE

Problems 34–38 refer to Fig. 6-44.

6-34 In Fig. 6-44,

a. how much current flows through M_1 when the Wheatstone bridge is balanced?

b. how much voltage exists between points C and D when the bridge is balanced?

Figure 6-44

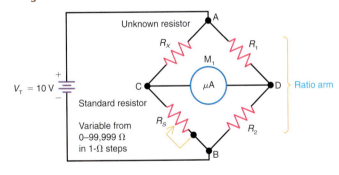

Unknown resistor
A
R_X
R_1
M_1
$V_T = 10$ V
C
μA
D ⟩ Ratio arm
Standard resistor
Variable from 0–99,999 Ω in 1-Ω steps
R_S
R_2
B

6-35 In Fig. 6-44, assume that the bridge is balanced when $R_1 = 1$ kΩ, $R_2 = 5$ kΩ, and $R_S = 34,080$ Ω. Determine

a. the value of the unknown resistor, R_X.

b. the voltages V_{CB} and V_{DB}.

c. the total current, I_T, flowing to and from the voltage source, V_T.

6-36 In reference to Problem 35, which direction (C to D or D to C) will electrons flow through M_1 if

a. R_S is reduced in value.

b. R_S is increased in value.

6-37 In Fig. 6-44, calculate the maximum unknown resistor, $R_{X(max)}$, that can be measured for the following ratio arm values:

a. $\dfrac{R_1}{R_2} = \dfrac{1}{1000}$

b. $\dfrac{R_1}{R_2} = \dfrac{1}{100}$

c. $\dfrac{R_1}{R_2} = \dfrac{1}{10}$

d. $\dfrac{R_1}{R_2} = \dfrac{1}{1}$

e. $\dfrac{R_1}{R_2} = \dfrac{10}{1}$

f. $\dfrac{R_1}{R_2} = \dfrac{100}{1}$

6-38 Assume that the same unknown resistor, R_X, is measured using different ratio arm fractions in Fig. 6-44. In each case, the standard resistor, R_S, is adjusted to provide the balanced condition. The values for each measurement are

a. $R_S = 123$ Ω and $\dfrac{R_1}{R_2} = \dfrac{1}{1}$.

b. $R_S = 1232$ Ω and $\dfrac{R_1}{R_2} = \dfrac{1}{10}$.

c. $R_S = 12{,}317$ Ω and $\dfrac{R_1}{R_2} = \dfrac{1}{100}$.

Calculate the value of the unknown resistor, R_X, for each measurement. Which ratio arm fraction provides the greatest accuracy?

PROBLEMS 39–41 REFER TO FIGURE 6–45.

6-39 In Fig. 6-45, to what value must R_3 be adjusted to provide zero volts across terminals C and D when the ambient temperature, T_A, is 25°C? (Note: R_0 is the resistance of the thermistor at an ambient temperature, T_A, of 25°C.)

Figure 6-45

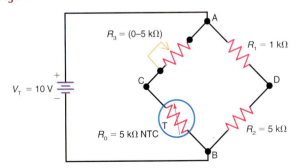

$R_3 = (0–5$ kΩ$)$
$R_1 = 1$ kΩ
A
$V_T = 10$ V
C
D
T
$R_0 = 5$ kΩ NTC
$R_2 = 5$ kΩ
B

6–40 In Fig. 6–45, assume that R_S is adjusted to provide zero volts across terminals C and D at an ambient temperature, T_A, of 25°C. What happens to the polarity of the output voltage, V_{CD}, when

   a. the ambient temperature, T_A, increases above 25°C?

   b. the ambient temperature, T_A, decreases below 25°C?

6–41 In Fig. 6–45, assume that R_3 has been adjusted to 850 Ω to provide zero volts across the output terminals C and D. Determine

   a. the resistance of the thermistor.

   b. whether the ambient temperature, T_A, has increased or decreased from 25°C.

SECTION 6–7 TROUBLESHOOTING: OPENS AND SHORTS IN SERIES–PARALLEL CIRCUITS

Figure 6–46 shows a series-parallel circuit with its normal operating voltages and currents.

6–42 ▐▐▐ **MultiSim** In Fig. 6–46, determine the voltages V_1, V_{AB}, V_3, V_{CD}, and V_5 for each of the following component troubles:

   a. R_4 is open.

   b. R_2 is shorted.

   c. R_3 is open.

   d. R_4 is shorted.

   e. R_2 is open.

   f. R_1 is open.

   g. R_1 is shorted.

Figure 6–46

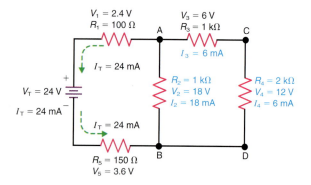

Critical Thinking

6–43 In Fig. 6–47, bulbs A and B each have an operating voltage of 28 V. If the wattage ratings for bulbs A and B are 1.12 W and 2.8 W, respectively, calculate (**a**) the required resistance of R_1; (**b**) the recommended wattage rating of R_1; (**c**) the total resistance R_T.

Figure 6–47 Circuit diagram for Critical Thinking Prob. 1.

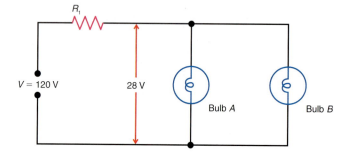

Figure 6–48 Circuit diagram for Critical Thinking Prob. 2.

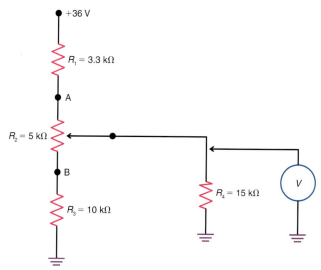

6–44 Refer to Fig. 6–48. How much voltage will be indicated by the voltmeter when the wiper arm of the linear potentiometer R_2 is set (**a**) to point A; (**b**) to point B; (**c**) midway between points A and B.

6–45 Explain how the temperature control circuit in Fig. 6–12 will be affected if the polarity of the applied voltage V_T is reversed.

Troubleshooting Challenge

Table 6–1 shows voltage measurements taken in Fig. 6–49. The first row shows the normal values that exist when the circuit is operating normally. Rows 2 to 13 are voltage measurements taken when one component in the circuit has failed. For each

row in Table 6–1, identify which component is defective and determine the type of defect that has occurred in the component.

Figure 6–49 Circuit diagram for troubleshooting challenge. Normal operating voltages are shown.

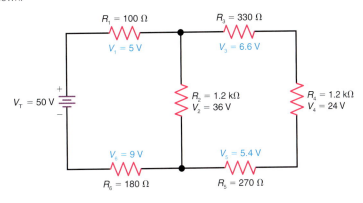

Table 6–1	Voltage Measurements for Troubleshooting Challenge						
	V_1	V_2	V_3	V_4	V_5	V_6	Defective Component
	VOLTS						
1 Normal values	5	36	6.6	24	5.4	9	None
2 Trouble 1	0	0	0	0	0	50	
3 Trouble 2	17.86	0	0	0	0	32.14	
4 Trouble 3	3.38	40.54	0	40.54	0	6.08	
5 Trouble 4	3.38	40.54	0	0	40.54	6.08	
6 Trouble 5	6.1	43.9	8.05	29.27	6.59	0	
7 Trouble 6	2.4	43.27	7.93	28.85	6.49	4.33	
8 Trouble 7	50	0	0	0	0	0	
9 Trouble 8	7.35	29.41	16.18	0	13.24	13.24	
10 Trouble 9	0	40	7.33	26.67	6	10	
11 Trouble 10	3.38	40.54	40.54	0	0	6.08	
12 Trouble 11	5.32	35.11	0	28.67	6.45	9.57	
13 Trouble 12	5.25	35.3	7.61	27.7	0	9.45	

Answers to Knowledge Check Problems

6–1 No, because the current through R_2 is the sum of the currents I_3 and I_4. Therefore, because I_2 is not equal to I_3 alone, R_2 and R_3 cannot be in series.

6–2 $I_1 = 800$ mA
$I_2 = 2.4$ A
$I_T = 3.2$ A

6–3 $I_2 = 480$ mA
$I_3 = 120$ mA
$I_T = 600$ mA

6–4 $R_T = 53\ \Omega$

6–5 $I_t = 3.64$ A

6–6 V_{CD} is negative.

6–7 V_{AB} increases.

Answers to Self-Reviews

6–1 a. $R = 1\ \text{k}\Omega$
b. $R = 0.5\ \text{k}\Omega$
c. $R_T = 1.5\ \text{k}\Omega$

6–2 a. 12 V
b. $I = 6$ A
c. $V_3 = 18$ V

6–3 a. $V_3 = 40$ V
b. $I = 8$ A
c. $V_1 = 4$ V

6–4 a. R_1
b. R_4
c. R_6

6–5 a. R_3
b. R_1
c. $I_2 = 4$ A
d. $V_3 = 60$ V

6–6 a. A and B are input; C and D are output.
b. Zero
c. $R_X = 71.35\ \Omega$
d. $99.99\ \Omega$
e. $500\ \Omega$

6–7 a. $I = 10$ A.
b. $I = 1.1$ A.
c. $V_{AB} = 12$ V
d. 0 V

Cumulative Summary (Chapters 1–6)

- The electron is the most basic particle of negative electricity; the proton is the most basic particle of positive electricity. Both have the same charge, but they have opposite polarities.

- A quantity of electrons is a negative charge; a deficiency of electrons is a positive charge. Like charges repel each other; unlike charges attract.

- Charge, Q, is measured in coulombs; 6.25×10^{18} electrons equals one coulomb of charge. Charge in motion is current. One coulomb of charge flowing past a given point each second equals one ampere of current.

- Potential difference, PD, is measured in volts. One volt produces one ampere of current against the opposition of one ohm of resistance.

- The main types of resistors are carbon-composition, carbon-film, metal-film, wire-wound, and surface-mount. Wire-wound resistors are used when the resistance must dissipate a lot of power, such as 5 W or more. Carbon-film and metal-film resistors are better than the older carbon-composition type because they have tighter tolerances, are less sensitive to temperature changes and aging, and generate less noise internally.

- Resistors having a power rating less than 2 W are often color-coded to indicate their resistance value. To review the color code, refer to Table 2–1 and Fig. 2–8.

- Surface-mount resistors (also called chip resistors) typically use a three-digit number printed on the body to indicate the resistance value in ohms. The first two digits indicate the first two digits in the numerical value of the resistance; the third digit is the multiplier. For example, a surface-mount resistor which is marked 103 has a resistance value of 10,000 Ω or 10 kΩ.

- A potentiometer is a variable resistor that has three terminals. It is used as a variable voltage divider. A rheostat is a variable resistor that has only two terminals. It is used to vary the current in a circuit.

- A thermistor is a resistor whose resistance changes with changes in operating temperature. Thermistors are available with either a positive or a negative temperature coefficient (NTC).

- The most common trouble with resistors is that they develop opens and thus have infinitely high resistance.

- The three forms of Ohm's law are $I = V/R$, $V = IR$, and $R = V/I$.

- The three power formulas are $P = VI$, $P = I^2R$, and $P = V^2/R$.

- The most common multiple and submultiples of the practical units are mega (M) for 10^6, micro (μ) for 10^{-6}, kilo (k) for 10^3, and milli (m) for 10^{-3}.

- For series resistances: (1) the current is the same in all resistances; (2) IR drops can be different with unequal resistances; (3) the applied voltage equals the sum of the series IR drops; (4) the total resistance equals the sum of the individual resistances; (5) an open circuit in one resistance results in no current through the entire series circuit.

- For parallel resistances: (1) the voltage is the same across all resistances; (2) the branch currents can be different with unequal resistances; (3) the total line current equals the sum of the parallel branch currents; (4) the combined equivalent resistance, R_{EQ}, of parallel branches is

less than the smallest resistance as determined by Formula (5-3); (5) an open circuit in one branch does not create an open in the other branches; (6) a short circuit across one branch short-circuits all branches.

- In series-parallel circuits, the resistances in one current path without any branch points are in series; all rules of series resistances apply. The resistances across the same two branch points are in

parallel; all rules of parallel resistances apply.

- A Wheatstone bridge has two input terminals and two output terminals. When the bridge is balanced, the voltage across the output terminals is 0 V. When the bridge is unbalanced, however, the output voltage may be either positive or negative. Balanced bridge circuits find many useful applications in electronics.

Cumulative Review Self-Test

Answers at back of book.

1. A carbon resistor is color-coded with brown, green, red, and gold stripes from left to right. Its value is (a) 1500 Ω ± 5%; (b) 6800 Ω ± 5%; (c) 10,000 Ω ± 10%; (d) 500,000 Ω ± 5%.

2. A metal-film resistor is color-coded with orange, orange, orange, red, and green stripes, reading from left to right. Its value is (a) 3.3 kΩ ± 5%; (b) 333 kΩ ± 5%; (c) 33.3 kΩ ± 0.5%; (d) 333 Ω ± 0.5%.

3. With 30 V applied across two equal resistors in series, 10 mA of current flows. Typical values for each resistor to be used here are (a) 10 Ω, 10 W; (b) 1500 Ω, ½ W; (c) 3000 Ω, 10 W; (d) 30 MΩ, 2 W.

4. In which of the following circuits will the voltage source produce the most current? (a) 10 V across a 10-Ω resistance; (b) 10 V across two 10-Ω resistances in series; (c) 10 V across two 10-Ω resistances in parallel; (d) 1000 V across a 1-MΩ resistance.

5. Three 120-V, 100-W bulbs are in parallel across a 120-V power line. If one bulb burns open: (a) the other two bulbs cannot light; (b) all three bulbs light; (c) the other two bulbs can light; (d) there is excessive current in the main line.

6. A circuit allows 1 mA of current to flow with 1 V applied. The conductance of the circuit equals (a) 0.002 Ω; (b) 0.005 μS; (c) 1000 μS; (d) 1 S.

7. If 2 A of current is allowed to accumulate charge for 5s, the resultant charge equals (a) 2 C; (b) 10 C; (c) 5 A; (d) 10 A.

8. A potential difference applied across a 1-MΩ resistor produces 1 mA of current. The applied voltage equals (a) 1 μV; (b) 1 mV; (c) 1 kV; (d) 1,000,000 V.

9. A string of two 1000-Ω resistances is in series with a parallel bank of two 1000-Ω resistances. The total resistance of the series-parallel circuit equals (a) 250 Ω; (b) 2500 Ω; (c) 3000 Ω; (d) 4000 Ω.

10. In the circuit of Question 9, one of the resistances in the series string opens. Then the current in the parallel bank (a) increases slightly in both branches; (b) equals zero in one branch but is maximum in the other branch; (c) is maximum in both branches; (d) equals zero in both branches.

11. With 100 V applied across a 10,000-Ω resistance, the power dissipation equals (a) 1 mW; (b) 1 W; (c) 100 W; (d) 1 kW.

12. A source of 10 V is applied across R_1, R_2, and R_3 in series, producing 1 A in the series circuit. R_1 equals 6 Ω and R_2 equals 2 Ω. Therefore, R_3 equals (a) 2 Ω; (b) 4 Ω; (c) 10 Ω; (d) 12 Ω.

13. A 5-V source and a 3-V source are connected with series-opposing polarities. The combined voltage across both sources equals (a) 5 V; (b) 3 V; (c) 2 V; (d) 8 V.

14. In a circuit with three parallel branches, if one branch opens, the main-line current will be (a) more; (b) less; (c) the same; (d) infinite.

15. A 10-Ω R_1 and a 20-Ω R_2 are in series with a 30-V source. If R_1 opens, the voltage drop across R_2 will be (a) zero; (b) 20 V; (c) 30 V; (d) infinite.

16. A voltage V_1 of 40 V is connected series-opposing with V_2 of 50 V. The total voltage across both components is: (a) 10 V; (b) 40 V; (c) 50 V; (d) 90 V.

17. Two series voltage drops V_1 and V_2 total 100 V for V_T. When V_1 is 60 V, then V_2 must equal: (a) 40 V; (b) 60 V; (c) 100 V; (d) 160 V.

18. Two parallel branch currents I_1 and I_2 total 100 mA for I_T. When I_1 is 60 mA, then I_2 must equal (a) 40 mA; (b) 60 mA; (c) 100 mA; (d) 160 mA.

19. A surface-mount resistor is marked 224. Its resistance is (a) 224 Ω; (b) 220 kΩ; (c) 224 kΩ; (d) 22 kΩ.

20. If a variable voltage is connected across a fixed resistance: (a) I and V will vary in direct proportion; (b) I and V will be inversely proportional; (c) I will remain constant as V is varied; (d) none of the above.

21. If a fixed value of voltage is connected across a variable resistance, (a) I will vary in direct proportion to R; (b) I will be inversely proportional to R; (c) I will remain constant as R is varied; (d) none of the above.

7

Voltage Dividers and Current Dividers

Any series circuit is a voltage divider in which the individual resistor voltage drops are proportional to the series resistance values. Similarly, any parallel circuit is a current divider in which the individual branch currents are inversely proportional to the branch resistance values. When parallel-connected loads are added to a series circuit, the circuit becomes a loaded voltage divider. Actually, a loaded voltage divider is just a practical application of a series-parallel circuit.

In a series circuit, it is possible to find the individual resistor voltage drops without knowing the series current. Likewise, it is possible to find the individual branch currents in a parallel circuit without knowing the value of the applied voltage. In this chapter, you will learn how to solve for the voltages in a series circuit and the currents in a parallel circuit using special formulas that provide shortcuts in the calculations. You will also learn how to design a loaded voltage divider that provides different load voltages and currents from a single supply voltage, V_T.

Objectives

After studying this chapter you should be able to

- *Calculate* the voltage drops in an unloaded voltage divider.

- *Explain* why resistor voltage drops are proportional to the resistor values in a series circuit.

- *Calculate* the branch currents in a parallel circuit.

- *Explain* why the branch currents are inversely proportional to the branch resistances in a parallel circuit.

- *Define* what is meant by the term *loaded voltage divider.*

- *Calculate* the voltage, current, and power values in a loaded voltage divider.

Outline

Important Terms

bleeder current

current divider

load currents

loaded voltage

voltage divider

voltage taps

7-1 Series Voltage Dividers

The current is the same in all resistances in a series circuit. Also, the voltage drops equal the product of I times R. Therefore, the IR voltages are proportional to the series resistances. A higher resistance has a greater IR voltage than a lower resistance in the same series circuit; equal resistances have the same amount of IR drop. If R_1 is double R_2, then V_1 will be double V_2.

The series string can be considered a *voltage divider*. Each resistance provides an IR drop V equal to its proportional part of the applied voltage. Stated as a formula,

$$V = \frac{R}{R_T} \times V_T \tag{7-1}$$

Example 7-1

Three 50-kΩ resistors R_1, R_2, and R_3 are in series across an applied voltage of 180 V. How much is the IR voltage drop across each resistor?

ANSWER The voltage drop across each R is 60 V. Since R_1, R_2, and R_3 are equal, each has one-third the total resistance of the circuit and one-third the total applied voltage. Using the formula,

$$V = \frac{R}{R_T} \times V_T = \frac{50 \text{ k}\Omega}{150 \text{ k}\Omega} \times 180 \text{ V}$$

$$= \frac{1}{3} \times 180 \text{ V}$$

$$= 60 \text{ V}$$

Note that R and R_T must be in the same units for the proportion. Then V is in the same units as V_T.

Typical Circuit

Figure 7-1 illustrates another example of a proportional voltage divider. Let the problem be to find the voltage across R_3. We can either calculate this voltage V_3 as IR_3 or determine its proportional part of the total applied voltage V_T. The answer is the same both ways. Note that R_T is $20 + 30 + 50 = 100 \text{ k}\Omega$.

Proportional Voltage Method

Using Formula (7-1), V_3 equals 20/100 of the 200 V applied for V_T because R_3 is 20 kΩ and R_T is 100 kΩ. Then V_3 is 20/100 of 200 or $\frac{1}{5}$ of 200, which is equal to 40 V. The calculations are

$$V_3 = \frac{R_3}{R_T} \times V_T = \frac{20}{100} \times 200 \text{ V}$$

$$V_3 = 40 \text{ V}$$

In the same way, V_2 is 60 V. The calculations are

$$V_2 = \frac{R_2}{R_T} \times V_T = \frac{30}{100} \times 200 \text{ V}$$

$$V_2 = 60 \text{ V}$$

MultiSim Figure 7-1 Series string of resistors as a proportional voltage divider. Each V_R is R/R_T fraction of the total source voltage V_T.

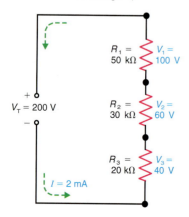

CALCULATOR

To do a problem like this on the calculator, you can divide R_3 by R_T first and then multiply by V_T. For the values here, to find V_3, the procedure can be as follows:

Punch in the number 20 for R_3.

Push the ⊕ key, then 100 for R_T and press the ⊗ key for the quotient of 0.2 on the display.

Next, punch in 200 for V_T and press the ⊜ key to display the quotient of 40 as the answer for V_3.

As another method, you can multiply R_3 by V_T first and then divide by R_T. The answers will be the same for either method.

Also, V_1 is 100 V. The calculations are

$$V_1 = \frac{R_1}{R_T} \times V_T = \frac{50}{100} \times 200 \text{ V}$$

$$V_1 = 100 \text{ V}$$

The sum of V_1, V_2, and V_3 in series is $100 + 60 + 40 = 200$ V, which is equal to V_T.

Method of *IR* Drops

If we want to solve for the current in Fig. 7–1, $I = V_T/R_T$ or 200 V/100 kΩ = 2 mA. This I flows through R_1, R_2, and R_3 in series. The *IR* drops are

$$V_1 = I \times R_1 = 2 \text{ mA} \times 50 \text{ k}\Omega = 100 \text{ V}$$
$$V_2 = I \times R_2 = 2 \text{ mA} \times 30 \text{ k}\Omega = 60 \text{ V}$$
$$V_3 = I \times R_3 = 2 \text{ mA} \times 20 \text{ k}\Omega = 40 \text{ V}$$

These voltages are the same values calculated by Formula (7–1) for proportional voltage dividers.

Two Voltage Drops in Series

For this case, it is not necessary to calculate both voltages. After you find one, subtract it from V_T to find the other.

As an example, assume that V_T is 48 V across two series resistors R_1 and R_2. If V_1 is 18 V, then V_2 must be $48 - 18 = 30$ V.

The Largest Series *R* Has the Most *V*

The fact that series voltage drops are proportional to the resistances means that a very small R in series with a much larger R has a negligible *IR* drop. An example is shown in Fig. 7–2a. Here the 1 kΩ of R_1 is in series with the much larger 999 kΩ of R_2. The V_T is 1000 V.

The voltages across R_1 and R_2 in Fig. 7–2a can be calculated by the voltage divider formula. Note that R_T is $1 + 999 = 1000$ kΩ.

$$V_1 = \frac{R_1}{R_T} \times V_T = \frac{1}{1000} \times 1000 \text{ V} = 1 \text{ V}$$

$$V_2 = \frac{R_2}{R_T} \times V_T = \frac{999}{1000} \times 1000 \text{ V} = 999 \text{ V}$$

The 999 V across R_2 is practically the entire applied voltage. Also, the very high series resistance dissipates almost all the power.

The current of 1 mA through R_1 and R_2 in Fig. 7–2a is determined almost entirely by the 999 kΩ of R_2. The I for R_T is 1000 V/1000 kΩ, which equals 1 mA. However, the 999 kΩ of R_2 alone would allow 1.001 mA of current, which differs very little from the original I of 1 mA.

Voltage Taps in a Series Voltage Divider

Consider the series voltage divider with **voltage taps** in Fig. 7–2b, where different voltages are available from the tap points A, B, and C. Note that the total resistance R_T is 20 kΩ, which can be found by adding the individual series resistance values. The voltage at each tap point is measured with respect to ground. The voltage at tap point C, designated V_{CG}, is the same as the voltage across R_4. The calculations for V_{CG} are as follows:

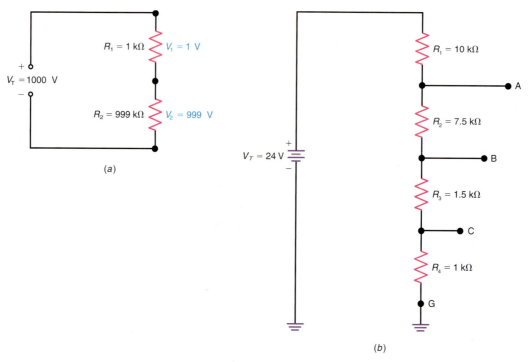

(a)

(b)

$$V_{CG} = \frac{R_4}{R_T} \times V_T$$

$$= \frac{1 \text{ k}\Omega}{20 \text{ k}\Omega} \times 24 \text{ V}$$

$$V_{CG} = 1.2 \text{ V}$$

The voltage at tap point B, designated V_{BG}, is the sum of the voltages across R_3 and R_4. The calculations for V_{BG} are

$$V_{BG} = \frac{R_3 + R_4}{R_T} \times V_T$$

$$= \frac{1.5 \text{ k}\Omega + 1 \text{ k}\Omega}{20 \text{ k}\Omega} \times 24 \text{ V}$$

$$V_{BG} = 3 \text{ V}$$

The voltage at tap point A, designated V_{AG}, is the sum of the voltage across R_2, R_3, and R_4. The calculations are

$$V_{AG} = \frac{R_2 + R_3 + R_4}{R_T} \times V_T$$

$$= \frac{7.5 \text{ k}\Omega + 1.5 \text{ k}\Omega + 1 \text{ k}\Omega}{20 \text{ k}\Omega} \times 24 \text{ V}$$

$$V_{AG} = 12 \text{ V}$$

Notice that the voltage V_{AG} equals 12 V, which is one-half of the applied voltage V_T. This makes sense, since $R_2 + R_3 + R_4$ make up 50% of the total resistance R_T. Similarly, since $R_3 + R_4$ make up 12.5% of the total resistance, the voltage V_{BG} will also be 12.5% of the applied voltage, which is 3 V in this case. The same analogy applies to V_{CG}.

Each tap voltage is positive because the negative terminal of the voltage source is grounded.

Advantage of the Voltage Divider Method

Using Formula (7–1), we can find the proportional voltage drops from V_T and the series resistances without knowing the amount of I. For odd values of R, calculating the I may be more troublesome than finding the proportional voltages directly. Also, in many cases, we can approximate the voltage division without the need for any written calculations.

■ *7–1 Knowledge Check*

Answer at end of chapter.

An applied voltage, V_T, of 15 V is across two resistors R_1 and R_2 in series. If R_2 is four times the value of R_1, how much is V_1?

■ *7–1 Self-Review*

Answers at end of chapter.

Refer to Fig. 7–1 for a to c.
a. How much is R_T?
b. What fraction of the applied voltage is V_3?
c. If each resistance is doubled in value, how much is V_1?
d. In Fig. 7–2b, how much is the voltage V_{BG} if resistors R_2 and R_3 are interchanged?

7–2 Current Divider with Two Parallel Resistances

It is often necessary to find the individual branch currents in a bank from the resistances and I_T, but without knowing the voltage across the bank. This problem can be solved by using the fact that currents divide inversely as the branch resistances. An example is shown in the *current divider* in Fig. 7–3. The formulas for the two branch currents are as follows.

$$I_1 = \frac{R_2}{R_1 + R_2} \times I_T$$

$$I_2 = \frac{R_1}{R_1 + R_2} \times I_T \qquad (7\text{–}2)$$

Figure 7–3 Current divider with two branch resistances. Each branch I is inversely proportional to its R. The smaller R has more I.

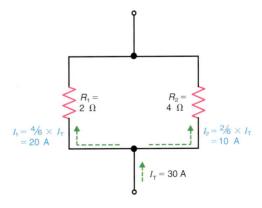

To use the calculator for a problem like this with current division between two branch resistances, as in Formula (7–2), there are several points to note. The numerator has the *R* of the branch opposite from the desired *I*. In adding R_1 and R_2, the parens keys Ⓘand Ⓘ should be used. The reason is that both terms in the denominator must be added before the division. The procedure for calculating I_1 in Fig. 7–3 can be as follows:

Punch in 4 for R_2.

Press the ⊕ key followed by the open parens key Ⓘ.

Punch in 2 ⊕ 4 for R_1 and R_2 followed by the close parens key Ⓘ. The sum of 6 will be displayed.

Press Ⓧ and 30, then press ⊜ to display the answer 20 for I_1.

Notice that the formula for each branch *I* has the opposite *R* in the numerator. The reason is that each branch current is inversely proportional to the branch resistance. The denominator is the same in both formulas, equal to the sum of the two branch resistances.

To calculate the currents in Fig. 7–3, with a 30-A I_T, a 2-Ω R_1, and a 4-Ω R_2,

$$I_1 = \frac{4}{2+4} \times 30$$

$$= \frac{4}{6} \times 30$$

$$I_1 = 20 \text{ A}$$

For the other branch,

$$I_2 = \frac{2}{2+4} \times 30$$

$$= \frac{2}{6} \times 30$$

$$I_2 = 10 \text{ A}$$

With all the resistances in the same units, the branch currents are in units of I_T. For instance, kilohms of *R* and milliamperes of *I* can be used.

Actually, it is not necessary to calculate both currents. After one *I* is calculated, the other can be found by subtracting from I_T.

Notice that the division of branch currents in a parallel bank is opposite from the voltage division of resistance in a series string. With series resistances, a higher resistance develops a higher *IR* voltage proportional to its *R*; with parallel branches, a lower resistance takes more branch current, equal to *V/R*.

In Fig. 7–3, the 20-A I_1 is double the 10-A I_2 because the 2-Ω R_1 is one-half the 4-Ω R_2. This is an inverse relationship between *I* and *R*.

The inverse relation between *I* and *R* in a parallel bank means that a very large *R* has little effect with a much smaller *R* in parallel. As an example, Fig. 7–4 shows a 999-kΩ R_2 in parallel with a 1-kΩ R_1 dividing the I_T of 1000 mA. The branch currents are calculated as follows:

$$I_1 = \frac{999}{1000} \times 1000 \text{ mA}$$

$$= 999 \text{ mA}$$

$$I_2 = \frac{1}{1000} \times 1000 \text{ mA}$$

$$= 1 \text{ mA}$$

The 999 mA for I_1 is almost the entire line current of 1000 mA because R_1 is so small compared with R_2. Also, the smallest branch *R* dissipates the most power because it has the most *I*.

The current divider Formula (7–2) can be used only for two branch resistances. The reason is the inverse relation between each branch *I* and its *R*. In comparison, the voltage divider Formula (7–1) can be used for any number of series resistances because of the direct proportion between each voltage drop *V* and its *R*.

For more branches, it is possible to combine the branches to work with only two divided currents at a time. However, a better method is to use parallel conductances, because *I* and *G* are directly proportional, as explained in the next section.

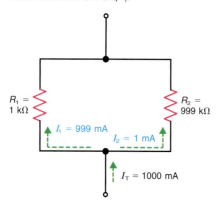

Figure 7–4 Example of a very large R_2 in parallel with a small R_1. For the branch currents, the small R_1 has almost the entire total line current, I_T.

■ **7-2 Knowledge Check**

Answers at end of chapter.

What are the values for I_1 and I_2 in Fig. 7–3 if $R_1 = 15\ \Omega$, $R_2 = 10\ \Omega$, and $I_T = 10$ A?

■ **7-2 Self-Review**

Answers at end of chapter.

Refer to Fig. 7–3.
a. What is the ratio of R_2 to R_1?
b. What is the ratio of I_2 to I_1?

7–3 Current Division by Parallel Conductances

Remember that the conductance G is $1/R$. Therefore, conductance and current are directly proportional. More conductance allows more current, for the same V. With any number of parallel branches, each branch current is

$$I = \frac{G}{G_T} \times I_T \tag{7–3}$$

where G is the conductance of one branch and G_T is the sum of all the parallel conductances. The unit for G is the siemens (S).

Note that Formula (7–3), for dividing branch currents in proportion to G, has the same form as Formula (7–1) for dividing series voltages in proportion to R. The reason is that both formulas specify a direct proportion.

Two Branches

As an example of using Formula (7–3), we can go back to Fig. 7–3 and find the branch currents with G instead of R. For the 2 Ω of R_1, the G_1 is $1/2 = 0.5$ S. The 4 Ω of R_2 has G_2 of $1/4 = 0.25$ S. Then G_T is $0.5 + 0.25 = 0.75$ S.

The I_T is 30A in Fig. 7–3. For the branch currents,

$$I_1 = \frac{G_1}{G_T} \times I_T = \frac{0.50}{0.75} \times 30 \text{ A}$$

$$I_1 = 20 \text{ A}$$

This 20 A is the same I_1 calculated before.

For the other branch, I_2 is $30 - 20 = 10$ A. Also, I_2 can be calculated as $0.25/0.75$ or $\frac{1}{3}$ of I_T for the same 10-A value.

Three Branches

A circuit with three branch currents is shown in Fig. 7–5. We can find G for the 10-Ω R_1, 2-Ω R_2, and 5-Ω R_3 as follows.

$$G_1 = \frac{1}{R_1} = \frac{1}{10\ \Omega} = 0.1 \text{ S}$$

$$G_2 = \frac{1}{R_2} = \frac{1}{2\ \Omega} = 0.5 \text{ S}$$

$$G_3 = \frac{1}{R_3} = \frac{1}{5\ \Omega} = 0.2 \text{ S}$$

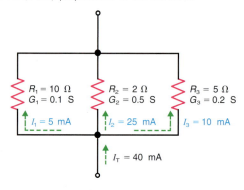

MultiSim **Figure 7–5** Current divider with branch conductances G_1, G_2, and G_3, each equal to $1/R$. Note that S is the siemens unit for conductance. With conductance values, each branch I is directly proportional to the branch G.

Remember that the siemens (S) unit is the reciprocal of the ohm (Ω) unit. The total conductance then is

$$G_T = G_1 + G_2 + G_3$$
$$= 0.1 + 0.5 + 0.2$$
$$G_T = 0.8 \text{ S}$$

The I_T is 40 mA in Fig. 7–5. To calculate the branch currents with Formula (7–3),

$$I_1 = 0.1/0.8 \times 40 \text{ mA} = 5 \text{ mA}$$
$$I_2 = 0.5/0.8 \times 40 \text{ mA} = 25 \text{ mA}$$
$$I_3 = 0.2/0.8 \times 40 \text{ mA} = 10 \text{ mA}$$

The sum is $5 + 25 + 10 = 40$ mA for I_T.

Although three branches are shown here, Formula (7–3) can be used to find the currents for any number of parallel conductances because of the direct proportion between I and G. The method of conductances is usually easier to use than the method of resistances for three or more branches.

■ 7–3 Knowledge Check

Answers at end of chapter.

In Fig. 7–5, assume $G_1 = 10$ mS, $G_2 = 6.67$ mS, $G_3 = 33.33$ mS, and $I_T = 150$ mA. Calculate I_1, I_2, and I_3.

■ 7–3 Self-Review

Answers at end of chapter.

Refer to Fig. 7–5.
a. What is the ratio of G_3 to G_1?
b. What is the ratio of I_3 to I_1?

7–4 Series Voltage Divider with Parallel Load Current

The voltage dividers shown so far illustrate just a series string without any branch currents. However, a voltage divider is often used to tap off part of the applied voltage for a load that needs less voltage than V_T. Then the added load is a parallel branch across part of the divider, as shown in Fig. 7–6. This example

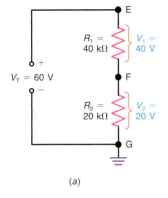

(a)

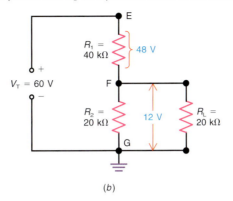

(b)

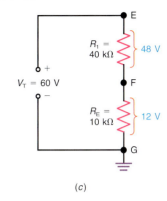

(c)

shows how the **loaded voltage** at the tap F is reduced below the potential it would have without the branch current for R_L.

Why the Loaded Voltage Decreases

We can start with Fig. 7–6a, which shows an R_1–R_2 voltage divider alone. Resistances R_1 and R_2 in series simply form a proportional divider across the 60-V source for V_T.

For the resistances, R_1 is 40 kΩ and R_2 is 20 kΩ, making R_T equal to 60 kΩ. Also, the current $I = V_T/R_T$, or 60 V/60 kΩ = 1 mA. For the divided voltages in Fig. 7–6a,

$$V_1 = \frac{40}{60} \times 60 \text{ V} = 40 \text{ V}$$

$$V_2 = \frac{20}{60} \times 60 \text{ V} = 20 \text{ V}$$

Note that $V_1 + V_2$ is 40 + 20 = 60 V, which is the total applied voltage.

However, in Fig. 7–6b, the 20-kΩ branch of R_L changes the equivalent resistance at tap F to ground. This change in the proportions of R changes the voltage division. Now the resistance from F to G is 10 kΩ, equal to the 20-kΩ R_2 and R_L in parallel. This equivalent bank resistance is shown as the 10-kΩ R_E in Fig. 7–6c.

Resistance R_1 is still the same 40 kΩ because it has no parallel branch. The new R_T for the divider in Fig. 7–6c is 40 kΩ + 10 kΩ = 50 kΩ. As a result, V_E from F to G in Fig. 7–6c becomes

$$V_E = \frac{R_E}{R_T} \times V_T = \frac{10}{50} \times 60 \text{ V}$$

$$V_E = 12 \text{ V}$$

Therefore, the voltage across the parallel R_2 and R_L in Fig. 7–6b is reduced to 12 V. This voltage is at the tap F for R_L.

Note that V_1 across R_1 increases to 48 V in Fig. 7–6c. Now V_1 is 40/50 × 60 V = 48 V. The V_1 increases here because there is more current through R_1.

The sum of $V_1 + V_E$ in Fig. 7–6c is 12 + 48 = 60 V. The IR drops still add to equal the applied voltage.

Path of Current for R_L

All current in the circuit must come from the source V_T. Trace the electron flow for R_L. It starts from the negative side of V_T, through R_L, to the tap at F, and returns through R_1 in the divider to the positive side of V_T. This current I_L goes through R_1 but not R_2.

Bleeder Current

In addition, both R_1 and R_2 have their own current from the source. This current through all the resistances in the divider is called the **bleeder current** I_B. The electron flow for I_B is from the negative side of V_T, through R_2 and R_1, and back to the positive side of V_T.

In summary, then, for the three resistances in Fig. 7–6b, note the following currents:

1. Resistance R_L has just its load current I_L.
2. Resistance R_2 has only the bleeder current I_B.
3. Resistance R_1 has both I_L and I_B.

Note that only R_1 is in the path for both the bleeder current and the load current.

■ 7–4 Knowledge Check

Answer at end of chapter.

In Fig. 7–6b, how much is the voltage, V_{FG}, if R_L equals 40 kΩ?

■ 7–4 Self-Review

Answers at end of chapter.

Refer to Fig. 7–6.
a. What is the proportion of R_2/R_T in Fig. 7–6a?
b. What is the proportion of R_E/R_T in Fig. 7–6c?

7–5 Design of a Loaded Voltage Divider

These principles can be applied to the design of a practical voltage divider, as shown in Fig. 7–7. This type of circuit is used for the output of a power supply in electronic equipment to supply different voltages at the taps, with different *load currents*. For instance, load D can represent the collector-emitter circuit for one or more power transistors that need +100 V for the collector supply. The tap at E can also be the 40-V collector supply for medium-power transistors. Finally, the 18-V tap at F can be for base-emitter bias current in the power transistors and collector voltage for smaller transistors.

Note the load specifications in Fig. 7–7. Load F needs 18 V from point F to chassis ground. When the 18 V is supplied by this part of the divider, a 36-mA branch current will flow through the load. Similarly, 40 V is needed at tap E for 54 mA of I_E in load E. Also, 100 V is available at D with a load current I_D of 180 mA. The total load current here is $36 + 54 + 180 = 270$ mA.

In addition, the bleeder current I_B through the entire divider is generally specified at about 10% of the load current. For the example here, I_B is taken as 30 mA to make a total line current I_T of $270 + 30 = 300$ mA from the source. Remember that the 30-mA I_B flows through R_1, R_2, and R_3.

Figure 7-7 Voltage divider for different voltages and currents from the source V_T. See text for design calculations to find the values of R_1, R_2, and R_3.

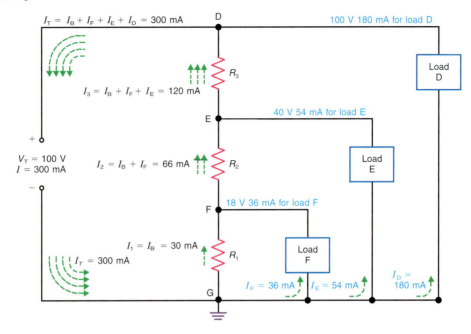

The design problem in Fig. 7–7 is to find the values of R_1, R_2, and R_3 needed to provide the specified voltages. Each R is calculated as its ratio of V/I. However, the question is what are the correct values of V and I to use for each part of the divider.

Find the Current in Each *R*

We start with R_1 because its current is only the 30-mA bleeder current I_B. No load current flows through R_1. Therefore I_1 through R_1 equals 30 mA.

The 36-mA current I_F for load F returns to the source through R_2 and R_3. Considering just R_2 now, its current is the I_F load current and the 30-mA bleeder current I_B. Therefore, I_2 through R_2 is 36 + 30 = 66 mA.

The 54-mA current I_E for load E returns to the source through R_3 alone. However, R_3 also has the 36-mA I_F and the 30-mA I_B. Therefore I_3 through R_3 is 54 + 36 + 30 = 120 mA. The values for I_1, I_2, and I_3 are given in Table 7–1.

Table 7-1	Design Values for Voltage Divider		
For Figure 7-7	Current, mA	Voltage, V	Resistance, Ω
R_1	30	18	600
R_2	66	22	333
R_3	120	60	500

Note that the load current I_D for load D at the top of the diagram does not flow through R_3 or any of the resistors in the divider. However, the I_D of 180 mA is the main load current through the source of applied voltage. The 120 mA of bleeder and load currents plus the 180-mA I_D load add to equal 300 mA for I_T in the main line of the power supply.

Calculate the Voltage across Each *R*

The voltages at the taps in Fig. 7–7 give the potential to chassis ground. But we need the voltage across the two ends of each *R*. For R_1, the voltage V_1 is the indicated 18 V to ground because one end of R_1 is grounded. However, the voltage across R_2 is the difference between the 40-V potential at point E and the 18 V at F. Therefore, V_2 is $40 - 18 = 22$ V. Similarly, V_3 is calculated as 100 V at point D minus the 40 V at E, or, V_3 is $100 - 40 = 60$ V. These values for V_1, V_2, and V_3 are summarized in Table 7–1.

Calculating Each *R*

Now we can calculate the resistance of R_1, R_2, and R_3 as each *V/I* ratio. For the values listed in Table 7–1,

$$R_1 = \frac{V_1}{I_1} = \frac{18\ V}{30\ mA} = 0.6\ k\Omega = 600\ \Omega$$

$$R_2 = \frac{V_2}{I_2} = \frac{22\ V}{66\ mA} = 0.333\ k\Omega = 333\ \Omega$$

$$R_3 = \frac{V_3}{I_3} = \frac{60\ V}{120\ mA} = 0.5\ k\Omega = 500\ \Omega$$

When these values are used for R_1, R_2, and R_3 and connected in a voltage divider across the source of 100 V, as in Fig. 7–7, each load will have the specified voltage at its rated current.

■ 7–5 Knowledge Check

Answers at end of chapter.

In Fig. 7–7, recalculate the required values of R_1, R_2, and R_3 if load E requires 34 mA at 40 V. (I_B is still 30 mA.)

■ 7–5 Self-Review

Answers at end of chapter.

Refer to Fig. 7–7.
a. How much is the bleeder current I_B through R_1, R_2, and R_3?
b. How much is the voltage for load E at tap E to ground?
c. How much is V_2 across R_2?
d. If load D opens, how much voltage will be measured at tap F to ground?

Summary

- In a series circuit, V_T is divided into IR voltage drops proportional to the resistances. Each $V_R = (R/R_T) \times V_T$, for any number of series resistances. The largest series R has the largest voltage drop.

- In a parallel circuit, I_T is divided into branch currents. Each I is inversely proportional to the branch R. The inverse division of branch currents is given by Formula (7–2) for only two resistances. The smaller branch R has the larger branch current.

- For any number of parallel branches, I_T is divided into branch currents directly proportional to each conductance G. Each $I = (G/G_T) \times I_T$.

- A series voltage divider is often tapped for a parallel load, as in Fig. 7–6. Then the voltage at the tap is reduced because of the load current.

- The design of a loaded voltage divider, as in Fig. 7–7, involves calculating each R. Find the I and potential difference V for each R. Then $R = V/I$.

Important Terms

- Bleeder Current - the current that flows through all resistors in a loaded voltage divider. The bleeder current, designated I_B, is generally specified at about 10% of the total load current.

- Current Divider - any parallel circuit is a current divider in which the individual branch currents are inversely proportional to the branch resistance values. With respect to conductances, the individual branch currents are directly proportional to the branch conductance values.

- Load Currents - the currents drawn by the electronic devices and/or components connected as loads in a loaded voltage divider.

- Loaded Voltage - the voltage at a point in a series voltage divider where a parallel load has been connected.

- Voltage Divider - any series circuit is a voltage divider in which the individual resistor voltage drops are proportional to the series resistance values.

- Voltage Taps - the points in a series voltage divider that provide different voltages with respect to ground.

Related Formulas

$$V = \frac{R}{R_T} \times V_T$$

For two resistors in parallel: $I_1 = \dfrac{R_2}{R_1 + R_2} \times I_T$ $\qquad I_2 = \dfrac{R_1}{R_1 + R_2} \times I_T$

$$I = \frac{G}{G_T} \times I_T$$

Self-Test

Answers at back of book.

1. **In a series circuit, the individual resistor voltage drops are**
   a. inversely proportional to the series resistance values.
   b. proportional to the series resistance values.
   c. unrelated to the series resistance values.
   d. none of the above

2. **In a parallel circuit, the individual branch currents are**
   a. not related to the branch resistance values.
   b. directly proportional to the branch resistance values.
   c. inversely proportional to the branch resistance values.
   d. none of the above

3. **Three resistors R_1, R_2, and R_3 are connected in series across an applied voltage, V_T, of 24 V. If R_2 is one-third the value of R_T, how much is V_2?**
   a. 8 V
   b. 16 V
   c. 4 V
   d. This is impossible to determine.

4. Two resistors R_1 and R_2 are in parallel. If R_1 is twice the value of R_2, how much is I_2 in R_2 if I_T equals 6 A?

   a. 1 A

   b. 2 A

   c. 3 A

   d. 4 A

5. Two resistors R_1 and R_2 are in parallel. If the conductance, G_1, of R_1 is twice the value of the conductance, G_2 of R_2, how much is I_2 if $I_T = 6$ A?

   a. 1 A

   b. 2 A

   c. 3 A

   d. 4 A

PROBLEMS 6–10 REFER TO FIGURE 7–8.

6. In Fig. 7–8, how much is I_1 in R_1?

   a. 400 mA

   b. 300 mA

   c. 100 mA

   d. 500 mA

Figure 7–8

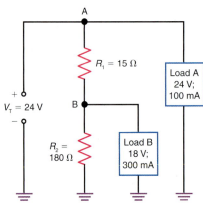

7. In Fig. 7–8, how much is the bleeder current, I_B?

   a. 500 mA

   b. 400 mA

   c. 100 mA

   d. 300 mA

8. In Fig. 7–8, how much is the total current, I_T?

   a. 500 mA

   b. 400 mA

   c. 100 mA

   d. 300 mA

9. In Fig. 7–8, what is the voltage, V_{BG}, if Load B becomes open?

   a. 18 V

   b. 19.2 V

   c. 6 V

   d. 22.15 V

10. In Fig. 7–8, what happens to the voltage, V_{BG}, if Load A becomes open?

    a. It increases.

    b. It decreases.

    c. It remains the same.

    d. This cannot be determined.

Questions

1. Define *series voltage divider.*

2. Define *parallel current divider.*

3. Give two differences between a series voltage divider and a parallel current divider.

4. Give three differences between Formula (7–2) for branch resistances and Formula (7–3) for branch conductances.

5. Define *bleeder current.*

6. What is the main difference between the circuits in Fig. 7–6a and b?

7. Referring to Fig. 7–1, why is V_1 series-aiding with V_2 and V_3 but in series opposition to V_T? Show the polarity of each IR drop.

8. Show the derivation of Formula (7–2) for each branch current in a parallel bank of two resistances. [Hint: The voltage across the bank is $I_T \times R_{EQ}$ and R_{EQ} is $R_1 R_2/(R_1 + R_2)$.]

Problems

SECTION 7–1 SERIES VOLTAGE DIVIDERS

7–1 A 100 Ω R_1 is in series with a 200-Ω R_2 and a 300-Ω R_3. The applied voltage, V_T, is 18 V. Calculate V_1, V_2, and V_3.

7–2 A 10-kΩ R_1 is in series with a 12-kΩ R_2, a 4.7-kΩ R_3, and a 3.3-kΩ R_4. The applied voltage, V_T, is 36 V. Calculate V_1, V_2, V_3, and V_4.

7–3 **MultiSim** In Fig. 7–9, calculate V_1, V_2, and V_3.

Figure 7–9

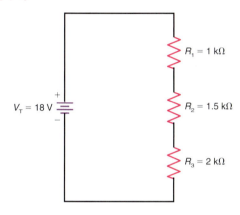

7–4 **MultiSim** In Fig. 7–9, recalculate V_1, V_2, and V_3 if $R_1 = 10\ \Omega$, $R_2 = 12\ \Omega$, $R_3 = 18\ \Omega$, and $V_T = 20$ V.

7–5 In Fig. 7–10, calculate V_1, V_2, and V_3. Note that resistor R_2 is three times the value of R_1 and resistor R_3 is two times the value of R_2.

Figure 7–10

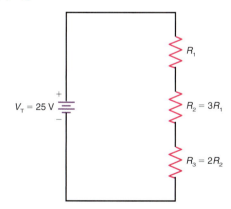

7–6 In Fig. 7–11, calculate
   a. V_1, V_2, and V_3.
   b. V_{AG}, V_{BG}, and V_{CG}.

Figure 7–11

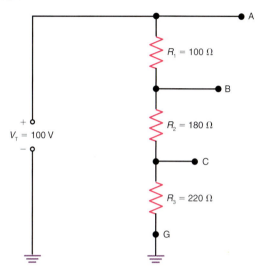

7–7 In Fig. 7–11, change R_1, R_2, R_3, and the applied voltage, V_T, to the following values: $R_1 = 9$ kΩ, $R_2 = 900\ \Omega$, $R_3 = 100\ \Omega$, and $V_T = 10$ V. Then, recalculate
   a. V_1, V_2, and V_3.
   b. V_{AG}, V_{BG}, and V_{CG}.

7–8 In Fig. 7–12, solve for
   a. V_1, V_2, V_3, and V_4.
   b. V_{AG}, V_{BG}, V_{CG}, and V_{DG}.

Figure 7–12

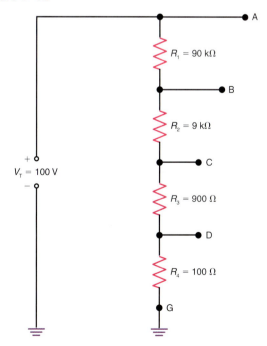

7-9 In Fig. 7–13, solve for
   a. V_1, V_2, V_3, and V_4.
   b. V_{AG}, V_{BG}, V_{CG}, and V_{DG}.

Figure 7–13

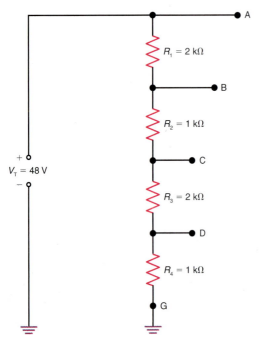

7-10 In Fig. 7–14, solve for
   a. V_1, V_2, V_3, V_4, and V_5.
   b. V_{AG}, V_{BG}, V_{CG}, and V_{DG}.

Figure 7–14

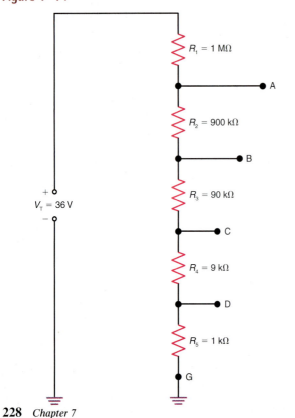

SECTION 7–2 CURRENT DIVIDER WITH TWO PARALLEL RESISTANCES

7-11 In Fig. 7–15, solve for I_1 and I_2.

Figure 7–15

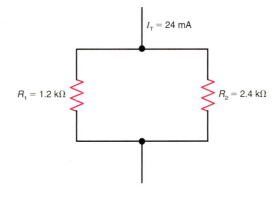

7-12 **MultiSim** In Fig. 7–16, solve for I_1 and I_2.

Figure 7–16

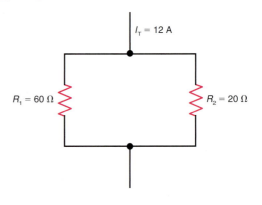

7-13 In Fig. 7–17, solve for I_1 and I_2.

Figure 7–17

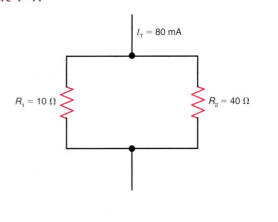

7–14 ‖ **MultiSim** In Fig. 7–18, solve for I_1 and I_2.

Figure 7–18

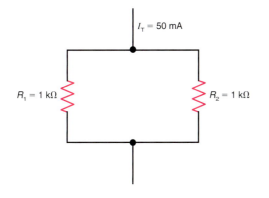

7–15 In Fig. 7–19, solve for I_1 and I_2.

Figure 7–19

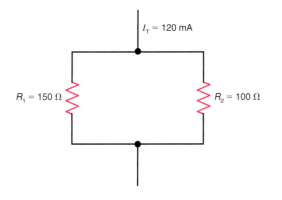

SECTION 7–3 CURRENT DIVISION BY PARALLEL CONDUCTANCES

7–16 In Fig. 7–20, solve for I_1, I_2, and I_3.

Figure 7–20

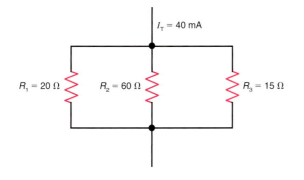

7–17 In Fig. 7–21, solve for I_1, I_2, and I_3.

Figure 7–21

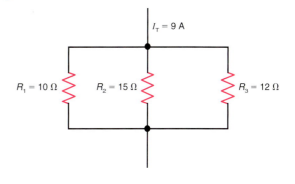

7–18 In Fig. 7–22, solve for I_1, I_2, and I_3.

Figure 7–22

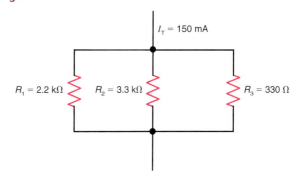

7–19 In Fig. 7–23, solve for I_1, I_2, and I_3.

Figure 7–23

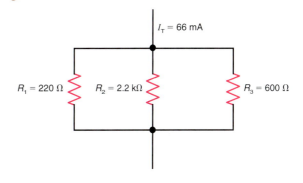

7–20 In Fig. 7–24, solve for I_1, I_2, I_3, and I_4.

Figure 7–24

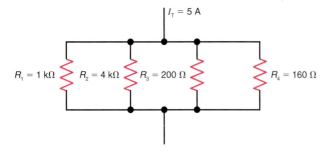

7–21 In Fig. 7–25, solve for I_1, I_2, I_3, and I_4.

Figure 7–25

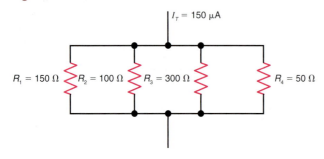

SECTION 7–4 SERIES VOLTAGE DIVIDER WITH PARALLEL LOAD CURRENT

7–22 In Fig. 7–26, calculate I_1, I_2, I_L, V_{BG}, and V_{AG} with
  a. S_1 open.
  b. S_1 closed.

Figure 7–26

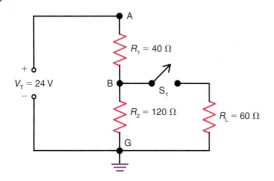

7–23 In Fig. 7–26, explain why the voltage, V_{BG}, decreases when the switch, S1 is closed.

7–24 In Fig. 7–27, calculate I_1, I_2, I_L, V_{BG}, and V_{AG} with
  a. S_1 open.
  b. S_1 closed.

Figure 7–27

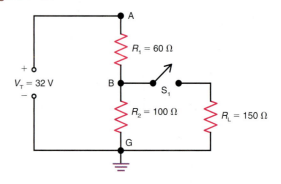

7–25 With S_1 closed in Fig. 7–27, which resistor has only the bleeder current, I_B, flowing through it?

7–26 In Fig. 7–28, calculate I_1, I_2, I_L, V_{BG}, and V_{AG} with
  a. S_1 open.
  b. S_1 closed.

Figure 7–28

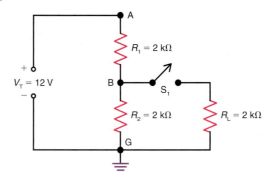

SECTION 7–5 DESIGN OF A LOADED VOLTAGE DIVIDER

7–27 If the bleeder current, I_B, is 10% of the total load current in Fig. 7–29, solve for the following:
  a. I_1, I_2, I_3, and I_T
  b. V_1, V_2, and V_3
  c. R_1, R_2, and R_3
  d. The power dissipated by R_1, R_2, and R_3.

Figure 7–29

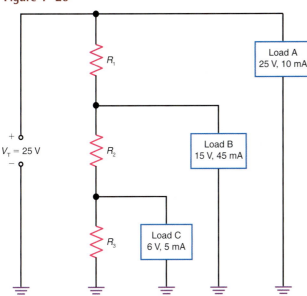

7–28 If the bleeder current, I_B, is 10% of the total load current in Fig. 7–30, solve for the following:

   a. I_1, I_2, I_3, and I_T
   b. V_1, V_2, and V_3
   c. R_1, R_2, and R_3
   d. The power dissipated by R_1, R_2, and R_3.

Figure 7–30

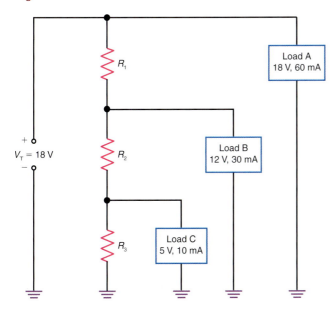

7–29 If the bleeder current, I_B, is 6 mA in Fig. 7–31, solve for the following:

   a. I_1, I_2, I_3, and I_T
   b. V_1, V_2, and V_3
   c. R_1, R_2, and R_3
   d. The power dissipated by R_1, R_2, and R_3.

Figure 7–31

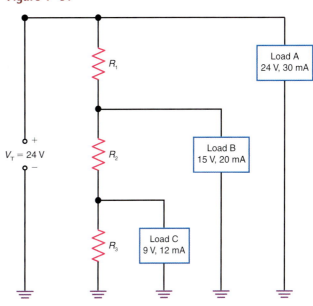

7–30 If the bleeder current, I_B, is 15mA in Fig. 7–32, solve for the following:

   a. I_1, I_2, I_3, and I_T
   b. V_1, V_2, and V_3
   c. R_1, R_2, and R_3
   d. The power dissipated by R_1, R_2, and R_3.

Figure 7–32

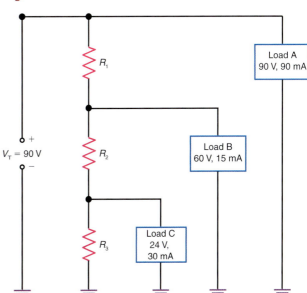

Critical Thinking

7–31 Refer to Fig. 7–33. Select values for R_1 and R_3 that will allow the output voltage to vary between 6 V and 15 V.

Figure 7–33

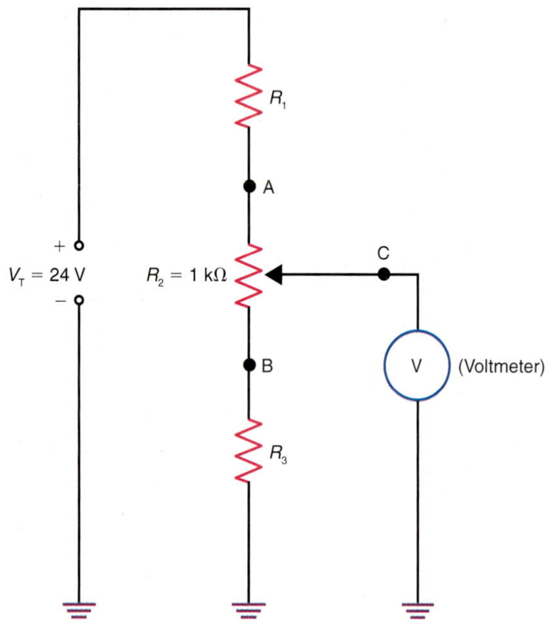

7–32 Design a loaded voltage divider, using a 25-V supply, to meet the following requirements: load A = 25 V, 10 mA; load B = 15 V, 45 mA; load C = 6 V, 5 mA; I_B = 10% of total load current. Draw the schematic diagram including all values.

7–33 Design a loaded voltage divider, using a 24-V supply, to meet the following requirements: load A = 18 V, 10 mA; load B = 12 V, 30 mA; load C = 5 V, 6 mA; I_B = 10% of total load current. Draw the schematic diagram including all values.

7–34 Design a loaded voltage divider, using a 25-V supply, to meet the following requirements: load A = 20 V, 25 mA; load B = 12 V, 10 mA; load C = -5 V, 10 mA; total current I_T = 40 mA. Draw the schematic diagram including all values.

Troubleshooting Challenge

Table 7–2 shows voltage measurements taken in Fig. 7–34. The first row shows the normal values when the circuit is operating properly. Rows 2 to 9 are voltage measurements taken when one component in the circuit has failed. For each row, identify which component is defective and determine the type of defect that has occurred in the component.

Table 7–3 shows voltage measurements taken in Fig. 7–35. The first row shows the normal values when the circuit is operating properly. Rows 2 to 9 are voltage measurements taken when one component in the circuit has failed. For each row, identify which component is defective and determine the type of defect that has occurred in the component.

Figure 7–34 Series voltage divider for Troubleshooting Challenge.

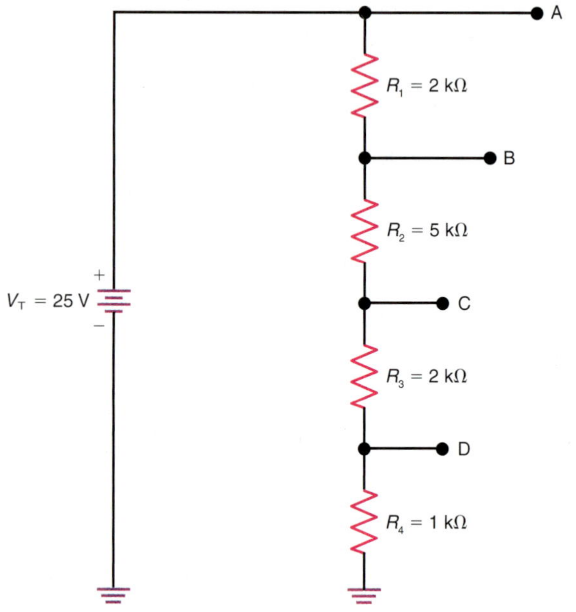

Table 7–2	Voltage Measurements Taken in Fig. 7–34				
	V_{AG}	V_{BG}	V_{CG}	V_{DG}	Defective Component
	VOLTS				
1 Normal values	25	20	7.5	2.5	None
2 Trouble 1	25	25	0	0	
3 Trouble 2	25	0	0	0	
4 Trouble 3	25	18.75	3.125	3.125	
5 Trouble 4	25	25	25	25	
6 Trouble 5	25	15	15	5	
7 Trouble 6	25	25	9.375	3.125	
8 Trouble 7	25	25	25	0	
9 Trouble 8	25	19.4	5.56	0	

Table 7–3	Voltage Measurements Taken in Fig. 7–35					
	V_A	V_B	V_C	V_D	Comments	Defective Component
	VOLTS					
1 Normal values	36	24	15	6	—	None
2 Trouble 1	36	32	0	0	—	
3 Trouble 2	36	16.94	0	0	R_2 Warm	
4 Trouble 3	36	25.52	18.23	0	—	
5 Trouble 4	36	0	0	0	R_1 Hot	
6 Trouble 5	36	0	0	0	—	
7 Trouble 6	36	27.87	23.23	9.29	—	
8 Trouble 7	36	23.25	13.4	0	—	
9 Trouble 8	36	36	22.5	9	—	

Figure 7–35 Loaded voltage divider for Troubleshooting Challenge.

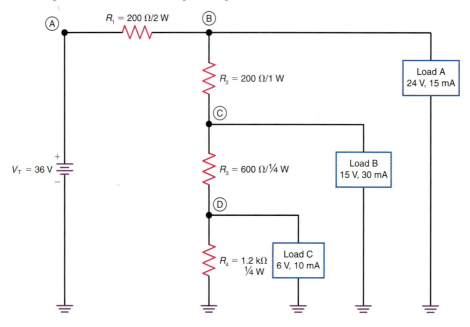

Answers to Knowledge Check Problems

7–1 3 V

7–2 $I_1 = 4$ A
$I_2 = 6$ A

7–3 $I_1 = 30$ mA
$I_2 = 20$ mA
$I_3 = 100$ mA

7–4 $V_{FG} = 15$ V

7–5 $R_1 = 600 \ \Omega$
$R_2 = 333 \ \Omega$
$R_3 = 600 \ \Omega$

Answers to Self-Reviews

7–1 **a.** $R_T = 100$ kΩ
b. $V_3 = (^2/_{10}) \times V_T$
c. $V_1 = 100$ V
d. $V_{BG} = 10.2$ V

7–2 **a.** 2 to 1
b. 1 to 2

7–3 **a.** 2 to 1
b. 2 to 1

7–4 **a.** $^1/_3$
b. $^1/_5$

7–5 **a.** $I_B = 30$ mA
b. $V_E = 40$ V
c. $V_2 = 22$ V
d. 18 V

8

Analog and Digital Multimeters

The digital multimeter (DMM) is the most common measuring instrument used by electronic technicians. All DMMs can measure voltage, current, and resistance, and some can even measure and test electronic components such as capacitors, diodes, and transistors. A DMM uses a numeric display to indicate the value of the measured quantity.

An analog multimeter uses a moving pointer and a printed scale. Like a DMM, an analog multimeter can measure voltage, current, and resistance. One disadvantage of an analog multimeter, however, is that the meter reading must be interpreted based on where the moving pointer rests along the printed scale. Although analog multimeters find somewhat limited use in electronics, there is great value in understanding their basic construction and operation. One of the main reasons for covering analog meters is that the concepts of series, parallel, and series-parallel circuits learned earlier are applied. In this chapter, therefore, you will be provided with a basic overview of the construction and operation of an analog meter as well as a concept called voltmeter loading. You will also learn about the main features of a DMM.

Objectives

After studying this chapter you should be able to

- *Explain* the difference between analog and digital meters.

- *Explain* the construction and operation of a moving-coil meter.

- *Calculate* the value of shunt resistance required to extend the current range of a basic moving-coil meter.

- *Calculate* the value of multiplier resistance required to make a basic moving-coil meter capable of measuring voltage.

- *Explain* the ohms-per-volt rating of a voltmeter.

- *Explain* what is meant by *voltmeter loading*.

- *Explain* how a basic moving-coil meter can be used with a battery to construct an ohmmeter.

- *List* the main features of a digital multimeter (DMM).

Outline

Important Terms

amp-clamp probe

analog multimeter

back-off ohmmeter scale

continuity testing

digital multimeter (DMM)

loading effect

multiplier resistor

ohms-per-volt (Ω/V) rating

shunt resistor

zero-ohms adjustment

8–1 Moving-Coil Meter

Figure 8–1 shows two types of multimeters used by electronic technicians. Figure 8–1*a* shows an analog volt-ohm-milliammeter (VOM), and Fig. 8–1*b* shows a digital multimeter (DMM). Both types are capable of measuring voltage, current, and resistance.

A moving-coil meter movement, shown in Fig. 8–2, is generally used in an analog VOM. The construction consists of a coil of fine wire wound on a drum mounted between the poles of a permanent magnet. When direct current flows in the coil, the magnetic field of the current reacts with the magnetic field of the permanent magnet.* The resultant force turns the drum with its pointer, winding up the restoring spring. When the current is removed, the pointer returns to zero. The amount of deflection indicates the amount of current in the coil. When the polarity is connected correctly, the pointer will read up-scale, to the right; the incorrect polarity forces the pointer off-scale, to the left. (It is interesting to note that the moving-coil arrangement is often called a D'Arsonval movement, after its inventor who patented this meter movement in 1881.)

The pointer deflection is directly proportional to the amount of current in the coil. If 100 μA is the current needed for full-scale deflection, 50 μA in the coil will produce half-scale deflection. The accuracy of the moving-coil meter mechanism is 0.1–2%.

Values of I_M

The full-scale deflection current I_M is the amount needed to deflect the pointer all the way to the right to the last mark on the printed scale. Typical values of I_M are from about 10 μA to 30 mA. In a VOM, the I_M is typically either 50 μA or 1 mA.

Figure 8–1 Typical multimeters used for measuring *V*, *I*, and *R*. (*a*) Analog VOM. (*b*) DMM.

(*a*)

(*b*)

---

* For more details on the interaction between two magnetic fields, see Chapter 14, Electromagnetic Induction.

Figure 8–2 Construction of a moving-coil meter. The diameter of the coil can be $\frac{1}{2}$ to 1 in.

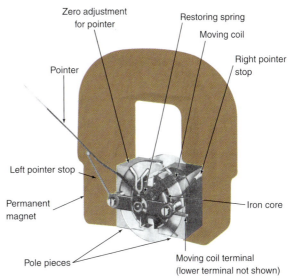

Refer to the analog VOM in Fig. 8–1a. The mirror along the scale is used to eliminate reading errors. The meter is read when the pointer and its mirror reflection appear as one. This eliminates the optical error called *parallax* caused by looking at the meter from the side.

Values of r_M

This is the internal resistance of the wire of the moving coil. Typical values range from 1.2 Ω for a 30-mA movement to 2000 Ω for a 50-μA movement. A movement with a smaller I_M has a higher r_M because many more turns of fine wire are needed. An average value of r_M for a 1-mA movement is about 50 Ω. Figure 8–3

Figure 8–3 Close-up view of a D'Arsonval moving-coil meter movement.

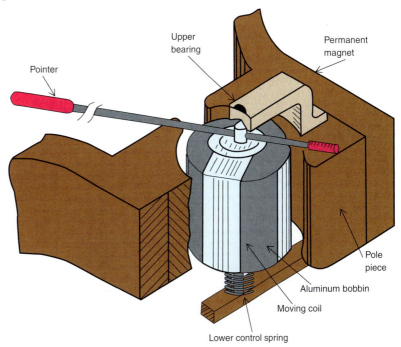

provides a close-up view of the basic components contained within a D'Arsonval meter movement. Notice that the moving coil is wound around a drum which rotates when direct current flows through the wire of the moving coil.

■ 8–1 Knowledge Check

Answers at end of chapter.

a. A moving-coil meter has an I_M value of 1 mA. How much current will produce one-quarter scale deflection?
b. What is the purpose of having a mirror along the scale of an analog VOM?

■ 8–1 Self-Review

Answers at end of chapter.

a. A D'Arsonval movement has an I_M value of 1 mA. How much is the deflection of the meter pointer if the current in the moving coil is 0.5 mA?
b. How much is the deflection in part a if the current in the moving coil is zero?

8–2 Meter Shunts

A *meter shunt* is a precision resistor connected across the meter movement for the purpose of shunting, or bypassing, a specific fraction of the circuit's current around the meter movement. The combination then provides a current meter with an extended range. The shunts are usually inside the meter case. However, the schematic symbol for the current meter usually does not show the shunt.

In current measurements, the parallel bank of the movement with its shunt is connected as a current meter in series in the circuit (Fig. 8–4). Note that the scale of a meter with an internal shunt is calibrated to take into account the current through both the shunt and the meter movement. Therefore, the scale reads total circuit current.

IIII MultiSim **Figure 8–4** Example of meter shunt R_S in bypassing current around the movement to extend the range from 1 to 2 mA. (*a*) Wiring diagram. (*b*) Schematic diagram showing the effect of the shunt. With $R_S = r_M$ the current range is doubled. (*c*) Circuit with 2-mA meter to read the current.

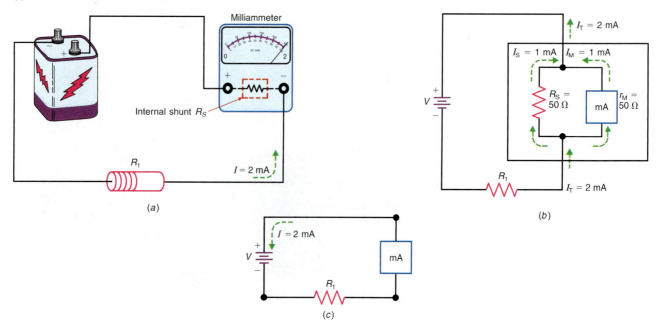

Resistance of the Meter Shunt

In Fig. 8–4b, the 1-mA movement has a resistance of 50 Ω, which is the resistance of the moving coil r_M. To double the range, the shunt resistance R_S is made equal to the 50 Ω of the movement. When the meter is connected in series in a circuit where the current is 2 mA, this total current into one terminal of the meter divides equally between the shunt and the meter movement. At the opposite meter terminal, these two branch currents combine to provide the 2 mA of the circuit current.

Inside the meter, the current is 1 mA through the shunt and 1 mA through the moving coil. Since it is a 1-mA movement, this current produces full-scale deflection. The scale is doubled, however, reading 2 mA, to account for the additional 1 mA through the shunt. Therefore, the scale reading indicates total current at the meter terminals, not just coil current. The movement with its shunt, then, is a 2-mA meter. Its internal resistance is $50 \times \frac{1}{2} = 25$ Ω.

Another example is shown in Fig. 8–5. In general, the shunt resistance for any range can be calculated with Ohm's law from the formula

$$R_S = \frac{V_M}{I_S} \tag{8–1}$$

where R_S is the resistance of the shunt and I_S is the current through it.

Voltage V_M is equal to $I_M \times r_M$. This is the voltage across both the shunt and the meter movement, which are in parallel.

Calculating I_S

This current through the shunt alone is the difference between the total current I_T through the meter and the divided current I_M through the movement or

$$I_S = I_T - I_M \tag{8–2}$$

Use the values of current for full-scale deflection, as these are known. In Fig. 8–5,

$$I_S = 5 - 1 = 4 \text{ mA}, \quad \text{or} \quad 0.004 \text{ A}$$

Calculating R_S

The complete procedure for using the formula $R_S = V_M/I_S$ can be as follows:

1. Find V_M. Calculate this for full-scale deflection as $I_M \times r_M$. In Fig. 8–5, with a 1-mA full-scale current through the 50-Ω movement,

$$V_M = 0.001 \times 50 = 0.05 \text{ V} \quad \text{or} \quad 50 \text{ mV}$$

2. Find I_S. For the values that are shown in Fig. 8–5,

$$I_S = 5 - 1 = 4 \text{ mA} = 0.004 \text{ A} \quad \text{or} \quad 4 \text{ mA}$$

Figure 8–5 Calculating the resistance of a meter shunt. R_S is equal to V_M/I_S. See text for calculations.

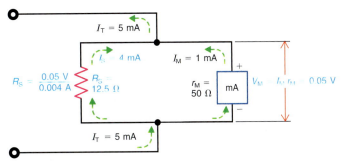

GOOD TO KNOW

When the insertion of a current meter reduces the circuit current below that which exists without the meter present, the effect is called current meter loading.

3. Divide V_M by I_S to find R_S. For the final result,

$$R_S = 0.05/0.004 = 12.5 \ \Omega$$

This shunt enables the 1-mA movement to be used for an extended range from 0–5 mA.

Note that R_S and r_M are inversely proportional to their full-scale currents. The 12.5 Ω for R_S equals one-fourth the 50 Ω of r_M because the shunt current of 4 mA is four times the 1 mA through the movement for full-scale deflection.

The shunts usually are precision wire-wound resistors. For very low values, a short wire of precise size can be used.

Since the moving-coil resistance, r_M, is in parallel with the shunt resistance, R_S, the resistance of a current meter can be calculated as $R_M = \dfrac{R_S \times r_M}{R_S + r_M}$. In general, a current meter should have very low resistance compared with the circuit where the current is being measured. As a general rule, the current meter's resistance should be no greater than $1/100$ of the circuit resistance. The higher the current range of a meter, the lower its shunt resistance, R_S, and in turn the overall meter resistance, R_M.

Example 8-1

A shunt extends the range of a 50-μA meter movement to 1 mA. How much is the current through the shunt at full-scale deflection?

ANSWER All currents must be in the same units for Formula (8–2). To avoid fractions, use 1000 μA for the 1-mA I_T. Then

$$I_S = I_T - I_M = 1000 \ \mu A - 50 \ \mu A$$
$$I_S = 950 \ \mu A$$

Example 8-2

A 50-μA meter movement has an r_M of 1000 Ω. What R_S is needed to extend the range to 500 μA?

ANSWER The shunt current I_S is $500 - 50$, or 450 μA. Then

$$R_S = \frac{V_M}{I_S}$$
$$= \frac{50 \times 10^{-6} \ A \times 10^3 \ \Omega}{450 \times 10^{-6} \ A} = \frac{50{,}000}{450}$$
$$R_S = 111.1 \ \Omega$$

■ 8–2A Knowledge Check

Answer at end of chapter.

A current meter should have very (low/high) internal resistance.

■ 8–2B Knowledge Check

Answer at end of chapter.

What shunt resistance, R_S, is needed to extend the range of a 1-mA, 50-Ω meter movement to a meter capable of measuring currents from 0–25 mA?

■ 8–2 Self-Review

Answers at end of chapter.

A 50-μA movement with a 900-Ω r_M has a shunt R_S for the range of 500 μA.
   a. How much is I_S?
   b. How much is V_M?
   c. What is the size of R_S?

8–3 Voltmeters

Although a meter movement responds only to current in the moving coil, it is commonly used for measuring voltage by the addition of a high resistance in series with the movement (Fig. 8–6). The series resistance must be much higher than the coil resistance to limit the current through the coil. The combination of the meter movement with this added series resistance then

┃┃┃ MultiSim **Figure 8–6** Multiplier resistor R_1 added in series with meter movement to form a voltmeter. (*a*) Resistance of R_1 allows 1 mA for full-scale deflection in 1-mA movement with 10 V applied. (*b*) Internal multiplier R_1 forms a voltmeter. The test leads can be connected across a potential difference to measure 0 to 10 V. (*c*) 10-V scale of voltmeter and corresponding 1-mA scale of meter movement.

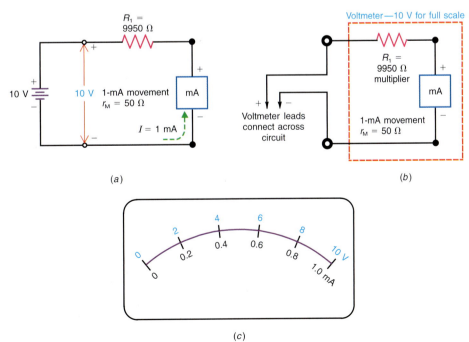

(a)

(b)

(c)

forms a voltmeter. The series resistor, called a *multiplier,* is usually connected inside the voltmeter case.

Since a voltmeter has high resistance, it must be connected in parallel to measure the potential difference across two points in a circuit. Otherwise, the high-resistance multiplier would add so much series resistance that the current in the circuit would be reduced to a very low value. Connected in parallel, though, the high resistance of the voltmeter is an advantage. The higher the voltmeter resistance, the smaller the effect of its parallel connection on the circuit being tested.

The circuit is not opened to connect the voltmeter in parallel. Because of this convenience, it is common practice to make voltmeter tests in troubleshooting. The voltage measurements apply the same way to an *IR* drop or a generated emf.

The correct polarity must be observed in using a dc voltmeter. Connect the negative voltmeter lead to the negative side of the potential difference being measured and the positive lead to the positive side.

Multiplier Resistance

Figure 8–6 illustrates how the meter movement and its multiplier R_1 form a voltmeter. With 10 V applied by the battery in Fig. 8–6a, there must be 10,000 Ω of resistance to limit the current to 1 mA for full-scale deflection of the meter movement. Since the movement has a 50-Ω resistance, 9950 Ω is added in series, resulting in a 10,000-Ω total resistance. Then I is 10 V/10 kΩ = 1 mA.

With 1 mA in the movement, the full-scale deflection can be calibrated as 10 V on the meter scale, as long as the 9950-Ω multiplier is included in series with the movement. It doesn't matter to which side of the movement the multiplier is connected.

If the battery is taken away, as in Fig. 8–6b, the movement with its multiplier forms a voltmeter that can indicate a potential difference of 0 to 10 V applied across its terminals. When the voltmeter leads are connected across a potential difference of 10 V in a dc circuit, the resulting 1-mA current through the meter movement produces full-scale deflection, and the reading is 10 V. In Fig. 8–6c, the 10-V scale is shown corresponding to the 1-mA range of the movement.

If the voltmeter is connected across a 5-V potential difference, the current in the movement is $\frac{1}{2}$ mA, the deflection is one-half of full scale, and the reading is 5 V. Zero voltage across the terminals means no current in the movement, and the voltmeter reads zero. In summary, then, any potential difference up to 10 V, whether an *IR* voltage drop or a generated emf, can be applied across the meter terminals. The meter will indicate less than 10 V in the same ratio that the meter current is less than 1 mA.

The resistance of a multiplier can be calculated from the formula

$$R_{\text{mult}} = \frac{\text{Full-scale } V}{\text{Full-scale } I} - r_{\text{M}} \tag{8–3}$$

Applying this formula to the example of R_1 in Fig. 8–6 gives

$$R_{\text{mult}} = \frac{10 \text{ V}}{0.001 \text{ A}} - 50 \ \Omega = 10,000 - 50$$
$$R_{\text{mult}} = 9950 \ \Omega \quad \text{or} \quad 9.95 \text{ k}\Omega$$

We can take another example for the same 10-V scale but with a 50-μA meter movement, which is commonly used. Now the multiplier resistance is much

Figure 8–7 A typical voltmeter circuit with multiplier resistors for different ranges.

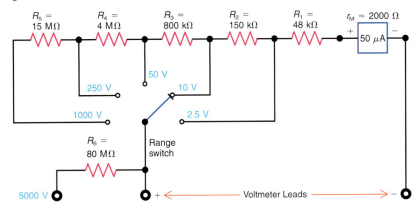

higher, though, because less I is needed for full-scale deflection. Let the resistance of the 50-μA movement be 2000 Ω. Then

$$R_{\text{mult}} = \frac{10 \text{ V}}{0.000\,050 \text{ A}} - 2000\ \Omega = 200{,}000 - 2000$$
$$R_{\text{mult}} = 198{,}000\ \Omega \quad \text{or} \quad 198 \text{ k}\Omega$$

Typical Multiple Voltmeter Circuit

An example of a voltmeter with multiple voltage ranges is shown in Fig. 8–7. Resistance R_1 is the series multiplier for the lowest voltage range of 2.5 V. When higher resistance is needed for the higher ranges, the switch adds the required series resistors.

The meter in Fig. 8–7 requires 50 μA for full-scale deflection. For the 2.5-V range, a series resistance of $2.5/(50 \times 10^{-6})$, or 50,000 Ω, is needed. Since r_M is 2000 Ω, the value of R_1 is $50{,}000 - 2000$, which equals 48,000 Ω or 48 kΩ.

For the 10-V range, a resistance of $10/(50 \times 10^{-6})$, or 200,000 Ω, is needed. Since $R_1 + r_M$ provides 50,000 Ω, R_2 is made 150,000 Ω, for a total of 200,000 Ω series resistance on the 10-V range. Similarly, additional resistors are switched in to increase the multiplier resistance for the higher voltage ranges. Note the separate jack and extra multiplier R_6 on the highest range for 5000 V. This method of adding series multipliers for higher voltage ranges is the circuit generally used in commercial multimeters.

Voltmeter Resistance

The high resistance of a voltmeter with a multiplier is essentially the value of the multiplier resistance. Since the multiplier is changed for each range, the voltmeter resistance changes.

Table 8–1 shows how the voltmeter resistance increases for higher ranges. The middle column lists the total internal resistance R_V, including R_{mult} and r_M, for the voltmeter circuit in Fig. 8–7. With a 50-μA movement, R_V increases from 50 kΩ on the 2.5-V range to 20 MΩ on the 1000-V range. Note that R_V has these values on each range whether you read full scale or not.

GOOD TO KNOW

The voltmeter resistance of an analog VOM can be measured by connecting the leads of a DMM to the leads of the analog VOM. With the DMM set to measure resistance and the VOM set to measure voltage, the DMM will indicate the voltmeter resistance, R_V. The higher the voltmeter range, the higher the resistance.

Table 8–1	A Voltmeter Using a 50-μA Movement	
Full-Scale Voltage V_F	$R_V = R_{mult} + r_M$	Ohms Per Volt $= R_V/V_F$
2.5	50 kΩ	20,000 Ω/V
10	200 kΩ	20,000 Ω/V
50	1 MΩ	20,000 Ω/V
250	5 MΩ	20,000 Ω/V
1000	20 MΩ	20,000 Ω/V

Ohms-per-Volt Rating

To indicate the voltmeter's resistance independently of the range, analog voltmeters are generally rated in ohms of resistance needed for 1 V of deflection. This value is the ohms-per-volt rating of the voltmeter. As an example, see the last column in Table 8–1. The values in the top row show that this meter needs 50,000 Ω R_V for 2.5 V of full-scale deflection. The resistance per 1 V of deflection then is 50,000/2.5, which equals 20,000 Ω/V.

The ohms-per-volt value is the same for all ranges because this characteristic is determined by the full-scale current I_M of the meter movement. To calculate the ohms-per-volt rating, take the reciprocal of I_M in ampere units. For example, a 1-mA movement results in 1/0.001 or 1000 Ω/V; a 50-μA movement allows 20,000 Ω/V, and a 20-μA movement allows 50,000 Ω/V. The ohms-per-volt rating is also called the *sensitivity* of the voltmeter.

A higher ohms-per-volt rating means higher voltmeter resistance R_V. R_V can be calculated as the product of the ohms-per-volt rating and the full-scale voltage of each range. For instance, across the second row in Table 8–1, on the 10-V range with a 20,000 Ω/V rating,

$$R_V = 10 \text{ V} \times \frac{20,000 \ \Omega}{\text{volts}}$$
$$R_V = 200,000 \ \Omega$$

Usually the ohms-per-volt rating of a voltmeter is printed on the meter face.

■ *8–3A Knowledge Check*

Answer at end of chapter.

In Fig. 8–6b, what value is needed for the multiplier resistor, R_1, to make a 0-50 V voltmeter?

■ *8–3B Knowledge Check*

Answer at end of chapter.

What is the Ω/V rating of a voltmeter whose R_V value is 2.5 MΩ on the 50-V range?

Answers at end of chapter.

Refer to Fig. 8–7.
a. Calculate the voltmeter resistance R_V on the 2.5-V range.
b. Calculate the voltmeter resistance R_V on the 50-V range.
c. Is the voltmeter multiplier resistor in series or parallel with the meter movement?
d. Is a voltmeter connected in series or parallel with the potential difference to be measured?
e. How much is the total R of a voltmeter with a sensitivity of 20,000 Ω/V on the 25-V scale?

8–4 Loading Effect of a Voltmeter

When the voltmeter resistance is not high enough, connecting it across a circuit can reduce the measured voltage, compared with the voltage present without the voltmeter. This effect is called *loading down* the circuit, since the measured voltage decreases because of the additional load current for the meter.

Loading Effect

Voltmeter loading can be appreciable in high-resistance circuits, as shown in Fig. 8–8. In Fig. 8–8a, without the voltmeter, R_1 and R_2 form a voltage divider across the applied voltage of 120 V. The two equal resistances of 100 kΩ each divide the applied voltage equally, with 60 V across each.

When the voltmeter in Fig. 8–8b is connected across R_2 to measure its potential difference, however, the voltage division changes. The voltmeter resistance R_V of 100 kΩ is the value for a 1000-ohms-per-volt meter on the 100-V range. Now the voltmeter in parallel with R_2 draws additional current, and the equivalent resistance between the measured points 1 and 2 is reduced from 100,000 to 50,000 Ω. This resistance is one-third the total circuit resistance, and the measured voltage across points 1 and 2 drops to 40 V, as shown in Fig. 8–8c.

As additional current drawn by the voltmeter flows through the series resistance R_1, this voltage goes up to 80 V.

Similarly, if the voltmeter were connected across R_1, this voltage would go down to 40 V, with the voltage across R_2 rising to 80 V. When the voltmeter is disconnected, the circuit returns to the condition in Fig. 8–8a, with 60 V across both R_1 and R_2.

||| **MultiSim** **Figure 8–8** How the loading effect of the voltmeter can reduce the voltage reading. (*a*) High-resistance series circuit without voltmeter. (*b*) Connecting voltmeter across one of the series resistances. (*c*) Reduced R and V between points 1 and 2 caused by the voltmeter as a parallel branch across R_2. The R_{2V} is the equivalent of R_2 and R_V in parallel.

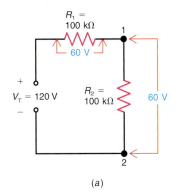

(a)

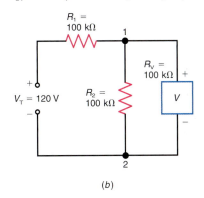

(b)

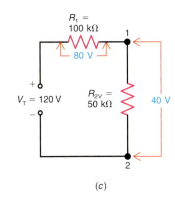

(c)

Figure 8–9 Negligible loading effect with a high-resistance voltmeter. (*a*) High-resistance series circuit without voltmeter, as in Fig. 8–8a. (*b*) Same voltages in circuit with voltmeter connected because R_V is so high.

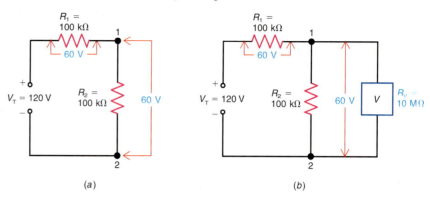

(*a*)　　　　　　　　　(*b*)

The loading effect is minimized by using a voltmeter with a resistance much greater than the resistance across which the voltage is measured. As shown in Fig. 8–9, with a voltmeter resistance of 10 MΩ, the loading effect is negligible. Because R_V is so high, it does not change the voltage division in the circuit. The 10 MΩ of the meter in parallel with the 100,000 Ω for R_2 results in an equivalent resistance practically equal to 100,000 Ω.

With multiple ranges on a VOM, the voltmeter resistance changes with the range selected. Higher ranges require more multiplier resistance, increasing the voltmeter resistance for less loading. As examples, a 20,000-ohms-per-volt meter on the 250-V range has an internal resistance R_V of 20,000 × 250, or 5 MΩ. However, on the 2.5-V range, the same meter has an R_V of 20,000 × 2.5, which is only 50,000 Ω.

On any one range, though, the voltmeter resistance is constant whether you read full-scale or less than full-scale deflection. The reason is that the multiplier resistance set by the range switch is the same for any reading on that range.

Correction for Loading Effect

The following formula can be used:

$$\overset{\textit{Actual reading}}{\underset{\downarrow}{}} + \overset{\textit{correction}}{\underset{\downarrow}{}}$$

$$V = V_M + \frac{R_1 R_2}{R_V(R_1 + R_2)} V_M \tag{8–4}$$

Voltage V is the corrected reading the voltmeter would show if it had infinitely high resistance. Voltage V_M is the actual voltage reading. Resistances R_1 and R_2 are the voltage-dividing resistances in the circuit without the voltmeter resistance R_V. As an example, in Fig. 8–8,

$$V = 40 \text{ V} + \frac{100 \text{ k}\Omega \times 100 \text{ k}\Omega}{100 \text{ k}\Omega \times 200 \text{ k}\Omega} \times 40 \text{ V} = 40 + \frac{1}{2} \times 40 = 40 + 20$$

$$V = 60 \text{ V}$$

The loading effect of a voltmeter causes too low a voltage reading because R_V is too low as a parallel resistance. This corresponds to the case of a current meter reading too low because R_M is too high as a series resistance. Both of these effects illustrate the general problem of trying to make any measurement without changing the circuit being measured.

Note that the digital multimeter (DMM) has practically no loading effect as a voltmeter. The input resistance is usually 10 MΩ or 20 MΩ, the same on all ranges.

Answer at end of chapter.

To eliminate the loading effect of a voltmeter, a voltmeter should have very (low/high) internal resistance.

■ *8-4 Self-Review*

Answers at end of chapter.

With the voltmeter across R_2 in Fig. 8–8b, what is the value for
a. V_1?
b. V_2?

8-5 Ohmmeters

An ohmmeter consists of an internal battery, the meter movement, and a current-limiting resistance, as illustrated in Fig. 8–10. For measuring resistance, the ohmmeter leads are connected across the external resistance to be measured. Power in the circuit being tested must be off. Then only the ohmmeter battery produces current for deflecting the meter movement. Since the amount of current through the meter depends on the external resistance, the scale can be calibrated in ohms.

The amount of deflection on the ohms scale indicates the measured resistance directly. The ohmmeter reads up-scale regardless of the polarity of the leads because the polarity of the internal battery determines the direction of current through the meter movement.

Series Ohmmeter Circuit

In Fig. 8–10a, the circuit has 1500 Ω for ($R_1 + r_M$). Then the 1.5-V cell produces 1 mA, deflecting the moving coil full scale. When these components are enclosed in a case, as in Fig. 8–10b, the series circuit forms an ohmmeter. Note that M indicates the meter movement.

If the leads are short-circuited together or connected across a short circuit, as in Fig. 8–10a, 1 mA flows. The meter movement is deflected full scale to the right. This ohmmeter reading is 0 Ω.

Figure 8–10   How meter movement M can be used as an ohmmeter with a 1.5-V battery. (*a*) Equivalent closed circuit with R_1 and the battery when ohmmeter leads are short-circuited for zero ohms of external R. (*b*) Internal ohmmeter circuit with test leads open, ready to measure an external resistance.

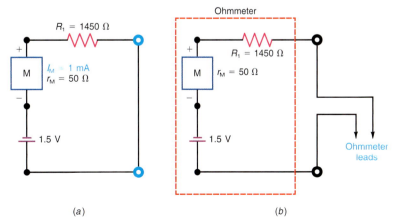

(a)   (b)

Table 8–2	Calibration of Ohmmeter in Fig. 8–10					
External R_X, Ω	Internal $R_i = R_1 + r_M$, Ω	$R_T = R_X + R_i$, Ω	$I = V/R_T$, mA	Deflection		Scale Reading, Ω
0	1500	1500	1	Full scale		0
750	1500	2250	$^2/_3 = 0.67$	$^2/_3$ scale		750
1500	1500	3000	$^1/_2 = 0.5$	$^1/_2$ scale		1500
3000	1500	4500	$^1/_3 = 0.33$	$^1/_3$ scale		3000
150,000	1500	151,500	0.01	$^1/_{100}$ scale		150,000
500,000	1500	501,500	0	None		∞

When the ohmmeter leads are open, not touching each other, the current is zero. The ohmmeter indicates infinitely high resistance or an open circuit across its terminals.

Therefore, the meter face can be marked zero ohms at the right for full-scale deflection and infinite ohms at the left for no deflection. In-between values of resistance result when less than 1 mA flows through the meter movement. The corresponding deflection on the ohm scale indicates how much resistance is across the ohmmeter terminals.

Back-Off Ohmmeter Scale

Table 8–2 and Fig. 8–11 illustrate the calibration of an ohmmeter scale in terms of meter current. The current equals V/R_T. Voltage V is the fixed applied voltage of 1.5 V supplied by the internal battery. Resistance R_T is the total resistance of R_X and the ohmmeter's internal resistance. Note that R_X is the external resistance to be measured.

Figure 8–11 Back-off ohmmeter scale with R readings increasing from right to left. (a) Series ohmmeter circuit for the unknown external resistor R_X to be measured. (b) Ohm scale has higher R readings to the left of the scale as more R_X decreases I_M. The R and I values are listed in Table 8–2.

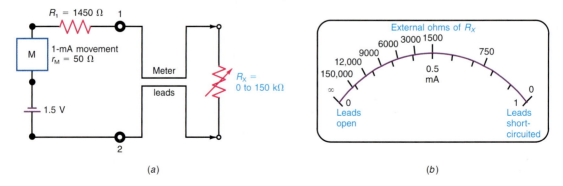

The ohmmeter's internal resistance R_i is constant at $50 + 1450$, or $1500 \, \Omega$ here. If R_X also equals $1500 \, \Omega$, for example, R_T equals $3000 \, \Omega$. The current then is $1.5 \, \text{V}/3000 \, \Omega$, or $0.5 \, \text{mA}$, resulting in half-scale deflection for the 1-mA movement. Therefore, the center of the ohm scale is marked for $1500 \, \Omega$. Similarly, the amount of current and meter deflection can be calculated for any value of the external resistance R_X.

Note that the ohm scale increases from right to left. This arrangement is called a *back-off scale*, with ohm values increasing to the left as the current backs off from full-scale deflection. The back-off scale is a characteristic of any ohmmeter where the internal battery is in series with the meter movement. Then more external R_X decreases the meter current.

A back-off ohmmeter scale is expanded at the right near zero ohms and crowded at the left near infinite ohms. This nonlinear scale results from the relation $I = V/R$ with V constant at $1.5 \, \text{V}$. Recall that with V constant, I and R are inversely related.

The highest resistance that can be indicated by the ohmmeter is about 100 times its total internal resistance. Therefore, the infinity mark on the ohms scale, or the "lazy eight" symbol ∞ for infinity, is only relative. It just means that the measured resistance is infinitely greater than the ohmmeter resistance.

It is important to note that the ohmmeter circuit in Fig. 8–11 is not entirely practical. The reason is that if the battery voltage, V_b, is not exactly $1.5 \, \text{V}$, the ohmmeter scale will not be calibrated correctly. Also, a single ohmmeter range is not practical when it is necessary to measure very small or large resistance values. Without going into the circuit detail, you should be aware of the fact that commercially available ohmmeters are designed to provide multiple ohmmeter ranges as well as compensation for a change in the battery voltage, V_b.

Multiple Ohmmeter Ranges

Commercial multimeters provide for resistance measurements from less than $1 \, \Omega$ up to many megohms in several ranges. The range switch in Fig. 8–12 shows the multiplying factors for the ohm scale. On the $R \times 1$ range, for low-resistance measurements, read the ohm scale directly. In the example here, the pointer indicates $12 \, \Omega$. When the range switch is on $R \times 100$, multiply the scale reading by 100; this reading would then be 12×100 or $1200 \, \Omega$. On the $R \times 10,000$ range, the pointer would indicate $120,000 \, \Omega$.

A multiplying factor, instead of full-scale resistance, is given for each ohm range because the highest resistance is infinite on all ohm ranges. This method for ohms should not be confused with full-scale values for voltage ranges. For the ohmmeter ranges, always multiply the scale reading by the $R \times$ factor. On voltage ranges, you may have to multiply or divide the scale reading to match the full-scale voltage with the value on the range switch.

Zero-Ohms Adjustment

To compensate for lower voltage output as the internal battery ages, an ohmmeter includes a variable resistor to calibrate the ohm scale. A back-off ohmmeter is always adjusted for zero ohms. With the test leads short-circuited, vary the ZERO OHMS control on the front panel of the meter until the pointer is exactly on zero at the right edge of the ohm scale. Then the ohm readings are correct for the entire scale.

This type of ohmmeter must be zeroed again every time you change the range because the internal circuit changes.

When the adjustment cannot deflect the pointer all the way to zero at the right edge, it usually means that the battery voltage is too low and it must

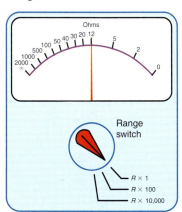

Figure 8–12 Multiple ohmmeter ranges with just one ohm scale. The ohm reading is multiplied by the factor set on the range switch.

be replaced. Usually, this trouble shows up first on the $R \times 1$ range, which takes the most current from the battery.

■ 8–5 Knowledge Check

Answer at end of chapter.

In Fig. 8–10, what value of external resistance, R_X, will produce three-fourths, full-scale deflection?

■ 8–5 Self-Review

Answers at end of chapter.

a. An ohmmeter reads 40 Ω on the $R \times 10$ range. How much is R_X?
b. A voltmeter reads 40 on the 300-V scale, but with the range switch on 30 V. How much is the measured voltage?

GOOD TO KNOW

When measuring an unknown value of current or voltage, always set the meter to its highest range and work your way down. The meter could be damaged if a very large value of current or voltage is attempted to be measured on the meter's lowest range setting.

8–6 Multimeters

Multimeters are also called *multitesters,* and they are used to measure voltage, current, or resistance. Table 8–3 compares the features of the main types of multimeters: first, the volt-ohm-milliammeter (VOM) in Fig. 8–13, and next the digital multimeter (DMM) in Fig. 8–14. The DMM is explained in more detail in the next section.

Beside its digital readout, an advantage of the DMM is its high input resistance R_V as a dc voltmeter. The R_V is usually 10 MΩ, the same on all ranges, which is high enough to prevent any loading effect by the voltmeter in most circuits. Some types have an R_V of 22 MΩ. Many modern DMMs are autoranging, that is, the internal circuitry selects the proper range for the meter and indicates the range as a readout.

For either a VOM or a DMM, it is important to have a low-voltage dc scale with resolution good enough to read 0.2 V or less. The range of 0.2 to 0.6 V, or 200 to 600 mV, is needed for measuring dc bias voltages in transistor circuits.

Low–Power Ohms (LPΩ)

Another feature needed for transistor measurements is an ohmmeter that does not have enough battery voltage to bias a semiconductor junction into the ON

Table 8–3	VOM Compared to DMM
VOM	**DMM**
Analog pointer reading	Digital readout
DC voltmeter R_V changes with range	R_V is 10 or 22 MΩ, the same on all ranges
Zero-ohms adjustment changed for each range	No zero-ohms adjustment
Ohm ranges up to $R \times 10,000$ Ω, as a multiplying factor	Ohm ranges up to 20 MΩ; each range is the maximum

Figure 8-13 Analog VOM that combines a function selector and range switch.

Figure 8-14 Portable digital multimeter (DMM).

or conducting state. The limit is 0.2 V or less. The purpose is to prevent any parallel conduction paths in the transistor amplifier circuit that can lower the ohmmeter reading.

Decibel Scale

Most analog multimeters have an ac voltage scale calibrated in decibel (dB) units, for measuring ac signals. The decibel is a logarithmic unit used for comparison of power levels or voltage levels. The mark of 0 dB on the scale indicates the reference level, which is usually 0.775 V for 1 mW across 600 Ω. Positive decibel values above the zero mark indicate ac voltages above the reference of 0.775 V; negative decibel values are less than the reference level.

Amp-Clamp Probe

The problem of opening a circuit to measure I can be eliminated by using a probe with a clamp that fits around the current-carrying wire. Its magnetic field is used to indicate the amount of current. An example is shown in Fig. 8–15. The clamp probe measures just ac amperes, generally for the 60-Hz ac power line.

High-Voltage Probe

An accessory probe can be used with a multimeter to measure dc voltages up to 30 kV. This probe is often referred to as a *high-voltage probe*. One application is measuring the high voltage of 20 to 30 kV at the anode of the color picture tube in a television receiver. The probe is just an external multiplier resistance for the dc voltmeter. The required R for a 30-kV probe is 580 MΩ with a 20-kΩ/V meter on the 1000-V range.

Figure 8–15 DMM with amp clamp accessory.

Figure 8–15 DMM with amp clamp accessory.

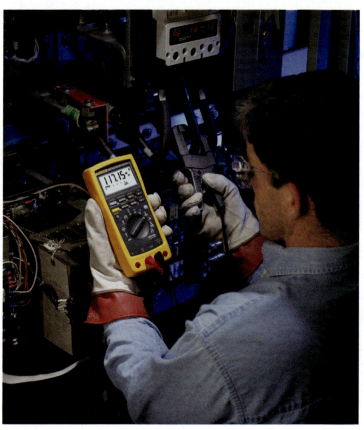

Figure 8–16 Typical digital multimeter (DMM).

◼ 8-6 Knowledge Check

Answer at end of chapter.

What is one of the main advantages of a DMM for measuring voltages?

◼ 8-6 Self-Review

Answers at end of chapter.

a. How much is R_V on the 1-V range for a VOM with a sensitivity of 20 kΩ/V?
b. If R_V is 10 MΩ for a DMM on the 100-V range, how much is R_V on the 200-mV range?
c. The low-power ohm function (LPΩ) does not require an internal battery. True or false?

8-7 Digital Multimeter (DMM)

The digital multimeter has become a very popular test instrument because the digital value of the measurement is displayed automatically with decimal point, polarity, and the unit for V, A, or Ω. Digital meters are generally easier to use because they eliminate the human error that often occurs in reading different scales on an analog meter with a pointer. Examples of the portable DMM are shown in Figs. 8–14 and 8–16.

The basis of the DMM operation is an analog-to-digital (A/D) converter circuit. It converts analog voltage values at the input to an equivalent binary form. These values are processed by digital circuits to be shown on a liquid-crystal display (LCD) as decimal values.

Voltage Measurements

The A/D converter requires a specific range of voltage input; typical values are -200 mV to $+200$ mV. For DMM input voltages that are higher, the voltage is divided down. When the input voltage is too low, it is increased by a dc amplifier circuit. The measured voltage can then be compared to a fixed reference voltage in the meter by a comparator circuit. Actually, all functions in the DMM depend on the voltage measurements by the converter and comparator circuits.

The input resistance of the DMM is in the range of 10 to 20 MΩ, shunted by 50 pF of capacitance. This R is high enough to eliminate the problem of voltmeter loading in most transistor circuits. Not only does the DMM have high input resistance, but the R is the same on all ranges.

With ac measurements, the ac input is converted to dc voltage for the A/D converter. The DMM has an internal diode rectifier that serves as an ac converter.

R Measurement

As an ohmmeter, the internal battery supplies I through the measured R for an IR drop measured by the DMM. The battery is usually the small 9-V type commonly used in portable equipment. A wide range of R values can be measured from a fraction of an ohm to more than 30 MΩ. Remember that power must be off in the circuit being tested with an ohmmeter.

A DMM ohmmeter usually has an open-circuit voltage across the meter leads, which is much too low to turn on a semiconductor junction. The result is low-power ohms operation.

I Measurements

To measure current, internal resistors provide a proportional IR voltage. The display shows the I values. Note that the DMM still must be connected as a series component in the circuit when current is measured.

Diode Test

The DMM usually has a setting for testing semiconductor diodes, either silicon or germanium. Current is supplied by the DMM for the diode to test the voltage across its junction. Normal values are 0.7 V for silicon and 0.3 V for germanium. A short-circuited junction will read 0 V. The voltage across an open diode reads much too high. Most diodes are silicon.

Resolution

This term for a DMM specifies how many places can be used to display the digits 0 to 9, regardless of the decimal point. For example, 9.99 V is a three-digit display; 9.999 V would be a four-digit display. Most portable units, however, compromise with a 3½-digit display. This means that the fourth digit at the left for the most significant place can only be a 1. If not, then the display has three digits. As examples, a 3½ digit display can show 19.99 V, but 29.99 V would be read as 30.0 V. Note that better resolution with more digits can be obtained with more expensive meters, especially the larger DMM units for bench mounting. Actually, though, 3½-digit resolution is enough for practically all measurements made in troubleshooting electronic equipment.

Range Overload

The DMM selector switch has specific ranges. Any value higher than the selected range is an overload. An indicator on the display warns that the value shown is not correct. Then a higher range is selected. Some units have an *auto-range function* that shifts the meter automatically to a higher range as soon as an overload is indicated.

Typical DMM

The unit in Fig. 8–16 can be used as an example. On the front panel, the two jacks at the bottom right are for the test leads. The lower jack is the common, with a black lead, used for all measurements. Above is the jack for the "hot" lead, usually red, used for the measurements of *V* and *R* either dc or ac values. The two jacks at the bottom left side are for the red lead when measuring either dc or ac *I*.

Consider each function of the large selector switch at the center in Fig. 8–16. The first position at the top, after the switch is turned clockwise from the off position, is used to measure ac volts, as indicated by the sine wave. No ranges are given as this meter has an autorange function. In operation, the meter has the ranges of 600 mV, 6 V, 60 V, 600 V, and, as a maximum, 1000 V.

If the autorange function is not desired, press the range button below the display to hold the range. Each touch of the button will change the range. Hold the button down to return to autorange operation.

The next two positions on the function switch are for dc volts. Polarity can be either positive or negative as indicated by the solid and dashed lines above the *V*. The ranges of dc voltages that can be measured are 6, 60, 600, and 1000 V as a maximum. For very low dc voltages, the mV switch setting should be used. Values below 600 mV can be measured on this range.

For an ohmmeter, the function switch is set to the position with the Ω symbol. The ohm values are from 0–50 MΩ in six ranges. Remember that power must be off in the circuit being measured, or the ohmmeter will read the wrong value. (Worse yet, the meter could be damaged.)

Next on the function switch is the position for testing semiconductor diodes, as shown by the diode symbol. The lines next to the symbol indicate that the meter produces a beep tone. Maximum diode test voltage is 2.4 V.

The last two positions on the function switch are for current measurements. The jacks at the lower left are used for larger or smaller current values.

In measuring ac values, either for *V* or *I*, the frequency range of the meter is limited to 45 to 1000 Hz, approximately. For amplitudes at higher frequencies, such as rf measurements, special meters are necessary. However, this meter can be used for *V* and *I* at the 60-Hz power-line frequency and the 400-Hz test frequency often used for audio equipment.

Analog Display

The bar at the bottom of the display in Fig. 8–16 is used only to show the relative magnitude of the input compared to the full-scale value of the range in use. This function is convenient when adjusting a circuit for a peak value or a minimum (null). The operation is comparable to watching the needle on a VOM for either a maximum or a null adjustment.

■ *8–7 Knowledge Check*

Answer at end of chapter.

Can a DMM with a 3½-digit display show a voltage of 49.72 V? Why or why not?

■ *8–7 Self-Review*

Answers at end of chapter.

a. The typical resistance of a DMM voltmeter is 10 MΩ. (True or False)
b. The ohmmeter on a portable DMM does not need an internal battery. (True or False)
c. A DMM voltmeter with 3½-digit resolution can display the value of 14.59 V. (True or False)

8–8 Meter Applications

Table 8–4 summarizes the main points to remember when using a voltmeter, ohmmeter, or milliammeter. These rules apply whether the meter is a single unit or one function on a multimeter. The voltage and current tests also apply to either dc or ac circuits.

To avoid excessive current through the meter, it is good practice to start on a high range when measuring an unknown value of voltage or current. It is very important not to make the mistake of connecting a current meter in parallel, because **usually this mistake ruins the meter.** The mistake of connecting a voltmeter in series does not damage the meter, but the reading will be wrong.

If the ohmmeter is connected to a circuit in which power is on, the meter can be damaged, beside giving the wrong reading. An ohmmeter has its own internal battery, and the power must be off in the circuit being tested. When *R* is tested with an ohmmeter, it may be necessary to disconnect one end of *R* from the circuit to eliminate parallel paths.

Connecting a Current Meter in the Circuit

In a series-parallel circuit, the current meter must be inserted in a branch to read branch current. In the main line, the meter reads the total current. These different

Table 8–4	Direct-Current Meters	
Voltmeter	**Milliammeter or Ammeter**	**Ohmmeter**
Power on in circuit	Power on in circuit	Power off in circuit
Connect in parallel	Connect in series	Connect in parallel
High internal *R*	Low internal *R*	Has internal battery
Has internal series multipliers; higher *R* for higher ranges	Has internal shunts; lower resistance for higher current ranges	Higher battery voltage and more sensitive meter for higher ohm ranges

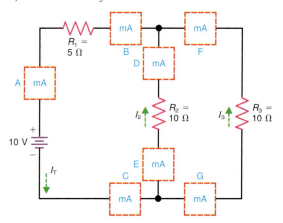

Figure 8–17 How to insert a current meter in different parts of a series-parallel circuit to read the desired current I. At point A, B, or C, the meter reads I_T; at D or E, the meter reads I_2; at F or G, the meter reads I_3.

connections are illustrated in Fig. 8–17. The meters are shown by dashed lines to illustrate the different points at which a meter could be connected to read the respective currents.

If the circuit is opened at point A to insert the meter in series in the main line here, the meter will read total line current I_T through R_1. A meter at B or C will read the same line current.

To read the branch current through R_2, this R must be disconnected from its junction with the main line at either end. A meter inserted at D or E, therefore, will read the R_2 branch current I_2. Similarly, a meter at F or G will read the R_3 branch current I_3.

Calculating *I* from Measured Voltage

The inconvenience of opening the circuit to measure current can often be eliminated by the use of Ohm's law. The voltage and resistance can be measured without opening the circuit, and the current calculated as V/R. In the example in Fig. 8–18, when the voltage across R_2 is 15 V and its resistance is 15 Ω, the current through R_2 must be 1 A. When values are checked during troubleshooting, if the voltage and resistance are normal, so is the current.

This technique can also be convenient for determining I in low-resistance circuits where the resistance of a microammeter may be too high. Instead of measuring I, measure V and R and calculate I as V/R.

Furthermore, if necessary, we can insert a known resistance R_S in series in the circuit, temporarily, just to measure V_S. Then I is calculated as V_S/R_S. The resistance of R_S, however, must be small enough to have little effect on R_T and I in the series circuit.

This technique is often used with oscilloscopes to produce a voltage waveform of IR which has the same waveform as the current in a resistor. The oscilloscope must be connected as a voltmeter because of its high input resistance.

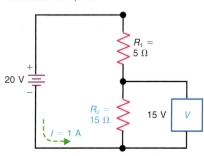

Figure 8–18 With 15 V measured across a known R of 15 Ω, the I can be calculated as V/R or $V/15 \, \Omega = 1$ A.

Checking Fuses

Turn the power off or remove the fuse from the circuit to check with an ohmmeter. A good fuse reads 0 Ω. A blown fuse is open, which reads infinity on the ohmmeter.

Figure 8–19 Voltage tests to localize an open circuit. (*a*) Normal circuit with voltages to chassis ground. (*b*) Reading of 0 V at point D shows R_3 is open.

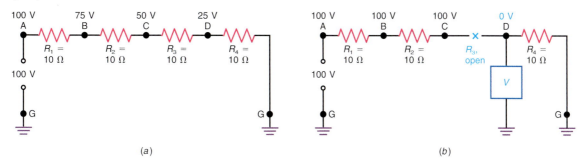

(*a*) (*b*)

A fuse can also be checked with the power on in the circuit by using a voltmeter. Connect the voltmeter across the two terminals of the fuse. A good fuse reads 0 V because there is practically no *IR* drop. With an open fuse, though, the voltmeter reading is equal to the full value of the applied voltage. Having the full applied voltage seems to be a good idea, but it should not be across the fuse.

Voltage Tests for an Open Circuit

Figure 8–19 shows four equal resistors in series with a 100-V source. A ground return is shown here because voltage measurements are usually made with respect to chassis or earth ground. Normally, each resistor would have an *IR* drop of 25 V. Then, at point B, the voltmeter to ground should read $100 - 25 = 75$ V. Also, the voltage at C should be 50 V, with 25 V at D, as shown in Fig. 8–19*a*.

However, the circuit in Fig. 8–19*b* has an open in R_3 toward the end of the series string of voltages to ground. Now when you measure at B, the reading is 100 V, equal to the applied voltage. This full voltage at B shows that the series circuit is open without any *IR* drop across R_1. The question is, however, which *R* has the open? Continue the voltage measurements to ground until you find 0 V. In this example, the open is in R_3 between the 100 V at C and 0 V at D.

The points that read the full applied voltage have a path back to the source of voltage. The first point that reads 0 V has no path back to the high side of the source. Therefore, the open circuit must be between points C and D in Fig. 8–19*b*.

■ 8-8 Knowledge Check

Answer at end of chapter.

In Fig. 8–19, how much voltage would be measured at point *B* if R_4 opens?

■ 8-8 Self-Review

Answers at end of chapter.

a. Which type of meter requires an internal battery?
b. How much is the normal voltage across a good fuse?
c. How much is the voltage across R_1 in Fig. 8–19*a*?
d. How much is the voltage across R_1 in Fig. 8–19*b*?

8–9 Checking Continuity with the Ohmmeter

A wire conductor that is continuous without a break has practically zero ohms of resistance. Therefore, the ohmmeter can be useful in testing for continuity. This test should be done on the lowest ohm range. There are many applications. A wire conductor can have an internal break which is not visible because of the insulated cover, or the wire can have a bad connection at the terminal. Checking for zero ohms between any two points along the conductor tests continuity. A break in the conducting path is evident from a reading of infinite resistance, showing an open circuit.

As another application of checking continuity, suppose that a cable of wires is harnessed together, as illustrated in Fig. 8–20, where the individual wires cannot be seen, but it is desired to find the conductor that connects to terminal A. This is done by checking continuity for each conductor to point A. The wire that has zero ohms to A is the one connected to this terminal. Often the individual wires are color-coded, but it may be necessary to check the continuity of each lead.

An additional technique that can be helpful is illustrated in Fig. 8–21. Here it is desired to check the continuity of the two-wire line, but its ends are too far apart for the ohmmeter leads to reach. The two conductors are temporarily short-circuited at one end, however, so that the continuity of both wires can be checked at the other end.

In summary, then, the ohmmeter is helpful in checking the continuity of any wire conductor. This includes resistance-wire heating elements, such as the wires in a toaster or the filament of an incandescent bulb. Their cold resistance is normally just a few ohms. Infinite resistance means that the wire element is open. Similarly, a good fuse has practically zero resistance. A burned-out fuse has infinite resistance, that is, it is open. Any coil for a transformer, solenoid, or motor will also have infinite resistance if the winding is open.

Figure 8–20 Continuity testing from point A to wire 3 shows that this wire is connected.

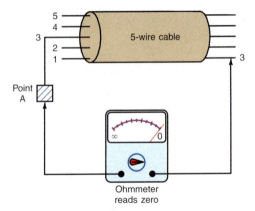

Figure 8–21 Temporary short circuit at one end of a long two-wire line to check continuity from the opposite end.

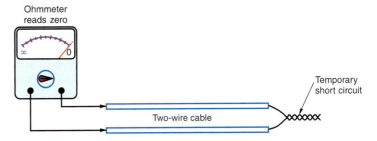

8–9 Knowledge Check

Answer at end of chapter.

An ohmmeter will read (zero/infinite) ohms if there is no continuity between two points.

8–9 Self-Review

Answers at end of chapter.

a. On a back-off ohmmeter, is zero ohms at the left or the right end of the scale?
b. What is the ohmmeter reading for an open circuit?

Summary

- Direct current in a moving-coil meter deflects the coil in proportion to the amount of current.

- A current meter is a low-resistance meter connected in series to read the amount of current in a circuit.

- A meter shunt R_S in parallel with the meter movement extends the range of a current meter [see Formula (8–1)].

- A voltmeter consists of a meter movement in series with a high-resistance multiplier. The voltmeter with its multiplier is connected across two points to measure the potential difference in volts. The multiplier R can be calculated from Formula (8–3).

- The ohms-per-volt rating of a voltmeter with series multipliers specifies the sensitivity on all voltage ranges. It equals the reciprocal of the full-scale deflection current of the meter. A typical value is 20,000 Ω/V for a voltmeter using a 50-μA movement. The higher the ohms-per-volt rating, the better.

- Voltmeter resistance R_V is higher for higher ranges because of higher-resistance multipliers. Multiply the ohms-per-volt rating by the voltage range to calculate the R_V for each range.

- An ohmmeter consists of an internal battery in series with the meter movement. Power must be off in the circuit being checked with an ohmmeter. The series ohmmeter has a back-off scale with zero ohms at the right edge and infinity at the left. Adjust for zero ohms with the leads short-circuited each time the ohms range is changed.

- The VOM is a portable multimeter that measures volts, ohms, and milliamperes.

- The digital multimeter generally has an input resistance of 10 MΩ on all dc voltage ranges.

- In checking wire conductors, the ohmmeter reads 0 Ω or very low R for normal continuity and infinite ohms for an open.

Important Terms

- Amp-Clamp Probe - a meter that can measure ac currents, generally from the 60-Hz ac power line, without breaking open the circuit. The probe of the meter is actually a clamp that fits around the current-carrying conductor.

- Analog Multimeter - a test instrument that is used to measure voltage, current, and resistance. An analog multimeter uses a moving pointer and a printed scale to display the value of the measured quantity.

- Back-Off Ohmmeter Scale - an ohmmeter scale that shows zero ohms (0 Ω) for full-scale deflection and infinite ohms (∞ Ω) for no deflection. As the name implies, the ohms of resistance increase from right to left on the scale as the pointer backs off from full-scale deflection.

- Continuity Testing - a resistance measurement that determines whether or not there is zero ohms of resistance (approximately) between two points, such as across the ends of a wire conductor.

- Digital Multimeter (DMM) - a popular test instrument that is used to measure voltage, current, and resistance. A DMM uses a numeric display to indicate directly the value of the measured quantity.

- Loading Effect - a term that describes the reduction in measured voltage when using a voltmeter to measure the voltage in a circuit. The term may also be applied to describe the reduction in current when using a current meter to measure the current in a circuit. The loading effect of a voltmeter occurs when the voltmeter resistance is not high enough. Conversely, the loading effect of a current meter occurs when the resistance of the current meter is too high.

- Multiplier Resistor - a large resistance in series with a moving-coil meter movement which allows the meter to measure voltages in a circuit.

- Ohms-Per-Volt (Ω/V) Rating - a voltmeter rating that specifies the ohms of resistance needed per 1 V of deflection. The Ω/V rating = $1/I_M$ or R_V/V_{Range}.

- Shunt Resistor - a resistor placed in parallel with a basic moving-coil meter movement to extend the current range beyond the I_M value of the meter movement.

- Zero-Ohms Adjustment - a control on an analog VOM which is adjusted for zero ohms with the ohmmeter leads shorted. This adjustment should be made each time the ohmmeter range is changed so that the ohmmeter scale remains calibrated.

Related Formulas

$$R_S = \frac{V_M}{I_S}$$

$$I_S = I_T - I_M$$

$$R_{mult} = \frac{\text{Full-scale } V}{\text{Full-scale } I} - r_M$$

$$V = V_M + \frac{R_1 R_2}{R_V(R_1 + R_2)} V_M$$

Self-Test

Answers at back of book.

1. **For a moving-coil meter movement, I_M is**
   a. the amount of current needed in the moving-coil to produce full-scale deflection of the meter's pointer.
   b. the value of current flowing in the moving-coil for any amount of pointer deflection.
   c. the amount of current required in the moving-coil to produce half-scale deflection of the meter's pointer.
   d. none of the above.

2. **For an analog VOM with a mirror along the printed scale,**
   a. the pointer deflection will be magnified by the mirror when measuring small values of voltage, current, and resistance.
   b. the meter should always be read by looking at the meter from the side.
   c. the meter is read when the pointer and its mirror reflection appear as one.
   d. both a and b.

3. **A current meter should have a**
   a. very high internal resistance.
   b. very low internal resistance.
   c. infinitely high internal resistance.
   d. none of the above.

4. **A voltmeter should have a**
   a. resistance of about 0 Ω.
   b. very low resistance.
   c. very high internal resistance.
   d. none of the above.

5. **Voltmeter loading is usually a problem when measuring voltages in**
   a. parallel circuits.
   b. low resistance circuits.
   c. a series circuit with low resistance values.
   d. high resistance circuits.

6. **To double the current range of a 50-μA, 2-kΩ moving-coil meter movement, the shunt resistance, R_S, should be**
   a. 2 kΩ.
   b. 1 kΩ.

   c. 18 kΩ.
   d. 50 kΩ.

7. **A voltmeter using a 20-μA meter movement has an Ω/V rating of**
   a. $\dfrac{20\ k\Omega}{V}$.
   b. $\dfrac{50\ k\Omega}{V}$.
   c. $\dfrac{1\ k\Omega}{V}$.
   d. $\dfrac{10\ M\Omega}{V}$.

8. **As the current range of an analog meter is increased, the overall meter resistance, R_M,**
   a. decreases.
   b. increases.
   c. stays the same.
   d. none of the above.

9. **As the voltage range of an analog VOM is increased, the total voltmeter resistance, R_V,**
   a. decreases.
   b. increases.
   c. stays the same.
   d. none of the above.

10. **An analog VOM has an Ω/V rating of 10 kΩ/V. What is the voltmeter resistance, R_V, if the voltmeter is set to the 25 V range?**
    a. 10 kΩ.
    b. 10 MΩ.
    c. 25 kΩ.
    d. 250 kΩ.

11. **What shunt resistance, R_S, is needed to make a 100-μA, 1-kΩ meter movement capable of measuring currents from 0–5 mA?**
    a. 25 Ω.
    b. 10.2 Ω.
    c. 20.41 Ω.
    d. 1 kΩ.

12. **For a 30-V range, a 50-μA, 2-kΩ meter movement needs a multiplier resistor of**
    a. 58 kΩ.
    b. 598 kΩ.

    c. 10 MΩ.
    d. 600 kΩ.

13. **When set to any of the voltage ranges, a typical DMM has an input resistance of**
    a. about 0 Ω.
    b. 20 kΩ.
    c. 10 MΩ.
    d. 1 kΩ.

14. **When using an ohmmeter to measure resistance in a circuit,**
    a. the power in the circuit being tested must be off.
    b. the power in the circuit being tested must be on.
    c. the power in the circuit being tested may be on or off.
    d. the power in the circuit being tested should be turned on after the leads are connected.

15. **Which of the following voltages cannot be displayed by a DMM with a $3\frac{1}{2}$-digit display?**
    a. 7.64 V.
    b. 13.5 V.
    c. 19.98 V.
    d. 29.98 V.

16. **What type of meter can be used to measure ac currents without breaking open the circuit?**
    a. an analog VOM.
    b. an amp-clamp probe.
    c. a DMM.
    d. There isn't such a meter.

17. **Which of the following measurements is usually the most inconvenient and time-consuming when troubleshooting?**
    a. resistance measurements.
    b. dc voltage measurements.
    c. current measurements.
    d. ac voltage measurements.

18. **An analog ohmmeter reads 18 on the $R \times 10$ k range. What is the value of the measured resistance?**
    a. 180 kΩ.
    b. 18 kΩ.
    c. 18 Ω.
    d. 180 Ω.

19. Which meter has a higher resistance, a DMM with 10 MΩ of resistance on all dc voltage ranges or an analog VOM with a 50 kΩ/V rating set to the 250-V range?

 a. the DMM.

 b. the analog VOM.

 c. They both have the same resistance.

 d. This is impossible to determine.

20. When using an ohmmeter to measure the continuity of a wire, the resistance should measure

 a. about 0 Ω if the wire is good.

 b. infinity if the wire is broken (open).

 c. very high resistance if the wire is good.

 d. both a and b.

Questions

1. (a) Why is a milliammeter connected in series in a circuit? (b) Why should the milliammeter have low resistance?

2. (a) Why is a voltmeter connected in parallel in a circuit? (b) Why should the voltmeter have high resistance?

3. A circuit has a battery across two resistances in series. (a) Draw a diagram showing how to connect a milliammeter in the correct polarity to read current through the junction of the two resistances. (b) Draw a diagram showing how to connect a voltmeter in the correct polarity to read the voltage across one resistance.

4. Explain briefly why a meter shunt equal to the resistance of the moving coil doubles the current range.

5. Describe how to adjust the ZERO OHMS control on a back-off ohmmeter.

6. What is meant by a 3½-digit display on a DMM?

7. Give two advantages of the DMM in Fig. 8–14 compared with the conventional VOM in Fig. 8–13.

8. What does the zero ohms control in the circuit of a back-off ohmmeter do?

9. State two precautions to be observed when you use a milliammeter.

10. State two precautions to be observed when you use an ohmmeter.

11. The resistance of a voltmeter R_V is 300 kΩ on the 300-V range when measuring 300 V. Why is R_V still 300 kΩ when measuring 250 V on the same range?

12. Give a typical value of voltmeter resistance for a DMM.

13. Would you rather use a DMM or VOM in troubleshooting? Why?

Problems

SECTION 8–2 METER SHUNTS

8–1 Calculate the value of the shunt resistance, R_S, needed to extend the range of the meter movement in Fig. 8–22 to (a) 2 mA; (b) 10 mA; (c) 25 mA; (d) 100 mA.

8–3 Calculate the value of the shunt resistance, R_S, needed to extend the range of the meter movement in Fig. 8–23 to (a) 100 µA; (b) 1 mA; (c) 5 mA; (d) 10 mA; (e) 50 mA; (f) 100 mA.

Figure 8–22

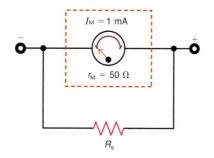

$I_M = 1$ mA

$r_M = 50$ Ω

R_s

Figure 8–23

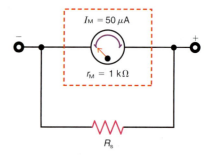

$I_M = 50$ µA

$r_M = 1$ kΩ

R_s

8–2 What is the resistance, R_M, of the meter (R_S in parallel with r_M) for each of the current ranges listed in Problem 8–1?

8–4 What is the resistance, R_M, of the meter (R_S in parallel with r_M) for each of the current ranges listed in Problem 8–3.

8–5 Refer to Fig. 8–24. (a) Calculate the values for the separate shunt resistances, R_{S1}, R_{S2}, and R_{S3}. (b) Calculate the resistance, R_M, of the meter (R_S in parallel with r_M) for each setting of the range switch.

Figure 8–24

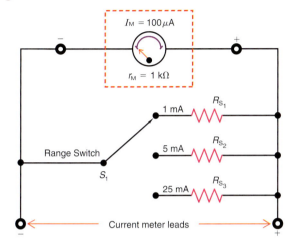

8–6 Repeat Problem 8–5 if the meter movement has the following characteristics: $I_M = 250\ \mu A$, $r_M = 2\ k\Omega$.

8–7 Why is it desirable for a current meter to have very low internal resistance?

SECTION 8–3 VOLTMETERS

8–8 Calculate the required multiplier resistance, R_{mult}, in Fig. 8–25 for each of the following voltage ranges: (a) 1 V; (b) 5 V; (c) 10 V; (d) 50 V; (e) 100 V; and (f) 500 V.

Figure 8–25

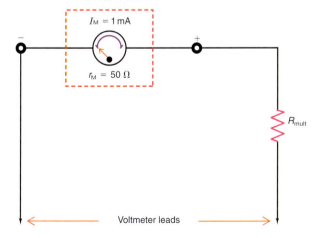

8–9 What is the Ω/V rating of the voltmeter in Problem 8–8?

8–10 Calculate the required multiplier resistance, R_{mult}, in Fig. 8–26 for each of the following voltage ranges: (a) 3 V; (b) 10 V; (c) 30 V; (d) 100 V; and (e) 300 V.

Figure 8–26

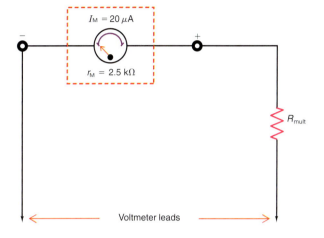

8–11 What is the Ω/V rating of the voltmeter in Problem 8–10?

8–12 Refer to Fig. 8–27. (a) Calculate the values for the multiplier resistors R_1, R_2, R_3, R_4, R_5, and R_6. (b) Calculate the total voltmeter resistance, R_V, for each setting of the range switch; (c) Determine the Ω/V rating of the voltmeter.

Figure 8–27

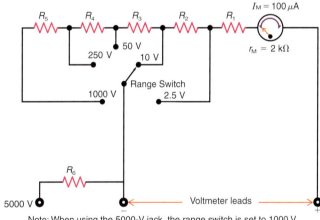

Note: When using the 5000-V jack, the range switch is set to 1000 V.

8-13 Refer to Fig. 8-28. (a) Calculate the values for the multiplier resistors R_1, R_2, R_3, R_4, R_5, and R_6; (b) Calculate the total voltmeter resistance, R_V, for each setting of the range switch; (c) Determine the Ω/V rating of the voltmeter.

Figure 8-28

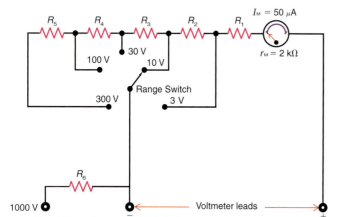

Note: When using the 1000 V jack, the range switch is set to 300 V.

8-14 A certain voltmeter has an Ω/V rating of 25 kΩ/V. Calculate the total voltmeter resistance, R_V, for the following voltmeter ranges: (a) 2.5 V; (b) 10 V; (c) 25 V; (d) 100 V; (e) 250 V; (f) 1000 V; (g) 5000 V.

8-15 Calculate the Ω/V rating of a voltmeter that uses a meter movement with an I_M value of (a) 1 mA; (b) 100 μA; (c) 50 μA; (d) 10 μA.

SECTION 8-4 LOADING EFFECT OF A VOLTMETER

8-16 Refer to Fig. 8-29. (a) Calculate the dc voltage that should exist across R_2 without the voltmeter present; (b) Calculate the dc voltage that would be measured across R_2 using a 10 kΩ/V analog voltmeter set to the 10-V range; (c) Calculate the dc voltage that would be measured across R_2 using a DMM having an R_V of 10 MΩ on all dc voltage ranges.

Figure 8-29

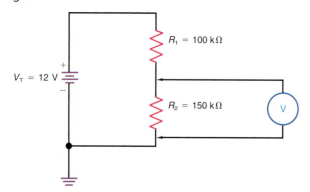

8-17 Repeat Problem 8-16 if $R_1 = 1$ kΩ and $R_2 = 1.5$ kΩ.

8-18 Refer to Fig. 8-30. (a) Calculate the dc voltage that should exist across R_2 without the voltmeter present; (b) Calculate the dc voltage that would be measured across R_2 using a 100 kΩ/V analog voltmeter set to the 10-V range; (c) Calculate the dc voltage that would be measured across R_2 using a DMM with an R_V of 10 MΩ on all dc voltage ranges.

Figure 8-30

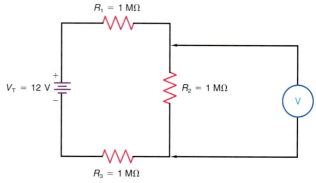

8-19 In Problem 8-18, which voltmeter produced a greater loading effect? Why?

8-20 In Fig. 8-31, determine (a) the voltmeter resistance, R_V (b) the corrected voltmeter reading using Formula 8-4.

Figure 8-31

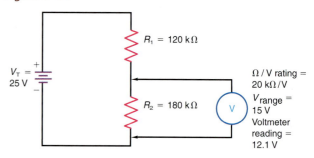

SECTION 8-5 OHMMETERS

8-21 Figure 8-32 shows a series ohmmeter and its corresponding meter face. How much is the external resistance, R_X, across the ohmmeter leads for (a) full-scale deflection; (b) three-fourths full-scale deflection; (c) one-half-scale deflection; (d) one-fourth full-scale deflection; (e) no deflection?

8-22 In Fig. 8-32, how much is the external resistance, R_X, for (a) four-fifths full-scale deflection; (b) two-thirds full-scale deflection; (c) three-fifths full-scale deflection; (d) two-fifths full-scale deflection; (e) one-third full-scale deflection; and (f) one-fifth full-scale deflection.

8-23 If the resistance values in Problems 8-21 and 8-22 were plotted on the scale of the meter face in Fig. 8-32, would the scale be linear or nonlinear? Why?

8-24 For the series ohmmeter in Fig. 8-32, is the orientation of the ohmmeter leads important when measuring the value of a resistor?

Figure 8-32

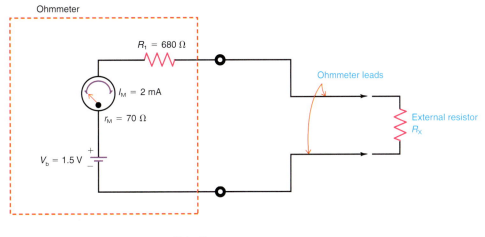

Ohmmeter

$R_1 = 680\ \Omega$

$I_M = 2\ \text{mA}$

$r_M = 70\ \Omega$

$V_b = 1.5\ \text{V}$

Ohmmeter leads

External resistor R_X

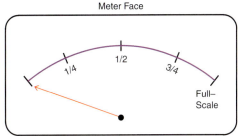

Meter Face

1/4   1/2   3/4

Full–
Scale

8-25 Why is the ohmmeter scale in Fig. 8-32 referred to as a back-off ohmmeter scale?

8-26 An analog ohmmeter has five range settings: R × 1, R × 10, R × 100, R × 1 k and R × 10 k. Determine the measured resistance for each ohmmeter reading listed below.

Ohmmeter Reading	Range Setting	Measured Resistance
2.4	R × 100	?
100	R × 10	?
8.6	R × 10 k	?
5.5	R × 1	?
100	R × 1 k	?
30	R × 100	?
500	R × 10 kΩ	?

8-27 Analog multimeters have a zero-ohm adjustment control for the ohmmeter portion of the meter. What purpose does it serve and how is it used?

SECTION 8-8 METER APPLICATIONS

8-28 On what range should you measure an unknown value of voltage or current? Why?

8-29 What might happen to an ohmmeter if it is connected across a resistor in a live circuit?

8-30 Why is one lead of a resistor disconnected from the circuit when measuring its resistance value?

8-31 Is a current meter connected in series or in parallel? Why?

8-32 How can the inconvenience of opening a circuit to measure current be eliminated in most cases?

8-33 What is the resistance of a
a. good fuse?
b. blown fuse?

8-34 In Fig. 8-33, list the voltages at points A, B, C, and D (with respect to ground) for each of the following situations.
a. All resistors normal
b. R_1 open
c. R_2 open
d. R_3 open
e. R_4 open

Figure 8-33

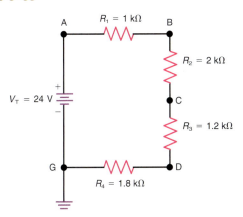

$R_1 = 1\ \text{k}\Omega$

$R_2 = 2\ \text{k}\Omega$

$V_T = 24\ \text{V}$

$R_3 = 1.2\ \text{k}\Omega$

$R_4 = 1.8\ \text{k}\Omega$

Critical Thinking

8-35 Figure 8-34 shows a universal-shunt current meter. Calculate the values for R_1, R_2, and R_3 that will provide current ranges of 2, 10, and 50 mA.

8-36 Design a series ohmmeter using a 2-kΩ, 50-μA meter movement and a 1.5-V battery. The center-scale ohm reading is to be 150 Ω.

8-37 The voltmeter across R_2 in Fig. 8-35 shows 20 V. If the voltmeter is set in the 30-V range, calculate the Ω/V rating of the meter.

Figure 8-34 Circuit diagram for Critical Thinking Prob. 8-35.

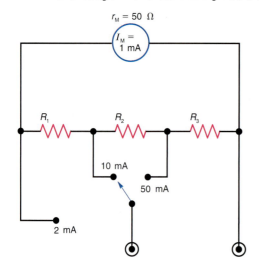

Figure 8-35 Circuit diagram for Critical Thinking Prob. 8-37.

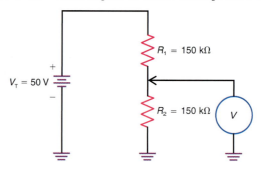

Answers to Knowledge Check Problems

8-1 **a.** 250 μA
   b. To eliminate the optical error called parallax.

8-2A Low

8-2B R_S needs to be 2.08 Ω approximately.

8-3A R_1 = 49.95 kΩ

8-3B Ω/V rating = 50 kΩ/V

8-4 High

8-5 R_X = 500 Ω

8-6 Its high resistance on all voltmeter ranges.

8-7 No, because the left-most digit can only be a 1 when four digits are displayed on a 3^1/$_2$-digit display.

8-8 100 V

8-9 Infinite ohms

Answers to Self-Reviews

8-1 **a.** ^{1}/$_2$ scale
   b. There is no deflection.

8-2 **a.** I_S = 450 μA
   b. V_M = 0.045 V
   c. R_S = 100 Ω

8-3 **a.** R_V = 50 kΩ
   b. R_V = 1 MΩ
   c. series
   d. parallel
   e. R = 500 kΩ

8-4 **a.** V_1 = 80 V
   b. V_2 = 40 V

8-5 **a.** R_X = 400 Ω
   b. V = 4 V

8-6 **a.** 20 kΩ
   b. 10 MΩ
   c. F

8-7 **a.** T
   b. F
   c. T

8-8 **a.** ohmmeter
   b. 0 V
   c. 25 V
   d. 0 V

8-9 **a.** right edge
   b. ∞ ohms

Cumulative Review Summary (Chapters 7–8)

- In a series voltage divider, the IR drop across each resistance is proportional to its R. A larger R has a larger voltage drop. Each $V = (R/R_T) \times V_T$. In this way, the series voltage drops can be calculated from V_T without I.

- In a parallel current divider, each branch current is inversely related to its R. A smaller R has more branch current. For only two resistances, we can use the inverse relation

$$I_1 = [R_2/(R_1 + R_2)] \times I_T$$

  In this way, the branch currents can be calculated from I_T without V.

- In a parallel current divider, each branch current is directly proportional to its conductance G. A larger G has more branch current. For any number of parallel resistances, each branch $I = (G/G_T) \times I_T$.

- A milliammeter or ammeter is a low-resistance meter connected in series in a circuit to measure current.

- Different current ranges are obtained by meter shunts in parallel with the meter.

- A voltmeter is a high-resistance meter connected across the voltage to be measured.

- Different voltage ranges are obtained by multipliers in series with the meter.

- An ohmmeter has an internal battery to indicate the resistance of a component across its two terminals with external power off.

- In making resistance tests, remember that $R = 0\ \Omega$ for continuity or a short circuit, but the resistance of an open circuit is infinitely high.

- Figure 8–1 shows a VOM and DMM. Both types can be used for voltage, current, and resistance measurements.

Cumulative Review Self-Test

Answers at back of book.

Answer True or False.

1. The internal R of a milliammeter must be low to have minimum effect on I in the circuit.

2. The internal R of a voltmeter must be high to have minimum current through the meter.

3. Power must be off when checking resistance in a circuit because the ohmmeter has its own internal battery.

4. In the series voltage divider in Fig. 8–19, the normal voltage from point B to ground is 75 V.

5. In Fig. 8–19, the normal voltage across R_1, between A and B, is 75 V.

6. The highest ohm range is best for checking continuity with an ohmmeter.

7. With four equal resistors in a series voltage divider with V_T of 44.4 V, each IR drop is 11.1 V.

8. With four equal resistors in parallel with I_T of 44.4 mA, each branch current is 11.1 mA.

9. Series voltage drops divide V_T in direct proportion to each series R.

10. Parallel currents divide I_T in direct proportion to each branch R.

11. The VOM cannot be used to measure current.

12. The DMM can be used as a high-resistance voltmeter.

chapter

9

Kirchhoff's Laws

Many types of circuits have components that are not in series, in parallel, or in series-parallel. For example, a circuit may have two voltages applied in different branches. Another example is an unbalanced bridge circuit. When the rules of series and parallel circuits cannot be applied, more general methods of analysis become necessary. These methods include the application of Kirchhoff's laws, as described in this chapter.

All circuits can be solved by Kirchhoff's laws because the laws do not depend on series or parallel connections. Although Kirchhoff's voltage and current laws were introduced briefly in Chaps. 4 and 5, respectively, this chapter takes a more in-depth approach to using Kirchhoff's laws for circuit analysis.

Kirchhoff's voltage and current laws were stated in 1847 by the German physicist Gustav R. Kirchhoff.

KVL: The algebraic sum of the voltage sources and IR voltage drops in any closed path must total zero.

KCL: At any point in a circuit, the algebraic sum of the currents directed into and out of a point must total zero.

These are the most precise statements of Kirchhoff's voltage and current laws. As you will see in this chapter, these statements do not conflict with the more general statements of Kirchhoff's laws used in earlier chapters.

Objectives

After studying this chapter you should be able to

- *State* Kirchhoff's current law.
- *State* Kirchhoff's voltage law.
- *Use* the method of branch currents to solve for all voltages and currents in a circuit containing two or more voltage sources in different branches.
- *Use* node-voltage analysis to solve for the unknown voltages and currents in a circuit containing two or more voltage sources in different branches.
- *Use* the method of mesh currents to solve for the unknown voltages and currents in a circuit containing two or more voltage sources in different branches.

Outline

Important Terms

Kirchhoff's current law (KCL)

Kirchhoff's voltage law (KVL)

loop

loop equation

mesh

mesh current

node

principal node

9–1 Kirchhoff's Current Law (KCL)

The algebraic sum of the currents entering and leaving any point in a circuit must equal zero. Or stated another way, *the algebraic sum of the currents into any point of the circuit must equal the algebraic sum of the currents out of that point.* Otherwise, charge would accumulate at the point, instead of having a conducting path. An *algebraic sum* means combining positive and negative values.

Algebraic Signs

In using Kirchhoff's laws to solve circuits, it is necessary to adopt conventions that determine the algebraic signs for current and voltage terms. A convenient system for currents is to *consider all currents into a branch point as positive and all currents directed away from that point as negative.*

As an example, in Fig. 9–1 we can write the currents as

$$I_A + I_B - I_C = 0$$

or

$$5\,A + 3\,A - 8\,A = 0$$

Currents I_A and I_B are positive terms because these currents flow into P, but I_C, directed out, is negative.

Figure 9–1 Current I_C out from point P equals 5 A + 3 A into P.

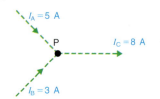

Current Equations

For a circuit application, refer to point C at the top of the diagram in Fig. 9–2. The 6-A I_T into point C divides into the 2-A I_3 and 4-A I_{4-5}, both directed out. Note that I_{4-5} is the current through R_4 and R_5. The algebraic equation is

$$I_T - I_3 - I_{4-5} = 0$$

Substituting the values for these currents,

$$6\,A - 2\,A - 4\,A = 0$$

For the opposite directions, refer to point D at the bottom of Fig. 9–2. Here the branch currents into D combine to equal the main-line current I_T

IIII MultiSim **Figure 9–2** Series-parallel circuit illustrating Kirchhoff's laws. See text for voltage and current equations.

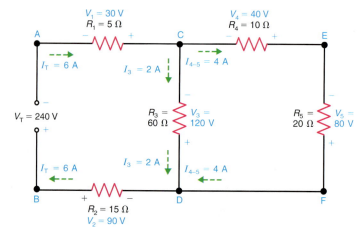

returning to the voltage source. Now I_T is directed out from D with I_3 and I_{4-5} directed in. The algebraic equation is

$$I_3 + I_{4-5} - I_T = 0$$

or

$$2\,A + 4\,A - 6\,A = 0$$

$I_{in} = I_{out}$

Note that at either point C or point D in Fig. 9–2, the sum of the 2-A and 4-A branch currents must equal the 6-A total line current. Therefore, Kirchhoff's current law can also be stated as $I_{in} = I_{out}$. For Fig. 9–2, the equations of current can be written:

At point C: $6\,A = 2\,A + 4\,A$
At point D: $2\,A + 4\,A = 6\,A$

Kirchhoff's current law is the basis for the practical rule in parallel circuits that the total line current must equal the sum of the branch currents.

Example 9–1

In Fig. 9–3, apply Kirchhoff's Current Law to solve for the unknown current, I_3.

Figure 9–3

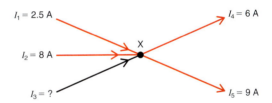

ANSWER In Fig. 9–3, the currents I_1, I_2, and I_3 flowing into point X are considered positive, whereas the currents I_4 and I_5 flowing away from point X are considered negative. Expressing the currents as an equation gives us,

$$I_1 + I_2 + I_3 - I_4 - I_5 = 0$$

or

$$I_1 + I_2 + I_3 = I_4 + I_5$$

Inserting the values from Fig. 9–3,

$$2.5\,A + 8\,A + I_3 = 6\,A + 9\,A$$

Solving for I_3 gives us

$$I_3 = 6\,A + 9\,A - 2.5\,A - 8\,A$$
$$= 4.5\,A$$

PIONEERS
IN ELECTRONICS

German physicist *Gustav Kirchhoff (1824–1887)* is best known for his statement of two basic laws of the behavior of current and voltage. Developed in 1847, these laws enable scientists to understand and therefore evaluate the behavior of networks.

■ *9–1 Knowledge Check*

> *Answer at end of chapter.*
> If $I_4 = 9.5$ A in Fig. 9–3, how much is I_3?

■ *9–1 Self-Review*

> *Answers at end of chapter.*
> a. With a 1-A I_1, 2-A I_2, and 3-A I_3 into a point, how much is I_{out}?
> b. If I_1 into a point is 3 A and I_3 out of that point is 7 A, how much is I_2 into that point?

9–2 Kirchhoff's Voltage Law (KVL)

The algebraic sum of the voltages around any closed path is zero. If you start from any point at one potential and come back to the same point and the same potential, the difference of potential must be zero.

Algebraic Signs

In determining the algebraic signs for voltage terms in a KVL equation, first mark the polarity of each voltage as shown in Fig. 9–2. A convenient system is to *go around any closed path and consider any voltage whose negative terminal is reached first as a negative term and any voltage whose positive terminal is reached first as a positive term.* This method applies to *IR* voltage drops and voltage sources. The direction can be clockwise or counterclockwise.

Remember that electrons flowing into a resistor make that end negative with respect to the other end. For a voltage source, the direction of electrons returning to the positive terminal is the normal direction for electron flow, which means that the source should be a positive term in the voltage equation.

When you go around the closed path and come back to the starting point, the algebraic sum of all the voltage terms must be zero. There cannot be any potential difference for one point.

If you do not come back to the start, then the algebraic sum is the voltage between the start and finish points.

You can follow any closed path because the voltage between any two points in a circuit is the same regardless of the path used in determining the potential difference.

Loop Equations

Any closed path is called a *loop*. A loop equation specifies the voltages around the loop.

Figure 9–2 has three loops. The outside loop, starting from point A at the top, through CEFDB, and back to A, includes the voltage drops V_1, V_4, V_5, and V_2 and the source V_T.

The inside loop ACDBA includes V_1, V_3, V_2, and V_T. The other inside loop, CEFDC with V_4, V_5, and V_3, does not include the voltage source.

Consider the voltage equation for the inside loop with V_T. In the clockwise direction starting from point A, the algebraic sum of the voltages is

$$-V_1 - V_3 - V_2 + V_T = 0$$

or

$$-30\,\text{V} - 120\,\text{V} - 90\,\text{V} + 240\,\text{V} = 0$$

Voltages V_1, V_3, and V_2 have negative signs, because the negative terminal for each of these voltages is reached first. However, the source V_T is a positive term because its plus terminal is reached first, going in the same direction.

For the opposite direction, going counterclockwise in the same loop from point B at the bottom, V_2, V_3, and V_1 have positive values and V_T is negative. Then

$$V_2 + V_3 + V_1 - V_T = 0$$

or

$$90\text{ V} + 120\text{ V} + 30\text{ V} - 240\text{ V} = 0$$

When we transpose the negative term of -240 V, the equation becomes

$$90\text{ V} + 120\text{ V} + 30\text{ V} = 240\text{ V}$$

This equation states that the sum of the voltage drops equals the applied voltage.

$$\Sigma V = V_T$$

The Greek letter Σ means "sum of." In either direction, for any loop, the sum of the IR voltage drops must equal the applied voltage V_T. In Fig. 9–2, for the inside loop with the source V_T, going counterclockwise from point B,

$$90\text{ V} + 120\text{ V} + 30\text{ V} = 240\text{ V}$$

This system does not contradict the rule for algebraic signs. If 240 V were on the left side of the equation, this term would have a negative sign.

Stating a loop equation as $\Sigma V = V_T$ eliminates the step of transposing the negative terms from one side to the other to make them positive. In this form, the loop equations show that Kirchhoff's voltage law is the basis for the practical rule in series circuits that the sum of the voltage drops must equal the applied voltage.

When a loop does not have any voltage source, the algebraic sum of the IR voltage drops alone must total zero. For instance, in Fig. 9–2, for the loop CEFDC without the source V_T, going clockwise from point C, the loop equation of voltages is

$$-V_4 - V_5 + V_3 = 0$$
$$-40\text{V} - 80\text{V} + 120\text{V} = 0$$
$$0 = 0$$

Notice that V_3 is positive now, because its plus terminal is reached first by going clockwise from D to C in this loop.

Example 9-2

In Fig. 9–4a, apply Kirchhoff's Voltage Law to solve for the voltages V_{AG} and V_{BG}.

ANSWER   In Fig. 9–4a, the voltage sources V_1 and V_2 are connected in a series-aiding fashion since they both force electrons to flow through the circuit in the same direction. The earth ground connection at the junction of V_1 and V_2 is used simply for a point of reference. The circuit is solved as follows:

$$V_T = V_1 + V_2$$
$$= 18\text{ V} + 18\text{ V}$$
$$= 36\text{ V}$$

Figure 9-4

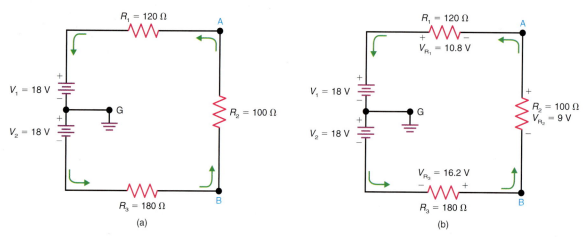

$$R_T = R_1 + R_2 + R_3$$
$$= 120\ \Omega + 100\ \Omega + 180\ \Omega$$
$$= 400\ \Omega$$

$$I = \frac{V_T}{R_T}$$
$$= \frac{36\ V}{400\ \Omega}$$
$$= 90\ mA$$

$$V_{R_1} = I \times R_1$$
$$= 90\ mA \times 120\ \Omega$$
$$= 10.8\ V$$

$$V_{R_2} = I \times R_2$$
$$= 90\ mA \times 100\ \Omega$$
$$= 9\ V$$

$$V_{R_3} = I \times R_3$$
$$= 90\ mA \times 180\ \Omega$$
$$= 16.2\ V$$

Figure 9–4b shows the voltage drops across each resistor. Notice that the polarity of each resistor voltage drop is negative at the end where the electrons enter the resistor and positive at the end where they leave.

Next, we can apply Kirchhoff's Voltage Law to determine if we have solved the circuit correctly. If we go counterclockwise (CCW) around the loop, starting and ending at the positive (+) terminal of V_1, we should obtain an algebraic sum of 0 V. The loop equation is written as

$$V_1 + V_2 - V_{R_3} - V_{R_2} - V_{R_1} = 0$$

Notice that the voltage sources V_1 and V_2 are considered positive terms in the equation because their positive (+) terminals were reached first when going around the loop. Similarly, the voltage drops V_{R_1}, V_{R_2}, and V_{R_3} are considered negative terms because the negative (−) end of each resistor's voltage drop is encountered first when going around the loop. Substituting the values from Fig. 9–4b gives us

$$18\ V + 18\ V - 16.2\ V - 9\ V - 10.8\ V = 0$$

It is important to realize that the sum of the resistor voltage drops must equal the applied voltage, V_T, which equals $V_1 + V_2$ or 36 V in this case. Expressed as an equation,

$$V_T = V_{R_1} + V_{R_2} + V_{R_3}$$
$$= 10.8\ V + 9\ V + 16.2\ V$$
$$= 36\ V$$

It is now possible to solve for the voltages V_{AG} and V_{BG} by applying Kirchhoff's Voltage Law. To do so, simply add the voltages algebraically between the start and finish points which are points A and G for V_{AG} and points B and G for V_{BG}. Using the values from Figure 9–4b,

$$V_{AG} = -V_{R_1} + V_1 \qquad \text{(CCW from A to G)}$$
$$= -10.8 \text{ V} + 18 \text{ V}$$
$$= 7.2 \text{ V}$$

Going clockwise (CW) from A to G produces the same result.

$$V_{AG} = V_{R_2} + V_{R_3} - V_2 \qquad \text{(CW from A to G)}$$
$$= 9 \text{ V} + 16.2 \text{ V} - 18 \text{ V}$$
$$= 7.2 \text{ V}$$

Since there are fewer voltages to add going counterclockwise from point A, it is the recommended solution for V_{AG}.

The voltage, V_{BG}, is found by using the same technique.

$$V_{BG} = V_{R_3} - V_2 \qquad \text{(CW from B to G)}$$
$$= 16.2 \text{ V} - 18 \text{ V}$$
$$= -1.8 \text{ V}$$

Going around the loop in the other direction gives us

$$V_{BG} = -V_{R_2} - V_{R_1} + V_1 \qquad \text{(CCW from B to G)}$$
$$= -9 \text{ V} - 10.8 \text{ V} + 18 \text{ V}$$
$$= -1.8 \text{ V}$$

Since there are fewer voltages to add going clockwise from point B, it is the recommended solution for V_{BG}.

■ 9–2 Knowledge Check

Answers at end of chapter.

Re-solve for V_{AG} and V_{BG} in Fig. 9–4 if $V_2 = 24$ V.

■ 9–2 Self-Review

Answers at end of chapter.

Refer to Fig. 9–2.
a. For partial loop CEFD, what is the total voltage across CD with −40 V for V_4 and −80 V for V_5?
b. For loop CEFDC, what is the total voltage with −40 V for V_4, −80 V for V_5, and including 120 V for V_3?

9–3 Method of Branch Currents

Now we can use Kirchhoff's laws to analyze the circuit in Fig. 9–5. The problem is to find the currents and voltages for the three resistors.

First, indicate current directions and mark the voltage polarity across each resistor consistent with the assumed current. Remember that electron flow in a resistor produces negative polarity where the current enters. In Fig. 9–5, we assume that the source V_1 produces electron flow from left to right through R_1, and V_2 produces electron flow from right to left through R_2.

The three different currents in R_1, R_2, and R_3 are indicated as I_1, I_2, and I_3. However, three unknowns would require three equations for the solution. From Kirchhoff's current law, $I_3 = I_1 + I_2$, as the current out of point C

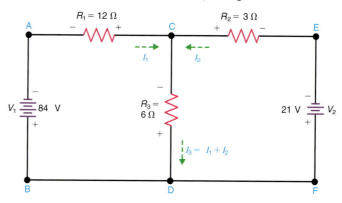

Figure 9–5 Application of Kirchhoff's laws to a circuit with two sources in different branches. See text for solution by finding the branch currents.

must equal the current in. The current through R_3, therefore, can be specified as $I_1 + I_2$.

With two unknowns, two independent equations are needed to solve for I_1 and I_2. These equations are obtained by writing two Kirchhoff's voltage law equations around two loops. There are three loops in Fig. 9–5, the outside loop and two inside loops, but we need only two. The inside loops are used for the solution here.

Writing the Loop Equations

For the loop with V_1, start at point B, at the bottom left, and go clockwise through V_1, V_{R_1}, and V_{R_3}. This equation for loop 1 is

$$84 - V_{R_1} - V_{R_3} = 0$$

For the loop with V_2, start at point F, at the lower right, and go counterclockwise through V_2, V_{R_2} and V_{R_3}. This equation for loop 2 is

$$21 - V_{R_2} - V_{R_3} = 0$$

Using the known values of R_1, R_2, and R_3 to specify the IR voltage drops,

$$V_{R_1} = I_1 R_1 = I_1 \times 12 = 12I_1$$
$$V_{R_2} = I_2 R_2 = I_2 \times 3 = 3I_2$$
$$V_{R_3} = (I_1 + I_2)R_3 = 6(I_1 + I_2)$$

Substituting these values in the voltage equation for loop 1,

$$84 - 12I_1 - 6(I_1 + I_2) = 0$$

Also, in loop 2,

$$21 - 3I_2 - 6(I_1 + I_2) = 0$$

Multiplying $(I_1 + I_2)$ by 6 and combining terms and transposing, the two equations are

$$-18I_1 - 6I_2 = -84$$
$$-6I_1 - 9I_2 = -21$$

Divide the top equation by -6 and the bottom equation by -3 to reduce the equations to their simplest terms and to have all positive terms. The two equations in their simplest form then become

$$3I_1 + I_2 = 14$$
$$2I_1 + 3I_2 = 7$$

GOOD TO KNOW

In Fig. 9-5, $I_3 = I_1 + I_2$ as shown. The current $I_1 = I_3 - I_2$ and the current $I_2 = I_3 - I_1$.

Solving for the Currents

These two equations in the two unknowns I_1 and I_2 contain the solution of the network. Note that the equations include every resistance in the circuit. Currents I_1 and I_2 can be calculated by any of the methods for the solution of simultaneous equations. Using the method of elimination, multiply the top equation by 3 to make the I_2 terms the same in both equations. Then

$$9I_1 + 3I_2 = 42$$
$$2I_1 + 3I_2 = 7$$

Subtract the bottom equation from the top equation, term by term, to eliminate I_2. Then, since the I_2 term becomes zero,

$$7I_1 = 35$$
$$I_1 = 5 \text{ A}$$

The 5-A I_1 is the current through R_1. Its direction is from A to C, as assumed, because the answer for I_1 is positive.

To calculate I_2, substitute 5 for I_1 in either of the two loop equations. Using the bottom equation for the substitution,

$$2(5) + 3I_2 = 7$$
$$3I_2 = 7 - 10$$
$$3I_2 = -3$$
$$I_2 = -1 \text{ A}$$

The negative sign for I_2 means that this current is opposite to the assumed direction. Therefore, I_2 flows through R_2 from C to E instead of from E to C as was previously assumed.

Why the Solution for I_2 Is Negative

In Fig. 9–5, I_2 was assumed to flow from E to C through R_2 because V_2 produces electron flow in this direction. However, the other voltage source V_1 produces electron flow through R_2 in the opposite direction from point C to E. This solution of -1 A for I_2 shows that the current through R_2 produced by V_1 is more than the current produced by V_2. The net result is 1 A through R_2 from C to E.

The actual direction of I_2 is shown in Fig. 9–6 with all the values for the solution of this circuit. Notice that the polarity of V_{R_2} is reversed from the assumed polarity in Fig. 9–5. Since the net electron flow through R_2 is actually from C to E, the end of R_2 at C is the negative end. However, the polarity of V_2 is the same in both diagrams because it is a voltage source that generates its own polarity.

Figure 9–6 Solution of circuit in Fig. 9-5 with all currents and voltages.

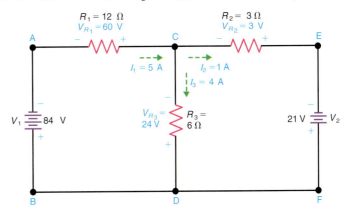

To calculate I_3 through R_3,

$$I_3 = I_1 + I_2 = 5 + (-1)$$
$$I_3 = 4 \text{ A}$$

The 4 A for I_3 is in the assumed direction from C to D. Although the negative sign for I_2 means only a reversed direction, its algebraic value of -1 must be used for substitution in the algebraic equations written for the assumed direction.

Calculating the Voltages

With all the currents known, the voltage across each resistor can be calculated as follows:

$$V_{R_1} = I_1 R_1 = 5 \times 12 = 60 \text{ V}$$
$$V_{R_2} = I_2 R_2 = 1 \times 3 = 3 \text{ V}$$
$$V_{R_3} = I_3 R_3 = 4 \times 6 = 24 \text{ V}$$

All currents are taken as positive, in the correct direction, to calculate the voltages. Then the polarity of each IR drop is determined from the actual direction of current, with electron flow into the negative end (see Fig. 9–6). Notice that V_{R_3} and V_{R_2} have opposing polarities in loop 2. Then the sum of $+3$ V and -24 V equals the -21 V of V_2.

Checking the Solution

As a summary of all answers for this problem, Fig. 9–6 shows the network with all currents and voltages. The polarity of each V is marked from the known directions. In checking the answers, we can see whether Kirchhoff's current and voltage laws are satisfied:

At point C: 5 A = 4 A + 1 A
At point D: 4 A + 1 A = 5 A

Around the loop with V_1 clockwise from B,

84 V − 60 V − 24 V = 0

Around the loop with V_2 counterclockwise from F,

21 V + 3 V − 24 V = 0

Note that the circuit has been solved using only the two Kirchhoff laws without any of the special rules for series and parallel circuits. Any circuit can be solved by applying Kirchhoff's laws for the voltages around a loop and the currents at a branch point.

■ *9–3 Knowledge Check*

Answer at end of chapter.

In the solution to Fig. 9–5, how was it determined that the assumed direction of the current, I_2, was incorrect?

■ *9–3 Self-Review*

Answers at end of chapter.

Refer to Fig. 9–6.
a. How much is the voltage around the partial loop CEFD?
b. How much is the voltage around loop CEFDC?

9–4 Node-Voltage Analysis

In the method of branch currents, these currents are used for specifying the voltage drops around the loops. Then loop equations are written to satisfy Kirchhoff's voltage law. Solving the loop equations, we can calculate the unknown branch currents.

Another method uses voltage drops to specify the currents at a branch point, also called a *node*. Then node equations of currents are written to satisfy Kirchhoff's current law. Solving the node equations, we can calculate the unknown node voltages. This method of node-voltage analysis often is shorter than the method of branch currents.

A node is simply a common connection for two or more components. A *principal node* has three or more connections. In effect, a principal node is a junction or branch point where currents can divide or combine. Therefore, we can always write an equation of currents at a principal node. In Fig. 9–7, points N and G are principal nodes.

However, one node must be the reference for specifying the voltage at any other node. In Fig. 9–7, point G connected to chassis ground is the reference node. Therefore, we need to write only one current equation for the other node N. In general, the number of current equations required to solve a circuit is one less than the number of principal nodes.

Writing the Node Equations

The circuit of Fig. 9–5, earlier solved by the method of branch currents, is redrawn in Fig. 9–7 to be solved now by node-voltage analysis. The problem here is to find the node voltage V_N from N to G. Once this voltage is known, all other voltages and currents can be determined.

The currents in and out of node N are specified as follows: I_1 is the only current through the 12-Ω R_1. Therefore, I_1 is V_{R_1}/R_1 or $V_{R_1}/12\ \Omega$. Similarly, I_2 is $V_{R_2}/3\ \Omega$. Finally, I_3 is $V_{R_3}/6\ \Omega$.

Note that V_{R_3} is the node voltage V_N that we are to calculate. Therefore, I_3 can also be stated as $V_N/6\ \Omega$. The equation of currents at node N is

$$I_1 + I_2 = I_3$$

or

$$\frac{V_{R_1}}{12} + \frac{V_{R_2}}{3} = \frac{V_N}{6}$$

Figure 9–7 Method of node-voltage analysis for the same circuit as in Fig. 9–5. See text for solution by finding V_N across R_3 from the principal node N to ground.

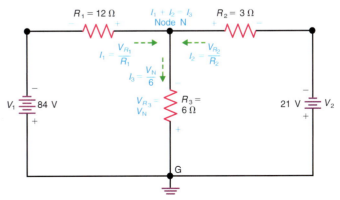

There are three unknowns here, but V_{R_1} and V_{R_2} can be specified in terms of V_N and the known values of V_1 and V_2. We can use Kirchhoff's voltage law because the applied voltage V must equal the algebraic sum of the voltage drops. For the loop with V_1 of 84 V,

$$V_{R_1} + V_N = 84 \quad \text{or} \quad V_{R_1} = 84 - V_N$$

For the loop with V_2 of 21 V,

$$V_{R_2} + V_N = 21 \quad \text{or} \quad V_{R_2} = 21 - V_N$$

Now substitute these values of V_{R_1} and V_{R_2} in the equation of currents:

$$I_1 + I_2 = I_3$$
$$\frac{V_{R_1}}{R_1} + \frac{V_{R_2}}{R_2} = \frac{V_{R_3}}{R_3}$$

Using the value of each V in terms of V_N,

$$\frac{84 - V_N}{12} + \frac{21 - V_N}{3} = \frac{V_N}{6}$$

This equation has only the one unknown, V_N. Clearing fractions by multiplying each term by 12, the equation is

$$(84 - V_N) + 4(21 - V_N) = 2\,V_N$$
$$84 - V_N + 84 - 4\,V_N = 2\,V_N$$
$$-7\,V_N = -168$$
$$V_N = 24 \text{ V}$$

This answer of 24 V for V_N is the same as that calculated for V_{R_3} by the method of branch currents. The positive value means that the direction of I_3 is correct, making V_N negative at the top of R_3 in Fig. 9–7.

Calculating All Voltages and Currents

The reason for finding the voltage at a node, rather than some other voltage, is the fact that a node voltage must be common to two loops. As a result, the node voltage can be used for calculating all voltages in the loops. In Fig. 9–7, with a V_N of 24 V, then V_{R_1} must be $84 - 24 = 60$ V. Also, I_1 is 60 V/12 Ω, which equals 5 A.

To find V_{R_2}, it must be $21 - 24$, which equals -3 V. The negative answer means that I_2 is opposite to the assumed direction and the polarity of V_{R_2} is the reverse of the signs shown across R_2 in Fig. 9–7. The correct directions are shown in the solution for the circuit in Fig. 9–6. The magnitude of I_2 is 3 V/3 Ω, which equals 1 A.

The following comparisons can be helpful in using node equations and loop equations. A node equation applies Kirchhoff's current law to the currents in and out of a node. However, the currents are specified as V/R so that the equation of currents can be solved to find a node voltage.

A loop equation applies Kirchhoff's voltage law to the voltages around a closed path. However, the voltages are specified as IR so that the equation of voltages can be solved to find a loop current. This procedure with voltage equations is used for the method of branch currents explained before with Fig. 9–5 and for the method of mesh currents to be described next with Fig. 9–8.

■ *9–4 Knowledge Check*

Answer at end of chapter.

In the solution to Fig. 9–7, how was it determined that the assumed direction of the current, I_2, was incorrect?

a. How many principal nodes does Fig. 9–7 have?
b. How many node equations are necessary to solve a circuit with three principal nodes?

9–5 Method of Mesh Currents

A mesh is the simplest possible closed path. The circuit in Fig. 9–8 has two meshes, ACDBA and CEFDC. The outside path ACEFDBA is a loop but not a mesh. Each mesh is like a single window frame. There is only one path without any branches.

A mesh current is assumed to flow around a mesh without dividing. In Fig. 9–8, the mesh current I_A flows through V_1, R_1, and R_3; mesh current I_B flows through V_2, R_2, and R_3. A resistance common to two meshes, such as R_3, has two mesh currents, which are I_A and I_B here.

The fact that a mesh current does not divide at a branch point is the difference between mesh currents and branch currents. A mesh current is an assumed current, and a branch current is the actual current. However, when the mesh currents are known, all individual currents and voltages can be determined.

As an example, Fig. 9–8, which has the same circuit as Fig. 9–5, will now be solved by using the assumed mesh currents I_A and I_B. The mesh equations are

$$18I_A - 6I_B = 84 \text{ V} \qquad \text{in mesh A}$$
$$-6I_A + 9I_B = -21 \text{ V} \qquad \text{in mesh B}$$

Writing the Mesh Equations

The number of meshes equals the number of mesh currents, which is the number of equations required. Here two equations are used for I_A and I_B in the two meshes.

The assumed current is usually taken in the same direction around each mesh to be consistent. Generally, the clockwise direction is used, as shown for I_A and I_B in Fig. 9–8.

In each mesh equation, the algebraic sum of the voltage drops equals the applied voltage.

The voltage drops are added going around a mesh in the same direction as its mesh current. Any voltage drop in a mesh produced by its own mesh

Figure 9–8 The same circuit as Fig. 9–5 analyzed as two meshes. See text for solution by calculating the assumed mesh currents I_A and I_B.

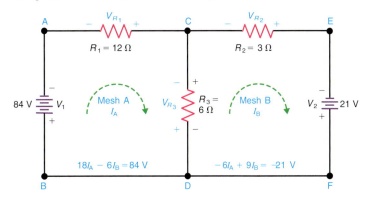

current is considered positive because it is added in the direction of the mesh current.

Since all the voltage drops of a mesh current in its own mesh must have the same positive sign, they can be written collectively as one voltage drop by adding all resistances in the mesh. For instance, in the first equation, for mesh A, the total resistance equals $12 + 6$, or $18\ \Omega$. Therefore, the voltage drop for I_A is $18\ I_A$ in mesh A.

In the second equation, for mesh B, the total resistance is $3 + 6$, or $9\ \Omega$, making the total voltage drop $9\ I_B$ for I_B in mesh B. You can add all resistances in a mesh for one R_T because they can be considered in series for the assumed mesh current.

Any resistance common to two meshes has two opposite mesh currents. In Fig. 9–8, I_A flows down and I_B is up through the common R_3, with both currents clockwise. As a result, a common resistance has two opposing voltage drops. One voltage is positive for the current of the mesh whose equation is being written. The opposing voltage is negative for the current of the adjacent mesh.

In mesh A, the common 6-Ω R_3 has opposing voltages $6I_A$ and $-6I_B$. The $6I_A$ of R_3 adds to the $12I_A$ of R_1 for the total positive voltage drop of $18I_A$ in mesh A. With the opposing voltage of $-6I_B$, then the equation for mesh A is $18I_A - 6I_B = 84$ V.

The same idea applies to mesh B. However, now the voltage $6I_B$ is positive because the equation is for mesh B. The $-6I_A$ voltage is negative here because I_A is for the adjacent mesh. The $6I_B$ adds to the $3I_B$ of R_2 for the total positive voltage drop of $9I_B$ in mesh B. With the opposing voltage of $-6I_A$, the equation for mesh B then is $-6I_A + 9I_B = -21$ V.

The algebraic sign of the source voltage in a mesh depends on its polarity. When the assumed mesh current flows into the positive terminal, as for V_1 in Fig. 9–8, it is considered positive for the right-hand side of the mesh equation. This direction of electron flow produces voltage drops that must add to equal the applied voltage.

With the mesh current into the negative terminal, as for V_2 in Fig. 9–8, it is considered negative. This is why V_2 is -21 V in the equation for mesh B. Then V_2 is actually a load for the larger applied voltage of V_1, instead of V_2 being the source. When a mesh has no source voltage, the algebraic sum of the voltage drops must equal zero.

These rules for the voltage source mean that the direction of electron flow is assumed for the mesh currents. Then electron flow is used to determine the polarity of the voltage drops. Note that considering the voltage source as a positive value with electron flow into the positive terminal corresponds to the normal flow of electron charges. If the solution for a mesh current comes out negative, the actual current for the mesh must be in the direction opposite from the assumed current flow.

Solving the Mesh Equations to Find the Mesh Currents

The two equations for the two meshes in Fig. 9–8 are

$$18I_A - 6I_B = 84$$
$$-6I_A + 9I_B = -21$$

These equations have the same coefficients as the voltage equations written for the branch currents, but the signs are different because the directions of the assumed mesh currents are not the same as those of the branch currents.

The solution will give the same answers for either method, but you must be consistent in algebraic signs. Use either the rules for meshes with mesh currents or the rules for loops with branch currents, but do not mix the two methods.

To eliminate I_B and solve for I_A, divide the first equation by 2 and the second equation by 3. Then

$$9I_A - 3I_B = 42$$
$$-2I_A + 3I_B = -7$$

Add the equations, term by term, to eliminate I_B. Then

$$7I_A = 35$$
$$I_A = 5 \text{ A}$$

To calculate I_B, substitute 5 for I_A in the second equation:

$$-2(5) + 3I_B = -7$$
$$3I_B = -7 + 10 = 3$$
$$I_B = 1 \text{ A}$$

The positive solutions mean that the electron flow for both I_A and I_B is actually clockwise, as assumed.

Finding the Branch Currents and Voltage Drops

Referring to Fig. 9–8, the 5-A I_A is the only current through R_1. Therefore, I_A and I_1 are the same. Then V_{R_1} across the 12-Ω R_1 is 5 × 12, or 60 V. The polarity of V_{R_1} is marked negative at the left, with the electron flow into this side.

Similarly, the 1-A I_B is the only current through R_2. The direction of this electron flow through R_2 is from left to right. Note that this value of 1 A for I_B clockwise is the same as −1 A for I_2, assumed in the opposite direction in Fig. 9–3. Then V_{R_2} across the 3-Ω R_2 is 1 × 3 or 3 V, with the left side negative.

The current I_3 through R_3, common to both meshes, consists of I_A and I_B. Then I_3 is 5 − 1 or 4 A. The currents are subtracted because I_A and I_B are in opposing directions through R_3. When all the mesh currents are taken one way, they will always be in opposite directions through any resistance common to two meshes.

The direction of the net 4-A I_3 through R_3 is downward, the same as I_A, because it is more than I_B. Then, V_{R_3} across the 6-Ω R_3 is 4 × 6 = 24 V, with the top negative.

The Set of Mesh Equations

The system for algebraic signs of the voltages in mesh equations is different from the method used with branch currents, but the end result is the same. The advantage of mesh currents is the pattern of algebraic signs for the voltages, without the need for tracing any branch currents. This feature is especially helpful in a more elaborate circuit, such as that in Fig. 9–9, that has three meshes. We can use Fig. 9–9 for more practice in writing mesh equations, without doing the numerical work of solving a set of three equations. Each R is 2 Ω.

In Fig. 9–9, the mesh currents are shown with solid arrows to indicate conventional current, which is a common way of analyzing these circuits. Also, the voltage sources V_1 and V_2 have the positive terminal at the top in the diagram. When the direction of conventional current is used, it is important to note that the voltage source is a positive value with mesh current into the negative terminal. This method corresponds to the normal flow of positive charges with conventional current.

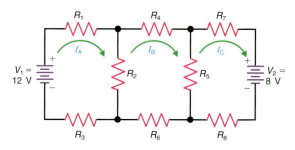

For the three mesh equations in Fig. 9–9,

In mesh A: $6I_A - 2I_B + 0 = -12$
In mesh B: $-2I_A + 8I_B - 2I_C = 0$
In mesh C: $0 - 2I_B + 6I_C = 8$

The zero term in equations A and C represents a missing mesh current. Only mesh B has all three mesh currents. However, note that mesh B has a zero term for the voltage source because it is the only mesh with only IR drops.

In summary, the only positive IR voltage in a mesh is for the R_T of each mesh current in its own mesh. All other voltage drops for any adjacent mesh current across a common resistance are always negative. This procedure for assigning algebraic signs to the voltage drops is the same whether the source voltage in the mesh is positive or negative. It also applies even if there is no voltage source in the mesh.

■ *9-5 Knowledge Check*

Answer at end of chapter.

What is the algebraic sum of the voltage drops in a mesh with no source voltage?

■ *9-5 Self-Review*

Answers at end of chapter.

a. **A network with four mesh currents needs four mesh equations for a solution. (True or False)**
b. **An R common to two meshes has opposing mesh currents. (True or False)**

Summary

■ Kirchhoff's voltage law states that the algebraic sum of all voltages around any closed path must equal zero. Stated another way, the sum of the voltage drops equals the applied voltage.

■ Kirchhoff's current law states that the algebraic sum of all currents directed in and out of any point in a circuit must equal zero. Stated another way, the current into a point equals the current out of that point.

■ A closed path is a loop. The method of using algebraic equations for the voltages around the loops to calculate the branch currents is illustrated in Fig. 9–5.

- A principal node is a branch point where currents divide or combine. The method of using algebraic equations for the currents at a node to calculate the node voltage is illustrated in Fig. 9–7.

- A mesh is the simplest possible loop. A mesh current is assumed to flow around the mesh without branching. The method of using algebraic equations for the voltages around the meshes to calculate the mesh currents is illustrated in Fig. 9–8.

Important Terms

- Kirchhoff's Current Law (KCL) - Kirchhoff's Current Law states that the algebraic sum of the currents entering and leaving any point in a circuit must equal zero.

- Kirchhoff's Voltage Law (KVL) - Kirchhoff's Voltage Law states that the algebraic sum of the voltages around any closed path must equal zero.

- Loop - another name for a closed path in a circuit.

- Loop Equation - an equation that specifies the voltages around a loop.

- Mesh - the simplest possible closed path within a circuit.

- Mesh Current - a current that is assumed to flow around a mesh without dividing.

- Node - a common connection for two or more components in a circuit where currents can combine or divide.

- Principal Node - a common connection for three or more components in a circuit where currents can combine or divide.

Self-Test

Answers at back of book.

1. **Kirchhoff's Current Law states that**

   a. the algebraic sum of the currents flowing into any point in a circuit must equal zero.

   b. the algebraic sum of the currents entering and leaving any point in a circuit must equal zero.

   c. the algebraic sum of the currents flowing away from any point in a circuit must equal zero.

   d. the algebraic sum of the currents around any closed path must equal zero.

2. **When applying Kirchhoff's Current Law,**

   a. consider all currents flowing into a branch point positive and all currents directed away from that point negative.

   b. consider all currents flowing into a branch point negative and all currents directed away from that point positive.

   c. remember that the total of all the currents entering a branch point must always be greater than the sum of the currents leaving that point.

   d. the algebraic sum of the currents entering and leaving a branch point does not necessarily have to be zero.

3. **If a 10-A I_1 and a 3-A I_2 flow into point X, how much current must flow away from point X?**

   a. 7A.

   b. 30A.

   c. 13A.

   d. This is impossible to determine.

4. **Three currents I_1, I_2, and I_3 flow into point X, whereas current I_4 flows away from point X. If $I_1 = 2.5$ A, $I_3 = 6$ A, and $I_4 = 18$ A, how much is current I_2?**

   a. 21.5 A.

   b. 14.5 A.

   c. 26.5 A.

   d. 9.5 A.

5. **When applying Kirchhoff's Voltage Law, a closed path is commonly referred to as a**

   a. node.

   b. principal node.

   c. loop.

   d. branch point.

6. **Kirchhoff's Voltage Law states that**

   a. the algebraic sum of the voltage sources and IR voltage drops in any closed path must total zero.

   b. the algebraic sum of the voltage sources and IR voltage drops around any closed path can never equal zero.

   c. the algebraic sum of all the currents flowing around any closed loop must equal zero.

   d. none of the above.

7. **When applying Kirchhoff's Voltage Law,**

   a. consider any voltage whose positive terminal is reached first as negative and any voltage whose negative terminal is reached first as positive.

   b. always consider all voltage sources as positive and all resistor voltage drops as negative.

   c. consider any voltage whose negative terminal is reached first as negative and any voltage whose positive terminal is reached first as positive.

   d. always consider all resistor voltage drops as positive and all voltage sources as negative.

8. **The algebraic sum of +40 V and −30 V is**

   a. −10 V.

   b. +10 V.

   c. +70 V.

   d. −70 V.

9. A principal node is

a. a closed path or loop where the algebraic sum of the voltages must equal zero.

b. the simplest possible closed path around a circuit.

c. a junction where branch currents can combine or divide.

d. none of the above.

10. How many equations are necessary to solve a circuit with two principal nodes?

a. 3.

b. 2.

c. 4.

d. 1.

11. The difference between a mesh current and a branch current is

a. a mesh current is an assumed current and a branch current is an actual current.

b. the direction of the currents themselves.

c. a mesh current does not divide at a branch point.

d. both a and c above.

12. Using the method of mesh currents, any resistance common to two meshes has

a. two opposing mesh currents.

b. one common mesh current.

c. zero current.

d. none of the above.

13. The fact that the sum of the resistor voltage drops equals the applied voltage in a series circuit is the basis for

a. Kirchhoff's Current Law.

b. node-voltage analysis.

c. Kirchhoff's Voltage Law.

d. the method of mesh currents.

14. The fact that the sum of the individual branch currents equals the total current in a parallel circuit is the basis for

a. Kirchhoff's Current Law.

b. node-voltage analysis.

c. Kirchhoff's Voltage Law.

d. the method of mesh currents.

15. If you do not go completely around the loop when applying Kirchhoff's Voltage Law, then

a. the algebraic sum of the voltages will always be positive.

b. the algebraic sum is the voltage between the start and finish points.

c. the algebraic sum of the voltages will always be negative.

d. the algebraic sum of the voltages cannot be determined.

Questions

1. State Kirchhoff's current law in two ways.

2. State Kirchhoff's voltage law in two ways.

3. What is the difference between a loop and a mesh?

4. What is the difference between a branch current and a mesh current?

5. Define principal node.

6. Define node voltage.

7. Use the values in Fig. 9–6 to show that the algebraic sum is zero for all voltages around the outside loop ACEFDBA.

8. Use the values in Fig. 9–6 to show that the algebraic sum is zero for all the currents into and out of node C and node D.

Problems

SECTION 9–1 KIRCHHOFF'S CURRENT LAW (KCL)

9–1 If a 5-A I_1 and a 10-A I_2 flow into point X, how much is the current, I_3, directed away from that point?

9–2 Applying Kirchhoff's Current Law, write an equation for the currents directed into and out of point X in Problem 9–1.

9–3 In Fig. 9–10, solve for the unknown current, I_3.

9–4 In Fig. 9–11, solve for the following unknown currents: I_3, I_5, and I_8.

9–5 Apply Kirchhoff's Current Law in Fig. 9–11 by writing an equation for the currents directed into and out of the following points:

a. Point X

b. Point Y

c. Point Z

Figure 9–10

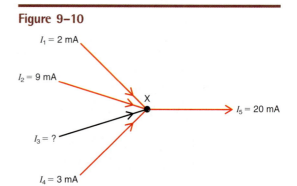

$I_1 = 2$ mA

$I_2 = 9$ mA

$I_3 = ?$

$I_4 = 3$ mA

X

$I_5 = 20$ mA

Figure 9-11

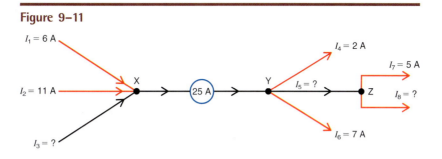

SECTION 9-2 KIRCHOFF'S VOLTAGE LAW (KVL)

9-6 **||| MultiSim** In Fig. 9-12,

a. Write a KVL equation for the loop CEFDC going clockwise from point C.

b. Write a KVL equation for the loop ACDBA going clockwise from point A.

c. Write a KVL equation for the loop ACEFDBA going clockwise from point A.

Figure 9-12

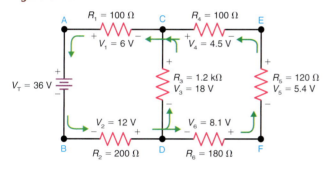

9-7 In Fig. 9-12,

a. Determine the voltage for the partial loop CEFD going clockwise from point C. How does your answer compare to the voltage drop across R_3?

b. Determine the voltage for the partial loop ACDB going clockwise from point A. How does your answer compare to the value of the applied voltage, V_T, across points A and B?

c. Determine the voltage for the partial loop ACEFDB going clockwise from point A. How does your answer compare to the value of the applied voltage, V_T, across points A and B?

d. Determine the voltage for the partial loop CDFE going counterclockwise from point C. How does your answer compare to the voltage drop across R_4?

9-8 In Fig. 9-13, solve for the voltages V_{AG} and V_{BG}. Indicate the proper polarity for each voltage.

9-9 In Fig. 9-14, solve for the voltages V_{AG} and V_{BG}. Indicate the proper polarity for each voltage.

9-10 In Fig. 9-15, solve for the voltages V_{AG} and V_{BG}. Indicate the proper polarity for each voltage.

Figure 9-13

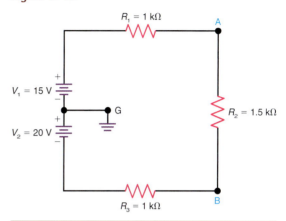

Figure 9-14

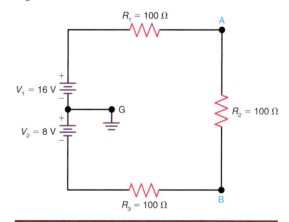

Figure 9-15

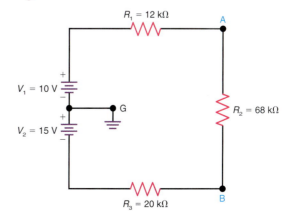

9–11 In Fig. 9–16, solve for the voltages V_{AG}, V_{BG}, V_{CG}, V_{DG}, and V_{AD}. Indicate the proper polarity for each voltage.

Figure 9–16

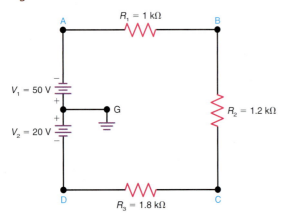

9–12 In Fig. 9–17, solve for the voltages V_{AG}, V_{BG}, and V_{CG}. Indicate the proper polarity for each voltage.

Figure 9–17

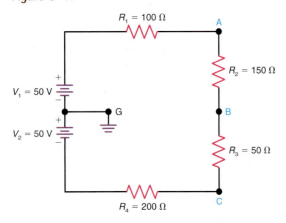

9–13 In Fig. 9–18, write a KVL equation for the loop ABDCA going counterclockwise from point A.

9–14 In Fig. 9–18, write a KVL equation for the loop EFDCE going clockwise from point E.

9–15 In Fig. 9–18, write a KVL equation for the loop ACEFDBA going clockwise from point A.

Figure 9–18

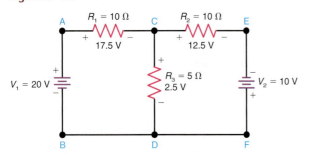

SECTION 9–3 METHOD OF BRANCH CURRENTS

9–16 Using the method of branch currents, solve for the unknown values of voltage and current in Fig. 9–19. To do this, complete steps a through m. The assumed direction of all currents is shown on the figure.

Figure 9–19

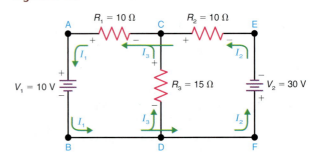

a. Using Kirchhoff's Current Law, write an equation for the currents I_1, I_2, and I_3 at point C.

b. Specify the current I_3 in terms of I_1 and I_2.

c. Write a KVL equation for the loop ABDCA, going counterclockwise from point A, using the terms V_1, V_{R_1}, and V_{R_3}. This loop will be called Loop 1.

d. Write a KVL equation for the loop FECDF, going counterclockwise from point F, using the terms V_2, V_{R_2}, and V_{R_3}. This loop will be called Loop 2.

e. Specify each resistor voltage drop as an IR product using actual resistor values for R_1, R_2, and R_3.

f. Rewrite the KVL equation for Loop 1 in part C using the IR voltage values for V_{R_1} and V_{R_3} specified in part (e).

g. Rewrite the KVL equation for Loop 2 in part D using the IR voltage values for V_{R_2} and V_{R_3} specified in part (e).

h. Reduce the Loop 1 and Loop 2 equations in parts f and g to their simplest possible form.

i. Solve for currents I_1 and I_2 using any of the methods for the solution of simultaneous equations. Next, solve for I_3.

j. In Fig. 9–19, were the assumed directions of all currents correct? How do you know?

k. Using the actual values of I_1, I_2, and I_3, calculate the individual resistor voltage drops.

l. Rewrite the KVL loop equations for both Loops 1 and 2 using actual voltage values. Go counterclockwise around both loops when adding voltages. (Be sure that the resistor voltage drops all have the correct polarity based on the actual directions for I_1, I_2, and I_3.)

m. Based on the actual directions for I_1, I_2, and I_3, write a KCL equation for the currents at point C.

9–17 Repeat Problem 9–16 for Fig. 9–20.

Figure 9-20

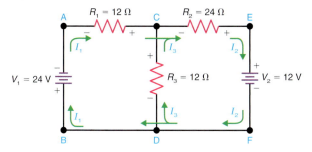

SECTION 9-4 NODE-VOLTAGE ANALYSIS

9-18 Using the method of node-voltage analysis, solve for all unknown values of voltage and current in Fig. 9-21. To do this, complete steps a through l. The assumed direction of all currents is shown on the figure.

Figure 9-21

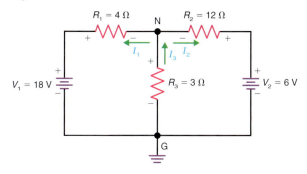

a. Using Kirchhoff's Current Law, write an equation for the currents I_1, I_2, and I_3 at the node-point, N.

b. Express the KCL equation in part a in terms of V_R/R and V_N/R.

c. Write a KVL equation for the loop containing V_1, V_{R_1}, and V_{R_3}.

d. Write a KVL equation for the loop containing V_2, V_{R_2} and V_{R_3}.

e. Specify V_{R_1} and V_{R_2} in terms of V_N and the known values of V_1 and V_2.

f. Using the values for V_{R_1} and V_{R_2} from part e, write a KCL equation for the currents at the node-point N.

g. Solve for the node voltage, V_N.

h. Using the equations from part e, solve for the voltage drops V_{R_1} and V_{R_2}.

i. In Fig. 9-21, were all the assumed directions of current correct? How do you know?

j. Calculate the currents I_1, I_2, and I_3.

k. Rewrite the KVL loop equation for both inner loops using actual voltage values. (Be sure that the resistor voltage drops all have the correct polarity based on the final answers for V_{R_1} and V_{R_2}.)

l. Based on the actual values for I_1, I_2, and I_3, write a KCL equation for the currents at the node-point, N.

9-19 Repeat Problem 9-18 for the circuit in Fig. 9-22.

Figure 9-22

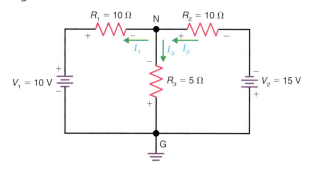

SECTION 9-5 METHOD OF MESH CURRENTS

9-20 Using the method of mesh currents, solve for all unknown values of voltage and current in Fig. 9-23. To do this, complete steps a through m.

Figure 9-23

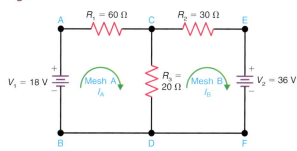

a. Identify the components through which the mesh current, I_A, flows.

b. Identify the components through which the mesh current, I_B, flows.

c. Which component has opposing mesh currents?

d. Write the mesh equation for mesh A.

e. Write the mesh equation for mesh B.

f. Solve for currents I_A and I_B using any of the methods for the solution of simultaneous equations.

g. Determine the values of currents I_1, I_2, and I_3.

h. Are the assumed directions of the mesh A and mesh B currents correct? How do you know?

i. What is the direction of the current, I_3, through R_3?

j. Solve for the voltage drops V_{R_1}, V_{R_2}, and V_{R_3}.

k. Using the final solutions for V_{R_1}, V_{R_2}, and V_{R_3}, write a KVL equation for the loop ACDBA going clockwise from point A.

l. Using the final solutions for V_{R_1}, V_{R_2}, and V_{R_3}, write a KVL equation for the loop EFDCE going clockwise from point E.

m. Using the final solutions (and directions) for I_1, I_2, and I_3, write a KCL equation for the currents at point C.

9-21 Repeat Problem 9-20 for Fig. 9-24.

Figure 9-24

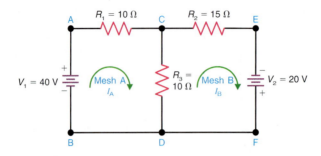

Critical Thinking

9-22 In Fig. 9-25, determine the values for R_1 and R_3 which will allow the output voltage to vary between $-5\ V$ and $+5\ V$.

9-23 Refer to Fig. 9-9. If all resistances are 10 Ω, calculate (a) I_A, I_B, and I_C (b) I_1, I_2, I_3, I_4, I_5, I_6, I_7, and I_8.

Figure 9-25 Circuit diagram for Critical Thinking Prob. 9-22.

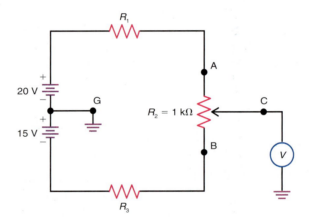

Answers to Knowledge Check Problems

9-1 $I_3 = 8\ A$

9-2 $V_{AG} = 5.4\ V$, $V_{BG} = -5.1\ V$

9-3 The solution for I_2 was negative which indicates that its assumed direction was incorrect.

9-4 The solution for V_{R_2} was negative which indicates the assumed direction of the current, I_2, was incorrect.

9-5 Zero

Answers to Self-Reviews

9-1 **a.** 6 A
 b. 4 A

9-2 **a.** $-120\ V$
 b. 0 V

9-3 **a.** $-24\ V$
 b. 0 V

9-4 **a.** two
 b. two

9-5 **a.** T
 b. T

10 Network Theorems

A network is a combination of components, such as resistances and voltage sources, interconnected to achieve a particular end result. However, networks generally need more than the rules of series and parallel circuits for analysis. Kirchhoff's laws can always be applied for any circuit connections. The network theorems, though, usually provide shorter methods for solving a circuit.

Some theorems enable us to convert a network into a simpler circuit, equivalent to the original. Then the equivalent circuit can be solved by the rules of series and parallel circuits. Other theorems enable us to convert a given circuit into a form that permits easier solutions.

Only the applications are given here, although all network theorems can be derived from Kirchhoff's laws. Note that resistance networks with batteries are shown as examples, but the theorems can also be applied to ac networks.

Objectives

After studying this chapter you should be able to

- *Apply* the superposition theorem to find the voltage across two points in a circuit containing more than one voltage source.

- *State* the requirements for applying the superposition theorem.

- *Determine* the Thevenin and Norton equivalent circuits with respect to any pair of terminals in a complex network.

- *Apply* Thevenin's and Norton's theorems in solving for an unknown voltage or current.

- *Convert* a Thevenin equivalent circuit to a Norton equivalent circuit and vice versa.

- *Apply* Millman's theorem to find a common voltage across any number of parallel branches.

- *Simplify* the analysis of a bridge circuit by using delta to wye conversion formulas.

Outline

Important Terms

active components

bilateral components

current source

linear component

Millman's theorem

Norton's theorem

passive components

superposition theorem

Thevenin's theorem

voltage source

10–1 Superposition Theorem

The superposition theorem is very useful because it extends the use of Ohm's law to circuits that have more than one source. In brief, we can calculate the effect of one source at a time and then superimpose the results of all sources. As a definition, the superposition theorem states: *In a network with two or more sources, the current or voltage for any component is the algebraic sum of the effects produced by each source acting separately.*

To use one source at a time, all other sources are "killed" temporarily. This means disabling the source so that it cannot generate voltage or current without changing the resistance of the circuit. A voltage source such as a battery is killed by assuming a short circuit across its potential difference. The internal resistance remains.

Voltage Divider with Two Sources

The problem in Fig. 10–1 is to find the voltage at P to chassis ground for the circuit in Fig. 10–1*a*. The method is to calculate the voltage at P contributed by each source separately, as in Fig. 10–1*b* and *c*, and then superimpose these voltages.

To find the effect of V_1 first, short-circuit V_2 as shown in Fig. 10–1*b*. Note that the bottom of R_1 then becomes connected to chassis ground because of the short circuit across V_2. As a result, R_2 and R_1 form a series voltage divider for the V_1 source.

Furthermore, the voltage across R_1 becomes the same as the voltage from P to ground. To find this V_{R_1} across R_1 as the contribution of the V_1 source, we use the voltage divider formula:

$$V_{R_1} = \frac{R_1}{R_1 + R_2} \times V_1 = \frac{60\ \text{k}\Omega}{30\ \text{k}\Omega + 60\ \text{k}\Omega} \times 24\ \text{V}$$

$$= \frac{60}{90} \times 24\ \text{V}$$

$$V_{R_1} = 16\ \text{V}$$

Next find the effect of V_2 alone, with V_1 short-circuited, as shown in Fig. 10–1*c*. Then point A at the top of R_2 becomes grounded. R_1 and R_2 form a series voltage divider again, but here the R_2 voltage is the voltage at P to ground.

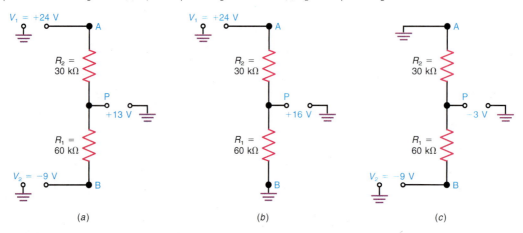

IIII MultiSim **Figure 10–1** Superposition theorem applied to a voltage divider with two sources V_1 and V_2. (*a*) Actual circuit with +13 V from point P to chassis ground. (*b*) V_1 alone producing +16 V at P. (*c*) V_2 alone producing −3 V at P.

(a) (b) (c)

With one side of R_2 grounded and the other side to point P, V_{R_2} is the voltage to calculate. Again we have a series divider, but this time for the negative voltage V_2. Using the voltage divider formula for V_{R_2} as the contribution of V_2 to the voltage at P,

$$V_{R_2} = \frac{R_2}{R_1 + R_2} \times V_2 = \frac{30 \text{ k}\Omega}{30 \text{ k}\Omega + 60 \text{ k}\Omega} \times -9 \text{ V}$$

$$= \frac{30}{90} \times -9 \text{ V}$$

$$V_{R_2} = -3 \text{ V}$$

This voltage is negative at P because V_2 is negative.

Finally, the total voltage at P is

$$V_P = V_{R_1} + V_{R_2} = 16 - 3$$
$$V_p = 13 \text{ V}$$

This algebraic sum is positive for the net V_P because the positive V_1 is larger than the negative V_2.

By superposition, therefore, this problem was reduced to two series voltage dividers. The same procedure can be used with more than two sources. Also, each voltage divider can have any number of series resistances. Note that in this case we were dealing with ideal voltage sources, that is, sources with zero internal resistance. If the source did have internal resistance, it would have been added in series with R_1 and R_2.

Requirements for Superposition

All components must be linear and bilateral to superimpose currents and voltages. *Linear* means that the current is proportional to the applied voltage. Then the currents calculated for different source voltages can be superimposed.

Bilateral means that the current is the same amount for opposite polarities of the source voltage. Then the values for opposite directions of current can be combined algebraically. Networks with resistors, capacitors, and air-core inductors are generally linear and bilateral. These are also *passive components,* that is, components that do not amplify or rectify. *Active components,* such as transistors, semiconductor diodes, and electron tubes, are never bilateral and often are not linear.

■ *10–1 Knowledge Check*

Answer at end of chapter.

In Fig. 10–1a what is the value of V_P if the polarity of V_1 is reversed?

■ *10–1 Self-Review*

Answers at end of chapter.

a. In Fig. 10–1b, which R is shown grounded at one end?
b. In Fig. 10–1c, which R is shown grounded at one end?

10–2 Thevenin's Theorem

Named after M. L. Thevenin, a French engineer, Thevenin's theorem is very useful in simplifying the process of solving for the unknown values of voltage and current in a network. By Thevenin's theorem, many sources and components, no

Figure 10–2 Any network in the block at the left can be reduced to the Thevenin equivalent series circuit at the right.

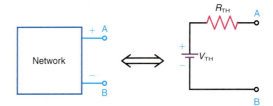

matter how they are interconnected, can be represented by an equivalent series circuit with respect to any pair of terminals in the network. In Fig. 10–2, imagine that the block at the left contains a network connected to terminals A and B. Thevenin's theorem states that the *entire* network connected to A and B can be replaced by a single voltage source V_{TH} in series with a single resistance R_{TH}, connected to the same two terminals.

Voltage V_{TH} is the open-circuit voltage across terminals A and B. This means finding the voltage that the network produces across the two terminals with an open circuit between A and B. The polarity of V_{TH} is such that it will produce current from A to B in the same direction as in the original network.

Resistance R_{TH} is the open-circuit resistance across terminals A and B, but with all sources killed. This means finding the resistance looking back into the network from terminals A and B. Although the terminals are open, an ohmmeter across AB would read the value of R_{TH} as the resistance of the remaining paths in the network without any sources operating.

Thevenizing a Circuit

As an example, refer to Fig. 10–3a, where we want to find the voltage V_L across the 2-Ω R_L and its current I_L. To use Thevenin's theorem, mentally disconnect R_L. The two open ends then become terminals A and B. Now we find the Thevenin equivalent of the remainder of the circuit that is still connected to A and B. In general, open the part of the circuit to be analyzed and "thevenize" the remainder of the circuit connected to the two open terminals.

Our only problem now is to find the value of the open-circuit voltage V_{TH} across AB and the equivalent resistance R_{TH}. The Thevenin equivalent always consists of a single voltage source in series with a single resistance, as in Fig. 10–3d.

The effect of opening R_L is shown in Fig. 10–3b. As a result, the 3-Ω R_1 and 6-Ω R_2 form a series voltage divider without R_L.

Furthermore, the voltage across R_2 now is the same as the open-circuit voltage across terminals A and B. Therefore V_{R_2} with R_L open is V_{AB}. This is the V_{TH} we need for the Thevenin equivalent circuit. Using the voltage divider formula,

$$V_{R_2} = \frac{6}{9} \times 36 \text{ V} = 24 \text{ V}$$

$$V_{R_2} = V_{AB} = V_{TH} = 24 \text{ V}$$

This voltage is positive at terminal A.

To find R_{TH}, the 2-Ω R_L is still disconnected. However, now the source V is short-circuited. So the circuit looks like Fig. 10–3c. The 3-Ω R_1 is now in parallel with the 6-Ω R_2 because both are connected across the same two points. This combined resistance is the product over the sum of R_1 and R_2.

$$R_{TH} = \frac{18}{9} = 2 \ \Omega$$

Again, we assume an ideal voltage source whose internal resistance is zero.

Figure 10–3 Application of Thevenin's theorem. (*a*) Actual circuit with terminals A and B across R_L. (*b*) Disconnect R_L to find that V_{AB} is 24 V. (*c*) Short-circuit V to find that R_{AB} is 2 Ω. (*d*) Thevenin equivalent circuit. (*e*) Reconnect R_L at terminals A and B to find that V_L is 12 V.

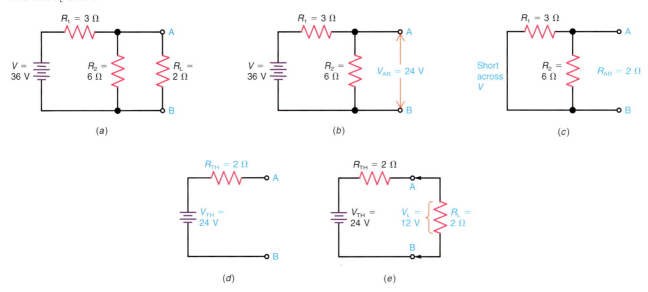

As shown in Fig. 10–3*d*, the Thevenin circuit to the left of terminals A and B then consists of the equivalent voltage V_{TH}, equal to 24 V, in series with the equivalent series resistance R_{TH}, equal to 2 Ω. This Thevenin equivalent applies for any value of R_L because R_L was disconnected. We are actually thevenizing the circuit that feeds the open AB terminals.

To find V_L and I_L, we can finally reconnect R_L to terminals A and B of the Thevenin equivalent circuit, as shown in Fig. 10–3*e*. Then R_L is in series with R_{TH} and V_{TH}. Using the voltage divider formula for the 2-Ω R_{TH} and 2-Ω R_L, $V_L = 1/2 \times 24$ V = 12 V. To find I_L as V_L/R_L, the value is 12 V/2 Ω, which equals 6 A.

These answers of 6 A for I_L and 12 V for V_L apply to R_L in both the original circuit in Fig. 10–3*a* and the equivalent circuit in Fig. 10–3*e*. Note that the 6-A I_L also flows through R_{TH}.

The same answers could be obtained by solving the series-parallel circuit in Fig. 10–3*a*, using Ohm's law. However, the advantage of thevenizing the circuit is that the effect of different values of R_L can be calculated easily. Suppose that R_L is changed to 4 Ω. In the Thevenin circuit, the new value of V_L would be 4/6 × 24 V = 16 V. The new I_L would be 16 V/4 Ω, which equals 4 A. If we used Ohm's law in the original circuit, a complete, new solution would be required each time R_L was changed.

Looking Back from Terminals A and B

The way we look at the resistance of a series-parallel circuit depends on where the source is connected. In general, we calculate the total resistance from the outside terminals of the circuit in toward the source as the reference.

When the source is short-circuited for thevenizing a circuit, terminals A and B become the reference. Looking back from A and B to calculate R_{TH}, the situation becomes reversed from the way the circuit was viewed to determine V_{TH}.

For R_{TH}, imagine that a source could be connected across AB, and calculate the total resistance working from the outside in toward terminals A and B. Actually, an ohmmeter placed across terminals A and B would read this resistance.

GOOD TO KNOW

The Thevenin equivalent circuit driving terminals A and B does not change even though the value of R_L may change.

Figure 10–4 Thevenizing the circuit of Fig. 10–3 but with a 4-Ω R_3 in series with the A terminal. (a) V_{AB} is still 24 V. (b) Now the R_{AB} is 2 + 4 = 6 Ω. (c) Thevenin equivalent circuit.

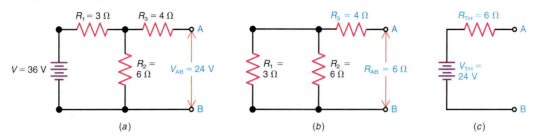

(a) (b) (c)

This idea of reversing the reference is illustrated in Fig. 10–4. The circuit in Fig. 10–4a has terminals A and B open, ready to be thevenized. This circuit is similar to that in Fig. 10–3 but with the 4-Ω R_3 inserted between R_2 and terminal A. The interesting point is that R_3 does not change the value of V_{AB} produced by the source V, but R_3 does increase the value of R_{TH}. When we look back from terminals A and B, the 4 Ω of R_3 is in series with 2 Ω to make R_{TH} 6 Ω, as shown for R_{AB} in Fig. 10–4b and R_{TH} in Fig. 10–4c.

Let us consider why V_{AB} is the same 24 V with or without R_3. Since R_3 is connected to the open terminal A, the source V cannot produce current in R_3. Therefore, R_3 has no IR drop. A voltmeter would read the same 24 V across R_2 and from A to B. Since V_{AB} equals 24 V, this is the value of V_{TH}.

Now consider why R_3 does change the value of R_{TH}. Remember that we must work from the outside in to calculate the total resistance. Then, A and B are like source terminals. As a result, the 3-Ω R_1 and 6-Ω R_2 are in parallel, for a combined resistance of 2 Ω. Furthermore, this 2 Ω is in series with the 4-Ω R_3 because R_3 is in the main line from terminals A and B. Then R_{TH} is 2 + 4 = 6 Ω. As shown in Fig. 10–4c, the Thevenin equivalent circuit consists of V_{TH} = 24 V and R_{TH} = 6 Ω.

■ *10–2 Knowledge Check*

Answers at end of chapter.

In Fig. 10–4a, what are the values for V_{TH} and R_{TH} if R_3 is changed to 10 Ω?

■ *10–2 Self-Review*

Answers at end of chapter.

a. For a Thevenin equivalent circuit, terminals A and B are open to find both V_{TH} and R_{TH}. (**True or False**)
b. For a Thevenin equivalent circuit, the source voltage is short-circuited only to find R_{TH}. (**True or False**)

10–3 Thevenizing a Circuit with Two Voltage Sources

The circuit in Fig. 10–5 has already been solved by Kirchhoff's laws, but we can use Thevenin's theorem to find the current I_3 through the middle resistance R_3. As shown in Fig. 10–5a, first mark the terminals A and B across R_3. In Fig. 10–5b, R_3 is disconnected. To calculate V_{TH}, find V_{AB} across the open terminals.

Figure 10–5 Thevenizing a circuit with two voltage sources V_1 and V_2. (a) Original circuit with terminals A and B across the middle resistor R_3. (b) Disconnect R_3 to find that V_{AB} is -33.6 V. (c) Short-circuit V_1 and V_2 to find that R_{AB} is 2.4 Ω. (d) Thevenin equivalent with R_L reconnected to terminals A and B.

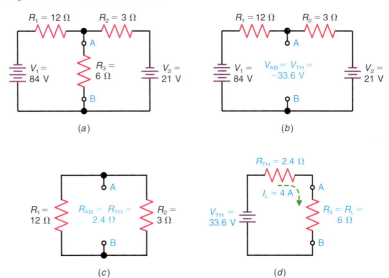

(a)   (b)

(c)   (d)

Superposition Method

With two sources, we can use superposition to calculate V_{AB}. First short-circuit V_2. Then the 84 V of V_1 is divided between R_1 and R_2. The voltage across R_2 is between terminals A and B. To calculate this divided voltage across R_2,

$$V_{R_2} = \frac{R_2}{R_{1-2}} \times V_1 = \frac{3}{15} \times (-84)$$

$$V_{R_2} = -16.8 \text{ V}$$

This is the only contribution of V_1 to V_{AB}. The polarity is negative at terminal A.

To find the voltage that V_2 produces between A and B, short-circuit V_1. Then the voltage across R_1 is connected from A to B. To calculate this divided voltage across R_1,

$$V_{R_1} = \frac{R_1}{R_{1-2}} \times V_2 = \frac{12}{15} \times (-21)$$

$$V_{R_1} = -16.8 \text{ V}$$

Both V_1 and V_2 produce -16.8 V across the AB terminals with the same polarity. Therefore, they are added.

The resultant value of $V_{AB} = -33.6$ V, shown in Fig. 10–5b, is the value of V_{TH}. The negative polarity means that terminal A is negative with respect to B.

To calculate R_{TH}, short-circuit the sources V_1 and V_2, as shown in Fig. 10–5c. Then the 12-Ω R_1 and 3-Ω R_2 are in parallel across terminals A and B. Their combined resistance is 36/15, or 2.4 Ω, which is the value of R_{TH}.

The final result is the Thevenin equivalent in Fig. 10–5d with an R_{TH} of 2.4 Ω and a V_{TH} of 33.6 V, negative toward terminal A.

GOOD TO KNOW

The polarity of V_{TH} is extremely critical because it allows us to determine the actual direction of I_3 through R_3.

To find the current through R_3, it is reconnected as a load resistance across terminals A and B. Then V_{TH} produces current through the total resistance of 2.4 Ω for R_{TH} and 6 Ω for R_3:

$$I_3 = \frac{V_{TH}}{R_{TH} + R_3} = \frac{33.6}{2.4 + 6} = \frac{33.6}{8.4} = 4 \text{ A}$$

This answer of 4 A for I_3 is the same value calculated before, using Kirchhoff's laws, in Fig. 9–5.

It should be noted that this circuit can be solved by superposition alone, without using Thevenin's theorem, if R_3 is not disconnected. However, opening terminals A and B for the Thevenin equivalent simplifies the superposition, as the circuit then has only series voltage dividers without any parallel current paths. In general, a circuit can often be simplified by disconnecting a component to open terminals A and B for Thevenin's theorem.

■ *10–3 Knowledge Check*

Answer at end of chapter.

In Fig. 10–5a, what is the value of V_{TH} if R_1 and R_2 are interchanged?

■ *10–3 Self-Review*

Answers at end of chapter.

In the Thevenin equivalent circuit in Fig. 10–5d,
a. How much is R_T?
b. How much is V_{R_L}?

10–4 Thevenizing a Bridge Circuit

As another example of Thevenin's theorem, we can find the current through the 2-Ω R_L at the center of the bridge circuit in Fig. 10–6a. When R_L is disconnected to open terminals A and B, the result is as shown in Fig. 10–6b. Notice how the circuit has become simpler because of the open. Instead of the unbalanced bridge in Fig. 10–6a which would require Kirchhoff's laws for a solution, the Thevenin equivalent in Fig. 10–6b consists of just two voltage dividers. Both the R_3–R_4 divider and the R_1–R_2 divider are across the same 30-V source.

Since the open terminal A is at the junction of R_3 and R_4, this divider can be used to find the potential at point A. Similarly, the potential at terminal B can be found from the R_1–R_2 divider. Then V_{AB} is the difference between the potentials at terminals A and B.

Note the voltages for the two dividers. In the divider with the 3-Ω R_3 and 6-Ω R_4, the bottom voltage V_{R_4} is 6/9 × 30 = 20 V. Then V_{R_3} at the top is 10 V because both must add up to equal the 30-V source. The polarities are marked negative at the top, the same as the source voltage V.

Similarly, in the divider with the 6-Ω R_1 and 4-Ω R_2, the bottom voltage V_{R_2} is 4/10 × 30 = 12 V. Then V_{R_1} at the top is 18 V because the two must add up to equal the 30-V source. The polarities are also negative at the top, the same as V.

Now we can determine the potentials at terminals A and B with respect to a common reference to find V_{AB}. Imagine that the positive side of the source V is connected to a chassis ground. Then we would use the bottom line in the diagram as our reference for voltages. Note that V_{R_4} at the bottom of the R_3–R_4 divider is the same as the potential of terminal A with respect to ground. This value is −20 V, with terminal A negative.

Figure 10–6 Thevenizing a bridge circuit. (*a*) Original circuit with terminals A and B across middle resistor R_L. (*b*) Disconnect R_L to find V_{AB} of −8 V. (*c*) With source V short-circuited, R_{AB} is 2 + 2.4 = 4.4 Ω. (*d*) Thevenin equivalent with R_L reconnected to terminals A and B.

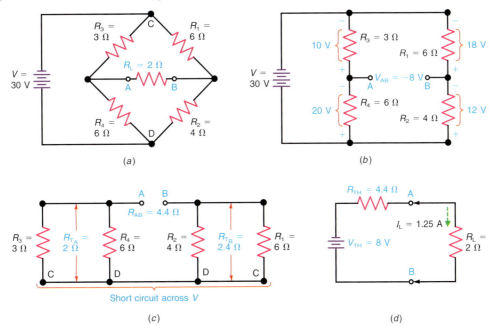

(*a*)　　　　(*b*)

(*c*)　　　　(*d*)

Similarly, V_{R_2} in the R_1–R_2 divider is the potential at B with respect to ground. This value is −12 V with terminal B negative. As a result, V_{AB} is the difference between the −20 V at A and the −12 V at B, both with respect to the common ground reference.

The potential difference V_{AB} then equals

$$V_{AB} = -20 - (-12) = -20 + 12 = -8 \text{ V}$$

Terminal A is 8 V more negative than B. Therefore, V_{TH} is 8 V, with the negative side toward terminal A, as shown in the Thevenin equivalent in Fig. 10–6*d*.

The potential difference V_{AB} can also be found as the difference between V_{R_3} and V_{R_1} in Fig. 10–6*b*. In this case, V_{R_3} is 10 V and V_{R_1} is 18 V, both positive with respect to the top line connected to the negative side of the source V. The potential difference between terminals A and B then is 10 − 18, which also equals −8 V. Note that V_{AB} must have the same value no matter which path is used to determine the voltage.

To find R_{TH}, the 30-V source is short-circuited while terminals A and B are still open. Then the circuit looks like Fig. 10–6*c*. Looking back from terminals A and B, the 3-Ω R_3 and 6-Ω R_4 are in parallel, for a combined resistance R_{T_A} of 18/9 or 2 Ω. The reason is that R_3 and R_4 are joined at terminal A, while their opposite ends are connected by the short circuit across the source V. Similarly, the 6-Ω R_1 and 4-Ω R_2 are in parallel for a combined resistance R_{T_B} of 24/10 = 2.4 Ω. Furthermore, the short circuit across the source now provides a path that connects R_{T_A} and R_{T_B} in series. The entire resistance is 2 + 2.4 = 4.4 Ω for R_{AB} or R_{TH}.

The Thevenin equivalent in Fig. 10–6*d* represents the bridge circuit feeding the open terminals A and B with 8 V for V_{TH} and 4.4 Ω for R_{TH}. Now connect the 2-Ω R_L to terminals A and B to calculate I_L. The current is

$$I_L = \frac{V_{TH}}{R_{TH} + R_L} = \frac{8}{4.4 + 2} = \frac{8}{6.4}$$

$$I_L = 1.25 \text{ A}$$

This 1.25 A is the current through the 2-Ω R_L at the center of the unbalanced bridge in Fig. 10–6a. Furthermore, the amount of I_L for any value of R_L in Fig. 10–6a can be calculated from the equivalent circuit in Fig. 10–6d.

■ **10–4 Knowledge Check**

Answers at end of chapter.

In Fig. 10–6, what are the values for V_{TH} and R_{TH} if $R_3 = 4\ \Omega$?

■ **10–4 Self-Review**

Answers at end of chapter.

In the Thevenin equivalent circuit in Fig. 10–6d,

a. How much is R_T?

b. How much is V_{R_L}?

10–5 Norton's Theorem

Named after E. L. Norton, a scientist with Bell Telephone Laboratories, Norton's theorem is used to simplify a network in terms of currents instead of voltages. In many cases, analyzing the division of currents may be easier than voltage analysis. For current analysis, therefore, Norton's theorem can be used to reduce a network to a simple parallel circuit with a current source. The idea of a *current source* is that it supplies a total line current to be divided among parallel branches, corresponding to a *voltage source* applying a total voltage to be divided among series components. This comparison is illustrated in Fig. 10–7.

Example of a Current Source

A source of electric energy supplying voltage is often shown with a series resistance that represents the internal resistance of the source, as in Fig. 10–7a. This method corresponds to showing an actual voltage source, such as a battery for dc circuits. However, the source may also be represented as a current source with a parallel resistance, as in Fig. 10–7b. Just as a voltage source is rated at, say, 10 V, a current source may be rated at 2 A. For the purpose of analyzing parallel branches, the concept of a current source may be more convenient than the concept of a voltage source.

If the current I in Fig. 10–7b is a 2-A source, it supplies 2 A no matter what is connected across the output terminals A and B. Without anything connected across A and B, all 2 A flows through the shunt R. When a load resistance R_L is connected across A and B, then the 2-A I divides according to the current division rules for parallel branches.

Remember that parallel currents divide inversely to branch resistances but directly with conductances. For this reason it may be preferable to consider the current source shunted by the conductance G, as shown in Fig. 10–7c. We can always convert between resistance and conductance because $1/R$ in ohms is equal to G in siemens.

The symbol for a current source is a circle with an arrow inside, as shown in Fig. 10–7b and c, to show the direction of current. This direction must be the same as the current produced by the polarity of the corresponding voltage source. Remember that a source produces electron flow out from the negative terminal.

An important difference between voltage and current sources is that a current source is killed by making it open, compared with short-circuiting a voltage source. Opening a current source kills its ability to supply current without affecting any parallel branches. A voltage source is short-circuited to kill its ability to supply voltage without affecting any series components.

Figure 10–7 General forms for a voltage source or current source connected to a load R_L across terminals A and B. (*a*) Voltage source V with series R. (*b*) Current source I with parallel R. (*c*) Current source I with parallel conductance G.

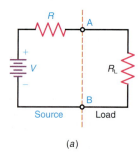

(a)

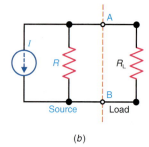

(b)

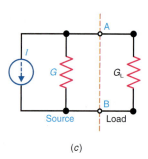

(c)

Figure 10–8 Any network in the block at the left can be reduced to the Norton equivalent parallel circuit at the right.

The Norton Equivalent Circuit

As illustrated in Fig. 10–8, Norton's theorem states that the entire network connected to terminals A and B can be replaced by a single current source I_N in parallel with a single resistance R_N. The value of I_N is equal to the short-circuit current through the AB terminals. This means finding the current that the network would produce through A and B with a short circuit across these two terminals.

The value of R_N is the resistance looking back from open terminals A and B. These terminals are not short-circuited for R_N but are open, as in calculating R_{TH} for Thevenin's theorem. Actually, the single resistor is the same for both the Norton and Thevenin equivalent circuits. In the Norton case, this value of R_{AB} is R_N in parallel with the current source; in the Thevenin case, it is R_{TH} in series with the voltage source.

Nortonizing a Circuit

As an example, let us recalculate the current I_L in Fig. 10–9a, which was solved before by Thevenin's theorem. The first step in applying Norton's theorem is to imagine a short circuit across terminals A and B, as shown in Fig. 10–9b. How much current is flowing in the short circuit? Note that a short circuit across AB short-circuits R_L and the parallel R_2. Then the only resistance in the circuit is the 3-Ω R_1 in series with the 36-V source, as shown in Fig. 10–9c. The short-circuit current, therefore, is

$$I_N = \frac{36 \text{ V}}{3 \text{ }\Omega} = 12 \text{ A}$$

This 12-A I_N is the total current available from the current source in the Norton equivalent in Fig. 10–9e.

To find R_N, remove the short circuit across A and B and consider the terminals open without R_L. Now the source V is considered short-circuited. As shown in Fig. 10–9d, the resistance seen looking back from terminals A and B is 6 Ω in parallel with 3 Ω, which equals 2 Ω for the value of R_N.

The resultant Norton equivalent is shown in Fig. 10–9e. It consists of a 12-A current source I_N shunted by the 2-Ω R_N. The arrow on the current source shows the direction of electron flow from terminal B to terminal A, as in the original circuit.

Finally, to calculate I_L, replace the 2-Ω R_L between terminals A and B, as shown in Fig. 10–9f. The current source still delivers 12 A, but now that current divides between the two branches of R_N and R_L. Since these two resistances are equal, the 12-A I_N divides into 6 A for each branch, and I_L is equal to 6 A. This value is the same current we calculated in Fig. 10–3, by Thevenin's theorem. Also, V_L can be calculated as $I_L R_L$, or 6 A × 2 Ω, which equals 12 V.

Looking at the Short-Circuit Current

In some cases, there may be a question of which current is I_N when terminals A and B are short-circuited. Imagine that a wire jumper is connected between

Figure 10–9 Same circuit as in Fig. 10-3, but solved by Norton's theorem. (*a*) Original circuit. (*b*) Short circuit across terminals A and B. (*c*) The short-circuit current I_N is 36/3 = 12 A. (*d*) Open terminals A and B but short-circuit V to find R_{AB} is 2 Ω, the same as R_{TH}. (*e*) Norton equivalent circuit. (*f*) R_L reconnected to terminals A and B to find that I_L is 6 A.

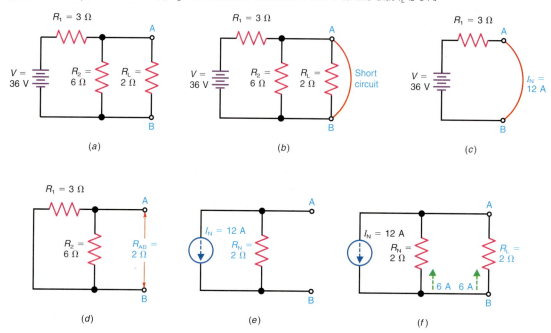

A and B to short-circuit these terminals. Then I_N must be the current that flows in this wire between terminals A and B.

Remember that any components directly across these two terminals are also short-circuited by the wire jumper. Then these parallel paths have no effect. However, any components in series with terminal A or terminal B are in series with the wire jumper. Therefore, the short-circuit current I_N also flows through the series components.

An example of a resistor in series with the short circuit across terminals A and B is shown in Fig. 10–10. The idea here is that the short-circuit I_N is a branch current, not the main-line current. Refer to Fig. 10–10*a*. Here the short circuit connects R_3 across R_2. Also, the short-circuit current I_N is now the same as the current I_3 through R_3. Note that I_3 is only a branch current.

To calculate I_3, the circuit is solved by Ohm's law. The parallel combination of R_2 with R_3 equals 72/18 or 4 Ω. The R_T is 4 + 4 = 8 Ω. As a result, the I_T from the source is 48 V/8 Ω = 6 A.

This I_T of 6 A in the main line divides into 4 A for R_2 and 2 A for R_3. The 2-A I_3 for R_3 flows through short-circuited terminals A and B. Therefore, this current of 2 A is the value of I_N.

To find R_N in Fig. 10–10*b*, the short circuit is removed from terminals A and B. Now the source V is short-circuited. Looking back from open terminals A and B, the 4-Ω R_1 is in parallel with the 6-Ω R_2. This combination is 24/10 = 2.4 Ω. The 2.4 Ω is in series with the 12-Ω R_3 to make R_{AB} = 2.4 + 12 = 14.4 Ω.

The final Norton equivalent is shown in Fig. 10–10*c*. Current I_N is 2 A because this branch current in the original circuit is the current that flows through R_3 and short-circuited terminals A and B. Resistance R_N is 14.4 Ω looking back from open terminals A and B with the source V short-circuited the same way as for R_{TH}.

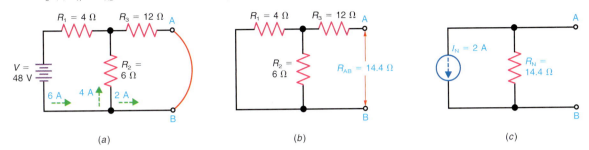

Figure 10–10 Nortonizing a circuit where the short-circuit current I_N is a branch current. (a) I_N is 2 A through short-circuited terminals A and B and R_3. (b) $R_N = R_{AB} = 14.4\ \Omega$. (c) Norton equivalent circuit.

(a)   (b)   (c)

■ **10–5 Knowledge Check**

Answers at end of chapter.

In Fig. 10–9*a*, what are the values for I_N and R_N if $R_1 = 6\ \Omega$?

■ **10–5 Self-Review**

Answers at end of chapter.

a. For a Norton equivalent circuit, terminals A and B are short-circuited to find I_N. (True or False)
b. For a Norton equivalent circuit, terminals A and B are open to find R_N. (True or False)

10–6 Thevenin–Norton Conversions

Thevenin's theorem says that any network can be represented by a voltage source and series resistance, and Norton's theorem says that the same network can be represented by a current source and shunt resistance. It must be possible, therefore, to convert directly from a Thevenin form to a Norton form and vice versa. Such conversions are often useful.

Norton from Thevenin

Consider the Thevenin equivalent circuit in Fig. 10–11*a*. What is its Norton equivalent? Just apply Norton's theorem, the same as for any other circuit. The short-circuit current through terminals A and B is

$$I_N = \frac{V_{TH}}{R_{TH}} = \frac{15\ V}{3\ \Omega} = 5\ A$$

Figure 10–11 Thevenin equivalent circuit in (a) corresponds to the Norton equivalent in (b).

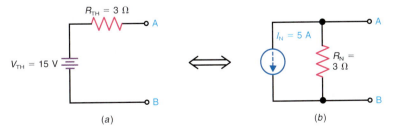

(a)   (b)

The resistance, looking back from open terminals A and B with the source V_{TH} short-circuited, is equal to the 3 Ω of R_{TH}. Therefore, the Norton equivalent consists of a current source that supplies the short-circuit current of 5 A, shunted by the same 3-Ω resistance that is in series in the Thevenin circuit. The results are shown in Fig. 10–11b.

Thevenin from Norton

For the opposite conversion, we can start with the Norton circuit of Fig. 10–11b and get back to the original Thevenin circuit. To do this, apply Thevenin's theorem, the same as for any other circuit. First, we find the Thevenin resistance by looking back from open terminals A and B. An important principle here, though, is that, although a voltage source is short-circuited to find R_{TH}, a current source is an open circuit. In general, a current source is killed by opening the path between its terminals. Therefore, we have just the 3-Ω R_N, in parallel with the infinite resistance of the open current source. The combined resistance then is 3 Ω.

In general, the resistance R_N always has the same value as R_{TH}. The only difference is that R_N is connected in parallel with I_N, but R_{TH} is in series with V_{TH}.

Now all that is required is to calculate the open-circuit voltage in Fig. 10–11b to find the equivalent V_{TH}. Note that with terminals A and B open, all current from the current source flows through the 3-Ω R_N. Then the open-circuit voltage across the terminals A and B is

$$I_N R_N = 5 \text{ A} \times 3 \text{ Ω} = 15 \text{ V} = V_{TH}$$

As a result, we have the original Thevenin circuit, which consists of the 15-V source V_{TH} in series with the 3-Ω R_{TH}.

Conversion Formulas

In summary, the following formulas can be used for these conversions:

Thevenin from Norton:

$$R_{TH} = R_N$$

$$V_{TH} = I_N \times R_N$$

Norton from Thevenin:

$$R_N = R_{TH}$$

$$I_N = V_{TH}/R_{TH}$$

Another example of these conversions is shown in Fig. 10–12.

Figure 10–12 Example of Thevenin-Norton conversions. (*a*) Original circuit, the same as in Figs. 10–3*a* and 10–9*a*. (*b*) Thevenin equivalent. (*c*) Norton equivalent.

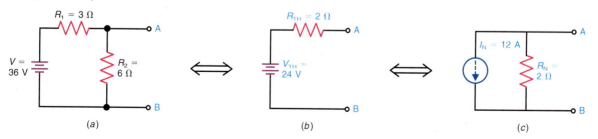

10-6A Knowledge Check

Answer at end of chapter.

If $V_{TH} = 24$ V and $R_{TH} = 8$ Ω in Fig. 10–11a, what are its Norton equivalent values in Fig. 10–11b?

10-6B Knowledge Check

Answer at end of chapter.

If $I_N = 2$ A and $R_N = 9$ Ω in Fig. 10–11b, what are its Thevenin equivalent values in Fig. 10–11a?

10-6 Self-Review

Answers at end of chapter.

a. In Thevenin-Norton conversions, resistances R_N and R_{TH} are equal. (True or False)
b. In Thevenin-Norton conversions, current I_N is V_{TH}/R_{TH}. (True or False)
c. In Thevenin-Norton conversions, voltage V_{TH} is $I_N \times R_N$. (True or False)

10–7 Conversion of Voltage and Current Sources

Norton conversion is a specific example of the general principle that any voltage source with its series resistance can be converted to an equivalent current source with the same resistance in parallel. In Fig. 10–13, the voltage source in Fig. 10–13a is equivalent to the current source in Fig. 10–13b. Just divide the source V by its series R to calculate the value of I for the equivalent current source shunted by the same R. Either source will supply the same current and voltage for any components connected across terminals A and B.

Conversion of voltage and current sources can often simplify circuits, especially those with two or more sources. Current sources are easier for parallel connections, where we can add or divide currents. Voltage sources are easier for series connections, where we can add or divide voltages.

Two Sources in Parallel Branches

In Fig. 10–14a, assume that the problem is to find I_3 through the middle resistor R_3. Note that V_1 with R_1 and V_2 with R_2 are branches in parallel with R_3. All three branches are connected across terminals A and B.

When we convert V_1 and V_2 to current sources in Fig. 10–14b, the circuit has all parallel branches. Current I_1 is $^{84}/_{12}$ or 7 A, and I_2 is $^{21}/_3$ which also happens to be 7 A. Current I_1 has its parallel R of 12 Ω, and I_2 has its parallel R of 3 Ω.

Figure 10–13 The voltage source in (a) corresponds to the current source in (b).

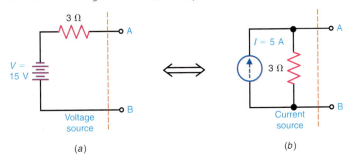

(a)   (b)

Figure 10–14 Converting two voltage sources in V_1 and V_2 in parallel branches to current sources I_1 and I_2 that can be combined. (a) Original circuit. (b) V_1 and V_2 converted to parallel current sources I_1 and I_2. (c) Equivalent circuit with one combined current source I_T.

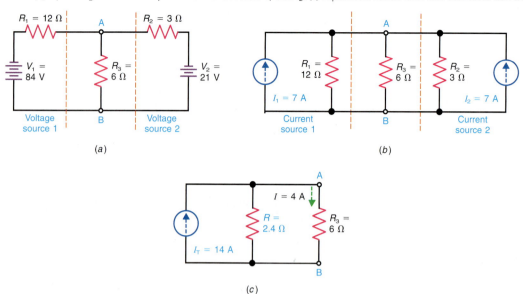

(a)

(b)

(c)

Furthermore, I_1 and I_2 can be combined for the one equivalent current source I_T in Fig. 10–14c. Since both sources produce current in the same direction through R_L, they are added for $I_T = 7 + 7 = 14$ A.

The shunt R for the 14-A combined source is the combined resistance of the 12-Ω R_1 and the 3-Ω R_2 in parallel. This R equals $^{36}/_{15}$ or 2.4 Ω, as shown in Fig. 10–14c.

To find I_L, we can use the current divider formula for the 6- and 2.4-Ω branches, into which the 14-A I_T from the current source was split. Then

$$I_L = \frac{2.4}{2.4 + 6} \times 14 = 4 \text{ A}$$

The voltage V_{R_3} across terminals A and B is $I_L R_L$, which equals $4 \times 6 = 24$ V. These are the same values calculated for V_{R_3} and I_3 by Kirchhoff's laws in Fig. 9–5 and by Thevenin's theorem in Fig. 10–5.

Two Sources in Series

Referring to Fig. 10–15, assume that the problem is to find the current I_L through the load resistance R_L between terminals A and B. This circuit has the two current sources I_1 and I_2 in series.

The problem here can be simplified by converting I_1 and I_2 to the series voltage sources V_1 and V_2 shown in Fig. 10–15b. The 2-A I_1 with its shunt 4-Ω R_1 is equivalent to 4×2, or 8 V, for V_1 with a 4-Ω series resistance. Similarly, the 5-A I_2 with its shunt 2-Ω R_2 is equivalent to 5×2, or 10 V, for V_2 with a 2-Ω series resistance. The polarities of V_1 and V_2 produce electron flow in the same direction as I_1 and I_2.

The series voltages can now be combined as in Fig. 10–15c. The 8 V of V_1 and 10 V of V_2 are added because they are series-aiding, resulting in the total V_T of 18 V, and, the resistances of 4 Ω for R_1 and 2 Ω for R_2 are added, for a combined R of 6 Ω. This is the series resistance of the 18-V source V_T connected across terminals A and B.

The total resistance of the circuit in Fig. 10–15c is R plus R_L, or $6 + 3 = 9\ \Omega$. With 18 V applied, $I_L = {}^{18}/_9 = 2$ A through R_L between terminals A and B.

Figure 10–15 Converting two current sources I_1 and I_2 in series to voltage sources V_1 and V_2 that can be combined. (*a*) Original circuit. (*b*) I_1 and I_2 converted to series voltage sources V_1 and V_2. (*c*) Equivalent circuit with one combined voltage source V_T.

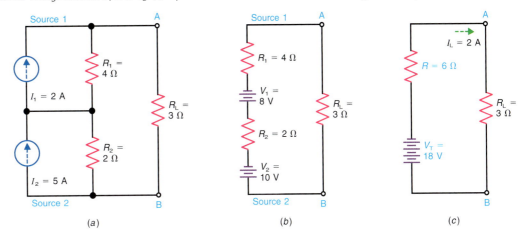

(*a*)   (*b*)   (*c*)

■ *10–7 Knowledge Check*

Answer at end of chapter.

In Fig. 10–14*c*, what is the value of the combined current source, I_T, if the polarity of V_1 is reversed in Fig. 10–14*a*?

■ *10–7 Self-Review*

Answers at end of chapter.

A voltage source has 21 V in series with 3 Ω. For the equivalent current source,

a. How much is *I*?
b. How much is the shunt *R*?

GOOD TO KNOW

When the values of *I* and *G* are known in any circuit, the voltage,

V, can be calculated as $V = \dfrac{I}{G}$.

10–8 Millman's Theorem

Millman's theorem provides a shortcut for finding the common voltage across any number of parallel branches with different voltage sources. A typical example is shown in Fig. 10–16. For all branches, the ends at point Y are connected to chassis ground. Furthermore, the opposite ends of all branches are also connected to the common point X. The voltage V_{XY}, therefore, is the common voltage across all branches.

Figure 10–16 Example of Millman's theorem to find V_{XY}, the common voltage across branches with separate voltage sources.

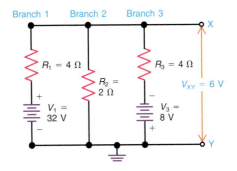

Finding the value of V_{XY} gives the net effect of all sources in determining the voltage at X with respect to chassis ground. To calculate this voltage,

$$V_{XY} = \frac{V_1/R_1 + V_2/R_2 + V_3/R_3}{1/R_1 + 1/R_2 + 1/R_3} \cdots \text{ etc.} \qquad (10\text{–}1)$$

This formula is derived by converting the voltage sources to current sources and combining the results. The numerator with V/R terms is the sum of the parallel current sources. The denominator with $1/R$ terms is the sum of the parallel conductances. The net V_{XY}, then, is the form of I/G or $I \times R$, which is in units of voltage.

Calculating V_{XY}

For the values in Fig. 10–16,

$$V_{XY} = \frac{32/4 + 0/2 - 8/4}{1/4 + 1/2 + 1/4}$$

$$= \frac{8 + 0 - 2}{1}$$

$$V_{XY} = 6 \text{ V}$$

Note that in branch 3, V_3 is considered negative because it would make point X negative. However, all resistances are positive. The positive answer for V_{XY} means that point X is positive with respect to Y.

In branch 2, V_2 is zero because this branch has no voltage source. However, R_2 is still used in the denominator.

This method can be used for any number of branches, but all must be in parallel without any series resistances between branches. In a branch with several resistances, they can be combined as one R_T. When a branch has more than one voltage source, the voltages can be combined algebraically for one V_T.

Applications of Millman's Theorem

In many cases, a circuit can be redrawn to show the parallel branches and their common voltage V_{XY}. Then with V_{XY} known, the entire circuit can be analyzed quickly. For instance, Fig. 10–17 has been solved before by other methods. For Millman's theorem, the common voltage V_{XY} across all branches is the same as V_3 across R_3. This voltage is calculated with Formula (10–1), as follows:

$$V_{XY} = \frac{-84/12 + 0/6 - 21/3}{1/12 + 1/6 + 1/3} = \frac{-7 + 0 - 7}{7/12}$$

$$= \frac{-14}{7/12} = -14 \times \frac{12}{7}$$

$$V_{XY} = -24 \text{ V} = V_3$$

Figure 10–17 The same circuit as in Fig. 9–5 for Kirchhoff's laws, but shown with parallel branches to calculate V_{XY} by Millman's theorem.

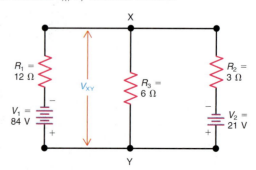

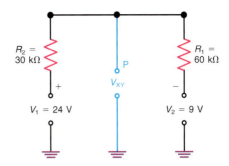

The negative sign means that point X is the negative side of V_{XY}.

With V_3 known to be 24 V across the 6-Ω R_3, then I_3 must be 24/6 = 4 A. Similarly, all voltages and currents in this circuit can then be calculated. (See Fig. 9–6 in Chap. 9.)

As another application, the example of superposition in Fig. 10–1 has been redrawn in Fig. 10–18 to show the parallel branches with a common voltage V_{XY} to be calculated by Millman's theorem. Then

$$V_{XY} = \frac{24 \text{ V}/30 \text{ k}\Omega - 9 \text{ V}/60 \text{ k}\Omega}{1/(30 \text{ k}\Omega) + 1/(60 \text{ k}\Omega)} = \frac{0.8 \text{ mA} - 0.15 \text{ mA}}{3/(60 \text{ k}\Omega)}$$

$$= 0.65 \times \frac{60}{3} = \frac{39}{3}$$

$$V_{XY} = 13 \text{ V} = V_P$$

The answer of 13 V from point P to ground, using Millman's theorem, is the same value calculated before by superposition.

■ *10-8 Knowledge Check*

> *Answer at end of chapter.*

> In Fig. 10–16, how much is the voltage, V_{XY}, if the polarity of V_1 is reversed?

■ *10-8 Self-Review*

> *Answers at end of chapter.*

> **For the example of Millman's theorem in Fig. 10–16,**
> a. **How much is V_{R_2}?**
> b. **How much is V_{R_3}?**

10–9 T or Y and π or Δ Connections

The circuit in Fig. 10–19 is called a T (tee) or Y (wye) network, as suggested by the shape. They are different names for the same network; the only difference is that the R_2 and R_3 legs are shown at an angle in the Y.

The circuit in Fig. 10–20 is called a π (pi) or Δ (delta) network because the shape is similar to these Greek letters. Both forms are the same network. Actually, the network can have the R_A arm shown at the top or bottom, as long as it is connected between R_B and R_C. In Fig. 10–20, R_A is at the top, as an inverted delta, to look like the π network.

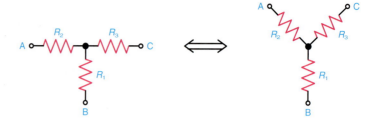

Figure 10–19 The form of a T or Y network.

The circuits in Figs. 10–19 and 10–20 are passive networks without any energy sources. They are also three-terminal networks with two pairs of connections for input and output voltages with one common. In Fig. 10–19, point B is the common terminal and point 2 is common in Fig. 10–20.

The Y and Δ forms are different ways to connect three resistors in a passive network. Note that resistors in the Y are labeled with subscripts 1, 2, and 3, whereas the Δ has subscripts A, B, and C to emphasize the different connections.

Conversion Formulas

In the analysis of networks, it is often helpful to convert a Δ to Y or vice versa. Either it may be difficult to visualize the circuit without the conversion, or the conversion makes the solution simpler. The formulas for these transformations are given here. All are derived from Kirchhoff's laws. Note that letters are used as subscripts for R_A, R_B, and R_C in the Δ, whereas the resistances are numbered R_1, R_2, and R_3 in the Y.

Conversions of Y to Δ or T to π:

$$R_A = \frac{R_1R_2 + R_2R_3 + R_3R_1}{R_1}$$

$$R_B = \frac{R_1R_2 + R_2R_3 + R_3R_1}{R_2} \tag{10–2}$$

$$R_C = \frac{R_1R_2 + R_2R_3 + R_3R_1}{R_3}$$

or

$$R_\Delta = \frac{\Sigma \text{ all cross products in Y}}{\text{opposite } R \text{ in Y}}$$

These formulas can be used to convert a Y network to an equivalent Δ or a T network to π. Both networks will have the same resistance across any pair of terminals.

Figure 10–20 The form of a π or Δ network.

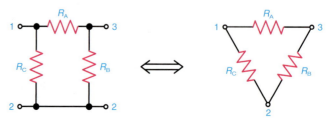

The three formulas have the same general form, indicated at the bottom as one basic rule. The symbol Σ is the Greek capital letter sigma, meaning "sum of."

For the opposite conversion,

Conversion of Δ to Y or π to T:

$$R_1 = \frac{R_B R_C}{R_A + R_B + R_C}$$

$$R_2 = \frac{R_C R_A}{R_A + R_B + R_C} \qquad (10\text{--}3)$$

$$R_3 = \frac{R_A R_B}{R_A + R_B + R_C}$$

or

$$R_Y = \frac{\text{product of two adjacent } R \text{ in } \Delta}{\Sigma \text{ all } R \text{ in } \Delta}$$

As an aid in using these formulas, the following scheme is useful. Place the Y inside the Δ, as shown in Fig. 10–21. Notice that the Δ has three closed sides, and the Y has three open arms. Also note how resistors can be considered opposite each other in the two networks. For instance, the open arm R_1 is opposite the closed side R_A, R_2 is opposite R_B, and R_3 is opposite R_C.

Furthermore, each resistor in an open arm has two adjacent resistors in the closed sides. For R_1, its adjacent resistors are R_B and R_C, also R_C and R_A are adjacent to R_2, and R_A and R_B are adjacent to R_3.

In the formulas for the Y-to-Δ conversion, each side of the delta is found by first taking all possible cross products of the arms of the wye, using two arms at a time. There are three such cross products. The sum of the three cross products is then divided by the opposite arm to find the value of each side of the delta. Note that the numerator remains the same, the sum of the three cross products. However, each side of the delta is calculated by dividing this sum by the opposite arm.

For the Δ-to-Y conversion, each arm of the wye is found by taking the product of the two adjacent sides in the delta and dividing by the sum of the three sides of the delta. The product of two adjacent resistors excludes the opposite resistor. The denominator for the sum of the three sides remains the same in the three formulas. However, each arm is calculated by dividing the sum into each cross product.

An Example of Conversion

The values shown for the equivalent Y and Δ in Fig. 10–21 are calculated as follows: Starting with 4, 6, and 10 Ω for sides R_A, R_B, and R_C, respectively, in the delta, the corresponding arms in the wye are

$$R_1 = \frac{R_B R_C}{R_A + R_B + R_C} = \frac{6 \times 10}{4 + 6 + 10} = \frac{60}{20} = 3 \ \Omega$$

$$R_2 = \frac{R_C R_A}{20} = \frac{10 \times 4}{20} = \frac{40}{20} = 2 \ \Omega$$

$$R_3 = \frac{R_A R_B}{20} = \frac{4 \times 6}{20} = \frac{24}{20} = 1.2 \ \Omega$$

As a check on these values, we can calculate the equivalent delta for this wye. Starting with values of 3, 2, and 1.2 Ω for R_1, R_2, and R_3, respectively, in the wye, the corresponding values in the delta are:

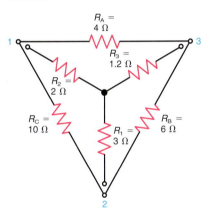

Figure 10–21 Conversion between Y and Δ networks. See text for conversion formulas.

$$R_A = \frac{R_1R_2 + R_2R_3 + R_3R_1}{R_1} = \frac{6 + 2.4 + 3.6}{3} = \frac{12}{3} = 4\,\Omega$$

$$R_B = \frac{12}{R_2} = \frac{12}{2} = 6\,\Omega$$

$$R_C = \frac{12}{R_3} = \frac{12}{1.2} = 10\,\Omega$$

These results show that the Y and Δ networks in Fig. 10–21 are equivalent to each other when they have the values obtained with the conversion formulas.

Note that the equivalent R values in the Y are less than those in the equivalent Δ network. The reason is that the Y has two legs between the terminals, whereas the Δ has only one.

Simplifying a Bridge Circuit

As an example of the use of such transformations, consider the bridge circuit of Fig. 10–22. The total current I_T from the battery is desired. Therefore, we must find the total resistance R_T.

One approach is to note that the bridge in Fig. 10–22a consists of two deltas connected between terminals P_1 and P_2. One of them can be replaced by

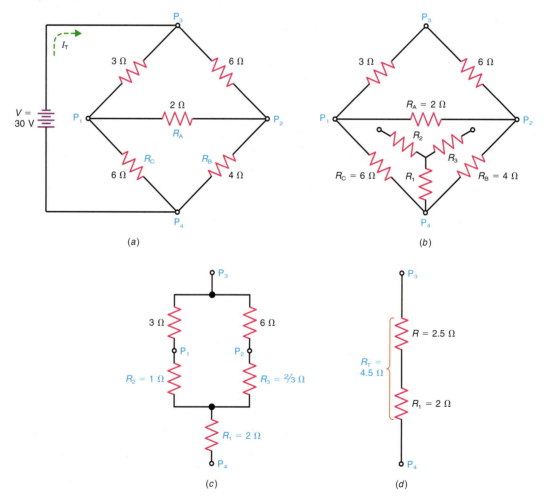

Figure 10–22 Solving a bridge circuit by Δ-to-Y conversion. (a) Original circuit. (b) How the Y of $R_1R_2R_3$ corresponds to the Δ of $R_AR_BR_C$. (c) The Y substituted for the Δ network. The result is a series-parallel circuit with the same R_T as the original bridge circuit. (d) R_T is 4.5 Ω between points P_3 and P_4.

(a)

(b)

(c)

(d)

an equivalent wye. We use the bottom delta with R_A across the top, in the same form as Fig. 10–21. We then replace this delta $R_A R_B R_C$ by an equivalent wye $R_1 R_2 R_3$, as shown in Fig. 10–22b. Using the conversion formulas,

$$R_1 = \frac{R_B R_C}{R_A + R_B + R_C} = \frac{24}{12} = 2 \ \Omega$$

$$R_2 = \frac{R_C R_A}{12} = \frac{12}{12} = 1 \ \Omega$$

$$R_3 = \frac{R_A R_B}{12} = \frac{8}{12} = \frac{2}{3} \ \Omega$$

We next use these values for R_1, R_2, and R_3 in an equivalent wye to replace the original delta. Then the resistances form the series-parallel circuit shown in Fig. 10–22c. The combined resistance of the two parallel branches here is 4×6.67 divided by 10.67, which equals 2.5 Ω for R. Adding this 2.5 Ω to the series R_1 of 2 Ω, the total resistance is 4.5 Ω in Fig. 10–22d.

This 4.5 Ω is the R_T for the entire bridge circuit between P_3 and P_4 connected to source V. Then I_T is 30 V/4.5 Ω, which equals 6.67 A supplied by the source.

Another approach to finding R_T for the bridge circuit in Fig. 10–22a is to recognize that the bridge also consists of two T or Y networks between terminals P_3 and P_4. One of them can be transformed into an equivalent delta. The result is another series-parallel circuit but with the same R_T of 4.5 Ω.

Balanced Networks

When all the R values are equal in a network, it is balanced. Then the conversion is simplified, as R in the wye network is one-third the R in the equivalent delta. As an example, for $R_A = R_B = R_C$ equal to 6 Ω in the delta, the equivalent wye has $R_1 = R_2 = R_3$ equal to 6/3 or 2 Ω. Or, converting the other way, for 2-Ω R values in a balanced wye, the equivalent delta network has each R equal to $3 \times 2 = 6$ Ω. This example is illustrated in Fig. 10–23.

■ *10–9 Knowledge Check*

Answer at end of chapter.

In Fig. 10–23a, assume $R_A = R_B = R_C = 36$ Ω. What are the equivalent values of R_1, R_2, and R_3 in Fig. 10–23b?

■ *10–9 Self-Review*

Answers at end of chapter.

In the standard form for conversion,
a. Which resistor in the Y is opposite R_A in the Δ?
b. Which two resistors in the Δ are adjacent to R_1, in the Y?

Figure 10–23 Equivalent balanced networks. (a) Delta form. (b) Wye form.

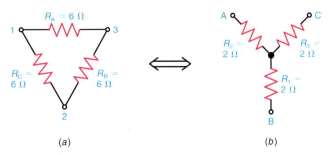

(a)

(b)

Summary

- *Superposition theorem.* In a linear, bilateral network having more than one source, the current and voltage in any part of the network can be found by adding algebraically the effect of each source separately. All other sources are temporarily killed by short-circuiting voltage sources and opening current sources.

- *Thevenin's theorem.* Any network with two open terminals A and B can be replaced by a single voltage source V_{TH} in series with a single resistance R_{TH} connected to terminals A and B. Voltage V_{TH} is the voltage produced by the network across terminals A and B. Resistance R_{TH} is the resistance across open terminals A and B with all voltage sources short-circuited.

- *Norton's theorem.* Any two-terminal network can be replaced by a single current source I_N in parallel with a single resistance R_N. The value of I_N is the current produced by the network through the short-circuited terminals. R_N is the resistance across the open terminals with all voltage sources short-circuited.

- *Millman's theorem.* The common voltage across parallel branches with different V sources can be determined with Formula (10-1).

- A voltage source V with its series R can be converted to an equivalent current source I with parallel R. Similarly, a current source I with a parallel R can be converted to a voltage source V with a series R. The value of I is V/R, or V is $I \times R$. The value of R is the same for both sources. However, R is in series with V but in parallel with I.

- The conversion between delta and wye networks is illustrated in Fig. 10–21. To convert from one network to the other, Formula (10-2) or (10-3) is used.

Important Terms

- Active Components - electronic components such as diodes and transistors that can rectify or amplify.

- Bilateral Components - electronic components that have the same current for opposite polarities of applied voltage.

- Current Source - a source that supplies a total current to be divided among parallel branches.

- Linear Component - an electronic component whose current is proportional to the applied voltage.

- Millman's Theorem - a theorem that provides a shortcut for finding the common voltage across any number of parallel branches with different voltage sources.

- Norton's Theorem - a theorem that states that an entire network connected to a pair of terminals can be replaced by a single current source, I_N, in parallel with a single resistance, R_N.

- Passive Components - electronic components that do not amplify or rectify.

- Superposition Theorem - a theorem that states that in a network with two or more sources, the current or voltage for any component is the algebraic sum of the effects produced by each source acting separately.

- Thevenin's Theorem - a theorem that states that an entire network connected to a pair of terminals can be replaced by a single voltage source, V_{TH}, in series with a single resistance, R_{TH}.

- Voltage Source - a source that supplies a total voltage to be divided among series components.

Related Formulas

Millman's theorem

$$V_{XY} = \frac{V_1/R_1 + V_2/R_2 + V_3/R_3}{1/R_1 + 1/R_2 + 1/R_3} \cdots \text{etc.}$$

Conversion of Y to Δ or T to π:

$$R_A = \frac{R_1 R_2 + R_2 R_3 + R_3 R_1}{R_1}$$

$$R_B = \frac{R_1 R_2 + R_2 R_3 + R_3 R_1}{R_2}$$

$$R_C = \frac{R_1 R_2 + R_2 R_3 + R_3 R_1}{R_3}$$

$$\text{or} \quad R_\Delta = \frac{\Sigma \text{ all cross products in Y}}{\text{opposite } R \text{ in Y}}$$

Conversion of Δ to Y or π to T.

$$R_1 = \frac{R_B R_C}{R_A + R_B + R_C}$$

$$R_2 = \frac{R_C R_A}{R_A + R_B + R_C}$$

$$R_3 = \frac{R_A R_B}{R_A + R_B + R_C}$$

or

$$R_Y = \frac{\text{product of two adjacent } R \text{ in } \Delta}{\Sigma \text{ all } R \text{ in } \Delta}$$

Self–Test

Answers at back of book.

1. A resistor is an example of a(n)

a. bilateral component.

b. active component.

c. passive component.

d. both a and c.

2. To apply the Superposition Theorem, all components must be

a. the active-type.

b. both linear and bilateral.

c. grounded.

d. both nonlinear and unidirectional.

3. When converting from a Norton-equivalent circuit to a Thevenin equivalent circuit or vice versa,

a. R_N and R_{TH} have the same value.

b. R_N will always be larger than R_{TH}.

c. I_N is short-circuited to find V_{TH}.

d. V_{TH} is short-circuited to find I_N.

4. When solving for the Thevenin equivalent resistance, R_{TH},

a. all voltage sources must be opened.

b. all voltage sources must be short-circuited.

c. all voltage sources must be converted to current sources.

d. none of the above.

5. Thevenin's Theorem states that an entire network connected to a pair of terminals can be replaced with

a. a single current source in parallel with a single resistance.

b. a single voltage source in parallel with a single resistance.

c. a single voltage source in series with a single resistance.

d. a single current source in series with a single resistance.

6. Norton's Theorem states that an entire network connected to a pair of terminals can be replaced with

a. a single current source in parallel with a single resistance.

b. a single voltage source in parallel with a single resistance.

c. a single voltage source in series with a single resistance.

d. a single current source in series with a single resistance.

7. With respect to terminals A and B in a complex network, the Thevenin voltage, V_{TH}, is

a. the voltage across terminals A and B when they are short-circuited.

b. the open-circuit voltage across terminals A and B.

c. the same as the voltage applied to the complex network.

d. none of the above.

8. A Norton equivalent circuit consists of a 100-μA current source, I_N, in parallel with a 10-kΩ resistance, R_N. If this circuit is converted into a Thevenin equivalent circuit, how much is V_{TH}?

a. 1 kV.

b. 10 V.

c. 1 V.

d. This is impossible to determine.

9. With respect to terminals A and B in a complex network, the Norton current, I_N, equals

a. the current flowing between terminals A and B when they are open.

b. the total current supplied by the applied voltage to the network.

c. zero when terminals A and B are short-circuited.

d. the current flowing between terminals A and B when they are short-circuited.

10. Which theorem provides a shortcut for finding the common voltage across any number of parallel branches with different voltage sources?

a. The Superposition Theorem.

b. Thevenin's Theorem.

c. Norton's Theorem.

d. Millman's Theorem.

Questions

1. State the superposition theorem, and discuss how to apply it.

2. State how to calculate V_{TH} and R_{TH} in Thevenin equivalent circuits.

3. State the method of calculating I_N and R_N for a Norton equivalent circuit.

4. How is a voltage source converted to a current source, and vice versa?

5. For what type of circuit is Millman's theorem used?

6. Draw a delta network and a wye network and give the six formulas needed to convert from one to the other.

Problems

SECTION 10-1 SUPERPOSITION THEOREM

10-1 **|||| MultiSim** In Fig. 10–24, use the Superposition Theorem to solve for the voltage, V_P, with respect to ground.

Figure 10–24

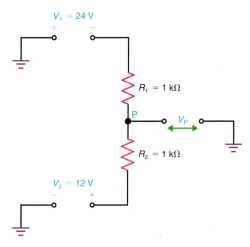

10-2 **|||| MultiSim** In Fig. 10–25, use the Superposition Theorem to solve for the voltage, V_P, with respect to ground.

Figure 10–25

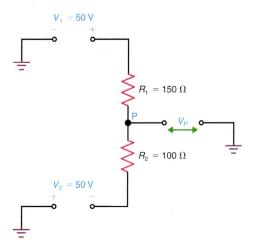

10-3 In Fig. 10–25, recalculate the voltage, V_P, if the resistors R_1 and R_2 are interchanged.

10-4 In Fig. 10–26, use the Superposition Theorem to solve for the voltage, V_{AB}.

10-5 In Fig. 10–26, recalculate the voltage, V_{AB}, if the polarity of V_2 is reversed.

Figure 10–26

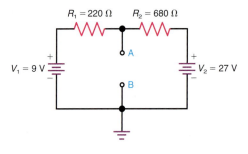

SECTION 10-2 THEVENIN'S THEOREM

10-6 **|||| MultiSim** In Fig. 10–27, draw the Thevenin equivalent circuit with respect to terminals A and B (mentally remove R_L).

Figure 10–27

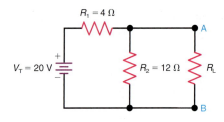

10-7 In Fig. 10–27, use the Thevenin equivalent circuit to calculate I_L and V_L for the following values of R_L; $R_L = 3\ \Omega$, $R_L = 6\ \Omega$, and $R_L = 12\ \Omega$.

10-8 **|||| MultiSim** In Fig. 10–28, draw the Thevenin equivalent circuit with respect to terminals A and B (mentally remove R_L).

Figure 10–28

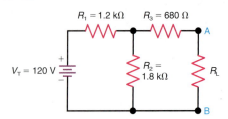

10-9 In Fig. 10–28, use the Thevenin equivalent circuit to calculate I_L and V_L for the following values of R_L: $R_L = 100\ \Omega$, $R_L = 1\ k\Omega$, and $R_L = 5.6\ k\Omega$.

10–10 In Fig. 10–29, draw the Thevenin equivalent circuit with respect to terminals A and B (mentally remove R_L).

Figure 10–29

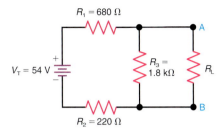

10–11 In Fig. 10–29, use the Thevenin equivalent circuit to calculate I_L and V_L for the following values of R_L: $R_L = 200\,\Omega$, $R_L = 1.2\,\text{k}\Omega$, and $R_L = 1.8\,\text{k}\Omega$.

10–12 In Fig. 10–30, draw the Thevenin equivalent circuit with respect to terminals A and B (mentally remove R_L).

Figure 10–30

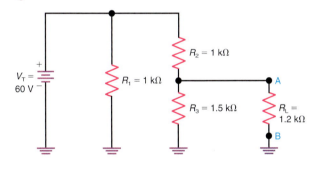

10–13 In Fig. 10–30, use the Thevenin equivalent circuit to solve for I_L and R_L.

SECTION 10-3 THEVENIZING A CIRCUIT WITH TWO VOLTAGE SOURCES

10–14 In Fig. 10–31, draw the Thevenin equivalent circuit with respect to terminals A and B (mentally remove R_3).

Figure 10–31

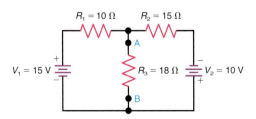

10–15 Using the Thevenin equivalent circuit for Fig. 10–31, calculate the values for I_3 and V_{R_3}.

10–16 In Fig. 10–32, draw the Thevenin equivalent circuit with respect to terminals A and B (mentally remove R_3).

Figure 10–32

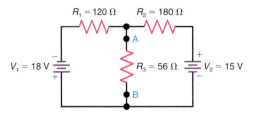

10–17 Using the Thevenin equivalent circuit for Fig. 10–32, calculate the values for I_3 and V_{R_3}.

10–18 With the polarity of V_1 reversed in Fig. 10–32, redraw the Thevenin equivalent circuit with respect to terminals A and B. Also, recalculate the new values for I_3 and V_{R_3}.

SECTION 10-4 THEVENIZING A BRIDGE CIRCUIT

10–19 In Fig. 10–33, draw the Thevenin equivalent circuit with respect to terminals A and B (mentally remove R_L).

Figure 10–33

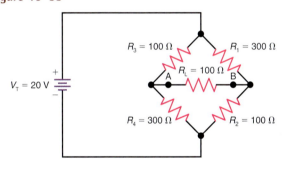

10–20 Using the Thevenin equivalent circuit for Fig. 10–33, calculate the values for I_L and V_L.

10–21 In Fig. 10–34, draw the Thevenin equivalent circuit with respect to terminals A and B (mentally remove R_L).

Figure 10–34

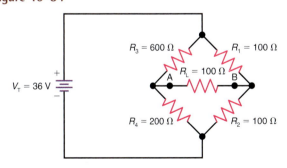

10–22 Using the Thevenin equivalent circuit for Fig. 10–34, calculate the values for I_L and V_L.

SECTION 10-5 NORTON'S THEOREM

10-23 IIII **MultiSim** In Fig. 10-35, draw the Norton equivalent circuit with respect to terminals A and B (mentally remove R_L).

Figure 10-35

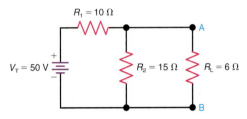

10-24 Using the Norton equivalent circuit for Fig. 10-35, calculate the values for I_L and V_L.

10-25 In Fig. 10-36, draw the Norton equivalent circuit with respect to terminals A and B (mentally remove R_L).

Figure 10-36

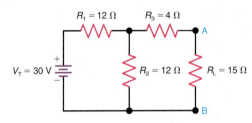

10-26 Using the Norton equivalent circuit for Fig. 10-36, calculate the values for I_L and V_L.

10-27 If R_3 is changed to 24 Ω in Fig. 10-36, redraw the Norton equivalent circuit with respect to terminals A and B. Also, recalculate the new values for I_L and V_L.

SECTION 10-6 THEVENIN-NORTON CONVERSIONS

10-28 Assume $V_{TH} = 15$ V and $R_{TH} = 5$ Ω for the Thevenin equivalent circuit in Fig. 10-37. What are the Norton equivalent values of I_N and R_N?

Figure 10-37

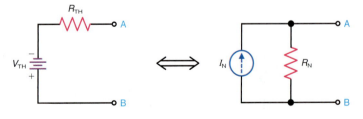

10-29 Assume $I_N = 20$ mA and $R_N = 1.2$ kΩ for the Norton equivalent circuit in Fig. 10-37. What are the Thevenin equivalent values of V_{TH} and R_{TH}?

10-30 Assume $I_N = 5$ mA and $R_N = 1.5$ kΩ for the Norton equivalent circuit in Fig. 10-37. What are the Thevenin equivalent values of V_{TH} and R_{TH}?

10-31 Assume $V_{TH} = 36$ V and $R_{TH} = 1.2$ kΩ for Thevenin equivalent circuit in Fig. 10-37. What are the Norton equivalent values of I_N and R_N?

SECTION 10-7 CONVERSION OF VOLTAGE AND CURRENT SOURCES

10-32 In Fig. 10-38,
a. convert voltage source 1 and voltage source 2 into equivalent current sources I_1 and I_2. Redraw the original circuit showing the current sources in place of V_1 and V_2.
b. combine the current sources I_1 and I_2 into one equivalent current source, I_T. Draw the equivalent circuit.
c. using the equivalent current source, I_T, calculate the values of I_3 and V_{R_3}.

Figure 10-38

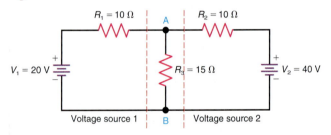

Voltage source 1 | B | Voltage source 2

10-33 In Fig. 10-39,
a. convert current source 1 and current source 2 into equivalent voltage sources V_1 and V_2. Redraw the original circuit showing the voltage sources in place of I_1 and I_2.
b. combine the voltage sources V_1 and V_2 into one equivalent voltage source, V_T. Draw the equivalent circuit.
c. using the equivalent voltage source, V_T, calculate the values of I_3 and V_{R_3}.

Figure 10-39

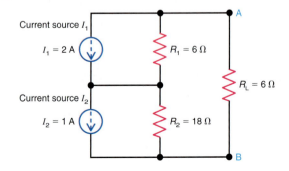

Current source I_1

Current source I_2

SECTION 10-8 MILLMAN'S THEOREM

10-34 In Fig. 10-40, apply Millman's Theorem to solve for the voltage, V_{XY}.

Figure 10-40

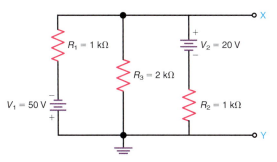

10-35 In Fig. 10-40, recalculate the voltage, V_{XY}, if the polarity of V_2 is reversed.

10-36 In Fig. 10-41, apply Millman's Theorem to solve for the voltage, V_{XY}.

Figure 10-41

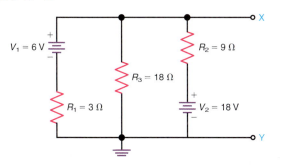

10-37 In Fig. 10-41, recalculate the voltage, V_{XY}, if the polarity of V_2 is reversed.

10-38 In Fig. 10-42, apply Millman's Theorem to solve for the voltage, V_{XY}.

Figure 10-42

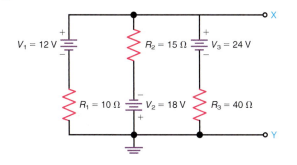

SECTION 10-9 T OR Y AND π OR Δ CONNECTIONS

10-39 Convert the T network in Fig. 10-43 into an equivalent π network.

Figure 10-43

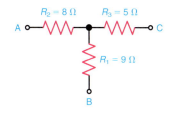

10-40 Convert the π network in Fig. 10-44 into an equivalent T network.

Figure 10-44

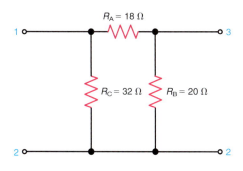

10-41 In Fig. 10-45, use delta-wye transformations to calculate both R_T and I_T.

Figure 10-45

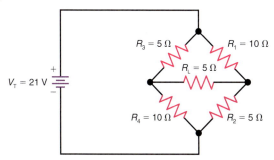

Critical Thinking

10-42 Thevenize the circuit driving terminals A and B in Fig. 10–46. Show the Thevenin equivalent circuit and calculate the values for I_L and V_L.

Figure 10–46 Circuit for Critical Thinking Problem 10-42.

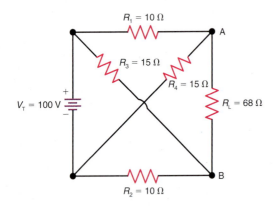

10-43 In Fig. 10–47, use the superposition theorem to solve for I_L and V_L.

10-44 In Fig. 10–47, show the Thevenin equivalent circuit driving terminals A and B.

Figure 10–47 Circuit for Critical Thinking Probs. 10–43 and 10–44.

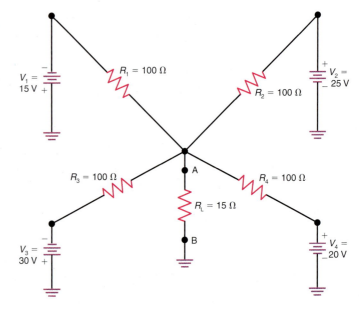

10-45 Refer to Fig. 10–34. Remove R_4 from the circuit and show the Thevenin equivalent circuit driving the open terminals. Also, calculate the value of I_4 and V_{R_4}.

Answers to Knowledge Check Problems

10-1 $V_P = -19$ V

10-2 $V_{TH} = 24$ V and $R_{TH} = 12\ \Omega$

10-3 $V_{TH} = -71.4$ V

10-4 $V_{TH} = -6$ V and $R_{TH} = 4.8\ \Omega$

10-5 $I_N = 6$ A and $R_N = 3\ \Omega$

10-6A $I_N = 3$ A and $R_N = 8\ \Omega$

10-6B $V_{TH} = 18$ V and $R_{TH} = 9\ \Omega$

10-7 $I_T = 0$ A

10-8 $V_{XY} = -10$ V

10-9 $R_1 = R_2 = R_3 = 12\ \Omega$

Answers to Self-Reviews

10-1 a. R_1
b. R_2

10-2 a. T
b. T

10-3 a. $8.4\ \Omega$
b. 24 V

10-4 a. $6.4\ \Omega$
b. 2.5 V

10-5 a. T
b. T

10-6 a. T
b. T
c. T

10-7 a. 7 A
b. $3\ \Omega$

10-8 a. 6 V
b. 14 V

10-9 a. R_1
b. R_B and R_C

Cumulative Review Summary (Chapters 9–10)

- Methods of applying Kirchhoff's laws include
  (a) equations of voltages using the branch currents in the loops to specify the voltages.
  (b) equations of currents at a node using the node voltage to specify the node currents.
  (c) equations of voltages using assumed mesh currents to specify the voltages.
- Methods of reducing a network to a simple equivalent circuit include
  (a) the superposition theorem using one source at a time.
  (b) Thevenin's theorem to convert the network to a series circuit with one voltage source.
  (c) Norton's theorem to convert the network to a parallel circuit with one current source.
  (d) Millman's theorem to find the common voltage across parallel branches with different sources.
  (e) delta (Δ) wye (Y) conversions to transform a network into a series-parallel circuit.

Cumulative Review Self-Test

Answers at back of book.

Answer True or False.

1. In Fig. 9–5, V_3 can be found by using Kirchhoff's laws with either branch currents or mesh currents.

2. In Fig. 9–5, V_3 can be found by superposition, thevenizing, or using Millman's theorem.

3. In Fig. 10–6, I_L cannot be found by delta-wye conversion because R_L disappears in the transformation.

4. In Fig. 10–6, I_L can be calculated with Kirchhoff's laws, using mesh currents for three meshes.

5. With superposition, we can use Ohm's law for circuits that have more than one source.

6. A Thevenin equivalent is a parallel circuit.

7. A Norton equivalent is a series circuit.

8. Either a Thevenin or a Norton equivalent of a network will produce the same current in any load across terminals A and B.

9. A Thevenin-to-Norton conversion means converting a voltage source to a current source.

10. The volt unit is equal to (volts/ohms) ÷ siemens.

11. A node voltage is a voltage between current nodes.

12. A π network can be converted to an equivalent T network.

13. A 10-V source with 10-Ω series R will supply 5 V to a 10-Ω load R_L.

14. A 10-A source with 10-Ω parallel R will supply 5 A to a 10-Ω load R_L.

15. Current sources in parallel can be added when they supply current in the same direction through R_L.

chapter 11

Conductors and Insulators

If you think of a wire as a water pipe for electricity, then it makes sense that the larger the diameter of a wire, the more current it can carry. The smaller the diameter of a wire, the less current it can carry. For a given length of wire, therefore, the resistance, R, decreases as its diameter and cross-sectional area increase. For a given diameter and cross-sectional area, however, the resistance of a wire increases with length. In general, conductors offer very little opposition or resistance to the flow of current.

An insulator is any material that resists or prevents the flow of electric charge, such as electrons. The resistance of an insulator is very high, typically several hundreds of megohms or more. An insulator provides the equivalent of an open circuit with practically infinite resistance and almost zero current.

In this chapter, you will be introduced to a variety of topics that includes wire conductors, insulators, connectors, mechanical switches and, fuses. All of these topics relate to the discussion of conductors and insulators since they either pass or prevent the flow of electricity, depending on their condition or state.

Objectives

After studying this chapter you should be able to

- *Explain* the main function of a conductor in an electric circuit.

- *Calculate* the cross-sectional area of round wire when the diameter is known.

- *List* the advantages of using stranded wire versus solid wire.

- *List* common types of connectors used with wire conductors.

- *Define* the terms *pole* and *throw* as they relate to switches.

- *Explain* how fast-acting and slow-blow fuses differ.

- *Calculate* the resistance of a wire conductor whose length, cross-sectional area, and specific resistance are known.

- *Explain* the meaning of *temperature coefficient of resistance*.

- *Explain* ion current and electron current.

- *Explain* why insulators are sometimes called *dielectrics*.

- *Explain* what is meant by the *corona effect*.

Outline

Important Terms

circuit breaker

circular mil (cmil)

corona effect

dielectric material

fuse

ionization current

pole

slow-blow fuse

specific resistance

switch

throw

temperature coefficient

wire gage

11–1 Function of the Conductor

In Fig. 11–1, the resistance of the two 10-ft lengths of copper-wire conductor is 0.08 Ω. This R is negligibly small compared with the 144-Ω R of the tungsten filament in the lightbulb. When the current of 0.833 A flows in the bulb and the series conductors, the IR voltage drop of the conductors is only 0.07 V, with 119.93 V across the bulb. Practically all the applied voltage is across the bulb filament. Since the bulb then has its rated voltage of 120 V, approximately, it will dissipate the rated power of 100 W and light with full brilliance.

The current in the wire conductors and the bulb is the same, since they are in series. However, the IR voltage drop in the conductor is practically zero because its R is almost zero.

Also, the I^2R power dissipated in the conductor is negligibly small, allowing the conductor to operate without becoming hot. Therefore, the conductor delivers energy from the source to the load with minimum loss by electron flow in the copper wires.

Although the resistance of wire conductors is very small, for some cases of high current, the resultant IR drop can be appreciable. For example, suppose that the 120-V power line is supplying 30 A of current to a load through two conductors, each of which has a resistance of 0.2 Ω. In this case, each conductor has an IR drop of 6 V, calculated as 30 A × 0.2 Ω = 6 V. With each conductor dropping 6 V, the load receives a voltage of only 108 V rather than the full 120 V. The lower-than-normal load voltage could result in the load not operating properly. Furthermore, the I^2R power dissipated in each conductor equals 180 W, calculated as 30^2 × 0.2 Ω = 180 W. The I^2R power loss of 180 W in the conductors is considered excessively high.

|||| MultiSim   Figure 11–1   The conductors should have minimum resistance to light the bulb with full brilliance. (*a*) Wiring diagram. (*b*) Schematic diagram. R_1 and R_2 represent the very small resistance of the wire conductors.

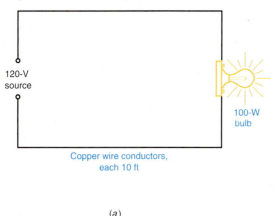

(a)

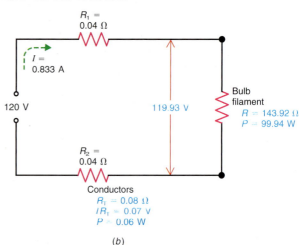

(b)

■ *11–1 Knowledge Check*

Answer at end of chapter.

A 15-A heater is connected to the 120-Vac power line through two 100-ft conductors each having a resistance of 0.5 Ω. How much voltage is available at the heater?

Answers at end of chapter.

Refer to Fig. 11–1.
a. How much is *R* for the 20 ft of copper wire?
b. How much is the *IR* voltage drop for the wire conductors?
c. The IR voltage in b is what percent of the applied voltage?

11–2 Standard Wire Gage Sizes

Table 11–1 lists the standard wire sizes in the system known as the American Wire Gage (AWG) or Brown and Sharpe (B&S) gage. The gage numbers specify the size of round wire in terms of its diameter and cross-sectional area. Note the following three points:

1. As the gage numbers increase from 1 to 40, the diameter and circular area decrease. Higher gage numbers indicate thinner wire sizes.
2. The circular area doubles for every three gage sizes. For example, No. 10 wire has approximately twice the area of No. 13 wire.
3. The higher the gage number and the thinner the wire, the greater the resistance of the wire for any given length.

In typical applications, hookup wire for electronic circuits with current of the order of milliamperes is generally about No. 22 gage. For this size, 0.5 to 1 A is the maximum current the wire can carry without excessive heating.

House wiring for circuits where the current is 5 to 15 A is usually No. 14 gage. Minimum sizes for house wiring are set by local electrical codes, which are usually guided by the National Electrical Code published by the National Fire Protection Association. A gage for measuring wire size is shown in Fig. 11–2.

Circular Mils

The cross-sectional area of round wire is measured in circular mils, abbreviated cmil. A mil is one-thousandth of an inch, or 0.001 in. One circular mil is the cross-sectional area of a wire with a diameter of 1 mil. The number of circular mils in any circular area is equal to the square of the diameter in mils or $cmil = d^2$ (mils).

Figure 11–2 American standard wire gage (AWG) for wire conductors.

Example 11–1

What is the area in circular mils of a wire with a diameter of 0.005 in.?

ANSWER We must convert the diameter to mils. Since 0.005 in. equals 5 mil,

$$\text{Circular mil area} = (5 \text{ mil})^2$$
$$\text{Area} = 25 \text{ cmil}$$

Note that the circular mil is a unit of area, obtained by squaring the diameter, whereas the mil is a linear unit of length, equal to one-thousandth of

| Table 11–1 | Copper–Wire Table |

Gage No.	Diameter, Mils	Area, Circular Mils	Ohms per 1000 ft of Copper Wire at 25°C*	Gage No.	Diameter, Mils	Area, Circular Mils	Ohms per 1000 ft of Copper Wire at 25°C*
1	289.3	83,690	0.1264	21	28.46	810.1	13.05
2	257.6	66,370	0.1593	22	25.35	642.4	16.46
3	229.4	52,640	0.2009	23	22.57	509.5	20.76
4	204.3	41,740	0.2533	24	20.10	404.0	26.17
5	181.9	33,100	0.3195	25	17.90	320.4	33.00
6	162.0	26,250	0.4028	26	15.94	254.1	41.62
7	144.3	20,820	0.5080	27	14.20	201.5	52.48
8	128.5	16,510	0.6405	28	12.64	159.8	66.17
9	114.4	13,090	0.8077	29	11.26	126.7	83.44
10	101.9	10,380	1.018	30	10.03	100.5	105.2
11	90.74	8234	1.284	31	8.928	79.70	132.7
12	80.81	6530	1.619	32	7.950	63.21	167.3
13	71.96	5178	2.042	33	7.080	50.13	211.0
14	64.08	4107	2.575	34	6.305	39.75	266.0
15	57.07	3257	3.247	35	5.615	31.52	335.0
16	50.82	2583	4.094	36	5.000	25.00	423.0
17	45.26	2048	5.163	37	4.453	19.83	533.4
18	40.30	1624	6.510	38	3.965	15.72	672.6
19	35.89	1288	8.210	39	3.531	12.47	848.1
20	31.96	1022	10.35	40	3.145	9.88	1069

* 20° to 25°C or 68° to 77°F is considered average room temperature.

an inch. Therefore, the circular-mil area increases as the square of the diameter. As illustrated in Fig. 11–3, doubling the diameter quadruples the area. Circular mils are convenient for round wire because the cross section is specified without using the formula πr^2 or $\pi d^2/4$ for the area of a circle.

Figure 11–3 Cross-sectional area for round wire. Doubling the diameter increases the circular area by four times.

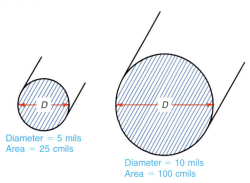

Diameter = 5 mils
Area = 25 cmils

Diameter = 10 mils
Area = 100 cmils

Figure 11–4 Types of wire conductors. (*a*) Solid wire. (*b*) Stranded wire. (*c*) Braided wire for very low *R*. (*d*) Coaxial cable. Note braided wire for shielding the inner conductor. (*e*) Twin-lead cable.

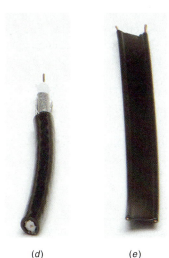

(*a*)   (*b*)   (*c*)

(*d*)   (*e*)

■ *11–2 Knowledge Check*

Answer at end of chapter.

If a wire has a circular area of 10,000 cmils, what is its diameter?

■ *11–2 Self-Review*

Answers at end of chapter.

a. How much is *R* for 1 ft of No. 22 wire?
b. What is the cross-sectional area in circular mils for wire with a diameter of 0.025 in.?
c. What is the wire gage size in Fig. 11–1?

11–3 Types of Wire Conductors

Most wire conductors are copper due to its low cost, although aluminum and silver are also used sometimes. The copper may be tinned with a thin coating of solder, which gives it a silvery appearance. Tinned wire is easier to solder for connections. The wire can be solid or stranded, as shown in Fig. 11–4*a* and *b*. Solid wire is made of only one conductor. If bent or flexed repeatedly, solid wire may break. Therefore solid wire is used in places where bending and flexing is not encountered. House wiring is a good example of the use of solid wire. Stranded wire is made up of several individual strands put together in a braid. Some uses for stranded wire include telephone cords, extension cords, and speaker wire, to name a few.

Stranded wire is flexible, easier to handle, and less likely to develop an open break. Sizes for stranded wire are equivalent to the sum of the areas for the individual strands. For instance, two strands of No. 30 wire correspond to solid No. 27 wire.

Example 11–2

A stranded wire is made up of 16 individual strands of No. 27 gage wire. What is its equivalent gage size in solid wire?

ANSWER The equivalent gage size in solid wire is determined by the total circular area of all individual strands. Referring to Table 11–1, the circular area for No. 27 gage wire is 201.5 cmils. Since there are 16 individual strands, the total circular area is calculated as follows:

$$\text{Total cmil Area} = 16 \text{ Strands} \times \frac{201.5 \text{ cmils}}{\text{strand}}$$

$$= 3224 \text{ cmils}$$

Referring to Table 11–1, we see that the circular area of 3224 cmils corresponds very closely to the cmil area of No. 15 gage wire. Therefore, 16 strands of No. 27 gage wire is roughly equivalent to No. 15 gage solid wire.

Very thin wire, such as No. 30, often has an insulating coating of enamel or shellac. It may look like copper, but the coating must be scraped off the ends to make a good connection. This type of wire is used for small coils.

Heavier wires generally are in an insulating sleeve, which may be rubber or one of many plastic materials. General-purpose wire for connecting electronic components is generally plastic-coated hookup wire of No. 20 gage. Hookup wire that is bare should be enclosed in a hollow insulating sleeve called *spaghetti*.

The braided conductor in Fig. 11–4c is used for very low resistance. It is wide for low R and thin for flexibility, and the braiding provides many strands. A common application is a grounding connection, which must have very low R.

Transmission Lines

Constant spacing between two conductors through the entire length provides a transmission line. Common examples are the coaxial cable in Fig. 11–4d and the twin lead in Fig. 11–4e.

Coaxial cable with an outside diameter of $\frac{1}{4}$ in. is generally used for the signals in cable television. In construction, there is an inner solid wire, insulated from metallic braid that serves as the other conductor. The entire assembly is covered by an outer plastic jacket. In operation, the inner conductor has the desired signal voltage with respect to ground, and the metallic braid is connected to ground to shield the inner conductor against interference. Coaxial cable, therefore, is a shielded type of transmission line.

With twin-lead wire, two conductors are embedded in plastic to provide constant spacing. This type of line is commonly used in television for connecting the antenna to the receiver. In this application, the spacing is $\frac{5}{8}$ in. between wires of No. 20 gage size, approximately. This line is not shielded.

Wire Cable

Two or more conductors in a common covering form a cable. Each wire is insulated from the others. Cables often consist of two, three, ten, or many more pairs of conductors, usually color-coded to help identify the conductors at both ends of a cable.

The ribbon cable in Fig. 11–5, has multiple conductors but not in pairs. This cable is used for multiple connections to a computer and associated equipment.

■ *11–3 Knowledge Check*

Answer at end of chapter.

A stranded wire consists of 16 strands of No. 30 gage wire. What is its equivalent gage size in solid wire?

Figure 11–5 Ribbon cable with multiple conductors.

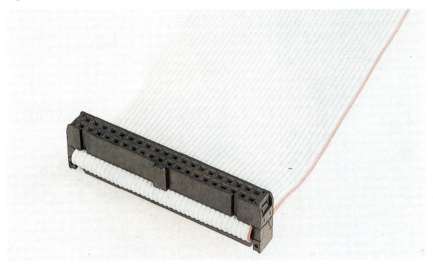

■ *11–3 Self-Review*

Answers at end of chapter.

a. The plastic coating on wire conductors has very high *R*. (True or False)
b. Coaxial cable is a shielded transmission line. (True or False)
c. With repeated bending and flexing, solid wire is more likely to develop an open than stranded wire. (True or False)

11–4 Connectors

Refer to Fig. 11–6 for different types. The spade lug in Fig. 11–6*a* is often used for screw-type terminals. The alligator clip in Fig. 11–6*b* is convenient for a temporary connection. Alligator clips come in small and large sizes. The banana pins in Fig. 11–6*c* have spring-type sides that make a tight connection. The terminal strip in Fig. 11–6*d* provides a block for multiple solder connections.

The RCA-type plug in Fig. 11–6*e* is commonly used for shielded cables with audio equipment. The inner conductor of the shielded cable is connected to the center pin of the plug, and the cable braid is connected to the shield. Both connections must be soldered.

The phone plug in Fig. 11–6*f* is still used in many applications but usually in a smaller size. The ring is insulated from the sleeve to provide for two connections. There may be a separate tip, ring, and sleeve for three connections. The sleeve is usually the ground side.

The plug in Fig. 11–6*g* is called an *F connector*. It is universally used in cable television because of its convenience. The center conductor of the coaxial cable serves as the center pin of the plug, so that no soldering is needed. Also, the shield on the plug is press-fit onto the braid of the cable underneath the plastic jacket.

Figure 11–6*h* shows a multiple pin connector having many conductors. This type of connector is often used to connect the components of a computer system, such as the monitor and the keyboard, to the computer.

Figure 11–6*i* shows a spring-loaded metal hook as a grabber for a temporary connection to a circuit. This type of connector is often used with the test leads of a VOM or a DMM.

Figure 11–6 Common types of connectors for wire conductors. (*a*) Spade lug. (*b*) Alligator clip. (*c*) Double banana-pin plug. (*d*) Terminal strip. (*e*) RCA-type plug for audio cables. (*f*) Phone plug. (*g*) F-type plug for cable TV. (*h*) Multiple-pin connector plug. (*i*) Spring-loaded metal hook as grabber for temporary connection in testing circuits.

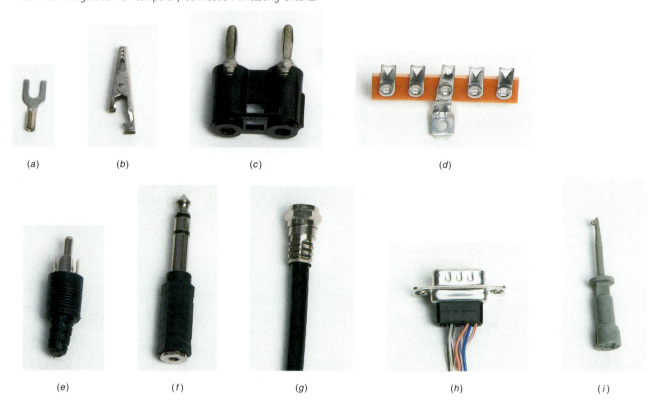

(*a*)   (*b*)   (*c*)   (*d*)

(*e*)   (*f*)   (*g*)   (*h*)   (*i*)

■ *11–4 Knowledge Check*

Answer at end of chapter.

In Fig. 11–6f, which part of the phone plug is grounded?

■ *11–4 Self-Review*

Answers at end of chapter.

a. The RCA-type plug is commonly used for shielded cables with audio equipment. (True or False)
b. The F-type connector is used with coaxial cable. (True or False)
c. The F-type connector can also be used with twin-lead line. (True or False)

GOOD TO KNOW

Many printed circuit boards are multilayer boards that contain several different layers of printed wiring. Printed circuit boards that have four or five layers of printed wiring are not uncommon.

11–5 Printed Wiring

Most electronic circuits are mounted on a plastic or fiberglass insulating board with printed wiring, as shown in Fig. 11–7. This is a printed-circuit (PC) or printed-wiring (PW) board. One side has the components, such as resistors, capacitors, coils, transistors, diodes, and integrated-circuit (IC) units. The other side has the conducting paths printed with silver or copper on the board, instead of using wires. On a double-sided board, the component side also has printed wiring. Sockets, small metal eyelets, or holes in the board are used to connect the components to the wiring.

With a bright light on one side, you can see through to the opposite side to trace the connections. However, the circuit may be drawn on the PC board.

Figure 11–7 Printed wiring board. (*a*) Component side with resistors, capacitors, transistors, and integrated circuits. (*b*) Side with printed wiring for the circuit.

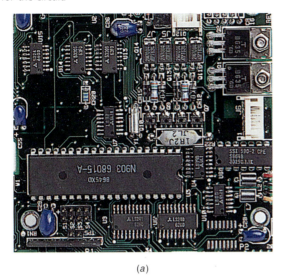

(*a*)

(*b*)

It is important not to use too much heat in soldering or desoldering. Otherwise the printed wiring can be lifted off the board. Use a small iron of about 25 to 30 W rating. When soldering semiconductor diodes and transistors, hold the lead with pliers or connect an alligator clip as a heat sink to conduct heat away from the semiconductor junction.

For desoldering, use a solder-sucker tool, with a soldering iron, to clean each terminal. Another method of removing solder is to use a copper wire braid that is impregnated with rosin flux. This copper wire braid, often called a desoldering braid, is excellent for attracting liquid or molten solder. Just put the desoldering braid on the solder joint and heat it until the solder runs up into the copper braid. The terminal must be clean enough to lift out the component easily without damaging the PC board. One advantage of using a desoldering braid versus a solder sucker tool is that the desoldering braid acts like a natural heat sink, thus reducing the risk of damaging the copper traces on the PC board.

A small crack in the printed wiring acts like an open circuit preventing current flow. Cracks can be repaired by soldering a short length of bare wire over the open circuit. If a larger section of printed wiring is open, or if the board is cracked, you can bridge the open circuit with a length of hookup wire soldered at two convenient end terminals of the printed wiring. In many electronic industries, special kits are available for replacing damaged or open traces on PC boards.

■ *11–5 Knowledge Check*

Answer at end of chapter.

When desoldering, list one advantage of using a desoldering braid versus a solder sucker tool.

■ *11–5 Self-Review*

Answers at end of chapter.

a. Which is the best size of iron to use to solder on a PC board: 25, 100, or 150 W?
b. How much is the resistance of a printed-wire conductor with a crack in the middle?

Conductors and Insulators **335**

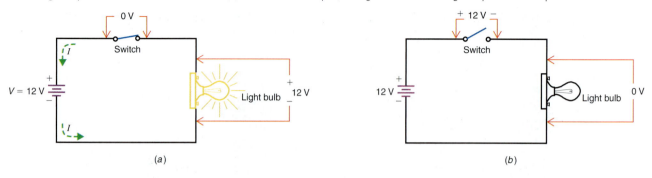

(a) (b)

Figure 11–9 Switches. (*a*) Single-pole, single-throw (SPST). (*b*) Single-pole, double-throw (SPDT). (*c*) Double-pole, single-throw (DPST). (*d*) Double-pole, double-throw (DPDT).

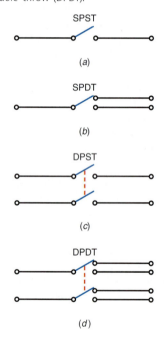

SPST

(a)

SPDT

(b)

DPST

(c)

DPDT

(d)

GOOD TO KNOW

Switches can have more than two poles and two throws. For example, a single-pole switch might have 3, 4, 5 or more throws. Similarly, there are triple pole, single-throw switches.

11–6 Switches

A switch is a component that allows us to control whether the current is ON or OFF in a circuit. A closed switch has practically zero resistance, whereas an open switch has nearly infinite resistance.

Figure 11–8 shows a switch in series with a voltage source and a lightbulb. With the switch closed, as in Fig. 11–8*a*, a complete path for current is provided and the light is ON. Since the switch has very low resistance when it is closed, all of the source voltage is across the load, with 0 V across the closed contacts of the switch. With the switch open, as in Fig. 11–8*b*, the path for current is interrupted and the bulb does not light. Since the switch has very high resistance when it is open, all of the source voltage is across the open switch contacts, with 0 V across the load.

Switch Ratings

All switches have a current rating and a voltage rating. The current rating corresponds to the maximum allowable current that the switch can carry when closed. The current rating is based on the physical size of the switch contacts as well as the type of metal used for the contacts. Many switches have gold- or silver-plated contacts to ensure very low resistance when closed.

The voltage rating of a switch corresponds to the maximum voltage that can safely be applied across the open contacts without internal arcing. The voltage rating does not apply when the switch is closed, since the voltage drop across the closed switch contacts is practically zero.

Switch Definitions

Toggle switches are usually described as having a certain number of poles and throws. For example, the switch in Fig. 11–8 is described as a *single-pole, single-throw* (SPST) switch. Other popular switch types include the *single-pole, double-throw* (SPDT), *double-pole, single-throw* (DPST), and *double-pole, double-throw* (DPDT). The schematic symbols for each type are shown in Fig. 11–9. Notice that the SPST switch has two connecting terminals, whereas the SPDT has three, the DPST has four, and the DPDT has six.

The term *pole* is defined as the number of completely isolated circuits that can be controlled by the switch. The term *throw* is defined as the number of closed contact positions that exist per pole. The SPST switch in Fig. 11–8 can control the current in only one circuit, and there is only one closed contact position, hence the name single-pole, single-throw.

Figure 11-10 Switch applications. (*a*) SPDT switch used to switch a 12-V source between one of two different loads. (*b*) DPST switch controlling two completely isolated circuits simultaneously. (*c*) DPDT switch used to reverse the polarity of voltage across a dc motor.

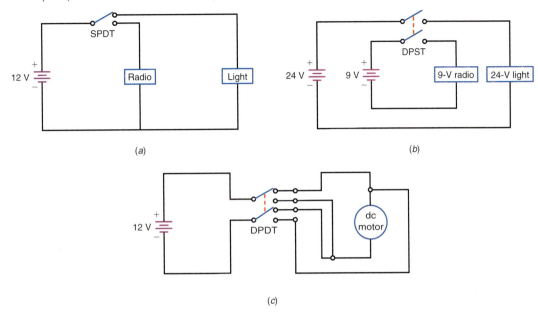

(*a*)　　　　　　　(*b*)

(*c*)

Figure 11-11 A variety of toggle switches.

Figure 11-12 Push-button switch schematic symbols. (*a*) Normally open (NO) push-button switch. (*b*) Normally closed (NC) push-button switch.

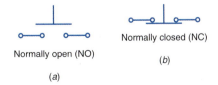

Normally closed (NC)

Normally open (NO)

(*b*)

(*a*)

Figure 11-10 shows a variety of switch applications. In Fig. 11-10*a*, a SPDT switch is being used to switch a 12-V dc source between one of two different loads. In Fig. 11-10*b*, a DPST switch is being used to control two completely separate circuits simultaneously. In Fig. 11-10*c*, a DPDT switch is being used to reverse the polarity of voltage across the terminals of a dc motor. (Reversing the polarity reverses the direction of the motor.) Note that the dashed lines shown between the poles in Figs. 11-10*b* and 11-10*c* indicate that both sets of contacts within the switch are opened and closed simultaneously.

Switch Types

Figure 11-11 shows a variety of toggle switches. Although the toggle switch is a very popular type of switch, several other types are found in electronic equipment. Additional types include push-button switches, rocker switches, slide switches, rotary switches, and DIP switches.

Push-button switches are often spring-loaded switches that are either normally open (NO) or normally closed (NC). Figure 11-12 shows the schematic symbols for both types. For the normally open switch in Fig. 11-12*a*, the switch contacts remain open until the push button is depressed. When the push button is depressed, the switch closes, allowing current to pass. The normally closed switch in Fig. 11-12*b* operates opposite the normally open switch in Fig. 11-12*a*. When the push button is depressed, the switch contacts open to interrupt current in the circuit. A typical push-button switch is shown in Fig. 11-13.

Figure 11-14 shows a DIP (dual-inline package) switch. It consists of eight miniature rocker switches, where each switch can be set separately. A DIP switch has pin connections that fit into a standard IC socket.

Figure 11-15 shows another type of switch known as a rotary switch. As shown, it consists of three wafers or decks mounted on a common shaft.

■ *11-6 Knowledge Check*

Answers at end of chapter.

What factors determine the current rating of a switch?

Figure 11–13 Typical push-button switch.

Figure 11–14 Dual-inline package (DIP) switch.

Figure 11–15 Rotary switch.

■ *11–6 Self-Review*

Answers at end of chapter.

a. How much is the *IR* voltage drop across a closed switch?
b. How many connections are there on an SPDT switch?
c. What is the resistance across the contacts of an open switch?
d. An SPST switch is rated at 10 A, 250 V. Should the switch be used to turn a 120-V, 1500-W heater ON and OFF?

11–7 Fuses

Many circuits have a fuse in series as a protection against an overload from a short circuit. Excessive current melts the fuse element, blowing the fuse and opening the series circuit. The purpose is to let the fuse blow before the components and wiring are damaged. The blown fuse can easily be replaced by a new one after the overload has been eliminated. A glass-cartridge fuse with holders is shown in Fig. 11–16. This is a type 3AG fuse with a diameter of ¼ in. and length of 1¼ in. *AG* is an abbreviation of "automobile glass," since that was one of the first applications of fuses in a glass holder to make the wire link visible. The schematic symbol for a fuse is ───⌢⌣─── as shown in Fig. 11–18a.

The metal fuse element may be made of aluminum, tin-coated copper, or nickel. Fuses are available with current ratings from ⅟₅₀₀ A to hundreds of amperes. The thinner the wire element in the fuse, the smaller its current rating. For example, a 2-in. length of No. 28 wire can serve as a 2-A fuse. As typical applications, the rating for fuses in each branch of older house wiring is often 15 A; the high-voltage circuit in a television receiver is usually protected by a ¼-A glass-cartridge fuse. For automobile fuses, the ratings are generally 10 to 30 A because of the higher currents needed with a 12-V source for a given amount of power.

Slow-Blow Fuses

These have coiled construction. They are designed to open only on a continued overload, such as a short circuit. The purpose of coiled construction is to prevent the fuse from blowing on a temporary current surge. As an example, a slow-blow fuse will hold a 400% overload in current for up to 2 s. Typical ratings are shown by the curves in Fig. 11–17. Circuits with an electric motor use slow-blow fuses because the starting current of a motor is much more than its running current.

Circuit Breakers

A circuit breaker can be used in place of a fuse to protect circuit components and wiring against the high current caused by a short circuit. It is constructed of a thin bimetallic strip that expands with heat and in turn trips open the circuit.

Figure 11–16 (*a*) Glass-cartridge fuse. (*b*) Fuse holder. (*c*) Panel-mounted fuse holder.

(*a*) (*b*) (*c*)

Figure 11–17 Chart showing percent of rated current vs. blowing time for fuses.

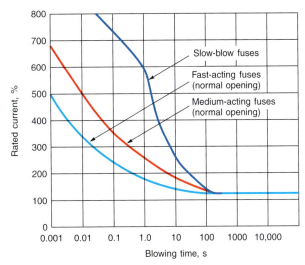

The advantage of a circuit breaker is that it can be reset once the bimetallic strip cools down and the short circuit has been removed. Because they can be reset, almost all new residential house wiring is protected by circuit breakers rather than fuses. The schematic symbol for a circuit breaker is often shown as ⌒.

Testing Fuses

In glass fuses, you can usually see whether the wire element inside is burned open. When measured with an ohmmeter, a good fuse has practically zero resistance. An open fuse reads infinite ohms. Power must be off or the fuse must be out of the circuit to test a fuse with an ohmmeter.

When you test with a voltmeter, a good fuse has zero volts across its two terminals (Fig. 11–18a). If you read appreciable voltage across the fuse, this means that it is open. In fact, the full applied voltage is across the open fuse in a series circuit, as shown in Fig. 11–18b. This is why fuses also have a voltage rating, which gives the maximum voltage without arcing in the open fuse.

Referring to Fig. 11–18, notice the results when measuring the voltages to ground at the two fuse terminals. In Fig. 11–18a, the voltage is the same 120 V at both ends because there is no voltage drop across the good fuse. In Fig. 11–18b, however, terminal B reads 0 V because this end is disconnected from V_T by the open fuse. These tests apply to either dc or ac voltages.

||| MultiSim **Figure 11–18** When a fuse opens, the applied voltage is across the fuse terminals. (a) Circuit closed with good fuse. Note schematic symbol for any type of fuse. (b) Fuse open. Voltage readings are explained in the text.

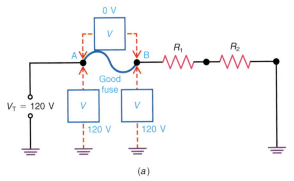

(a)

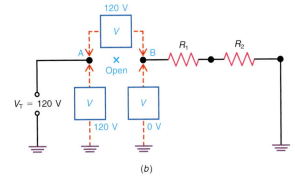

(b)

Answer at end of chapter.

In Fig. 11–18b, how much is the voltage across either R_1 or R_2?

■ 11–7 *Self-Review*

Answers at end of chapter.

a. How much is the resistance of a good fuse?
b. How much is the *IR* voltage drop across a good fuse?

11–8 Wire Resistance

The longer a wire, the higher its resistance. More work must be done to make electrons drift from one end to the other. However, the greater the diameter of the wire, the less the resistance, since there are more free electrons in the cross-sectional area. As a formula,

$$R = \rho \frac{l}{A} \tag{11–1}$$

where R is the total resistance, l the length, A the cross-sectional area, and ρ the specific resistance or *resistivity* of the conductor. The factor ρ then enables the resistance of different materials to be compared according to their nature without regard to different lengths or areas. Higher values of ρ mean more resistance. Note that ρ is the Greek letter *rho*.

Specific Resistance

Table 11–2 lists resistance values for different metals having the standard wire size of a 1-ft length with a cross-sectional area of 1 cmil. This rating is the *specific resistance* of the metal, in circular-mil ohms per foot. Since silver, copper, gold, and aluminum are the best conductors, they have the lowest values of specific resistance. Tungsten and iron have much higher resistance.

> **GOOD TO KNOW**
>
> The resistance of a wire conductor is directly proportional to its length and inversely proportional to its cross-sectional area.

Example 11–3

How much is the resistance of 100 ft of No. 20 gage copper wire?

ANSWER Note that from Table 11–1, the cross-sectional area for No. 20 gage wire is 1022 cmil; from Table 11–2, the ρ for copper is 10.4. Using Formula (11–1) gives

$$R = \rho \frac{l}{A}$$

$$= 10.4 \frac{\text{cmil} \cdot \Omega}{\text{ft}} \times \frac{100 \text{ ft}}{1022 \text{ cmil}}$$

$$R = 1.02 \ \Omega$$

All units cancel except the ohms for R. Note that 1.02 Ω for 100 ft is approximately one-tenth the resistance of 10.35 Ω for 1000 ft of No. 20 copper

Table 11-2 | Properties of Conducting Materials*

Material	Description and Symbol	Specific Resistance (ρ) at 20°C, cmil · Ω/ft	Temperature Coefficient per °C, α	Melting Point, °C
Aluminum	Element (Al)	17	0.004	660
Carbon	Element (C)	†	−0.0003	3000
Constantan	Alloy, 55% Cu, 45% Ni	295	0 (average)	1210
Copper	Element (Cu)	10.4	0.004	1083
Gold	Element (Au)	14	0.004	1063
Iron	Element (Fe)	58	0.006	1535
Manganin	Alloy, 84% Cu, 12% Mn, 4% Ni	270	0 (average)	910
Nichrome	Alloy 65% Ni, 23% Fe, 12% Cr	676	0.0002	1350
Nickel	Element (Ni)	52	0.005	1452
Silver	Element (Ag)	9.8	0.004	961
Steel	Alloy, 99.5% Fe, 0.5% C	100	0.003	1480
Tungsten	Element (W)	33.8	0.005	3370

* Listings approximate only, since precise values depend on exact composition of material.
† Carbon has about 2500 to 7500 times the resistance of copper. Graphite is a form of carbon.

wire listed in Table 11–1, showing that the resistance is proportional to length. Also note that a wire that is three gage sizes higher has half the circular area and double the resistance for the same wire length.

Example 11-4

How much is the resistance of a 100-ft length of No. 23 gage copper wire?

ANSWER

$$R = \rho \frac{l}{A}$$
$$= 10.4 \frac{\text{cmil} \cdot \Omega}{\text{ft}} \times \frac{100 \text{ ft}}{509.5 \text{ cmil}}$$
$$R = 2.04 \ \Omega$$

Table 11–3	Comparison of Resistivities	
Material	ρ, $\Omega \cdot$ cm, at 25°C	**Description**
Silver	1.6×10^{-6}	Conductor
Germanium	55	Semiconductor
Silicon	55,000	Semiconductor
Mica	2×10^{12}	Insulator

Units of Ohm-Centimeters for ρ

Except for wire conductors, specific resistances are usually compared for the standard size of a 1-cm cube. Then ρ is specified in $\Omega \cdot$ cm for the unit cross-sectional area of 1 cm^2.

As an example, pure germanium has $\rho = 55 \, \Omega \cdot$ cm, as listed in Table 11–3. This value means that R is 55 Ω for a cube with a cross-sectional area of 1 cm^2 and length of 1 cm.

For other sizes, use Formula (11–1) with l in cm and A in cm^2. Then all units of size cancel to give R in ohms.

Example 11–5

How much is the resistance of a slab of germanium 0.2 cm long with a cross-sectional area of 1 cm^2?

ANSWER

$$R = \rho \frac{l}{A}$$

$$= 55 \, \Omega \cdot \text{cm} \times \frac{0.2 \text{ cm}}{1 \text{ cm}^2}$$

$$R = 11 \, \Omega$$

The same size slab of silicon would have R of 11,000 Ω. Note from Table 11–3 that ρ is 1000 times more for silicon than for germanium.

Types of Resistance Wire

For applications in heating elements, such as a toaster, an incandescent light-bulb, or a heater, it is necessary to use wire that has more resistance than good conductors like silver, copper, or aluminum. Higher resistance is preferable so that the required amount of I^2R power dissipated as heat in the wire can be obtained without excessive current. Typical materials for resistance wire are the elements tungsten, nickel, or iron and alloys* such as manganin, Nichrome, and

---

* An *alloy* is a fusion of elements without chemical action between them. Metals are commonly alloyed to alter their physical characteristics.

constantan. These types are generally called *resistance wire* because R is greater than that of copper wire for the same length.

■ *11–8 Knowledge Check*

Answer at end of chapter

What is the resistance of 300 ft of No. 26 gage aluminum wire?

■ *11–8 Self-Review*

Answers at end of chapter.

a. **Does Nichrome wire have less or more resistance than copper wire?**
b. **For 100 ft of No. 14 gage copper wire, R is 0.26 Ω. How much is R for 1000 ft?**

11–9 Temperature Coefficient of Resistance

This factor with the symbol alpha (α) states how much the resistance changes for a change in temperature. A positive value for α means that R increases with temperature; with a negative α, R decreases; zero for α means that R is constant. Some typical values of α for metals and for carbon are listed in Table 11–2 in the fourth column.

Positive α

All metals in their pure form, such as copper and tungsten, have positive temperature coefficients. The α for tungsten, for example, is 0.005. Although α is not exactly constant, an increase in wire resistance caused by a rise in temperature can be calculated approximately from the formula

$$R_t = R_0 + R_0(\alpha\Delta t) \tag{11–2}$$

where R_0 is the resistance at 20°C, R_t is the higher resistance at the higher temperature, and Δt is the temperature rise above 20°C.

Example 11–6

A tungsten wire has a 14-Ω R at 20°C. Calculate its resistance at 120°C.

ANSWER The temperature rise Δt here is 100°C; α is 0.005. Substituting in Formula (11–2),

$$R_t = 14 + 14(0.005 \times 100)$$
$$= 14 + 7$$
$$R_t = 21 \ \Omega$$

The added resistance of 7 Ω increases the wire resistance by 50% because of the 100°C rise in temperature.

In practical terms, a positive α means that heat increases R in wire conductors. Then I is reduced for a specified applied voltage.

Negative α

Note that carbon has a negative temperature coefficient. In general, α is negative for all semiconductors, including germanium and silicon. Also, all electrolyte solutions, such as sulfuric acid and water, have a negative α.

A negative value of α means less resistance at higher temperatures. The resistance of semiconductor diodes and transistors, therefore, can be reduced appreciably when they become hot with normal load current.

Zero α

This means that R is constant with changes in temperature. The metal alloys constantan and manganin, for example, have a value of zero for α. They can be used for precision wire-wound resistors that do not change resistance when the temperature increases.

Hot Resistance

Because resistance wire is made of tungsten, Nichrome, iron, or nickel, there is usually a big difference in the amount of resistance the wire has when hot in normal operation and when cold without its normal load current. The reason is that the resistance increases with higher temperatures, since these materials have a positive temperature coefficient, as shown in Table 11–2.

As an example, the tungsten filament of a 100-W, 120-V incandescent bulb has a current of 0.833 A when the bulb lights with normal brilliance at its rated power, since $I = P/V$. By Ohm's law, the hot resistance is V/I, or 120 V/0.833 A, which equals 144 Ω. If, however, the filament resistance is measured with an ohmmeter when the bulb is not lit, the cold resistance is only about 10 Ω.

The Nichrome heater elements in appliances and the tungsten heaters in vacuum tubes also become several hundred degrees hotter in normal operation. In these cases, only the cold resistance can be measured with an ohmmeter. The hot resistance must be calculated from voltage and current measurements with the normal value of load current. As a practical rule, the cold resistance is generally about one-tenth the hot resistance. In troubleshooting, however, the approach is usually just to check whether the heater element is open. Then it reads infinite ohms on the ohmmeter.

Superconductivity

The effect opposite to hot resistance occurs when cooling a metal down to very low temperatures to reduce its resistance. Near absolute zero, 0 K or $-273°C$, some metals abruptly lose practically all their resistance. As an example, when cooled by liquid helium, the metal tin becomes superconductive at 3.7 K. Tremendous currents can be produced, resulting in very strong electromagnetic fields. Such work at very low temperatures, near absolute zero, is called *cryogenics*.

New types of ceramic materials have been developed and are stimulating great interest in superconductivity because they provide zero resistance at temperatures much above absolute zero. One type is a ceramic pellet, with a 1-in. diameter, that includes yttrium, barium, copper, and oxygen atoms. The superconductivity occurs at a temperature of 93 K, equal to $-160°C$. This value is still far below room temperature, but the cooling can be done with liquid nitrogen, which is much cheaper than liquid helium. As research continues, it is likely that new materials will be discovered that are superconductive at even higher temperatures.

Do semiconductor materials such as germanium and silicon have a positive or negative temperature coefficient?

a. Metal conductors have more R at higher temperatures. (True or False)
b. Tungsten can be used for resistance wire. (True or False)
c. A superconductive material has practically zero resistance. (True or False)

GOOD TO KNOW

A fluorescent light is a good example of an electric current in a gas.

Figure 11-19 Formation of ions. (*a*) Normal sodium (Na) atom. (*b*) Positively charged ion indicated as Na⁺, missing one free electron.

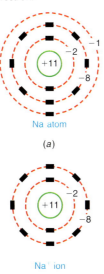

11-10 Ion Current in Liquids and Gases

We usually think of metal wire for a conductor, but there are other possibilities. Liquids such as salt water or dilute sulfuric acid can also allow the movement of electric charges. For gases, consider the neon glow lamp, in which neon serves as a conductor.

The mechanism may be different for conduction in metal wire, liquids, or gases, but in any case, the current is a motion of charges. Furthermore, either positive or negative charges can be the carriers that provide electric current. The amount of current is Q/T. For one coulomb of charge per second, the current is one ampere.

In solid materials such as metals, the atoms are not free to move among each other. Therefore, conduction of electricity must take place by the drift of free electrons. Each atom remains neutral, neither gaining nor losing charge, but the metals are good conductors because they have plenty of free electrons that can be forced to drift through the solid substance.

In liquids and gases, however, each atom can move freely among all the other atoms because the substance is not solid. As a result, the atoms can easily take on electrons or lose electrons, particularly the valence electrons in the outside shell. The result is an atom that is no longer electrically neutral. Adding one or more electrons produces a negative charge; the loss of one or more electrons results in a positive charge. The charged atoms are called *ions*. Such charged particles are commonly formed in liquids and gases.

The Ion

An ion is an atom, or group of atoms, that has a net electric charge, either positive or negative, resulting from a loss or gain of electrons. In Fig. 11-19*a*, the sodium atom is neutral, with 11 positive charges in the nucleus balanced by 11 electrons in the outside shells. This atom has only one electron in the shell farthest from the nucleus. When the sodium is in solution, this one electron can easily leave the atom. The reason may be another atom close by that needs one electron to have a stable ring of eight electrons in its outside shell. Notice that if the sodium atom loses one valence electron, the atom will still have an outside ring of eight electrons, as shown in Fig. 11-19*b*. This sodium atom now is a positive ion, with a charge equal to one proton.

Current of Ions

Just as in electron flow, opposite ion charges are attracted to each other, and like charges repel. The resultant motion of ions provides electric current. In liquids and gases, therefore, conduction of electricity results mainly from the movement of ions. This motion of ion charges is called *ionization current*. Since an ion includes the nucleus of the atom, the ion charge is much heavier than an electron charge and moves with less velocity. We can say that ion charges are less mobile than electron charges.

The direction of ionization current can be the same as that of electron flow or the opposite. When negative ions move, they are attracted to the positive terminal of an applied voltage in the same direction as electron flow. However, when positive ions move, this ionization current is in the opposite direction, toward the negative terminal of an applied voltage.

For either direction, though, the amount of ionization current is determined by the rate at which the charge moves. If 3 C of positive ion charges move past a given point per second, the current is 3 A, the same as 3 C of negative ions or 3 C of electron charges.

Ionization in Liquids

Ions are usually formed in liquids when salts or acids are dissolved in water. Salt water is a good conductor because of ionization, but pure distilled water is an insulator. In addition some metals immersed in acids or alkaline solutions ionize. Liquids that are good conductors because of ionization are called *electrolytes*. In general, electrolytes have a negative value of α, as more ionization at higher temperatures lowers the resistance.

Ionization in Gases

Gases have a minimum striking or ionization potential, which is the lowest applied voltage that will ionize the gas. Before ionization, the gas is an insulator, but the ionization current makes the ionized gas have a low resistance. An ionized gas usually glows. Argon, for instance, emits blue light when the gas is ionized. Ionized neon gas glows red. The amount of voltage needed to reach the striking potential varies with different gases and depends on the gas pressure. For example, a neon glow lamp for use as a night light ionizes at approximately 70 V.

Ionic Bonds

The sodium ion in Fig. 11–19b has a charge of 1+ because it is missing one electron. If such positive ions are placed near negative ions with a charge of 1−, there will be an electrical attraction to form an ionic bond.

A common example is the combination of sodium (Na) ions and chlorine (Cl) ions to form table salt (NaCl), as shown in Fig. 11–20. Notice that the one outer electron of the Na atom can fit into the seven-electron shell of the Cl atom. When these two elements are combined, the Na atom gives up one electron to form a positive ion, with a stable L shell having eight electrons; also, the Cl atom adds this one electron to form a negative ion, with a stable M shell

Figure 11–20 Ionic bond between atoms of sodium (Na) and chlorine (Cl) to form a molecule of sodium chloride (NaCl).

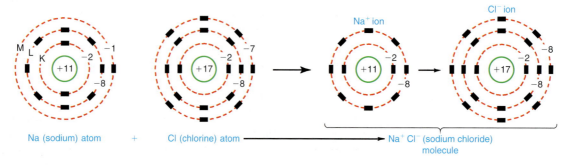

having eight electrons. The two opposite types of ions are bound in NaCl because of the strong attractive force between opposite charges close together.

The ions in NaCl can separate in water to make salt water a conductor of electricity; pure water is not a conductor of electricity. When current flows in salt water, then, the moving charges must be ions, another example of ionization current.

■ 11–10 Knowledge Check

Answer at end of chapter.

If a neutral atom loses one valence electron, it becomes a (positive/negative) ion?

■ 11–10 Self-Review

Answers at end of chapter.

a. **How much is I for 2 C/s of positive ion charges?**
b. **Which have the greatest mobility: positive ions, negative ions, or electrons?**
c. **A dielectric material is a good conductor of electricity. (True or False)**

11–11 Insulators

Substances that have very high resistance, of the order of many megohms, are classed as insulators. With such high resistance, an insulator cannot conduct appreciable current when voltage is applied. As a result, insulators can have either of two functions. One is to isolate conductors to eliminate conduction between them. The other is to store an electric charge when voltage is applied.

An insulator maintains its charge because electrons cannot flow to neutralize the charge. The insulators are commonly called *dielectric materials,* which means that they can store a charge.

Among the best insulators, or dielectrics, are air, vacuum, rubber, wax, shellac, glass, mica, porcelain, oil, dry paper, textile fibers, and plastics such as Bakelite, Formica, and polystyrene. Pure water is a good insulator, but saltwater is not. Moist earth is a fairly good conductor, and dry, sandy earth is an insulator.

For any insulator, a high enough voltage can be applied to break down the internal structure of the material, forcing the dielectric to conduct. This dielectric breakdown is usually the result of an arc, which ruptures the physical structure of the material, making it useless as an insulator. Table 11–4 compares several insulators in terms of dielectric strength, which is the voltage breakdown rating. The higher the dielectric strength, the better the insulator, since it is less likely to break down at a high value of applied voltage. The breakdown voltages in Table 11–4 are approximate values for the standard thickness of 1 mil, or 0.001 in. More thickness allows a higher breakdown-voltage rating. Note that the value of 20 V/mil for air or vacuum is the same as 20 kV/in.

Insulator Discharge Current

An insulator in contact with a voltage source stores charge, producing a potential on the insulator. The charge tends to remain on the insulator, but it can be discharged by one of the following methods:

1. **Conduction through a conducting path.** For instance, a wire across the charged insulator provides a discharge path. Then the discharged dielectric has no potential.

Table 11–4	Voltage Breakdown of Insulators		
Material	**Dielectric Strength, V/mil**	**Material**	**Dielectric Strength, V/mil**
Air or vacuum	20	Paraffin wax	200–300
Bakelite	300–550	Phenol, molded	300–700
Fiber	150–180	Polystyrene	500–760
Glass	335–2000	Porcelain	40–150
Mica	600–1500	Rubber, hard	450
Paper	1250	Shellac	900
Paraffin oil	380		

2. Brush discharge. As an example, high voltage on a sharp pointed wire can discharge through the surrounding atmosphere by ionization of the air molecules. This may be visible in the dark as a bluish or reddish glow, called the *corona effect*.
3. Spark discharge. This is a result of breakdown in the insulator because of a high potential difference that ruptures the dielectric. The current that flows across the insulator at the instant of breakdown causes the spark.

A corona is undesirable because it reduces the potential by brush discharge into the surrounding air. In addition, the corona often indicates the beginning of a spark discharge. A potential of the order of kilovolts is usually necessary for a corona because the breakdown voltage for air is approximately 20 kV/in. To reduce the corona effect, conductors that have high voltage should be smooth, rounded, and thick. This equalizes the potential difference from all points on the conductor to the surrounding air. Any sharp point can have a more intense field, making it more susceptible to a corona and eventual spark discharge.

■ *11–11 Knowledge Check*

Answer at end of chapter.

Under what condition will an insulator conduct electricity?

■ *11–11 Self-Review*

Answers at end of chapter.

a. Which has a higher voltage breakdown rating, air or mica?
b. Can 30 kV arc across an air gap of 1 in.?

11–12 Troubleshooting Hints for Wires and Connectors

For all types of electronic equipment, a common problem is an open circuit in the wire conductors, the connectors, and the switch contacts.

You can check continuity of conductors, both wires and printed wiring, with an ohmmeter. A good conductor reads 0 Ω for continuity. An open reads infinite ohms.

A connector can also be checked for continuity between the wire and the connector itself. Also, the connector may be tarnished, oxide coated, or rusted. Then it must be cleaned with either fine sandpaper or emery cloth. Sometimes, it helps just to pull out the plug and reinsert it to make sure of tight connections.

With a plug connector for cable, make sure the wires have continuity to the plug. Except for the F-type connector, most plugs require careful soldering to the center pin.

A switch with dirty or pitted contacts can produce intermittent operation. In most cases, the switch cannot be disassembled for cleaning. Therefore, the switch must be replaced with a new one.

■ *11–12 Knowledge Check*

> *Answer at end of chapter.*
>
> **An intermittent switch is usually replaced with a new one because it is not practical to clean it. (True or False)**

■ *11–12 Self-Review*

> *Answers at end of chapter.*
>
> a. **Printed wiring cannot be checked for continuity with an ohmmeter. (True or False)**
> b. **A tarnished or rusty connection has higher than normal resistance. (True or False)**

Summary

- A conductor has very low resistance. All metals are good conductors; the best are silver, copper, and aluminum. Copper is generally used for wire conductors due to its lower cost.

- The sizes for copper wire are specified by the American Wire Gage. Higher gage numbers mean thinner wire. Typical sizes are No. 22 gage hookup wire for electronic circuits and No. 12 and No. 14 for house wiring.

- The cross-sectional area of round wire is measured in circular mils. One mil is 0.001 in. The area in circular mils equals the diameter in mils squared.

- The resistance R of a conductor can be found using the formula $R = \rho(l/A)$, where ρ is the specific resistance, l is the length of the conductor, and A is the cross-sectional area of the conductor. Wire resistance increases directly with length l, but decreases inversely with cross-sectional area A.

- The voltage drop across a closed switch in a series circuit is zero volts. When open, the switch has the applied voltage across it.

- A fuse protects circuit components and wiring against overload; excessive current melts the fuse element to open the entire series circuit. A good fuse has very low resistance and practically zero voltage across it.

- Ionization in liquids and gases produces atoms that are not electrically neutral. These are ions. Negative ions have an excess of electrons; positive ions have a deficiency of electrons. In liquids and gases, electric current is a result of the movement of ions.

- The resistance of pure metals increases with temperature. For semiconductors and liquid electrolytes, resistance decreases at higher temperatures.

- An insulator has very high resistance. Common insulating materials are air, vacuum, rubber, paper, glass, porcelain, shellac, and plastics. Insulators are also called *dielectrics*.

- Superconductors have practically no resistance.

- Common circuit troubles are an open in wire conductors; dirty contacts in switches; and dirt, oxides, and corrosion on connectors and terminals.

Important Terms

- **Circuit Breaker** - a device used to protect the components and wiring in a circuit in the event of a short circuit. It is constructed of a thin bimetallic strip that expands with heat and in turn trips open the circuit. The circuit breaker can be reset after the bimetallic strip cools down and the short circuit is removed.

- **Circular Mil (cmil)** - a unit that specifies the cross-sectional area of round wire. 1 cmil is the cross-sectional area of a wire with a diameter, d, of 1 mil, where 1 mil = 0.001 inches. For round wire the cmil area is equal to the square of the diameter, d, in mils.

- **Corona Effect** - a visible blue or red glow caused by ionization of the air molecules when the high voltage on a sharp, pointed wire discharges into the atmosphere.

- **Dielectric Material** - another name used in conjunction with insulating materials. An insulator is commonly referred to as a dielectric because it can hold or store an electric charge.

- **Fuse** - a device used to protect the components and wiring in a circuit in the event of a short circuit. The fuse element is made of either aluminum, tin-coated copper, or nickel. Excessive current melts the fuse element which blows the fuse.

- **Ionization Current** - a current from the movement of ion charges in a liquid or gas.

- **Pole** - the number of completely isolated circuits that can be controlled by a switch.

- **Slow-Blow Fuse** - a type of fuse that can handle a temporary surge current which exceeds the current rating of the fuse. This type of fuse has an element with a coiled construction and is designed to open only on a continued overload such as a short circuit.

- **Specific Resistance** - the resistance of a metal conductor whose cross-sectional area is 1 cmil and whose length is 1 ft. The specific resistance, designated ρ, is specified in cmil $\cdot$ Ω/ft.

- **Switch** - a component that controls whether the current is on or off in a circuit.

- **Throw** - the number of closed contact positions that exist per pole on a switch.

- **Temperature Coefficient** - a factor that indicates how much the resistance of a material changes with temperature. A positive temperature coefficient means that the resistance increases with temperature, whereas a negative temperature coefficient means that the resistance decreases with temperature.

- **Wire Gauge** - a number assigned to a specific size of round wire in terms of its diameter and cross-sectional area. The American Wire Gage (AWG) system provides a table of all wire sizes which includes the gage size, the diameter, d, in mils and the area, A, in circular mils (cmils).

Related Formulas

$$R = \rho \frac{l}{A}$$

$$R_t = R_0 + R_0(\alpha \Delta t)$$

Self-Test

Answers at back of book.

1. **A closed switch has a resistance of approximately**
   a. infinity.
   b. zero ohms.
   c. 1 MΩ.
   d. none of the above.

2. **An open fuse has a resistance that approaches**
   a. infinity.
   b. zero ohms.
   c. 1 to 2 Ω.
   d. none of the above.

3. **How many connecting terminals does an SPDT switch have?**
   a. 2.
   b. 6.
   c. 3.
   d. 4.

4. **The voltage drop across a closed switch equals**
   a. the applied voltage.
   b. zero volts.
   c. infinity.
   d. none of the above.

5. **For round wire, as the gage numbers increase from 1 to 40**
   a. the diameter and circular area increase.
   b. the wire resistance decreases for a specific length and type.
   c. the diameter increases but the circular area remains constant.
   d. the diameter and circular area decrease.

6. **The circular area of round wire, doubles for**
   a. every 2 gage sizes.
   b. every 3 gage sizes.
   c. each successive gage size.
   d. every 10 gage sizes.

7. Which has more resistance, a 100-ft length of No. 12 gage copper wire or a 100-ft length of No. 12 gage aluminum wire?

a. The 100-ft length of No. 12 gage aluminum wire.

b. The 100-ft length of No. 12 gage copper wire.

c. They both have exactly the same resistance.

d. This is impossible to determine.

8. In their pure form, all metals have a

a. negative temperature coefficient.

b. temperature coefficient of zero.

c. positive temperature coefficient.

d. very high resistance.

9. The current rating of a switch corresponds to the maximum current the switch can safely handle when it is

a. open.

b. either open or closed.

c. closed.

d. none of the above.

10. How much is the resistance of a 2000-ft length of No. 20 gage aluminum wire?

a. less than 1 Ω.

b. 20.35 Ω.

c. 3.33 kΩ.

d. 33.27 Ω.

11. How many completely isolated circuits can be controlled by a DPST switch?

a. 1.

b. 2.

c. 3.

d. 4.

12. Which of the following metals is the best conductor of electricity?

a. steel.

b. aluminum.

c. silver.

d. gold.

13. What is the area in circular mils (cmils) of a wire whose diameter, d, is 0.01 inches?

a. 0.001 cmil.

b. 10 cmil.

c. 1 cmil.

d. 100 cmil.

14. The term "pole" as it relates to switches is defined as

a. the number of completely isolated circuits that can be controlled by the switch.

b. the number of closed contact positions that the switch has.

c. the number of connecting terminals the switch has.

d. none of the above.

15. The motion of ion charges in a liquid or gas is called

a. the corona effect.

b. hole flow.

c. superconductivity.

d. ionization current.

Questions

1. Name three good metal conductors in order of resistance. Describe at least one application.

2. Name four insulators. Give one application.

3. Name two semiconductors.

4. Name two types of resistance wire. Give one application.

5. What is meant by the "dielectric strength of an insulator"?

6. Why does ionization occur more readily in liquids and gases, compared with solid metals? Give an example of ionization current.

7. Define the following: ion, ionic bond, and electrolyte.

8. Draw a circuit with two bulbs, a battery, and an SPDT switch that determines which bulb lights.

9. Why is it not possible to measure the hot resistance of a filament with an ohmmeter?

10. Give one way in which negative ion charges are similar to electron charges and one way in which they are different.

11. Define the following abbreviations for switches: SPST, SPDT, DPST, DPDT, NO, and NC.

12. Give two common circuit troubles with conductors and connector plugs.

Problems

SECTION 11–1 FUNCTION OF THE CONDUCTOR

11–1 In Fig. 11–21, an 8-Ω heater is connected to the 120-Vac power line by two 50-ft lengths of copper wire. If each 50-ft length of wire has a resistance of 0.08 Ω, then calculate the following:

a. The total length of copper wire that connects the 8-Ω heater to the 120-Vac power line.

b. The total resistance, R_T, of the circuit.

c. The current, I, in the circuit.

d. The voltage drop across each 50-ft length of copper wire.

e. The voltage across the 8-Ω heater.

f. The $I^2 R$ power loss in each 50-ft length of copper wire.

g. The power dissipated by the 8-Ω heater.

h. The total power, P_T, supplied to the circuit by the 120-Vac power line.

i. The percentage of the total power, P_T, dissipated by the 8-Ω heater.

Figure 11–21

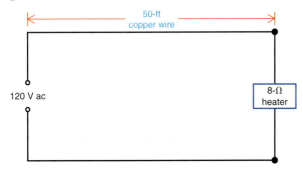

11–2 In Fig. 11–21, recalculate the values in parts a through i in Problem 11–1 if the 8-Ω heater is replaced with a 24-Ω fan.

SECTION 11–2 STANDARD WIRE GAGE SIZES

11–3 What is the area in circular mils for a wire if its diameter, d, equals

a. 0.005 in.

b. 0.021 in.

c. 0.032 in.

d. 0.05 in.

e. 0.1 in.

f. 0.2 in.

11–4 What is the approximate AWG size of a wire whose diameter, d, equals 0.072 in.?

11–5 Using Table 11–1, determine the resistance of a 1000-ft length of copper wire for the following gage sizes:

a. No. 10 gage

b. No. 13 gage

c. No. 16 gage

d. No. 24 gage

11–6 Which would you expect to have more resistance, a 1000-ft length of No. 14 gage copper wire or a 1000-ft length of No. 12 gage copper wire?

11–7 Which would you expect to have more resistance, a 1000-ft length of No. 23 gage copper wire or a 100-ft length of No. 23 gage copper wire?

SECTION 11–3 TYPES OF WIRE CONDUCTORS

11–8 A stranded wire consists of 41 strands of No. 30 gage copper wire. What is its equivalent gage size in solid wire?

11–9 If an extension cord is made up of 65 strands of No. 28 gage copper wire, what is its equivalent gage size in solid wire?

11–10 How many strands of No. 36 gage wire does it take to make a stranded wire whose equivalent gage size is No. 16?

11–11 What is the gage size of the individual strands in a No. 10 gage stranded wire if there are eight strands?

SECTION 11–6 SWITCHES

11–12 With the switch, S_1, closed in Fig. 11–22,

a. how much is the voltage across the switch?

b. how much is the voltage across the lamp?

c. will the lamp light?

d. what is the current, I, in the circuit based on the specifications of the lamp?

Figure 11–22

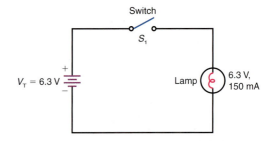

11–13 With the switch, S_1, open in Fig. 11–22,

a. how much is the voltage across the switch?

b. how much is the voltage across the lamp?

c. will the lamp light?

d. what is the current, I, in the circuit based on the specifications of the lamp?

11–14 Draw a schematic diagram showing how a SPDT switch can be used to supply a resistive heating element with either 6V or 12V.

11–15 Draw a schematic diagram showing how a DPDT switch can be used to

a. allow a stereo receiver to switch between two different speakers.

b. reverse the polarity of voltage across a dc motor to reverse its direction.

11–16 An SPST switch is rated at 10 A/250 V. Can this switch be used to control a 120-V, 1000 W appliance?

SECTION 11–8 WIRE RESISTANCE

11–17 Calculate the resistance of the following conductors:

a. 250 ft of No. 20 gage copper wire.

b. 250 ft of No. 20 gage aluminum wire.

c. 250 ft of No. 20 gage steel wire.

11–18 Calculate the resistance of the following conductors:

a. 100 ft of No. 14 gage copper wire.

b. 200 ft of No. 14 gage copper wire.

11–19 Calculate the resistance of the following conductors:

a. 100 ft of No. 15 gage copper wire.

b. 100 ft of No. 18 gage copper wire.

11–20 How much is the resistance of a slab of silicon 0.1 cm long with a cross-sectional area of 1 cm^2?

11–21 What is the resistance for each conductor of a 50-ft extension cord made of No. 14 gage copper wire?

11–22 A 100-ft extension cord uses No. 14 gage copper wire for each of its conductors. If the extension cord is used to connect a 10-A load to the 120-Vac power line, how much voltage is available at the load?

11–23 What is the smallest gage size copper wire that will limit the conductor voltage drop to 5 V when 120 Vac is supplied to a 6-A load? The total line length of both conductors is 200 ft.

SECTION 11–9 TEMPERATURE COEFFICIENT OF RESISTANCE

11–24 A tungsten wire has a resistance, R, of 20 Ω at 20°C. Calculate its resistance at 70°C.

11–25 A steel wire has a resistance of 10 Ω at 20°C. Calculate its resistance at 100°C.

11–26 A nickel wire has a resistance of 150 Ω at 20°C. Calculate its resistance at 250°C.

11–27 An aluminum wire has a resistance of 100 Ω at 20°C. Calculate its resistance at 120°C.

11–28 The resistance of a nichrome wire is 1 kΩ at 20°C. Calculate its resistance at 220°C.

11–29 The resistance of a steel wire is 10 Ω at 20°C. How much is its resistance at −30°C?

11–30 A No. 14 gage aluminum wire has 8.4 Ω of resistance at −20°C. What is its resistance at 50°C?

Critical Thinking

11–31 Use two switches having the appropriate number of poles and throws to build a partial decade resistance box. The resistance is to be adjustable in 1-Ω and 10-Ω steps from 0 to 99 Ω across two terminals identified as A and B. Draw the circuit showing all resistance values and switch connections.

11–32 Show how two SPDT switches can be wired to turn ON and OFF a light from two different locations. The voltage source is a 12-V battery.

Answers to Knowledge Check Problems

11–1 105 V

11–2 0.1 inches or 100 mils

11–3 No. 18 gage

11–4 the sleeve

11–5 The desoldering braid acts like a natural heat sink thus preventing damage to the PC board.

11–6 the physical size of the switch contacts as well as the type of metal used for the contacts.

11–7 0 V across both R_1 and R_2

11–8 approx. 20 Ω

11–9 negative

11–10 positive ion

11–11 when the voltage across the insulator is high enough to break down the internal structure of the material forcing it to conduct

11–12 True

Answers to Self-Reviews

11–1 a. 0.08 Ω
b. 0.07 V approx.
c. 0.06% approx.

11–2 a. 0.01646 Ω
b. 625 cmil
c. No. 16

11–3 a. T
b. T
c. T

11–4 a. T
b. T
c. F

11–5 a. 25 W
b. infinite ohms

11–6 a. zero
b. three
c. infinite
d. no

11–7 a. zero
b. zero

11–8 a. more
b. 2.6 Ω

11–9 a. T
b. T
c. T

11–10 a. $I = 2A$
b. electrons
c. F

11–11 a. mica
b. yes

11–12 a. F
b. T

12 Batteries

A *battery* is a group of cells that generate energy from an internal chemical reaction. The cell itself consists of two different conducting materials as the electrodes that are immersed in an electrolyte. The chemical reaction between the electrodes and the electrolyte results in a separation of electric charges as ions and free electrons. Then the two electrodes have a difference of potential that provides a voltage output from the cell.

The main types are the alkaline cell with an output of 1.5 V and the lead-sulfuric acid wet cell with 2.1 V for its output. A common battery is the 9-V flat battery. It has six cells connected in series internally for an output of $6 \times 1.5 = 9$ V. Dry batteries are used to power many different types of portable electronic equipment.

The lead-sulfuric acid cell is the type used in most automobiles. Six cells are connected in series internally for a 12-V output.

A battery provides a source of steady dc voltage of fixed polarity and is a good example of a generator or energy source. The battery supplies voltage for a circuit as the load to produce the desired load current. An important factor is the internal resistance r_i of the source, which affects the output voltage when a load is connected. A low r_i means that the source can maintain a constant output voltage for different values of load current. For the opposite case, a high r_i makes the output voltage drop, but a constant value of load current can be maintained.

Objectives

After studying this chapter you should be able to

- *Explain* the difference between primary and secondary cells.

- *Define* what is meant by the *internal resistance of a cell.*

- *List* several different types of voltaic cells.

- *Explain* how cells can be connected to increase either the current capacity or voltage output of a battery.

- *Explain* why the terminal voltage of a battery drops with more load current.

- *Explain* the difference between voltage sources and current sources.

- *Explain* the concept of maximum power transfer.

Outline

Important Terms

ampere-hour (A-H) rating
battery
charging
constant-current generator
constant-voltage generator
discharging

float charging
fuel cell
hydrometer
internal resistance, r_i
open-circuit voltage
primary cell

secondary cell
specific gravity
storage cell
voltaic cell

12–1 Introduction to Batteries

We rely on batteries to power an almost unlimited number of electronic products available today. For example, batteries are used in cars, personal computers (PCs), handheld radios, laptops, MP3 players, and cell phones, to name just a few of the more common applications. Batteries are available in a wide variety of shapes and sizes and have many different voltage and current ratings. The different sizes and ratings are necessary to meet the needs of the vast number of applications. Regardless of the application, however, all batteries are made up of a combination of individual voltaic cells. Together, the cells provide a steady dc voltage at the output terminals of the battery. The voltage output and current rating of a battery are determined by several factors, including the type of elements used for the electrodes, the physical size of the electrodes, and the type of electrolyte.

As you know, some batteries become exhausted with use and cannot be recharged. Others can be recharged hundreds or even thousands of times before they are no longer able to produce or maintain the rated output voltage. Whether a battery is rechargeable or not is determined by the type of cells that make up the battery. There are two types, primary cells and secondary cells.

Primary Cells

This type cannot be recharged. After it has delivered its rated capacity, the primary cell must be discarded because the internal chemical reaction cannot be restored. Figure 12–1 shows a variety of dry cells and batteries, all of which are of the primary type. In Table 12–1 several different cells are listed by name. Each of the cells is listed as either the primary or the secondary type. Notice the open circuit voltage for each of the cell types listed.

Secondary Cells

This type can be recharged because the chemical action is reversible. When it supplies current to a load resistance, the cell is *discharging* because the current tends to neutralize the separated charges at the electrodes. For the opposite case, the current can be reversed to re-form the electrodes as the chemical action is reversed. This action is *charging* the cell. The charging current must be supplied by an external dc voltage source, with the cell serving as a load resistance. The

Figure 12–1 Typical dry cells and batteries. These primary types cannot be recharged.

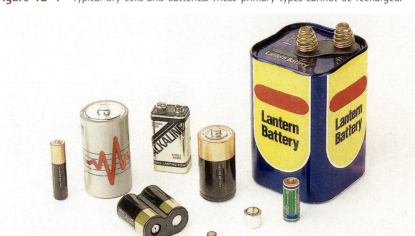

Table 12-1	Cell Types and Open-Circuit Voltage	
Cell Name	**Type**	**Nominal Open-Circuit* Voltage, Vdc**
Carbon-zinc	Primary	1.5
Zinc chloride	Primary	1.5
Manganese dioxide (alkaline)	Primary or secondary	1.5
Mercuric oxide	Primary	1.35
Silver oxide	Primary	1.5
Lithium	Primary	3.0
Lead-acid	Secondary	2.1
Nickel-cadmium	Secondary	1.2
Nickel-metal-hydride	Secondary	1.2
Nickel-iron (Edison cell)	Secondary	1.2
Nickel-zinc	Secondary	1.6
Solar	Secondary	0.5

* Open-circuit *V* is the terminal voltage without a load.

GOOD TO KNOW

A battery is continually doing the work of separating the positive and negative charges within itself. This is true even when the battery is not in use. Because of this, batteries made up of primary cells can become exhausted before they ever leave the store. Therefore, always be sure to buy batteries that are fresh from the manufacturer.

Figure 12–2 Example of a 12-V auto battery using six lead-acid cells in series. This is a secondary type, which can be recharged.

discharging and recharging is called *cycling* of the cell. Since a secondary cell can be recharged, it is also called a *storage cell*. The most common type is the lead-acid cell generally used in automotive batteries (Fig. 12–2). In addition, the list in Table 12–1 indicates which are secondary cells.

Dry Cells

What we call a *dry cell* really has a moist electrolyte. However, the electrolyte cannot be spilled and the cell can operate in any position.

Sealed Rechargeable Cells

This type is a secondary cell that can be recharged, but it has a sealed electrolyte that cannot be refilled. These cells are capable of charge and discharge in any position.

■ *12-1 Knowledge Check*

> *Answer at end of chapter.*
>
> **What is the term used to describe the discharging and recharging of a cell?**

a. How much is the output voltage of a carbon-zinc cell?
b. How much is the output voltage of a lead-acid cell?
c. Which type can be recharged, a primary or a secondary cell?

12-2 The Voltaic Cell

When two different conducting materials are immersed in an electrolyte, as illustrated in Fig. 12–3a, the chemical action of forming a new solution results in the separation of charges. This device for converting chemical energy into electric energy is a voltaic cell. It is also called a *galvanic cell,* named after Luigi Galvani (1737–1798).

In Fig. 12–3a, the charged conductors in the electrolyte are the electrodes or plates of the cell. They are the terminals that connect the voltage output to an external circuit, as shown in Fig. 12–3b. Then the potential difference resulting from the separated charges enables the cell to function as a source of applied voltage. The voltage across the cell's terminals forces current to flow in the circuit to light the bulb.

Current Outside the Cell

Electrons from the negative terminal of the cell flow through the external circuit with R_L and return to the positive terminal. The chemical action in the cell separates charges continuously to maintain the terminal voltage that produces current in the circuit.

The current tends to neutralize the charges generated in the cell. For this reason, the process of producing load current is considered discharging of the cell. However, the internal chemical reaction continues to maintain the separation of charges that produces the output voltage.

Current Inside the Cell

The current through the electrolyte is a motion of ion charges. Notice in Fig. 12–3b that the current inside the cell flows from the positive terminal to the negative terminal. This action represents the work being done by the chemical reaction to generate the voltage across the output terminals.

Figure 12–3 How a voltaic cell converts chemical energy into electrical energy. (a) Electrodes or plates in liquid electrolyte solution. (b) Schematic of a circuit with a voltaic cell as a dc voltage source V to produce current in load R_L, which is the lightbulb.

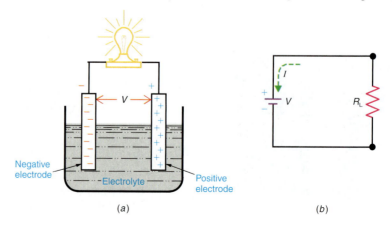

(a)   (b)

The negative terminal in Fig. 12–3a is considered the anode of the cell because it forms positive ions in the electrolyte. The opposite terminal of the cell is its cathode.

Internal Resistance

Any practical voltage source has internal resistance, indicated as r_i, which limits the current it can deliver. For a chemical cell, as in Fig. 12–3, the r_i is mainly the resistance of the electrolyte. For a good cell, r_i is very low, with typical values less than 1 Ω. As the cell deteriorates, though, r_i increases, preventing the cell from producing its normal terminal voltage when load current is flowing because the internal voltage drop across r_i opposes the output terminal voltage. This is why you can often measure the normal voltage of a dry cell with a voltmeter, which drains very little current, but the terminal voltage drops when the load is connected.

The voltage output of a cell depends on the elements used for the electrodes and the electrolyte. The current rating depends mostly on the size. Larger batteries can supply more current. Dry cells are generally rated up to 250 mA, and the lead-acid wet cell can supply current up to 300 A or more. Note that a smaller r_i allows a higher current rating.

Electromotive Series

The fact that the voltage output of a cell depends on its elements can be seen from Table 12–2. This list, called the *electrochemical series* or *electromotive series,* gives the relative activity in forming ion charges for some of the chemical elements. The potential for each element is the voltage with respect to hydrogen

Table 12–2	Electromotive Series of Elements
Element	**Potential, V**
Lithium	−2.96
Magnesium	−2.40
Aluminum	−1.70
Zinc	−0.76
Cadmium	−0.40
Nickel	−0.23
Lead	−0.13
Hydrogen (reference)	0.00
Copper	+0.35
Mercury	+0.80
Silver	+0.80
Gold	+1.36

as a zero reference. The difference between the potentials for two different elements indicates the voltage of an ideal cell using these electrodes. Note that other factors, such as the electrolyte, cost, stability, and long life, are important for the construction of commercial batteries.

■ *12–2 Knowledge Check*

Answer at end of chapter.

The current rating of a battery or cell depends mostly on what factor?

■ *12–2 Self-Review*

Answers at end of chapter.

a. The negative terminal of a chemical cell has a charge of excess electrons. (True or False)
b. The internal resistance of a cell limits the amount of output current. (True or False)
c. Two electrodes of the same metal provide the highest voltage output. (True or False)

12–3 Common Types of Primary Cells

In this section, you will be introduced to several different types of primary cells in use today.

Carbon–Zinc

The carbon-zinc dry cell is a very common type because of its low cost. It is also called the *Leclanché cell,* named after its inventor. Examples are shown in Fig. 12–1, and Fig. 12–4 illustrates the internal construction of the D-size round cell. The voltage output of the carbon-zinc cell is 1.4 to 1.6 V, with a nominal value of 1.5 V. The suggested current range is up to 150 mA for the D size, which has a height of $2\frac{1}{4}$ in. and volume of 3.18 in.^{3}. The C, AA, and AAA sizes are smaller, with lower current ratings.

Figure 12–4 Cutaway view of carbon-zinc dry cell. This is size D with a height of $2\frac{1}{4}$ in.

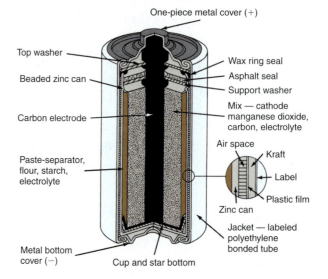

One-piece metal cover (+)

Top washer

Beaded zinc can

Carbon electrode

Paste-separator, flour, starch, electrolyte

Metal bottom cover (−)

Cup and star bottom

Wax ring seal

Asphalt seal

Support washer

Mix — cathode manganese dioxide, carbon, electrolyte

Air space

Kraft

Label

Plastic film

Zinc can

Jacket — labeled polyethylene bonded tube

The electrochemical system consists of a zinc anode and a manganese dioxide cathode in a moist electrolyte. The electrolyte is a combination of ammonium chloride and zinc chloride dissolved in water. For the round-cell construction, a carbon rod is used down the center, as shown in Fig. 12–4. The rod is chemically inert. However, it serves as a current collector for the positive terminal at the top. The path for current inside the cell includes the carbon rod as the positive terminal, the manganese dioxide, the electrolyte, and the zinc can which is the negative electrode. The carbon rod also prevents leakage of the electrolyte but is porous to allow the escape of gases which accumulate in the cell.

In operation of the cell, the ammonia releases hydrogen gas which collects around the carbon electrode. This reaction is called *polarization,* and it can reduce the voltage output. However, the manganese dioxide releases oxygen, which combines with the hydrogen to form water. The manganese dioxide functions as a *depolarizer.* Powdered carbon is also added to the depolarizer to improve conductivity and retain moisture.

Carbon-zinc dry cells are generally designed for an operating temperature of 70°F. Higher temperatures will enable the cell to provide greater output. However, temperatures of 125°F or more will cause rapid deterioration of the cell.

The chemical efficiency of the carbon-zinc cell increases with less current drain. Stated another way, the application should allow for the largest battery possible, within practical limits. In addition, performance of the cell is generally better with intermittent operation. The reason is that the cell can recuperate between discharges, probably by depolarization.

As an example of longer life with intermittent operation, a carbon-zinc D cell may operate for only a few hours with a continuous drain at its rated current. Yet the same cell could be used for a few months or even a year with intermittent operation of less than 1 hour at a time with smaller values of current.

Alkaline Cell

Another popular type is the manganese-zinc cell shown in Fig. 12–5, which has an alkaline electrolyte. It is available as either a primary or a secondary cell,

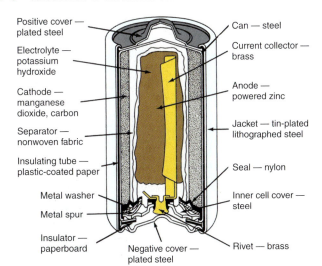

Figure 12–5 Construction of the alkaline cell.

Positive cover — plated steel

Electrolyte — potassium hydroxide

Cathode — manganese dioxide, carbon

Separator — nonwoven fabric

Insulating tube — plastic-coated paper

Metal washer

Metal spur

Insulator — paperboard

Negative cover — plated steel

Can — steel

Current collector — brass

Anode — powered zinc

Jacket — tin-plated lithographed steel

Seal — nylon

Inner cell cover — steel

Rivet — brass

but the primary type is more common. The output is the same 1.5 V as that of a carbon-zinc cell, but the alkaline cell lasts much longer.

The electrochemical system consists of a powdered zinc anode and a manganese dioxide cathode in an alkaline electrolyte. The electrolyte is potassium hydroxide, which is the main difference between the alkaline and Leclanché cells. Hydroxide compounds are alkaline with negative hydroxyl (OH) ions, whereas an acid electrolyte has positive hydrogen (H) ions. The voltage output from the alkaline cell is 1.5 V.

The alkaline cell has many applications because of its ability to work at high efficiency with continuous, high discharge rates. Depending on the application, an alkaline cell can provide up to seven times the service of a carbon-zinc cell. As examples, in a portable CD player, an alkaline cell will normally have twice the service life of a general-purpose carbon-zinc cell; in toys, the alkaline cell typically provides about seven times more service.

The outstanding performance of the alkaline cell is due to its low internal resistance. Its r_i is low because of the dense cathode material, the large surface area of the anode in contact with the electrolyte, and the high conductivity of the electrolyte. In addition, alkaline cells perform satisfactorily at low temperatures.

Zinc Chloride Cells

This type is actually a modified carbon-zinc cell whose construction is illustrated in Fig. 12–4. However, the electrolyte contains only zinc chloride. The zinc chloride cell is often referred to as a *heavy duty* type. It can normally deliver more current over a longer period of time than the Leclanché cell. Another difference is that the chemical reaction in the zinc chloride cell consumes water along with the chemically active materials, so that the cell is nearly dry at the end of its useful life. As a result, liquid leakage is not a problem.

Additional Types of Primary Cells

The miniature button construction shown in Fig. 12–6 is often used for the mercury cell and the silver oxide cell. The cell diameter is ⅜ to 1 in.

Mercury Cell

The electrochemical system consists of a zinc anode, a mercury compound for the cathode and an electrolyte of potassium or sodium hydroxide. Mercury cells are available as flat, round cylinders and miniature button shapes. Note, though, that some round mercury cells have the top button as the negative terminal and the bottom terminal positive. The open-circuit voltage is 1.35 V when the

---

Figure 12–6 Construction of miniature button type of primary cell. Diameter is ⅜ to 1 in. Note the chemical symbols AgO_2 for silver oxide, HgO for mercuric oxide, and MnO_2 for manganese dioxide.

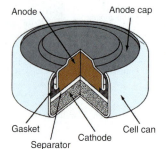

• Anodes are a gelled mixture of amalgamated zinc powder and electrolyte.

• Cathodes
  Silver cells: AgO_2, MnO_2, and conductor
  Mercury cells: HgO and conductor (may contain MnO_2)
  Manganese dioxide cells: MNO_2 and conductor

cathode is mercuric oxide (HgO) and 1.4 V or more with mercuric oxide/manganese dioxide. The 1.35-V type is more common.

The mercury cell is used where a relatively flat discharge characteristic is required with high current density. Its internal resistance is low and essentially constant. These cells perform well at elevated temperatures, up to 130°F continuously or 200°F for short periods. One drawback of the mercury cell is its relatively high cost compared with a carbon-zinc cell. Mercury cells are becoming increasingly unavailable due to the hazards associated with proper disposal after use.

Silver Oxide Cell

The electrochemical system consists of a zinc anode, a cathode of silver oxide (AgO_2) with small amounts of manganese dioxide, and an electrolyte of potassium or sodium hydroxide. It is commonly available in the miniature button shape shown in Fig. 12–6. The open-circuit voltage is 1.6 V, but the nominal output with a load is considered 1.5 V. Typical applications include hearing aids, cameras, and electronic watches, which use very little current.

Summary of the Most Common Types of Dry Cells

The most common types of dry cells include carbon-zinc, zinc chloride (heavy-duty), and manganese-zinc (alkaline). It should be noted that, the alkaline cell is better for heavy-duty use than the zinc-chloride type. They are available in the round, cylinder types, listed in Table 12–3, for the D, C, AA, and AAA sizes. The small button cells generally use either mercury or silver oxide. All these dry cells are the primary type and cannot be recharged. Each has an output of 1.5 V except for the 1.35-V mercury cell.

Any dry cell loses its ability to produce output voltage even when it is not being used. The shelf life is about 2 years for the alkaline type, but much less with the carbon-zinc cell, especially for small sizes and partially used cells. The reasons are self-discharge within the cell and loss of moisture from the electrolyte. Therefore, dry cells should be used fresh from the manufacturer. It is worth noting, however, that the shelf life of dry cells is steadily increasing due to recent advances in battery technology.

Note that shelf life can be extended by storing the cell at low temperatures, about 40 to 50°F. Even temperatures below freezing will not harm the cell. However, the cell should be allowed to return to normal room temperature before being used, preferably in its original packaging, to avoid condensation.

The alkaline type of dry cell is probably the most cost-efficient. It costs more but lasts much longer, besides having a longer shelf life. Compared with

Table 12–3	Sizes for Popular Types of Dry Cells*	
Size	Height, in.	Diameter, in.
D	$2\frac{1}{4}$	$1\frac{1}{4}$
C	$1\frac{3}{4}$	1
AA	$1\frac{7}{8}$	$\frac{9}{16}$
AAA	$1\frac{3}{4}$	$\frac{3}{8}$

* Cylinder shape shown in Fig. 12-1.

Figure 12–7 Lithium battery.

size-D batteries, the alkaline type can last about 10 times longer than the carbon-zinc type in continuous operation, or about seven times longer for typical intermittent operation. The zinc chloride heavy-duty type can last two or three times longer than the general-purpose carbon-zinc cell. For low-current applications of about 10 mA or less, however, there is not much difference in battery life.

Lithium Cell

The lithium cell is a relatively new primary cell. However, its high output voltage, long shelf life, low weight, and small volume make the lithium cell an excellent choice for special applications. The open-circuit output voltage is either 2.9 V or 3.7 V, depending on the electrolyte. Note the high potential of lithium in the electromotive list of elements shown before in Table 12–2. Figure 12–7 shows an example of a lithium battery with a 6-V output.

A lithium cell can provide at least 10 times more energy than the equivalent carbon-zinc cell. However, lithium is a very active chemical element. Many of the problems in construction have been solved, though, especially for small cells delivering low current. One interesting application is a lithium cell as the dc power source for a cardiac pacemaker. The long service life is important for this use.

Two forms of lithium cells are in widespread use, the lithium-sulfur dioxide ($LiSO_2$) type and the lithium-thionyl chloride type. Output is approximately 3 V.

In the $LiSO_2$ cell, the sulfur dioxide is kept in a liquid state by using a high-pressure container and an organic liquid solvent, usually methyl cyanide. One problem is safe encapsulation of toxic vapor if the container should be punctured or cracked. This problem can be significant for safe disposal of the cells when they are discarded after use.

The shelf life of the lithium cell, 10 years or more, is much longer than that of other types.

■ *12–3 Knowledge Check*

Answer at end of chapter.

Which cell can provide the same current over a longer period of time, the C-size or D-size cell?

■ *12–3 Self-Review*

Answers at end of chapter.

a. **Which has longer shelf life, the alkaline or carbon-zinc cell?**
b. **For the same application, which will provide a longer service life, an alkaline or carbon-zinc dry cell?**
c. **Which size cell is larger, C or AA?**
d. **What type of cell is typically used in watches, hearing aids, cameras, etc?**

Figure 12–8 Common 12-V lead-acid battery used in automobiles.

12–4 Lead-Acid Wet Cell

Where high load current is necessary, the lead-acid cell is the type most commonly used. The electrolyte is a dilute solution of sulfuric acid (H_2SO_4). In the application of battery power to start the engine in an automobile, for example, the load current to the starter motor is typically 200 to 400 A. One cell has a nominal output of 2.1 V, but lead-acid cells are often used in a series combination of three for a 6-V battery and six for a 12-V battery. Examples are shown in Figs. 12–2 and 12–8.

The lead-acid type is a secondary cell or storage cell, which can be recharged. The charge and discharge cycle can be repeated many times to restore the output voltage, as long as the cell is in good physical condition. However, heat with excessive charge and discharge currents shortens the useful life to about 3 to 5 years for an automobile battery. Of the different types of secondary cells, the lead-acid type has the highest output voltage, which allows fewer cells for a specified battery voltage.

Construction

Inside a lead-acid battery, the positive and negative electrodes consist of a group of plates welded to a connecting strap. The plates are immersed in the electrolyte, consisting of eight parts of water to three parts of concentrated sulfuric acid. Each plate is a grid or framework, made of a lead-antimony alloy. This construction enables the active material, which is lead oxide, to be pasted into the grid. In manufacture of the cell, a forming charge produces the positive and negative electrodes. In the forming process, the active material in the positive plate is changed to lead peroxide (PbO_2). The negative electrode is spongy lead (Pb).

Automobile batteries are usually shipped dry from the manufacturer. The electrolyte is put in at installation, and then the battery is charged to form the plates. With maintenance-free batteries, little or no water need be added in normal service. Some types are sealed, except for a pressure vent, without provision for adding water.

Chemical Action

Sulfuric acid is a combination of hydrogen and sulfate ions. When the cell discharges, lead peroxide from the positive electrode combines with hydrogen ions to form water and with sulfate ions to form lead sulfate. The lead sulfate is also produced by combining lead on the negative plate with sulfate ions. Therefore, the net result of discharge is to produce more water, which dilutes the electrolyte, and to form lead sulfate on the plates.

As discharge continues, the sulfate fills the pores of the grids, retarding circulation of acid in the active material. Lead sulfate is the powder often seen on the outside terminals of old batteries. When the combination of weak electrolyte and sulfation on the plate lowers the output of the battery, charging is necessary.

On charge, the external dc source reverses the current in the battery. The reversed direction of ions flowing in the electrolyte results in a reversal of the chemical reactions. Now the lead sulfate on the positive plate reacts with water and sulfate ions to produce lead peroxide and sulfuric acid. This action re-forms the positive plate and makes the electrolyte stronger by adding sulfuric acid. At the same time, charging enables the lead sulfate on the negative plate to react with hydrogen ions; this also forms sulfuric acid while re-forming lead on the negative electrode.

As a result, the charging current can restore the cell to full output, with lead peroxide on the positive plates, spongy lead on the negative plate, and the required concentration of sulfuric acid in the electrolyte. The chemical equation for the lead-acid cell is

$$Pb + PbO_2 + 2H_2SO_4 \underset{\text{Discharge}}{\overset{\text{Charge}}{\rightleftarrows}} 2\,PbSO_4 + 2H_2O$$

On discharge, the Pb and PbO_2 combine with the SO_4 ions at the left side of the equation to form lead sulfate ($PbSO_4$) and water (H_2O) on the right side of the equation.

On charge, with reverse current through the electrolyte, the chemical action is reversed. Then the Pb ions from the lead sulfate on the right side of the equation re-form the lead and lead peroxide electrodes. Also, the SO_4 ions combine with H_2 ions from the water to produce more sulfuric acid on the left side of the equation.

Current Ratings

Lead-acid batteries are generally rated in terms of the amount of discharge current they can supply for a specified period of time. The output voltage must be maintained above a minimum level, which is 1.5 to 1.8 V per cell. A common rating is ampere-hours (A · h) based on a specific discharge time, which is often 8 h. Typical values for automobile batteries are 100 to 300 A · h.

As an example, a 200-A · h battery can supply a load current of 200/8 or 25 A, based on an 8-h discharge. The battery can supply less current for a longer time or more current for a shorter time. Automobile batteries may be rated for "cold cranking power," which is related to the job of starting the engine. A typical rating is 450 A for 30 s at a temperature of 0°F.

Note that the ampere-hour unit specifies coulombs of charge. For instance, 200 A · h corresponds to 200 A × 3600 s (1 h = 3600 s). This equals 720,000 A · s, or coulombs. One ampere-second is equal to one coulomb. Then the charge equals 720,000 or 7.2×10^5 C. To put this much charge back into the battery would require 20 h with a charging current of 10 A.

The ratings for lead-acid batteries are given for a temperature range of 77 to 80°F. Higher temperatures increase the chemical reaction, but operation above 110°F shortens the battery life.

Low temperatures reduce the current capacity and voltage output. The ampere-hour capacity is reduced approximately 0.75% for each decrease of 1°F below the normal temperature rating. At 0°F, the available output is only 60% of the ampere-hour battery rating. In cold weather, therefore, it is very important to have an automobile battery up to full charge. In addition, the electrolyte freezes more easily when diluted by water in the discharged condition.

Specific Gravity

The state of discharge for a lead-acid cell is generally checked by measuring the specific gravity of the electrolyte. Specific gravity is a ratio comparing the weight of a substance with the weight of water. For instance, concentrated sulfuric acid is 1.835 times as heavy as water for the same volume. Therefore, its specific gravity equals 1.835. The specific gravity of water is 1, since it is the reference.

In a fully charged automotive cell, the mixture of sulfuric acid and water results in a specific gravity of 1.280 at room temperatures of 70 to 80°F. As the cell discharges, more water is formed, lowering the specific gravity. When the specific gravity is below about 1.145, the cell is considered completely discharged.

Specific-gravity readings are taken with a battery hydrometer, such as the one in Fig. 12–9. With this type of hydrometer, the state of charge of a cell within the battery is indicated by the number of floating disks. For example, one floating disk indicates the cell is at 25% of full charge. Two floating disks indicate the cell is at 50% of full charge. Similarly, three floating disks indicate 75% of full charge, whereas four floating disks indicate the cell is at 100% of full charge. The number of floating disks is directly correlated with the value of the specific gravity. As the specific gravity increases, more disks will float. Note that all cells within the battery must be tested for full charge.

The importance of the specific gravity can be seen from the fact that the open-circuit voltage of the lead-acid cell is approximately equal to

$$V = \text{specific gravity} + 0.84$$

For the specific gravity of 1.280, the voltage is 1.280 + 0.84 = 2.12 V, as an example. These values are for a fully charged battery.

Figure 12–9 Hydrometer to check specific gravity of lead-acid battery.

Figure 12-10 Reversed directions for charge and discharge currents of a battery. The r_i is internal resistance. (*a*) The V_B of the battery discharges to supply the load current for R_L. (*b*) The battery is the load resistance for V_G, which is an external source of charging voltage.

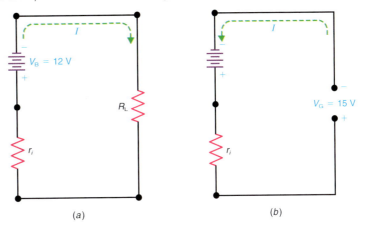

(a)   (b)

Charging the Lead-Acid Battery

The requirements are illustrated in Fig. 12–10. An external dc voltage source is necessary to produce current in one direction. Also, the charging voltage must be more than the battery emf. Approximately 2.5 V per cell is enough to overcome the cell emf so that the charging voltage can produce current opposite to the direction of the discharge current.

Note that the reversal of current is obtained by connecting the battery V_B and charging source V_G with + to + and − to −, as shown in Fig. 12–10*b*. The charging current is reversed because the battery effectively becomes a load resistance for V_G when it is higher than V_B. In this example, the net voltage available to produce a charging current is $15 - 12 = 3$ V.

A commercial charger for automobile batteries is shown in Fig. 12–11. This unit can also be used to test batteries and jump-start cars. The charger is essentially a dc power supply, rectifying input from an ac power line to provide dc output for charging batteries.

Float charging refers to a method in which the charger and the battery are always connected to each other to supply current to the load. In Fig. 12–12, the charger provides current for the load and the current necessary to keep the battery fully charged. The battery here is an auxiliary source for dc power.

It may be of interest to note that an automobile battery is in a floating-charge circuit. The battery charger is an ac generator or alternator with rectifier diodes, driven by a belt from the engine. When you start the car, the battery supplies the cranking power. Once the engine is running, the alternator charges the battery. It is not necessary for the car to be moving. A voltage regulator is used in this system to maintain the output at approximately 13 to 15 V.

Figure 12-11 Charger for auto batteries.

Figure 12-12 Circuit for battery in float-charge application.

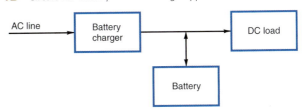

■ *12–4 Knowledge Check*

Answer at end of chapter.

What is the open-circuit voltage of a lead-acid cell whose specific gravity measures 1.06?

■ *12–4 Self-Review*

Answers at end of chapter.

a. How many lead-acid cells in series are needed for a 12-V battery?
b. A battery is rated for 120 A · h at an 8-h rate at 77°F. How much discharge current can it supply for 8 h?
c. Which of the following is the specific gravity reading for a fully charged lead acid cell: 1.080, 1.180, or 1.280?

12–5 Additional Types of Secondary Cells

A secondary cell is a storage cell that can be recharged by reversing the internal chemical reaction. A primary cell must be discarded after it has been completely discharged. The lead-acid cell is the most common type of storage cell. However, other types of secondary cells are available. Some of these are described next.

Nickel-Cadmium (NiCd) Cell

This type is popular because of its ability to deliver high current and to be cycled many times for recharging. Also, the cell can be stored for a long time, even when discharged, without any damage. The NiCd cell is available in both sealed and nonsealed designs, but the sealed construction shown in Fig. 12–13 is common. Nominal output voltage is 1.2 V per cell. Applications include portable power tools, alarm systems, and portable radio or video equipment.

Figure 12–13 Examples of nickel-cadmium cells. The output voltage for each is 1.2 V.

The chemical equation for the NiCd cell can be written as follows:

$$2\,Ni(OH)_3 + Cd \underset{\text{Discharge}}{\overset{\text{Charge}}{\rightleftarrows}} 2\,Ni(OH)_2 + Cd(OH)_2$$

The electrolyte is potassium hydroxide (KOH), but it does not appear in the chemical equation because the function of this electrolyte is to act as a conductor for the transfer of hydroxyl (OH) ions. Therefore, unlike the lead-acid cell, the specific gravity of the electrolyte in the NiCd cell does not change with the state of charge.

The NiCd cell is a true storage cell with a reversible chemical reaction for recharging that can be cycled up to 1000 times. Maximum charging current is equal to the 10-h discharge rate. Note that a new NiCd battery may need charging before use. A disadvantage of NiCd batteries is that they can develop a memory whereby they won't accept full discharge if they are routinely discharged to the same level, then charged. For this reason, it is a good idea to discharge them to different levels and occassionally to discharge them completely before recharging to erase the memory.

Nickel–Metal–Hydride (NiMH) Cell

Nickel-metal-hydride (NiMH) cells are currently finding widespread application in those high-end portable electrical and electronic products where battery performance parameters, notably run-time, are of major concern. NiMH cells are an extension of the proven, sealed, NiCd cells discussed previously. A NiMH cell has about 40% more capacity than a comparably sized NiCd cell, however. In other words, for a given weight and volume, a NiMH cell has a higher A-H rating than a NiCd cell.

With the exception of the negative electrode, NiMH cells use the same general types of components as a sealed NiCd cell. As a result, the nominal output voltage of a NiMH cell is 1.2 V, the same as the NiCd cell. In addition to having higher A-H ratings compared to NiCd cells, NiMH cells also do not suffer nearly as much from the memory effect. As a result of the advantages offered by the NiMH cell, they are finding widespread use in the power-tool market where additional operating time and higher power are of major importance.

The disadvantage of NiMH cells versus NiCd cells is their much higher cost. Also, NiMH cells self-discharge much more rapidly during storage or non-use than NiCd cells. Furthermore, NiMH cells cannot be cycled as many times as their NiCd counterparts. NiMH cells are continually being improved and it is foreseeable that they will overcome, at least to a large degree, the disadvantages listed here.

Nickel–Iron (Edison) Cell

Developed by Thomas Edison, this cell was once used extensively in industrial truck and railway applications. However, it has been replaced almost entirely by the lead-acid battery. New methods of construction with less weight, though, are making this cell a possible alternative in some applications.

The Edison cell has a positive plate of nickel oxide, a negative plate of iron, and an electrolyte of potassium hydroxide in water with a small amount of lithium hydroxide added. The chemical reaction is reversible for recharging. Nominal output is 1.2 V per cell.

Nickel–Zinc Cell

This type has been used in limited railway applications. There has been renewed interest in it for use in electric cars because of its high energy density. However, one drawback is its limited cycle life for recharging. Nominal output is 1.6 V per cell.

Fuel Cells

A fuel cell is an electrochemical device that converts hydrogen and oxygen into water and in the process produces electricity. A single fuel cell is a piece of plastic between a couple of pieces of carbon plates that are sandwiched between two end plates acting as electrodes. These plates have channels that distribute the fuel and oxygen. As long as the reactants—pure hydrogen and oxygen—are supplied to the fuel cell, it will continually produce electricity. A conventional battery has all of its chemicals stored inside and it converts the chemical energy into electrical energy. This means that a battery will eventually go dead or deteriorate to a point where it is no longer useful and must be discarded. Chemicals constantly flow into a fuel cell so it never goes dead. (This assumes of course that there is always a flow of chemicals into the cell.) Most fuel cells in use today use hydrogen and oxygen. However, research is being done on a new type of fuel cell that uses methanol and oxygen. This type of fuel cell is in the early stages of development, however. A fuel cell provides a dc voltage at its output and can be used to power motors, lights, and other electrical devices. Note that fuel cells are used extensively in the space program for dc power. Fuel cells are very efficient and can provide hundreds of kilowatts of power.

Several different types of fuel cells are available today, typically classified by the type of electrolyte that they use. The proton exchange membrane fuel cell (PEMFC) is one of the most promising. This is the type that is likely to be used to power cars, buses, and maybe even your house in the future.

Solar Cells

This type converts the sun's light energy directly into electric energy. The cells are made of semiconductor materials, which generate voltage output with light input. Silicon, with an output of 0.5 V per cell, is mainly used now. Research is continuing, however, on other materials, such as cadmium sulfide and gallium arsenide, that might provide more output. In practice, the cells are arranged in modules that are assembled into a large solar array for the required power.

In most applications, the solar cells are used in combination with a lead-acid cell specifically designed for this use. When there is sunlight, the solar cells charge the battery and supply power to the load. When there is no light, the battery supplies the required power.

■ *12–5 Knowledge Check*

Answer at end of chapter.

Which type of cell suffers less from the memory effect, NiMH or NiCd?

■ *12–5 Self-Review*

Answers at end of chapter.

a. The NiCd cell is a primary type. (True or False)
b. The output of the NiCd cell is 1.2 V. (True or False)
c. NiMH cells cannot be cycled as many times as NiCd cells. (True or False)
d. The output of a solar cell is typically 0.5 V. (True or False)

12–6 Series-Connected and Parallel-Connected Cells

An applied voltage higher than the voltage of one cell can be obtained by connecting cells in series. The total voltage available across the battery of cells is equal to the sum of the individual values for each cell. Parallel cells have the same voltage as one cell but have more current capacity. The combination of cells is called a *battery*.

Series Connections

Figure 12–14 shows series-aiding connections for three dry cells. Here the three 1.5-V cells in series provide a total battery voltage of 4.5 V. Notice that the two end terminals, A and B, are left open to serve as the plus and minus terminals of the battery. These terminals are used to connect the battery to the load circuit, as shown in Fig. 12–14c.

In the lead-acid battery in Fig. 12–2, short, heavy metal straps connect the cells in series. The current capacity of a battery with cells in series is the same as that for one cell because the same current flows through all series cells.

Parallel Connections

For more current capacity, the battery has cells in parallel, as shown in Fig. 12–15. All positive terminals are strapped together, as are all the negative terminals. Any point on the positive side can be the plus terminal of the battery, and any point on the negative side can be the negative terminal.

The parallel connection is equivalent to increasing the size of the electrodes and electrolyte, which increases the current capacity. The voltage output of the battery, however, is the same as that for one cell.

Figure 12–14 Cells connected in series for higher voltage. Current rating is the same as for one cell. (*a*) Wiring. (*b*) Schematic symbol for battery with three series cells. (*c*) Battery connected to load resistance R_L.

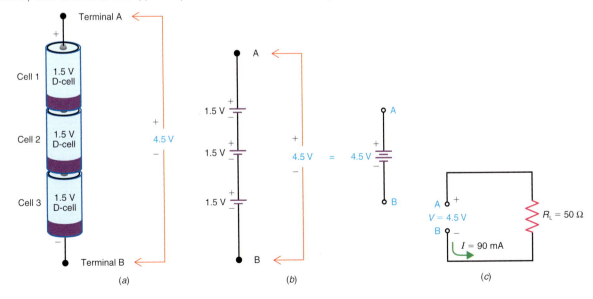

Figure 12-15 Cells connected in parallel for higher current rating. (*a*) Wiring. (*b*) Schematic symbol for battery with three parallel cells. (*c*) Battery connected to load resistance R_L.

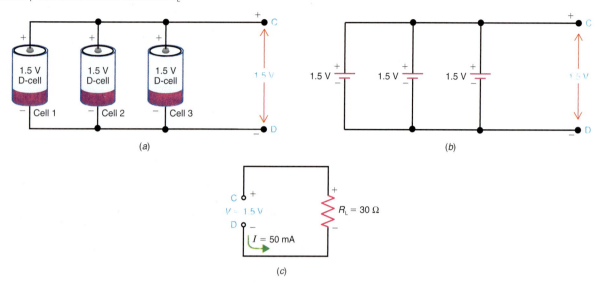

(*a*)

(*b*)

(*c*)

Identical cells in parallel supply equal parts of the load current. For example, with three identical parallel cells producing a load current of 300 mA, each cell has a drain of 100 mA. Bad cells should not be connected in parallel with good cells, however, since the cells in good condition will supply more current, which may overload the good cells. In addition, a cell with lower output voltage will act as a load resistance, draining excessive current from the cells that have higher output voltage.

Series-Parallel Connections

To provide a higher output voltage and more current capacity, cells can be connected in series-parallel combinations. Figure 12–16 shows four D-cells connected in series-parallel to form a battery that has a 3-V output with a current capacity of ½ A. Two of the 1.5-V cells in series provide 3 V total output voltage. This series string has a current capacity of ¼ A, however, assuming this current rating for one cell.

To double the current capacity, another string is connected in parallel. The two strings in parallel have the same 3-V output as one string, but with a current capacity of ½ A instead of the ¼ A for one string.

■ *12–6 Knowledge Check*

Answer at end of chapter.

In Fig. 12–14, what is the current rating of the battery if each cell has a current rating of 150 mA?

■ *12–6 Self-Review*

Answers at end of chapter.

a. How many carbon-zinc cells in series are required to obtain a 9-V dc output? How many lead-acid cells are required to obtain 12.6 Vdc?
b. How many identical cells in parallel would be required to double the current rating of a single cell?

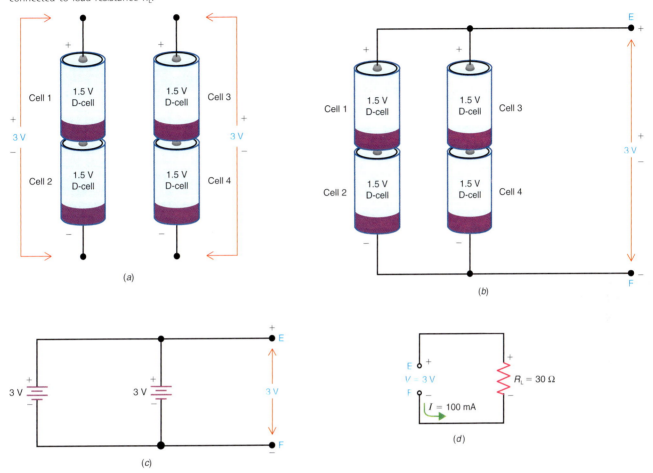

|||MultiSim **Figure 12–16** Cells connected in series-parallel combinations. (a) Wiring two 3-V strings, each with two 1.5-V cells in series. (b) Wiring two 3-V strings in parallel. (c) Schematic symbol for the battery in (b) with output of 3 V. (d) Equivalent battery connected to load resistance R_L.

c. How many cells rated 1.5 Vdc 300 mA would be required in a series-parallel combination that would provide a rating of 900 mA at 6 Vdc?

12–7 Current Drain Depends on Load Resistance

It is important to note that the current rating of batteries, or any voltage source, is only a guide to typical values permissible for normal service life. The actual amount of current produced when the battery is connected to a load resistance is equal to $I = V/R$ by Ohm's law.

Figure 12–17 illustrates three different cases of using the applied voltage of 1.5 V from a dry cell. In Fig. 12–17a, the load resistance R_1 is 7.5 Ω. Then I is 1.5/7.5 = ⅕ A or 200 mA.

A No. 6 carbon-zinc cell with a 1500 mA · h rating could supply this load of 200 mA continuously for about 7.5 h at a temperature of 70°F before dropping to an end voltage of 1.2 V. If an end voltage of 1.0 V could be used, the same load would be served for a longer period of time.

Figure 12–17 An example of how current drain from a battery used as a voltage source depends on the R of the load resistance. Different values of I are shown for the same V of 1.5 V. (a) V/R_1 equals I of 200 mA. (b) V/R_2 equals I of 10 mA. (c) V/R_3 equals I of 600 mA.

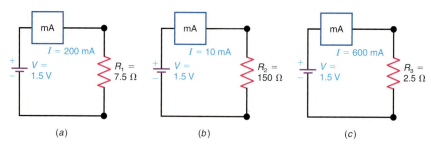

(a)   (b)   (c)

In Fig. 12–17b, a larger load resistance R_2 is used. The value of 150 Ω limits the current to 1.5/150 = 0.01 A or 10 mA. Again using the No. 6 carbon-zinc cell at 70°F, the load could be served continuously for 150 h with an end voltage of 1.2 V. The two principles here are

1. The cell delivers less current with higher resistance in the load circuit.
2. The cell can deliver a smaller load current for a longer time.

In Fig. 12–17c, the load resistance R_3 is reduced to 2.5 Ω. Then I is 1.5/2.5 = 0.6 A or 600 mA. The No. 6 cell could serve this load continuously for only 2.5 h for an end voltage of 1.2 V. The cell could deliver even more load current, but for a shorter time. The relationship between current and time is not linear. For any one example, though, the amount of current is determined by the circuit, not by the current rating of the battery.

■ *12-7 Knowledge Check*

Answer at end of chapter.

For which circuit in Fig. 12–17 will the battery provide the shortest service life?

■ *12-7 Self-Review*

Answers at end of chapter.

a. **A cell rated at 250 mA will produce this current for any value of R_L. (True or False)**
b. **A higher value of R_L allows the cell to operate at normal voltage for a longer time. (True or False)**

12–8 Internal Resistance of a Generator

Any source that produces voltage output continuously is a generator. It may be a cell separating charges by chemical action or a rotary generator converting motion and magnetism into voltage output, for common examples. In any case, all generators have internal resistance, which is labeled r_i in Fig. 12–18.

The internal resistance, r_i, is important when a generator supplies load current because its internal voltage drop, Ir_i, subtracts from the generated emf, resulting in lower voltage across the output terminals. Physically, r_i may be the resistance of the wire in a rotary generator, or r_i is the resistance of the electrolyte

Figure 12–18 Internal resistance r_i is in series with the generator voltage V_G. (a) Physical arrangement for a voltage cell. (b) Schematic symbol for r_i. (c) Equivalent circuit of r_i in series with V_G.

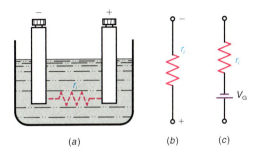

(a)   (b)   (c)

between electrodes in a chemical cell. More generally, the internal resistance r_i is the opposition to load current inside the generator.

Since any current in the generator must flow through the internal resistance, r_i is in series with the generated voltage, as shown in Fig. 12–18c. It may be of interest to note that, with just one load resistance connected across a generator, they are in series with each other because R_L is in series with r_i.

If there is a short circuit across the generator, its r_i prevents the current from becoming infinitely high. As an example, if a 1.5-V cell is temporarily short-circuited, the short-circuit current I_{sc} could be about 15 A. Then r_i is V/I_{sc}, which equals 1.5/15, or 0.1 Ω for the internal resistance. These are typical values for a carbon-zinc D-size cell. (The value of r_i would be lower for a D-size alkaline cell.)

Why Terminal Voltage Drops with More Load Current

Figure 12–19 illustrates how the output of a 100-V source can drop to 90 V because of the internal 10-V drop across r_i. In Fig. 12–19a, the voltage across the output terminals is equal to the 100 V of V_G because there is no load current in an open circuit. With no current, the voltage drop across r_i is zero. Then the full generated voltage is available across the output terminals. This value is the generated emf, *open-circuit voltage,* or *no-load voltage.*

We cannot connect the test leads inside the source to measure V_G. However, measuring this no-load voltage without any load current provides a

⦀ MultiSim **Figure 12–19** Example of how an internal voltage drop decreases voltage at the output terminal of the generator. (a) Open-circuit voltage output equals V_G of 100 V because there is no load current. (b) Terminal voltage V_L between points A and B is reduced to 90 V because of 10-V drop across 100-Ω r_i with 0.1-A I_L.

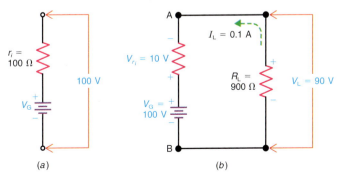

(a)   (b)

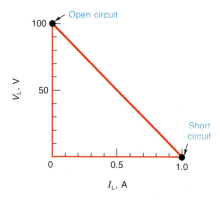

Figure 12–20 How terminal voltage V_L drops with more load current. The graph is plotted for values in Table 12–4.

method of determining the internally generated emf. We can assume that the voltmeter draws practically no current because of its very high resistance.

In Fig. 12–19*b* with a load, however, current of 0.1 A flows to produce a drop of 10 V across the 100 Ω of r_i. Note that R_T is 900 + 100 = 1000 Ω. Then I_L equals 100/1000, which is 0.1 A.

As a result, the voltage output V_L equals 100 − 10 = 90 V. This terminal voltage or load voltage is available across the output terminals when the generator is in a closed circuit with load current. The 10-V internal drop is subtracted from V_G because they are series-opposing voltages.

The graph in Fig. 12–20 shows how the terminal voltage V_L drops with increasing load current I_L. The reason is the greater internal voltage drop across r_i, as shown by the calculated values listed in Table 12–4. For this example, V_G is 100 V and r_i is 100 Ω.

Across the top row, infinite ohms for R_L means an open circuit. Then I_L is zero, there is no internal drop V_i, and V_L is the same 100 V as V_G.

Across the bottom row, zero ohms for R_L means a short circuit. Then the short-circuit current of 1 A results in zero output voltage because the entire generator voltage is dropped across the internal resistance. Or we can say that with a short circuit of zero ohms across the load, the current is limited to V_G/r_i.

The lower the internal resistance of a generator, the better it is in producing full output voltage when supplying current for a load. For example, the very low r_i, about 0.01 Ω, for a 12-V lead-acid battery, is the reason it can supply high values of load current and maintain its output voltage.

For the opposite case, a higher r_i means that the terminal voltage of a generator is much less with load current. As an example, an old dry battery with r_i of 500 Ω would appear normal when measured by a voltmeter but be useless because of low voltage when normal load current flows in an actual circuit.

How to Measure r_i

The internal resistance of any generator can be measured indirectly by determining how much the output voltage drops for a specified amount of load current. The difference between the no-load voltage and the load voltage is the amount of internal voltage drop $I_L r_i$. Dividing by I_L gives the value of r_i. As a formula,

$$r_i = \frac{V_{NL} - V_L}{I_L} \tag{12–1}$$

Table 12–4	How V_L Drops with More I_L (for Fig. 12–20)					
V_G, V	r_i, Ω	R_L, Ω	$R_T = R_L + r_i$, Ω	$I_L = V_G/R_T$, A	$V_i = I_L r_i$, V	$V_L = V_G - V_i$, V
100	100	∞	∞	0	0	100
100	100	900	1000	0.1	10	90
100	100	600	700	0.143	14.3	85.7
100	100	300	400	0.25	25	75
100	100	100	200	0.5	50	50
100	100	0	100	1.0	100	0

Example 12-1

Calculate r_i if the output of a generator drops from 100 V with zero load current to 80 V when $I_L = 2$ A.

ANSWER

$$r_i = \frac{100 - 80}{2}$$
$$= \frac{20}{2}$$
$$r_i = 10 \ \Omega$$

A convenient technique for measuring r_i is to use a variable load resistance R_L. Vary R_L until the load voltage is one-half the no-load voltage. This value of R_L is also the value of r_i, since they must be equal to divide the generator voltage equally. For the same 100-V generator with the 10-Ω r_i used in Example 12–1, if a 10-Ω R_L were used, the load voltage would be 50 V, equal to one-half the no-load voltage.

You can solve this circuit by Ohm's law to see that I_L is 5 A with 20 Ω for the combined R_T. Then the two voltage drops of 50 V each add to equal the 100 V of the generator.

■ 12–8 Knowledge Check

Answer at end of chapter.

The output voltage of a battery drops from 12 V with no load to 11.1 V when supplying 1.8 A of current. Calculate r_i.

■ 12–8 Self-Review

Answers at end of chapter.

a. For formula (12–1), V_L must be more than V_{NL}. (True or False)
b. For formula (12–1), when V_L is one-half V_{NL}, the r_i is equal to R_L. (True or False)
c. The generator's internal resistance, r_i, is in series with the load. (True or False)
d. More load current produces a larger voltage drop across r_i. (True or False)

12–9 Constant-Voltage and Constant-Current Sources

A generator with very low internal resistance is considered a constant-voltage source. Then the output voltage remains essentially the same when the load current changes. This idea is illustrated in Fig. 12–21a for a 6-V lead-acid battery with an r_i of 0.005 Ω. If the load current varies over the wide range of 1 to 100 A, the internal Ir_i drop across 0.005 Ω is less than 0.5 V for any of these values.

Figure 12–21 Constant-voltage generator with low r_i. The V_L stays approximately the same 6 V as I varies with R_L. (a) Circuit. (b) Graph for V_L.

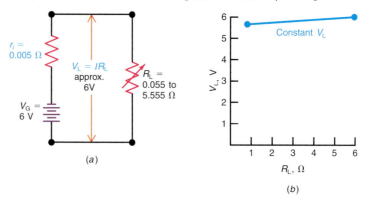

Constant-Current Generator

It has very high resistance, compared with the external load resistance, resulting in constant current, although the output voltage varies.

The constant-current generator shown in Fig. 12–22, has such high resistance, with an r_i of 0.9 MΩ, that it is the main factor determining how much current can be produced by V_G. Here R_L varies in a 3:1 range from 50 to 150 kΩ. Since the current is determined by the total resistance of R_L and r_i in series, however, I is essentially constant at 1.05 to 0.95 mA, or approximately 1 mA. This relatively constant I is shown by the graph in Fig. 12–22b.

Note that the terminal voltage V_L varies in approximately the same 3:1 range as R_L. Also, the output voltage is much less than the generator voltage because of the high internal resistance compared with R_L. This is a necessary condition, however, in a circuit with a constant-current generator.

A common practice is to insert a series resistance to keep the current constant, as shown in Fig. 12–23a. Resistance R_1 must be very high compared with R_L. In this example, I_L is 50 μA with 50 V applied, and R_T is practically equal to the 1 MΩ of R_1. The value of R_L can vary over a range as great as 10:1 without changing R_T or I appreciably.

A circuit with an equivalent constant-current source is shown in Fig. 12–23b. Note the arrow symbol for a current source. As far as R_L is concerned, its terminals A and B can be considered as receiving either 50 V in series with 1 MΩ or 50 μA in a shunt with 1 MΩ.

Figure 12–22 Constant-current generator with high r_i. The I stays approximately the same 1 mA as V_L varies with R_L. (a) Circuit. (b) Graph for I.

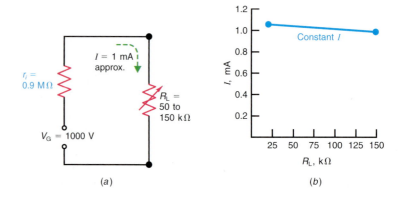

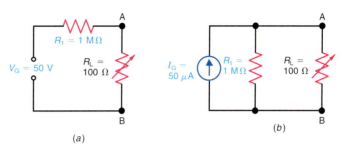

Figure 12–23 Voltage source in (*a*) equivalent to current source in (*b*) for load resistance R_L across terminals A and B.

(*a*)

(*b*)

■ *12–9 Knowledge Check*

Answer at end of chapter.

In Fig. 12–21A, what is the load voltage, V_L, if $R_L = 10\ k\Omega$?

■ *12–9 Self-Review*

Answers at end of chapter.

Is the internal resistance high or low for
a. a constant-voltage source?
b. a constant-current source?

GOOD TO KNOW

To get the maximum radiated power from an antenna in a communications system, the radiation resistance of the antenna must match the output resistance of the radio transmitter. This is just one of many instances where it is crictical that $r_i = R_L$.

12–10 Matching a Load Resistance to the Generator r_i

In the diagram in Fig. 12–24, when R_L equals r_i, the load and generator are matched. The matching is significant because the generator then produces maximum power in R_L, as verified by the values listed in Table 12–5.

Maximum Power in R_L

When R_L is $100\ \Omega$ to match the $100\ \Omega$ of r_i, maximum power is transferred from the generator to the load. With higher resistance for R_L, the output voltage V_L is higher, but the current is reduced. Lower resistance for R_L allows more current, but V_L is less. When r_i and R_L both equal $100\ \Omega$, this combination of current and voltage produces the maximum power of 100 W across R_L.

With generators that have very low resistance, however, matching is often impractical. For example, if a 6-V lead-acid battery with a 0.003-Ω internal resistance were connected to a 0.003-Ω load resistance, the battery could be damaged by excessive current as high as 1000 A.

Maximum Voltage Across R_L

If maximum voltage, rather than power, is desired, the load should have as high a resistance as possible. Note that R_L and r_i form a voltage divider for the generator voltage, as illustrated in Fig. 12–24*b*. The values for IR_L listed in Table 12–5 show how the output voltage V_L increases with higher values of R_L.

Maximum Efficiency

Note also that the efficiency increases as R_L increases because there is less current, resulting in less power lost in r_i. When R_L equals r_i, the efficiency is only 50%, since one-half the total generated power is dissipated in r_i, the

Figure 12–24 Circuit for varying R_L to match r_i. (a) Schematic diagram. (b) Equivalent voltage divider for voltage output across R_L. (c) Graph of power output P_L for different values of R_L. All values are listed in Table 12–5.

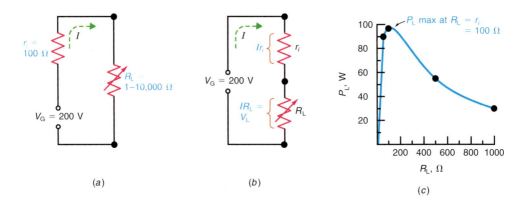

(a)   (b)   (c)

internal resistance of the generator. In conclusion, then, matching the load and generator resistances is desirable when the load requires maximum power rather than maximum voltage or efficiency, assuming that the match does not result in excessive current.

■ *12–10 Knowledge Check*

   Answer at end of chapter.

   If a 10-V generator has an r_i of 20 Ω, how much power will be delivered to the load if $R_L = r_i$?

■ *12–10 Self-Review*

   Answers at end of chapter.

   a.   When $R_L = r_i$, the P_L is maximum. (True or False)
   b.   The V_L is maximum when R_L is maximum. (True or False)

Table 12–5	Effect of Load Resistance on Generator Output*						
	R_L, Ω	$I = V_G/R_T$, A	Ir_i, V	IR_L, V	P_L, W	P_i, W	P_T, W	Efficiency = P_L/P_T, %
	1	1.98	198	2	4	392	396	1
	50	1.33	133	67	89	178	267	33
$R_L = r_i \rightarrow$	100	1	100	100	100	100	200	50
	500	0.33	33	167	55	11	66	83
	1,000	0.18	18	180	32	3.24	35.24	91
	10,000	0.02	2	198	4	0.04	4.04	99

* Values calculated approximately for circuit in Fig. 12–24, with $V_G = 200$ V and $r_i = 100$ Ω.

Summary

- A voltaic cell consists of two different conductors as electrodes immersed in an electrolyte. The voltage output depends only on the chemicals in the cell. The current capacity increases with larger sizes. A primary cell cannot be recharged. A secondary or storage cell can be recharged.

- A battery is a group of cells in series or in parallel. With cells in series, the voltages add, but the current capacity is the same as that of one cell. With cells in parallel, the voltage output is the same as that of one cell, but the total current capacity is the sum of the individual values.

- The carbon-zinc dry cell is a common type of primary cell. Zinc is the negative electrode; carbon is the positive electrode. Its output voltage is approximately 1.5 V.

- The lead-acid cell is the most common form of storage battery. The positive electrode is lead peroxide; spongy lead is the negative electrode. Both are in a dilute solution of sulfuric acid as the electrolyte. The voltage output is approximately 2.1 V per cell.

- To charge a lead-acid battery, connect it to a dc voltage equal to approximately 2.5 V per cell. Connecting the positive terminal of the battery to the positive side of the charging source and the negative terminal to the negative side results in charging current through the battery.

- The nickel-cadmium cell is rechargeable and has an output of 1.2 V.

- A constant-voltage generator has very low internal resistance.

- A constant-current generator has very high internal resistance.

- Any generator has an internal resistance r_i. With load current I_L, the internal $I_L r_i$ drop reduces the voltage across the output terminals. When I_L makes the terminal voltage drop to one-half the no-load voltage, the external R_L equals the internal r_i.

- Matching a load to a generator means making the R_L equal to the generator's r_i. The result is maximum power delivered to the load from the generator.

Important Terms

- Ampere-Hour (A-H) Rating - a common rating for batteries that indicates how much load current a battery can supply during a specified discharge time. For example, a battery with a 100 A-H rating can deliver 1 A for 100 h, 2 A for 50 h, 4 A for 25 h, etc.

- Battery - a device containing a group of individual voltaic cells that provides a constant or steady dc voltage at its output terminals.

- Charging - the process of reversing the current, and thus the chemical action, in a cell or battery to re-form the electrodes and the electrolyte.

- Constant-Current Generator - a generator whose internal resistance is very high compared with the load resistance. Because its internal resistance is so high, it can supply constant current to a load whose resistance value varies over a wide range.

- Constant-Voltage Generator - a generator whose internal resistance is very low compared with the load resistance. Because its internal resistance is so low, it can supply constant voltage to a load whose resistance value varies over a wide range.

- Discharging - the process of neutralizing the separated charges on the electrodes of a cell or battery as a result of supplying current to a load resistance.

- Float Charging - a method of charging in which the charger and the battery are always connected to each other to supply current to the load. With this method, the charger provides the current for the load and the current necessary to keep the battery fully charged.

- Fuel Cell - an electrochemical device that converts hydrogen and oxygen into water and produces electricity. A fuel cell provides a steady dc output voltage that can power motors, lights, or other appliances. Unlike a regular battery, however, a fuel cell has chemicals constantly flowing into it so it never goes dead.

- Hydrometer - a device used to check the state of charge of a cell within a lead-acid battery.

- Internal Resistance, r_i - the resistance inside a voltage source that limits the amount of current it can deliver to a load.

- Open-Circuit Voltage - the voltage across the output terminals of a voltage source when no load is present.

- Primary Cell - a type of voltaic cell that cannot be recharged because the internal chemical reaction to restore the electrodes is not possible.

- Secondary Cell - a type of voltaic cell that can be recharged because the internal chemical reaction to restore the electrodes is possible.

- Specific Gravity - the ratio of the weight of a volume of a substance to that of water.

- Storage Cell - another name for a secondary cell.

- Voltaic Cell - a device that converts chemical energy into electric energy. The output voltage of a voltaic cell depends on the type of elements used for the electrodes and the type of electrolyte.

Related Formulas

$$r_i = \frac{V_{NL} - V_L}{I_L}$$

$V = $ Specific gravity $+ 0.84$

Self-Test

1. Which of the following cells is not a primary cell?
   a. Carbon-zinc.
   b. Alkaline.
   c. Zinc-chloride.
   d. Lead-acid.

2. The dc output voltage of a C-size alkaline cell is
   a. 1.2 V.
   b. 1.5 V.
   c. 2.1 V.
   d. about 3 V.

3. Which of the following cells is a secondary cell?
   a. Silver oxide.
   b. Lead-acid.
   c. Nickel-cadmium.
   d. both b and c.

4. What happens to the internal resistance, r_i, of a voltaic cell as the cell deteriorates?
   a. It increases.
   b. It decreases.
   c. It stays the same.
   d. It usually disappears.

5. The dc output voltage of a lead-acid cell is
   a. 1.35 V.
   b. 1.5 V.
   c. 2.1 V.
   d. about 12 V.

6. Cells are connected in series to
   a. increase the current capacity.
   b. increase the voltage output.
   c. decrease the voltage output.
   d. decrease the internal resistance.

7. Cells are connected in parallel to
   a. increase the current capacity.
   b. increase the voltage output.
   c. decrease the voltage output.
   d. decrease the current capacity.

8. Five D-size alkaline cells in series have a combined voltage of
   a. 1.5 V.
   b. 5.0 V.
   c. 7.5 V.
   d. 11.0 V.

9. The main difference between a primary cell and a secondary cell is that
   a. a primary cell can be recharged and a secondary cell cannot.
   b. a secondary cell can be recharged and a primary cell cannot.
   c. a primary cell has an unlimited shelf life and a secondary cell does not.
   d. primary cells produce a dc voltage and secondary cells produce an ac voltage.

10. A constant-voltage source
    a. has very high internal resistance.
    b. supplies constant-current to any load resistance.
    c. has very low internal resistance.
    d. none of the above.

11. A constant-current source
    a. has very low internal resistance.
    b. supplies constant current to a wide range of load resistances.
    c. has very high internal resistance.
    d. both b and c.

12. The output voltage of a battery drops from 6.0 V with no load to 5.4 V with a load current of 50 mA. How much is the internal resistance, r_i?
    a. 12 Ω.
    b. 108 Ω.
    c. 120 Ω.
    d. This is impossible to determine.

13. Maximum power is transferred from a generator to a load when
    a. $R_L = r_i$.
    b. R_L is maximum.
    c. R_L is minimum.
    d. R_L is 10 or more times the value of r_i.

14. What is the efficiency of power transfer for the matched load condition?
    a. 100%.
    b. 0%.
    c. 50%.
    d. This is impossible to determine.

15. The internal resistance of a battery
    a. cannot be measured with an ohmmeter.
    b. can be measured with an ohmmeter.
    c. can be measured indirectly by determining how much the output voltage drops for a given load current.
    d. both a and c.

Questions

1. (a) What is the advantage of connecting cells in series? (b) What is connected to the end terminals of the series cells?

2. (a) What is the advantage of connecting cells in parallel? (b) Why can the load be connected across any one of the parallel cells?

3. How many cells are necessary in a battery to double the voltage and current ratings of a single cell? Show the wiring diagram.

4. Draw a diagram showing two 12-V lead-acid batteries being charged by a 15-V source.

5. Why is a generator with very low internal resistance called a constant-voltage source?

6. Why does discharge current lower the specific gravity in a lead-acid cell?

7. Would you consider the lead-acid battery a constant-current source or a constant-voltage source? Why?

8. List five types of chemical cells, giving two features of each.

9. Referring to Fig. 12–21b, draw the corresponding graph that shows how I varies with R_L.

10. Referring to Fig. 12–22b, draw the corresponding graph that shows how V_L varies with R_L.

11. Referring to Fig. 12–24c, draw the corresponding graph that shows how V_L varies with R_L.

Problems

SECTION 12–6 SERIES-CONNECTED AND PARALLEL-CONNECTED CELLS

In Problems 12–1 to 12–5, assume that each individual cell is identical and that the current capacity for each cell is not being exceeded for the load conditions presented.

12–1 In Fig. 12–25, solve for the load voltage, V_L, the load current, I_L, and the current supplied by each cell in the battery.

Figure 12–25

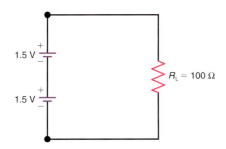

12–2 Repeat Problem 12–1 for the circuit in Fig. 12–26.

Figure 12–26

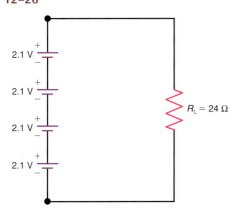

12–3 Repeat Problem 12–1 for the circuit in Fig. 12–27.

Figure 12–27

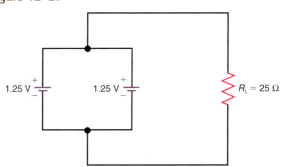

12–4 Repeat Problem 12–1 for the circuit in Fig. 12–28.

Figure 12–28

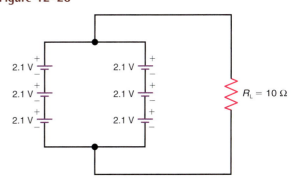

12-5 Repeat Problem 12-1 for the circuit in Fig. 12-29.

Figure 12-29

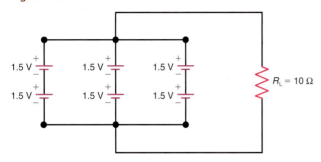

SECTION 12-8 INTERNAL RESISTANCE OF A GENERATOR

12-6 With no load, the output voltage of a battery is 9 V. If the output voltage drops to 8.5 V when supplying 50 mA of current to a load, how much is its internal resistance?

12-7 The output voltage of a battery drops from 6 V with no load to 5.2 V with a load current of 400 mA. Calculate the internal resistance, r_i.

12-8 **MultiSim** A 9-V battery has an internal resistance of 0.6 Ω. How much current flows from the 9 V battery in the event of a short circuit?

12-9 A 1.5-V "AA" alkaline cell develops a terminal voltage of 1.35 V while delivering 25 mA to a load resistance. Calculate r_i.

12-10 Refer to Fig. 12-30. With S_1 in Position 1, $V = 50$ V. With S_1 in Position 2, $V = 37.5$ V. Calculate r_i.

Figure 12-30

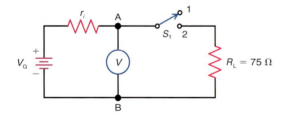

12-11 A generator has an open-circuit voltage of 18 V. Its terminal voltage drops to 15 V when a 75-Ω load is connected. Calculate r_i.

SECTION 12-9 CONSTANT-VOLTAGE AND CONSTANT-CURRENT SOURCES

12-12 Refer to Fig. 12-31. If $r_i = 0.01$ Ω, calculate I_L and V_L for the following values of load resistance:
  a. $R_L = 1$ Ω
  b. $R_L = 5$ Ω
  c. $R_L = 10$ Ω
  d. $R_L = 100$ Ω

Figure 12-31

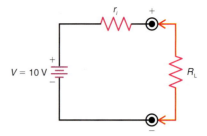

12-13 Refer to Fig. 12-31. If $r_i = 10$ MΩ, calculate I_L and V_L for the following values of load resistance:
  a. $R_L = 0$ Ω
  b. $R_L = 100$ Ω
  c. $R_L = 1$ kΩ
  d. $R_L = 100$ kΩ

12-14 Redraw the circuit in Fig. 12-32 using the symbol for a current source.

Figure 12-32

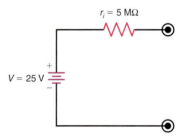

SECTION 12-10 MATCHING A LOAD RESISTANCE TO THE GENERATOR, r_i

12-15 Refer to Fig. 12-33. Calculate I_L, V_L, P_L, P_T, and % efficiency for the following values of R_L:
  a. $R_L = 10$ Ω
  b. $R_L = 25$ Ω
  c. $R_L = 50$ Ω
  d. $R_L = 75$ Ω
  e. $R_L = 100$ Ω

Figure 12-33

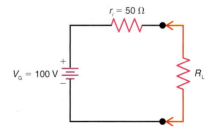

12–16 In Problem 12–15, what value of R_L provides

    a. the highest load voltage, V_L?

    b. the smallest voltage drop across the 50-Ω r_i?

    c. the maximum transfer of power?

    d. the maximum efficiency?

12–17 In Fig. 12–34,

    a. what value of R_L will provide maximum transfer of power from generator to load?

    b. what is the load power for the matched load condition?

    c. what percentage of the total power is delivered to R_L when $R_L = r_i$?

Figure 12–34

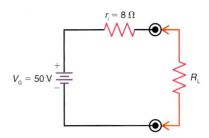

Critical Thinking

12–18 In Fig. 12–35, calculate (**a**) V_L, (**b**) I_L, and (**c**) the current supplied to R_L by each separate voltage source.

12–19 In Fig. 12–36, calculate (**a**) the value of R_L for which the maximum transfer of power occurs; (**b**) the maximum power delivered to R_L.

Figure 12–35   Circuit diagram for Critical Thinking Prob. 12–18.

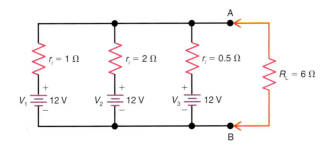

Figure 12–36   Circuit diagram for Critical Thinking Prob. 12–19.

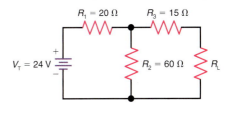

Answers to Knowledge Check Problems

12–1 Cycling

12–2 Its physical size

12–3 D-Size cell

12–4 1.9 V

12–5 NiMH

12–6 150 mA

12–7 Figure 12–17c

12–8 $r_i = 0.5\ \Omega$

12–9 6 V

12–10 1.25 W

Answers to Self-Reviews

12–1 a. 1.5 V
    b. 2.1 V
    c. secondary

12–2 a. T
    b. T
    c. F

12–3 a. alkaline
    b. alkaline
    c. size C
    d. silver oxide

12–4 a. six
    b. 15 A
    c. 1.280

12–5 a. F
    b. T
    c. T
    d. T

12–6 a. six, six
    b. two
    c. twelve

12–7 a. F
    b. T

12–8 a. F
    b. T
    c. T
    d. T

12–9 a. low
    b. high

12–10 a. T
    b. T

Cumulative Review Summary (Chapters 11–12)

- A conductor is a material whose resistance is very low. Some examples of good conductors are silver, copper, and aluminum; copper is generally used for wire. An insulator is a material whose resistance is very high. Some examples of good insulators include air, mica, rubber, porcelain, and plastics.

- The gage sizes for copper wires are listed in Table 11–1. As the gage sizes increase from 1 to 40, the diameter and circular area decrease. Higher gage numbers correspond to thinner wire.

- For switches, the term *pole* refers to the number of completely isolated circuits that can be controlled by the switch. The term *throw* refers to the number of closed-contact positions that exist per pole. A switch can have any number of poles and throws.

- A good fuse has very low resistance, with an *IR* voltage of practically zero. An open fuse has nearly infinite resistance. If an open fuse exists in a series circuit or in the main line of a parallel circuit, its voltage drop equals the applied voltage.

- The resistance of a wire is directly proportional to its length and inversely proportional to its cross-sectional area.

- All metals in their purest form have positive temperature coefficients, which means that their resistance increases with an increase in temperature. Carbon has a negative temperature coefficient, which means that its resistance decreases as the temperature increases.

- An ion is an atom that has either gained or lost electrons. A negative ion is an atom with more electrons than protons. Conversely, a positive ion is an atom with more protons than electrons. Ions can move to provide electric current in liquids and gases. The motion of ions is called *ionization current.*

- A battery is a combination of individual voltaic cells. A primary cell cannot be recharged, whereas a secondary cell can be recharged several times. The main types of cells for batteries include alkaline, silver oxide, nickel cadmium, lithium, and lead-acid.

- With individual cells in series, the total battery voltage equals the sum of the individual cell voltages. This assumes that the cells are connected in a series-aiding manner. The current rating of the series-aiding cells is the same as that for the cell with the lowest current rating.

- With individual cells in parallel, the voltage is the same as that across one cell. However, the current rating of the combination equals the sum of the individual current-rating values. Only cells that have the same voltage should be connected in parallel.

- All types of dc and ac generators have an internal resistance r_i. The value of r_i may be the resistance of the electrolyte in a battery or the wire in a rotary generator.

- When a generator supplies current to a load, the terminal voltage drops because some voltage is dropped across the internal resistance r_i.

- Matching a load to a generator means making R_L equal to r_i. When $R_L = r_i$, maximum power is delivered from the generator to the load.

- A constant-voltage source has very low internal r_i, whereas a constant-current source has very high internal r_i.

Cumulative Review Self-Test

Answers at back of book.

1. Which of the following is the best conductor of electricity? (a) carbon; (b) silicon; (c) rubber; (d) copper.

2. Which of the following wires has the largest cross-sectional area? (a) No. 28 gage; (b) No. 23 gage; (c) No. 12 gage; (d) No. 16 gage.

3. The filament of a lightbulb measures 2.5 Ω when cold. With 120 V applied across the filament, the bulb dissipates 75 W of power. What is the hot resistance of the bulb? (a) 192 Ω; (b) 0.625 Ω; (c) 2.5 Ω; (d) 47 Ω.

4. A DPST switch has how many terminal connections for soldering? (a) 3; (b) 1; (c) 4; (d) 6.

5. Which of the following materials has a negative temperature coefficient? (a) steel; (b) carbon; (c) tungsten; (d) nichrome.

6. The *IR* voltage across a good fuse equals (a) the applied voltage; (b) one-half the applied voltage; (c) infinity; (d) zero.

7. A battery has a no-load voltage of 9 V. Its terminal voltage drops to 8.25 V when a load current of 200 mA is drawn from the battery. The internal resistance r_i equals (a) 0.375 Ω; (b) 3.75 Ω; (c) 41.25 Ω; (d) 4.5 Ω.

8. When $R_L = r_i$, (a) maximum voltage is across R_L; (b) maximum power is delivered to R_L; (c) the efficiency is 100%; (d) the minimum power is delivered to R_L.

9. A constant-current source has (a) very high internal resistance; (b) constant output voltage; (c) very low internal resistance; (d) output voltage that is always zero.

10. Cells can be connected in series-parallel to (a) increase the voltage above that of a single cell; (b) increase the current capacity above that of a single cell; (c) reduce the voltage and current rating below that of a single cell; (d) both (a) and (b).

13 Magnetism

The phenomenon known as magnetism was first discovered by the ancient Greeks in about 100 BC. Then it was observed that a peculiar stone had the property of attracting small fragments of iron to itself. The peculiar stone was called a lodestone, and the power of attraction it possessed was called magnetism. Any material possessing the property of magnetism is a magnet. Every magnet has both a north (N) pole and a south (S) pole. Just as "like" electric charges repel each other and "unlike" charges attract, "like" magnetic poles repel each other and "unlike" poles attract. The discovery of natural magnets led to the invention of the compass, which is a direction-finding device. Since the earth itself is a huge natural magnet, a freely suspended magnet will align itself with the magnetic North and South Poles of the earth.

Every magnet has invisible magnetic field lines that extend outward from the magnetic poles. The number of magnetic field lines and their concentration can be measured with special test equipment. In this chapter, you will be introduced to the basic units for magnetic fields. You will also learn about the different types of magnets and how magnetic materials are classified.

Objectives

After studying this chapter you should be able to

- *Describe* the magnetic field surrounding a magnet.
- *Define* the units of *magnetic flux* and *flux density.*
- *Convert* between magnetic units.
- *Describe* how an iron bar is magnetized by induction.
- *Define* the term *relative permeability.*
- *Explain* the difference between a bar magnet and an electromagnet.
- *List* the three classifications of magnetic materials.
- *Explain* the electrical and magnetic properties of ferrites.
- *Describe* the Hall effect.

Outline

Important Terms

Curie temperature
diamagnetic
electromagnet
ferrite
ferromagnetic
flux density (B)

gauss (G)
Hall effect
induction
magnetic flux ϕ
maxwell (Mx)
paramagnetic

permanent magnet
relative permeability (μ_r)
tesla (T)
toroid
weber (Wb)

Figure 13–1 Poles of a magnet.
(*a*) Electromagnet (EM) produced by
current from a battery. (*b*) Permanent
magnet (PM) without any external source
of current.

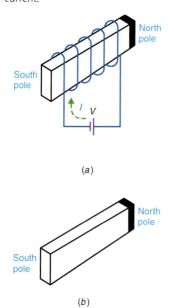

(a)

(b)

13–1 The Magnetic Field

As shown in Fig. 13–1 and Fig. 13–2, the north and south poles of a magnet are the points of concentration of magnetic strength. The practical effects of this ferromagnetism result from the magnetic field of force between the two poles at opposite ends of the magnet. Although the magnetic field is invisible, evidence of its force can be seen when small iron filings are sprinkled on a glass or paper sheet placed over a bar magnet (Fig. 13–2*a*). Each iron filing becomes a small bar magnet. If the sheet is tapped gently to overcome friction so that the filings can move, they become aligned by the magnetic field.

Many filings cling to the ends of the magnet, showing that the magnetic field is strongest at the poles. The field exists in all directions but decreases in strength with increasing distance from the poles of the magnet.

Field Lines

To visualize the magnetic field without iron filings, we show the field as lines of force, as in Fig. 13–2*b*. The direction of the lines outside the magnet shows the path a north pole would follow in the field, repelled away from the north pole of the magnet and attracted to its south pole. Although we cannot actually have a unit north pole by itself, the field can be explored by noting how the north pole on a small compass needle moves.

The magnet can be considered the generator of an external magnetic field, provided by the two opposite magnetic poles at the ends. This idea corresponds to the two opposite terminals on a battery as the source of an external electric field provided by opposite charges.

Magnetic field lines are unaffected by nonmagnetic materials such as air, vacuum, paper, glass, wood, or plastics. When these materials are placed in the magnetic field of a magnet, the field lines are the same as though the material were not there.

However, the magnetic field lines become concentrated when a magnetic substance such as iron is placed in the field. Inside the iron, the field lines are more dense, compared with the field in air.

North and South Magnetic Poles

The earth itself is a huge natural magnet, with its greatest strength at the North and South Poles. Because of the earth's magnetic poles, if a small bar magnet is suspended so that it can turn easily, one end will always point north. This end of the bar magnet is defined as the *north-seeking pole,* as shown in

Figure 13–2 Magnetic field of force around a bar magnet. (*a*) Field outlined by iron filings. (*b*) Field indicated by lines of force.

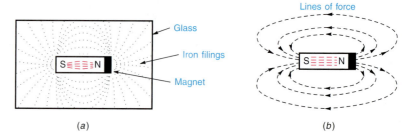

(a)

(b)

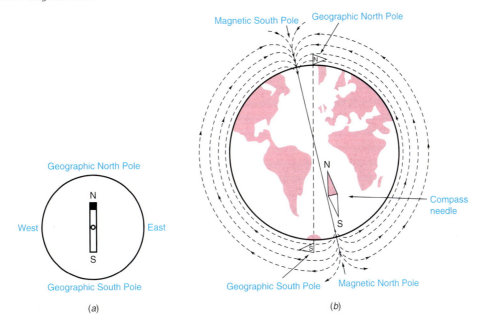

(a)

(b)

Fig. 13–3*a*. The opposite end is the *south-seeking pole*. When polarity is indicated on a magnet, the north-seeking end is the north pole (N) and the opposite end is the south pole (S). It is important to note that the earth's geographic North Pole has south magnetic polarity and the geographic South Pole has north magnetic polarity. This is shown in Fig. 13–3*b*.

Similar to the force between electric charges is the force between magnetic poles causing attraction of opposite poles and repulsion between similar poles:

1. A north pole (N) and a south pole (S) tend to attract each other.
2. A north pole (N) tends to repel another north pole (N), and a south pole (S) tends to repel another south pole (S).

These forces are illustrated by the fields of iron filings between opposite poles in Fig. 13–4*a* and between similar poles in Fig. 13–4*b*.

■ *13–1 Knowledge Check*

Answer at end of chapter.

What is the magnetic polarity of the South Geographic Pole?

■ *13–1 Self-Review*

Answers at end of chapter.

a. On a magnet, the north-seeking pole is labeled *N*. (True or False)
b. Like poles have a force of repulsion. (True or False)

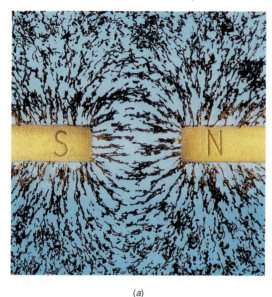
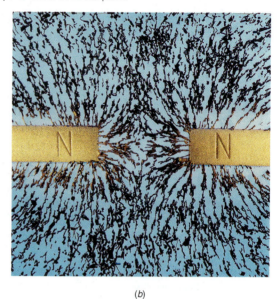

(*a*)　　　　　　　　　　　　　　　(*b*)

13–2 Magnetic Flux ϕ

The entire group of magnetic field lines, which can be considered flowing outward from the north pole of a magnet, is called *magnetic flux*. Its symbol is the Greek letter ϕ (phi). A strong magnetic field has more lines of force and more flux than a weak magnetic field.

The Maxwell

One maxwell (Mx) unit equals one magnetic field line. In Fig. 13–5, as an example, the flux illustrated is 6 Mx because there are six field lines flowing in or out for each pole. A 1-lb magnet can provide a magnetic flux ϕ of about 5000 Mx. This unit is named after James Clerk Maxwell (1831–1879), an important Scottish mathematical physicist, who contributed much to electrical and field theory.

PIONEERS IN ELECTRONICS

Physicist *James Clerk Maxwell (1831–1879)* unified scientific theories of electricity and magnetism into a unified theory of the electromagnetic field. In 1865, Maxwell proved that electromagnetic phenomena travel in waves at the speed of light. In 1873, he went on to state that light itself is an electromagnetic wave.

||| **MultiSim** **Figure 13–5** Total flux ϕ is six lines or 6 Mx. Flux density B at point P is two lines per square centimeter or 2 G.

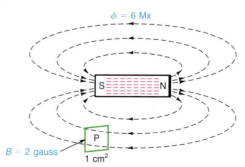

$\phi = 6$ Mx

$B = 2$ gauss

1 cm²

The Weber

This is a larger unit of magnetic flux. One weber (Wb) equals 1×10^8 lines or maxwells. Since the weber is a large unit for typical fields, the microweber unit can be used. Then $1 \,\mu\text{Wb} = 10^{-6}$ Wb. This unit is named after Wilhelm Weber (1804–1890), a German physicist.

To convert microwebers to lines or maxwells, multiply by the conversion factor 10^8 lines per weber, as follows:

$$1 \,\mu\text{Wb} = 1 \times 10^{-6} \text{ Wb} \times 10^8 \, \frac{\text{lines}}{\text{Wb}}$$

$$= 1 \times 10^2 \text{ lines}$$

$$1\mu\text{Wb} = 100 \text{ lines or Mx}$$

Note that the conversion is arranged to make the weber units cancel, since we want maxwell units in the answer.

Even the microweber unit is larger than the maxwell unit. For the same 1-lb magnet, a magnetic flux of 5000 Mx corresponds to $50 \,\mu\text{Wb}$. The calculations for this conversion of units are

$$\frac{5000 \text{ Mx}}{100 \text{ Mx}/\mu\text{Wb}} = 50 \,\mu\text{Wb}$$

Note that the maxwell units cancel. Also, the $1/\mu\text{Wb}$ becomes inverted from the denominator to μWb in the numerator.

Conversion Between Units

Converting from maxwells (Mx) to webers (Wb) or vice versa, is easier if you use the following conversion formulas:

$$\#\text{Wb} = \#\text{Mx} \times \frac{1 \text{ Wb}}{1 \times 10^8 \text{ Mx}}$$

$$\#\text{Mx} = \#\text{Wb} \times \frac{1 \times 10^8 \text{ Mx}}{1 \text{ Wb}}$$

Example 13–1

Make the following conversions: (a) 25,000 Mx to Wb; (b) 0.005 Wb to Mx.

ANSWER

$$\text{(a)} \quad \#\text{Wb} = \#\text{Mx} \times \frac{1 \text{ Wb}}{1 \times 10^8 \text{ Mx}}$$

$$= 25{,}000 \text{ Mx} \times \frac{1 \text{ Wb}}{1 \times 10^8 \text{ Mx}}$$

$$\#\text{Wb} = 250 \times 10^{-6} \text{ Wb or } 250 \,\mu\text{Wb}$$

$$\text{(b)} \quad \#\text{Mx} = \#\text{Wb} \times \frac{1 \times 10^8 \text{ Mx}}{1 \text{ Wb}}$$

$$= 0.005 \text{ Wb} \times \frac{1 \times 10^8 \text{ Mx}}{1 \text{ Wb}}$$

$$\#\text{Mx} = 5.0 \times 10^5 \text{ Mx}$$

PIONEERS IN ELECTRONICS

Systems of Magnetic Units

The basic units in metric form can be defined in two ways:

1. The centimeter-gram-second system defines small units. This is the cgs system.
2. The meter-kilogram-second system is for larger units of a more practical size. This is the mks system.

Furthermore, the Système Internationale (SI) units provide a worldwide standard in mks dimensions. They are practical values based on the ampere of current.

For magnetic flux ϕ, the maxwell (Mx) is a cgs unit, and the weber (Wb) is an mks or SI unit. The SI units are preferred for science and engineering, but the cgs units are still used in many practical applications of magnetism.

■ *13–2 Knowledge Check*

> *Answer at end of chapter.*
>
> Convert a magnetic flux of 1×10^{-8} Wb to Mx units.

■ *13–2 Self-Review*

> *Answers at end of chapter.*
>
> The value of 2000 magnetic lines is how much flux in
> a. maxwell units?
> b. microweber units?

13–3 Flux Density B

As shown in Fig. 13–5, the *flux density* is the number of magnetic field lines per unit area of a section perpendicular to the direction of flux. As a formula,

$$B = \frac{\phi}{A} \qquad\qquad (13\text{–}1)$$

where ϕ is the flux through an area A and the flux density is B.

The Gauss

In the cgs system, this unit is one line per square centimeter, or 1 Mx/cm^2. As an example, in Fig. 13–5, the total flux ϕ is six lines, or 6 Mx. At point P in this field, however, the flux density B is 2 G because there are two lines per square centimeter. The flux density is higher close to the poles, where the flux lines are more crowded.

As an example of flux density, B for a 1-lb magnet would be 1000 G at the poles. This unit is named after Karl F. Gauss (1777–1855), a German mathematician.

Example 13-2

With a flux of 10,000 Mx through a perpendicular area of 5 cm², what is the flux density in gauss?

ANSWER

$$B = \frac{\phi}{A} = \frac{10{,}000 \text{ Mx}}{5 \text{ cm}^2} = 2000 \frac{\text{Mx}}{\text{cm}^2}$$

$$B = 2000 \text{ G}$$

As typical values, B for the earth's magnetic field can be about 0.2 G; a large laboratory magnet produces B of 50,000 G. Since the gauss is so small, kilogauss units are often used, where $1 \text{ kG} = 10^3 \text{ G}$.

The Tesla

In SI, the unit of flux density B is webers per square meter (Wb/m^2). One weber per square meter is called a *tesla*, abbreviated T. This unit is named for Nikola Tesla (1857–1943), a Yugoslav-born American inventor in electricity and magnetism.
    When converting between cgs and mks units, note that

$$1 \text{ m} = 100 \text{ cm} \quad \text{or} \quad 1 \times 10^2 \text{ cm}$$
$$1 \text{ m}^2 = 10{,}000 \text{ cm}^2 \quad \text{or} \quad 10^4 \text{ cm}^2$$

These conversions are from the larger m (meter) and m^2 (square meter) to the smaller units of cm (centimeter) and cm^2 (square centimeter). To go the opposite way,

$$1 \text{ cm} = 0.01 \text{ m} \quad \text{or} \quad 1 \times 10^{-2} \text{ m}$$
$$1 \text{ cm}^2 = 0.0001 \text{ m}^2 \quad \text{or} \quad 1 \times 10^{-4} \text{ m}^2$$

As an example, 5 cm^2 is equal to 0.0005 m^2 or 5×10^{-4} m^2. The calculations for the conversion are

$$5 \text{ cm}^2 \times \frac{0.0001 \text{ m}^2}{\text{cm}^2} = 0.0005 \text{ m}^2$$

In powers of 10, the conversion is

$$5 \text{ cm}^2 \times \frac{1 \times 10^{-4} \text{ m}^2}{\text{cm}^2} = 5 \times 10^{-4} \text{ m}^2$$

In both cases, note that the units of cm^2 cancel to leave m^2 as the desired unit.

Example 13-3

With a flux of 400 μWb through an area of 0.0005 m^2, what is the flux density B in tesla units?

ANSWER

$$B = \frac{\phi}{A} = \frac{400 \times 10^{-6} \text{ Wb}}{5 \times 10^{-4} \text{ m}^2}$$
$$= \frac{400}{5} \times 10^{-2}$$
$$= 80 \times 10^{-2} \text{ Wb/m}^2$$
$$B = 0.80 \text{ T}$$

The tesla is a larger unit than the gauss, as $1\,\text{T} = 1 \times 10^4\,\text{G}$.

For example, the flux density of 20,000 G is equal to 2 T. The calculations for this conversion are

$$\frac{20{,}000\,\text{G}}{1 \times 10^4\,\text{G/T}} = \frac{2 \times 10^4\,\text{T}}{1 \times 10^4} = 2\,\text{T}$$

Note that the G units cancel to leave T units for the desired answer. Also, the $1/\text{T}$ in the denominator becomes inverted to T units in the numerator.

Conversion Between Units

Converting from teslas (T) to gauss (G), or vice versa, is easier if you use the following conversion formulas:

$$\#\text{G} = \#\text{T} \times \frac{1 \times 10^4\,\text{G}}{1\,\text{T}}$$

$$\#\text{T} = \#\text{G} \times \frac{1\,\text{T}}{1 \times 10^4\,\text{G}}$$

Example 13-4

Make the following conversions: (a) 0.003 T to G; (b) 15,000 G to T.

ANSWER

(a) $\#\text{G} = \#\text{T} \times \dfrac{1 \times 10^4\,\text{G}}{\text{T}}$

$\qquad\ = 0.003\,\text{T} \times \dfrac{1 \times 10^4\,\text{G}}{\text{T}}$

$\#\text{G} = 30\,\text{G}$

(b) $\#\text{T} = \#\text{G} \times \dfrac{1\,\text{T}}{1 \times 10^4\,\text{G}}$

$\qquad\ = 15{,}000\,\text{G} \times \dfrac{1\,\text{T}}{1 \times 10^4\,\text{G}}$

$\#\text{T} = 1.5\,\text{T}$

Comparison of Flux and Flux Density

Remember that the flux ϕ includes total area, whereas the flux density B is for a specified unit area. The difference between ϕ and B is illustrated in Fig. 13–6 with cgs units. The total area A here is 9 cm², equal to 3 cm × 3 cm. For one unit box of 1 cm², 16 lines are shown. Therefore, the flux density B is 16 lines or maxwells per square centimeter, which equals 16 G. The total area includes nine of these boxes. Therefore, the total flux ϕ is 144 lines or maxwells, equal to 9 × 16 for $B \times A$.

For the opposite case, if the total flux ϕ is given as 144 lines or maxwells, the flux density is found by dividing 144 by 9 cm². This division of 144/9 equals 16 lines or maxwells per square centimeter, which is 16 G.

Figure 13–6 Comparison of total flux ϕ and flux density B. The total area of 9 cm^2 has 144 lines or 144 Mx. For 1 cm^2, the flux density is 144/9 = 16 G.

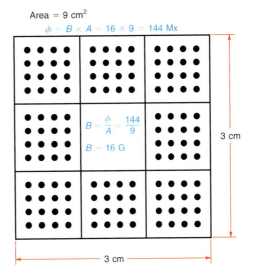

Area = 9 cm^2

$\phi = B \times A = 16 \times 9 = 144$ Mx

$B = \dfrac{\phi}{A} = \dfrac{144}{9}$

$B = 16$ G

3 cm

3 cm

■ **13–3 Knowledge Check**

Answer at end of chapter.

Convert a flux density of 1×10^{-4} T to gauss units.

■ **13–3 Self-Review**

Answers at end of chapter.

a. The ϕ is 9000 Mx through 3 cm^2. How much is B in gauss units?
b. How much is B in tesla units for ϕ of 90 μWb through 0.0003 m^2?

13–4 Induction by the Magnetic Field

The electric or magnetic effect of one body on another without any physical contact between them is called *induction*. For instance, a permanent magnet can induce an unmagnetized iron bar to become a magnet without the two touching. The iron bar then becomes a magnet, as shown in Fig. 13–7. What happens is that the magnetic lines of force generated by the permanent magnet make the internal molecular magnets in the iron bar line up in the same direction, instead of the random directions in unmagnetized iron. The magnetized iron bar then has magnetic poles at the ends, as a result of magnetic induction.

||| **MultiSim** **Figure 13–7** Magnetizing an iron bar by induction.

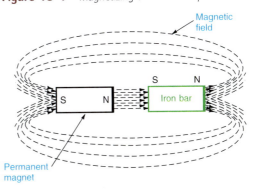

Magnetic field

S   N

S   N   Iron bar

Permanent magnet

Note that the induced poles in the iron have polarity opposite from the poles of the magnet. Since opposite poles attract, the iron bar will be attracted. Any magnet attracts to itself all magnetic materials by induction.

Although the two bars in Fig. 13–7 are not touching, the iron bar is in the magnetic flux of the permanent magnet. It is the invisible magnetic field that links the two magnets, enabling one to affect the other. Actually, this idea of magnetic flux extending outward from the magnetic poles is the basis for many inductive effects in ac circuits. More generally, the magnetic field between magnetic poles and the electric field between electric charges form the basis for wireless radio transmission and reception.

Polarity of Induced Poles

Note that the north pole of the permanent magnet in Fig. 13–7 induces an opposite south pole at this end of the iron bar. If the permanent magnet were reversed, its south pole would induce a north pole. The closest induced pole will always be of opposite polarity. This is the reason why either end of a magnet can attract another magnetic material to itself. No matter which pole is used, it will induce an opposite pole, and opposite poles are attracted.

Relative Permeability

Soft iron, as an example, is very effective in concentrating magnetic field lines by induction in the iron. This ability to concentrate magnetic flux is called *permeability*. Any material that is easily magnetized has high permeability, therefore, because the field lines are concentrated by induction.

Numerical values of permeability for different materials compared with air or vacuum can be assigned. For example, if the flux density in air is 1 G but an iron core in the same position in the same field has a flux density of 200 G, the relative permeability of the iron core equals 200/1, or 200.

The symbol for relative permeability is μ_r (mu), where the subscript r indicates relative permeability. Typical values for μ_r are 100 to 9000 for iron and steel. There are no units because μ_r is a comparison of two flux densities and the units cancel. The symbol K_m may also be used for relative permeability to indicate this characteristic of a material for a magnetic field, corresponding to K_e for an electric field.

■ *13–4 Knowledge Check*

> *Answer at end of chapter.*
>
> **What is the unit of relative permeability?**

■ *13–4 Self-Review*

> *Answers at end of chapter.*
>
> a. Induced poles always have polarity opposite from the inducing poles. (True or False)
> b. The relative permeability of air or vacuum is approximately 300. (True or False)

Figure 13–8 The horseshoe magnet in (*a*) has a smaller air gap than the bar magnet in (*b*).

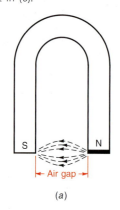

(a)

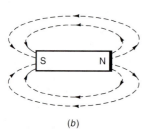

(b)

13–5 Air Gap of a Magnet

As shown in Fig. 13–8, the air space between the poles of a magnet is its air gap. The shorter the air gap, the stronger the field in the gap for a given pole strength. Since air is not magnetic and cannot concentrate magnetic lines, a larger air gap provides additional space for the magnetic lines to spread out.

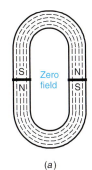

(a)

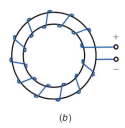

(b)

Figure 13–10 Storing permanent magnets in a closed loop, with opposite poles touching. (a) Four bar magnets. (b) Two bar magnets. (c) Horseshoe magnet with iron keeper across air gap.

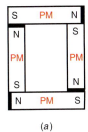

(a)

(b)

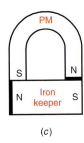

(c)

Referring to Fig. 13–8a, note that the horseshoe magnet has more crowded magnetic lines in the air gap, compared with the widely separated lines around the bar magnet in Fig. 13–8b. Actually, the horseshoe magnet can be considered a bar magnet bent around to place the opposite poles closer. Then the magnetic lines of the poles reinforce each other in the air gap. The purpose of a short air gap is to concentrate the magnetic field outside the magnet for maximum induction in a magnetic material placed in the gap.

Ring Magnet without Air Gap

When it is desired to concentrate magnetic lines within a magnet, however, the magnet can be formed as a closed magnetic loop. This method is illustrated in Fig. 13–9a by the two permanent horseshoe magnets placed in a closed loop with opposite poles touching. Since the loop has no open ends, there can be no air gap and no poles. The north and south poles of each magnet cancel as opposite poles touch.

Each magnet has its magnetic lines inside, plus the magnetic lines of the other magnet, but outside the magnets, the lines cancel because they are in opposite directions. The effect of the closed magnetic loop, therefore, is maximum concentration of magnetic lines in the magnet with minimum lines outside.

The same effect of a closed magnetic loop is obtained with the *toroid* or ring magnet in Fig. 13–9b, made in the form of a doughnut. Iron is often used for the core. This type of electromagnet has maximum strength in the iron ring and little flux outside. As a result, the toroidal magnet is less sensitive to induction from external magnetic fields and, conversely, has little magnetic effect outside the coil.

Note that, even if the winding is over only a small part of the ring, practically all the flux is in the iron core because its permeability is so much greater than that of air. The small part of the field in the air is called *leakage flux.*

Keeper for a Magnet

The principle of the closed magnetic ring is used to protect permanent magnets in storage. In Fig. 13–10a, four permanent-magnet bars are in a closed loop, while Fig. 13–10b shows a stacked pair. Additional even pairs can be stacked this way, with opposite poles touching. The closed loop in Fig. 13–10c shows one permanent horseshoe magnet with a soft-iron *keeper* across the air gap. The keeper maintains the strength of the permanent magnet as it becomes magnetized by induction to form a closed loop. Then any external magnetic field is concentrated in the closed loop without inducing opposite poles in the permanent magnet. If permanent magnets are not stored this way, the polarity can be reversed with induced poles produced by a strong external field from a dc source; an alternating field can demagnetize the magnet.

■ *13–5 Knowledge Check*

Answer at end of chapter.

Where is the magnetic field most concentrated in a toroid?

■ *13–5 Self-Review*

Answers at end of chapter.

a. A short air gap has a stronger field than a large air gap for the same magnetizing force. (True or False)
b. A toroid is made in the form of a doughnut. (True or False)

13-6 Types of Magnets

The two broad classes are permanent magnets and electromagnets. An electromagnet needs current from an external source to maintain its magnetic field. With a permanent magnet, not only is its magnetic field present without any external current, but the magnet can maintain its strength indefinitely. Sharp mechanical shock as well as extreme heat, however, can cause demagnetization.

Electromagnets

Current in a wire conductor has an associated magnetic field. If the wire is wrapped in the form of a coil, as in Fig. 13–11, the current and its magnetic field become concentrated in a smaller space, resulting in a stronger field. With the length much greater than its width, the coil is called a *solenoid*. It acts like a bar magnet, with opposite poles at the ends.

More current and more turns make a stronger magnetic field. Also, the iron core concentrates magnetic lines inside the coil. Soft iron is generally used for the core because it is easily magnetized and demagnetized.

The coil in Fig. 13–11, with the switch closed and current in the coil, is an electromagnet that can pick up the steel nail shown. If the switch is opened, the magnetic field is reduced to zero, and the nail will drop off. This ability of an electromagnet to provide a strong magnetic force of attraction that can be turned on or off easily has many applications in lifting magnets, buzzers, bells or chimes, and relays. A *relay* is a switch with contacts that are opened or closed by an electromagnet.

Another common application is magnetic tape recording. The tape is coated with fine particles of iron oxide. The recording head is a coil that produces a magnetic field in proportion to the current. As the tape passes through the air gap of the head, small areas of the coating become magnetized by induction. On playback, the moving magnetic tape produces variations in electric current.

Permanent Magnets

These are made of hard magnetic materials, such as cobalt steel, magnetized by induction in the manufacturing process. A very strong field is needed for induction in these materials. When the magnetizing field is removed, however, residual induction makes the material a permanent magnet. A common PM material is *alnico*, a commercial alloy of aluminum, nickel, and iron, with cobalt, copper, and titanium added to produce about 12 grades. The Alnico V grade is often used for PM loudspeakers (Fig. 13–12). In this application, a typical size of PM slug for a steady magnetic field is a few ounces to about 5 lb, with a flux ϕ of 500 to 25,000 lines or maxwells. One advantage of a PM loudspeaker is that only two connecting leads are needed for the voice coil because the steady magnetic field of the PM slug is obtained without any field-coil winding.

Commercial permanent magnets will last indefinitely if they are not subjected to high temperatures, physical shock, or a strong demagnetizing field. If the magnet becomes hot, however, the molecular structure can be rearranged, resulting in loss of magnetism that is not recovered after cooling. The point at which a magnetic material loses its ferromagnetic properties is the *Curie temperature*. For iron, this temperature is about 800°C, when the relative permeability drops to unity. A permanent magnet does not become exhausted with use because its magnetic properties are determined by the structure of the internal atoms and molecules.

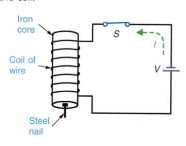

Figure 13–11 Electromagnet holding a nail when switch *S* is closed for current in the coil.

Iron core

Coil of wire

S

I

V

Steel nail

Figure 13–12 Example of a PM loudspeaker.

Classification of Magnetic Materials

When we consider materials simply as either magnetic or nonmagnetic, this division is based on the strong magnetic properties of iron. However, weak magnetic materials can be important in some applications. For this reason, a more exact classification includes the following three groups:

1. *Ferromagnetic materials.* These include iron, steel, nickel, cobalt, and commercial alloys such as alnico and Permalloy. They become strongly magnetized in the same direction as the magnetizing field, with high values of permeability from 50 to 5000. Permalloy has a μ_r of 100,000 but is easily saturated at relatively low values of flux density.
2. *Paramagnetic materials.* These include aluminum, platinum, manganese, and chromium. Their permeability is slightly more than 1. They become weakly magnetized in the same direction as the magnetizing field.
3. *Diamagnetic materials.* These include bismuth, antimony, copper, zinc, mercury, gold, and silver. Their permeability is less than 1. They become weakly magnetized but in the direction opposite from the magnetizing field.

The basis of all magnetic effects is the magnetic field associated with electric charges in motion. Within the atom, the motion of its orbital electrons generates a magnetic field. There are two kinds of electron motion in the atom. First is the electron revolving in its orbit. This motion provides a diamagnetic effect. However, this magnetic effect is weak because thermal agitation at normal room temperature results in random directions of motion that neutralize each other.

More effective is the magnetic effect from the motion of each electron spinning on its own axis. The spinning electron serves as a tiny permanent magnet. Opposite spins provide opposite polarities. Two electrons spinning in opposite directions form a pair, neutralizing the magnetic fields. In the atoms of ferromagnetic materials, however, there are many unpaired electrons with spins in the same direction, resulting in a strong magnetic effect.

In terms of molecular structure, iron atoms are grouped in microscopically small arrangements called *domains*. Each domain is an elementary *dipole magnet*, with two opposite poles. In crystal form, the iron atoms have domains parallel to the axes of the crystal. Still, the domains can point in different directions because of the different axes. When the material becomes magnetized by an external magnetic field, though, the domains become aligned in the same direction. With PM materials, the alignment remains after the external field is removed.

■ *13-6 Knowledge Check*

> *Answer at end of chapter.*
>
> Which of the following materials is classified as ferromagnetic: steel, aluminum, or copper?

■ *13-6 Self-Review*

> *Answers at end of chapter.*
>
> a. An electromagnet needs current to maintain its magnetic field. (True or False)
> b. A relay coil is an electromagnet. (True or False)
> c. Iron is a diamagnetic material. (True or False)

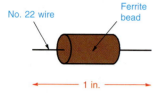

13-7 Ferrites

Ferrite is the name for nonmetallic materials that have the ferromagnetic properties of iron. Ferrites have very high permeability, like iron. However, a ferrite is a nonconducting ceramic material, whereas iron is a conductor. The permeability of ferrites is in the range of 50 to 3000. The specific resistance is $10^5 \ \Omega \cdot$ cm, which makes a ferrite an insulator.

A common application is a ferrite core, usually adjustable, in the coils of RF transformers. The ferrite core is much more efficient than iron when the current alternates at high frequency. The reason is that less I^2R power is lost by eddy currents in the core because of its very high resistance.

A ferrite core is used in small coils and transformers for signal frequencies up to 20 MHz, approximately. The high permeability means that the transformer can be very small. However, ferrites are easily saturated at low values of magnetizing current. This disadvantage means that ferrites are not used for power transformers.

Another application is in ferrite beads (Fig. 13–13). A bare wire is used as a string for one or more beads. The bead concentrates the magnetic field of the current in the wire. This construction serves as a simple, economical RF choke, instead of a coil. The purpose of the choke is to reduce the current just for an undesired radio frequency.

■ *13–7 Knowledge Check*

> *Answer at end of chapter.*
>
> **Are ferrites good conductors or insulators of electricity?**

■ *13–7 Self-Review*

> *Answers at end of chapter.*
>
> a. **Which has more R, ferrites or soft iron?**
> b. **Which has more I^2R losses, an insulator or a conductor?**

13–8 Magnetic Shielding

The idea of preventing one component from affecting another through their common electric or magnetic field is called *shielding*. Examples are the braided copper-wire shield around the inner conductor of a coaxial cable, a metal shield can that encloses an RF coil, or a shield of magnetic material enclosing a cathode-ray tube.

The problem in shielding is to prevent one component from inducing an effect in the shielded component. The shielding materials are always metals, but there is a difference between using good conductors with low resistance, such as copper and aluminum, and using good magnetic materials such as soft iron.

A good conductor is best for two shielding functions. One is to prevent induction of static electric charges. The other is to shield against the induction of a varying magnetic field. For static charges, the shield provides opposite induced charges, which prevent induction inside the shield. For a varying magnetic field, the shield has induced currents that oppose the inducing field. Then there is little net field strength to produce induction inside the shield.

The best shield for a steady magnetic field is a good magnetic material of high permeability. A steady field is produced by a permanent magnet, a coil with steady direct current, or the earth's magnetic field. A magnetic shield of high permeability concentrates the magnetic flux. Then there is little flux to

PIONEERS IN ELECTRONICS

In 1879, *Edwin Hall* was a graduate student at Johns Hopkins University when he discovered the Hall effect. When a wire carrying a current is placed in an applied magnetic field, a voltage across the wire is created that is proportional to the strength of the magnetic field. This effect is at the heart of a number of technologies such as antilock brake sensors and some computer keyboards.

Figure 13–14 The Hall effect. Hall voltage V_H generated across the element is proportional to the perpendicular flux density B.

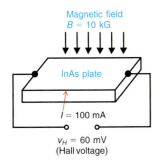

Magnetic field
B = 10 kG

InAs plate

I = 100 mA

v_H = 60 mV
(Hall voltage)

induce poles in a component inside the shield. The shield can be considered a short circuit for the lines of magnetic flux.

■ *13–8 Knowledge Check*

Answer at end of chapter.

What type of metal is used to shield a component against a steady magnetic field?

■ *13–8 Self-Review*

Answers at end of chapter.

a. A magnetic material with high permeability is a good shield for a steady magnetic field. (True or False)
b. A conductor is a good shield against a varying magnetic field. (True or False)

13–9 The Hall Effect

In 1879, E. H. Hall observed that a small voltage is generated across a conductor carrying current in an external magnetic field. The Hall voltage was very small with typical conductors, and little use was made of this effect. However, with the development of semiconductors, larger values of Hall voltage can be generated. The semiconductor material indium arsenide (InAs) is generally used. As illustrated in Fig. 13–14, the InAs element inserted in a magnetic field can generate 60 mV with B equal to 10 kG and an I of 100 mA. The applied flux must be perpendicular to the direction of the current. With current in the direction of the length of conductor, the generated voltage is developed across the width.

The amount of Hall voltage V_H is directly proportional to the value of flux density B. This means that values of B can be measured by V_H. As an example, the gaussmeter in Fig. 13–15 uses an InAs probe in the magnetic field to generate a proportional Hall voltage V_H. This value of V_H is then read by the meter, which is calibrated in gauss. The original calibration is made in terms of a reference magnet with a specified flux density.

■ *13–9 Knowledge Check*

Answer at end of chapter.

In Fig. 13–14, the applied flux must be (parallel/perpendicular) to the direction of the current.

■ *13–9 Self-Review*

Answers at end of chapter.

a. In Fig. 13–14, how much is the generated Hall voltage?
b. Does the gaussmeter in Fig. 13–15 measure flux or flux density?

Figure 13–15 A gaussmeter to measure flux density, with a probe containing an indium arsenide element.

Summary

- Iron, nickel, and cobalt are common examples of magnetic materials. Air, paper, wood, and plastics are nonmagnetic.

- The pole of a magnet that seeks the geographic North Pole of the earth is called a *north pole*; the opposite pole is a *south pole*.

- Opposite magnetic poles are attracted; similar poles repel.

- An electromagnet needs current from an external source to provide a magnetic field. Permanent magnets retain their magnetism indefinitely.

- Any magnet has an invisible field of force outside the magnet, indicated by magnetic field lines. Their direction is from the north to the south pole.

- The open ends of a magnet where it meets a nonmagnetic material provide magnetic poles. At opposite open ends, the poles have opposite polarity.

- A magnet with an air gap has opposite poles with magnetic lines of force across the gap. A closed magnetic ring has no poles.

- Magnetic induction enables the field of a magnet to induce magnetic poles in a magnetic material without touching.

- Permeability is the ability to concentrate magnetic flux. A good magnetic material has high permeability.

- Magnetic shielding means isolating a component from a magnetic field. The best shield against a steady magnetic field is a material with high permeability.

- The Hall voltage is a small voltage generated across the width of a conductor carrying current through its length, when magnetic flux is applied perpendicular to the current. This effect is generally used in the gaussmeter to measure flux density.

- Table 13–1 summarizes the units of magnetic flux ϕ and flux density B.

Table 13–1	Magnetic Flux ϕ and Flux Density B		
Name	Symbol	CGS Units	MKS or SI Units
Flux, or total lines	$\phi = B \times$ area	1 maxwell (Mx) = 1 line	1 weber (Wb) = 10^8 Mx
Flux density, or lines per unit area	$B = \dfrac{\phi}{\text{area}}$	1 gauss (G) = $\dfrac{1\text{ Mx}}{\text{cm}^2}$	1 tesla (T) = $\dfrac{1\text{ Wb}}{\text{m}^2}$

Important Terms

- Curie Temperature - the temperature at which a magnetic material loses its ferromagnetic properties.

- Diamagnetic - a classification of materials that become weakly magnetized but in the direction opposite to the magnetizing field. Diamagnetic materials have a permeability less than 1. Examples include antimony, bismuth, copper, gold, mercury, silver and zinc.

- Electromagnet - a magnet that requires an electric current flowing in the turns of a coil to create a magnetic field. With no current in the coil, there is no magnetic field.

- Ferrite - a nonmetallic material that has the ferromagnetic properties of iron.

- Ferromagnetic - a classification of materials that become strongly magnetized in the same direction as the magnetizing field. Ferromagnetic materials have high values of permeability in the range of 50 to 5,000 or even higher. Examples include iron, steel, nickel, and cobalt.

- Flux Density (B) - the number of magnetic field lines per unit area of a section perpendicular to the direction of flux.

- Gauss (G) - the cgs unit of flux density. $1\text{ G} = \dfrac{1\text{ Mx}}{\text{cm}^2}$.

- Hall Effect - the effect that describes a small voltage generated across the width of a conductor that is carrying current in an external magnetic field. To develop the Hall voltage, the current in the conductor and the external flux must be at right angles to each other.

- Induction - the electric or magnetic effect of one body on another without any physical contact between them.

- Magnetic Flux (ϕ) - another name used to describe magnetic field lines.

- Maxwell (Mx) - the cgs unit of magnetic flux. 1 Mx = 1 magnetic field line.

- Paramagnetic - a classification of materials that become weakly magnetized in the same direction as the magnetizing field. Their permeability is slightly more than 1. Examples include aluminum, platinum, manganese, and chromium.

- Permanent Magnet - a hard magnetic material such as cobalt steel which is magnetized by induction in the manufacturing process. A permanent magnet retains its magnetic properties indefinitely as long as it is not subjected to very high temperatures, physical shock, or a strong demagnetizing field.

- Relative Permeability (μ_r) - the ability of a material to concentrate magnetic flux. Mathematically, relative permeability, designated μ_r, is the ratio of the flux density (B) in a material such as iron and the flux density, B, in air. There are no units for μ_r because it compares two flux densities and the units cancel.

- Tesla (T) - the SI unit of flux density.

$$1\ T = \frac{1\ Wb}{m^2}$$

- Toroid - an electromagnet wound in the form of a doughnut. It has no magnetic poles and the maximum strength of the magnetic field is concentrated in its iron core.

- Weber (Wb) - the SI unit of magnet flux. 1 Wb = 1×10^8 Mx or lines.

Related Formulas

$$B = \frac{\phi}{A}$$

$$\#Mx = \#Wb \times \frac{1 \times 10^8\ Mx}{1\ Wb}$$

$$\#Wb = \#Mx \times \frac{1\ Wb}{1 \times 10^8\ Mx}$$

$$\#G = \#T \times \frac{1 \times 10^4\ G}{1\ T}$$

$$\#T = \#G \times \frac{1\ T}{1 \times 10^4\ G}$$

Self-Test

Answers at back of book.

1. The maxwell (Mx) is a unit of
   a. flux density.
   b. permeability.
   c. magnetic flux.
   d. field intensity.

2. With bar magnets,
   a. like poles attract each other and unlike poles repel each other.
   b. unlike poles attract each other and like poles repel each other.
   c. there are no north or south poles on the ends of the magnet.
   d. none of the above.

3. The tesla (T) is a unit of
   a. flux density.
   b. magnetic flux.
   c. permeability.
   d. magnetomotive force.

4. 1 maxwell (Mx) is equal to
   a. 1×10^8 Wb.
   b. $\frac{1\ Wb}{m^2}$.
   c. 1×10^4 G.
   d. One magnetic field line.

5. 1 Wb is equal to
   a. 1×10^8 Mx.
   b. one magnetic field line.
   c. $\frac{1\ Mx}{cm^2}$.
   d. 1×10^4 kG.

6. The electric or magnetic effect of one body on another without any physical contact between them is called
   a. its permeability.
   b. induction.
   c. the Hall effect.
   d. hysteresis.

7. A commercial permanent magnet will last indefinitely if it is not subjected to
   a. a strong demagnetizing field.
   b. physical shock.
   c. high temperatures.
   d. all of the above.

8. What is the name for a nonmetallic material that has the ferromagnetic properties of iron?
   a. lodestone.
   b. toroid.
   c. ferrite.
   d. solenoid.

9. One tesla (T) is equal to
   a. $\frac{1\ Mx}{m^2}$.
   b. $\frac{1\ Mx}{cm^2}$.
   c. $\frac{1\ Wb}{m^2}$.
   d. $\frac{1\ Wb}{cm^2}$.

10. The ability of a material to concentrate magnetic flux is called its
    a. induction.
    b. permeability.
    c. Hall effect.
    d. diamagnetic.

11. If the north (N) pole of a permanent magnet is placed near a piece of soft iron, what is the polarity of the nearest induced pole?
    a. south (S) pole.
    b. north (N) pole.
    c. It could be either a north (N) or a south (S) pole.
    d. This is impossible to determine.

12. A magnet that requires current in a coil to create the magnetic field is called a(n)

  a. permanent magnet.

  b. electromagnet.

  c. solenoid.

  d. either b or c.

13. The point at which a magnetic material loses its ferromagnetic properties is called the

  a. melting point.

  b. freezing point.

  c. Curie temperature.

  d. leakage point.

14. A material that becomes strongly magnetized in the same direction as the magnetizing field is classified as

  a. diamagnetic.

  b. ferromagnetic.

  c. paramagnetic.

  d. toroidal.

15. Which of the following materials are nonmagnetic?

  a. air.

  b. wood.

  c. glass.

  d. all of the above.

16. The gauss (G) is a unit of

  a. flux density.

  b. magnetic flux.

  c. permeability.

  d. none of the above.

17. One gauss (G) is equal to

  a. $\dfrac{1 \text{ Mx}}{\text{m}^2}$.

  b. $\dfrac{1 \text{ Wb}}{\text{cm}^2}$.

  c. $\dfrac{1 \text{ Mx}}{\text{cm}^2}$.

  d. $\dfrac{1 \text{ Wb}}{\text{m}}$.

18. 1 μWb equals

  a. 1×10^8 Mx.

  b. 10,000 Mx.

  c. 1×10^{-8} Mx.

  d. 100 Mx.

19. A toroid

  a. is an electromagnet.

  b. has no magnetic poles.

  c. uses iron for the core around which the coil is wound.

  d. all of the above.

20. When a small voltage is generated across the width of a conductor carrying current in an external magnetic field, the effect is called

  a. the Doppler effect.

  b. the Miller effect.

  c. the Hall effect.

  d. the Schultz effect.

21. The weber (Wb) is a unit of

  a. magnetic flux.

  b. flux density.

  c. permeability.

  d. none of the above.

22. The flux density in the iron-core of an electromagnet is 0.25 T. When the iron-core is removed, the flux density drops to 62.5×10^{-6} T. What is the relative permeability of the iron core?

  a. $\mu_r = 4$.

  b. $\mu_r = 250$.

  c. $\mu_r = 4000$.

  d. This is impossible to determine.

23. What is the flux density, B, for a magnetic flux of 500 Mx through an area of 10 cm²?

  a. 50×10^{-3} T.

  b. 50 G.

  c. 5000 G.

  d. both a and b.

24. The geographic North Pole of the earth has

  a. no magnetic polarity.

  b. south magnetic polarity.

  c. north magnetic polarity.

  d. none of the above.

25. With an electromagnet,

  a. more current and more coil turns mean a stronger magnetic field.

  b. less current and fewer coil turns mean a stronger magnetic field.

  c. if there is no current in the coil, there is no magnetic field.

  d. both a and c.

Questions

1. Name two magnetic materials and three nonmagnetic materials.

2. Explain the difference between a permanent magnet and an electromagnet.

3. Draw a horseshoe magnet and its magnetic field. Label the magnetic poles, indicate the air gap, and show the direction of flux.

4. Define: relative permeability, shielding, induction, Hall voltage.

5. Give the symbols, cgs units, and SI units for magnetic flux and for flux density.

6. How are the north and south poles of a bar magnet determined with a magnetic compass?

7. Referring to Fig. 13–11, why can either end of the magnet pick up the nail?

8. What is the difference between flux ϕ and flux density B?

Problems

SECTION 13–2 MAGNETIC FLUX ϕ

13–1 Define (a) the maxwell (Mx) unit of magnetic flux ϕ; (b) the weber (Wb) unit of magnetic flux, ϕ.

13–2 Make the following conversions:
  a. 0.001 Wb to Mx
  b. 0.05 Wb to Mx
  c. 15×10^{-4} Wb to Mx
  d. 1×10^{-8} Wb to Mx

13–3 Make the following conversions:
  a. 1000 Mx to Wb
  b. 10,000 Mx to Wb
  c. 1 Mx to Wb
  d. 100 Mx to Wb

13–4 Make the following conversions:
  a. 0.0002 Wb to Mx
  b. 5500 Mx to Wb
  c. 70 Mx to Wb
  d. 30×10^{-6} Wb to Mx

13–5 Make the following conversions:
  a. 0.00004 Wb to Mx
  b. 225 Mx to Wb
  c. 80,000 Mx to Wb
  d. 650×10^{-6} Wb to Mx

13–6 A permanent magnet has a magnetic flux of 12,000 μWb. How many magnetic field lines does this correspond to?

13–7 An electromagnet produces a magnetic flux of 900 μWb. How many magnetic field lines does this correspond to?

13–8 A permanent magnet has a magnetic flux of 50,000 Mx. How many Webers (Wb) of magnetic flux does this correspond to?

SECTION 13–3 FLUX DENSITY B

13–9 Define (a) The gauss (G) unit of flux density, B; (b) the tesla (T) unit of flux density, B.

13–10 Make the following conversions:
  a. 2.5 T to G
  b. 0.05 T to G
  c. 1×10^{-4} T to G
  d. 0.1 T to G

13–11 Make the following conversions:
  a. 4000 G to T
  b. 800,000 G to T
  c. 600 G to T
  d. 10,000 G to T

13–12 Make the following conversions:
  a. 0.004 T to G
  b. 1000 G to T
  c. 1×10^5 G to T
  d. 10 T to G

13–13 Make the following conversions:
  a. 0.0905 T to G
  b. 100 T to G
  c. 75,000 G to T
  d. 1.75×10^6 G to T

13–14 With a flux of 250 Mx through an area of 2 cm^2, what is the flux density in gauss units?

13–15 A flux of 500 μWb exists in an area, A, of 0.01 m^2. What is the flux density in tesla units?

13–16 Calculate the flux density, in teslas, for a flux, ϕ, of 400 μWb in an area of 0.005 m^2?

13–17 Calculate the flux density in gauss units for a flux, ϕ, of 200 μWb in an area of 5×10^{-4} m^2?

13–18 With a magnetic flux, ϕ, of 30,000 Mx through a perpendicular area of 6 cm^2, what is the flux density in gauss units?

13–19 How much is the flux density in teslas for a flux, ϕ, of 160 μWb through an area of 0.0012 m^2?

13–20 With a flux, ϕ, of 2000 μWb through an area of 0.0004 m^2, what is the flux density in gauss units?

13–21 For a flux density of 30 kG at the north pole of a magnet through a cross-sectional area of 8 cm^2, how much is the total flux in maxwells?

13–22 The flux density in an iron-core is 5×10^{-3} T. If the area of the core is 10 cm^2, calculate the total number of magnetic flux lines in the core.

13–23 The flux density in an iron core is 5 T. If the area of the core is 40 cm^2, calculate the magnetic flux in weber units.

13–24 The flux density in an iron core is 80 kG. If the area of the core is 0.2 m^2, calculate the magnetic flux in weber units.

13–25 If the flux density in 0.05 m^2 is 2000 G, how many magnetic field lines are there?

Critical Thinking

13–26 A flux ϕ of 25 μWb exists in an area of 0.25 in^2. What is the flux density B in (a) gauss units; (b) teslas?

13–27 At the north pole of an electromagnet, the flux density B equals 5 T. If the area A equals 0.125 in^2, determine the total number of flux lines ϕ in (a) maxwells; (b) webers.

Answers to Knowledge Check Problems

13–1 North pole

13–2 1×10^{-8} Wb = 1 Mx

13–3 1×10^{-4} T = 1 G

13–4 μ_r has no units

13–5 in its iron core

13–6 steel

13–7 insulators

13–8 one which is a good magnetic material with high permeability

13–9 perpendicular

Answers to Self-Reviews

13–1 a. T
b. T

13–2 a. 2000 Mx
b. 20 μWb

13–3 a. 3000 G
b. 0.3 T

13–4 a. T
b. F

13–5 a. T
b. T

13–6 a. T
b. T
c. F

13–7 a. ferrites
b. conductor

13–8 a. T
b. T

13–9 a. 60 mV
b. flux density

14 Electromagnetism

A magnetic field is always associated with an electric current. Therefore, the units of measure for the strength or intensity of a magnetic field are based on the electric current that produces the field. For an electromagnet, the strength and intensity of the magnetic field depend on the amount of current flow and the number of coil turns in a given length. The electromagnet acts like a bar magnet with opposite magnetic poles at its ends.

When a conductor passes through a magnetic field, the work put into this action forces free electrons to move along the length of the conductor. The rate at which the conductor moves through the magnetic field and how many field lines are cut determines the amount of induced current and/or voltage. In this chapter, you will learn about the units and laws of electromagnetism. You will also learn about an electromechanical device known as a relay. As you will see, a relay uses an electromagnet to open or close one or more sets of switch contacts.

Objectives

After studying this chapter you should be able to

- Define the terms *magnetomotive force* and *field intensity* and list the units of each.

- *Explain* the *B-H* magnetization curve.

- *Define* the term *saturation* as it relates to a magnetic core.

- *Explain* what is meant by *magnetic hysteresis*.

- *Describe* the magnetic field of an electric current in a straight conductor.

- *Determine* the magnetic polarity of a solenoid using the left-hand rule.

- *Explain* the concept of motor action.

- *Explain* how an induced voltage can be developed across the ends of a conductor that passes through a magnetic field.

- *State* Lenz's law.

- Using Faraday's law, *calculate* the induced voltage across a conductor being passed through a magnetic field.

- *Explain* the basic construction and operation of an electromechanical relay.

- *List and explain* some important relay ratings.

Outline

Important Terms

ampere-turn (A · t)

ampere-turns/meter (A · t /m)

B-H magnetization curve

degaussing

Faraday's law

field intensity (*H*)

holding current

hysteresis

left-hand rule

Lenz's law

magnetomotive force (mmf)

motor action

pickup current

saturation

14–1 Ampere-turns of Magnetomotive Force (mmf)

The strength of the magnetic field of a coil magnet depends on how much current flows in the turns of the coil. The more current, the stronger the magnetic field. Also, more turns in a specific length concentrate the field. The coil serves as a bar magnet with opposite poles at the ends, providing a magnetic field proportional to the ampere-turns. As a formula,

$$\text{Ampere-turns} = I \times N = \text{mmf} \qquad (14\text{--}1)$$

where I is the current in amperes multiplied by the number of turns N. The quantity IN specifies the amount of *magnetizing force* or *magnetic potential*, which is the *magnetomotive force (mmf)*.

The practical unit is the ampere-turn. The SI abbreviation for ampere-turn is A, the same as for the ampere, since the number of turns in a coil usually is constant but the current can be varied. However, for clarity, we shall use the abbreviation A · t.

As shown in Fig. 14–1, a solenoid with five turns and 2 amperes has the same magnetizing force as one with 10 turns and 1 ampere, as the product of the amperes and turns is 10 for both cases. With thinner wire, more turns can be placed in a given space. The amount of current is determined by the resistance of the wire and the source voltage. The number of ampere-turns necessary depends on the magnetic field strength required.

||||MultiSim **Figure 14–1** Two examples of equal ampere-turns for the same mmf. (a) IN is $2 \times 5 = 10$. (b) IN is $1 \times 10 = 10$.

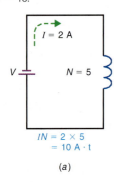

(a)

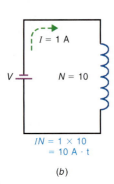

(b)

Example 14–1

Calculate the ampere-turns of mmf for a coil with 2000 turns and a 5-mA current.

ANSWER $\text{mmf} = I \times N = 2000 \times 5 \times 10^{-3}$
$$= 10 \text{ A} \cdot \text{t}$$

Example 14–2

A coil with 4 A is to provide a magnetizing force of 600 A · t. How many turns are necessary?

ANSWER $N = \dfrac{\text{A} \cdot \text{t}}{I} = \dfrac{600}{4}$
$$= 150 \text{ turns}$$

Example 14-3

A coil with 400 turns must provide 800 A · t of magnetizing force. How much current is necessary?

ANSWER $\quad I = \dfrac{A \cdot t}{N} = \dfrac{800}{400}$

$$= 2 \text{ A}$$

Example 14-4

The wire in a solenoid of 250 turns has a resistance of 3 Ω. (a) How much is the current when the coil is connected to a 6-V battery? (b) Calculate the ampere-turns of mmf.

ANSWER

a. $I = \dfrac{V}{R} = \dfrac{6 \text{ V}}{3 \text{ }\Omega}$

$$= 2 \text{ A}$$

b. mmf $= I \times N = 2 \text{ A} \times 250 \text{ t}$
$$= 500 \text{ A} \cdot \text{t}$$

The ampere-turn A · t, is an SI unit. It is calculated as *IN* with the current in amperes.

The cgs unit of mmf is the *gilbert,** abbreviated Gb. One ampere-turn equals 1.26 Gb. The number 1.26 is approximately $4\pi/10$, derived from the surface area of a sphere, which is $4\pi r^2$.

To convert *IN* to gilberts, multiply the ampere-turns by the constant conversion factor 1.26 Gb/1 A · t. As an example, 1000 A · t is the same mmf as 1260 Gb. The calculations are

$$1000 \text{ A} \cdot \text{t} \times 1.26 \frac{\text{Gb}}{1 \text{ A} \cdot \text{t}} = 1260 \text{ Gb}$$

Note that the units of A · t cancel in the conversion.

■ *14-1 Knowledge Check*

> **Answer at end of chapter.**
>
> **A coil with 300 turns has a current of 50 mA. How much is the mmf in gilberts (Gb)?**

---

* William Gilbert (1540–1603) was an English scientist who investigated the magnetism of the earth.

a. If the mmf is 243 A · t, and *I* is doubled from 2 to 4 A with the same number of turns, how much is the new mmf?
b. Convert 500 A · t to gilberts.

14–2 Field Intensity (*H*)

The ampere-turns of mmf specify the magnetizing force, but the intensity of the magnetic field depends on the length of the coil. At any point in space, a specific value of ampere-turns must produce less field intensity for a long coil than for a short coil that concentrates the same mmf. Specifically, the field intensity *H* in mks units is

$$H = \frac{\text{ampere-turns of mmf}}{l \text{ meters}} \qquad \text{(14–2)}$$

This formula is for a solenoid. The field intensity *H* is at the center of an air core. For an iron core, *H* is the intensity through the entire core. By means of units for *H*, the magnetic field intensity can be specified for either electromagnets or permanent magnets, since both provide the same kind of magnetic field.

The length in Formula (14–2) is between poles. In Fig. 14–2*a*, the length is 1 m between the poles at the ends of the coil. In Fig. 14–2*b*, *l* is also 1 m between the ends of the iron core. In Fig. 14–2*c*, though, *l* is 2 m between the poles at the ends of the iron core, although the winding is only 1 m long.

The examples in Fig. 14–2 illustrate the following comparisons:

1. In all three cases, the mmf is 1000 A · t for the same value of *IN*.
2. In Fig. 14–2*a* and *b*, *H* equals 1000 A · t/m. In *a*, this *H* is the intensity at the center of the air core; in *b*, this *H* is the intensity through the entire iron core.
3. In Fig. 14–2*c*, because *l* is 2 m, *H* is 1000/2, or 500 A · t/m. This *H* is the intensity in the entire iron core.

Units For *H*

The field intensity is basically mmf per unit of length. In practical units, *H* is ampere-turns per meter. The cgs unit for *H* is the *oersted,** abbreviated Oe, which equals one gilbert of mmf per centimeter.

Figure 14–2 Relation between ampere-turns of mmf and the resultant field intensity *H* for different cores. Note that *H* = mmf/length. (*a*) Intensity *H* is 1000 A · t/m with an air core. (*b*) *H* = 1000 A · t/m in an iron core of the same length as the coil. (*c*) *H* is 1000/2 = 500 A · t/m in an iron core twice as long as the coil.

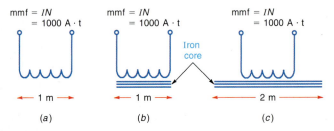

* H. C. Oersted (1777–1851), a Danish physicist, discovered electromagnetism.

Conversion of Units

To convert SI units of A · t/m to cgs units of Oe, multiply by the conversion factor 0.0126 Oe per 1 A · t/m. As an example, 1000 A · t/m is the same H as 12.6 Oe. The calculations are

$$1000 \, \frac{A \cdot t}{m} \times 0.0126 \, \frac{Oe}{1 \, A \cdot t/m} = 12.6 \, Oe$$

Note that the units of A · t and m cancel. The m in the conversion factor becomes inverted to the numerator.

Permeability (μ)

Whether we say H is 1000 A · t/m or 12.6 Oe, these units specify how much field intensity is available to produce magnetic flux. However, the amount of flux produced by H depends on the material in the field. A good magnetic material with high relative permeability can concentrate flux and produce a large value of flux density B for a specified H. These factors are related by the formula:

$$B = \mu \times H \tag{14–3}$$

or

$$\mu = \frac{B}{H} \tag{14–4}$$

Using SI units, B is the flux density in webers per square meter, or teslas; H is the field intensity in ampere-turns per meter. In the cgs system, the units are gauss for B and oersted for H. The factor μ is the absolute permeability, not referred to any other material, in units of B/H.

In the cgs system, the units of gauss for B and oersteds for H have been defined to give μ the value of 1 G/Oe, for vacuum, air, or space. This simplification means that B and H have the same numerical values in air and in vacuum. For instance, the field intensity H of 12.6 Oe produces a flux density of 12.6 G in air.

Furthermore, the values of relative permeability μ_r are the same as those for absolute permeability in B/H units in the cgs system. The reason is that μ is 1 for air or vacuum, used as the reference for the comparison. As an example, if μ_r for an iron sample is 600, the absolute μ is also 600 G/Oe.

In SI, however, the permeability of air or vacuum is not 1. This value is $4\pi \times 10^{-7}$, or 1.26×10^{-6}, with the symbol μ_0. Therefore, values of relative permeability μ_r must be multiplied by 1.26×10^{-6} for μ_0 to calculate μ as B/H in SI units.

For an example of $\mu_r = 100$, the SI value of μ can be calculated as follows:

$$\mu = \mu_r \times \mu_0$$
$$= 100 \times 1.26 \times 10^{-6} \, \frac{T}{A \cdot t/m}$$
$$\mu = 126 \times 10^{-6} \, \frac{T}{A \cdot t/m}$$

GOOD TO KNOW

The permeability of a material is similar in many respects to the conductivity in electric circuits.

Example 14–5

A magnetic material has a μ_r of 500. Calculate the absolute μ as B/H (a) in cgs units and (b) in SI units.

ANSWER

a. $\mu = \mu_r \times \mu_0$ in cgs units. Then

$$= 500 \times 1 \frac{G}{Oe}$$

$$= 500 \frac{G}{Oe}$$

b. $\mu = \mu_r \times \mu_0$ in SI units. Then

$$= 500 \times 1.26 \times 10^{-6} \frac{T}{A \cdot t/m}$$

$$= 630 \times 10^{-6} \frac{T}{A \cdot t/m}$$

Example 14–6

For this example of $\mu = 630 \times 10^{-6}$ in SI units, calculate the flux density B that will be produced by the field intensity H equal to 1000 A · t/m.

ANSWER $B = \mu H$

$$= \left(630 \times 10^{-6} \frac{T}{A \cdot t/m} \right) \left(1000 \frac{A \cdot t}{m} \right)$$

$$= 630 \times 10^{-3}\ T$$

$$= 0.63\ T$$

Note that the ampere-turns and meter units cancel, leaving only the tesla unit for the flux density B.

■ *14–2 Knowledge Check*

Answer at end of chapter.

Two coils of different lengths have the same mmf. Which coil, the longer or shorter, will have a higher value of H?

■ *14–2 Self-Review*

Answers at end of chapter.

a. What are the values of μ_r for air, vacuum, and space?
b. An iron core has 200 times more flux density than air for the same field intensity H. How much is μ_r?
c. An iron core produces 200 G of flux density for 1 Oe of field intensity H. How much is μ?

d. A coil with an mmf of 25 A · t is 0.1 m long. How much is the field intensity H?

e. Convert 500 $\dfrac{A \cdot t}{m}$ to oersted units.

14–3 *B-H* Magnetization Curve

The *B-H* curve in Fig. 14–3 is often used to show how much flux density B results from increasing the amount of field intensity H. This curve is for soft iron, plotted for the values in Table 14–1, but similar curves can be obtained for all magnetic materials.

Calculating *H* and *B*

The values in Table 14–1 are calculated as follows:

1. The current I in the coil equals V/R. For a 10-Ω coil resistance with 20 V applied, I is 2 A, as listed in the top row of Table 14–1. Increasing values of V produce more current in the coil.

2. The ampere-turns IN of magnetizing force increase with more current. Since the turns are constant at 100, the values of IN increase from 200 for 2 A in the top row to 1000 for 10 A in the bottom row.

3. The field intensity H increases with higher IN. The values of H are in mks units of ampere-turns per meter. These values equal $IN/0.2$ because the length is 0.2 m. Therefore, each IN is divided by 0.2, or multiplied by 5, for the corresponding values of H. Since H increases in the same proportion as I, sometimes the horizontal axis on a B-H curve is given only in amperes, instead of in H units.

4. The flux density B depends on the field intensity H and the permeability of the iron. The values of B in the last column are obtained by multiplying $\mu \times H$. However, for SI units, the values of μ_r listed must be multiplied by 1.26×10^{-6} to obtain $\mu \times H$ in teslas.

Figure 14–3 *B-H* magnetization curve for soft iron. No values are shown near zero, where μ may vary with previous magnetization.

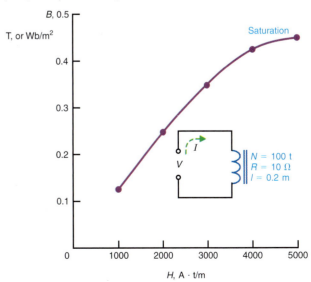

Table 14–1		B-H Values for Fig. 14–3						
V, Volts	R, Ω	$I = V/R$, Amperes	N, Turns	mmf, A · t	l, m	H, A · t/m	μ_r	$B = \mu \times H$, T
20	10	2	100	200	0.2	1000	100	0.126
40	10	4	100	400	0.2	2000	100	0.252
60	10	6	100	600	0.2	3000	100	0.378
80	10	8	100	800	0.2	4000	85	0.428
100	10	10	100	1000	0.2	5000	70	0.441

Saturation

Note that the permeability decreases for the highest values of H. With less μ, the iron core cannot provide proportional increases in B for increasing values of H. In Fig. 14–3, for values of H above 4000 A · t/m, approximately, the values of B increase at a much slower rate, making the curve relatively flat at the top. The effect of little change in flux density when the field intensity increases is called *saturation*.

Iron becomes saturated with magnetic lines of induction. After most of the molecular dipoles and the magnetic domains are aligned by the magnetizing force, very little additional induction can be produced. When the value of μ is specified for a magnetic material, it is usually the highest value before saturation.

■ *14–3 Knowledge Check*

Answer at end of chapter.

In Fig. 14–3, what is the value of B for an H of 2500 $\dfrac{\text{A} \cdot \text{t}}{\text{m}}$?

■ *14–3 Self-Review*

Answers at end of chapter.

Refer to Fig. 14–3.
a. How much is B in tesla units for 1500 A · t/m?
b. What value of H starts to produce saturation?

14–4 Magnetic Hysteresis

Hysteresis means "lagging behind." With respect to the magnetic flux in an iron core of an electromagnet, the flux lags the increases or decreases in magnetizing force. Hysteresis results because the magnetic dipoles are not perfectly elastic. Once aligned by an external magnetizing force, the dipoles do not return exactly to their original positions when the force is removed. The effect is the same as if the dipoles were forced to move against internal friction between molecules. Furthermore, if the magnetizing force is reversed in direction by reversal of the current in an electromagnet, the flux produced in the opposite direction lags behind the reversed magnetizing force.

Hysteresis Loss

When the magnetizing force reverses thousands or millions of times per second, as with rapidly reversing alternating current, hysteresis can cause a considerable loss of energy. A large part of the magnetizing force is then used to overcome the internal friction of the molecular dipoles. The work done by the magnetizing force against this internal friction produces heat. This energy wasted in heat as the molecular dipoles lag the magnetizing force is called the *hysteresis loss*. For steel and other hard magnetic materials, hysteresis losses are much higher than in soft magnetic materials like iron.

When the magnetizing force varies at a slow rate, hysteresis losses can be considered negligible. An example is an electromagnet with direct current that is simply turned on and off or the magnetizing force of an alternating current that reverses 60 times per second or less. The faster the magnetizing force changes, however, the greater the hysteresis effect.

Hysteresis Loop

To show the hysteresis characteristics of a magnetic material, its values of flux density B are plotted for a periodically reversing magnetizing force. See Fig. 14–4. This curve is the hysteresis loop of the material. The larger the area enclosed by the curve, the greater the hysteresis loss. The hysteresis loop is actually a *B-H* curve with an ac magnetizing force.

Values of flux density B are indicated on the vertical axis. The units can be gauss or teslas.

The horizontal axis indicates values of field intensity H. On this axis, the units can be oersteds, ampere-turns per meter, ampere-turns, or magnetizing current because all factors are constant except I.

Opposite directions of current result in opposite directions of $+H$ and $-H$ for the field lines. Similarly, opposite polarities are indicated for flux density as $+B$ or $-B$.

The current starts from zero at the center, when the material is unmagnetized. Then positive H values increase B to saturation at $+B_{max}$. Next H

Figure 14–4 Hysteresis loop for magnetic materials. This graph is a *B-H* curve like Fig. 14–3, but H alternates in polarity with alternating current.

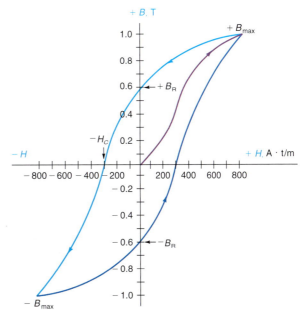

decreases to zero, but B drops to the value B_R, instead of to zero, because of hysteresis. When H becomes negative, B drops to zero and continues to $-B_{max}$, which is saturation in the opposite direction from $+B_{max}$ because of the reversed magnetizing current.

Then, as the $-H$ values decrease, the flux density is reduced to $-B_R$. Finally, the loop is completed; positive values of H produce saturation at B_{max} again. The curve does not return to the zero origin at the center because of hysteresis. As the magnetizing force periodically reverses, the values of flux density are repeated to trace out the hysteresis loop.

The value of either $+B_R$ or $-B_R$, which is the flux density remaining after the magnetizing force has been reduced to zero, is the *residual induction* of a magnetic material, also called its *retentivity*. In Fig. 14–4, the residual induction is 0.6 T, in either the positive or the negative direction.

The value of $-H_C$, which equals the magnetizing force that must be applied in the reverse direction to reduce the flux density to zero, is the *coercive force* of the material. In Fig. 14–4, the coercive force $-H_C$ is 300 A $\cdot$ t/m.

Demagnetization

To demagnetize a magnetic material completely, the residual induction B_R must be reduced to zero. This usually cannot be accomplished by a reversed dc magnetizing force because the material then would become magnetized with opposite polarity. The practical way is to magnetize and demagnetize the material with a continuously decreasing hysteresis loop. This can be done with a magnetic field produced by alternating current. Then, as the magnetic field and the material are moved away from each other or the current amplitude is reduced, the hysteresis loop becomes smaller and smaller. Finally, with the weakest field, the loop collapses practically to zero, resulting in zero residual induction.

This method of demagnetization is also called *degaussing*. One application is degaussing the metal electrodes in a color picture tube with a deguassing coil providing alternating current from the power line. Another example is erasing the recorded signal on magnetic tape by demagnetizing with an ac bias current. The average level of the erase current is zero, and its frequency is much higher than the recorded signal.

■ *14–4 Knowledge Check*

> *Answer at end of chapter.*
>
> In Fig. 14–4, what is the value of the coercive force, $-H_C$?

■ *14–4 Self-Review*

> *Answers at end of chapter.*
> a. Hysteresis loss increases with higher frequencies. (True or False)
> b. Degaussing is done with alternating current. (True or False)

14–5 Magnetic Field Around an Electric Current

In Fig. 14–5, the iron filings aligned in concentric rings around the conductor show the magnetic field of the current in the wire. The iron filings are dense next to the conductor, showing that the field is strongest at this point. Furthermore, the field strength decreases inversely as the square of the distance

Figure 14–5 How iron filings can be used to show the invisible magnetic field around the electric current in a wire conductor.

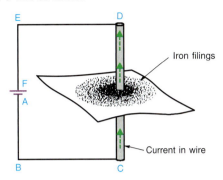

from the conductor. It is important to note the following two factors about the magnetic lines of force:

1. The magnetic lines are circular because the field is symmetrical with respect to the wire in the center.
2. The magnetic field with circular lines of force is in a plane perpendicular to the current in the wire.

From points C to D in the wire, the circular magnetic field is in the horizontal plane because the wire is vertical. Also, the vertical conductor between points EF and AB has the associated magnetic field in the horizontal plane. Where the conductor is horizontal, as from B to C and D to E, the magnetic field is in a vertical plane.

These two requirements of a circular magnetic field in a perpendicular plane apply to any charge in motion. Whether electron flow or motion of positive charges is considered, the associated magnetic field must be at right angles to the direction of current.

In addition, the current need not be in a wire conductor. As an example, the beam of moving electrons in the vacuum of a cathode-ray tube has an associated magnetic field. In all cases, the magnetic field has circular lines of force in a plane perpendicular to the direction of motion of the electric charges.

Clockwise and Counterclockwise Fields

With circular lines of force, the magnetic field would tend to move a magnetic pole in a circular path. Therefore, the direction of the lines must be considered either clockwise or counterclockwise. This idea is illustrated in Fig. 14–6, showing how a north pole would move in the circular field.

Figure 14–6 Rule for determining direction of circular field around a straight conductor. Field is counterclockwise for direction of electron flow shown here. Circular field is clockwise for reversed direction of electron flow.

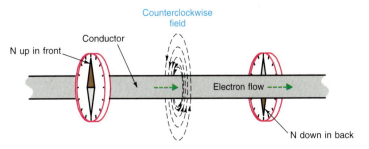

The directions are tested with a magnetic compass needle. When the compass is in front of the wire, the north pole on the needle points up. On the opposite side, the compass points down. If the compass were placed at the top, its needle would point toward the back of the wire; below the wire, the compass would point forward. When all these directions are combined, the result is the circular magnetic field shown with counterclockwise lines of force. (The counterclockwise direction of the magnetic field assumes that you are looking into the end of the wire, in the same direction as electron flow.)

Instead of testing every conductor with a magnetic compass, however, we can use the following rule for straight conductors to determine the circular direction of the magnetic field: *If you grasp the conductor with your left hand so that the thumb points in the direction of electron flow, your fingers will encircle the conductor in the same direction as the circular magnetic field lines.* In Fig. 14–6, the direction of electron flow is from left to right. Facing this way, you can assume that the circular magnetic flux in a perpendicular plane has lines of force in the counterclockwise direction.

The opposite direction of electron flow produces a reversed field. Then the magnetic lines of force rotate clockwise. If the charges were moving from right to left in Fig. 14–6, the associated magnetic field would be in the opposite direction with clockwise lines of force.

Fields Aiding or Canceling

When the magnetic lines of two fields are in the same direction, the lines of force aid each other, making the field stronger. When magnetic lines are in opposite directions, the fields cancel.

In Fig. 14–7, the fields are shown for two conductors with opposite directions of electron flow. The dot in the middle of the field at the left indicates the tip of an arrowhead to show current up from the paper. The cross symbolizes the back of an arrow to indicate electron flow into the paper.

Notice that the magnetic lines *between the conductors* are in the same direction, although one field is clockwise and the other counterclockwise. Therefore, the fields aid here, making a stronger total field. On either side of the conductors, the two fields are opposite in direction and tend to cancel each other. The net result, then, is to strengthen the field in the space between the conductors.

■ *14–5 Knowledge Check*

Answer at end of chapter.

Imagine a vertical wire on this page. If electrons flow upward through the conductor, what is the direction of the circular magnetic field, clockwise or counterclockwise, when looking down at the page?

|||| **MultiSim**   **Figure 14–7**   Magnetic fields aiding between parallel conductors with opposite directions of current.

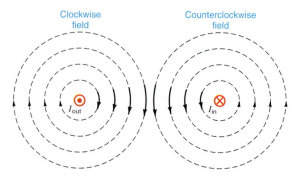

a. **Magnetic field lines around a conductor are circular in a perpendicular cross section of the conductor. (True or False)**
b. **In Fig. 14–7, the field is strongest between the conductors. (True or False)**

14–6 Magnetic Polarity of a Coil

Bending a straight conductor into a loop, as shown in Fig. 14–8, has two effects. First, the magnetic field lines are more dense inside the loop. The total number of lines is the same as those for the straight conductor, but the lines inside the loop are concentrated in a smaller space. Furthermore, all lines inside the loop are aiding in the same direction. This makes the loop field effectively the same as a bar magnet with opposite poles at opposite faces of the loop.

Solenoid as a Bar Magnet

A coil of wire conductor with more than one turn is generally called a *solenoid*. An ideal solenoid, however, has a length much greater than its diameter. Like a single loop, the solenoid concentrates the magnetic field inside the coil and provides opposite magnetic poles at the ends. These effects are multiplied, however, by the number of turns as the magnetic field lines aid each other in the same direction inside the coil. Outside the coil, the field corresponds to a bar magnet with north and south poles at opposite ends, as illustrated in Fig. 14–9.

Magnetic Polarity

To determine the magnetic polarity of a solenoid, use the *left-hand rule* illustrated in Fig. 14–10: *If the coil is grasped with the fingers of the left hand curled around the coil in the direction of electron flow, the thumb points to the north pole of the coil.* The left hand is used here because the current is electron flow.

The solenoid acts like a bar magnet, whether or not it has an iron core. Adding an iron core increases the flux density inside the coil. In addition, the field strength is uniform for the entire length of the core. The polarity is the same, however, for air-core and iron-core coils.

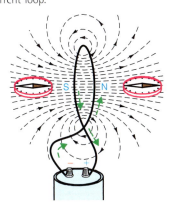

Figure 14–8 Magnetic poles of a current loop.

GOOD TO KNOW

The magnetic polarity of a solenoid can be verified with a compass.

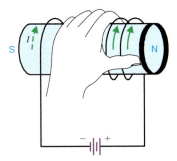

Figure 14–10 Left-hand rule for north pole of a coil with current *I*. The *I* is electron flow.

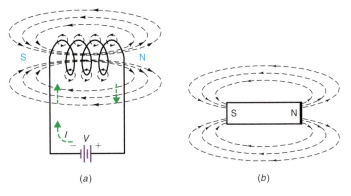

Figure 14–9 Magnetic poles of a solenoid. (*a*) Coil winding. (*b*) Equivalent bar magnet.

(*a*)   (*b*)

Figure 14-11 Examples for determining the magnetic polarity of a coil with direct current *I*. The *I* is electron flow. The polarities are reversed in (*a*) and (*b*) because the battery is reversed to reverse the direction of current. Also, (*d*) is the opposite of (*c*) because of the reversed winding.

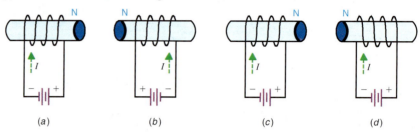

(*a*)　　　　　(*b*)　　　　　(*c*)　　　　　(*d*)

The magnetic polarity depends on the direction of current flow and the direction of winding. The current is determined by the connections to the voltage source. Electron flow is from the negative side of the voltage source, through the coil, and back to the positive terminal.

The direction of winding can be over and under, starting from one end of the coil, or under and over with respect to the same starting point. Reversing either the direction of winding or the direction of current reverses the magnetic poles of the solenoid. See Fig. 14–11. When both are reversed, though, the polarity is the same.

■ *14-6 Knowledge Check*

Answer at end of chapter.

List the two factors that determine the magnetic polarity of a coil.

■ *14-6 Self-Review*

Answers at end of chapter.

a. **In Fig. 14–9, if the battery is reversed, will the north pole be at the left or the right?**

b. **If one end of a solenoid is a north pole, is the opposite end a north or a south pole?**

14-7 Motor Action Between Two Magnetic Fields

The physical motion from the forces of magnetic fields is called *motor action.* One example is the simple attraction or repulsion between bar magnets.

We know that like poles repel and unlike poles attract. It can also be considered that fields in the same direction repel and opposite fields attract.

Consider the repulsion between two north poles, illustrated in Fig. 14–12. Similar poles have fields in the same direction. Therefore, the similar fields of the two like poles repel each other.

A more fundamental reason for motor action, however, is the fact that the force in a magnetic field tends to produce motion from a stronger field toward a weaker field. In Fig. 14–12, note that the field intensity is greatest in the space between the two north poles. Here the field lines of similar poles in both magnets reinforce in the same direction. Farther away the field intensity is less, for essentially one magnet only. As a result, there is a difference in field

Figure 14–12 Repulsion between similar poles of two bar magnets. The motion is from the stronger field to the weaker field.

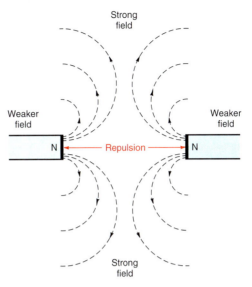

strength, providing a net force that tends to produce motion. The direction of motion is always toward the weaker field.

To remember the directions, we can consider that the stronger field moves to the weaker field, tending to equalize field intensity. Otherwise, the motion would make the strong field stronger and the weak field weaker. This must be impossible because then the magnetic field would multiply its own strength without any work added.

Force on a Straight Conductor in a Magnetic Field

Current in a conductor has its associated magnetic field. When this conductor is placed in another magnetic field from a separate source, the two fields can react to produce motor action. The conductor must be perpendicular to the magnetic field, however, as shown in Fig. 14–13. This way, the perpendicular

IIII MultiSim **Figure 14–13** Motor action of current in a straight conductor when it is in an external magnetic field. The H_I is the circular field of the current. The H_M indicates field lines between the north and south poles of the external magnet.

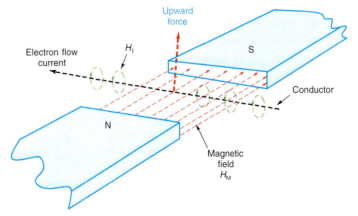

magnetic field produced by the current is in the same plane as the external magnetic field.

Unless the two fields are in the same plane, they cannot affect each other. In the same plane, however, lines of force in the same direction reinforce to make a stronger field, whereas lines in the opposite direction cancel and result in a weaker field.

To summarize these directions:

1. When the conductor is at 90°, or perpendicular to the external field, the reaction between the two magnetic fields is maximum.
2. When the conductor is at 0°, or parallel to the external field, there is no effect between them.
3. When the conductor is at an angle between 0 and 90°, only the perpendicular component is effective.

In Fig. 14–13, electrons flow in the wire conductor in the plane of the paper from the bottom to the top of the page. This flow provides the counterclockwise field H_I around the wire in a perpendicular plane cutting through the paper. The external field H_M has lines of force from left to right in the plane of the paper. Then lines of force in the two fields are parallel above and below the wire.

Below the conductor, its field lines are left to right in the same direction as the external field. Therefore, these lines reinforce to produce a stronger field. Above the conductor, the lines of the two fields are in opposite directions, causing a weaker field. As a result, the net force of the stronger field makes the conductor move upward out of the page toward the weaker field.

If electrons flow in the reverse direction in the conductor or if the external field is reversed, the motor action will be in the opposite direction. Reversing both the field and the current, however, results in the same direction of motion.

Rotation of a Conductor Loop in a Magnetic Field

When a loop of wire is in the magnetic field, opposite sides of the loop have current in opposite directions. Then the associated magnetic fields are opposite. The resulting forces are upward on one side of the loop and downward on the other side, making it rotate. This effect of a force in producing rotation is called *torque.*

The principle of motor action between magnetic fields producing rotational torque is the basis of all electric motors. The moving-coil meter described in Sec. 8–1 is a similar application. Since torque is proportional to current, the amount of rotation indicates how much current flows through the coil.

■ *14–7 Knowledge Check*

Answer at end of chapter.

In Fig. 14–13, which way will the conductor move if the polarity of the magnetic field is reversed?

■ *14–7 Self-Review*

Answers at end of chapter.

a. In Fig. 14–12, the field is strongest between the two north poles. (True or False)
b. In Fig. 14–13, if both the magnetic field and the current are reversed, the motion will still be upward. (True or False)

Just as electrons in motion provide an associated magnetic field, when magnetic flux moves, the motion of magnetic lines cutting across a conductor forces free electrons in the conductor to move, producing current. This action is called *induction* because there is no physical connection between the magnet and the conductor. The induced current is a result of generator action as the mechanical work put into moving the magnetic field is converted into electric energy when current flows in the conductor.

Referring to Fig. 14–14, let the conductor AB be placed at right angles to the flux in the air gap of the horseshoe magnet. Then, when the magnet is moved up or down, its flux cuts across the conductor. The action of magnetic flux cutting across the conductor generates current. The fact that current flows is indicated by the microammeter.

When the magnet is moved downward, current flows in the direction shown. If the magnet is moved upward, current will flow in the opposite direction. Without motion, there is no current.

Direction of Motion

Motion is necessary for the flux lines of the magnetic field to cut across the conductor. This cutting can be accomplished by motion of either the field or the conductor. When the conductor is moved upward or downward, it cuts across the flux. The generator action is the same as moving the field, except that the relative motion is opposite. Moving the conductor upward, for instance, corresponds to moving the magnet downward.

Conductor Perpendicular to External Flux

To have electromagnetic induction, the conductor and the magnetic lines of flux must be perpendicular to each other. Then the motion makes the flux cut through the cross-sectional area of the conductor. As shown in Fig. 14–14, the conductor is at right angles to the lines of force in the field *H*.

The reason the conductor must be perpendicular is to make its induced current have an associated magnetic field in the same plane as the external flux. If the field of the induced current does not react with the external field, there can be no induced current.

How Induced Current Is Generated

The induced current can be considered the result of motor action between the external field *H* and the magnetic field of free electrons in every cross-sectional area of the wire. Without an external field, the free electrons move at random without any specific direction, and they have no net magnetic field. When the conductor is in the magnetic field *H*, there still is no induction without relative motion, since the magnetic fields for the free electrons are not disturbed. When the field or conductor moves, however, there must be a reaction opposing the motion. The reaction is a flow of free electrons resulting from motor action on the electrons.

Referring to Fig. 14–14, for example, the induced current must flow in the direction shown because the field is moved downward, pulling the magnet away from the conductor. The induced current of electrons then has a clockwise field; lines of force aid *H* above the conductor and cancel *H* below. When motor action between the two magnetic fields tends to move the conductor toward the weaker field, the conductor will be forced downward, staying with the magnet to oppose the work of pulling the magnet away from the conductor.

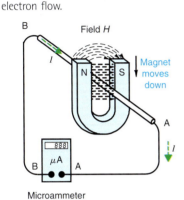

Figure 14–14 Induced current produced by magnetic flux cutting across a conductor. Direction of *I* here is for electron flow.

GOOD TO KNOW

In Fig. 14–14, extend the thumb, forefinger, and middle finger of the left hand at right angles to each other. With the forefinger pointing in the direction of magnetic flux, and the thumb in the direction the conductor is moving, the middle finger will point in the direction of the induced current. This is called the left-hand generator rule.

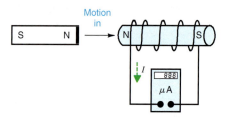

The effect of electromagnetic induction is increased when a coil is used for the conductor. Then the turns concentrate more conductor length in a smaller area. As illustrated in Fig. 14–15, moving the magnet into the coil enables the flux to cut across many turns of conductors.

Lenz's Law

Lenz's law is the basic principle for determining the direction of an induced voltage or current. Based on the principle of conservation of energy, the law simply states that the direction of the induced current must be such that its own magnetic field will oppose the action that produced the induced current.

In Fig. 14–15, for example, the induced current has the direction that produces a north pole at the left to oppose the motion by repulsion of the north pole being moved in. This is why it takes some work to push the permanent magnet into the coil. The work expended in moving the permanent magnet is the source of energy for the current induced in the coil.

Using Lenz's law, we can start with the fact that the left end of the coil in Fig. 14–15 must be a north pole to oppose the motion. Then the direction of the induced current is determined by the left-hand rule for electron flow. If the fingers coil around the direction of electron flow shown, under and over the winding, the thumb will point to the left for the north pole.

For the opposite case, suppose that the north pole of the permanent magnet in Fig. 14–15 is moved away from the coil. Then the induced pole at the left end of the coil must be a south pole by Lenz's law. The induced south pole will attract the north pole to oppose the motion of the magnet being moved away. For a south pole at the left end of the coil, then, the electron flow will be reversed from the direction shown in Fig. 14–15. We could generate an alternating current in the coil by moving the magnet periodically in and out.

■ *14–8 Knowledge Check*

> *Answer at end of chapter.*
>
> In Fig. 14–14, what is the direction of the induced current if the magnet is moved up?

■ *14–8 Self-Review*

> *Answers at end of chapter.*
>
> Refer to Fig. 14–15.

a. If the north end of the magnet is moved away from the coil, will its left side be north or south?
b. If the south end of the magnet is moved in, will the left end of the coil be north or south?
c. Referring to Fig. 14–14, if the conductor is moved up, instead of moving the magnet down, will the induced current flow in the same direction?

PIONEERS
IN ELECTRONICS

Russian physicist *Heinrich Friedrich Emil Lenz (1804–1865)* demonstrated that an increase in temperature increases the resistance of a metal. In 1834 he formulated Lenz's law, which states that the current induced by a change flows so as to oppose the effect producing the change. In 1838, he demonstrated the "Peltier Effect" with reversing current by using bismuth-antimony rods.

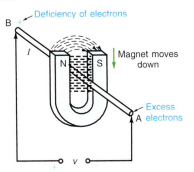

14–9 Generating an Induced Voltage

Consider a magnetic flux cutting a conductor that is not in a closed circuit, as shown in Fig. 14–16. The motion of flux across the conductor forces free electrons to move, but in an open circuit, the displaced electrons produce opposite electric charges at the two open ends.

For the directions shown, free electrons in the conductor are forced to move to point A. Since the end is open, electrons accumulate here. Point A then develops a negative potential.

At the same time, point B loses electrons and becomes charged positively. The result is a potential difference across the two ends, provided by the separation of electric charges in the conductor.

The potential difference is an electromotive force (emf), generated by the work of cutting across the flux. You can measure this potential difference with a voltmeter. However, a conductor cannot store electric charge. Therefore, the voltage is present only while the motion of flux cutting across the conductor is producing the induced voltage.

Induced Voltage across a Coil

For a coil, as in Fig. 14–17a, the induced emf is increased by the number of turns. Each turn cut by flux adds to the induced voltage, since each turn cut forces free electrons to accumulate at the negative end of the coil with a deficiency of electrons at the positive end.

The polarity of the induced voltage follows from the direction of induced current. The end of the conductor to which the electrons go and at which they accumulate is the negative side of the induced voltage. The opposite end, with a deficiency of electrons, is the positive side. The total emf across the coil is the sum of the induced voltages, since all the turns are in series.

Furthermore, the total induced voltage acts in series with the coil, as illustrated by the equivalent circuit in Fig. 14–17b, showing the induced voltage as a separate generator. This generator represents a voltage source with a potential difference resulting from the separation of charges produced by electromagnetic induction. The source v then can produce current in an external load circuit connected across the negative and positive terminals, as shown in Fig. 14–17c.

The induced voltage is in series with the coil because current produced by the generated emf must flow through all the turns. An induced voltage of 10 V, for example, with R_L equal to 5 Ω, results in a current of 2 A, which flows through the coil, the equivalent generator v, and the load resistance R_L.

Figure 14–17 Voltage induced across coil cut by magnetic flux. (a) Motion of flux generating voltage across coil. (b) Induced voltage acts in series with the coil. (c) The induced voltage is a source that can produce current in an external load resistor R_L connected across the coil.

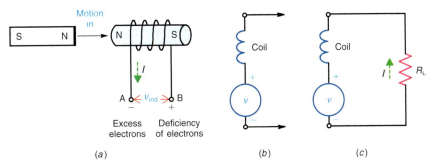

The direction of current in Fig. 14–17c shows electron flow around the circuit. Outside the source v, the electrons move from its negative terminal, through R_L, and back to the positive terminal of v because of its potential difference.

Inside the generator, however, the electron flow is from the + terminal to the − terminal. This direction of electron flow results from the fact that the left end of the coil in Fig. 14–17a must be a north pole, by Lenz's law, to oppose the north pole being moved in.

Notice how motors and generators are similar in using the motion of a magnetic field, but with opposite applications. In a motor, current is supplied so that an associated magnetic field can react with the external flux to produce motion of the conductor. In a generator, motion must be supplied so that the flux and conductor can cut across each other to induce voltage across the ends of the conductor.

Faraday's Law of Induced Voltage

The voltage induced by magnetic flux cutting the turns of a coil depends on the number of turns and how fast the flux moves across the conductor. Either the flux or the conductor can move. Specifically, the amount of induced voltage is determined by the following three factors:

1. *Amount of flux.* The more magnetic lines of force that cut across the conductor, the higher the amount of induced voltage.
2. *Number of turns.* The more turns in a coil, the higher the induced voltage. The v_{ind} is the sum of all individual voltages generated in each turn in series.
3. *Time rate of cutting.* The faster the flux cuts a conductor, the higher the induced voltage. Then more lines of force cut the conductor within a specific period of time.

These factors are fundamental in many applications. Any conductor with current will have voltage induced in it by a change in current and its associated magnetic flux.

The amount of induced voltage can be calculated by Faraday's law:

$$v_{\text{ind}} = N \frac{d\phi \text{ (webers)}}{dt \text{ (seconds)}} \tag{14–5}$$

where N is the number of turns and $d\phi/dt$ specifies how fast the flux ϕ cuts across the conductor. With $d\phi/dt$ in webers per second, the induced voltage is in volts.

As an example, suppose that magnetic flux cuts across 300 turns at the rate of 2 Wb/s.

To calculate the induced voltage,

$$v_{\text{ind}} = N \frac{d\phi}{dt}$$
$$= 300 \times 2$$
$$v_{\text{ind}} = 600 \text{ V}$$

It is assumed that all flux links all turns, which is true for an iron core.

Rate of Change

The symbol d in $d\phi$ and dt is an abbreviation for *change*. The $d\phi$ means a change in the flux ϕ, and dt means a change in time. In mathematics, dt represents an infinitesimally small change in time, but in this book we are using the d to mean rate of change in general. The results are exactly the same for the practical changes used here because the rate of change is constant.

Figure 14–18 Graphs of induced voltage produced by magnetic flux changes in a coil. (*a*) Linear increase of flux ϕ. (*b*) Constant rate of change for $d\phi/dt$ at 2 Wb/s. (*c*) Constant induced voltage of 600 V for a coil with 300 turns.

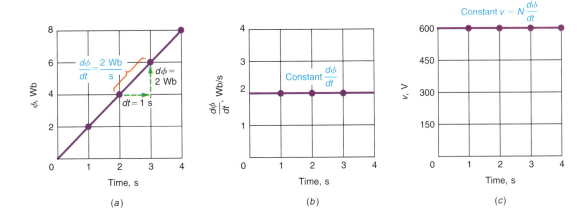

As an example, if the flux ϕ is 4 Wb one time but then changes to 6 Wb, the change in flux $d\phi$ is 2 Wb. The same idea applies to a decrease as well as an increase. If the flux changed from 6 to 4 Wb, $d\phi$ would still be 2 Wb. However, an increase is usually considered a change in the positive direction, with an upward slope, whereas a decrease has a negative slope downward.

Similarly, dt means a change in time. If we consider the flux at a time 2 s after the start and at a later time 3 s after the start, the change in time is 3 − 2, or 1 s for dt. Time always increases in the positive direction.

Combining the two factors of $d\phi$ and dt, we can say that for magnetic flux increasing by 2 Wb in 1 s, $d\phi/dt$ equals 2/1, or 2 Wb/s. This states the rate of change of the magnetic flux.

As another example, suppose that the flux increases by 2 Wb in 0.5 s. Then

$$\frac{d\phi}{dt} = \frac{2 \text{ Wb}}{0.5 \text{ s}} = 4 \text{ Wb/s}$$

Analysis of Induced Voltage as $N(d\phi/dt)$

This fundamental concept of voltage induced by a change in flux is illustrated by the graphs in Fig. 14–18, for the values listed in Table 14–2. The linear rise in Fig. 14–18*a* shows values of flux ϕ increasing at a uniform rate. In this case, the curve goes up 2 Wb for every 1-s interval. The slope of this curve, then,

Table 14–2	Induced–Voltage Calculations for Fig. 14–18					
ϕ, Wb	$d\phi$, Wb	t, s	dt, s	$d\phi/dt$, Wb/s	N, TURNS	$N(d\phi/dt)$, v
2	2	1	1	2	300	600
4	2	2	1	2	300	600
6	2	3	1	2	300	600
8	2	4	1	2	300	600

equal to $d\phi/dt$, is 2 Wb/s. Note that, although ϕ increases, the rate of change is constant because the linear rise has a constant slope.

For induced voltage, only the $d\phi/dt$ factor is important, not the actual value of flux. To emphasize this basic concept, the graph in Fig. 14–18b shows the $d\phi/dt$ values alone. This graph is a straight horizontal line for the constant value of 2 Wb/s.

The induced-voltage graph in Fig. 14–18c is also a straight horizontal line. Since $v_{ind} = N(d\phi/dt)$, the graph of induced voltage is the $d\phi/dt$ values multiplied by the number of turns. The result is a constant 600 V, with 300 turns cut by flux changing at the constant rate of 2 Wb/s.

The example illustrated here can be different in several ways without changing the basic fact that the induced voltage is equal to $N(d\phi/dt)$. First, the number of turns or the $d\phi/dt$ values can be greater or less than the values assumed here. More turns provide more induced voltage, whereas fewer turns mean less voltage. Similarly, a higher value for $d\phi/dt$ results in more induced voltage.

Note that two factors are included in $d\phi/dt$. Its value can be increased by a higher value of $d\phi$ or a smaller value of dt. As an example, the value of 2 Wb/s for $d\phi/dt$ can be doubled either by increasing $d\phi$ to 4 Wb or reducing dt to 0.5 s. Then $d\phi/dt$ is 4/1 or 2/0.5, which equals 4 Wb/s in either case. The same flux changing within a shorter time means a faster rate of flux cutting the conductor, resulting in a higher value of $d\phi/dt$ and more induced voltage.

For the opposite case, a smaller value of $d\phi/dt$, with less flux or a slower rate of change, results in a lower value of induced voltage. As $d\phi/dt$ decreases, the induced voltage will reverse polarity.

Finally, note that the $d\phi/dt$ graph in Fig. 14–18b has the constant value of 2 Wb/s because the flux is increasing at a linear rate. However, the flux need not have a uniform rate of change. Then the $d\phi/dt$ values will not be constant. In any case, though, the values of $d\phi/dt$ at all instants will determine the values of the induced voltage equal to $N(d\phi/dt)$.

Polarity of the Induced Voltage

The polarity is determined by Lenz's law. Any induced voltage has the polarity that opposes the change causing the induction. Sometimes this fact is indicated by using a negative sign for v_{ind} in Formula (14–5). However, the absolute polarity depends on whether the flux is increasing or decreasing, the method of winding, and which end of the coil is the reference.

When all these factors are considered, v_{ind} has polarity such that the current it produces and the associated magnetic field oppose the change in flux producing the induced voltage. If the external flux increases, the magnetic field of the induced current will be in the opposite direction. If the external field decreases, the magnetic field of the induced current will be in the same direction as the external field to oppose the change by sustaining the flux. In short, the induced voltage has polarity that opposes the change.

■ *14–9 Knowledge Check*

Answer at end of chapter.

Are the windings of a coil and its resistance in series or parallel with the induced voltage?

■ *14–9 Self-Review*

Answers at end of chapter.

a. **The magnetic flux of 8 Wb changes to 10 Wb in 1 s. How much is $d\phi/dt$?**
b. **The flux of 8 μWb changes to 10 μWb in 1 μs. How much is $d\phi/dt$?**

Figure 14–19 Schematic symbols commonly used to represent relay contacts. (*a*) Symbols used to represent normally open (NO) contacts. (*b*) Symbols used to represent normally closed (NC) contacts.

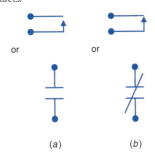

or   or

(*a*)   (*b*)

GOOD TO KNOW

Many relays are packaged in a hermetically sealed enclosure, which is a type of enclosure that isolates the relay, namely the switching contacts, from the environment.

14–10 Relays

A *relay* is an electromechanical device that operates by electromagnetic induction. It uses either an ac- or a dc-actuated electromagnet to open or close one or more sets of contacts. Relay contacts that are open when the relay is not energized are called *normally open* (NO) contacts. Conversely, relay contacts that are closed when the relay is not energized are called *normally closed* (NC) contacts. Relay contacts are held in their resting or normal position either by a spring or by some type of gravity-actuated mechanism. In most cases, an adjustment of the spring tension is provided to set the restraining force on the normally open and normally closed contacts to some desired level based on predetermined circuit conditions.

Figure 14–19 shows the schematic symbols that are commonly used to represent relay contacts. Figure 14–19*a* shows the symbols used to represent normally open contacts, and Fig. 14–19*b* shows the symbols used to represent normally closed contacts. When normally open contacts close, they are said to *make*, whereas when normally closed contacts open they are said to *break*. Like mechanical switches, the switching contacts of a relay can have any number of poles and throws.

Figure 14–20 shows the basic parts of an SPDT armature relay. Terminal connections 1 and 2 provide connection to the electromagnet (relay coil), and terminal connections 3, 4, and 5 provide connections to the SPDT relay contacts which open or close when the relay is energized. A relay is said to be *energized* when NO contacts close and NC contacts open. The movable arm of an electromechanical relay is called the *armature*. The armature is magnetic and has contacts that make or break with other contacts when the relay is energized. For example, when terminals 1 and 2 in Fig. 14–20 are connected to a dc source, current flows in the relay coil and an electromagnet is formed. If there is sufficient current in the relay coil, contacts 3 and 4 close (make) and contacts 4 and 5 open (break). The armature is attracted whether the electromagnet produces a north or a south pole on the end adjacent to the armature. Figure 14–21 is a photo of a typical relay.

Relay Specifications

Manufacturers of electromechanical relays always supply a specification sheet for each of their relays. The specification sheet contains voltage and current ratings for both the relay coil and its switch contacts. The specification sheet also includes information regarding the location of the relay coil and switching contact terminals. And finally, the specification sheet will indicate whether the

Figure 14–20 Basic parts of an SPDT armature relay. Terminal connections 1 and 2 provide connection to the electromagnet, and terminal connections 3, 4, and 5 provide connections to the SPDT relay contacts which open or close when the relay is energized.

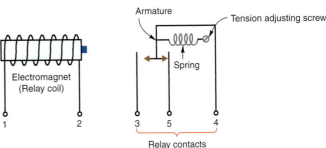

Figure 14–21 Typical relay.

relay can be energized from either an ac or a dc source. The following is an explanation of a relay's most important ratings.

Pickup voltage. The minimum amount of relay coil voltage necessary to energize or operate the relay.

Pickup current. The minimum amount of relay coil current necessary to energize or operate the relay.

Holding current. The minimum amount of current required to keep a relay energized or operating. (The holding current is less than the pickup current.)

Dropout voltage. The maximum relay coil voltage at which the relay is no longer energized.

Contact voltage rating. The maximum voltage the relay contacts can switch safely.

Contact current rating. The maximum current the relay contacts can switch safely.

Contact voltage drop. The voltage drop across the closed contacts of a relay when operating.

Insulation resistance. The resistance measured across the relay contacts in the open position.

Relay Applications

Figure 14–22 shows schematic diagrams for two relay systems. The diagram in Fig. 14–22a represents an open-circuit system. With the control switch S_1 open, the SPST relay contacts are open and the load is inoperative. Closing S_1 energizes the relay. This closes the NO relay contacts and makes the load operative.

||| MultiSim **Figure 14–22** Schematic diagrams for two relay systems. (a) Open-circuit system. (b) Closed-circuit system.

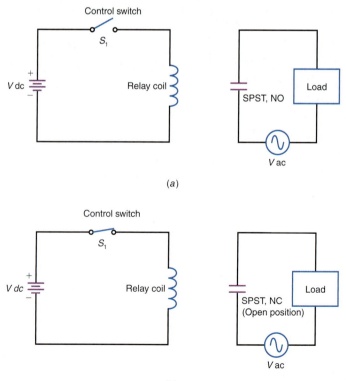

Figure 14–22b represents a closed-circuit system. In this case, the relay is energized by the control switch S_1, which is closed during normal operation. With the relay energized, the normally closed relay contacts are open and the load is inoperative. When it is desired to operate the load, the control switch S_1 is opened. This returns the relay contacts to their normally closed position, thereby activating the load.

It is important to note that a relay can be energized using a low-voltage, low-power source. However, the relay contacts can be used to control a circuit whose load consumes much more power at a much higher voltage than the relay coil circuit. In fact, one of the main advantages of using a relay is its ability to switch or control very high power loads with a relatively low amount of input power. In remote-control applications, a relay can control a high power load a long distance away much more efficiently than a mechanical switch can. When a mechanical switch is used to control a high power load a long distance away, the I^2R power loss in the conductors carrying current to the load can become excessive. Critical thinking problems 14–25 and 14–26 illustrate the advantages of using a relay to control a high power load a long distance away.

Common Relay Troubles

If a relay coil develops an open, the relay cannot be energized. The reason is simple. With an open relay coil, the current is zero and no magnetic field is set up by the electromagnet to attract the armature. An ohmmeter can be used to check for the proper relay coil resistance. An open relay coil measures infinite (∞) resistance. Since it is usually not practical to repair an open relay coil, the entire relay must be replaced.

A common problem with relays is dirty switch contacts. The switch contacts develop a thin carbon coating after extended use from arcing across the contact terminals when they are opened and closed. Dirty switch contacts usually produce intermittent operation of the load being controlled—for example, a motor. In some cases, the relay contacts may chatter (vibrate) if they are dirty.

One final point: The manufacturer of a relay usually indicates its life expectancy in terms of the number of times the relay can be energized (operated). A typical value is 5 million operations.

■ *14–10 Knowledge Check*

Answer at end of chapter.

What is the name given to the moveable arm of an electromechanical relay?

■ *14–10 Self-Review*

Answers at end of chapter.

a. A relay is energized if NC contacts are opened. (True or False)
b. The pickup current is the minimum relay coil current required to keep a relay energized. (True or False)
c. The voltage drop across a set of closed relay contacts carrying 1 A of current is very low. (True or False)
d. An open relay coil measures 0 Ω with an ohmmeter. (True or False)

Summary

- For an electromagnet, the strength of the magnetic field depends on how much current flows in the turns of the coil. The coil serves as a bar magnet with opposite poles at the ends.

- The strength of an electromagnet is specified in SI units of ampere-turns. The product of amperes (A) and turns (t) indicates the magnetomotive force (mmf) of the coil. The cgs unit of mmf is the gilbert (Gb).

- The field intensity, H, of a coil specifies the mmf per unit length. The units of H are A · t/m and the oersted.

- The permeability of a material indicates its ability to concentrate magnetic flux.

- Demagnetization of a magnetic material is also known as degaussing.

- Current in a straight conductor has an associated magnetic field with circular lines of force in a plane perpendicular to the conductor. The direction of the circular field is counterclockwise when you look along the conductor in the direction of electron flow.

- The left-hand rule for determining the polarity of an electromagnet says that when your fingers curl around the turns in the direction of electron flow, the thumb points to the north pole.

- Motor action is the motion that results from the net force of two fields that can aid or cancel each other. The direction of the resultant force is always from the stronger field to the weaker field.

- Generator action refers to induced voltage. For N turns, $v_{ind} = N(d\phi/dt)$, where $d\phi/dt$ stands for the change in flux (ϕ) in time (t). The change is given in webers per second. There must be a change in flux to produce induced voltage.

- Lenz's law states that the direction of an induced current must be such that its own magnetic field will oppose the action that produced the induced current.

- The switching contacts of an electromechanical relay may be either normally open (NO) or normally closed (NC). The contacts are held in their normal or resting positions by springs or some gravity-actuated mechanism.

- The movable arm on a relay is called the *armature*. The armature is magnetic and has contacts that open or close with other contacts when the relay is energized.

- The pickup current of a relay is the minimum amount of relay coil current that will energize the relay. The holding current is the minimum relay coil current required to keep a relay energized.

Important Terms

- Ampere-Turn (A · t) - the SI unit of magnetomotive force (mmf).

- Ampere-Turns/Meter (A · t/m) - the SI unit of field intensity, H.

- B-H Magnetization Curve - a graph showing how the flux density, B, in teslas, increases with the field intensity, H, in ampere-turns/meter.

- Degaussing - a method of demagnetizing a material by using an alternating current. The method involves magnetizing and demagnetizing a material with a diminishing magnetic field until the material has practically zero residual induction.

- Faraday's Law - a law for determining the amount of induced voltage in a conductor. The amount of induced voltage, v_{ind}, is calculated as

$$v_{ind} = N\frac{d\phi}{dt}.$$

- Field Intensity (H) - the amount of mmf per unit length. The units for field intensity are A · t/m and oersted.

- Holding Current - the minimum amount of relay-coil current required to keep a relay energized or operating.

- Hysteresis - hysteresis means lagging behind. With respect to the magnetic flux in an iron core of an electromagnet, the flux lags behind the increases and decreases in magnetizing force.

- Left-Hand Rule - if a coil is grasped with the fingers of the left hand curled around the coil in the direction of electron flow, the thumb points to the north pole of the coil.

- Lenz's Law - Lenz's law states that the direction of the induced current in a conductor must be such that its own magnetic field will oppose the action that produced the induced current.

- Magnetomotive Force (mmf) - a measure of the strength of a magnetic field based on the amount of current flowing in the turns of a coil. The units of mmf are the ampere-turn (A · t) and the gilbert (Gb).

- Motor Action - a motion that results from the net force of two magnetic fields that can aid or cancel each other. The direction of the resultant force is always from a stronger field to a weaker field.

- Pickup Current - the minimum amount of relay-coil current necessary to energize or operate a relay.

- Saturation - the point in a magnetic material, such as an iron core, where further increases in field intensity produce no further increases in flux density.

Related Formulas

Ampere-turns $= I \times N =$ mmf

$H = \dfrac{\text{ampere-turns of mmf}}{l \text{ meters}}$

$B = \mu \times H$

$\mu = \dfrac{B}{H}$

$V_{\text{ind}} = N \dfrac{d\phi}{dt}$

Self-Test

1. A current of 20 mA flowing through a coil with 500 turns produces an mmf of
   a. 100 A · t.
   b. 1 A · t.
   c. 10 A · t.
   d. 7.93 A · t.

2. A coil with 1000 turns must provide an mmf of 50 A · t. The required current is
   a. 5 mA.
   b. 0.5 A.
   c. 50 μA.
   d. 50 mA.

3. The left-hand rule for solenoids states that
   a. if the fingers of the left-hand encircle the coil in the same direction as electron flow, the thumb points in the direction of the north pole.
   b. if the thumb of the left-hand points in the direction of current flow, the fingers point toward the north pole.
   c. if the fingers of the left-hand encircle the coil in the same direction as electron flow, the thumb points in the direction of the south pole.
   d. if the thumb of the right-hand points in the direction of electron flow, the fingers will point in the direction of the north pole.

4. The physical motion resulting from the forces of two magnetic fields is called
   a. Lenz's law.
   b. motor action.
   c. the left-hand rule for coils.
   d. integration.

5. Motor action always tends to produce motion from
   a. a stronger field toward a weaker field.
   b. a weaker field toward a stronger field.
   c. a north pole toward a south pole.
   d. none of the above.

6. A conductor will have an induced current or voltage only when there is
   a. a stationary magnetic field.
   b. a stationary conductor.
   c. relative motion between the wire and magnetic field.
   d. both a and b.

7. The polarity of an induced voltage is determined by
   a. motor action.
   b. Lenz's law.
   c. the number of turns in the coil.
   d. the amount of current in the coil.

8. For a relay, the pickup current is defined as
   a. the maximum current rating of the relay coil.
   b. the minimum relay-coil current required to keep a relay energized.
   c. the minimum relay-coil current required to energize a relay.
   d. the minimum current in the switching contacts.

9. The moveable arm of an attraction-type relay is called the
   a. contacts.
   b. relay coil.
   c. terminal.
   d. armature.

10. For a conductor being moved through a magnetic field, the amount of induced voltage is determined by
    a. the rate at which the conductor cuts the magnetic flux.
    b. the number of magnetic flux lines which are cut by the conductor.
    c. the time of day during which the conductor is moved through the field.
    d. both a and b.

11. Degaussing is done with
    a. strong permanent magnets.
    b. alternating current.
    c. static electricity.
    d. direct current.

12. Hysteresis losses
    a. increase with higher frequencies.
    b. decrease with higher frequencies.
    c. are greater with direct current.
    d. increase with lower frequencies.

13. The saturation of an iron core occurs when
    a. all of the molecular dipoles and magnetic domains are aligned by the magnetizing force.
    b. the coil is way too long.
    c. the flux density cannot be increased in the core when the field intensity is increased.
    d. both a and c.

14. The unit of field intensity is the
    a. oersted.
    b. gilbert.
    c. A · t/m.
    d. both a and c.

15. For a single conductor carrying an alternating current, the associated magnetic field is
    a. only on the top side.
    b. parallel to the direction of current.
    c. at right angles to the direction of current.
    d. only on the bottom side.

16. A coil with 200 mA of current has an mmf of 80 A · t. How many turns does the coil have?
    a. 4000 turns.
    b. 400 turns.
    c. 40 turns.
    d. 16 turns.

17. The magnetic field surrounding a solenoid is
    a. like that of a permanent magnet.
    b. unable to develop north and south poles.
    c. one without magnetic flux lines.
    d. unlike that of a permanent magnet.

18. For a relay, the holding current is defined as
    a. the maximum current the relay contacts can handle.
    b. the minimum amount of relay-coil current required to keep a relay energized.
    c. the minimum amount of relay-coil current required to energize a relay.
    d. the maximum current required to operate a relay.

19. A vertical wire with electron flow into this page has an associated magnetic field which is
    a. clockwise.
    b. counterclockwise.
    c. parallel to the wire.
    d. none of the above.

20. How much is the induced voltage when a magnetic flux cuts across 150 turns at the rate of 5 Wb/s?
    a. 7.5 kV.
    b. 75 V.
    c. 750 V.
    d. 750 mV.

Questions

1. Draw a diagram showing two conductors connecting a battery to a load resistance through a closed switch. (a) Show the magnetic field of the current in the negative side of the line and in the positive side. (b) Where do the two fields aid? Where do they oppose?

2. State the rule for determining the magnetic polarity of a solenoid. (a) How can the polarity be reversed? (b) Why are there no magnetic poles when the current through the coil is zero?

3. Why does the motor action between two magnetic fields result in motion toward the weaker field?

4. Why does current in a conductor perpendicular to this page have a magnetic field in the plane of the paper?

5. Why must the conductor and the external field be perpendicular to each other to have motor action or to generate induced voltage?

6. Explain briefly how either motor action or generator action can be obtained with the same conductor in a magnetic field.

7. Assume that a conductor being cut by the flux of an expanding magnetic field has 10 V induced with the top end positive. Now analyze the effect of the following changes: (a) The magnetic flux continues to expand, but at a slower rate. How does this affect the amount of induced voltage and its polarity? (b) The magnetic flux is constant, neither increasing nor decreasing. How much is the induced voltage? (c) The magnetic flux contracts, cutting across the conductor with the opposite direction of motion. How does this affect the polarity of the induced voltage?

8. Redraw the graph in Fig. 14–18c for 500 turns with all other factors the same.

9. Redraw the circuit with the coil and battery in Fig. 14–10, showing two different ways to reverse the magnetic polarity.

10. Referring to Fig. 14–18, suppose that the flux decreases from 8 Wb to zero at the same rate as the increase. Tabulate all values as in Table 14–2 and draw the three graphs corresponding to those in Fig. 14–18.

11. Assume that you have a relay whose pickup and holding current values are unknown. Explain how you can determine their values experimentally.

12. List two factors that determine the strength of an electromagnet.

13. What is meant by magnetic hysteresis?

14. What is meant by the saturation of an iron core?

Problems

SECTION 14–1 AMPERE-TURNS OF MAGNETOMOTIVE FORCE (mmf)

14–1 What is (a) the cgs unit of mmf? (b) the SI unit of mmf?

14–2 Calculate the ampere-turns of mmf for a coil with the following values:
   a. $I = 10$ mA, $N = 150$ turns
   b. $I = 15$ mA, $N = 100$ turns
   c. $I = 2$ mA, $N = 5000$ turns
   d. $I = 100$ μA, $N = 3000$ turns

14–3 Calculate the ampere-turns of mmf for a coil with the following values:
   a. $I = 5$ mA, $N = 4000$ turns
   b. $I = 40$ mA, $N = 50$ turns
   c. $I = 250$ mA, $N = 40$ turns
   d. $I = 600$ mA, $N = 300$ turns

14–4 Calculate the current required in a coil to provide an mmf of 2 A $\cdot$ t if the number of turns equals
   a. 50.
   b. 500.
   c. 100.
   d. 2000.

14–5 Calculate the number of turns required in a coil to provide an mmf of 100 A $\cdot$ t if the current equals
   a. $I = 100$ mA.
   b. $I = 25$ mA.
   c. $I = 40$ mA.
   d. $I = 2$ A.

14–6 Convert the following values of mmf to gilberts (Gb):
   a. 100 A $\cdot$ t
   b. 30 A $\cdot$ t
   c. 500 A $\cdot$ t

14–7 Convert the following values of mmf to ampere-turns (A $\cdot$ t):
   a. 126 Gb
   b. 37.8 Gb
   c. 630 Gb

SECTION 14–2 FIELD INTENSITY (*H*)

14–8 What is (a) the cgs unit of field intensity? (b) the SI unit of field intensity?

14–9 Calculate the field intensity, H, in ampere-turns per meter, for each of the following cases:
   a. mmf = 100 A $\cdot$ t, $l = 0.2$ m
   b. mmf = 25 A $\cdot$ t, $l = 0.25$ m
   c. mmf = 4 A $\cdot$ t, $l = 0.08$ m
   d. mmf = 20 A $\cdot$ t, $l = 0.1$ m

14–10 Calculate the field intensity, *H*, in ampere-turns per meter, for each of the following cases:

   a. $I = 40$ mA, N = 500 turns, $l = 0.2$ m
   b. $I = 100$ mA, N = 1000 turns, $l = 0.5$ m
   c. $I = 60$ mA, N = 600 turns, $l = 0.25$ m
   d. $I = 10$ mA, N = 300 turns, $l = 0.075$ m

14–11 Convert the following values of field intensity to oersteds:
   a. 50 A $\cdot$ t/m
   b. 150 A $\cdot$ t/m

14–12 Convert the following values of field intensity to A $\cdot$ t/m.
   a. 0.63 oersteds
   b. 1.89 oersteds

14–13 Calculate the absolute permeability, μ, of a material if its relative permeability, μ_r, equals
   a. 10.
   b. 50.
   c. 100.
   d. 500.
   e. 1000.

14–14 A coil with an iron core has a field intensity, *H*, of 50 A $\cdot$ t/m. If the relative permeability, μ_r, equals 300, calculate the flux density, *B*, in teslas.

14–15 Calculate the relative permeability, μ_r, of an iron core when a field intensity, *H*, of 750 A $\cdot$ t/m produces a flux density, *B*, of 0.126 T.

14–16 Calculate the field intensity, *H*, of an electromagnet if the flux density, *B*, equals 0.504 teslas and the relative permeability of the core is 200.

SECTION 14–3 *B-H* MAGNETIZATION CURVE

14–17 Referring to the *B-H* curve in Figure 14–3, calculate the absolute permeability, μ, in SI units for the iron core at a field intensity, *H*, of
   a. 3000 A $\cdot$ t/m.
   b. 5000 A $\cdot$ t/m.

SECTION 14–9 GENERATING AN INDUCED VOLTAGE

14–18 A magnetic field cuts across a coil of 500 turns at the rate of 100 μWb/s. Calculate v_{ind}.

14–19 A magnetic field cuts across a coil of 400 turns at the rate of 0.02 Wb/s. Calculate v_{ind}.

14–20 A magnetic flux of 300 Mx cuts across a coil of 1500 turns in 200 μs. Calculate v_{ind}.

14–21 The magnetic flux surrounding a coil changes from 1000 to 6000 Mx in 5 μs. If the coil has 200 turns, how much is the induced voltage?

14–22 A coil has an induced voltage of 1 kV when the rate of flux change is 0.5 Wb/s. How many turns are in the coil?

Critical Thinking

14-23 Derive the value of 1.26×10^{-6} T/(A · t/m) for μ_0 from $\mu = B/H$.

14-24 What is the relative permeability (μ_r) of a piece of soft iron whose permeability (μ) equals 3.0×10^{-3} T/(A · t/m)?

14-25 Refer to Fig. 14–23a. Calculate **(a)** the total wire resistance R_W of the No. 12-gage copper wires; **(b)** the total resistance R_T of the circuit; **(c)** the voltage available across the load R_L; **(d)** the I^2R power loss in the wire conductors; **(e)** the load power P_L; **(f)** the total power P_T consumed by the circuit; **(g)** the percent efficiency of the system calculated as $(P_L/P_T) \times 100$.

14-26 Refer to Fig. 14–23b. Calculate **(a)** the total wire resistance R_W of the No. 20-gage copper wires; **(b)** the total

resistance R_T of the relay coil circuit; **(c)** the voltage across the relay coil; **(d)** the I^2R power loss in the No. 20–gage copper wires in the relay coil circuit; **(e)** the total wire resistance R_W of the 10-ft length of No. 12-gage copper wires that connect the 16-Ω load R_L to the 240-V_{ac} power line; **(f)** the voltage available across the load R_L; **(g)** the I^2R power loss in the 10-ft length of the No. 12-gage copper wire; **(h)** the load power P_L; **(i)** the total power P_T consumed by the load side of the circuit; **(j)** the percent efficiency of the system calculated as $(P_L/P_T) \times 100$.

14-27 Explain the advantage of using a relay rather than an ordinary mechanical switch when controlling a high power load a long distance away. Use your solutions from Critical Thinking Probs. 14–25 and 14–26 to support your answer.

Figure 14–23 Circuit diagram for Critical Thinking Probs. 14–25 and 14–26. (a) Mechanical switch controlling a high power load a long distance away. (b) Relay controlling a high power load a long distance away.

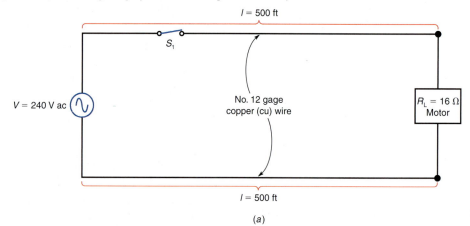

(a)

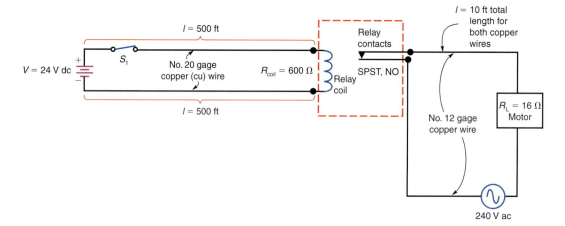

(b)

Answers to Knowledge Check Problems

14-1 18.9 Gb

14-2 the shorter coil

14-3 0.315 T

14-4 $-300 \dfrac{A \cdot t}{m}$

14-5 clockwise

14-6 the direction of electron flow and the direction of the coil windings

14-7 downward

14-8 electron flow is from A to B

14-9 in series

14-10 armature

Answers to Self-Reviews

14-1 a. 486 A · t
b. 630 Gb

14-2 a. 1
b. 200
c. 200 G/Oe
d. $250 \dfrac{A \cdot t}{m}$
e. 6.3 oersteds

14-3 a. 0.189 T
b. 4000 A · t/m approx.

14-4 a. T
b. T

14-5 a. T
b. T

14-6 a. left
b. south

14-7 a. T
b. T

14-8 a. south
b. south
c. yes

14-9 a. 2 Wb//s
b. 2 Wb/s

14-10 a. T
b. F
c. T
d. F

chapter 15

Alternating Voltage and Current

This chapter begins the analysis of alternating voltage, as used for the 120-Vac power line. An alternating voltage is one that continuously varies in amplitude and periodically reverses in polarity. One cycle includes two alternations in polarity. The number of cycles per second is the frequency measured in hertz (Hz). Every ac voltage has both amplitude variations and polarity reversals. The amplitude values and rate of polarity reversal, however, vary from one ac waveform to the next. This chapter covers the theory, the terminology, and the measurements of alternating voltage and current.

Outline

Objectives

After studying this chapter you should be able to

- *Describe* how a sine wave of alternating voltage is generated.
- *Calculate* the instantaneous value of a sine wave of alternating voltage or current.
- *Define* the following values for a sine wave: peak, peak-to-peak, root-mean-square, and average.
- *Calculate* the rms, average, and peak-to-peak values of a sine wave when the peak value is known.
- *Define* frequency and period and list the units of each.
- *Calculate* the wavelength when the frequency is known.
- *Explain* the concept of phase angles.
- *Describe* the makeup of a nonsinusoidal waveform.
- *Define* the term *harmonics*.
- *Outline* the basics of residential house wiring.

Important Terms

alternation

average value

cycle

decade

effective value

form factor

frequency

generator

harmonic frequency

hertz (Hz)

motor

nonsinusoidal waveform

octave

peak value

period

phase angle

phasor

quadrature phase

radian

root-mean-square (rms) value

sine wave

wavelength

15–1 Alternating Current Applications

Figure 15–1 shows the output from an ac voltage generator with the reversals between positive and negative polarities and the variations in amplitude. In Fig. 15–1a, the waveform shown simulates an ac voltage as it would appear on the screen of an oscilloscope, which is an important test instrument for ac voltages. The oscilloscope shows a picture of any ac voltage connected to its input terminals. It also indicates the amplitude. The details of how to use the oscilloscope for ac voltage measurements is explained in App. E.

In Fig. 15–1b, the graph of the ac waveform shows how the output from the generator in Fig. 15–1c varies with respect to time. Assume that this graph shows *V* at terminal 2 with respect to terminal 1. Then the voltage at terminal 1 corresponds to the zero axis in the graph as the reference level. At terminal 2, the output voltage has positive amplitude variations from zero up to the peak value and down to zero. All these voltage values are with respect to terminal 1. After a half-cycle, the voltage at terminal 2 becomes negative, still with respect to the other terminal. Then the same voltage variations are repeated at terminal 2, but they have negative polarity compared to the reference level. Note that if we take the voltage at terminal 1 with terminal 2 as the reference, the waveform in Fig. 15–1b would have the same shape but be inverted in polarity. The negative half-cycle would come first, but it does not matter which is first or second.

The characteristic of varying values is the reason that ac circuits have so many uses. For instance, a transformer can operate only with alternating current to step up or step down an ac voltage. The reason is that the changing current produces changes in its associated magnetic field. This application is just an example of inductance *L* in ac circuits, where the changing magnetic flux of a varying current can produce induced voltage. The details of inductance are explained in Chaps. 19, 20, and 21.

A similar but opposite effect in ac circuits is capacitance *C*. The capacitance is important with the changing electric field of a varying voltage. Just as *L*

|||| MultiSim **Figure 15–1** Waveform of ac power-line voltage with frequency of 60 Hz. Two cycles are shown. (*a*) Oscilloscope display. (*b*) Details of waveform and alternating polarities. (*c*) Symbol for an ac voltage source.

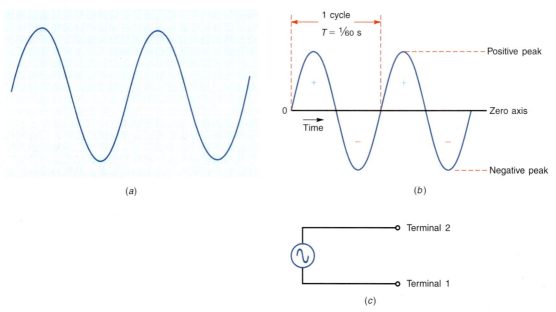

(*a*)

(*b*)

(*c*)

has an effect with alternating current, C has an effect that depends on alternating voltage. The details of capacitance are explained in Chaps. 16, 17, and 18.

The L and C are additional factors, beside resistance R, in the operation of ac circuits. Note that R is the same for either a dc or an ac circuit. However, the effects of L and C depend on having an ac source. The rate at which the ac variations occur, which determines the frequency, allows a greater or lesser reaction by L and C. Therefore, the effect is different for different frequencies. One important application is a resonant circuit with L and C that is tuned to a particular frequency. Tuning in radio and television stations is an application of resonance in an LC circuit.

In general, electronic circuits are combinations of R, L, and C, with both direct current and alternating current. Audio, video, and radio signals are ac voltages and currents. However, amplifiers that use transistors and integrated circuits need dc voltages to conduct any current at all. The resulting output of an amplifier circuit, therefore, consists of direct current with a superimposed ac signal.

■ *15–1 Knowledge Check*

Answer at end of chapter.

Which test instrument is used for viewing ac voltages?

■ *15–1 Self-Review*

Answers at end of chapter.

a. **An ac voltage varies in magnitude and reverses in polarity. (True or False)**
b. **A transformer can operate with either ac or a steady dc input. (True or False)**
c. **Inductance L and capacitance C are important factors in ac circuits. (True or False)**

Figure 15–2 A loop rotating in a magnetic field to produce induced voltage v with alternating polarities. (*a*) Loop conductors moving parallel to magnetic field results in zero voltage. (*b*) Loop conductors cutting across magnetic field produce maximum induced voltage.

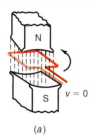
(a)

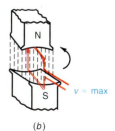
(b)

15–2 Alternating-Voltage Generator

An alternating voltage is a voltage that continuously varies in magnitude and periodically reverses in polarity. In Fig. 15–1, the variations up and down on the waveform show the changes in magnitude. The zero axis is a horizontal line across the center. Then voltages above the center have positive polarity, and values below center are negative.

Figure 15–2 shows how such a voltage waveform is produced by a rotary generator. The conductor loop rotates through the magnetic field to generate the induced ac voltage across its open terminals. The magnetic flux shown here is vertical, with lines of force in the plane of the paper.

In Fig. 15–2a, the loop is in its horizontal starting position in a plane perpendicular to the paper. When the loop rotates counterclockwise, the two longer conductors move around a circle. Note that in the flat position shown, the two long conductors of the loop move vertically up or down but parallel to the vertical flux lines. In this position, motion of the loop does not induce a voltage because the conductors are not cutting across the flux.

When the loop rotates through the upright position in Fig. 15–2b, however, the conductors cut across the flux, producing maximum induced voltage. The shorter connecting wires in the loop do not have any appreciable voltage induced in them.

Each of the longer conductors has opposite polarity of induced voltage because the conductor at the top is moving to the left while the bottom conductor is moving to the right. The amount of voltage varies from zero to maximum as the loop moves from a flat position to upright, where it can cut

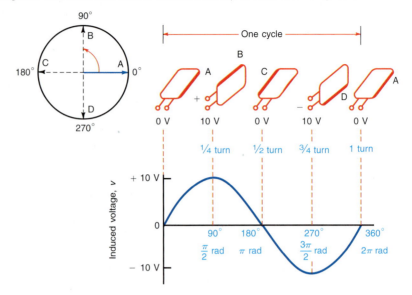

Figure 15–3 One cycle of alternating voltage generated by a rotating loop. The magnetic field, not shown here, is directed from top to bottom, as in Fig. 15–2.

across the flux. Also, the polarity at the terminals of the loop reverses as the motion of each conductor reverses during each half-revolution.

With one revolution of the loop in a complete circle back to the starting position, therefore, the induced voltage provides a potential difference v across the loop, varying in the same way as the wave of voltage shown in Fig. 15–1. If the loop rotates at the speed of 60 revolutions per second, the ac voltage has a frequency of 60 Hz.

The Cycle

One complete revolution of the loop around the circle is a *cycle*. In Fig. 15–3, the generator loop is shown in its position at each quarter-turn during one complete cycle. The corresponding wave of induced voltage also goes through one cycle. Although not shown, the magnetic field is from top to bottom of the page, as in Fig. 15–2.

At position A in Fig. 15–3, the loop is flat and moves parallel to the magnetic field, so that the induced voltage is zero. Counterclockwise rotation of the loop moves the dark conductor to the top at position B, where it cuts across the field to produce maximum induced voltage. The polarity of the induced voltage here makes the open end of the dark conductor positive. This conductor at the top is cutting across the flux from right to left. At the same time, the opposite conductor below is moving from left to right, causing its induced voltage to have opposite polarity. Therefore, maximum induced voltage is produced at this time across the two open ends of the loop. Now the top conductor is positive with respect to the bottom conductor.

In the graph of induced voltage values below the loop in Fig. 15–3, the polarity of the dark conductor is shown with respect to the other conductor. Positive voltage is shown above the zero axis in the graph. As the dark conductor rotates from its starting position parallel to the flux toward the top position, where it cuts maximum flux, more and more induced voltage is produced with positive polarity.

When the loop rotates through the next quarter-turn, it returns to the flat position shown in C, where it cannot cut across flux. Therefore, the induced voltage values shown in the graph decrease from the maximum value to zero at the half-turn, just as the voltage was zero at the start. The half-cycle of revolution is called an *alternation*.

The next quarter-turn of the loop moves it to the position shown at D in Fig. 15–3, where the loop cuts across the flux again for maximum induced voltage. Note, however, that here the dark conductor is moving left to right at the bottom of the loop. This motion is reversed from the direction it had when it was at the top, moving right to left. Because the direction of motion is reversed during the second half-revolution, the induced voltage has opposite polarity with the dark conductor negative. This polarity is shown as negative voltage below the zero axis. The maximum value of induced voltage at the third quarter-turn is the same as at the first quarter-turn but with opposite polarity.

When the loop completes the last quarter-turn in the cycle, the induced voltage returns to zero as the loop returns to its flat position at A, the same as at the start. This cycle of values of induced voltage is repeated as the loop continues to rotate with one complete cycle of voltage values, as shown, for each circle of revolution.

Note that zero at the start and zero after the half-turn of an alternation are not the same. At the start, the voltage is zero because the loop is flat, but the dark conductor is moving upward in the direction that produces positive voltage. After one half-cycle, the voltage is zero with the loop flat, but the dark conductor is moving downward in the direction that produces negative voltage. After one complete cycle, the loop and its corresponding waveform of induced voltage are the same as at the start. *A cycle can be defined, therefore, as including the variations between two successive points having the same value and varying in the same direction.*

Angular Measure

Because the cycle of voltage in Fig. 15–3 corresponds to rotation of the loop around a circle, it is convenient to consider parts of the cycle in angles. The complete circle includes 360°. One half-cycle, or one alternation, is 180° of revolution. A quarter-turn is 90°. The circle next to the loop positions in Fig. 15–3 illustrates the angular rotation of the dark conductor as it rotates counterclockwise from 0 to 90 to 180° for one half-cycle, then to 270°, and returning to 360° to complete the cycle. Therefore, one cycle corresponds to 360°.

Radian Measure

In angular measure it is convenient to use a specific unit angle called the *radian* (abbreviated rad), which is an angle equal to 57.3°. Its convenience is due to the fact that a radian is the angular part of the circle that includes an arc equal to the radius r of the circle, as shown in Fig. 15–4. The circumference around the circle equals $2\pi r$. A circle includes 2π rad, then, as each radian angle includes one length r of the circumference. Therefore, one cycle equals 2π rad.

As shown in the graph in Fig. 15–3, divisions of the cycle can be indicated by angles in either degrees or radians. The comparison between degrees and radians can be summarized as follows:

Zero degrees is also zero radians
$360° = 2\pi$ rad
$180° = \frac{1}{2} \times 2\pi$ rad $= \pi$ rad
$90° = \frac{1}{2} \times \pi$ rad $= \pi/2$ rad
$270° = 180° + 90°$ or π rad $+ \pi/2$ rad $= 3\pi/2$ rad

The constant 2π in circular measure is numerically equal to 6.2832. This is double the value of 3.1416 for π. The Greek letter π (pi) is used to represent the ratio of the circumference to the diameter for any circle, which always has the numerical value of 3.1416. The fact that 2π rad is 360° can be shown as $2 \times 3.1416 \times 57.3° = 360°$ for a complete cycle.

Figure 15–4 One radian (rad) is the angle equal to 57.3°. The complete circle of 360° includes 2π rad.

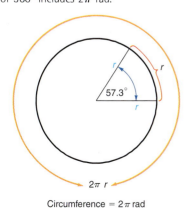

Circumference $= 2\pi$ rad

Answer at end of chapter.

What is the numerical value of π?

■ *15–2 Self-Review*

Answers at end of chapter.

Refer to Fig. 15–3.
a. How much is the induced voltage at $\pi/2$ rad?
b. How many degrees are in a complete cycle?

15–3 The Sine Wave

The voltage waveform in Figs. 15–1 and 15–3 is called a *sine wave, sinusoidal wave,* or *sinusoid* because the amount of induced voltage is proportional to the sine of the angle of rotation in the circular motion producing the voltage. The sine is a trigonometric function of an angle; it is equal to the ratio of the opposite side to the hypotenuse in a right triangle. This numerical ratio increases from zero for 0° to a maximum value of 1 for 90° as the side opposite the angle becomes larger.

The voltage waveform produced by the circular motion of the loop is a sine wave because the induced voltage increases to a maximum at 90°, when the loop is vertical, in the same way that the sine of the angle of rotation increases to a maximum at 90°. The induced voltage and sine of the angle correspond for the full 360° of the cycle. Table 15–1 lists the numerical values of the sine for several important angles to illustrate the specific characteristics of a sine wave.

Table 15–1		Values in a Sine Wave	
Angle θ			
Degrees	**Radians**	**Sin θ**	**Loop Voltage**
0	0	0	Zero
30	$\dfrac{\pi}{6}$	0.500	50% of maximum
45	$\dfrac{\pi}{4}$	0.707	70.7% of maximum
60	$\dfrac{\pi}{3}$	0.866	86.6% of maximum
90	$\dfrac{\pi}{2}$	1.000	Positive maximum value
180	π	0	Zero
270	$\dfrac{3\pi}{2}$	−1.000	Negative maximum value
360	2π	0	Zero

Notice that the sine wave reaches one-half its maximum value in $30°$, which is only one-third of $90°$. This fact means that the sine wave has a sharper slope of changing values when the wave is near the zero axis, compared with more gradual changes near the maximum value.

The instantaneous value of a sine-wave voltage for any angle of rotation is expressed by the formula

$$v = V_M \sin \theta \qquad\qquad (15\text{--}1)$$

where θ (Greek letter *theta*) is the angle, sin is the abbreviation for its sine, V_M is the maximum voltage value, and v is the instantaneous value of voltage at angle θ.

Example 15-1

A sine wave of voltage varies from zero to a maximum of 100 V. How much is the voltage at the instant of $30°$ of the cycle? $45°$? $90°$? $270°$?

ANSWER $\quad v = V_M \sin \theta = 100 \sin \theta$

At $30°$: $\quad v = V_M \sin 30° = 100 \times 0.5$
$\qquad\qquad\quad = 50$ V

At $45°$: $\quad v = V_M \sin 45° = 100 \times 0.707$
$\qquad\qquad\quad = 70.7$ V

At $90°$: $\quad v = V_M \sin 90° = 100 \times 1$
$\qquad\qquad\quad = 100$ V

At $270°$: $\quad v = V_M \sin 270° = 100 \times -1$
$\qquad\qquad\quad = -100$ V

The value of -100 V at $270°$ is the same as that at $90°$ but with opposite polarity.

To do the problems in Example 15–1, you must either refer to a table of trigonometric functions or use a scientific calculator that has trig functions.

Between zero at $0°$ and maximum at $90°$, the amplitudes of a sine wave increase exactly as the sine value of the angle of rotation. These values are for the first quadrant in the circle, that is, 0 to $90°$. From 90 to $180°$ in the second quadrant, the values decrease as a mirror image of the first $90°$. The values in the third and fourth quadrants, from 180 to $360°$, are exactly the same as 0 to $180°$ but with opposite sign. At $360°$, the waveform is back to $0°$ to repeat its values every $360°$.

In summary, the characteristics of the sine-wave ac waveform are

1. The cycle includes $360°$ or 2π rad.
2. The polarity reverses each half-cycle.
3. The maximum values are at 90 and $270°$.
4. The zero values are at 0 and $180°$.
5. The waveform changes its values fastest when it crosses the zero axis.
6. The waveform changes its values slowest when it is at its maximum value. The values must stop increasing before they can decrease.

A perfect example of the sine-wave ac waveform is the 60-Hz power-line voltage in Fig. 15–1.

■ 15–3 Knowledge Check

Answer at end of chapter.

The instantaneous voltage of a sine wave at 60° equals 75 V. What is the peak or maximum voltage at 90°?

■ 15–3 Self-Review

Answers at end of chapter.

A sine-wave voltage has a peak value of 170 V. What is its value at
a. 30°?
b. 45°?
c. 90°?

15–4 Alternating Current

When a sine wave of alternating voltage is connected across a load resistance, the current that flows in the circuit is also a sine wave. In Fig. 15–5, let the sine-wave voltage at the left in the diagram be applied across R of 100 Ω. The resulting sine wave of alternating current is shown at the right in the diagram. Note that the frequency is the same for v and i.

During the first alternation of v in Fig. 15–5, terminal 1 is positive with respect to terminal 2. Since the direction of electron flow is from the negative side of v, through R, and back to the positive side of v, current flows in the direction indicated by arrow A for the first half-cycle. This direction is taken as the positive direction of current in the graph for i, corresponding to positive values of v.

The amount of current is equal to v/R. If several instantaneous values are taken, when v is zero, i is zero; when v is 50 V, i equals 50 V/100, or 0.5 A; when v is 100 V, i equals 100 V/100, or 1 A. For all values of applied voltage with positive polarity, therefore, the current is in one direction, increasing to its maximum value and decreasing to zero, just like the voltage.

In the next half-cycle, the polarity of the alternating voltage reverses. Then terminal 1 is negative with respect to terminal 2. With reversed voltage polarity, current flows in the opposite direction. Electron flow is from terminal 1 of the voltage source, which is now the negative side, through R, and back to terminal 2. This direction of current, as indicated by arrow B in Fig. 15–5, is negative.

Figure 15–5 A sine wave of alternating voltage applied across R produces a sine wave of alternating current in the circuit. (*a*) Waveform of applied voltage. (*b*) AC circuit. Note the symbol for sine-wave generator V. (*c*) Waveform of current in the circuit.

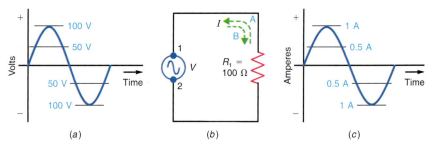

(a)   (b)   (c)

The negative values of i in the graph have the same numerical values as the positive values in the first half-cycle, corresponding to the reversed values of applied voltage. As a result, the alternating current in the circuit has sine-wave variations corresponding exactly to the sine-wave alternating voltage.

Only the waveforms for v and i can be compared. There is no comparison between relative values because the current and voltage are different quantities.

It is important to note that the negative half-cycle of applied voltage is just as useful as the positive half-cycle in producing current. The only difference is that the reversed polarity of voltage produces the opposite direction of current.

Furthermore, the negative half-cycle of current is just as effective as the positive values when heating the filament to light a bulb. With positive values, electrons flow through the filament in one direction. Negative values produce electron flow in the opposite direction. In both cases, electrons flow from the negative side of the voltage source, through the filament, and return to the positive side of the source. For either direction, the current heats the filament. The direction does not matter, since it is the motion of electrons against resistance that produces power dissipation. In short, resistance R has the same effect in reducing I for either direct current or alternating current.

■ 15–4 Knowledge Check

Answer at end of chapter.

In Fig. 15–5, what is the value of current in the circuit 30° into the ac cycle if $R = 5\ k\Omega$?

■ 15–4 Self-Review

Answers at end of chapter.

Refer to Fig. 15–5.
a. **When v is 70.7 V, how much is i?**
b. **How much is i at 30°?**

15–5 Voltage and Current Values for a Sine Wave

Since an alternating sine wave of voltage or current has many instantaneous values through the cycle, it is convenient to define specific magnitudes to compare one wave with another. The peak, average, and root-mean-square (rms) values can be specified, as indicated in Fig. 15–6. These values can be used for either current or voltage.

Peak Value

This is the maximum value V_M or I_M. For example, specifying that a sine wave has a peak value of 170 V states the highest value the sine wave reaches. All other values during the cycle follow a sine wave. The peak value applies to either the positive or the negative peak.

To include both peak amplitudes, the *peak-to-peak* (p-p) *value* may be specified. For the same example, the peak-to-peak value is 340 V, double the peak value of 170 V, since the positive and negative peaks are symmetrical. Note that the two opposite peak values cannot occur at the same time. Furthermore, in some waveforms, the two peaks are not equal.

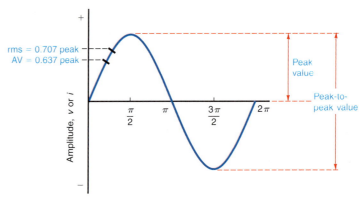

Average Value

This is an arithmetic average of all values in a sine wave for one alternation, or half-cycle. The half-cycle is used for the average because over a full cycle the average value is zero, which is useless for comparison. If the sine values for all angles up to 180° for one alternation are added and then divided by the number of values, this average equals 0.637. These calculations are shown in Table 15–2.

Since the peak value of the sine function is 1 and the average equals 0.637, then

$$\text{Average value} = 0.637 \times \text{peak value} \qquad (15\text{–}2)$$

With a peak of 170 V, for example, the average value is 0.637 × 170 V, which equals approximately 108 V.

Root-Mean-Square, or Effective, Value

The most common method of specifying the amount of a sine wave of voltage or current is by relating it to the dc voltage and current that will produce the same heating effect. This is called its *root-mean-square* value, abbreviated rms. The formula is

$$\text{rms value} = 0.707 \times \text{peak value} \qquad (15\text{–}3)$$

or

$$V_{\text{rms}} = 0.707 V_{\text{max}}$$

and

$$I_{\text{rms}} = 0.707 I_{\text{max}}$$

With a peak of 170 V, for example, the rms value is 0.707 × 170, or 120 V, approximately. This is the voltage of the commercial ac power line, which is always given in rms value.

It is often necessary to convert from rms to peak value. This can be done by inverting Formula (15–3), as follows:

$$\text{Peak} = \frac{1}{0.707} \times \text{rms} = 1.414 \times \text{rms} \qquad (15\text{–}4)$$

or

$$V_{\text{max}} = 1.414 V_{\text{rms}}$$

Table 15–2	Derivation of Average and rms Values for a Sine-Wave Alternation		
Interval	Angle θ	Sin θ	(Sin $\theta)^2$
1	15°	0.26	0.07
2	30°	0.50	0.25
3	45°	0.71	0.50
4	60°	0.87	0.75
5	75°	0.97	0.93
6	90°	1.00	1.00
7*	105°	0.97	0.93
8	120°	0.87	0.75
9	135°	0.71	0.50
10	150°	0.50	0.25
11	165°	0.26	0.07
12	180°	0.00	0.00
	Total	7.62	6.00

Average voltage:
$$\frac{7.62}{12} = 0.635^{\dagger}$$

rms value:
$$\sqrt{6/12} = \sqrt{0.5} = 0.707$$

* For angles between 90 and 180°, $sin\ \theta = sin\ (180° - \theta)$.
† More intervals and precise values are needed to get the exact average of 0.637.

and

$$I_{max} = 1.414 I_{rms}$$

Dividing by 0.707 is the same as multiplying by 1.414.

For example, commercial power-line voltage with an rms value of 120 V has a peak value of 120 × 1.414, which equals 170 V, approximately. Its peak-to-peak value is 2 × 170, or 340 V, which is double the peak value. As a formula,

Peak-to-peak value = 2.828 × rms value   (15–5)

The factor 0.707 for the rms value is derived as the square root of the average (mean) of all the squares of the sine values. If we take the sine for each angle in the cycle, square each value, add all the squares, divide by the number of values added to obtain the average square, and then take the square root of this mean value, the answer is 0.707. These calculations are shown in Table 15–2 for one alternation from 0 to 180°. The results are the same for the opposite alternation.

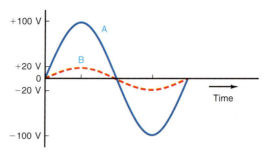

The advantage of the rms value derived in terms of the squares of the voltage or current values is that it provides a measure based on the ability of the sine wave to produce power, which is I^2R or V^2/R. As a result, the rms value of an alternating sine wave corresponds to the same amount of direct current or voltage in heating power. An alternating voltage with an rms value of 120 V, for instance, is just as effective in heating the filament of a light bulb as 120 V from a steady dc voltage source. For this reason, the rms value is also called the *effective* value.

Unless indicated otherwise, all sine-wave ac measurements are in rms values. The capital letters V and I are used, corresponding to the symbols for dc values. As an example, $V = 120$ V for ac power-line voltage.

The ratio of the rms to average values is the *form factor*. For a sine wave, this ratio is $0.707/0.637 = 1.11$.

Note that sine waves can have different amplitudes but still follow the sinusoidal waveform. Figure 15–7 compares a low-amplitude voltage with a high-amplitude voltage. Although different in amplitude, they are both sine waves. In each wave, the rms value = 0.707 × peak value.

■ 15–5 Knowledge Check

Answer at end of chapter.

What is the rms value of a sine wave whose average value equals 7.644 V?

■ 15–5 Self-Review

Answers at end of chapter.

a. **Convert 170 V peak to rms value.**
b. **Convert 10 V rms to peak value.**
c. **Convert 1 V rms to peak-to-peak value.**

||| MultiSim   Figure 15–8

Number of cycles per second is the frequency in hertz (Hz) units. (*a*) $f = 1$ Hz. (*b*) $f = 4$ Hz.

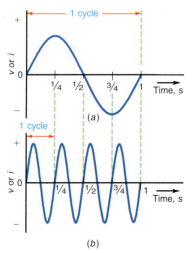

15–6 Frequency

The number of cycles per second is the *frequency*, with the symbol f. In Fig. 15–3, if the loop rotates through 60 complete revolutions, or cycles, during 1 s, the frequency of the generated voltage is 60 cps, or 60 Hz. You see only one cycle of the sine waveform, instead of 60 cycles, because the time interval shown here is $\frac{1}{60}$ s. Note that the factor of time is involved. More cycles per second means a higher frequency and less time for one cycle, as illustrated in Fig. 15–8. Then the changes in values are faster for higher frequencies.

A complete cycle is measured between two successive points that have the same value and direction. In Fig. 15–8, the cycle is between successive points where the waveform is zero and ready to increase in the positive direction. Or the cycle can be measured between successive peaks.

On the time scale of 1 s, waveform *a* goes through one cycle; waveform *b* has much faster variations, with four complete cycles during 1 s. Both waveforms are sine waves, even though each has a different frequency.

In comparing sine waves, the amplitude has no relation to frequency. Two waveforms can have the same frequency with different amplitudes (Fig. 15–7), the same amplitude but different frequencies (Fig. 15–8), or different amplitudes and frequencies. The amplitude indicates the amount of voltage or current, and the frequency indicates the rate of change of amplitude variations in cycles per second.

Frequency Units

The unit called the *hertz* (Hz), named after Heinrich Hertz, is used for cycles per second. Then 60 cps = 60 Hz. All metric prefixes can be used. As examples

$$1 \text{ kilocycle per second} = 1 \times 10^3 \text{ Hz} = 1 \text{ kHz}$$
$$1 \text{ megacycle per second} = 1 \times 10^6 \text{ Hz} = 1 \text{ MHz}$$
$$1 \text{ gigacycle per second} = 1 \times 10^9 \text{ Hz} = 1 \text{ GHz}$$

Audio and Radio Frequencies

The entire frequency range of alternating voltage or current from 1 Hz to many megahertz can be considered in two broad groups: audio frequencies (af) and radio frequencies (rf). *Audio* is a Latin word meaning "I hear." The audio range includes frequencies that can be heard as sound waves by the human ear. This range of audible frequencies is approximately 16 to 16,000 Hz.

The higher the frequency, the higher the pitch or tone of the sound. High audio frequencies, about 3000 Hz and above, provide *treble* tone. Low audio frequencies, about 300 Hz and below, provide *bass* tone.

Loudness is determined by amplitude. The greater the amplitude of the af variation, the louder its corresponding sound.

Alternating current and voltage above the audio range provide rf variations, since electrical variations of high frequency can be transmitted by electromagnetic radio waves. Examples of frequency allocations are given in Table 15–3.

Sonic and Ultrasonic Frequencies

These terms refer to sound waves, which are variations in pressure generated by mechanical vibrations, rather than electrical variations. The velocity of sound waves through dry air at 20°C equals 1130 ft/s. Sound waves above the audible range of frequencies are called *ultrasonic* waves. The range of frequencies for ultrasonic applications, therefore, is from 16,000 Hz up to several megahertz.

PIONEERS IN ELECTRONICS

In 1887 German physicist *Heinrich Hertz* (*1857–1894*) proved that electricity could be transmitted in electromagnetic waves. In his honor, the hertz (Hz) is now the standard unit for the measurement of frequency. One Hz equals one complete cycle per second.

Table 15–3	Examples of Common Frequencies
Frequency	**Use**
60 Hz	AC power line (US)
50–15,000 Hz	Audio equipment
535–1605 kHz*	AM radio broadcast band
54–60 MHz	TV channel 2
88–108 MHz	FM radio broadcast band

* Expanded to 1705 kHz in 1991.

Sound waves in the audible range of frequencies below 16,000 Hz can be considered *sonic* or sound frequencies. The term *audio* is reserved for electrical variations that can be heard when converted to sound waves.

■ *15–6 Knowledge Check*

> *Answer at end of chapter.*
>
> A frequency of 1.49 MHz corresponds to how many kilohertz?

■ *15–6 Self-Review*

> *Answers at end of chapter.*
>
> a. What is the frequency of the bottom waveform in Fig. 15–8?
> b. Convert 1605 kHz to megahertz.

15–7 Period

The amount of time it takes for one cycle is called the *period*. Its symbol is T for time. With a frequency of 60 Hz, as an example, the time for one cycle is $\frac{1}{60}$ s. Therefore, the period is $\frac{1}{60}$ s. Frequency and period are reciprocals of each other:

$$T = \frac{1}{f} \quad \text{or} \quad f = \frac{1}{T} \tag{15–6}$$

The higher the frequency, the shorter the period. In Fig. 15–8*a*, the period for the wave with a frequency of 1 Hz is 1 s, and the higher frequency wave of 4 Hz in Fig. 15–8*b* has a period of $\frac{1}{4}$ s for a complete cycle.

Units of Time

The second is the basic unit of time, but for higher frequencies and shorter periods, smaller units of time are convenient. Those used most often are:

$$T = 1 \text{ millisecond} = 1 \text{ ms} = 1 \times 10^{-3} \text{ s}$$
$$T = 1 \text{ microsecond} = 1 \text{ } \mu\text{s} = 1 \times 10^{-6} \text{ s}$$
$$T = 1 \text{ nanosecond} = 1 \text{ ns} = 1 \times 10^{-9} \text{ s}$$

These units of time for a period are reciprocals of the corresponding units for frequency. The reciprocal of frequency in kilohertz gives the period T in milliseconds; the reciprocal of megahertz is microseconds; the reciprocal of gigahertz is nanoseconds.

GOOD TO KNOW

An oscilloscope can measure the period and frequency of an ac waveform.

Example 15-2

An alternating current varies through one complete cycle in $\frac{1}{1000}$ s. Calculate the period and frequency.

ANSWER $T = \dfrac{1}{1000} \text{ s}$

$$f = \frac{1}{T} = \frac{1}{{}^1/_{1000}}$$

$$= \frac{1000}{1} = 1000$$

$$= 1000 \text{ Hz or 1 kHz}$$

Example 15-3

Calculate the period for the two frequencies of 1 MHz and 2 MHz.

ANSWER

a. For 1 MHz,

$$T = \frac{1}{f} = \frac{1}{1 \times 10^6}$$
$$= 1 \times 10^{-6}\,s = 1\,\mu s$$

b. For 2 MHz,

$$T = \frac{1}{f} = \frac{1}{2 \times 10^6}$$
$$= 0.5 \times 10^{-6}\,s = 0.5\,\mu s$$

To do these problems on a calculator, you need the reciprocal key, usually marked $\boxed{1/x}$. Keep the powers of 10 separate and remember that the reciprocal has the same exponent with opposite sign. With f of 2×10^6, for $1/f$ just punch in 2 and then press $\boxed{2^{nd}F}$ and the $\boxed{1/x}$ key to see 0.5 as the reciprocal. The 10^6 for f becomes 10^{-6} for T so that the answer is 0.5×10^{-6} s or 0.5 μs.

■ **15–7 Knowledge Check**

Answer at end of chapter.

What is the period, T, of a 5–kHz sine wave?

■ **15–7 Self-Review**

Answers at end of chapter.

a. $T = \frac{1}{400}$ s. Calculate f.
b. $f = 400$ Hz. Calculate T.

15–8 Wavelength

When a periodic variation is considered with respect to distance, one cycle includes the *wavelength*, which is the length of one complete wave or cycle (Fig. 15–9). For example, when a radio wave is transmitted, variations in the electromagnetic field travel through space. Also, with sound waves, the variations in air pressure corresponding to the sound wave move through air. In these applications, the distance traveled by the wave in one cycle is the wavelength. The wavelength depends upon the frequency of the variation and its velocity of transmission:

$$\lambda = \frac{\text{velocity}}{\text{frequency}} \tag{15–7}$$

where λ (the Greek letter lambda) is the symbol for one complete wavelength.

Figure 15–9 Wavelength λ is the distance traveled by the wave in one cycle.

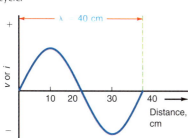

Wavelength of Radio Waves

The velocity of electromagnetic radio waves in air or vacuum is 186,000 mi/s, or 3×10^{10} cm/s, which is the speed of light. Therefore,

$$\lambda(\text{cm}) = \frac{3 \times 10^{10} \text{ cm/s}}{f(\text{Hz})} \qquad (15\text{–}8)$$

Note that the higher the frequency, the shorter the wavelength. For instance, the short-wave radio broadcast band of 5.95 to 26.1 MHz includes frequencies higher than the standard AM radio broadcast band of 535 to 1605 kHz.

Example 15–4

Calculate λ for a radio wave with f of 30 GHz.

ANSWER

$$\lambda = \frac{3 \times 10^{10} \text{ cm/s}}{30 \times 10^9 \text{ Hz}} = \frac{3}{30} \times 10 \text{ cm}$$
$$= 0.1 \times 10$$
$$= 1 \text{ cm}$$

Such short wavelengths are called *microwaves*. This range includes λ of 1 m or less for frequencies of 300 MHz or more.

Example 15–5

The length of a TV antenna is $\lambda/2$ for radio waves with f of 60 MHz. What is the antenna length in centimeters and feet?

ANSWER

a. $\lambda = \dfrac{3 \times 10^{10} \text{ cm/s}}{60 \times 10^6 \text{ Hz}} = \dfrac{1}{20} \times 10^4 \text{ cm}$

$\qquad = 0.05 \times 10^4$

$\qquad = 500 \text{ cm}$

Then, $\lambda/2 = {}^{500}\!/_2 = 250 \text{ cm}$.

b. Since 2.54 cm = 1 in.,

$\lambda/2 = \dfrac{250 \text{ cm}}{2.54 \text{ cm/in.}} = 98.4 \text{ in.}$

$\qquad = \dfrac{98.4 \text{ in}}{12 \text{ in./ft}} = 8.2 \text{ ft}$

Example 15-6

For the 6-m band used in amateur radio, what is the corresponding frequency?

ANSWER The formula $\lambda = v/f$ can be inverted

$$f = \frac{v}{\lambda}$$

Then

$$f = \frac{3 \times 10 \text{ cm/s}}{6 \text{ m}} = \frac{3 \times 10^{10} \text{ cm/s}}{6 \times 10^2 \text{ cm}}$$

$$= \frac{3}{6} \times 10^8 = 0.5 \times 10^8 \text{ Hz}$$

$$= 50 \times 10^6 \text{ Hz} \quad \text{or} \quad 50 \text{ MHz}$$

Wavelength of Sound Waves

The velocity of sound waves is much lower than that of radio waves because sound waves result from mechanical vibrations rather than electrical variations. In average conditions, the velocity of sound waves in air equals 1130 ft/s. To calculate the wavelength, therefore,

$$\lambda = \frac{1130 \text{ ft/s}}{f \text{ Hz}} \tag{15-9}$$

This formula can also be used for ultrasonic waves. Although their frequencies are too high to be audible, ultrasonic waves are still sound waves rather than radio waves.

Example 15-7

What is the wavelength of the sound waves produced by a loudspeaker at a frequency of 100 Hz?

ANSWER

$$\lambda = \frac{1130 \text{ ft/s}}{100 \text{ Hz}}$$
$$\lambda = 11.3 \text{ ft}$$

Example 15-8

For ultrasonic waves at a frequency of 34.44 kHz, calculate the wavelength in feet and in centimeters.

ANSWER

$$\lambda = \frac{1130}{34.44 \times 10^3}$$
$$= 32.8 \times 10^{-3} \text{ ft}$$
$$= 0.0328 \text{ ft}$$

To convert to inches,

$$0.0328 \text{ ft} \times 12 = 0.3936 \text{ in.}$$

To convert to centimeters,

$$0.3936 \text{ in.} \times 2.54 = 1 \text{ cm} \quad \text{approximately}$$

Note that the 34.44-kHz sound waves in this example have the same wavelength (1 cm) as the 30-GHz radio waves in Example 15–4. The reason is that radio waves have a much higher velocity than sound waves.

■ *15–8 Knowledge Check*

Answer at end of chapter.

What is the wavelength of a sound wave whose frequency is 60 Hz?

■ *15–8 Self-Review*

Answers at end of chapter.

a. The higher the frequency, the shorter the wavelength λ. (True or False)
b. The higher the frequency, the longer the period T. (True or False)
c. The velocity of propagation for radio waves in free space is 3×10^{10} cm/s. (True or False)

15–9 Phase Angle

Referring back to Fig. 15–3, suppose that the generator started its cycle at point B, where maximum voltage output is produced, instead of starting at the point of zero output. If we compare the two cases, the two output voltage waves would be as shown in Fig. 15–10. Each is the same waveform of alternating voltage, but wave B starts at maximum, and wave A starts at zero. The complete cycle of wave B through 360° takes it back to the maximum value from which it started. Wave A starts and finishes its cycle at zero. With respect to time, therefore, wave B is ahead of wave A in values of generated voltage. The amount it leads in time equals one quarter-revolution, which is 90°. This angular difference is the phase angle between waves B and A. Wave B leads wave A by the phase angle of 90°.

The 90° phase angle between waves B and A is maintained throughout the complete cycle and in all successive cycles, as long as they both have the same frequency. At any instant, wave B has the value that A will have 90° later. For instance, at 180° wave A is at zero, but B is already at its negative maximum value, where wave A will be later at 270°.

To compare the phase angle between two waves, they must have the same frequency. Otherwise, the relative phase keeps changing. Also, they must have sine-wave variations because this is the only kind of waveform that is

Figure 15–10 Two sine-wave voltages 90° out of phase. (*a*) Wave B leads wave A by 90°. (*b*) Corresponding phasors V_B and V_A for the two sine-wave voltages with phase angle $\theta = 90°$. The right angle shows quadrature phase.

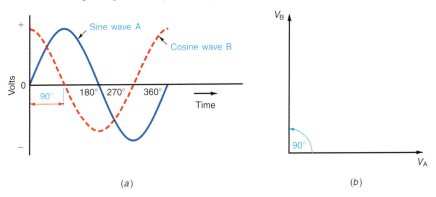

(*a*)   (*b*)

measured in angular units of time. The amplitudes can be different for the two waves, although they are shown the same here. We can compare the phases of two voltages, two currents, or a current with a voltage.

The 90° Phase Angle

The two waves in Fig. 15–10 represent a sine wave and a cosine wave 90° out of phase with each other. The 90° phase angle means that one has its maximum amplitude when the other is at zero value. Wave A starts at zero, corresponding to the sine of 0°, has its peak amplitude at 90 and 270°, and is back to zero after one cycle of 360°. Wave B starts at its peak value, corresponding to the cosine of 0°, has its zero value at 90 and 270°, and is back to the peak value after one cycle of 360°.

However, wave B can also be considered a sine wave that starts 90° before wave A in time. This phase angle of 90° for current and voltage waveforms has many applications in sine-wave ac circuits with inductance or capacitance.

The sine and cosine waveforms have the same variations but displaced by 90°. Both waveforms are called *sinusoids*. The 90° angle is called *quadrature phase*.

Phase–Angle Diagrams

To compare phases of alternating currents and voltages, it is much more convenient to use phasor diagrams corresponding to the voltage and current waveforms, as shown in Fig. 15–10*b*. The arrows here represent the phasor quantities corresponding to the generator voltage.

A phasor is a quantity that has magnitude and direction. The length of the arrow indicates the magnitude of the alternating voltage in rms, peak, or any ac value, as long as the same measure is used for all phasors. The angle of the arrow with respect to the horizontal axis indicates the phase angle.

The terms *phasor* and *vector* are used for a quantity that has direction, requiring an angle to specify the value completely. However, a vector quantity has direction in space, whereas a phasor quantity varies in time. As an example of a vector, a mechanical force can be represented by a vector arrow at a specific angle, with respect to either the horizontal or the vertical direction.

For phasor arrows, the angles shown represent differences in time. One sinusoid is chosen as the reference. Then the timing of the variations in another sinusoid can be compared to the reference by means of the angle between the phasor arrows.

Figure 15–11 Leading and lagging phase angles for 90°. (*a*) When phasor V_A is the horizontal reference, phasor V_B leads by 90°. (*b*) When phasor V_B is the horizontal reference, phasor V_A lags by –90°.

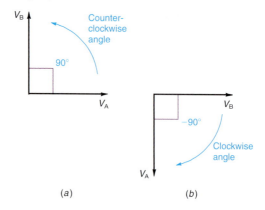

(*a*)   (*b*)

The phasor corresponds to the entire cycle of voltage, but is shown only at one angle, such as the starting point, since the complete cycle is known to be a sine wave. Without the extra details of a whole cycle, phasors represent the alternating voltage or current in a compact form that is easier for comparing phase angles.

In Fig. 15–10*b*, for instance, the phasor V_A represents the voltage wave A with a phase angle of 0°. This angle can be considered the plane of the loop in the rotary generator where it starts with zero output voltage. The phasor V_B is vertical to show the phase angle of 90° for this voltage wave, corresponding to the vertical generator loop at the start of its cycle. The angle between the two phasors is the phase angle.

The symbol for a phase angle is θ (the Greek letter theta). In Fig. 15–10, as an example, $\theta = 90°$.

Phase–Angle Reference

The phase angle of one wave can be specified only with respect to another as reference. How the phasors are drawn to show the phase angle depends on which phase is chosen as the reference. Generally, the reference phasor is horizontal, corresponding to 0°. Two possibilities are shown in Fig. 15–11. In Fig. 15–11*a*, the voltage wave A or its phasor V_A is the reference. Then the phasor V_B is 90° counterclockwise. This method is standard practice, using counterclockwise rotation as the positive direction for angles. Also, a leading angle is positive. In this case, then, V_B is 90° counterclockwise from the reference V_A to show that wave B leads wave A by 90°.

However, wave B is shown as the reference in Fig. 15–11*b*. Now V_B is the horizontal phasor. To have the same phase angle, V_A must be 90° clockwise, or –90° from V_B. This arrangement shows that negative angles, clockwise from the 0° reference, are used to show lagging phase angles. The reference determines whether the phase angle is considered leading or lagging in time.

The phase is not actually changed by the method of showing it. In Fig. 15–11, V_A and V_B are 90° out of phase, and V_B leads V_A by 90° in time. There is no fundamental difference whether we say V_B is ahead of V_A by +90° or V_A is behind V_B by –90°.

Two waves and their corresponding phasors can be out of phase by any angle, either less or more than 90°. For instance, a phase angle of 60° is shown in Fig. 15–12. For the waveforms in Fig. 15–12*a*, wave D is behind C by 60° in time. For the phasors in Fig. 15–12*b*, this lag is shown by the phase angle of –60°.

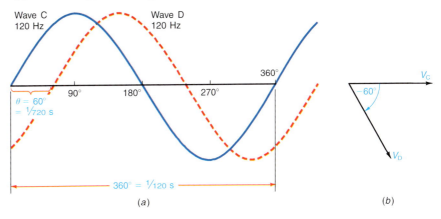

Figure 15–12 Phase angle of 60° is the time for 60/360 or 1/6 of the cycle. (a) Waveforms. (b) Phasor diagram.

(a)   (b)

Figure 15–13 Two waveforms in phase, or the phase angle is 0°. (a) Waveforms. (b) Phasor diagram.

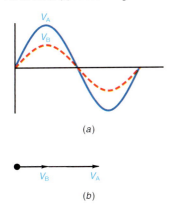

(a)

(b)

Figure 15–14 Two waveforms out of phase or in opposite phase with phase angle of 180°. (a) Waveforms. (b) Phasor diagram.

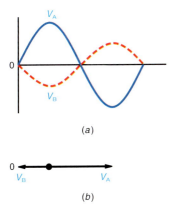

(a)

(b)

In-Phase Waveforms

A phase angle of 0° means that the two waves are in phase (Fig. 15–13).

Out-of-Phase Waveforms

An angle of 180° means opposite phase, or that the two waveforms are exactly out of phase (Fig. 15–14). Then the amplitudes are opposing.

■ *15-9 Knowledge Check*

   Answer at end of chapter.

   Which varies with respect to time, a phasor or a vector?

■ *15-9 Self-Review*

   Answers at end of chapter.

   Give the phase angle in
   a. **Fig. 15–10.**
   b. **Fig. 15–12.**
   c. **Fig. 15–13.**

15–10 The Time Factor in Frequency and Phase

It is important to remember that the waveforms we are showing are graphs drawn on paper. The physical factors represented are variations in amplitude, usually on the vertical scale, with respect to equal intervals on the horizontal scale, which can represent either distance or time. To show wavelength, as in Fig. 15–9, the cycles of amplitude variations are plotted against distance or length. To show frequency, the cycles of amplitude variations are shown with respect to time in angular measure. The angle of 360° represents the time for one cycle, or the period T.

As an example of how frequency involves time, a waveform with stable frequency is actually used in electronic equipment as a clock reference for very small units of time. Assume a voltage waveform with the frequency of 10 MHz. The period T is 0.1 μs. Therefore, every cycle is repeated at 0.1-μs intervals. When each cycle of voltage variations is used to indicate time, then, the result is effectively a clock that measures 0.1-μs units. Even smaller units of time can be measured with higher frequencies. In everyday applications, an electric clock connected to the power line keeps correct time because it is controlled by the exact frequency of 60 Hz.

Furthermore, the phase angle between two waves of the same frequency indicates a specific difference in time. As an example, Fig. 15–12 shows a phase angle of 60°, with wave C leading wave D. Both have the same frequency of 120 Hz. The period T for each wave then is $\frac{1}{120}$ s. Since 60° is one-sixth of the complete cycle of 360°, this phase angle represents one-sixth of the complete period of $\frac{1}{120}$ s. If we multiply $\frac{1}{6} \times \frac{1}{120}$, the answer is $\frac{1}{720}$ s for the time corresponding to the phase angle of 60°. If we consider wave D lagging wave C by 60°, this lag is a time delay of $\frac{1}{720}$ s.

More generally, the time for a phase angle θ can be calculated as

$$ t = \frac{\theta}{360} \times \frac{1}{f} \qquad (15\text{--}10) $$

where f is in Hz, θ is in degrees, and t is in seconds.

The formula gives the time of the phase angle as its proportional part of the total period of one cycle. For the example of θ equal to 60° with f at 120 Hz,

$$ t = \frac{\theta}{360} \times \frac{1}{f} $$
$$ = \frac{60}{360} \times \frac{1}{120} = \frac{1}{6} \times \frac{1}{120} $$
$$ = \frac{1}{720} \text{ s} $$

■ **15–10 Knowledge Check**

Answer at end of chapter.

For two 60-Hz ac voltage waveforms, how much time does a phase angle difference of 45° correspond to?

■ **15–10 Self-Review**

Answers at end of chapter.

a. In Fig. 15–12, how much time corresponds to 180°?
b. For two waves with a frequency of 1 MHz, how much time is the phase angle of 36°?

15–11 Alternating Current Circuits with Resistance

An ac circuit has an ac voltage source. Note the symbol in Fig. 15–15 used for any source of sine-wave alternating voltage. This voltage connected across an external load resistance produces alternating current of the same waveform, frequency, and phase as the applied voltage.

The amount of current equals V/R by Ohm's law. When V is an rms value, I is also an rms value. For any instantaneous value of V during the cycle, the value of I is for the corresponding instant.

In an ac circuit with only resistance, the current variations are in phase with the applied voltage, as shown in Fig. 15–15b. This in-phase relationship between V and I means that such an ac circuit can be analyzed by the same methods used for dc circuits, since there is no phase angle to consider. Circuit

GOOD TO KNOW

When t and f are known, θ can be calculated as $\theta = \frac{t}{T} \times 360°$ where $T = \frac{1}{f}$.

Figure 15–15 An ac circuit with resistance R alone. (a) Schematic diagram. (b) Waveforms.

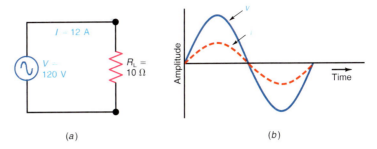

(a)   (b)

components that have R alone include resistors, the filaments of lightbulbs, and heating elements.

The calculations in ac circuits are generally in rms values, unless noted otherwise. In Fig. 15–15a, for example, the 120 V applied across the 10-Ω R_L produces rms current of 12 A. The calculations are

$$I = \frac{V}{R_L} = \frac{120 \text{ V}}{10 \ \Omega} = 12 \text{ A}$$

Furthermore, the rms power dissipation is I^2R, or

$$P = 144 \times 10 = 1440 \text{ W}$$

Series AC Circuit with R

In Fig. 15–16, R_T is 30 Ω, equal to the sum of 10 Ω for R_1 plus 20 Ω for R_2. The current in the series circuit is

$$I = \frac{V_T}{R_T} = \frac{120 \text{ V}}{30 \ \Omega} = 4 \text{ A}$$

The 4-A current is the same in all parts of the series circuit. This principle applies for either an ac or a dc source.

Next, we can calculate the series voltage drops in Fig. 15–16. With 4 A through the 10-Ω R_1, its IR voltage drop is

$$V_1 = I \times R_1 = 4 \text{ A} \times 10 \ \Omega = 40 \text{ V}$$

The same 4 A through the 20-Ω R_2 produces an IR voltage drop of 80 V. The calculations are

$$V_2 = I \times R_2 = 4 \text{ A} \times 20 \ \Omega = 80 \text{ V}$$

Note that the sum of 40 V for V_1 and 80 V for V_2 in series equals the 120 V applied.

Parallel AC Circuit with R

In Fig. 15–17, the 10-Ω R_1 and 20-Ω R_2 are in parallel across the 120-V ac source. Therefore, the voltage across the parallel branches is the same as the applied voltage.

Each branch current, then, is equal to 120 V divided by the branch resistance. The branch current for the 10-Ω R_1 is

$$I_1 = \frac{120 \text{ V}}{10 \ \Omega} = 12 \text{ A}$$

Figure 15–16 Series ac circuit with resistance only.

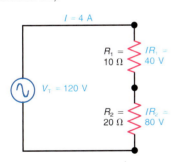

Figure 15–17 Parallel ac circuit with resistance only.

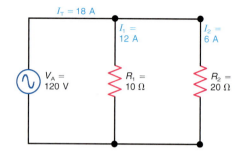

The same 120 V is across the 20-Ω branch with R_2. Its branch current is

$$I_2 = \frac{120\ \text{V}}{20\ \Omega} = 6\ \text{A}$$

The total line current I_T is $12 + 6 = 18$ A, or the sum of the branch currents.

Series–Parallel AC Circuit with R

See Fig. 15–18. The 20-Ω R_2 and 20-Ω R_3 are in parallel, for an equivalent bank resistance of 20/2 or 10 Ω. This 10-Ω bank is in series with the 20-Ω R_1 in the main line and totals 30 Ω for R_T across the 120-V source. Therefore, the main line current produced by the 120-V source is

$$I_T = \frac{V_T}{R_T} = \frac{120\ \text{V}}{30\ \Omega} = 4\ \text{A}$$

The voltage drop across R_1 in the main line is calculated as

$$V_1 = I_T \times R_1 = 4\ \text{A} \times 20\ \Omega = 80\ \text{V}$$

Subtracting this 80-V drop from the 120 V of the source, the remaining 40 V is across the bank of R_2 and R_3 in parallel. Since the branch resistances are equal, the 4-A I_T divides equally, 2 A in R_2 and 2 A in R_3. The branch currents can be calculated as

$$I_2 = \frac{40\ \text{V}}{20\ \Omega} = 2\ \text{A}$$
$$I_3 = \frac{40\ \text{V}}{20\ \Omega} = 2\ \text{A}$$

Note that the 2 A for I_2 and 2 A for I_3 in parallel branches add to equal the 4-A current in the main line.

Figure 15–18 Series-parallel ac circuit with resistance only.

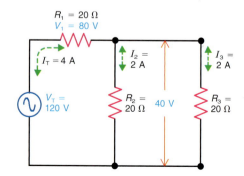

■ *15–11 Knowledge Check*

    Answer at end of chapter.

    In Fig. 15–18, how much is the peak-to-peak voltage across R_3?

■ *15–11 Self-Review*

    Answers at end of chapter.

    Calculate R_T in
  a. Fig. 15–16.
  b. Fig. 15–17.
  c. Fig. 15–18.

15–12 Nonsinusoidal AC Waveforms

The sine wave is the basic waveform for ac variations for several reasons. This waveform is produced by a rotary generator; the output is proportional to the angle of rotation. In addition, electronic oscillator circuits with inductance and capacitance naturally produce sine-wave variations.

    Because of its derivation from circular motion, any sine wave can be analyzed in terms of angular measure, either in degrees from 0 to 360° or in radians from 0 to 2π rad.

    Another feature of a sine wave is its basic simplicity; the rate of change of the amplitude variations corresponds to a cosine wave that is similar but 90° out of phase. The sine wave is the only waveform that has this characteristic of a rate of change with the same waveform as the original changes in amplitude.

    In many electronic applications, however, other waveshapes are important. Any waveform that is not a sine or cosine wave is a *nonsinusoidal waveform*. Common examples are the square wave and sawtooth wave in Fig. 15–19.

    For either voltage or current nonsinusoidal waveforms, there are important differences and similarities to consider. Note the following comparisons with sine waves.

1. In all cases, the cycle is measured between two points having the same amplitude and varying in the same direction. The period is the time for one cycle. In Fig. 15–19, T for any of the waveforms is 4 μs and the corresponding frequency is $1/T$, equal to 0.25 MHz.
2. Peak amplitude is measured from the zero axis to the maximum positive or negative value. However, peak-to-peak amplitude is better for measuring nonsinusoidal waveshapes because they can have unsymmetrical peaks, as in Fig. 15–19d. For all waveforms shown here, though, the peak-to-peak (p–p) amplitude is 20 V.
3. The rms value 0.707 of maximum applies only to sine waves because this factor is derived from the sine values in the angular measure used only for the sine waveform.
4. Phase angles apply only to sine waves because angular measure is used only for sine waves. Note that the horizontal axis for time is divided into angles for the sine wave in Fig. 15–19a, but there are no angles shown for the nonsinusoidal waveshapes.
5. All waveforms represent ac voltages. Positive values are shown above the zero axis, and negative values below the axis.

    The sawtooth wave in Fig. 15–19b represents a voltage that slowly increases to its peak value with a uniform or linear rate of change and then drops sharply to its starting value. This waveform is also called a *ramp voltage*. It is also often referred to as a *time base* because of its constant rate of change.

Figure 15–19 Comparison of sine wave with nonsinusoidal waveforms. Two cycles shown. (a) Sine wave. (b) Sawtooth wave. (c) Symmetrical square wave. (d) Unsymmetrical rectangular wave or pulse waveform.

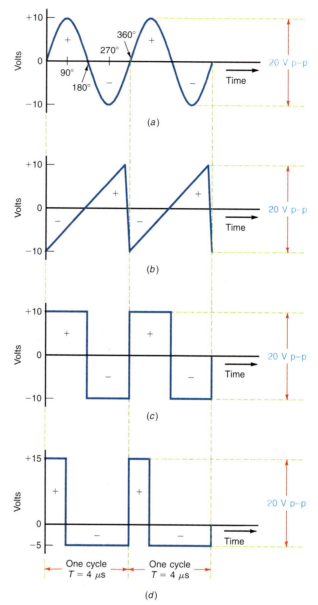

Note that one complete cycle includes a slow rise and a fast drop in voltage. In this example, the period T for a complete cycle is 4 μs. Therefore, these sawtooth cycles are repeated at the frequency of 0.25 MHz. The sawtooth waveform of voltage or current is often used for horizontal deflection of the electron beam in the cathode-ray tube (CRT) for oscilloscopes and TV receivers.

The square wave in Fig. 15–19c represents a switching voltage. First, the 10-V peak is instantaneously applied in positive polarity. This voltage remains on for 2 μs, which is one half-cycle. Then the voltage is instantaneously switched to −10 V for another 2 μs. The complete cycle then takes 4 μs, and the frequency is 0.25 MHz.

The rectangular waveshape in Fig. 15–19d is similar, but the positive and negative half-cycles are not symmetrical either in amplitude or in time. However, the frequency is the same 0.25 MHz and the peak-to-peak amplitude is the same 20 V, as in all the waveshapes. This waveform shows pulses of voltage or current, repeated at a regular rate.

■ *15-12 Knowledge Check*

Answer at end of chapter.

The waveform in Fig. 15–19*c* has an rms value of 7.07 V. (True or False)

■ *15-12 Self-Review*

Answers at end of chapter.

a. In Fig. 15–19*c*, for how much time is the waveform at +10 V?
b. In Fig. 15–19*d*, what voltage is the positive peak amplitude?

15-13 Harmonic Frequencies

Consider a repetitive nonsinusoidal waveform, such as a 100-Hz square wave. Its fundamental rate of repetition is 100 Hz. Exact multiples of the fundamental frequency are called *harmonic frequencies.* The second harmonic is 200 Hz, the third harmonic is 300 Hz, etc. Even multiples are even harmonics, and odd multiples are odd harmonics.

Harmonics are useful in analyzing distorted sine waves or nonsinusoidal waveforms. Such waveforms consist of a pure sine wave at the fundamental frequency plus harmonic frequency components. For example, Fig. 15–20 illustrates how a square wave corresponds to a fundamental sine wave with odd harmonics. Typical audio waveforms include odd and even harmonics. The harmonic components make one source of sound different from another with the same fundamental frequency.

A common unit for frequency multiples is the *octave,* which is a range of 2:1. Doubling the frequency range—from 100 to 200 Hz, from 200 to 400 Hz, and from 400 to 800 Hz, as examples—raises the frequency by one octave. The reason for this name is that an octave in music includes eight consecutive tones, for double the frequency. One-half the frequency is an octave lower.

Another unit for representing frequency multiples is the decade. A decade corresponds to a 10:1 range in frequencies such as 100 Hz to 1 kHz and 30 kHz to 300 kHz.

■ *15-13 Knowledge Check*

Answer at end of chapter.

What is the second even harmonic of 150 Hz?

■ *15-13 Self-Review*

Answers at end of chapter.

a. What frequency is the fourth harmonic of 12 MHz?
b. Give the frequency one octave above 220 Hz.

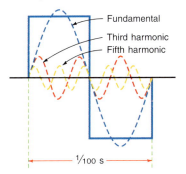

Figure 15–20 Fundamental and harmonic frequencies for an example of a 100-Hz square wave.

15-14 The 60-Hz AC Power Line

Practically all homes in the United States are supplied alternating voltage between 115 and 125 V rms at a frequency of 60 Hz. This is a sine-wave voltage produced by a rotary generator. The electricity is distributed by high-voltage power lines from a generating station and reduced to the lower voltages used in the home. Here the incoming voltage is wired to all wall outlets and electrical equipment in parallel. The 120-V source of commercial electricity is the 60-Hz *power line* or the *mains,* indicating that it is the main line for all parallel branches.

Advantages

The incoming electric service to residences is normally given as 120 V rms. With an rms value of 120 V, the ac power is equivalent to 120-Vdc power in heating effect. If the value were higher, there would be more danger of a fatal electric shock. Lower voltages would be less efficient in supplying power.

Higher voltage can supply electric power with less I^2R loss, since the same power is produced with less I. Note that the I^2R power loss increases as the square of the current. For applications where large amounts of power are used, such as central air-conditioners and clothes dryers, a line voltage of 240 V is often used.

The advantage of ac over dc power is greater efficiency in distribution from the generating station. Alternating voltages can easily be stepped up by a transformer with very little loss, but a transformer cannot operate on direct current because it needs the varying magnetic field produced by an ac voltage.

With a transformer, the alternating voltage at the generating station can be stepped up to values as high as 500 kV for high-voltage distribution lines. These high-voltage lines supply large amounts of power with much less current and less I^2R loss, compared with a 120-V line. In the home, the lower voltage required is supplied by a step-down transformer. The step-up and step-down characteristics of a transformer refer to the ratio of voltages across the input and output connections.

The 60-Hz frequency is convenient for commercial ac power. Much lower frequencies would require much bigger transformers because larger windings would be necessary. Also, too low a frequency for alternating current in a lamp could cause the light to flicker. For the opposite case, too high a frequency results in excessive iron-core heating in the transformer because of eddy currents and hysteresis losses. Based on these factors, 60 Hz is the frequency of the ac power line in the United States. However, the frequency of the ac power mains in England and most European countries is 50 Hz.

The 60-Hz Frequency Reference

All power companies in the United States, except those in Texas, are interconnected in a grid that maintains the ac power-line frequency between 59.98 and 60.02 Hz. The frequency is compared with the time standard provided by the Bureau of Standards radio station WWV at Fort Collins, Colorado. As a result, the 60-Hz power-line frequency is maintained accurately to ±0.033%. This accuracy makes the power-line voltage a good secondary standard for checking frequencies based on 60 Hz.

Residential Wiring

At the electrical service entrance (where power enters a house), most homes have the three-wire power lines illustrated in Fig. 15–21. The three wires, including the grounded neutral, can be used for either 240 or 120 V single phase. The 240 V at the residence is stepped down from the high-voltage distribution lines.

Note the color coding for the wiring in Fig. 15–21. The grounded neutral is white. Each high side can use any color except white or green, but usually black* or red is used. White is reserved for the neutral wire, and green or bare wire is reserved for grounding.

Figure 15–21 Three-wire, single-phase power lines that can provide either 240 or 120 V.

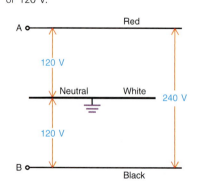

---

* Note that in electronic equipment, black is the color-coded wiring used for chassis-ground returns. However, in electric power work, black wire is used for high-side connections.

From either the red or black high side to the neutral, 120 V is available for separate branch circuits to the lights and outlets. Across the red and black wires, 240 V is available for high-power appliances. This three-wire service with a grounded neutral is called the *Edison system.*

The electrical service is commonly rated for 100 A. At 240 V, then, the power available is $100 \times 240 = 24{,}000$ W, or 24 kW.

The main wires to the service entrance are generally No. 4 to 8 gage. Sizes 6 and heavier are always stranded wire. The 120-V branch circuits, usually rated at 15 A or 20 A, use No. 12 or 14 gage wire. Each branch has its own fuse or circuit breaker. A main switch is usually included to cut off all power from the service entrance.

The neutral wire is grounded at the service entrance to a water pipe or a metal rod driven into the earth, which is *ground.* All 120-V branches must have one side connected to the grounded neutral. White wire is used for these connections. In addition, all metal boxes for outlets, switches, and lights must have a continuous ground to each other and to the neutral. The wire cable usually has a bare wire for grounding boxes.

Cables commonly used are armored sheath with the trade name BX and nonmetallic flexible cable with the trade name Romex. Each has two or more wires for the neutral, high-side connections, and grounding. Both cables contain an extra bare wire for grounding. Rules and regulations for residential wiring are governed by local electrical codes. These are usually based on the National Electrical Code published by the National Fire Protection Association.

Grounding

In ac power distribution systems, grounding is the practice of connecting one side of the power line to earth or ground. The purpose is safety in two ways. First is protection against dangerous electric shock. Also, the power distribution lines are protected against excessively high voltage, particularly from lightning. If the system is struck by lightning, excessive current in the grounding system will energize a cutout device to deenergize the lines.

The grounding in the power distribution system means that it is especially important to have grounding for the electric wiring at the residence. For instance, suppose that an electric appliance such as a clothes dryer does not have its metal case grounded. An accidental short circuit in the equipment can connect the metal frame to the "hot" side of the ac power line. Then the frame has voltage with respect to earth ground. If somebody touches the frame and has a return to ground, the result is a dangerous electric shock. With the case grounded, however, the accidental short circuit blows the fuse or circuit breaker to cut off the power.

In normal operation, the electric circuits function the same way with or without the ground, but grounding is an important safety precaution. Figure 15–22 shows two types of plug connectors for the ac power line that help provide protection because they are polarized with respect to the ground connections. Although an ac voltage does not have any fixed polarity, the plugs ensure grounding of the chassis or frame of the equipment connected to the power line. In Fig. 15–22a, the plug has two blades for the 120-V line, but the wider blade will fit only the side of the outlet that is connected to the neutral wire. This wiring is standard practice. For the three-prong plug in Fig. 15–22b, the rounded pin is for a separate grounding wire, usually color-coded green.

In some cases, there may be leakage of current from the "hot" side of the power line to ground. A leakage current of 5 mA or more is considered

Figure 15–22 Plug connectors polarized for ground connection to an ac power line. (*a*) Wider blade connects to neutral. (*b*) Rounded pin connects to ground.

(*a*)

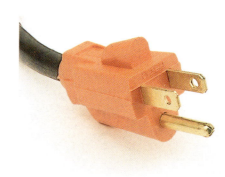

(*b*)

Figure 15–23 Ground-fault circuit interrupter (GFCI).

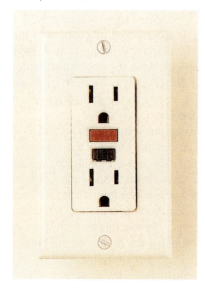

dangerous. The ground-fault circuit interrupter (GCFI) shown in Fig. 15–23 is a device that can sense excessive leakage current and open the circuit as a protection against shock hazard.

It may be of interest to note that with high-fidelity audio equipment, the lack of proper grounding can cause a hum heard in the sound. The hum is usually not a safety problem, but it still is undesirable.

■ *15–14 Knowledge Check*

Answer at end of chapter.

In standard house wiring, the white neutral wire has no current because it is grounded. (True or False)

■ *15–14 Self-Review*

Answers at end of chapter.

a. The 120 V of the ac power line is a peak-to-peak value. (True or False)
b. The frequency of the ac power-line voltage is 60 Hz ± 0.033%. (True or False)
c. In Fig. 15–21, the voltage between the black and white wires is 120 V. (True or False)
d. The color code for grounding wires is green. (True or False)

15–15 Motors and Generators

A generator converts mechanical energy into electric energy; a motor does the opposite, converting electricity into rotary motion. The main parts in the assembly of motors and generators are essentially the same (Fig. 15–24).

Armature

In a generator, the armature connects to the external circuit to provide the generator output voltage. In a motor, it connects to the electrical source that drives the motor. The armature is often constructed in the form of a drum, using

Figure 15-24 Main parts of a dc motor.

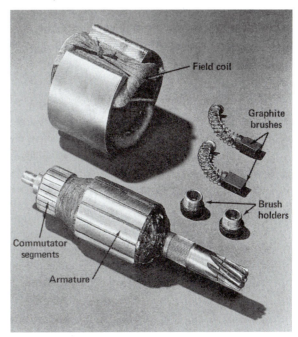

many conductor loops for increased output. In Fig. 15–24, the rotating armature is the *rotor* part of the assembly.

Field Winding

This electromagnet provides the flux cut by the rotor. In a motor, current for the field is produced by the same source that supplies the armature. In a generator, the field current may be obtained from a separate exciter source, or from its own armature output. Residual magnetism in the iron yoke of the field allows this *self-excited generator* to start.

The field coil may be connected in series with the armature in parallel, or in a series-parallel *compound winding*. When the field winding is stationary, it is the *stator* part of the assembly.

Slip Rings

In an ac machine, two or more slip rings or *collector rings* connect the rotating loop to the stationary wire leads for the external circuit.

Brushes

These graphite connectors are spring-mounted to brush against the spinning rings on the rotor. The stationary external leads are connected to the brushes for connection to the rotating loop. Constant rubbing slowly wears down the brushes, and they must be replaced after they are worn.

Commutator

A dc machine has a commutator ring instead of slip rings. As shown in Fig. 15–24, the commutator ring has segments, one pair for each loop in the armature. Each of the commutator segments is insulated from the others by mica.

The commutator converts the ac machine to dc operation. In a generator, the commutator segments reverse the loop connections to the brushes every half-cycle to maintain a constant polarity of output voltage. For a dc motor, the commutator segments allow the dc source to produce torque in one direction.

Brushes are necessary with a commutator ring. The two stationary brushes contact opposite segments on the rotating commutator. Graphite brushes are used for very low resistance.

Alternating Current Induction Motor

This type, for alternating current only, does not have any brushes. The stator is connected directly to the ac source. Then alternating current in the stator winding induces current in the rotor without any physical connection between them. The magnetic field of the current induced in the rotor reacts with the stator field to produce rotation. Alternating-current induction motors are economical and rugged without any troublesome brush arcing.

With a single-phase source, however, a starting torque must be provided for an ac induction motor. One method uses a starting capacitor in series with a separate starting coil. The capacitor supplies an out-of-phase current just for starting and then is switched out. Another method of starting uses shaded poles. A solid copper ring on the main field pole makes the magnetic field unsymmetrical to allow starting.

The rotor of an ac induction motor may be wire-wound or the squirrel-cage type. This rotor is constructed with a frame of metal bars.

Universal Motor

This type operates on either alternating or direct current because the field and armature are in series. Its construction is like that of a dc motor with the rotating armature connected to a commutator and brushes. The universal motor is commonly used for small machines such as portable drills and food mixers.

Alternators

Alternating current generators are alternators. For large power requirements, the alternator usually has a rotating field, and the armature is the stator.

■ *15–15 Knowledge Check*

Answer at end of chapter.

Which converts mechanical energy into electrical energy, a motor or a generator?

■ *15–15 Self-Review*

Answers at end of chapter.

a. **In Fig. 15–24, the commutator segments are on the armature. (True or False)**
b. **Motor brushes are made of graphite because of its very low resistance. (True or False)**
c. **A starting capacitor is used with dc motors that have small brushes. (True or False)**

15–16 Three–Phase AC Power

In an alternator with three generator windings equally spaced around the circle, the windings produce output voltages 120° out of phase with each other. The three-phase output is illustrated by the sine-wave voltages in Fig. 15–25*a* and the corresponding phasors in Fig. 15–25*b*. The advantage of three-phase ac voltage

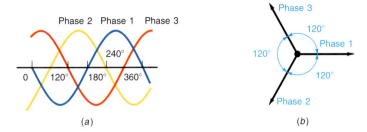

Figure 15–25 Three-phase alternating voltage or current with 120° between each phase. (a) Sine waves. (b) Phasor diagram.

is more efficient distribution of power. Also, ac induction motors are self-starting with three-phase alternating current. Finally, the ac ripple is easier to filter in the rectified output of a dc power supply.

In Fig. 15–26a, the three windings are in the form of a Y, also called *wye* or *star* connections. All three coils are joined at one end, and the opposite ends are for the output terminals A, B, and C. Note that any pair of terminals is across two coils in series. Each coil has 120 V. The voltage output across any two output terminals is 120 × 1.73 = 208 V, because of the 120° phase angle.

In Fig. 15–26b, the three windings are connected in the form of a *delta* (Δ). Any pair of terminals is across one generator winding. The output then is 120 V. However, the other coils are in a parallel branch. Therefore, the current capacity of the line is increased by the factor 1.73.

In Fig. 15–27, the center point of the Y is used for a fourth line, as the neutral wire in the three-phase power distribution system. This way, power

Figure 15–26 Types of connections for three-phase power. (a) Wye or Y. (b) Delta or Δ.

Figure 15–27 Y connections to a four-wire line with neutral.

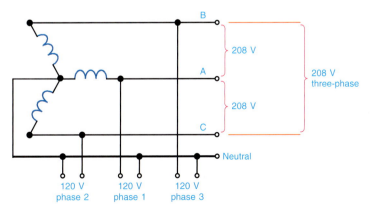

is available at either 208 V three phase or 120 V single phase. Note that the three-phase voltage is 208 V, not the 240 V in the Edison single-phase system. From terminal A, B, or C to the neutral line in Fig. 15–27, the output is 120 V across one coil. This 120-V single-phase power is used in conventional lighting circuits. However, across terminals AB, BC, or CA, without the neutral, the output is 208 V for three-phase induction motors or other circuits that need three-phase power. Although illustrated here for the 120-V, 60-Hz power line, note that three-phase connections are commonly used for higher voltages.

■ *15–16 Knowledge Check*

> *Answer at end of chapter.*
>
> **What is the advantage of using three-phase ac power?**

■ *15–16 Self-Review*

> *Answers at end of chapter.*
>
> a. **What is the angle between three-phase voltages?**
> b. **For the Y in Fig. 15–26a, how much is V_{AC} or V_{AB}?**

Summary

- Alternating voltage varies continuously in magnitude and periodically reverses in polarity. When alternating voltage is applied across a load resistance, the result is alternating current in the circuit.

- A complete set of values repeated periodically is one cycle of the ac waveform. The cycle can be measured from any one point on the wave to the next successive point having the same value and varying in the same direction. One cycle includes 360° in angular measure, or 2π rad.

- The rms value of a sine wave is 0.707 × peak value.

- The peak amplitude, at 90° and 270° in the cycle, is 1.414 × rms value.

- The peak-to-peak value is double the peak amplitude, or 2.828 × rms for a symmetrical ac waveform.

- The average value is 0.637 × peak value.

- The frequency equals the number of cycles per second. One cps is 1 Hz. The audio-frequency (af) range is 16 to 16,000 Hz. Higher frequencies up to 300,000 MHz are radio frequencies.

- The amount of time for one cycle is the period T. The period and frequency are reciprocals: $T = 1/f$, or $f = 1/T$. The higher the frequency, the shorter the period.

- Wavelength λ is the distance a wave travels in one cycle. The higher the frequency, the shorter the wavelength. The wavelength also depends on the velocity at which the wave travels: $\lambda = v/f$, where v is velocity of the wave and f is the frequency.

- Phase angle is the angular difference in time between corresponding values in the cycles for two waveforms of the same frequency.

- When one sine wave has its maximum value while the other is at zero, the two waves are 90° out of phase. Two waveforms with a zero phase angle between them are in phase; a 180° phase angle means opposite phase.

- The length of a phasor arrow indicates amplitude, and the angle corresponds to the phase. A leading phase is shown by counterclockwise angles.

- Sine-wave alternating voltage V applied across a load resistance R produces alternating current I in the circuit. The current has the same waveform, frequency, and phase as the applied voltage because of the resistive load. The amount of $I = V/R$.

- The sawtooth wave and square wave are two common examples of nonsinusoidal waveforms. The amplitudes of these waves are usually measured in peak-to-peak value.

- Harmonic frequencies are exact multiples of the fundamental frequency.

- The ac voltage used in residences range from 115 to 125 V rms with a frequency of 60 Hz. The nominal voltage is usually given as 120 V.

- For residential wiring, the three-wire single-phase Edison system shown in Fig. 15–21 is used to provide either 120 or 240 V.

- In a motor, the rotating armature connects to the power line. The stator field coils provide the magnetic flux cut by the armature as it is forced to rotate. A generator has the opposite effect: it converts mechanical energy into electrical output.

- A dc motor has commutator segments contacted by graphite brushes for the external connections to the power source. An ac induction motor does not have brushes.

- In three-phase power, each phase angle is 120°. For the Y connections in Fig. 15–26a, each pair of output terminals has an output of 120 × 1.73 = 208 V. This voltage is known as the line-to-line voltage.

Important Terms

- Alternation - one-half cycle of revolution of a conductor loop rotating through a magnetic field. This corresponds to one-half cycle of alternating voltage or current.

- Average Value - the arithmetic average of all values in a sine wave for one alternation. Average value = 0.637 × peak value.

- Cycle - one complete revolution of a conductor loop rotating through a magnetic field. For any ac waveform, a cycle can be defined to include the variations between two successive points having the same value and varying in the same direction.

- Decade - a unit for representing a 10:1 range in frequencies.

- Effective Value - another name for an rms value.

- Form Factor - the ratio of the rms to average values. For a sine wave, $\frac{\text{rms}}{\text{avg}} = 1.11$.

- Frequency - the number of cycles a waveform completes each second.

- Generator - a machine or device that converts mechanical energy into electrical energy.

- Harmonic Frequency - a frequency that is an exact multiple of the fundamental frequency.

- Hertz (Hz) - the basic unit of frequency. 1 Hz = 1 cycle per second.

- Motor - a machine or device that converts electrical energy into mechanical energy.

- Nonsinusoidal Waveform - any waveform that is not a sine wave or a cosine wave.

- Octave - a unit for representing a 2:1 range in frequencies.

- Peak Value - the maximum amplitude of a sine wave.

- Period - the amount of time it takes to complete one cycle of alternating voltage or current. The symbol for the period is T for time. The unit for T is the second(s).

- Phase Angle - the angular difference between two sinusoidal waveforms or phasors.

- Phasor - a line representing the magnitude and direction of a quantity, such as voltage or current, with respect to time.

- Quadrature Phase - a phase angle of 90°.

- Radian - an angle equal to approximately 57.3°.

- Root-Mean-Square (rms) Value - the value of a sine wave that corresponds to the same amount of direct current or voltage in heating power. Unless indicated otherwise, all sine wave ac measurements are in rms values. rms value = 0.707 × peak value.

- Sine Wave - a waveform whose value is proportional to the sine of the angle of rotation in the circular motion producing the induced voltage or current.

- Wavelength - the distance a waveform travels through space to complete one cycle.

Related Formulas

$v = V_M \sin \theta$

Average value = 0.637 × peak value

rms value = 0.707 × peak value

$\text{Peak} = \dfrac{1}{0.707} \times \text{rms} = 1.414 \times \text{rms}$

Peak-to-peak value = 2.828 × rms value

$T = \dfrac{1}{f} \quad \text{or} \quad f = \dfrac{1}{T}$

$\lambda = \dfrac{\text{velocity}}{\text{frequency}}$

$\lambda \text{ (cm)} = \dfrac{3 \times 10^{10} \text{ cm/s}}{f(\text{Hz})} \quad \text{(radio wave)}$

$\lambda = \dfrac{1130 \text{ ft/s}}{f(\text{Hz})} \quad \text{(sound wave)}$

$t = \dfrac{\theta}{360} \times \dfrac{1}{f}$

Self-Test

Answers at back of book.

1. **An alternating voltage is one that**
   a. varies continuously in magnitude.
   b. reverses periodically in polarity.
   c. never varies in amplitude.
   d. both a and b.

2. **One complete revolution of a conductor loop through a magnetic field is called a(n)**
   a. octave.
   b. decade.
   c. cycle.
   d. alternation.

3. **For a sine wave, one-half cycle is often called a(n)**
   a. alternation.
   b. harmonic.
   c. octave.
   d. period.

4. **One cycle includes**
   a. 180°.
   b. 360°.
   c. 2π rad.
   d. both b and c.

5. **In the the United States, the frequency of the ac power-line voltage is**
   a. 120 Hz.
   b. 60 Hz.
   c. 50 Hz.
   d. 100 Hz.

6. **For a sine wave, the number of complete cycles per second is called the**
   a. period.
   b. wavelength.
   c. frequency.
   d. phase angle.

7. **A sine wave of alternating voltage has its maximum values at**
   a. 90° and 270°.
   b. 0° and 180°.
   c. 180° and 360°.
   d. 30° and 150°.

8. **To compare the phase angle between two waveforms, both must have**
   a. the same amplitude.
   b. the same frequency.
   c. different frequencies.
   d. both a and b.

9. **A 2-kHz sine wave has a period, T, of**
   a. 0.5 μs.
   b. 50 μs.
   c. 500 μs.
   d. 2 ms.

10. **If a sine wave has a period, T, of 40 μs, its frequency, f, equals**
    a. 25 kHz.
    b. 250 Hz.
    c. 40 kHz.
    d. 2.5 kHz.

11. **What is the wavelength of a radio wave whose frequency is 15 MHz?**
    a. 20 meters.
    b. 15 meters.
    c. 0.753 ft.
    d. 2000 meters.

12. **The value of alternating current or voltage that has the same heating effect as a corresponding dc value is known as the**
    a. peak value.
    b. average value.
    c. rms value.
    d. peak-to-peak value.

13. **The wavelength of a 500-Hz sound wave is**
    a. 60 km.
    b. 2.26 ft.
    c. 4.52 ft.
    d. 0.226 ft.

14. **In residential house wiring, the hot wire is usually color-coded**
    a. white.
    b. green.
    c. black or red.
    d. as a bare copper wire.

15. **A sine wave with a peak value of 20 V has an rms value of**
    a. 28.28 V.
    b. 14.14 V.
    c. 12.74 V.
    d. 56.6 V.

16. **A sine wave whose rms voltage is 25.2 V has a peak value of approximately**
    a. 17.8 V.
    b. 16 V.
    c. 50.4 V.
    d. 35.6 V.

17. **The unit of frequency is the**
    a. hertz.
    b. maxwell.
    c. radian.
    d. second.

18. **For an ac waveform, the period, T, refers to**
    a. the number of complete cycles per second.
    b. the length of time required to complete one cycle.
    c. the time it takes for the waveform to reach its peak value.
    d. none of the above.

19. **The wavelength of a radio wave is**
    a. inversely proportional to its frequency.
    b. directly proportional to its frequency.
    c. inversely proportional to its amplitude.
    d. unrelated to its frequency.

20. **Exact multiples of the fundamental frequency are called**
    a. ultrasonic frequencies.
    b. harmonic frequencies.
    c. treble frequencies.
    d. resonant frequencies.

21. **Raising the frequency of 500 Hz by two octaves corresponds to a frequency of**
    a. 2 kHz.
    b. 1 kHz.
    c. 4 kHz.
    d. 250 Hz.

22. In residential house wiring, the neutral wire is always color-coded
    a. black.
    b. bare copper.
    c. green.
    d. white.

23. The second harmonic of 7 MHz is
    a. 3.5 MHz.
    b. 28 MHz.
    c. 14 MHz.
    d. 7 MHz.

24. A sine wave has a peak voltage of 170 V. What is the instantaneous voltage at an angle of 45°?
    a. 240 V.
    b. 85 V.
    c. 0 V.
    d. 120 V.

25. Unless indicated otherwise, all sine wave ac measurements are in
    a. peak-to-peak values.
    b. peak values.
    c. rms values.
    d. average values.

Questions

1. (a) Define an alternating voltage. (b) Define an alternating current. (c) Why does ac voltage applied across a load resistance produce alternating current in the circuit?

2. (a) State two characteristics of a sine wave of voltage. (b) Why does the rms value of 0.707 × peak value apply just to sine waves?

3. Draw two cycles of an ac sawtooth voltage waveform with a peak-to-peak amplitude of 40 V. Do the same for a square wave.

4. Give the angle in degrees and radians for each of the following: one cycle, one half-cycle, one quarter-cycle, three quarter-cycles.

5. The peak value of a sine wave is 1 V. How much is its average value? rms value? Effective value? Peak-to-peak value?

6. State the following ranges in hertz: (a) audio frequencies; (b) radio frequencies; (c) standard AM radio broadcast band; (d) FM broadcast band; (e) VHF band; (f) microwave band.

7. Make a graph with two waves, one with a frequency of 500 kHz and the other with 1000 kHz. Mark the horizontal axis in time, and label each wave.

8. Draw the sine waves and phasor diagrams to show (a) two waves 180° out of phase; (b) two waves 90° out of phase.

9. Give the voltage value for the 60-Hz ac line voltage with an rms value of 120 V at each of the following times in a cycle: 0°, 30°, 45°, 90°, 180°, 270°, 360°.

10. (a) The phase angle of 90° equals how many radians? (b) For two sine waves 90° out of phase with each other, compare their amplitudes at 0°, 90°, 180°, 270°, and 360°.

11. Tabulate the sine and cosine values every 30° from 0 to 360° and draw the corresponding sine wave and cosine wave.

12. Draw a graph of the values for $(\sin \theta)^2$ plotted against θ for every 30° from 0 to 360°.

13. Why is the wavelength of an ultrasonic wave at 34.44 kHz the same 1 cm as for the much higher frequency radio wave at 30 GHz?

14. Draw the sine waves and phasors to show wave V_1 leading wave V_2 by 45°.

15. Why are amplitudes for nonsinusoidal waveforms generally measured in peak-to-peak values, rather than rms or average value?

16. Define harmonic frequencies, giving numerical values.

17. Define one octave, with an example of numerical values.

18. Which do you consider more important for applications of alternating current—polarity reversals or variations in value?

19. Define the following parts in the assembly of motors: (a) armature rotor; (b) field stator; (c) collector rings; (d) commutator segments.

20. Show diagrams of Y and Δ connections for three-phase ac power.

Problems

SECTION 15–2 ALTERNATING-VOLTAGE GENERATOR

15–1 For a sine wave of alternating voltage, how many degrees are included in
   a. ¼ cycle?
   b. ½ cycle?
   c. ¾ cycle?
   d. 1 complete cycle?

15–2 For a sine wave of alternating voltage, how many radians are included in
   a. ¼ cycle?
   b. ½ cycle?
   c. ¾ cycle?
   d. 1 complete cycle?

15–3 At what angle does a sine wave of alternating voltage

    a. reach its maximum positive value?

    b. reach its maximum negative value?

    c. cross the zero axis?

15–4 One radian corresponds to how many degrees?

SECTION 15–3 THE SINE WAVE

15–5 The peak value of a sine wave equals 20 V. Calculate the instantaneous voltage of the sine wave for the phase angles listed.

    a. 30°

    b. 45°

    c. 60°

    d. 75°

    e. 120°

    f. 210°

    g. 300°

15–6 The peak value of a sine wave equals 100 mV. Calculate the instantaneous voltage of the sine wave for the phase angles listed.

    a. 15°

    b. 50°

    c. 90°

    d. 150°

    e. 180°

    f. 240°

    g. 330°

15–7 A sine wave of alternating voltage has an instantaneous value of 45 V at an angle of 60°. Determine the peak value of the sine wave.

SECTION 15–4 ALTERNATING CURRENT

15–8 In Fig. 15–28, the sine wave of applied voltage has a peak or maximum value of 10 V, as shown. Calculate the instantaneous value of current for the phase angles listed.

    a. 30°

    b. 60°

    c. 90°

    d. 120°

    e. 150°

    f. 180°

    g. 210°

    h. 240°

    i. 270°

    j. 300°

    k. 330°

15–9 In Fig. 15–28, do electrons flow clockwise or counterclockwise in the circuit during

    a. the positive alternation?

    b. the negative alternation?

Note: During the positive alternation, terminal 1 is positive with respect to terminal 2.

Figure 15–28

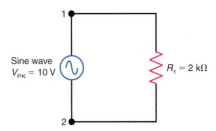

SECTION 15–5 VOLTAGE AND CURRENT VALUES FOR A SINE WAVE

15–10 If the sine wave in Fig. 15–29 has a peak value of 15 V, then calculate

    a. the peak-to-peak value.

    b. the rms value.

    c. the average value.

Figure 15–29

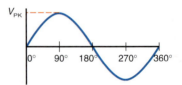

15–11 If the sine wave in Fig. 15–29 has a peak value of 50 V, then calculate

    a. the peak-to-peak value.

    b. the rms value.

    c. the average value.

15–12 If the sine wave in Fig. 15–29 has an rms value of 60 V, then calculate

    a. the peak value.

    b. the peak-to-peak value.

    c. the average value.

15–13 If the sine wave in Fig. 15–29 has an rms value of 40 V, then calculate

    a. the peak value.

    b. the peak-to-peak value.

    c. the average value.

15–14 If the sine wave of alternating voltage in Fig. 15–30 has a peak value of 25 V, then calculate

    a. the peak current value.

    b. the peak-to-peak current value.

    c. the rms current value.

    d. the average current value.

Figure 15–30

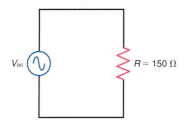

15–15 If the sine wave of alternating voltage in Fig. 15–30 has an rms value of 7.07 V, then calculate

    a. the rms current value.

    b. the peak current value.

    c. the peak-to-peak current value.

    d. the average current value.

15–16 Convert the following values into rms values:

    a. 32 V peak

    b. 18 V peak-to-peak

    c. 90.09 V average

    d. 120 mA peak-to-peak

15–17 Convert the following values into peak values:

    a. 12 V rms

    b. 72 V average

    c. 50 V peak-to-peak

    d. 750 mV rms

SECTION 15–6 FREQUENCY

15–18 What is the frequency, f, of a sine wave that completes

    a. 10 cycles per second?

    b. 500 cycles per second?

    c. 50,000 cycles per second?

    d. 2,000,000 cycles per second?

15–19 How many cycles per second (cps) do the following frequencies correspond to?

    a. 2 kHz

    b. 15 MHz

    c. 10 kHz

    d. 5 GHz

SECTION 15–7 PERIOD

15–20 Calculate the period, T, for the following sine wave frequencies:

    a. 50 Hz

    b. 100 Hz

    c. 500 Hz

    d. 1 kHz

15–21 Calculate the period, T, for the following sine wave frequencies:

    a. 2 kHz

    b. 4 kHz

    c. 200 kHz

    d. 2 MHz

15–22 Calculate the frequency, f, of a sine wave whose period, T, is

    a. 40 μs.

    b. 50 μs.

    c. 2.5 ms.

    d. 16.67 ms.

15–23 Calculate the frequency, f, of a sine wave whose period, T, is

    a. 5 ms.

    b. 10 μs.

    c. 500 ns.

    d. 33.33 μs.

15–24 For a 5-kHz sine wave, how long does it take for

    a. $^1/_4$ cycle?

    b. $^1/_2$ cycle?

    c. $^3/_4$ cycle?

    d. 1 full cycle?

SECTION 15–8 WAVELENGTH

15–25 What is the velocity of an electromagnetic radio wave in

    a. miles per second (mi/s).

    b. centimeters per sec (cm/s).

    c. meters per sec (m/s).

15–26 What is the velocity in ft/s of a sound wave produced by mechanical vibrations?

15–27 What is the wavelength in cm of an electromagnetic radio wave whose frequency is

    a. 3.75 MHz?

    b. 7.5 MHz?

    c. 15 MHz?

    d. 20 MHz?

15–28 Convert the wavelengths in Prob. 15–27 into meters (m).

15–29 What is the wavelength in meters of an electromagnetic radio wave whose frequency is 150 MHz?

15–30 What is the wavelength in ft of a sound wave whose frequency is

    a. 50 Hz?

    b. 200 Hz?

    c. 750 Hz?

    d. 2 kHz?

    e. 4 kHz?

    f. 10 kHz?

15–31 What is the frequency of an electromagnetic radio wave whose wavelength is

    a. 160 m?

    b. 10 m?

    c. 17 m?

    d. 11 m?

15-32 What is the frequency of a sound wave whose wavelength is

   a. 4.52 ft?

   b. 1.13 ft?

   c. 3.39 ft?

   d. 0.226 ft?

SECTION 15–9 PHASE ANGLE

15-33 Describe the difference between a sine wave and a cosine wave.

15-34 Two voltage waveforms of the same amplitude, V_X and V_Y, are 45° out of phase with each other, with V_Y lagging V_X. Draw the phasors representing these voltage waveforms if

   a. V_X is used as the reference phasor.

   b. V_Y is used as the reference phasor.

SECTION 15–10 THE TIME FACTOR IN FREQUENCY AND PHASE

15-35 For two waveforms with a frequency of 1 kHz, how much time corresponds to a phase angle difference of

   a. 30°?

   b. 45°?

   c. 60°?

   d. 90°?

15-36 For two waveforms with a frequency of 50 kHz, how much time corresponds to a phase angle difference of

   a. 15°?

   b. 36°?

   c. 60°?

   d. 150°?

SECTION 15–11 ALTERNATING CURRENT CIRCUITS WITH RESISTANCE

15-37 In Fig. 15–31, solve for the following values: R_T, I, V_1, V_2, P_1, P_2, and P_T.

Figure 15–31

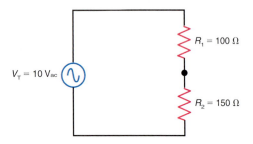

15-38 In Fig. 15–32, solve for the following values: I_1, I_2, I_T, R_{EQ}, P_1, P_2, and P_T.

Figure 15–32

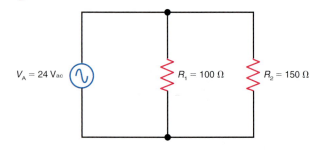

15-39 In Fig. 15–33, solve for the following values: R_T, I_T, V_1, V_2, V_3, I_2, I_3, P_1, P_2, P_3, and P_T.

Figure 15–33

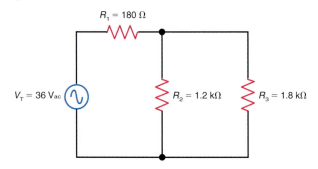

15-40 In Fig. 15–33, find the following values:

   a. the peak-to-peak current through R_1

   b. the average voltage across R_2

   c. the peak voltage across R_3

   d. the average current through R_3

SECTION 15–12 NONSINUSOIDAL AC WAVEFORMS

15-41 Determine the peak-to-peak voltage and frequency for the waveform in

   a. Fig. 15–34a

   b. Fig. 15–34b

   c. Fig. 15–34c

Figure 15-34

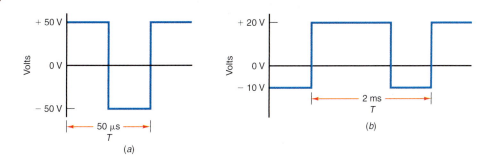

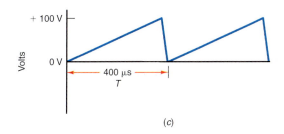

(c)

SECTION 15-13 HARMONIC FREQUENCIES

15-42 List the first four harmonics of a 3.8-MHz radio signal.

15-43 List the first seven harmonics of a 1-kHz sine wave. Label each harmonic as either an even or odd harmonic.

15-44 Raising the frequency of 250 Hz by one octave corresponds to what frequency?

15-45 Lowering the frequency of 3 kHz by two octaves corresponds to what frequency?

15-46 Raising the frequency of 300 Hz by three octaves corresponds to what frequency?

15-47 What is the frequency two decades above 1 kHz?

SECTION 15-14 THE 60-Hz AC POWER LINE

15-48 What is the frequency of the ac power line in most European countries?

15-49 What device or component is used to step up or step down an ac voltage in the distribution of ac power to our homes and industries?

15-50 What is the main reason for using extremely high voltages, such as 500 kV, on the distribution lines for ac power?

Critical Thinking

15-51 The electrical length of an antenna is to be one-half wavelength long at a frequency f of 7.2 MHz. Calculate the length of the antenna in: **(a)** feet; **(b)** centimeters.

15-52 A transmission line has a length l of 7.5 m. What is its electrical wavelength at 10 MHz?

15-53 The total length of an antenna is 120 ft. At what frequency is the antenna one-half wavelength long?

15-54 A cosine wave of current has an instantaneous amplitude of 45 mA at $\theta = \pi/3$ rad. Calculate the waveform's instantaneous amplitude at $\theta = 3\pi/2$ rad.

Answers to Knowledge Check Problems

15-1 Oscilloscope	**15-4** 10 mA	**15-7** 200 μs
15-2 3.1416	**15-5** 8.48 V	**15-8** 18.83 ft
15-3 86.6 V	**15-6** 1490	**15-9** Phasor

15-10 2.083 ms

15-11 113.12 V

15-12 False

15-13 600 Hz

15-14 False

15-15 Generator

15-16 More efficient distribution of power

Answers to Self-Reviews

15-1 a. T
    b. F
    c. T

15-2 a. 10 V
    b. 360°

15-3 a. 85 V
    b. 120 V
    c. 170 V

15-4 a. 0.707 A
    b. 0.5 A

15-5 a. 120 V rms
    b. 14.14 V peak
    c. 2.83 V p-p

15-6 a. 4 Hz
    b. 1.605 MHz

15-7 a. 400 Hz
    b. $1/400$ S

15-8 a. T
    b. F
    c. T

15-9 a. 90°
    b. 60°
    c. 0°

15-10 a. $1/240$ S
    b. 0.1 μs

15-11 a. 30 Ω
    b. 6.67 Ω
    c. 30 Ω

15-12 a. 2 μs
    b. 15 V

15-13 a. 48 MHz
    b. 440 Hz

15-14 a. F
    b. T
    c. T
    d. T

15-15 a. T
    b. T
    c. F

15-16 a. 120°
    b. 208 V

Cumulative Review Summary (Chapters 13–15)

- Iron, nickel, and cobalt are magnetic materials. Magnets have a north pole and a south pole at opposite ends. Opposite poles attract; like poles repel.

- A magnet has an invisible, external magnetic field. This magnetic flux is indicated by field lines. The direction of field lines outside the magnet is from a north pole to a south pole.

- An electromagnet has an iron core that becomes magnetized when current flows in the coil winding.

- Magnetic units are defined in Table 13-1.

- Continuous magnetization and demagnetization of an iron core by alternating current causes hysteresis losses that increase with higher frequencies.

- Current in a conductor has an associated magnetic field with circular lines of force in a plane perpendicular to the wire.

- Motor action results from the net force of two fields that can aid or cancel. The direction of the resultant force is from the stronger field to the weaker.

- The motion of flux cutting across a perpendicular conductor generates an induced voltage.

- Faraday's law of induced voltage states that $v = N \, d\phi/dt$.

- Lenz's law states that an induced voltage must have the polarity that opposes the change causing the induction.

- Alternating voltage varies in magnitude and reverses in polarity.

- One cycle includes the values between points having the same value and varying in the same direction. The cycle includes 360°, or 2π rad.

- Frequency f equals cycles per second (cps). One cps = 1 Hz.

- Period T is the time for one cycle. It equals 1/f. When f is in cycles per second, T is in seconds.

- Wavelength λ is the distance a wave travels in one cycle. $\lambda = v/f$.

- The rms, or effective value, of a sine wave equals 0.707 × peak value. Or the peak value equals 1.414 × rms value. The average value equals 0.637 × peak value.

- Phase angle θ is the angular difference between corresponding values in the cycles for two sine waves of the same frequency. The angular difference can be expressed in time based on the frequency of the waves.

- Phasors, similar to vectors, indicate the amplitude and phase angle of alternating voltage or current. The length of the phasor is the amplitude, and the angle is the phase.

- The square wave and sawtooth wave are common examples of nonsinusoidal waveforms.

- Direct current motors generally use commutator segments with graphite brushes. Alternating current motors are usually the induction type without brushes.

- House wiring uses three-wire, single-phase power with a frequency of 60 Hz. The voltages for house wiring are 120 V to the grounded neutral and 240 V across the two high sides.

- Three-phase ac power has three legs 120° out of phase. A Y connection with 120 V across each phase has 208 V available across each two legs.

Cumulative Review Self-Test

Answers at back of book.

1. Which of the following statements is true? (a) Alnico is commonly used for electromagnets. (b) Paper cannot affect magnetic flux because it is not a magnetic material. (c) Iron is generally used for permanent magnets. (d) Ferrites have lower permeability than air or vacuum.

2. Hysteresis losses (a) are caused by high-frequency alternating current in a coil with an iron core; (b) generally increase with direct current in a coil; (c) are especially important for permanent magnets that have a steady magnetic field; (d) cannot be produced in an iron core because it is a conductor.

3. A magnetic flux of 25,000 lines through an area of 5 cm^2 results in (a) 5 lines of flux; (b) 5000 Mx of flux; (c) flux density of 5000 G; (d) flux density corresponding to 25,000 A.

4. If 10 V is applied across a relay coil with 100 turns having 2 Ω of resistance, the total force producing magnetic flux in the circuit is (a) 10 Mx; (b) 50 G; (c) 100 Oe; (d) 500 A · t.

5. The ac power-line voltage of 120 V rms has a peak value of (a) 100 V; (b) 170 V; (c) 240 V; (d) 338 V.

6. Which of the following can produce the most induced voltage? (a) 1-A direct current; (b) 50-A direct current; (c) 1-A 60-Hz alternating current; (d) 1-A 400-Hz alternating current.

7. Which of the following has the highest frequency? (a) $T = \frac{1}{1000}$ s; (b) $T = \frac{1}{60}$ s; (c) $T = 1$ s; (d) $T = 2$ s.

8. Two waves of the same frequency are opposite in phase when the phase angle between them is (a) 0°; (b) 90°; (c) 360°; (d) π rad.

9. A 120-V, 60-Hz power-line voltage is applied across a 120-Ω resistor. The current equals (a) 1 A, peak value; (b) 120 A, peak value; (c) 1 A, rms value; (d) 5 A, rms value.

10. When an alternating voltage reverses in polarity, the current it produces (a) reverses in direction; (b) has a steady dc value; (c) has a phase angle of 180°; (d) alternates at 1.4 times the frequency of the applied voltage.

16 Capacitance

Capacitance is the ability of a dielectric to hold or store an electric charge. The more charge stored for a given voltage, the higher the capacitance. The symbol for capacitance is C, and the unit is the farad (F), named after Michael Faraday.

A capacitor consists of an insulator (also called a *dielectric*) between two conductors. The conductors make it possible to apply voltage across the insulator. Different types of capacitors are manufactured for specific values of C. They are named according to the dielectric. Common types are air, ceramic, mica, paper, film, and electrolytic capacitors. Capacitors used in electronic circuits are small and economical.

The most important property of a capacitor is its ability to block a steady dc voltage while passing ac signals. The higher the frequency, the less the opposition to ac voltage.

Capacitors are a common source of troubles because they can have either an open at the conductors or a short circuit through the dielectric. These troubles are described here, including the method of checking a capacitor with an ohmmeter, even though a capacitor is actually an insulator.

Objectives

After studying this chapter you should be able to

- *Describe* how charge is stored in the dielectric of a capacitor.
- *Describe* how a capacitor charges and discharges.
- *Define* the farad unit of capacitance.
- *List* the physical factors affecting the capacitance of a capacitor.
- *List* several types of capacitors and the characteristics of each.
- *Explain* how an electrolytic capacitor is constructed.
- *Explain* how capacitors are coded.
- *Calculate* the total capacitance of parallel-connected capacitors.
- *Calculate* the equivalent capacitance of series-connected capacitors.
- *Calculate* the energy stored in a capacitor.
- *Define* the terms *leakage, dielectric absorption,* and *equivalent series resistance* as they relate to capacitors.
- *Describe* how an ohmmeter can be used to test a capacitor.

Outline

Important Terms

capacitance (C)

capacitor

charging

condenser

dielectric absorption

dielectric constant, K_ϵ

dielectric material

dielectric strength

discharging

electric field

equivalent series resistance (ESR)

farad (F) unit

ganged capacitors

leakage current

leakage resistance

microfarad (μF)

nanofarad (nF)

picofarad (pF)

relative permittivity, ϵ_r

16–1 How Charge Is Stored in a Dielectric

It is possible for dielectric materials such as air or paper to hold an electric charge because free electrons cannot flow through an insulator. However, the charge must be applied by some source. In Fig. 16–1a, the battery can charge the capacitor shown. With the dielectric contacting the two conductors connected to the potential difference V, electrons from the voltage source accumulate on the side of the capacitor connected to the negative terminal of V. The opposite side of the capacitor connected to the positive terminal of V loses electrons.

As a result, the excess of electrons produces a negative charge on one side of the capacitor, and the opposite side has a positive charge. As an example, if 6.25×10^{18} electrons are accumulated, the negative charge equals 1 coulomb (C). The charge on only one plate need be considered because the number of electrons accumulated on one plate is exactly the same as the number taken from the opposite plate.

What the voltage source does is simply redistribute some electrons from one side of the capacitor to the other. This process is called *charging* the capacitor. Charging continues until the potential difference across the capacitor is equal to the applied voltage. Without any series resistance, the charging is instantaneous. Practically, however, there is always some series resistance. This charging current is transient, or temporary; it flows only until the capacitor is charged to the applied voltage. Then there is no current in the circuit.

The result is a device for storing charge in the dielectric. Storage means that the charge remains even after the voltage source is disconnected. The measure of how much charge can be stored is the capacitance C. More charge stored for a given amount of applied voltage means more capacitance. Components made to provide a specified amount of capacitance are called *capacitors,* or by their old name *condensers.*

Electrically, then, capacitance is the ability to store charge. A capacitor consists simply of two conductors separated by an insulator. For example, Fig. 16–1b shows a variable capacitor using air for the dielectric between the metal plates. There are many types with different dielectric materials, including paper, mica, and ceramics, but the schematic symbols shown in Fig. 16–1c apply to all capacitors.

Figure 16–1 Capacitance stores the charge in the dielectric between two conductors. (a) Structure. (b) Air-dielectric variable capacitor. Length is 2 in. (c) Schematic symbols for fixed and variable capacitors.

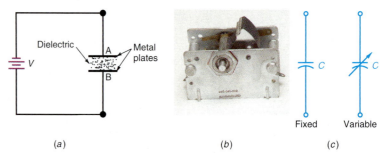

Electric Field in the Dielectric

Any voltage has a field of electric lines of force between the opposite electric charges. The electric field corresponds to the magnetic lines of force of the magnetic field associated with electric current. What a capacitor does is concentrate the electric field in the dielectric between the plates. This concentration corresponds to a magnetic field concentrated in the turns of a coil. The only function of the capacitor plates and wire conductors is to connect the voltage source V across the dielectric. Then the electric field is concentrated in the capacitor, instead of being spread out in all directions.

Electrostatic Induction

The capacitor has opposite charges because of electrostatic induction by the electric field. Electrons that accumulate on the negative side of the capacitor provide electric lines of force that repel electrons from the opposite side. When this side loses electrons, it becomes positively charged. The opposite charges induced by an electric field correspond to the opposite poles induced in magnetic materials by a magnetic field.

■ 16-1 Knowledge Check

Answer at end of chapter.

In Fig. 16-1a how much current flows in the circuit once the capacitor voltage equals the applied voltage, V?

■ 16-1 Self-Review

Answers at end of chapter.

a. In a capacitor, is the electric charge stored in the dielectric or on the metal plates?
b. What is the unit of capacitance?

16-2 Charging and Discharging a Capacitor

Charging and discharging are the two main effects of capacitors. Applied voltage puts charge in the capacitor. The accumulation of charge results in a buildup of potential difference across the capacitor plates. When the capacitor voltage equals the applied voltage, there is no more charging. The charge remains in the capacitor, with or without the applied voltage connected.

The capacitor discharges when a conducting path is provided across the plates, without any applied voltage. Actually, it is necessary only that the capacitor voltage be more than the applied voltage. Then the capacitor can serve as a voltage source, temporarily, to produce discharge current in the discharge path. The capacitor discharge continues until the capacitor voltage drops to zero or is equal to the applied voltage.

Applying the Charge

In Fig. 16-2a, the capacitor is neutral with no charge because it has not been connected to any source of applied voltage and there is no electrostatic field in the dielectric. Closing the switch in Fig. 16-2b, however, allows the negative battery terminal to repel free electrons in the conductor to plate A. At the same time, the positive terminal attracts free electrons from plate B. The side of the dielectric at plate A accumulates electrons because they cannot flow through the insulator, and plate B has an equal surplus of protons.

Remember that opposite charges have an associated potential difference, which is the voltage across the capacitor. The charging process continues until the capacitor voltage equals the battery voltage, which is 10 V in this example. Then no further charging is possible because the applied voltage cannot make free electrons flow in the conductors.

Note that the potential difference across the charged capacitor is 10 V between plates A and B. There is no potential difference from each plate to its battery terminal, however, which is why the capacitor stops charging.

Storing the Charge

The negative and positive charges on opposite plates have an associated electric field through the dielectric, as shown by the dotted lines in Figs. 16–2b and 16–2c. The direction of these electric lines of force is shown repelling electrons from plate B, making this side positive. The effect of electric lines of force through the dielectric results in storage of the charge. The electric field distorts the molecular structure so that the dielectric is no longer neutral. The dielectric is actually stressed by the invisible force of the electric field. As evidence, the dielectric can be ruptured by a very intense field with high voltage across the capacitor.

The result of the electric field, then, is that the dielectric has charge supplied by the voltage source. Since the dielectric is an insulator that cannot conduct, the charge remains in the capacitor even after the voltage source is removed, as illustrated in Fig. 16–2c. You can now take this charged capacitor by itself out of the circuit, and it still has 10 V across the two terminals.

Discharging

The action of neutralizing the charge by connecting a conducting path across the dielectric is called *discharging* the capacitor. In Fig. 16–2d, the wire between plates A and B is a low-resistance path for discharge current. With the stored charge in the dielectric providing the potential difference, 10 V is available to produce discharge current. The negative plate repels electrons, which are attracted to the positive plate through the wire, until the positive and negative charges are neutralized. Then there is no net charge. The capacitor is completely discharged, the voltage across it equals zero, and there is no discharge current. Now the capacitor is in the same uncharged condition as in Fig. 16–2a. It can be charged again, however, by a source of applied voltage.

IIII MultiSim **Figure 16–2** Storing electric charge in a capacitance. (*a*) Capacitor without any charge. (*b*) Battery charges capacitor to applied voltage of 10 V. (*c*) Stored charge remains in the capacitor, providing 10 V without the battery. (*d*) Discharging the capacitor.

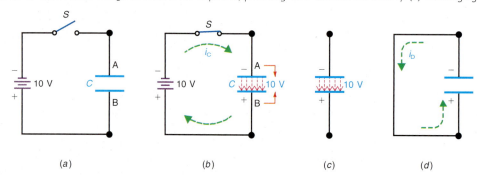

Nature of the Capacitance

A capacitor can store the amount of charge necessary to provide a potential difference equal to the charging voltage. If 100 V were applied in Fig. 16–2, the capacitor would charge to 100 V.

The capacitor charges to the applied voltage because it takes on more charge when the capacitor voltage is less. As soon as the capacitor voltage equals the applied voltage, no more charging current can flow. *Note that any charge or discharge current flows through the conducting wires to the plates but not through the dielectric.*

Charge and Discharge Currents

In Fig. 16–2b, i_C is in the opposite direction from i_D in Fig. 16–2d. In both cases, the current shown is electron flow. However, i_C is charging current to the capacitor and i_D is discharge current from the capacitor. The charge and discharge currents must always be in opposite directions. In Fig. 16–2b, the negative plate of C accumulates electrons from the voltage source. In Fig. 16–2d, the charged capacitor is a voltage source to produce electron flow around the discharge path.

More charge and discharge current result from a higher value of C for a given amount of voltage. Also, more V produces more charge and discharge current with a given amount of capacitance. However, the value of C does not change with the voltage because the amount of C depends on the physical construction of the capacitor.

■ *16–2 Knowledge Check*

Answer at end of chapter.

For any capacitor that is being charged, the charge current flows through the dielectric from plate to plate until the capacitor is fully charged. (True or False)

■ *16–2 Self-Review*

Answers at end of chapter.

Refer to Fig. 16–2.
a. If the applied voltage were 14.5 V, how much would the voltage be across C after it is charged?
b. How much is the voltage across C after it is completely discharged?
c. Can a capacitor be charged again after it is discharged?

PIONEERS IN ELECTRONICS

The unit of measure for capacitance, the farad (F), was named for *Michael Faraday (1791–1867),* an English chemist and physicist who discovered the principle of induction (1 F is the unit of capacitance that will store 1 coulomb (C) of charge when 1 volt (V) is applied).

16–3 The Farad Unit of Capacitance

With more charging voltage, the electric field is stronger and more charge is stored in the dielectric. The amount of charge Q stored in the capacitance is therefore proportional to the applied voltage. Also, a larger capacitance can store more charge. These relations are summarized by the formula

$$Q = CV \text{ coulombs} \tag{16-1}$$

where Q is the charge stored in the dielectric in coulombs (C), V is the voltage across the plates of the capacitor, and C is the capacitance in farads.

The C is a physical constant, indicating the capacitance in terms of the amount of charge that can be stored for a given amount of charging voltage. When one coulomb is stored in the dielectric with a potential difference of one volt, the capacitance is one *farad.*

Practical capacitors have sizes in millionths of a farad, or smaller. The reason is that typical capacitors store charge of microcoulombs or less. Therefore, the common units are

$$1 \text{ microfarad} = 1 \ \mu\text{F} = 1 \times 10^{-6} \text{ F}$$
$$1 \text{ nanofarad} = 1 \text{ nF} = 1 \times 10^{-9} \text{ F}$$
$$1 \text{ picofarad} = 1 \text{ pF} = 1 \times 10^{-12} \text{ F}$$

Although traditionally it has not been used, the nanofarad unit of capacitance is gaining acceptance in the electronics industry.

Example 16–1

How much charge is stored in a 2-μF capacitor connected across a 50-V supply?

ANSWER $Q = CV = 2 \times 10^{-6} \times 50$
$= 100 \times 10^{-6} \text{ C}$

Example 16–2

How much charge is stored in a 40-μF capacitor connected across a 50-V supply?

ANSWER $Q = CV = 40 \times 10^{-6} \times 50$
$= 2000 \times 10^{-6} \text{ C}$

Note that the larger capacitor stores more charge for the same voltage, in accordance with the definition of capacitance as the ability to store charge.

The factors in $Q = CV$ can be inverted to

$$C = \frac{Q}{V} \tag{16–2}$$

or

$$V = \frac{Q}{C} \tag{16–3}$$

For all three formulas, the basic units are volts for V, coulombs for Q, and farads for C. Note that the formula $C = Q/V$ actually defines one farad of capacitance as one coulomb of charge stored for one volt of potential difference. The letter C (in italic type) is the symbol for capacitance. The same letter C (in Roman type) is the abbreviation for the coulomb unit of charge. The difference between C and C will be made clearer in the examples that follow.

Example 16-3

A constant current of 2 μA charges a capacitor for 20 s. How much charge is stored? Remember $I = Q/t$ or $Q = I \times t$.

ANSWER $Q = I \times t$
$$= 2 \times 10^{-6} \times 20$$
$$= 40 \times 10^{-6} \text{ or } 40 \ \mu C$$

Example 16-4

The voltage across the charged capacitor in Example 16–3 is 20 V. Calculate C.

ANSWER $C = \dfrac{Q}{V} = \dfrac{40 \times 10^{-6}}{20} = 2 \times 10^{-6}$
$$= 2 \ \mu F$$

Example 16-5

A constant current of 5 mA charges a 10-μF capacitor for 1 s. How much is the voltage across the capacitor?

ANSWER Find the stored charge first:

$$Q = I \times t = 5 \times 10^{-3} \times 1$$
$$= 5 \times 10^{-3} \text{ C or 5 mC}$$
$$V = \frac{Q}{C} = \frac{5 \times 10^{-3}}{10 \times 10^{-6}} = 0.5 \times 10^{3}$$
$$= 500 \text{ V}$$

Larger Plate Area Increases Capacitance

As illustrated in Fig. 16–3, when the area of each plate is doubled, the capacitance in Fig. 16–3b stores twice the charge of Fig. 16–3a. The potential difference in both cases is still 10 V. This voltage produces a given strength of electric field. A larger plate area, however, means that more of the dielectric surface can contact each plate, allowing more lines of force through the dielectric between the plates and less flux leakage outside the dielectric. Then the field can store more charge in the dielectric. The result of larger plate area is more charge stored for the same applied voltage, which means that the capacitance is larger.

Figure 16–3 Increasing stored charge and capacitance by increasing the plate area and decreasing the distance between plates. (a) Capacitance of 1 μF. (b) A 2-μF capacitance with twice the plate area and the same distance. (c) A 2-μF capacitance with one-half the distance and the same plate area.

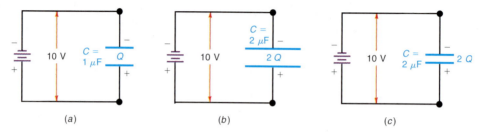

(a)   (b)   (c)

Thinner Dielectric Increases Capacitance

As illustrated in Fig. 16–3c, when the distance between plates is reduced by one-half, the capacitance stores twice the charge of Fig. 16–3a. The potential difference is still 10 V, but its electric field has greater flux density in the thinner dielectric. Then the field between opposite plates can store more charge in the dielectric. With less distance between the plates, the stored charge is greater for the same applied voltage, which means that the capacitance is greater.

Dielectric Constant K_ϵ

This indicates the ability of an insulator to concentrate electric flux. Its numerical value is specified as the ratio of flux in the insulator compared with the flux in air or vacuum. The dielectric constant of air or vacuum is 1, since it is the reference.

Mica, for example, has an average dielectric constant of 6, which means that it can provide a density of electric flux six times as great as that of air or vacuum for the same applied voltage and equal size. Insulators generally have a dielectric constant K_ϵ greater than 1, as listed in Table 16–1. Higher values of K_ϵ allow greater values of capacitance.

Note that the aluminum oxide and tantalum oxide listed in Table 16–1 are used for the dielectric in electrolytic capacitors. Also, plastic film is often used instead of paper for the rolled-foil type of capacitor.

The dielectric constant for an insulator is actually its *relative permittivity*. The symbol ϵ_r, or K_ϵ, indicates the ability to concentrate electric flux. This factor corresponds to relative permeability, with the symbol μ_r or K_m, for magnetic flux. Both ϵ_r and μ_r are pure numbers without units as they are just ratios.*

These physical factors for a parallel-plate capacitor are summarized by the formula

$$C = K_\epsilon \times \frac{A}{d} \times 8.85 \times 10^{-12} \text{ F} \qquad (16-4)$$

where A is the area in square meters of either plate, d is the distance in meters between plates, K_ϵ is the dielectric constant, or relative permittivity, as listed in Table 16–1, and C is capacitance in farads. The constant factor 8.85×10^{-12} is the absolute permittivity of air or vacuum, in SI, since the farad is an SI unit.

> **GOOD TO KNOW**
>
> The capacitance of a capacitor is determined only by its physical construction and not by external circuit parameters such as frequency, voltage, etc.

---

* The absolute permittivity ϵ_0 is 8.854×10^{-12} F/m in SI units for electric flux in air or vacuum. This value corresponds to an absolute permeability μ_0 of $4\pi \times 10^{-7}$ H/m in SI units for magnetic flux in air or a vacuum.

Table 16–1	Dielectric Materials*	
Material	Dielectric Constant K_ϵ	Dielectric Strength, V/MIL
Air or vacuum	1	20
Aluminum oxide	7	
Ceramics	80–1200	600–1250
Glass	8	335–2000
Mica	3–8	600–1500
Oil	2–5	275
Paper	2–6	1250
Plastic film	2–3	
Tantalum oxide	25	

* Exact values depend on the specific composition of different types.

Example 16-6

Calculate C for two plates, each with an area 2 m^2, separated by 1 cm, or 10^{-2} m, with a dielectric of air.

ANSWER Substituting in Formula (16–4),

$$C = 1 \times \frac{2}{10^{-2}} \times 8.85 \times 10^{-12} \text{ F}$$
$$= 200 \times 8.85 \times 10^{-12}$$
$$= 1770 \times 10^{-12} \text{ F or } 1770 \text{ pF}$$

This value means that the capacitor can store 1770×10^{-12} C of charge with 1 V. Note the relatively small capacitance, in picofarad units, with the extremely large plates of 2 m^2, which is really the size of a tabletop or a desktop.

    If the dielectric used is paper with a dielectric constant of 6, then C will be six times greater. Also, if the spacing between plates is reduced by one-half to 0.5 cm, the capacitance will be doubled. Note that practical capacitors for electronic circuits are much smaller than this parallel-plate capacitor. They use a very thin dielectric with a high dielectric constant, and the plate area can be concentrated in a small space.

Dielectric Strength

Table 16–1 also lists breakdown-voltage ratings for typical dielectrics. *Dielectric strength* is the ability of a dielectric to withstand a potential difference without arcing across the insulator. This voltage rating is important because rupture of the insulator provides a conducting path through the dielectric. Then it cannot store charge because the capacitor has been short-circuited. Since the breakdown voltage increases with greater thickness, capacitors with higher voltage ratings have more distance between plates. This increased distance reduces the capacitance, however, all other factors remaining the same.

■ *16–3 Knowledge Check*

Answer at end of chapter.

What is the value of capacitance in Example 16–6 if a glass dielectric is used?

■ *16–3 Self-Review*

Answers at end of chapter.

a. A capacitor charged to 100 V has 1000 μC of charge. How much is C?
b. A mica capacitor and ceramic capacitor have the same physical dimensions. Which has more C?

16–4 Typical Capacitors

Commercial capacitors are generally classified according to the dielectric. Most common are air, mica, paper, plastic film, and ceramic capacitors, plus the electrolytic type. Electrolytic capacitors use a molecular-thin oxide film as the dielectric, resulting in large capacitance values in little space. These types are compared in Table 16–2 and discussed in the sections that follow.

Except for electrolytic capacitors, capacitors can be connected to a circuit without regard to polarity, since either side can be the more positive plate. Electrolytic capacitors are marked to indicate the side that must be connected to the positive or negative side of the circuit. *Note that the polarity of the charging source determines the polarity of the capacitor voltage.* Failure to observe the correct polarity can damage the dielectric and lead to the complete destruction of the capacitor.

Table 16–2	Types of Capacitors		
Dielectric	**Construction**	**Capacitance**	**Breakdown, V**
Air	Meshed plates	10–400 pF	400 (0.02-in. air gap)
Ceramic	Tubular	0.5–1600 pF	500–20,000
	Disk	1 pF to 1 μF	
Electrolytic	Aluminum	1–6800 μF	10–450
	Tantalum	0.047 to 330 μF	6–50
Mica	Stacked sheets	10–5000 pF	500–20,000
Paper	Rolled foil	0.001–1 μF	200–1600
Plastic film	Foil or metallized	100 pF to 100 μF	50–600

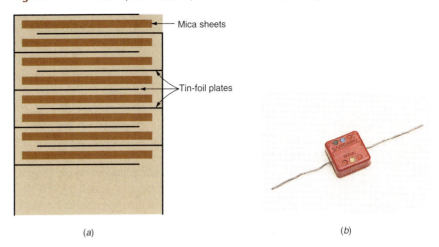

Mica sheets

Tin-foil plates

(*a*)

(*b*)

Mica Capacitors

Thin mica sheets as the dielectric are stacked between tinfoil sections for the conducting plates to provide the required capacitance. Alternate strips of tinfoil are connected and brought out as one terminal for one set of plates, and the opposite terminal connects to the other set of interlaced plates. The construction is shown in Fig. 16–4*a*. The entire unit is generally in a molded Bakelite case. Mica capacitors are often used for small capacitance values of about 10 to 5000 pF; their length is ¾ in. or less with about ⅛-in. thickness. A typical mica capacitor is shown in Fig. 16–4*b*.

Paper Capacitors

In this construction shown in Fig. 16–5*a*, two rolls of tinfoil conductor separated by a paper dielectric are rolled into a compact cylinder. Each outside lead connects to its roll of tinfoil as a plate. The entire cylinder is generally placed in a cardboard container coated with wax or encased in plastic. Paper capacitors are often used for medium capacitance values of 0.001 to 1.0 μF, approximately. The size of a 0.05-μF capacitor is typically 1 in. long and ⅜-in. in diameter. A paper capacitor is shown in Fig. 16–5*b*.

A black or a white band at one end of a paper capacitor indicates the lead connected to the outside foil. This lead should be used for the ground or low-potential side of the circuit to take advantage of shielding by the outside foil. There is no required polarity, however, since the capacitance is the same no matter which side is grounded. Also note that in the schematic symbol for *C*, the curved line usually indicates the low-potential side of the capacitor.

Figure 16–5 Paper capacitor. (*a*) Physical construction. (*b*) Example of a paper capacitor.

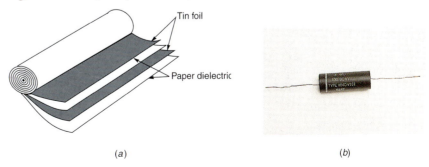

Tin foil

Paper dielectric

(*a*)

(*b*)

Film Capacitors

Film capacitors are constructed much like paper capacitors except that the paper dielectric is replaced with a plastic film such as polypropylene, polystyrene, polycarbonate, or polyethelene terepthalate (Mylar). There are two main types of film capacitors: the foil type and the metallized type. The foil type uses sheets of metal foil, such as aluminum or tin, for its conductive plates. The metallized type is constructed by depositing (spraying) a thin layer of metal, such as aluminum or zinc, on the plastic film. The sprayed-on metal serves as the plates of the capacitor. The advantage of the metallized type over the foil type is that the metallized type is much smaller for a given capacitance value and breakdown voltage rating. The reason is that the metallized type has much thinner plates because they are sprayed on. Another advantage of the metallized type is that it is self-healing. This means that if the dielectric is punctured because its breakdown voltage rating is exceeded, the capacitor is not damaged permanently. Instead, the capacitor heals itself. This is not true of the foil type.

Film capacitors are very temperature-stable and are therefore used frequently in circuits that require very stable capacitance values. Some examples are radio-frequency oscillators and timer circuits. Film capacitors are available with values ranging from about 100 pF to 100 μF. Figure 16–6 shows a typical film capacitor.

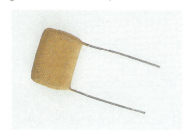

Figure 16–6   Film capacitor.

Ceramic Capacitors

The ceramic materials used in ceramic capacitors are made from earth fired under extreme heat. With titanium dioxide or one of several types of silicates, very high values of dielectric constant K_ϵ can be obtained. Most ceramic capacitors come in disk form, as shown in Fig. 16–7. In the disk form, silver is deposited on both sides of the ceramic dielectric to form the capacitor plates. Ceramic capacitors are available with values of 1pF (or less) up to about 1 μF. The wide range of values is possible because the dielectric constant K_ϵ can be tailored to provide almost any desired value of capacitance.

Note that ceramic capacitors are also available in forms other than disks. Some ceramic capacitors are available with axial leads and use a color code similar to that of a resistor.

Figure 16–7   Ceramic disk capacitor.

Surface-Mount Capacitors

Like resistors, capacitors are also available as surface-mounted components. Surface-mounted capacitors are often called *chip capacitors.* Chip capacitors are constructed by placing a ceramic dielectric material between layers of conductive film which form the capacitor plates. The capacitance is determined by the dielectric constant K_ϵ and the physical area of the plates. Chip capacitors are available in many sizes. A common size is 0.125 in. long by 0.063 in. wide in various thicknesses. Another common size is 0.080 in. long by 0.050 in. wide in various thicknesses. Figure 16–8 shows two sizes of chip capacitors. Like chip resistors, chip capacitors have their end electrodes soldered directly to the copper traces of the printed-circuit board. Chip capacitors are available with values ranging from a fraction of a picofarad up to several microfarads.

Variable Capacitors

Figure 16–1*b* shows a variable air capacitor. In this construction, the fixed metal plates connected together form the *stator.* The movable plates connected together on the shaft form the *rotor.* Capacitance is varied by rotating the shaft to make the rotor plates mesh with the stator plates. They do not touch, however,

Figure 16-8 Chip capacitors.

since air is the dielectric. Full mesh is maximum capacitance. Moving the rotor completely out of mesh provides minimum capacitance.

A common application is the tuning capacitor in radio receivers. When you tune to different stations, the capacitance varies as the rotor moves in or out of mesh. Combined with an inductance, the variable capacitance then tunes the receiver to a different resonant frequency for each station. Usually two or three capacitor sections are *ganged* on one common shaft.

Temperature Coefficient

Ceramic capacitors are often used for temperature compensation to increase or decrease capacitance with a rise in temperature. The temperature coefficient is given in parts per million (ppm) per degree Celsius, with a reference of 25°C. As an example, a negative 750-ppm unit is stated as N750. A positive temperature coefficient of the same value would be stated as P750. Units that do not change in capacitance are labeled NPO.

Capacitance Tolerance

Ceramic disk capacitors for general applications usually have a tolerance of ±20%. For closer tolerances, mica or film capacitors are used. These have tolerance values of ±2 to 20%. Silver-plated mica capacitors are available with a tolerance of ±1%.

The tolerance may be less on the minus side to make sure that there is enough capacitance, particularly with electrolytic capacitors, which have a wide tolerance. For instance, a 20-μF electrolytic with a tolerance of -10%, $+50\%$ may have a capacitance of 18 to 30 μF. However, the exact capacitance value is not critical in most applications of capacitors for filtering, ac coupling, and bypassing.

Voltage Rating of Capacitors

This rating specifies the maximum potential difference that can be applied across the plates without puncturing the dielectric. Usually the voltage rating is for temperatures up to about 60°C. Higher temperatures result in a lower voltage rating. Voltage ratings for general-purpose paper, mica, and ceramic capacitors are typically 200 to 500 V. Ceramic capacitors with ratings of 1 to 20 kV are also available.

Electrolytic capacitors are typically available in 16-, 35-, and 50-V ratings. For applications where a lower voltage rating is permissible, more capacitance can be obtained in a smaller size.

The potential difference across the capacitor depends on the applied voltage and is not necessarily equal to the voltage rating. A voltage rating higher than the potential difference applied across the capacitor provides a safety factor for long life in service. However, the actual capacitor voltage of electrolytic capacitors should be close to the rated voltage to produce the oxide film that provides the specified capacitance.

The voltage ratings are for dc voltage applied. The breakdown rating is lower for ac voltage because of the internal heat produced by continuous charge and discharge.

Capacitor Applications

In most electronic circuits, a capacitor has dc voltage applied, combined with a much smaller ac signal voltage. The usual function of the capacitor is to block the dc voltage but pass the ac signal voltage by means of the charge and discharge current. These applications include coupling, bypassing, and filtering of ac signals.

■ 16–4 Knowledge Check

Answer at end of chapter.

What type of capacitor might be destroyed if the voltage polarity across its plates is incorrect?

■ 16–4 Self-Review

Answers at end of chapter.

a. An electrolytic capacitor must be connected in the correct polarity. (True or False)
b. The potential difference across a capacitor is always equal to its maximum voltage rating. (True or False)
c. Ceramic and paper capacitors generally have less C than electrolytic capacitors. (True or False)
d. The letters NPO indicate zero temperature coefficient. (True or False)

16–5 Electrolytic Capacitors

Electrolytic capacitors are commonly used for C values ranging from about 1 to 6800 μF because electrolytics provide the most capacitance in the smallest space with least cost.

Construction

Figure 16–9 shows the aluminum-foil type. The two aluminum electrodes are in an electrolyte of borax, phosphate, or carbonate. Between the two aluminum strips, absorbent gauze soaks up electrolyte to provide the required electrolysis that produces an oxide film. This type is considered a wet electrolytic, but it can be mounted in any position.

Figure 16–9 Construction of aluminum electrolytic capacitor. (*a*) Internal electrodes. (*b*) Foil rolled into cartridge. (*c*) Typical capacitor with multiple sections.

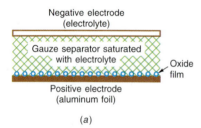

(*a*)

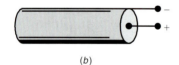

(*b*)

(*c*)

When dc voltage is applied to form the capacitance in manufacture, the electrolytic action accumulates a molecular-thin layer of aluminum oxide at the junction between the positive aluminum foil and the electrolyte. The oxide film is an insulator. As a result, capacitance is formed between the positive aluminum electrode and the electrolyte in the gauze separator. The negative aluminum electrode simply provides a connection to the electrolyte. Usually, the metal can itself is the negative terminal of the capacitor, as shown in Fig. 16–9c.

Because of the extremely thin dielectric film, very large C values can be obtained. The area is increased by using long strips of aluminum foil and gauze, which are rolled into a compact cylinder with very high capacitance. For example, an electrolytic capacitor the same size as a 0.1-μF paper capacitor, but rated at 10 V breakdown, may have 1000 μF of capacitance or more. Higher voltage ratings, up to 450 V, are available, with typical C values up to about 6800 μF. The very high C values usually have lower voltage ratings.

Polarity

Electrolytic capacitors are used in circuits that have a combination of dc voltage and ac voltage. The dc voltage maintains the required polarity across the electrolytic capacitor to form the oxide film. A common application is for electrolytic filter capacitors to eliminate the 60- or 120-Hz ac ripple in a dc power supply. Another use is for audio coupling capacitors in transistor amplifiers. In both applications, for filtering or coupling, electrolytics are needed for large C with a low-frequency ac component, whereas the circuit has a dc component for the required voltage polarity. Incidentally, the difference between filtering out an ac component or coupling it into a circuit is only a question of parallel or series connections. The filter capacitors for a power supply are typically 100 to 1000 μF. Audio capacitors are usually 10 to 47 μF.

If the electrolytic is connected in opposite polarity, the reversed electrolysis forms gas in the capacitor. It becomes hot and may explode. This is a possibility only with electrolytic capacitors.

Leakage Current

The disadvantage of electrolytics, in addition to the required polarization, is their relatively high leakage current compared with other capacitors, since the oxide film is not a perfect insulator. The problem with leakage current in a capacitor is that it allows part of the dc component to be coupled into the next circuit along with the ac component. In newer electrolytic capacitors, the leakage current is quite small. Section 16–10 takes a closer look at leakage current in capacitors.

Nonpolarized Electrolytics

This type is available for applications in circuits without any dc polarizing voltage, as in a 60-Hz ac power line. One application is the starting capacitor for ac motors. A nonpolarized electrolytic actually contains two capacitors, connected internally in series-opposing polarity.

Tantalum Capacitors

This is another form of electrolytic capacitor, using tantalum (Ta) instead of aluminum. Titanium (Ti) is also used. Typical tantalum capacitors are shown in Fig. 16–10. They feature

1. larger C in a smaller size
2. longer shelf life
3. less leakage current

Figure 16–10 Tantalum capacitors.

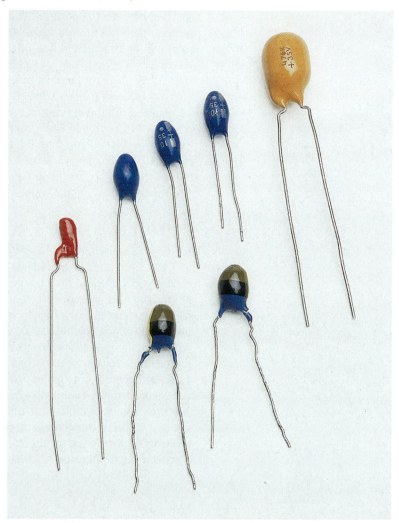

However, tantalum electrolytics cost more than the aluminum type. Construction of tantalum capacitors include the wet-foil type and a solid chip or slug. The solid tantalum is processed in manufacture to have an oxide film as the dielectric. Referring back to Table 16–1, note that tantalum oxide has a dielectric constant of 25, compared with 7 for aluminum oxide.

■ 16–5 Knowledge Check

Answer at end of chapter.

What is the dielectric material in an aluminum electrolytic capacitor?

■ 16–5 Self-Review

Answers at end of chapter.

a. The rating of 1000 μF at 25 V is probably for an electrolytic capacitor. (True or False)
b. Electrolytic capacitors allow more leakage current than mica capacitors. (True or False)
c. Tantalum capacitors have a longer shelf life than aluminum electrolytics. (True or False)

16–6 Capacitor Coding

The value of a capacitor is always specified in either microfarad or picofarad units of capacitance. This is true for all types of capacitors. As a general rule, if a capacitor (other than an electrolytic capacitor) is marked using a whole number such as 33, 220, or 680, the capacitance C is in picofarads (pF). Conversely, if a capacitor is labeled using a decimal fraction such as 0.1, 0.047, or 0.0082, the capacitance C is in microfarads (μF). There are a variety of ways in which a manufacturer may indicate the value of a capacitor. What follows is an explanation of the most frequently encountered coding systems.

Film-Type Capacitors

Figure 16–11 shows a popular coding system for film-type capacitors. The first two numbers on the capacitor indicate the first two digits in the numerical value of the capacitance. The third number is the *multiplier*, indicating by what factor the first two digits must be multiplied. The letter at the far right indicates the capacitor's tolerance. In this coding system, the capacitance is always in

||| MultiSim **Figure 16–11** Film capacitor coding system.

Film-Type Capacitors

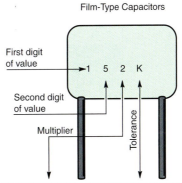

Multiplier		Tolerance of Capacitor		
For the Number	Multiplier	Letter	10 pF or Less	Over 10 pF
0 1	1 10	B C	±0.1 pF ±0.25 pF	
2 3	100 1,000	D F	±0.5 pF ±1.0 pF	±1%
4 5	10,000 100,000	G H	±2.0 pF	±2% ±3%
8	0.01	J K		±5% ±10%
9	0.1	M		±20%

Examples:
    152K = 15 × 100 = 1500 pF or 0.0015 μF, ±10%
    759J = 75 × 0.1 = 7.5 pF, ±5%

Note: The letter R may be used at times to signify a decimal point, as in 2R2 = 2.2 (pF or μF).

picofarad units. The capacitor's breakdown voltage rating is usually printed on the body directly below the coded value of capacitance.

Example 16-7

Determine the value of capacitance for the film capacitors in Fig. 16–12a and Fig. 16–12b.

Figure 16-12 Film capacitors for Example 16-7.

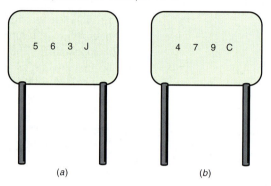

(a)   (b)

ANSWER   In Fig. 16–12a, the first two numbers are 5 and 6, respectively, for 56 as the first two digits in the numerical value of the capacitance. The third number, 3, indicates a multiplier of 1000, or $56 \times 1000 = 56,000$ pF. The letter J indicates a capacitor tolerance of $\pm 5\%$.

   In Fig. 16–12b, the first two numbers are 4 and 7, respectively, for 47 as the first two digits in the numerical value of the capacitance. The third number, 9, indicates a fractional multiplier of 0.1, or $47 \times 0.1 = 4.7$ pF. The letter C indicates a capacitor tolerance of ± 0.25 pF.

Ceramic Disk Capacitors

Figure 16–13 shows how most ceramic disk capacitors are marked to indicate their capacitance. As you can see, the capacitance is expressed either as a whole number or as a decimal fraction. The type of coding system used depends on the manufacturer. Ceramic disk capacitors are often used for coupling and bypassing ac signals, where it is allowable to have a wide or lopsided tolerance.

Example 16-8

In Fig. 16–14, determine (a) the capacitance value and tolerance; (b) the temperature-range identification information.

ANSWER   (a) Since the capacitance is expressed as a decimal fraction, its value is in microfarads. In this case, $C = 0.047$ μF. The letter Z, to the right of 0.047, indicates a capacitor tolerance of $+80\%$, -20%. Notice that the actual capacitance value can be as much as 80% above its coded value but only 20% below its coded value.

(b) The alphanumeric code, Z5V, printed below the capacitance value, provides additional capacitor information. Referring to Fig. 16–13, note that the letter Z and number 5 indicate the low and high temperatures of +10°C and +85°C, respectively. The letter V indicates that the maximum capacitance change over the specified temperature range (+10°C to +85°C) is +22%, −82%. For temperature changes less than the range indicated, the percent change in capacitance will be less than that indicated.

Figure 16–13 Ceramic disk capacitor coding system.

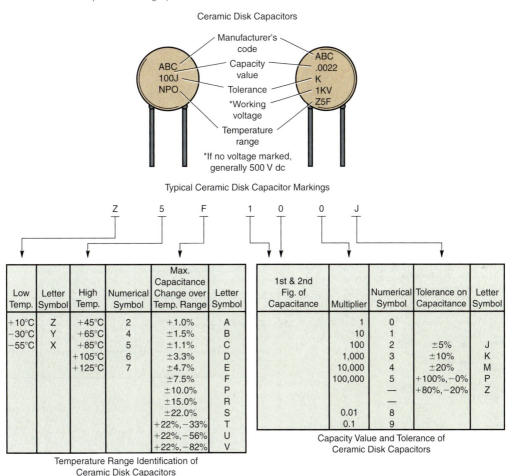

Figure 16–14 Ceramic disk capacitor for Example 16-8.

Mica Capacitors

Mica capacitors are coded using colored dots to indicate the capacitance value in picofarads. Three different coding systems are shown in Fig. 16–16. The color code is best understood through an example.

Example 16-9

Determine the capacitance and tolerance for the capacitor in Fig. 16–15 (refer to Fig. 16–16).

Figure 16–15 Mica capacitor for Example 16-9.

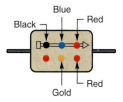

ANSWER   The dots in the top row are read from left to right in the direction of the arrow. In the bottom row, they are read in the reverse order from right to left. The first dot at the left in the top row is black, indicating a mica capacitor. The next two color dots are blue and red, for 62 as the first two digits in the numerical value of the capacitance. The next dot, at the far right in the bottom row, is red, indicating a multiplier of 100. Therefore, $C = 62 \times 100 = 6200$ pF. The next dot is gold, indicating a capacitor tolerance of $\pm 5\%$.

Figure 16–16   Three different coding systems used for mica capacitors.

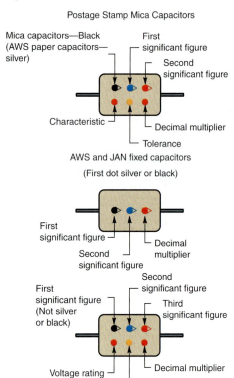

Color	Significant Figure	Multiplier	Tolerance (%)	Voltage Rating
Black	0	1	—	—
Brown	1	10	1	100
Red	2	100	2	200
Orange	3	1,000	3	300
Yellow	4	10,000	4	400
Green	5	100,000	5	500
Blue	6	1,000,000	6	600
Violet	7	10,000,000	7	700
Gray	8	100,000,000	8	800
White	9	1,000,000,000	9	900
Gold	—	0.1	5	1,000
Silver	—	0.01	10	2,000
No color	—	—	20	500

Chip Capacitors

Before determining the capacitance value of a chip capacitor, make sure it is a capacitor and not a resistor. Chip capacitors have the following identifiable features:

1. The body is one solid color, such as off-white, beige, gray, tan, or brown.
2. The end electrodes completely enclose the end of the part.

Figure 16–17 Chip capacitor coding system.

Value (33 Value Symbols)—Upper and Lowercase Letters					Multiplier
A-1.0	H-2.0	b-3.5	f-5.0	X-7.5	0 = × 1.0
B-1.1	J-2.2	P-3.6	T-5.1	t-8.0	1 = × 10
C-1.2	K-2.4	Q-3.9	U-5.6	Y-8.2	2 = × 100
D-1.3	a-2.5	d-4.0	m-6.0	y-9.0	3 = × 1,000
E-1.5	L-2.7	R-4.3	V-6.2	Z-9.1	4 = × 10,000
F-1.6	M-3.0	e-4.5	W-6.8		5 = × 100,000
G-1.8	N-3.3	S-4.7	n-7.0		etc.

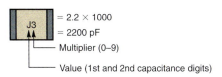

J3 = 2.2 × 1000
   = 2200 pF

Multiplier (0–9)

Value (1st and 2nd capacitance digits)

Three popular coding systems are used by manufacturers of chip capacitors. In all three systems, the values represented are in picofarads. One system, shown in Fig. 16–17, uses a two-place system in which a letter indicates the first and second digits of the capacitance value and a number indicates the multiplier (0 to 9). Thirty-three symbols are used to represent the two significant figures. The symbols used include 24 uppercase letters and 9 lowercase letters. In Fig. 16–17, note that J3 represents 2200 pF.

Another system, shown in Fig. 16–18, also uses two places. In this case, however, values below 100 pF are indicated using two numbers from which the capacitance value is read directly. Values above 100 pF are indicated by a letter and a number as before. In this system, only 24 uppercase letters are used. Also note that the alphanumeric codes in this system are 10 times higher than those in the system shown in Fig. 16–17.

Figure 16–18 Chip capacitor coding system.

Alternate Two-Place Code
• Values below 100 pF—Value read directly

• Values 100 pF and above—Letter/number code

A1 = 10 × 10
   = 100 pF

N3 = 33 × 1000
   = 33000 pF
   = 0.033 μF

Multiplier (1–9)

Value (1st and 2nd significant digits)

05 = 5 pF

82 = 82 pF

Value (24 Value Symbols)—Uppercase Letters Only					Multiplier
A-10	F-16	L-27	R-43	W-68	1 = × 10
B-11	G-18	M-30	S-47	X-75	2 = × 100
C-12	H-20	N-33	T-51	Y-82	3 = × 1,000
D-13	J-22	P-36	U-56	Z-91	4 = × 10,000
E-15	K-24	Q-39	V-62		5 = × 100,000 etc.

Figure 16-19 Chip capacitor coding system.

Standard Single-Place Code

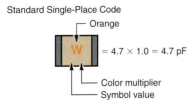

Orange

W = 4.7 × 1.0 = 4.7 pF

Color multiplier
Symbol value

Examples: R (Green) = 3.3 × 100 = 330 pF
7 (Blue)   = 8.2 × 1000 = 8200 pF

Value (24 Value Symbols)—Uppercase Letters and Numerals					Multiplier (Color)
A-1.0	H-1.6	N-2.7	V-4.3	3-6.8	Orange = × 1.0
B-1.1	I-1.8	O-3.0	W-4.7	4-7.5	Black = × 10
C-1.2	J-2.0	R-3.3	X-5.1	7-8.2	Green = × 100
D-1.3	K-2.2	S-3.6	Y-5.6	9-9.1	Blue = × 1,000
E-1.5	L-2.4	T-3.9	Z-6.2		Violet = × 10,000
					Red = × 100,000

Figure 16–19 shows yet another system, in which a single letter or number is used to designate the first two digits in the capacitance value. The multiplier is determined by the color of the letter. In the example shown, an orange-colored W represents a capacitance C of 4.7 pF.

Note that other coding systems are used for chip capacitors; these systems are not covered here. However, the three coding systems shown in this section are the most common systems presently in use. Also note that some chip capacitors found on printed-circuit boards are not marked or coded. When this is the case, the only way to determine the capacitance value is to check it with a capacitance tester.

Tantalum Capacitors

Tantalum capacitors are frequently coded to indicate their capacitance in picofarads. Figure 16–20 shows how to interpret this system.

Figure 16-20 Tantalum capacitor coding system.

Dipped Tantalum Capacitors

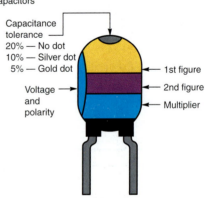

Capacitance tolerance
20% — No dot
10% — Silver dot
5% — Gold dot

Voltage and polarity

1st figure
2nd figure
Multiplier

Color	Rated Voltage	Capacitance in Picofarads		Multiplier
		1st Figure	2nd Figure	
Black	4	0	0	—
Brown	6	1	1	—
Red	10	2	2	—
Orange	15	3	3	—
Yellow	20	4	4	10,000
Green	25	5	5	100,000
Blue	35	6	6	1,000,000
Violet	50	7	7	10,000,000
Gray	—	8	8	—
White	3	9	9	—

Figure 16–21 Tantalum capacitor for Example 16–10.

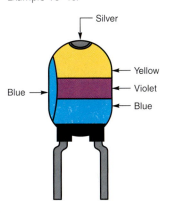

Silver

Yellow

Blue

Violet

Blue

Example 16–10

For the tantalum capacitor shown in Fig. 16–21, determine the capacitance C in both pF and μF units. Also, determine the voltage rating and tolerance.

ANSWER Moving from top to bottom, the first two color bands are yellow and violet, which represent the digits 4 and 7, respectively. The third color band is blue, indicating a multiplier of 1,000,000. Therefore the capacitance C is 47 × 1,000,000 = 47,000,000 pF, or 47 μF. The blue color at the left indicates a voltage rating of 35 V. And, finally, the silver dot at the very top indicates a tolerance of ±10%.

■ *16–6 Knowledge Check*

> *Answer at end of chapter.*
>
> **What is the tolerance of a plastic-film capacitor whose coded value is 102 F?**

■ *16–6 Self-Review*

> *Answers at end of chapter.*
>
> a. **A ceramic disk capacitor that is marked .01 has a capacitance of 0.01 pF. (True or False)**
> b. **A film capacitor that is marked 224 has a capacitance of 220,000 pF. (True or False)**
> c. **A chip capacitor has a green letter E marked on it. Its capacitance is 150 pF. (True or False)**
> d. **A ceramic disk capacitor is marked .001P. Its tolerance is +100%, −0%. (True or False)**

Figure 16–22 Capacitances in parallel.

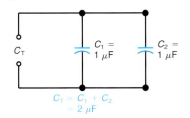

C_T

$C_1 = 1\ \mu F$

$C_2 = 1\ \mu F$

$C_T = C_1 + C_2$
$= 2\ \mu F$

16–7 Parallel Capacitances

Connecting capacitances in parallel is equivalent to adding the plate areas. Therefore, the total capacitance is the sum of the individual capacitances. As illustrated in Fig. 16–22,

$$C_T = C_1 + C_2 + \cdots + \text{etc.} \tag{16–5}$$

A 10-μF capacitor in parallel with a 5-μF capacitor, for example, provides a 15-μF capacitance for the parallel combination. The voltage is the same across the parallel capacitors. Note that adding parallel capacitances is opposite to inductances in parallel and resistances in parallel.

■ *16–7 Knowledge Check*

> *Answer at end of chapter.*
>
> **In Fig. 16–22, how much is C_T if $C_2 = 3\ \mu F$?**

Answers at end of chapter.

a. How much is C_T for 0.01 μF in parallel with 0.02 μF?
b. What C must be connected in parallel with 100 pF to make C_T 250 pF?

16-8 Series Capacitances

Connecting capacitances in series is equivalent to increasing the thickness of the dielectric. Therefore, the combined capacitance is less than the smallest individual value. As shown in Fig. 16-23, the combined equivalent capacitance is calculated by the reciprocal formula:

$$C_{EQ} = \frac{1}{\dfrac{1}{C_1} + \dfrac{1}{C_2} + \dfrac{1}{C_3} + \cdots + \text{etc.}} \qquad (16\text{-}6)$$

Any of the shortcut calculations for the reciprocal formula apply. For example, the combined capacitance of two equal capacitances of 10 μF in series is 5 μF.

Capacitors are used in series to provide a higher working voltage rating for the combination. For instance, each of three equal capacitances in series has one-third the applied voltage.

Division of Voltage across Unequal Capacitances

In series, the voltage across each C is inversely proportional to its capacitance, as illustrated in Fig. 16-24. The smaller capacitance has the larger proportion of the applied voltage. The reason is that the series capacitances all have the same charge because they are in one current path. With equal charge, a smaller capacitance has a greater potential difference.

We can consider the amount of charge in the series capacitors in Fig. 16-24. Let the charging current be 600 μA flowing for 1 s. The charge Q equals $I \times t$ or 600 μC. Both C_1 and C_2 have Q equal to 600 μC because they are in the same series path for charging current.

Although the charge is the same in C_1 and C_2, they have different voltages because of different capacitance values. For each capacitor, $V = Q/C$. For the two capacitors in Fig. 16-24, then,

$$V_1 = \frac{Q}{C_1} = \frac{600\ \mu C}{1\ \mu F} = 600\ V$$

$$V_2 = \frac{Q}{C_2} = \frac{600\ \mu C}{2\ \mu F} = 300\ V$$

GOOD TO KNOW

For two capacitors, C_1 and C_2 in series, the individual capacitor voltages can be calculated using the following equations:

$$V_{C_1} = \frac{C_2}{C_1 + C_2} \times V_T.$$

$$V_{C_2} = \frac{C_1}{C_1 + C_2} \times V_T.$$

||| **MultiSim** **Figure 16-23** Capacitances in series.

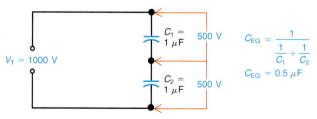

Figure 16–24 With series capacitors, the smaller C has more voltage for the same charge.

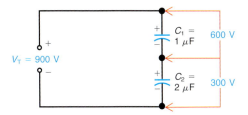

Charging Current for Series Capacitances

The charging current is the same in all parts of the series path, including the junction between C_1 and C_2, even though this point is separated from the source voltage by two insulators. At the junction, the current is the resultant of electrons repelled by the negative plate of C_2 and attracted by the positive plate of C_1. The amount of current in the circuit is determined by the equivalent capacitance of C_1 and C_2 in series. In Fig. 16–24, the equivalent capacitance is $\frac{2}{3}$ μF.

■ *16–8 Knowledge Check*

    Answer at end of chapter.

    In Fig. 16–24, how much charge is stored by both capacitors in series?

■ *16–8 Self-Review*

    Answers at end of chapter.

  a. How much is C_{EQ} for two 0.2-μF capacitors in series?
  b. With 50 V applied across both, how much is V_C across each capacitor?
  c. How much is C_{EQ} for 100 pF in series with 50 pF?

16–9 Energy Stored in Electrostatic Field of Capacitance

The electrostatic field of the charge stored in a dielectric has electric energy supplied by the voltage source that charges C. This energy is stored in the dielectric. The proof is the fact that the capacitance can produce discharge current when the voltage source is removed. The electric energy stored is

$$\text{Energy} = \mathscr{E} = \tfrac{1}{2}\, CV^2 \text{ joules} \tag{16–7}$$

where C is the capacitance in farads, V is the voltage across the capacitor, and $\mathscr{E}$ is the electric energy in joules. For example, a 1-μF capacitor charged to 400 V has stored energy equal to

$$\mathscr{E} = \tfrac{1}{2}\, CV^2 = \frac{1 \times 10^{-6} \times (4 \times 10^2)^2}{2}$$

$$= \frac{1 \times 10^{-6} \times (16 \times 10^4)}{2} = 8 \times 10^{-2}$$

$$= 0.08 \text{ J}$$

This 0.08 J of energy is supplied by the voltage source that charges the capacitor to 400 V. When the charging circuit is opened, the stored energy remains as charge in the dielectric. With a closed path provided for discharge, the entire 0.08 J is available to produce discharge current. As the capacitor discharges, the energy is used in producing discharge current. When the capacitor is completely discharged, the stored energy is zero.

The stored energy is the reason that a charged capacitor can produce an electric shock, even when not connected in a circuit. When you touch the two leads of the charged capacitor, its voltage produces discharge current through your body. Stored energy greater than 1 J can be dangerous from a capacitor charged to a voltage high enough to produce an electric shock.

Example 16-11

The high-voltage circuit for a color picture tube can have 30 kV across 500 pF of C. Calculate the stored energy.

ANSWER

$$\mathscr{E} = \frac{1}{2}\, CV^2 = \frac{500 \times 10^{-12} \times (30 \times 10^3)^2}{2}$$
$$= 250 \times 10^{-12} \times 900 \times 10^6$$
$$= 225 \times 10^{-3}$$
$$= 0.225 \text{ J}$$

■ *16-9 Knowledge Check*

Answer at end of chapter.

How much energy is stored by a 1000-μF capacitor charged to 50 V?

■ *16-9 Self-Review*

Answers at end of chapter.

a. The stored energy in C increases with more V. (True or False)
b. The stored energy decreases with less C. (True or False)

16-10 Measuring and Testing Capacitors

A *capacitance meter* is a piece of test equipment specifically designed to measure the capacitance value of capacitors. Although capacitance meters can be purchased as stand-alone units, many handheld and benchtop digital multimeters (DMMs) are capable of measuring a wide range of capacitance values. For example, the benchtop DMM shown in Fig. 16–25 has five capacitance ranges: 2 nF, 20 nF, 200 nF, 2000 nF, and 20 μF. To measure the value of a capacitor using this meter, insert the leads of the capacitor into the capacitance socket, labeled CX, located in the upper right-hand corner of the meter. Next, depress the CX (capacitance) button and select the desired capacitance range.

Figure 16–25 Typical DMM with capacitance measurement capability.

Capacitance socket

CX   Range settings

The meter will display the measured capacitance value. For best accuracy, always select the lowest range setting that still displays the measured capacitance value. Note that the polarity markings next to the capacitance socket need to be observed when electrolytic capacitors are inserted. For nonelectrolytic capacitors, lead polarity does not matter. Before inserting any capacitor in the socket, it must be fully discharged to avoid damage to the meter.

Recall from Sec. 16–6 that capacitors are always coded in either microfarad or picofarad units but never in nanofarad units. Although this is standard industry practice, you will nevertheless encounter the nanofarad unit of capacitance when you use meters capable of measuring capacitance, such as that shown in Fig. 16–25. Therefore, it is important to know how to convert between the nanofarad unit and either microfarad or picofarad units. To convert from nanofarad units to picofarad units, simply move the decimal point three places to the right. For example, $33 \text{ nF} = 33 \times 10^{-9} \text{ F} = 33,000 \times 10^{-12} \text{ F} = 33,000 \text{ pF}$. To convert from nanofarads to microfarads, move the decimal point three places to the left. For another example, $470 \text{ nF} = 470 \times 10^{-9} \text{ F} = 0.47 \times 10^{-6} \text{ F} = 0.47 \text{ } \mu\text{F}$. When using meters having nanofarad capacitance ranges, you will need to make these conversions to compare the measured value of capacitance with the coded value.

Example 16-12

Suppose a film capacitor, coded 393J, is being measured using the meter shown in Fig. 16–25. If the meter reads 37.6 on the 200-nF range, (a) What is the capacitance value in picofarad units? (b) Is the measured capacitance value within its specified tolerance?

ANSWER   The capacitor code, 393J, corresponds to a capacitance value of 39,000 pF ±5%. (a) A reading of 37.6 on the 200-nF range corresponds to a capacitance of 37.6 nF. To convert 37.6 nF to picofarad units, move the decimal point three places to the right. This gives an answer of 37,600 pF. (b) The acceptable capacitance range is calculated as follows: $39,000 \text{ pF} \times 0.05 = \pm 1950 \text{ pF}$. Therefore, the measured value of capacitance can range anywhere from 37,050 pF to 40,950 pF and still be considered within tolerance. Note that in nanofarad units, this corresponds to a range of 37.05 to 40.95 nF. Since the measured value of 37.6 nF falls within this range, the measured capacitance value is within tolerance.

Figure 16–26 Leakage resistance R_ℓ of a capacitor.

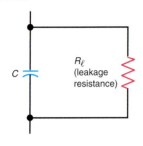

Figure 16–27 Capacitor-inductor analyzer.

Leakage Resistance of a Capacitor

Consider a capacitor charged by a dc voltage source. After the charging voltage is removed, a perfect capacitor would hold its charge indefinitely. Because there is no such thing as a perfect insulator, however, the charge stored in the capacitor will eventually leak or bleed off, thus neutralizing the capacitor. There are three leakage paths through which the capacitor might discharge: (1) leakage through the dielectric, (2) leakage across the insulated case or body between the capacitor leads, and (3) leakage through the air surrounding the capacitor. For paper, film, mica, and ceramic, the leakage current is very slight, or inversely, the leakage resistance is very high. The combination of all leakage paths can be represented as a single parallel resistance R_ℓ across the capacitor plates, as shown in Fig. 16–26. For paper, film, mica, and ceramic capacitors, the leakage resistance R_ℓ is typically 100,000 MΩ or more. The leakage resistance is much less for larger capacitors, such as electrolytics however, with a typical value of R_ℓ ranging from about 500 kΩ up to 10 MΩ. In general, the larger the capacitance of a capacitor, the lower its leakage resistance. Note that the leakage current in capacitors is fairly temperature-sensitive. The higher the temperature, the greater the leakage current (because of lower leakage resistance).

The leakage resistance of a capacitor can be measured with a DMM or an analog ohmmeter, but this is not the best way to test a capacitor for leakage. The best way is to measure the leakage current in the capacitor while the rated working voltage is applied across the capacitor plates. A capacitor is much more likely to show leakage when the dielectric is under stress from the applied voltage. In fact, a capacitor may not show any leakage at all until the dielectric is under stress from the applied voltage. To measure the value of a capacitor and test it for leakage, technicians often use a capacitor-inductor analyzer like that shown in Fig. 16–27. This analyzer allows the user to apply the rated working voltage to the capacitor while testing for leakage. The amount of leakage acceptable depends on the type of capacitor. Most nonpolarized capacitors should have no leakage at all, whereas electrolytics will almost always show some. Pull-out charts showing the maximum allowable leakage for the most common electrolytic capacitors are usually provided with a capacitor-inductor analyzer.

Dielectric Absorption

Dielectric absorption is the inability of a capacitor to completely discharge to zero. It is sometimes referred to as *battery action* or *capacitor memory* and is due to the dielectric of the capacitor retaining a charge after it is supposedly discharged. The effect of dielectric absorption is that it reduces the capacitance value of the capacitor. All capacitors have at least some dielectric absorption, but electrolytics have the highest amount. Dielectric absorption has an undesirable effect on circuit operation if it becomes excessive. The dielectric absorption of a capacitor can be checked using the capacitor-inductor analyzer in Fig. 16–27. Note that there is no way to test for dielectric absorption with an ohmmeter.

Equivalent Series Resistance (ESR)

With ac voltage applied to a capacitor, the continuous charge, discharge, and reverse charging action cannot be followed instantaneously in the dielectric. This corresponds to hysteresis in magnetic materials. With a high-frequency charging voltage applied to the capacitor, there may be a difference between the amount of ac voltage applied to the capacitor and the actual ac voltage across the dielectric. The difference, or loss, can be attributed to the effects of hysteresis in the dielectric. As you might expect, dielectric hysteresis losses increase with frequency.

All losses in a capacitor can be represented as a resistor either in series or in parallel with an ideal capacitor. For example, the losses from dielectric hysteresis can be represented as a single resistor in series with the capacitor as shown in Fig. 16–28a. The other resistor shown in series with the capacitor represents the resistance of the capacitor leads and plates. It also includes any resistance at the point where the capacitor leads are bonded to the metal plates. As before, the leakage resistance R_ℓ is shown directly in parallel with the capacitor. Collectively, the resistances shown in Fig. 16–28a can be lumped into one equivalent series resistance (ESR) as shown in Fig. 16–28b. This is an accurate and convenient way to represent all losses in a capacitor. Ideally, the ESR of a capacitor should be zero. For paper, film, ceramic, and mica capacitors, the ESR value is approximately zero. For electrolytics, however, the ESR may be several ohms or more depending on the way they are constructed. Note that ESR is most often a problem in capacitors used in high-frequency filtering applications. For example, most computers use switching power supplies to power the computer. These power supplies require capacitors for filtering high frequencies. In these applications, a high ESR interferes with the normal filtering action of the capacitor and therefore causes improper circuit operation. In some cases, the power dissipated by the ESR may cause the capacitor to overheat.

The ESR of a capacitor cannot be checked with an ohmmeter because the ESR is in series with the very high resistance of the dielectric. To check a capacitor for ESR, you must use a capacitor-inductor analyzer like that shown in Fig. 16–27. Pull-out charts showing the maximum allowable ESR for different types of capacitors are usually provided with the analyzer.

◾ 16–10 Knowledge Check

Answer at end of chapter.

A plastic-film capacitor is coded 103K. Indicate the coded value in pF, nF, and μF units.

◾ 16–10 Self-Review

Answers at end of chapter.

a. A 150-nF capacitor is the same as a 0.15-μF capacitor. (True or False)
b. It is best to test a capacitor for leakage with the rated working voltage applied. (True or False)
c. Ideally, the ESR of an electrolytic capacitor should be infinite. (True or False)
d. Dielectric absorption in a capacitor can be detected with an ohmmeter. (True or False)

Figure 16–28 Resistances representing losses in a capacitor. (*a*) Series and parallel resistance represents capacitor losses. (*b*) Equivalent series resistance (ESR) represents the total losses in a capacitor.

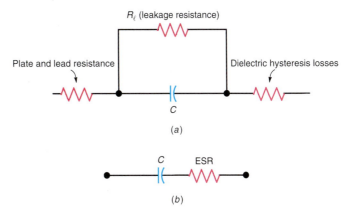

16–11 Troubles in Capacitors

Capacitors can become open or short-circuited. In either case, the capacitor is useless because it cannot store charge. A leaky capacitor is equivalent to a partial short circuit where the dielectric gradually loses its insulating properties under the stress of applied voltage, thus lowering its resistance. A good capacitor has very high resistance of the order of several megohms; a short-circuited capacitor has zero ohms resistance, or continuity; the resistance of a leaky capacitor is lower than normal. Capacitor-inductor analyzers, like that shown in Fig. 16–27, should be used to test a capacitor. However, if a capacitor-inductor analyzer is not available, an ohmmeter (preferably analog) may be able to identify the problem. What follows is a general procedure for testing capacitors using an analog ohmmeter.

Checking Capacitors with an Ohmmeter

A capacitor usually can be checked with an ohmmeter. The highest ohm range, such as $R \times 1 \, M\Omega$, is preferable. Also, disconnect one side of the capacitor from the circuit to eliminate any parallel resistance paths that can lower the resistance. Keep your fingers off the connections, since body resistance lowers the reading.

As shown in Fig. 16–29, the ohmmeter leads are connected across the capacitor. For a good capacitor, the meter pointer moves quickly toward the low-resistance side of the scale and then slowly recedes toward infinity. When the pointer stops moving, the reading is the dielectric resistance of the capacitor, which is normally very high. For paper, film, mica, and ceramic capacitors, the resistance is usually so high that the needle of the meter rests on the infinity mark (∞). However, electrolytic capacitors will usually measure a much lower resistance of about 500 kΩ to 10 MΩ. In all cases, discharge the capacitor before checking with the ohmmeter.

When the ohmmeter is initially connected, its battery charges the capacitor. This charging current is the reason the meter pointer moves away from infinity, since more current through the ohmmeter means less resistance. Maximum current flows at the first instant of charge. Then the charging current decreases as the capacitor voltage increases toward the applied voltage; therefore, the needle pointer slowly moves toward infinite resistance. Finally, the capacitor is completely charged to the ohmmeter battery voltage, the charging current is zero, and the ohmmeter reads just the small leakage current through the dielectric. This charging effect, called *capacitor action,* shows that the capacitor can store charge, indicating a normal capacitor. Note that both the rise and the fall of the meter readings are caused by charging. The capacitor discharges when the meter leads are reversed.

Ohmmeter Readings

Troubles in a capacitor are indicated as follows:

1. If an ohmmeter reading immediately goes practically to zero and stays there, the capacitor is short-circuited.
2. If a capacitor shows charging, but the final resistance reading is appreciably less than normal, the capacitor is leaky. Such capacitors are particularly troublesome in high-resistance circuits. When checking electrolytics, reverse the ohmmeter leads and take the higher of the two readings.
3. If a capacitor shows no charging action but reads very high resistance, it may be open. Some precautions must be remembered, however,

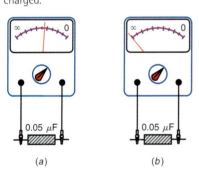

Figure 16–29 Checking a capacitor with an ohmmeter. The *R* scale is shown right to left, as on a VOM. Use the highest ohms range. (*a*) Capacitor action as needle is moved by the charging current from the battery in the ohmmeter. (*b*) Practically infinite leakage resistance reading after the capacitor has been charged.

since very high resistance is a normal condition for capacitors. Reverse the ohmmeter leads to discharge the capacitor, and check it again. In addition, remember that capacitance values of 100 pF, or less, normally have very little charging current for the low battery voltage of the ohmmeter.

Short-Circuited Capacitors

In normal service, capacitors can become short-circuited because the dielectric deteriorates with age, usually over a period of years under the stress of charging voltage, especially at higher temperatures. This effect is more common with paper and electrolytic capacitors. The capacitor may become leaky gradually, indicating a partial short circuit, or the dielectric may be punctured, causing a short circuit.

Open Capacitors

In addition to the possibility of an open connection in any type of capacitor, electrolytics develop high resistance in the electrolyte with age, particularly at high temperatures. After service of a few years, if the electrolyte dries up, the capacitor will be partially open. Much of the capacitor action is gone, and the capacitor should be replaced.

Leaky Capacitors

A leaky capacitor reads R less than normal with an ohmmeter. However, dc voltage tests are more definite. In a circuit, the dc voltage at one terminal of the capacitor should not affect the dc voltage at the other terminal.

Shelf Life

Except for electrolytics, capacitors do not deteriorate with age while stored, since there is no applied voltage. Electrolytic capacitors, however, like dry cells, should be used fresh from the manufacturer because the wet electrolyte may dry out over a period of time.

Capacitor Value Change

All capacitors can change value over time, but some are more prone to change than others. Ceramic capacitors often change value by 10 to 15% during the first year, as the ceramic material relaxes. Electrolytics change value from simply sitting because the electrolytic solution dries out.

Replacing Capacitors

Approximately the same C and V ratings should be used when installing a new capacitor. Except for tuning capacitors, the C value is usually not critical. Also, a higher voltage rating can be used. An important exception, however, is the electrolytic capacitor. Then the ratings should be close to the original values for two reasons. First, the specified voltage is needed to form the internal oxide film that provides the required capacitance. Also, too much C may allow excessive charging current in the circuit that charges the capacitor. Remember that electrolytics generally have large values of capacitance.

■ *16–11 Knowledge Check*

Answer at end of chapter.

When tested with an analog ohmmeter, which capacitor, a 100 pF or 0.33 μF, will show more capacitor action?

a. What is the ohmmeter reading for a shorted capacitor?
b. Does capacitor action with an ohmmeter show that the capacitor is good or bad?
c. Which type of capacitor is more likely to develop trouble, a mica or an electrolytic?

Summary

■ A capacitor consists of two conductors separated by an insulator, or dielectric. Its ability to store charge is the capacitance *C*. Applying voltage to store charge is called *charging the capacitor;* short-circuiting the two leads or terminals of the capacitor to neutralize the charge is called *discharging the capacitor.* Schematic symbols for *C* are summarized in Fig. 16–30.

Figure 16–30 Schematic symbols for types of *C.* (*a*) Fixed type with air, paper, plastic film, mica, or ceramic dielectric. (*b*) Electrolytic type, which has polarity. (*c*) Variable. (*d*) Ganged variable capacitors on one shaft.

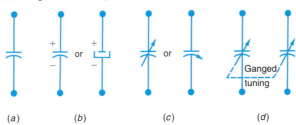

(*a*)   (*b*)   (*c*)   (*d*)

■ The unit of capacitance is the farad. One farad of capacitance stores one coulomb of charge with one volt applied. Practical capacitors have much smaller capacitance values from 1 pF to 1000 μF. A capacitance of 1 pF is 1×10^{-12} F; 1 μF $= 1 \times 10^{-6}$ F; and 1 nF $= 1 \times 10^{-9}$ F.

■ $Q = CV$, where Q is the charge in coulombs, C the capacitance in farads, and V is the potential difference across the capacitor in volts.

■ Capacitance increases with larger plate area and less distance between plates.

■ The ratio of charge stored in different insulators to the charge stored in air is the dielectric constant K_{ϵ} of the

material. Air or vacuum has a dielectric constant of 1.

■ The most common types of commercial capacitors are air, plastic film, paper, mica, ceramic, and electrolytic. Electrolytics are the only capacitors that require observing polarity when connecting to a circuit. The different types are compared in Table 16–2.

■ Capacitors are coded to indicate their capacitance in either microfarads (μF) or picofarads (pF).

■ For parallel capacitors, $C_T = C_1 + C_2 + C_3 + \cdots +$ etc.

■ For series capacitors,

$$C_{EQ} = \cfrac{1}{\cfrac{1}{C_1} + \cfrac{1}{C_2} + \cfrac{1}{C_3} + \cdots + \text{etc.}}$$

■ The electric field of a capacitance has stored energy $\mathcal{E} = \frac{1}{2} CV^2$, where V is volts, C is in farads, and electric energy is in joules.

■ When checked with an analog ohmmeter, a good capacitor shows charging current, and then the ohmmeter reading steadies at the leakage resistance. All types except electrolytics normally have very high leakage resistance such as 100,000 MΩ or more. Electrolytics have more leakage current; a typical leakage resistance is about 500 kΩ to 10 MΩ?

Important Terms

■ Capacitance (*C*) - the ability of a dielectric to hold or store an electric charge. The more charge stored for a given voltage, the greater the capacitance.

■ Capacitor - a component that can store electric charge. A capacitor consists of two metal plates separated by an insulator. Capacitors are named according to the type of

dielectric used. Common capacitor types include air, ceramic, plastic film, mica, paper, and aluminum electrolytic.

- Charging - increasing the amount of charge stored in a capacitor. The accumulation of stored charge results in a buildup of voltage across the capacitor.
- Condenser - another (older) name for a capacitor.
- Dielectric Absorption - the inability of a capacitor to discharge completely to zero. Dielectric absorption is sometimes called battery action or capacitor memory.
- Dielectric Constant, K_ϵ - a factor that indicates the ability of an insulator to concentrate electric flux, also known as relative permittivity, ϵ_r.
- Dielectric Material - another name for an insulator.
- Dielectric Strength - the ability of a dielectric to withstand a potential difference without internal arcing.

- Discharging - the action of neutralizing the charge stored in a capacitor by connecting a conducting path across the capacitor leads.
- Electric Field - the invisible lines of force between opposite electric charges.
- Equivalent Series Resistance (ESR) - a resistance in series with an ideal capacitor that collectively represents all losses in a capacitor. Ideally, the ESR of a capacitor should be zero.
- Farad (F) Unit - the basic unit of capacitance. $1\ F = \dfrac{1\ C}{1\ V}$
- Ganged Capacitors - two or three capacitor sections on one common shaft which can be rotated.

- Leakage Current - the current that flows through the dielectric of a capacitor when voltage is applied across the capacitor plates.
- Leakage Resistance - a resistance in parallel with a capacitor that represents all leakage paths through which a capacitor can discharge.
- Microfarad - a small unit of capacitance equal to 1×10^{-6} F.
- Nanofarad - a small unit of capacitance equal to 1×10^{-9} F.
- Picofarad - a small unit of capacitance equal to 1×10^{-12} F.
- Relative Permittivity, ϵ_r - a factor that indicates the ability of an insulator to concentrate electric flux, also known as the dielectric constant, K_ϵ.

Related Formulas

$Q = CV$ coulombs

$C = \dfrac{Q}{V}$

$V = \dfrac{Q}{C}$

$C = K_\epsilon \times \dfrac{A}{d} \times 8.85 \times 10^{-12}$ F

$C_T = C_1 + C_2 + \cdots + \text{etc. (parallel capacitors)}$

$C_{EQ} = \dfrac{1}{\dfrac{1}{C_1} + \dfrac{1}{C_2} + \dfrac{1}{C_3} + \cdots + \text{etc.}}$ (series capacitors)

Energy $= \mathscr{E} = \frac{1}{2}\, CV^2$ joules

Self-Test

Answers at back of book.

1. **In general, a capacitor is a component that can**
   a. pass a dc current.
   b. store an electric charge.
   c. act as a bar magnet.
   d. step up or step down an ac voltage.

2. **The basic unit of capacitance is the**
   a. farad.
   b. henry.
   c. tesla.
   d. ohm.

3. **Which of the following factors affect the capacitance of a capacitor?**
   a. the area, A, of the plates.
   b. the distance, d, between the plates.
   c. the type of dielectric used.
   d. all of the above.

4. **How much charge in coulombs is stored by a 50-μF capacitor with 20 V across its plates?**
   a. $Q = 100\ \mu$C.
   b. $Q = 2.5\ \mu$C.
   c. $Q = 1$ mC.
   d. $Q = 1\ \mu$C.

5. **A capacitor consists of**
   a. two insulators separated by a conductor.
   b. a coil of wire wound on an iron core.
   c. two conductors separated by an insulator.
   d. none of the above.

6. **A capacitance of 82,000 pF is the same as**
   a. 0.082 μF.
   b. 82 μF.
   c. 82 nF.
   d. both a and c.

7. A 47μF-capacitor has a stored charge of 2.35 mC. What is the voltage across the capacitor plates?

a. 50 V.

b. 110 V approx.

c. 5 V.

d. 100 V.

8. Which of the following types of capacitors typically has the highest leakage current?

a. plastic-film.

b. electrolytic.

c. mica.

d. air-variable.

9. One of the main applications of a capacitor is to

a. block ac and pass dc.

b. block both dc and ac.

c. block dc and pass ac.

d. pass both dc and ac.

10. When checked with an ohmmeter, a shorted capacitor will measure

a. infinite ohms.

b. zero ohms.

c. somewhere in the range of 1 to 10 MΩ.

d. none of the above.

11. The equivalent capacitance, C_{EQ}, of a 10-μF and a 40-μF capacitor in series is

a. 50 μF.

b. 125 μF.

c. 8 μF.

d. 400 μF.

12. A 0.33-μF capacitor is in parallel with a 0.15-μF and a 220,000-pF capacitor. What is the total capacitance, C_T?

a. 0.7 μF.

b. 0.007 μF.

c. 0.07 μF.

d. 7 nF.

13. A 5-μF capacitor, C_1, and a 15-μF capacitor, C_2, are connected in series. If the charge stored in C_1 equals 90 μC, what is the voltage across the capacitor C_2?

a. 18 V.

b. 12 V.

c. 9 V.

d. 6 V.

14. A plastic-film capacitor, whose coded value is 333M, measures 0.025 μF when tested with a capacitor-inductor analyzer. The measured capacitance is

a. well within tolerance.

b. barely within tolerance.

c. slightly out of tolerance.

d. right on the money.

15. Capacitors are never coded in

a. nanofarad units.

b. microfarad units.

c. picofarad units.

d. both b and c.

16. Which type of capacitor could explode if the polarity of voltage across its plates is incorrect?

a. air-variable.

b. mica.

c. ceramic disk.

d. aluminum electrolytic.

17. The voltage rating of a capacitor is not affected by

a. the area of the plates.

b. the distance between the plates.

c. the type of dielectric used.

d. both b and c.

18. The leakage resistance of a capacitor is typically represented as a(n)

a. resistance in series with the capacitor plates.

b. electric field between the capacitor plates.

c. resistance in parallel with the capacitor plates.

d. closed switch across the dielectric material.

19. A 2200-μF capacitor with a voltage rating of 35 V is most likely a(n)

a. electrolytic capacitor.

b. air-variable capacitor.

c. mica capacitor.

d. paper capacitor.

20. A capacitor that can store 100 μC of charge with 10 V across its plates has a capacitance value of

a. 0.01 μF.

b. 10 μF.

c. 10 nF.

d. 100 mF.

21. Calculate the permissible capacitance range of a ceramic disk capacitor whose coded value is .068Z.

a. 0.0544 μF to 0.1224 μF.

b. 0.0136 μF to 0.0816 μF.

c. 0.0136 μF to 0.1224 μF.

d. 0.0544 pF to 0.1224 pF.

22. The equivalent series resistance (ESR) of a capacitor should ideally be

a. infinite.

b. as high as possible.

c. around 100 kΩ or so.

d. zero.

23. The charge and discharge current of a capacitor flows

a. through the dielectric.

b. to and from the capacitor plates.

c. through the dielectric only until the capacitor is fully charged.

d. straight through the dielectric from one plate to the other.

24. Capacitance increases with

a. larger plate area and greater distance between the plates.

b. smaller plate area and greater distance between the plates.

c. larger plate area and less distance between the plates.

d. higher values of applied voltage.

25. Two 0.02-μF, 500-V capacitors in series have an equivalent capacitance and breakdown voltage rating of

a. 0.04 μF, 1 kV.

b. 0.01 μF, 250 V.

c. 0.01 μF, 500 V.

d. 0.01 μF, 1 kV.

Questions

1. Define capacitance with respect to physical structure and electrical function. Explain how a two-wire conductor has capacitance.

2. (a) What is meant by a dielectric material? (b) Name five common dielectric materials. (c) Define dielectric flux.

3. Explain briefly how to charge a capacitor. How is a charged capacitor discharged?

4. Define 1-F of capacitance. Convert the following into farads using powers of 10: (a) 50 pF; (b) 0.001 μF; (c) 0.047 μF; (d) 0.01 μF; (e) 10 μF.

5. State the effect on capacitance of (a) larger plate area; (b) thinner dielectric; (c) higher value of dielectric constant.

6. Give one reason for your choice of the type of capacitor to be used in the following applications: (a) 80-μF capacitance for a circuit where one side is positive and the applied voltage never exceeds 150 V; (b) 1.5-pF capacitance for an rf circuit where the required voltage rating is less than 500 V; (c) 5-μF capacitance for an audio circuit where the required voltage rating is less than 25 V.

7. Give the capacitance value of six-dot mica capacitors color-coded as follows: (a) Black, red, green, brown, black, black. (b) White, green, brown, black, silver, brown. (c) Brown, green, black, red, gold, blue.

8. Draw a diagram showing the fewest number of 400-V, 2-μF capacitors needed for a combination rated at 800 V with 2-μF total capacitance.

9. Suppose you are given two identical uncharged capacitors. One is charged to 50 V and connected across the uncharged capacitor. Why will the voltage across both capacitors then be 25 V?

10. Describe briefly how you would check a 0.05-μF capacitor with an ohmmeter. State the ohmmeter indications when the capacitor is good, short-circuited, or open.

11. Define the following: (a) Leakage resistance. (b) Dielectric absorption. (c) Equivalent series resistance.

12. Give two comparisons between the electric field in a capacitor and the magnetic field in a coil.

13. Give three types of troubles in capacitors.

14. When a capacitor discharges, why is its discharge current in the direction opposite from the charging current?

15. Compare the features of aluminum and tantalum electrolytic capacitors.

16. Why can plastic film be used instead of paper for capacitors?

17. What two factors determine the breakdown voltage rating of a capacitor?

Problems

SECTION 16–3 THE FARAD UNIT OF CAPACITANCE

16–1 Calculate the amount of charge, Q, stored by a capacitor if
    a. $C = 10\ \mu$F and $V = 5$ V.
    b. $C = 1\ \mu$F and $V = 25$ V.
    c. $C = 0.01\ \mu$F and $V = 150$ V.
    d. $C = 0.22\ \mu$F and $V = 50$ V.
    e. $C = 680$ pF and $V = 200$ V.
    f. $C = 47$ pF and $V = 3$ kV.

16–2 How much charge, Q, is stored by a 0.05-μF capacitor if the voltage across the plates equals
    a. 10 V?
    b. 40 V?
    c. 300 V?
    d. 500 V?
    e. 1 kV?

16–3 How much voltage exists across the plates of a 200-μF capacitor if a constant current of 5 mA charges it for
    a. 100 ms?
    b. 250 ms?
    c. 0.5 s?
    d. 2 s?
    e. 3 s?

16–4 Determine the voltage, V, across a capacitor if
    a. $Q = 2.5\ \mu$C and $C = 0.01\ \mu$F.
    b. $Q = 49.5$ nC and $C = 330$ pF.
    c. $Q = 10$ mC and $C = 1{,}000\ \mu$F.
    d. $Q = 500\ \mu$C and $C = 0.5\ \mu$F.
    e. $Q = 188$ nC and $C = 0.0047\ \mu$F.
    f. $Q = 75$ nC and $C = 0.015\ \mu$F.

16-5 Determine the capacitance, C, of a capacitor if
   a. $Q = 15\ \mu C$ and $V = 1\ V$.
   b. $Q = 15\ \mu C$ and $V = 30\ V$.
   c. $Q = 100\ \mu C$ and $V = 25\ V$.
   d. $Q = 3.3\ \mu C$ and $V = 15\ V$.
   e. $Q = 0.12\ \mu C$ and $V = 120\ V$.
   f. $Q = 100\ \mu C$ and $V = 2.5k\ V$.

16-6 List the physical factors that affect the capacitance, C, of a capacitor.

16-7 Calculate the capacitance, C, of a capacitor for each set of physical characteristics listed.
   a. $A = 0.1\ cm^2$, $d = 0.005\ cm$, $K_\epsilon = 1$
   b. $A = 0.05\ cm^2$, $d = 0.001\ cm$, $K_\epsilon = 500$
   c. $A = 0.1\ cm^2$, $d = 1 \times 10^{-5}\ cm$, $K_\epsilon = 50$
   d. $A = 1\ cm^2$, $d = 5 \times 10^{-6}\ cm$, $K_\epsilon = 6$

SECTION 16-6 CAPACITOR CODING

16-8 Determine the capacitance and tolerance of each of the capacitors shown in Fig. 16-31.

Figure 16-31

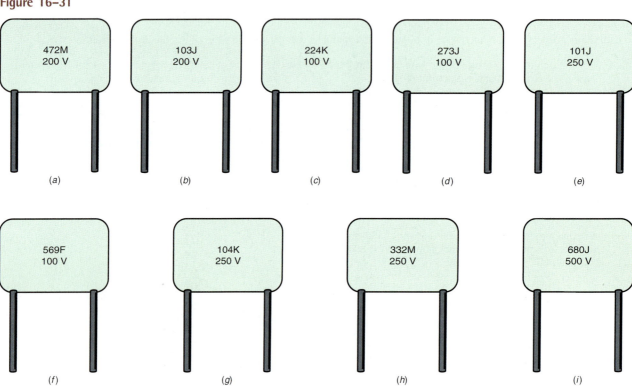

(a) 472M 200 V
(b) 103J 200 V
(c) 224K 100 V
(d) 273J 100 V
(e) 101J 250 V
(f) 569F 100 V
(g) 104K 250 V
(h) 332M 250 V
(i) 680J 500 V

16-9 Determine the capacitance and tolerance of each of the capacitors shown in Fig. 16-32.

Figure 16-32

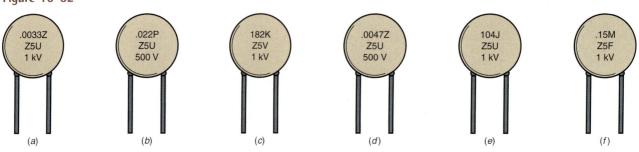

(a) .0033Z Z5U 1 kV
(b) .022P Z5U 500 V
(c) 182K Z5V 1 kV
(d) .0047Z Z5U 500 V
(e) 104J Z5U 1 kV
(f) .15M Z5F 1 kV

16-10 Determine the capacitance and tolerance of each of the capacitors shown in Fig. 16–33.

Figure 16–33

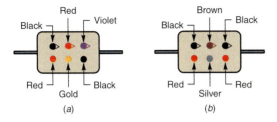

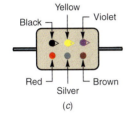

(c)

16-11 Determine the capacitance of each chip capacitor in Fig. 16–34. Use the coding scheme in Fig. 16–17.

Figure 16–34

(a)   (b)   (c)   (d)

16-12 Determine the capacitance of each chip capacitor in Fig. 16–35. Use the coding scheme in Fig. 16–18.

Figure 16–35

(a)   (b)   (c)   (d)

16-13 Determine the capacitance of each chip capacitor in Fig. 16–36.

Figure 16–36

(a)   (b)   (c)   (d)

16-14 Determine the capacitance and tolerance of each capacitor in Fig. 16–37.

Figure 16–37

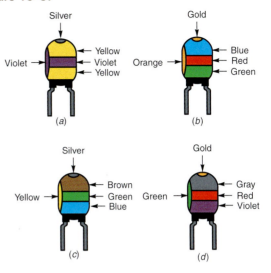

16-15 Determine the permissible capacitance range of the capacitors in
   a. Fig. 16–31*a*.
   b. Fig. 16–31*d*.
   c. Fig. 16–31*f*.
   d. Fig. 16–32*c*.
   e. Fig. 16–32*d*.

16-16 Explain the alphanumeric code, Z5U, for the capacitor in Fig. 16–32*b*.

SECTION 16–7 PARALLEL CAPACITANCES

16-17 A 5-μF and 15-μF capacitor are in parallel. How much is C_T?

16-18 A 0.1-μF, 0.27-μF, and a 0.01-μF capacitor are in parallel. How much is C_T?

16-19 A 150-pF, 330-pF, and a 0.001-μF capacitor are in parallel. How much is C_T?

16-20 In Fig. 16–38,
   a. how much voltage is across each individual capacitor?
   b. how much charge is stored by C_1?
   c. how much charge is stored by C_2?
   d. how much charge is stored by C_3?
   e. what is the total charge stored by all capacitors?
   f. how much is C_T?

Figure 16–38

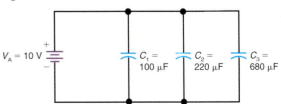

SECTION 16–8 SERIES CAPACITANCES

16–21 A 0.1-μF and 0.4-μF capacitor are in series. How much is the equivalent capacitance, C_{EQ}?

16–22 A 1500-pF and a 0.001-μF capacitor are in series. How much is the equivalent capacitance, C_{EQ}?

16–23 A 0.082-μF, 0.047-μF, and a 0.012 μF capacitor are in series. How much is the equivalent capacitance, C_{EQ}?

16–24 In Fig. 16–39, assume a charging current of 180 μA flows for 1 s. Solve for

   a. C_{EQ}.

   b. the charge stored by C_1, C_2, and C_3.

   c. the voltage across C_1, C_2, and C_3.

   d. the total charge stored by all capacitors.

Figure 16–39

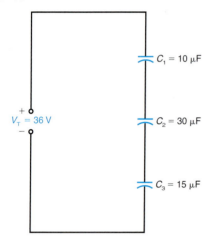

16–25 In Fig. 16–40, assume a charging current of 2.4 mA flows for 1 ms. Solve for

   a. C_{EQ}.

   b. the charge stored by C_1, C_2, and C_3.

   c. the voltage across C_1, C_2, and C_3.

   d. the total charge stored by all capacitors.

Figure 16–40

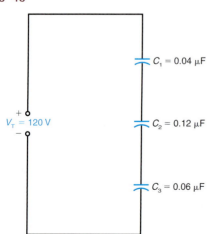

16–26 How much capacitance must be connected in series with a 120-pF capacitor to obtain an equivalent capacitance, C_{EQ}, of 100 pF.

SECTION 16–9 ENERGY STORED IN ELECTROSTATIC FIELD OF CAPACITANCE

16–27 How much energy is stored by a 100-μF capacitor which is charged to

   a. 5 V?

   b. 10 V?

   c. 50 V?

16–28 How much energy is stored by a 0.027-μF capacitor which is charged to

   a. 20 V?

   b. 100 V?

   c. 500 V?

16–29 Calculate the energy stored by each capacitor in Fig. 16–39.

SECTION 16–10 MEASURING AND TESTING CAPACITORS

16–30 Make the following conversions:

   a. 0.047 μF to pF

   b. 0.0015 μF to pF

   c. 390,000 pF to μF

   d. 1000 pF to μF

16–31 Make the following conversions:

   a. 15 nf to pF

   b. 1 nF to pF

   c. 680 nF to pF

   d. 33,000 pF to nF

   e. 1,000,000 pF to nF

   f. 560,000 pF to nF

16–32 A plastic-film capacitor has a coded value of 154K. If the measured value of capacitance is 0.160 μF, is the capacitance value within tolerance?

16–33 A ceramic disk capacitor is coded 102Z. If the measured value of capacitance is 680 pF, is the capacitance within tolerance?

16–34 A plastic-film capacitor has a coded value of 229B. If the measured value of capacitance is 2.05 pF, is the capacitance within tolerance?

SECTION 16–11 TROUBLES IN CAPACITORS

16–35 What is the ohmmeter reading for a(n)

   a. shorted capacitor.

   b. open capacitor.

   c. leaky capacitor.

16–36 Describe the effect of connecting a 0.47-μF capacitor to the leads of an analog ohmmeter set to the $R \times$ 10K range.

Critical Thinking

16–37 Three capacitors in series have a combined equivalent capacitance C_{EQ} of 1.6 nF. If $C_1 = 4C_2$ and $C_3 = 20C_1$, calculate the values for C_1, C_2, and C_3.

16–38 A 100-pF ceramic capacitor has a temperature coefficient T_C of N500. Calculate its capacitance at (a) 75°C; (b) 125°C; (c) −25°C.

16–39 (a) Calculate the energy stored by a 100-μF capacitor charged to 100 V. (b) If this capacitor is now connected across another 100-μF capacitor which is uncharged, calculate the total energy stored by both capacitors. (c) Is the energy stored by both capacitors in part (b) less than the energy stored by the single capacitor in part (a)? If yes, where did the energy go?

Answers to Knowledge Check Problems

16–1 None, $I = 0$ A

16–2 False

16–3 $C = 14.16$ nF

16–4 Electrolytic capacitor

16–5 A thin oxide film

16–6 $\pm 1\%$

16–7 $C_T = 4\ \mu F$

16–8 $Q_T = 600\ \mu C$

16–9 1.25 J

16–10 10,000 pF, 10 nF, and 0.01 μF

16–11 0.33 μF

Answers to Self-Reviews

16–1 a. dielectric
b. farad

16–2 a. 14.5 V
b. 0 V
c. Yes

16–3 a. 10 μF
b. ceramic

16–4 a. T
b. F
c. T
d. T

16–5 a. T
b. T
c. T

16–6 a. F
b. T
c. T
d. T

16–7 a. 0.03 μF
b. 150 pF

16–8 a. 0.1 μF
b. 25 V
c. 33.3 pF

16–9 a. T
b. T

16–10 a. T
b. T
c. F
d. F

16–11 a. 0 Ω
b. good
c. electrolytic

Capacitive Reactance

When a capacitor charges and discharges with a varying voltage applied, alternating current can flow. Although there cannot be any current through the dielectric of the capacitor, its charge and discharge produce alternating current in the circuit connected to the capacitor plates. The amount of I that results from the applied sine-wave V depends on the capacitor's capacitive reactance. The symbol for capacitive reactance is X_C, and its unit is the ohm. The X in X_C indicates reactance, whereas the subscript C specifies capacitive reactance.

The amount of X_C is a V/I ratio but it can also be calculated as $X_C = 1/(2\pi fC)$ in terms of the value of the capacitance and the frequency of the varying V and I. With f and C in the units of the hertz and farad, X_C is in units of ohms. The reciprocal relation in $1/(2\pi fC)$ means that the ohms of X_C decrease for higher frequencies and with more C because more charge and discharge current results either with more capacitance or faster changes in the applied voltage.

Objectives

After studying this chapter you should be able to

- *Explain* how alternating current can flow in a capacitive circuit.
- *Calculate* the reactance of a capacitor when the frequency and capacitance are known.
- *Calculate* the total capacitive reactance of series-connected capacitors.
- *Calculate* the equivalent capacitive reactance of parallel-connected capacitors.
- *Explain* how Ohm's law can be applied to capacitive reactance.
- *Calculate* the capacitive current when the capacitance and rate of voltage change are known.

Outline

Important Terms

capacitive reactance, X_C

charging current

discharge current

inversely proportional

phase angle

Current in a capacitive circuit. (*a*) The
4-μF capacitor allows enough current *I* to
light the bulb brightly. (*b*) Less current
with a smaller capacitor causes dim light.
(*c*) The bulb cannot light with dc voltage
applied because a capacitor blocks direct
current.

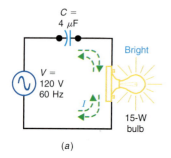

(*a*)

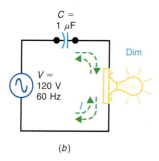

(*b*)

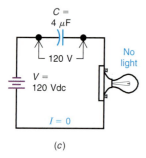

(*c*)

GOOD TO KNOW

For a capacitor, the charge and
discharge current flows to and
from the plates but not through
the dielectric.

17–1 Alternating Current in a Capacitive Circuit

The fact that current flows with ac voltage applied is demonstrated in Fig. 17–1, where the bulb lights in Fig. 17–1*a* and *b* because of the capacitor charge and discharge current. There is no current through the dielectric, which is an insulator. While the capacitor is being charged by increasing applied voltage, however, the charging current flows in one direction in the conductors to the plates. While the capacitor is discharging, when the applied voltage decreases, the discharge current flows in the reverse direction. With alternating voltage applied, the capacitor alternately charges and discharges.

First the capacitor is charged in one polarity, and then it discharges; next the capacitor is charged in the opposite polarity, and then it discharges again. The cycles of charge and discharge current provide alternating current in the circuit at the same frequency as the applied voltage. This is the current that lights the bulb.

In Fig. 17–1*a*, the 4-μF capacitor provides enough alternating current to light the bulb brightly. In Fig. 17–1*b*, the 1-μF capacitor has less charge and discharge current because of the smaller capacitance, and the light is not so bright. Therefore, the smaller capacitor has more opposition to alternating current as less current flows with the same applied voltage, that is, it has more reactance for less capacitance.

In Fig. 17–1*c*, the steady dc voltage will charge the capacitor to 120 V. Because the applied voltage does not change, though, the capacitor will just stay charged. Since the potential difference of 120 V across the charged capacitor is a voltage drop opposing the applied voltage, no current can flow. Therefore, the bulb cannot light. The bulb may flicker on for an instant because charging current flows when voltage is applied, but this current is only temporary until the capacitor is charged. Then the capacitor has the applied voltage of 120 V, but there is zero voltage across the bulb.

As a result, the capacitor is said to *block* direct current or voltage. In other words, after the capacitor has been charged by a steady dc voltage, there is no current in the dc circuit. All the applied dc voltage is across the charged capacitor with zero voltage across any series resistance.

In summary, then, this demonstration shows the following points:

1. Alternating current flows in a capacitive circuit with ac voltage applied.
2. A smaller capacitance allows less current, which means more X_C with more ohms of opposition.
3. Lower frequencies for the applied voltage result in less current and more X_C. With a steady dc voltage source, which corresponds to a frequency of zero, the opposition of the capacitor is infinite and there is no current. In this case, the capacitor is effectively an open circuit.

These effects have almost unlimited applications in practical circuits because X_C depends on frequency. A very common use of a capacitor is to provide little opposition for ac voltage but to block any dc voltage. Another example is to use X_C for less opposition to a high-frequency alternating current, compared with lower frequencies.

Capacitive Current

The reason that a capacitor allows current to flow in an ac circuit is the alternate charge and discharge. If we insert an ammeter in the circuit, as shown in Fig. 17–2, the ac meter will read the amount of charge and discharge current.

Figure 17–2 Capacitive reactance X_C is the ratio V_C/I_C.

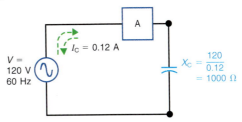

In this example, I_C is 0.12 A. This current is the same in the voltage source, the connecting leads, and the plates of the capacitor. However, there is no current through the insulator between the plates of the capacitor.

Values for X_C

When we consider the ratio of V_C/I_C for the ohms of opposition to the sine-wave current, this value is 120/0.12, which equals 1000 Ω. This 1000 Ω is what we call X_C, to indicate how much current can be produced by sine-wave voltage applied to a capacitor. In terms of current, $X_C = V_C/I_C$. In terms of frequency and capacitance, $X_C = 1/(2\pi fC)$.

The X_C value depends on the amount of capacitance and the frequency of the applied voltage. If C in Fig. 17–2 were increased, it could take on more charge for more charging current and then produce more discharge current. Then X_C is less for more capacitance. Also, if the frequency in Fig. 17–2 were increased, the capacitor could charge and discharge faster to produce more current. This action also means that V_C/I_C would be less with more current for the same applied voltage. Therefore, X_C is less for higher frequencies. Reactance X_C can have almost any value from practically zero to almost infinite ohms.

GOOD TO KNOW

Capacitive reactance, X_C, is a measure of a capacitor's opposition to the flow of alternating current. The unit of X_C is the ohm (Ω). X_C applies only to sine waves.

■ *17–1 Knowledge Check*

  Answer at end of chapter.

  In Fig. 17–2, how much is X_C if I_C = 60 mA?

■ *17–1 Self-Review*

  Answers at end of chapter.

  a. **Which has more reactance, a 0.1- or a 0.5-µF capacitor, at the same frequency?**
  b. **Which allows more charge and discharge current, a 0.1- or a 0.5-µF capacitor, at the same frequency?**

17–2 The Amount of X_C Equals $1/(2\pi fC)$

The effects of frequency and capacitance are included in the formula for calculating ohms of reactance. The f is in hertz units and the C is in farads for X_C in ohms. As an example, we can calculate X_C for C of 2.65 µF and f of 60 Hz. Then

$$X_C = \frac{1}{2\pi fC} \qquad \text{(17–1)}$$

$$= \frac{1}{2\pi \times 60 \times 2.65 \times 10^{-6}} = \frac{1}{6.28 \times 159 \times 10^{-6}}$$

$$= 0.00100 \times 10^{6}$$

$$= 1000 \ \Omega$$

Note the following factors in the formula $X_C = \dfrac{1}{2\pi fC}$.

1. The constant factor 2π is always $2 \times 3.14 = 6.28$. It indicates the circular motion from which a sine wave is derived. Therefore, the formula $X_C = \dfrac{1}{2\pi fC}$ applies only to sine-wave ac circuits. The 2π is actually 2π rad or $360°$ for a complete circle or cycle.

2. The frequency, f, is a time element. A higher frequency means that the voltage varies at a faster rate. A faster voltage change can produce more charge and discharge current for a given value of capacitance, C. The result is less X_C.

3. The capacitance, C, indicates the physical factors of the capacitor that determine how much charge and discharge current it can produce for a given change in voltage.

4. Capacitive reactance, X_C, is measured in ohms corresponding to the $\dfrac{V_C}{I_C}$ ratio for sine-wave ac circuits. The X_C value determines how much current C allows for a given value of applied voltage.

Example 17-1

How much is X_C for (a) 0.1 μF of C at 1400 Hz? (b) 1 μF of C at the same frequency?

ANSWER

a. $X_C = \dfrac{1}{2\pi fC} = \dfrac{1}{6.28 \times 1400 \times 0.1 \times 10^{-6}}$

$= \dfrac{1}{6.28 \times 140 \times 10^{-6}} = 0.00114 \times 10^6$

$= 1140 \ \Omega$

b. At the same frequency, with ten times more C, X_C is one-tenth or $1140/10$, which equals $114 \ \Omega$.

Example 17-2

How much is the X_C of a 47-pF value of C at (a) 1 MHz? (b) 10 MHz?

ANSWER

a. $X_C = \dfrac{1}{2\pi fC} = \dfrac{1}{6.28 \times 47 \times 10^{-12} \times 1 \times 10^6}$

$$= \frac{1}{295.16 \times 10^{-6}} = 0.003388 \times 10^{6}$$
$$= 3388 \ \Omega$$

b. At 10 times the frequency,

$$X_C = \frac{3388}{10} = 338 \ \Omega.$$

Note that X_C in Example 17–2b is one-tenth the value in Example 17–2a because f is 10 times greater.

X_C Is Inversely Proportional to Capacitance

This statement means that X_C increases as capacitance is decreased. In Fig. 17–3, when C is reduced by a factor of 1/10 from 1.0 to 0.1 μF, then X_C increases 10 times from 1000 to 10,000 Ω. Also, decreasing C by one-half from 0.2 to 0.1 μF doubles X_C from 5000 to 10,000 Ω.

This inverse relation between C and X_C is illustrated by the graph in Fig. 17–3. Note that values of X_C increase downward on the graph, indicating negative reactance that is opposite from inductive reactance. (Inductive reactance is covered in chapter 20.) With C increasing to the right, the decreasing values of X_C approach the zero axis of the graph.

X_C Is Inversely Proportional to Frequency

Figure 17–4 illustrates the inverse relationship between X_C and f. With f increasing to the right in the graph from 0.1 to 1 MHz, the value of X_C for the 159-pF capacitor decreases from 10,000 to 1000 Ω as the X_C curve comes closer to the zero axis.

The graphs are nonlinear because of the inverse relation between X_C and f or C. At one end, the curves approach infinitely high reactance for zero capacitance or zero frequency. At the other end, the curves approach zero reactance for infinitely high capacitance or frequency.

||| MultiSim **Figure 17–3** A table of values and a graph to show that capacitive reactance X_C decreases with higher values of C. Frequency is constant at 159 Hz.

X_C increases as C decreases

C, μF	$X_C* = \frac{1}{2\pi f C}$, Ω
1.0	1000
0.5	2000
0.2	5000
0.1	10,000

*For f = 159 Hz

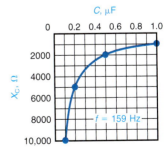

Figure 17–4 A table of values and a graph to show that capacitive reactance X_C decreases with higher frequencies. C is constant at 159 pF.

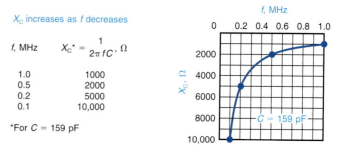

X_C increases as f decreases

f, MHz	$X_C^* = \dfrac{1}{2\pi fC}$, Ω
1.0	1000
0.5	2000
0.2	5000
0.1	10,000

*For $C = 159$ pF

Calculating C from Its Reactance

In some applications, it may be necessary to find the value of capacitance required for a desired amount of X_C. For this case, the reactance formula can be inverted to

$$C = \frac{1}{2\pi fX_C} \qquad (17\text{–}2)$$

The value of 6.28 for 2π is still used. The only change from Formula (17–1) is that the C and X_C values are inverted between denominator and numerator on the left and right sides of the equation.

Example 17–3

What C is needed for X_C of 100 Ω at 3.4 MHz?

ANSWER

$$C = \frac{1}{2\pi fX_C} = \frac{1}{6.28 \times 3.4 \times 10^6 \times 100}$$
$$= \frac{1}{628 \times 3.4 \times 10^6}$$
$$= 0.000468 \times 10^{-6}\ \text{F} = 0.000468\ \mu\text{F} \quad \text{or} \quad 468\ \text{pF}$$

A practical size for this capacitor would be 470 pF. The application is to have low reactance at the specified frequency of 3.4 MHz.

Calculating Frequency from the Reactance

Another use is to find the frequency at which a capacitor has a specified amount of X_C. Again, the reactance formula can be inverted to the form shown in Formula (17–3).

$$f = \frac{1}{2\pi CX_C} \qquad (17\text{–}3)$$

The following example illustrates the use of this formula.

Example 17-4

At what frequency will a 10-μF capacitor have X_C equal to 100 Ω?

ANSWER

$$f = \frac{1}{2\pi C X_C} = \frac{1}{6.28 \times 10 \times 10^{-6} \times 100}$$

$$= \frac{1}{6280 \times 10^{-6}}$$

$$= 0.000159 \times 10^6$$

$$= 159 \text{ Hz}$$

This application is a capacitor for low reactance at audio frequencies.

Summary of X_C Formulas

Formula (17–1) is the basic form for calculating X_C when f and C are known values. As another possibility, the value of X_C can be measured as V_C/I_C.

With X_C known, the value of C can be calculated for a specified f by Formula (17–2), or f can be calculated with a known value of C by using Formula (17–3).

■ *17–2 Knowledge Check*

Answer at end of chapter.

What is the value of X_C for a 0.1 μF capacitor at dc?

■ *17–2 Self-Review*

Answers at end of chapter.

The X_C for a capacitor is 400 Ω at 8 MHz.
a. **How much is X_C at 16 MHz?**
b. **How much is X_C at 4 MHz?**
c. **Is a smaller or larger C needed for less X_C?**

17-3 Series or Parallel Capacitive Reactances

Because capacitive reactance is an opposition in ohms, series or parallel reactances are combined in the same way as resistances. As shown in Fig. 17–5a, series capacitive reactances are added arithmetically.
Series capacitive reactance:

$$X_{C_T} = X_{C_1} + X_{C_2} + \cdots + \text{etc.} \tag{17-4}$$

For parallel reactances, the combined reactance is calculated by the reciprocal formula, as shown in Fig. 17–5b.
Parallel capacitive reactance:

$$X_{C_{EQ}} = \frac{1}{\dfrac{1}{X_{C_1}} + \dfrac{1}{X_{C_2}} + \dfrac{1}{X_{C_3}} + \cdots + \text{etc.}} \tag{17-5}$$

Figure 17–5 Reactances alone combine like resistances. (*a*) Addition of series reactances. (*b*) Two reactances in parallel equal their product divided by their sum.

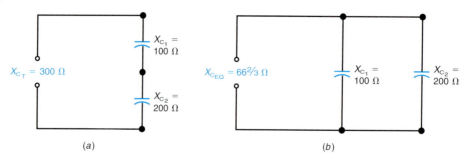

(*a*)   (*b*)

In Fig. 17–5*b*, the parallel combination of 100 and 200 Ω is 66⅔ Ω for $X_{C_{EQ}}$. The combined parallel reactance is less than the lowest branch reactance. Any shortcuts for combining parallel resistances also apply to parallel reactances.

Combining capacitive reactances is opposite to the way capacitances are combined. The two procedures are compatible, however, because capacitive reactance is inversely proportional to capacitance. The general case is that ohms of opposition add in series but combine by the reciprocal formula in parallel.

■ 17–3 Knowledge Check

Answer at end of chapter.

What is the combined reactance of three 150-Ω X_C values in (a) series, (b) parallel?

■ 17–3 Self-Review

Answers at end of chapter.

a. How much is X_{C_T} for a 200-Ω X_{C_1} in series with a 300-Ω X_{C_2}?
b. How much is $X_{C_{EQ}}$ for a 200-Ω X_{C_1} in parallel with a 300-Ω X_{C_2}?

17–4 Ohm's Law Applied to X_C

The current in an ac circuit with X_C alone is equal to the applied voltage divided by the ohms of X_C. Three examples with X_C are illustrated in Fig. 17–6. In Fig. 17–6*a*, there is just one reactance of 100 Ω. The current *I* then is equal to V/X_C, or 100 V/100 Ω, which is 1 A.

For the series circuit in Fig. 17–6*b*, the total reactance, equal to the sum of the series reactances, is 300 Ω. Then the current is 100 V/300 Ω, which equals ⅓ A. Furthermore, the voltage across each reactance is equal to its IX_C product. The sum of these series voltage drops equals the applied voltage.

For the parallel circuit in Fig. 17–6*c*, each parallel reactance has its individual branch current, equal to the applied voltage divided by the branch reactance. The applied voltage is the same across both reactances, since all are in parallel. In addition, the total line current of 1½ A is equal to the sum of the individual branch currents of 1 and ½ A each. Because the applied voltage is an rms value, all calculated currents and voltage drops in Fig. 17–6 are also rms values.

■ 17–4 Knowledge Check

Answer at end of chapter.

In Fig. 17–6b, what are the values for *I*, V_1, and V_2 if X_{C_2} = 400 Ω?

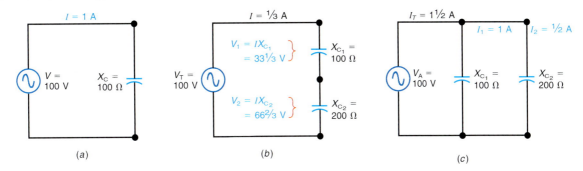

(a) (b) (c)

■ 17–4 Self-Review

Answers at end of chapter.

a. In Fig. 17–6b, how much is X_{C_T}?
b. In Fig. 17–6c, how much is $X_{C_{EQ}}$?

17–5 Applications of Capacitive Reactance

The general use of X_C is to block direct current but provide low reactance for alternating current. In this way, a varying ac component can be separated from a steady direct current. Furthermore, a capacitor can have less reactance for alternating current of high frequencies, compared with lower frequencies.

Note the following difference in ohms of R and X_C. Ohms of R remain the same for dc circuits or ac circuits, whereas X_C depends on the frequency.

If 100 Ω is taken as a desired value of X_C, capacitor values can be calculated for different frequencies, as listed in Table 17–1. The C values indicate typical capacitor sizes for different frequency applications. Note that the required C becomes smaller for higher frequencies.

Table 17–1	Capacitance Values for a Reactance of 100 Ω	
C (Approx.)	Frequency	Remarks
27 μF	60 Hz	Power-line and low audio frequency
1.6 μF	1000 Hz	Audio frequency
0.16 μF	10,000 Hz	Audio frequency
1600 pF	1000 kHz (RF)	AM radio
160 pF	10 MHz (HF)	Short-wave radio
16 pF	100 MHz (VHF)	FM radio

The 100 Ω of reactance for Table 17–1 is taken as a low X_C in common applications of C as a coupling capacitor, bypass capacitor, or filter capacitor for ac variations. For all these functions, the X_C must be low compared with the resistance in the circuit. Typical values of C, then, are 16 to 1600 pF for rf signals and 0.16 to 27 μF for af signals. The power-line frequency of 60 Hz, which is a low audio frequency, requires C values of about 27 μF or more.

■ 17–5 Knowledge Check

Answer at end of chapter.

What value of capacitance will provide an X_C of 150 Ω at 20 kHz?

■ 17–5 Self-Review

Answers at end of chapter.

A capacitor C has 100 Ω X_C at 60 Hz.
a. How much is X_C at 120 Hz?
b. How much is X_C at 6 Hz?

17–6 Sine-Wave Charge and Discharge Current

In Fig. 17–7, sine-wave voltage applied across a capacitor produces alternating charge and discharge current. The action is considered for each quarter-cycle. Note that the voltage v_C across the capacitor is the same as the applied voltage v_A at all times because they are in parallel. The values of current i, however, depend on the charge and discharge of C. When v_A is increasing, it charges C to keep v_C at the same voltage as v_A; when v_A is decreasing, C discharges to maintain v_C at the same voltage as v_A. When v_A is not changing, there is no charge or discharge current.

During the first quarter-cycle in Fig. 17–7a, v_A is positive and increasing, charging C in the polarity shown. The electron flow is from the negative terminal of the source voltage, producing charging current in the direction indicated by the arrow for i. Next, when the applied voltage decreases during the second quarter-cycle, v_C also decreases by discharging. The discharge current is from the negative plate of C through the source and back to the positive plate. Note that the direction of discharge current in Fig. 17–7b is opposite that of the charge current in Fig. 17–7a.

For the third quarter-cycle in Fig. 17–7c, the applied voltage v_A increases again but in the negative direction. Now C charges again but in reversed polarity. Here the charging current is in the direction opposite from the charge current in Fig. 17–7a but in the same direction as the discharge current in Fig. 17–7b. Finally,

Figure 17–7 Capacitive charge and discharge currents. (a) Voltage V_A increases positive to charge C. (b) The C discharges as V_A decreases. (c) Voltage V_A increases negative to charge C in opposite polarity. (d) The C discharges as reversed V_A decreases.

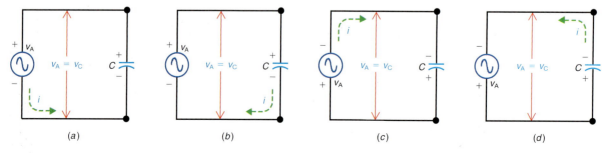

the negative applied voltage decreases during the final quarter-cycle in Fig. 17–7d. As a result, C discharges. This discharge current is opposite to the charge current in Fig. 17–7c but in the same direction as the charge current in Fig. 17–7a.

For the sine wave of applied voltage, therefore, the capacitor provides a cycle of alternating charge and discharge current. Notice that capacitive current flows for either charge or discharge, whenever the voltage changes, for either an increase or a decrease. Also, i and v have the same frequency.

Calculating the Values of i_C

The greater the voltage change, the greater the amount of capacitive current. Furthermore, a larger capacitor can allow more charge current when the applied voltage increases and can produce more discharge current. Because of these factors the amount of capacitive current can be calculated as

$$i_C = C\frac{dv}{dt} \tag{17-6}$$

where i is in amperes, C is in farads, and dv/dt is in volts per second. As an example, suppose that the voltage across a 240-pF capacitor changes by 25 V in 1 μs. The amount of capacitive current then is

$$i_C = C\frac{dv}{dt} = 240 \times 10^{-12} \times \frac{25}{1 \times 10^{-6}}$$
$$= 240 \times 25 \times 10^{-6} = 6000 \times 10^{-6}$$
$$= 6 \times 10^{-3} \text{ A or } 6 \text{ mA}$$

Notice how Formula (17–6) is similar to the capacitor charge formula $Q = CV$. When the voltage changes, this dv/dt factor produces a change in the charge Q. When the charge moves, this dq/dt change is the current i_C. Therefore, dq/dt or i_C is proportional to dv/dt. With the constant factor C, then, i_C becomes equal to $C(dv/dt)$.

By means of Formula (17–6), then, i_C can be calculated to find the instantaneous value of charge or discharge current when the voltage changes across a capacitor.

Example 17–5

Calculate the instantaneous value of charging current i_C produced by a 6-μF C when its potential difference is increased by 50 V in 1 s.

ANSWER

$$i_C = C\frac{dv}{dt} = 6 \times 10^{-6} \times \frac{50}{1}$$
$$= 300 \text{ } \mu\text{A}$$

Example 17–6

Calculate i_C for the same C as in Example 17–5 when its potential difference is decreased by 50 V in 1 s.

ANSWER For the same $C(dv/dt)$, i_C is the same 300 μA. However, this 300 μA is discharge current, which flows in the direction opposite from i_C on charge. If desired, the i_C for discharge current can be considered negative, or –300 μA.

Example 17-7

Calculate i_C produced by a 250-pF capacitor for a change of 50 V in 1 μs.

ANSWER

$$i_C = C \frac{dv}{dt}$$

$$= 250 \times 10^{-12} \times \frac{50}{1 \times 10^{-6}}$$

$$= 12{,}500 \times 10^{-6} \text{ A or } 12{,}500 \ \mu\text{A or } 12.5 \text{ mA}$$

Table 17–2		Values for $i_C = C(dv/dt)$ Curves in Fig. 17–8					
Time		**dt**		dv, V	dv/dt, V/μs	C, pF	$i_c = C(dv/dt)$, mA
θ	μs	θ	μs				
30°	2	30°	2	50	25	240	6
60°	4	30°	2	36.6	18.3	240	4.4
90°	6	30°	2	13.4	6.7	240	1.6
120°	8	30°	2	−13.4	−6.7	240	−1.6
150°	10	30°	2	−36.6	−18.3	240	−4.4
180°	12	30°	2	−50	−25	240	−6
210°	14	30°	2	−50	−25	240	−6
240°	16	30°	2	−36.6	−18.3	240	−4.4
270°	18	30°	2	−13.4	−6.7	240	−1.6
300°	20	30°	2	13.4	6.7	240	1.6
330°	22	30°	2	36.6	18.3	240	4.4
360°	24	30°	2	50	25	240	6

Notice that more i_C is produced in Example 17–7, although C is smaller than in Example 17–6, because dv/dt is a much faster voltage change.

Waveshapes of v_C and i_C

More details of capacitive circuits can be analyzed by plotting the values calculated in Table 17–2. Figure 17–8 shows the waveshapes representing these values. Figure 17–8a shows a sine wave of voltage v_C across a 240-pF capacitance C. Since the capacitive current i_C depends on the rate of change of voltage, rather than on the absolute value of v, the curve in Fig. 17–8b shows how much the voltage changes. In this curve, the dv/dt values are plotted for every 30° of the cycle.

Figure 17–8c shows the actual capacitive current i_C. This i_C curve is similar to the dv/dt curve because i_C equals the constant C multiplied by dv/dt.

90° Phase Angle

The i_C curve at the bottom of Fig. 17–8 has its zero values when the v_C curve at the top is at maximum. This comparison shows that the curves are 90° out

Figure 17–8 Waveshapes of capacitive circuits. (a) Waveshape of sine-wave voltage at top. (b) Changes in voltage below causing (c) current i_C charge and discharge waveshape. Values plotted are those given in Table 17–2.

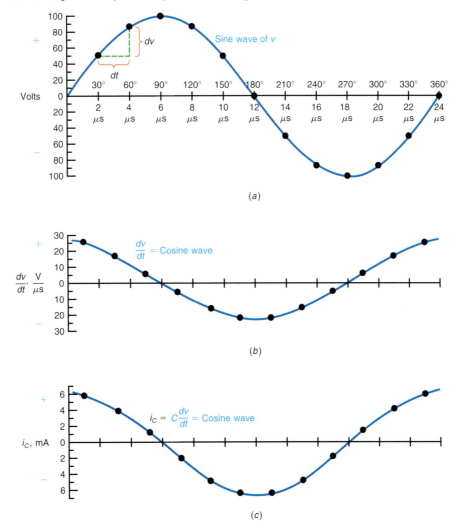

of phase because i_C is a cosine wave of current for the sine wave of voltage v_C. The 90° phase difference results from the fact that i_C depends on the dv/dt rate of change, rather than on v itself. More details of this 90° phase angle for capacitance are explained in the next chapter.

For each of the curves, the period T is 24 μs. Therefore, the frequency is $1/T$ or $\frac{1}{24}$, which equals 41.67 kHz. Each curve has the same frequency, although there is a 90° phase difference between i and v.

Ohms of X_C

The ratio of v_C/i_C specifies the capacitive reactance in ohms. For this comparison, we use the actual value of v_C, which has a peak of 100 V. The rate-of-change factor is included in i_C. Although the peak of i_C at 6 mA is 90° ahead of the peak of v_C at 100 V, we can compare these two peak values. Then v_C/i_C is 100/0.006, which equals 16,667 Ω,

This X_C is only an approximate value because i_C cannot be determined exactly for the large dt changes every 30°. If we used smaller intervals of time, the peak i_C would be 6.28 mA with X_C then 15,900 Ω, the same as $1/(2\pi fC)$ with a 240-pF C and a frequency of 41.67 kHz.

■ *17–6 Knowledge Check*

Answer at end of chapter.

For a capacitor, what is the phase relation between v_C and i_C?

■ *17–6 Self-Review*

Answers at end of chapter.

Refer to the curves in Fig. 17–8.
a. **At what angle does v have its maximum positive value?**
b. **At what angle does dv/dt have its maximum positive value?**
c. **What is the phase angle difference between v_C and i_C?**

Summary

■ Capacitive reactance, indicated by X_C, is the opposition of a capacitance to the flow of sine-wave alternating current.

■ Reactance X_C is measured in ohms because it limits the current to the value V/X_C. With V in volts and X_C in ohms, I is in amperes.

■ $X_C = 1/(2\pi fC)$. With f in hertz and C in farads, X_C is in ohms.

■ For the same value of capacitance, X_C decreases when the frequency increases.

■ For the same frequency, X_C decreases when the capacitance increases.

■ With X_C and f known, the capacitance $C = 1/(2\pi fX_C)$.

■ With X_C and C known, the frequency $f = 1/(2\pi CX_C)$.

■ The total X_C of capacitive reactances in series equals the sum of the individual values, as for series resistances. The series reactances have the same current. The voltage across each reactance is IX_C.

■ The combined reactance of parallel capacitive reactances, is calculated by the reciprocal formula, as for parallel resistances. Each branch current is V/X_C. The total line current is the sum of the individual branch currents.

■ Table 17–3 summarizes the differences between C and X_C.

Table 17-3	Comparison of Capacitance and Capacitive Reactance	
Capacitance	**Capacitive Reactance**	
Symbol is C	Symbol is X_C	
Measured in farad units	Measured in ohm units	
Depends on construction of capacitor	Depends on frequency of sine-wave voltage	
$C = i_C/(dv/dt)$ or Q/V	$X_C = v_C/i_C$ or $1/(2\pi fC)$	

Important Terms

- Capacitive Reactance, X_C - a measure of a capacitor's opposition to the flow of alternating current. X_C is measured in ohms. $X_C = \dfrac{1}{2\pi fC}$ or $X_C = \dfrac{V_C}{I_C}$. X_C applies only to sine-wave ac circuits.

- Charging Current - the current that flows to and from the plates of a capacitor as the charge stored by the dielectric increases.

- Discharge Current - the current that flows to and from the plates of a capacitor as the charge stored by the dielectric decreases. The discharge current of a capacitor is opposite in direction to the charging current.

- Inversely Proportional - the same as a reciprocal relation; as the value in the denominator increases the resultant quotient decreases. In the formula $X_C = \dfrac{1}{2\pi fC}$, X_C is inversely proportional to both f and C. This means that as f and C increase, X_C decreases.

- Phase Angle - the angular difference or displacement between two waveforms. For a capacitor, the charge and discharge current, i_C, reaches its maximum value 90° ahead of the capacitor voltage, v_C. As a result, the charge and discharge current, i_C, is said to lead the capacitor voltage, v_C, by a phase angle of 90°.

Related Formulas

$X_C = \dfrac{1}{2\pi fC}$

$C = \dfrac{1}{2\pi fX_C}$

$f = \dfrac{1}{2\pi CX_C}$

$X_{C_T} = X_{C_1} + X_{C_2} + \cdots + \text{etc.}$

(Series capacitive reactances)

$X_{C_{EQ}} = \dfrac{1}{\dfrac{1}{X_{C_1}} + \dfrac{1}{X_{C_2}} + \dfrac{1}{X_{C_3}} + \cdots + \text{etc.}}$ (Parallel capacitive reactances)

$i_C = C\dfrac{dv}{dt}$

$X_C = \dfrac{V_C}{I_C}$

Self-Test

Answers at back of book.

1. The capacitive reactance, X_C, of a capacitor is
   a. inversely proportional to frequency.
   b. unaffected by frequency.
   c. directly proportional to frequency.
   d. directly proportional to capacitance.

2. The charge and discharge current of a capacitor flows
   a. through the dielectric.
   b. only when a dc voltage is applied.
   c. to and from the plates.
   d. both a and b.

3. For direct current (dc), a capacitor acts like a(n)
   a. closed switch.
   b. open.
   c. short.
   d. small resistance.

4. At the same frequency, a larger capacitance provides
   a. more charge and discharge current.
   b. less charge and discharge current.
   c. less capacitive reactance, X_C.
   d. both a and c.

5. How much is the capacitance, C, of a capacitor that draws 4.8 mA of current from a 12-Vac generator? The frequency of the ac generator is 636.6 Hz.
   a. 0.01 μF.
   b. 0.1 μF.
   c. 0.001 μF.
   d. 100 pF.

6. At what frequency does a 0.015-μF capacitor have an X_C value of 2 kΩ?
   a. 5.3 MHz.
   b. 5.3 Hz.
   c. 5.3 kHz.
   d. 106 kHz.

7. What is the capacitive reactance, X_C, of a 330-pF capacitor at a frequency of 1 MHz?
   a. 482 Ω.
   b. 48.2 Ω.
   c. 1 kΩ.
   d. 482 MΩ.

8. What is the instantaneous value of charging current, i_C, of a 10-μF capacitor if the voltage across the capacitor plates changes at the rate of 250 V per second?
   a. 250 μA.
   b. 2.5 A.
   c. 2.5 μA.
   d. 2.5 mA.

9. For a capacitor, the charge and discharge current, i_C,
   a. lags the capacitor voltage, v_C, by a phase angle of 90°.
   b. leads the capacitor voltage, v_C, by a phase angle of 90°.
   c. is in phase with the capacitor voltage, v_C.
   d. none of the above.

10. Two 1-kΩ X_C values in series have a total capacitive reactance of
   a. 1.414 kΩ.
   b. 500 Ω.
   c. 2 kΩ.
   d. 707 Ω.

11. Two 5-kΩ X_C values in parallel have an equivalent capacitive reactance of
   a. 7.07 kΩ.
   b. 2.5 kΩ.
   c. 10 kΩ.
   d. 3.53 kΩ.

12. For any capacitor,
   a. the stored charge increases with more capacitor voltage.
   b. the charge and discharge currents are in opposite directions.
   c. i_C leads v_C by 90°.
   d. all of the above.

13. The unit of capacitive reactance, X_C, is the
   a. ohm.
   b. farad.
   c. hertz.
   d. radian.

14. The main difference between resistance, R, and capacitive reactance, X_C, is that
   a. X_C is the same for both dc and ac, whereas R depends on frequency.
   b. R is the same for both dc and ac, whereas X_C depends on frequency.
   c. R is measured in ohms and X_C is measured in farads.
   d. none of the above.

15. A very common use for a capacitor is to
   a. block any dc voltage but provide very little opposition to an ac voltage.
   b. block both dc and ac voltages.
   c. pass both dc and ac voltages.
   d. none of the above.

Questions

1. Why is capacitive reactance measured in ohms? State two differences between capacitance and capacitive reactance.

2. Explain briefly why the bulb lights in Fig. 17–1a but not in Fig. 17–1c.

3. Explain briefly what is meant by two factors being inversely proportional. How does this apply to X_C and C? X_C and f?

4. In comparing X_C and R, give two differences and one similarity.

5. Why are the waves in Fig. 17–8a and b considered to be 90° out of phase, but the waves in Fig. 17–8b and c have the same phase?

6. Referring to Fig. 17–3, how does this graph show an inverse relation between X_C and C?

542 *Chapter 17*

7. Referring to Fig. 17–4, how does this graph show an inverse relation between X_C and f?

8. Referring to Fig. 17–8, draw three similar curves but for a sine wave of voltage with a period $T = 12\ \mu s$ for the full cycle. Use the same C of 240 pF. Compare the value of X_C obtained as $1/(2\pi fC)$ and v_C/i_C.

9. (a) What is the relationship between charge q and current i? (b) How is this comparison similar to the relation between the two formulas $Q = CV$ and $i_c = C(dv/dt)$?

Problems

SECTION 17–1 ALTERNATING CURRENT IN A CAPACITIVE CIRCUIT

17–1 With the switch, S_1, closed in Fig. 17–9, how much is
  a. the current, I, in the circuit?
  b. the dc voltage across the 12–V lamp?
  c. the dc voltage across the capacitor?

Figure 17–9

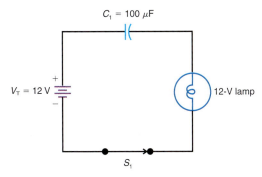

17–2 In Fig. 17–9 explain why the bulb will light for just an instant when S_1 is initially closed.

17–3 In Fig. 17–10, the capacitor and the lightbulb draw 400 mA from the 120-Vac source. How much current flows
  a. to and from the terminals of the 120 Vac source?
  b. through the lightbulb?
  c. to and from the plates of the capacitor?
  d. through the connecting wires?
  e. through the dielectric of the capacitor?

Figure 17–10

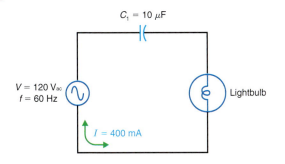

17–4 In Fig. 17–11, calculate the capacitive reactance, X_C, for the following values of Vac and I?
  a. Vac = 10 V and I = 20 mA
  b. Vac = 24 V and I = 8 mA
  c. Vac = 15 V and I = 300 μA
  d. Vac = 100 V and I = 50 μA

Figure 17–11

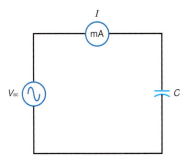

17–5 In Fig. 17–11, list three factors that can affect the amount of charge and discharge current flowing in the circuit.

SECTION 17–2 THE AMOUNT OF X_C EQUALS $\dfrac{1}{2\pi fC}$

17–6 Calculate the capacitive reactance, X_C, of a 0.1-μF capacitor at the following frequencies:
  a. f = 10 Hz
  b. f = 50 Hz
  c. f = 200 Hz
  d. f = 10 kHz

17–7 Calculate the capacitive reactance, X_C, of a 10-μF capacitor at the following frequencies:
  a. f = 60 Hz
  b. f = 120 Hz
  c. f = 500 Hz
  d. f = 1 kHz

17–8 What value of capacitance will provide an X_C of 1 kΩ at the following frequencies?
  a. f = 318.3 Hz
  b. f = 1.591 kHz
  c. f = 3.183 kHz
  d. f = 6.366 kHz

17–9 At what frequency will a 0.047-μF capacitor provide an X_C value of

    a. 100 kΩ?

    b. 5 kΩ?

    c. 1.5 kΩ?

    d. 50 Ω?

17–10 How much is the capacitance of a capacitor that draws 2 mA of current from a 10-V ac generator whose frequency is 3.183 kHz?

17–11 At what frequency will a 820-pF capacitance have an X_C value of 250 Ω?

17–12 A 0.01-μF capacitor draws 50 mA of current when connected directly across a 50-Vac source. What is the value of current drawn by the capacitor when

    a. the frequency is doubled?

    b. the frequency is decreased by one-half?

    c. the capacitance is doubled to 0.02 μF?

    d. the capacitance is reduced by one-half to 0.005 μF?

17–13 A capacitor has an X_C value of 10 kΩ at a given frequency. What is the new value of X_C when the frequency is

    a. cut in half?

    b. doubled?

    c. quadrupled?

    d. increased by a factor of 10?

17–14 Calculate the capacitive reactance, X_C, for the following capacitance and frequency values:

    a. $C = 0.47\ \mu$F, $f = 1$ kHz

    b. $C = 100\ \mu$F, $f = 120$ Hz

    c. $C = 250$ pF, $f = 1$ MHz

    d. $C = 0.0022\ \mu$F, $f = 50$ kHz

17–15 Determine the capacitance value for the following frequency and X_C values:

    a. $X_C = 1$ kΩ, $f = 3.183$ kHz

    b. $X_C = 200\ \Omega$, $f = 63.66$ kHz

    c. $X_C = 25$ kΩ, $f = 1.592$ kHz

    d. $X_C = 1$ MΩ, $f = 100$ Hz

17–16 Determine the frequency for the following capacitance and X_C values:

    a. $C = 0.05\ \mu$F, $X_C = 4$ kΩ

    b. $C = 0.1\ \mu$F, $X_C = 1.591$ kΩ

    c. $C = 0.0082\ \mu$F, $X_C = 6.366$ kΩ

    d. $C = 50\ \mu$F, $X_C = 100\ \Omega$

SECTION 17–3 SERIES OR PARALLEL CAPACITIVE REACTANCES

17–17 How much is the total capacitive reactance, X_{C_T}, for the following series capacitive reactances:

    a. $X_{C_1} = 1$ kΩ, $X_{C_2} = 1.5$ kΩ, $X_{C_3} = 2.5$ kΩ

    b. $X_{C_1} = 500\ \Omega$, $X_{C_2} = 1$ kΩ, $X_{C_3} = 1.5$ kΩ

    c. $X_{C_1} = 20$ kΩ, $X_{C_2} = 10$ kΩ, $X_{C_3} = 120$ kΩ

    d. $X_{C_1} = 340\ \Omega$, $X_{C_2} = 570\ \Omega$, $X_{C_3} = 2.09$ kΩ

17–18 What is the equivalent capacitive reactance, $X_{C_{EQ}}$, for the following parallel capacitive reactances:

    a. $X_{C_1} = 100\ \Omega$ and $X_{C_2} = 400\ \Omega$

    b. $X_{C_1} = 1.2$ kΩ and $X_{C_2} = 1.8$ kΩ

    c. $X_{C_1} = 15\ \Omega$, $X_{C_2} = 6\ \Omega$, $X_{C_3} = 10\ \Omega$

    d. $X_{C_1} = 2.5$ kΩ, $X_{C_2} = 10$ kΩ, $X_{C_3} = 2$ kΩ, $X_{C_4} = 1$ kΩ

SECTION 17–4 OHM'S LAW APPLIED TO X_C

17–19 In Fig. 17–12, calculate the current, I.

Figure 17–12

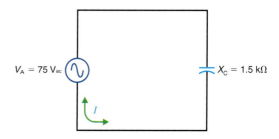

$V_A = 75$ V$_{ac}$ $X_C = 1.5$ kΩ

I

17–20 In Fig. 17–12, what happens to the current, I, when the frequency of the applied voltage

    a. decreases.

    b. increases.

17–21 In Fig. 17–13, solve for

    a. X_{C_T}.

    b. I.

    c. V_{C_1}, V_{C_2}, and V_{C_3}.

Figure 17–13

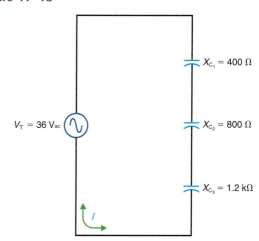

$X_{C_1} = 400\ \Omega$

$X_{C_2} = 800\ \Omega$

$V_T = 36$ V$_{ac}$

$X_{C_3} = 1.2$ kΩ

I

17–22 In Fig. 17–14, solve for

    a. X_{C_1}, X_{C_2}, and X_{C_3}.

    b. X_{C_T}.

    c. I.

    d. V_{C_1}, V_{C_2}, and V_{C_3}.

    e. C_{EQ}.

Figure 17–14

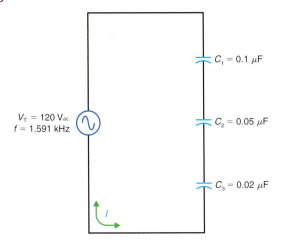

$C_1 = 0.1\ \mu F$

$V_T = 120\ V_{ac}$
$f = 1.591\ kHz$

$C_2 = 0.05\ \mu F$

$C_3 = 0.02\ \mu F$

I

17–23 In Fig. 17–13, solve for C_1, C_2, C_3, and C_{EQ} if the applied voltage has a frequency of 318.3 Hz.

17–24 In Fig. 17–15, solve for
  a. I_{C_1}, I_{C_2}, and I_{C_3}.
  b. I_T.
  c. $X_{C_{EQ}}$.

Figure 17–15

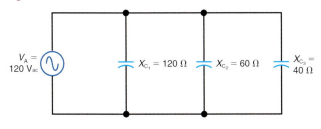

$V_A =$
$120\ V_{ac}$

$X_{C_1} = 120\ \Omega$

$X_{C_2} = 60\ \Omega$

$X_{C_3} =$
$40\ \Omega$

17–25 In Fig. 17–16, solve for
  a. X_{C_1}, X_{C_2}, and X_{C_3}.
  b. I_{C_1}, I_{C_2}, and I_{C_3}.
  c. I_T.
  d. $X_{C_{EQ}}$.
  e. C_T.

Figure 17–16

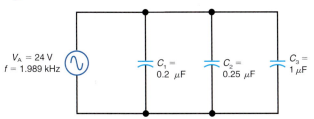

$V_A = 24\ V$
$f = 1.989\ kHz$

$C_1 =$
$0.2\ \mu F$

$C_2 =$
$0.25\ \mu F$

$C_3 =$
$1\ \mu F$

17–26 In Fig. 17–15, solve for C_1, C_2, C_3, and C_T if the frequency of the applied voltage is 6.366 kHz.

SECTION 17–5 APPLICATIONS OF CAPACITIVE REACTANCE

17–27 Calculate the value of capacitance, C, required to produce an X_C value of 500 Ω at the following frequencies:
  a. $f = 100$ Hz
  b. $f = 2$ kHz
  c. $f = 50$ kHz
  d. $f = 10$ MHz

SECTION 17–6 SINE–WAVE CHARGE AND DISCHARGE CURRENT

17–28 Calculate the instantaneous charging current, i_C, for a 0.33-μF capacitor if the voltage across the capacitor plates changes at the rate of 10 V/1 ms.

17–29 Calculate the instantaneous charging current, i_C, for a 0.01-μF capacitor if the voltage across the capacitor plates changes at the rate of
  a. 100 V/s
  b. 100 V/ms
  c. 50 V/μs

17–30 What is the instantaneous discharge current, i_C, for a 100-μF capacitor if the voltage across the capacitor plates decreases at the rate of
  a. 10 V/s
  b. 1 V/ms
  c. 50 V/ms

17–31 For a capacitor, what is the phase relationship between the charge and discharge current, i_C, and the capacitor voltage, v_C? Explain your answer.

17–32 A capacitor has a discharge current, i_C, of 15 mA when the voltage across its plates decreases at the rate of 150 V/μs. Calculate C.

17–33 What rate of voltage change, $\dfrac{dv}{dt}$, will produce a charging current of 25 mA in a 0.01-μF capacitor? Express your answer in volts per second.

Critical Thinking

17–34 Explain an experimental procedure for determining the value of an unmarked capacitor. (Assume that a capacitance meter is not available.)

17–35 In Fig. 17–17, calculate X_{C_1}, X_{C_1}, X_{C_2}, C_1, C_3, V_{C_1}, V_{C_2}, V_{C_3}, I_2, and I_3.

Figure 17–17 Circuit for Critical Thinking Prob. 17–35.

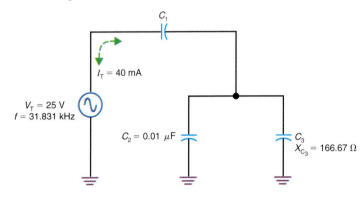

Answers to Knowledge Check Problems

17–1 $X_C = 2\ k\Omega$

17–2 X_C is infinite for dc

17–3 a. 450 Ω
　　　b. 50 Ω

17–4 $I = 200$ mA, $V_1 = 20$ V, and $V_2 = 80$ V

17–5 $C = 0.053\ \mu F$

17–6 i_C and v_C are 90° out of phase with each other.

Answers to Self-Reviews

17–1 a. 0.1 μF
　　　b. 0.5 μF

17–2 a. 200 Ω
　　　b. 800 Ω
　　　c. larger

17–3 a. 500 Ω
　　　b. 120 Ω

17–4 a. 300 Ω
　　　b. 66.7 Ω

17–5 a. 50 Ω
　　　b. 1000 Ω

17–6 a. 90°
　　　b. 0 or 360°
　　　c. 90°

18 Capacitive Circuits

This chapter analyzes circuits that combine capacitive reactance X_C and resistance R. The main questions are, how do we combine the ohms of opposition, how much current flows, and what is the phase angle? Although both X_C and R are measured in ohms, they have different characteristics. Specifically, X_C decreases with more C and higher frequencies for sine-wave ac voltage applied, whereas R is the same for dc and ac circuits. Furthermore, the phase angle for the voltage across X_C is at $-90°$ measured in the clockwise direction with i_C as the reference at $0°$.

In addition, the practical application of a coupling capacitor shows how a low value of X_C can be used to pass the desired ac signal variations, while blocking the steady dc level of a fluctuating dc voltage. In a coupling circuit with C and R in series, the ac component is across R for the output voltage but the dc component across C is not present across the output terminals.

Finally, the general case of capacitive charge and discharge current produced when the applied voltage changes is shown with nonsinusoidal voltage variations. In this case, we compare the waveshapes of v_C and i_C. Remember that the $-90°$ angle for an IX_C voltage applies only to sine waves.

Objectives

After studying this chapter you should be able to

- *Explain* why the current leads the voltage by 90° for a capacitor.

- *Define* the term *impedance.*

- *Calculate* the total impedance and phase angle of a series *RC* circuit.

- *Describe* the operation and application of an *RC* phase-shifter circuit.

- *Calculate* the total current, equivalent impedance, and phase angle of a parallel *RC* circuit.

- *Explain* how a capacitor can couple some ac frequencies but not others.

- *Calculate* the individual capacitor voltage drops for capacitors in series.

- *Calculate* the capacitive current that flows with nonsinusoidal waveforms.

Outline

Important Terms

arctangent (arctan)
capacitive voltage divider
coupling capacitor, C_C

impedance, Z
phase angle, θ
phasor triangle

RC phase-shifter
tangent (tan)

18–1 Sine Wave v_C Lags i_C by $90°$

For a sine wave of applied voltage, a capacitor provides a cycle of alternating charge and discharge current, as shown in Fig. 18–1a. In Fig. 18–1b, the waveshape of this charge and discharge current i_C is compared with the voltage v_C.

Examining the v_C and i_C Waveforms

In Fig. 18–1b, note that the instantaneous value of i_C is zero when v_C is at its maximum value. At either its positive or its negative peak, v_C is not changing. For one instant at both peaks, therefore, the voltage must have a static value before changing its direction. Then v is not changing and C is not charging or discharging. The result is zero current at this time.

Also note that i_C is maximum when v_C is zero. When v_C crosses the zero axis, i_C has its maximum value because then the voltage is changing most rapidly.

Therefore, i_C and v_C are $90°$ out of phase, since the maximum value of one corresponds to the zero value of the other; i_C leads v_C because i_C has its maximum value a quarter-cycle before the time that v_C reaches its peak. The phasors in Fig. 18–1c show i_C leading v_C by the counterclockwise angle of $90°$. Here v_C is the horizontal phasor for the reference angle of $0°$. In Fig. 18–1d, however, the current i_C is the horizontal phasor for reference. Since i_C must be $90°$ leading, v_C is shown lagging by the clockwise angle of $−90°$. In series circuits, the current i_C is the reference, and then the voltage v_C can be considered to lag i_C by $90°$.

Why i_C Leads v_C by $90°$

The $90°$ phase angle results because i_C depends on the rate of change of v_C. In other words, i_C has the phase of dv/dt, not the phase of v. As shown previously in Fig. 17–8 for a sine wave of v_C, the capacitive charge and discharge current is a cosine wave. This $90°$ phase between v_C and i_C is true in any sine-wave ac circuit, whether C is in series or parallel and whether C is alone or combined with other components. We can always say that for any X_C, its current and voltage are $90°$ out of phase.

Capacitive Current Is the Same in a Series Circuit

The leading phase angle of capacitive current is only with respect to the voltage across the capacitor, which does not change the fact that the current is the same

Figure 18–1 Capacitive current i_C leads v_C by $90°$. (a) Circuit with sine wave V_A across C. (b) Waveshapes of i_C $90°$ ahead of v_C. (c) Phasor diagram of i_C leading the horizontal reference v_C by a counterclockwise angle of $90°$. (d) Phasor diagram with i_C as the reference phasor to show v_C lagging i_C by an angle of $−90°$.

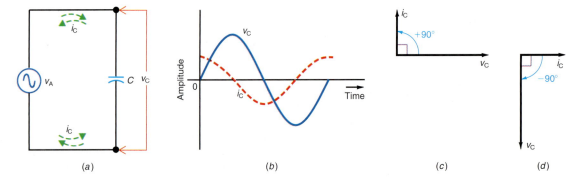

in all parts of a series circuit. In Fig. 18–1a, for instance, the current in the generator, the connecting wires, and both plates of the capacitor must be the same because they are all in the same path.

Capacitive Voltage Is the Same Across Parallel Branches

In Fig. 18–1a, the voltage is the same across the generator and C because they are in parallel. There cannot be any lag or lead in time between these two parallel voltages. At any instant, whatever the voltage value is across the generator at that time, the voltage across C is the same. With respect to the series current, however, both v_A and v_C are 90° out of phase with i_C.

The Frequency Is the Same for v_C and i_C

Although v_C lags i_C by 90°, both waves have the same frequency. For example, if the frequency of the sine wave v_C in Fig. 18–1b is 100 Hz, this is also the frequency of i_C.

■ *18–1 Knowledge Check*

> *Answer at end of chapter.*
>
> **In Fig. 18–1b, what is the value of i_C when v_C is at its positive peak?**

■ *18–1 Self-Review*

> *Answers at end of chapter.*
>
> **Refer to Fig. 18–1.**
> a. **What is the phase angle between v_A and v_C?**
> b. **What is the phase angle between v_C and i_C?**
> c. **Does v_C lead or lag i_C?**

18–2 X_C and R in Series

When a capacitor and a resistor are connected in series, as in Fig. 18–2a, the current I is limited by both X_C and R. The current I is the same in both X_C and R since they are in series. However, each component has its own series voltage drop, equal to IR for the resistance and IX_C for the capacitive reactance.

Note the following points about a circuit that combines both X_C and R in series, like that in Fig. 18–2a.

1. The current is labeled I, rather than I_C, because I flows through all series components.
2. The voltage across X_C, labeled V_C, can be considered an IX_C voltage drop, just as we use V_R for an IR voltage drop.
3. The current I through X_C must lead V_C by 90° because this is the phase angle between the voltage and current for a capacitor.
4. The current I through R and its IR voltage drop are in phase. There is no reactance to sine-wave alternating current in any resistance. Therefore, I and IR have a phase angle of 0°.

It is important to note that the values of I and V may be in rms, peak, peak-to-peak, or instantaneous, as long as the same measure is applied to the entire circuit. Peak values will be used here for convenience in comparing waveforms.

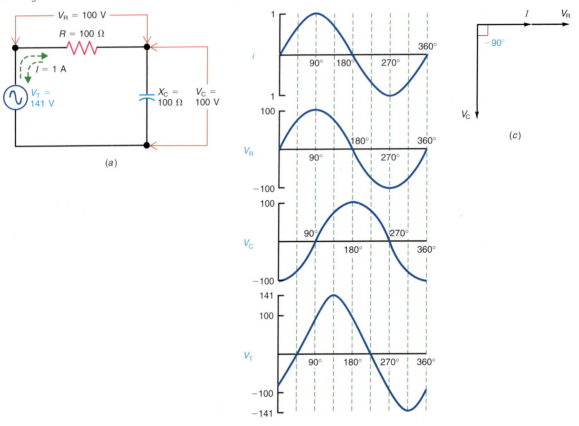

Phase Comparisons

Note the following points about a circuit containing series resistance and reactance:

1. The voltage V_C is 90° out of phase with I.
2. However, V_R and I are in phase.
3. If I is used as the reference, V_C is 90° out of phase with V_R.

Specifically, V_C lags V_R by 90° just as the voltage V_C lags the current I by 90°. The phase relationships between I, V_R, V_C, and V_T are shown by the waveforms in Fig. 18–2*b*. Figure 18–2*c* shows the phasors representing I, V_R, and V_C.

Combining V_R and V_C

As shown in Fig. 18–2*b*, when the voltage wave V_R is combined with the voltage wave V_C, the result is the voltage wave of the applied voltage V_T. The voltage drops, V_R and V_C, must add to equal the applied voltage V_T. The 100-V peak values for V_R and V_C total 141 V, however, instead of 200 V, because of the 90° phase difference.

Consider some instantaneous values in Fig. 18–2*b*, to see why the 100-V peak V_R and 100-V peak V_C cannot be added arithmetically. When V_R is at its maximum of 100 V, for instance, V_C is at zero. The total voltage V_T at

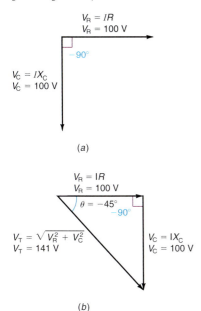

(a)

(b)

this instant, then, is 100 V. Similarly, when V_C is at its maximum of 100 V, V_R is at zero and the total voltage V_T is again 100 V.

Actually, V_T reaches its maximum of 141 V when V_C and V_R are each at 70.7 V. When series voltage drops that are out of phase are combined, therefore, they cannot be added without taking the phase difference into account.

Phasor Voltage Triangle

Instead of combining waveforms that are out of phase, as in Fig. 18–2*b*, we can add them more quickly by using their equivalent phasors, as shown in Fig. 18–3. The phasors in Fig. 18–3*a* show the 90° phase angle without any addition. The method in Fig. 18–3*b* is to add the tail of one phasor to the arrowhead of the other, using the angle required to show their relative phase. Note that voltages V_R and V_C are at right angles to each other because they are 90° out of phase. Note also that the phasor for V_C is downward at an angle of −90° from the phasor for V_R. Here V_R is used as the reference phasor because it has the same phase as the series current I, which is the same everywhere in the circuit. The phasor V_T, extending from the tail of the V_R phasor to the arrowhead of the V_C phasor, represents the applied voltage V_T, which is the phasor sum of V_R and V_C. Since V_R and V_C form a right angle, the resultant phasor V_T is the hypotenuse of a right triangle. The hypotenuse is the side opposite the 90° angle.

From the geometry of a right triangle, the Pythagorean theorem states that the hypotenuse is equal to the square root of the sum of the squares of the sides. For the voltage triangle in Fig. 18–3*b*, therefore, the resultant is

$$V_T = \sqrt{V_R^2 + V_C^2} \qquad \text{(18–1)}$$

where V_T is the phasor sum of the two voltages V_R and V_C 90° out of phase.

This formula is for V_R and V_C when they are in series, since they are 90° out of phase. All voltages must be expressed in the same units. When V_T is an rms value, V_R and V_C must also be rms values. For the voltage triangle in Fig. 18–3*b*,

$$V_T = \sqrt{100^2 + 100^2} = \sqrt{10,000 + 10,000}$$
$$= \sqrt{20,000}$$
$$= 141 \text{ V}$$

■ *18–2 Knowledge Check*

Answer at end of chapter.

In a series *RC* circuit, $V_R = 48$ V and $V_C = 20$ V. How much is V_T?

■ *18–2 Self-Review*

Answers at end of chapter.

a. In a series circuit with X_C and *R*, what is the phase angle between *I* and V_R?
b. What is the phase angle between V_R and V_C?
c. In a series circuit with X_C and *R*, does the series current *I* lead or lag the applied voltage V_T?

CALCULATOR

To do a problem like this on a calculator, remember that the square root sign is a sign of grouping. All terms within the group must be added before you take the square root. Also, each term must be squared individually before adding for the sum. Specifically, for this problem:

■ Punch in 100 and push the $\boxed{x^2}$ key for 10,000 as the square.

■ Next, punch the $\boxed{+}$ key and then punch in 100 and $\boxed{x^2}$. Press the $\boxed{=}$ key. The display should read 20,000.

■ Press $\boxed{\sqrt{}}$ to read the answer of 141.421.

In some calculators, either the $\boxed{x^2}$ or the $\boxed{\sqrt{}}$ key must be preceded by the second function key $\boxed{2^{nd}F}$.

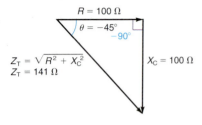

$R = 100 \ \Omega$

$\theta = -45°$

$-90°$

$Z_T = \sqrt{R^2 + X_C^2}$
$Z_T = 141 \ \Omega$

$X_C = 100 \ \Omega$

18–3 Impedance Z Triangle

A triangle of R and X_C in series corresponds to the voltage triangle, as shown in Fig. 18–4. It is similar to the voltage triangle in Fig. 18–3b, but the common factor I cancels because the series current I is the same in X_C and R. The resultant of the phasor addition of X_C and R is their total opposition in ohms, called *impedance*, with the symbol Z_T. The Z takes into account the 90° phase relation between R and X_C.

For the impedance triangle of a series circuit with capacitive reactance X_C and resistance R,

$$Z_T = \sqrt{R^2 + X_C^2} \tag{18–2}$$

where R, X_C, and Z_T are all in ohms. For the phasor triangle in Fig. 18–4,

$$Z_T = \sqrt{100^2 + 100^2} = \sqrt{10{,}000 + 10{,}000}$$
$$= \sqrt{20{,}000}$$
$$Z_T = 141 \ \Omega$$

This is the total impedance Z_T in Fig. 18–2a.

Note that the applied voltage V_T of 141 V divided by the total impedance of 141 Ω results in 1 A of current in the series circuit. The IR voltage V_R is 1 A × 100 Ω or 100 V; the IX_C voltage is also 1 A × 100 Ω or 100 V. The series IR and IX_C voltage drops of 100 V each are added using phasors to equal the applied voltage V_T of 141 V. Finally, the applied voltage equals IZ_T or 1 A × 141 Ω, which is 141 V.

Summarizing the similar phasor triangles for voltage and ohms in a series RC circuit,

1. The phasor for R, IR, or V_R is used as a reference at 0°.
2. The phasor for X_C, IX_C, or V_C is at −90°.
3. The phasor for Z_T, IZ_T, or V_T has the phase angle θ of the complete circuit.

Phase Angle with Series X_C and R

The angle between the applied voltage V_T and the series current I is the phase angle of the circuit. Its symbol is θ (theta). In Fig. 18–3b, the phase angle between V_T and IR is −45°. Since IR and I have the same phase, the angle is also −45° between V_T and I.

In the corresponding impedance triangle in Fig. 18–4, the angle between Z_T and R is also equal to the phase angle. Therefore, the phase angle can be calculated from the impedance triangle of a series RC circuit by the formula

$$\tan \theta_Z = -\frac{X_C}{R} \tag{18–3}$$

The tangent (tan) is a trigonometric function of an angle, equal to the ratio of the opposite side to the adjacent side of a triangle. In this impedance triangle, X_C is the opposite side and R is the adjacent side of the angle. We use the subscript Z for θ to show that θ_Z is found from the impedance triangle for a series circuit. To calculate this phase angle,

$$\tan \theta_Z = -\frac{X_C}{R} = -\frac{100}{100} = -1$$

The angle that has the tangent value of −1 is −45° in this example. The numerical values of the trigonometric functions can be found from a table or by using a scientific calculator. Note that the phase angle of −45° is halfway between 0° and −90° because R and X_C are equal.

CALCULATOR

To do the trigonometry in Example 18-1 with a calculator, keep in mind the following points:

- The ratio of $-X_C/R$ specifies the angle's tangent function as a numerical value, but this is not the angle θ in degrees. Finding X_C/R is a division problem.

- The angle θ itself is an inverse function of $\tan \theta$ that is indicated as arctan θ or $\tan^{-1}\theta$. A scientific calculator can give the trigonometric functions directly from the value of an angle, or inversely show the angle from its trigonometric functions.

- As a check on your values, note that $\tan \theta = -1$, $\tan^{-1}$ (arctan θ) is $-45°$. Tangent values less than -1 must be for angles smaller than $-45°$; angles more than $-45°$ must have tangent values higher than -1.

For the values in Example 18-1 specifically, punch in -40 for X_C, press the $\oplus$ key, punch in 30 for R, and press the $\ominus$ key for the ratio of -1.33 on the display. This value is $\tan \theta$. Although it is on the display, push the $\boxed{\text{TAN}^{-1}}$ key, and the answer of $-53.1°$ will appear for the angle. Use of the $\boxed{\text{TAN}^{-1}}$ key is usually preceded by pressing the second function key, $\boxed{2^{\text{nd}}\text{F}}$.

Example 18-1

If a 30-Ω R and a 40-Ω X_C are in series with 100 V applied, find the following: Z_T, I, V_R, V_C, and θ_Z. What is the phase angle between V_C and V_R with respect to I? Prove that the sum of the series voltage drops equals the applied voltage V_T.

ANSWER

$$Z_T = \sqrt{R^2 + X_C^2} = \sqrt{30^2 + 40^2}$$
$$= \sqrt{900 + 1600}$$
$$= \sqrt{2500}$$
$$= 50 \ \Omega$$
$$I = \frac{V_T}{Z_T} = \frac{100 \text{ V}}{50 \ \Omega} = 2 \text{ A}$$
$$V_R = IR = 2\text{A} \times 30 \ \Omega = 60 \text{ V}$$
$$V_C = IX_C = 2\text{A} \times 40 \ \Omega = 80 \text{ V}$$
$$\tan \theta_Z = -\frac{X_C}{R} = -\frac{40}{30} = -1.333$$
$$\theta_Z = -53.1°$$

Therefore, V_T lags I by 53.1°. Furthermore, I and V_R are in phase, and V_C lags I by 90°. Finally,

$$V_T = \sqrt{V_R^2 + V_C^2} = \sqrt{60^2 + 80^2} = \sqrt{3600 + 6400}$$
$$= \sqrt{10,000}$$
$$= 100 \text{ V}$$

Note that the phasor sum of the voltage drops equals the applied voltage V_T.

Series Combinations of X_C and R

In series, the higher the X_C compared with R, the more capacitive the circuit. There is more voltage drop across the capacitive reactance X_C, and the phase angle increases toward $-90°$. The series X_C always makes the series current I lead the applied voltage V_T. With all X_C and no R, the entire applied voltage V_T is across X_C and θ equals $-90°$.

Several combinations of X_C and R in series are listed in Table 18-1 with their resultant impedance values and phase angle. Note that a ratio of 10:1, or more, for X_C/R means that the circuit is practically all capacitive. The phase

GOOD TO KNOW

For a series RC circuit, when $X_C \geq 10 \ R$, $Z_T \cong X_C$. When $R \geq 10 \ X_C$, $Z_T \cong R$.

Table 18-1	Series R and X_C Combinations		
R, Ω	X_c, Ω	Z_T, Ω (Approx.)	Phase Angle θ_Z
1	10	$\sqrt{101} = 10$	$-84.3°$
10	10	$\sqrt{200} = 14$	$-45°$
10	1	$\sqrt{101} = 10$	$-5.7°$

Note: θ_Z is the phase angle of Z_T or V_T with respect to the reference phasor I in series circuits.

angle of $-84.3°$ is almost $-90°$, and the total impedance Z_T is approximately equal to X_C. The voltage drop across X_C in the series circuit is then practically equal to the applied voltage V_T with almost none across R.

At the opposite extreme, when R is 10 times more than X_C, the series circuit is mainly resistive. The phase angle of $-5.7°$ then means that the current is almost in phase with the applied voltage V_T; Z_T is approximately equal to R, and the voltage drop across R is practically equal to the applied voltage V_T with almost none across X_C.

When X_C and R equal each other, the resultant impedance Z_T is 1.41 times either one. The phase angle then is $-45°$, halfway between $0°$ for resistance alone and $-90°$ for capacitive reactance alone.

■ *18–3 Knowledge Check*

Answer at end of chapter.

A series RC circuit consists of a 1-kΩ R and a 2-kΩ X_C. How much is Z_T?

■ *18–3 Self-Review*

Answers at end of chapter.

a. How much is Z_T for a 20-Ω R in series with a 20-Ω X_C?
b. How much is V_T for 20 V across R and 20 V across X_C in series?
c. What is the phase angle θ_Z of this circuit?

18–4 *RC* Phase-Shifter Circuit

Figure 18–5 shows an application of X_C and R in series to provide a desired phase shift in the output V_R compared with the input V_T. The R can be varied up to 100 kΩ to change the phase angle. The C is 0.05 μF here for the 60-Hz ac power-line voltage, but a smaller C would be used for a higher frequency. The capacitor must have an appreciable value of reactance for the phase shift.

For the circuit in Fig. 18–5a, assume that R is set for 50 kΩ at its middle value. The reactance of the 0.05-μF capacitor at 60 Hz is approximately 53 kΩ. For these values of X_C and R, the phase angle of the circuit is $-46.7°$. This angle has a tangent of $-53/50 = -1.06$.

The phasor triangle in Fig. 18–5b shows that IR or V_R is out of phase with V_T by the leading angle of $46.7°$. Note that V_C is always $90°$ lagging V_R in a series circuit. The angle between V_C and V_T then becomes $90° - 46.7° = 43.3°$.

GOOD TO KNOW

In Fig. 18–5a, another *RC* phase shifting network could be added at the output of the first one to provide an even greater range in over-all phase shift.

▌▌▌ **MultiSim** **Figure 18–5** An *RC* phase-shifter circuit. (*a*) Schematic diagram. (*b*) Phasor triangle with *IR*, or *V*$_R$, as the horizontal reference. *V*$_R$ leads *V*$_T$ by 46.7° with *R* set at 50 kΩ. (*c*) Phasors shown with *V*$_T$ as the horizontal reference.

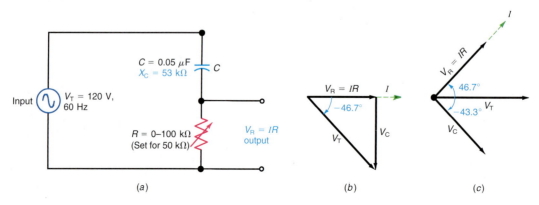

This circuit provides a phase-shifted voltage V_R at the output with respect to the input. For this reason, the phasors are redrawn in Fig. 18–5c to show the voltages with the input V_T as the horizontal reference. The conclusion, then, is that the output voltage across R leads the input V_T by 46.7°, whereas V_C lags V_T by 43.3°.

Now let R be varied for a higher value at 90 kΩ, while X_C stays the same. The phase angle becomes −30.5°. This angle has a tangent of $-53/90 = -0.59$. As a result, V_R leads V_T by 30.5°, and V_C lags V_T by 59.5°.

For the opposite case, let R be reduced to 10 kΩ. Then the phase angle becomes −79.3°. This angle has the tangent $-53/10 = -5.3$. Then V_R leads V_T by 79.3° and V_C lags V_T by 10.7°. Notice that the phase angle between V_R and V_T becomes larger as the series circuit becomes more capacitive with less resistance.

A practical application for this circuit is providing a voltage of variable phase to set the conduction time of semiconductors in power-control circuits. In this case, the output voltage is taken across the capacitor C. This provides a lagging phase angle with respect to the input voltage V_T. As R is varied from 0 Ω to 100 kΩ, the phase angle between V_C and V_T increases from 0° to about −62°. If R were changed so that it varied from 0 to 1 MΩ, the phase angle between V_C and V_T would vary between 0° and −90° approximately.

■ 18–4 Knowledge Check

Answer at end of chapter.

In Fig. 18–5 what is the approximate phase angle between V_C and V_T when $R = 1$ kΩ?

■ 18–4 Self-Review

Answers at end of chapter.

In Fig. 18–5, give the phase angle between
a. V_R and V_T.
b. V_R and V_C.
c. V_C and V_T.

18–5 X_C and R in Parallel

For parallel circuits with X_C and R, the 90° phase angle must be considered for each of the branch currents. Remember that any series circuit has different voltage drops but one common current. A parallel circuit has different branch currents but one common voltage.

In the parallel circuit in Fig. 18–6a, the applied voltage V_A is the same across X_C, R, and the generator, since they are all in parallel. There cannot be any phase difference between these voltages. Each branch, however, has its own individual current. For the resistive branch, $I_R = V_A/R$; for the capacitive branch, $I_C = V_A/X_C$.

The resistive branch current I_R is in phase with the generator voltage V_A. The capacitive branch current I_C leads V_A, however, because the charge and discharge current of a capacitor leads the capacitor voltage by 90°. The waveforms for V_A, I_R, I_C, and I_T in Fig. 18–6a are shown in Fig. 18–6b. The individual branch currents I_R and I_C must add to equal the total current I_T. The 10-A peak values for I_R and I_C total 14.14 A, however, instead of 20 A, because of the 90° phase difference.

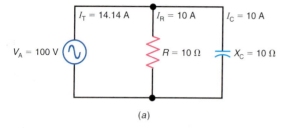

(a)

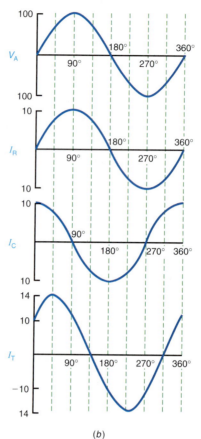

(b)

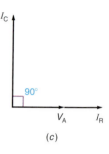

(c)

Consider some instantaneous values in Fig. 18–6b to see why the 10-A peak for I_R and 10-A peak for I_C cannot be added arithmetically. When I_C is at its maximum of 10 A, for instance, I_R is at zero. The total for I_T at this instant then is 10 A. Similarly, when I_R is at its maximum of 10 A, I_C is at zero and the total current I_T at this instant is also 10 A.

Actually, I_T has its maximum of 14.14 A when I_R and I_C are each 7.07 A. When branch currents that are out of phase are combined, therefore, they cannot be added without taking the phase difference into account.

Figure 18–6c shows the phasors representing V_A, I_R, and I_C. Notice that I_C leads V_A and I_R by 90°. In this case, the applied voltage V_A is used as the reference phasor since it is the same across both branches.

Phasor Current Triangle

Figure 18–7   Phasor triangle of capacitive and resistive branch currents 90° out of phase in a parallel circuit to find the resultant I_T.

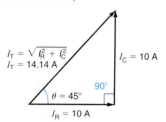

Figure 18–7 shows the phasor current triangle for the parallel RC circuit in Fig. 18–6a. Note that the resistive branch current I_R is used as the reference phasor since V_A and I_R are in phase. The capacitive branch current I_C is drawn upward at an angle of +90° since I_C leads V_A and thus I_R by 90°. The sum of the I_R and I_C phasors is indicated by the phasor for I_T, which connects the tail of the I_R phasor to the tip of the I_C phasor. The I_T phasor is the hypotenuse of the right triangle. The phase angle between I_T and I_R represents the phase angle of the circuit. Peak values are shown here for convenience, but rms and peak-to-peak values could also be used.

Using the Pythagorean theorem, the total current I_T could be calculated by taking the square root of the sum of the squares of the sides. For the current triangle in Fig. 18–7 therefore, the resultant I_T is

$$I_T = \sqrt{I_R^2 + I_C^2} \qquad (18\text{–}4)$$

For the values in Fig. 18–6,

$$I_T = \sqrt{10^2 + 10^2} = \sqrt{100 + 100}$$
$$= \sqrt{200}$$
$$= 14.14 \text{ A}$$

Impedance of X_C and R in Parallel

A practical approach to the problem of calculating the total or equivalent impedance of X_C and R in parallel is to calculate the total line current I_T and divide the applied voltage V_A by this value.

$$Z_{EQ} = \frac{V_A}{I_T} \qquad (18\text{–}5)$$

For the circuit in Fig. 18–6a, V_A is 100 V, and the total current I_T, obtained as the phasor sum of I_R and I_C, is 14.14 A. Therefore, we can calculate the equivalent impedance Z_{EQ} as

$$Z_{EQ} = \frac{V_A}{I_T} = \frac{100 \text{ V}}{14.14 \text{ A}}$$
$$= 7.07 \text{ }\Omega$$

This impedance, the combined opposition in ohms across the generator, is equal to the 10-Ω resistance in parallel with the 10-Ω X_C.

Note that the impedance Z_{EQ} for equal values of R and X_C in parallel is not one-half but instead equals 70.7% of either one. Still, the value of Z_{EQ} will always be less than the lowest ohm value in the parallel branches.

For the general case of calculating the Z_{EQ} of X_C and R in parallel, any number can be assumed for the applied voltage V_A because, in the calculations for Z_{EQ} in terms of the branch currents, the value of V_A cancels. A good value to assume for V_A is the value of either R or X_C, whichever is the larger number. This way, there are no fractions smaller than that in the calculation of the branch currents.

Phase Angle in Parallel Circuits

In Fig. 18–7, the phase angle θ is 45° because R and X_C are equal, resulting in equal branch currents. The phase angle is between the total current I_T and the generator voltage V_A. However, V_A and I_R are in phase. Therefore θ is also between I_T and I_R.

Using the tangent formula to find θ from the current triangle in Fig. 18–7 gives

$$\tan \theta_I = \frac{I_C}{I_R} \qquad (18\text{–}6)$$

The phase angle is positive because the I_C phasor is upward, leading V_A by 90°. This direction is opposite from the lagging phasor of series X_C. The effect of X_C is no different, however. Only the reference is changed for the phase angle.

Note that the phasor triangle of branch currents for parallel circuits gives θ_I as the angle of I_T with respect to the generator voltage V_A. This phase angle for I_T is labeled θ_I with respect to the applied voltage. For the phasor triangle of voltages in a series circuit, the phase angle for Z_T and V_T is labeled θ_Z with respect to the series current.

Example 18-2

A 30-mA I_R is in parallel with another branch current of 40 mA for I_C. The applied voltage V_A is 72 V. Calculate I_T, Z_{EQ}, and θ_I.

ANSWER This problem can be calculated in mA units for I and kΩ for Z without powers of 10.

$$I_T = \sqrt{I_R^2 + I_C^2} = \sqrt{(30)^2 + (40)^2}$$
$$= \sqrt{900 + 1600} = \sqrt{2500}$$
$$= 50 \text{ mA}$$
$$Z_{EQ} = \frac{V_A}{I_T} = \frac{72 \text{ V}}{50 \text{ mA}}$$
$$= 1.44 \text{ k}\Omega$$
$$\tan \theta_I = \frac{I_C}{I_R} = \frac{40}{30} = 1.333$$
$$= \arctan (1.333)$$
$$\theta_I = 53.1°$$

Parallel Combinations of X_C and R

In Table 18-2, when X_C is 10 times R, the parallel circuit is practically resistive because there is little leading capacitive current in the main line. The small value of I_C results from the high reactance of shunt X_C. Then the total impedance of the parallel circuit is approximately equal to the resistance, since the high value of X_C in a parallel branch has little effect. The phase angle of 5.7° is practically 0° because almost all of the line current is resistive.

As X_C becomes smaller, it provides more leading capacitive current in the main line. When X_C is 1/10 R, practically all of the line current is the I_C component. Then, the parallel circuit is practically all capacitive with a total impedance practically equal to X_C. The phase angle of 84.3° is almost 90° because the line current is mostly capacitive. Note that these conditions are opposite to the case of X_C and R in series. With X_C and R equal, their branch currents are equal and the phase angle is 45°.

Table 18-2	Parallel Resistance and Capacitance Combinations*					
R, Ω	X_C, Ω	I_R, A	I_C, A	I_T, A (Approx.)	Z_{EQ}, Ω (Approx.)	Phase Angle θ_I
1	10	10	1	$\sqrt{101} = 10$	1	5.7°
10	10	1	1	$\sqrt{2} = 1.4$	7.07	45°
10	1	1	10	$\sqrt{101} = 10$	1	84.3°

* $V_A = 10$ V. Note that θ_I is the phase angle of I_T with respect to the reference V_A in parallel circuits.

As additional comparisons between series and parallel *RC* circuits, remember that

1. The series voltage drops V_R and V_C have individual values that are 90° out of phase. Therefore, V_R and V_C are added by phasors to equal the applied voltage V_T. The negative phase angle $-\theta_Z$ is between V_T and the common series current I. More series X_C allows more V_C to make the circuit more capacitive with a larger negative phase angle for V_T with respect to I.
2. The parallel branch currents I_R and I_C have individual values that are 90° out of phase. Therefore, I_R and I_C are added by phasors to equal I_T, which is the main-line current. The positive phase angle θ_I is between the line current I_T and the common parallel voltage V_A. Less parallel X_C allows more I_C to make the circuit more capacitive with a larger positive phase angle for I_T with respect to V_A.

■ *18–5 Knowledge Check*

> *Answer at end of chapter.*
>
> In a parallel *RC* circuit where $R = X_C$, does I_T lead or lag V_A?

■ *18–5 Self-Review*

> *Answers at end of chapter.*
>
> a. How much is I_T for branch currents I_R of 2 A and I_C of 2 A?
> b. Find the phase angle θ_I between I_T and V_A.

18–6 RF and AF Coupling Capacitors

In Fig. 18–8, C_C is used in the application of a coupling capacitor. Its low reactance allows developing practically all the ac signal voltage of the generator across R. Very little of the ac voltage is across C_C.

The coupling capacitor is used for this application because it provides more reactance at lower frequencies, resulting in less ac voltage coupled across R and more across C_C. For dc voltage, all voltage is across C with none across R, since the capacitor blocks direct current. As a result, the output signal voltage across R includes the desired higher frequencies but not direct current or very low frequencies. This application of C_C, therefore, is called *ac coupling*.

The dividing line for C_C to be a coupling capacitor at a specific frequency can be taken as X_C one-tenth or less of the series R. Then the series RC circuit is primarily resistive. Practically all the voltage drop of the ac generator is across R, with little across C. In addition, the phase angle is almost 0°.

Typical values of a coupling capacitor for audio or radio frequencies can be calculated if we assume a series resistance of 16,000 Ω. Then X_C must be 1600 Ω or less. Typical values for C_C are listed in Table 18–3. At 100 Hz, a coupling capacitor must be 1 μF to provide 1600 Ω of reactance. Higher frequencies allow a smaller value of C_C for a coupling capacitor having the same reactance. At 100 MHz in the VHF range, the required capacitance is only 1 pF.

Note that the C_C values are calculated for each frequency as a lower limit. At higher frequencies, the same size C_C will have less reactance than one-tenth of R, which improves coupling.

Choosing a Coupling Capacitor for a Circuit

As an example of using these calculations, suppose that we have the problem of determining C_C for an audio amplifier. This application also illustrates the

Figure 18–8 Series circuit for *RC* coupling. Small X_C compared with R allows practically all the applied voltage to be developed across R for the output, with little across C.

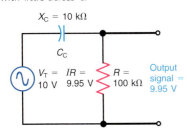

$X_C = 10$ kΩ

C_C

$V_T = 10$ V $IR = 9.95$ V $R = 100$ kΩ   Output signal = 9.95 V

Table 18–3	Coupling Capacitors with a Reactance of 1600 Ω*	
f	C_C	Remarks
100 Hz	1 μF	Low audio frequencies
1000 Hz	0.1 μF	Audio frequencies
10 kHz	0.01 μF	Audio frequencies
1000 kHz	100 pF	Radio frequencies
100 MHz	1 pF	Very high frequencies

* For an X_C one-tenth of a series R of 16,000 Ω

relatively large capacitance needed with low series resistance. The C is to be a coupling capacitor for audio frequencies of 50 Hz and up with a series R of 4000 Ω. Then the required X_C is 4000/10, or 400 Ω. To find C at 50 Hz,

$$C = \frac{1}{2\pi f X_C} = \frac{1}{6.28 \times 50 \times 400}$$
$$= \frac{1}{125,600} = 0.0000079$$
$$= 7.9 \times 10^{-6} \quad \text{or} \quad 7.9 \; \mu F$$

A 10-μF electrolytic capacitor would be a good choice for this application. The slightly higher capacitance value is better for coupling. The voltage rating should exceed the actual voltage across the capacitor in the circuit. Although electrolytic capacitors have a slight leakage current, they can be used for coupling capacitors in this application because of the low series resistance.

■ *18–6 Knowledge Check*

> *Answer at end of chapter.*
>
> In Fig. 18–8, calculate C_C if R = 2 kΩ and f = 5 kHz.

■ *18–6 Self-Review*

> *Answers at end of chapter.*
>
> a. The X_C of a coupling capacitor is 70 Ω at 200 Hz. How much is its X_C at 400 Hz?
> b. From Table 18–3, what C would be needed for 1600 Ω of X_C at 50 MHz?

18–7 Capacitive Voltage Dividers

When capacitors are connected in series across a voltage source, the series capacitors serve as a voltage divider. Each capacitor has part of the applied voltage, and the sum of all the series voltage drops equals the source voltage.

The amount of voltage across each is inversely proportional to its capacitance. For instance, with 2 μF in series with 1 μF, the smaller capacitor has double the voltage of the larger capacitor. Assuming 120 V applied,

Figure 18–9 Series capacitors divide V_T inversely proportional to each C. The smaller C has more V. (a) An ac divider with more X_C for the smaller C. (b) A dc divider.

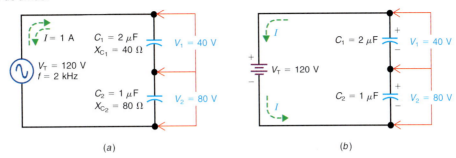

(a)   (b)

one-third of this, or 40 V, is across the 2-μF capacitor, and two-thirds, or 80 V, is across the 1-μF capacitor.

The two series voltage drops of 40 and 80 V add to equal the applied voltage of 120 V. The phasor addition is the same as the arithmetic sum of the two voltages because they are in phase. When voltages are out of phase with each other, arithmetic addition is not possible and phasor addition becomes necessary.

AC Divider

With sine-wave alternating current, the voltage division between series capacitors can be calculated on the basis of reactance. In Fig. 18–9a, the total reactance is 120 Ω across the 120-V source. The current in the series circuit is 1 A. This current is the same for X_{C_1} and X_{C_2} in series. Therefore, the IX_C voltage across C_1 is 40 V with 80 V across C_2.

The voltage division is proportional to the series reactances, as it is to series resistances. However, reactance is inversely proportional to capacitance. As a result, the smaller capacitance has more reactance and a greater part of the applied voltage.

DC Divider

In Fig. 18–9b, both C_1 and C_2 will be charged by the battery. The voltage across the series combination of C_1 and C_2 must equal V_T. When charging current flows, electrons repelled from the negative battery terminal accumulate on the negative plate of C_2, repelling electrons from its positive plate. These electrons flow through the conductor to the negative plate of C_1. As the positive battery terminal attracts electrons, the charging current from the positive plate of C_1 returns to the positive side of the dc source. Then C_1 and C_2 become charged in the polarity shown.

Since C_1 and C_2 are in the same series path for charging current, both have the same amount of charge. However, the potential difference provided by the equal charges is inversely proportional to capacitance. The reason is that $Q = CV$, or $V = Q/C$. Therefore, the 1-μF capacitor has double the voltage of the 2-μF capacitor with the same charge in both.

If you measure across C_1 with a dc voltmeter, the meter reads 40 V. Across C_2, the dc voltage is 80 V. The measurement from the negative side of C_2 to the positive side of C_1 is the same as the applied battery voltage of 120 V.

If the meter is connected from the positive side of C_2 to the negative plate of C_1, however, the voltage is zero. These plates have the same potential because they are joined by a conductor of zero resistance.

GOOD TO KNOW

Connecting a dc voltmeter across either C_1 or C_2 in Fig. 18–9b will cause the capacitor to discharge through the resistance of the meter. It is best to use a DMM with a high internal resistance so that the amount of discharge is minimal. The voltmeter reading should be taken immediately after it is connected across the capacitor.

The polarity marks at the junction between C_1 and C_2 indicate the voltage at this point with respect to the opposite plate of each capacitor. This junction is positive compared with the opposite plate of C_2 with a surplus of electrons. However, the same point is negative compared with the opposite plate of C_1, which has a deficiency of electrons.

In general, the following formula can be used for capacitances in series as a voltage divider:

$$V_C = \frac{C_{EQ}}{C} \times V_T \qquad (18\text{-}7)$$

Note that C_{EQ} is in the numerator, since it must be less than the smallest individual C with series capacitances. For the divider examples in Fig. 18–9a and b,

$$V_1 = \frac{C_{EQ}}{C_1} \times 120 = \frac{2/3}{2} \times 120 = 40 \text{ V}$$

$$V_2 = \frac{C_{EQ}}{C_2} \times 120 = \frac{2/3}{1} \times 120 = 80 \text{ V}$$

This method applies to series capacitances as dividers for either dc or ac voltage, as long as there is no series resistance. Note that the case of capacitive dc dividers also applies to pulse circuits. Furthermore, bleeder resistors may be used across each of the capacitors to ensure more exact division.

■ 18–7 Knowledge Check

Answer at end of chapter.

In Fig. 18–9b, what is the voltage across C_1 if $C_1 = 50 \ \mu F$ and $C_2 = 150 \ \mu F$? $V_T = 120$ V.

■ 18–7 Self-Review

Answers at end of chapter.

a. Capacitance C_1 of 10 pF and C_2 of 90 pF are across 20 kV. Calculate the amount of V_1 and V_2.
b. In Fig. 18–9a, how much is X_{C_T}?

18–8 The General Case of Capacitive Current i_C

The capacitive charge and discharge current i_C is always equal to $C(dv/dt)$. A sine wave of voltage variations for v_C produces a cosine wave of current i. This means that v_C and i_C have the same waveform, but they are 90° out of phase.

It is usually convenient to use X_C for calculations in sine-wave circuits. Since X_C is $1/(2\pi f C)$, the factors that determine the amount of charge and discharge current are included in f and C. Then I_C equals V_C/X_C. Or, if I_C is known, V_C can be calculated as $I_C \times X_C$.

With a nonsinusoidal waveform for voltage v_C, the concept of reactance cannot be used. Reactance X_C applies only to sine waves. Then i_C must be determined as $C(dv/dt)$. An example is illustrated in Fig. 18–10 to show the change of waveform here, instead of the change of phase angle in sine-wave circuits.

Note that the sawtooth waveform of voltage v_C corresponds to a rectangular waveform of current. The linear rise of the sawtooth wave produces a constant amount of charging current i_C because the rate of change is constant

GOOD TO KNOW

For a capacitor, the value of charge or discharge current is zero when dv/dt is zero. When dv/dt is a constant value the charge or discharge current is also a constant value.

Figure 18–10 Waveshape of i_C equal to $C(dv/dt)$. (a) Sawtooth waveform of V_C. (b) Rectangular current waveform of i_C resulting from the uniform rate of change in the sawtooth waveform of voltage.

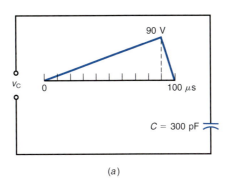

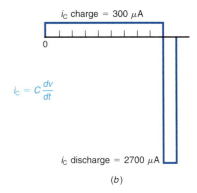

$$i_C = C\frac{dv}{dt}$$

(a)

(b)

for the charging voltage. When the capacitor discharges, v_C drops sharply. Then the discharge current is in the direction opposite from the charge current. Also, the discharge current has a much larger value because of the faster rate of change in v_C.

■ 18–8 Knowledge Check

Answer at end of chapter.

In Fig. 18–10, why is the discharge current greater than the charge current?

■ 18–8 Self-Review

Answers at end of chapter.

a. **In Fig. 18–10a, how much is dv/dt in V/s for the sawtooth rise from 0 to 90 V in 90 μs?**
b. **How much is the charge current i_C, as $C(dv/dt)$ for this dv/dt?**

Summary

■ In a sine-wave ac circuit, the voltage across a capacitance lags its charge and discharge current by 90°.

■ Therefore, capacitive reactance X_C is a phasor quantity out of phase with its series resistance by −90° because $i_C = C(dv/dt)$. This fundamental fact is the basis of all the following relations.

■ The combination of X_C and R in series is their total impedance Z_T. These three types of ohms of opposition to current are compared in Table 18–4.

■ The opposite characteristics for series and parallel circuits with X_C and R are summarized in Table 18–5.

■ Two or more capacitors in series across a voltage source serve as a voltage divider. The smallest C has the largest part of the applied voltage.

■ A coupling capacitor has X_C less than its series resistance by a factor of one-tenth or less to provide practically all the ac applied voltage across R with little across C.

■ In sine-wave circuits, $I_C = V_C/X_C$. Then I_C is out of phase with V_C by 90°.

■ For a circuit with X_C and R in series, $\tan \theta_Z = -(X_C/R)$, and in parallel, $\tan \theta_I = I_C/I_R$. See Table 18–5.

■ When the voltage is not a sine wave, $i_C = C(dv/dt)$. Then the waveshape of i_C is different from that of the voltage.

Table 18–4	Comparison of R, X_C, and Z	
R	$X_C = 1/(2\pi fC)$	$Z_T = \sqrt{R^2 + X_C^2}$
Ohm unit	Ohm unit	Ohm unit
IR voltage in phase with I	IX_C voltage lags I_C by 90°	IZ_T is the applied voltage
Same ohm value for all f	Ohm value decreases for higher f	Becomes more resistive with more f Becomes more capacitive with less f

Table 18–5	Series and Parallel RC Circuits
X_C and R in Series	X_C and R in Parallel
I the same in X_C and R	V the same across X_C and R
$V_T = \sqrt{V_R^2 + V_C^2}$	$I_T = \sqrt{I_R^2 + I_C^2}$
$Z_T = \sqrt{R^2 + X_C^2}$	$Z_{EQ} = \dfrac{V_A}{I_T}$
V_C lags V_R by 90°	I_C leads I_R by 90°
$\tan \theta_Z = -\dfrac{X_C}{R}$; θ_Z increases as X_C increases, resulting in more V_C	$\tan \theta_I = \dfrac{I_C}{I_R}$; θ_I decreases as X_C increases, resulting in less I_C

Important Terms

■ Arctangent (arctan) - an inverse trigonometric function that specifies the angle, θ, corresponding to a given tangent (tan) value.

■ Capacitive Voltage Divider - a voltage divider that consists of series-connected capacitors. The amount of voltage across each capacitor is inversely proportional to its capacitance value.

■ Coupling Capacitor, C_C - a capacitor that is selected to pass ac signals above a specified frequency from one point in a circuit to another. The dividing line for calculating the coupling capacitance, C_C, is to make X_C one-tenth the value of the series R. The value of C_C is calculated for a specified frequency as a lower limit.

■ Impedance, Z - the total opposition to the flow of current in a sine-wave ac circuit. In an RC circuit, the impedance, Z, takes into account the 90° phase relation between X_C and R. Impedance, Z, is measured in ohms.

■ Phase Angle, θ - the angle between the generator voltage and current in a sine-wave ac circuit.

■ Phasor Triangle - a right triangle which represents the phasor sum of two quantities 90° out of phase with each other.

■ RC Phase-Shifter - an application of a series RC circuit in which the output across either R or C provides a desired phase shift with respect to the input voltage. RC phase-shifter circuits are commonly used to control the conduction time of semiconductors in power-control circuits.

■ Tangent (tan) - a trigonometric function of an angle, equal to the ratio of the opposite side to the adjacent side of a triangle.

Related Formulas

Series *RC* Circuits

$$V_T = \sqrt{V_R^2 + V_C^2}$$

$$Z_T = \sqrt{R^2 + X_C^2}$$

$$\tan \theta_Z = -\frac{X_C}{R}$$

Parallel *RC* Circuits

$$I_T = \sqrt{I_R^2 + I_C^2}$$

$$Z_{EQ} = \frac{V_A}{I_T}$$

$$\tan \theta_I = \frac{I_C}{I_R}$$

Series Capacitors

$$V_C = \frac{C_{EQ}}{C} \times V_T$$

Self-Test

Answers at back of book.

1. For a capacitor in a sine-wave ac circuit,
 a. V_C lags i_C by 90°.
 b. i_C leads V_C by 90°.
 c. i_C and V_C have the same frequency.
 d. all of the above.

2. In a series *RC* circuit,
 a. V_C leads V_R by 90°.
 b. V_C and I are in phase.
 c. V_C lags V_R by 90°.
 d. both b and c.

3. In a series *RC* circuit where $V_C = 15$ V and $V_R = 20$ V, how much is the total voltage, V_T?
 a. 35 V.
 b. 25 V.
 c. 625 V.
 d. 5 V.

4. A 10-Ω resistor is in parallel with a capacitive reactance of 10 Ω. The combined equivalent impedance, Z_{EQ}, of this combination is
 a. 7.07 Ω.
 b. 20 Ω.
 c. 14.14 Ω.
 d. 5 Ω.

5. In a parallel *RC* circuit,
 a. I_C lags I_R by 90°.
 b. I_R and I_C are in phase.
 c. I_C leads I_R by 90°.
 d. I_R leads I_C by 90°.

6. In a parallel *RC* circuit where $I_R = 8$ A and $I_C = 10$ A, how much is the total current, I_T?
 a. 2 A.
 b. 12.81 A.
 c. 18 A.
 d. 164 A.

7. In a series *RC* circuit where $R = X_C$, the phase angle, θ_Z, is
 a. +45°.
 b. −90°.
 c. 0°.
 d. −45°.

8. A 10-μF capacitor, C_1, and a 15-μF capacitor, C_2, are connected in series with a 12-V dc source. How much voltage is across C_2?
 a. 4.8 V.
 b. 7.2 V.
 c. 12 V.
 d. 0 V.

9. The dividing line for a coupling capacitor at a specific frequency can be taken as
 a. X_C ten or more times the series resistance.
 b. X_C equal to R.
 c. X_C one-tenth or less the series resistance.
 d. none of the above.

10. A 100-Ω resistance is in series with a capacitive reactance of 75 Ω. The total impedance, Z_T, is
 a. 125 Ω.
 b. 25 Ω.
 c. 175 Ω.
 d. 15.625 kΩ.

11. In a series *RC* circuit,
 a. V_C and V_R are in phase.
 b. V_T and I are always in phase.
 c. V_R and I are in phase.
 d. V_R leads I by 90°.

12. In a parallel *RC* circuit,
 a. V_A and I_R are in phase.
 b. V_A and I_C are in phase.
 c. I_C and I_R are in phase.
 d. V_A and I_R are 90° out of phase.

13. When the frequency of the applied voltage increases in a parallel *RC* circuit,
 a. the phase angle, θ_I, increases.
 b. Z_{EQ} increases.
 c. Z_{EQ} decreases.
 d. both a and c.

14. When the frequency of the applied voltage increases in a series *RC* circuit,
 a. the phase angle, θ, becomes more negative.
 b. Z_T increases.
 c. Z_T decreases.
 d. both a and c.

15. Capacitive reactance, X_C,
 a. applies only to nonsinusoidal waveforms or dc.
 b. applies only to sine waves.
 c. applies to either sinusoidal or nonsinusoidal waveforms.
 d. is directly proportional to frequency.

Questions

1. (a) Why does a capacitor charge when the applied voltage increases? (b) Why does the capacitor discharge when the applied voltage decreases?

2. A sine wave of voltage V is applied across a capacitor C. (a) Draw the schematic diagram. (b) Draw the sine waves of voltage and current out of phase by 90°. (c) Draw a phasor diagram showing the phase angle of $-90°$ between V and I.

3. Why will a circuit with R and X_C in series be less capacitive as the frequency of the applied voltage is increased?

4. Define the following: coupling capacitor, sawtooth voltage, capacitive voltage divider.

5. State two troubles possible in coupling capacitors and describe briefly how you would check the capacitor with an ohmmeter.

6. Explain the function of R and C in an RC coupling circuit.

7. Explain briefly why a capacitor can block dc voltage.

8. What is the waveshape of i_C for a sine-wave v_C?

9. Explain why the impedance Z_{EQ} of a parallel RC circuit decreases as the frequency increases.

10. Explain why θ_z in a series RC circuit increases (becomes more negative) as frequency decreases.

Problems

SECTION 18–1 SINE WAVE v_C LAGS i_C BY 90°

18–1 In Fig. 18–11, what is the
    a. peak value of the capacitor voltage, V_C?
    b. peak value of the charge and discharge current, i_C?
    c. frequency of the charge and discharge current?
    d. phase relationship between V_C and i_C?

Figure 18–11

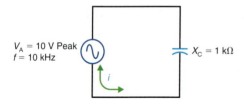

18–2 In Fig. 18–11, what is the value of the capacitor current, i_C, at the instant when V_C equals:
    a. its positive peak of $+10$ V?
    b. 0 V?
    c. its negative peak of -10 V?

18–3 In Fig. 18–11, draw the phasors representing V_C and i_C using
    a. V_C as the reference phasor.
    b. i_C as the reference phasor.

SECTION 18–2 X_C AND R IN SERIES

18–4 In Fig. 18–12, how much current, I, is flowing
    a. through the 30 Ω resistor, R?
    b. through the 40-Ω capacitive reactance, X_C?
    c. to and from the terminals of the applied voltage, V_T?

Figure 18–12

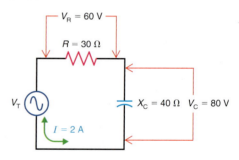

18–5 In Fig. 18–12, what is the phase relationship between
    a. I and V_R?
    b. I and V_C?
    c. V_C and V_R?

18–6 In Fig. 18–12, how much is the applied voltage, V_T?

18–7 Draw the phasor voltage triangle for the circuit in Fig. 18–12. (Use V_R as the reference phasor.)

18–8 In Fig. 18–13, solve for
    a. the resistor voltage, V_R.
    b. the capacitor voltage, V_C.
    c. the total voltage, V_T.

Figure 18–13

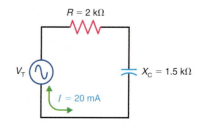

18-9 In Fig. 18-14, solve for
  a. the resistor voltage, V_R.
  b. the capacitor voltage, V_C.
  c. the total voltage, V_T.

Figure 18-14

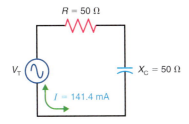

18-10 In a series RC circuit, solve for the applied voltage, V_T if
  a. $V_R = 40$ V and $V_C = 40$ V.
  b. $V_R = 10$ V and $V_C = 5$ V.
  c. $V_R = 48$ V and $V_C = 72$ V.
  d. $V_R = 12$ V and $V_C = 18$ V.

SECTION 18-3 IMPEDANCE Z TRIANGLE

18-11 In Fig. 18-15, solve for Z_T, I, V_C, V_R, and θ_Z.

Figure 18-15

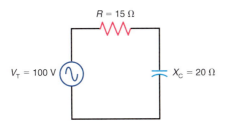

18-12 Draw the impedance triangle for the circuit in Fig. 18-15 (Use R as the reference phasor.)

18-13 In Fig. 18-16, solve for Z_T, I, V_C, V_R, and θ_Z.

Figure 18-16

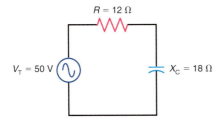

18-14 In Fig. 18-17, solve for Z_T, I, V_C, V_R, and θ_Z.

Figure 18-17

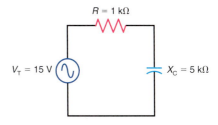

18-15 In Fig. 18-18, solve for Z_T, I, V_C, V_R, and θ_Z.

Figure 18-18

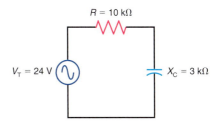

18-16 In Fig. 18-19, solve for Z_T, I, V_C, V_R, and θ_Z for the following circuit values:
  a. $X_C = 30\ \Omega$, $R = 40\ \Omega$, and $V_T = 50$ V
  b. $X_C = 200\ \Omega$, $R = 200\ \Omega$, and $V_T = 56.56$ V
  c. $X_C = 10\ \Omega$, $R = 100\ \Omega$, and $V_T = 10$ V
  d. $X_C = 100\ \Omega$, $R = 10\ \Omega$, and $V_T = 10$ V

Figure 18-19

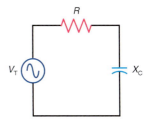

18-17 In Fig. 18-20, solve for X_C, Z_T, I, V_C, V_R, and θ_Z.

Figure 18-20

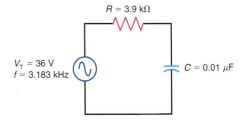

Capacitive Circuits **569**

18–18 In Fig. 18–20, what happens to each of the following quantities if the frequency of the applied voltage increases?

  a. X_C
  b. Z_T
  c. I
  d. V_C
  e. V_R
  f. θ_Z

18–19 Repeat Problem 18–18 if the frequency of the applied voltage decreases.

SECTION 18–4 RC PHASE-SHIFTER CIRCUIT

18–20 With R set to 50 kΩ in Fig. 18–21, solve for X_C, Z_T, I, V_R, V_C, and θ_Z.

Figure 18–21

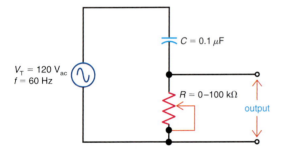

18–21 With R set to 50 kΩ in Fig. 18–21, what is the phase relationship between

  a. V_T and V_R
  b. V_T and V_C

18–22 Draw the phasors for V_R, V_C, and V_T in Fig. 18–21 with R set at 50 kΩ. Use V_T as the reference phasor.

18–23 With R set at 1 kΩ in Fig. 18–21, solve for

  a. Z_T, I, V_R, V_C, and θ_Z.
  b. the phase relationship between V_T and V_R.
  c. the phase relationship between V_T and V_C.

18–24 With R set at 100 kΩ in Fig. 18–21, solve for

  a. Z_T, I, V_R, V_C, and θ_Z.
  b. the phase relationship between V_T and V_R.
  c. the phase relationship between V_T and V_C.

SECTION 18–5 X_C AND R IN PARALLEL

18–25 In Fig. 18–22, how much voltage is across

  a. the 40-Ω resistor, R?
  b. the 30-Ω capacitive reactance, X_C?

Figure 18–22

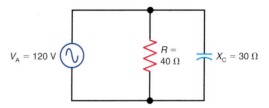

18–26 In Fig. 18–22, what is the phase relationship between

  a. V_A and I_R?
  b. V_A and I_C?
  c. I_C and I_R?

18–27 In Fig. 18–22, solve for I_R, I_C, I_T, Z_{EQ}, and θ_I.

18–28 Draw the phasor current triangle for the circuit in Fig. 18–22. (Use I_R as the reference phasor.)

18–29 In Fig. 18–23, solve for I_R, I_C, I_T, Z_{EQ}, and θ_I.

Figure 18–23

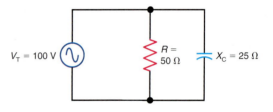

18–30 In Fig. 18–24, solve for I_R, I_C, I_T, Z_{EQ}, and θ_I.

Figure 18–24

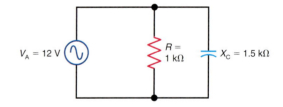

18–31 In Fig. 18–25, solve for I_R, I_C, I_T, Z_{EQ}, and θ_I.

Figure 18–25

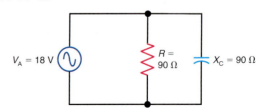

18–32 In Fig. 18–26, solve for I_R, I_C, I_T, Z_{EQ}, and θ_I.

Figure 18–26

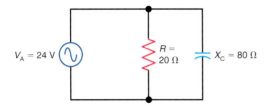

18–33 In Fig. 18–27, solve for I_R, I_C, I_T, Z_{EQ}, and θ_I for the following circuit values:

    a. $R = 50 \ \Omega$, $X_C = 50 \ \Omega$, and $V_A = 50$ V

    b. $R = 10 \ \Omega$, $X_C = 100 \ \Omega$, and $V_A = 20$ V

    c. $R = 100 \ \Omega$, $X_C = 10 \ \Omega$, and $V_A = 20$ V

Figure 18–27

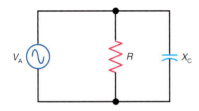

18–34 In Fig. 18–27, how much is Z_{EQ} if $R = 60 \ \Omega$ and $X_C = 80 \ \Omega$?

18–35 In Fig. 18–28, solve for X_C, I_R, I_C, I_T, Z_{EQ}, and θ_I.

Figure 18–28

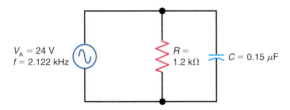

18–36 In Fig. 18–28, what happens to each of the following quantities if the frequency of the applied voltage increases?

    a. I_R

    b. I_C

    c. I_T

    d. Z_{EQ}

    e. θ_I

18–37 Repeat Problem 18–36 if the frequency of the applied voltage decreases.

SECTION 18–6 RF AND AF COUPLING CAPACITORS

18–38 In Fig. 18–29, calculate the minimum coupling capacitance, C_C, in series with the 1-kΩ resistance, R, if the frequency of the applied voltage is

    a. 159.1 Hz.

    b. 1591 Hz.

    c. 15.91 kHz.

Figure 18–29

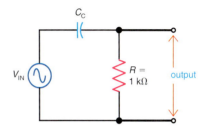

18–39 In Fig. 18–29, assume that $C_C = 0.047 \ \mu$F and $R = 1 \ k\Omega$, as shown. For these values, what is the lowest frequency of the applied voltage that will provide an X_C of 100 Ω? At this frequency, what is the phase angle, θ_Z?

SECTION 18–7 CAPACITIVE VOLTAGE DIVIDERS

18–40 In Fig. 18–30, calculate the following:

    a. X_{C_1}, X_{C_2}, X_{C_3}, X_{C_4}, and X_{C_T}.

    b. I.

    c. V_{C_1}, V_{C_2}, V_{C_3}, and V_{C_4}.

Figure 18–30

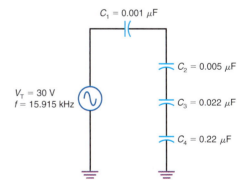

18-41 In Fig. 18-31, calculate V_{C_1}, V_{C_2}, and V_{C_3}.

Figure 18-31

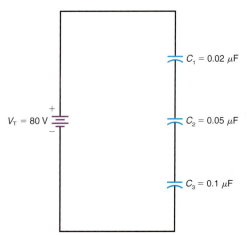

Figure 18-32

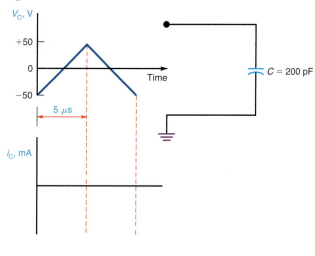

SECTION 18-8 THE GENERAL CASE OF CAPACITIVE CURRENT, i_C

18-42 For the waveshape of capacitor voltage, V_C, in Fig. 18-32, show the corresponding charge and discharge current, i_C, with values for a 200-pF capacitance.

18-43 In Fig. 18-33, show the corresponding charge and discharge current for the waveshape of capacitor voltage shown.

Figure 18-33

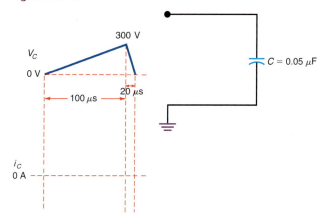

Critical Thinking

18-44 In Fig. 18-34, calculate X_C, Z_T, I, f, V_T, and V_R.

Figure 18-34 Circuit for Critical Thinking Prob. 18-44.

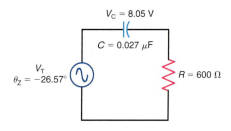

Figure 18-35 Circuit for Critical Thinking Prob. 18-45.

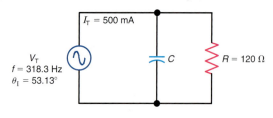

18-45 In Fig. 18-35, calculate I_C, I_R, V_T, X_C, C, and Z_{EQ}.

18-46 In Fig. 18-36, calculate I_C, I_R, I_T, X_C, R, and C.

Figure 18-36 Circuit for Critical Thinking Prob. 18-46.

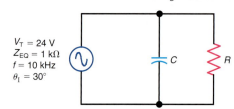

Answers to Knowledge Check Problems

18-1 $i_C = 0$

18-2 $V_T = 52$ V

18-3 $Z_T = 2.236$ kΩ

18-4 $0°$

18-5 I_T leads V_A by $45°$

18-6 $C_C = 0.159$ μF

18-7 $V_{C_1} = 90$ V

18-8 Because $\dfrac{dv}{dt}$ is greater when C is discharging.

Answers to Self-Reviews

18-1 a. $0°$
    b. $90°$
    c. lag

18-2 a. $0°$
    b. $90°$
    c. lead

18-3 a. 28.28 Ω
    b. 28.28 V
    c. $\theta_Z = -45°$

18-4 a. $46.7°$
    b. $90°$
    c. $43.3°$

18-5 a. 2.828 A
    b. $\theta_I = 45°$

18-6 a. 35 Ω
    b. 2 pF

18-7 a. $V_1 = 18$ kV
       $V_2 = 2$ kV
    b. $X_{C_T} = 120$ Ω

18-8 a. $dv/dt = 1 \times 10^6$ V/s
    b. $i_C = 300$ μA

Cumulative Review Summary (Chapters 16–18)

■ A capacitor consists of two conductors separated by an insulator, which is a dielectric material. When voltage is applied to the conductors, charge is stored in the dielectric. One coulomb of charge stored with 1 volt applied corresponds to 1 farad of capacitance C. The common units of capacitance are microfarads (1 μF $= 10^{-6}$ F) or picofarads (1 pF $= 10^{-12}$ F).

■ Capacitance increases with plate area and larger values of dielectric constant but decreases with increased distance between plates.

■ The most common types of capacitors are air, film, paper, mica, ceramic disk, surface-mount (chip), and electrolytic. Electrolytics must be connected in the correct polarity. The capacitance coding systems for film, ceramic disk, mica, and tantalum capacitors are illustrated in Figs. 16–11, 16–13, 16–16, and 16–20, respectively. The capacitance coding systems used with chip capacitors are illustrated in Figs. 16–17, 16–18, and 16–19.

■ The total capacitance of parallel capacitors is the sum of individual values; the combined capacitance of series capacitors is found by the reciprocal formula. These rules are opposite from the formulas used for resistors in series or parallel.

■ When a good capacitor is checked with an ohmmeter, it shows charging current, and then the ohmmeter reads a very high value of ohms equal to the insulation resistance. A short-circuited capacitor reads zero ohms; an open capacitor does not show any charging current.

■ $X_C = 1/(2\pi fC)$ Ω, where f is in hertz, C is in farads, and X_C is in ohms. The higher the frequency and the greater the capacitance, the smaller X_C.

■ The total X_C of capacitive reactances in series equals the sum of the individual values, just as for series resistances. The series reactances have the same current. The voltage across each X_C equals IX_C.

■ With parallel capacitive reactances, the combined reactance is calculated using the reciprocal formula, as for parallel resistances. Each branch current equals V_A/X_C. The total current is the sum of the individual branch currents.

■ A common application of X_C is in af or rf coupling capacitors, which have low reactance for higher frequencies but more reactance for lower frequencies.

■ Reactance X_C is a phasor quantity in which the voltage across the capacitor lags $90°$ behind its charge and discharge current.

■ In a series RC circuit, R and X_C are added by phasors because the voltage drops are $90°$ out of phase. Therefore, the total impedance $Z_T = \sqrt{R^2 + X_C^2}$; the current $I = V_T/Z_T$.

■ For parallel RC circuits, the resistive and capacitive branch currents are added by phasors, $I_T = \sqrt{I_R^2 + I_C^2}$; the impedance $Z_{EQ} = V_A/I_T$.

■ Capacitive charge or discharge current i_C is equal to $C(dv/dt)$ for any waveshape of v_C.

■ For a series capacitor, the amount of voltage drop is inversely proportional to its capacitance. The smaller the capacitance, the larger the voltage drop.

Cumulative Review Self-Test

Answers at back of book.

Answer True or False.

1. A capacitor can store charge because it has a dielectric between two conductors.

2. With 100-V applied, a 0.01-μF capacitor stores 1 μC of charge.

3. The smaller the capacitance, the higher the potential difference across it for a given amount of charge stored in the capacitor.

4. A 250-pF capacitance equals 250×10^{-12} F.

5. The thinner the dielectric, the greater the capacitance and the lower the breakdown voltage rating for a capacitor.

6. Larger plate area increases capacitance.

7. Capacitors in series provide less capacitance but a higher breakdown voltage rating for the combination.

8. Capacitors in parallel increase the total capacitance with the same voltage rating.

9. Two 0.01-μF capacitors in parallel have a total C of 0.005 μF.

10. A good 0.1-μF film capacitor will show charging current and read 500 MΩ or more on an ohmmeter.

11. If the capacitance is doubled, the reactance is halved.

12. If the frequency is doubled, the capacitive reactance is doubled.

13. The reactance of a 0.1-μF capacitor at 60 Hz is approximately 60 Ω.

14. In a series RC circuit, the voltage across X_C lags 90° behind the current.

15. The phase angle of a series RC circuit can be any angle between 0° and −90°, depending on the ratio of X_C to R.

16. In a parallel RC circuit, the voltage across X_C lags 90° behind its capacitive branch current.

17. In a parallel circuit of two resistances with 1 A in each branch, the total line current equals 1.414 A.

18. A 1000-Ω X_C in parallel with a 1000-Ω R has a combined Z of 707 Ω.

19. A 1000-Ω X_C in series with a 1000-Ω R has a total Z of 1414 Ω.

20. Neglecting its sign, the phase angle is 45° for both circuits in Probs. 18 and 19.

21. The total impedance of a 1-MΩ R in series with a 5-Ω X_C is approximately 1 MΩ with a phase angle of 0°.

22. The combined impedance of a 5-Ω R in parallel with a 1–MΩ X_C is approximately 5 Ω with a phase angle of 0°.

23. Both resistance and impedance are measured in ohms.

24. The impedance Z of an RC circuit can change with frequency because the circuit includes reactance.

25. Capacitors in series have the same charge and discharge current.

26. Capacitors in parallel have the same voltage.

27. The phasor combination of a 30-Ω R in series with a 40-Ω X_C equals 70 Ω impedance.

28. A film capacitor coded 103 has a value of 0.001 μF.

29. Capacitive current can be considered leading current in a series circuit.

30. In a series RC circuit, the higher the value of X_C, the greater its voltage drop compared with the IR drop.

31. Electrolytic capacitors typically have more leakage current than plastic-film capacitors.

32. A 0.04-μF capacitor in series with a 0.01-μF capacitor has an equivalent capacitance, C_{EQ} of 0.008 μF.

33. A shorted capacitor measures 0 Ω.

34. An open capacitor measures infinite ohms.

35. The X_C of a capacitor is inversely proportional to both f and C.

36. The equivalent series resistance, ESR, of a capacitor can be measured with an ohmmeter.

37. In an RC coupling circuit, the output is taken across C.

38. The equivalent impedance, Z_{EQ}, of a parallel RC circuit will decrease if the frequency of the applied voltage increases.

39. Electrolytic capacitors usually have lower breakdown voltage ratings than mica, film, and ceramic capacitors.

40. A 10-μF and a 5-μF capacitor are in series with a dc voltage source. The 10-μF capacitor will have the larger voltage drop.

chapter 19

Inductance

Inductance is the ability of a conductor to produce induced voltage when the current varies. A long wire has more inductance than a short wire, since more conductor length cut by magnetic flux produces more induced voltage. Similarly, a coil has more inductance than the equivalent length of straight wire because the coil concentrates magnetic flux. Components manufactured to have a definite value of inductance are coils of wire, called *inductors*. The symbol for inductance is L, and the unit is the henry (H).

The wire for a coil can be wound around a hollow, insulating tube, or the coil can be the wire itself. This type is an air-core coil because the magnetic field of the current in the coil is in air. With another basic type, the wire is wound on an iron core to concentrate the magnetic flux for more inductance.

Air-core coils are used in rf circuits because higher frequencies need less L for the required inductive effect. Iron-core inductors are used in the audio-frequency range, especially in the ac power-line frequency of 60 Hz and for lower frequencies in general.

Objectives

After studying this chapter you should be able to

- *Explain* the concept of self-inductance.

- *Define* the henry unit of inductance and *define* mutual inductance.

- *Calculate* the inductance when the induced voltage and rate of current change are known.

- *List* the physical factors affecting the inductance of an inductor.

- *Calculate* the induced voltage across an inductor, given the inductance and rate of current change.

- *Explain* how induced voltage opposes a change in current.

- *Describe* how a transformer works and *list* important transformer ratings.

- *Calculate* the currents, voltages, and impedances of a transformer circuit.

- *Identify* the different types of transformer cores.

- *Calculate* the total inductance of series-connected inductors.

- *Calculate* the equivalent inductance of parallel-connected inductors.

- *List* some common troubles with inductors.

Outline

Important Terms

autotransformer
coefficient of coupling, k
counter emf (cemf)
eddy current
efficiency
ferrite core
henry (H)
impedance matching

inductance, L
leakage flux
Lenz's law
mutual inductance, L_M
phasing dots
reflected impedance
series-aiding
series-opposing

stray capacitance
stray inductance
transformer
turns ratio
variac
volt-ampere (VA)

19–1 Induction by Alternating Current

Induced voltage is the result of flux cutting across a conductor. This action can be produced by physical motion of either the magnetic field or the conductor. When the current in a conductor varies in amplitude, however, the variations of current and its associated magnetic field are equivalent to motion of the flux. As the current increases in value, the magnetic field expands outward from the conductor. When the current decreases, the field collapses into the conductor. As the field expands and collapses with changes of current, the flux is effectively in motion. Therefore, a varying current can produce induced voltage without the need for motion of the conductor.

Figure 19–1 illustrates the changes in the magnetic field of a sine wave of alternating current. Since the alternating current varies in amplitude and reverses in direction, its magnetic field has the same variations. At point A, the current is zero and there is no flux. At B, the positive direction of current provides some field lines taken here in the counterclockwise direction. Point C has maximum current and maximum counterclockwise flux.

At D there is less flux than at C. Now the field is collapsing because of reduced current. At E, with zero current, there is no magnetic flux. The field can be considered as having collapsed into the wire.

The next half-cycle of current allows the field to expand and collapse again, but the directions are reversed. When the flux expands at points F and G, the field lines are clockwise, corresponding to current in the negative direction. From G to H and I, this clockwise field collapses into the wire.

The result of an expanding and collapsing field, then, is the same as that of a field in motion. This moving flux cuts across the conductor that is providing the current, producing induced voltage in the wire itself. Furthermore, any other conductor in the field, whether or not carrying current, also is cut by the varying flux and has induced voltage.

It is important to note that induction by a varying current results from the change in current, not the current value itself. The current must change to provide motion of the flux. A steady direct current of 1000 A, as an example of a large current, cannot produce any induced voltage as long as the current value is constant. A current of 1 μA changing to 2 μA, however, does induce voltage. Also, the faster the current changes, the higher the induced voltage because when the flux moves at a higher speed, it can induce more voltage.

Since inductance is a measure of induced voltage, the amount of inductance has an important effect in any circuit in which the current changes.

Figure 19–1 The magnetic field of an alternating current is effectively in motion as it expands and contracts with current variations.

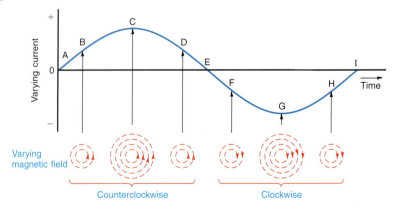

The inductance is an additional characteristic of a circuit beside its resistance. The characteristics of inductance are important in

1. *AC circuits.* Here the current is continuously changing and producing induced voltage. Lower frequencies of alternating current require more inductance to produce the same amount of induced voltage as a higher frequency current. The current can have any waveform, as long as the amplitude is changing.
2. *DC circuits in which the current changes in value.* It is not necessary for the current to reverse direction. One example is a dc circuit turned on or off. When the direct current is changing between zero and its steady value, the inductance affects the circuit at the time of switching. This effect of a sudden change is called the circuit's *transient response.* A steady direct current that does not change in value is not affected by inductance, however, because there can be no induced voltage without a change in current.

■ *19–1 Knowledge Check*

Answer at end of chapter.

Which can produce more induced voltage in a conductor, a dc current of 2 A or a sine wave with a peak-to-peak value of 1 mA?

■ *19–1 Self-Review*

Answers at end of chapter.

a. **For the same number of turns and frequency, which has more inductance, a coil with an iron core or one without an iron core?**
b. **In Fig. 19–1, are the changes of current faster at time B or C?**

19–2 Self-Inductance *L*

The ability of a conductor to induce voltage in itself when the current changes is its *self-inductance* or simply *inductance.* The symbol for inductance is *L*, for linkages of the magnetic flux, and its unit is the *henry* (H). This unit is named after Joseph Henry (1797–1878).

Definition of the Henry Unit

As illustrated in Fig. 19–2, 1 henry is the amount of inductance that allows 1 volt to be induced when the current changes at the rate of one ampere per second. The formula is

$$L = \frac{v_L}{di/dt} \qquad\qquad (19\text{–}1)$$

where v_L is in volts and di/dt is the current change in amperes per second.

Figure 19–2 When a current change of 1 A/s induces 1 V across *L*, its inductance equals 1 H.

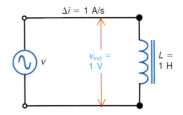

Again the symbol d is used to indicate an infinitesimally small change in current with time. The factor di/dt for the current variation with respect to time specifies how fast the current's magnetic flux is cutting the conductor to produce v_L.

Example 19–1

The current in an inductor changes from 12 to 16 A in 1 s. How much is the di/dt rate of current change in amperes per second?

ANSWER The di is the difference between 16 and 12, or 4 A in 1 s. Then

$$\frac{di}{dt} = 4 \text{ A/s}$$

Example 19–2

The current in an inductor changes by 50 mA in 2 μs. How much is the di/dt rate of current change in amperes per second?

ANSWER

$$\frac{di}{dt} = \frac{50 \times 10^{-3}}{2 \times 10^{-6}} = 25 \times 10^3$$
$$= 25,000 \text{ A/s}$$

Example 19–3

How much is the inductance of a coil that induces 40 V when its current changes at the rate of 4 A/s?

ANSWER

$$L = \frac{v_L}{di/dt} = \frac{40}{4}$$
$$= 10 \text{ H}$$

Example 19-4

How much is the inductance of a coil that induces 1000 V when its current changes at the rate of 50 mA in 2 μs?

ANSWER For this example, the $1/dt$ factor in the denominator of Formula (19–1) can be inverted to the numerator.

$$L = \frac{v_L}{di/dt} = \frac{v_L \times dt}{di}$$
$$= \frac{1 \times 10^3 \times 2 \times 10^{-6}}{50 \times 10^{-3}}$$
$$= \frac{2 \times 10^{-3}}{50 \times 10^{-3}} = \frac{2}{50}$$
$$= 0.04 \text{ H or } 40 \text{ mH}$$

Notice that the smaller inductance in Example 19–4 produces much more v_L than the inductance in Example 19–3. The very fast current change in Example 19–4 is equivalent to 25,000 A/s.

Inductance of Coils

In terms of physical construction, the inductance depends on how a coil is wound. Note the following factors.

1. A greater number of turns N increases L because more voltage can be induced. L increases in proportion to N^2. Double the number of turns in the same area and length increases the inductance four times.
2. More area A enclosed by each turn increases L. This means that a coil with larger turns has more inductance. The L increases in direct proportion to A and as the square of the diameter of each turn.
3. The L increases with the permeability of the core. For an air core, μ_r is 1. With a magnetic core, L is increased by the μ_r factor because the magnetic flux is concentrated in the coil.
4. The L decreases with more length for the same number of turns because the magnetic field is less concentrated.

These physical characteristics of a coil are illustrated in Fig. 19–3. For a long coil, where the length is at least 10 times the diameter, the inductance can be calculated from the formula

$$L = \mu_r \times \frac{N^2 \times A}{l} \times 1.26 \times 10^{-6} \text{ H} \qquad (19\text{–}2)$$

GOOD TO KNOW

The inductance of a coil with a magnetic core will vary with the amount of current (both dc and ac) that passes through the coil. Too much current will saturate the magnetic core, thus reducing its permeability, μ_r, and in turn the inductance, L. Most iron-core inductors (also known as chokes) have an inductance rating at a predetermined value of direct current. As an example, an iron-core choke may have the following rating: 8.5 H @ 50 mA.

Figure 19–3 Physical factors for inductance L of a coil. See text for calculating L.

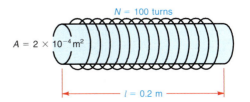

N = 100 turns

$A = 2 \times 10^{-4} \text{ m}^2$

$l = 0.2$ m

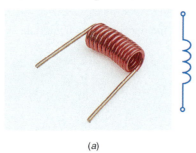

(*a*)

(*b*)

GOOD TO KNOW

A steady dc current cannot induce a voltage in a coil because the magnetic flux is stationary.

where L is in henrys, l is in meters, and A is in square meters. The constant factor 1.26×10^{-6} is the absolute permeability of air or vacuum in SI units to calculate L in henrys.

For the air-core coil in Fig. 19–3,

$$L = 1 \times \frac{10^4 \times 2 \times 10^{-4}}{0.2} \times 1.26 \times 10^{-6}$$
$$= 12.6 \times 10^{-6}\,\text{H} = 12.6\,\mu\text{H}$$

This value means that the coil can produce a self-induced voltage of 12.6 μV when its current changes at the rate of 1 A/s because $v_L = L(di/dt)$. Furthermore, if the coil has an iron core with $\mu_r = 100$, then L will be 100 times greater.

Typical Coil Inductance Values

Air-core coils for rf applications have L values in millihenrys (mH) and microhenrys (μH). A typical air-core rf inductor (called a *choke*) is shown with its schematic symbol in Fig. 19–4*a*. Note that

$$1\,\text{mH} = 1 \times 10^{-3}\,\text{H}$$
$$1\,\mu\text{H} = 1 \times 10^{-6}\,\text{H}$$

For example, an rf coil for the radio broadcast band of 535 to 1605 kHz may have an inductance L of 250 μH, or 0.250 mH.

Iron-core inductors for the 60-Hz power line and for audio frequencies have inductance values of about 1 to 25 H. An iron-core choke is shown in Fig. 19–4*b*.

■ 19–2 Knowledge Check

Answer at end of chapter.

In Fig. 19–3, how much is L if $N = 200$ turns?

■ 19–2 Self-Review

Answers at end of chapter.

a. **A coil induces 2 V with di/dt of 1 A/s. How much is L?**
b. **A coil has L of 8 mH with 125 turns. If the number of turns is doubled, how much will L be?**

19–3 Self-Induced Voltage v_L

The self-induced voltage across an inductance L produced by a change in current di/dt can be stated as

$$v_L = L\frac{di}{dt} \tag{19–3}$$

where v_L is in volts, L in henrys, and di/dt in amperes per second. This formula is an inverted version of Formula (19–1), which defines inductance.

Actually, both versions are based on Formula (14–1): $v = N(d\phi/dt)$ for magnetism. This gives the voltage in terms of the amount of magnetic flux cut by a conductor per second. When the magnetic flux associated with the current varies the same as i, then Formula (19–3) gives the same results for calculating induced voltage. Remember also that the induced voltage across the coil is actually the result of inducing electrons to move in the conductor, so that there is also an induced current. In using Formula (19–3) to calculate v_L, multiply L by the di/dt factor.

Example 19-5

How much is the self-induced voltage across a 4-H inductance produced by a current change of 12 A/s?

ANSWER

$$v_L = L\frac{di}{dt} = 4 \times 12$$
$$= 48 \text{ V}$$

Example 19-6

The current through a 200-mH L changes from 0 to 100 mA in 2 μs. How much is v_L?

ANSWER

$$v_L = L\frac{di}{dt}$$
$$= 200 \times 10^{-3} \times \frac{100 \times 10^{-3}}{2 \times 10^{-6}}$$
$$= 10,000 \text{ V or } 10 \text{ kV}$$

Note the high voltage induced in the 200-mH inductance because of the fast change in current.

The induced voltage is an actual voltage that can be measured, although v_L is produced only while the current is changing. When di/dt is present for only a short time, v_L is in the form of a voltage pulse. For a sine-wave current, which is always changing, v_L is a sinusoidal voltage 90° out of phase with i_L.

■ 19-3 Knowledge Check

Answer at end of chapter.

For a 50-mH inductor, how much is v_L if $di/dt = 50$ mA/1 μs?

■ 19-3 Self-Review

Answers at end of chapter.

a. If L is 2 H and di/dt is 1 A/s, how much is v_L?
b. For the same coil, the di/dt is increased to 100 A/s. How much is v_L?

19–4 How v_L Opposes a Change in Current

By Lenz's law, the induced voltage v_L must produce current with a magnetic field that opposes the change of current that induces v_L. The polarity of v_L, therefore, depends on the direction of the current variation di. When di increases, v_L has polarity that opposes the increase in current; when di decreases, v_L has opposite polarity to oppose the decrease in current.

In both cases, the change in current is opposed by the induced voltage. Otherwise, v_L could increase to an unlimited amount without the need to add any work. *Inductance, therefore, is the characteristic that opposes any change in current.* This is the reason that an induced voltage is often called a *counter emf* or *back emf*.

More details of applying Lenz's law to determine the polarity of v_L in a circuit are shown in Fig. 19–5. Note the directions carefully. In Fig. 19–5a, the electron flow is into the top of the coil. This current is increasing. By Lenz's law, v_L must have the polarity needed to oppose the increase. The induced voltage shown with the top side negative opposes the increase in current. The reason is that this polarity of v_L can produce current in the opposite direction, from minus to plus in the external circuit. Note that for this opposing current, v_L is the generator. This action tends to keep the current from increasing.

In Fig. 19–5b, the source is still producing electron flow into the top of the coil, but i is decreasing because the source voltage is decreasing. By Lenz's law, v_L must have the polarity needed to oppose the decrease in current. The induced voltage shown with the top side positive now opposes the decrease. The reason is that this polarity of v_L can produce current in the same direction, tending to keep the current from decreasing.

In Fig. 19–5c, the voltage source reverses polarity to produce current in the opposite direction, with electron flow into the bottom of the coil. The current in this reversed direction is now increasing. The polarity of v_L

Figure 19–5 Determining the polarity of v_L that opposes the change in i. (*a*) The i is increasing, and v_L has the polarity that produces an opposing current. (*b*) The i is decreasing, and v_L produces an aiding current. (*c*) The i is increasing but is flowing in the opposite direction. (*d*) The same direction of i as in (*c*) but with decreasing values.

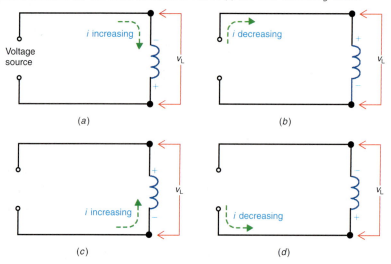

must oppose the increase. As shown, now the bottom of the coil is made negative by v_L to produce current opposing the source current. Finally, in Fig. 19–5d, the reversed current is decreasing. This decrease is opposed by the polarity shown for v_L to keep the current flowing in the same direction as the source current.

Notice that the polarity of v_L reverses for either a reversal of direction for i or a reversal of change in di between increasing or decreasing values. When both the direction of the current and the direction of change are reversed, as in a comparison of Fig. 19–5a and d, the polarity of v_L remains unchanged.

Sometimes the formulas for induced voltage are written with minus signs to indicate that v_L opposes the change, as specified by Lenz's law. However, the negative sign is omitted here so that the actual polarity of the self-induced voltage can be determined in typical circuits.

In summary, Lenz's law states that the reaction v_L opposes its cause, which is the change in i. When i is increasing, v_L produces an opposing current. For the opposite case when i is decreasing, v_L produces an aiding current.

■ 19–4 Knowledge Check

Answer at end of chapter.

What is another name for an induced voltage?

■ 19–4 Self-Review

Answers at end of chapter.

a. In Fig. 19–5a and b, the v_L has opposite polarities. (True or False)
b. In Fig. 19–5b and c, the polarity of v_L is the same. (True or False)

19–5 Mutual Inductance L_M

When the current in an inductor changes, the varying flux can cut across any other inductor nearby, producing induced voltage in both inductors. In Fig. 19–6, the coil L_1 is connected to a generator that produces varying current in the turns. The winding L_2 is not connected to L_1, but the turns are linked by the magnetic field. A varying current in L_1, therefore, induces voltage across L_1 and across L_2. If all flux of the current in L_1 links all turns of the coil L_2, each turn in L_2 will have the same amount of induced voltage as each turn in L_1. Furthermore, the induced voltage v_{L_2} can produce current in a load resistance connected across L_2.

Figure 19–6 Mutual inductance L_M between L_1 and L_2 linked by magnetic flux.

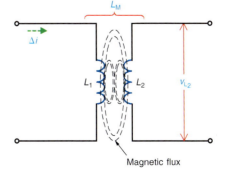

GOOD TO KNOW

In Fig. 19–6 the induced voltage across L_2 can be determined if L_1, L_M and V_{L_1} are known. The formula is: $V_{L_2} = \dfrac{L_M}{L_1} \times V_{L_1}$.

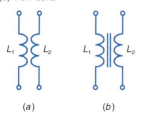

When the induced voltage produces current in L_2, its varying magnetic field induces voltage in L_1. The two coils, L_1 and L_2, have mutual inductance, therefore, because current in one can induce voltage in the other.

The unit of mutual inductance is the henry, and the symbol is L_M. *Two coils have L_M of 1 H when a current change of 1 A/s in one coil induces 1 V in the other coil.*

The schematic symbol for two coils with mutual inductance is shown in Fig. 19–7*a* for an air core and in Fig. 19–7*b* for an iron core. Iron increases the mutual inductance, since it concentrates magnetic flux. Any magnetic lines that do not link the two coils result in *leakage flux.*

Coefficient of Coupling

The fraction of total flux from one coil linking another coil is the coefficient of coupling k between the two coils. As examples, if all the flux of L_1 in Fig. 19–6 links L_2, then k equals 1, or unity coupling; if half the flux of one coil links the other, k equals 0.5. Specifically, the coefficient of coupling is

$$k = \frac{\text{flux linkages between } L_1 \text{ and } L_2}{\text{flux produced by } L_1}$$

There are no units for k, because it is a ratio of two values of magnetic flux. The value of k is generally stated as a decimal fraction, like 0.5, rather than as a percent.

The coefficient of coupling is increased by placing the coils close together, possibly with one wound on top of the other, by placing them parallel rather than perpendicular to each other, or by winding the coils on a common iron core. Several examples are shown in Fig. 19–8.

A high value of k, called *tight coupling,* allows the current in one coil to induce more voltage in the other coil. *Loose coupling,* with a low value of k, has the opposite effect. In the extreme case of zero coefficient of coupling, there is no mutual inductance. Two coils may be placed perpendicular to each other and far apart for essentially zero coupling to minimize interaction between the coils.

Air-core coils wound on one form have values of k equal to 0.05 to 0.3, approximately, corresponding to 5 to 30% linkage. Coils on a common iron core can be considered to have practically unity coupling, with k equal to 1. As shown in Fig. 19–8*c*, for both windings L_1 and L_2, practically all magnetic flux is in the common iron core. Mutual inductance is also called *mutual coupling.*

Figure 19–8 Examples of coupling between two coils linked by L_M. (*a*) L_1 or L_2 on paper or plastic form with air core; k is 0.1. (*b*) L_1 wound over L_2 for tighter coupling; k is 0.3. (*c*) L_1 and L_2 on the same iron core; k is 1. (*d*) Zero coupling between perpendicular air-core coils.

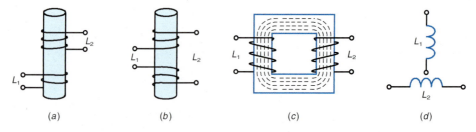

(a)   (b)   (c)   (d)

Example 19-7

A coil L_1 produces 80 μWb of magnetic flux. Of this total flux, 60 μWb are linked with L_2. How much is k between L_1 and L_2?

ANSWER

$$k = \frac{60\ \mu\text{Wb}}{80\ \mu\text{Wb}}$$
$$= 0.75$$

Example 19-8

A 10-H inductance L_1 on an iron core produces 4 Wb of magnetic flux. Another coil L_2 is on the same core. How much is k between L_1 and L_2?

ANSWER   Unity or 1. All coils on a common iron core have practically perfect coupling.

Calculating L_M

Mutual inductance increases with higher values for the primary and secondary inductances and tighter coupling:

$$L_M = k\sqrt{L_1 \times L_2} \tag{19-4}$$

where L_1 and L_2 are the self-inductance values of the two coils, k is the coefficient of coupling, and L_M is the mutual inductance linking L_1 and L_2, in the same units as L_1 and L_2. The k factor is needed to indicate the flux linkages between the two coils.

As an example, suppose that $L_1 = 2$ H and $L_2 = 8$ H, with both coils on an iron core for unity coupling. Then the mutual inductance is

$$L_M = 1\sqrt{2 \times 8} = \sqrt{16} = 4\ \text{H}$$

The value of 4 H for L_M in this example means that when the current changes at the rate of 1 A/s in either coil, it will induce 4 V in the other coil.

Example 19-9

Two 400-mH coils L_1 and L_2 have a coefficient of coupling k equal to 0.2. Calculate L_M.

$$L_M = k\sqrt{L_1 \times L_2}$$
$$= 0.2\sqrt{400 \times 10^{-3} \times 400 \times 10^{-3}}$$
$$= 0.2 \times 400 \times 10^{-3}$$
$$= 80 \times 10^{-3} \text{ H or } 80 \text{ mH}$$

Example 19-10

If the two coils in Example 19–9 had a mutual inductance L_M of 40 mH, how much would k be?

ANSWER   Formula (19–4) can be inverted to find k.

$$k = \frac{L_M}{\sqrt{L_1 \times L_2}}$$
$$= \frac{40 \times 10^{-3}}{\sqrt{400 \times 10^{-3} \times 400 \times 10^{-3}}}$$
$$= \frac{40 \times 10^{-3}}{400 \times 10^{-3}}$$
$$= 0.1$$

Notice that the same two coils have one-half the mutual inductance L_M because the coefficient of coupling k is 0.1 instead of 0.2.

■ 19–5 Knowledge Check

Answer at end of chapter.

Two 50-mH coils L_1 and L_2, have a coefficient of coupling, k, equal to 0.5. Calculate L_M.

■ 19–5 Self-Review

Answers at end of chapter.

a. All flux from the current in L_1 links L_2. How much is the coefficient of coupling k?
b. Mutual inductance L_M is 9 mH with k of 0.2. If k is doubled to 0.4, how much will L_M be?

19–6 Transformers

The transformer is an important application of mutual inductance. As shown in Fig. 19–9, a transformer has a primary winding inductance L_P connected to a voltage source that produces alternating current, and the secondary winding inductance L_S is connected across the load resistance R_L. The purpose of the transformer is to transfer power from the primary, where the generator is

Figure 19–9 Iron-core transformer with a 1:10 turn ratio. Primary current I_P induces secondary voltage V_S, which produces current in secondary load R_L.

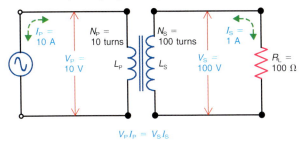

$$V_P I_P = V_S I_S$$

CALCULATOR

To do Example 19–9 on a calculator that does not have an (EXP) key, multiply $L_1 \times L_2$, take the square root of the product, and multiply by k. Keep the powers of 10 separate. Specifically, punch in 400 for L_1, push the $(\times)$ key, punch in 400 for L_2, and push the $(=)$ key for the product, 16,000. Press the $(\sqrt{})$ key, which is sometimes the $(2^{nd}F)$ of the (x^2) key to get 400. While it is on the display, push the $(\times)$ key, punch in 0.2, and press the $(=)$ key for the answer of 80. For the powers of 10, $10^{-3} \times 10^{-3} = 10^{-6}$, and the square root is equal to 10^{-3} for the unit of millihenry in the answer.

For Example 19–10, the formula is L_M divided by $\sqrt{L_1 \times L_2}$. Specifically, punch in 40 for the value in the numerator, press the $(\div)$ key, then the $(\,(\,)$ key, multiply 400×400, and press the $(\,)\,)$ key, followed by the $(\sqrt{})$ and $(=)$ keys. The display will read 0.1. The powers of 10 cancel with 10^{-3} in the numerator and denominator. Also, there are no units for k, since the units of L cancel.

connected, to the secondary, where the induced secondary voltage can produce current in the load resistance that is connected across L_S.

Although the primary and secondary are not physically connected to each other, power in the primary is coupled into the secondary by the magnetic field linking the two windings. The transformer is used to provide power for the load resistance R_L, instead of connecting R_L directly across the generator, whenever the load requires an ac voltage higher or lower than the generator voltage. By having more or fewer turns in L_S, compared with L_P, the transformer can step up or step down the generator voltage to provide the required amount of secondary voltage. Typical transformers are shown in Figs. 19–10 and 19–11. Note that a steady dc voltage cannot be stepped up or down by a transformer because a steady current cannot produce induced voltage.

Turns Ratio

The ratio of the number of turns in the primary to the number in the secondary is the turns ratio of the transformer:

$$\text{Turns ratio} = \frac{N_P}{N_S} \qquad (19\text{–}5)$$

where N_P = number of turns in the primary and N_S = number of turns in the secondary. For example, 500 turns in the primary and 50 turns in the secondary provide a turns ratio of 500/50, or 10:1, which is stated as "ten-to-one."

Voltage Ratio

With unity coupling between primary and secondary, the voltage induced in each turn of the secondary is the same as the self-induced voltage of each turn

Figure 19–11 Iron-core power transformer.

Figure 19–10 (a) Air-core rf transformer. Height is 2 in. (b) Color code and typical dc resistance of windings.

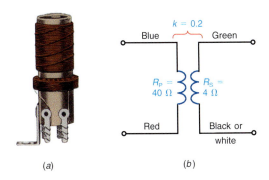

(a)   (b)

in the primary. Therefore, the voltage ratio is in the same proportion as the turns ratio:

$$\frac{V_P}{V_S} = \frac{N_P}{N_S} \tag{19–6}$$

When the secondary has more turns than the primary, the secondary voltage is higher than the primary voltage and the primary voltage is said to be stepped up. This principle is illustrated in Fig. 19–9 with a step-up ratio of 10/100, or 1:10. When the secondary has fewer turns, the voltage is stepped down.

In either case, the ratio is in terms of the primary voltage, which may be stepped up or down in the secondary winding.

These calculations apply only to iron-core transformers with unity coupling. Air-core transformers for rf circuits (as shown in Fig. 19–10a) are generally tuned to resonance. In this case, the resonance factor is considered instead of the turns ratio.

Example 19–11

A power transformer has 100 turns for N_P and 600 turns for N_S. What is the turns ratio? How much is the secondary voltage V_S if the primary voltage V_P is 120 V?

ANSWER The turns ratio is 100/600, or 1:6. Therefore, V_P is stepped up by the factor 6, making V_S equal to 6 × 120, or 720 V.

Example 19–12

A power transformer has 100 turns for N_P and 5 turns for N_S. What is the turns ratio? How much is the secondary voltage V_S with a primary voltage of 120 V?

ANSWER The turns ratio is 100/5, or 20:1. The secondary voltage is stepped down by a factor of $^1/_{20}$, making V_S equal to 120/20, or 6 V.

Secondary Current

By Ohm's law, the amount of secondary current equals the secondary voltage divided by the resistance in the secondary circuit. In Fig. 19–9, with a value of 100 Ω for R_L and negligible coil resistance assumed,

$$I_S = \frac{V_S}{R_L} = \frac{100 \text{ V}}{100 \text{ }\Omega} = 1 \text{ A}$$

Power in the Secondary

The power dissipated by R_L in the secondary is $I_S^2 \times R_L$ or $V_S \times I_S$, which equals 100 W in this example. The calculations are

$$P = I_S^2 \times R_L = 1 \times 100 = 100 \text{ W}$$
$$P = V_S \times I_S = 100 \times 1 = 100 \text{ W}$$

It is important to note that power used by the secondary load, such as R_L in Fig. 19–9, is supplied by the generator in the primary. How the load in the secondary draws power from the generator in the primary can be explained as follows.

With current in the secondary winding, its magnetic field opposes the varying flux of the primary current. The generator must then produce more primary current to maintain the self-induced voltage across L_P and the secondary voltage developed in L_S by mutual induction. If the secondary current doubles, for instance, because the load resistance is reduced by one-half, the primary current will also double in value to provide the required power for the secondary. Therefore, the effect of the secondary-load power on the generator is the same as though R_L were in the primary, except that the voltage for R_L in the secondary is stepped up or down by the turns ratio.

Current Ratio

With zero losses assumed for the transformer, the power in the secondary equals the power in the primary:

$$V_S I_S \times V_P I_P \tag{19–7}$$

or

$$\frac{I_S}{I_P} = \frac{V_P}{V_S} \tag{19–8}$$

The current ratio is the inverse of the voltage ratio, that is, voltage step-up in the secondary means current step-down, and vice versa. The secondary does not generate power but takes it from the primary. Therefore, the current step-up or step-down is in terms of the secondary current I_S, which is determined by the load resistance across the secondary voltage. These points are illustrated by the following two examples.

Example 19–13

A transformer with a 1:6 turns ratio has 720 V across 7200 Ω in the secondary. (a) How much is I_S? (b) Calculate the value of I_P.

ANSWER

a. $I_S = \dfrac{V_S}{R_L} = \dfrac{720 \text{ V}}{7200 \ \Omega}$

  $= 0.1$ A

b. With a turns ratio of 1:6, the current ratio is 6:1. Therefore,

  $I_P = 6 \times I_S = 6 \times 0.1$

  $= 0.6$ A

Example 19-14

A transformer with a 20:1 voltage step-down ratio has 6 V across 0.6 Ω in the secondary. (a) How much is I_S? (b) How much is I_P?

ANSWER

a. $I_S = \dfrac{V_S}{R_L} = \dfrac{6\ V}{0.6\ \Omega}$

$= 10\ A$

b. $I_P = \frac{1}{20} \times I_S = \frac{1}{20} \times 10$

$= 0.5\ A$

As an aid in these calculations, remember that the side with the higher voltage has the lower current. The primary and secondary V and I are in the same proportion as the number of turns in the primary and secondary.

Total Secondary Power Equals Primary Power

Figure 19–12 illustrates a power transformer with two secondary windings L_1 and L_2. There can be one, two, or more secondary windings with unity coupling to the primary as long as all the windings are on the same iron core. Each secondary winding has induced voltage in proportion to its turns ratio with the primary winding, which is connected across the 120 V source.

The secondary winding L_1 has a voltage step-up of 6:1, providing 720 V. The 7200-Ω load resistance R_1, across L_1, allows the 720 V to produce 0.1 A for I_1 in this secondary circuit. The power here is 720 V $\times$ 0.1 A = 72 W.

The other secondary winding L_2 provides voltage step-down with the ratio 20:1, resulting in 6 V across R_2. The 0.6-Ω load resistance in this circuit allows 10 A for I_2. Therefore, the power here is 6 V $\times$ 10 A, or 60 W. Since the windings have separate connections, each can have its individual values of voltage and current.

The total power used in the secondary circuits is supplied by the primary. In this example, the total secondary power is 132 W, equal to 72 W for P_1 and 60 W for P_2. The power supplied by the 120-V source in the primary then is 72 + 60 = 132 W.

Figure 19–12 Total power used by two secondary loads R_1 and R_2 is equal to the power supplied by the source in the primary.

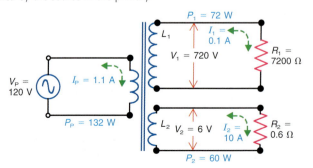

The primary current I_P equals the primary power P_P divided by the primary voltage V_P. This is 132 W divided by 120 V, which equals 1.1 A for the primary current. The same value can be calculated as the sum of 0.6 A of primary current providing power for L_1 plus 0.5 A of primary current for L_2, resulting in the total of 1.1 A as the value of I_P.

This example shows how to analyze a loaded power transformer. The main idea is that the primary current depends on the secondary load. The calculations can be summarized as follows:

1. Calculate V_S from the turns ratio and V_P.
2. Use V_S to calculate I_S: $I_S = V_S/R_L$.
3. Use I_S to calculate P_S: $P_S = V_S \times I_S$.
4. Use P_S to find P_P: $P_P = P_S$.
5. Finally, I_P can be calculated: $I_P = P_P/V_P$.

With more than one secondary, calculate each I_S and P_S. Then add all P_S values for the total secondary power, which equals the primary power.

Autotransformers

As illustrated in Fig. 19–13, an autotransformer consists of one continuous coil with a tapped connection such as terminal 2 between the ends at terminals 1 and 3. In Fig. 19–13a, the autotransformer steps up the generator voltage. Voltage V_P between 1 and 2 is connected across part of the total turns, and V_S is induced across all the turns. With six times the turns for the secondary voltage, V_S also is six times V_P.

In Fig. 19–13b, the autotransformer steps down the primary voltage connected across the entire coil. Then the secondary voltage is taken across less than the total turns.

The winding that connects to the voltage source to supply power is the primary, and the secondary is across the load resistance R_L. The turns ratio and voltage ratio apply the same way as in a conventional transformer having an isolated secondary winding.

Autotransformers are used often because they are compact and efficient and usually cost less since they have only one winding. Note that the autotransformer in Fig. 19–13 has only three leads, compared with four leads for the transformer in Fig. 19–9 with an isolated secondary.

Isolation of the Secondary

In a transformer with a separate winding for L_S, as in Fig. 19–9, the secondary load is not connected directly to the ac power line in the primary. This isolation

GOOD TO KNOW

Up until the invention of solid-state power control devices, autotransformers were used in theatrical stage lighting.

Figure 19–13 Autotransformer with tap at terminal 2 for 10 turns of the complete 60-turn winding. (a) V_P between terminals 1 and 2 stepped up across 1 and 3. (b) V_P between terminals 1 and 3 stepped down across 1 and 2.

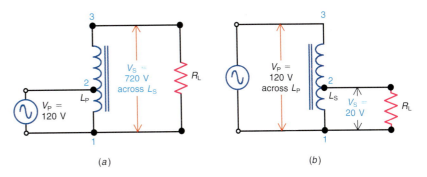

(a)   (b)

is an advantage in reducing the chance of electric shock. With an autotransformer, as in Fig. 19–13, the secondary is not isolated. Another advantage of an isolated secondary is that any direct current in the primary is blocked from the secondary. Sometimes a transformer with a 1:1 turns ratio is used for isolation from the ac power line.

Transformer Efficiency

Efficiency is defined as the ratio of power out to power in. Stated as a formula,

$$\% \text{ Efficiency} = \frac{P_{\text{out}}}{P_{\text{in}}} \times 100 \tag{19–9}$$

For example, when the power out in watts equals one-half the power in, the efficiency is one-half, which equals $0.5 \times 100\%$, or 50%. In a transformer, power out is secondary power, and power in is primary power.

Assuming zero losses in the transformer, power out equals power in and the efficiency is 100%. Actual power transformers, however, have an efficiency slightly less than 100%. The efficiency is approximately 80 to 90% for transformers that have high power ratings. Transformers for higher power are more efficient because they require heavier wire, which has less resistance. In a transformer that is less than 100% efficient, the primary supplies more than the secondary power. The primary power that is lost is dissipated as heat in the transformer, resulting from I^2R in the conductors and certain losses in the core material. The R of the primary winding is generally about 10 Ω or less for power transformers.

■ *19–6 Knowledge Check*

    Answer at end of chapter.

    A transformer has a turns ratio, $\dfrac{N_P}{N_S}$ of $\dfrac{2.5}{1}$. If $I_S = 750$ mA, how much is I_P?

■ *19–6 Self-Review*

    Answers at end of chapter.

    a. A transformer connected to the 120-V ac power line has a turns ratio of 1:2. Calculate the stepped-up V_S.
    b. A V_S of 240 V is connected across a 2400-Ω R_L. Calculate I_S.
    c. An autotransformer has an isolated secondary. (True or False)
    d. With more I_S for the secondary load, does the I_P increase or decrease?

19–7 Transformer Ratings

Like most other components, transformers have voltage, current, and power ratings that must not be exceeded. Exceeding any of these ratings will usually destroy the transformer. What follows is a brief description of the most important transformer ratings.

Voltage Ratings

Manufacturers of transformers always specify the voltage rating of the primary and secondary windings. Under no circumstances should the primary voltage rating be exceeded. In many cases, the rated primary and secondary voltages are printed on the transformer. For example, consider the transformer shown in Fig. 19–14a. Its rated primary voltage is 120 V, and its secondary voltage is

Figure 19–14 Transformer with primary and secondary voltage ratings. (*a*) Top black leads are primary leads. Yellow and black leads on bottom are secondary leads. (*b*) Schematic symbol.

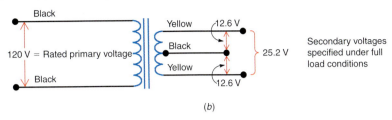

Black

Yellow ⌒12.6 V

120 V = Rated primary voltage

Black

Yellow

Black

12.6 V

25.2 V

Secondary voltages specified under full load conditions

(*b*)

(*a*)

specified as 12.6–0–12.6, which indicates that the secondary is center tapped. The notation 12.6–0–12.6 indicates that 12.6 V is available between the center tap connection and either outside secondary lead. The total secondary voltage available is 2 × 12.6 V or 25.2 V. In Fig. 19–14*a*, the black leads coming out of the top of the transformer provide connection to the primary winding. The two yellow leads coming out of the bottom of the transformer provide connection to the outer leads of the secondary winding. The bottom middle black lead connects to the center tap on the secondary winding.

Note that manufacturers may specify the secondary voltages of a transformer differently. For example, the secondary in Fig. 19–14*a* may be specified as 25.2 V CT, where CT indicates a center-tapped secondary. Another way to specify the secondary voltage in Fig. 19–14*a* would be 12.6 V each side of center.

Regardless of how the secondary voltage of a transformer is specified, the rated value is always specified under full load conditions with the rated primary voltage applied. A transformer is considered fully loaded when the rated current is drawn from the secondary. When unloaded, the secondary voltage will measure a value that is approximately 5 to 10% higher than its rated value. Let's use the transformer in Fig. 19–14*a* as an example. It has a rated secondary current of 2 A. If 120 V is connected to the primary and no load is connected to the secondary, each half of the secondary will measure somewhere between 13.2 and 13.9 V approximately. However, with the rated current of 2 A drawn from the secondary, each half of the secondary will measure approximately 12.6 V.

Figure 19–14*b* shows the schematic diagram for the transformer in Fig. 19–14*a*. Notice that the colors of each lead are identified for clarity.

As you already know, transformers can have more than one secondary winding. They can also have more than one primary winding. The purpose is to allow using the transformer with more than one value of primary voltage. Figure 19–15 shows a transformer with two separate primaries and a single secondary. This transformer can be wired to work with a primary voltage of either 120 or 240 V. For either value of primary voltage, the secondary voltage is 24 V. Figure 19–15*a* shows the individual primary windings with phasing dots to identify those leads with the same instantaneous polarity. Figure 19–15*b* shows how to connect the primary windings to 240 V. Notice the connections of the leads with the phasing dots. With this connection, each half of the primary voltage is in the proper phase to provide a series-aiding connection of the induced voltages. Furthermore, the series connection of the primary windings provides a turns ratio N_P/N_S of 10:1, thus allowing a secondary voltage of 24 V. Figure 19–15*c* shows how to connect the primaries to 120 V. Again, notice the connection of the leads with the phasing dots. When the primary windings are

Figure 19–15 Transformer with multiple primary windings. (*a*) Phasing dots show primary leads with same instantaneous polarity. (*b*) Primary windings connected in series to work with a primary voltage of 240 V; N_P/N_S = 10:1. (*c*) Primary windings connected in parallel to work with a primary voltage of 120 V; N_P/N_S = 5:1.

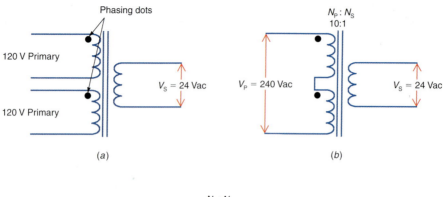

(a)   (b)

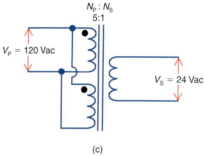

(c)

in parallel, the total primary current I_P is divided evenly between the windings. The parallel connection also provides a turns ratio N_P/N_S of 5:1, thus allowing a secondary voltage of 24 V.

Figure 19–16 shows a transformer that can operate with a primary voltage of either 120 or 440 V. In this case, only one of the primary windings is used with a given primary voltage. For example, if 120 V is applied to the lower primary, the upper primary winding is not used. Conversely, if 440 V is applied to the upper primary, the lower primary winding is not used.

Current Ratings

Manufacturers of transformers usually specify current ratings only for the secondary windings. The reason is quite simple. If the secondary current is not exceeded, there is no possible way the primary current can be exceeded. If the secondary current exceeds its rated value, excessive I^2R losses will result in the secondary winding. This will cause the secondary, and perhaps the primary, to

Figure 19–16 Transformer that has two primaries, which are used separately and never together.

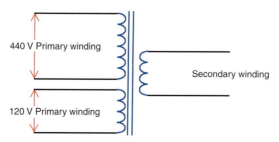

overheat, thus eventually destroying the transformer. The IR voltage drop across the secondary windings is the reason that the secondary voltage decreases as the load current increases.

Example 19-15

In Fig. 19–14b, calculate the primary current I_P if the secondary current I_S equals its rated value of 2 A.

ANSWER Rearrange Formula (19–8) and solve for the primary current I_P.

$$I_P = \frac{V_S}{V_P} \times I_S$$

$$= \frac{25.2 \text{ V}}{120 \text{ V}} \times 2 \text{ A}$$

$$= 0.42 \text{ A} \quad \text{or} \quad 420 \text{ mA}$$

Power Ratings

The power rating of a transformer is the amount of power the transformer can deliver to a resistive load. The power rating is specified in volt-amperes (VA) rather than watts (W) because the power is not actually dissipated by the transformer. The product VA is called *apparent power,* since it is the power that is *apparently* used by the transformer. The unit of apparent power is VA because the watt unit is reserved for the dissipation of power in a resistance.

Assume that a power transformer whose primary and secondary voltage ratings are 120 and 25 V, respectively, has a power rating of 125 VA. What does this mean? It means that the product of the transformer's primary, or secondary, voltage and current must not exceed 125 VA. If it does, the transformer will overheat and be destroyed. The maximum allowable secondary current for this transformer can be calculated as

$$I_{S(max)} = \frac{125 \text{ VA}}{25 \text{ V}}$$

$$I_{S(max)} = 5 \text{ A}$$

The maximum allowable primary current can be calculated as

$$I_{P(max)} = \frac{125 \text{ VA}}{120 \text{ V}}$$

$$I_{P(max)} = 1.04 \text{ A}$$

With multiple secondary windings, the VA rating of each individual secondary may be given without any mention of the primary VA rating. In this case, the sum of all secondary VA ratings must be divided by the rated primary voltage to determine the maximum allowable primary current.

In summary, you will never overload a transformer or exceed any of its maximum ratings if you obey two fundamental rules:

1. Never apply more than the rated voltage to the primary.
2. Never draw more than the rated current from the secondary.

Frequency Ratings

All transformers have a frequency rating which must be adhered to. Typical frequency ratings for power transformers are 50, 60, and 400 Hz. A power transformer with a frequency rating of 400 Hz cannot be used at 50 or 60 Hz, because it will overheat. However, many power transformers are designed to operate at either 50 or 60 Hz, because many types of equipment may be sold in both Europe and the United States, where the power-line frequencies are 50 and 60 Hz, respectively. Power transformers with a 400-Hz rating are often used in aircraft because these transformers are much smaller and lighter than 50- or 60-Hz transformers having the same power rating.

■ 19–7 Knowledge Check

Answer at end of chapter.

How is the power rating of a transformer usually specified?

■ 19–7 Self-Review

Answers at end of chapter.

a. The measured voltage across an unloaded secondary is usually 5 to 10% higher than its rated value. (True or False)
b. The current rating of a transformer is usually specified only for the secondary windings. (True or False)
c. A power rating of 300 VA for a transformer means that the transformer secondary must be able to dissipate this amount of power. (True or False)

19–8 Impedance Transformation

Transformers can be used to change or transform a secondary load impedance to a new value as seen by the primary. The secondary load impedance is said to be reflected back into the primary and is therefore called a *reflected impedance*. The reflected impedance of the secondary may be stepped up or down in accordance with the square of the transformer turns ratio.

By manipulating the relationships between the currents, voltages, and turns ratio in a transformer, an equation for the reflected impedance can be developed. This relationship is

$$Z_P = \left(\frac{N_P}{N_S}\right)^2 \times Z_S \qquad (19\text{--}10)$$

where Z_P = primary impedance and Z_S = secondary impedance (see Fig.19–17). If the turns ratio N_P/N_S is greater than 1, Z_S will be stepped up in value. Conversely, if the turns ratio N_P/N_S is less than 1, Z_S will be stepped down in value. It should be noted that the term *impedance* is used rather loosely here, since the primary and secondary impedances may be purely resistive. In the discussions and examples that follow, Z_P and Z_S will be assumed to be purely resistive. The concept of reflected impedance has several practical applications in electronics.

To find the required turns ratio when the impedance ratio is known, rearrange Formula (19–10) as follows:

$$\frac{N_P}{N_S} = \sqrt{\frac{Z_P}{Z_S}} \qquad (19\text{--}11)$$

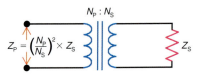

||| MultiSim   Figure 19–17   The secondary load impedance Z_S is reflected back into the primary as a new value that is proportional to the square of the turns ratio, N_P/N_S.

GOOD TO KNOW

If the secondary impedance, Z_S, of a transformer is capacitive or inductive in nature, the reflected impedance will also be capacitive or inductive in nature.

Example 19-16

Determine the primary impedance Z_P for the transformer circuit in Fig. 19–18.

Figure 19–18 Circuit for Example 19–16.

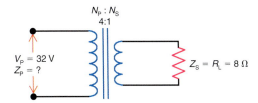

ANSWER Use Formula (19–10). Since $Z_S = R_L$, we have

$$Z_P = \left(\frac{N_P}{N_S}\right)^2 \times R_L$$

$$= \left(\frac{4}{1}\right)^2 \times 8\ \Omega$$

$$= 16 \times 8\ \Omega$$

$$= 128\ \Omega$$

The value of 128 Ω obtained for Z_P using Formula (19–10) can be verified as follows.

$$V_S = \frac{N_S}{N_P} \times V_P$$

$$= \frac{1}{4} \times 32\ \text{V}$$

$$= 8\ \text{V}$$

$$I_S = \frac{V_S}{R_L}$$

$$= \frac{8\ \text{V}}{8\ \Omega}$$

$$= 1\ \text{A}$$

$$I_P = \frac{V_S}{V_P} \times I_S$$

$$= \frac{8\ \text{V}}{32\ \text{V}} \times 1\ \text{A}$$

$$= 0.25\ \text{A}$$

And finally,

$$Z_P = \frac{V_P}{I_P}$$

$$= \frac{32\ \text{V}}{0.25\ \text{A}}$$

$$= 128\ \Omega$$

Example 19-17

In Fig. 19–19, calculate the turns ratio N_P/N_S which will produce a reflected primary impedance Z_P of (a) 75 Ω; (b) 600 Ω.

Figure 19-19 Circuit for Example 19-17.

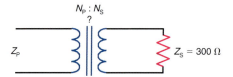

ANSWER (a) Use Formula (19–11).

$$\frac{N_P}{N_S} = \sqrt{\frac{Z_P}{Z_S}}$$

$$= \sqrt{\frac{75\ \Omega}{300\ \Omega}}$$

$$= \sqrt{\frac{1}{4}}$$

$$= \frac{1}{2}$$

(b)
$$\frac{N_P}{N_S} = \sqrt{\frac{Z_P}{Z_S}}$$

$$= \sqrt{\frac{600\ \Omega}{300\ \Omega}}$$

$$= \sqrt{\frac{2}{1}}$$

$$= \frac{1.414}{1}$$

Impedance Matching for Maximum Power Transfer

Transformers are used when it is necessary to achieve maximum transfer of power from a generator to a load when the generator and load impedances are not the same. This application of a transformer is called *impedance matching*.

As an example, consider the amplifier and load in Fig. 19–20a. Notice that the internal resistance r_i of the amplifier is 200 Ω and the load R_L is 8 Ω. If the amplifier and load are connected directly, as in Fig. 19–20b, the load receives 1.85 W of power, which is calculated as

$$P_L = \left(\frac{V_G}{r_i + R_L}\right)^2 \times R_L$$

$$= \left(\frac{100\ V}{200\ \Omega + 8\ \Omega}\right)^2 \times 8\ \Omega$$

$$= 1.85\ W$$

Figure 19–20 Transferring power from an amplifier to a load R_L. (a) Amplifier has $r_i = 200\ \Omega$ and $R_L = 8\ \Omega$. (b) Connecting the amplifier directly to R_L. (c) Using a transformer to make the 8-Ω R_L appear like 200 Ω in the primary.

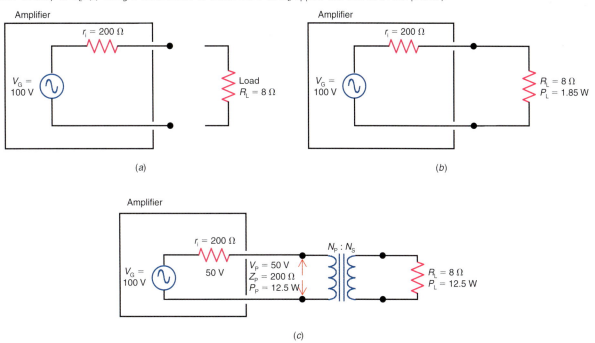

(a)

(b)

(c)

To increase the power delivered to the load, a transformer can be used between the amplifier and load. This is shown in Fig. 19–20c. We know that to transfer maximum power from the amplifier to the load, R_L must be transformed to a value equaling 200 Ω in the primary. With Z_P equaling r_i, maximum power will be delivered from the amplifier to the primary. Since the primary power P_P must equal the secondary power P_S, maximum power will also be delivered to the load R_L. In Fig. 19–20c, the turns ratio that provides a Z_P of 200 Ω can be calculated as

$$\frac{N_P}{N_S} = \sqrt{\frac{Z_P}{Z_S}}$$

$$= \sqrt{\frac{200\ \Omega}{8\ \Omega}}$$

$$= \frac{5}{1}$$

With r_i and Z_P equal, the power delivered to the primary can be calculated as

$$P_P = \left(\frac{V_G}{r_i + Z_P}\right)^2 \times Z_P$$

$$= \left(\frac{100\ \text{V}}{400\ \Omega}\right)^2 \times 200\ \Omega$$

$$= 12.5\ \text{W}$$

Since $P_P = P_S$, the load R_L also receives 12.5 W of power. As proof, calculate the secondary voltage.

$$V_S = \frac{N_S}{N_P} \times V_P$$

$$= \frac{1}{5} \times 50\ \text{V}$$

$$= 10\ \text{V}$$

Inductance **601**

(Notice that V_P is $\frac{1}{2}$ V_G, since r_i and Z_P divide V_G evenly.) Next, calculate the load power P_L.

$$P_L = \frac{V_S{}^2}{R_L}$$
$$= \frac{10^2 \text{ V}}{8 \text{ } \Omega}$$
$$= 12.5 \text{ W}$$

Notice how the transformer has been used as an impedance matching device to obtain the maximum transfer of power from the amplifier to the load. Compare the power dissipated by R_L in Fig. 19–20*b* to that in Fig. 19–20*c*. There is a big difference between the load power of 1.85 W in Fig. 19–20*b* and the load power of 12.5 W in Fig. 19–20*c*.

■ 19–8 Knowledge Check

Answer at end of chapter.

A transformer has a turns ratio, N_P/N_S of 3.16:1. If $Z_S = 8$ Ω, how much is Z_P?

■ 19–8 Self-Review

Answers at end of chapter.

a. The turns ratio of a transformer will not affect the primary impedance Z_P. (True or False)
b. When the turns ratio N_P/N_S is greater than 1, the primary impedance Z_P is less than the value of Z_S. (True or False)
c. If the turns ratio N_P/N_S of a transformer is 2/1 and $Z_S = 50$ Ω, the primary impedance $Z_P = 200$ Ω. (True or False)

19–9 Core Losses

The fact that the magnetic core can become warm, or even hot, shows that some of the energy supplied to the coil is used up in the core as heat. The two main effects are eddy-current losses and hysteresis losses.

Eddy Currents

In any inductance with an iron core, alternating current induces voltage in the core itself. Since it is a conductor, the iron core has current produced by the induced voltage. This current is called an *eddy current* because it flows in a circular path through the cross section of the core, as illustrated in Fig. 19–21.

The eddy currents represent wasted power dissipated as heat in the core. Note in Fig. 19–21 that the eddy-current flux opposes the coil flux, so that more current is required in the coil to maintain its magnetic field. The higher the frequency of the alternating current in the inductance, the greater the eddy-current loss.

Eddy currents can be induced in any conductor near a coil with alternating current, not only in its core. For instance, a coil has eddy-current losses in a metal cover. In fact, the technique of induction heating is an application of heat resulting from induced eddy currents.

RF Shielding

The reason that a coil may have a metal cover, usually copper or aluminum, is to provide a shield against the varying flux of rf current. In this case, the shielding

Figure 19–21 Cross-sectional view of iron core showing eddy currents.

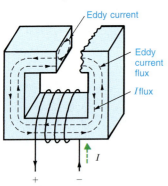

Figure 19–22 Laminated iron core.
(*a*) Shell-type construction. (*b*) E- and
I-shaped laminations. (*c*) Symbol for iron
core.

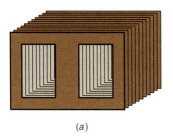

(*a*)

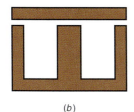

(*b*)

(*c*)

Figure 19–23 rf coils with ferrite
core. Width of coil is $^1/_2$ in. (*a*) Variable *L*
from 1 to 3 mH. (*b*) Tuning coil.

(*a*)   (*b*)

effect depends on using a good conductor for the eddy currents produced by the
varying flux, rather than magnetic materials used for shielding against static
magnetic flux.

The shield cover not only isolates the coil from external varying magnetic fields but also minimizes the effect of the coil's rf current for external circuits. The reason that the shield helps both ways is the same, as the induced
eddy currents have a field that opposes the field that is inducing the current.
Note that the clearance between the sides of the coil and the metal should be
equal to or greater than the coil radius to minimize the effect of the shield in
reducing the inductance.

Hysteresis Losses

Another loss factor present in magnetic cores is hysteresis, although hysteresis
losses are not as great as eddy-current losses. The hysteresis losses result from
the additional power needed to reverse the magnetic field in magnetic materials
in the presence of alternating current. The greater the frequency, the more
hysteresis losses.

Air-Core Coils

Note that air has practically no losses from eddy currents or hysteresis. However,
the inductance for small coils with an air core is limited to low values in the
microhenry or millihenry range.

■ *19–9 Knowledge Check*

    Answer at end of chapter.

    What are the two main losses in the core of a transformer?

■ *19–9 Self-Review*

    Answers at end of chapter.

    a. **Which has greater eddy-current losses, an iron core or an air core?**
    b. **Which produces more hysteresis losses, 60 Hz or 60 MHz?**

19–10 Types of Cores

To minimize losses while maintaining high flux density, the core can be made
of laminated steel layers insulated from each other. Insulated powdered-iron
granules and ferrite materials can also be used. These core types are illustrated
in Figs. 19–22 and 19–23. The purpose is to reduce the amount of eddy currents.
The type of steel itself can help reduce hysteresis losses.

Laminated Core

Figure 19–22*a* shows a shell-type core formed with a group of individual laminations. Each laminated section is insulated by a very thin coating of iron oxide,
silicon steel, or varnish. The insulating material increases the resistance in the
cross section of the core to reduce the eddy currents but allows a low-reluctance
path for high flux density around the core. Transformers for audio frequencies
and 60-Hz power are generally made with a laminated iron core.

Powdered-Iron Core

Powdered iron is generally used to reduce eddy currents in the iron core of an
inductance for radio frequencies. It consists of individual insulated granules
pressed into one solid form called a *slug*.

Ferrite Core

Ferrites are synthetic ceramic materials that are ferromagnetic. They provide high values of flux density, like iron, but have the advantage of being insulators. Therefore, a ferrite core can be used for high frequencies with minimum eddy-current losses.

This core is usually a slug that can move in or out of the coil to vary L, as in Fig. 19–23a. In Fig. 19–23b, the core has a hole to fit a plastic alignment tool for tuning the coil. Maximum L results with the slug in the coil.

■ 19–10 Knowledge Check

Answer at end of chapter.

Is ferrite material a conductor or an insulator?

■ 19–10 Self-Review

Answers at end of chapter.

a. An iron core provides a coefficient of coupling k of unity or 1. (True or False)
b. A laminated iron core reduces eddy-current losses. (True or False)
c. Ferrites have less eddy-current losses than iron. (True or False)

19–11 Variable Inductance

The inductance of a coil can be varied by one of the methods illustrated in Fig. 19–24. In Fig. 19–24a, more or fewer turns can be used by connection to one of the taps on the coil. Also, in Fig. 19–24b, a slider contacts the coil to vary the number of turns used. These methods are for large coils.

Figure 19–24c shows the schematic symbol for a coil with a slug of powdered iron or ferrite. The dotted lines indicate that the core is not solid iron. The arrow shows that the slug is variable. Usually, an arrow at the top means that the adjustment is at the top of the coil. An arrow at the bottom, pointing down, shows that the adjustment is at the bottom.

The symbol in Fig. 19–24d is a *variometer,* which is an arrangement for varying the position of one coil within the other. The total inductance of the series-aiding coils is minimum when they are perpendicular.

For any method of varying L, the coil with an arrow in Fig. 19–24e can be used. However, an adjustable slug is usually shown as in Fig. 19–24c.

A practical application of variable inductance is the *Variac.* The Variac is an autotransformer with a variable tap to change the turns ratio. The output

GOOD TO KNOW

A Variac is a common piece of test equipment used by technicians. It allows the technician to increase the ac voltage slowly while monitoring the operation of the equipment being repaired.

Figure 19–24 Methods of varying inductance. (*a*) Tapped coil. (*b*) Slider contact. (*c*) Adjustable slug. (*d*) Variometer. (*e*) Symbol for variable *L*.

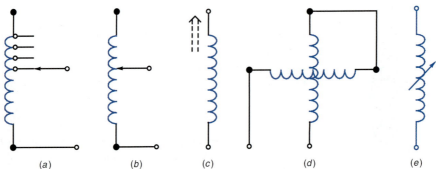

(a)　　(b)　　(c)　　(d)　　(e)

Figure 19–25 Variac with isolated output.

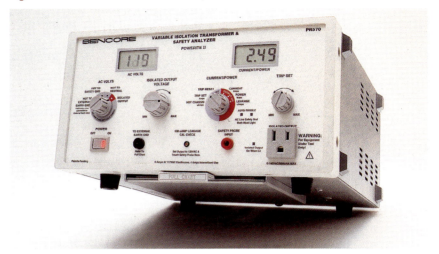

voltage in the secondary can be varied from 0 to approximately 140 V, with input from the 120-V, 60-Hz power line. One use is to test equipment with voltage above or below the normal line voltage.

The Variac is plugged into the power line, and the equipment to be tested is plugged into the Variac. Note that the power rating of the Variac should be equal to or more than the power used by the equipment being tested. Figure 19–25 shows a Variac with an isolated output.

■ *19–11 Knowledge Check*

    Answer at end of chapter.

    What is one use for the Variac shown in Fig. 19–25?

■ *19–11 Self-Review*

    Answers at end of chapter.

    a. A Variac is a transformer with a variable secondary voltage. (True or False)

    b. Figure 19–24*c* shows a ferrite or powdered iron core. (True or False)

19–12 Inductances in Series or Parallel

Figure 19–26 Inductances L_1 and L_2 in series without mutual coupling.

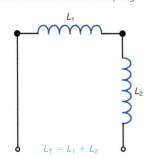

$$L_T = L_1 + L_2$$

As shown in Fig. 19–26, the total inductance of coils connected in series is the sum of the individual L values, as for series R. Since the series coils have the same current, the total induced voltage is a result of the total number of turns. Therefore, total series inductance is,

$$L_T = L_1 + L_2 + L_3 + \cdots + \text{etc.} \tag{19–12}$$

where L_T is in the same units of inductance as L_1, L_2, and L_3. This formula assumes no mutual induction between the coils.

Example 19-18

Inductance L_1 in Fig. 19–26 is 5 mH and L_2 is 10 mH. How much is L_T?

ANSWER $L_T = 5 \text{ mH} + 10 \text{ mH} = 15 \text{ mH}$.

With coils connected in parallel, the equivalent inductance is calculated from the reciprocal formula

$$L_{EQ} = \cfrac{1}{\cfrac{1}{L_1} + \cfrac{1}{L_2} + \cfrac{1}{L_3} + \cdots + \text{etc.}} \qquad (19\text{–}13)$$

Again, no mutual induction is assumed, as illustrated in Fig. 19–27.

Example 19-19

Inductances L_1 and L_2 in Fig. 19–27 are each 8 mH. How much is L_{EQ}?

ANSWER

$$L_{EQ} = \cfrac{1}{\frac{1}{8} + \frac{1}{8}}$$
$$= 4 \text{ mH}$$

Figure 19–27 Inductances L_1 and L_2 in parallel without mutual coupling.

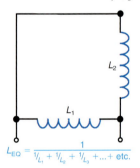

$$L_{EQ} = \frac{1}{\frac{1}{L_1} + \frac{1}{L_2} + \frac{1}{L_3} + \ldots + \text{etc.}}$$

All shortcuts for calculating parallel R can be used with parallel L, since both are based on the reciprocal formula. In this example, L_{EQ} is $\frac{1}{2} \times 8 = 4$ mH.

Series Coils with L_M

This depends on the amount of mutual coupling and on whether the coils are connected series-aiding or series-opposing. *Series-aiding* means that the common current produces the same direction of magnetic field for the two coils. The *series-opposing* connection results in opposite fields.

The coupling depends on the coil connections and direction of winding. Reversing either one reverses the field. Inductances L_1 and L_2 with the same direction of winding are connected series-aiding in Fig. 19–28a. However, they are series-opposing in Fig. 19–28b because L_1 is connected to the opposite end of L_2. To calculate the total inductance of two coils that are series-connected and have mutual inductance,

$$L_T = L_1 + L_2 \pm 2L_M \qquad (19\text{–}14)$$

The mutual inductance L_M is plus, increasing the total inductance, when the coils are series-aiding, or minus when they are series-opposing to reduce the total inductance.

Note the phasing dots above the coils in Fig. 19–28. Coils with phasing dots at the same end have the same direction of winding. When current enters the dotted ends for two coils, their fields are aiding and L_M has the same sense as L.

Figure 19–28 Inductances L_1 and L_2 in series but with mutual coupling L_M. (a) Aiding magnetic fields. (b) Opposing magnetic fields.

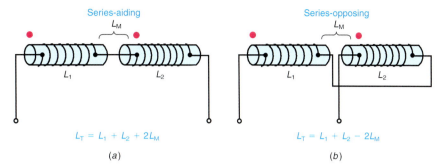

Series-aiding

L_M

L_1 L_2

$L_T = L_1 + L_2 + 2L_M$

(a)

Series-opposing

L_M

L_1 L_2

$L_T = L_1 + L_2 - 2L_M$

(b)

How to Measure L_M

Formula (19–14) provides a method of determining the mutual inductance between two coils L_1 and L_2 of known inductance. First, the total inductance is measured for the series-aiding connection. Let this be L_{T_a}. Then the connections to one coil are reversed to measure the total inductance for the series-opposing coils. Let this be L_{T_o}. Then

$$L_M = \frac{L_{T_a} - L_{T_o}}{4} \qquad\qquad (19\text{–}15)$$

When the mutual inductance is known, the coefficient of coupling k can be calculated from the fact that $L_M = k\sqrt{L_1 L_2}$.

Example 19–20

Two series coils, each with an L of 250 μH, have a total inductance of 550 μH connected series-aiding and 450 μH series-opposing. (a) How much is the mutual inductance L_M between the two coils? (b) How much is the coupling coefficient k?

ANSWER

a. $L_M = \dfrac{L_{T_a} - L_{T_o}}{4}$

$\quad = \dfrac{550 - 450}{4} = \dfrac{100}{4}$

$\quad = 25\ \mu\text{H}$

b. $L_M = k\sqrt{L_1 L_2}$, or

$\quad k = \dfrac{L_M}{\sqrt{L_1 L_2}} = \dfrac{25}{\sqrt{250 \times 250}}$

$\quad\quad = \dfrac{25}{250} = \dfrac{1}{10}$

$\quad\quad = 0.1$

Coils may also be in parallel with mutual coupling. However, the inverse relations with parallel connections and the question of aiding or opposing fields make this case complicated. Actually, it would hardly ever be used.

■ *19–12 Knowledge Check*

Answers at end of chapter.

Assuming no mutual coupling, how much is the combined inductance of a 30-mH and 50-mH inductor in (a) series (b) parallel?

■ *19–12 Self-Review*

Answers at end of chapter.

a. A 500-μH coil and a 1-mH coil are in series without L_M. Calculate L_T.
b. The same coils are in parallel without L_M. Calculate L_{EQ}.

19–13 Energy in a Magnetic Field of Inductance

The magnetic flux of the current in an inductance has electric energy supplied by the voltage source producing the current. The energy is stored in the magnetic field, since it can do the work of producing induced voltage when the flux moves. The amount of electric energy stored is

$$\text{Energy} = \mathscr{E} = \tfrac{1}{2}LI^2 \qquad\qquad (19\text{–}16)$$

The factor of $\tfrac{1}{2}$ gives the average result of I in producing energy. With L in henrys and I in amperes, the energy is in watt-seconds, or *joules*. For a 10-H L with a 3-A I, the electric energy stored in the magnetic field equals

$$\text{Energy} = \tfrac{1}{2}LI^2 = \frac{10 \times 9}{2} = 45 \text{ J}$$

This 45 J of energy is supplied by the voltage source that produces 3 A in the inductance. When the circuit is opened, the magnetic field collapses. The energy in the collapsing magnetic field is returned to the circuit in the form of induced voltage, which tends to keep the current flowing.

The entire 45 J is available for the work of inducing voltage, since no energy is dissipated by the magnetic field. With resistance in the circuit, however, the I^2R loss with induced current dissipates all energy after a period of time.

CALCULATOR

To do Example 19–21 on a calculator, first square the *I*, multiply by *L*, and divide by 2. Specifically, punch in 1.2 and push the $\boxed{x^2}$ key for 1.44. While this is on the display, push the $\boxed{\times}$ key, punch in 0.4, and push the $\boxed{=}$ key for 0.576 on the display. Now push the $\boxed{\div}$ key, punch in 2, and then push the $\boxed{=}$ key to get 0.288 as the answer.

Example 19–21

A current of 1.2 A flows in a coil with an inductance of 0.4 H. How much energy is stored in the magnetic field?

ANSWER

$$\text{Energy} = \frac{LI^2}{2} = \frac{0.4 \times 1.44}{2}$$
$$= 0.288 \text{ J}$$

■ *19–13 Knowledge Check*

Answer at end of chapter.

For an inductor carrying current, where is the energy stored?

■ *19–13 Self-Review*

Answers at end of chapter.

a. What is the unit of electric energy stored in a magnetic field?
b. Does a 4-H coil store more or less energy than a 2-H coil for the same current?

19–14 Stray Capacitive and Inductive Effects

Stray capacitive and inductive effects can occur in all circuits with all types of components. A capacitor has a small amount of inductance in the conductors. A coil has some capacitance between windings. A resistor has a small amount of inductance and capacitance. After all, physically a capacitance is simply an insulator between two conductors having a difference of potential. An inductance is basically a conductor carrying current.

Actually, though, these stray effects are usually quite small, compared with the concentrated or lumped values of capacitance and inductance. Typical values of stray capacitance may be 1 to 10 pF, whereas stray inductance is usually a fraction of 1 μH. For very high radio frequencies, however, when small values of L and C must be used, the stray effects become important. As another example, any wire cable has capacitance between the conductors.

A practical case of problems caused by stray L and C is a long cable used for rf signals. If the cable is rolled in a coil to save space, a serious change in the electrical characteristics of the line will take place. Specifically, for twin-lead or coaxial cable feeding the antenna input to a television receiver, the line should not be coiled because the added L or C can affect the signal. Any excess line should be cut off, leaving the little slack that may be needed. This precaution is not so important with audio cables.

Stray Circuit Capacitance

The wiring and components in a circuit have capacitance to the metal chassis. This stray capacitance C_S is typically 5 to 10 pF. To reduce C_S, the wiring should be short with the leads and components placed high off the chassis. Sometimes, for very high frequencies, stray capacitance is included as part of the circuit design. Then changing the placement of components or wiring affects the circuit operation. Such critical *lead dress* is usually specified in the manufacturer's service notes.

Stray Inductance

Although practical inductors are generally made as coils, all conductors have inductance. The amount of L is $v_L/(di/dt)$, as with any inductance producing induced voltage when the current changes. The inductance of any wiring not included in the conventional inductors can be considered stray inductance. In most cases, stray inductance is very small; typical values are less than 1 μH. For high radio frequencies, though, even a small L can have an appreciable inductive effect.

One source of stray inductance is connecting leads. A wire 0.04 in. in diameter and 4 in. long has an L of approximately 0.1 μH. At low frequencies, this inductance is negligible. However, consider the case of rf current, where i varies from 0 to 20-mA peak value, in the short time of 0.025 μs, for a quarter-cycle of a 10-MHz sine wave. Then v_L equals 80 mV, which is an appreciable inductive effect. This is one reason that connecting leads must be very short in rf circuits.

As another example, wire-wound resistors can have appreciable inductance when wound as a straight coil. This is why carbon resistors are preferred for minimum stray inductance in rf circuits. However, noninductive wire-wound resistors can also be used. These are wound so that adjacent turns have current in opposite directions and the magnetic fields oppose each other to cancel the inductance. Another application of this technique is twisting a pair of connecting leads to reduce the inductive effect.

Inductance of a Capacitor

Capacitors with coiled construction, particularly paper and electrolytic capacitors, have some internal inductance. The larger the capacitor, the greater its series inductance. Mica and ceramic capacitors have very little inductance, however, which is why they are generally used for radio frequencies.

For use above audio frequencies, the rolled-foil type of capacitor must have noninductive construction. This means that the start and finish of the foil winding must not be the terminals of the capacitor. Instead, the foil windings are offset. Then one terminal can contact all layers of one foil at one edge, and the opposite edge of the other foil contacts the second terminal. Most rolled-foil capacitors, including the paper and film types, are constructed this way.

Distributed Capacitance of a Coil

As illustrated in Fig. 19–29, a coil has distributed capacitance C_d between turns. Note that each turn is a conductor separated from the next turn by an insulator, which is the definition of capacitance. Furthermore, the potential of each turn is different from the next, providing part of the total voltage as a potential difference to charge C_d. The result then is the equivalent circuit shown for an rf coil. The L is the inductance and R_e its internal effective ac resistance in series with L, and the total distributed capacitance C_d for all turns is across the entire coil.

Special methods for minimum C_d include *space-wound* coils, where the turns are spaced far apart; the honeycomb or *universal* winding, with the turns crossing each other at right angles; and the *bank winding,* with separate sections called *pies.* These windings are for rf coils. In audio and power transformers, a grounded conductor shield, called a *Faraday screen,* is often placed between windings to reduce capacitive coupling.

Reactive Effects in Resistors

As illustrated by the high-frequency equivalent circuit in Fig. 19–30, a resistor can include a small amount of inductance and capacitance. The inductance of carbon-composition resistors is usually negligible. However, approximately 0.5 pF of capacitance across the ends may have an effect, particularly with large resistances used for high radio frequencies. Wire-wound resistors definitely have enough inductance to be evident at radio frequencies. However, special resistors are available with double windings in a noninductive method based on cancellation of opposing magnetic fields.

Capacitance of an Open Circuit

An open switch or a break in a conducting wire has capacitance C_O across the open. The reason is that the open consists of an insulator between two conductors. With a voltage source in the circuit, C_O charges to the applied voltage. Because of the small C_O, of the order of picofarads, the capacitance charges to the source voltage in a short time. This charging of C_O is the reason that an

Figure 19–29 Equivalent circuit of an rf coil. (*a*) Distributed capacitance C_d between turns of wire. (*b*) Equivalent circuit.

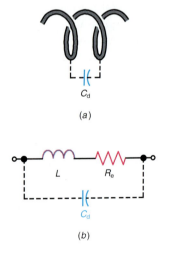

(a)

(b)

Figure 19–30 High-frequency equivalent circuit of a resistor.

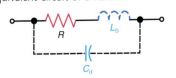

open series circuit has the applied voltage across the open terminals. After a momentary flow of charging current, C_O charges to the applied voltage and stores the charge needed to maintain this voltage.

■ *19–14 Knowledge Check*

Answer at end of chapter.

What type of resistor has an appreciable amount of inductance?

■ *19–14 Self-Review*

Answers at end of chapter.

a. A two-wire cable has distributed C between the conductors. (True or False)
b. A coil has distributed C between the turns. (True or False)
c. Stray inductance and stray capacitance are most likely to be a problem at high frequencies. (True or False)

19–15 Measuring and Testing Inductors

Although many DMMs are capable of measuring the value of a capacitor, few are capable of measuring the value of an inductor. Therefore, when it is necessary to measure the value of an inductor, you may want to use a capacitor-inductor analyzer like that shown earlier in Chap. 16. The capacitor-inductor analyzer can also test the quality (Q) of the inductor by using something called a *ringing test*.

Another test instrument that is capable of measuring inductance L, capacitance C, and resistance R, is an LCR meter. A typical LCR meter is shown in Fig. 19–31. Although this is a handy piece of test equipment, most LCR meters are not capable of measuring anything except the value of a component. Note, however, that some LCR meters are capable of making a few additional tests besides measuring the component value.

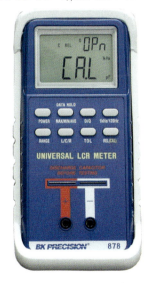

Figure 19–31 Typical LCR meter.

Inductor Coding

Inductors may or may not be coded to indicate their inductance value in henrys (H), millihenrys (mH), or microhenrys (μH). Some very small inductors used in rf circuits may consist of five or six turns of bare wire and therefore cannot be coded. Larger inductors, such as chokes used in the audio frequency range, normally have their inductance values printed on them. Some inductors use a coding system similar to that used with film capacitors. In this case, a three-digit code is used to indicate the inductance value in microhenrys. For example, an inductor may be coded 103; this is interpreted as follows: The first two digits (1 and 0) represent the first and second digits in the inductance value. The last digit (3), called the *multiplier digit,* tells how many zeros to add after the first two digits. In this case, 103 corresponds to an inductance of 10,000 μH. Some manufacturers put the multiplier digit first instead of last. For example, for an inductor coded 210, the second and third digits represent the first and second digits of the inductance value, and the first digit tells how many zeros to add. In this case, the code 210 corresponds to an inductance value of 1000 μH. Usually the three-digit codes include no tolerance rating. Sometimes inductors have their value printed on the body, sometimes they have colored stripes, and sometimes they are not coded at all. Confusing you say? Absolutely! Sometimes, the only sure way to determine the value of an inductor is to measure its value. Because no standardization is in place for the coding of inductors, no further coverage of the topic is provided here.

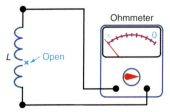

Ohmmeter

L Open

Figure 19–33 The internal dc resistance r_i of a coil is in series with its inductance L.

L

r_i

Troubles in Coils

The most common trouble in coils is an open winding. As illustrated in Fig. 19–32, an ohmmeter connected across the coil reads infinite resistance for the open circuit. It does not matter whether the coil has an air core or an iron core. Since the coil is open, it cannot conduct current and therefore has no inductance because it cannot produce induced voltage. When the resistance is checked, the coil should be disconnected from the external circuit to eliminate any parallel paths that could affect the resistance readings.

Direct Current Resistance of a Coil

A coil has dc resistance equal to the resistance of the wire used in the winding. The amount of resistance is less with heavier wire and fewer turns. For rf coils with inductance values up to several millihenrys, requiring 10 to 100 turns of fine wire, the dc resistance is 1 to 20 Ω, approximately. Inductors for 60 Hz and audio frequencies with several hundred turns may have resistance values of 10 to 500 Ω, depending on the wire size.

As shown in Fig. 19–33, the dc resistance and inductance of a coil are in series, since the same current that induces voltage in the turns must overcome the resistance of the wire. Although resistance has no function in producing induced voltage, it is useful to know the dc coil resistance because if it is normal, usually the inductance can also be assumed to have its normal value.

Open Coil

An open winding has infinite resistance, as indicated by an ohmmeter reading. With a transformer that has four leads or more, check the resistance across the two leads for the primary, across the two leads for the secondary, and across any other pairs of leads for additional secondary windings. For an autotransformer with three leads, check the resistance from one lead to each of the other two.

When the open circuit is inside the winding, it is usually not practical to repair the coil, and the entire unit is replaced. In some cases, an open connection at the terminals can be resoldered.

Value Change

The value of an inductor can change over time because of core breakage, windings relaxing, or shorted turns. Note that a coil whose inductance value is changed may check okay with an ohmmeter. To check the value of an inductor, use either a capacitor-inductor analyzer or an LCR meter.

Open Primary Winding

When the primary of a transformer is open, no primary current can flow, and no voltage is induced in any of the secondary windings.

Open Secondary Winding

When the secondary of a transformer is open, it cannot supply power to any load resistance across the open winding. Furthermore, with no current in the secondary, the primary current is also practically zero, as though the primary winding were open. The only primary current needed is the small magnetizing current to sustain the field producing induced voltage across the secondary without any load. If the transformer has several secondary windings, however, an open winding in one secondary does not affect transformer operation for the secondary circuits that are normal.

Short across Secondary Winding

In this case, excessive primary current flows, as though it were short-circuited, often burning out the primary winding. The reason is that the large secondary current has a strong field that opposes the flux of the self-induced voltage across the primary, making it draw more current from the generator.

■ 19–15 Knowledge Check

Answer at end of chapter.

What is the inductance of a small coil that is coded 102?

■ 19–15 Self-Review

Answers at end of chapter.

a. The normal R of a coil is 18 Ω. How much will an ohmmeter read if the coil is open?
b. The primary of a 1:3 step-up autotransformer is connected to a 120-Vac power line. How much will the secondary voltage be if the primary is open?
c. Are the dc resistance and inductance of a coil in series or in parallel?

Summary

- Varying current induces voltage in a conductor, since the expanding and collapsing field of the current is equivalent to flux in motion.

- Lenz's law states that the induced voltage produces I that opposes the change in current causing the induction. Inductance, therefore, tends to keep the current from changing.

- The ability of a conductor to produce induced voltage across itself when the current varies is its self-inductance, or inductance. The symbol is L, and the unit of inductance is the henry. One henry of inductance allows 1 V to be induced when the current changes at the rate of 1 A/s. For smaller units, 1 mH $= 1 \times 10^{-3}$ H and 1 μH $= 1 \times 10^{-6}$ H.

- To calculate self-induced voltage, $v_L = L(di/dt)$, with v in volts, L in henrys, and di/dt in amperes per second.

- Mutual inductance is the ability of varying current in one conductor to induce voltage in another conductor nearby. Its symbol is L_M, measured in henrys. $L_M = k\sqrt{L_1L_2}$, where k is the

coefficient of coupling between conductors.

- A transformer consists of two or more windings with mutual inductance. The primary winding connects to the source voltage; the load resistance is connected across the secondary winding. A separate winding is an isolated secondary. The transformer is used to step up or step down ac voltage.

- An autotransformer is a tapped coil, used to step up or step down the primary voltage. There are three leads with one connection common to both the primary and the secondary.

- A transformer with an iron core has essentially unity coupling. Therefore, the voltage ratio is the same as the turns ratio: $V_P/V_S = N_P/N_S$.

- Assuming 100% efficiency for an iron-core power transformer, the power supplied to the primary equals the power used in the secondary.

- The voltage rating of a transformer's secondary is always specified under full load conditions with the rated primary voltage applied. The measured

voltage across an unloaded secondary is usually 5 to 10% higher than its rated value.

- The current or power rating of a transformer is usually specified only for the secondary windings.

- Transformers can be used to reflect a secondary load impedance back into the primary as a new value that is either larger or smaller than its actual value. The primary impedance Z_P can be determined using Formula (19–10).

- The impedance transforming properties of a transformer make it possible to obtain maximum transfer of power from a generator to a load when the generator and load impedances are not equal. The required turns ratio can be determined using Formula (19–11).

- Eddy currents are induced in the iron core of an inductance, causing wasted power that heats the core. Eddy-current losses increase with higher frequencies of alternating current. To reduce eddy currents, the iron core is laminated. Powdered-iron and ferrite cores have minimum eddy-current losses at radio frequencies. Hysteresis also causes power loss.

■ With no mutual coupling, series inductances are added like series resistances. The equivalent inductance of parallel inductances is calculated by the reciprocal formula, as for parallel resistances.

■ The magnetic field of an inductance has stored energy $\mathscr{E} = \frac{1}{2} LI^2$. With I in amperes and L in henrys, energy $\mathscr{E}$ is in joules.

■ In addition to its inductance, a coil has dc resistance equal to the resistance of the wire in the coil. An open coil has infinitely high resistance.

■ An open primary in a transformer results in no induced voltage in any of the secondary windings.

■ Figure 19–34 summarizes the main types of inductors, or coils, with their schematic symbols.

■ The characteristics of inductance and capacitance are compared in Table 19-1.

■ Stray inductance can be considered the inductance of any wiring not included in conventional inductors. Stray capacitance can be considered the capacitance of any two conductors separated from each other by an insulator and not included in conventional capacitors.

Figure 19–34 Summary of types of inductors. (*a*) Air-core coil. (*b*) Iron-core coil. (*c*) Adjustable ferrite core. (*d*) Air-core transformer. (*e*) Variable L_P and L_S. (*f*) Iron-core transformer. (*g*) Autotransformer.

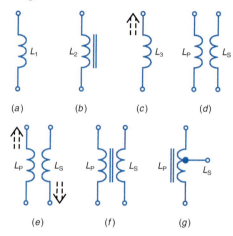

Table 19–1	Comparison of Capacitance and Inductance	
Capacitance		**Inductance**
Symbol is C		Symbol is L
Unit is the farad (F)		Unit is the henry (H)
Needs dielectric as insulator		Needs conductor for circuit path
More plate area allows more C		More turns allow more L
Dielectric can concentrate electric field for more C		Core can concentrate magnetic field for more L
$C_{EQ} = \dfrac{1}{\dfrac{1}{C_1} + \dfrac{1}{C_2} + \cdots + \text{etc.}}$ in series		$L_T = L_1 + L_2$ in series
$C_T = C_1 + C_2$ in parallel		$L_{EQ} = \dfrac{1}{1/L_1 + 1/L_2}$ in parallel

Important Terms

- Autotransformer - a transformer made of one continuous coil with a tapped connection between the end terminals. An autotransformer has only three leads and provides no isolation between the primary and secondary.

- Coefficient of Coupling, k - the fraction of total flux from one coil linking another coil nearby.

- Counter emf (cemf) - a term used to describe the effect of an induced voltage in opposing a change in current.

- Eddy Current - the current that flows in a circular path through the cross section of the iron core in a transformer.

- Efficiency - the ratio of power output to power input. In a transformer, power out is secondary power and power in is primary power.

- Ferrite Core - a type of core that has high flux density, like iron, but is an insulator. A ferrite core used in a coil has minimum eddy current losses due to its high resistance.

- Henry (H) Unit - the basic unit of inductance. 1 H is the amount of inductance that produces 1 volt of induced voltage when the current varies at the rate of 1 A/s.

- Impedance Matching - an application of a transformer in which the secondary load resistance is stepped up or down to provide maximum

transfer of power from the generator to the load.

- Inductance, L - the ability of a conductor to produce an induced voltage in itself when the current changes. Inductance can also be defined as the characteristic that opposes any change in current.

- Leakage Flux - any magnetic field lines that do not link two coils which are close to each other.

- Lenz's Law - Lenz's Law states that the polarity of an induced voltage must be such that it opposes the current which produces the induced voltage.

- Mutual Inductance, L_M - the ability of one coil to induce a voltage in another coil nearby. Two coils have a mutual inductance, L_M, of 1 H when a current change of 1 A/s in one coil induces 1 V in the other coil.

- Phasing Dots - dots on the primary and secondary leads of a transformer schematic symbol that identify those leads having the same instantaneous polarity.

- Reflected Impedance - the term used to describe the transformation of a secondary load resistance to a new value as seen by the primary.

- Series-Aiding - a connection of coils in which the coil current produces the same direction of magnetic field for both coils.

- Series-Opposing - a connection of coils in which the coil current

produces opposing magnetic fields for each coil.

- Stray Capacitance - a very small capacitance that exists between any two conductors separated by an insulator. For example, the capacitance can be between two wires in a wiring harness or between a single wire and a metal chassis.

- Stray Inductance - the small inductance of any length of conductor or component lead. The effects of both stray inductance and stray capacitance are most noticeable at very high frequencies.

- Transformer - a device that uses the concept of mutual inductance to step up or step down an alternating voltage.

- Turns Ratio - the ratio of the number of turns in the primary to the number of turns in the secondary of a transformer.

- Variac - a piece of test equipment which provides a variable output voltage. The Variac is plugged into a 120-Vac power line and the equipment under test is plugged into the Variac. Most Variacs available today have an isolated output.

- Volt-Ampere (VA) - the unit of apparent power that specifies the power rating of a transformer. The product VA is called the apparent power because it is the power that is apparently used by the transformer.

Related Formulas

$$L = \frac{v_L}{di/dt}$$

$$L = \mu_r \times \frac{N^2 \times A}{l} \times 1.26 \times 10^{-6}\,\text{H}$$

$$v_L = L\frac{di}{dt}$$

$$L_M = k\sqrt{L_1 \times L_2}$$

$$\text{Turns ratio} = \frac{N_P}{N_S}$$

$$\frac{V_P}{V_S} = \frac{N_P}{N_S}$$

$$V_S I_S = V_P I_P$$

$$\frac{I_S}{I_P} = \frac{V_P}{V_S}$$

$$\%\ \text{Efficiency} = \frac{P_{out}}{P_{in}} \times 100$$

$$Z_P = \left(\frac{N_P}{N_S}\right)^2 \times Z_S$$

$$\frac{N_P}{N_S} = \sqrt{\frac{Z_P}{Z_S}}$$

Series connection with no L_M

$$L_T = L_1 + L_2 + L_3 + \cdots + \text{etc.}$$

Series connection with L_M

$$L_T = L_1 + L_2 \pm 2L_M$$

Parallel connection: (No L_M)

$$L_{EQ} = \cfrac{1}{\cfrac{1}{L_1} + \cfrac{1}{L_2} + \cfrac{1}{L_3} + \cdots + etc.}$$

$$L_M = \frac{L_{T_a} - L_{T_o}}{4}$$

$$\text{Energy} = \mathscr{E} = \tfrac{1}{2}LI^2$$

Self-Test

Answers at back of book.

1. **The unit of inductance is the**
   a. henry.
   b. farad.
   c. ohm.
   d. volt-ampere.

2. **Which of the following factors affect the inductance, L, of an inductor?**
   a. number of turns.
   b. area enclosed by each turn.
   c. permeability of the core.
   d. all of the above.

3. **A transformer cannot be used to**
   a. step up or down an ac voltage.
   b. step up or down a dc voltage.
   c. match impedances.
   d. transfer power from primary to secondary.

4. **The interaction between two inductors physically close together is called**
   a. counter emf.
   b. self inductance.
   c. mutual inductance.
   d. hysteresis.

5. **If the secondary current in a step-down transformer increases, the primary current will**
   a. not change.
   b. increase.
   c. decrease.
   d. drop a little.

6. **Inductance can be defined as the characteristic that**
   a. opposes a change in current.
   b. opposes a change in voltage.
   c. aids or enhances any change in current.
   d. stores electric charge.

7. **If the number of turns in a coil is doubled in the same length and area, the inductance, L, will**
   a. double.
   b. quadruple.
   c. stay the same.
   d. be cut in half.

8. **An open coil has**
   a. zero resistance and zero inductance.
   b. infinite inductance and zero resistance.
   c. normal inductance but infinite resistance.
   d. infinite resistance and zero inductance.

9. **Two 10-H inductors are connected in series-aiding and have a mutual inductance, L_M, of 0.75 H. The total inductance, L_T, of this combination is**
   a. 18.5 H.
   b. 20.75 H.
   c. 21.5 H.
   d. 19.25 H.

10. **How much is the self-induced voltage, V_L, across a 100–mH inductor produced by a current change of 50,000 A/s?**
    a. 5 kV.
    b. 50 V.
    c. 5 MV.
    d. 500 kV.

11. **The measured voltage across an unloaded secondary of a transformer is usually**
    a. the same as the rated secondary voltage.
    b. 5 to 10% higher than the rated secondary voltage.
    c. 50% higher than the rated secondary voltage.
    d. 5 to 10% lower than the rated secondary voltage.

12. **A laminated iron-core transformer has reduced eddy current losses because**
    a. the laminations are stacked vertically.
    b. more wire can be used with less dc resistance.
    c. the magnetic flux is in the air gap of the core.
    d. the laminations are insulated from each other.

13. **How much is the inductance of a coil that induces 50 V when its current changes at the rate of 500 A/s?**
    a. 100 mH.
    b. 1 H.
    c. 100 μH.
    d. 10 μH.

14. **A 100 mH inductor is in parallel with a 150-mH and a 120-mH inductor. Assuming no mutual inductance between coils, how much is L_{EQ}?**
    a. 400 mH.
    b. 370 mH.
    c. 40 mH.
    d. 80 mH.

15. **A 400-μH coil is in series with a 1.2-mH coil without mutual inductance. How much is L_T?**
    a. 401.2 μH.
    b. 300 μH.
    c. 160 μH.
    d. 1.6 mH.

16. **A step-down transformer has a turns ratio, $\dfrac{N_P}{N_S}$, of 4:1. If the primary voltage, V_P, is 120 Vac, how much is the secondary voltage, V_S?**
    a. 480 Vac.
    b. 120 Vac.
    c. 30 Vac.
    d. This is impossible to determine.

17. If an iron-core transformer has a turns ratio, $\dfrac{N_P}{N_s}$, of 3:1 and $Z_s = 16\ \Omega$, how much is Z_P?

a. $48\ \Omega$.

b. $144\ \Omega$.

c. $1.78\ \Omega$.

d. $288\ \Omega$.

18. How much is the induced voltage, V_L, across a 5-H inductor carrying a steady dc current of 200 mA?

a. 0 V.

b. 1 V.

c. 100 kV.

d. 120 Vac.

19. The secondary current, I_s, in an iron-core transformer equals 1.8 A. If the turns ratio, $\dfrac{N_P}{N_s}$, equals 3:1, how much is the primary current, I_P?

a. $I_P = 1.8$ A.

b. $I_P = 600$ mA.

c. $I_P = 5.4$ A.

d. none of the above.

20. For a coil, the dc resistance, r_i, and inductance, L, are

a. in parallel.

b. infinite.

c. the same thing.

d. in series.

Questions

1. Define 1 H of self-inductance and 1 H of mutual inductance.

2. State Lenz's law in terms of induced voltage produced by varying current.

3. Refer to Fig. 19–5. Explain why the polarity of v_L is the same for the examples in Fig. 19–5a and d.

4. Make a schematic diagram showing the primary and secondary of an iron-core transformer with a 1:6 voltage step-up ratio (a) using an autotransformer; (b) using a transformer with isolated secondary winding. Then (c) with 100 turns in the primary, how many turns are in the secondary for both cases?

5. Define the following: coefficient of coupling, transformer efficiency, stray inductance, and eddy-current losses.

6. Why are eddy-current losses reduced with the following cores: (a) laminated; (b) powdered iron; (c) ferrite?

7. Why is a good conductor used for an rf shield?

8. Show two methods of providing a variable inductance.

9. (a) Why will the primary of a power transformer have excessive current if the secondary is short-circuited? (b) Why is there no voltage across the secondary if the primary is open?

10. (a) Describe briefly how to check a coil for an open winding with an ohmmeter. Which ohmmeter range should be used? (b) Which leads will be checked on an autotransformer with one secondary and a transformer with two isolated secondary windings?

11. Derive the formula $L_M = (L_{T_a} - L_{T_o})/4$ from the fact that $L_{T_a} = L_1 + L_2 + 2L_M$ and $L_{T_o} = L_1 + L_2 - 2L_M$.

12. Explain how a transformer with a 1:1 turns ratio and an isolated secondary can be used to reduce the chance of electric shock from the 120-Vac power line.

13. Explain the terms *stray inductance* and *stray capacitance* and give an example of each.

Problems

SECTION 19–1 INDUCTION BY ALTERNATING CURRENT

19–1 Which can induce more voltage in a conductor, a steady dc current of 10 A or a small current change of 1 to 2 mA?

19–2 Examine the sine wave of alternating current in Fig. 19–1. Identify the points on the waveform (using the letters A–I) where the rate of current change, $\dfrac{di}{dt}$, is

a. greatest.

b. zero.

19–3 Which will induce more voltage across a conductor, a low-frequency alternating current or a high-frequency alternating current?

SECTION 19–2 SELF-INDUCTANCE L

19–4 Convert the following current changes, $\frac{di}{dt}$, to amperes per second:

    a. 0 to 3 A in 2 s.

    b. 0 to 50 mA in 5 μs.

    c. 100 to 150 mA in 5 ms.

    d. 150 to 100 mA in 20 μs.

    e. 30 to 35 mA in 1 μs.

    f. 80 to 96 mA in 0.4 μs.

    g. 10 to 11 A in 1 s.

19–5 How much inductance, L, will be required to produce an induced voltage, V_L, of 15 V for each of the $\frac{di}{dt}$ values listed in Prob. 19–4?

19–6 How much is the inductance, L, of a coil that induces 75 V when the current changes at the rate of 2500 A/s?

19–7 How much is the inductance, L, of a coil that induces 20 V when the current changes at the rate of 400 A/s?

19–8 Calculate the inductance, L, for the following long coils: (NOTE: 1 m = 100 cm and 1 m^2 = 10,000 cm^2)

    a. air core, 20 turns, area 3.14 cm^2, length 25 cm

    b. same coil as (a) with ferrite core having a μ_r of 5000

    c. air core, 200 turns, area 3.14 cm^2, length 25 cm

    d. air core, 20 turns, area 3.14 cm^2, length 50 cm

    e. iron core with μ_r of 2000, 100 turns, area 5 cm^2, length 10 cm

19–9 Recalculate the inductance, L, in Prob. 19–8a if the number of turns is doubled to 40.

19–10 What is another name for an rf inductor?

SECTION 19–3 SELF-INDUCED VOLTAGE V_L

19–11 How much is the self-induced voltage across a 5-H inductance produced by a current change of 100 to 200 mA in 1 ms.

19–12 How much is the self-induced voltage across a 33-mH inductance when the current changes at the rate of 1500 A/s?

19–13 How much is the self-induced voltage across a 100-mH inductor for the following values of $\frac{di}{dt}$:

    a. 100 A/s

    b. 200 A/s

    c. 50 A/s

    d. 1000 A/s

SECTION 19–5 MUTUAL INDUCTANCE L_M

19–14 A coil, L_1, produces 200 μWb of magnetic flux. A nearby coil, L_2, is linked with L_1 by 50 μWb of magnetic flux. What is the coefficient of coupling, k, between L_1 and L_2?

19–15 A coil, L_1, produces 40 μWb of magnetic flux. A coil, L_2, nearby, is linked with L_1 by 30 μWb of magnetic flux. What is the value of k?

19–16 Two 50-mH coils, L_1 and L_2, have a coefficient of coupling, k, equal to 0.6. Calculate L_M.

19–17 Two inductors, L_1 and L_2, have a coefficient of coupling, k, equal to 0.5. L_1 = 100 mH and L_2 = 150 mH. Calculate L_M.

19–18 What is the assumed value of k for an iron-core transformer?

SECTION 19–6 TRANSFORMERS

19–19 In Fig. 19–35, solve for

    a. the secondary voltage, V_S.

    b. the secondary current, I_S.

    c. the secondary power, P_{sec}.

    d. the primary power, P_{pri}.

    e. the primary current, I_P.

Figure 19–35

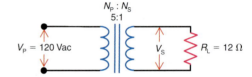

19–20 Repeat Prob. 19–19 if $N_P:N_S$ = 10:1.

19–21 In Fig. 19–36, solve for

    a. V_{S_1} (Secondary 1 Voltage).

    b. V_{S_2} (Secondary 2 Voltage).

    c. I_{S_1} (Secondary 1 Current).

    d. I_{S_2} (Secondary 2 Current).

    e. P_{Sec_1}.

    f. P_{Sec_2}.

    g. P_{pri}.

    h. I_P.

Figure 19–36

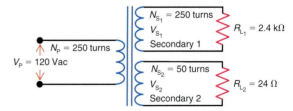

19–22 In Fig. 19–36, calculate the primary current, I_P, if R_{L_1} opens.

19–23 In Fig. 19–37, solve for

    a. the turns ratio $\frac{N_P}{N_S}$.

    b. the secondary current, I_S.

    c. the primary current, I_P.

Figure 19–37

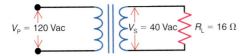

19–24 In Fig. 19–38, what turns ratio, $\dfrac{N_P}{N_S}$, is required to obtain a secondary voltage of

  a. 60 Vac?

  b. 600 Vac?

  c. 420 Vac?

  d. 24 Vac?

  e. 12.6 Vac?

Figure 19–38

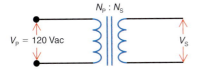

19–25 A transformer delivers 400 W to a load connected to its secondary. If the input power to the primary is 500 W, what is the efficiency of the transformer?

19–26 Explain the advantages of a transformer having an isolated secondary.

SECTION 19–7 TRANSFORMER RATINGS

19–27 How is the power rating of a transformer specified?

19–28 To avoid overloading a transformer, what two rules should be observed?

19–29 What is the purpose of phasing dots on the schematic symbol of a transformer?

19–30 Assume that a 6-Ω load is connected to the secondary in Fig. 19–15b and c. How much is the current in each individual primary winding in

  a. Fig. 19–15b?

  b. Fig. 19–15c?

19–31 Refer to Fig. 19–39. Calculate the following:

  a. V_{sec_1}

  b. V_{sec_2}

  c. the maximum allowable current in secondary 1

  d. the maximum allowable current in secondary 2

  e. the maximum allowable primary current

Figure 19–39

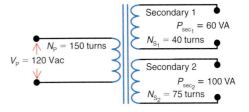

19–32 Refer to the transformer in Fig. 19–40. How much voltage would a DMM measure across the following secondary leads if the secondary current is 2 A?

  a. V_{AB}

  b. V_{AC}

  c. V_{BC}

Figure 19–40

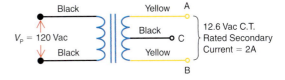

19–33 How much is the primary current, I_P, in Fig. 19–40 if the secondary current is 2 A?

19–34 Repeat Prob. 19–32 if the secondary is unloaded.

SECTION 19–8 IMPEDANCE TRANSFORMATION

19–35 In Fig. 19–41, calculate the primary impedance, Z_P, for a turns ratio $\dfrac{N_P}{N_S}$ of

  a. 2:1

  b. 1:2

  c. 11.18:1

  d. 10:1

  e. 1:3.16

Figure 19–41

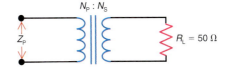

19–36 In Fig. 19–42, calculate the required turns ratio $\frac{N_P}{N_S}$ for

    a. $Z_P = 10\text{ k}\Omega$ and $R_L = 75\ \Omega$

    b. $Z_P = 100\ \Omega$ and $R_L = 25\ \Omega$

    c. $Z_P = 100\ \Omega$ and $R_L = 10\text{ k}\Omega$

    d. $Z_P = 1\text{ k}\Omega$ and $R_L = 200\ \Omega$

    e. $Z_P = 50\ \Omega$ and $R_L = 600\ \Omega$

    f. $Z_P = 200\ \Omega$ and $R_L = 10\ \Omega$

Figure 19–42

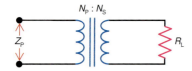

19–37 In Fig. 19–43, what turns ratio, $\frac{N_P}{N_S}$, will provide maximum transfer of power from the amplifier to the 4-Ω speaker?

Figure 19–43

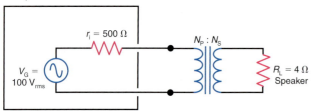

19–38 Using your answer from Prob. 19–37, calculate

    a. the primary impedance, Z_P.

    b. the power delivered to the 4-Ω speaker.

    c. the primary power.

SECTION 19–12 INDUCTANCES IN SERIES OR PARALLEL

19–39 Calculate the total inductance, L_T, for the following combinations of series inductors. Assume no mutual induction.

    a. $L_1 = 5\text{ mH}$ and $L_2 = 15\text{ mH}$

    b. $L_1 = 12\text{ mH}$ and $L_2 = 6\text{ mH}$

    c. $L_1 = 220\ \mu\text{H}$, $L_2 = 330\ \mu\text{H}$, and $L_3 = 450\ \mu\text{H}$

    d. $L_1 = 1\text{ mH}$, $L_2 = 500\ \mu\text{H}$, $L_3 = 2.5\text{ mH}$, and $L_4 = 6\text{ mH}$

19–40 Assuming that the inductor combinations listed in Prob. 19–39 are in parallel rather than series, calculate the equivalent inductance, L_{EQ}. Assume no mutual induction.

19–41 A 100-mH and a 300-mH inductor are connected in series-aiding and have a mutual inductance, L_M, of 130 mH. What is the total inductance, L_T?

19–42 If the inductors in Prob. 19–41 are connected in a series-opposing arrangement, how much is L_T?

19–43 A 20-mH and a 40-mH inductor have a coefficient of coupling, k, of 0.4. Calculate L_T if the inductors are

    a. series-aiding.

    b. series-opposing.

19–44 Two 100-mH inductors in series have a total inductance, L_T, of 100 mH when connected in a series-opposing arrangement and 300 mH when connected in a series-aiding arrangement. Calculate

    a. L_M

    b. k

SECTION 19–13 ENERGY IN A MAGNETIC FIELD OF INDUCTANCE

19–45 Calculate the energy in joules stored by a magnetic field created by 90 mA of current in a 60-mH inductor.

19–46 Calculate the energy in joules stored by a magnetic field created by 200 mA in a 5-H inductor.

19–47 A current of 3 A flows in a coil with an inductance of 150 mH. How much energy is stored in the magnetic field?

Critical Thinking

19–48 Derive the formula:
$$Z_P = \left(\frac{N_P}{N_S}\right)^2 \times Z_S$$

19–49 Calculate the primary impedance Z_P in Fig. 19–44.

Figure 19–44   Circuit for Critical Thinking Prob. 19–49.

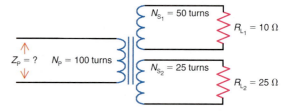

19-50 In Fig. 19–45, calculate the impedance Z_P across primary leads: (**a**) 1 and 3; (**b**) 1 and 2. (Note: Terminal 2 is a center-tap connection on the transformer primary. Also, the turns ratio of 4:1 is specified using leads 1 and 3 of the primary.)

19-51 Refer to Fig. 19–36. If the transformer has an efficiency of 80 percent, calculate the primary current I_P.

Figure 19–45 Circuit for Critical Thinking Prob. 19-50.

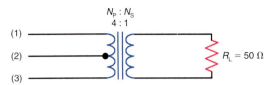

Answers to Knowledge Check Problems

19-1 A sine wave with a peak-to-peak value of 1 mA

19-2 $L = 50.4\ \mu H$

19-3 $v_L = 2.5$ kV

19-4 Counter emf

19-5 $L_M = 25$ mH

19-6 $I_P = 300$ mA

19-7 In volt-amperes (VA)

19-8 $Z_P = 80\ \Omega$

19-9 Eddy-current losses and hysteresis losses.

19-10 Insulator

19-11 To test electronic equipment with a voltage that is either less than or greater than the normal ac power-line voltage

19-12 **a.** $L_T = 80$ mH
b. $L_{EQ} = 18.75$ mH

19-13 In the magnetic field

19-14 A wire-wound resistor

19-15 1000 μH

Answers to Self-Reviews

19-1 a. coil with an iron core
b. time B

19-2 a. $L = 2$ H
b. $L = 32$ mH

19-3 a. $v_L = 2$ V
b. $v_L = 200$ V

19-4 a. T
b. T

19-5 a. $k = 1$
b. $L_M = 18$ mH

19-6 a. $V_S = 240$ V
b. $I_S = 0.1$ A

c. F
d. increase

19-7 a. T
b. T
c. F

19-8 a. F
b. F
c. T

19-9 a. iron core
b. 60 MHz

19-10 a. T
b. T
c. T

19-11 a. T
b. T

19-12 a. $L_T = 1.5$ mH
b. $L_{EQ} = 0.33$ mH

19-13 a. joule
b. more

19-14 a. T
b. T
c. T

19-15 a. infinite ohms
b. 120 V
c. series

When alternating current flows in an inductance L, the amount of current is much less than the dc resistance alone would allow. The reason is that the current variations induce a voltage across L that opposes the applied voltage. This additional opposition of an inductance to sine-wave alternating current is specified by the amount of its inductive reactance X_L. It is an opposition to current, measured in ohms. The X_L is the ohms of opposition, therefore, that an inductance L has for sine-wave current.

The amount of X_L equals $2\pi f L$ ohms, with f in hertz and L in henrys. Note that the opposition in ohms of X_L increases for higher frequencies and more inductance. The constant factor 2π indicates sine-wave variations.

The requirements for X_L correspond to what is needed to produce induced voltage. There must be variations in current and its associated magnetic flux. For a steady direct current without any changes in current, X_L is zero. However, with sine-wave alternating current, X_L is the best way to analyze the effect of L.

Objectives

After studying this chapter you should be able to

- *Explain* how inductive reactance reduces the amount of alternating current.

- *Calculate* the reactance of an inductor when the frequency and inductance are known.

- *Calculate* the total reactance of series connected inductors.

- *Calculate* the equivalent reactance of parallel-connected inductors.

- *Explain* how Ohm's law can be applied to inductive reactance.

- *Describe* the waveshape of induced voltage produced by sine-wave alternating current.

Outline

Important Terms

inductive reactance, X_L phase angle   proportional

Illustrating the effect of inductive reactance X_L in reducing the amount of sine-wave alternating current. (*a*) Bulb lights with 2.4 A. (*b*) Inserting an X_L of 1000 Ω reduces I to 0.12 A, and the bulb cannot light. (*c*) With direct current, the coil has no inductive reactance, and the bulb lights.

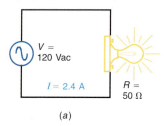

(a)

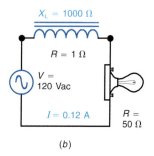

(b)

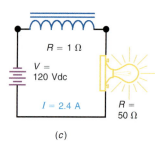

(c)

20–1 How X_L Reduces the Amount of I

Figure 20–1 illustrates the effect of X_L in reducing the alternating current for a lightbulb. The more ohms of X_L, the less current flows. When X_L reduces I to a very small value, the bulb cannot light.

In Fig. 20–1*a*, there is no inductance, and the ac voltage source produces a 2.4-A current to light the bulb with full brilliance. This 2.4-A I results from 120 V applied across the 50-Ω R of the bulb's filament.

In Fig. 20–1*b*, however, a coil is connected in series with the bulb. The coil has a dc resistance of only 1 Ω, which is negligible, but the reactance of the inductance is 1000 Ω. This 1000-Ω X_L is a measure of the coil's reaction to sine-wave current in producing a self-induced voltage that opposes the applied voltage and reduces the current. Now I is 120 V/1000 Ω, approximately, which equals 0.12 A. This I is not enough to light the bulb.

Although the dc resistance is only 1 Ω, the X_L of 1000 Ω for the coil limits the amount of alternating current to such a low value that the bulb cannot light. This X_L of 1000 Ω for a 60-Hz current can be obtained with an inductance L of approximately 2.65 H.

In Fig. 20–1*c*, the coil is also in series with the bulb, but the applied battery voltage produces a steady value of direct current. Without any current variations, the coil cannot induce any voltage and, therefore, it has no reactance. The amount of direct current, then, is practically the same as though the dc voltage source were connected directly across the bulb, and it lights with full brilliance. In this case, the coil is only a length of wire because there is no induced voltage without current variations. The dc resistance is the resistance of the wire in the coil.

In summary, we can draw the following conclusions:

1. An inductance can have appreciable X_L in ac circuits to reduce the amount of current. Furthermore, the higher the frequency of the alternating current, and the greater the inductance, the higher the X_L opposition.
2. There is no X_L for steady direct current. In this case, the coil is a resistance equal to the resistance of the wire.

These effects have almost unlimited applications in practical circuits. Consider how useful ohms of X_L can be for different kinds of current, compared with resistance, which always has the same ohms of opposition. One example is to use X_L where it is desired to have high ohms of opposition to alternating current but little opposition to direct current. Another example is to use X_L for more opposition to a high-frequency alternating current, compared with lower frequencies.

X_L Is an Inductive Effect

An inductance can have X_L to reduce the amount of alternating current because self-induced voltage is produced to oppose the applied voltage. In Fig. 20–2, V_L is the voltage across L, induced by the variations in sine-wave current produced by the applied voltage V_A.

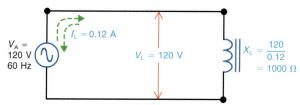

The two voltages V_A and V_L are the same because they are in par.. However, the current I_L is the amount that allows the self-induced voltage v_L to be equal to V_A. In this example, I is 0.12 A. This value of a 60-Hz current in the inductance produces a V_L of 120 V.

The Reactance Is a V/I Ratio

The V/I ratio for the ohms of opposition to the sine-wave current is 120/0.12, which equals 1000 Ω. This 1000 Ω is what we call X_L, to indicate how much current can be produced by sine-wave voltage across an inductance. The ohms of X_L can be almost any amount, but the 1000 Ω here is a typical example.

The Effect of L and f on X_L

The X_L value depends on the amount of inductance and on the frequency of the alternating current. If L in Fig. 20–2 were increased, it could induce the same 120 V for V_L with less current. Then the ratio of V_L/I_L would be greater, meaning more X_L for more inductance.

Also, if the frequency were increased in Fig. 20–2, the current variations would be faster with a higher frequency. Then the same L could produce the 120 V for V_L with less current. For this condition also, the V_L/I_L ratio would be greater because of the smaller current, indicating more X_L for a higher frequency.

■ *20–1 Knowledge Check*

Answer at end of chapter.

For an inductor, does X_L increase or decrease with an increase in frequency?

■ *20–1 Self-Review*

Answers at end of chapter.

a. **For the dc circuit in Fig. 20–1c, how much is X_L?**
b. **For the ac circuit in Fig. 20–1b, how much is the V/I ratio for X_L?**

20–2 $X_L = 2\pi fL$

The formula $X_L = 2\pi fL$ includes the effects of frequency and inductance for calculating the inductive reactance. The frequency is in hertz, and L is in henrys for an X_L in ohms. As an example, we can calculate X_L for an inductance of 2.65 H at the frequency of 60 Hz:

$$X_L = 2\pi fL \tag{20–1}$$
$$= 6.28 \times 60 \times 2.65$$
$$X_L = 1000 \ \Omega$$

Note the following factors in the formula $X_L = 2\pi fL$.

1. The constant factor 2π is always $2 \times 3.14 = 6.28$. It indicates the circular motion from which a sine wave is derived. Therefore, this formula applies only to sine-wave ac circuits. The 2π is actually 2π rad or 360° for a complete circle or cycle.
2. The frequency f is a time element. Higher frequency means that the current varies at a faster rate. A faster current change can produce more self-induced voltage across a given inductance. The result is more X_L.

X_L increases as f increases

Frequency, Hz	$X_L = 2\pi fL$, Ω
0	0
100	200
200	400
300	600
400	800

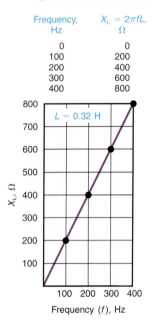

3. The inductance L indicates the physical factors of the coil that determine how much voltage it can induce for a given current change.
4. Inductive reactance X_L is in ohms, corresponding to a V_L/I_L ratio for sine-wave ac circuits, to determine how much current L allows for a given applied voltage.

Stating X_L as V_L/I_L and as $2\pi fL$ are two ways of specifying the same value of ohms. The $2\pi fL$ formula gives the effect of L and f on the X_L. The V_L/I_L ratio gives the result of $2\pi fL$ in reducing the amount of I.

The formula $2\pi fL$ shows that X_L is proportional to frequency. When f is doubled, for instance, X_L is doubled. This linear increase in inductive reactance with frequency is illustrated in Fig. 20–3.

The reactance formula also shows that X_L is proportional to the inductance. When the value of henrys for L is doubled, the ohms of X_L is also doubled. This linear increase of inductive reactance with inductance is illustrated in Fig. 20–4.

Example 20-1

How much is X_L of a 6-mH L at 41.67 kHz?

ANSWER

$$X_L = 2\pi fL$$
$$= 6.28 \times 41.67 \times 10^3 \times 6 \times 10^{-3}$$
$$= 1570 \ \Omega$$

Figure 20-4 Graph of values to show linear increase of X_L for higher values of inductance L. The frequency is constant at 100 Hz.

X_L increases as L increases

Inductance, H	$X_L = 2\pi fL$, Ω
0	0
0.32	200
0.64	400
0.96	600
1.28	800

Example 20-2

Calculate the X_L of (a) a 10-H L at 60 Hz and (b) a 5-H L at 60 Hz.

ANSWER

a. For a 10-H L,
$$X_L = 2\pi fL = 6.28 \times 60 \times 10$$
$$= 3768 \ \Omega$$

b. For a 5-H L,
$$X_L = \tfrac{1}{2} \times 3768 = 1884 \ \Omega$$

Example 20-3

Calculate the X_L of a 250-μH coil at (a) 1 MHz and (b) 10 MHz.

ANSWER

a. At 1 MHz,
$$X_L = 2\pi fL = 6.28 \times 1 \times 10^6 \times 250 \times 10^{-6}$$
$$= 1570 \ \Omega$$

b. At 10 MHz,
$$X_L = 10 \times 1570 = 15{,}700 \ \Omega$$

CALCULATOR

To do a problem like Example 20–1 with a calculator requires continued multiplication. Multiply all the factors and then press the $\bigodot$ key only at the end. If the calculator does not have an $\boxed{EXP}$ (exponential) function key, do the powers of 10 separately without the calculator. Specifically, for this example with $2\pi \times 6 \times 10^{-3} \times 41.67 \times 10^3$, the 10^3 and 10^{-3} cancel. Then calculate $2\pi \times 6 \times 41.67$ as factors. To save time in the calculation, 2π can be memorized as 6.28, since it occurs in many ac formulas. For the multiplication, punch in 6.28 for 2π and then push the $\bigotimes$ key, punch in 6 and push the $\bigotimes$ key again, punch in 41.67, and push the $\bigodot$ key for the total product of 1570 as the final answer. It is not necessary to use the $\bigodot$ key until the last step for the final product. The factors can be multiplied in any order.

The last two examples illustrate the fact that X_L is proportional to frequency and inductance. In Example 20–2b, X_L is one–half the value in Example 20–2a because the inductance is one-half. In Example 20–3b, the X_L is 10 times more than in Example 20–3a because the frequency is 10 times higher.

Finding L from X_L

Not only can X_L be calculated from f and L, but if any two factors are known, the third can be found. Very often X_L can be determined from voltage and current measurements. With the frequency known, L can be calculated as

$$L = \frac{X_L}{2\pi f} \tag{20-2}$$

This formula has the factors inverted from Formula (20–1). Use the basic units with ohms for X_L and hertz for f to calculate L in henrys.

It should be noted that Formula (20–2) can also be stated as

$$L = \frac{1}{2\pi f} \times X_L$$

This form is easier to use with a calculator because $1/2\pi f$ can be found as a reciprocal value and then multiplied by X_L.

The following problems illustrate how to find X_L from V and I measurements and using X_L to determine L with Formula (20–2).

Example 20-4

A coil with negligible resistance has 62.8 V across it with 0.01 A of current. How much is X_L?

ANSWER

$$X_L = \frac{V_L}{I_L} = \frac{62.8 \ \text{V}}{0.01 \ \text{A}}$$
$$= 6280 \ \Omega$$

Example 20-5

Calculate L of the coil in Example 20–4 when the frequency is 1000 Hz.

ANSWER

$$L = \frac{X_L}{2\pi f} = \frac{6280}{6.28 \times 1000}$$
$$= 1 \text{ H}$$

Example 20-6

Calculate L of a coil that has 15,700 Ω of X_L at 12 MHz.

ANSWER

$$L = \frac{X_L}{2\pi f} = \frac{1}{2\pi f} \times X_L$$
$$= \frac{1}{6.28 \times 12 \times 10^6} \times 15{,}700$$
$$= 0.0133 \times 10^{-6} \times 15{,}700$$
$$= 208.8 \times 10^{-6} \text{ H} \quad \text{or} \quad 208.8 \text{ } \mu\text{H}$$

Finding f from X_L

For a third version of the inductive reactance formula,

$$f = \frac{X_L}{2\pi L} \tag{20–3}$$

Use the basic units of ohms for X_L and henrys for L to calculate the frequency in hertz.

Formula 20–3 can also be stated as

$$f = \frac{1}{2\pi L} \times X_L$$

This form is easier to use with a calculator. Find the reciprocal value and multiply by X_L, as explained before in Example 20–6.

Example 20-7

At what frequency will an inductance of 1 H have a reactance of 1000 Ω?

ANSWER

$$f = \frac{1}{2\pi L} \times X_{\text{L}} = \frac{1}{6.28 \times 1} \times 1000$$
$$= 0.159 \times 1000$$
$$= 159 \text{ Hz}$$

■ *20-2 Knowledge Check*

Answer at end of chapter.

How much is the inductance, *L*, of a coil that draws 8 mA of current from a 12-Vac source whose frequency is 2 kHz?

■ *20-2 Self-Review*

Answers at end of chapter.

Calculate X_{L} for the following:
a. *L* is 1 H and *f* is 100 Hz.
b. *L* is 0.5 H and *f* is 100 Hz.
c. *L* is 1 H and *f* is 1000 Hz.

20-3 Series or Parallel Inductive Reactances

Since reactance is an opposition in ohms, the values of X_{L} in series or in parallel are combined the same way as ohms of resistance. With series reactances, the total is the sum of the individual values, as shown in Fig. 20-5a. For example, the series reactances of 100 and 200 Ω add to equal 300 Ω of X_{L} across both reactances. Therefore, in series,

$$X_{L_{\text{T}}} = X_{L_1} + X_{L_2} + X_{L_3} + \cdots + \text{etc.} \qquad (20\text{-}4)$$

The combined reactance of parallel reactances is calculated by the reciprocal formula. As shown in Fig. 20-5b, in parallel

$$X_{L_{\text{EQ}}} = \frac{1}{\dfrac{1}{X_{L_1}} + \dfrac{1}{X_{L_2}} + \dfrac{1}{X_{L_3}} + \cdots + \text{etc.}} \qquad (20\text{-}5)$$

The combined parallel reactance will be less than the lowest branch reactance. Any shortcuts for calculating parallel resistances also apply to parallel reactances. For instance, the combined reactance of two equal reactances in parallel is one-half either reactance.

Figure 20–5 Combining ohms of X_L for inductive reactances. (a) X_{L_1} and X_{L_2} in series. (b) X_{L_1} and X_{L_2} in parallel.

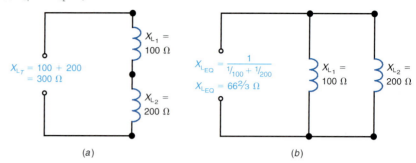

(a)   (b)

■ 20–3 Knowledge Check

Answers at end of chapter.

How much is the combined reactance for two 500-Ω X_L values in (a) series (b) parallel?

■ 20–3 Self-Review

Answers at end of chapter.

a. An X_L of 200 Ω is in series with a 300-Ω X_L. How much is the total X_{L_T}?

b. An X_L of 200 Ω is in parallel with a 300-Ω X_L. How much is the combined $X_{L_{EQ}}$?

20–4 Ohm's Law Applied to X_L

The amount of current in an ac circuit with only inductive reactance is equal to the applied voltage divided by X_L. Three examples are given in Fig. 20–6. No dc resistance is indicated, since it is assumed to be practically zero for the coils shown. In Fig. 20–6a, there is one reactance of 100 Ω. Then I equals V/X_L, or 100 V/100 Ω, which is 1 A.

In Fig. 20–6b, the total reactance is the sum of the two individual series reactances of 100 Ω each, for a total of 200 Ω. The current, calculated as V/X_{L_T}, then equals 100 V/200 Ω, which is 0.5 A. This current is the same in both series reactances. Therefore, the voltage across each reactance equals its IX_L product. This is 0.5 A × 100 Ω, or 50 V across each X_L.

Figure 20–6 Circuit calculations with V, I, and ohms of reactance X_L. (a) One reactance. (b) Two series reactances. (c) Two parallel reactances.

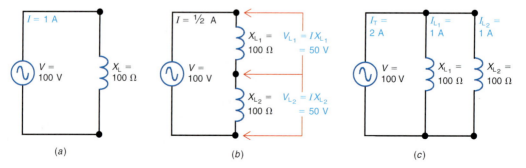

(a)   (b)   (c)

In Fig. 20–6c, each parallel reactance has its individual branch current, equal to the applied voltage divided by the branch reactance. Then each branch current equals 100 V/100 Ω, which is 1 A. The voltage is the same across both reactances, equal to the generator voltage, since they are all in parallel.

The total line current of 2 A is the sum of the two individual 1-A branch currents. With the rms value for the applied voltage, all calculated values of currents and voltage drops in Fig. 20–6 are also rms values.

■ *20-4 Knowledge Check*

Answers at end of chapter.

If $X_{L_1} = 300\ \Omega$ in Fig. 20–6b, recalculate X_{L_T}, I, V_{L_1}, and V_{L_2}.

■ *20-4 Self-Review*

Answers at end of chapter.

a. In Fig. 20–6b, how much is the I through both X_{L_1} and X_{L_2}?
b. In Fig. 20–6c, how much is the V across both X_{L_1} and X_{L_2}?

20–5 Applications of X_L for Different Frequencies

The general use of inductance is to provide minimum reactance for relatively low frequencies but more for higher frequencies. In this way, the current in an ac circuit can be reduced for higher frequencies because of more X_L. There are many circuits in which voltages of different frequencies are applied to produce current with different frequencies. Then, the general effect of X_L is to allow the most current for direct current and low frequencies, with less current for higher frequencies, as X_L increases.

Compare this frequency factor for ohms of X_L with ohms of resistance. The X_L increases with frequency, but R has the same effect in limiting direct current or alternating current of any frequency.

If 1000 Ω is taken as a suitable value of X_L for many applications, typical inductances can be calculated for different frequencies. These are listed in Table 20–1.

Table 20–1	Values of Inductance L for X_L of 1000 Ω	
L^* (Approx.)	Frequency	Remarks
2.65 H	60 Hz	Power-line frequency and low audio frequency
160 mH	1000 Hz	Medium audio frequency
16 mH	10,000 Hz	High audio frequency
160 μH	1000 kHz (RF)	In radio broadcast band
16 μH	10 MHz (HF)	In short-wave radio band
1.6 μH	100 MHz (VHF)	In FM broadcast band

* Calculated as $L = 1000/(2\pi f)$.

At 60 Hz, for example, the inductance L in the top row of Table 20–1 is 2.65 H for 1000 Ω of X_L. The calculations are

$$L = \frac{X_L}{2\pi f} = \frac{1000}{2\pi \times 60}$$
$$= \frac{1000}{377}$$
$$= 2.65 \text{ H}$$

For this case, the inductance has practically no reactance for direct current or for very low frequencies below 60 Hz. However, above 60 Hz, the inductive reactance increases to more than 1000 Ω.

To summarize, the effects of increasing frequencies for this 2.65-H inductance are as follows:

Inductive reactance X_L is zero for 0 Hz which corresponds to a steady direct current.

Inductive reactance X_L is less than 1000 Ω for frequencies below 60 Hz.

Inductive reactance X_L equals 1000 Ω at 60 Hz.

Inductive reactance X_L is more than 1000 Ω for frequencies above 60 Hz.

Note that the smaller inductances at the bottom of the first column still have the same X_L of 1000 Ω as the frequency is increased. Typical rf coils, for instance, have an inductance value of the order of 100 to 300 μH. For the very high frequency (VHF) range, only several microhenrys of inductance are needed for an X_L of 1000 Ω.

It is necessary to use smaller inductance values as the frequency is increased because a coil that is too large can have excessive losses at high frequencies. With iron-core coils, particularly, the hysteresis and eddy-current losses increase with frequency.

■ *20–5 Knowledge Check*

Answer at end of chapter.

What is the main difference between inductive reactance, X_L, and resistance, R?

■ *20–5 Self-Review*

Answers at end of chapter.

Refer to Table 20–1.
a. **Which frequency requires the smallest L for 1000 Ω of X_L?**
b. **How much would X_L be for a 1.6-μH L at 200 MHz?**

20–6 Waveshape of v_L Induced by Sine-Wave Current

More details of inductive circuits can be analyzed by means of the waveshapes in Fig. 20–7, plotted for the calculated values in Table 20–2. The top curve shows a sine wave of current i_L flowing through a 6-mH inductance L. Since induced voltage depends on the rate of change of current rather than on the absolute value of i, the curve in Fig. 20–7b shows how much the current changes. In this curve, the di/dt values are plotted for the current changes every 30° of the cycle. The bottom curve shows the actual induced voltage v_L. This v_L curve is similar to the di/dt curve because v_L equals the constant factor L multiplied by di/dt. Note that di/dt indicates infinitely small changes in i and t.

Figure 20–7 Waveshapes in inductive circuits. (*a*) Sine-wave current *i*; (*b*) changes in current with time *di/dt*; (*c*) induced voltage v_L.

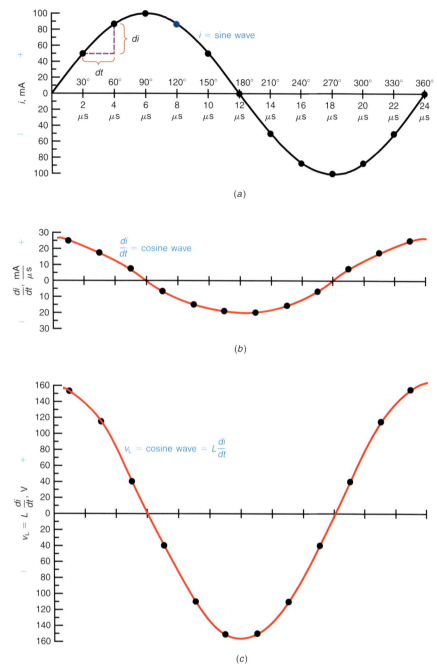

(a)

(b)

(c)

90° Phase Angle

The v_L curve at the bottom of Fig. 20–7 has its zero values when the i_L curve at the top is at maximum. This comparison shows that the curves are 90° out of phase. The v_L is a cosine wave of voltage for the sine wave of current i_L.

The 90° phase difference results from the fact that v_L depends on the *di/dt* rate of change, rather than on *i* itself. More details of this 90° phase angle between v_L and i_L for inductance are explained in the next chapter.

Table 20-2				Values for $V_L = L(di/dt)$ Curves in Fig. 20-7			
Time		**dt**					
θ	μs	θ	μs	di, mA	di/dt, mA/μs	L, mH	$v_L = L(di/dt)$, V
30°	2	30°	2	50	25	6	150
60°	4	30°	2	36.6	18.3	6	109.8
90°	6	30°	2	13.4	6.7	6	40.2
120°	8	30°	2	−13.4	−6.7	6	−40.2
150°	10	30°	2	−36.6	−18.3	6	−109.8
180°	12	30°	2	−50	−25	6	−150
210°	14	30°	2	−50	−25	6	−150
240°	16	30°	2	−36.6	−18.3	6	−109.8
270°	18	30°	2	−13.4	−6.7	6	−40.2
300°	20	30°	2	13.4	6.7	6	40.2
330°	22	30°	2	36.6	18.3	6	109.8
360°	24	30°	2	50	25	6	150

Frequency

For each of the curves, the period T is 24 μs. Therefore, the frequency is $1/T$ or $\frac{1}{24}$ μs, which equals 41.67 kHz. Each curve has the same frequency.

Ohms of X_L

The ratio of v_L/i_L specifies the inductive reactance in ohms. For this comparison, we use the actual value of i_L, which has a peak value of 100 mA. The rate-of-change factor is included in the induced voltage v_L. Although the peak of v_L at 150 V is 90° before the peak of i_L at 100 mA, we can compare these two peak values. Then v_L/i_L is 150/0.1, which equals 1500 Ω.

This X_L is only approximate because v_L cannot be determined exactly for the large dt changes every 30°. If we used smaller intervals of time, the peak v_L would be 157 V. Then X_L would be 1570 Ω, the same as $2\pi fL$ Ω with a 6-mH L and a frequency of 41.67 kHz. This is the same X_L problem as Example 20–1.

The Tabulated Values from 0° to 90°

The numerical values in Table 20–2 are calculated as follows: The i curve is a sine wave. This means that it rises to one-half its peak value in 30°, to 0.866 of the peak in 60°, and the peak value is at 90°.

In the di/dt curve, the changes in i are plotted. For the first 30°, the di is 50 mA; the dt change is 2 μs. Then di/dt is 50/2 or 25 mA/μs. This point is plotted between 0° and 30° to indicate that 25 mA/μs is the rate of change of current for the 2-μs interval between 0° and 30°. If smaller intervals were used, the di/dt values could be determined more accurately.

During the next 2-μs interval from 30° to 60°, the current increases from 50 to 86.6 mA. The change of current during this time is 86.6 − 50, which equals 36.6 mA. The time is the same 2 μs for all the intervals. Then di/dt for the next plotted point is 36.6/2, or 18.3.

For the final 2-μs change before i reaches its peak at 100 mA, the di value is 100 − 86.6, or 13.4 mA, and the di/dt value is 6.7. All of these values are listed in Table 20–2.

Notice that the di/dt curve in Fig. 20–7b has its peak at the zero value of the i curve and the peak i values correspond to zero on the di/dt curves. These conditions result because the sine wave of i has its sharpest slope at the zero values. The rate of change is greatest when the i curve is going through the zero axis. The i curve flattens near the peaks and has a zero rate of change exactly at the peak. The curve must stop going up before it can come down. In summary, then, the di/dt curve and the i curve are 90° out of phase with each other.

The v_L curve follows the di/dt curve exactly, as $v_L = L(di/dt)$. The phase of the v_L curve is exactly the same as that of the di/dt curve, 90° out of phase with the i curve. For the first plotted point,

$$v_L = L\frac{di}{dt} = 6 \times 10^{-3} \times \frac{50 \times 10^{-3}}{2 \times 10^{-6}}$$
$$= 150 \text{ V}$$

The other v_L values are calculated the same way, multiplying the constant factor of 6 mH by the di/dt value for each 2-μs interval.

90° to 180°

In this quarter-cycle, the sine wave of i decreases from its peak of 100 mA at 90° to zero at 180°. This decrease is considered a negative value for di, as the slope is negative going downward. Physically, the decrease in current means that its associated magnetic flux is collapsing, compared with the expanding flux as the current increases. The opposite motion of the collapsing flux must make v_L of opposite polarity, compared with the induced voltage polarity for increasing flux. This is why the di values are negative from 90° to 180°. The di/dt values are also negative, and the v_L values are negative.

180° to 270°

In this quarter-cycle, the current increases in the reverse direction. If the magnetic flux is considered counterclockwise around the conductor with $+i$ values, the flux is in the reversed clockwise direction with $-i$ values. Any induced voltage produced by expanding flux in one direction will have opposite polarity from voltage induced by expanding flux in the opposite direction. This is why the di values are considered negative from 180° to 270°, as in the second quarter-cycle, compared with the positive di values from 0° to 90°. Actually, increasing negative values and decreasing positive values are changing in the same direction. This is why v_L is negative for both the second and third quarter-cycles.

Figure 20–8 How a 90° phase angle for the V_L applies in a complex circuit with more than one inductance. The current I_1 lags V_{L_1} by 90°, I_2 lags V_{L_2} by 90°, and I_3 lags V_{L_3} by 90°.

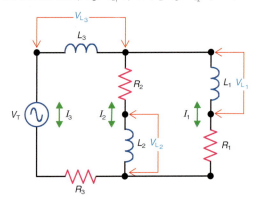

270° to 360°

In the last quarter-cycle, the negative i values are decreasing. Now the effect on polarity is like two negatives making a positive. The current and its magnetic flux have the negative direction. But the flux is collapsing, which induces opposite voltage from increasing flux. Therefore, the di values from 270° to 360° are positive, as are the di/dt values and the induced voltages v_L.

The same action is repeated for each cycle of sine-wave current. Then the current i_L and the induced voltage v_L are 90° out of phase. The reason is that v_L depends on di/dt, not on i alone.

Application of the 90° Phase Angle in a Circuit

The phase angle of 90° between V_L and I will always apply for any L with sine-wave current. Remember, though, that the specific comparison is only between the induced voltage across any one coil and the current flowing in its turns. To emphasize this important principle, Fig. 20–8 shows an ac circuit with a few coils and resistors. The details of this complex circuit are not to be analyzed now. However, for each L in the circuit, the V_L is 90° out of phase with its I. The I lags V_L by 90°, or V_L leads I. For the three coils in Fig. 20–8,

Current I_1 lags V_{L_1} by 90°.
Current I_2 lags V_{L_2} by 90°.
Current I_3 lags V_{L_3} by 90°.
Note that I_3 is also I_T for the series-parallel circuit.

■ *20–6 Knowledge Check*

Answer at end of chapter.

For an inductor, why does v_L lead i by 90°?

■ *20–6 Self-Review*

Answers at end of chapter.

Refer to Fig. 20–7.
a. **At what angle does i have its maximum positive value?**
b. **At what angle does v_L have its maximum positive value?**
c. **What is the phase angle difference between the waveforms for i and v_L?**

Summary

- Inductive reactance is the opposition of an inductance to the flow of sine-wave alternating current. The symbol for inductive reactance is X_L.

- Reactance X_L is measured in ohms because it limits the current to the value $I = V/X_L$. With V in volts and X_L in ohms, I is in amperes.

- $X_L = 2\pi fL$, where f is in hertz, L is in henrys, and X_L is in ohms.

- With a constant L, X_L increases proportionately with higher frequencies.

- At a constant frequency, X_L increases proportionately with higher inductances.

- With X_L and f known, the inductance $L = X_L/(2\pi f)$.

- With X_L and L known, the frequency $f = X_L/(2\pi L)$.

- The total X_L of reactances in series is the sum of the individual values, as for series resistances. Series reactances have the same current. The voltage across each inductive reactance is IX_L.

- The equivalent reactance of parallel reactances is calculated by the reciprocal formula, as for parallel resistances. Each branch current is V/X_L. The total line current is the sum of the individual branch currents.

- Table 20–3 summarizes the differences between L and X_L.

- Table 20–4, compares X_L and R.

- Table 20–5, summarizes the differences between capacitive reactance and inductive reactance.

Table 20–3	Comparison of Inductance and Inductive Reactance	
Inductance		**Inductive Reactance**
Symbol is L		Symbol is X_L
Measured in henry units		Measured in ohm units
Depends on construction of coil		Depends on frequency and inductance
$L = v_L/(di/dt)$, in H units		$X_L = v_L/i_L$ or $2\pi fL$, in Ω units

Table 20–4	Comparison of X_L and R	
X_L		**R**
Ohm unit		Ohm unit
Increases for higher frequencies		Same for all frequencies
Current lags voltage by 90° ($\theta = 90°$)		Current in phase with voltage ($\theta = 0°$)

Table 20-5	Comparison of Capacitive and Inductive Reactances	
X_C, Ω		X_L, Ω
Decreases with more capacitance C		Increases with more inductance L
Decreases with increase in frequency f		Increases with increase in frequency f
Allows less current at lower frequencies; blocks direct current.		Allows more current at lower frequencies; passes direct current.

Important Terms

■ Inductive Reactance, X_L - a measure of an inductor's opposition to the flow of alternating current. X_L is measured in ohms and is calculated as $X_L = 2\pi fL$ or $X_L = \dfrac{V_L}{I_L}$. X_L applies only to sine-wave alternating current.

■ Phase Angle - the angular difference or displacement between two waveforms. For an inductor, the induced voltage, V_L, reaches its maximum value 90° ahead of the inductor current, i_L. As a result, the induced voltage, V_L, across an inductor is said to lead the inductor current, i_L, by a phase angle of 90°.

■ Proportional - a mathematical term used to describe the relationship between two quantities. For example, in the formula $X_L = 2\pi fL$, X_L is directly proportional to both the frequency, f, and the inductance, L. The term proportional means that if either f or L is doubled X_L will double. Similarly, if either f or L is reduced by one-half, X_L will be reduced by one-half. In other words, X_L will increase or decrease in direct proportion to either f or L.

Related Formulas

$X_L = 2\pi fL$

$L = \dfrac{X_L}{2\pi f}$

$f = \dfrac{X_L}{2\pi L}$

$X_L = \dfrac{V_L}{I_L}$

$X_{L_T} = X_{L_1} + X_{L_2} + X_{L_3} + \cdots + \text{etc.}$ (Series inductors)

$X_{L_{EQ}} = \dfrac{1}{\frac{1}{X_{L_1}} + \frac{1}{X_{L_2}} + \frac{1}{X_{L_3}} + \cdots + \text{etc.}}$ (Parallel inductors)

Self-Test

Answers at back of book.

1. **The unit of inductive reactance, X_L, is the**
   a. henry.
   b. ohm.
   c. farad.
   d. hertz.

2. **The inductive reactance, X_L, of an inductor is**
   a. inversely proportional to frequency.
   b. unaffected by frequency.
   c. directly proportional to frequency.
   d. inversely proportional to inductance.

3. **For an inductor, the induced voltage, V_L**
   a. leads the inductor current, i_L, by 90°.
   b. lags the inductor current, i_L, by 90°.
   c. is in phase with the inductor current, i_L.
   d. none of the above.

4. For a steady dc current, the X_L of an inductor is
   a. infinite.
   b. extremely high.
   c. usually about 10 kΩ.
   d. 0 Ω.

5. What is the inductive reactance, X_L, of a 100-mH coil at a frequency of 3.183 kHz?
   a. 2 kΩ.
   b. 200 Ω.
   c. 1 MΩ.
   d. 4 Ω.

6. At what frequency does a 60-mH inductor have an X_L value of 1 kΩ?
   a. 377 Hz.
   b. 265 kHz.
   c. 2.65 kHz.
   d. 15.9 kHz.

7. What value of inductance will provide an X_L of 500 Ω at a frequency of 159.15 kHz?
   a. 5 H.
   b. 500 μH.
   c. 500 mH.
   d. 750 μH.

8. Two inductors, L_1 and L_2, are in series. If X_{L_1} = 4 kΩ and X_{L_2} = 2 kΩ, how much is X_{L_T}?
   a. 6 kΩ.
   b. 1.33 kΩ.
   c. 4.47 kΩ.
   d. 2 kΩ.

9. Two inductors, L_1 and L_2, are in parallel. If X_{L_1} = 1 kΩ and X_{L_2} = 1 kΩ, how much is $X_{L_{EQ}}$?
   a. 707 Ω.
   b. 2 kΩ.
   c. 1.414 kΩ.
   d. 500 Ω.

10. How much is the inductance of a coil that draws 25 mA of current from a 24-V ac source whose frequency is 1 kHz?
    a. 63.7 μH.
    b. 152.8 mH.
    c. 6.37 H.
    d. 15.28 mH.

Questions

1. Explain briefly why X_L limits the amount of alternating current.

2. Give two differences and one similarity between X_L and R.

3. Explain why X_L increases with higher frequencies and more inductance.

4. Give two differences between the inductance L of a coil and its reactance X_L.

5. Why are the waves in Fig. 20–7a and b considered 90° out of phase, whereas the waves in Fig. 20–7b and c have the same phase?

6. Referring to Fig. 20–3, how does this graph show a linear proportion between X_L and frequency?

7. Referring to Fig. 20–4, how does this graph show a linear proportion between X_L and L?

8. Referring to Fig. 20–3, tabulate the values of L that would be needed for each frequency listed but for an X_L of 2000 Ω. (Do not include 0 Hz.)

9. (a) Draw the circuit for a 40-Ω R across a 120-V, 60-Hz source. (b) Draw the circuit for a 40-Ω X_L across a 120-V, 60-Hz source. (c) Why is I equal to 3 A for both circuits? (d) Give two differences between the circuits.

10. Why are coils for rf applications generally smaller than af coils?

Problems

SECTION 20–1 HOW X_L REDUCES THE AMOUNT OF I

20–1 How much is the inductive reactance, X_L, of a coil for a steady dc current?

20–2 List two factors that determine the amount of inductive reactance a coil will have.

20–3 In Fig. 20–9, how much dc current will be indicated by the ammeter, M_1, with S_1 in position 1?

Figure 20–9

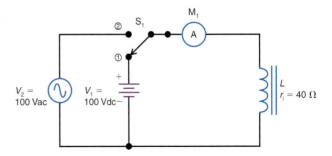

20-4 In Fig. 20-9, how much inductive reactance, X_L, does the coil have with S_1 in position 1? Explain your answer.

20-5 In Fig. 20-9, the ammeter, M_1, reads an ac current of 25 mA with S_1 in position 2.

    a. Why is there less current in the circuit with S_1 in position 2 compared to position 1?

    b. How much is the inductive reactance, X_L, of the coil? (Ignore the effect of the coil resistance, r_i.)

20-6 In Fig. 20-10, how much is the inductive reactance, X_L, for each of the following values of Vac and I?

    a. Vac = 10 V and I = 2 mA

    b. Vac = 50 V and I = 20 μA

    c. Vac = 12 V and I = 15 mA

    d. Vac = 6 V and I = 40 μA

    e. Vac = 120 V and I = 400 mA

SECTION 20-2 $X_L = 2\pi fL$

Figure 20-10

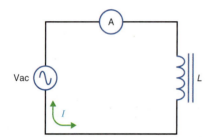

20-7 Calculate the inductive reactance, X_L, of a 100-mH inductor at the following frequencies:

    a. f = 60 Hz

    b. f = 120 Hz

    c. f = 1.592 kHz

    d. f = 10 kHz

20-8 Calculate the inductive reactance, X_L, of a 50-μH coil at the following frequencies:

    a. f = 60 Hz

    b. f = 10 kHz

    c. f = 500 kHz

    d. f = 3.8 MHz

20-9 What value of inductance, L, will provide an X_L value of 1 kΩ at the following frequencies?

    a. f = 318.3 Hz

    b. f = 1.591 kHz

    c. f = 5 kHz

    d. f = 6.36 kHz

20-10 At what frequency will a 30-mH inductor provide an X_L value of

    a. 50 Ω?

    b. 200 Ω?

    c. 1 kΩ?

    d. 40 kΩ?

20-11 How much is the inductance of a coil that draws 15 mA from a 24-Vac source whose frequency is 1 kHz?

20-12 At what frequency will a stray inductance of 0.25 μH have an X_L value of 100 Ω?

20-13 A 25-mH coil draws 2mA of current from a 10-Vac source. What is the value of current drawn by the inductor when

    a. the frequency is doubled?

    b. the frequency is reduced by one-half?

    c. the inductance is doubled to 50 mH?

    d. the inductance is reduced by one-half to 12.5 mH?

20-14 A coil has an inductive reactance, X_L, of 10 kΩ at a given frequency. What is the value of X_L when the frequency is

    a. cut in half?

    b. doubled?

    c. quadrupled?

    d. increased by a factor 10?

20-15 Calculate the inductive reactance, X_L, for the following inductance and frequency values:

    a. L = 7 H, f = 60 Hz

    b. L = 25 μH, f = 7 MHz

    c. L = 500 mH, f = 318.31 Hz

    d. L = 1 mH, f = 159.2 kHz

20-16 Determine the inductance value for the following frequency and X_L values:

    a. X_L = 50 Ω, f = 15.91 kHz

    b. X_L = 2k Ω, f = 5 kHz

    c. X_L = 10 Ω, f = 795.7 kHz

    d. X_L = 4k Ω, f = 6 kHz

20-17 Determine the frequency for the following inductance and X_L values:

    a. L = 80 mH, X_L = 1 kΩ

    b. L = 60 μH, X_L = 200 Ω

    c. L = 5 H, X_L = 100 kΩ

    d. L = 150 mH, X_L = 7.5 kΩ

SECTION 20-3 SERIES OR PARALLEL INDUCTIVE REACTANCES

20-18 How much is the total inductive reactance, X_{L_T}, for the following series inductive reactances:

    a. X_{L_1} = 250 Ω and X_{L_2} = 1.5 kΩ

    b. X_{L_1} = 200 Ω, X_{L_2} = 400 Ω and X_{L_3} = 800 Ω

    c. X_{L_1} = 10 kΩ, X_{L_2} = 30 kΩ and X_{L_3} = 15 kΩ

    d. X_{L_1} = 1.8 kΩ, X_{L_2} = 2.2 kΩ and X_{L_3} = 1 kΩ

20-19 What is the equivalent inductive reactance, $X_{L_{EQ}}$, for the following parallel inductive reactances?

    a. X_{L_1} = 1.2 kΩ and X_{L_2} = 1.8 kΩ

    b. X_{L_1} = 1.5 kΩ and X_{L_2} = 1 kΩ

    c. X_{L_1} = 1.2 kΩ, X_{L_2} = 400 Ω and X_{L_3} = 300 Ω

    d. X_{L_1} = 1 kΩ, X_{L_2} = 4 kΩ, X_{L_3} = 800 Ω, and X_{L_4} = 200 Ω

SECTION 20-4 OHM'S LAW APPLIED TO X_L

20-20 In Fig. 20-11, calculate the current, I.

Figure 20-11

$V = 120$ Vac $X_L = 1.5$ kΩ

20-21 In Fig. 20-11, what happens to the current, I, when the frequency of the applied voltage

    a. decreases?

    b. increases?

20-22 In Fig. 20-12, solve for

    a. X_{L_1}.

    b. I.

    c. V_{L_1}, V_{L_2}, and V_{L_3}.

Figure 20-12

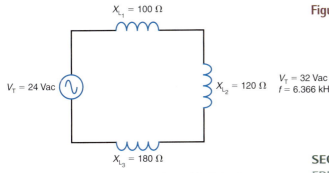

$X_{L_1} = 100$ Ω

$V_T = 24$ Vac $X_{L_2} = 120$ Ω

$X_{L_3} = 180$ Ω

20-23 In Fig. 20-12, solve for L_1, L_2, L_3, and L_T if the frequency of the applied voltage is 1.591 kHz.

20-24 In Fig. 20-13, solve for

    a. X_{L_1}, X_{L_2}, and X_{L_3}.

    b. X_{L_T}.

    c. I.

    d. V_{L_1}, V_{L_2}, and V_{L_3}.

    e. L_T.

Figure 20-13

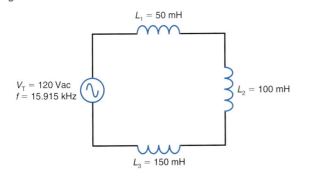

$L_1 = 50$ mH

$V_T = 120$ Vac
$f = 15.915$ kHz $L_2 = 100$ mH

$L_3 = 150$ mH

20-25 In Fig. 20-14, solve for

    a. I_{L_1}, I_{L_2}, and I_{L_3}.

    b. I_T.

    c. $X_{L_{EQ}}$.

Figure 20-14

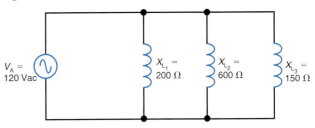

$V_A = 120$ Vac $X_{L_1} = 200$ Ω $X_{L_2} = 600$ Ω $X_{L_3} = 150$ Ω

20-26 In Fig. 20-14, solve for L_1, L_2, L_3, and L_T if the frequency of the applied voltage is 6.366 kHz.

20-27 In Fig. 20-15, solve for

    a. X_{L_1}, X_{L_2}, and X_{L_3}.

    b. I_{L_1}, I_{L_2}, and I_{L_3}.

    c. I_T.

    d. $X_{L_{EQ}}$.

    e. L_{EQ}.

Figure 20-15

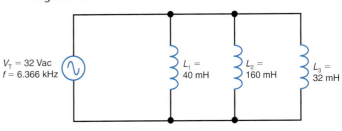

$V_T = 32$ Vac
$f = 6.366$ kHz $L_1 = 40$ mH $L_2 = 160$ mH $L_3 = 32$ mH

SECTION 20-5 APPLICATIONS OF X_L FOR DIFFERENT FREQUENCIES

20-28 Calculate the value of inductance, L, required to produce an X_L value of 500 Ω at the following frequencies:

    a. $f = 250$ Hz

    b. $f = 636.6$ Hz

    c. $f = 3.183$ kHz

    d. $f = 7.957$ kHz

SECTION 20-6 WAVESHAPE OF V_L INDUCED BY SINE-WAVE CURRENT

20-29 For an inductor, what is the phase relationship between the induced voltage, V_L, and the inductor current, i_L? Explain your answer.

20-30 For a sine wave of alternating current flowing through an inductor, at what angles in the cycle will the induced voltage be

    a. maximum?

    b. zero?

Critical Thinking

20–31 In Fig. 20–16, calculate L_1, L_2, L_3, L_T, X_{L_1}, X_{L_2}, X_{L_3}, V_{L_1}, V_{L_3}, I_{L_2}, and I_{L_3}.

20–32 Two inductors in series without L_M have a total inductance L_T of 120 μH. If $L_1/L_2 = 1/20$, what are the values for L_1 and L_2?

20–33 Three inductors in parallel have an equivalent inductance L_{EQ} of 7.5 mH. If $L_2 = 3\ L_3$ and $L_3 = 4\ L_1$, calculate L_1, L_2, and L_3.

Figure 20–16 Circuit for Critical Thinking Prob. 20–31.

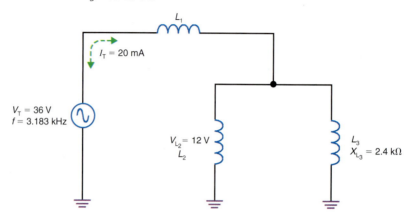

Answers to Knowledge Check Problems

20–1 X_L increases as the frequency increases.

20–2 $L = 119.4$ mH

20–3 a. $X_{L_T} = 1$ kΩ

     b. $X_{L_{EQ}} = 250\ \Omega$

20–4 $X_{L_T} = 400\ \Omega$, $I = 250$ mA, $V_{L_1} = 75$ V, and $V_{L_2} = 25$ V

20–5 Inductive reactance, X_L, is dependent on frequency, whereas resistance is not.

20–6 Because V_L is dependent on the rate of current change, di/dt, rather than on the actual value of current itself.

Answers to Self-Reviews

20–1 a. $0\ \Omega$

     b. $1000\ \Omega$

20–2 a. $X_L = 628\ \Omega$

     b. $X_L = 314\ \Omega$

     c. $X_L = 6280\ \Omega$

20–3 a. $X_{L_T} = 500\ \Omega$

     b. $X_{L_{EQ}} = 120\ \Omega$

20–4 a. 0.5 A

     b. 100 V

20–5 a. 100 MHz

     b. $2000\ \Omega$

20–6 a. 90°

     b. 0° or 360°

     c. 90°

21 Inductive Circuits

This chapter analyzes circuits that combine inductive reactance X_L and resistance R. The main questions are, how do we combine the ohms of opposition, how much current flows, and what is the phase angle? Although X_L and R are both measured in ohms, they have different characteristics. Specifically, X_L increases with more L and higher frequencies, when sine-wave ac voltage is applied, whereas R is the same for dc or ac circuits. Furthermore, the phase angle for the voltage across X_L is at 90° with respect to the current through L.

In addition, the practical application of using a coil as a choke to reduce the current for a specific frequency is explained here. For a circuit with L and R in series, the X_L can be high for an undesired ac signal frequency, whereas R is the same for either direct current or alternating current.

Finally, the general case of induced voltage produced across L is shown with nonsinusoidal current variations. In this case, we compare the waveshapes of i_L and v_L instead of their phase. Remember that the 90° angle for an IX_L voltage applies only to sine waves.

With nonsinusoidal waveforms, such as pulses of current or voltage, the circuit can be analyzed in terms of its L/R time constant, as explained in Chapter 22.

Objectives

After studying this chapter you should be able to

- *Explain* why the voltage leads the current by 90° for an inductor.
- *Calculate* the total impedance and phase angle of a series RL circuit.
- *Calculate* the total current, equivalent impedance, and phase angle of a parallel RL circuit.
- *Define* what is meant by the Q of a coil.
- *Explain* how an inductor can be used to pass some ac frequencies but block others.
- *Calculate* the induced voltage that is produced by a nonsinusoidal current.

Outline

Important Terms

ac effective resistance, R_e

arctangent (arctan)

choke

impedance, Z

phase angle, θ

phasor triangle

Q of a coil

skin effect

tangent (tan)

21–1 Sine Wave i_L Lags v_L by 90°

When sine-wave variations of current produce an induced voltage, the current lags its induced voltage by exactly 90°, as shown in Fig. 21–1. The inductive circuit in Fig. 21–1a has the current and voltage waveshapes shown in Fig. 21–1b. The phasors in Fig. 21–1c show the 90° phase angle between i_L and v_L. Therefore, we can say that i_L lags v_L by 90°, or v_L leads i_L by 90°.

This 90° phase relationship between i_L and v_L is true in any sine-wave ac circuit, whether L is in series or parallel and whether L is alone or combined with other components. We can always say that the voltage across any X_L is 90° out of phase with the current through it.

Why the Phase Angle Is 90°

This results because v_L depends on the rate of change of i_L. As previously shown in Fig. 20–7 for a sine wave of i_L, the induced voltage is a cosine wave. In other words, v_L has the phase of di/dt, not the phase of i.

Why i_L Lags v_L

The 90° difference can be measured between any two points having the same value on the i_L and v_L waves. A convenient point is the positive peak value. Note that the i_L wave does not have its positive peak until 90° after the v_L wave. Therefore, i_L lags v_L by 90°. This 90° lag is in time. The time lag equals one quarter-cycle, which is one-quarter of the time for a complete cycle.

Inductive Current Is the Same in a Series Circuit

The time delay and resultant phase angle for the current in an inductance apply only with respect to the voltage across the inductance. This condition does not change the fact that the current is the same in all parts of a series circuit. In Fig. 21–1a, the current in the generator, the connecting wires, and L must be the same because they are in series. Whatever the current value is at any instant, it is the same in all series components. The time lag is between current and voltage.

Inductive Voltage Is the Same across Parallel Branches

In Fig. 21–1a, the voltage across the generator and the voltage across L are the same because they are in parallel. There cannot be any lag or lead in time between these two parallel voltages. Whatever the voltage value is across the generator at any instant, the voltage across L is the same. The parallel voltage v_A or v_L is 90° out of phase with the current.

Figure 21–1   (a) Circuit with inductance L. (b) Sine wave of i_L lags v_L by 90°. (c) Phasor diagram.

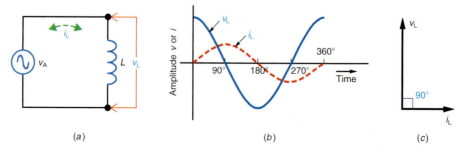

(a)   (b)   (c)

The voltage across L in this circuit is determined by the applied voltage, since they must be the same. The inductive effect here is to make the current have values that produce $L(di/dt)$ equal to the parallel voltage.

The Frequency Is the Same for i_L and v_L

Although i_L lags v_L by 90°, both waves have the same frequency. The i_L wave reaches its peak values 90° later than the v_L wave, but the complete cycles of variations are repeated at the same rate. As an example, if the frequency of the sine wave v_L in Fig. 21–1b is 100 Hz, this is also the frequency for i_L.

■ 21-1 Knowledge Check

Answer at end of chapter.

For an inductor, why does v_L lead i_L by 90°?

■ 21-1 Self-Review

Answers at end of chapter.

Refer to Fig. 21–1.
a. What is the phase angle between v_A and v_L?
b. What is the phase angle between v_L and i_L?
c. Does i_L lead or lag v_L?

21-2 X_L and R in Series

When a coil has series resistance, the current is limited by both X_L and R. This current I is the same in X_L and R, since they are in series. Each has its own series voltage drop, equal to IR for the resistance and IX_L for the reactance.

Note the following points about a circuit that combines series X_L and R, as in Fig. 21–2:

1. The current is labeled I, rather than I_L, because I flows through all series components.

⫿⫿⫿ MultiSim   Figure 21–2   Inductive reactance X_L and resistance R in series. (a) Circuit. (b) Waveforms of current and voltage. (c) Phasor diagram.

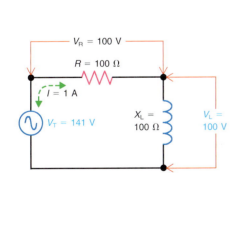

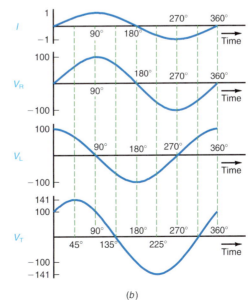

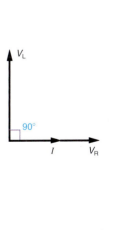

(a)   (b)   (c)

2. The voltage across X_L, labeled V_L, can be considered an IX_L voltage drop, just as we use V_R for an IR voltage drop.
3. The current I through X_L must lag V_L by 90° because this is the phase angle between current through an inductance and its self-induced voltage.
4. The current I through R and its IR voltage drop are in phase. There is no reactance to sine-wave current in any resistance. Therefore, I and IR have a phase angle of 0°.

Resistance R can be either the internal resistance of the coil or an external series resistance. The I and V values may be rms, peak, or instantaneous, as long as the same measure is applied to all. Peak values are used here for convenience in comparing waveforms.

Phase Comparisons

Note the following:

1. Voltage V_L is 90° out of phase with I.
2. However, V_R and I are in phase.
3. If I is used as the reference, V_L is 90° out of phase with V_R.

Specifically, V_R lags V_L by 90°, just as the current I lags V_L. These phase relations are shown by the waveforms in Fig. 21–2b and the phasors in Fig. 21–2c.

Combining V_R and V_L

As shown in Fig. 21–2b, when the V_R voltage wave is combined with the V_L voltage wave, the result is the voltage wave for the applied generator voltage V_T. The voltage drops must add to equal the applied voltage. The 100-V peak values for V_R and for V_L total 141 V, however, instead of 200 V, because of the 90° phase difference.

Consider some instantaneous values to see why the 100-V peak V_R and 100-V peak V_L cannot be added arithmetically. When V_R is at its maximum value of 100 V, for instance, V_L is at zero. The total for V_T then is 100 V. Similarly, when V_L is at its maximum value of 100 V, then V_R is zero and the total V_T is also 100 V.

Actually, V_T has its maximum value of 141 V when V_L and V_R are each 70.7 V. When series voltage drops that are out of phase are combined, therefore, they cannot be added without taking the phase difference into account.

Phasor-Voltage Triangle

Instead of combining waveforms that are out of phase, we can add them more quickly by using their equivalent phasors, as shown in Fig. 21–3. The phasors in Fig. 21–3a show only the 90° angle without any addition. The method in Fig. 21–3b is to add the tail of one phasor to the arrowhead of the other, using the angle required to show their relative phase. Voltages V_R and V_L are at right angles because they are 90° out of phase. The sum of the phasors is a resultant phasor from the start of one to the end of the other. Since the V_R and V_L phasors form a right angle, the resultant phasor is the hypotenuse of a right triangle. The hypotenuse is the side opposite the 90° angle.

From the geometry of a right triangle, the Pythagorean theorem states that the hypotenuse is equal to the square root of the sum of the squares of the sides. For the voltage triangle in Fig. 21–3b, therefore, the resultant is

$$V_T = \sqrt{V_R^2 + V_L^2} \tag{21–1}$$

where V_T is the phasor sum of the two voltages V_R and V_L 90° out of phase.

CALCULATOR

To do a problem like this on the calculator, remember that the square root sign is a sign of grouping. All terms within the group must be added before you take the square root. Also, each term must be squared individually before adding for the sum. Specifically for this problem:

- Punch in 100 and push the $\boxed{x^2}$ button for 10,000 as the square. Press $\boxed{+}$.
- Next punch in 100 and $\boxed{x^2}$. Press $\boxed{=}$. The display should read 20,000.
- Press $\boxed{\sqrt{}}$ to read the answer 141.421.

In some calculators, either the $\boxed{x^2}$ or the $\boxed{\sqrt{}}$ key must be preceded by the second function key $\boxed{2^{nd}F}$.

$$V_L = IX_L = 100\ V \qquad 90° \qquad V_R = IR = 100\ V$$

$$(a)$$

$$V_T = \sqrt{V_R^2 + V_L^2}$$
$$V_T = 141\ V \qquad V_L = IX_L = 100\ V$$
$$\theta = 45° \qquad 90°$$
$$V_R = IR = 100\ V$$

$$(b)$$

This formula is for V_R and V_L when they are in series, since then they are 90° out of phase. All voltages must be in the same units. When V_T is an rms value, V_R and V_L are also rms values. For the example in Fig. 21–3,

$$V_T = \sqrt{100^2 + 100^2} = \sqrt{10{,}000 + 10{,}000}$$
$$= \sqrt{20{,}000}$$
$$= 141\ V$$

■ *21-2 Knowledge Check*

Answer at end of chapter.

In a series *RL* circuit, how much is V_T if V_R = 20 V and V_L = 15 V?

■ *21-2 Self-Review*

Answers at end of chapter.

a.   **In a series circuit with X_L and R, what is the phase angle between I and V_R?**
b.   **What is the phase angle between V_R and V_L?**

21–3 Impedance Z Triangle

A triangle of R and X_L in series corresponds to a voltage triangle, as shown in Fig. 21–4. It is similar to the voltage triangle in Fig. 21–3, but the common factor I cancels because the current is the same in X_L and R. The resultant of the phasor addition of R and X_L is their total opposition in ohms, called *impedance*, with the symbol Z_T.* The Z takes into account the 90° phase relation between R and X_L.

For the impedance triangle of a series circuit with reactance and resistance,

$$Z_T = \sqrt{R^2 + X_L^2} \qquad\qquad (21\text{-}2)$$

where R, X_L, and Z_T are all in ohms. For the example in Fig. 21–4,

$$Z_T = \sqrt{100^2 + 100^2} = \sqrt{10{,}000 + 10{,}000}$$
$$= \sqrt{20{,}000}$$
$$= 141\ \Omega$$

Figure 21–4   Addition of R and X_L 90° out of phase in series circuit, to find the resultant impedance Z_T.

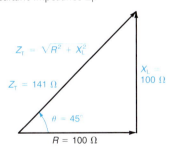

$$Z_T = \sqrt{R^2 + X_L^2}$$
$$Z_T = 141\ \Omega$$
$$X_L = 100\ \Omega$$
$$\theta = 45°$$
$$R = 100\ \Omega$$

---

* Although Z_T is a passive component, we consider it a phasor here because it determines the phase angle of V and I.

Note that the applied voltage of 141 V divided by the total impedance of 141 Ω results in 1 A of current in the series circuit. The IR voltage is 1×100, or 100 V; the IX_L voltage is also 1×100, or 100 V. The total of the series IR and IX_L drops of 100 V each, added by phasors, equals the applied voltage of 141 V. Finally, the applied voltage equals IZ, or 1×141, which is 141 V.

Summarizing the similar phasor triangles for volts and ohms in a series circuit,

1. The phasor for R, IR, or V_R is used as a reference at 0°.
2. The phasor for X_L, IX_L, or V_L is at 90°.
3. The phasor for Z, IZ, or V_T has the phase angle θ of the complete circuit.

Phase Angle with Series X_L

The angle between the generator voltage and its current is the phase angle of the circuit. Its symbol is θ (theta). In Fig. 21–3, the phase angle between V_T and IR is 45°. Since IR and I have the same phase, the angle is also 45° between V_T and I.

In the corresponding impedance triangle in Fig. 21–4, the angle between Z_T and R is also equal to the phase angle. Therefore, the phase angle can be calculated from the impedance triangle of a series circuit by the formula

$$\tan \theta_Z = \frac{X_L}{R} \qquad (21\text{--}3)$$

The tangent (tan) is a trigonometric function of any angle, equal to the ratio of the opposite side to the adjacent side of a right triangle. In this impedance triangle, X_L is the opposite side and R is the adjacent side of the angle. We use the subscript z for θ to show that θ_Z is found from the impedance triangle for a series circuit. To calculate this phase angle,

$$\tan \theta_Z = \frac{X_L}{R} = \frac{100}{100} = 1$$

The angle whose tangent is equal to 1 is 45°. Therefore, the phase angle is 45° in this example. The numerical values of the trigonometric functions can be found from a table or from a scientific calculator.

Note that the phase angle of 45° is halfway between 0° and 90° because R and X_L are equal.

Example 21-1

If a 30-Ω R and a 40-Ω X_L are in series with 100 V applied, find the following: Z_T, I, V_R, V_L, and θ_Z. What is the phase angle between V_L and V_R with respect to I? Prove that the sum of the series voltage drops equals the applied voltage V_T.

ANSWER

$$\begin{aligned} Z_T &= \sqrt{R^2 + X_L^2} = \sqrt{900 + 1600} \\ &= \sqrt{2500} \\ &= 50\ \Omega \end{aligned}$$

$$I = \frac{V_T}{Z_T} = \frac{100}{50} = 2 \text{ A}$$

$$V_R = IR = 2 \times 30 = 60 \text{ V}$$

$$V_L = IX_L = 2 \times 40 = 80 \text{ V}$$

$$\tan \theta_Z = \frac{X_L}{R} = \frac{40}{30} = \frac{4}{3} = 1.33$$

$$\theta_Z = 53.1°$$

Therefore, I lags V_T by 53.1°. Furthermore, I and V_R are in phase, and I lags V_L by 90°. Finally,

$$V_T = \sqrt{V_R^2 + V_L^2} = \sqrt{60^2 + 80^2} = \sqrt{3600 + 6400}$$
$$= \sqrt{10,000}$$
$$= 100 \text{ V}$$

Note that the phasor sum of the voltage drops equals the applied voltage.

Series Combinations of X_L and R

In a series circuit, the higher the value of X_L compared with R, the more inductive the circuit. This means that there is more voltage drop across the inductive reactance and the phase angle increases toward 90°. The series current lags the applied generator voltage. With all X_L and no R, the entire applied voltage is across X_L, and θ_Z equals 90°.

Several combinations of X_L and R in series are listed in Table 21–1 with their resultant impedance and phase angles. Note that a ratio of 10:1 or more for X_L/R means that the circuit is practically all inductive. The phase angle of 84.3° is only slightly less than 90° for the ratio of 10:1, and the total impedance Z_T is approximately equal to X_L. The voltage drop across X_L in the series circuit will be practically equal to the applied voltage, with almost none across R.

At the opposite extreme, when R is 10 times as large as X_L, the series circuit is mainly resistive. The phase angle of 5.7°, then, means that the current is almost in phase with the applied voltage, the total impedance Z_T is approximately equal to R, and the voltage drop across R is practically equal to the applied voltage, with almost none across X_L.

When X_L and R equal each other, their resultant impedance Z_T is 1.41 times the value of either one. The phase angle then is 45°, halfway between 0° for resistance alone and 90° for inductive reactance alone.

Table 21–1	Series R and X_L Combinations		
R, Ω	X_L, Ω	Z_T, Ω (Approx.)	Impedance Angle θ_Z
1	10	$\sqrt{101} = 10$	84.3°
10	10	$\sqrt{200} = 14.1$	45°
10	1	$\sqrt{101} = 10$	5.7°

Note: θ_Z is the angle of Z_T with respect to the reference I in a series circuit.

■ *21–3 Knowledge Check*

Answers at end of chapter.

For a series RL circuit, how much are Z_T and θ_Z if $X_L = 150\ \Omega$ and $R = 112.5\ \Omega$?

■ *21–3 Self-Review*

Answers at end of chapter.

a. How much is Z_T for a 20-Ω R in series with a 20-Ω X_L?
b. How much is V_T for 20 V across R and 20 V across X_L in series?
c. What is the phase angle of the circuit in **a** and **b**?

21–4 X_L and R in Parallel

For parallel circuits with X_L and R, the 90° phase angle must be considered for each of the branch currents, instead of the voltage drops. Remember that any series circuit has different voltage drops but one common current. A parallel circuit has different branch currents but one common voltage.

In the parallel circuit in Fig. 21–5*a*, the applied voltage V_A is the same across X_L, R, and the generator, since they are all in parallel. There cannot be any phase difference between these voltages. Each branch, however, has its individual current. For the resistive branch, $I_R = V_A/R$; in the inductive branch, $I_L = V_A/X_L$.

The resistive branch current I_R is in phase with the generator voltage V_A. The inductive branch current I_L lags V_A, however, because the current in an inductance lags the voltage across it by 90°.

The total line current, therefore, consists of I_R and I_L, which are 90° out of phase with each other. The phasor sum of I_R and I_L equals the total line current I_T. These phase relations are shown by the waveforms in Fig. 21–5*b*, and the phasors in Fig. 21–5*c*. Either way, the phasor sum of 10 A for I_R and 10 A for I_L is equal to 14.14 A for I_T.

|||| MultiSim **Figure 21–5** Inductive reactance X_L and R in parallel. (*a*) Circuit. (*b*) Waveforms of applied voltage and branch currents. (*c*) Phasor diagram.

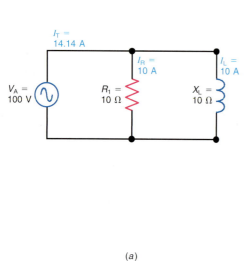

(a)

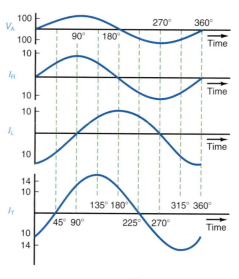

(b)

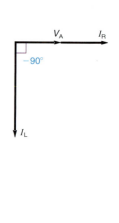

(c)

Both methods illustrate the general principle that quadrature components must be combined by phasor addition. The branch currents are added by phasors here because they are the factors that are 90° out of phase in a parallel circuit. This method is similar to combining voltage drops 90° out of phase in a series circuit.

Phasor Current Triangle

Note that the phasor diagram in Fig. 21–5c has the applied voltage V_A of the generator as the reference phasor because V_A is the same throughout the parallel circuit.

The phasor for I_L is down, compared with up for an X_L phasor. Here the parallel branch current I_L lags the parallel voltage reference V_A. In a series circuit, the X_L voltage leads the series current reference I. For this reason, the I_L phasor is shown with a negative 90° angle. The −90° means that the current I_L lags the reference phasor V_A.

The phasor addition of the branch currents in a parallel circuit can be calculated by the phasor triangle for currents shown in Fig. 21–6. Peak values are used for convenience in this example, but when the applied voltage is an rms value, the calculated currents are also in rms values. To calculate the total line current,

$$I_T = \sqrt{I_R^2 + I_L^2} \tag{21–4}$$

For the values in Fig. 21–6,

$$\begin{aligned} I_T &= \sqrt{10^2 + 10^2} = \sqrt{100 + 100} \\ &= \sqrt{200} \\ &= 14.14 \text{ A} \end{aligned}$$

Impedance of X_L and R in Parallel

A practical approach to the problem of calculating the total impedance of X_L and R in parallel is to calculate the total line current I_T and divide this value into the applied voltage V_A:

$$Z_{EQ} = \frac{V_A}{I_T} \tag{21–5}$$

For example, in Fig. 21–5, V_A is 100 V and the resultant I_T, obtained as the phasor sum of the resistive and reactive branch currents, is equal to 14.14 A. Therefore, we calculate the impedance as

$$\begin{aligned} Z_{EQ} &= \frac{V_A}{I_T} = \frac{100 \text{ V}}{14.14 \text{ A}} \\ &= 7.07 \ \Omega \end{aligned}$$

This impedance is the combined opposition in ohms across the generator, equal to the resistance of 10 Ω in parallel with the reactance of 10 Ω.

Note that the impedance for equal values of R and X_L in parallel is not one-half but equals 70.7% of either one. Still, the combined value of ohms must be less than the lowest ohms value in the parallel branches.

For the general case of calculating the impedance of X_L and R in parallel, any number can be assumed for the applied voltage because the value of V_A cancels in the calculations for Z in terms of the branch currents. A good value to assume for V_A is the value of either R or X_L, whichever is the higher number. This way, there are no fractions smaller than 1 in the calculation of the branch currents.

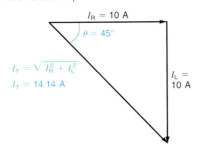

Figure 21–6 Phasor triangle of inductive and resistive branch currents 90° out of phase in a parallel circuit to find resultant I_T.

$I_R = 10$ A

$\theta = 45°$

$I_T = \sqrt{I_R^2 + I_L^2}$

$I_T = 14.14$ A

$I_L = 10$ A

Example 21-2

What is the total Z of a 600-Ω R in parallel with a 300-Ω X_L? Assume 600 V for the applied voltage. Then

ANSWER

$$I_R = \frac{600\ \text{V}}{600\ \Omega} = 1\ \text{A}$$

$$I_L = \frac{600\ \text{V}}{300\ \Omega} = 2\ \text{A}$$

$$I_T = \sqrt{I_R^2 + I_L^2}$$
$$= \sqrt{1 + 4} = \sqrt{5}$$
$$= 2.24\ \text{A}$$

Then, dividing the assumed value of 600 V for the applied voltage by the total line current gives

$$Z_{EQ} = \frac{V_A}{I_T} = \frac{600\ \text{V}}{2.24\ \text{A}}$$
$$= 268\ \Omega$$

The combined impedance of a 600-Ω R in parallel with a 300-Ω X_L is equal to 268 Ω, no matter how much the applied voltage is.

Phase Angle with Parallel X_L and R

In a parallel circuit, the phase angle is between the line current I_T and the common voltage V_A applied across all branches. However, the resistive branch current I_R has the same phase as V_A. Therefore, the phase of I_R can be substituted for the phase of V_A. This is shown in Fig. 21–5c. The triangle of currents is shown in Fig. 21–6. To find θ_I from the branch currents, use the tangent formula:

$$\tan \theta_I = -\frac{I_L}{I_R} \qquad (21\text{–}6)$$

We use the subscript I for θ to show that θ_I is found from the triangle of branch currents in a parallel circuit. In Fig. 21–6, θ_I is $-45°$ because I_L and I_R are equal. Then $\tan \theta_I = -1$.

The negative sign is used for this current ratio because I_L is lagging at $-90°$, compared with I_R. The phase angle of $-45°$ here means that I_T lags I_R and V_A by 45°.

Note that the phasor triangle of branch currents gives θ_I as the angle of I_T with respect to the generator voltage V_A. This phase angle for I_T is with respect to the applied voltage as the reference at 0°. For the phasor triangle of voltages in a series circuit, the phase angle θ_Z for Z_T and V_T is with respect to the series current as the reference phasor at 0°.

Parallel Combinations of X_L and R

Several combinations of X_L and R in parallel are listed in Table 21–2. When X_L is 10 times R, the parallel circuit is practically resistive because there is little inductive current in the line. The small value of I_L results from the high X_L. The total impedance of the parallel circuit is approximately equal to the resistance,

Table 21–2 | Parallel Resistance and Inductance Combinations*

R, Ω	X_L, Ω	I_R, A	I_L, A	I_T, A (Approx.)	$Z_{EQ} = V_A/I_T$, Ω	Phase Angle θ_I
1	10	10	1	$\sqrt{101} = 10$	1	$-5.7°$
10	10	1	1	$\sqrt{2} = 1.4$	7.07	$-45°$
10	1	1	10	$\sqrt{101} = 10$	1	$-84.3°$

* $V_A = 10$ V. Note that θ_I is the angle of I_T with respect to the reference V_A in parallel circuits.

then, since the high value of X_L in a parallel branch has little effect. The phase angle of $-5.7°$ is practically 0° because almost all of the line current is resistive.

As X_L becomes smaller, it provides more inductive current in the main line. When X_L is $\frac{1}{10} R$, practically all of the line current is the I_L component. Then the parallel circuit is practically all inductive, with a total impedance practically equal to X_L. The phase angle of $-84.3°$ is almost $-90°$ because the line current is mostly inductive. Note that these conditions are opposite from those of X_L and R in series.

When X_L and R are equal, their branch currents are equal and the phase angle is $-45°$. All these phase angles are negative for parallel I_L and I_R.

As additional comparisons between series and parallel circuits, remember that

1. The series voltage drops V_R and V_L have individual values that are 90° out of phase. Therefore, V_R and V_L are added by phasors to equal the applied voltage V_T. The phase angle θ_Z is between V_T and the common series current I. More series X_L allows more V_L to make the circuit more inductive, with a larger positive phase angle for V_T with respect to I.

2. The parallel branch currents I_R and I_L have individual values that are 90° out of phase. Therefore, I_R and I_L are added by phasors to equal I_T, which is the main-line current. The negative phase angle $-\theta_I$ is between the line current I_T and the common parallel voltage V_A. Less parallel X_L allows more I_L to make the circuit more inductive, with a larger negative phase angle for I_T with respect to V_A.

■ 21-4 Knowledge Check

Answers at end of chapter.

For a parallel RL circuit, how much are Z_{EQ} and θ_I if $X_L = 400$ Ω and $R = 300$ Ω?

■ 21-4 Self-Review

Answers at end of chapter.

a. How much is I_T for a branch current I_R of 2 A and I_L of 2 A?
b. Find the phase angle θ_I.

21-5 Q of a Coil

The ability of a coil to produce self-induced voltage is indicated by X_L, since it includes the factors of frequency and inductance. However, a coil has internal resistance equal to the resistance of the wire in the coil. This internal r_i of the coil

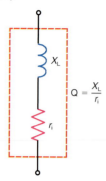

Figure 21–7 The Q of a coil depends on its inductive reactance X_L and resistance r_i.

$$Q = \frac{X_L}{r_i}$$

reduces the current, which means less ability to produce induced voltage. Combining these two factors of X_L and r_i, the *quality* or *merit* of a coil is indicated by

$$Q = \frac{X_L}{r_i} = \frac{2\pi f L}{r_i} \qquad (21\text{–}7)$$

As shown in Fig. 21–7, the internal r_i is in series with X_L.

As an example, a coil with X_L of 500 Ω and r_i of 5 Ω has a Q of $500/5 = 100$. The Q is a numerical value without any units, since the ohms cancel in the ratio of reactance to resistance. This Q of 100 means that the X_L of the coil is 100 times more than its r_i.

The Q of coils may range in value from less than 10 for a low-Q coil up to 1000 for a very high Q. Radio-frequency (rf) coils generally have Qs of about 30 to 300.

At low frequencies, r_i is just the dc resistance of the wire in the coil. However, for rf coils, the losses increase with higher frequencies and the effective r_i increases. The increased resistance results from eddy currents and other losses.

Because of these losses, the Q of a coil does not increase without limit as X_L increases for higher frequencies. Generally, Q can increase by a factor of about 2 for higher frequencies, within the range for which the coil is designed. The highest Q for rf coils generally results from an inductance value that provides an X_L of about 1000 Ω at the operating frequency.

More fundamentally, Q is defined as the ratio of reactive power in the inductance to the real power dissipated in the resistance. Then

$$Q = \frac{P_L}{P_{r_i}} = \frac{I^2 X_L}{I^2 r_i} = \frac{X_L}{r_i} = \frac{2\pi f L}{r_i}$$

which is the same as Formula (21–7).

Skin Effect

Radio-frequency current tends to flow at the surface of a conductor at very high frequencies, with little current in the solid core at the center. This skin effect results from the fact that current in the center of the wire encounters slightly more inductance because of the magnetic flux concentrated in the metal, compared with the edges, where part of the flux is in air. For this reason, conductors for VHF currents are often made of hollow tubing. The skin effect increases the effective resistance because a smaller cross-sectional area is used for the current path in the conductor.

AC Effective Resistance

When the power and current applied to a coil are measured for rf applied voltage, the $I^2 R$ loss corresponds to a much higher resistance than the dc resistance measured with an ohmmeter. This higher resistance is the ac effective resistance R_e. Although it is a result of high-frequency alternating current, R_e is not a reactance; R_e is a resistive component because it draws in-phase current from the ac voltage source.

The factors that make the R_e of a coil more than its dc resistance include skin effect, eddy currents, and hysteresis losses. Air-core coils have low losses but are limited to small values of inductance.

For a magnetic core in rf coils, a powdered-iron or ferrite slug is generally used. In a powdered-iron slug, the granules of iron are insulated from each other to reduce eddy currents. Ferrite materials have small eddy-current losses because they are insulators, although magnetic. A ferrite core is easily

GOOD TO KNOW

The Q of a parallel RL circuit is calculated as $Q = \dfrac{R}{X_L}$, assuming the series resistance of the coil is negligible. The Q formula for parallel RL circuits is derived as follows: $Q = \dfrac{P_L}{P_R} = \dfrac{V_A{}^2/X_L}{V_A{}^2/R} = \dfrac{R}{X_L}$.
Note: R is the resistance in parallel with X_L.

Figure 21–8 Ferrite coil antenna for a radio receiver.

saturated. Therefore, its use must be limited to coils with low values of current. A common application is the ferrite-core antenna coil in Fig. 21–8.

To reduce the R_e for small rf coils, stranded wire can be made with separate strands insulated from each other and braided so that each strand is as much on the outer surface as all other strands. This is called *litzendraht* or *litz wire*.

As an example of the total effect of ac losses, assume that an air-core rf coil of 50-μH inductance has a dc resistance of 1 Ω measured with the battery in an ohmmeter. However, in an ac circuit with a 2-MHz current, the effective coil resistance R_e can increase to 12 Ω. The increased resistance reduces the Q of the coil.

Actually, the Q can be used to determine the effective ac resistance. Since Q is X_L/R_e, then R_e equals X_L/Q. For this 50-μH L at 2 MHz, its X_L, equal to $2\pi f L$, is 628 Ω. The Q of the coil can be measured on a Q meter, which operates on the principle of resonance. Let the measured Q be 50. Then $R_e = 628/50$, equal to 12.6 Ω.

In general, the lower the internal resistance of a coil, the higher its Q.

Example 21–3

An air-core coil has an X_L of 700 Ω and an R_e of 2 Ω. Calculate the value of Q for this coil.

ANSWER

$$Q = \frac{X_L}{R_e} = \frac{700}{2}$$
$$= 350$$

Example 21-4

A 200-μH coil has a Q of 40 at 0.5 MHz. Find R_e.

ANSWER

$$R_e = \frac{X_L}{Q} = \frac{2\pi f L}{Q}$$

$$= \frac{2\pi \times 0.5 \times 10^6 \times 200 \times 10^{-6}}{40}$$

$$= \frac{628}{40}$$

$$= 15.7 \; \Omega$$

The Q of a Capacitor

The quality Q of a capacitor in terms of minimum loss is often indicated by its power factor. The lower the numerical value of the power factor, the better the quality of the capacitor. Since the losses are in the dielectric, the power factor of the capacitor is essentially the power factor of the dielectric, independent of capacitance value or voltage rating. At radio frequencies, approximate values of power factor are 0.000 for air or vacuum, 0.0004 for mica, about 0.01 for paper, and 0.0001 to 0.03 for ceramics.

The reciprocal of the power factor can be considered the Q of the capacitor, similar to the idea of the Q of a coil. For instance, a power factor of 0.001 corresponds to a Q of 1000. A higher Q therefore means better quality for the capacitor. If the leakage resistance R_l is known, the Q can be calculated as $Q = 2\pi f R_l C$. Capacitors have Qs that are much higher than those of inductors. The Q of capacitors typically ranges into the thousands, depending on design.

■ *21-5 Knowledge Check*

   Answer at end of chapter.

   Why doesn't the Q of a coil increase without limit as X_L increases at higher frequencies?

■ *21-5 Self-Review*

   Answers at end of chapter.

   a. A 200-μH coil with an 8-Ω internal R_e has an X_L of 600 Ω. Calculate the Q.
   b. A coil with a Q of 50 has a 500-Ω X_L at 4 MHz. Calculate its internal R_e.

21-6 AF and RF Chokes

Inductance has the useful characteristic of providing more ohms of reactance at higher frequencies. Resistance has the same opposition at all frequencies and for direct current. The skin effect for L at very high frequencies is not being considered here. These characteristics of L and R are applied to the circuit in

GOOD TO KNOW

Always remember that an $\frac{X_L}{R}$ ratio

of 10:1 is considered the dividing

line for calculating the value of

the choke inductance, L. This 10:1

ratio of $\frac{X_L}{R}$ should exist for the

lowest frequency intended to be

blocked from the output.

Figure 21–9 Coil used as a choke with X_L at least $10 \times R$. Note that R is an external resistor; V_L across L is practically all of the applied voltage with very little V_R. (a) Circuit with X_L and R in series. (b) Input and output voltages.

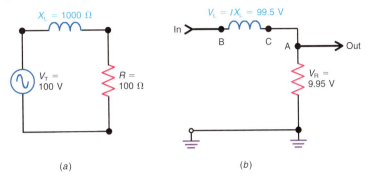

(a)　　　　　(b)

Fig. 21–9 where X_L is much greater than R for the frequency of the ac source V_T. The result is that L has practically all the voltage drop in this series circuit with very little of the applied voltage across R.

The inductance L is used here as a *choke*. Therefore, a choke is an inductance in series with an external R to prevent the ac signal voltage from developing any appreciable output across R at the frequency of the source.

The dividing line in calculations for a choke can be taken as X_L 10 or more times the series R. Then the circuit is primarily inductive. Practically all the ac voltage drop is across L, with little across R. This case also results in θ of practically 90°, but the phase angle is not related to the action of X_L as a choke.

Figure 21–9b illustrates how a choke is used to prevent ac voltage in the input from developing voltage in the output for the next circuit. Note that the output here is V_R from point A to earth ground. Practically all ac input voltage is across X_L between points B and C. However, this voltage is not coupled out because neither B nor C is grounded.

The desired output across R could be direct current from the input side without any ac component. Then X_L has no effect on the steady dc component. Practically all dc voltage would be across R for the output, but the ac voltage would be just across X_L. The same idea applies to passing an af signal through to R, while blocking an rf signal as IX_L across the choke because of more X_L at the higher frequency.

Calculations for a Choke

Typical values for audio or radio frequencies can be calculated if we assume a series resistance of 100 Ω as an example. Then X_L must be at least 1000 Ω. As listed in Table 21–3, at 100 Hz the relatively large inductance of 1.6 H provides 1000 Ω of X_L. Higher frequencies allow a smaller value of L for a choke with the same reactance. At 100 MHz in the VHF range, the choke is only 1.6 μH.

Some typical chokes are shown in Fig. 21–10. The iron-core choke in Fig. 21–10a is for audio frequencies. The air-core choke in Fig. 21–10b is for radio frequencies. The rf choke in Fig. 21–10c has color coding, which is often used for small coils. The color values are the same as for resistors, except that the values of L are given in microhenrys. As an example, a coil with yellow, red, and black stripes or dots is 42 μH.

Note that inductors are also available as surface-mount components. There are basically two body styles: completely encased and open. The encased

Table 21–3	Typical Chokes for a Reactance of 1000 Ω*	
F	L	Remarks
100 Hz	1.6 H	Low audio frequency
1000 Hz	0.16 H	Audio frequency
10 kHz	16 mH	Audio frequency
1000 kHz	0.16 mH	Radio frequency
100 MHz	1.6 μH	Very high radio frequency

* For an X_L that is 10 times a series R of 100 Ω.

Figure 21–10 Typical chokes. (a) Choke for 60 Hz with 8-H inductance and r_i of 350 Ω. Width is 2 in. (b) RF choke with 5 mH of inductance and r_i of 50 Ω. Length is 1 in. (c) Small rf choke encapsulated in plastic with leads for printed-circuit board; $L = 42$ μH. Width is 3/4 in.

(a)

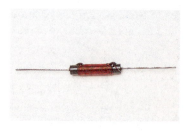

(b)

(c)

body style looks like a thick capacitor with a black body. The open body style inductor is easy to identify because the coil is visible. The value of a surface-mount inductor, if marked, is usually represented using the same three-digit system used for resistors, with the value displayed in microhenrys (μH).

Choosing a Choke for a Circuit

As an example of using these calculations, suppose that we have the problem of determining what kind of coil to use as a choke for the following application. The L is to be an rf choke in series with an external R of 300 Ω, with a current of 90 mA and a frequency of 0.2 MHz. Then X_L must be at least 10 × 300 = 3000 Ω. At f of 0.2 MHz,

$$L = \frac{X_L}{2\pi f} = \frac{3000}{2\pi \times 0.2 \times 10^6} = \frac{3 \times 10^3}{1.256 \times 10^6}$$

$$= \frac{3}{1.256} \times 10^{-3}$$

$$= 2.4 \text{ mH}$$

A typical and easily available commercial size is 2.5 mH, with a current rating of 115 mA and an internal resistance of 20 Ω, similar to the rf choke in Fig. 21–10b. Note that the higher current rating is suitable. Also, the internal resistance is negligible compared with the external R. An inductance a little higher than the calculated value will provide more X_L, which is better for a choke.

■ *21–6 Knowledge Check*

Answers at end of chapter.

Calculate the minimum inductance, L, for a choke in series with a resistance of 2.7 kΩ if the lowest frequency of the input signal is (a) 15 kHz (b) 2 MHz.

■ *21–6 Self-Review*

Answers at end of chapter.

a. How much is the minimum X_L for a choke in series with R of 80 Ω?
b. If X_L is 800 Ω at 3 MHz, how much will X_L be at 6 MHz for the same coil?

21–7 The General Case of Inductive Voltage

The voltage across any inductance in any circuit is always equal to $L(di/dt)$. This formula gives the instantaneous values of v_L based on the self-induced voltage produced by a change in magnetic flux from a change in current.

A sine waveform of current i produces a cosine waveform for the induced voltage v_L, equal to $L(di/dt)$. This means that v_L has the same waveform as i, but v_L and i are 90° out of phase for sine-wave variations.

The inductive voltage can be calculated as IX_L in sine-wave ac circuits. Since X_L is $2\pi fL$, the factors that determine the induced voltage are included in the frequency and inductance. Usually, it is more convenient to work with IX_L for the inductive voltage in sine-wave ac circuits, instead of $L(di/dt)$.

However, with a nonsinusoidal current waveform, the concept of reactance cannot be used. The X_L applies only to sine waves. Then v_L must be calculated as $L(di/dt)$, which applies for any inductive voltage.

An example is illustrated in Fig. 21–11a for sawtooth current. The sawtooth rise is a uniform or linear increase of current from zero to 90 mA in this example. The sharp drop in current is from 90 mA to zero. Note that the rise is relatively slow; it takes 90 μs. This is nine times longer than the fast drop in 10 μs.

The complete period of one cycle of this sawtooth wave is 100 μs. A cycle includes the rise of i to the peak value and its drop back to the starting value.

The Slope of I

The slope of any curve is a measure of how much it changes vertically for each horizontal unit. In Fig. 21–11a, the increase in current has a constant slope. Here i increases 90 mA in 90 μs, or 10 mA for every 10 μs of time. Then di/dt is constant at 10 mA/10 μs for the entire rise time of the sawtooth waveform. Actually, di/dt is the slope of the i curve. The constant di/dt is why the v_L waveform has a constant value of voltage during the linear rise of i. Remember that the amount of induced voltage depends on the change in current with time.

The drop in i is also linear but much faster. During this time, the slope is 90 mA/10 μs for di/dt.

Figure 21–11 Rectangular waveshape of v_L produced by sawtooth current through inductance L. (a) Waveform of current i. (b) Induced voltage v_L equal to $L(di/dt)$.

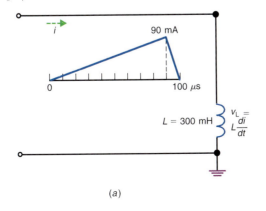

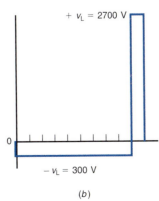

(a)

(b)

The Polarity of v_L

In Fig. 21–11, apply Lenz's law to indicate that v_L opposes the change in current. With electron flow into the top of L, the v_L is negative to oppose an increase in current. This polarity opposes the direction of electron flow shown for the current i produced by the source. For the rise time, then, the induced voltage here is labeled $-v_L$.

During the drop in current, the induced voltage has opposite polarity, which is labeled $+v_L$. These voltage polarities are for the top of L with respect to earth ground.

Calculations for v_L

The values of induced voltage across the 300-mH L are calculated as follows.
For the sawtooth rise:

$$-v_L = L\frac{di}{dt}$$

$$= 300 \times 10^{-3} \times \frac{10 \times 10^{-3}}{10 \times 10^{-6}}$$

$$= 300 \text{ V}$$

For the sawtooth drop:

$$+v_L = L\frac{di}{dt}$$

$$= 300 \times 10^{-3} \times \frac{90 \times 10^{-3}}{10 \times 10^{-6}}$$

$$= 2700 \text{ V}$$

The decrease in current produces nine times more voltage because the sharp drop in i is nine times faster than the relatively slow rise.

Remember that the di/dt factor can be very large, even with small currents, when the time is short. For instance, a current change of 1 mA in 1 μs is equivalent to the very high di/dt value of 1000 A/s.

An interesting feature of the inductive waveshapes in Fig. 21–11 is that they are the same as the capacitive waveshapes shown before in Fig. 18–10, but with current and voltage waveshapes interchanged. This comparison follows from the fact that both v_L and i_C depend on the rate of change. Then i_C is $C(dv/dt)$, and v_L is $L(di/dt)$.

It is important to note that v_L and i_L have different waveshapes with nonsinusoidal current. In this case, we compare the waveshapes instead of the phase angle. Common examples of nonsinusoidal waveshapes for either v or i are the sawtooth waveform, square wave, and rectangular pulses. For a sine wave, the $L(di/dt)$ effects result in a cosine wave, as shown before, in Fig. 20–7.

■ 21–7 Knowledge Check

Answer at end of chapter.

In Fig. 21–11a, how much is V_L if i increases from 0 to 120 mA in 90 μs?

■ 21–7 Self-Review

Answers at end of chapter.

Refer to Fig. 21–11.
a. How much is di/dt in amperes per second for the sawtooth rise of i?
b. How much is di/dt in amperes per second for the drop in i?

Summary

- In a sine-wave ac circuit, the current through an inductance lags 90° behind the voltage across the inductance because $v_L = L(di/dt)$. This fundamental fact is the basis of all the following relations.

- Therefore, inductive reactance X_L is a phasor quantity 90° out of phase with R. The phasor combination of X_L and R is their impedance Z_T.

- These three types of opposition to current are compared in Table 21–4.

- The phase angle θ is the angle between the applied voltage and its current.

- The opposite characteristics for series and parallel circuits with X_L and R are summarized in Table 21–5.

- The Q of a coil is X_L/r_i, where r_i is the coil's internal resistance.

- A choke is an inductance with X_L greater than the series R by a factor of 10 or more.

- In sine-wave circuits, $V_L = IX_L$. Then V_L is out of phase with I by an angle of 90°.

- For a circuit with X_L and R in series, $\tan \theta_Z = X_L/R$. When the components are in parallel, $\tan \theta_I = -(I_L/I_R)$. See Table 21–5.

- When the current is not a sine wave, $v_L = L(di/dt)$. Then the waveshape of V_L is different from the waveshape of I.

- Inductors are available as surface-mount components. Surface-mount inductors are available in both completely encased and open body styles.

Table 21–4	Comparison of R, X_L, and Z_T	
R	$X_L = 2\pi fL$	$Z_T = \sqrt{R^2 + X_L^2}$
Ohm unit	Ohm unit	Ohm unit
IR voltage in phase with I	IX_L voltage leads I by 90°	IZ is applied voltage; it leads line I by $\theta°$
Same for all frequencies	Increases as frequency increases	Increases with X_L at higher frequencies

Table 21–5	Series and Parallel RL Circuits	
X_L and R in Series	**X_L and R in Parallel**
I the same in X_L and R	V_A the same across X_L and R
$V_T = \sqrt{V_R^2 + V_L^2}$	$I_T = \sqrt{I_R^2 + I_L^2}$
$Z_T = \sqrt{R^2 + X_L^2}$	$Z_{EQ} = \dfrac{V_A}{I_T}$
V_L leads V_R by 90°	I_L lags I_R by 90°
$\tan \theta_z = \dfrac{X_L}{R}$	$\tan \theta_I = -\dfrac{I_L}{I_R}$
The θ_Z increases with more X_L, which means more V_L, thus making the circuit more inductive	The $-\theta_I$ decreases with more X_L, which means less I_L, thus making the circuit less inductive

Important Terms

- **AC Effective Resistance, R_e** - the resistance of a coil for higher frequency alternating current. The value of R_e is more than the dc resistance of the coil because it includes the losses associated with high frequency alternating current in a coil. These losses include skin effect, eddy currents, and hysteresis losses.

- **Arctangent (arctan)** - an inverse trigonometric function that specifies the angle, θ, corresponding to a given tangent (tan) value.

- **Choke** - a name used for a coil when its application is to appreciably reduce the amount of ac voltage that is developed across a series resistor, R. The dividing line for calculating the choke inductance is to make X_L 10 or more times larger than the series R at a specific frequency as a lower limit.

- **Impedance, Z** - the total opposition to the flow of current in a sine-wave ac circuit. In an RL circuit, the impedance, Z, takes into account the 90° phase relation between X_L and R. Impedance is measured in ohms.

- **Phase Angle, θ** - the angle between the applied voltage and current in a sine-wave ac circuit.

- **Phasor Triangle** - a right triangle that represents the phasor sum of two quantities 90° out of phase with each other.

- **Q of a Coil** - the quality or figure of merit for a coil. More specifically, the Q of a coil is the ratio of reactive power in the inductance to the real power dissipated in the coil's resistance,
$$Q = \frac{X_L}{r_i}.$$

- **Skin Effect** - a term used to describe current flowing on the outer surface of a conductor at very high frequencies. The skin effect causes the effective resistance of a coil to increase at higher frequencies since the effect is the same as reducing the cmil area of the wire.

- **Tangent (tan)** - a trigonometric function of an angle, equal to the ratio of the opposite side to the adjacent side of a right triangle.

Related Formulas

Series RL circuit

$$V_T = \sqrt{V_R^2 + V_L^2}$$
$$Z_T = \sqrt{R^2 + X_L^2}$$
$$\tan \theta_z = \frac{X_L}{R}$$

Parallel RL circuit

$$I_T = \sqrt{I_R^2 + I_L^2}$$
$$Z_{EQ} = \frac{V_A}{I_T}$$
$$\tan \theta_I = -\frac{I_L}{I_R}$$

Q of a coil

$$Q = \frac{X_L}{r_i} = \frac{2\pi fL}{r_i}$$

Self-Test

Answers at back of book.

1. **Inductive reactance, X_L,**
   a. applies only to nonsinusoidal waveforms or dc.
   b. applies only to sine waves.
   c. applies to either sinusoidal or nonsinusoidal waveforms.
   d. is inversely proportional to frequency.

2. **For an inductor in a sine wave ac circuit,**
   a. V_L leads i_L by 90°.
   b. V_L lags i_L by 90°.
   c. V_L and i_L are in phase.
   d. none of the above.

3. **In a series RL circuit,**
   a. V_L lags V_R by 90°.
   b. V_L leads V_R by 90°.
   c. V_R and I are in phase.
   d. both b and c.

4. **In a series RL circuit where $V_L = 9$ V and $V_R = 12$ V, how much is the total voltage, V_T?**
   a. 21 V.
   b. 225 V.
   c. 15 V.
   d. 3 V.

5. **A 50-Ω resistor is in parallel with an inductive reactance, X_L, of 50 Ω. The combined equivalent impedance, Z_{EQ} of this combination is**
   a. 70.7 Ω.
   b. 100 Ω.

   c. 35.36 Ω.
   d. 25 Ω.

6. **In a parallel RL circuit,**
   a. I_L lags I_R by 90°.
   b. I_L leads I_R by 90°.
   c. I_L and I_R are in phase.
   d. I_R lags I_L by 90°.

7. **In a parallel RL circuit, where $I_R = 1.2$ A and $I_L = 1.6$ A, how much is the total current, I_T?**
   a. 2.8 A.
   b. 2 A.
   c. 4 A.
   d. 400 mA.

8. In a series *RL* circuit where $X_L = R$, the phase angle, θ_Z, is
   a. −45°.
   b. 0°.
   c. +90°.
   d. +45°.

9. In a parallel *RL* circuit,
   a. V_A and I_L are in phase.
   b. I_L and I_R are in phase.
   c. V_A and I_R are in phase.
   d. V_A and I_R are 90° out of phase.

10. A 1-kΩ resistance is in series with an inductive reactance, X_L, of 2 kΩ. The total impedance, Z_T, is
    a. 2.24 kΩ.
    b. 3 kΩ.
    c. 1 kΩ.
    d. 5 MΩ.

11. When the frequency of the applied voltage decreases in a parallel *RL* circuit,
    a. the phase angle, θ_I, becomes less negative.
    b. Z_{EQ} increases.
    c. Z_{EQ} decreases.
    d. both a and b.

12. When the frequency of the applied voltage increases in a series *RL* circuit,
    a. θ_Z increases.
    b. Z_T decreases.
    c. Z_T increases.
    d. both a and c.

13. The dividing line for calculating the value of a choke inductance is to make
    a. X_L 10 or more times larger than the series *R*.
    b. X_L one-tenth or less than the series *R*.

    c. X_L equal to *R*.
    d. *R* 10 or more times larger than the series X_L.

14. The *Q* of a coil is affected by
    a. frequency.
    b. the resistance of the coil.
    c. skin effect.
    d. all of the above.

15. If the current through a 300-mH coil increases at the linear rate of 50 mA per 10 μs, how much is the induced voltage, V_L?
    a. 1.5 V.
    b. 1.5 kV.
    c. This is impossible to determine because X_L is unknown.
    d. This is impossible to determine because V_L also increases at a linear rate.

Questions

1. What characteristic of the current in an inductance determines the amount of induced voltage? State briefly why.

2. Draw a schematic diagram showing an inductance connected across a sine-wave voltage source and indicate the current and voltage that are 90° out of phase with one another.

3. Why is the voltage across a resistance in phase with the current through the resistance?

4. (a) Draw the sine waveforms for two voltages 90° out of phase, each with a peak value of 100 V. (b) Why does their phasor sum equal 141 V and not 200 V? (c) When will the sum of two 100-V drops in series equal 200 V?

5. (a) Define the phase angle of a sine-wave ac circuit. (b) State the formula for the phase angle in a circuit with X_L and *R* in series.

6. Define the following: (a) *Q* of a coil; (b) ac effective resistance; (c) rf choke; (d) sawtooth current.

7. Why do all waveshapes in Fig. 21–2*b* have the same frequency?

8. Describe how to check the trouble of an open choke with an ohmmeter.

9. Redraw the circuit and graph in Fig. 21–11 for a sawtooth current with a peak of 30 mA.

10. Why is the R_e of a coil considered resistance rather than reactance?

11. Why are rf chokes usually smaller than af chokes?

12. What is the waveshape of v_L for a sine-wave i_L?

Problems

SECTION 21–1 SINE WAVE i_L LAGS V_L BY 90°

21–1 In Fig. 21–12, what is the
   a. peak value of the inductor voltage, V_L?
   b. peak value of the inductor current, i_L?
   c. frequency of the inductor current, i_L?
   d. phase relationship between V_L and i_L?

Figure 21–12

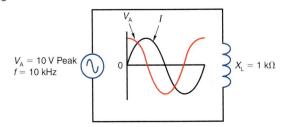

$V_A = 10$ V Peak
$f = 10$ kHz
$X_L = 1$ kΩ

21–2 In Fig. 21–12, what is the value of the induced voltage, V_L, when i_L is at

    a. 0 mA?

    b. its positive peak of 10 mA?

    c. its negative peak of 10 mA?

21–3 In Fig. 21–12, draw the phasors representing V_L and i_L using

    a. i_L as the reference phasor.

    b. V_L as the reference phasor.

SECTION 21–2 X_L AND R IN SERIES

21–4 In Fig. 21–13, how much current, I, is flowing

    a. through the 15-Ω resistor, R?

    b. through the 20-Ω inductive reactance, X_L?

    c. to and from the terminals of the applied voltage, V_T?

Figure 21–13

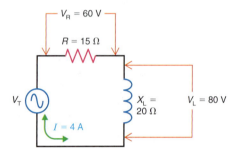

21–5 In Fig. 21–13, what is the phase relationship between

    a. I and V_R?

    b. I and V_L?

    c. V_L and V_R?

21–6 In Fig. 21–13, how much is the applied voltage, V_T?

21–7 Draw the phasor voltage triangle for the circuit in Fig. 21–13. (Use V_R as the reference phasor.)

21–8 In Fig. 21–14, solve for

    a. the resistor voltage, V_R.

    b. the inductor voltage, V_L.

    c. the total voltage, V_T.

Figure 21–14

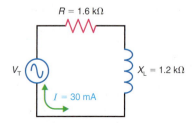

21–9 In Fig. 21–15, solve for

    a. the resistor voltage, V_R.

    b. the inductor voltage, V_L.

    c. the total voltage, V_T.

Figure 21–15

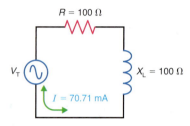

21–10 In a series RL circuit, solve for the applied voltage, V_T, if

    a. $V_R = 12$ V and $V_L = 6$ V.

    b. $V_R = 25$ V and $V_L = 40$ V.

    c. $V_R = 9$ V and $V_L = 16$ V.

    d. $V_R = 40$ V and $V_L = 40$ V.

SECTION 21–3 IMPEDANCE Z TRIANGLE

21–11 In Fig. 21–16, solve for Z_T, I, V_L, V_R, and θ_Z.

Figure 21–16

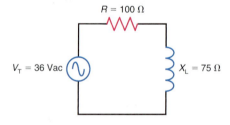

21–12 Draw the impedance triangle for the circuit in Fig. 21–16. (Use R as the reference phasor.)

21–13 In Fig. 21–17, solve for Z_T, I, V_L, V_R, and θ_Z.

Figure 21–17

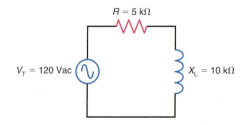

21–14 In Fig. 21–18, solve for Z_T, I, V_L, V_R, and θ_Z.

Figure 21–18

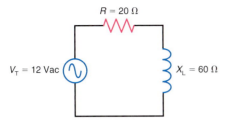

$R = 20\ \Omega$
$V_T = 12\ \text{Vac}$
$X_L = 60\ \Omega$

21–15 In Fig. 21–19, solve for Z_T, I, V_L, V_R, and θ_Z.

Figure 21–19

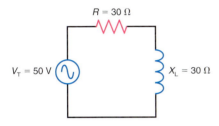

$R = 30\ \Omega$
$V_T = 50\ \text{V}$
$X_L = 30\ \Omega$

21–16 In Fig. 21–20, solve for Z_T, I, V_L, V_R, and θ_Z for the following circuit values:

a. $X_L = 30\ \Omega$, $R = 40\ \Omega$, and $V_T = 50\ \text{V}$
b. $X_L = 50\ \Omega$, $R = 50\ \Omega$, and $V_T = 141.4\ \text{V}$
c. $X_L = 10\ \Omega$, $R = 100\ \Omega$, and $V_T = 10\ \text{V}$
d. $X_L = 100\ \Omega$, $R = 10\ \Omega$, and $V_T = 10\ \text{V}$

Figure 21–20

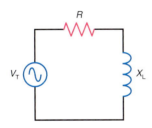

R
V_T
X_L

21–17 In Fig. 21–21, solve for X_L, Z_T, I, V_R, V_L, and θ_Z.

Figure 21–21

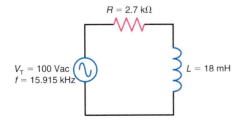

$R = 2.7\ \text{k}\Omega$
$V_T = 100\ \text{Vac}$
$f = 15.915\ \text{kHz}$
$L = 18\ \text{mH}$

21–18 In Fig. 21–21, what happens to each of the following quantities if the frequency of the applied voltage increases?

a. X_L
b. Z_T
c. I
d. V_R
e. V_L
f. θ_Z

21–19 Repeat Prob. 21–18 if the frequency of the applied voltage decreases.

SECTION 21–4 X_L AND R IN PARALLEL

21–20 In Fig. 21–22, how much voltage is across

a. the 30-Ω resistor, R?
b. the 40-Ω inductive reactance, X_L?

Figure 21–22

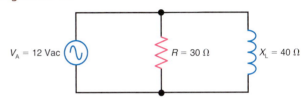

$V_A = 12\ \text{Vac}$
$R = 30\ \Omega$
$X_L = 40\ \Omega$

21–21 In Fig. 21–22, what is the phase relationship between

a. V_A and I_R?
b. V_A and I_L?
c. I_L and I_R?

21–22 In Fig. 21–22, solve for I_R, I_L, I_T, Z_{EQ}, and θ_I.

21–23 Draw the phasor current triangle for the circuit in Fig. 21–22. (Use I_R as the reference phasor.)

21–24 In Fig. 21–23, solve for I_R, I_L, I_T, Z_{EQ}, and θ_I.

Figure 21–23

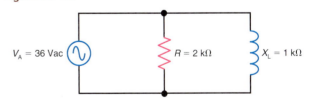

$V_A = 36\ \text{Vac}$
$R = 2\ \text{k}\Omega$
$X_L = 1\ \text{k}\Omega$

21–25 In Fig. 21–24, solve for I_R, I_L, I_T, Z_{EQ}, and θ_I.

Figure 21–24

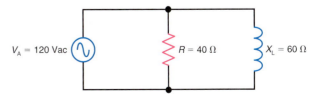

$V_A = 120\ \text{Vac}$
$R = 40\ \Omega$
$X_L = 60\ \Omega$

Inductive Circuits **667**

21-26 In Fig. 21-25, solve for I_R, I_L, I_T, Z_{EQ}, and θ_I.

Figure 21-25

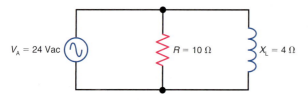

21-27 In Fig. 21-26, solve for I_R, I_L, I_T, Z_{EQ}, and θ_I.

Figure 21-26

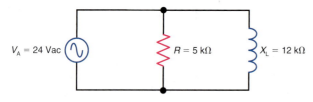

21-28 In Fig. 21-27, solve for I_R, I_L, I_T, Z_{EQ}, and θ_I for the following circuit values?
  a. $R = 50\ \Omega$, $X_L = 50\ \Omega$, and $V_A = 50\ V$
  b. $R = 10\ \Omega$, $X_L = 100\ \Omega$, and $V_A = 20\ V$
  c. $R = 100\ \Omega$, $X_L = 10\ \Omega$, and $V_A = 20\ V$

Figure 21-27

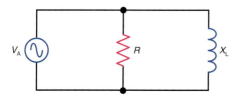

21-29 In Fig. 21-27, how much is Z_{EQ} if $R = 320\ \Omega$ and $X_L = 240\ \Omega$?

21-30 In Fig. 21-28, solve for X_L, I_R, I_L, I_T, Z_{EQ}, and θ_I.

Figure 21-28

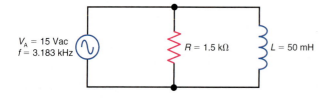

21-31 In Fig. 21-28, what happens to each of the following quantities if the frequency of the applied voltage increases?
  a. I_R
  b. I_L

  c. I_T
  d. Z_{EQ}
  e. θ_I

21-32 Repeat Prob. 21-31 if the frequency of the applied voltage decreases.

SECTION 21-5 Q OF A COIL

21-33 For the inductor shown in Fig. 21-29, calculate the Q for the following frequencies:
  a. $f = 500\ Hz$
  b. $f = 1\ kHz$
  c. $f = 1.592\ kHz$
  d. $f = 10\ kHz$

Figure 21-29

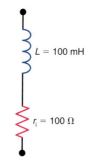

21-34 Why can't the Q of a coil increase without limit as the value of X_L increases for higher frequencies?

21-35 Calculate the ac effective resistance, R_e, of a 350-μH inductor whose Q equals 35 at 1.5 MHz.

21-36 Recalculate the value of R_e in Prob. 21-35 if the value of Q decreases to 25 at 5 MHz.

SECTION 21-6 AF AND RF CHOKES

21-37 In Fig. 21-30, calculate the required value of the choke inductance, L, at the following frequencies:
  a. $f = 500\ Hz$
  b. $f = 2.5\ kHz$
  c. $f = 200\ kHz$
  d. $f = 1\ MHz$

Figure 21-30

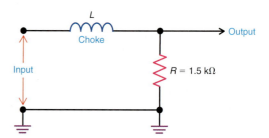

21–38 If $L = 50$ mH in Fig. 21–30, then what is the lowest frequency at which L will serve as a choke?

21–39 In Fig. 21–30 assume that the input voltage equals 10 V peak-to-peak for all frequencies. If $L = 150$ mH, then calculate V_{out} for the following frequencies:

a. 159.2 Hz

b. 1.592 kHz

c. 15.92 kHz

SECTION 21–7 THE GENERAL CASE OF INDUCTIVE VOLTAGE

21–40 In Fig. 21–31, draw the waveform of induced voltage, V_L, across the 8-mH inductor for the triangular current waveform shown.

21–41 In Fig. 21–32, draw the waveform of induced voltage, V_L, across the 250-mH inductor for the sawtooth current waveform shown.

Figure 21–31

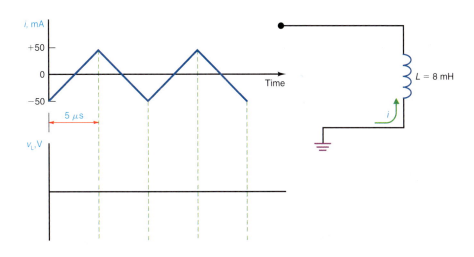

Figure 21–32

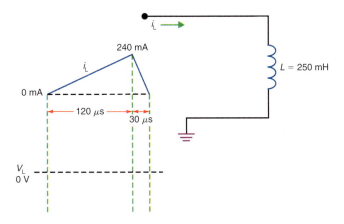

Critical Thinking

21–42 In Fig. 21–33, calculate X_L, R, L, I, V_L, and V_R.

21–43 In Fig. 21–34, calculate I_T, I_R, I_L, X_L, R, and L.

Figure 21–33 Circuit for Critical Thinking Prob. 21–42.

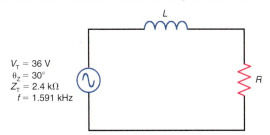

Figure 21–34 Circuit for Critical Thinking Prob. 21–43.

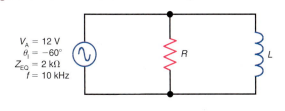

21-44 In Fig. 21-35, calculate V_R, V_{L_1}, X_{L_1}, X_{L_2}, I, Z_T, L_1, L_2, and θ_Z.

Figure 21-35 Circuit for Critical Thinking Prob. 21-44.

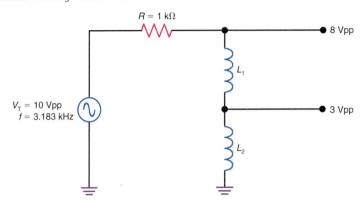

Answers to Knowledge Check Problems

21-1 Because the value of V_L is dependent on the rate of current change, di/dt, rather than on the actual value of current itself.

21-2 $V_T = 25$ V

21-3 $Z_T = 187.5\ \Omega$ and $\theta_Z = 53.13°$

21-4 $Z_{EQ} = 240\ \Omega$ and $\theta_I = -36.87°$

21-5 Because the coil has additional losses with higher frequency alternating current. These losses, which include skin effect, eddy currents, and hysteresis losses, cause the ac effective resistance, R_e, to increase with frequency, thus limiting Q.

21-6 a. $L = 286.5$ mH
b. $L = 2.15$ mH

21-7 $-V_L = 400$ V

Answers to Self-Reviews

21-1 a. 0°
b. 90°
c. lag

21-2 a. 0°
b. 90°

21-3 a. 28.28 Ω
b. 28.28 V
c. $\theta_Z = 45°$

21-4 a. $I_T = 2.828$ A
b. $\theta_I = -45°$

21-5 a. $Q = 75$
b. $R_e = 10\ \Omega$

21-6 a. $X_L = 800\ \Omega$
b. $X_L = 1600\ \Omega$

21-7 a. $di/dt = 1000$ A/s
b. $di/dt = 9000$ A/s

RC and *L/R* Time Constants

Many applications of inductance are for sine-wave ac circuits, but any time the current changes, *L* has the effect of producing induced voltage. Examples of nonsinusoidal waveshapes include dc voltages that are switched on or off, square waves, sawtooth waves, and rectangular pulses. For capacitance, also, many applications are for sine waves, but whenever the voltage changes, *C* produces charge or discharge current.

With nonsinusoidal voltage and current, the effect of *L* or *C* is to produce a change in waveshape. This effect can be analyzed by means of the time constant for capacitive and inductive circuits. The time constant is the time for a change of 63.2% in the current through *L* or the voltage across *C*.

Actually, *RC* circuits are more common than *RL* circuits because capacitors are smaller and more economical and do not have strong magnetic fields.

Objectives

After studying this chapter you should be able to

- *Define* the term *transient response*.

- *Define* the term *time constant*.

- *Calculate* the time constant of a circuit containing resistance and inductance.

- *Explain* the effect of producing a high voltage when opening an *RL* circuit.

- *Calculate* the time constant of a circuit containing resistance and capacitance.

- *Explain* how capacitance opposes a change in voltage.

- *List* the criteria for proper differentiation and integration.

- *Explain* why a long time constant is required for an *RC* coupling circuit.

- *Use* the universal time constant graph to solve for voltage and current values in an *RC* or *RL* circuit that is charging or discharging.

- *Explain* the difference between time constants and reactance.

Outline

Important Terms

differentiator

integrator

long time constant

short time constant

steady-state

time constant

transient response

universal time-constant graph

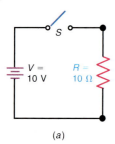

(a)

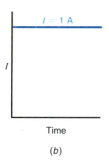

(b)

22–1 Response of Resistance Alone

To emphasize the special features of L and C, the circuit in Fig. 22–1a illustrates how an ordinary resistive circuit behaves. When the switch is closed, the battery supplies 10 V across the 10-Ω R and the resultant I is 1 A. The graph in Fig. 22–1b shows that I changes from 0 to 1 A instantly when the switch is closed. If the applied voltage is changed to 5 V, the current will change instantly to 0.5 A. If the switch is opened, I will immediately drop to zero.

Resistance has only opposition to current; there is no reaction to a change because R has no concentrated magnetic field to oppose a change in I, like inductance, and no electric field to store charge that opposes a change in V, like capacitance.

■ *22–1 Knowledge Check*

Answer at end of chapter.

In Fig. 22–1, does the current drop to zero immediately after the switch is opened?

■ *22–1 Self-Review*

Answers at end of chapter.

a. **Resistance R does not produce induced voltage for a change in I. (True or False)**
b. **Resistance R does not produce charge or discharge current for a change in V. (True or False)**

22–2 *L/R* Time Constant

Consider the circuit in Fig. 22–2, where L is in series with R. When S is closed, the current changes as I increases from zero. Eventually, I will reach the steady value of 1 A, equal to the battery voltage of 10 V divided by the circuit resistance of 10 Ω. While the current is building up from 0 to 1 A, however, I is changing and the inductance opposes the change. The action of the RL circuit during this time is its *transient response,* which means that a temporary

GOOD TO KNOW

Theoretically, the current, I, in Fig. 22-2 never reaches its steady-state value of 1 A with the switch closed.

IIII MultiSim **Figure 22–2** Transient response of circuit with R and inductance L. When the switch is closed, I rises from zero to the steady-state value of 1 A. (a) Circuit with time constant L/R of 1 H/10 Ω = 0.1 s. (b) Graph of I during five time constants. Compare with graph in Fig. 22–1b.

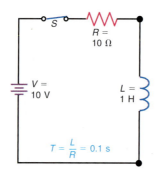

(a)

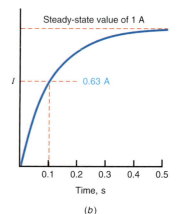

(b)

condition exists only until the steady-state current of 1 A is reached. Similarly, when S is opened, the transient response of the *RL* circuit opposes the decay of current toward the steady-state value of zero.

The transient response is measured in terms of the ratio L/R, which is the time constant of an inductive circuit. To calculate the time constant,

$$T = \frac{L}{R} \qquad (22\text{–}1)$$

where T is the time constant in seconds, L is the inductance in henrys, and R is the resistance in ohms. The resistance in series with L is either the coil resistance, an external resistance, or both in series. In Fig. 22–2,

$$T = \frac{L}{R} = \frac{1}{10} = 0.1 \text{ s}$$

Specifically, the time constant is a measure of how long it takes the current to change by 63.2%, or approximately 63%. In Fig. 22–2, the current increases from 0 to 0.63 A, which is 63% of the steady-state value, in a period of 0.1 s, which is one time constant. In a period of five time constants, the current is practically equal to its steady-state value of 1 A.

The reason why L/R equals time can be illustrated as follows: Since induced voltage $V = L(di/dt)$, by transposing terms, L has the dimensions of $V \times T/I$. Dividing L by R results in $V \times T/IR$. As the IR and V factors cancel, T remains to indicate the dimension of time for the ratio L/R.

Example 22–1

What is the time constant of a 20-H coil having 100 Ω of series resistance?

ANSWER

$$T = \frac{L}{R} = \frac{20 \text{ H}}{100 \text{ }\Omega}$$
$$= 0.2 \text{ s}$$

Example 22–2

An applied dc voltage of 10 V will produce a steady-state current of 100 mA in the 100-Ω coil of Example 22–1. How much is the current after 0.2 s? After 1 s?

ANSWER Since 0.2 s is one time constant, I is 63% of 100 mA, which equals 63 mA. After five time constants, or 1 s (0.2 s × 5), the current will reach its steady-state value of 100 mA and remain at this value as long as the applied voltage stays at 10 V.

Example 22-3

If a 1-MΩ R is added in series with the coil of Example 22–1, how much will the time constant be for the higher resistance RL circuit?

ANSWER

$$T = \frac{L}{R} = \frac{20 \text{ H}}{1{,}000{,}000 \text{ }\Omega}$$
$$= 20 \times 10^{-6} \text{ s}$$
$$= 20 \text{ } \mu\text{s}$$

The L/R time constant becomes longer with larger values of L. More series R, however, makes the time constant shorter. With more series resistance, the circuit is less inductive and more resistive.

■ *22-2 Knowledge Check*

> ***Answer at end of chapter.***
>
> **In Fig. 22–2a, which component opposes the buildup of current with S_1 closed?**

■ *22-2 Self-Review*

> ***Answers at end of chapter.***
>
> a. **Calculate the time constant for 2 H in series with 100 Ω.**
> b. **Calculate the time constant for 2 H in series with 4000 Ω.**

22-3 High Voltage Produced by Opening an *RL* Circuit

When an inductive circuit is opened, the time constant for current decay becomes very short because L/R becomes smaller with the high resistance of the open circuit. Then the current drops toward zero much faster than the rise of current when the switch is closed. The result is a high value of self-induced voltage V_L across a coil whenever an RL circuit is opened. This high voltage can be much greater than the applied voltage.

There is no gain in energy, though, because the high-voltage peak exists only for the short time the current is decreasing at a very fast rate at the start of the decay. Then, as I decays at a slower rate, the value of V_L is reduced. After the current has dropped to zero, there is no voltage across L.

This effect can be demonstrated by a neon bulb connected across a coil, as shown in Fig. 22–3. The neon bulb requires 90 V for ionization, at which time it glows. The source here is only 8 V, but when the switch is opened, the self-induced voltage is high enough to light the bulb for an instant. The sharp voltage pulse or spike is more than 90 V just after the switch is opened, when I drops very fast at the start of the decay in current.

Note that the 100-Ω R_1 is the internal resistance of the 2-H coil. This resistance is in series with L whether S is closed or open. The 4-kΩ R_2 across the switch is in the circuit only when S is opened, to have a specific resistance

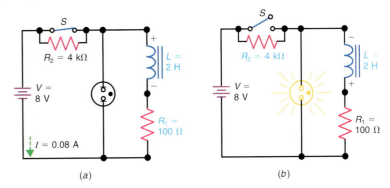

Figure 22–3 Demonstration of high voltage produced by opening inductive circuit. (*a*) With switch closed, 8 V applied cannot light the 90-V neon bulb. (*b*) When the switch is opened, the short L/R time constant results in high V_L, which lights the bulb.

(a)   (b)

across the open switch. Since R_2 is much more than R_1, the L/R time constant is much shorter with the switch open.

Closing the Circuit

In Fig. 22–3*a*, the switch is closed to allow current in L and to store energy in the magnetic field. Since R_2 is short-circuited by the switch, the 100-Ω R_1 is the only resistance. The steady-state I is $V/R_1 = 8/100 = 0.08$ A. This value of I is reached after five time constants.

One time constant is $L/R = 2/100 = 0.02$ s. Five time constants equal $5 \times 0.02 = 0.1$ s. Therefore, I is 0.08 A after 0.1 s, or 100 ms. The energy stored in the magnetic field is 64×10^{-4} J, equal to $\frac{1}{2}LI^2$.

Opening the Circuit

When the switch is opened in Fig. 22–3*b*, R_2 is in series with L, making the total resistance 4100 Ω, or approximately 4 kΩ. The result is a much shorter time constant for current decay. Then L/R is 2/4000, or 0.5 ms. The current decays practically to zero in five time constants, or 2.5 ms.

This rapid drop in current results in a magnetic field collapsing at a fast rate, inducing a high voltage across L. The peak v_L in this example is 320 V. Then v_L serves as the voltage source for the bulb connected across the coil. As a result, the neon bulb becomes ionized, and it lights for an instant. One problem is arcing produced when an inductive circuit is opened. Arcing can destroy contact points and under certain conditions cause fires or explosions.

Calculating the Peak of v_L

The value of 320 V for the peak induced voltage when S is opened in Fig. 22–3 can be determined as follows: With the switch closed, I is 0.08 A in all parts of the series circuit. The instant S is opened, R_2 is added in series with L and R_1. The energy stored in the magnetic field maintains I at 0.08 A for an instant before the current decays. With 0.08 A in the 4-kΩ R_2, its potential difference is $0.08 \times 4000 = 320$ V. The collapsing magnetic field induces this 320-V pulse to allow an I of 0.08 A at the instant the switch is opened.

The *di/dt* for v_L

The required rate of change in current is 160 A/s for the v_L of 320 V induced by the L of 2 H. Since $v_L = L(di/dt)$, this formula can be transposed to specify *di/dt* as equal to v_L/L. Then *di/dt* corresponds to 320 V/2 H, or 160 A/s. This value is the actual *di/dt* at the start of the decay in current when the switch is opened in Fig. 22–3*b*, as a result of the short time constant.*

Applications of Inductive Voltage Pulses

There are many uses for the high voltage generated by opening an inductive circuit. One example is the high voltage produced for the ignition system in an automobile. Here the circuit of the battery in series with a high-inductance spark coil is opened by the breaker points of the distributor to produce the high voltage needed for each spark plug. When an inductive circuit is opened very rapidly, 10,000 V can easily be produced.

■ *22–3 Knowledge Check*

> *Answer at end of chapter.*
>
> If R_2 is changed to 15 kΩ in Fig. 22–3*a*, how much is the peak inductor voltage, V_L, when the switch is opened?

■ *22–3 Self-Review*

> *Answers at end of chapter.*
>
> a. Is the *L/R* time constant longer or shorter in Fig. 22–3 when *S* is opened?
> b. Which produces more v_L, a faster *di/dt* or a slower *di/dt*?

22–4 *RC* Time Constant

The transient response of capacitive circuits is measured in terms of the product $R \times C$. To calculate the time constant,

$$T = R \times C \tag{22–2}$$

where *R* is in ohms, *C* is in farads, and *T* is in seconds. In Fig. 22–4, for example, with an *R* of 3 MΩ and a *C* of 1 μF,

$$T = 3 \times 10^6 \times 1 \times 10^{-6}$$
$$= 3 \text{ s}$$

Note that the 10^6 for megohms and the 10^{-6} for microfarads cancel. Therefore, multiplying the units of MΩ × μF gives the *RC* product in seconds.

Common combinations of units for the *RC* time constant are

$$\text{MΩ} \times μ\text{F} = \text{s}$$
$$\text{kΩ} \times μ\text{F} = \text{ms}$$
$$\text{MΩ} \times p\text{F} = μ\text{s}$$

The reason that the *RC* product is expressed in units of time can be illustrated as follows: $C = Q/V$. The charge *Q* is the product of $I \times T$. The factor *V* is *IR*. Therefore, *RC* is equivalent to $(R \times Q)/V$, or $(R \times IT)/IR$. Since *I* and *R* cancel, *T* remains to indicate the dimension of time.

---

* The *di/dt* value can be calculated from the slope at the start of decay, shown by the dashed line for curve *b* in Fig. 22–9.

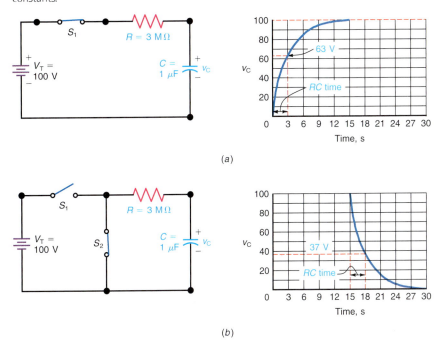

III MultiSim **Figure 22–4** Details of how a capacitor charges and discharges in an *RC* circuit. (*a*) With S_1 closed, *C* charges through *R* to 63% of V_T in one *RC* time constant of 3 s and is almost completely charged in five time constants. (*b*) With S_1 opened to disconnect the battery and S_2 closed for *C* to discharge through *R*, V_C drops to 37% of its initial voltage in one time constant of 3 s and is almost completely discharged in five time constants.

The Time Constant Indicates the Rate of Charge or Discharge

RC specifies the time it takes *C* to charge to 63% of the charging voltage. Similarly, *RC* specifies the time it takes *C* to discharge 63% of the way down to the value equal to 37% of the initial voltage across *C* at the start of discharge.

In Fig. 22–4*a*, for example, the time constant on charge is 3 s. Therefore, in 3 s, *C* charges to 63% of the 100 V applied, reaching 63 V in *RC* time. After five time constants, which is 15 s here, *C* is almost completely charged to the full 100 V applied. If *C* discharges after being charged to 100 V, then *C* will discharge down to 36.8 V or approximately 37 V in 3 s. After five time constants, *C* discharges to zero.

A shorter time constant allows the capacitor to charge or discharge faster. If the *RC* product in Fig. 22–4 is 1 s, then *C* will charge to 63 V in 1 s instead of 3 s. Also, v_C will reach the full applied voltage of 100 V in 5 s instead of 15 s. Charging to the same voltage in less time means a faster charge.

On discharge, also, the shorter time constant will allow *C* to discharge from 100 to 37 V in 1 s instead of 3 s. Also, v_C will be down to zero in 5 s instead of 15 s.

For the opposite case, a longer time constant means slower charge or discharge of the capacitor. More *R* or *C* results in a longer time constant.

RC Applications

Several examples are given here to illustrate how the time constant can be applied to *RC* circuits.

Example 22-4

What is the time constant of a 0.01-μF capacitor in series with a 1-MΩ resistance?

ANSWER

$$T = R \times C = 1 \times 10^6 \times 0.01 \times 10^{-6}$$
$$= 0.01 \text{ s}$$

The time constant in Example 22-4 is for charging or discharging, assuming the series resistance is the same for charge or discharge.

Example 22-5

With a dc voltage of 300 V applied, how much is the voltage across C in Example 22-4 after 0.01 s of charging? After 0.05 s? After 2 hours? After 2 days?

ANSWER Since 0.01 s is one time constant, the voltage across C then is 63% of 300 V, which equals 189 V. After five time constants, or 0.05 s, C will be charged practically to the applied voltage of 300 V. After 2 hours or 2 days, C will still be charged to 300 V if the applied voltage is still connected.

Example 22-6

If the capacitor in Example 22-5 is allowed to charge to 300 V and then discharged, how much is the capacitor voltage 0.01 s after the start of discharge? The series resistance is the same on discharge as on charge.

ANSWER In one time constant, C discharges to 37% of its initial voltage, or 0.37×300 V, which equals 111 V.

Example 22-7

Assume the capacitor in Example 22-5 is discharging after being charged to 200 V. How much will the voltage across C be 0.01 s after the beginning of discharge? The series resistance is the same on discharge as on charge.

ANSWER In one time constant, C discharges to 37% of its initial voltage, or 0.37×200, which equals 74 V.

Example 22–7 shows that the capacitor can charge or discharge from any voltage value. The rate at which it charges or discharges is determined by RC, counting from the time the charge or discharge starts.

Example 22-8

If a 1-MΩ resistance is added in series with the capacitor and resistor in Example 22–4, how much will the time constant be?

ANSWER Now the series resistance is 2 MΩ. Therefore, RC is 2×0.01, or 0.02 s.

The RC time constant becomes longer with larger values of R and C. More capacitance means that the capacitor can store more charge. Therefore, it takes longer to store the charge needed to provide a potential difference equal to 63% of the applied voltage. More resistance reduces the charging current, requiring more time to charge the capacitor.

Note that the RC time constant only specifies a rate. The actual amount of voltage across C depends on the amount of applied voltage as well as on the RC time constant.

A capacitor takes on charge whenever its voltage is less than the applied voltage. The charging continues at the RC rate until the capacitor is completely charged, or the voltage is disconnected.

A capacitor discharges whenever its voltage is more than the applied voltage. The discharge continues at the RC rate until the capacitor is completely discharged, the capacitor voltage equals the applied voltage, or the load is disconnected.

To summarize these two important principles:

1. Capacitor C charges when the net charging voltage is more than v_C.
2. Capacitor C discharges when v_C is more than the net charging voltage.

The net charging voltage equals the difference between v_C and the applied voltage.

■ 22-4 Knowledge Check

Answer at end of chapter.

An RC time constant equals 10 ms. If $R = 100$ kΩ, what is C?

■ 22-4 Self-Review

Answers at end of chapter.

a. How much is the RC time constant for 470 pF in series with 2 MΩ on charge?
b. How much is the RC time constant for 470 pF in series with 1 kΩ on discharge?

22–5 *RC* Charge and Discharge Curves

In Fig. 22–4, the rise is shown in the *RC* charge curve because the charging is fastest at the start and then tapers off as *C* takes on additional charge at a slower rate. As *C* charges, its potential difference increases. Then the difference in voltage between V_T and v_C is reduced. Less potential difference reduces the current that puts the charge in *C*. The more *C* charges, the more slowly it takes on additional charge.

Similarly, on discharge, *C* loses its charge at a declining rate. At the start of discharge, v_C has its highest value and can produce maximum discharge current. As the discharge continues, v_C goes down and there is less discharge current. The more *C* discharges, the more slowly it loses the remainder of its charge.

Charge and Discharge Current

There is often the question of how current can flow in a capacitive circuit with a battery as the dc source. The answer is that current flows any time there is a change in voltage. When V_T is connected, the applied voltage changes from zero. Then charging current flows to charge *C* to the applied voltage. After v_C equals V_T, there is no net charging voltage and *I* is zero.

Similarly, *C* can produce discharge current any time v_C is greater than V_T. When V_T is disconnected, v_C can discharge down to zero, producing discharge current in the direction opposite from the charging current. After v_C equals zero, there is no current.

Capacitance Opposes Voltage Changes Across Itself

This ability corresponds to the ability of inductance to oppose a change in current. When the applied voltage in an *RC* circuit increases, the voltage across the capacitance cannot increase until the charging current has stored enough charge in *C*. The increase in applied voltage is present across the resistance in series with *C* until the capacitor has charged to the higher applied voltage. When the applied voltage decreases, the voltage across the capacitor cannot go down immediately because the series resistance limits the discharge current.

The voltage across the capacitance in an *RC* circuit, therefore, cannot follow instantaneously the changes in applied voltage. As a result, the capacitance is able to oppose changes in voltage across itself. The instantaneous variations in V_T are present across the series resistance, however, since the series voltage drops must add to equal the applied voltage at all times.

■ *22–5 Knowledge Check*

Answer at end of chapter.

In Fig. 22–4, which component, *R* or *C*, controls the amount of charge and discharge current?

■ *22–5 Self-Review*

Answers at end of chapter.

a. From the curve in Fig. 22–4*a*, how much is v_C after 3 s of charge?
b. From the curve in Fig. 22–4*b*, how much is v_C after 3 s of discharge?

GOOD TO KNOW

The voltage across a capacitor cannot change instantaneously.

22–6 High Current Produced by Short-Circuiting an *RC* Circuit

A capacitor can be charged slowly by a small charging current through a high resistance and then be discharged quickly through a low resistance to obtain a momentary surge, or pulse, of discharge current. This idea corresponds to the pulse of high voltage obtained by opening an inductive circuit.

The circuit in Fig. 22–5 illustrates the application of a battery-capacitor (BC) unit to fire a flashbulb for cameras. The flashbulb needs 5 A to ignite, but this is too much load current for the small 15-V battery, which has a rating of 30 mA for normal load current. Instead of using the bulb as a load for the battery, though, the 100-μF capacitor is charged by the battery through the 3-kΩ R in Fig. 22–5a, and then the capacitor is discharged through the bulb in Fig. 22–5b.

Charging the Capacitor

In Fig. 22–5a, S_1 is closed to charge C through the 3-kΩ R without the bulb. The time constant of the RC charging circuit is 0.3 s.

After five time constants, or 1.5 s, C is charged to the 15 V of the battery. The peak charging current, at the first instant of charge, is V/R or 15 V/3 kΩ, which equals 5 mA. This value is an easy load current for the battery.

Discharging the Capacitor

In Fig. 22–5b, v_C is 15 V without the battery. Now S_2 is closed, and C discharges through the 3-Ω resistance of the bulb. The time constant for discharge with the lower r of the bulb is $3 \times 100 \times 10^{-6}$, which equals 300 μs. At the first instant of discharge, when v_C is 15 V, the peak discharge current is 15/3, which equals 5 A. This current is enough to fire the bulb.

Energy Stored in C

When the 100-μF C is charged to 15 V by the battery, the energy stored in the electric field is $CV^2/2$, which equals 0.01 J, approximately. This energy is available to maintain v_C at 15 V for an instant when the switch is closed. The result is the 5-A I through the 3-Ω r of the bulb at the start of the decay. Then v_C and i_C drop to zero in five time constants.

Figure 22–5 Demonstration of high current produced by discharging a charged capacitor through a low resistance. (a) When S_1 is closed, C charges to 15 V through 3 kΩ. (b) Without the battery, S_2 is closed to allow V_C to produce the peak discharge current of 5 A through the 3-Ω bulb. V_C in (b) is across the same C used in (a).

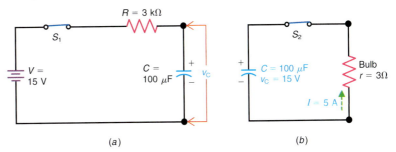

(a)   (b)

The dv/dt for i_C

The required rate of change in voltage is 0.05×10^6 V/s for the discharge current i_C of 5 A produced by the C of 100 μF. Since $i_C = C(dv/dt)$, this formula can be transposed to specify dv/dt as equal to i_C/C. Then dv/dt corresponds to 5 A/100 μF, or 0.05×10^6 V/s. This value is the actual dv/dt at the start of discharge when the switch is closed in Fig. 22–5b. The dv/dt is high because of the short RC time constant.*

■ *22-6 Knowledge Check*

    Answer at end of chapter.

    In Fig. 22–5, does the value of C affect the peak discharge current?

■ *22-6 Self-Review*

    Answers at end of chapter.

    a. Is the RC time constant longer or shorter in Fig. 22–5b compared with Fig. 22–5a?

    b. Which produces more i_C, a faster dv/dt or a slower dv/dt?

22–7 RC Waveshapes

The voltage and current waveshapes in the RC circuit in Fig. 22–6 show when a capacitor is allowed to charge through a resistance for RC time and then discharge through the same resistance for the same amount of time. Note that this particular case is not typical of practical RC circuits, but the waveshapes show some useful details about the voltage and current for charging and discharging. The RC time constant here equals 0.1 s to simplify the calculations.

Square Wave of Applied Voltage

The idea of closing S_1 to apply 100 V and then opening it to disconnect V_T at a regular rate corresponds to a square wave of applied voltage, as shown by the waveform in Fig. 22–6a. When S_1 is closed for charge, S_2 is open; when S_1 is open, S_2 is closed for discharge. Here the voltage is on for the RC time of 0.1 s and off for the same time of 0.1 s. The period of the square wave is 0.2 s, and f is 1/0.2 s, which equals 5 Hz for the frequency.

Capacitor Voltage v_C

As shown in Fig. 22–6b, the capacitor charges to 63 V, equal to 63% of the charging voltage, in the RC time of 0.1 s. Then the capacitor discharges because the applied V_T drops to zero. As a result, v_C drops to 37% of 63 V, or 23.3 V in RC time.

    The next charge cycle begins with v_C at 23.3 V. The net charging voltage now is $100 - 23.3 = 76.7$ V. The capacitor voltage increases by 63% of 76.7 V, or 48.3 V. When 48.3 V is added to 23.3 V, v_C rises to 71.6 V. On discharge, after 0.3 s, v_C drops to 37% of 71.6 V, or to 26.5 V.

Charge and Discharge Current

As shown in Fig. 22–6c, the current i has its positive peak at the start of charge and its negative peak at the start of discharge. On charge, i is calculated as the

---

* See footnote on p. 678.

Figure 22–6 Waveshapes for the charge and discharge of an RC circuit in RC time. Circuit on top with S_1 and S_2 provides the square wave of applied voltage.

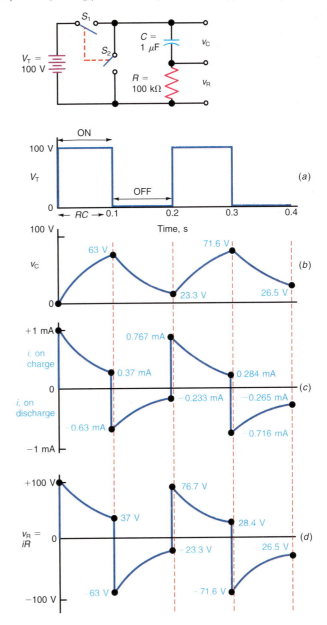

net charging voltage, which is $(V_T - v_C)$, divided by R. On discharge, i always equals v_C/R.

At the start of charge, i is maximum because the net charging voltage is maximum before C charges. Similarly, the peak i for discharge occurs at the start, when v_C is maximum before C discharges.

Note that i is an ac waveform around the zero axis, since the charge and discharge currents are in opposite directions. We are arbitrarily taking the charging current as positive values for i.

Resistor Voltage v_R

This waveshape in Fig. 22–6d follows the waveshape of current because v_R is $i \times R$. Because of the opposite directions of charge and discharge current, the iR waveshape is an ac voltage.

GOOD TO KNOW

In Fig. 22–6*a*, the average dc value of V_T is 50 V. The dc value of any waveform is its average height over one full cycle.

Note that on charge, v_R must always be equal to $V_T - v_C$ because of the series circuit.

On discharge, v_R has the same values as v_C because they are in parallel, without V_T. Then S_2 is closed to connect R across C.

Why the i_C Waveshape Is Important

The v_C waveshape of capacitor voltage in Fig. 22–6 shows the charge and discharge directly, but the i_C waveshape is very interesting. First, the voltage waveshape across R is the same as the i_C waveshape. Also, whether C is charging or discharging, the i_C waveshape is the same except for the reversed polarity. We can see the i_C waveshape as the voltage across R. It generally is better to connect an oscilloscope for voltage waveshapes across R, especially with one side grounded.

Finally, we can tell what v_C is from the v_R waveshape. The reason is that at any instant, V_T must equal the sum of v_R and v_C. Therefore v_C is equal to $V_T - v_R$, when V_T is charging C. When C is discharging, there is no V_T. Then v_R is the same as v_C.

■ *22–7 Knowledge Check*

Answer at end of chapter.

In Fig. 22–6, do R and C have the same voltage during charge or discharge?

■ *22–7 Self-Review*

Answers at end of chapter.

Refer to the waveforms in Fig. 22–6.
a. **When v_C is 63 V, how much is v_R?**
b. **When v_R is 76.7 V, how much is v_C?**

22–8 Long and Short Time Constants

Useful waveshapes can be obtained by using RC circuits with the required time constant. In practical applications, RC circuits are used more than RL circuits because almost any value of an RC time constant can be obtained easily. With coils, the internal series resistance cannot be short-circuited and the distributed capacitance often causes resonance effects.

Long RC Time

GOOD TO KNOW

Inductors and resistors can also be used to obtain either a differentiated or integrated output. For an RL differentiator the output is taken across the inductor. Conversely, for an RL integrator the output is taken across the resistor.

Whether an RC time constant is long or short depends on the pulse width of the applied voltage. We can arbitrarily define a long time constant as at least five times longer than the pulse width, in time, for the applied voltage. As a result, C takes on very little charge. The time constant is too long for v_C to rise appreciably before the applied voltage drops to zero and C must discharge. On discharge also, with a long time constant, C discharges very little before the applied voltage rises to make C charge again.

Short RC Time

A short time constant is defined as no more than one-fifth the pulse width, in time, for the applied voltage V_T. Then V_T is applied for a period of at least five time constants, allowing C to become completely charged. After C is charged, v_C remains at the value of V_T while the voltage is applied. When V_T drops to zero, C discharges completely in five time constants and remains at zero while there is no applied voltage. On the next cycle, C charges and discharges completely again.

Differentiation

The voltage across R in an RC circuit is called a *differentiated output* because v_R can change instantaneously. A short time constant is always used for differentiating circuits to provide sharp pulses of v_R.

Integration

The voltage across C is called an *integrated output* because it must accumulate over a period of time. A medium or long time constant is always used for integrating circuits.

■ *22–8 Knowledge Check*

Answer at end of chapter.

Does an integrator require a long or a short time constant?

■ *22–8 Self-Review*

Answers at end of chapter.

a. Voltage V_T is on for 0.4 s and off for 0.4 s. RC is 6 ms for charge and discharge. Is this a long or short RC time constant?
b. Voltage V_T is on for 2 μs and off for 2 μs. RC is 6 ms for charge and discharge. Is this a long or short RC time constant?

22–9 Charge and Discharge with a Short RC Time Constant

Usually, the time constant is made much shorter or longer than a factor of 5 to obtain better waveshapes. In Fig. 22–7, RC is 0.1 ms. The frequency of the square wave is 25 Hz, with a period of 0.04 s, or 40 ms. One-half this period is the time when V_T is applied. Therefore, the applied voltage is on for 20 ms and off for 20 ms. The RC time constant of 0.1 ms is shorter than the pulse width of 20 ms by a factor of $\frac{1}{200}$. Note that the time axis of all waveshapes is calibrated in seconds for the period of V_T, not in RC time constants.

Square Wave of V_T Is Across C

The waveshape of v_C in Fig. 22–7b is the same as the square wave of applied voltage because the short time constant allows C to charge or discharge completely very soon after V_T is applied or removed. The charge or discharge time of five time constants is much less than the pulse width.

Sharp Pulses of i

The waveshape of i shows sharp peaks for the charge or discharge current. Each current peak is $V_T/R = 1$ mA, decaying to zero in five RC time constants. These pulses coincide with the leading and trailing edges of the square wave of V_T.

Actually, the pulses are much sharper than shown. They are not to scale horizontally to indicate the charge and discharge action. Also, v_C is actually a square wave, like the applied voltage, but with slightly rounded corners for the charge and discharge.

Charge and discharge of an *RC* circuit with a short time constant. Note that the waveshape of V_R in (*d*) has sharp voltage peaks for the leading and trailing edges of the square-wave applied voltage.

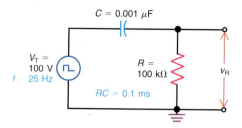

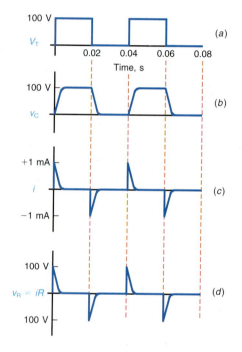

Sharp Pulses of v_R

The waveshape of voltage across the resistor follows the current waveshape because $v_R = iR$. Each current pulse of 1 mA across the 100-kΩ *R* results in a voltage pulse of 100 V.

More fundamentally, the peaks of v_R equal the applied voltage V_T before *C* charges. Then v_R drops to zero as v_C rises to the value of V_T.

On discharge, $v_R = v_C$, which is 100 V at the start of discharge. Then the pulse drops to zero in five time constants. The pulses of v_R in Fig. 22–7 are useful as timing pulses that match the edges of the square-wave applied voltage V_T. Either the positive or the negative pulses can be used.

The *RC* circuit in Fig. 22–7*a* is a good example of an *RC* differentiator. With the *RC* time constant much shorter than the pulse width of V_T, the voltage V_R follows instantaneously the changes in the applied voltage. Keep in mind that a differentiator must have a short time constant with respect to the pulse width of V_T to provide good differentiation. For best results, an *RC* differentiator should have a time constant which is one-tenth or less of the pulse width of V_T.

Answer at end of chapter.

In Fig. 22–7, how long does it take for the capacitor voltage to reach 100 V after V_T switches to 100 V?

Answers at end of chapter.

Refer to Fig. 22–7.
a. Is the time constant here short or long?
b. Is the square wave of applied voltage across C or R?

22–10 Long Time Constant for an *RC* Coupling Circuit

The *RC* circuit in Fig. 22–8 is the same as that in Fig. 22–7, but now the *RC* time constant is long because of the higher frequency of the applied voltage. Specifically, the *RC* time of 0.1 ms is 200 times longer than the 0.5-μs pulse width of V_T with a frequency of 1 MHz. Note that the time axis is calibrated in microseconds for the period of V_T, not in *RC* time constants.

Very Little of V_T Is Across C

The waveshape of v_C in Fig. 22–8b shows very little voltage rise because of the long time constant. During the 0.5 μs when V_T is applied, C charges to only $\frac{1}{200}$ of the charging voltage. On discharge, also, v_C drops very little.

Square Wave of i

The waveshape of i stays close to the 1-mA peak at the start of charging. The reason is that v_C does not increase much, allowing V_T to maintain the charging current. On discharge, the reverse i for discharge current is very small because v_C is low.

Square Wave of V_T Is Across R

The waveshape of v_R is the same square wave as i because $v_R = iR$. The waveshapes of i and v_R are essentially the same as the square-wave V_T applied. They are not shown to scale vertically to indicate the slight charge and discharge action.

Eventually, v_C will climb to the average dc value of 50 V, i will vary ± 0.5 mA above and below zero, and v_R will vary ± 50 V above and below zero. This application is an *RC* coupling circuit to block the average value of the varying dc voltage V_T as the capacitive voltage v_C, and v_R provides an ac voltage output having the same variations as V_T.

If the output is taken across C rather than R in Fig. 22–8a, the circuit is classified as an *RC* integrator. In Fig. 22–8b, it can be seen that C combines or integrates its original voltage with the new change in voltage. Eventually, however, the voltage across C will reach a steady-state value of 50 V after the input waveform has been applied for approximately five *RC* time constants. Keep in mind that an integrator must have a long time constant with respect to the pulse width of V_T to provide good integration. For best results, an *RC* integrator should have a time constant which is 10 or more times longer than the pulse width of V_T.

GOOD TO KNOW

An *RC* integrator can also be described as a circuit whose output is proportional to the charge being stored.

Figure 22–8 Charge and discharge of an *RC* circuit with a long time constant. Note that the waveshape of V_R in (*d*) has the same waveform as the applied voltage.

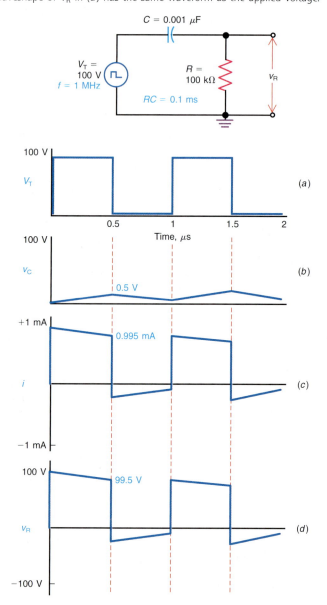

22–10 Knowledge Check

Answer at end of chapter.

In Fig. 22–8, what is the average dc voltage to which *C* eventually charges?

22–10 Self-Review

Answers at end of chapter.

Refer to Fig. 22–8.
a. Is the *RC* time constant here short or long?
b. Is the square wave of applied voltage across *R* or *C*?

22–11 Advanced Time Constant Analysis

We can determine transient voltage and current values for any amount of time with the curves in Fig. 22–9. The rising curve *a* shows how v_C builds up as *C* charges in an *RC* circuit; the same curve applies to i_L, increasing in the inductance for an *RL* circuit. The decreasing curve *b* shows how v_C drops as *C* discharges or i_L decays in an inductance.

Note that the horizontal axis is in units of time constants rather than absolute time. Suppose that the time constant of an *RC* circuit is 5 μs. Therefore, one *RC* time unit = 5 μs, two *RC* units = 10 μs, three *RC* units = 15 μs, four *RC* units = 20 μs, and five *RC* units = 25 μs.

As an example, to find v_C after 10 μs of charging, we can take the value of curve *a* in Fig. 22–9 at two *RC*. This point is at 86% amplitude. Therefore, we can say that in this *RC* circuit with a time constant of 5 μs, v_C charges to 86% of the applied V_T after 10 μs. Similarly, some important values that can be read from the curve are listed in Table 22–1.

If we consider curve *a* in Fig. 22–9 as an *RC* charge curve, v_C adds 63% of the net charging voltage for each additional unit of one time constant, although it may not appear so. For instance, in the second interval of *RC* time, v_C adds 63% of the net charging voltage, which is 0.37 V_T. Then 0.63 × 0.37 equals 0.23, which is added to 0.63 to give 0.86, or 86%, as the total charge from the start.

Slope at *t* = 0

The curves in Fig. 22–9 can be considered approximately linear for the first 20% of change. In 0.1 time constant, for instance, the change in amplitude is 10%; in 0.2 time constant, the change is 20%. The dashed lines in Fig. 22–9 show that if this constant slope continued, the result would be 100% charge in one time constant. This does not happen, though, because the change is opposed by the energy stored in *L* and *C*. However, at the first instant of rise or decay, at *t* = 0, the change in v_C or i_L can be calculated from the dotted slope line.

Figure 22–9 Universal time-constant chart for *RC* and *RL* circuits. The rise or fall changes by 63% in one time constant.

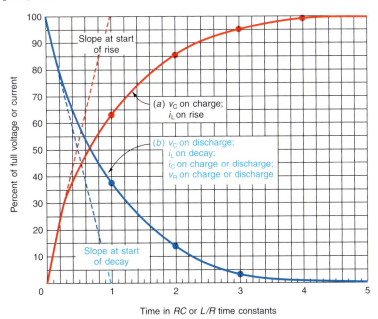

Table 22–1	Time Constant Factors
Factor	**Amplitude**
0.2 time constant	20%
0.5 time constant	40%
0.7 time constant	50%
1 time constant	63%
2 time constants	86%
3 time constants	96%
4 time constants	98%
5 time constants	99%

Equation of the Decay Curve

The rising curve a in Fig. 22–9 may seem more interesting because it describes the buildup of v_C or i_L, but the decaying curve b is more useful. For RC circuits, curve b can be applied to

1. v_C on discharge
2. i and v_R on charge or discharge

If we use curve b for the voltage in RC circuits, the equation of this decay curve can be written as

$$v = V \times \epsilon^{-t/RC} \tag{22–3}$$

where V is the voltage at the start of decay and v is the instantaneous voltage after the time t. Specifically, v can be v_R on charge and discharge or v_C only on discharge.

The constant ϵ is the base 2.718 for natural logarithms. The negative exponent $-t/RC$ indicates a declining exponential or logarithmic curve. The value of t/RC is the ratio of actual time of decline t to the RC time constant.

This equation can be converted to common logarithms for easier calculations. Since the natural base ϵ is 2.718, its logarithm to base 10 equals 0.434. Therefore, the equation becomes

$$v = \text{antilog}\left(\log V - 0.434 \times \frac{t}{RC}\right) \tag{22–4}$$

Calculations for v_R

As an example, let us calculate v_R dropping from 100 V, after RC time. Then the factor t/RC is 1. Substituting these values,

$$
\begin{aligned}
v_R &= \text{antilog}(\log 100 - 0.434 \times 1) \\
&= \text{antilog}(2 - 0.434) \\
&= \text{antilog } 1.566 \\
&= 37 \text{ V}
\end{aligned}
$$

Figure 22-10 How v_C and v_R add to
equal the applied voltage V_T of 100 V.
(*a*) Zero time at the start of charging.
(*b*) After one *RC* time constant. (*c*) After
two *RC* time constants. (*d*) After five or
more *RC* time constants.

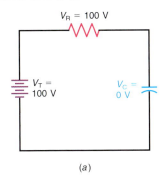

(*a*)

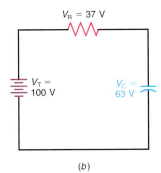

(*b*)

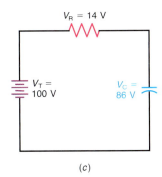

(*c*)

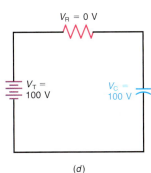

(*d*)

All these logs are to base 10. Note that log 100 is taken first so that 0.434 can be subtracted from 2 before the antilog of the difference is found. The antilog of 1.566 is 37.

We can also use V_R to find V_C, which is $V_T - V_R$. Then $100 - 37 = 63$ V for V_C. These answers agree with the fact that in one time constant, V_R drops 63% and V_C rises 63%.

Figure 22-10 illustrates how the voltages across R and C in series must add to equal the applied voltage V_T. The four examples with 100 V applied are

1. At time zero, at the start of charging, V_R is 100 V and V_C is 0 V. Then $100 + 0 = 100$ V.
2. After one time constant, V_R is 37 V and V_C is 63 V. Then $37 + 63 = 100$ V.
3. After two time constants, V_R is 14 V and V_C is 86 V. Then $14 + 86 = 100$ V.
4. After five time constants, V_R is 0 V and V_C is 100 V, approximately. Then $0 + 100 = 100$ V

It should be emphasized that Formulas (22-3) and (22-4) can be used to calculate any decaying value on curve *b* in Fig. 22-9. These applications for an *RC* circuit include V_R on charge or discharge, i on charge or discharge, and V_C only on discharge. For an *RC* circuit in which C is charging, Formula (22-5) can be used to calculate the capacitor voltage v_C at any point along curve *a* in Fig. 22-9:

$$v_C = V(1 - \epsilon^{-t/RC}) \tag{22-5}$$

In Formula (22-5), V represents the maximum voltage to which C can charge, whereas v_C is the instantaneous capacitor voltage after time t. Formula (22-5) is derived from the fact that v_C must equal $V_T - V_R$ while C is charging.

Example 22-9

An *RC* circuit has a time constant of 3 s. The capacitor is charged to 40 V. Then C is discharged. After 6 s of discharge, how much is V_R?

ANSWER Note that 6 s is twice the *RC* time of 3 s. Then $t/RC = 2$.

$$\begin{aligned}
V_R &= \text{antilog}(\log 40 - 0.434 \times 2) \\
&= \text{antilog}(1.602 - 0.868) \\
&= \text{antilog}(0.734) \\
&= 5.42 \text{ V}
\end{aligned}$$

Note that in two *RC* time constants, the v_R is down to approximately 14% of its initial voltage, a drop of about 86%.

Calculations for *t*

Furthermore, Formula (22-4), can be transposed to find the time t for a specific voltage decay. Then

$$t = 2.3 \, RC \log \frac{V}{v} \tag{22-6}$$

where V is the higher voltage at the start and v is the lower voltage at the finish. The factor 2.3 is $1/0.434$.

As an example, let RC be 1 s. How long will it take for v_R to drop from 100 to 50 V? The required time for this decay is

$$t = 2.3 \times 1 \times \log \frac{100}{50} = 2.3 \times 1 \times \log 2$$
$$= 2.3 \times 1 \times 0.3$$
$$= 0.7 \text{ s} \quad \text{approximately}$$

This answer agrees with the fact that a drop of 50% takes 0.7 time constant. Formula (22–6) can also be used to calculate the time for any decay of v_C or v_R.

Formula (22–6) cannot be used for a rise in v_C. However, if you convert this rise to an equivalent drop in v_R, the calculated time is the same for both cases.

Example 22–10

An RC circuit has an R of 10 kΩ and a C of 0.05 μF. The applied voltage for charging is 36 V. (a) Calculate the time constant. (b) How long will it take C to charge to 24 V?

ANSWER

a. RC is 10 kΩ $\times$ 0.05 μF = 0.5 ms or 0.5×10^{-3} s.
b. The v_C rises to 24 V while v_R drops from 36 to 12 V. Then

$$t = 2.3 \, RC \log \frac{V}{v}$$
$$= 2.3 \times 0.5 \times 10^{-3} \times \log \frac{36}{12}$$
$$= 2.3 \times 0.5 \times 10^{-3} \times 0.477$$
$$= 0.549 \times 10^{-3} \text{ s} \quad \text{or} \quad 0.549 \text{ ms}$$

GOOD TO KNOW

When a capacitor charges from an initial voltage other than zero, the capacitor voltage can be determined at any time, t, with the use of the following equation:

$$v_C = (V_F - V_i)(1 - \epsilon^{-t/RC}) + V_i$$

where V_F and V_i represent the final and initial voltages respectively. The quantity $(V_F - V_i)$ represents the net charging voltage.

■ *22–11 Knowledge Check*

Answer at end of chapter.

How many time constants correspond to a 50% change?

■ *22–11 Self-Review*

Answers at end of chapter.

For the universal curves in Fig. 22–9.
a. Curve a applies to v_C on charge. (True or False)
b. Curve b applies to v_C on discharge. (True or False)
c. Curve b applies to v_R when C charges or discharges. (True or False)

22–12 Comparison of Reactance and Time Constant

The formula for capacitive reactance includes the factor of time in terms of frequency as $X_C = 1/(2\pi fC)$. Therefore, X_C and the RC time constant are both measures of the reaction of C to a change in voltage. The reactance X_C is a special case but a very important one that applies only to sine waves. The RC time constant can be applied to square waves and rectangular pulses.

Phase Angle of Reactance

The capacitive charge and discharge current i_C is always equal to $C(dv/dt)$. A sine wave of voltage variations for v_C produces a cosine wave of current i_C. This means that v_C and i_C are both sinusoids, but 90° out of phase.

In this case, it is usually more convenient to use X_C for calculations in sine-wave ac circuits to determine Z, I, and the phase angle θ. Then $I_C = V_C/X_C$. Moreover, if I_C is known, $V_C = I_C \times X_C$. The phase angle of the circuit depends on the amount of X_C compared with the resistance R.

Changes in Waveshape

With nonsinusoidal voltage applied, X_C cannot be used. Then i_C must be calculated as $C(dv/dt)$. In this comparison of i_C and v_C, their waveshapes can be different, instead of the change in phase angle for sine waves. The waveshapes of v_C and i_C depend on the RC time constant.

Coupling Capacitors

If we consider the application of a coupling capacitor, X_C must be one-tenth or less of its series R at the desired frequency. This condition is equivalent to having an RC time constant that is long compared with the period of one cycle. In terms of X_C, the C has little IX_C voltage, with practically all the applied voltage across the series R. In terms of a long RC time constant, C cannot take on much charge. Practically all the applied voltage is developed as $v_R = iR$ across the series resistance by the charge and discharge current. These comparisons are summarized in Table 22–2.

Inductive Circuits

Similar comparisons can be made between $X_L = 2\pi fL$ for sine waves and the L/R time constant. The voltage across any inductance is $v_L = L(di/dt)$.

Table 22–2	Comparison of Reactance X_C and RC Time Constant	
Sine-Wave Voltage	**Nonsinusoidal Voltage**	
Examples are 60-Hz power line, af signal voltage, rf signal voltage	Examples are dc circuit turned on and off, square waves, rectangular pulses	
Reactance $X_C = \dfrac{1}{2\pi fC}$	Time constant $T = RC$	
Larger C results in smaller reactance X_C	Larger C results in longer time constant	
Higher frequency results in smaller X_C	Shorter pulse width corresponds to longer time constant	
$I_C = \dfrac{V_C}{X_C}$	$i_C = C\dfrac{dv}{dt}$	
X_C makes I_C and V_C 90° out of phase	Waveshape changes between i_C and v_C	

Sine-wave variations for i_L produce a cosine wave of voltage v_L, 90° out of phase.

In this case, X_L can be used to determine Z, I, and the phase angle θ. Then $I_L = V_L/X_L$. Furthermore, if I_L is known, $V_L = I_L \times X_L$. The phase angle of the circuit depends on the amount of X_L compared with R.

With nonsinusoidal voltage, however, X_L cannot be used. Then v_L must be calculated as $L(di/dt)$. In this comparison, i_L and v_L can have different waveshapes, depending on the L/R time constant.

Choke Coils

For this application, the idea is to have almost all the applied ac voltage across L. The condition of X_L being at least 10 times R corresponds to a long time constant. The high value of X_L means that practically all the applied ac voltage is across X_L as IX_L, with little IR voltage.

The long L/R time constant means that i_L cannot rise appreciably, resulting in little v_R voltage across the resistor. The waveform for i_L and v_R in an inductive circuit corresponds to v_C in a capacitive circuit.

When Do We Use the Time Constant?

In electronic circuits, the time constant is useful in analyzing the effect of L or C on the waveshape of nonsinusoidal voltages, particularly rectangular pulses. Another application is the transient response when a dc voltage is turned on or off. The 63% change in one time constant is a natural characteristic of v or i, where the magnitude of one is proportional to the rate of change of the other.

When Do We Use Reactance?

X_L and X_C are generally used for sine-wave V or I. We can determine Z, I, voltage drops, and phase angles. The phase angle of 90° is a natural characteristic of a cosine wave when its magnitude is proportional to the rate of change in a sine wave.

■ *22–12 Knowledge Check*

Answer at end of chapter.

When a nonsinusoidal voltage is applied to an *RC* circuit, do V_C and i_C have the same waveshape?

■ *22–12 Self-Review*

Answers at end of chapter.

a. Does an *RC* coupling circuit have a small or large X_C compared with R?
b. Does an *RC* coupling circuit have a long or short time constant for the frequency of the applied voltage?

Summary

- The transient response of an inductive circuit with nonsinusoidal current is indicated by the time constant L/R. With L in henrys and R in ohms, T is the time in seconds for the current i_L to change by 63%. In five time constants, i_L reaches the steady value of V_T/R.

- At the instant an inductive circuit is opened, high voltage is generated across L because of the fast current decay with a short time constant. The induced voltage $v_L = L(di/dt)$. The di is the change in i_L.

- The transient response of a capacitive circuit with nonsinusoidal voltage is indicated by the time constant RC. With C in farads and R in ohms, T is the time in seconds for the voltage across the capacitor v_C to change by 63%. In five time constants, v_C reaches the steady value of V_T.

- At the instant a charged capacitor is discharged through a low resistance, a high value of discharge current can be produced. The discharge current $i_C = C(dv/dt)$ can be large because of the fast discharge with a short time constant. The dv is the change in v_C.

- The waveshapes of v_C and i_L correspond, as both rise relatively slowly to the steady-state value.

- Also, i_C and v_L correspond because they are waveforms that can change instantaneously.

- The resistor voltage $v_R = iR$ for both RC and RL circuits.

- A short time constant is one-fifth or less of the pulse width, in time, for the applied voltage.

- A long time constant is greater than the pulse width, in time, for the applied voltage by a factor of 5 or more.

- An RC circuit with a short time constant produces sharp voltage spikes for v_R at the leading and trailing edges of a square-wave of applied voltage. The waveshape of voltage V_T is across the capacitor as v_C. See Fig. 22–7.

- An RC circuit with a long time constant allows v_R to be essentially the same as the variations in applied voltage V_T, and the average dc value of V_T is blocked as v_C. See Fig. 22–8.

- The universal rise and decay curves in Fig. 22–9 can be used for current or voltage in RC and RL circuits for any time up to five time constants.

- A differentiator is a circuit whose output voltage is proportional to the change in applied voltage.

- An integrator is a circuit whose output combines, or integrates, its original voltage with the new change in voltage.

- The concept of reactance is useful for sine-wave ac circuits with L and C.

- The time constant method is used with L or C to analyze nonsinusoidal waveforms.

Important Terms

- Differentiator - a circuit whose output voltage is proportional to the change in applied voltage. To provide good differentiation, the time constant of a circuit must be short with respect to the pulse width of the applied voltage.

- Integrator - a circuit whose output combines or integrates its original voltage with the new change in voltage. For best integration, the time constant of a circuit must be long with respect to the pulse width of the applied voltage.

- Long Time Constant - a long time constant is arbitrarily defined as one that is five or more times longer than the pulse width of the applied voltage.

- Short Time Constant - a short time constant is arbitrarily defined as one that is one-fifth or less the time of the pulse width of the applied voltage.

- Steady-State - a term to describe the final condition of a circuit after it has passed through its initial transitional state.

- Time Constant - a measure of how long it takes for a 63.2% change to occur.

- Transient Response - a term to describe the transitional state of a circuit when power is first applied or removed.

- Universal Time-Constant Graph - a graph which shows the percent change in voltage or current in an RC or RL circuit with respect to the number of time constants that have elapsed.

Related Formulas

$$T = \frac{L}{R}$$

$$T = R \times C$$

$$v = V \times \epsilon^{-t/RC}$$

$$v = \text{antilog}\left(\log V - 0.434 \times \frac{t}{RC}\right)$$

$$v_C = V(1 - \epsilon^{-t/RC})$$

$$t = 2.3 \, RC \, \log\frac{V}{v}$$

Self-Test

1. **What is the time constant of the circuit in Fig. 22–11 with S_1 closed?**

   a. 250 μs.

   b. 31.6 μs.

   c. 50 μs.

   d. 5 ms.

Figure 22–11

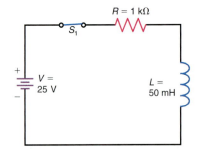

2. **With S_1 closed in Fig. 22–11, what is the eventual steady-state value of current?**

   a. 15.8 mA.

   b. 12.5 mA.

   c. 0 mA.

   d. 25 mA.

3. **In Fig. 22–11, how long does it take the current, I, to reach its steady-state value after S_1 is closed?**

   a. 50 μs.

   b. 250 μs.

   c. 500 μs.

   d. This is impossible to determine.

4. **In Fig. 22–11, how much is the resistor voltage at the very first instant ($t = 0$ s) S_1 is closed?**

   a. 0 V.

   b. 25 V.

   c. 15.8 V.

   d. 9.2 V.

5. **In Fig. 22–11, what is the value of the resistor voltage exactly one time constant after S_1 is closed?**

   a. 15.8 V.

   b. 9.2 V.

   c. 6.32 V.

   d. 21.5 V.

6. **If a 2-MΩ resistor is placed across the switch, S_1, in Fig. 22–11, how much is the peak inductor voltage, V_L, when S_1 is opened?**

   a. 0 V.

   b. 25 V.

   c. 50 kV.

   d. This is impossible to determine.

7. **In Fig. 22–11, what is the value of the current 35 μs after S_1 is closed?**

   a. approximately 20 mA.

   b. approximately 12.5 mA.

   c. 15.8 mA.

   d. 20 mA.

8. **With S_1 closed in Fig. 22–11, the length of one time constant could be increased by**

   a. decreasing L.

   b. decreasing R.

   c. increasing L.

   d. both b and c.

9. **In Fig. 22–11, what is the value of the inductor voltage five time constants after S_1 is closed?**

   a. 50 kV.

   b. 25 V.

   c. 0 V.

   d. 9.2 V.

10. **In Fig. 22–11, how much is the resistor voltage exactly 100 μs after S_1 is closed?**

    a. 12 V.

    b. 21.6 V.

    c. 3.4 V.

    d. 15.8 V.

11. **In Fig. 22–12, what is the time constant of the circuit with S_1 in Position 1?**

    a. 2 s.

    b. 5 s.

    c. 10 s.

    d. 1 s.

12. **In Fig. 22–12, what is the time constant of the circuit with S_1 in Position 2?**

    a. 2 s.

    b. 5 s.

    c. 10 s.

    d. 1 s.

Figure 22–12

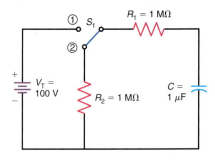

13. **In Fig. 22–12, how long will it take for the voltage across C to reach 100 V after S_1 is placed in Position 1?**

    a. 1 s.

    b. 2 s.

    c. 10 s.

    d. 5 s.

14. **In Fig. 22–12, how much voltage is across resistor, R_1, at the first instant the switch is moved from Position 2 to Position 1? (Assume that C was completely discharged with S_1 in Position 2.)**

    a. 100 V.

    b. 63.2 V.

    c. 0 V.

    d. 36.8 V.

15. **In Fig. 22–12, assume that C is fully charged to 100 V with S_1 in Position 1. How long will it take for C to discharge fully if S_1 is moved to Position 2?**

    a. 1 s.

    b. 5 s.

    c. 10 s.

    d. 2 s.

16. **In Fig. 22–12, assume that C is completely discharged while in Position 2. What is the voltage across C exactly 1s after S_1 is moved to Position 1?**

    a. 50 v.

    b. 63.2 V.

    c. 36.8 V.

    d. 100 V.

17. In Fig. 22–12, assume that C is completely discharged while in Position 2. What is the voltage across R_1 exactly two time constants after S_1 is moved to Position 1?

   a. 37 V.

   b. 13.5 V.

   c. 50 V.

   d. 86 V.

18. In Fig. 22–12, what is the steady-state value of current with S_1 in Position 1?

   a. 100 μA.

   b. 50 μA.

   c. 1 A.

   d. 0 μA.

19. In Fig. 22–12, assume that C is fully charged to 100 V with S_1 in Position 1. What is the value of the capacitor voltage 3 s after S_1 is moved to Position 2?

   a. 77.7 V.

   b. 0 V.

   c. 22.3 V.

   d. 36.8 V.

20. In Fig. 22–12, assume that C is charging with S_1 in Position 1. At the instant the capacitor voltage reaches 75 V, S_1 is moved to Position 2. What is the approximate value of the capacitor voltage 0.7 time constants after S_1 is moved to Position 2?

   a. 75 V.

   b. 27.6 V.

   c. 50 V.

   d. 37.5 V.

21. For best results, an RC coupling circuit should have a(n)

   a. short time constant.

   b. medium time constant.

   c. long time constant.

   d. zero time constant.

22. A differentiator is a circuit whose

   a. output combines its original voltage with the new change in voltage.

   b. output is always one-half of V_{in}.

   c. time constant is long with the output across C.

   d. output is proportional to the change in applied voltage.

23. An integrator is a circuit whose

   a. output combines its original voltage with the new change in voltage.

   b. output is always equal to V_{in}.

   c. output is proportional to the change in applied voltage.

   d. time constant is short with the output across R.

24. The time constant of an RL circuit is 47μs. If $L = 4.7$ mH, calculate R.

   a. $R = 10$ kΩ.

   b. $R = 100\ \Omega$.

   c. $R = 10$ MΩ.

   d. $R = 1$ kΩ.

25. The time constant of an RC circuit is 330 μs. If $R = 1$ kΩ, calculate C.

   a. $C = 0.33\ \mu$F.

   b. $C = 0.033\ \mu$F.

   c. $C = 3.3\ \mu$F.

   d. $C = 330$ pF.

Questions

1. Give the formula, with units, for calculating the time constant of an RL circuit.

2. Give the formula, with units, for calculating the time constant of an RC circuit.

3. Redraw the RL circuit and graph in Fig. 22–2 for a 2-H L and a 100-Ω R.

4. Redraw the graphs in Fig. 22–4 to fit the circuit in Fig. 22–5 with a 100-μF C. Use a 3000-Ω R for charge but a 3-Ω R for discharge.

5. List two comparisons of RC and RL circuits for nonsinusoidal voltage.

6. List two comparisons between RC circuits with nonsinusoidal voltage and sine-wave voltage applied.

7. Define the following: (a) a long time constant; (b) a short time constant; (c) an RC differentiating circuit; (d) an RC integrating circuit.

8. Redraw the horizontal time axis of the universal curve in Fig. 22–9, calibrated in absolute time units of milliseconds for an RC circuit with a time constant equal to 2.3 ms.

9. Redraw the circuit and graphs in Fig. 22–7 with everything the same except that R is 20 kΩ, making the RC time constant shorter.

10. Redraw the circuit and graphs in Fig. 22–8 with everything the same except that R is 500 kΩ, making the RC time constant longer.

11. Invert the equation $T = RC$, in two forms, to find R or C from the time constant.

12. Show three types of nonsinusoidal waveforms.

13. Give an application in electronic circuits for an RC circuit with a long time constant and with a short time constant.

14. Why can arcing voltage be a problem with coils used in switching circuits?

Problems

SECTION 22–1 RESPONSE OF RESISTANCE ALONE

22–1 In Fig. 22–13, how long does it take for the current, I, to reach its steady-state value after S_1 is closed?

Figure 22–13

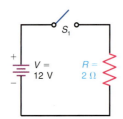

22–2 In Fig. 22–13, what is the current with S_1 closed?

22–3 Explain how the resistor in Fig. 22–13 reacts to the closing or opening of S_1.

SECTION 22–2 L/R TIME CONSTANT

22–4 In Fig. 22–14,

   a. what is the time constant of the circuit with S_1 closed?

   b. what is the eventual steady-state current with S_1 closed?

   c. what is the value of the circuit current at the first instant S_1 is closed? ($t = 0$ s)

   d. what is the value of the circuit current exactly one time constant after S_1 is closed?

   e. how long after S_1 is closed will it take before the circuit current reaches its steady-state value?

Figure 22–14

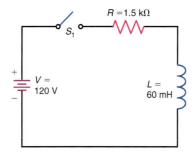

22–5 Repeat Prob. 22–4 if $L = 100$ mH and $R = 500\ \Omega$.

22–6 Calculate the time constant for an inductive circuit with the following values:

   a. $L = 10$ H, $R = 1$ kΩ

   b. $L = 500$ mH, $R = 2$ kΩ

   c. $L = 250\ \mu$H, $R = 50\ \Omega$

   d. $L = 15$ mH, $R = 7.5$ kΩ

22–7 List two ways to

   a. increase the time constant of an inductive circuit.

   b. decrease the time constant of an inductive circuit.

SECTION 22–3 HIGH VOLTAGE PRODUCED BY OPENING AN *RL* CIRCUIT

22–8 Assume that the switch, S_1, in Fig. 22–14 has been closed for more than five L/R time constants. If a 1-MΩ resistor is placed across the terminals of the switch, calculate

   a. the approximate time constant of the circuit with S_1 open.

   b. the peak inductor voltage, V_L, when S_1 is opened.

   c. the di/dt value at the instant S_1 is opened.

   d. how long it takes for the current to decay to zero after S_1 is opened (approximately).

22–9 Without a resistor across S_1 in Fig. 22–14, is it possible to calculate the time constant of the circuit with the switch open? Also, what effect will probably occur inside the switch when it is opened?

SECTION 22–4 *RC* TIME CONSTANT

22–10 In Fig. 22–15, what is the time constant of the circuit with the switch, S_1, in position

   a. 1?

   b. 2?

Figure 22–15

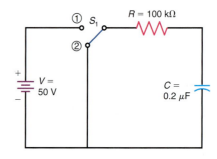

22–11 Assume that the capacitor in Fig. 22–15 is fully discharged with S_1 in Position 2. How much is the capacitor voltage, V_C,

   a. exactly one time constant after S_1 is moved to Position 1?

   b. five time constants after S_1 is moved to Position 1?

   c. 1 week after S_1 is moved to Position 1?

22-12 Assume that the capacitor in Fig. 22–15 is fully charged with S_1 in Position 1. How much is the capacitor voltage, V_C,

    a. exactly one time constant after S_1 is moved to Position 2?

    b. five time constants after S_1 is moved to Position 2?

    c. 1 week after S_1 is moved to Position 2?

22-13 Calculate the time constant of a capacitive circuit with the following values:

    a. $R = 1 M\,\Omega$, $C = 1\ \mu F$

    b. $R = 150\ \Omega$, $C = 0.01\ \mu F$

    c. $R = 330\ k\Omega$, $C = 270\ pF$

    d. $R = 5\ k\Omega$, $C = 40\ \mu F$

22-14 List two ways to

    a. increase the time constant of a capacitive circuit.

    b. decrease the time constant of a capacitive circuit.

22-15 Assume that the capacitor in Fig. 22–15 is discharging from 50 V with S_1 in Position 2. At the instant the capacitor voltage reaches 25 V, S_1 is moved back to Position 1. What is

    a. the net charging voltage at the first instant S_1 is put back in Position 1?

    b. the value of the capacitor voltage exactly one time constant after S_1 is moved back to Position 1?

    c. the value of the capacitor voltage five time constants after S_1 is moved back to Position 1?

22-16 Assume that the capacitor in Fig. 22–15 is charging from 0 V with S_1 in Position 1. At the instant the capacitor voltage reaches 35 V, S_1 is moved back to Position 2. What is

    a. the value of the capacitor voltage exactly one time constant after S_1 is moved back to Position 2?

    b. the value of the capacitor voltage five time constants after S_1 is moved back to Position 2?

SECTION 22–5 *RC* CHARGE AND DISCHARGE CURVES

22-17 Assume that the capacitor in Fig. 22–15 is fully discharged with S_1 in Position 2. What is

    a. the value of the charging current at the first instant S_1 is moved to Position 1?

    b. the value of the charging current five time constants after S_1 is moved to Position 1?

    c. the value of the resistor voltage exactly one time constant after S_1 is moved to Position 1?

    d. the value of the charging current exactly one time constant after S_1 is moved to Position 1?

22-18 Assume that the capacitor in Fig. 22–15 is fully charged to 50 V with S_1 in Position 1. What is the value of the discharge current

    a. at the first instant S_1 is moved to Position 2?

    b. exactly one time constant after S_1 is moved to Position 2?

    c. five time constants after S_1 is moved to Position 2?

SECTION 22–6 HIGH CURRENT PRODUCED BY SHORT-CIRCUITING AN *RC* CIRCUIT

22-19 In Fig. 22–16, what is the *RC* time constant with S_1 in Position

    a. 1?

    b. 2?

Figure 22–16

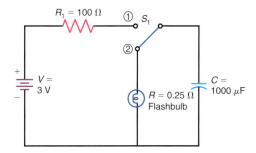

22-20 In Fig. 22–16, how long will it take the capacitor voltage to

    a. reach 3 V after S_1 is moved to Position 1?

    b. discharge to 0 V after S_1 is moved to Position 2?

22-21 Assume that the capacitor in Fig. 22–16 is fully discharged with S_1 in Position 2. At the first instant S_1 is moved to Position 1, how much is

    a. the voltage across the capacitor?

    b. the voltage across the resistor?

    c. the initial charging current?

22-22 Assume that the capacitor in Fig. 22–16 is fully charged with S_1 in Position 1. At the first instant S_1 is moved to Position 2, what is

    a. the dc voltage across the flashbulb?

    b. the initial value of the discharge current?

    c. the initial rate of voltage change, dv/dt?

22-23 How much energy is stored by the capacitor in Fig. 22–16 if it is fully charged to 3 V?

SECTION 22–7 *RC* WAVESHAPES

22-24 For the circuit in Fig. 22–17,

    a. calculate the *RC* time constant.

    b. draw the capacitor voltage waveform and include voltage values at times t_0, t_1, t_2, t_3, and t_4.

    c. draw the resistor voltage waveform and include voltage values at times t_0, t_1, t_2, t_3, and t_4.

    d. draw the charge and discharge current waveform and include current values at times t_0, t_1, t_2, t_3, and t_4.

Figure 22–17

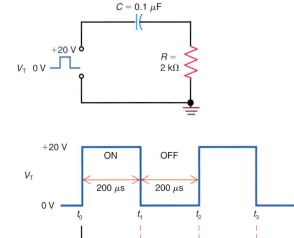

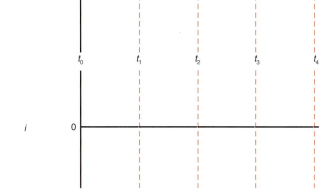

SECTION 22–8 LONG AND SHORT TIME CONSTANTS

22–25 In Fig. 22–17, is the time constant of the circuit considered long or short with respect to the pulse width of the applied voltage, V_T, if the resistance, R, is

a. increased to 10 kΩ?

b. decreased to 400 Ω?

22–26 For an RC circuit used as a differentiator,

    a. across which component is the output taken?

    b. should the time constant be long or short with respect to the pulse width of the applied voltage?

22–27 For an RC circuit used as an integrator,

    a. across which component is the output taken?

    b. should the time constant be long or short with respect to the pulse width of the applied voltage?

SECTION 22–9 CHARGE AND DISCHARGE WITH A SHORT RC TIME CONSTANT

22–28 For the circuit in Fig. 22–18,

    a. calculate the RC time constant.

    b. draw the capacitor voltage waveform and include voltage values at times t_0, t_1, t_2, t_3, and t_4.

    c. draw the resistor voltage waveform and include voltage values at times t_0, t_1, t_2, t_3, and t_4.

    d. specify the ratio of the pulse width of the applied voltage to the RC time constant.

Figure 22–18

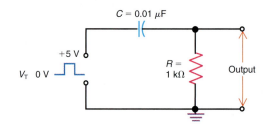

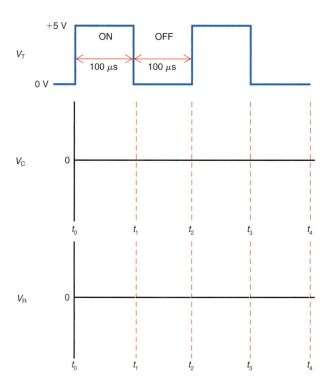

SECTION 22–10 LONG TIME CONSTANT FOR AN *RC* COUPLING CIRCUIT

22–29 Assume that the resistance, *R*, in Fig. 22–18 is increased to 100 kΩ but the frequency of the applied voltage, V_T, remains the same. Determine

    a. the new *RC* time constant of the circuit.

    b. the ratio of the pulse width of the applied voltage to the *RC* time constant.

    c. the approximate capacitor and resistor voltage waveforms, assuming that the input voltage has been applied for longer than five *RC* time constants.

SECTION 22–11 ADVANCED TIME CONSTANT ANALYSIS

22–30 What is the time constant of the circuit in Fig. 22–19?

Figure 22–19

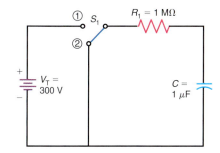

22–31 Assume that *C* in Fig. 22–19 is completely discharged with S_1 in Position 2. If S_1 is moved to Position 1, how much is the capacitor voltage at the following time intervals?

    a. *t* = 0 s

    b. *t* = 0.7 s

    c. *t* = 1 s

    d. *t* = 1.5 s

    e. *t* = 2 s

    f. *t* = 2.5 s

    g. *t* = 3.5 s

22–32 Assume that *C* in Fig. 22–19 is fully charged with S_1 in Position 1. If S_1 is moved to Position 2, how much is the resistor voltage at the following time intervals?

    a. *t* = 0 s

    b. *t* = 0.7 s

    c. *t* = 1 s

    d. *t* = 1.5 s

    e. *t* = 2 s

    f. *t* = 2.5 s

    g. *t* = 3.5 s

22–33 Assume that *C* in Fig. 22–19 is completely discharged with S_1 in Position 2. If S_1 is moved back to Position 1, how long will it take for the capacitor voltage to reach

    a. 90 V?

    b. 150 V?

    c. 200 V?

    d. 240 V?

    e. 270 V?

22–34 Assume that the capacitor in Fig. 22–19 is discharging from 300 V with S_1 in Position 2. At the instant V_C reaches 150 V, S_1 is moved back to Position 1. What is the value of the capacitor voltage 1.25 s later?

22–35 What is the time constant of the circuit in Fig. 22–20?

Figure 22–20

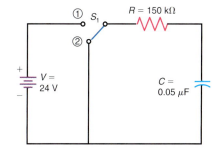

22–36 Assume that *C* in Fig. 22–20 is completely discharged with S_1 in Position 2. If S_1 is moved back to Position 1, how long will it take for the capacitor voltage to reach

    a. 3 V?

    b. 6 V?

    c. 15 V?

    d. 20 V?

22–37 Assume that *C* in Fig. 22–20 is completely discharged with S_1 in Position 2. If S_1 is moved back to Position 1, how much is the resistor voltage at the following time intervals?

    a. *t* = 0 s

    b. *t* = 4.5 ms

    c. *t* = 10 ms

    d. *t* = 15 ms

    e. *t* = 25 ms

22–38 Assume that *C* in Fig. 22–20 is fully charged with S_1 in Position 1. If S_1 is moved to Position 2, how long will it take the capacitor to discharge to

    a. 4 V?

    b. 8 V?

    c. 12 V?

    d. 18 V?

22–39 When analyzing a sine-wave ac circuit containing resistance and capacitance or resistance and inductance, do we use the concepts involving reactance or time constants?

22–40 When analyzing a circuit having a square-wave input voltage, should we use the concepts involving reactance or time constants?

22–41 Should an *RC* coupling circuit have a long or short time constant with respect to the period of the ac input voltage?

Critical Thinking

22–42 Refer to Fig. 22–21. **(a)** If S_1 is closed long enough for the capacitor *C* to become fully charged, what voltage is across *C*? **(b)** With *C* fully charged, how long will it take *C* to discharge fully when S_1 is opened?

is initially closed? **(c)** What is V_C 415.8 μs after S_1 is initially closed? **(d)** What is V_C 1.5 ms after S_1 is initially closed?

22–44 Refer to Fig. 22–22. Assume that *C* is allowed to charge fully and then the polarity of V_T is suddenly reversed. What is the capacitor voltage v_C for the following time intervals after the reversal of V_T: **(a)** 0 s; **(b)** 6.93 ms; **(c)** 10 ms; **(d)** 15 ms; **(e)** 30 ms?

Figure 22–21 Circuit for Critical Thinking Probs. 22–42 and 22–43.

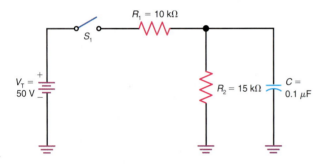

Figure 22–22 Circuit for Critical Thinking Prob. 22–44.

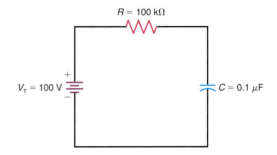

22–43 Refer to Fig. 22–21. **(a)** How long will it take *C* to fully charge after S_1 is closed? **(b)** What is V_C 1 ms after S_1

Answers to Knowledge Check Problems

22–1 Yes

22–2 The inductor, *L*

22–3 $V_L = 1.2$ kV

22–4 $C = 0.1 \ \mu$F

22–5 The resistor, *R*

22–6 No

22–7 Discharge

22–8 Long

22–9 0.5 ms

22–10 50 V

22–11 Approximately 0.7 time constants

22–12 No

Answers to Self-Reviews

22–1 **a.** T
 b. T

22–2 **a.** 0.02 s
 b. 0.5 ms

22–3 **a.** shorter
 b. faster

22–4 **a.** 940 μs
 b. 470 ns

22–5 **a.** 63 V
 b. 37 V

22–6 **a.** shorter
 b. faster

22–7 **a.** $v_R = 37$ V
 b. $v_C = 23.3$ V

22–8 **a.** short
 b. long

22–9 **a.** short
 b. across *C*

22–10 **a.** long
 b. across *R*

22–11 **a.** T
 b. T
 c. T

22–12 **a.** small X_C
 b. long time constant

Cumulative Review Summary (Chapters 19–22)

■ The ability of a conductor to produce induced voltage across itself when the current changes is its self-inductance, or inductance. The symbol is L, and the unit is the henry. One henry allows 1 V to be induced when the current changes at the rate of 1 A/s.

■ The polarity of the induced voltage always opposes the change in current that is causing the induced voltage. This is Lenz's law.

■ Mutual inductance is the ability of varying current in one coil to induce voltage in another coil nearby, without any connection between them. Its symbol is L_M, and the unit is also the henry.

■ A transformer consists of two or more windings with mutual inductance. The primary connects to the source voltage, the secondary to the load. With an iron core, the voltage ratio between primary and secondary equals the turns ratio.

■ The efficiency of a transformer equals the ratio of power output from the secondary to power input to the primary $\times$ 100.

■ Eddy currents are induced in the iron core of an inductance, causing I^2R losses that increase with higher frequencies. Laminated iron, powdered-iron, or ferrite cores have minimum eddy-current losses. Hysteresis also increases the losses.

■ Series inductances without mutual coupling add like series resistances. The combined inductance of parallel inductances is calculated by the reciprocal formula, as for parallel resistances.

■ Inductive reactance X_L equals $2\pi fL\ \Omega$, where f is in hertz and L is in henrys. Reactance X_L increases with more inductance and higher frequencies.

■ A common application of X_L is an af or rf choke, which has high reactance for one group of frequencies but less reactance for lower frequencies.

■ Reactance X_L is a phasor quantity whose current lags 90° behind its induced voltage. In series circuits, R and X_L are added by phasors because their voltage drops are 90° out of phase. In parallel circuits, the resistive and inductive branch currents are 90° out of phase.

■ Impedance Z, in ohms, is the total opposition of an ac circuit with resistance and <u>reactance</u>. For series circuits, $Z_T = \sqrt{R^2 + X_L^2}$ and $I = V_T/Z_T$. For parallel circuits, $I_T = \sqrt{I_R^2 + I_L^2}$ and $Z_{EQ} = V_A/I_T$.

■ The Q of a coil is X_L/r_i.

■ Energy stored by an inductance is $\frac{1}{2}LI^2$, where I is in amperes, L is in henrys, and the energy is in joules.

■ The voltage across L is always equal to $L(di/dt)$ for any waveshape of current.

■ The transient response of a circuit refers to the temporary condition that exists until the circuit's current or voltage reaches its steady-state value. The transient response of a circuit is measured in time constants, where one time constant is defined as the length of time during which a 63.2% change in current or voltage occurs.

■ For an inductive circuit, one time constant is the time in seconds for the current to change by 63.2%. For inductive circuits, one time constant equals L/R, that is, $T = L/R$, where L is in henrys, R is in ohms, and T is in seconds. The current reaches its steady-state value after five L/R time constants have elapsed.

■ In a capacitive circuit, one time constant is the time in seconds for the capacitor voltage to change by 63.2%. For capacitive circuits, one time constant equals RC, that is, $T = RC$, where R is in ohms, C is in farads, and T is in seconds. The capacitor voltage reaches its steady-state value after five RC time constants have elapsed.

■ When the input voltage to an inductive or capacitive circuit is nonsinusoidal, time constants rather than reactances are used to determine the circuit's voltage and current values.

■ Whether an L/R or RC time constant is considered short or long depends on its relationship to the pulse width of the applied voltage. In general, a short time constant is considered one that is one-fifth or less the time of the pulse width of the applied voltage. Conversely, a long time constant is generally considered one that is five or more times longer than the pulse width of the applied voltage.

■ To calculate the voltage across a capacitor during charge, use curve a in Fig. 22–9 or use Formula (22–5). To calculate the voltage across a resistor during charge, use curve b in Fig. 22–9 or Formula (22–3). To calculate the voltage across a capacitor or resistor during discharge, use curve b in Fig. 22–9 or Formula (22–3).

Cumulative Review Self-Test

Answers at back of book.

1. A coil induces 200 mV when the current changes at the rate of 1 A/s. The inductance L is (a) 1 mH; (b) 2 mH; (c) 200 mH; (d) 100 mH.

2. Alternating current in an inductance produces maximum induced voltage when the current has its (a) maximum value; (b) maximum change in magnetic flux; (c) minimum change in magnetic flux; (d) rms value of 0.707 $\times$ peak.

3. An iron-core transformer connected to a 120-V, 60-Hz power line has a turns ratio of 1:20. The voltage across the secondary equals (a) 20 V; (b) 60 V; (c) 120 V; (d) 2400 V.

4. Two 250-mH chokes in series have a total inductance of (a) 60 mH; (b) 125 mH; (c) 250 mH; (d) 500 mH.

5. Which of the following will have minimum eddy-current losses? (a) Solid iron core; (b) laminated iron core; (c) powdered-iron core; (d) air core.

6. Which of the following will have maximum inductive reactance? (a) 2-H inductance at 60 Hz; (b) 2-mH inductance at 60 kHz; (c) 5-mH inductance at 60 kHz; (d) 5-mH inductance at 100 kHz.

7. A 100-Ω R is in series with 100 Ω of X_L. The total impedance Z equals (a) 70.7 Ω; (b) 100 Ω; (c) 141 Ω; (d) 200 Ω.

8. A 100-Ω R is in parallel with 100 Ω of X_L. The total impedance Z equals (a) 70.7 Ω; (b) 100 Ω; (c) 141 Ω; (d) 200 Ω.

9. If two waves have the frequency of 1000 Hz and one is at the maximum value when the other is at zero, the phase angle between them is (a) 0°; (b) 90°; (c) 180°; (d) 360°.

10. If an ohmmeter check on a 50-μH choke reads 3 Ω, the coil is probably (a) open; (b) defective; (c) normal; (d) partially open.

11. An inductive circuit with $L = 100$ mH and $R = 10$ kΩ has a time constant of (a) 1 μs; (b) 100 μs; (c) 10 μs; (d) 1000 μs.

12. A capacitive circuit with $R = 1.5$ kΩ and $C = 0.01$ μF has a time constant of (a) 15 μs; (b) 1.5 μs; (c) 150 μs; (d) 150 s.

13. With respect to the pulse width of the applied voltage, the time constant of an RC integrator should be (a) short; (b) the same as the pulse width of V_T; (c) long; (d) shorter than the pulse width of V_T.

14. With respect to the pulse width of the applied voltage, the time constant of an RC differentiator should be (a) long; (b) the same as the pulse width of V_T; (c) longer than the pulse width of V_T; (d) short.

15. The current rating of a transformer is usually specified for (a) the primary windings only; (b) the secondary windings only; (c) both the primary and secondary windings; (d) the core only.

16. The secondary of a transformer is connected to a 15-Ω resistor. If the turns ratio $N_P/N_S = 3:1$, the primary impedance Z_P equals (a) 135 Ω; (b) 45 Ω; (c) 5 Ω; (d) none of these.

23 Alternating Current Circuits

This chapter shows how to analyze sine-wave ac circuits that have R, X_L, and X_C. How do we combine these three types of ohms of opposition, how much current flows, and what is the phase angle? These questions are answered for both series and parallel circuits.

The problems are simplified by the fact that in series circuits X_L is at $90°$ and X_C is at $-90°$, which are opposite phase angles. Then all of one reactance can be canceled by part of the other reactance, resulting in only a single net reactance.

Similarly, in parallel circuits, I_L and I_C have opposite phase angles. These phasor currents oppose each other and result in a single net reactive line current.

Finally, the idea of how ac power and dc power can differ because of ac reactance is explained. Also, types of ac current meters, including the wattmeter, are described.

Objectives

After studying this chapter you should be able to

- *Explain* why opposite reactances in series cancel.

- *Determine* the total impedance and phase angle of a series circuit containing resistance, capacitance, and inductance.

- *Determine* the total current, equivalent impedance, and phase angle of a parallel circuit containing resistance, capacitance, and inductance.

- *Define* the terms *real power, apparent power, volt-ampere reactive,* and *power factor.*

- *Calculate* the power factor of a circuit.

Outline

Important Terms

apparent power
double subscripts
power factor (PF)

real power
volt-ampere (VA)

volt-ampere reactive (VAR)
wattmeter

23–1 AC Circuits with Resistance but No Reactance

Combinations of series and parallel resistances are shown in Fig. 23–1. In Fig. 23–1a and b, all voltages and currents throughout the resistive circuit are in phase. There is no reactance to cause a lead or lag in either current or voltage.

Series Resistances

For the circuit in Fig. 23–1a, with two 50-Ω resistances in series across the 100-V source, the calculations are as follows:

$$R_T = R_1 + R_2 = 50 + 50 = 100\ \Omega$$
$$I = \frac{V_T}{R_T} = \frac{100}{100} = 1\ \text{A}$$
$$V_1 = IR_1 = 1 \times 50 = 50\ \text{V}$$
$$V_2 = IR_2 = 1 \times 50 = 50\ \text{V}$$

Note that the series resistances R_1 and R_2 serve as a voltage divider, as in dc circuits. Each R has one-half the applied voltage for one-half the total series resistance.

The voltage drops V_1 and V_2 are both in phase with the series current I, which is the common reference. Also, I is in phase with the applied voltage V_T because there is no reactance.

Parallel Resistances

For the circuit in Fig. 23–1b, with two 50-Ω resistances in parallel across the 100-V source, the calculations are

$$I_1 = \frac{V_A}{R_1} = \frac{100}{50} = 2\ \text{A}$$
$$I_2 = \frac{V_A}{R_2} = \frac{100}{50} = 2\ \text{A}$$
$$I_T = I_1 + I_2 = 2 + 2 = 4\ \text{A}$$

With a total current of 4 A in the main line from the 100-V source, the combined parallel resistance is 25 Ω. This R_{EQ} equals 100 V/4 A for the two 50-Ω branches.

Each branch current has the same phase as that of the applied voltage. Voltage V_A is the reference because it is common to both branches.

Figure 23–1 Alternating-current circuits with resistance but no reactance. (a) Resistances R_1 and R_2 in series. (b) Resistances R_1 and R_2 in parallel.

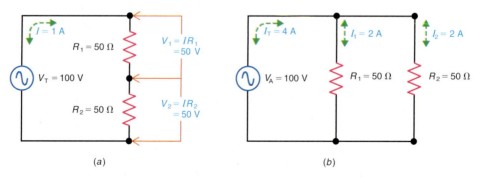

(a)   (b)

■ *23–1 Knowledge Check*

Answer at end of chapter.

In Fig. 23–1a, do the values for I, V_1, and V_2 change if $V_T = 100$ V dc?

■ *23–1 Self-Review*

Answers at end of chapter.

a. In Fig. 23–1a, what is the phase angle between V_T and I?
b. In Fig. 23–1b, what is the phase angle between I_T and V_A?

23–2 Circuits with X_L Alone

The circuits with X_L in Figs. 23–2 and 23–3 correspond to the series and parallel circuits in Fig. 23–1, with ohms of X_L equal to R values. Since the applied voltage is the same, the values of current correspond because ohms of X_L are just as effective as ohms of R in limiting the current or producing a voltage drop.

Although X_L is a phasor quantity with a 90° phase angle, all ohms of opposition are the same kind of reactance in this example. Therefore, without any R or X_C, the series ohms of X_L can be combined directly. Similarly, the parallel I_L currents can be added.

Figure 23–2 Series circuit with X_L alone. (a) Schematic diagram. (b) Phasor diagram of voltages and series current.

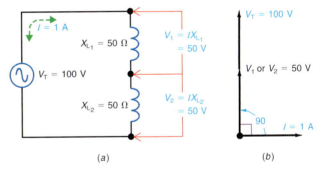

(a) (b)

Figure 23–3 Parallel circuit with X_L alone. (a) Schematic diagram. (b) Phasor diagram of branch and total line currents and applied voltage.

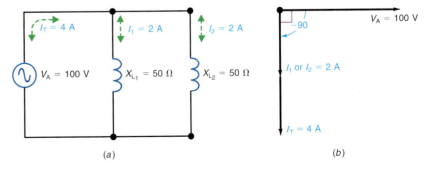

(a) (b)

X_L Values in Series

For Fig. 23–2*a*, the calculations are

$$X_{L_T} = X_{L_1} + X_{L_2} = 50 + 50 = 100 \ \Omega$$
$$I = \frac{V_T}{X_{L_T}} = \frac{100}{100} = 1 \ A$$
$$V_1 = IX_{L_1} = 1 \times 50 = 50 \ V$$
$$V_2 = IX_{L_2} = 1 \times 50 = 50 \ V$$

Note that the two series voltage drops of 50 V each add to equal the total applied voltage of 100 V.

With regard to the phase angle for the inductive reactance, the voltage across any X_L always leads the current through it by 90°. In Fig. 23–2*b*, I is the reference phasor because it is common to all series components. Therefore, the voltage phasors for V_1 and V_2 across either reactance, or V_T across both reactances, are shown leading I by 90°.

I_L Values in Parallel

For Fig. 23–3*a* the calculations are

$$I_1 = \frac{V_A}{X_{L_1}} = \frac{100}{50} = 2 \ A$$
$$I_2 = \frac{V_A}{X_{L_2}} = \frac{100}{50} = 2 \ A$$
$$I_T = I_1 + I_2 = 2 + 2 = 4 \ A$$

These two branch currents can be added because both have the same phase. This angle is 90° lagging the voltage reference phasor, as shown in Fig. 23–3*b*.

Since the voltage V_A is common to the branches, this voltage is across X_{L_1} and X_{L_2}. Therefore V_A is the reference phasor for parallel circuits.

Note that there is no fundamental change between Fig. 23–2*b*, which shows each X_L voltage leading its current by 90°, and Fig. 23–3*b*, showing each X_L current lagging its voltage by −90°. The phase angle between the inductive current and voltage is still the same 90°.

■ *23–2 Knowledge Check*

   Answers at end of chapter.

   For a practical inductor, does V_L always lead I by 90°? Why?

■ *23–2 Self-Review*

   Answers at end of chapter.

   a. In Fig. 23–2, what is the phase angle of V_T with respect to I?
   b. In Fig. 23–3, what is the phase angle of I_T with respect to V_A?

23–3 Circuits with X_C Alone

Again, reactances are shown in Figs. 23–4 and 23–5 but with X_C values of 50 Ω. Since there is no R or X_L, the series ohms of X_C can be combined directly. Also, the parallel I_C currents can be added.

X_C Values in Series

For Fig. 23–4*a*, the calculations for V_1 and V_2 are the same as before. These two series voltage drops of 50 V each add to equal the total applied voltage.

Figure 23–4 Series circuit with X_C alone. (a) Schematic diagram. (b) Phasor diagram of voltages and series current.

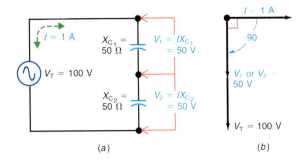

(a)   (b)

Figure 23–5 Parallel circuit with X_C alone. (a) Schematic diagram. (b) Phasor diagram of branch and total line currents and applied voltage.

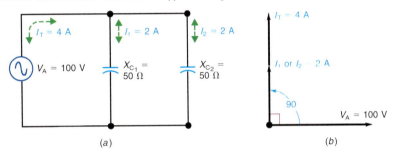

(a)   (b)

With regard to the phase angle for the capacitive reactance, the voltage across any X_C always lags its capacitive charge and discharge current I by 90°. For the series circuit in Fig. 23–4, I is the reference phasor. The capacitive current leads by 90°, or we can say that each voltage lags I by $-90°$.

I_C Values in Parallel

For Fig. 23–5, V_A is the reference phasor. The calculations for I_1 and I_2 are the same as before. However, now each of the capacitive branch currents or the I_T leads V_A by 90°.

23–3 Knowledge Check

Answers at end of chapter.

If $V_T = 100$ V dc Fig. 23–4A, what are the values for I, V_1, and V_2?

23–3 Self-Review

Answers at end of chapter.

a. In Fig. 23–4, what is the phase angle of V_T with respect to I?
b. In Fig. 23–5, what is the phase angle of I_T with respect to V_A?

23–4 Opposite Reactances Cancel

In a circuit with both X_L and X_C, the opposite phase angles enable one to offset the effect of the other. For X_L and X_C in series, the net reactance is the difference between the two series reactances, resulting in less reactance than in either one. In parallel circuits, the net reactive current is the difference between the I_L and I_C branch currents, resulting in less total line current than in either branch current.

X_L and X_C in Series

For the example in Fig. 23–6, the series combination of a 60-Ω X_L and a 40-Ω X_C in Fig. 23–6a and b is equivalent to the net reactance of the 20-Ω X_L shown in Fig. 23–6c. Then, with 20 Ω as the net reactance across the 120-V source, the current is 6 A. This current lags the applied voltage V_T by 90° because the net reactance is inductive.

For the two series reactances in Fig. 23–6a, the current is the same through both X_L and X_C. Therefore, the voltage drops can be calculated as

$$V_L \text{ or } IX_L = 6 \text{ A} \times 60 \text{ } \Omega = 360 \text{ V}$$
$$V_C \text{ or } IX_C = 6 \text{ A} \times 40 \text{ } \Omega = 240 \text{ V}$$

Note that each individual reactive voltage drop can be more than the applied voltage. The phasor sum of the series voltage drops still is 120 V, however, equal to the applied voltage because the IX_L and IX_C voltages are opposite. The IX_L voltage leads the series current by 90°; the IX_C voltage lags the same current by 90°. Therefore, IX_L and IX_C are 180° out of phase with each other, which means that they are of opposite polarity and offset each other. Then the total voltage across the two in series is 360 V minus 240 V, which equals the applied voltage of 120 V.

If the values in Fig. 23–6 were reversed, with an X_C of 60 Ω and an X_L of 40 Ω, the net reactance would be a 20-Ω X_C. The current would be 6 A again but with a lagging phase angle of −90° for the capacitive voltage. The IX_C voltage would then be greater at 360 V than an IX_L value of 240 V, but the difference would still equal the applied voltage of 120 V.

X_L and X_C in Parallel

In Fig. 23–7, the 60-Ω X_L and 40-Ω X_C are in parallel across the 120-V source. Then the 60-Ω X_L branch current I_L is 2 A, and the 40-Ω X_C branch current I_C is 3 A. The X_C branch has more current because its reactance is less than X_L.

||| MultiSim   Figure 23–6   When X_L and X_C are in series, their ohms of reactance subtract. (a) Series circuit with 60-Ω X_L and 40-Ω X_C. (b) Phasor diagram. (c) Equivalent circuit with net value of 20 Ω of X_L for the total reactance.

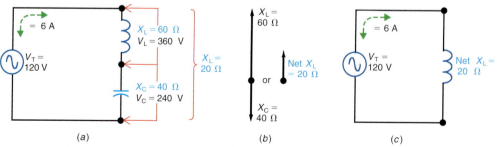

(a)   (b)   (c)

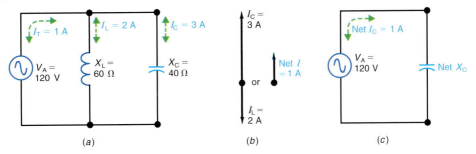

(a)   (b)   (c)

||| **MultiSim** **Figure 23–8**

Impedance Z_T of series circuit with resistance and reactance. (a) Circuit with R, X_L, and X_C in series. (b) Equivalent circuit with one net reactance. (c) Phasor diagram. The voltage triangle of phasors is equivalent to an impedance triangle for series circuits.

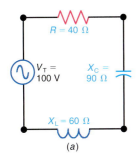

(a)

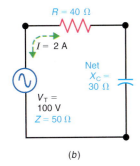

(b)

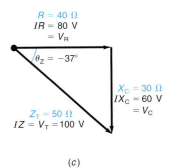

(c)

In terms of phase angle, I_L lags the parallel voltage V_A by 90°, and I_C leads the same voltage by 90°. Therefore, the opposite reactive branch currents are 180° out of phase with each other. The net line current then is the difference between 3 A for I_C and 2 A for I_L, which equals the net value of 1 A. This resultant current leads V_A by 90° because it is capacitive current.

If the values in Fig. 23–7 were reversed, with an X_C of 60 Ω and an X_L of 40 Ω, I_L would be larger. The I_L would then equal 3 A, with an I_C of 2 A. The net line current would be 1 A again but inductive with a net I_L.

■ *23-4 Knowledge Check*

    Answer at end of chapter.

    In Fig. 23–7c, what is the value of the net X_C?

■ *23-4 Self-Review*

    Answers at end of chapter.

    a. In Fig. 23–6, how much is the net X_L?
    b. In Fig. 23–7, how much is the net I_C?

23–5 Series Reactance and Resistance

In the case of series reactance and resistance, the resistive and reactive effects must be combined by phasors. For series circuits, the ohms of opposition are added to find Z_T. First add all the series resistances for one total R. Also, combine all series reactances, adding all X_Ls and all X_Cs and finding the net X by subtraction. The result is one net reactance. It may be either capacitive or inductive, depending on which kind of reactance is larger. Then the total R and net X can be added by phasors to find the total ohms of opposition in the entire series circuit.

Magnitude of Z_T

After the total R and net reactance X are found, they can be combined by the formula

$$Z_T = \sqrt{R^2 + X^2} \qquad (23\text{–}1)$$

The circuit's total impedance Z_T is the phasor sum of the series resistance and reactance. Whether the net X is at +90° for X_L or −90° for X_C does not matter in calculating the magnitude of Z_T.

An example is illustrated in Fig. 23–8. Here the net series reactance in Fig. 23–8b is a 30-Ω X_C. This value is equal to a 60-Ω X_L subtracted from a

90-Ω X_C, as shown in Fig. 23–8a. The net 30-Ω X_C in Fig. 23–8b is in series with a 40-Ω R. Therefore,

$$Z_T = \sqrt{R^2 + X^2} = \sqrt{(40)^2 + (30)^2} = \sqrt{1600 + 900}$$
$$= \sqrt{2500}$$
$$= 50 \ \Omega$$

$I = V/Z_T$

The current is 100 V/50 Ω in this example, or 2 A. This value is the magnitude without considering the phase angle.

Series Voltage Drops

All series components have the same 2-A current. Therefore, the individual drops in Fig. 23–8a are

$$V_R = IR = 2 \times 40 = 80 \text{ V}$$
$$V_C = IX_C = 2 \times 90 = 180 \text{ V}$$
$$V_L = IX_L = 2 \times 60 = 120 \text{ V}$$

Since IX_C and IX_L are voltages of opposite polarity, the net reactive voltage is 180 V minus 120 V, which equals 60 V. The phasor sum of IR at 80 V and the net reactive voltage IX of 60 V equals the applied voltage V_T of 100 V.

Angle of Z_T

The impedance angle of the series circuit is the angle whose tangent equals X/R. This angle is negative for X_C but positive for X_L.

In this example, X is the net reactance of 30 Ω for X_C and R is 40 Ω. Then tan $\theta_Z = -0.75$, and θ_Z is $-37°$, approximately.

The negative angle for Z indicates a net capacitive reactance for the series circuit. If the values of X_L and X_C were reversed, θ_Z would be $+37°$, instead of $-37°$, because of the net X_L. However, the magnitude of Z would still be the same.

Example 23–1

A 27-Ω R is in series with 54 Ω of X_L and 27 Ω of X_C. The applied voltage V_T is 50 mV. Calculate Z_T, I, and θ_Z.

ANSWER The net X_L is 27 Ω. Then

$$Z_T = \sqrt{R^2 + X_L^2} = \sqrt{(27)^2 + (27)^2}$$
$$= \sqrt{729 + 729} = \sqrt{1458}$$
$$= 38.18 \ \Omega$$
$$I = \frac{V_T}{Z_T} = \frac{50 \text{ mV}}{38.18 \ \Omega}$$
$$= 1.31 \text{ mA}$$
$$\tan \theta_Z = \frac{X}{R} = \frac{27 \ \Omega}{27 \ \Omega}$$
$$= 1$$
$$\theta_Z = \arctan (1)$$
$$= 45°$$

Figure 23–9 Series circuit with more components than Fig. 23–8 but the same Z_T, I, and θ_Z.

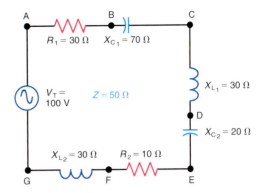

In general, when the series resistance and reactance are equal, Z_T is 1.414 times either value. Here, Z_T is $1.414 \times 27 = 38.18\ \Omega$. Also, $\tan \theta$ must be 1 and the angle is $45°$ for equal sides in a right triangle. To find Z_T on a calculator, see the procedure described on page 648 for the square root of the sum of two squares.

More Series Components

Figure 23–9 shows how to combine any number of series resistances and reactances. Here the total series R of $40\ \Omega$ is the sum of $30\ \Omega$ for R_1 and $10\ \Omega$ for R_2. Note that the order of connection does not matter, since the current is the same in all series components.

The total series X_C is $90\ \Omega$, equal to the sum of $70\ \Omega$ for X_{C_1} and $20\ \Omega$ for X_{C_2}. Similarly, the total series X_L is $60\ \Omega$. This value is equal to the sum of $30\ \Omega$ for X_{L_1} and $30\ \Omega$ for X_{L_2}.

The net reactance X equals $30\ \Omega$, which is $90\ \Omega$ of X_C minus $60\ \Omega$ of X_L. Since X_C is larger than X_L, the net reactance is capacitive. The circuit in Fig. 23–9 is equivalent to Fig. 23–8, therefore, since a $40\text{-}\Omega$ R is in series with a net X_C of $30\ \Omega$.

Double-Subscript Notation

This method for specifying ac and *dc* voltages is useful to indicate the polarity or phase. For instance, in Fig. 23–9 the voltage across R_2 can be taken as either V_{EF} or V_{FE}. With opposite subscripts, these two voltages are $180°$ out of phase. In using double subscripts, note that the first letter in the subscript is the point of measurement with respect to the second letter.

■ *23–5 Knowledge Check*

Answer at end of chapter.

In Fig. 23–9, what is the magnitude of the voltage across points C and E?

■ *23–5 Self-Review*

Answers at end of chapter.

a. In Fig. 23–8, how much is the net reactance?
b. In Fig. 23–9, how much is the net reactance?
c. In Fig. 23–9, give the phase difference between V_{CD} and V_{DC}.

23–6 Parallel Reactance and Resistance

In parallel circuits, the branch currents for resistance and reactance are added by phasors. Then the total line current is found by the formula

$$I_T = \sqrt{I_R^2 + I_X^2}$$

(23–2)

Calculating I_T

As an example, Fig. 23–10a, shows a circuit with three branches. Since the voltage across all the parallel branches is the applied 100 V, the individual branch currents are

$$I_R = \frac{V_A}{R} = \frac{100 \text{ V}}{25 \text{ }\Omega} = 4 \text{ A}$$

$$I_L = \frac{V_A}{X_L} = \frac{100 \text{ V}}{25 \text{ }\Omega} = 4 \text{ A}$$

$$I_C = \frac{V_A}{X_C} = \frac{100 \text{ V}}{100 \text{ }\Omega} = 1 \text{ A}$$

The net reactive branch current I_X is 3 A, then, equal to the difference between the 4-A I_L and the 1-A I_C, as shown in Fig. 23–10b.

The next step is to calculate I_T as the phasor sum of I_R and I_X. Then

$$I_T = \sqrt{I_R^2 + I_X^2} = \sqrt{4^2 + 3^2} = \sqrt{16 + 9} = \sqrt{25}$$
$$= 5 \text{ A}$$

The phasor diagram for I_T is shown in Fig. 23–10c.

$Z_{EQ} = V_A/I_T$

This gives the total impedance of a parallel circuit. In this example, Z_{EQ} is 100 V/5 A, which equals 20 Ω. This value is the equivalent impedance of all three branches in parallel across the source.

Figure 23–10 Total line current I_T of parallel circuit with resistance and reactance. (a) Parallel branches with I_R, I_C, and I_L. (b) Equivalent circuit with net I_X. (c) Phasor diagram.

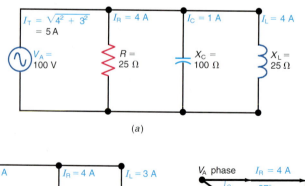

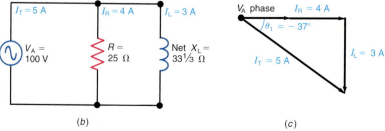

Phase Angle

The phase angle of the parallel circuit is found from the branch currents. Now θ is the angle whose tangent equals I_X/I_R.

For this example, I_X is the net inductive current of 3-A I_L. Also, I_R is 4 A. These phasors are shown in Fig. 23–10c. Then θ is a negative angle with a tangent of -0.75. This phase angle is approximately $-37°$.

The negative angle for I_T indicates lagging inductive current. The value of $-37°$ is the phase angle of I_T with respect to the voltage reference V_A.

When Z_{EQ} is calculated as V_A/I_T for a parallel circuit, the phase angle is the same value as for I_T but with opposite sign. In this example, Z_{EQ} is 20 Ω with a phase angle of $+37°$, for an I_T of 5 A with an angle of $-37°$. We can consider that Z_{EQ} has the phase angle of the voltage source with respect to I_T.

Example 23-2

The following branch currents are supplied from a 50-mV source: $I_R = 1.8$ mA; $I_L = 2.8$ mA; $I_C = 1$ mA. Calculate I_T, Z_{EQ}, and θ_I.

ANSWER The net I_X is 1.8 mA. Then

$$I_T = \sqrt{I_R^2 + I_X^2} = \sqrt{(1.8)^2 + (1.8)^2}$$
$$= \sqrt{3.24 + 3.24} = \sqrt{6.48}$$
$$= 2.55 \text{ mA}$$

$$Z_{EQ} = \frac{V_A}{I_T} = \frac{50 \text{ mV}}{2.55 \text{ mA}}$$
$$= 19.61 \text{ Ω}$$

$$\tan \theta_I = -\frac{I_L}{I_R} = -\frac{1.8 \text{ mA}}{1.8 \text{ mA}}$$
$$= -1$$

$$\theta_I = \arctan (1)$$
$$= -45°$$

Note that with equal branch currents, I_T is $1.414 \times 1.8 = 2.55$ mA. Also, the phase angle θ_I is negative for inductive branch current.

More Parallel Branches

Figure 23–11 shows how any number of parallel resistances and reactances can be combined. The total resistive branch current I_R of 4 A is the sum of 2 A each for the R_1 branch and the R_2 branch. Note that the order of connection does not matter, since the parallel branch currents add in the main line. Effectively, two 50-Ω resistances in parallel are equivalent to one 25-Ω resistance.

Similarly, the total inductive branch current I_L is 4 A, equal to 3 A for I_{L_1} and 1 A for I_{L_2}. Also, the total capacitive branch current I_C is 1 A, equal to ½ A each for I_{C_1} and I_{C_2}.

The net reactive branch current I_X is 3 A, then, equal to a 4-A I_L minus a 1-A I_C. Since I_L is larger, the net current is inductive.

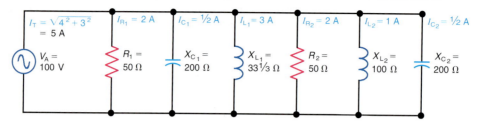

Therefore, the circuit in Fig. 23–11 is equivalent to the circuit in Fig. 23–10. Both have a 4-A resistive current I_R and a 3-A net reactive current I_X. These values added by phasors make a total of 5 A for I_T in the main line.

■ 23-6 Knowledge Check

Answer at end of chapter.

How much is I_T in Fig. 23–11 if a third capacitor with an X_C of 16.67 Ω is added to the circuit?

■ 23-6 Self-Review

Answers at end of chapter.

a. In Fig. 23–10, what is the net reactive branch current?
b. In Fig. 23–11, what is the net reactive branch current?

23-7 Series-Parallel Reactance and Resistance

Figure 23–12 shows how a series-parallel circuit can be reduced to a series circuit with just one reactance and one resistance. The method is straightforward as long as resistance and reactance are not combined in one parallel bank or series string.

Working backward toward the generator from the outside branch in Fig. 23–12a, we have an X_{L_1} and an X_{L_2} of 100 Ω each in series, which total 200 Ω. This string in Fig. 23–12a is equivalent to X_{L_5} in Fig. 23–12b.

In the other branch, the net reactance of X_{L_3} and X_C is equal to 600 Ω minus 400 Ω. This is equivalent to the 200 Ω of X_{L_4} in Fig. 23–12b. The X_{L_4} and X_{L_5} of 200 Ω each in parallel are combined for an X_L of 100 Ω.

In Fig. 23–12c, the 100-Ω X_L is in series with the 100-Ω R_{1-2}. This value is for R_1 and R_2 in parallel.

The triangle diagram for the equivalent circuit in Fig. 23–12d shows the total impedance Z of 141 Ω for a 100-Ω R in series with a 100-Ω X_L.

With a 141-Ω impedance across the applied V_T of 100 V, the current in the generator is 0.7 A. The phase angle θ is 45° for this circuit.*

■ 23-7 Knowledge Check

Answer at end of chapter.

In Fig. 23–12, how much is Z_T if $X_{L_3} = 1$ kΩ?

---

* More complicated ac circuits with series-parallel impedances are analyzed with complex numbers, as explained in Chap. 24.

Figure 23–12 Reducing an ac series-parallel circuit with R, X_L, and X_C to a series circuit with one net resistance and one net reactance. (*a*) Actual circuit. (*b*) Simplified arrangement. (*c*) Equivalent series circuit. (*d*) Impedance triangle with phase angle.

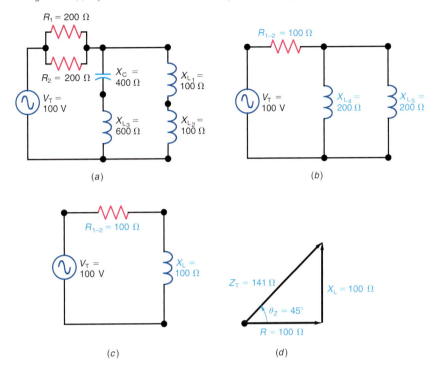

(a)   (b)

(c)   (d)

■ 23-7 Self-Review

Answers at end of chapter.

Refer to Fig. 23–12.
a. How much is $X_{L_1} + X_{L_2}$?
b. How much is $X_{L_3} - X_C$?
c. How much is X_{L_4} in parallel with X_{L_5}?

23–8 Real Power

In an ac circuit with reactance, the current I supplied by the generator either leads or lags the generator voltage V. Then the product VI is not the real power produced by the generator, since the instantaneous voltage may have a high value while at the same time the current is near zero, or vice versa. The real power, in watts, however, can always be calculated as I^2R, where R is the total resistive component of the circuit, because current and voltage are in phase in a resistance. To find the corresponding value of power as VI, this product must be multiplied by the cosine of the phase angle θ. Then

$$\text{Real power} = P = I^2R \qquad (23\text{–}3)$$

or

$$\text{Real power} = P = VI \cos \theta \qquad (23\text{–}4)$$

where V and I are in rms values, and P, the real power, is in watts. Multiplying VI by the cosine of the phase angle provides the resistive component for real power equal to I^2R.

For example, the ac circuit in Fig. 23–13 has 2 A through a 100-Ω R in series with the X_L of 173 Ω. Therefore,

$$P = I^2R = 4 \times 100 = 400 \text{ W}$$

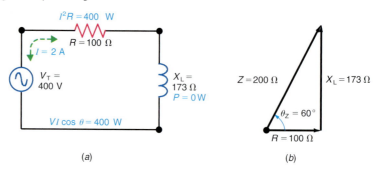

Figure 23–13 Real power, *P*, in a series circuit. (*a*) Schematic diagram. (*b*) Impedance triangle with phase angle.

(a)

(b)

Furthermore, in this circuit, the phase angle is 60° with a cosine of 0.5. The applied voltage is 400 V. Therefore,

$$P = VI \cos \theta = 400 \times 2 \times 0.5 = 400 \text{ W}$$

In both examples, the real power is the same 400 W because this is the amount of power supplied by the generator and dissipated in the resistance. Either formula can be used for calculating the real power, depending on which is more convenient.

Real power can be considered resistive power that is dissipated as heat. A reactance does not dissipate power but stores energy in an electric or magnetic field.

Power Factor

Because it indicates the resistive component, $\cos \theta$ is the power factor of the circuit, converting the *VI* product to real power. The power factor formulas are

For series circuits:

$$\text{Power factor} = PF = \cos \theta = \frac{R}{Z} \qquad \text{(23–5)}$$

For parallel circuits:

$$\text{Power factor} = \cos \theta = \frac{I_R}{I_T} \qquad \text{(23–6)}$$

In Fig. 23–13, as an example of a series circuit, we use *R* and *Z* for the calculations:

$$PF = \cos \theta = \frac{R}{Z} = \frac{100 \ \Omega}{200 \ \Omega} = 0.5$$

For the parallel circuit in Fig. 23–10, we use the resistive current I_R and the I_T:

$$PF = \cos \theta = \frac{I_R}{I_T} \quad \frac{4 \text{ A}}{5 \text{ A}} = 0.8$$

The power factor is not an angular measure but a numerical ratio with a value between 0 and 1, equal to the cosine of the phase angle.

With all resistance and zero reactance, *R* and *Z* are the same for a series circuit, or I_R and I_T are the same for a parallel circuit, and the ratio is 1. Therefore, unity power factor means a resistive circuit. At the opposite extreme, all reactance with zero resistance makes the power factor zero, which means that the circuit is all reactive. The power factor is frequently given in percent so that unity power factor is 100%. To convert from decimal *PF* to percent *PF*, just multiply by 100.

GOOD TO KNOW

The apparent power is the power which is "apparently" used by the circuit before the phase angle between *V* and *I* is considered.

Apparent Power

When *V* and *I* are out of phase because of reactance, the product of $V \times I$ is called *apparent power*. The unit is *volt-amperes* (VA) instead of watts, since the watt is reserved for real power.

For the example in Fig. 23–13, with 400 V and the 2-A *I*, 60° out of phase, the apparent power is *VI*, or $400 \times 2 = 800$ VA. Note that apparent power is the *VI* product alone, without considering the power factor cos θ.

The power factor can be calculated as the ratio of real power to apparent power because this ratio equals cos θ. As an example, in Fig. 23–13, the real power is 400 W, and the apparent power is 800 VA. The ratio of 400/800, then, is 0.5 for the power factor, the same as cos 60°.

The VAR

This is an abbreviation for volt-ampere reactive. Specifically, VARs are volt-amperes at the angle of 90°.

In general, for any phase angle θ between *V* and *I*, multiplying *VI* by sin θ gives the vertical component at 90° for the value of the VARs. In Fig. 23–13, the value of *VI* sin 60° is $800 \times 0.866 = 692.8$ VAR.

Note that the factor sin θ for the VARs gives the vertical or reactive component of the apparent power *VI*. However, multiplying *VI* by cos θ as the power factor gives the horizontal or resistive component of the real power.

Correcting the Power Factor

In commercial use, the power factor should be close to unity for efficient distribution of electric power. However, the inductive load of motors may result in a power factor of 0.7, as an example, for the phase angle of 45°. To correct for this lagging inductive component of the current in the main line, a capacitor can be connected across the line to draw leading current from the source. To bring the power factor up to 1.0, that is, unity PF, the value of capacitance is calculated to take the same amount of volt-amperes as the VARs of the load.

■ *23–8 Knowledge Check*

Answer at end of chapter.

The apparent power in a circuit is 2.5 kVA. If the power factor is 0.55, how much real power is dissipated by the circuit?

■ *23–8 Self-Review*

Answers at end of chapter.

a. What is the unit of real power?
b. What is the unit of apparent power?
c. Is I^2R real or apparent power?

23–9 AC Meters

The D'Arsonval moving-coil type of meter movement will not read if it is used in an ac circuit because the ac wave changes polarity too rapidly. Since the two opposite polarities cancel, an alternating current cannot deflect the meter movement either up-scale or down-scale. An ac meter must produce deflection of the meter pointer up-scale regardless of polarity. This deflection is accomplished by one of the following three methods for nonelectronic ac meters.

1. *Thermal type.* In this method, the heating effect of the current, which is independent of polarity, is used to provide meter deflection. Two examples are the thermocouple type and hot-wire meter.

2. *Electromagnetic type.* In this method, the relative magnetic polarity is maintained constant although the current reverses. Examples are the iron-vane meter, dynamometer, and wattmeter.

3. *Rectifier type.* The rectifier changes the ac input to dc output for the meter, which is usually a D'Arsonval movement. This type is the most common for ac voltmeters generally used for audio and radio frequencies.

All analog ac meters (meters with scales and pointers) have scales calibrated in rms values, unless noted otherwise on the meter.

A thermocouple consists of two dissimilar metals joined together at one end but open at the opposite side. Heat at the short-circuited junction produces a small dc voltage across the open ends, which are connected to a dc meter movement. In the hot-wire meter, current heats a wire to make it expand, and this motion is converted into meter deflection. Both types are used as ac meters for radio frequencies.

The iron-vane meter and dynamometer have very low sensitivity compared with a D'Arsonval movement. They are used in power circuits for either direct current or 60-Hz alternating current.

■ 23–9 Knowledge Check

Answer at end of chapter.

What is the most common type of ac voltmeter for measuring af and rf voltages?

■ 23–9 Self-Review

Answers at end of chapter.

a. **The iron-vane meter can read alternating current. (True or False)**
b. **The D'Arsonval meter movement works with direct current only. (True or False)**

Figure 23–14 Schematic of voltage and current coils of an analog wattmeter.

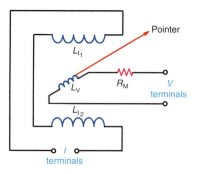

GOOD TO KNOW

Four circuit connections are needed when using the wattmeter in Fig. 23–14.

23–10 Wattmeters

The wattmeter uses fixed coils to measure current in a circuit, and the movable coil measures voltage (Fig. 23–14). The deflection, then, is proportional to power. Either dc power or real ac power can be read directly by the wattmeter.

In Fig. 23–14, the coils L_{I_1} and L_{I_2} in series are heavy stationary coils serving as an ammeter to measure current. The two *I* terminals are connected in one side of the line in series with the load. The movable coil L_V and its multiplier resistance R_M are used as a voltmeter with the *V* terminals connected across the line in parallel with the load. Then the current in the fixed coils is proportional to *I*, and the current in the movable coil is proportional to *V*. As a result, the deflection is proportional to *V* and *I*.

Furthermore, it is the *VI* product for each instant that produces deflection. For instance, if the *V* value is high when the *I* value is low for a phase angle close to 90°, there will be little deflection. The meter deflection is proportional to the watts of real power, therefore, regardless of the power factor in ac circuits. The wattmeter is commonly used to measure power from the 60-Hz power line. For radio frequencies, however, power is generally measured in terms of heat transfer.

Can a wattmeter measure the real power in an ac circuit if the power factor is 0.75?

■ 23-10 *Self-Review*

Answers at end of chapter.

a. Does a wattmeter measure real or apparent power?
b. In Fig. 23–14, does the movable coil of a wattmeter measure *V* or *I*?

23-11 Summary of Types of Ohms in AC Circuits

The differences in R, X_L, X_C, and Z_T are listed in Table 23–1, but the following general features should also be noted. Ohms of opposition limit the amount of current in dc circuits or ac circuits. Resistance R is the same for either case. However, ac circuits can have ohms of reactance because of the variations in alternating current or voltage. Reactance X_L is the reactance of an inductance with sine-wave changes in current. Reactance X_C is the reactance of a capacitor with sine-wave changes in voltage.

Both X_L and X_C are measured in ohms, like R, but reactance has a 90° phase angle, whereas the phase angle for resistance is 0°. A circuit with steady direct current cannot have any reactance.

Ohms of X_L or X_C are opposite because X_L has a phase angle of $+90°$ and X_C has an angle of $-90°$. Any individual X_L or X_C always has a phase angle that is exactly 90°.

Ohms of impedance Z result from the phasor combination of resistance and reactance. In fact, Z can be considered the general form of any ohms of opposition in ac circuits.

Impedance can have any phase angle, depending on the relative amounts of R and X. When Z consists mostly of R with little reactance, the phase angle of Z is close to 0°. With R and X equal, the phase angle of Z is 45°. Whether the angle is positive or negative depends on whether the net reactance

Table 23-1	Types of Ohms in AC Circuits			
	Resistance R, Ω	**Inductive Reactance X_L, Ω**	**Capacitive Reactance X_C, Ω**	**Impedance Z_T, Ω**
Definition	In-phase opposition to alternating or direct current	90° leading opposition to alternating current	90° lagging opposition to alternating current	Combination of resistance and reactance $Z_T = \sqrt{R^2 + X^2}$
Effect of frequency	Same for all frequencies	Increases with higher frequencies	Decreases at higher frequencies	X_L component increases, but X_C decreases at higher frequencies
Phase angle	0°	I_L lags V_L by 90°	I_C leads V_C by 90°	$\tan \theta_Z = \pm X/R$ in series, $\tan \theta_I = \pm I_X/I_R$ in parallel

is inductive or capacitive. When Z consists mainly of X with little R, the phase angle of Z is close to 90°.

The phase angle is θ_Z for Z or V_T with respect to the common I in a series circuit. With parallel branch currents, θ_I is for I_T in the main line with respect to the common voltage.

■ 23–11 Knowledge Check

Answers at end of chapter.

A series ac circuit has an impedance, Z, of 1 kΩ. If $\theta_Z = 45°$, calculate R and X.

■ 23–11 Self-Review

Answers at end of chapter.

a. Which of the following does not change with frequency: Z, X_L, X_C, or R?
b. Which has lagging current: R, X_L, or X_C?
c. Which has leading current: R, X_L, or X_C?

23–12 Summary of Types of Phasors in AC Circuits

Phasors for ohms, volts, and amperes are shown in Fig. 23–15. Note the similarities and differences.

Figure 23–15 Summary of phasor relations in ac circuits. (*a*) Series R and X_L. (*b*) Series R and X_C. (*c*) Parallel branches with I_R and I_C. (*d*) Parallel branches with I_R and I_L.

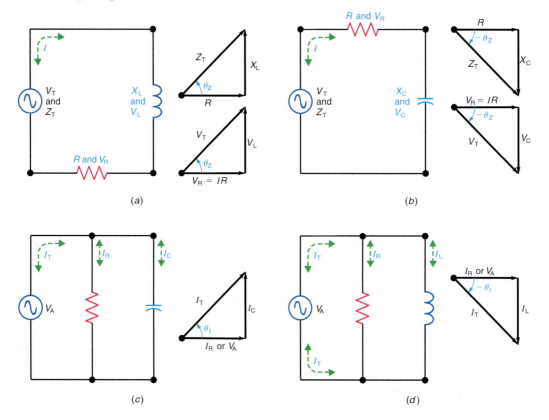

(a)   (b)

(c)   (d)

Series Components

In series circuits, ohms and voltage drops have similar phasors. The reason is the common I for all series components. Therefore,

V_R or IR has the same phase as R.
V_L or IX_L has the same phase as X_L.
V_C or IX_C has the same phase as X_C.

Resistance

The R, V_R, and I_R always have the same phase angle because there is no phase shift in a resistance. This applies to R in either a series or a parallel circuit.

Reactance

Reactances X_L and X_C are 90° phasors in opposite directions. The X_L or V_L has an angle of $+90°$ with an upward phasor, and the X_C or V_C has an angle of $-90°$ with a downward phasor.

Reactive Branch Currents

The phasor of a parallel branch current is opposite from its reactance. Therefore, I_C is upward at $+90°$, opposite from X_C downward at $-90°$. Also, I_L is downward at $-90°$, opposite from X_L upward at $+90°$.

In short, I_C and I_L are opposite each other, and both are opposite from their corresponding reactances.

Angle θ_Z

The phasor resultant for ohms of reactance and resistance is Z. The phase angle θ for Z can be any angle between 0° and 90°. In a series circuit θ_Z for Z is the same as θ for V_T with respect to the common current I.

Angle θ_I

The phasor resultant of branch currents is the total line current I_T. The phase angle of I_T can be any angle between 0° and 90°. In a parallel circuit, θ_I is the angle of I_T with respect to the applied voltage V_A.

Such phasor combinations are necessary in sine-wave ac circuits to take into account the effect of reactance. Phasors can be analyzed either graphically, as in Fig. 23–15, or by the shorter technique of complex numbers, with a j operator that corresponds to the 90° phasor. Complex numbers are explained in the next chapter.

Circuit Phase Angle θ

The phase angle for all types of sine-wave ac circuits is usually considered the angle between the current I from the source and its applied voltage as the reference. This angle can be labeled θ, without any subscript. No special identification is necessary because θ is the phase angle of the circuit. Then there are only the two possibilities shown in Fig. 23–16. In Fig. 23–16a, the θ is a counterclockwise angle for a positive value, which means that I leads V. The leading I is in a circuit with series X_C or with I_C in a parallel branch. In Fig. 23–16b, the phase angle is clockwise for $-\theta$, which means that I lags V. The lagging I is produced in a circuit with series X_L or with I_L in a parallel branch.

Note that, in general, θ is the same as θ_I in parallel branch currents. However, θ has a sign opposite from θ_Z with series reactances.

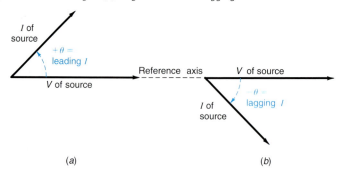

(a)   (b)

23–12 Knowledge Check

Answer at end of chapter.

For any type of ac circuit, what is the phase relationship between V_R and I?

23–12 Self-Review

Answers at end of chapter.

a. Of the following phasors, which two are 180° apart: V_L, V_C, or V_R?
b. Of the following phasors, which two are out of phase by 90°: I_R, I_T, or I_L?

Summary

- In ac circuits with resistance alone, the circuit is analyzed the same way as dc circuits, generally with rms ac values. Without any reactance, the phase angle between V and I is zero.

- When capacitive reactances alone are combined, the X_C values are added in series and combined by the reciprocal formula in parallel, just like ohms of resistance. Similarly, ohms of X_L alone can be added in series or combined by the reciprocal formula in parallel, just like ohms of resistance.

- Since X_C and X_L are opposite reactances, they offset each other. In series, ohms of X_C and X_L can be subtracted. In parallel, the capacitive

and inductive branch currents I_C and I_L can be subtracted.

- In ac circuits, R, X_L, and X_C can be reduced to one equivalent resistance and one net reactance.

- In series, the total R and net X at 90° are combined as $Z_T = \sqrt{R^2 + X^2}$. The phase angle of the series R and X is the angle with tangent $\pm X/R$. To find I, first we calculate Z_T and then divide into V_T.

- For parallel branches, the total I_R and net reactive I_X at 90° are combined as $I_T = \sqrt{I_R^2 + I_X^2}$. The phase angle of the parallel R and X is the angle with tangent $\pm I_X/I_R$. To find Z_{EQ} first we calculate I_T and then divide into V_A.

- The quantities R, X_L, X_C, and Z in ac circuits all are ohms of opposition. The differences with respect to frequency and phase angle are summarized in Table 23–1.

- The phase relations for resistance and reactance are summarized in Fig. 23–15.

- In ac circuits with reactance, the real power P in watts equals I^2R, or $VI \cos \theta$, where θ is the phase angle. The real power is the power dissipated as heat in resistance. Cos θ is the power factor of the circuit.

- The wattmeter measures real ac power or dc power.

Important Terms

- **Apparent Power** - the power that is apparently consumed by an ac circuit. Apparent power is calculated as $V \times I$ without considering the phase angle. The unit of apparent power is the volt-ampere (VA) since the watt unit is reserved for real power.

- **Double-Subscripts** - a notational system used to specify the dc and ac voltages in a circuit. The first letter in the subscript indicates the point of measurement, whereas the second letter indicates the point of reference. With double-subscript notation, the polarity or phase of a voltage can be indicated.

- **Power Factor (PF)** - a numerical ratio between 0 and 1 that specifies the ratio of real to apparent power in an ac circuit. For any ac circuit, the power factor is equal to the cosine of the phase angle.

- **Real Power** - the actual power dissipated as heat in the resistance of an ac circuit. The unit of real power is the watt (W). Real power can be calculated as I^2R where R is the resistive component of the circuit or as $V \times I \times \cos \theta$ where θ is the phase angle of the circuit.

- **Volt-ampere (VA)** - the unit of apparent power.

- **Volt-ampere Reactive (VAR)** - the volt-amperes at the angle of 90°.

- **Wattmeter** - a test instrument used to measure the real power in watts.

Related Formulas

$$Z_T = \sqrt{R^2 + X^2}$$
$$I_T = \sqrt{I_R^2 + I_X^2}$$
Real power $= P = I^2R$
Real power $= P = VI \cos \theta$

Power factor $= PF = \cos \theta = \dfrac{R}{Z}$ (Series circuits)

Power factor $= PF = \cos \theta = \dfrac{I_R}{I_T}$ (Parallel circuits)

Self-Test

Answers at back of book.

1. **In an ac circuit with only series resistances,**
   a. V_T and I are in phase.
   b. $R_T = R_1 + R_2 + R_3 \cdots + $ etc.
   c. each voltage drop is in phase with the series current.
   d. all of the above.

2. **In an ac circuit with only parallel inductors,**
   a. I_T lags V_A by 90°.
   b. V_A lags I_T by 90°.
   c. V_A and I_T are in phase.
   d. none of the above.

3. **A series circuit contains 150 Ω of X_L and 250 Ω of X_C. What is the net reactance?**
   a. 400 Ω, X_L.
   b. 400 Ω, X_C.
   c. 100 Ω, X_C.
   d. 291.5 Ω, X_C.

4. **What is the power factor (PF) of a purely resistive ac circuit?**
   a. 0.
   b. 1.

c. 0.707.
   d. Without values, this is impossible to determine.

5. **The unit of apparent power is the**
   a. volt-ampere (VA).
   b. watt (W).
   c. volt-ampere reactive (VAR).
   d. joule (J).

6. **A 15-Ω resistance is in series with 50 Ω of X_L and 30 Ω of X_C. If the applied voltage equals 50 V, how much real power is dissipated by the circuit?**
   a. 60 W.
   b. 100 W.
   c. 100 VA.
   d. 4.16 W.

7. **An ac circuit has a 100-Ω R, a 300-Ω X_L, and a 200-Ω X_C all in series. What is the phase angle of the circuit?**
   a. 78.7°.
   b. 45°.
   c. −90°.
   d. 56.3°.

8. **A 10-Ω resistor is in parallel with an X_L of 10 Ω. If the applied voltage is 120 V, what is the power factor of the circuit?**
   a. 0.
   b. 0.5.
   c. 1.
   d. 0.707.

9. **An ac circuit has an 80-Ω R, 20-Ω X_L, and a 40-Ω X_C in parallel. If the applied voltage is 24 Vac, what is the phase angle of the circuit?**
   a. −26.6°.
   b. 45°.
   c. −63.4°.
   d. −51.3°.

10. **In an ac circuit with only series capacitors,**
   a. V_T leads I by 90°.
   b. V_T lags I by 90°.
   c. each capacitor voltage drop leads I by 90°.
   d. both a and c.

11. A 10-Ω R is in parallel with a 15-Ω X_L. The applied voltage is 120 Vac. How much is the apparent power in the circuit?
   a. 2.4 kW.
   b. 1.44 kVA.
   c. 1.44 kW.
   d. 1.73 kVA.

12. The unit of real power is the
   a. watt (W).
   b. volt-ampere (VA).
   c. joule (J).
   d. volt-ampere reactive (VAR).

13. In a parallel ac circuit with X_L and X_C,
   a. I_L and I_C are 90° out of phase.
   b. I_L and I_C are in phase.
   c. I_L and I_C are 180° out of phase.
   d. X_L and X_C are 90° out of phase.

14. In a series RLC circuit,
   a. X_L and X_C are 180° out of phase.
   b. I_L and I_C are 180° out of phase.
   c. X_L and X_C are 90° out of phase.
   d. X_L and X_C are in phase.

15. A parallel ac circuit with 120 Vac applied has a total current, I_T, of 5 A. If the phase angle of the circuit is −53.13°, how much real power is dissipated by the circuit?
   a. 600 VA.
   b. 480 W.
   c. 360 W.
   d. 3.6 kVA.

Questions

1. Why can series or parallel resistances be combined in ac circuits the same way as in dc circuits?

2. (a) Why do X_L and X_C reactances in series offset each other? (b) With X_L and X_C reactances in parallel, why can their branch currents be subtracted?

3. Give one difference in electrical characteristics comparing R and X_C, R and Z, X_C and C, X_L and L.

4. Name three types of ac meters.

5. Make a diagram showing a resistance R_1 in series with the load resistance R_L, with a wattmeter connected to measure the power in R_L.

6. Make a phasor diagram for the circuit in Fig. 23–8a showing the phase of the voltage drops IR, IX_C, and IX_L with respect to the reference phase of the common current I.

7. Explain briefly why the two opposite phasors at +90° for X_L and −90° for I_L both follow the principle that

any self-induced voltage leads the current through the coil by 90°.

8. Why is it that a reactance phasor is always at exactly 90° but an impedance phasor can be less than 90°?

9. Why must the impedance of a series circuit be more than either its X or its R?

10. Why must I_T in a parallel circuit be more than either I_R or I_X?

11. Compare real power and apparent power.

12. Define power factor.

13. Make a phasor diagram showing the opposite direction of positive and negative angles.

14. In Fig. 23–15, which circuit has leading current with a positive phase angle θ where I from the source leads the V applied by the source?

Problems

SECTION 23–1 AC CIRCUITS WITH RESISTANCE BUT NO REACTANCE

23–1 In Fig. 23–17, solve for R_T, I, V_1, and V_2.

Figure 23–17

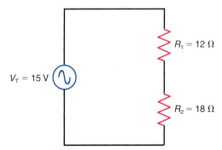

23–2 In Fig. 23–17, what is the phase relationship between
   a. V_T and I?
   b. V_1 and I?
   c. V_2 and I?

23–3 In Fig. 23–18, solve for I_1, I_2, I_T, and R_{EQ}.

23–4 In Fig. 23–18, what is the phase relationship between
   a. V_A and I_1?
   b. V_A and I_2?
   c. V_A and I_T?

Figure 23-18

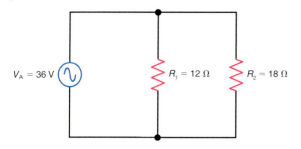

SECTION 23-3 CIRCUITS WITH X_C ALONE

23-9 In Fig. 23-21, solve for X_{CT}, I, V_1, and V_2.

Figure 23-21

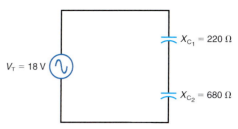

SECTION 23-2 CIRCUITS WITH X_L ALONE

23-5 In Fig. 23-19, solve for X_{L_1}, I, V_1, and V_2.

Figure 23-19

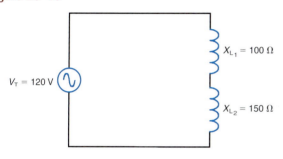

23-6 In Fig. 23-19, what is the phase relationship between
   a. V_T and I?
   b. V_1 and I?
   c. V_2 and I?

23-7 In Fig. 23-20, solve for I_1, I_2, I_T, and $X_{L_{EQ}}$.

Figure 23-20

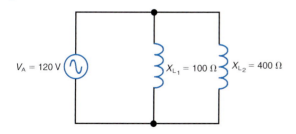

23-8 In Fig. 23-20, what is the phase relationship between
   a. V_A and I_1?
   b. V_A and I_2?
   c. V_A and I_T?

23-10 In Fig. 23-21, what is the phase relationship between
   a. V_T and I?
   b. V_1 and I?
   c. V_2 and I?

23-11 In Fig. 23-22, solve for I_1, I_2, I_T, and $X_{C_{EQ}}$.

Figure 23-22

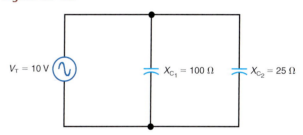

23-12 In Fig. 23-22, what is the phase relationship between
   a. V_A and I_1?
   b. V_A and I_2?
   c. V_A and I_T?

SECTION 23-4 OPPOSITE REACTANCES CANCEL

23-13 In Fig. 23-23, solve for
   a. the net reactance, X.
   b. the current, I.
   c. the inductor voltage, V_L.
   d. the capacitor voltage, V_C.

Figure 23-23

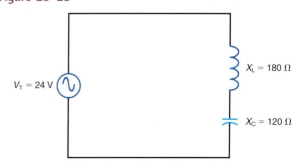

23-14 In Fig. 23–23, what is the phase relationship between
   a. X_L and X_C?
   b. V_L and I?
   c. V_C and I?
   d. V_T and I?
   e. V_L and V_C?
   f. V_T and I if the values of X_L and X_C are interchanged?

23-15 In Fig. 23–24, solve for
   a. the inductive branch current, I_L.
   b. the capacitive branch current, I_C.
   c. the net line current, I_T.
   d. the net reactance, X.

Figure 23–24

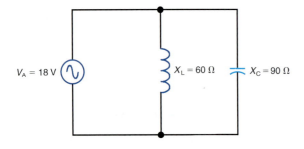

$V_A = 18\ V$ $X_L = 60\ \Omega$ $X_C = 90\ \Omega$

23-16 In Fig. 23–24, what is the phase relationship between
   a. I_L and V_A?
   b. I_C and V_A?
   c. I_L and I_C?
   d. V_A and I_T?
   e. V_A and I_T if the values of X_L and X_C are interchanged?

SECTION 23–5 SERIES REACTANCE AND RESISTANCE

23-17 In Fig. 23–25, solve for
   a. the net reactance, X.
   b. Z_T
   c. I
   d. V_R
   e. V_L
   f. V_C
   g. θ_Z

Figure 23–25

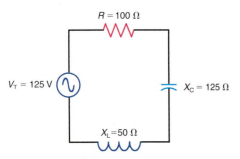

$R = 100\ \Omega$
$V_T = 125\ V$ $X_C = 125\ \Omega$
$X_L = 50\ \Omega$

23-18 In Fig. 23–25, what is the phase relationship between
   a. X_L and X_C?
   b. V_L and I?
   c. V_C and I?
   d. V_R and I?
   e. V_L and V_C?
   f. V_T and I?
   g. V_T and V_R?
   h. V_T and I if the values of X_L and X_C are interchanged?

23-19 Repeat Prob. 23–17 for the circuit in Fig. 23–26.

Figure 23–26

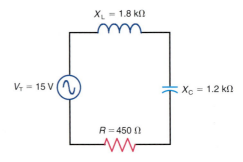

$X_L = 1.8\ k\Omega$
$V_T = 15\ V$ $X_C = 1.2\ k\Omega$
$R = 450\ \Omega$

23-20 In Fig. 23–27, solve for
   a. the net reactance, X.
   b. Z_T.
   c. I.
   d. V_{R_1} and V_{R_2}.
   e. V_{L_1} and V_{L_2}.
   f. V_{C_1} and V_{C_2}.
   g. θ_Z.

Figure 23–27

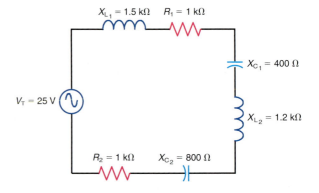

$X_{L_1} = 1.5\ k\Omega$ $R_1 = 1\ k\Omega$
$X_{C_1} = 400\ \Omega$
$V_T = 25\ V$
$X_{L_2} = 1.2\ k\Omega$
$R_2 = 1\ k\Omega$ $X_{C_2} = 800\ \Omega$

SECTION 23–6 PARALLEL REACTANCE AND RESISTANCE

23-21 In Fig. 23–28, solve for
   a. I_R.
   b. I_C.

c. I_L.

d. the net reactive branch current, I_X.

e. I_T.

f. Z_{EQ}.

g. θ_I.

Figure 23–28

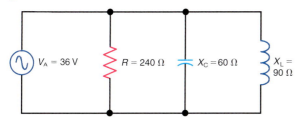

23–22 In Fig. 23–28, what is the phase relationship between

a. V_A and I_R?

b. V_A and I_C?

c. V_A and I_L?

d. I_L and I_C?

e. V_A and I_T?

f. V_A and I_T if the values of X_L and X_C are interchanged?

23–23 Repeat Prob. 23–21 for the circuit in Fig. 23–29.

Figure 23–29

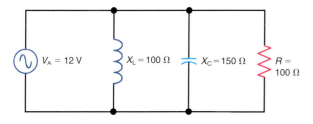

23–24 In Fig. 23–30, solve for

a. I_{R_1} and I_{R_2}.

b. I_{L_1} and I_{L_2}.

c. I_{C_1} and I_{C_2}.

d. the net reactive branch current, I_X.

e. I_T.

f. Z_{EQ}.

g. θ_I.

SECTION 23–7 SERIES-PARALLEL REACTANCE AND RESISTANCE

23–25 In Fig. 23–31, solve for

a. Z_T.

b. I_T.

c. V_{R_1}.

d. V_{C_1}, V_{C_2}, and V_{C_3}.

e. V_{L_1} and V_{L_2}.

f. θ_Z.

Figure 23–31

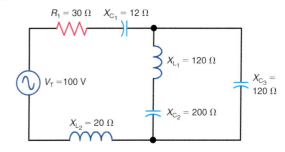

SECTION 23–8 REAL POWER

23–26 Determine the real power, apparent power, and power factor (PF) for each of the following circuits:

a. Fig. 23–17

b. Fig. 23–19

c. Fig. 23–22

d. Fig. 23–24

23–27 Determine the real power, apparent power, and power factor (PF) for each of the following circuits:

a. Fig. 23–25

b. Fig. 23–26

c. Fig. 23–28

d. Fig. 23–29

23–28 Calculate the real power, apparent power, and power factor for each of the following circuit conditions:

a. A parallel RLC circuit with $V_A = 120$ V, $I_T = 5$ A, and $\theta_I = -45°$.

b. A parallel RLC circuit with $V_A = 240$ V, $I_T = 18$ A, and $\theta_I = -26.56°$.

c. A parallel RLC circuit with $V_A = 100$ V, $I_T = 3$ A, and $\theta_I = 78°$.

d. A parallel RLC circuit with $V_A = 120$ V, $I_T = 8$ A, and $\theta_I = 56°$.

Figure 23–30

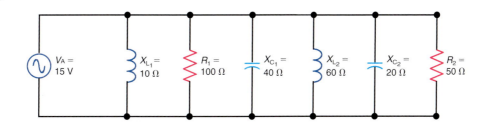

Critical Thinking

23-29 In Fig. 23-32, what value of L will produce a circuit power factor of 0.8?

23-30 In Fig. 23-33, what value of C in parallel with R and L will produce a power factor of 0.8?

Figure 23-32 Circuit for Critical Thinking Prob. 23-29.

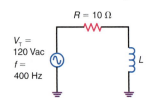

Figure 23-33 Circuit for Critical Thinking Prob. 23-30.

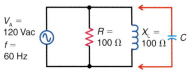

Answers to Knowledge Check Problems

23-1 No

23-2 No, because the internal series resistance of a practical inductor causes V_L to lead I by less than 90°.

23-3 $I = 0$ A, $V_1 = 50$ V, and $V_2 = 50$ V

23-4 $X_C = 120$ Ω

23-5 20 V

23-6 $I_T = 5$ A

23-7 $Z_T = 180.3$ Ω

23-8 1.375 kW

23-9 The rectifier type

23-10 Yes

23-11 $R = 707$ Ω and $X = X_L = 707$ Ω

23-12 0°

Answers to Self-Reviews

23-1 a. 0°
b. 0°

23-2 a. 90°
b. −90°

23-3 a. −90°
b. 90°

23-4 a. 20 Ω
b. 1 A

23-5 a. $X_C = 30$ Ω
b. $X_C = 30$ Ω
c. 180°

23-6 a. $I_L = 3$ A
b. $I_L = 3$ A

23-7 a. 200 Ω
b. 200 Ω
c. 100 Ω

23-8 a. watt
b. volt-ampere
c. real

23-9 a. T
b. T

23-10 a. real power
b. V

23-11 a. R
b. X_L
c. X_C

23-12 a. V_L and V_C
b. I_R and I_L

chapter 24

Complex Numbers for AC Circuits

Complex numbers refer to a numerical system that includes the phase angle of a quantity with its magnitude. Therefore, complex numbers are useful in ac circuits when the reactance of X_L or X_C makes it necessary to consider the phase angle. For instance, complex notation explains why θ_Z is negative with X_C and θ_I is negative with I_L.

Any type of ac circuit can be analyzed with complex numbers. They are especially convenient for solving series-parallel circuits that have both resistance and reactance in one or more branches. Although graphical analysis with phasor arrows can be used, the method of complex numbers is probably the best way to analyze ac circuits with series-parallel impedances.

Objectives

After studying this chapter you should be able to

- *Explain* the *j* operator.
- *Define* a complex number.
- *Add, subtract, multiply,* and *divide* complex numbers.
- *Explain* the difference between the rectangular and polar forms of a complex number.
- *Convert* a complex number from polar to rectangular form and vice versa.
- *Explain* how to use complex numbers to solve series and parallel ac circuits containing resistance, capacitance, and inductance.

Outline

Important Terms

admittance, *Y*

complex number

imaginary number

j operator

polar form

real number

rectangular form

susceptance, *B*

24–1 Positive and Negative Numbers

Our common use of numbers as either positive or negative represents only two special cases. In their more general form, numbers have both quantity and phase angle. In Fig. 24–1, positive and negative numbers are shown corresponding to the phase angles of 0° and 180°, respectively.

For example, the numbers 2, 4, and 6 represent units along the horizontal or *x* axis, extending toward the right along the line of zero phase angle. Therefore, positive numbers represent units having the phase angle of 0°, or this phase angle corresponds to the factor of +1. To indicate 6 units with zero phase angle, then, 6 is multiplied by +1 as a factor for the positive number 6. The + sign is often omitted, as it is assumed unless indicated otherwise.

In the opposite direction, negative numbers correspond to 180°, or this phase angle corresponds to the factor of −1. Actually, −6 represents the same quantity as 6 but rotated through the phase angle of 180°. The angle of rotation is the *operator* for the number. The operator for −1 is 180°; the operator for +1 is 0°.

■ 24–1 Knowledge Check

Answers at end of chapter.

What phase angle corresponds to the factor
a.   +1?
b.   −1?

■ 24–1 Self-Review

Answers at end of chapter.

a.   **What is the angle for the number +5?**
b.   **What is the angle for the number −5?**

24–2 The *j* Operator

The operator for a number can be any angle between 0° and 360°. Since the angle of 90° is important in ac circuits, the factor *j* is used to indicate 90°. See Fig. 24–2. Here, the number 5 means 5 units at 0°, the number −5 is at 180°, and *j*5 indicates the number 5 at the 90° angle.

Figure 24–1   Positive and negative numbers.

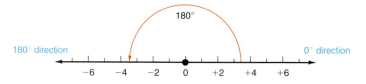

Figure 24–2   The *j* axis at 90° from the horizontal real axis.

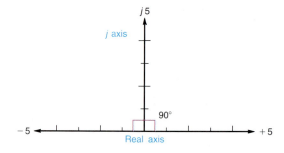

The j is usually written before the number. The reason is that the j sign is a 90° operator, just as the + sign is a 0° operator and the − sign is a 180° operator. Any quantity at right angles to the zero axis, or 90° counterclockwise, is on the +j axis.

In mathematics, numbers on the horizontal axis are *real numbers*, including positive and negative values. Numbers on the j axis are called *imaginary numbers* because they are not on the real axis. In mathematics, the abbreviation i is used to indicate imaginary numbers. In electricity, however, j is used to avoid confusion with i as the symbol for current. Furthermore, there is nothing imaginary about electrical quantities on the j axis. An electric shock from j500 V is just as dangerous as 500 V positive or negative.

More features of the j operator are shown in Fig. 24–3. The angle of 180° corresponds to the j operation of 90° repeated twice. This angular rotation is indicated by the factor j^2. Note that the j operation multiplies itself, instead of adding.

Since j^2 means 180°, which corresponds to the factor of −1, we can say that j^2 is the same as −1. In short, the operator j^2 for a number means to multiply by −1. For instance, $j^2 8$ is −8.

Furthermore, the angle of 270° is the same as −90°, which corresponds to the operator −j. These characteristics of the j operator are summarized as follows:

$$0° = 1$$
$$90° = j$$
$$180° = j^2 = -1$$
$$270° = j^3 = j^2 \times j = -1 \times j = -j$$
$$360° = \text{same as } 0°$$

As examples, the number 4 or −4 represents 4 units on the real horizontal axis; j4 means 4 units with a leading phase angle of 90°; −j4 means 4 units with a lagging phase angle of −90°.

Figure 24–3 The j operator indicates 90° rotation from the real axis; the −j operator is −90°; j^2 operation is 180° rotation back to the real axis in a negative direction.

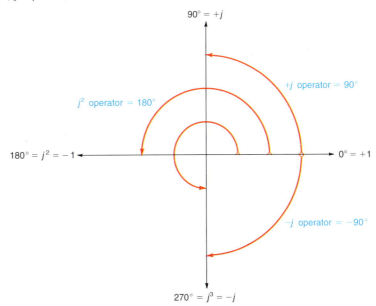

■ *24-2 Knowledge Check*

Answer at end of chapter.

$j^2 \times (-1) = ?$

■ *24-2 Self-Review*

Answers at end of chapter.

a. What is the angle for the operator j?
b. What is the angle for the operator $-j$?

GOOD TO KNOW

When phasors are added, the tail of one phasor is placed at the arrowhead of the other. The resultant phasor extends from the tail of the first phasor to the arrowhead of the second.

24–3 Definition of a Complex Number

The combination of a real and an imaginary term is called a *complex number*. Usually, the real number is written first. As an example, $3 + j4$ is a complex number including 3 units on the real axis added to 4 units 90° out of phase on the j axis. Complex numbers must be added as phasors.

Phasors for complex numbers shown in Fig. 24–4 are typical examples. The $+j$ phasor is up for 90°; the $-j$ phasor is down for −90°. The phasors are shown with the end of one joined to the start of the next, to indicate addition. Graphically, the sum is the hypotenuse of the right triangle formed by the two phasors. Since a number like $3 + j4$ specifies the phasors in rectangular coordinates, this system is the *rectangular form* of complex numbers.

Be careful to distinguish a number like $j2$, where 2 is a coefficient, from j^2, where 2 is the exponent. The number $j2$ means 2 units up on the j axis of 90°. However, j^2 is the operator of −1, which is on the real axis in the negative direction.

Another comparison to note is between $j3$ and j^3. The number $j3$ is 3 units up on the j axis, and j^3 is the same as the $-j$ operator, which is down on the −90° axis.

Also note that either the real term or the j term can be the larger of the two. When the j term is larger, the angle is more than 45°; when the j term is smaller, the angle is less than 45°. If the j term and the real term are equal, the angle is 45°.

■ *24-3 Knowledge Check*

Answer at end of chapter.

When a complex number is expressed in rectangular form, which term is written first? What is the second term?

Figure 24–4 Phasors corresponding to real terms and imaginary (j) terms, in rectangular coordinates.

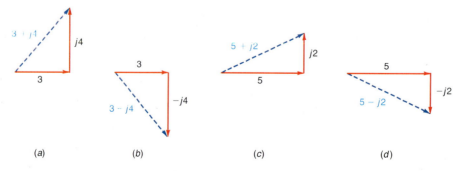

(a)   (b)   (c)   (d)

Answers at end of chapter.

a. For 7 + *j*6, the 6 is at 90° leading the 7. (True or False)
b. For 7 − *j*6, the 6 is at −90° lagging the 7. (True or False)

24–4 How Complex Numbers Are Applied to AC Circuits

Applications of complex numbers are a question of using a real term for 0°, +*j* for 90°, and −*j* for −90°, to denote phase angles. Figure 24–5 illustrates the following rules:

An *angle of* 0° or a real number without any *j* operator is used for resistance *R*. For instance, 3 Ω of *R* is stated as 3 Ω.

An *angle of* 90° or +*j* is used for inductive reactance X_L. For instance, a 4-Ω X_L is *j*4 Ω. This rule always applies to X_L, whether it is in series or parallel with *R*. The reason is the fact that IX_L represents voltage across an inductance, which always leads the current in the inductance by 90°. The +*j* is also used for V_L.

An *angle of* −90° or −*j* is used for X_C. For instance, a 4-Ω X_C is −*j*4 Ω. This rule always applies to X_C, whether it is in series or parallel with *R*. The reason is that IX_C is the voltage across a capacitor, which always lags the capacitor's charge and discharge current by −90°. The −*j* is also used for V_C.

With reactive branch currents, the sign for *j* is reversed, compared with reactive ohms, because of the opposite phase angle. In Fig. 24–6*a* and *b*, −*j*

Figure 24–5 Rectangular form of complex numbers for impedances. (*a*) Reactance X_L is +*j*. (*b*) Reactance X_C is −*j*.

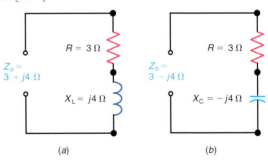

(*a*)   (*b*)

Figure 24–6 Rectangular form of complex numbers for branch currents. (*a*) Current I_L is −*j*. (*b*) Current I_C is +*j*.

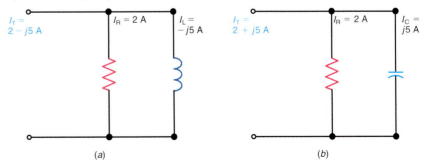

(*a*)   (*b*)

is used for inductive branch current I_L, and $+j$ is used for capacitive branch current I_C.

■ 24-4 Knowledge Check

Answer at end of chapter.

Write 10 mA of inductive current, I_L, using the j operator.

■ 24-4 Self-Review

Answers at end of chapter.

a. Write 3 kΩ of X_L using the j operator.
b. Write 5 mA of I_L using the j operator.

24-5 Impedance in Complex Form

The rectangular form of complex numbers is a convenient way to state the impedance of series resistance and reactance. In Fig. 24–5a, the impedance is $3 + j4$ because Z_a is the phasor sum of a 3-Ω R in series with $j4$ Ω for X_L. Similarly, Z_b is $3 - j4$ for a 3-Ω R in series with $-j4$ Ω for X_C. The minus sign in Z_b results from adding the negative term for $-j$, that is, $3 + (-j4) = 3 - j4$.

For a 4-kΩ R and a 2-kΩ X_L in series: $Z_T = 4000 + j2000$ Ω
For a 3-kΩ R and a 9-kΩ X_C in series: $Z_T = 3000 - j9000$ Ω
For $R = 0$ and a 7-Ω X_L in series: $Z_T = 0 + j7$ Ω
For a 12-Ω R and $X = 0$ in series: $Z_T = 12 + j0$

Note the general form of stating $Z = R \pm jX$. If one term is zero, substitute 0 for this term to keep Z in its general form. This procedure is not required, but there is usually less confusion when the same form is used for all types of Z.

The advantage of this method is that multiple impedances written as complex numbers can then be calculated as follows:

For series impedances:

$$Z_T = Z_1 + Z_2 + Z_3 + \cdots + \text{etc.} \tag{24-1}$$

For parallel impedances:

$$\frac{1}{Z_T} = \frac{1}{Z_1} + \frac{1}{Z_2} + \frac{1}{Z_3} + \cdots + \text{etc.} \tag{24-2}$$

For two parallel impedances:

$$Z_T = \frac{Z_1 \times Z_2}{Z_1 + Z_2} \tag{24-3}$$

Examples are shown in Fig. 24–7. The circuit in Fig. 24–7a is a series combination of resistances and reactances. Combining the real terms and j terms separately, $Z_T = 12 + j4$. The calculations are $3 + 9 = 12$ Ω for R and $j6$ added to $-j2$ equals $j4$ for the net X.

The parallel circuit in Fig. 24–7b shows that X_L is $+j$ and X_C is $-j$, even though they are in parallel branches, because they are reactances, not currents.

So far, these types of circuits can be analyzed with or without complex numbers. For the series-parallel circuit in Fig. 24–7c, however, the notation of complex numbers is necessary to state the complex impedance Z_T, consisting of branches with reactance and resistance in one or more of the branches. Impedance Z_T is stated here in its form as a complex impedance. To calculate

Figure 24–7 Reactance X_L is a $+j$ term and X_C is a $-j$ term whether in series or parallel. (a) Series circuit. (b) Parallel branches. (c) Complex branch impedances Z_1 and Z_2 in parallel.

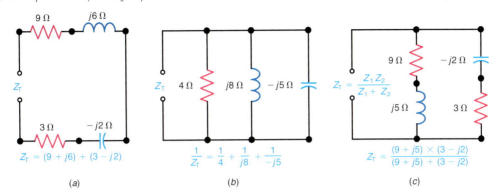

(a) (b) (c)

Z_T, some of the rules described in the next section must be used for combining complex numbers.

■ 24–5 Knowledge Check

Answer at end of chapter.

A 10-Ω R is in series with 20 Ω of X_L and 30 Ω of X_C. Express the total impedance, Z_T, in rectangular form.

■ 24–5 Self-Review

Answers at end of chapter.

Write the following impedances in complex form:
a. X_L of 7 Ω in series with R of 4 Ω.
b. X_C of 7 Ω in series with zero R.

24–6 Operations with Complex Numbers

Real numbers and j terms cannot be combined directly because they are 90° out of phase. The following rules apply:

For Addition or Subtraction

Add or subtract the real and j terms separately:

$$(9 + j5) + (3 + j2) = 9 + 3 + j5 + j2$$

$$= 12 + j7$$

$$(9 + j5) + (3 - j2) = 9 + 3 + j5 - j2$$

$$= 12 + j3$$

$$(9 + j5) + (3 - j8) = 9 + 3 + j5 - j8$$

$$= 12 - j3$$

The answer should be in the form of $R \pm jX$, where R is the algebraic sum of all the real or resistive terms and X is the algebraic sum of all the imaginary or reactive terms.

To Multiply or Divide a *j* Term by a Real Number

Multiply or divide the numbers. The answer is still a *j* term. Note the algebraic signs in the following examples. If both factors have the same sign, either + or −, the answer is +; if one factor is negative, the answer is negative.

$$4 \times j3 = j12 \qquad\qquad j12 \div 4 = j3$$
$$j5 \times 6 = j30 \qquad\qquad j30 \div 6 = j5$$
$$j5 \times (-6) = -j30 \qquad\qquad -j30 \div (-6) = j5$$
$$-j5 \times 6 = -j30 \qquad\qquad -j30 \div 6 = -j5$$
$$-j5 \times (-6) = j30 \qquad\qquad j30 \div (-6) = -j5$$

To Multiply or Divide a Real Number by a Real Number

Just multiply or divide the real numbers, as in arithmetic. There is no *j* operation. The answer is still a real number.

To Multiply a *j* Term by a *j* Term

Multiply the numbers and the *j* coefficients to produce a j^2 term. The answer is a real term because j^2 is −1, which is on the real axis. Multiplying two *j* terms shifts the number 90° from the *j* axis to the real axis of 180°. As examples,

$$j4 \times j3 = j^2 12 = (-1)(12)$$
$$= -12$$
$$j4 \times (-j3) = -j^2 12 = -(-1)(12)$$
$$= 12$$

To Divide a *j* Term by a *j* Term

Divide the *j* coefficients to produce a real number; the *j* factors cancel. For instance:

$$j12 \div j4 = 3 \qquad\qquad -j12 \div j4 = -3$$
$$j30 \div j5 = 6 \qquad\qquad j30 \div (-j6) = -5$$
$$j15 \div j3 = 5 \qquad\qquad -j15 \div (-j3) = 5$$

To Multiply Complex Numbers

Follow the rules of algebra for multiplying two factors, each having two terms:

$$(9 + j5) \times (3 - j2) = 27 - j18 + j15 - j^2 10$$
$$= 27 - j3 - (-1)10$$
$$= 27 - j3 + 10$$
$$= 37 - j3$$

Note that $-j^2 10$ equals +10 because the operator j^2 is −1 and −(−1)10 becomes +10.

To Divide Complex Numbers

This process becomes more involved because division of a real number by an imaginary number is not possible. Therefore, the denominator must first be converted to a real number without any *j* term.

Converting the denominator to a real number without any *j* term is called *rationalization* of the fraction. To do this, multiply both numerator and denominator by the *conjugate* of the denominator. Conjugate complex numbers have equal terms but opposite signs for the *j* term. For instance, $(1 + j2)$ has the conjugate $(1 - j2)$.

Rationalization is permissible because the value of a fraction is not changed when both numerator and denominator are multiplied by the same factor. This procedure is the same as multiplying by 1. In the following example of division with rationalization, the denominator $(1 + j2)$ has the conjugate $(1 - j2)$:

$$\frac{4 - j1}{1 + j2} = \frac{4 - j1}{1 + j2} \times \frac{(1 - j2)}{(1 - j2)} = \frac{4 - j8 - j1 + j^2 2}{1 - j2 + j2 - j^2 4}$$

$$= \frac{4 - j9 - 2}{1 + 4}$$

$$= \frac{2 - j9}{5}$$

$$= 0.4 - j1.8$$

As a result of the rationalization, $4 - j1$ has been divided by $1 + j2$ to find the quotient that is equal to $0.4 - j1.8$.

Note that the product of a complex number and its conjugate always equals the sum of the squares of the numbers in each term. As another example, the product of $(2 + j3)$ and its conjugate $(2 - j3)$ must be $4 + 9$, which equals 13. Simple numerical examples of division and multiplication are given here because when the required calculations become too long, it is easier to divide and multiply complex numbers in polar form, as explained soon in Sec. 24–8.

■ *24–6 Knowledge Check*

Answer at end of chapter.

$$\frac{10 + j10}{5 - j5} = ?$$

■ *24–6 Self-Review*

Answers at end of chapter.

a. $(2 + j3) + (3 + j4) = ?$
b. $(2 + j3) \times 2 = ?$

GOOD TO KNOW

Learn how to use the polar-to-rectangular and rectangular-to-polar conversion keys on your calculator.

24–7 Magnitude and Angle of a Complex Number

In electrical terms the complex impedance $(4 + j3)$ means $4 \, \Omega$ of resistance and $3 \, \Omega$ of inductive reactance with a leading phase angle of $90°$. See Fig. 24–8a. The magnitude of Z is the resultant, equal to $\sqrt{16 + 9} = \sqrt{25} = 5 \, \Omega$. Finding the square root of the sum of the squares is vector or phasor addition of two terms in quadrature, $90°$ out of phase.

The phase angle of the resultant is the angle whose tangent is 0.75. This angle equals $37°$. Therefore, $4 + j3 = 5 \; \underline{/37°}$.

When calculating the tangent ratio, note that the j term is the numerator and the real term is the denominator because the tangent of an angle is the ratio of the opposite side to the adjacent side. For a negative j term, the tangent is negative, which means a negative angle.

Note the following definitions: $(4 + j3)$ is the complex number in rectangular coordinates. The real term is 4. The imaginary term is $j3$. The resultant 5 is the magnitude, absolute value, or modulus of the complex number. Its phase angle or argument is $37°$. The resultant value by itself can be written as $|5|$; the vertical lines indicate that it is the magnitude without the phase angle.

Figure 24–8 Magnitude and angle of a complex number. (a) Rectangular form. (b) Polar form.

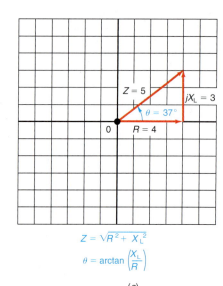

$$Z = \sqrt{R^2 + X_L^2}$$

$$\theta = \arctan\left(\frac{X_L}{R}\right)$$

(a)

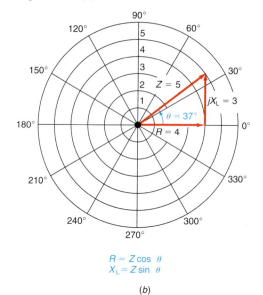

$$R = Z \cos \theta$$
$$X_L = Z \sin \theta$$

(b)

The magnitude is the value a meter would read. For instance, with a current of 5 $\underline{/37°}$ A in a circuit, an ammeter reads 5 A. As additional examples,

$$2 + j4 = \sqrt{4 + 16} \; \underline{/\arctan 2} = 4.47 \; \underline{/63°}$$

$$4 + j2 = \sqrt{16 + 4} \; \underline{/\arctan 0.5} = 4.47 \; \underline{/26.5°}$$

$$8 + j6 = \sqrt{64 + 36} \; \underline{/\arctan 0.75} = 10 \; \underline{/37°}$$

$$8 - j6 = \sqrt{64 + 36} \; \underline{/\arctan -0.75} = 10 \; \underline{/-37°}$$

$$4 + j4 = \sqrt{16 + 16} \; \underline{/\arctan 1} = 5.66 \; \underline{/45°}$$

$$4 - j4 = \sqrt{16 + 16} \; \underline{/\arctan -1} = 5.66 \; \underline{/-45°}$$

Note that arctan 0.75 in the third example means the angle with a tangent equal to 0.75. This value is $^6/_8$ or $^3/_4$ for the ratio of the opposite side to the adjacent side. The arctan can also be indicated as $\tan^{-1} 0.75$. In either case, this angle has 0.75 for its tangent, which makes the angle 36.87°.

   Many scientific calculators have keys that can convert from rectangular coordinates to the magnitude-phase angle form (called *polar coordinates*) directly. See your calculator manual for the particular steps used. If your calculator does not have these keys, the problem can be done in two separate parts: (1) the magnitude as the square root of the sum of two squares and (2) the angle as the arctan equal to the *j* term divided by the real term.

■ **24–7 Knowledge Check**

Answer at end of chapter.

Find the resultant magnitude and phase angle for the complex number 20 + *j*15.

■ **24–7 Self-Review**

Answers at end of chapter.

For the complex impedance 10 + *j*10 Ω,
a. calculate the magnitude.
b. calculate the phase angle.

24–8 Polar Form of Complex Numbers

Calculating the magnitude and phase angle of a complex number is actually converting to an angular form in polar coordinates. As shown in Fig. 24–8, the rectangular form $4 + j3$ is equal to $5\ \underline{/37°}$ in polar form. In polar coordinates, the distance from the center is the magnitude of the phasor Z. Its phase angle θ is counterclockwise from the 0° axis.

To convert any complex number to polar form,

1. find the magnitude by phasor addition of the j term and real term.
2. find the angle whose tangent is the j term divided by the real term. As examples,

$$2 + j4 = 4.47\ \underline{/63°}$$
$$4 + j2 = 4.47\ \underline{/26.5°}$$
$$8 + j6 = 10\ \underline{/37°}$$
$$8 - j6 = 10\ \underline{/-37°}$$
$$4 + j4 = 5.66\ \underline{/45°}$$
$$4 - j4 = 5.66\ \underline{/-45°}$$

These examples are the same as those given before for finding the magnitude and phase angle of a complex number.

The magnitude in polar form must be more than either term in rectangular form, but less than their arithmetic sum. For instance, in $8 + j6 = 10\ \underline{/37°}$ the magnitude of 10 is more than 8 or 6 but less than their sum of 14.

Applied to ac circuits with resistance for the real term and reactance for the j term, then, the polar form of a complex number states the resultant impedance and its phase angle. Note the following cases for an impedance where either the resistance or the reactance is zero:

$$0 + j5 = 5\ \underline{/90°}$$
$$0 - j5 = 5\ \underline{/-90°}$$
$$5 + j0 = 5\ \underline{/0°}$$

The polar form is much more convenient for multiplying or dividing complex numbers. The reason is that multiplication in polar form merely involves multiplying the magnitudes and adding the angles. Division involves dividing the magnitudes and subtracting the angles. The following rules apply.

For Multiplication

Multiply the magnitudes but add the angles algebraically:

$$24\ \underline{/40°} \times 2\ \underline{/30°} = 24 \times 2\ \underline{/40° + 30°} = 48\ \underline{/+70°}$$
$$24\ \underline{/40°} \times (-2\ \underline{/30°}) = -48\ \underline{/+70°}$$
$$12\ \underline{/-20°} \times 3\ \underline{/-50°} = 36\ \underline{/-70°}$$
$$12\ \underline{/-20°} \times 4\ \underline{/5°} = 48\ \underline{/-15°}$$

When you multiply by a real number, just multiply the magnitudes:

$$4 \times 2\ \underline{/30°} = 8\ \underline{/30°}$$
$$4 \times 2\ \underline{/-30°} = 8\ \underline{/-30°}$$
$$-4 \times 2\ \underline{/30°} = -8\ \underline{/30°}$$
$$-4 \times (-2\ \underline{/30°}) = 8\ \underline{/30°}$$

This rule follows from the fact that a real number has an angle of 0°. When you add 0° to any angle, the sum equals the same angle.

For Division

Divide the magnitudes and subtract the angles algebraically:

$$24 \underline{/40°} \div 2 \underline{/30°} = 24 \div 2 \underline{/40° - 30°} = 12 \underline{/10°}$$

$$12 \underline{/20°} \div 3 \underline{/50°} = 4 \underline{/-30°}$$

$$12 \underline{/-20°} \div 4 \underline{/50°} = 3 \underline{/-70°}$$

To divide by a real number, just divide the magnitudes:

$$12 \underline{/30°} \div 2 = 6 \underline{/30°}$$

$$12 \underline{/-30°} \div 2 = 6 \underline{/-30°}$$

This rule is also a special case that follows from the fact that a real number has a phase angle of 0°. When you subtract 0° from any angle, the remainder equals the same angle.

For the opposite case, however, when you divide a real number by a complex number, the angle of the denominator changes its sign in the answer in the numerator. This rule still follows the procedure of subtracting angles for division, since a real number has a phase angle of 0°. As examples,

$$\frac{10}{5 \underline{/30°}} = \frac{10 \underline{/0°}}{5 \underline{/30°}} = 2 \underline{/0° - 30°}$$

$$= 2 \underline{/-30°}$$

$$\frac{10}{5 \underline{/-30°}} = \frac{10 \underline{/0°}}{5 \underline{/-30°}} = 2 \underline{/0° - (-30°)}$$

$$= 2 \underline{/+30°}$$

Stated another way, we can say that the reciprocal of an angle is the same angle but with opposite sign. Note that this operation is similar to working with powers of 10. Angles and powers of 10 follow the general rules of exponents.

■ 24-8 Knowledge Check

Answer at end of chapter.

Is it quicker to divide complex numbers in rectangular form or polar form?

■ 24-8 Self-Review

Answers at end of chapter.

a. $6 \underline{/20°} \times 2 \underline{/30°} = ?$

b. $6 \underline{/20°} \div 2 \underline{/30°} = ?$

24-9 Converting Polar to Rectangular Form

Complex numbers in polar form are convenient for multiplication and division, but they cannot be added or subtracted if their angles are different because the real and imaginary parts that make up the magnitude are different. When complex numbers in polar form are to be added or subtracted, therefore, they must be converted into rectangular form.

Consider the impedance $Z \underline{/\theta}$ in polar form. Its value is the hypotenuse of a right triangle with sides formed by the real term and j term in rectangular coordinates. See Fig. 24–9. Therefore, the polar form can be converted to

Conversion to rectangular form
can be done fast with a calculator.
Again, some scientific calculators
contain conversion keys that make
going from polar coordinates to
rectangular coordinates a simple
four-key procedure. Check your
calculator manual for the exact
procedure. If your calculator does
not have this capability, use the
following routine. Punch in the
value of the angle θ in degrees.
Make sure that the correct sign is
used and the calculator is set to
handle angles in degrees. Find
cos θ or sin θ and multiply by
the magnitude for each term.
Remember to use cos θ for the real
term and sin θ for the j term. For
the example of 100 /30°, punch in
the number 30 and press the (COS)
key for 0.866 as cos θ. While it is
on the display, press the ⊗ key,
punch in 100, and press the ⊜ key
for the answer of 86.6 as the real
term. Clear the display for the
next operation with sin θ. Punch
in 30, push the (SIN) key for 0.5 as
sin θ, press the ⊗ key, punch in
100, and push the ⊜ key for the
answer of 50 as the j term.

Figure 24–9   Converting the polar form of $Z\,/\theta$ to the rectangular form of $R \pm jX$. (*a*) The positive angle θ in the first quadrant has a $+j$ term. (*b*) The negative angle $-\theta$ in the fourth quadrant has a $-j$ term.

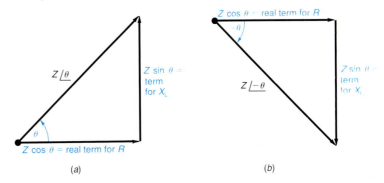

(a)   (b)

rectangular form by finding the horizontal and vertical sides of the right triangle. Specifically,

$$\text{Real term for } R = Z \cos \theta$$
$$j \text{ term for } X = Z \sin \theta$$

In Fig. 24–9*a*, assume that $Z\,/\theta$ in polar form is $5\,/37°$. The sine of 37° is 0.6 and its cosine is 0.8.

To convert to rectangular form,

$$R = Z \cos \theta = 5 \times 0.8 = 4$$
$$X = Z \sin \theta = 5 \times 0.6 = 3$$

Therefore,

$$5\,/37° = 4 + j3$$

This example is the same as the illustration in Fig. 24–8. The $+$ sign for the j term means that it is X_L, not X_C.

In Fig. 24–9*b*, the values are the same, but the j term is negative when θ is negative. The negative angle has a negative j term because the opposite side is in the fourth quadrant, where the sine is negative. However, the real term is still positive because the cosine is positive.

Note that R for cos θ is the horizontal component, which is an adjacent side of the angle. The X for sin θ is the vertical component, which is opposite the angle. The $+X$ is X_L; the $-X$ is X_C.

These rules apply for angles in the first or fourth quadrant, from 0 to 90° or from 0 to −90°. As examples,

$$14.14\,/45° = 14.14 \cos 45° + 14.14 \sin 45° = 10 + j10$$
$$14.14\,/-45° = 14.14 \cos (-45°) + 14.14 \sin (-45°) = 10 + j(-10)$$
$$= 10 - j10$$
$$10\,/90° = 0 + j10$$
$$10\,/-90° = 0 - j10$$
$$100\,/30° = 86.6 + j50$$
$$100\,/-30° = 86.6 - j50$$
$$100\,/60° = 50 + j86.6$$
$$100\,/-60° = 50 - j86.6$$

When going from one form to the other, keep in mind whether the angle is smaller or greater than 45° and whether the j term is smaller or larger than the real term. For angles between 0 and 45°, the opposite side, which is the *j* term, must be smaller than the real term. For angles between 45 and 90°, the *j* term must be larger than the real term.

To summarize how complex numbers are used in ac circuits in rectangular and polar form:

1. For addition or subtraction, complex numbers must be in rectangular form. This procedure applies to the addition of impedances in a series circuit. If the series impedances are in rectangular form, combine all the real terms and the *j* terms separately. If the series impedances are in polar form, they must be converted to rectangular form to be added.
2. For multiplication and division, complex numbers are generally used in polar form because the calculations are faster. If the complex number is in rectangular form, convert to polar form. With the complex number available in both forms, you can quickly add or subtract in rectangular form and multiply or divide in polar form. Sample problems showing how to apply these methods in ac circuits are given in the following sections.

■ *24–9 Knowledge Check*

Answer at end of chapter.

Can complex numbers in polar form be added or subtracted?

■ *24–9 Self-Review*

Answers at end of chapter.

Convert to rectangular form.
a. $14.14 \, \underline{/45°}$.
b. $14.14 \, \underline{/-45°}$.

24–10 Complex Numbers in Series AC Circuits

Refer to Fig. 24–10. Although a circuit like this with only series resistances and reactances can be solved graphically with phasor arrows, the complex numbers show more details of the phase angles.

Z_T in Rectangular Form

The total Z_T in Fig. 24–10*a* is the sum of the impedances:

$$Z_T = 2 + j4 + 4 - j12$$
$$= 6 - j8$$

The total series impedance then is $6 - j8$. Actually, this amounts to adding all series resistances for the real term and finding the algebraic sum of all series reactances for the *j* term.

Z_T in Polar Form

We can convert Z_T from rectangular to polar form as follows:

$$Z_T = 6 - j8$$
$$= \sqrt{36 + 64} \, \underline{/\arctan -8/6}$$
$$= \sqrt{100} \, \underline{/\arctan -1.33}$$
$$= 10 \, \underline{/-53°} \; \Omega$$

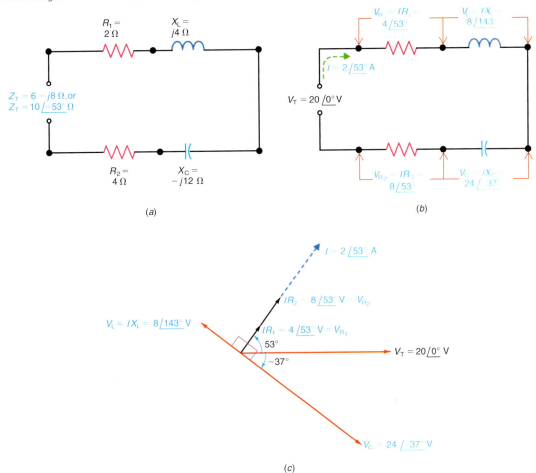

The angle of $-53°$ for Z_T means that the applied voltage and the current are $53°$ out of phase. Specifically, this angle is θ_Z.

Calculating I

The reason for the polar form is to divide the applied voltage V_T by Z_T to calculate the current I. See Fig. 24–10b. Note that the V_T of 20 V is a real number without any j term. Therefore, the applied voltage is $20 \angle 0°$. This angle of $0°$ for V_T makes it the reference phase for the following calculations. We can find the current as

$$I = \frac{V_T}{Z_T} = \frac{20 \angle 0°}{10 \angle -53°} = 2 \angle 0° - (-53°)$$
$$= 2 \angle 53° \text{ A}$$

Note that Z_T has a negative angle of $-53°$ but the sign changes to $+53°$ for I because of the division into a quantity with the angle of $0°$. In general, the reciprocal of an angle in polar form is the same angle with opposite sign.

Phase Angle of the Circuit

The fact that I has an angle of $+53°$ means that it leads V_T. The positive angle for I shows that the series circuit is capacitive with leading current. This angle is more than 45° because the net reactance is more than the total resistance, resulting in a tangent greater than 1.

Finding Each Voltage Drop

To calculate the voltage drops around the circuit, each resistance or reactance can be multiplied by I:

$$V_{R_1} = IR_1 = 2\underline{/53°} \times 2\underline{/0°} = 4\underline{/53°} \text{ V}$$
$$V_L = IX_L = 2\underline{/53°} \times 4\underline{/90°} = 8\underline{/143°} \text{ V}$$
$$V_C = IX_C = 2\underline{/53°} \times 12\underline{/-90°} = 24\underline{/-37°} \text{ V}$$
$$V_{R_2} = IR_2 = 2\underline{/53°} \times 4\underline{/0°} = 8\underline{/53°} \text{ V}$$

Phase Angle of Each Voltage

The phasors for these voltages are in Fig. 24–10c. They show the phase angles using the applied voltage V_T as the zero reference phase.

The angle of 53° for V_{R_1} and V_{R_2} shows that the voltage across a resistance has the same phase as I. These voltages lead V_T by 53° because of the leading current.

The angle of −37° for V_C means that it lags the generator voltage V_T by this much. However, this voltage across X_C still lags the current by 90°, which is the difference between 53° and −37°.

The angle of 143° for V_L in the second quadrant is still 90°, leading the current at 53°, because 143° − 53° = 90°. With respect to the generator voltage V_T, though, the phase angle of V_L is 143°.

Total Voltage V_T Equals the Phasor Sum of the Series Voltage Drops

If we want to add the voltage drops around the circuit and find out whether they equal the applied voltage, each V must be converted to rectangular form. Then these values can be added. In rectangular form, then, the individual voltages are

$$V_{R_1} = 4\underline{/53°} = 2.408 + j3.196 \text{ V}$$
$$V_L = 8\underline{/143°} = -6.392 + j4.816 \text{ V}$$
$$V_C = 24\underline{/-37°} = 19.176 - j14.448 \text{ V}$$
$$V_{R_2} = 8\underline{/53°} = \underline{4.816 + j6.392} \text{ V}$$
$$\text{Total } V = 20.008 - j0.044 \text{ V}$$

or converting to polar form,

$$V_T = 20 \underline{/0°} \text{ V} \quad \text{approximately}$$

Note that for $8\underline{/143°}$ in the second quadrant, the cosine is negative for a negative real term but the sine is positive for a positive j term.

■ *24–10 Knowledge Check*

Answer at end of chapter.

In Fig. 24–10, what is the phase angle of V_L with respect to V_C?

■ *24–10 Self-Review*

Answers at end of chapter.

Refer to Fig. 24–10.
a. What is the phase angle of I with reference to V_T?
b. What is the phase angle of V_L with reference to V_T?
c. What is the phase angle of V_L with reference to V_R?

24-11 Complex Numbers in Parallel AC Circuits

A useful application is converting a parallel circuit to an equivalent series circuit. See Fig. 24–11, with a 10-Ω X_L in parallel with a 10-Ω R. In complex notation, R is $10 + j0$ and X_L is $0 + j10$. Their combined parallel impedance Z_T equals the product divided by the sum. For Fig. 24–11a, then,

$$Z_T = \frac{(10 + j0) \times (0 + j10)}{(10 + j0) + (0 + j10)} = \frac{10 \times j10}{10 + j10} = \frac{j100}{10 + j10}$$

$$= \frac{j10}{1 + j1}$$

Converting to polar form for division,

$$Z_T = \frac{j100}{10 + j10} = \frac{100\ \underline{/90°}}{14.14\ \underline{/+45°}} = 7.07\ \underline{/45°}$$

Converting the Z_T of 7.07 $\underline{/45°}$ into rectangular form to see its resistive and reactive components,

$$\text{Real term} = 7.07 \cos 45°$$
$$= 7.07 \times 0.707 = 5$$
$$j \text{ term} = 7.07 \sin 45°$$
$$= 7.07 \times 0.707 = 5$$

Therefore,

$$Z_T = 7.07\ \underline{/45°} \quad \text{in polar form}$$
$$Z_T = 5 + j5 \quad \text{in rectangular form}$$

The rectangular form of Z_T means that a 5-Ω R in series with a 5-Ω X_L is the equivalent of 10-Ω R in parallel with 10-Ω X_L, as shown in Fig. 24–11b.

Admittance Y and Susceptance B

In parallel circuits, it is usually easier to add branch currents than to combine reciprocal impedances. For this reason, branch conductance G is often used instead of branch resistance, where $G = 1/R$. Similarly, reciprocal terms can be defined for complex impedances. The two main types are *admittance Y*, which

---

Figure 24–11 Complex numbers used for a parallel ac circuit to convert a parallel bank to an equivalent series impedance.

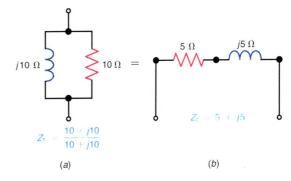

(a)　　　(b)

is the reciprocal of impedance, and *susceptance B*, which is the reciprocal of reactance. These reciprocals can be summarized as follows:

$$\text{Conductance} = G = \frac{1}{R} \text{ S} \tag{24–4}$$

$$\text{Susceptance} = B = \frac{1}{\pm X} \text{ S} \tag{24–5}$$

$$\text{Admittance} = Y = \frac{1}{Z} \text{ S} \tag{24–6}$$

With R, X, and Z in units of ohms, the reciprocals G, B, and Y are in siemens (S) units.

The phase angle for B or Y is the same as that of the current. Therefore, the sign is opposite from the angle of X or Z because of the reciprocal relation. An inductive branch has susceptance $-jB$, whereas a capacitive branch has susceptance $+jB$, with the same angle as a branch current.

For parallel branches of conductance and susceptance, the total admittance $Y_T = G \pm jB$. For the two branches in Fig. 24–11a, as an example, G is 0.1 and B is also 0.1.

In rectangular form:

$$Y_T = 0.1 - j0.1 \text{ S}$$

In polar form:

$$Y_T = 0.14 \, \underline{/-45°} \text{ S}$$

This value for Y_T is the same as I_T with 1 V applied across Z_T of 7.07 $\underline{/45°}$ Ω .

As another example, suppose that a parallel circuit has 4 Ω for R in one branch and $-j4$ Ω for X_C in the other branch. In rectangular form, then, Y_T is $0.25 + j0.25$ S. Also, the polar form is $Y_T = 0.35 \, \underline{/45°}$ S.

■ 24–11 Knowledge Check

Answer at end of chapter.

What is the impedance, Z_T, for a parallel ac circuit whose admittance, Y_T, is 0.05 S $- j0.08$ S?

■ 24–11 Self-Review

Answers at end of chapter.

a. A Z of $3 + j4$ Ω is in parallel with an R of 2 Ω. State Z_T in rectangular form.
b. Do the same as in *a* for X_C instead of X_L.

24–12 Combining Two Complex Branch Impedances

A common application is a circuit with two branches Z_1 and Z_2, where each is a complex impedance with both reactance and resistance. A circuit such as that in Fig. 24–12 can be solved only graphically or by complex numbers. Actually, using complex numbers is the shortest method.

The procedure here is to find Z_T as the product divided by the sum of Z_1 and Z_2. A good way to start is to state each branch impedance in both rectangular and polar forms. Then Z_1 and Z_2 are ready for addition, multiplication, and division. The solution of this circuit is as follows:

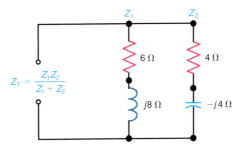

$$Z_1 = 6 + j8 = 10\ \underline{/53°}$$
$$Z_2 = 4 - j4 = 5.66\ \underline{/-45°}$$

The combined impedance is

$$Z_T = \frac{Z_1 \times Z_2}{Z_1 + Z_2}$$

Use the polar form of Z_1 and Z_2 to multiply, but add in rectangular form:

$$Z_T = \frac{10\ \underline{/53°} \times 5.66\ \underline{/-45°}}{6 + j8 + 4 - j4}$$

$$= \frac{56.6\ \underline{/8°}}{10 + j4}$$

Converting the denominator to polar form for easier division,

$$10 + j4 = 10.8\ \underline{/22°}$$

Then

$$Z_T = \frac{56.6\ \underline{/8°}}{10.8\ \underline{/22°}} = 5.24\ \underline{/-14°}\ \Omega$$

We can convert Z_T into rectangular form. The R component is $5.24 \times \cos(-14°)$ or $5.24 \times 0.97 = 5.08$. Note that $\cos \theta$ is positive in the first and fourth quadrants. The j component equals $5.24 \times \sin(-14°)$ or $5.24 \times (-0.242) = -1.127$. In rectangular form, then,

$$Z_T = 5.08 - j1.27$$

Therefore, this series-parallel circuit combination is equivalent to $5.08\ \Omega$ of R in series with $1.27\ \Omega$ of X. Notice that the minus j term means that the circuit is capacitive. This problem can also be done in rectangular form by rationalizing the fraction for Z_T.

■ 24–12 Knowledge Check

Answer at end of chapter.

In Fig. 24–12, express Y_T in polar form.

■ 24–12 Self-Review

Answers at end of chapter.

Refer to Fig. 24–12.
a. Add $(6 + j8) + (4 - j4)$ for the sum of Z_1 and Z_2.
b. Multiply $10\ \underline{/53°} \times 5.66\ \underline{/-45°}$ for the product of Z_1 and Z_2.

24–13 Combining Complex Branch Currents

Figure 24–13 gives an example of finding I_T for two branch currents. The branch currents can just be added in rectangular form for the total I_T of parallel branches. This method corresponds to adding series impedances in rectangular form to find Z_T. The rectangular form is necessary for the addition of phasors.

Adding the branch currents in Fig. 24–13,

$$I_T = I_1 + I_2$$
$$= (6 + j6) + (3 - j4)$$
$$= 9 + j2 \text{ A}$$

Note that I_1 has $+j$ for the $+90°$ of capacitive current, and I_2 has $-j$ for inductive current. These current phasors have signs opposite from their reactance phasors.

In polar form, the I_T of $9 + j2$ A is calculated as the phasor sum of the branch currents.

$$I_T = \sqrt{9^2 + 2^2} = \sqrt{85}$$
$$= 9.22 \text{ A}$$
$$\tan \theta = \frac{2}{9} = 0.222$$
$$\theta_1 = \arctan (0.22)$$
$$= 12.53°$$

Therefore, I_T is $9 + j2$ A in rectangular form or $9.22 \underline{/12.53°}$ A in polar form. The complex currents for any number of branches can be added in rectangular form.

■ *24–13 Knowledge Check*

Answer at end of chapter.

In Fig. 24–13 how much is I_T if $I_2 = 6 - j10$ A?

■ *24–13 Self-Review*

Answers at end of chapter.

a. Find I_T in rectangular form for I_1 of $0 + j2$ A and I_2 of $4 + j3$ A.
b. Find I_T in rectangular form for I_1 of $6 + j7$ A and I_2 of $3 - j9$ A.

Figure 24–13   Finding I_T for two branch currents in parallel.

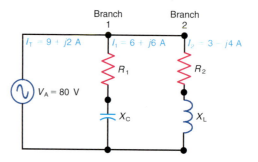

24–14 Parallel Circuit with Three Complex Branches

Because the circuit in Fig. 24–14 has more than two complex impedances in parallel, use the method of branch currents. There will be several conversions between rectangular and polar form, since addition must be in rectangular form, but division is easier in polar form. The sequence of calculations is

1. Convert each branch impedance to polar form. This is necessary for dividing into the applied voltage V_A to calculate the individual branch currents. If V_A is not given, any convenient value can be assumed. Note that V_A has a phase angle of $0°$ because it is the reference.
2. Convert the individual branch currents from polar to rectangular form so that they can be added for the total line current. This step is necessary because the resistive and reactive components must be added separately.
3. Convert the total line current from rectangular to polar form for dividing into the applied voltage to calculate Z_T.
4. The total impedance can remain in polar form with its magnitude and phase angle or can be converted to rectangular form for its resistive and reactive components.

These steps are used in the following calculations to solve the circuit in Fig. 24–14. All the values are in A, V, or Ω units.

Branch Impedances

Each Z is converted from rectangular form to polar form:

$$Z_1 = 50 - j50 = 70.7 \; \underline{/-45°}$$
$$Z_2 = 40 + j30 = 50 \; \underline{/+37°}$$
$$Z_3 = 30 + j40 = 50 \; \underline{/+53°}$$

Figure 24–14 Finding Z_T for any three complex impedances in parallel. See text for solution by means of branch currents.

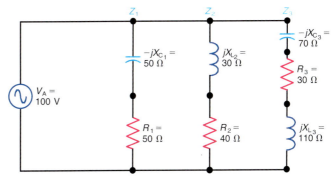

Branch Currents

Each I is calculated as V_A divided by Z in polar form:

$$I_1 = \frac{V_A}{Z_1} = \frac{100 \angle 0°}{70.7 \angle -45°} = 1.414 \angle +45° = 1 + j1$$

$$I_2 = \frac{V_A}{Z_2} = \frac{100 \angle 0°}{50 \angle +37°} = 2.00 \angle -37° = 1.6 - j1.2$$

$$I_3 = \frac{V_A}{Z_3} = \frac{100 \angle 0°}{50 \angle +53°} = 2.00 \angle -53° = 1.2 - j1.6$$

The polar form of each I is converted to rectangular form for addition of the branch currents.

Total Line Current

In rectangular form,

$$
\begin{aligned}
I_T &= I_1 + I_2 + I_3 \\
&= (1 + j1) + (1.6 - j1.2) + (1.2 - j1.6) \\
&= 1 + 1.6 + 1.2 + j1 - j1.2 - j1.6 \\
&= 3.8 - j1.8
\end{aligned}
$$

Converting $3.8 - j1.8$ into polar form,

$$I_T = 4.2 \angle -25.4°$$

Total Impedance

In polar form,

$$Z_T = \frac{V_A}{I_T} = \frac{100 \angle 0°}{4.2 \angle -25.4°}$$

$$= 23.8 \angle +25.4° \ \Omega$$

Converting $23.8 \angle +25.4°$ into rectangular form,

$$Z_T = 21.5 + j10.2 \ \Omega$$

Therefore, the complex ac circuit in Fig. 24–14 is equivalent to the combination of 21.5 Ω of R in series with 10.2 Ω of X_L. The circuit is inductive.

This problem can also be done by combining Z_1 and Z_2 in parallel as $Z_1 Z_2/(Z_1 + Z_2)$. Then combine this value with Z_3 in parallel to find the total Z_T of the three branches.

■ *24–14 Knowledge Check*

Answer at end of chapter.

In Fig. 24–14, does the value of V_A affect the total impedance, Z_T?

■ *24–14 Self-Review*

Answers at end of chapter.

Refer to Fig. 24–14.
a. State Z_2 in rectangular form for branch 2.
b. State Z_2 in polar form.
c. Find I_2.

Summary

- In complex numbers, resistance R is a real term and reactance is a j term. Thus, an 8-Ω R is 8; an 8-Ω X_L is $j8$; an 8-Ω X_C is $-j8$. The general form of a complex impedance with series resistance and reactance, then, is $Z_T = R \pm jX$, in rectangular form.

- The same notation can be used for series voltages where $V = V_R \pm jV_X$.

- For branch currents $I_T = I_R \pm jI_X$, but the reactive branch currents have signs opposite from impedances. Capacitive branch current is jI_C, and inductive branch current is $-jI_L$.

- The complex branch currents are added in rectangular form for any number of branches to find I_T.

- To convert from rectangular to polar form: $R \pm jX = Z_T \underline{/\theta}$. The angle is θ_Z.

- The magnitude of Z_T is $\sqrt{R^2 + X^2}$. Also, θ_Z is the angle with $\tan = X/R$.

- To convert from polar to rectangular form, $Z_T \underline{/\theta_Z} = R \pm jX$, where R is $Z_T \cos \theta_Z$ and the j term is $Z_T \sin \theta_Z$. A positive angle has a positive j term; a negative angle has a negative j term. Also, the angle is more than 45° for a j term larger than the real term; the angle is less than 45° for a j term smaller than the real term.

- The rectangular form must be used for addition or subtraction of complex numbers.

- The polar form is usually more convenient in multiplying and dividing complex numbers. For multiplication, multiply the magnitudes and add the angles; for division, divide the magnitudes and subtract the angles.

- To find the total impedance Z_T of a series circuit, add all resistances for the real term and find the algebraic sum of the reactances for the j term. The result is $Z_T = R \pm jX$. Then convert Z_T to polar form for dividing into the applied voltage to calculate the current.

- To find the total impedance Z_T of two complex branch impedances Z_1 and Z_2 in parallel, Z_T can be calculated as $Z_1 Z_2 / (Z_1 + Z_2)$.

Important Terms

- Admittance, Y - the reciprocal of impedance or $Y = 1/Z$. The unit of admittance is siemens (S).

- Complex Number - the combination of a real and an imaginary term. Complex numbers can be expressed in either rectangular or polar form.

- Imaginary Number - a number on the j axis is called an imaginary number because it is not on the real axis. (Any quantity at right angles to the zero axis, or at 90°, is considered on the j axis.)

- j Operator - the j operator indicates a phase angle of either plus or minus 90°. For example, $+j1$ kΩ indicates 1 kΩ of inductive reactance, X_L, on

the $+j$ axis of 90°. Similarly, $-j1$ kΩ indicates 1 kΩ of capacitive reactance, X_C, on the $-j$ axis of $-90°$.

- Polar Form - the form of a complex number that specifies its magnitude and phase angle. The general form of a complex number expressed in polar form is $r\angle\theta$ where r is the magnitude of the resultant phasor and θ is the phase angle with respect to the horizontal or real axis.

- Real Number - any number on the horizontal axis, either positive (0°) or negative (180°).

- Rectangular Form - the form of a complex number that specifies the real and imaginary terms individually. The general form of a complex number specified in rectangular form is $a \pm jb$ where a is the real term and $\pm jb$ is the imaginary term at $\pm90°$.

- Susceptance, B - the reciprocal of reactance or $B = 1/\pm X$. The unit of susceptance is siemens (S).

Related Formulas

$Z_T = Z_1 + Z_2 + Z_3 + \cdots + $ etc. (Series impedances)

$\dfrac{1}{Z_T} = \dfrac{1}{Z_1} + \dfrac{1}{Z_2} + \dfrac{1}{Z_3} + \cdots + $ etc. (Parallel impedances)

$Z_T = \dfrac{Z_1 Z_2}{Z_1 + Z_2}$ (Two parallel impedances)

Conductance $= G = \dfrac{1}{R}$ S

Susceptance $= B = \dfrac{1}{\pm X}$ S

Admittance $= Y = \dfrac{1}{Z}$ S

Self-Test

1. **Numbers on the horizontal axis are called**
   a. imaginary numbers.
   b. conjugate numbers.
   c. real numbers.
   d. complex numbers.

2. **Numbers on the plus or minus j axis are called**
   a. imaginary numbers.
   b. conjugate numbers.
   c. real numbers.
   d. complex numbers.

3. **A value of $-j500 \, \Omega$ represents**
   a. $500 \, \Omega$ of inductive reactance.
   b. $500 \, \Omega$ of capacitive reactance.
   c. $500 \, \Omega$ of resistance.
   d. $500 \, \Omega$ of conductance.

4. **An inductive reactance of $20 \, \Omega$ can be expressed as**
   a. $+20 \, \Omega$.
   b. $j^2 20 \, \Omega$.
   c. $-j20 \, \Omega$.
   d. $+j20 \, \Omega$.

5. **A series ac circuit consists of $10 \, \Omega$ of resistance and $15 \, \Omega$ of inductive reactance. What is the impedance of this circuit when expressed in polar form?**
   a. $15 \, \Omega + j10 \, \Omega$.
   b. $18\angle56.3° \, \Omega$.
   c. $18\angle33.7° \, \Omega$.
   d. $25\angle56.3° \, \Omega$.

6. **An ac circuit has an impedance, Z, of $50\angle-36.87° \, \Omega$. What is the impedance of this circuit when expressed in rectangular form?**

   a. $40 \, \Omega - j30 \, \Omega$.
   b. $40 \, \Omega + j30 \, \Omega$.
   c. $30 \, \Omega + j40 \, \Omega$.
   d. $30 \, \Omega - j40 \, \Omega$.

7. **When adding or subtracting complex numbers, all numbers must be in**
   a. polar form.
   b. scientific notation.
   c. rectangular form.
   d. none of the above.

8. **When multiplying complex numbers in polar form,**
   a. multiply the magnitudes and subtract the phase angles.
   b. multiply the magnitudes and add the phase angles.
   c. multiply the angles and add the magnitudes.
   d. multiply both the magnitudes and phase angles.

9. **When dividing complex numbers in polar form,**
   a. divide the magnitudes and subtract the phase angles.
   b. divide the magnitudes and add the phase angles.
   c. divide the phase angles and subtract the magnitudes.
   d. divide both the magnitudes and phase angles.

10. **What is the admittance, Y, of a parallel branch whose impedance is $200\angle-63.43° \, \Omega$?**
    a. $5\angle-63.43° \, \text{mS}$.
    b. $5\angle26.57° \, \text{mS}$.
    c. $200\angle63.43 \, \text{mS}$.
    d. $5\angle63.43 \, \text{mS}$.

11. **In complex numbers, j^2 corresponds to**
    a. $180°$.
    b. -1.
    c. $-90°$.
    d. both a and b.

12. **Susceptance, B, is**
    a. the reciprocal of impedance.
    b. the reciprocal of reactance.
    c. the reciprocal of resistance.
    d. the same as conductance.

13. **A branch current of $+j250 \, \text{mA}$ represents**
    a. $250 \, \text{mA}$ of inductive current.
    b. $250 \, \text{mA}$ of resistive current.
    c. $250 \, \text{mA}$ of capacitive current.
    d. $250 \, \text{mA}$ of in-phase current.

14. **A parallel ac circuit has an admittance, Y_T, of $6 \, \text{mS} + j8 \, \text{mS}$. What is the impedance, Z, in polar form?**
    a. $10\angle45° \, \text{k}\Omega$.
    b. $100\angle53.13° \, \Omega$.
    c. $100\angle-53.13° \, \Omega$.
    d. $14\angle53.13° \, \Omega$.

15. **What is the resistance, R, of an ac circuit whose impedance, Z, is $300\angle53.13° \, \Omega$?**
    a. $240 \, \Omega$.
    b. $180 \, \Omega$.
    c. $270 \, \Omega$.
    d. $60 \, \Omega$.

Questions

1. Give the mathematical operator for the angles of $0°$, $90°$, $180°$, $270°$, and $360°$.

2. Define the sine, cosine, and tangent functions of an angle.

3. Compare the following combinations: resistance R and conductance G; reactance X and susceptance B; impedance Z and admittance Y.

4. What are the units for admittance Y and susceptance B?

5. Why do Z_T and I_T for a circuit have angles with opposite signs?

Problems

SECTION 24–1 POSITIVE AND NEGATIVE NUMBERS

24–1 What is the phase angle for
  a. positive numbers on the horizontal or x axis?
  b. negative numbers on the horizontal or x axis?

24–2 What factor corresponds to a phase angle of
  a. 0°?
  b. 180°?

SECTION 24–2 THE j OPERATOR

24–3 What is the name of the axis at right angles to the real or horizontal axis?

24–4 What is the phase angle for numbers on the
  a. $+j$ axis?
  b. $-j$ axis?

24–5 What is the name given to numbers on the
  a. horizontal axis?
  b. j axis?

24–6 List the phase angle for each of the following factors:
  a. $+1$
  b. -1
  c. $+j$
  d. $-j$
  e. j^2
  f. j^3

24–7 What do the following numbers mean?
  a. $j25$
  b. $-j36$

SECTION 24–3 DEFINITION OF A COMPLEX NUMBER

24–8 What is the definition of a complex number?

24–9 In what form is the complex number 100 Ω + j400 Ω?

24–10 For the complex number 8 + j6, identify the real and imaginary terms.

24–11 In each of the following examples, identify when the phase angle is less than 45°, greater than 45°, or equal to 45°:
  a. $3 + j5$
  b. $180 + j60$
  c. $40 - j40$
  d. $100 - j120$
  e. $40 + j30$

SECTION 24–4 HOW COMPLEX NUMBERS ARE APPLIED TO AC CIRCUITS

24–12 Is a resistance value considered a real or imaginary number?

24–13 What is the phase angle of a positive real number?

24–14 Express the following quantities using the j operator.
  a. 50 Ω of X_L
  b. 100 Ω of X_C
  c. V_L of 25 V
  d. V_C of 15 V
  e. 4 A of I_L
  f. 600 mA of I_C

SECTION 24–5 IMPEDANCE IN COMPLEX FORM

24–15 Express the following impedances in rectangular form.
  a. 10 Ω of R in series with 20 Ω of X_L.
  b. 10 Ω of X_L in series with 15 Ω of R.
  c. 0 Ω of R in series with 1 kΩ of X_C.
  d. 1.5 kΩ of R in series with 2 kΩ of X_C.
  e. 150 Ω of R in series with 0 Ω of X.
  f. 75 Ω of R in series with 75 Ω of X_C.

24–16 In the following examples, combine the real terms and j terms separately, and express the resultant values in rectangular form.
  a. 40 Ω of R in series with 30 Ω of X_C and 60 Ω of X_L.
  b. 500 Ω of R in series with 150 Ω of X_L and 600 Ω of X_C.
  c. 1 kΩ of X_C in series with 2 kΩ of X_L, 3 kΩ of R, and another 2 kΩ of R.

SECTION 24–6 OPERATIONS WITH COMPLEX NUMBERS

24–17 Add the following complex numbers:
  a. $(6 + j9) + (9 + j6)$
  b. $(25 + j10) + (15 - j30)$
  c. $(0 + j100) + (200 + j50)$
  d. $(50 - j40) + (40 - j10)$
  e. $(12 + j0) + (24 - j48)$

24–18 Multiply or divide the following j terms and real numbers:
  a. $j10 \times 5$
  b. $-j60 \times (-4)$
  c. $-j8 \times 9$
  d. $j4 \times (-8)$
  e. $j100 \div 20$
  f. $-j600 \div 6$
  g. $-j400 \div (-20)$
  h. $j16 \div (-8)$

24–19 Multiply or divide the following j terms.
  a. $j8 \times j9$
  b. $-j12 \times j5$
  c. $-j7 \times (-j4)$
  d. $j3 \times j8$
  e. $j12 \div j6$
  f. $-j100 \div j8$
  g. $-j250 \div (-j10)$
  h. $j1000 \div (-j40)$

24–20 Multiply the following complex numbers:

    a. $(3 + j5) \times (4 + j3)$

    b. $(6 - j8) \times (8 + j6)$

    c. $(12 + j3) \times (5 + j9)$

    d. $(4 - j2) \times (8 - j12)$

24–21 Divide the following complex numbers:

    a. $(15 - j3) \div (10 + j4)$

    b. $(6 + j3) \div (24 - j8)$

    c. $(10 + j2) \div (20 - j4)$

    d. $(2 - j6) \times (4 - j4)$

SECTION 24–7 MAGNITUDE AND ANGLE OF A COMPLEX NUMBER

24–22 Calculate the resultant magnitude and phase angle for each of the following complex numbers expressed in rectangular form:

    a. $5 - j8$

    b. $10 + j15$

    c. $100 + j50$

    d. $20 - j35$

    e. $150 + j200$

    f. $75 - j75$

    g. $0 + j100$

    h. $100 + j0$

    i. $10 - j40$

    j. $2000 - j6000$

SECTION 24–8 POLAR FORM OF COMPLEX NUMBERS

24–23 Convert the following complex numbers, written in rectangular form, into polar form:

    a. $10 + j10$

    b. $8 - j10$

    c. $12 + j18$

    d. $140 - j55$

24–24 Multiply the following complex numbers expressed in polar form:

    a. $50\angle30° \times 2\angle-65°$

    b. $3\angle-15° \times 5\angle-40°$

    c. $9\angle20° \times 8\angle30°$

    d. $15\angle-70° \times 4\angle10°$

    e. $2 \times 150\angle-45°$

    f. $1000\angle-90° \times 0.5\angle90°$

    g. $40\angle25° \times 1.5$

24–25 Divide the following complex numbers expressed in polar form:

    a. $48\angle-80° \div 16\angle45°$

    b. $120\angle60° \div 24\angle-90°$

    c. $172\angle-45° \div 43\angle-45°$

    d. $210\angle22° \div 45\angle-44°$

    e. $180\angle75° \div 6$

    f. $750\angle80° \div 30$

    g. $2500\angle50° \div 200$

SECTION 24–9 CONVERTING POLAR TO RECTANGULAR FORM

24–26 Convert the following numbers expressed in polar form into rectangular form:

    a. $50\angle45°$

    b. $100\angle60°$

    c. $250\angle-53.13°$

    d. $1000\angle-30°$

    e. $12\angle0°$

    f. $180\angle-78.5°$

    g. $5\angle36.87°$

    h. $25\angle15°$

    i. $45\angle100°$

    j. $60\angle-90°$

    k. $100\angle53.13°$

    l. $40\angle90°$

SECTION 24–10 COMPLEX NUMBERS IN SERIES AC CIRCUITS

24–27 In Fig. 24–15, state

    a. Z_T in rectangular form.

    b. Z_T in polar form.

    c. I in polar form.

    d. V_R in polar form.

    e. V_L in polar form.

    f. V_C in polar form.

Figure 24–15

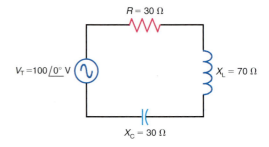

24–28 Repeat Prob. 24–27 for Fig. 24–16.

Figure 24–16

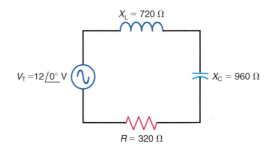

SECTION 24–11 COMPLEX NUMBERS IN PARALLEL AC CIRCUITS

24–29 In Fig. 24–17, state the total impedance, Z_T, in both polar and rectangular form. Use Formula 24–3 to solve for Z_T.

Figure 24–17

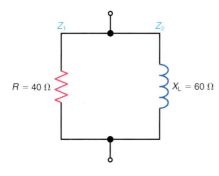

$R = 40\ \Omega$ $X_L = 60\ \Omega$

24–30 In Fig. 24–17, state the total admittance, Y_T, in both polar and rectangular form. Using the polar form of Y_T, solve for Z_T.

24–31 In Fig. 24–18, state the total admittance, Y_T, in both rectangular and polar form. Solve for Z_T from Y_T.

Figure 24–18

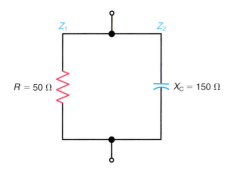

$R = 50\ \Omega$ $X_C = 150\ \Omega$

24–32 In Fig. 24–18, solve for the total impedance, Z_T, using Formula 24–3. State Z_T in both polar and rectangular form.

24–33 Draw the equivalent series circuit for the circuit in Fig.
  a. 24–17.
  b. 24–18.

SECTION 24–12 COMBINING TWO COMPLEX BRANCH IMPEDANCES

24–34 In Fig. 24–19,
  a. state Z_1 in both rectangular and polar form.
  b. state Z_2 in both rectangular and polar form.
  c. state Z_T in both rectangular and polar form.

Figure 24–19

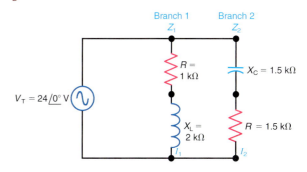

Branch 1 — Z_1 Branch 2 — Z_2

$V_T = 24\ \underline{/0°}\ V$

$R = 1\ k\Omega$ $X_C = 1.5\ k\Omega$

$X_L = 2\ k\Omega$ $R = 1.5\ k\Omega$

I_1 I_2

24–35 Repeat Prob. 24–34 for Fig. 24–20.

Figure 24–20

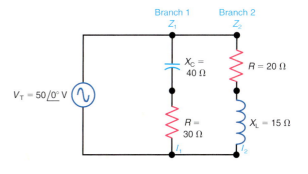

Branch 1 — Z_1 Branch 2 — Z_2

$V_T = 50\ \underline{/0°}\ V$

$X_C = 40\ \Omega$ $R = 20\ \Omega$

$R = 30\ \Omega$ $X_L = 15\ \Omega$

I_1 I_2

SECTION 24–13 COMBINING COMPLEX BRANCH CURRENTS

24–36 In Fig. 24–19,
  a. state the branch current, I_1, in both polar and rectangular form.
  b. state the branch current, I_2, in both polar and rectangular form.
  c. state the total current, I_T, in both polar and rectangular form.

24–37 Repeat Prob. 24–36 for Fig. 24–20.

SECTION 24–14 PARALLEL CIRCUIT WITH THREE COMPLEX BRANCHES

24-38 In Fig. 24–21,

    a. state Z_1 in polar form.

    b. state Z_2 in polar form.

    c. state Z_3 in polar form.

    d. state I_1 in polar and rectangular form.

    e. state I_2 in polar and rectangular form.

    f. state I_3 in polar and rectangular form.

    g. state I_T in polar and rectangular form.

    h. state Z_T in polar and rectangular form.

Figure 24–21

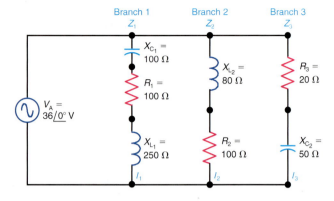

Critical Thinking

24-39 In Fig. 24–22, calculate the input voltage V_{in}, in polar form.

Figure 24–22 Circuit for Critical Thinking Prob. 24–39.

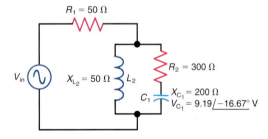

Answers to Knowledge Check Problems

24-1 a. $0°$

    b. $180°$

24-2 $+1$

24-3 The real term is written first followed by the imaginary (j) term.

24-4 $-j10$ mA

24-5 $10\ \Omega - j10\ \Omega$

24-6 $0 + j2$

24-7 $25∠36.87°$

24-8 Polar form

24-9 No

24-10 V_L and V_C are $180°$ out of phase.

24-11 $Z_T = 10.6∠58°\ \Omega$

24-12 $Y_T = 190.8∠14°$ mS

24-13 $I_T = 12 - j4$ A or $12.65∠-18.44°$ A

24-14 No

Answers to Self-Reviews

24-1 a. $0°$
b. $180°$

24-2 a. $90°$
b. -90 or $270°$

24-3 a. T
b. T

24-4 a. $j3 \text{ k}\Omega$
b. $-j5 \text{ mA}$

24-5 a. $4 + j7$
b. $0 - j7$

24-6 a. $5 + j7$
b. $4 + j6$

24-7 a. $14.14 \ \Omega$
b. $45°$

24-8 a. $12 \ \underline{/50°}$
b. $3 \ \underline{/-10°}$

24-9 a. $10 + j10$
b. $10 - j10$

24-10 a. $53°$
b. $143°$
c. $90°$

24-11 a. $(6 + j8)/(5 + j4)$
b. $(6 - j8)/(5 - j4)$

24-12 a. $10 + j4$
b. $56.6 \ \underline{/8°}$

24-13 a. $4 + j5 \text{ A}$
b. $9 - j2 \text{ A}$

24-14 a. $40 + j30$
b. $50 \ \underline{/37°} \ \Omega$
c. $2 \ \underline{/-37°} \text{ A}$

Cumulative Review Summary (Chapters 23–24)

- Reactances X_C and X_L are opposite. In series, the ohms of X_C and X_L cancel. In parallel, the branch currents I_C and I_L cancel.

- As a result, circuits with R, X_C, and X_L can be reduced to one net reactance X and one equivalent R.

- In series circuits, the net X is added with the total R by phasors for the impedance: $Z_T = \sqrt{R^2 + X^2}$. Then $I = V_T/Z_T$.

- For branch currents in parallel circuits, the net I_X is added with I_R by phasors for the total line current: $I_T = \sqrt{I_R^2 + I_X^2}$. Then $Z_{EQ} = V/I_T$.

- The characteristics of ohms of R, X_C, X_L, and Z in ac circuits are compared in Table 23–1.

- In ac circuits with reactance, the real power in watts equals I^2R. This value equals $VI \cos \theta$, where θ is the phase angle of the circuit and $\cos \theta$ is the power factor.

- The wattmeter uses an ac meter movement to read V and I at the same time, measuring watts of real power.

- In complex numbers, R is a real term at $0°$ and reactance is a $\pm j$ term at $\pm 90°$. In rectangular form, $Z_T = R \pm jX$. For example, $10 \ \Omega$ of R in series with $10 \ \Omega$ of X_L is $10 + j10 \ \Omega$.

- The polar form of $10 + j10 \ \Omega$ is $14 \ \underline{/45°} \ \Omega$. The angle of $45°$ is arctan X/R. The magnitude of 14 is $\sqrt{R^2 + X^2}$.

- The rectangular form of complex numbers must be used for addition and subtraction. Add or subtract the real terms and the j terms separately.

- The polar form of complex numbers is easier for multiplication and division. For multiplication, multiply the magnitudes and add the angles. For division, divide the magnitudes and subtract the angle of the divisor.

- In double-subscript notation for a voltage, such as V_{BE}, the first letter in the subscript is the point of measurement with respect to the second letter. So V_{BE} is the base voltage with respect to the emitter in a transistor.

Cumulative Review Self-Test

Answers at back of book.

Fill in the numerical answer.

1. An ac circuit with $100 \ \Omega \ R_1$ in series with $200 \ \Omega \ R_2$ has R_T of _____ Ω.

2. With $100 \ \Omega \ X_{L_1}$ in series with $200 \ \Omega \ X_{L_2}$, the total X_L is _____ Ω.

3. For $200 \ \Omega \ X_{C_1}$ in series with $100 \ \Omega \ X_{C_2}$, the total X_C is _____ Ω.

4. Two X_C branches of $500 \ \Omega$ each in parallel have combined X_C of _____ Ω.

5. Two X_L branches of $500 \ \Omega$ each in parallel have combined X_L of _____ Ω.

6. A $500\text{-}\Omega \ X_L$ is in series with a $300\text{-}\Omega \ X_C$. The net X_L is _____ Ω.

7. For $500 \ \Omega \ X_C$ in series with $300 \ \Omega \ X_{L_1}$, the net X_C is _____ Ω.

8. A $10\text{-}\Omega \ X_L$ is in series with a $10\text{-}\Omega \ R$. The total Z_T is _____ Ω.

9. With a $10\text{-}\Omega \ X_C$ in series with a $10\text{-}\Omega \ R$, the total Z_T is _____ Ω.

10. With 14 V applied across $14 \ \Omega \ Z_T$, the I is _____ A.

11. For $10 \ \Omega \ X_L$ and $10 \ \Omega \ R$ in series, the phase angle θ is _____ degrees.

12. For 10-Ω X_C and 10-Ω R in series, the phase angle θ is _____ degrees.

13. A 10-Ω X_L and a 10-Ω R are in parallel across 10 V. The amount of each branch I is _____ A.

14. In question 13, the total line current I_T equals _____ A.

15. In questions 13 and 14, Z_T of the parallel branches equals _____ Ω.

16. With 120 V, an I of 10 A, and θ of 60°, a wattmeter reads _____ W.

17. The Z of $4 + j4$ Ω converted to polar form is _____ Ω.

18. The impedance value of $8 \underline{/40°} / 2 \underline{/30°}$ is equal to _____ Ω.

Answer True/False.

19. In an ac circuit with X_C and R in series, if the frequency is raised, the current will increase.

20. In an ac circuit with X_L and R in series, if the frequency is increased, the current will be reduced.

21. The volt-ampere is a unit of apparent power.

22. The polar form of complex numbers is best for adding impedance values.

25 Resonance

This chapter explains how X_L and X_C can be combined to favor one particular frequency, the resonant frequency to which the LC circuit is tuned. The resonance effect occurs when the inductive and capacitive reactances are equal.

In radio frequency (rf) circuits, the main application of resonance is for tuning to an ac signal of the desired frequency. Applications of resonance include tuning in communication receivers, transmitters, and electronic equipment in general.

Tuning by means of the resonant effect provides a practical application of selectivity. The resonant circuit can be operated to select a particular frequency for the output with many different frequencies at the input.

Objectives

After studying this chapter you should be able to

- *Define* the term *resonance.*

- *List* four characteristics of a series resonant circuit.

- *List* three characteristics of a parallel resonant circuit.

- *Explain* how the resonant frequency formula is derived.

- *Calculate* the Q of a series or parallel resonant circuit.

- *Calculate* the equivalent impedance of a parallel resonant circuit.

- *Explain* what is meant by the *bandwidth* of a resonant circuit.

- *Calculate* the bandwidth of a series or parallel resonant circuit.

- *Explain* the effect of varying L or C in tuning an LC circuit.

- *Calculate* L or C for a resonant circuit.

Outline

Important Terms

antiresonance	flywheel effect	resonant frequency
bandwidth	half-power points	tank circuit
damping	Q of a resonant circuit	tuning

25–1 The Resonance Effect

Inductive reactance increases as the frequency is increased, but capacitive reactance decreases with higher frequencies. Because of these opposite characteristics, for any *LC* combination, there must be a frequency at which the X_L equals the X_C because one increases while the other decreases. This case of equal and opposite reactances is called *resonance,* and the ac circuit is then a *resonant circuit.*

Any *LC* circuit can be resonant. It all depends on the frequency. At the resonant frequency, an *LC* combination provides the resonance effect. Off the resonant frequency, either below or above, the *LC* combination is just another ac circuit.

The frequency at which the opposite reactances are equal is the *resonant frequency.* This frequency can be calculated as $f_r = 1/(2\pi\sqrt{LC})$, where L is the inductance in henrys, C is the capacitance in farads, and f_r is the resonant frequency in hertz that makes $X_L = X_C$.

In general, we can say that large values of L and C provide a relatively low resonant frequency. Smaller values of L and C allow higher values for f_r. The resonance effect is most useful for radio frequencies, where the required values of microhenrys for L and picofarads for C are easily obtained.

The most common application of resonance in rf circuits is called *tuning.* In this use, the *LC* circuit provides maximum voltage output at the resonant frequency, compared with the amount of output at any other frequency either below or above resonance. This idea is illustrated in Fig. 25–1, where the *LC* circuit resonant at 1000 kHz magnifies the effect of this particular frequency. The result is maximum output at 1000 kHz, compared with lower or higher frequencies.

Tuning in radio and television receivers is an application of resonance. When you tune a radio to one station, the *LC* circuits are tuned to resonance for that particular carrier frequency. Also, when you tune a television receiver to a particular channel, the *LC* circuits are tuned to resonance for that station. There are almost unlimited uses for resonance in ac circuits.

■ *25–1 Knowledge Check*

Answer at end of chapter.

The capacitor in an *LC* circuit has an X_C value of 500 Ω at its resonant frequency. How much is X_L at f_r?

■ *25–1 Self-Review*

Answers at end of chapter.

Refer to Fig. 25–1.
a. **Give the resonant frequency.**
b. **Give the frequency that has maximum output.**

Figure 25–1 *LC* circuit resonant at f_r of 1000 kHz to provide maximum output at this frequency.

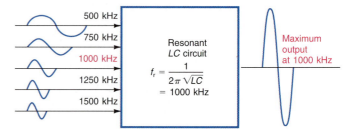

25–2 Series Resonance

When the frequency of the applied voltage is 1000 kHz in the series ac circuit in Fig. 25–2a, the reactance of the 239-μH inductance equals 1500 Ω. At the same frequency, the reactance of the 106-pF capacitance also is 1500 Ω. Therefore, this LC combination is resonant at 1000 kHz. This is f_r because the inductive reactance and capacitive reactance are equal at this frequency.

In a series ac circuit, inductive reactance leads by 90°, compared with the zero reference angle of the resistance, and capacitive reactance lags by 90°. Therefore, X_L and X_C are 180° out of phase. The opposite reactances cancel each other completely when they are equal.

Figure 25–2b shows X_L and X_C equal, resulting in a net reactance of zero ohms. The only opposition to current, then, is the coil resistance r_S, which limits how low the series resistance in the circuit can be. With zero reactance and just the low value of series resistance, the generator voltage produces the greatest amount of current in the series LC circuit at the resonant frequency. The series resistance should be as small as possible for a sharp increase in current at resonance.

Maximum Current at Series Resonance

The main characteristic of series resonance is the resonant rise of current to its maximum value of V_T/r_S at the resonant frequency. For the circuit in Fig. 25–2a, the maximum current at series resonance is 30 μA, equal to 300 μV/10 Ω. At any other frequency, either below or above the resonant frequency, there is less current in the circuit.

This resonant rise of current to 30 μA at 1000 kHz is shown in Fig. 25–3. In Fig. 25–3a, the amount of current is shown as the amplitude of individual cycles of the alternating current produced in the circuit by the ac generator voltage. Whether the amplitude of one ac cycle is considered in terms of peak, rms, or average value, the amount of current is greatest at the resonant frequency. In Fig. 25–3b, the current amplitudes are plotted on a graph for frequencies at and near the resonant frequency, producing a typical *response curve* for a series resonant circuit. The response curve in Fig. 25–3b can be considered an outline of the increasing and decreasing amplitudes of the individual cycles shown in Fig. 25–3a.

The response curve of the series resonant circuit shows that the current is small below resonance, rises to its maximum value at the resonant frequency,

IIII MultiSim Figure 25–2   Series resonance. (a) Schematic diagram of series r_S, L, and C. (b) Graph to show that reactances X_C and X_L are equal and opposite at the resonant frequency f_r. Inductive reactance is shown up for jX_L and capacitive reactance is down for $-jX_C$.

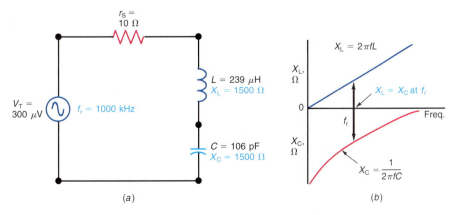

Figure 25–3 Graphs showing maximum current at resonance for the series circuit in Fig. 25–2. (*a*) Amplitudes of individual cycles. (*b*) Response curve to show the amount of *I* below and above resonance. Values of *I* are in Table 25–1.

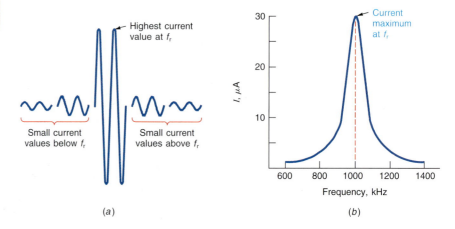

(a)

(b)

and then drops off to small values above resonance. To prove this fact, Table 25–1 lists the calculated values of impedance and current in the circuit of Fig. 25–2 at the resonant frequency of 1000 kHz and at two frequencies below and two frequencies above resonance.

Below resonance, at 600 kHz, X_C is more than X_L and there is appreciable net reactance, which limits the current to a relatively low value. At the higher frequency of 800 kHz, X_C decreases and X_L increases, making the two reactances closer to the same value. The net reactance is then smaller, allowing more current.

At the resonant frequency, X_L and X_C are equal, the net reactance is zero, and the current has its maximum value equal to V_T/r_S.

Above resonance at 1200 and 1400 kHz, X_L is greater than X_C, providing net reactance that limits the current to values much smaller than at resonance. In summary,

1. Below the resonant frequency, X_L is small, but X_C has high values that limit the amount of current.

Table 25–1	Series-Resonance Calculations for the Circuit in Fig. 25–2*							
Frequency, kHz	$X_L = 2\pi fL$, Ω	$X_C = 1/(2\pi fC)$, Ω	Net Reactance, Ω		Z_T, Ω†	$I = V_T/Z_T$, μA†	$V_L = IX_L$, μV	$V_C = IX_C$, μV
			$X_C - X_L$	$X_L - X_C$				
600	900	2500	1600		1600	0.19	171	475
800	1200	1875	675		675	0.44	528	825
$f_r \rightarrow$ 1000	1500	1500	0	0	10	30	45,000	45,000
1200	1800	1250		550	550	0.55	990	688
1400	2100	1070		1030	1030	0.29	609	310

* $L = 239$ μH, $C = 106$ pF, $V_T = 300$ μV, $r_S = 10$ Ω.
† Z_T and *I* calculated without r_S when its resistance is very small compared with the net X_L or X_C. Z_T and *I* are resistive at f_r.

2. Above the resonant frequency, X_C is small, but X_L has high values that limit the amount of current.

3. At the resonant frequency, X_L equals X_C, and they cancel to allow maximum current.

Minimum Impedance at Series Resonance

Since reactances cancel at the resonant frequency, the impedance of the series circuit is minimum, equal to just the low value of series resistance. This minimum impedance at resonance is resistive, resulting in zero phase angle. At resonance, therefore, the resonant current is in phase with the generator voltage.

Resonant Rise in Voltage Across Series *L* or *C*

The maximum current in a series *LC* circuit at resonance is useful because it produces maximum voltage across either X_L or X_C at the resonant frequency. As a result, the series resonant circuit can select one frequency by providing much more voltage output at the resonant frequency, compared with frequencies above and below resonance. Figure 25–4 illustrates the resonant rise in voltage across the capacitance in a series ac circuit. At the resonant frequency of 1000 kHz, the voltage across *C* rises to 45,000 μV, and the input voltage is only 300 μV.

In Table 25–1, the voltage across *C* is calculated as IX_C, and across *L* as IX_L. Below the resonant frequency, X_C has a higher value than at resonance, but the current is small. Similarly, above the resonant frequency, X_L is higher than at resonance, but the current has a low value because of inductive reactance. At resonance, although X_L and X_C cancel each other to allow maximum current, each reactance by itself has an appreciable value. Since the current is the same in all parts of a series circuit, the maximum current at resonance produces maximum voltage IX_C across *C* and an equal IX_L voltage across *L* for the resonant frequency.

Although the voltage across X_C and X_L is reactive, it is an actual voltage that can be measured. In Fig. 25–5, the voltage drops around the series resonant circuit are 45,000 μV across *C*, 45,000 μV across *L*, and 300 μV across r_S. The voltage across the resistance is equal to and in phase with the generator voltage.

Across the series combination of both *L* and *C*, the voltage is zero because the two series voltage drops are equal and opposite. To use the resonant rise of voltage, therefore, the output must be connected across either *L* or *C* alone. We can consider the V_L and V_C voltages similar to the idea of two batteries connected in series opposition. Together, the resultant is zero for equal and opposite voltages, but each battery still has its own potential difference.

In summary, the main characteristics of a series resonant circuit are

1. The current *I* is maximum at the resonant frequency f_r.

2. The current *I* is in phase with the generator voltage, or the phase angle of the circuit is 0°.

III MultiSim **Figure 25–4** Series circuit selects frequency by producing maximum IX_C voltage output across *C* at resonance.

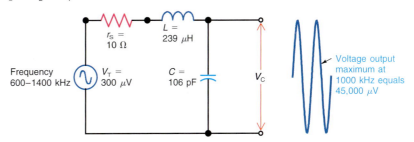

Figure 25–5 Voltage drops around series resonant circuit.

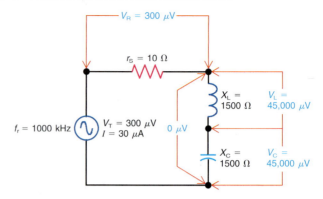

3. The voltage is maximum across either L or C alone.
4. The impedance is minimum at f_r, equal only to the low r_S.

■ *25–2 Knowledge Check*

Answer at end of chapter.

In a series resonant circuit, do the values of X_L and X_C affect the value of I at f_r?

■ *25–2 Self-Review*

Answers at end of chapter.

For series resonance,
a. X_L and X_C are maximum. (True or False)
b. X_L and X_C are equal. (True or False)
c. current I is maximum. (True or False)

25–3 Parallel Resonance

When L and C are in parallel, as shown in Fig. 25–6 and X_L equals X_C, the reactive branch currents are equal and opposite at resonance. Then they cancel each other to produce minimum current in the main line. Since the line current is minimum, the impedance is maximum. These relations are based on r_S being very small compared with X_L at resonance. In this case, the branch currents are practically equal when X_L and X_C are equal.

Minimum Line Current at Parallel Resonance

To show how the current in the main line dips to its minimum value when the parallel LC circuit is resonant, Table 25–2 lists the values of branch currents and the total line current for the circuit in Fig. 25–6.

With L and C the same as in the series circuit of Fig. 25–2, X_L and X_C have the same values at the same frequencies. Since L, C, and the generator are in parallel, the voltage applied across the branches equals the generator voltage of 300 μV. Therefore, each reactive branch current is calculated as 300 μV divided by the reactance of the branch.

The values in the top row of Table 25–2 are obtained as follows: At 600 kHz, the capacitive branch current equals 300 μV/2500 Ω, or 0.12 μA. The inductive branch current at this frequency is 300 μV/900 Ω, or 0.33 μA.

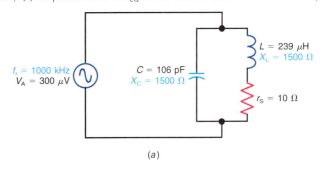

|||| MultiSim **Figure 25–6** Parallel resonant circuit. (*a*) Schematic diagram of *L* and *C* in parallel branches. (*b*) Response curve of I_T shows that the line current dips to a minimum at f_r. (*c*) Response curve of Z_{EQ} shows that it rises to a maximum at f_r.

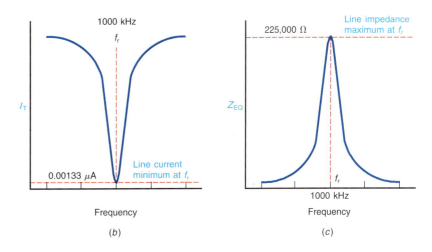

(*b*)   (*c*)

Table 25–2	Parallel-Resonance Calculations for the Circuit in Fig. 25–6*								
Frequency, kHz	$X_C = 1/(2\pi fC)$, Ω	$X_L = 2\pi fL$, Ω	$I_C = V/X_C$, μA	$I_L = V/X_L$, $\mu A\dagger$	Net Reactive Line Current, μA		I_T, $\mu A\dagger$	$Z_{EQ} = V_A/I_T$, $\Omega\dagger$
					$I_L - I_C$	$I_C - I_L$		
600	2500	900	0.12	0.33	0.21		0.21	1400
800	1875	1200	0.16	0.25	0.09		0.09	3333
$f_r \rightarrow$ 1000	1500	1500	0.20	0.20	0	0	0.00133	225,000‡
1200	1250	1800	0.24	0.17		0.07	0.07	3800
1400	1070	2100	0.28	0.14		0.14	0.14	2143

* $L = 239\ \mu H$, $C = 106\ pF$, $V_T = 300\ \mu V$, $r_S = 10\ \Omega$.
† Z_{EQ} and I calculated without r_S when its resistance is very small compared with the net X_L or X_C. Z_{EQ} and I are resistive at f_r.
‡ At resonance, Z_{EQ} is calculated by Formula (25–7). Z_{EQ} and I_T are resistive at f_r.

Since this is a parallel ac circuit, the capacitive current leads by 90°, whereas the inductive current lags by 90°, compared with the reference angle of the generator voltage, which is applied across the parallel branches. Therefore, the opposite currents are 180° out of phase. The net current in the line, then, is the difference between 0.33 and 0.12, which equals 0.21 μA.

Following this procedure, the calculations show that as the frequency is increased toward resonance, the capacitive branch current increases because of the lower value of X_C and the inductive branch current decreases with higher values of X_L. As a result, there is less net line current as the two branch currents become more nearly equal.

At the resonant frequency of 1000 kHz, both reactances are 1500 Ω, and the reactive branch currents are both 0.20 μA, canceling each other completely.

Above the resonant frequency, there is more current in the capacitive branch than in the inductive branch, and the net line current increases above its minimum value at resonance.

The dip in I_T to its minimum value at f_r is shown by the graph in Fig. 25–6b. At parallel resonance, I_T is minimum and Z_{EQ} is maximum.

The in-phase current due to r_S in the inductive branch can be ignored off-resonance because it is so small compared with the reactive line current. At the resonant frequency when the reactive currents cancel, however, the resistive component is the entire line current. Its value at resonance equals 0.00133 μA in this example. This small resistive current is the minimum value of the line current at parallel resonance.

Maximum Line Impedance at Parallel Resonance

The minimum line current resulting from parallel resonance is useful because it corresponds to maximum impedance in the line across the generator. Therefore, an impedance that has a high value for just one frequency but a low impedance for other frequencies, either below or above resonance, can be obtained by using a parallel LC circuit resonant at the desired frequency. This is another method of selecting one frequency by resonance. The response curve in Fig. 25–6c shows how the impedance rises to a maximum for parallel resonance.

The main application of parallel resonance is the use of an LC tuned circuit as the load impedance Z_L in the output circuit of rf amplifiers. Because of the high impedance, then, the gain of the amplifier is maximum at f_r. The voltage gain of an amplifier is directly proportional to Z_L. The advantage of a resonant LC circuit is that Z is maximum only for an ac signal at the resonant frequency. Also, L has practically no dc resistance, which means practically no dc voltage drop.

Referring to Table 25–2, the total impedance of the parallel ac circuit is calculated as the generator voltage divided by the total line current. At 600 kHz, for example, Z_{EQ} equals 300 μV/0.21 μA, or 1400 Ω. At 800 kHz, the impedance is higher because there is less line current.

At the resonant frequency of 1000 kHz, the line current is at its minimum of 0.00133 μA. Then the impedance is maximum and is equal to 300 μV/0.00133 μA, or 225,000 Ω.

Above 1000 kHz, the line current increases, and the impedance decreases from its maximum.

How the line current can be very low even though the reactive branch currents are appreciable is illustrated in Fig. 25–7. In Fig. 25–7a, the resistive component of the total line current is shown as though it were a separate branch drawing an amount of resistive current from the generator in the main line equal to the current resulting from the coil resistance. Each reactive branch current has its value equal to the generator voltage divided by the reactance. Since they are equal and of opposite phase, however, in any part of the circuit where both

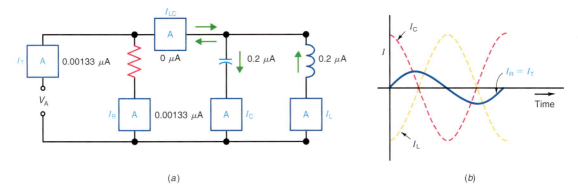

(a)   (b)

reactive currents are present, the net amount of electron flow in one direction at any instant corresponds to zero current. The graph in Fig. 25–7b shows how equal and opposite currents for I_L and I_C cancel.

If a meter is inserted in series with the main line to indicate total line current I_T, it dips sharply to the minimum value of line current at the resonant frequency. With minimum current in the line, the impedance across the line is maximum at the resonant frequency. The maximum impedance at parallel resonance corresponds to a high value of resistance, without reactance, since the line current is then resistive with zero phase angle.

In summary, the main characteristics of a parallel resonant circuit are

1. The line current I_T is minimum at the resonant frequency.
2. The current I_T is in phase with the generator voltage V_A, or the phase angle of the circuit is $0°$.
3. The impedance Z_{EQ}, equal to V_A/I_T, is maximum at f_r because of the minimum I_T.

The *LC* Tank Circuit

Note that the individual branch currents are appreciable at resonance, although I_T is minimum. For the example in Table 25–2, at f_r, either the I_L or the I_C equals 0.2 μA. This current is greater than the I_C values below f_r or the I_L values above f_r.

The branch currents cancel in the main line because I_C is at $90°$ with respect to the source V_A while I_L is at $-90°$, making them opposite with respect to each other.

However, inside the *LC* circuit, I_L and I_C do not cancel because they are in separate branches. Then I_L and I_C provide a circulating current in the *LC* circuit, which equals 0.2 μA in this example. For this reason, a parallel resonant *LC* circuit is often called a *tank circuit*.

Because of the energy stored by L and C, the circulating tank current can provide full sine waves of current and voltage output when the input is only a pulse. The sine-wave output is always at the natural resonant frequency of the *LC* tank circuit. This ability of the *LC* circuit to supply complete sine waves is called the *flywheel effect*. Also, the process of producing sine waves after a pulse of energy has been applied is called *ringing* of the *LC* circuit.

■ *25-3 Knowledge Check*

Answer at end of chapter.

In Fig. 25–6, why can't I_L and I_C be exactly equal at f_r?

Answers at end of chapter.

For parallel resonance,
a. currents I_L and I_C are maximum. (True or False)
b. currents I_L and I_C are equal. (True or False)
c. current I_T is minimum. (True or False)

25-4 Resonant Frequency $f_r = 1/(2\pi\sqrt{LC})$

The formula for the resonant frequency is derived from $X_L = X_C$. Using f_r to indicate the resonant frequency in the formulas for X_L and X_C,

$$2\pi f_r L = \frac{1}{2\pi f_r C}$$

Inverting the factor f_r gives

$$2\pi L (f_r)^2 = \frac{1}{2\pi C}$$

Inverting the factor $2\pi L$ gives

$$(f_r)^2 = \frac{1}{(2\pi)^2 LC}$$

The square root of both sides is then

$$f_r = \frac{1}{2\pi\sqrt{LC}} \tag{25-1}$$

where L is in henrys, C is in farads, and the resonant frequency f_r is in hertz (Hz). For example, to find the resonant frequency of the LC combination in Fig. 25–2, the values of 239×10^{-6} and 106×10^{-12} are substituted for L and C. Then,

$$f_r = \frac{1}{2\pi\sqrt{LC}} = \frac{1}{2\pi\sqrt{239 \times 10^{-6} \times 106 \times 10^{-12}}}$$

$$= \frac{1}{6.28\sqrt{25,334 \times 10^{-18}}} = \frac{1}{6.28 \times 159.2 \times 10^{-9}} = \frac{1}{1000 \times 10^{-9}}$$

$$= 1 \times 10^6 \text{ Hz} = 1 \text{ MHz} = 1000 \text{ kHz}$$

For any series or parallel LC circuit, the f_r equal to $1/(2\pi\sqrt{LC})$ is the resonant frequency that makes the inductive and capacitive reactances equal.

How the f_r Varies with L and C

It is important to note that higher values of L and C result in lower values of f_r. Either L or C, or both, can be varied. An LC circuit can be resonant at any frequency from a few hertz to many megahertz.

As examples, an LC combination with the relatively large values of an 8-H inductance and a 20-μF capacitance is resonant at the low audio frequency of 12.6 Hz. For a much higher frequency in the rf range, a small inductance of 2 μH will resonate with the small capacitance of 3 pF at an f_r of 64.9 MHz. These examples are solved in the next two problems for more practice with the resonant frequency formula. Such calculations are often used in practical applications of tuned circuits. Probably the most important feature of any LC

CALCULATOR

To do this problem on a calculator, keep in mind the following points. If your calculator does not have an exponential (EXP) key, work with the powers of 10 separately, without the calculator. For multiplication, add the exponents. The square root has one-half the exponent, but be sure that the exponent is an even number before dividing by 2. The reciprocal has the same exponent but with opposite sign.

For the example just solved with 239 μH for L and 106 pF for C, first punch in 239, push the $\otimes$ key, punch in 106, and then press the $\ominus$ key for the product of 25,334. While this number is on the display, push the $\sqrt{}$ key for 159.2. Keep this display, press the $\otimes$ key, punch in 6.28 for 2π, and push the $\ominus$ key for the total product of approximately 1000 in the denominator.

The powers of 10 in the denominator are $10^{-6} \times 10^{-12} = 10^{-18}$. Its square root is 10^{-9}.

While 1000 for the denominator is on the display, press the $(1/x)$ key for the reciprocal, equal to 0.001. The reciprocal of 10^{-9} is 10^9. The answer for f_r then is 0.001×10^9, which equals 1×10^6.

combination is its resonant frequency, especially in rf circuits. The applications of resonance are mainly for radio frequencies.

Example 25-1

Calculate the resonant frequency for an 8-H inductance and a 20-μF capacitance.

ANSWER

$$f_r = \frac{1}{2\pi\sqrt{LC}}$$

$$= \frac{1}{2\pi\sqrt{8 \times 20 \times 10^{-6}}}$$

$$= \frac{1}{6.28\sqrt{160 \times 10^{-6}}}$$

$$= \frac{1}{6.28 \times 12.65 \times 10^{-3}}$$

$$= \frac{1}{79.44 \times 10^{-3}}$$

$$= 0.0126 \times 10^{3}$$

$$= 12.6 \text{ Hz} \qquad \text{(approx.)}$$

Example 25-2

Calculate the resonant frequency for a 2-μH inductance and a 3-pF capacitance.

ANSWER

$$f_r = \frac{1}{2\pi\sqrt{LC}}$$

$$= \frac{1}{2\pi\sqrt{2 \times 10^{-6} \times 3 \times 10^{-12}}}$$

$$= \frac{1}{6.28\sqrt{6 \times 10^{-18}}}$$

$$= \frac{1}{6.28 \times 2.45 \times 10^{-9}}$$

$$= \frac{1}{15.4 \times 10^{-9}} = 0.065 \times 10^{9}$$

$$= 65 \times 10^{6} \text{ Hz} = 65 \text{ MHz}$$

Specifically, because of the square root in the denominator of Formula (25–1), the f_r decreases inversely as the square root of L or C. For instance, if L or C is quadrupled, the f_r is reduced by one-half. The $\frac{1}{2}$ is equal to the square root of $\frac{1}{4}$.

As a numerical example, suppose that f_r is 6 MHz with particular values of L and C. If either L or C is made four times larger, then f_r will be reduced to 3 MHz.

Or, to take the opposite case of doubling the frequency from 6 MHz to 12 MHz, the following can be done:

1. Use one-fourth the L with the same C.
2. Use one-fourth the C with the same L.
3. Reduce both L and C by one-half.
4. Use any new combination of L and C whose product will be one-fourth the original product of L and C.

LC Product Determines f_r

There are any number of LC combinations that can be resonant at one frequency. With more L, then less C can be used for the same f_r. Or less L can be used with more C. Table 25–3 lists five possible combinations of L and C resonant at 1000 kHz, as an example of one f_r. The resonant frequency is the same 1000 kHz here for all five combinations. When either L or C is increased by a factor of 10 or 2, the other is decreased by the same factor, resulting in a constant value for the LC product.

The reactance at resonance changes with different combinations of L and C, but in all five cases, X_L and X_C are equal to each other at 1000 kHz. This is the resonant frequency determined by the value of the LC product in $f_r = 1/(2\pi\sqrt{LC})$.

Measuring L or C by Resonance

Of the three factors L, C, and f_r in the resonant-frequency formula, any one can be calculated when the other two are known. The resonant frequency of the LC combination can be found experimentally by determining the frequency that produces the resonant response in an LC combination. With a known value of either L or C, and the resonant frequency determined, the third factor can be calculated.

Table 25–3		LC Combinations Resonant at 1000 kHz		
L, μH	C, pF	$L \times C$ LC Product	X_L, Ω at 1000 kHz	X_C, Ω at 1000 kHz
23.9	1060	25,334	150	150
119.5	212	25,334	750	750
239	106	25,334	1500	1500
478	53	25,334	3000	3000
2390	10.6	25,334	15,000	15,000

This method is commonly used for measuring inductance or capacitance. A test instrument for this purpose is the Q meter, which also measures the Q of a coil.

Calculating C from f_r

The C can be taken out of the square root sign or radical in the resonance formula, as follows:

$$f_r = \frac{1}{2\pi\sqrt{LC}}$$

Squaring both sides to eliminate the radical gives

$$f_r^2 = \frac{1}{(2\pi)^2 LC}$$

Inverting C and f_r^2 gives

$$C = \frac{1}{4\pi^2 f_r^2 L} \qquad (25\text{--}2)$$

where f_r is in hertz, C is in farads, and L is in henrys.

Calculating L from f_r

Similarly, the resonance formula can be transposed to find L. Then

$$L = \frac{1}{4\pi^2 f_r^2 C} \qquad (25\text{--}3)$$

With Formula (25–3), L is determined by f_r with a known value of C. Similarly, C is determined from Formula (25–2) by f_r with a known value of L.

Example 25-3

What value of C resonates with a 239-μH L at 1000 kHz?

ANSWER

$$C = \frac{1}{4\pi^2 f_r^2 L}$$

$$= \frac{1}{4\pi^2 (1000 \times 10^3)^2 239 \times 10^{-6}}$$

$$= \frac{1}{39.48 \times 1 \times 10^6 \times 239}$$

$$= \frac{1}{9435.75 \times 10^6}$$

$$= 0.000106 \times 10^{-6}\,\text{F} = 106\,\text{pF}$$

Note that 39.48 is a constant for $4\pi^2$.

Example 25-4

What value of L resonates with a 106-pF C at 1000 kHz, equal to 1 MHz?

ANSWER

$$L = \frac{1}{4\pi^2 f_r^2 C}$$

$$= \frac{1}{39.48 \times 1 \times 10^{12} \times 106 \times 10^{-12}}$$

$$= \frac{1}{4184.88}$$

$$= 0.000239 \text{ H} = 239 \; \mu\text{H}$$

Note that 10^{12} and 10^{-12} in the denominator cancel each other. Also, 1×10^{12} is the square of 1×10^6, or 1 MHz.

The values in Examples 25–3 and 25–4 are from the LC circuit illustrated in Fig. 25–2 for series resonance and Fig. 25–6 for parallel resonance.

■ *25–4 Knowledge Check*

Answer at end of chapter.

If the value of the inductance in a resonant circuit is quadrupled, what happens to the resonant frequency?

■ *25–4 Self-Review*

Answers at end of chapter.

a.   To increase f_r, must C be increased or decreased?
b.   If C is increased from 100 to 400 pF, L must be decreased from 800 μH to what value for the same f_r?
c.   Give the constant value for $4\pi^2$.

25–5 Q Magnification Factor of a Resonant Circuit

The quality, or *figure of merit,* of the resonant circuit, in sharpness of resonance, is indicated by the factor Q. In general, the higher the ratio of the reactance at resonance to the series resistance, the higher the Q and the sharper the resonance effect.

Q of Series Circuit

In a series resonant circuit, we can calculate Q from the following formula:

$$Q = \frac{X_L}{r_S} \tag{25–4}$$

where Q is the figure of merit, X_L is the inductive reactance in ohms at the resonant frequency, and r_S is the resistance in ohms in series with X_L. For the series resonant circuit in Fig. 25–2,

$$Q = \frac{1500\ \Omega}{10\ \Omega} = 150$$

The Q is a numerical factor without any units, because it is a ratio of reactance to resistance and the ohms cancel. Since the series resistance limits the amount of current at resonance, the lower the resistance, the sharper the increase to maximum current at the resonant frequency, and the higher the Q. Also, a higher value of reactance at resonance allows the maximum current to produce higher voltage for the output.

The Q has the same value if it is calculated with X_C instead of X_L, since they are equal at resonance. However, the Q of the circuit is generally considered in terms of X_L, because usually the coil has the series resistance of the circuit. In this case, the Q of the coil and the Q of the series resonant circuit are the same. If extra resistance is added, the Q of the circuit will be less than the Q of the coil. The highest possible Q for the circuit is the Q of the coil.

The value of 150 can be considered a high Q. Typical values are 50 to 250, approximately. Less than 10 is a low Q; more than 300 is a very high Q.

Higher *L/C* Ratio Can Provide Higher *Q*

As shown before in Table 25–3, different combinations of L and C can be resonant at the same frequency. However, the amount of reactance at resonance is different. More X_L can be obtained with a higher L and lower C for resonance, although X_L and X_C must be equal at the resonant frequency. Therefore, both X_L and X_C are higher with a higher L/C ratio for resonance.

More X_L can allow a higher Q if the ac resistance does not increase as much as the reactance. An approximate rule for typical rf coils is that maximum Q can be obtained when X_L is about 1000 Ω. In many cases, though, the minimum C is limited by stray capacitance in the circuit.

Q Rise in Voltage Across Series *L* or *C*

The Q of the resonant circuit can be considered a magnification factor that determines how much the voltage across L or C is increased by the resonant rise of current in a series circuit. Specifically, the voltage output at series resonance is Q times the generator voltage:

$$V_L = V_C = Q \times V_{gen} \tag{25–5}$$

In Fig. 25–4, for example, the generator voltage is 300 μV and Q is 150. The resonant rise of voltage across either L or C then equals 300 μV $\times$ 150, or 45,000 μV. Note that this is the same value calculated in Table 25–1 for V_C or V_L at resonance.

How to Measure *Q* in a Series Resonant Circuit

The fundamental nature of Q for a series resonant circuit is seen from the fact that the Q can be determined experimentally by measuring the Q rise in voltage across either L or C and comparing this voltage with the generator voltage. As a formula,

$$Q = \frac{V_{out}}{V_{in}} \tag{25–6}$$

GOOD TO KNOW

Formula 25–5 is derived as follows: $V_L = I \times X_L$ where $I = \dfrac{V_{Gen}}{r_S}$ at the resonant frequency,

f_r. Therefore, $V_L = \dfrac{V_{Gen}}{r_S} \times X_L$ or

$V_L = \dfrac{X_L}{r_S} \times V_{Gen}$. Since $Q = \dfrac{X_L}{r_S}$

we have $V_L = Q \times V_{Gen}$.

Since $X_L = X_C$ at f_r, formula 25–5 can be used to find both V_L and V_C.

where V_{out} is the ac voltage measured across the coil or capacitor and V_{in} is the generator voltage.

Referring to Fig. 25–5, suppose that you measure with an ac voltmeter across L or C and this voltage equals 45,000 μV at the resonant frequency. Also, measure the generator input of 300 μV. Then

$$Q = \frac{V_{out}}{V_{in}}$$
$$= \frac{45{,}000 \; \mu V}{300 \; \mu V}$$
$$= 150$$

This method is better than the X_L/r_S formula for determining Q because r_S is the ac resistance of the coil, which is not so easily measured. Remember that the coil's ac resistance can be more than double the dc resistance measured with an ohmmeter. In fact, measuring Q with Formula (25–6) makes it possible to calculate the ac resistance. These points are illustrated in the following examples.

Example 25–5

A series circuit resonant at 0.4 MHz develops 100 mV across a 250-μH L with a 2-mV input. Calculate Q.

ANSWER

$$Q = \frac{V_{out}}{V_{in}} = \frac{100 \; mV}{2 \; mV}$$
$$= 50$$

Example 25–6

What is the ac resistance of the coil in the preceding example?

ANSWER The Q of the coil is 50. We need to know the reactance of this 250-μH coil at the frequency of 0.4 MHz. Then,

$$X_L = 2\pi fL = 6.28 \times 0.4 \times 10^6 \times 250 \times 10^{-6}$$
$$= 628 \; \Omega$$

Also, $Q = \dfrac{X_L}{r_S}$ or $r_S = \dfrac{X_L}{Q}$

$$r_S = \frac{628 \; \Omega}{50}$$
$$= 12.56 \; \Omega$$

Figure 25–8 The *Q* of a parallel resonant circuit in terms of X_L and its series resistance r_S.

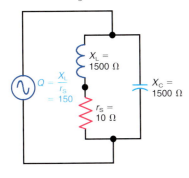

Q of Parallel Circuit

In a parallel resonant circuit where r_S is very small compared with X_L, the Q also equals X_L/r_S. Note that r_S is still the resistance of the coil in series with X_L (see Fig. 25–8). The Q of the coil determines the Q of the parallel circuit here because it is less than the Q of the capacitive branch. Capacitors used in tuned circuits generally have a very high Q because of their low losses. In Fig. 25–8, the Q is 1500 Ω/10 Ω, or 150, the same as the series resonant circuit with the same values.

This example assumes that the generator resistance is very high and that there is no other resistance branch shunting the tuned circuit. Then the Q of the parallel resonant circuit is the same as the Q of the coil. Actually, shunt resistance can lower the Q of a parallel resonant circuit, as analyzed in Sec. 25–10.

Q Rise in Impedance Across a Parallel Resonant Circuit

For parallel resonance, the Q magnification factor determines by how much the impedance across the parallel LC circuit is increased because of the minimum line current. Specifically, the impedance across the parallel resonant circuit is Q times the inductive reactance at the resonant frequency:

$$Z_{EQ} = Q \times X_L \tag{25–7}$$

Referring back to the parallel resonant circuit in Fig. 25–6 as an example, X_L is 1500 Ω and Q is 150. The result is a rise of impedance to the maximum value of 150 × 1500 Ω, or 225,000 Ω, at the resonant frequency.

Since the line current equals V_A/Z_{EQ}, the minimum line current is 300 μV/225,000 Ω, which equals 0.00133 μA.

At f_r, the minimum line current is $1/Q$ of either branch current. In Fig. 25–7, I_L or I_C is 0.2 μA and Q is 150. Therefore, I_T is 0.2/150, or 0.00133 μA, which is the same answer as V_A/Z_{EQ}. Or, stated another way, the circulating tank current is Q times the minimum I_T.

How to Measure Z_{EQ} of a Parallel Resonant Circuit

Formula (25–7) for Z_{EQ} is also useful in its inverted form as $Q = Z_{EQ}/X_L$. We can measure Z_{EQ} by the method illustrated in Fig. 25–9. Then Q can be calculated from the value of Z_{EQ} and the inductive reactance of the coil.

GOOD TO KNOW

Another formula which can be used to calculate Z_{EQ} at f_r is:

$Z_{EQ} = (1 + Q^2)r_S$, where $Q = \dfrac{X_L}{r_S}$.

Figure 25–9 How to measure Z_{EQ} of a parallel resonant circuit. Adjust R_1 to make its V_R equal to V_{LC}. Then $Z_{EQ} = R_1$.

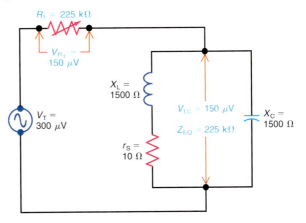

To measure Z_{EQ}, first tune the LC circuit to resonance. Then adjust R_1 in Fig. 25–9 to the resistance that makes its ac voltage equal to the ac voltage across the tuned circuit. With equal voltages, the Z_{EQ} must have the same value as R_1.

For the example here, which corresponds to the parallel resonance shown in Figs. 25–6 and 25–8, Z_{EQ} is equal to 225,000 Ω. This high value is a result of parallel resonance. The X_L is 1500 Ω. Therefore, to determine Q, the calculations are

$$Q = \frac{Z_{EQ}}{X_L} = \frac{225,000}{1500} = 150$$

Example 25-7

In Fig. 25–9, assume that with a 4-mV ac input signal for V_T, the voltage across R_1 is 2 mV when R_1 is 225 kΩ. Determine Z_{EQ} and Q.

ANSWER   Because they divide V_T equally, Z_{EQ} is 225 kΩ, the same as R_1. The amount of input voltage does not matter, as the voltage division determines the relative proportions between R_1 and Z_{EQ}. With 225 kΩ for Z_{EQ} and 1.5 kΩ for X_L, the Q is 225/1.5, or $Q = 150$.

Example 25-8

A parallel LC circuit tuned to 200 kHz with a 350-μH L has a measured Z_{EQ} of 17,600 Ω. Calculate Q.

ANSWER   First, calculate X_L as $2\pi fL$ at f_r:

$$X_L = 2\pi \times 200 \times 10^3 \times 350 \times 10^{-6} = 440 \ \Omega$$

Then,

$$Q = \frac{Z_{EQ}}{X_L} = \frac{17,600}{440}$$

$$= 40$$

■ 25–5 *Knowledge Check*

Answer at end of chapter.

Give an example of a low Q and a high Q.

■ 25–5 *Self-Review*

Answers at end of chapter.

a. In a series resonant circuit, V_L is 300 mV with an input of 3 mV. Calculate Q.
b. In a parallel resonant circuit, X_L is 500 Ω. With a Q of 50, calculate Z_{EQ}.

25–6 Bandwidth of a Resonant Circuit

When we say that an *LC* circuit is resonant at one frequency, this is true for the maximum resonance effect. However, other frequencies close to f_r also are effective. For series resonance, frequencies just below and above f_r produce increased current, but a little less than the value at resonance. Similarly, for parallel resonance, frequencies close to f_r can provide high impedance, although a little less than the maximum Z_{EQ}.

Therefore, any resonant frequency has an associated band of frequencies that provide resonance effects. How wide the band is depends on the Q of the resonant circuit. Actually, it is practically impossible to have an *LC* circuit with a resonant effect at only one frequency. The width of the resonant band of frequencies centered around f_r is called the *bandwidth* of the tuned circuit.

Measurement of Bandwidth

The group of frequencies with a response 70.7% of maximum, or more, is generally considered the bandwidth of the tuned circuit, as shown in Fig. 25–10*b*. The resonant response here is increasing current for the series circuit in Fig. 25–10*a*. Therefore, the bandwidth is measured between the two frequencies f_1 and f_2 producing 70.7% of the maximum current at f_r.

For a parallel circuit, the resonant response is increasing impedance Z_{EQ}. Then the bandwidth is measured between the two frequencies allowing 70.7% of the maximum Z_{EQ} at f_r.

The bandwidth indicated on the response curve in Fig. 25–10*b* equals 20 kHz. This is the difference between f_2 at 60 kHz and f_1 at 40 kHz, both with 70.7% response.

Compared with the maximum current of 100 mA for f_r at 50 kHz, f_1 below resonance and f_2 above resonance each allows a rise to 70.7 mA. All frequencies in this band 20 kHz wide allow 70.7 mA, or more, as the resonant response in this example.

Bandwidth Equals f_r/Q

Sharp resonance with high Q means narrow bandwidth. The lower the Q, the broader the resonant response and the greater the bandwidth.

Figure 25–10 Bandwidth of a tuned *LC* circuit. (*a*) Series circuit with input of 0 to 100 kHz. (*b*) Response curve with bandwidth Δf equal to 20 kHz between f_1 and f_2.

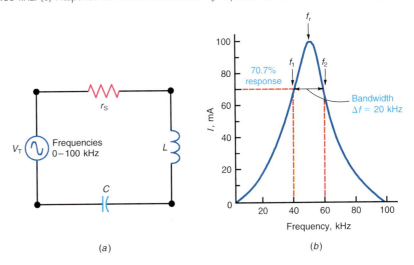

Also, the higher the resonant frequency, the greater the range of frequency values included in the bandwidth for a given sharpness of resonance. Therefore, the bandwidth of a resonant circuit depends on the factors f_r and Q. The formula is

$$f_2 - f_1 = \Delta f = \frac{f_r}{Q} \tag{25–8}$$

where Δf is the total bandwidth in the same units as the resonant frequency f_r. The bandwidth Δf can also be abbreviated BW.

For example, a series circuit resonant at 800 kHz with a Q of 100 has a bandwidth of 800/100, or 8 kHz. Then the I is 70.7% of maximum, or more, for all frequencies for a band 8 kHz wide. This frequency band is centered around 800 kHz, from 796 to 804 kHz.

With a parallel resonant circuit having a Q higher than 10, Formula (25-8) also can be used for calculating the bandwidth of frequencies that provide 70.7% or more of the maximum Z_{EQ}. However, the formula cannot be used for parallel resonant circuits with low Q, as the resonance curve then becomes unsymmetrical.

High Q Means Narrow Bandwidth

The effect for different values of Q is illustrated in Fig. 25–11. Note that a higher Q for the same resonant frequency results in less bandwidth. The slope is sharper for the sides or *skirts* of the response curve, in addition to its greater amplitude.

High Q is generally desirable for more output from the resonant circuit. However, it must have enough bandwidth to include the desired range of signal frequencies.

The Edge Frequencies

Both f_1 and f_2 are separated from f_r by one-half of the total bandwidth. For the top curve in Fig. 25–11, as an example, with a Q of 80, Δf is ±5 kHz centered around 800 kHz for f_r. To determine the edge frequencies,

Figure 25–11 Higher Q provides a sharper resonant response. Amplitude is I for series resonance or Z_{EQ} for parallel resonance. Bandwidth at half-power frequencies is Δf.

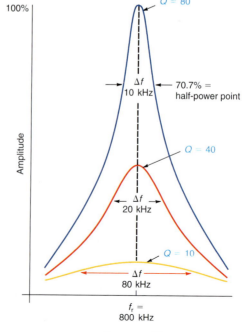

$$f_1 = f_r - \frac{\Delta f}{2} = 800 - 5 = 795 \text{ kHz}$$

$$f_2 = f_r + \frac{\Delta f}{2} = 800 + 5 = 805 \text{ kHz}$$

These examples assume that the resonance curve is symmetrical. This is true for a high-Q parallel resonant circuit and a series resonant circuit with any Q.

Example 25-9

An LC circuit resonant at 2000 kHz has a Q of 100. Find the total bandwidth Δf and the edge frequencies f_1 and f_2.

ANSWER

$$\Delta f = \frac{f_r}{Q} = \frac{2000 \text{ kHz}}{100} = 20 \text{ kHz}$$

$$f_1 = f_r - \frac{\Delta f}{2} = 2000 - 10 = 1990 \text{ kHz}$$

$$f_2 = f_r + \frac{\Delta f}{2} = 2000 + 10 = 2010 \text{ kHz}$$

Example 25-10

Repeat Example 25-9 for an f_r equal to 6000 kHz and the same Q of 100.

ANSWER

$$f = \frac{f_r}{Q} = \frac{6000 \text{ kHz}}{100} = 60 \text{ kHz}$$

$$f_1 = 6000 - 30 = 5970 \text{ kHz}$$

$$f_2 = 6000 + 30 = 6030 \text{ kHz}$$

Notice that Δf is three times as wide as Δf in Example 25-9 for the same Q because f_r is three times higher.

Half-Power Points

It is simply for convenience in calculations that the bandwidth is defined between the two frequencies having 70.7% response. At each of these frequencies, the net capacitive or inductive reactance equals the resistance. Then the total impedance of the series reactance and resistance is 1.4 times greater than R. With this much more impedance, the current is reduced to $1/1.414$, or 0.707, of its maximum value.

Furthermore, the relative current or voltage value of 70.7% corresponds to 50% in power, since power is $I^2 R$ or V^2/R and the square of 0.707 equals 0.50.

Therefore, the bandwidth between frequencies having 70.7% response in current or voltage is also the bandwidth in terms of half-power points. Formula (25–8) is derived for Δf between the points with 70.7% response on the resonance curve.

Measuring Bandwidth to Calculate Q

The half-power frequencies f_1 and f_2 can be determined experimentally. For series resonance, find the two frequencies at which the current is 70.7% of maximum I, or for parallel resonance, find the two frequencies that make the impedance 70.7% of the maximum Z_{EQ}. The following method uses the circuit in Fig. 25–9 for measuring Z_{EQ}, but with different values to determine its bandwidth and Q:

1. Tune the circuit to resonance and determine its maximum Z_{EQ} at f_r. In this example, assume that Z_{EQ} is 10,000 Ω at the resonant frequency of 200 kHz.
2. Keep the same amount of input voltage, but change its frequency slightly below f_r to determine the frequency f_1 that results in a Z_1 equal to 70.7% of Z_{EQ}. The required value here is 0.707 $\times$ 10,000, or 7070 Ω, for Z_1 at f_1. Assume that this frequency f_1 is determined to be 195 kHz.
3. Similarly, find the frequency f_2 above f_r that results in the impedance Z_2 of 7070 Ω. Assume that f_2 is 205 kHz.
4. The total bandwidth between the half-power frequencies equals $f_2 - f_1$ or 205 − 195. Then the value of $\Delta f = 10$ kHz.
5. Then $Q = f_r / \Delta f$ or 200 kHz/10 kHz = 20 for the calculated value of Q.

In this way, measuring the bandwidth makes it possible to determine Q. With Δf and f_r, Q can be determined for either parallel or series resonance.

■ *25–6 Knowledge Check*

Answer at end of chapter.

If a parallel resonant circuit has a tank impedance of 400 kΩ at f_r, how much is its impedance at the edge frequencies?

■ *25–6 Self-Review*

Answers at end of chapter.

a. **An *LC* circuit with f_r of 10 MHz has a Q of 40. Calculate the half-power bandwidth.**
b. **For an f_r of 500 kHz and bandwidth Δf of 10 kHz, calculate Q.**

25–7 Tuning

Tuning means obtaining resonance at different frequencies by varying either L or C. As illustrated in Fig. 25–12, the variable capacitance C can be adjusted to tune the series *LC* circuit to resonance at any one of the five different frequencies. Each of the voltages V_1 to V_5 indicates an ac input with a specific frequency. Which one is selected for maximum output is determined by the resonant frequency of the *LC* circuit.

When C is set to 424 pF, for example, the resonant frequency of the *LC* circuit is 500 kHz for f_{r_1}. The input voltage whose frequency is 500 kHz then produces a resonant rise of current which results in maximum output voltage across C. At other frequencies, such as 707 kHz, the voltage output is less than the input. With C at 424 pF, therefore, the *LC* circuit tuned to 500 kHz selects this frequency by providing much more voltage output than other frequencies.

Figure 25–12 Tuning a series *LC* circuit. (*a*) Input voltages at different frequencies. (*b*) Relative response for each frequency when *C* is varied (not to scale).

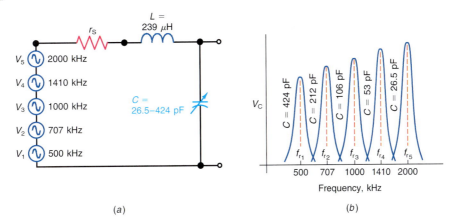

(*a*) (*b*)

Suppose that we want maximum output for the ac input voltage that has the frequency of 707 kHz. Then *C* is set at 212 pF to make the *LC* circuit resonant at 707 kHz for f_{r_2}. Similarly, the tuned circuit can resonate at a different frequency for each input voltage. In this way, the *LC* circuit is tuned to select the desired frequency.

The variable capacitance *C* can be set at the values listed in Table 25–4 to tune the *LC* circuit to different frequencies. Only five frequencies are listed here, but any one capacitance value between 26.5 and 424 pF can tune the 239-μH coil to resonance at any frequency in the range of 500 to 2000 kHz. Note that a parallel resonant circuit also can be tuned by varying *C* or *L*.

Tuning Ratio

When an *LC* circuit is tuned, the change in resonant frequency is inversely proportional to the square root of the change in *L* or *C*. Referring to Table 25–4, notice that when *C* is decreased to one-fourth, from 424 to 106 pF, the resonant frequency doubles from 500 to 1000 kHz, or the frequency is increased by the factor $1/\sqrt{1/4}$, which equals 2.

Suppose that we want to tune through the whole frequency range of 500 to 2000 kHz. This is a tuning ratio of 4:1 for the highest to the lowest

Table 25–4	Tuning *LC* Circuit by Varying *C*	
L, μH	*C*, pF	f_r, kHz
239	424	500
239	212	707
239	106	1000
239	53	1410
239	26.5	2000

Figure 25–13 Application of tuning an *LC* circuit through the AM radio band.

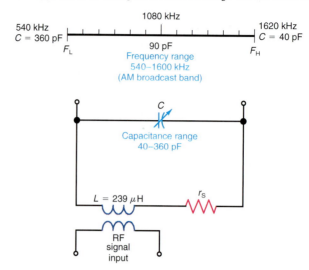

frequency. Then the capacitance must be varied from 424 to 26.5 pF, which is a 16:1 capacitance ratio.

Radio Tuning Dial

Figure 25–13 illustrates a typical application of resonant circuits in tuning a receiver to the carrier frequency of a desired station in the AM broadcast band. The tuning is done by the air capacitor *C*, which can be varied from 360 pF with the plates completely in mesh to 40 pF out of mesh. The fixed plates form the *stator,* whereas the *rotor* has plates that move in and out.

Note that the lowest frequency F_L at 540 kHz is tuned with the highest *C* at 360 pF. Resonance at the highest frequency F_H at 1620 kHz results from the lowest *C* at 40 pF.

The capacitance range of 40 to 360 pF tunes through the frequency range from 1620 kHz down to 540 kHz. Frequency F_L is one-third F_H because the maximum *C* is nine times the minimum *C*.

The same idea applies to tuning through the commercial FM broadcast band of 88 to 108 MHz with smaller values of *L* and *C*. Also, television receivers are tuned to a specific broadcast channel by resonance at the desired frequencies.

For electronic tuning, the *C* is varied by a *varactor.* This is a semiconductor diode that varies in capacitance when its voltage is changed.

■ 25–7 Knowledge Check

Answer at end of chapter.

If *C* is adjusted to 53 pF in Fig. 25–12, will the signal whose frequency is 707 kHz provide any significant output?

■ 25–7 Self-Review

Answers at end of chapter.

a. **When a tuning capacitor is completely in mesh, is the *LC* circuit tuned to the highest or lowest frequency in the band?**
b. **A tuning ratio of 2:1 in frequency requires what ratio of variable *L* or *C*?**

25-8 Mistuning

Suppose that a series *LC* circuit is tuned to 1000 kHz, but the frequency of the input voltage is 17 kHz, completely off-resonance. The circuit could provide a *Q* rise in output voltage for current having the frequency of 1000 kHz, but there is no input voltage and therefore no current at this frequency.

The input voltage produces current that has a frequency of 17 kHz. This frequency cannot produce a resonant rise in current, however, because the current is limited by the net reactance. When the frequency of the input voltage and the resonant frequency of the *LC* circuit are not the same, therefore, the mistuned circuit has very little output compared with the *Q* rise in voltage at resonance.

Similarly, when a parallel circuit is mistuned, it does not have a high value of impedance. Furthermore, the net reactance off-resonance makes the *LC* circuit either inductive or capacitive.

Series Circuit Off-Resonance

When the frequency of the input voltage is lower than the resonant frequency of a series *LC* circuit, the capacitive reactance is greater than the inductive reactance. As a result, there is more voltage across the capacitive reactance than across the inductive reactance. The series *LC* circuit is capacitive below resonance, therefore, with capacitive current leading the generator voltage.

Above the resonant frequency, the inductive reactance is greater than the capacitive reactance. As a result, the circuit is inductive above resonance with inductive current that lags the generator voltage. In both cases, there is much less output voltage than at resonance.

Parallel Circuit Off-Resonance

With a parallel *LC* circuit, the smaller amount of inductive reactance below resonance results in more inductive branch current than capacitive branch current. The net line current is inductive, therefore, making the parallel *LC* circuit inductive below resonance, as the line current lags the generator voltage.

Above the resonant frequency, the net line current is capacitive because of the higher value of capacitive branch current. Then the parallel *LC* circuit is capacitive with line current leading the generator voltage. In both cases, the total impedance of the parallel circuit is much less than the maximum impedance at resonance. Note that the capacitive and inductive effects off resonance are opposite for series and parallel *LC* circuits.

■ *25-8 Knowledge Check*

Answer at end of chapter.

Is the resonant frequency the only frequency where $\theta = 0°$ in a series *LC* circuit?

■ *25-8 Self-Review*

Answers at end of chapter.

a. Is a series resonant circuit inductive or capacitive below resonance?
b. Is a parallel resonant circuit inductive or capacitive below resonance?

25–9 Analysis of Parallel Resonant Circuits

Parallel resonance is more complex than series resonance because the reactive branch currents are not exactly equal when X_L equals X_C. The reason is that the coil has its series resistance r_S in the X_L branch, whereas the capacitor has only X_C in its branch.

For high-Q circuits, we consider r_S negligible. In low-Q circuits, however, the inductive branch must be analyzed as a complex impedance with X_L and r_S in series. This impedance is in parallel with X_C, as shown in Fig. 25–14. The total impedance Z_{EQ} can then be calculated by using complex numbers, as explained in Chap. 24.

Figure 25–14 General method of calculating Z_{EQ} for a parallel resonant circuit as $(Z_1 \times Z_2)/(Z_1 + Z_2)$ with complex numbers.

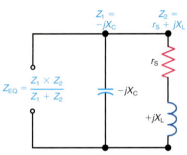

High-Q Circuit

We can apply the general method in Fig. 25–14 to the parallel resonant circuit shown before in Fig. 25–6 to see whether Z_{EQ} is 225,000 Ω. In this example, X_L and X_C are 1500 Ω and r_S is 10 Ω. The calculations are

$$Z_{EQ} = \frac{Z_1 \times Z_2}{Z_1 + Z_2} = \frac{-j1500 \times (j1500 + 10)}{-j1500 + j1500 + 10}$$

$$= \frac{-j^2 2.25 \times 10^6 - j15,000}{10} = -j^2 2.25 \times 10^5 - j1500$$

$$= 225,000 - j1500 = 225,000 \underline{/0°}\ \Omega$$

Note that $-j^2$ is $+1$. Also, the reactive $j1500$ Ω is negligible compared with the resistive 225,000 Ω. This answer for Z_{EQ} is the same as $Q \times X_L$, or 150×1500, because of the high Q with negligibly small r_S.

Low-Q Circuit

We can consider a Q less than 10 as low. For the same circuit in Fig. 25–6, if r_S is 300 Ω with an X_L of 1500 Ω, the Q will be 1500/300, which equals 5. For this case of appreciable r_S, the branch currents cannot be equal when X_L and X_C are equal because then the inductive branch will have more impedance and less current.

With a low-Q circuit, Z_{EQ} must be calculated in terms of the branch impedances. For this example, the calculations are simpler with all impedances stated in kilohms:

$$Z_{EQ} = \frac{Z_1 \times Z_2}{Z_1 + Z_2} = \frac{-j1.5 \times (j1.5 + 0.3)}{-j1.5 + j1.5 + 0.3} = \frac{-j^2 2.25 - j0.45}{0.3}$$

$$= 7.5 - j1.5\ \Omega = 7.65 \underline{/-11.3°}\ k\Omega = 7650 \underline{/-11.3°}\ \Omega$$

The phase angle θ is not zero because the reactive branch currents are unequal, even though X_L and X_C are equal. The appreciable value of r_S in the X_L branch makes this branch current smaller than I_C in the X_C branch.

Criteria for Parallel Resonance

The frequency f_r that makes $X_L = X_C$ is always $1/(2\pi\sqrt{LC})$. However, for low-Q circuits, f_r does not necessarily provide the desired resonance effect. The three main criteria for parallel resonance are

1. zero phase angle and unity power factor.
2. maximum impedance and minimum line current.
3. $X_L = X_C$. This is resonance at $f_r = 1/(2\pi\sqrt{LC})$.

GOOD TO KNOW

When $Q < 10$ in a parallel resonant circuit, Z_{EQ} is maximum and I_T is minimum at a frequency slightly less than f_r.

These three effects do not occur at the same frequency in parallel circuits that have low Q. The condition for unity power factor is often called *antiresonance* in a parallel LC circuit to distinguish it from the case of equal X_L and X_C.

Note that when Q is 10 or higher, though, the parallel branch currents are practically equal when $X_L = X_C$. Then at $f_r = 1/(2\pi\sqrt{LC})$, the line current is minimum with zero phase angle, and the impedance is maximum.

For a series resonant circuit, there are no parallel branches to consider. Therefore, the current is maximum at exactly f_r, whether the Q is high or low.

■ 25–9 Knowledge Check

Answer at end of chapter.

In a low-Q parallel resonant circuit, is Z_{EQ} maximum at f_r?

■ 25–9 Self-Review

Answers at end of chapter.

a. Is a Q of 8 a high or low value?
b. With this Q, will the I_L be more or less than I_C in the parallel branches when $X_L = X_C$?

25–10 Damping of Parallel Resonant Circuits

In Fig. 25–15a, the shunt R_P across L and C is a damping resistance because it lowers the Q of the tuned circuit. The R_P may represent the resistance of the external source driving the parallel resonant circuit, or R_P can be an actual resistor added for lower Q and greater bandwidth. Using the parallel R_P to reduce Q is better than increasing the series resistance r_s because the resonant response is more symmetrical with shunt damping.

Figure 25–15 The Q of a parallel resonant circuit in terms of coil resistance r_S and parallel damping resistor R_P. See Formula (25–10) for calculating Q. (a) Parallel R_P but negligible r_S. (b) Series r_S but no R_P branch. (c) Both R_P and r_S.

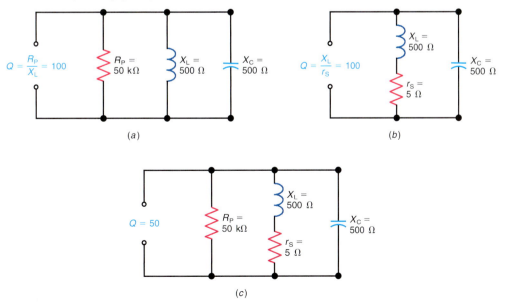

The effect of varying the parallel R_P is opposite from that of the series r_S. A lower value of R_P lowers the Q and reduces the sharpness of resonance. Remember that less resistance in a parallel branch allows more current. This resistive branch current cannot be canceled at resonance by the reactive currents. Therefore, the resonant dip to minimum line current is less sharp with more resistive line current. Specifically, when Q is determined by parallel resistance

$$Q = \frac{R_P}{X_L} \tag{25-9}$$

This relationship with shunt R_P is the reciprocal of the Q formula with series r_S. Reducing R_P decreases Q, but reducing r_S increases Q. The damping can be done by series r_S, parallel R_P, or both.

Parallel R_P Without r_S

In Fig. 25–15a, Q is determined only by the R_P, as no series r_S is shown. We can consider that r_S is zero or very small. Then the Q of the coil is infinite or high enough to be greater than the damped Q of the tuned circuit, by a factor of 10 or more. The Q of the damped resonant circuit here is $R_P/X_L = 50,000/500 = 100$.

Series r_S Without R_P

In Fig. 25–15b, Q is determined only by the coil resistance r_S, as no shunt damping resistance is used. Then $Q = X_L/r_S = 500/5 = 100$. This is the Q of the coil, which is also the Q of the parallel resonant circuit without shunt damping.

Conversion of r_S or R_P

For the circuits in both Fig. 25–15a and b, Q is 100 because the 50,000-Ω R_P is equivalent to the 5-Ω r_S as a damping resistance. One value can be converted to the other. Specifically,

$$r_S = \frac{X_L^2}{R_P}$$

or

$$R_P = \frac{X_L^2}{r_S}$$

In this example, r_S equals $250,000/50,000 = 5\ \Omega$, or R_P is $250,000/5 = 50,000\ \Omega$.

Damping with Both r_S and R_P

Figure 25–15c shows the general case of damping where both r_S and R_P must be considered. Then the Q of the circuit can be calculated as

$$Q = \frac{X_L}{r_S + X_L^2/R_P} \tag{25-10}$$

For the values in Fig. 25–15c,

$$Q = \frac{500}{5 + 250,000/50,000} = \frac{500}{5 + 5} = \frac{500}{10}$$
$$= 50$$

The Q is lower here compared with Fig. 25–15a or b because this circuit has both series and shunt damping.

Note that for an r_S of zero, Formula (25–10) can be inverted and simplified to $Q = R_P/X_L$. This is the same as Formula (25–9) for shunt damping alone.

For the opposite case where R_P is infinite, that is, an open circuit, Formula (25–10) reduces to X_L/r_S. This is the same as Formula (25–4) without shunt damping.

■ *25–10 Knowledge Check*

Answer at end of chapter.

Does the formula $R_P = \dfrac{X_L^2}{r_S}$ provide the same answer as

$Z_{EQ} = Q \times X_L$?

■ *25–10 Self-Review*

Answers at end of chapter.

a. A parallel resonant circuit has an X_L of 1000 Ω and an r_S of 20 Ω, without any shunt damping. Calculate Q.
b. A parallel resonant circuit has an X_L of 1000 Ω, negligible r_S, and shunt R_P of 50 kΩ. Calculate Q.
c. How much is Z_{EQ} at f_r for the circuits in (a) and (b)?

25–11 Choosing L and C for a Resonant Circuit

The following example illustrates how resonance is an application of X_L and X_C. Suppose that we have the problem of determining the inductance and capacitance for a circuit resonant at 159 kHz. First, we need a known value for either L or C to calculate the other. Which one to choose depends on the application. In some cases, particularly at very high frequencies, C must be the minimum possible value, which might be about 10 pF. At medium frequencies, though, we can choose L for the general case when an X_L of 1000 Ω is desirable and can be obtained. Then the inductance of the required L, equal to $X_L/2\pi f$, is 0.001 H or 1 mH, for the inductive reactance of 1000 Ω.

For resonance at 159 kHz with a 1-mH L, the required C is 0.001 μF or 1000 pF. This value of C can be calculated for an X_C of 1000 Ω, equal to X_L at the f_r of 159 kHz, or from Formula (25–2). In either case, if you substitute 1×10^{-9} F for C and 1×10^{-3} H for L in the resonant frequency formula, f_r will be 159 kHz.

This combination is resonant at 159 kHz whether L and C are in series or parallel. In series, the resonant effect produces maximum current and maximum voltage across L or C at 159 kHz. The effect is desirable for the input circuit of an rf amplifier tuned to f_r because of the maximum signal. In parallel, the resonant effect at 159 kHz is minimum line current and maximum impedance across the generator. This effect is desirable for the output circuit of an rf amplifier, as the gain is maximum at f_r because of the high Z.

If we assume that the 1-mH coil used for L has an internal resistance of 20 Ω, the Q of the coil is 1000 Ω/20 Ω, which equals 50. This value is also the Q of the series resonant circuit. If there is no shunt damping resistance across the parallel LC circuit, its Q is also 50. With a Q of 50, the bandwidth of the resonant circuit is 159 kHz/50, which equals 3.18 kHz for Δf.

■ 25-11 Knowledge Check

Answer at end of chapter.

For medium frequencies, what value of X_L is usually chosen at f_r?

■ 25-11 Self-Review

Answers at end of chapter.

a. What is f_r for 1000 pF of C and 1 mH of L?
b. What is f_r for 250 pF of C and 1 mH of L?

Summary

- Series and parallel resonance are compared in Table 25–5. The main difference is that series resonance produces maximum current and very low impedance at f_r, but with parallel resonance, the line current is minimum to provide very high impedance. Remember that these formulas for parallel resonance are very close approximations that can be used for circuits with Q higher than 10. For series resonance, the formulas apply whether the Q is high or low.

Table 25–5	Comparison of Series and Parallel Resonance	
Series Resonance	**Parallel Resonance (high Q)**	
$f_r = \dfrac{1}{2\pi\sqrt{LC}}$	$f_r = \dfrac{1}{2\pi\sqrt{LC}}$	
I maximum at f_r with θ of 0°	I_T minimum at f_r with θ of 0°	
Impedance Z minimum at f_r	Impedance Z maximum at f_r	
$Q = X_L/r_S$, or	$Q = X_L/r_S$, or	
$Q = V_{out}/V_{in}$	$Q = Z_{max}/X_L$	
Q rise in voltage $= Q \times V_{gen}$	Q rise in impedance $= Q \times X_L$	
Bandwidth $\Delta f = f_r/Q$	Bandwidth $\Delta f = f_r/Q$	
Circuit capacitive below f_r, but inductive above f_r	Circuit inductive below f_r, but capacitive above f_r	
Needs low-resistance source for low r_S, high Q, and sharp tuning	Needs high-resistance source for high R_P, high Q, and sharp tuning	
Source is inside LC circuit	Source is outside LC circuit	

Important Terms

- Antiresonance - a term to describe the condition of unity power factor in a parallel LC circuit. The term antiresonance is used to distinguish it from the case of equal X_L and X_C values in a series LC circuit.

- Bandwidth - the width of the resonant band of frequencies centered around the resonant frequency of an LC circuit.

- Damping - a technique for reducing the Q of a resonant circuit to increase the bandwidth. For a parallel resonant circuit, damping is typically accomplished by adding a parallel resistor across the tank circuit.

- Flywheel Effect - the effect that reproduces complete sine waves in a parallel LC tank circuit when the input is only a pulse.

- Half-Power Points - the frequencies above and below the resonant frequency which have a current or voltage value equal to 70.7% of its value at resonance.

- Q of a Resonant Circuit - a measure of the sharpness of a resonant circuit's response curve. The higher the ratio of the reactance at resonance to the series resistance, the higher the Q and the sharper the resonant effect.

- Resonant Frequency - the frequency at which the inductive reactance, X_L, and the capacitive reactance, X_C, of an LC circuit are equal.

- Tank Circuit - another name for a parallel resonant LC circuit.

- Tuning - a means of obtaining resonance at different frequencies by varying either L or C in an LC circuit.

Related Formulas

$$f_r = \frac{1}{2\pi\sqrt{LC}}$$

$$C = \frac{1}{4\pi^2 f_r^2 L}$$

$$L = \frac{1}{4\pi^2 f_r^2 C}$$

$$Q = \frac{X_L}{r_S} \text{ (series resonant circuit or parallel resonant circuit with no } R_P\text{)}$$

$$V_L = V_C = Q \times V_{gen}$$

$$Q = \frac{V_{out}}{V_{in}}$$

$$Z_{EQ} = Q \times X_L \text{ (parallel resonant circuit)}$$

$$f_2 - f_1 = \Delta f = \frac{f_r}{Q}$$

$$Q = \frac{R_P}{X_L} \text{ (} Q \text{ for parallel resonant circuit without series } r_S\text{)}$$

$$Q = \frac{X_L}{r_S + X_L^2/R_P} \text{ (} Q \text{ for parallel resonant circuit with both } r_S \text{ and } R_P\text{)}$$

Self-Test

Answers at back of book.

1. **The resonant frequency of an LC circuit is the frequency where**
   a. $X_L = 0\ \Omega$ and $X_C = 0\ \Omega$.
   b. $X_L = X_C$.
   c. X_L and r_S of the coil are equal.
   d. X_L and X_C are in phase.

2. **The impedance of a series LC circuit at resonance is**
   a. maximum.
   b. nearly infinite.
   c. minimum.
   d. both a and b.

3. **The total line current, I_T, of a parallel LC circuit at resonance is**
   a. minimum.
   b. maximum.

   c. equal to I_L and I_C.
   d. Q times larger than I_L or I_C.

4. **The current at resonance in a series LC circuit is**
   a. zero.
   b. minimum.
   c. different in each component.
   d. maximum.

5. **The impedance of a parallel LC circuit at resonance is**
   a. zero.
   b. maximum.
   c. minimum.
   d. equal to the r_S of the coil.

6. **The phase angle of an LC circuit at resonance is**
   a. 0°.
   b. +90°.
   c. 180°.
   d. −90°.

7. **Below resonance, a series LC circuit appears**
   a. inductive.
   b. resistive.
   c. capacitive.
   d. none of the above.

8. Above resonance, a parallel *LC* circuit appears
   a. inductive.
   b. resistive.
   c. capacitive.
   d. none of the above.

9. A parallel *LC* circuit has a resonant frequency of 3.75 MHz and a *Q* of 125. What is the bandwidth?
   a. 15 kHz.
   b. 30 kHz.
   c. 60 kHz.
   d. none of the above.

10. What is the resonant frequency of an *LC* circuit with the following values: $L = 100\ \mu H$ and $C = 63.3$ pF?
    a. $f_r = 1$ MHz.
    b. $f_r = 8$ MHz.
    c. $f_r = 2$ MHz.
    d. $f_r = 20$ MHz.

11. What value of capacitance is needed to provide a resonant frequency of 1 MHz if *L* equals $50\ \mu H$?
    a. 506.6 pF.
    b. $506.6\ \mu F$.
    c. $0.001\ \mu F$.
    d. $0.0016\ \mu F$.

12. When either *L* or *C* is increased, the resonant frequency of an *LC* circuit
    a. decreases.
    b. increases.
    c. doesn't change.
    d. This is impossible to determine.

13. A series *LC* circuit has a *Q* of 100 at resonance. If $V_{in} = 5$ mV$_{pp}$, how much is the voltage across *C*?
    a. $50\ \mu V_{pp}$.
    b. 5 mV$_{pp}$.
    c. 50 mV$_{pp}$.
    d. 500 mV$_{pp}$.

14. In a low *Q* parallel resonant circuit, when $X_L = X_C$,
    a. $I_L = I_C$.
    b. I_L is less than I_C.
    c. I_C is less than I_L.
    d. I_L is more than I_C.

15. To double the resonant frequency of an *LC* circuit with a fixed value of *L*, the capacitance, *C*, must be
    a. doubled.
    b. quadrupled.
    c. reduced by one-half.
    d. reduced by one-quarter.

16. A higher *Q* for a resonant circuit provides a
    a. dampened response curve.
    b. wider bandwidth.
    c. narrower bandwidth.
    d. none of the above.

17. The current at the resonant frequency of a series *LC* circuit is 10mA$_{pp}$. What is the value of current at the half-power points?
    a. 7.07 mA$_{pp}$.
    b. 14.14 mA$_{pp}$.
    c. 5 mA$_{pp}$.
    d. 10 mA$_{pp}$.

18. The *Q* of a parallel resonant circuit can be lowered by
    a. placing a resistor in parallel with the tank.
    b. adding more resistance in series with the coil.
    c. decreasing the value of *L* or *C*.
    d. both a and b.

19. The ability of an *LC* circuit to supply complete sine waves when the input to the tank is only a pulse is called
    a. tuning.
    b. the flywheel effect.
    c. antiresonance.
    d. its *Q*.

20. Which of the following can provide a higher *Q*?
    a. a higher *L/C* ratio.
    b. a lower *L/C* ratio.
    c. more resistance in series with the coil.
    d. either b or c.

Questions

1. (a) State two characteristics of series resonance. (b) With a microammeter measuring current in the series *LC* circuit of Fig. 25–2, describe the meter readings for the different frequencies from 600 to 1400 kHz.

2. (a) State two characteristics of parallel resonance. (b) With a microammeter measuring current in the main line for the parallel *LC* circuit in Fig. 25–6a, describe the meter readings for frequencies from 600 to 1400 kHz.

3. State the *Q* formula for the following *LC* circuits: (a) series resonant; (b) parallel resonant, with series resistance r_S in the inductive branch; (c) parallel resonant with zero series resistance but shunt R_P.

4. Explain briefly why a parallel *LC* circuit is inductive but a series *LC* circuit is capacitive below f_r.

5. What is the effect on *Q* and bandwidth of a parallel resonant circuit if its shunt damping resistance is decreased from 50,000 to 10,000 Ω?

6. Describe briefly how you would use an ac meter to measure the bandwidth of a series resonant circuit to calculate the circuit *Q*.

7. Why is a low-resistance generator good for high *Q* in series resonance, but a high-resistance generator is needed for high *Q* in parallel resonance?

8. Referring to Fig. 25–13, why is it that the middle frequency of 1080 kHz does not correspond to the middle capacitance value of 200 pF?

9. (a) Give three criteria for parallel resonance. (b) Why is the antiresonant frequency f_a different from f_r with a low-Q circuit? (c) Why are they the same for a high-Q circuit?

10. Show how Formula (25–10) reduces to R_P/X_L when r_S is zero.

11. (a) Specify the edge frequencies f_1 and f_2 for each of the three response curves in Fig. 25–11. (b) Why does lower Q allow more bandwidth?

12. (a) Why does maximum Z for a parallel resonant circuit correspond to minimum line current? (b) Why does zero phase angle for a resonant circuit correspond to unity power factor?

13. Explain how manual tuning of an LC circuit can be done with a capacitor or a coil.

14. What is meant by *electronic tuning*?

15. Suppose it is desired to tune an LC circuit from 540 to 1600 kHz by varying either L or C. Explain how the bandwidth Δf is affected by (a) varying L to tune the LC circuit; (b) varying C to tune the LC circuit.

Problems

SECTION 25–1 THE RESONANCE EFFECT

25–1 Define what is meant by a resonant circuit.

25–2 What is the main application of resonance?

25–3 If an inductor in a resonant LC circuit has an X_L value of 1 kΩ, how much is X_C?

SECTION 25–2 SERIES RESONANCE

25–4 List the main characteristics of a series resonant circuit.

25–5 Figure 25–16 shows a series resonant circuit with the values of X_L and X_C at f_r. Calculate the
  a. net reactance, X.
  b. total impedance, Z_T.
  c. current, I.
  d. phase angle, θ.
  e. voltage across L.
  f. voltage across C.
  g. voltage across r_S.

25–8 In Fig. 25–16, assume that the frequency of the applied voltage increases slightly above f_r and $X_L = 2.02$ kΩ, and $X_C = 1.98$ kΩ. Calculate the
  a. net reactance, X.
  b. total impedance, Z_T.
  c. current, I.
  d. phase angle, θ.
  e. voltage across L.
  f. voltage across C.
  g. voltage across r_S.

SECTION 25–3 PARALLEL RESONANCE

25–9 List the main characteristics of a parallel resonant circuit.

25–10 Figure 25–17 shows a parallel resonant circuit with the same values for X_L, X_C, and r_S as in Fig. 25–16. With an applied voltage of 10 V and a total line current, I_T, of 100 μA, calculate the
  a. inductive current, I_L (Ignore r_S).
  b. capacitive current, I_C.
  c. net reactive branch current, I_X.
  d. equivalent impedance, Z_{EQ}, of the tank circuit.

Figure 25–16

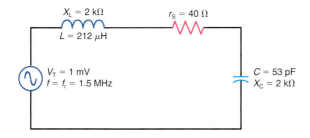

25–6 In Fig. 25–16, what happens to Z_T and I if the frequency of the applied voltage increases or decreases from the resonant frequency, f_r? Explain your answer.

25–7 In Fig. 25–16, why is the phase angle, θ, 0° at f_r?

Figure 25–17

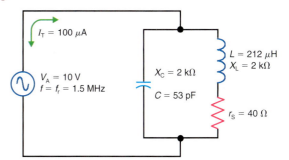

25–11 In Fig. 25–17, is X_L, X_C, or r_S responsible for the line current, I_T, of 100 μA at f_r?

25–12 In Fig. 25–17, what happens to Z_{EQ} and I_T as the frequency of the applied voltage increases or decreases from the resonant frequency, f_r? Explain your answer.

SECTION 25–4 RESONANT FREQUENCY $f_r = 1/2\pi \sqrt{LC}$

25–13 Calculate the resonant frequency, f_r, of an LC circuit with the following values:

   a. $L = 100\ \mu$H and $C = 40.53$ pF

   b. $L = 250\ \mu$H and $C = 633.25$ pF

   c. $L = 40\ \mu$H and $C = 70.36$ pF

   d. $L = 50\ \mu$H and $C = 20.26$ pF

25–14 What value of inductance, L, must be connected in series with a 50-pF capacitance to obtain an f_r of 3.8 MHz?

25–15 What value of capacitance, C, must be connected in parallel with a 100-μH inductance to obtain an f_r of 1.9 MHz?

25–16 In Fig. 25–18, what is the range of resonant frequencies as C is varied from 40 to 400 pF?

Figure 25–18

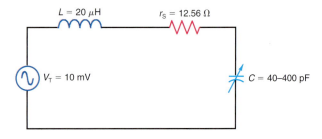

25–17 With C set at 50.67 pF in Fig. 25–18, solve for the following:

   a. f_r

   b. X_L and X_C at f_r

   c. Z_T at f_r

   d. I at f_r

   e. V_L and V_C at f_r

   f. θ_Z at f_r

25–18 In Fig. 25–18, what value of C will provide an f_r of 2.5 MHz?

25–19 If C is set at 360 pF in Fig. 25–18, what is the resonant frequency, f_r? What value of C will double the resonant frequency?

SECTION 25–5 Q MAGNIFICATION FACTOR OF A RESONANT CIRCUIT

25–20 What is the Q of the series resonant circuit in Fig. 25–18 with C set at

   a. 202.7 pF?

   b. 50.67 pF?

25–21 A series resonant circuit has the following values: $L = 50\ \mu$H, $C = 506.6$ pF, $r_S = 3.14\ \Omega$, and $V_{in} = 10$ mV. Calculate the following:

   a. f_r

   b. Q

   c. V_L and V_C

25–22 In reference to Prob. 25–21, assume that L is doubled to 100 μH and C is reduced by one-half to 253.3 pF. If r_S and V_{in} remain the same, recalculate the following:

   a. f_r

   b. Q

   c. V_L and V_C

25–23 What is the Q of a series resonant circuit if the output voltage across the capacitor is 15 V_{pp} with an input voltage of 50 mV_{pp}?

25–24 Explain why the Q of a resonant circuit cannot increase without limit as X_L increases for higher frequencies.

25–25 In Fig. 25–19, solve for the following:

   a. f_r

   b. X_L and X_C at f_r

   c. I_L and I_C at f_r

   d. Q

   e. Z_{EQ} at f_r

   f. I_T

Figure 25–19

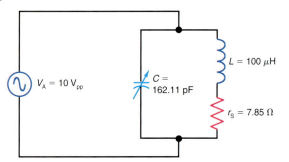

25–26 For the parallel resonant circuit in Fig. 25–17, prove that $Z_{EQ} = 100$ kΩ, as calculated earlier in Prob. 25–10(d).

25–27 The equivalent impedance, Z_{EQ}, of an LC tank circuit measures 150 kΩ using the experimental technique shown in Fig. 25–9. If the resonant frequency is 2.5 MHz and $L = 50\ \mu$H, calculate the Q of the resonant circuit.

SECTION 25–6 BANDWIDTH OF A RESONANT CIRCUIT

25–28 With C set at 50.67 pF in Fig. 25–18, calculate the following:

   a. the bandwidth, Δf

   b. the edge frequencies f_1 and f_2

   c. the current, I, at f_r, f_1, and f_2

25–29 In Fig. 25–19, calculate the following:

    a. the bandwidth, Δf

    b. the edge frequencies f_1 and f_2

    c. the equivalent impedance, Z_{EQ} at f_r, f_1, and f_2

25–30 Does a higher Q correspond to a wider or narrower bandwidth?

25–31 In Fig. 25–20, calculate the following:

    a. f_r

    b. X_L and X_C at f_r

    c. Z_T at f_r

    d. I at f_r

    e. Q

    f. V_L and V_C at f_r

    g. θ_Z at f_r

    h. Δf, f_1, and f_2

    i. I at f_1 and f_2

Figure 25–20

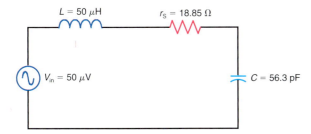

25–32 In Fig. 25–20, calculate the following values at the edge frequency, f_1:

    a. X_L

    b. X_C

    c. the net reactance, X

    d. Z_T

    e. I

    f. θ_Z

25–33 Using the values from Probs. 25–31 and 25–32, compare Z_T and I at f_1 and f_r.

25–34 In Fig. 25–21, calculate the following:

    a. f_r

    b. X_L and X_C at f_r

    c. I_L and I_C at f_r

    d. Q

    e. Z_{EQ}

    f. I_T

    g. θ_I

    h. Δf, f_1, and f_2

Figure 25–21

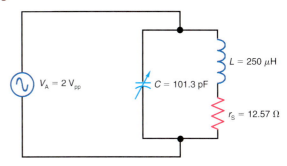

25–35 In Fig. 25–21, how much is Z_{EQ} at f_1 and f_2? How much is I_T?

SECTION 25–7 TUNING

25–36 In Fig. 25–18, calculate

    a. the capacitance tuning ratio.

    b. the ratio of the highest to lowest resonant frequency when C is varied from its lowest to its highest value.

25–37 In Fig. 25–18, is the voltage across the capacitor the same for all resonant frequencies across the tuning range? Why or why not?

SECTION 25–8 MISTUNING

25–38 Does a series LC circuit appear capacitive or inductive when the frequency of the input voltage is

    a. lower than its resonant frequency?

    b. higher than its resonant frequency?

25–39 Does a parallel LC circuit appear capacitive or inductive when the frequency of the input voltage is

    a. lower than its resonant frequency?

    b. higher than its resonant frequency?

SECTION 25–9 ANALYSIS OF PARALLEL RESONANT CIRCUITS

25–40 Is a Q of 5 considered a low or a high Q?

25–41 In Fig. 25–21, is the Q at f_r considered a low or a high Q?

25–42 If r_S is increased to 392.8 Ω in Fig. 25–21, calculate the following values:

    a. f_r

    b. Q

    c. Z_{EQ}

    d. I_T

25–43 In Prob. 25–42, calculate I_L and I_C at f_r. Are they equal? If not, why?

25–44 In reference to Prob. 25–42c, calculate Z_{EQ} as $Q \times X_L$. Does this answer agree with the original value obtained in Prob. 25–42c? If not, why?

25-45 In reference to Prob. 25–42, do you think Z_{EQ} is maximum at f_r or is Z_{EQ} maximum above or below f_r?

SECTION 25–10 DAMPING OF PARALLEL RESONANT CIRCUITS

25-46 In Fig. 25–19, calculate the Q and bandwidth, Δf, if a 100-kΩ resistor is placed in parallel with the tank circuit.

25-47 In Fig. 25–21, calculate the Q and bandwidth, Δf, if a 2-MΩ resistor is placed in parallel with the tank circuit.

25-48 In Fig. 25–19, convert the series resistance, r_S, to an equivalent parallel resistance, R_P.

25-49 Repeat Prob. 25–48 for Fig. 25–21.

SECTION 25–11 CHOOSING L AND C FOR A RESONANT CIRCUIT

25-50 Assume that it is desired to have an X_L value of 1.5 kΩ at the resonant frequency of 2 MHz. What are the required values of L and C?

Critical Thinking

25-51 Prove that

$$X_L = \sqrt{L/C}$$

for an LC circuit at f_r.

25-52 Suppose you are an engineer designing a coil to be used in a resonant LC circuit. Besides obtaining the required inductance L, your main concern is to reduce the skin effect so as to obtain as high a Q as possible for the LC circuit. List three design techniques that would reduce or minimize the skin effect in the coil windings.

Answers to Knowledge Check Problems

25-1 $X_L = X_C = 500 \ \Omega$

25-2 No

25-3 Because r_S in the inductive branch causes its branch impedance to be slightly greater than X_C at f_r. Therefore, I_L is slightly less than I_C at f_r which prevents I_L and I_C from canceling completely.

25-4 The resonant frequency, f_r, is cut in half.

25-5 A Q of 5 is an example of a low Q, whereas a Q of 100 is an example of a high Q.

25-6 283 kΩ

25-7 No

25-8 Yes

25-9 No, slightly below f_r.

25-10 Yes

25-11 1 kΩ

Answers to Self-Reviews

25-1 a. 1000 kHz
b. 1000 kHz

25-2 a. F
b. T
c. T

25-3 a. F
b. T
c. T

25-4 a. decreased
b. 200 μH
c. 39.48

25-5 a. $Q = 100$
b. $Z_{EQ} = 25$ kΩ

25-6 a. $\Delta f = 0.25$ MHz
b. $Q = 50$

25-7 a. lowest
b. 1:4

25-8 a. capacitive
b. inductive

25-9 a. low
b. less

25-10 a. $Q = 50$
b. $Q = 50$
c. $Z_{EQ} = 50$ kΩ

25-11 a. $f_r = 159$ kHz
b. $f_r = 318$ kHz

26 Filters

A filter separates different components that are mixed together. For instance, a mechanical filter can separate particles from liquid or small particles from large particles. An electrical filter can separate different frequency components.

Generally, inductors and capacitors are used for filtering because of their opposite frequency characteristics. Inductive reactance X_L increases but capacitive reactance X_C decreases with higher frequencies. In addition, their filtering action depends on whether L and C are in series or in parallel with the load.

The amount of attenuation offered by a filter is usually specified in decibels (dB). The decibel is a logarithmic expression that compares two power levels. The frequency response of a filter is usually drawn as a graph of frequency versus decibel attenuation.

The most common filtering applications are separating audio from radio frequencies, or vice versa, and separating ac variations from the average dc level. There are many of these applications in electronic circuits.

Objectives

After studying this chapter you should be able to

- *State* the difference between a low-pass and a high-pass filter.

- *Explain* what is meant by *pulsating direct current*.

- *Explain* how a transformer acts as a high-pass filter.

- *Explain* how an *RC* coupling circuit couples alternating current but blocks direct current.

- *Explain* the function of a bypass capacitor.

- *Calculate* the cutoff frequency, output voltage, and phase angle of basic *RL* and *RC* filters.

- *Explain* the operation of band-pass and band-stop filters.

- *Explain* why log-log graph paper or semilog graph paper is used to plot a frequency response.

- *Define* the term *decibel*.

- *Explain* how resonant circuits can be used as band-pass or band-stop filters.

- *Describe* the function of a power-line filter and a television antenna filter.

Outline

Important Terms

attenuation
band-pass filter
band-stop filter
bypass capacitor

crystal filter
cutoff frequency
decade

decibel (dB)
fluctuating DC
high-pass filter

low-pass filter
octave
pulsating DC

26–1 Examples of Filtering

Electronic circuits often have currents of different frequencies corresponding to voltages of different frequencies because a source produces current with the same frequency as the applied voltage. As examples, the ac signal applied to an audio circuit can have high and low audio frequencies; an rf circuit can have a wide range of radio frequencies at its input; the audio detector in a radio has both radio frequencies and audio frequencies in the output. Finally, the rectifier in a power supply produces dc output with an ac ripple superimposed on the average dc level.

In such applications where the current has different frequency components, it is usually necessary either to favor or to reject one frequency or a band of frequencies. Then an electrical filter is used to separate higher or lower frequencies.

The electrical filter can pass the higher frequency component to the load resistance, which is the case of a high-pass filter, or a low-pass filter can be used to favor the lower frequencies. In Fig. 26–1a, the high-pass filter allows 10 kHz to produce output, while rejecting or attenuating the lower frequency of 100 Hz. In Fig. 26–1b, the filtering action is reversed to pass the lower frequency of 100 Hz, while attenuating 10 kHz. These examples are for high and low audio frequencies.

For the case of audio frequencies mixed with radio frequencies, a low-pass filter allows the audio frequencies in the output, whereas a high-pass filter allows passing the radio frequencies to the output.

Figure 26–1 Function of electrical filters. (a) High-pass filter couples higher frequencies to the load. (b) Low-pass filter couples lower frequencies to the load.

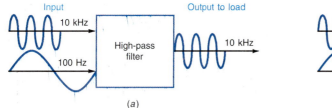

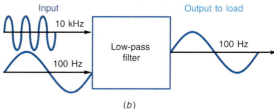

(a)   (b)

■ *26–1 Knowledge Check*

Answer at end of chapter.

What type of filter would be used to pass all frequencies below 30 MHz but eliminate all frequencies above 30 MHz?

■ *26–1 Self-Review*

Answers at end of chapter.

A high-pass filter will pass which of the following:
a. **100 Hz or 500 kHz.**
b. **60 Hz or a steady dc level.**

26–2 Direct Current Combined with Alternating Current

Current that varies in amplitude but does not reverse in polarity is considered *pulsating* or *fluctuating* direct current. It is not a steady direct current because its value fluctuates. However, it is not alternating current because the polarity remains the same, either positive or negative. The same idea applies to voltages.

Figure 26–2 An example of a pulsating or fluctuating direct current and voltage. (*a*) Circuit. (*b*) Graph of voltage across R_L. This V equals V_B of the battery plus V_A of the ac source with a frequency of 1000 Hz.

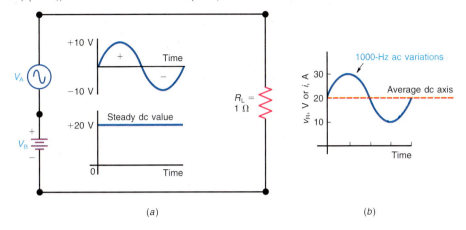

(*a*)   (*b*)

Figure 26–2 illustrates how a circuit can have pulsating direct current or voltage. Here, the steady dc voltage of the battery V_B is in series with the ac voltage V_A. Since the two series generators add, the voltage across R_L is the sum of the two applied voltages, as shown by the waveshape of v_R in Fig. 26–2*b*.

If values are taken at opposite peaks of the ac variation, when V_A is at $+10$ V, it adds to the $+20$ V of the battery to provide $+30$ V across R_L; when the ac voltage is -10 V, it bucks the battery voltage of $+20$ V to provide $+10$ V across R_L. When the ac voltage is at zero, the voltage across R_L equals the battery voltage of $+20$ V.

The combined voltage v_R then consists of the ac variations fluctuating above and below the battery voltage as the axis, instead of the zero axis for ac voltage. The result is a pulsating dc voltage, since it is fluctuating but always has positive polarity with respect to zero.

The pulsating direct current i through R_L has the same waveform, fluctuating above and below the steady dc level of 20 A. The i and v values are the same because R_L is 1 Ω.

Another example is shown in Fig. 26–3. If a 100-Ω R_L is connected across 120 V, 60 Hz, as in Fig. 26–3*a*, the current in R_L will be V/R_L. This is an ac sine wave with an rms value of 120/100 or 1.2 A.

Figure 26–3 A combination of ac and dc voltage to provide fluctuating dc voltage across R_L. (*a*) An ac source alone. (*b*) A dc source alone. (*c*) The ac source and dc source in series for the fluctuating voltage across R_L.

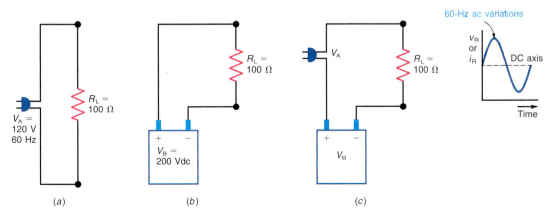

(*a*)   (*b*)   (*c*)

Also, if you connect the same R_L across the 200-Vdc source in Fig. 26–3b, instead of using the ac source, the steady direct current in R_L will be 200/100, or 2 A. The battery source voltage and its current are considered steady dc values because there are no variations.

However, suppose that the ac source V_A and dc source V_B are connected in series with R_L, as in Fig. 26–3c. What will happen to the current and voltage for R_L? Will V_A or V_B supply the current? The answer is that both sources will. Each voltage source produces current as though the other were not there, assuming the sources have negligibly small internal impedance. The result then is the fluctuating dc voltage or current shown, with the ac variations of V_A superimposed on the average dc level of V_B.

DC and AC Components

The pulsating dc voltage v_R in Fig. 26–3c is the original ac voltage V_A with its axis shifted to a dc level by the battery voltage V_B. In effect, a dc component has been inserted into the ac variations. This effect is called *dc insertion*.

Referring back to Fig. 26–2, if you measure across R_L with a dc voltmeter, it will read the dc level of 20 V. An ac-coupled oscilloscope[*] will show only the peak-to-peak variations of ±10 V.

It is convenient, therefore, to consider the pulsating or fluctuating voltage and current in two parts. One is the steady dc component, which is the axis or average level of the variations; the other is the ac component, consisting of the variations above and below the dc axis. Here the dc level for V_T is +20 V, and the ac component equals 10 V peak or 7.07 V rms value. The ac component is also called ac *ripple*.

Note that with respect to the dc level, the fluctuations represent alternating voltage or current that actually reverses in polarity. For example, the change of v_R from +20 to +10 V is a decrease in positive voltage compared with zero. However, compared with the dc level of +20 V, the value of +10 V is 10 V more negative than the axis.

Typical Examples of DC Level with AC Component

As a common application, transistors and ICs always have fluctuating dc voltage or current when used for amplifying an ac signal. The transistor or IC amplifier needs steady dc voltages to operate. The signal input is an ac variation, usually with a dc axis to establish the desired operating level. The amplified output is also an ac variation superimposed on a dc supply voltage that supplies the required power output. Therefore, the input and output circuits have fluctuating dc voltage.

The examples in Fig. 26–4 illustrate two possibilities in terms of polarities with respect to chassis ground. In Fig. 26–4a, the waveform is always positive, as in the previous examples. This example could apply to the collector voltage on an npn transistor amplifier. Note the specific values. The average dc axis is the steady dc level. The positive peak equals the dc level plus the peak ac value. The minimum point equals the dc level minus the peak ac value. The peak-to-peak value of the ac component and its rms value are the same as that of the ac signal alone. However, it is better to subtract the minimum from the maximum for the peak-to-peak value in case the waveform is unsymmetrical.

---

[*] See App. E for an explanation of how to use the oscilloscope.

Figure 26–4 Typical examples of a dc voltage with an ac component. (*a*) Positive fluctuating dc values because of a large positive dc component. (*b*) Negative fluctuating dc values because of a large negative dc component.

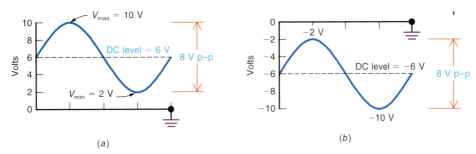

(*a*)   (*b*)

In Fig. 27–4*b*, all values are negative. Notice that here the positive peak of the ac component subtracts from the dc level because of opposite polarities. Now the negative peak adds to the negative dc level to provide a maximum point of negative voltage.

Separating the AC Component

In many applications, the circuit has pulsating dc voltage, but only the ac component is desired. Then the ac component can be passed to the load, while the steady dc component is blocked, either by transformer coupling or by capacitive coupling. A transformer with a separate secondary winding isolates or blocks steady direct current in the primary. A capacitor isolates or blocks a steady dc voltage.

■ *26–2 Knowledge Check*

Answer at end of chapter.

In Fig 26–2, what is the peak and rms value of the ac component?

■ *26–2 Self-Review*

Answers at end of chapter.

For the fluctuating dc waveform in Fig. 26–4*a*, specify the following voltages:
a. average dc level.
b. maximum and minimum values.
c. peak-to-peak of ac component.
d. peak and rms of ac component.

26–3 Transformer Coupling

Remember that a transformer produces induced secondary voltage just for variations in primary current. With pulsating direct current in the primary, the secondary has output voltage, therefore, only for the ac variations. The steady dc component in the primary has no effect in the secondary.

In Fig. 26–5, the pulsating dc voltage in the primary produces pulsating primary current. The dc axis corresponds to a steady value of primary current that has a constant magnetic field, but only when the field changes, can secondary voltage be induced. Therefore, only the fluctuations in the primary can produce output in the secondary. Since there is no output for the steady primary current, this dc level corresponds to the zero level for the ac output in the secondary.

Figure 26–5 Transformer coupling blocks the dc component. With fluctuating direct current in the primary L_P, only the ac component produces induced voltage in the secondary L_S.

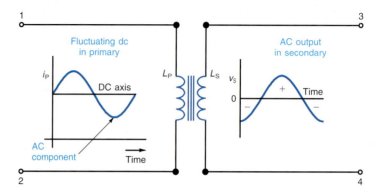

When the primary current increases above the steady level, this increase produces one polarity for the secondary voltage as the field expands; when the primary current decreases below the steady level, the secondary voltage has reverse polarity as the field contracts. The result in the secondary is an ac variation having opposite polarities with respect to the zero level.

The phase of the ac secondary voltage may be as shown or 180° opposite, depending on the connections and direction of the windings. Also, the ac secondary output may be more or less than the ac component in the primary, depending on the turns ratio. This ability to isolate the steady dc component in the primary while providing ac output in the secondary applies to all iron-core and air-core transformers with a separate secondary winding.

■ *26–3 Knowledge Check*

Answer at end of chapter.

Does a transformer block or pass dc?

■ *26–3 Self-Review*

Answers at end of chapter.

a. **Is transformer coupling an example of a high-pass or low-pass filter?**
b. **In Fig. 26–5, what is the level of v_s for the average dc level of i_p?**

26–4 Capacitive Coupling

Capacitive coupling is probably the most common type of coupling in amplifier circuits. The coupling connects the output of one circuit to the input of the next. The requirements are to include all frequencies in the desired signal, while rejecting undesired components. Usually, the dc component must be blocked from the input to ac amplifiers. The purpose is to maintain a specific dc level for the amplifier operation.

In Fig. 26–6, the pulsating dc voltage across input terminals 1 and 2 is applied to the RC coupling circuit. Capacitance C_C will charge to the steady dc level, which is the average charging voltage. The steady dc component is blocked, therefore, since it cannot produce voltage across R. However, the ac component is developed across R, between the output terminals 3 and 4, because the ac voltage allows C to produce charge and discharge current through R. Note

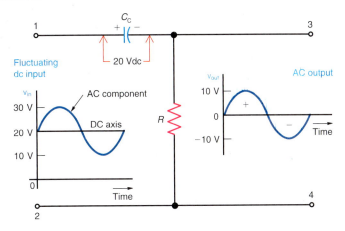

Figure 26–6 The *RC* coupling blocks the dc component. With fluctuating dc voltage applied, only the ac component produces charge and discharge current for the output voltage across *R*.

that the zero axis of the ac voltage output corresponds to the average level of the pulsating dc voltage input.

The DC Component Across *C*

The voltage across C_C is the steady dc component of the input voltage because the variations of the ac component are symmetrical above and below the average level. Furthermore, the series resistance is the same for charge and discharge. As a result, any increase in charging voltage above the average level is counteracted by an equal discharge below the average.

In Fig. 26–6, for example, when v_{in} increases from 20 to 30 V, this effect on charging C_C is nullified by the discharge when v_{in} decreases from 20 to 10 V. At all times, however, v_{in} has a positive value that charges C_C in the polarity shown.

The net result is that only the average level is effective in charging C_C, since the variations from the axis neutralize each other. After a period of time, depending on the *RC* time constant, C_C will charge to the average value of the pulsating dc voltage applied, which is 20 V here.

The AC Component Across *R*

Although C_C is charged to the average dc level, when the pulsating input voltage varies above and below this level, the charge and discharge current produces *IR* voltage corresponding to the fluctuations of the input. When v_{in} increases above the average level, C_C takes on charge, producing charging current through *R*. Even though the charging current may be too small to affect the voltage across C_C appreciably, the *IR* drop across a large value of resistance can be practically equal to the ac component of the input voltage. In summary, a long *RC* time constant is needed for good coupling.

If the polarity is considered, in Fig. 26–6, the charging current produced by an increase of v_{in} produces electron flow from the low side of *R* to the top, adding electrons to the negative side of C_C. The voltage at the top of *R* is then positive with respect to the line below.

When v_{in} decreases below the average level, *C* loses charge. The discharge current then is in the opposite direction through *R*. The result is negative polarity for the ac voltage output across *R*.

Filters **813**

When the input voltage is at its average level, there is no charge or discharge current, resulting in zero voltage across R. The zero level in the ac voltage across R corresponds to the average level of the pulsating dc voltage applied to the RC circuit.

The end result is that with positive pulsating dc voltage applied, the values above the average produce the positive half-cycle of the ac voltage across R; the values below the average produce the negative half-cycle. Only this ac voltage across R is coupled to the next circuit, as terminals 3 and 4 provide the output from the RC coupling circuit.

It is important to note that there is practically no phase shift. This rule applies to all RC coupling circuits, since R must be 10 or more times X_C. Then the reactance is negligible compared with the series resistance, and the phase angle of less than $5.7°$ is practically zero.

Voltages around the *RC* Coupling Circuit

If you measure the fluctuating dc voltage across the input terminals 1 and 2 in Fig. 26–6 with a dc voltmeter, it will read the average dc level of 20 V. If you connect an ac-coupled oscilloscope across the same two points, it will show only the fluctuating ac component. These voltage variations have a peak value of 10 V, a peak-to-peak value of 20 V, or an rms value of $0.707 \times 10 = 7.07$ V.

Across points 1 and 3 for V_C in Fig. 26–6, a dc voltmeter reads the steady dc value of 20 V. An ac voltmeter across points 1 and 3 reads practically zero.

However, an ac voltmeter across the output R between points 3 and 4 will read the ac voltage of 7 V, approximately, for V_R. Furthermore, a dc voltmeter across R reads zero. The dc component of the input voltage is across C_C but is blocked from the output across R.

Typical Coupling Capacitors

Common values of rf and af coupling capacitors for different sizes of series R are listed in Table 26–1. In all cases, the coupling capacitor blocks the steady dc component of the input voltage, and the ac component is passed to the resistance.

Table 26–1	Typical Audio Frequency and Radio Frequency Coupling Capacitors*			
	Values of C_C			
Frequency	$R = 1.6$ kΩ	$R = 16$ kΩ	$R = 160$ kΩ	Frequency Band
100 Hz	10 μF	1 μF	0.1 μF	Audio frequency
1000 Hz	1 μF	0.1 μF	0.01 μF	Audio frequency
10 kHz	0.1 μF	0.01 μF	0.001 μF	Audio frequency
100 kHz	0.01 μF	0.001 μF	100 pF	Radio frequency
1 MHz	0.001 μF	100 pF	10 pF	Radio frequency
10 MHz	100 pF	10 pF	1 pF	Radio frequency
100 MHz	10 pF	1 pF	0.1 pF	Very high frequency

* For coupling circuit in Fig. 26–6; $X_{C_C} = \frac{1}{10} R$.

The size of C_C required depends on the frequency of the ac component. At each frequency listed at the left in Table 26–1, the values of capacitance in the horizontal row have an X_C equal to one-tenth the resistance value for each column. The R increases from 1.6 to 16 to 160 kΩ for the three columns, allowing smaller values of C_C. Typical audio coupling capacitors, then, are about 0.1 to 10 μF, depending on the lowest audio frequency to be coupled and the size of the series resistance. Typical rf coupling capacitors are about 1 to 100 pF.

Values of C_C more than about 1 μF are usually electrolytic capacitors, which must be connected in the correct polarity. These can be very small; many are ½ in. long with a low voltage rating of 6 to 25 V for transistor circuits. The small leakage current of electrolytic capacitors is not a serious problem in this application because of the low voltage and small series resistance of transistor coupling circuits.

26–4 Knowledge Check

Answer at end of chapter.

Is a long or short time constant required for an RC coupling circuit?

26–4 Self-Review

Answers at end of chapter.

a. In Fig. 26–6, what is the level of v_{out} across R corresponding to the average dc level of v_{in}?
b. Which of the following is a typical audio coupling capacitor with a 1-kΩ R: 1 pF; 0.001 μF; or 5 μF?

26–5 Bypass Capacitors

A bypass is a path around a component. In circuits, the bypass is a parallel or shunt path. Capacitors are often used in parallel with resistance to bypass the ac component of a pulsating dc voltage. The result, then, is steady dc voltage across the RC parallel combination, if the bypass capacitance is large enough to have little reactance at the lowest frequency of the ac variations.

As illustrated in Fig. 26–7, the capacitance C_1 in parallel with R_1 is an ac bypass capacitor for R_1. For any frequency at which X_{C_1} is one-tenth of R_1, or less, the ac component is bypassed around R_1 through the low reactance in the shunt path. The result is practically zero ac voltage across the bypass capacitor because of its low reactance.

Figure 26–7   Low reactance of bypass capacitor C_1 short-circuits R_1 for an ac component of fluctuating dc input voltage.

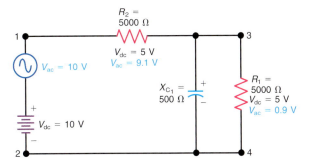

Since the voltage is the same across R_1 and C_1 because they are in parallel, there is also no ac voltage across R_1 for the frequency at which C_1 is a bypass capacitor. We can say that R is bypassed for the frequency at which X_C is one-tenth of R. The bypassing also applies to higher frequencies where X_C is less than one-tenth of R. Then the ac voltage across the bypass capacitor is even closer to zero because of its lower reactance.

Bypassing the AC Component of a Pulsating DC Voltage

The voltages in Fig. 26–7 are calculated by considering the effect of C_1 separately for V_{dc} and for V_{ac}. For direct current, C_1 is practically an open circuit. Then its reactance is so high compared with the 5000-Ω R_1 that X_{C_1} can be ignored as a parallel branch. Therefore, R_1 can be considered a voltage divider in series with R_2. Since R_1 and R_2 are equal, each has 5 V, equal to one-half V_{dc}. Although this dc voltage division depends on R_1 and R_2, the dc voltage across C_1 is the same 5 V as across its parallel R_1.

For the ac component of the applied voltage, however, the bypass capacitor has very low reactance. In fact, X_{C_1} must be one-tenth of R_1, or less. Then the 5000-Ω R_1 is so high compared with the low value of X_{C_1} that R_1 can be ignored as a parallel branch. Therefore, the 500-Ω X_{C_1} can be considered a voltage divider in series with R_2.

With an X_{C_1} of 500 Ω, this value in series with the 5000-Ω R_2 allows approximately one-eleventh of V_{ac} across C_1. This ac voltage, equal to 0.9 V here, is the same across R_1 and C_1 in parallel. The remainder of the ac applied voltage, approximately equal to 9.1 V, is across R_2. In summary, then, the bypass capacitor provides an ac short circuit across its shunt resistance, so that little or no ac voltage can be developed without affecting the dc voltages.

Measuring voltages around the circuit in Fig. 26–7, a dc voltmeter reads 5 V across R_1 and 5 V across R_2. An ac voltmeter across R_2 reads 9.1 V, which is almost all of the ac input voltage. Across the bypass capacitor C_1, the ac voltage is only 0.9 V.

Typical sizes of rf and af bypass capacitors are listed in Table 26–2. The values of C have been calculated at different frequencies for an X_C one-tenth

Table 26–2	Typical Audio Frequency and Radio Frequency Bypass Capacitors*			
	Values of C			
Frequency	**R = 16 kΩ**	**R = 1.6 kΩ**	**R = 160 Ω**	**Frequency Band**
100 Hz	1 μF	10 μF	100 μF	Audio frequency
1000 Hz	0.1 μF	1 μF	10 μF	Audio frequency
10 kHz	0.01 μF	0.1 μF	1 μF	Audio frequency
100 kHz	0.001 μF	0.01 μF	0.1 μF	Radio frequency
1 MHz	100 pF	0.001 μF	0.01 μF	Radio frequency
10 MHz	10 pF	100 pF	0.001 μF	Radio frequency
100 MHz	1 pF	10 pF	100 pF	Very high frequency

* For RC bypass circuit in Fig. 26–7; $X_{C_1} = {}^1\!/_{10}R$.

Figure 26–8 Capacitor C_1 bypasses R_1 for radio frequencies but not for audio frequencies.

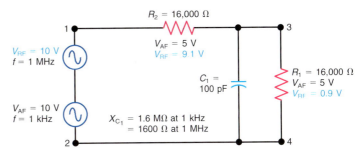

the shunt resistance given in each column. The R decreases for the three columns, from 16 kΩ to 1.6 kΩ and 160 Ω. Note that smaller values of R require larger values of C for bypassing. Also, when X_C equals one-tenth of R at one frequency, X_C will be even less for higher frequencies, improving the bypassing action. Therefore, the size of bypass capacitors should be considered on the basis of the lowest frequency to be bypassed.

Note that the applications of coupling and bypassing for C are really the same, except that C_C is in series with R and the bypass C is in parallel with R. In both cases X_C must be one-tenth or less of R. Then C_C couples the ac signal to R, or the shunt bypass short-circuits R for the ac signal.

Bypassing Radio Frequencies but not Audio Frequencies

See Fig. 26–8. At the audio frequency of 1000 Hz, C_1 has a reactance of 1.6 MΩ. This reactance is so much higher than R_1 that the impedance of the parallel combination is essentially equal to the 16,000 Ω of R_1. Then R_1 and R_2 serve as a voltage divider for the applied af voltage of 10 V. Each of the equal resistances has one-half the applied voltage, equal to the 5 V across R_2 and 5 V across R_1. This 5 V at 1000 Hz is also present across C_1, since it is in parallel with R_1.

For the rf voltage at 1 MHz, however, the reactance of the bypass capacitor is only 1600 Ω. This is one-tenth of R_1. Then X_{C_1} and R_1 in parallel have a combined impedance approximately equal to 1600 Ω.

Now, with a 1600-Ω impedance for the R_1C_1 bank in series with the 16,000 Ω of R_2, the voltage across R_1 and C_1 is one-eleventh the applied rf voltage. Then there is 0.9 V across the lower impedance of R_1 and C_1 with 9.1 V across the larger resistance of R_2. As a result, the rf component of the applied voltage can be considered bypassed. The capacitor C_1 is the rf bypass across R_1.

■ 26–5 Knowledge Check

Answer at end of chapter.

Should the size of a bypass capacitor be based on the lowest or highest frequency to be bypassed?

■ 26–5 Self-Review

Answers at end of chapter.

a. In Fig. 26–8, is C_1 an af or rf bypass?
b. Which of the following is a typical audio bypass capacitor across a 1-kΩ R: 1 pF; 0.001 μF; or 5 μF?

26–6 Filter Circuits

In terms of their function, filters can be classified as either low-pass or high-pass. A low-pass filter allows the lower frequency components of the applied voltage to develop output voltage across the load resistance, whereas the higher frequency components are attenuated, or reduced, in the output. A high-pass filter does the opposite, allowing the higher frequency components of the applied voltage to develop voltage across the output load resistance.

An RC coupling circuit is an example of a high-pass filter because the ac component of the input voltage is developed across R while the dc voltage is blocked by the series capacitor. Furthermore, with higher frequencies in the ac component, more ac voltage is coupled. For the opposite case, a bypass capacitor is an example of a low-pass filter. The higher frequencies are bypassed, but the lower the frequency, the less the bypassing action. Then lower frequencies can develop output voltage across the shunt bypass capacitor.

To make the filtering more selective in terms of which frequencies are passed to produce output voltage across the load, filter circuits generally combine inductance and capacitance. Since inductive reactance increases with higher frequencies and capacitive reactance decreases, the two opposite effects improve the filtering action.

With combinations of L and C, filters are named to correspond to the circuit configuration. Most common types of filters are the L, T, and π. Any one of the three can function as either a low-pass filter or a high-pass filter.

The reactance X_L of either low-pass or high-pass filters with L and C increases with higher frequencies, while X_C decreases. The frequency characteristics of X_L and X_C cannot be changed. However, the circuit connections are opposite to reverse the filtering action.

In general, high-pass filters use

1. Coupling capacitance C in series with the load. Then X_C can be low for high frequencies to be passed to R_L, while low frequencies are blocked.

2. Choke inductance L in parallel across R_L. Then the shunt X_L can be high for high frequencies to prevent a short circuit across R_L, while low frequencies are bypassed.

The opposite characteristics for low-pass filters are

1. Inductance L in series with the load. The high X_L for high frequencies can serve as a choke, while low frequencies can be passed to R_L.

2. Bypass capacitance C in parallel across R_L. Then high frequencies are bypassed by a small X_C, while low frequencies are not affected by the shunt path.

The ability of any filter to reduce the amplitude of undesired frequencies is called the *attenuation* of the filter. The frequency at which the attenuation reduces the output to 70.7% is the *cutoff frequency*, usually designated f_c.

■ *26–6 Knowledge Check*

Answer at end of chapter.

The input to a low-pass filter is 10 V at the cutoff frequency. How much is V_{out}?

■ *26–6 Self-Review*

Answers at end of chapter.

a. Does high-pass filtering or low-pass filtering require series C?
b. Which filtering requires parallel C?

26-7 Low-Pass Filters

Figure 26–9 illustrates low-pass circuits from a single filter element with a shunt bypass capacitor in Fig. 26–9a or a series choke in b, to the more elaborate combinations of an L-type filter in c, a T type in d, and a π type in e and f. With an applied input voltage having different frequency components, the low-pass filter action results in maximum low-frequency voltage across R_L, while most of the high-frequency voltage is developed across the series choke or resistance.

In Fig. 26–9a, the shunt capacitor C bypasses R_L at high frequencies. In Fig. 26–9b, the choke L acts as a voltage divider in series with R_L. Since L has maximum reactance for the highest frequencies, this component of the input voltage is developed across L with little across R_L. At lower frequencies, L has low reactance, and most of the input voltage can be developed across R_L.

In Fig. 26–9c, the use of both the series choke and the bypass capacitor improves the filtering by providing a sharper cutoff between the low frequencies that can develop voltage across R_L and the higher frequencies stopped from the load by producing maximum voltage across L. Similarly, the T-type circuit in Fig. 26–9d and the π-type circuits in e and f improve filtering.

Using the series resistance in Fig. 26–9f instead of a choke provides an economical π filter in less space.

Passband and Stopband

As illustrated in Fig. 26–10, a low-pass filter attenuates frequencies above the cutoff frequency f_c of 15 kHz in this example. Any component of the input voltage having a frequency lower than 15 kHz can produce output voltage across the load. These frequencies are in the *passband*. Frequencies of 15 kHz or more are in the *stopband*. The sharpness of filtering between the passband and the stopband depends on the type of circuit. In general, the more L and C components, the sharper the response of the filter. Therefore, π and T types are better filters than the L type and the bypass or choke alone.

Figure 26–9   Low-pass filter circuits. (a) Bypass capacitor C in parallel with R_L. (b) Choke L in series with R_L. (c) Inverted-L type with choke and bypass capacitor. (d) The T type with two chokes and one bypass capacitor. (e) The π type with one choke and bypass capacitors at both ends. (f) The π type with a series resistor instead of a choke.

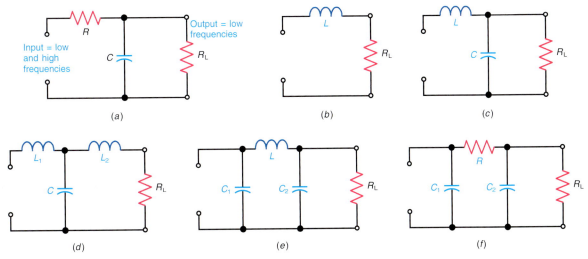

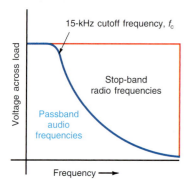

Figure 26–10 The response of a low-pass filter with cutoff at 15 kHz. The filter passes the audio signal but attenuates radio frequencies.

The response curve in Fig. 26–10 is illustrated for the application of a low-pass filter attenuating rf voltages while passing audio frequencies to the load. This is necessary when the input voltage has rf and af components but only the audio voltage is desired for the af circuits that follow the filter.

A good example is filtering the audio output of the detector circuit in a radio receiver, after the rf-modulated carrier signal has been rectified. Another common application of low-pass filtering is separating the steady dc component of a pulsating dc input from the higher frequency 60-Hz ac component, as in the pulsating dc output of the rectifier in a power supply.

Circuit Variations

The choice between the T-type filter with a series input choke and the π type with a shunt input capacitor depends on the internal resistance of the generator supplying input voltage to the filter. A low-resistance generator needs the T filter so that the choke can provide high series impedance for the bypass capacitor. Otherwise, the bypass capacitor must have extremely large values to short-circuit the low-resistance generator at high frequencies.

The π filter is more suitable with a high-resistance generator when the input capacitor can be effective as a bypass. For the same reasons, the L filter can have the shunt bypass either in the input for a high-resistance generator or across the output for a low-resistance generator.

In all filter circuits, the series choke can be connected either in the high side of the line, as in Fig. 26–9, or in series in the opposite side of the line, without having any effect on the filtering action. Also, the series components can be connected in both sides of the line for a *balanced filter* circuit.

Passive and Active Filters

All circuits here are passive filters, as they use only capacitors, inductors, and resistors, which are passive components. An active filter, however, uses an operational amplifier (op amp) on an IC chip, with R and C. The purpose is to eliminate the need for inductance L. This feature is important in filters for audio frequencies when large coils would be necessary.

26–7 Knowledge Check

Answer at end of chapter.

In Fig. 26–9, which filter, the inverted L type or π type, provides a sharper cutoff between the low and high frequencies?

26–7 Self-Review

Answers at end of chapter.

a. Which diagrams in Fig. 26–9 show a π-type filter?
b. Does the response curve in Fig. 26–10 show low-pass or high-pass filtering?

26–8 High–Pass Filters

As illustrated in Fig. 26–11, the high-pass filter passes to the load all frequencies higher than the cutoff frequency f_c, whereas lower frequencies cannot develop appreciable voltage across the load. The graph in Fig. 26–11a shows the response of a high-pass filter with a stopband of 0 to 50 Hz. Above the cutoff frequency of 50 Hz, the higher audio frequencies in the passband can produce af voltage across the output load resistance.

Figure 26–11 High-pass filters. (*a*) The response curve for an audio frequency filter cutting off at 50 Hz. (*b*) An *RC* coupling circuit. (*c*) Inverted L type. (*d*) The T type. (*e*) The π type.

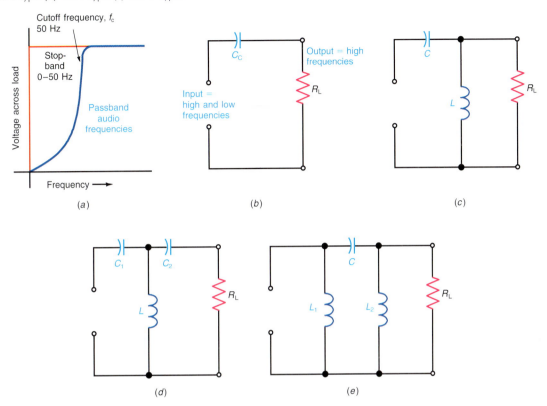

(*a*)   (*b*)   (*c*)

(*d*)   (*e*)

The high-pass filtering action results from using C_C as a coupling capacitor in series with the load, as in Fig. 26–11*b*. The *L*, *T*, and π types use the inductance for a high-reactance choke across the line. In this way, the higher frequency components of the input voltage can develop very little voltage across the series capacitance, allowing most of this voltage to be produced across R_L. The inductance across the line has higher reactance with increasing frequencies, allowing the shunt impedance to be no lower than the value of R_L.

For low frequencies, however, R_L is effectively short-circuited by the low inductive reactance across the line. Also, C_C has high reactance and develops most of the voltage at low frequencies, stopping these frequencies from developing voltage across the load.

■ 26–8 Knowledge Check

Answer at end of chapter.

In Fig. 26–11*a*, is the frequency of 10 kHz in the stopband or passband?

■ 26–8 Self-Review

Answers at end of chapter.

a. Which diagram in Fig. 26–11 shows a T-type filter?
b. Does the response curve in Fig. 26–11*a* show high-pass or low-pass filtering?

26–9 Analyzing Filter Circuits

Any low-pass or high-pass filter can be thought of as a frequency-dependent voltage divider, since the amount of output voltage is a function of frequency. Special formulas can be used to calculate the output voltage for any frequency of the applied voltage. What follows is a more mathematical approach in analyzing the operation of the most basic low-pass and high-pass filter circuits.

RC Low–Pass Filter

Figure 26–12a shows a simple RC low-pass filter, and Fig. 26–12b shows how its output voltage V_{out} varies with frequency. Let's examine how the RC low-pass filter responds when $f = 0$ Hz (dc) and $f = \infty$ Hz. At $f = 0$ Hz, the capacitor C has infinite capacitive reactance X_C, calculated as

$$X_C = \frac{1}{2\pi f C}$$

$$= \frac{1}{2 \times \pi \times 0 \text{ Hz} \times 0.01 \ \mu F}$$

$$= \infty \ \Omega$$

Figure 26–13a shows the equivalent circuit for this condition. Notice that C appears as an open. Since all of the input voltage appears across the open in a series circuit, V_{out} must equal V_{in} when $f = 0$ Hz.

At the other extreme, consider the circuit when the frequency f is very high or infinitely high. Then $X_C = 0 \ \Omega$, calculated as

$$X_C = \frac{1}{2\pi f C}$$

$$= \frac{1}{2 \times \pi \times \infty \text{ Hz} \times 0.01 \ \mu F}$$

$$= 0 \ \Omega$$

Figure 26–13b shows the equivalent circuit for this condition. Notice that C appears as a short. Since the voltage across a short is zero, the output voltage for very high frequencies must be zero.

When the frequency of the input voltage is somewhere between zero and infinity, the output voltage can be determined by using Formula (26–1):

$$V_{out} = \frac{X_C}{Z_T} \times V_{in} \tag{26–1}$$

where

$$Z_T = \sqrt{R^2 + X_C^2}$$

||| MultiSim **Figure 26–12** RC low-pass filter. (a) Circuit. (b) Graph of V_{out} versus frequency.

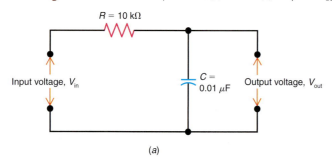

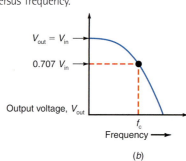

(a)

(b)

Figure 26–13 *RC* low-pass equivalent circuits. (*a*) Equivalent circuit for *f* = 0 Hz. (*b*) Equivalent circuit for very high frequencies, or *f* = ∞ Hz.

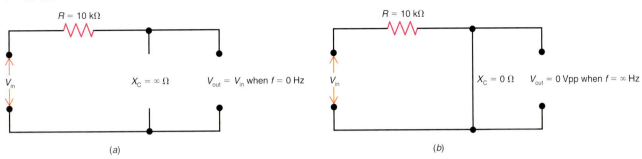

(*a*)   (*b*)

At very low frequencies, where X_C approaches infinity, V_{out} is approximately equal to V_{in}. This is true because the ratio X_C/Z_T approaches one as X_C and Z_T become approximately the same value. At very high frequencies, where X_C approaches zero, the ratio X_C/Z_T becomes very small, and V_{out} is approximately zero.

With respect to the input voltage V_{in}, the phase angle θ of the output voltage V_{out} can be calculated as

$$\theta = \arctan\left(-\frac{R}{X_C}\right) \qquad (26\text{--}2)$$

At very low frequencies, X_C is very large and θ is approximately 0°. At very high frequencies, however, X_C is nearly zero and θ approaches −90°.

The frequency where $X_C = R$ is the *cutoff frequency*, designated f_c. At f_c, the series current I is at 70.7% of its maximum value because the total impedance Z_T is 1.41 times larger than the resistance of R. The formula for the cutoff frequency f_c of an *RC* low-pass filter is derived as follows. Because $X_C = R$ at f_c,

$$\frac{1}{2\pi f_c C} = R$$

Solving for f_c gives

$$f_c = \frac{1}{2\pi RC} \qquad (26\text{--}3)$$

The response curve in Fig. 26–12*b* shows that $V_{out} = 0.707V_{in}$ at the cutoff frequency f_c.

Example 26-1

In Fig. 26–12*a*, calculate (a) the cutoff frequency f_c; (b) V_{out} at f_c; (c) θ at f_c. (Assume V_{in} = 10 V_{pp} for all frequencies.)

ANSWER

a. To calculate f_c, use Formula (26–3):

$$f_c = \frac{1}{2\pi RC}$$

$$= \frac{1}{2 \times \pi \times 10\ k\Omega \times 0.01\ \mu F}$$

$$= 1.592\ kHz$$

b. To calculate V_{out} at f_c, use Formula (26–1). First, however, calculate X_C and Z_T at f_c:

$$X_C = \frac{1}{2\pi f_c C}$$

$$= \frac{1}{2 \times \pi \times 1.592 \text{ kHz} \times 0.01 \text{ } \mu\text{F}}$$

$$= 10 \text{ k}\Omega$$

$$Z_T = \sqrt{R^2 + X_C^2}$$

$$= \sqrt{10^2 \text{ k}\Omega + 10^2 \text{ k}\Omega}$$

$$= 14.14 \text{ k}\Omega$$

Next,

$$V_{out} = \frac{X_C}{Z_T} \times V_{in}$$

$$= \frac{10 \text{ k}\Omega}{14.14 \text{ k}\Omega} \times 10 \text{ V}_{pp}$$

$$= 7.07 \text{ V}_{pp}.$$

c. To calculate θ, use Formula (26–2):

$$\theta = \arctan\left(-\frac{R}{X_C}\right)$$

$$= \arctan\left(-\frac{10 \text{ k}\Omega}{10 \text{ k}\Omega}\right)$$

$$= \arctan(-1)$$

$$= -45°$$

The phase angle of $-45°$ tells us that V_{out} lags V_{in} by $45°$ at the cutoff frequency f_c.

RL Low-Pass Filter

Figure 26–14a shows a simple RL low-pass filter, and Fig. 26–14b shows how its output voltage V_{out} varies with frequency. For the analysis that follows, it is assumed that the coil's dc resistance r_s is negligible in comparison with the series resistance R.

Figure 26–15a shows the equivalent circuit when $f = 0$ Hz (dc). Notice that the inductor L acts as a short, since X_L must equal 0 Ω when $f = 0$ Hz. As a result, $V_{out} = V_{in}$ at very low frequencies and for direct current (0 Hz). At very high frequencies, X_L approaches infinity and the equivalent circuit appears

Figure 26–14 RL low-pass filter. (a) Circuit. (b) Graph of V_{out} versus frequency.

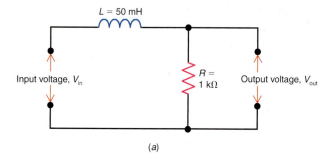

(a)

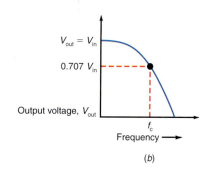

(b)

Figure 26–15 *RL* low-pass equivalent circuits. (*a*) Equivalent circuit for *f* = 0 Hz. (*b*) Equivalent circuit for very high frequencies, or *f* = ∞ Hz.

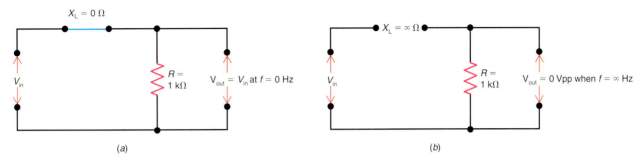

(*a*)   (*b*)

as in Fig. 26–15*b*. Since *L* is basically equivalent to an open at very high frequencies, all of the input voltage will be dropped across *L* rather than *R*. Therefore, V_{out} = 0 V_{pp} at very high frequencies.

To calculate the output voltage at any frequency in Fig. 26–14*a*, use Formula (26–4):

$$V_{out} = \frac{R}{Z_T} \times V_{in} \tag{26–4}$$

where

$$Z_T = \sqrt{R^2 + X_L^2}$$

At very low frequencies, where X_L is very small, V_{out} is approximately equal to V_{in}. This is true because the ratio R/Z_T approaches one as Z_T and R become approximately the same value. At very high frequencies, the output voltage is approximately zero, because the ratio R/Z_T becomes very small as X_L and thus Z_T approach infinity.

The phase angle θ between V_{in} and V_{out} can be determined using Formula (26–5):

$$\theta = \arctan\left(-\frac{X_L}{R}\right) \tag{26–5}$$

At very low frequencies, X_L approaches zero and θ is approximately 0°. At very high frequencies, X_L approaches infinity and θ is approximately −90°.

The frequency at which $X_L = R$ is the cutoff frequency f_c. At f_c, the series current *I* is at 70.7% of its maximum value, since $Z_T = 1.41R$ when $X_L = R$. The formula for the cutoff frequency of an *RL* low-pass filter is derived as follows. Since $X_L = R$ at f_c,

$$2\pi f_c L = R$$

Solving for f_c gives

$$f_c = \frac{R}{2\pi L} \tag{26–6}$$

The response curve in Fig. 26–14*b* shows that $V_{out} = 0.707V_{in}$ at the cutoff frequency f_c.

Example 26–2

In Fig. 26–14*a*, calculate (a) the cutoff frequency f_c; (b) V_{out} at 1 kHz; (c) θ at 1 kHz. (Assume V_{in} = 10 V_{pp} for all frequencies.)

ANSWER

a. To calculate f_c, use Formula (26–6):

$$f_c = \frac{R}{2\pi L}$$

$$= \frac{1\ k\Omega}{2 \times \pi \times 50\ mH}$$

$$= 3.183\ kHz$$

b. To calculate V_{out} at 1 kHz, use Formula (26–4). First, however, calculate X_L and Z_T at 1 kHz:

$$X_L = 2\pi f L$$

$$= 2 \times \pi \times 1\ kHz \times 50\ mH$$

$$= 314\ \Omega$$

$$Z_T = \sqrt{R^2 + X_L^{\,2}}$$

$$= \sqrt{1^2\ k\Omega + 314^2\ \Omega}$$

$$= 1.05\ k\Omega$$

Next,

$$V_{out} = \frac{R}{Z_T} \times V_{in}$$

$$= \frac{1\ k\Omega}{1.05\ k\Omega} \times 10\ V_{pp}$$

$$= 9.52\ V_{pp}$$

Notice that $V_{out} \cong V_{in}$, since 1 kHz is in the passband of the low-pass filter.

c. To calculate θ at 1 kHz, use Formula (26–5). Recall that $X_L = 314\ \Omega$ at 1 kHz:

$$\theta = \arctan\left(-\frac{X_L}{R}\right)$$

$$= \arctan\left(-\frac{314\ \Omega}{1\ k\Omega}\right)$$

$$= \arctan(-0.314)$$

$$= -17.4°$$

The phase angle of $-17.4°$ tells us that V_{out} lags V_{in} by 17.4° at a frequency of 1 kHz.

RC High–Pass Filter

Figure 26–16*a* shows an *RC* high-pass filter. Notice that the output is taken across the resistor *R* rather than across the capacitor *C*. Figure 26–16*b* shows how the output voltage varies with frequency. To calculate the output voltage V_{out} at any frequency, use Formula (26–7):

$$V_{out} = \frac{R}{Z_T} \times V_{in} \tag{26–7}$$

where

$$Z_T = \sqrt{R^2 + X_C^{\,2}}$$

At very low frequencies, the output voltage approaches zero because the ratio R/Z_T becomes very small as X_C and thus Z_T approach infinity. At very

GOOD TO KNOW

Another way to calculate the output voltage of an *RC* or *RL* high-pass filter is:

$$V_{out} = \frac{V_{in}}{\sqrt{1 + (f_c/f)^2}}$$

where *f* represents any frequency.

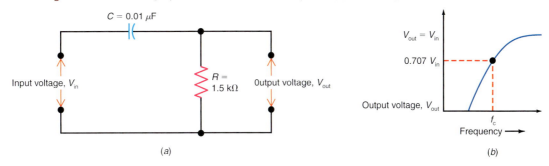

(*a*)   (*b*)

high frequencies, V_{out} is approximately equal to V_{in}, because the ratio R/Z_T approaches one as Z_T and R become approximately the same value.

The phase angle of V_{out} with respect to V_{in} for an *RC* high-pass filter can be calculated using Formula (26–8):

$$\theta = \arctan\left(\frac{X_C}{R}\right) \tag{26–8}$$

At very low frequencies where X_C is very large, θ is approximately 90°. At very high frequencies where X_C approaches zero, θ is approximately 0°.

To calculate the cutoff frequency f_c for an *RC* high-pass filter, use Formula (26–3). Although this formula is used to calculate f_c for an *RC* low-pass filter, it can also be used to calculate f_c for an *RC* high-pass filter. The reason is that, in both circuits, $X_C = R$ at the cutoff frequency. In Fig. 26–16*b*, notice that $V_{out} = 0.707V_{in}$ at f_c.

RL High-Pass Filter

An *RL* high-pass filter is shown in Fig. 26–17*a*, and its response curve is shown in Fig. 26–17*b*. In Fig. 26–17*a*, notice that the output is taken across the inductor *L* rather than across the resistance *R*.

To calculate the output voltage V_{out} at any frequency, use Formula (26–9):

$$V_{out} = \frac{X_L}{Z_T} \times V_{in} \tag{26–9}$$

where

$$Z_T = \sqrt{R^2 + X_L^2}$$

At very low frequencies, where X_L is very small, V_{out} is approximately zero. At very high frequencies, $V_{out} = V_{in}$ because the ratio X_L/Z_T is approximately one.

Figure 26–17   *RL* high-pass filter. (*a*) Circuit. (*b*) Graph of V_{out} versus frequency.

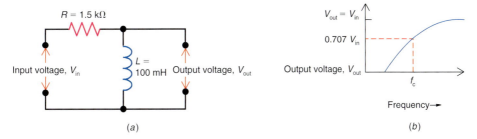

(*a*)   (*b*)

The phase angle θ of the output voltage V_{out} with respect to the input voltage V_{in} is

$$\theta = \arctan\left(\frac{R}{X_L}\right) \tag{26–10}$$

At very low frequencies, θ approaches 90° because the ratio R/X_L becomes very large when X_L approaches zero. At very high frequencies, θ approaches 0° because the ratio R/X_L becomes approximately zero as X_L approaches infinity. To calculate the cutoff frequency of an RL high-pass filter, use Formula (26–6).

Example 26-3

Calculate the cutoff frequency for (a) the RC high-pass filter in Fig. 26–16a; (b) the RL high-pass filter in Fig. 26–17a.

ANSWER

a. Use Formula (26–3):

$$f_c = \frac{1}{2\pi RC}$$

$$= \frac{1}{2 \times \pi \times 1.5 \text{ k}\Omega \times 0.01 \text{ }\mu\text{F}}$$

$$= 10.61 \text{ kHz}$$

b. Use Formula (26–6):

$$f_c = \frac{R}{2\pi L}$$

$$= \frac{1.5 \text{ k}\Omega}{2 \times \pi \times 100 \text{ mH}}$$

$$= 2.39 \text{ kHz}$$

RC Band-Pass Filter

A high-pass filter can be combined with a low-pass filter when it is desired to pass only a certain band of frequencies. This type of filter is called a *band-pass filter*. Figure 26–18a shows an RC band-pass filter, and Fig. 26–18b shows how its output voltage varies with frequency. In Fig. 26–18a, R_1 and C_1 constitute

||| MultiSim **Figure 26–18** *RC* bandpass filter. (*a*) Circuit. (*b*) Graph of V_{out} versus frequency.

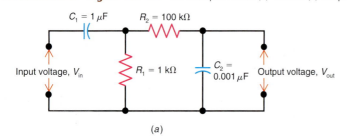

(a)

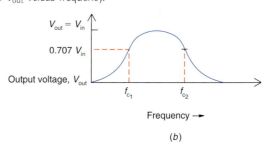

(b)

the high-pass filter, and R_2 and C_2 constitute the low-pass filter. To ensure that the low-pass filter does not load the high-pass filter, R_2 is usually 10 or more times larger than the resistance of R_1. The cutoff frequency of the high-pass filter is designated f_{c_1}, and the cutoff frequency of the low-pass filter is designated f_{c_2}. These two frequencies can be found on the response curve in Fig. 26–18b. To calculate the values for f_{c_1} and f_{c_2}, use the formulas given earlier for individual RC low-pass and RC high-pass filter circuits.

Example 26-4

In Fig. 26–18a, calculate the cutoff frequencies f_{c_1} and f_{c_2}.

ANSWER Calculate f_{c_1} for the high-pass filter consisting of R_1 and C_1:

$$f_{c_1} = \frac{1}{2\pi R_1 C_1}$$

$$= \frac{1}{2 \times \pi \times 1\ k\Omega \times 1\ \mu F}$$

$$= 159\ Hz$$

Next, calculate f_{c_2}.

$$f_{c_2} = \frac{1}{2\pi R_2 C_2}$$

$$= \frac{1}{2 \times \pi \times 100\ k\Omega \times 0.001\ \mu F}$$

$$= 1.59\ kHz$$

The frequencies below 159 Hz and above 1.59 kHz are severely attenuated, whereas those between 159 Hz and 1.59 kHz are effectively passed from the input to the output.

RC Band-Stop Filter

A high-pass filter can also be combined with a low-pass filter when it is desired to block or severely attenuate a certain band of frequencies. Such a filter is called a *band-stop* or *notch filter*. Figure 26–19a shows an RC band-stop filter,

Figure 26–19 Notch filter. (a) Circuit. (b) Graph of V_{out} versus frequency.

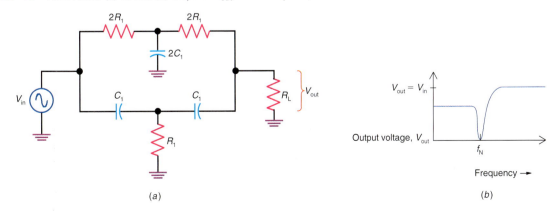

(a)

(b)

and Fig. 26–19b shows how its output voltage varies with frequency. In Fig. 26–19a, the components identified as $2R_1$ and $2C_1$ constitute the low-pass filter section, and the components identified as R_1 and C_1 constitute the high-pass filter section. Notice that the individual filters are in parallel. The frequency of maximum attenuation is called the *notch frequency*, identified as f_N in Fig. 26–19b. Notice that the maximum value of V_{out} below f_N is less than the maximum value of V_{out} above f_N. The reason for this is that the series resistances $(2R_1)$ in the low-pass filter provide greater circuit losses than the series capacitors (C_1) in the high-pass filter.

To calculate the notch frequency f_N in Fig. 26–19a, use Formula (26–11):

$$f_N = \frac{1}{4\pi R_1 C_1} \tag{26–11}$$

Example 26-5

Calculate the notch frequency f_N in Fig. 26–19a if $R_1 = 1\ k\Omega$ and $C_1 = 0.01\ \mu F$. Also, calculate the required values for $2R_1$ and $2C_1$ in the low-pass filter.

ANSWER Use Formula (26–11):

$$f_N = \frac{1}{4\pi R_1 C_1}$$
$$= \frac{1}{4 \times \pi \times 1\ k\Omega \times 0.01\ \mu F}$$
$$= 7.96\ kHz$$
$$2R_1 = 2 \times 1\ k\Omega$$
$$= 2\ k\Omega$$
$$2C_1 = 2 \times 0.01\ \mu F$$
$$= 0.02\ \mu F$$

■ *26-9 Knowledge Check*

Answer at end of chapter.

At the output of a low-pass filter, what is the approximate phase angle for signal frequencies well below the cutoff frequency?

■ *26-9 Self-Review*

Answers at end of chapter.

a. Increasing the capacitance C in Fig. 26–12a raises the cutoff frequency f_c. (True or False)
b. Decreasing the inductance L in Fig. 26–14a raises the cutoff frequency f_c. (True or False)
c. Increasing the value of C_2 in Fig. 26–18a reduces the passband. (True or False)
d. In Fig. 26–17a, V_{out} is approximately zero at very low frequencies. (True or False)

26-10 Decibels and Frequency Response Curves

In analyzing filters, the decibel (dB) unit is often used to describe the amount of attenuation offered by the filter. In basic terms, the *decibel* is a logarithmic expression that compares two power levels. Expressed mathematically,

$$N_{dB} = 10 \log \frac{P_{out}}{P_{in}} \tag{26-12}$$

where

N_{dB} = gain or loss in decibels
P_{in} = input power
P_{out} = output power

If the ratio P_{out}/P_{in} is greater than one, the N_{dB} value is positive, indicating an increase in power from input to output. If the ratio P_{out}/P_{in} is less than one, the N_{dB} value is negative, indicating a loss or reduction in power from input to output. A reduction in power, corresponding to a negative N_{dB} value, is referred to as *attenuation*.

Example 26-6

A certain amplifier has an input power of 1 W and an output power of 100 W. Calculate the dB power gain of the amplifier.

ANSWER   Use Formula (26-12):

$$N_{dB} = 10 \log \frac{P_{out}}{P_{in}}$$
$$= 10 \log \frac{100 \text{ W}}{1 \text{ W}}$$
$$= 10 \times 2$$
$$= 20 \text{ dB}$$

Example 26-7

The input power to a filter is 100 mW, and the output power is 5 mW. Calculate the attenuation, in decibels, offered by the filter.

ANSWER

$$N_{dB} = 10 \log \frac{P_{out}}{P_{in}}$$
$$= 10 \log \frac{5 \text{ mW}}{100 \text{ mW}}$$
$$= 10 \times (-1.3)$$
$$= -13 \text{ dB}$$

The power gain or loss in decibels can also be computed from a voltage ratio if the measurements are made across equal resistances.

$$N_{dB} = 20 \log \frac{V_{out}}{V_{in}} \qquad (26\text{–}13)$$

where

$$
\begin{aligned}
N_{dB} &= \text{gain or loss in decibels} \\
V_{in} &= \text{input voltage} \\
V_{out} &= \text{output voltage}
\end{aligned}
$$

The N_{dB} values of the passive filters discussed in this chapter can never be positive because V_{out} can never be greater than V_{in}.

Consider the RC low-pass filter in Fig. 26–20. The cutoff frequency f_c for this circuit is 1.592 kHz, as determined by Formula (26–1). Recall that the formula for V_{out} at any frequency is

$$V_{out} = \frac{X_C}{Z_T} \times V_{in}$$

Dividing both sides of the equation by V_{in} gives

$$\frac{V_{out}}{V_{in}} = \frac{X_C}{Z_T}$$

Substituting X_C/Z_T for V_{out}/V_{in} in Formula (26–13) gives

$$N_{dB} = 20 \log \frac{X_C}{Z_T}$$

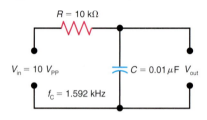

Figure 26–20 *RC* low-pass filter.

$R = 10\ \text{k}\Omega$

$V_{in} = 10\ V_{PP}$ $C = 0.01\ \mu F$ V_{out}

$f_c = 1.592\ \text{kHz}$

Example 26–8

In Fig. 26–20, calculate the attenuation, in decibels, at the following frequencies: (a) 0 Hz; (b) 1.592 kHz; (c) 15.92 kHz. (Assume that $V_{in} = 10\ V_{pp}$ at all frequencies.)

ANSWER

a. At 0 Hz, $V_{out} = V_{in} = 10\ V_{pp}$, since the capacitor C appears as an open. Therefore,

$$
\begin{aligned}
N_{dB} &= 20 \log \frac{V_{out}}{V_{in}} \\
&= 20 \log \frac{10\ V_{pp}}{10\ V_{pp}} \\
&= 20 \log 1 \\
&= 20 \times 0 \\
&= 0\ \text{dB}
\end{aligned}
$$

b. Since 1.592 kHz is the cutoff frequency f_c, V_{out} will be $0.707 \times V_{in}$ or $7.07\ V_{pp}$. Therefore,

$$
\begin{aligned}
N_{dB} &= 20 \log \frac{V_{out}}{V_{in}} \\
&= 20 \log \frac{7.07\ V_{pp}}{10\ V_{pp}} \\
&= 20 \log 0.707
\end{aligned}
$$

$$= 20 \times (-0.15)$$
$$= -3 \text{ dB}$$

c. To calculate N_{dB} at 15.92 kHz, X_C and Z_T must first be determined.

$$X_C = \frac{1}{2\pi f C}$$
$$= \frac{1}{2 \times \pi \times 15.92 \text{ kHz} \times 0.01 \text{ } \mu\text{F}}$$
$$= 1 \text{ k}\Omega$$
$$Z_T = \sqrt{R^2 + X_C^2}$$
$$= \sqrt{10^2 \text{ k}\Omega + 1^2 \text{ k}\Omega}$$
$$= 10.05 \text{ k}\Omega$$

Next,

$$N_{dB} = 20 \log \frac{X_C}{Z_T}$$
$$= 20 \log \frac{1 \text{ k}\Omega}{10.05 \text{ k}\Omega}$$
$$= 20 \log 0.0995$$
$$= 20(-1)$$
$$= -20 \text{ dB}$$

In Example 26–8, notice that N_{dB} is 0 dB at a frequency of 0 Hz, which is in the filter's passband. This may seem unusual, but the 0-dB value simply indicates that there is no attenuation at this frequency. For an ideal passive filter, $N_{dB} = 0$ dB in the passband. As another point of interest from Example 26–8, N_{dB} is −3 dB at the cutoff frequency of 1.592 kHz. Since $V_{out} = 0.707 \, V_{in}$ at f_c for any passive filter, N_{dB} is always −3 dB at the cutoff frequency of a passive filter.

The N_{dB} value of loss can be determined for any filter if the values of V_{in} and V_{out} are known. Figure 26–21 shows the basic RC and RL low-pass and high-pass filters. The formula for calculating the N_{dB} attenuation is provided for each filter.

Frequency Response Curves

The frequency response of a filter is typically shown by plotting its gain (or loss) versus frequency on logarithmic graph paper. The two types of logarithmic graph paper are log-log and semilog. On *semilog graph paper,* the divisions along one axis are spaced logarithmically, and the other axis has conventional linear spacing between divisions. On *log-log graph paper,* both axes have logarithmic spacing between divisions. Logarithmic spacing results in a scale that expands the display of smaller values and compresses the display of larger values. On logarithmic graph paper, a 2-to-1 range of frequencies is called an *octave,* and a 10-to-1 range of values is called a *decade.*

One advantage of logarithmic spacing is that a larger range of values can be shown in one plot without losing resolution in the smaller values. For example, if frequencies between 10 Hz and 100 kHz were plotted on 100 divisions of linear graph paper, each division would represent approximately 1000 Hz and it would be impossible to plot values in the decade between 10 Hz and 100 Hz. On the other hand, by using logarithmic graph paper, the decade between 10 Hz and 100 Hz would occupy the same space on the graph as the decade between 10 kHz and 100 kHz.

Figure 26–21 *RC* and *RL* filter circuits, showing formulas for calculating decibel attenuation.

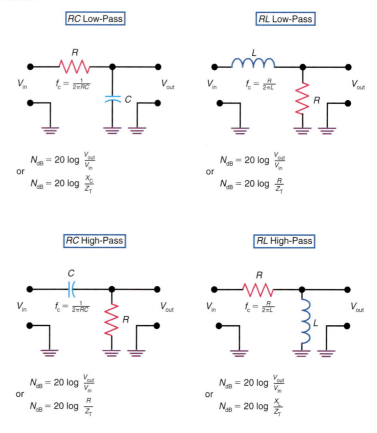

RC Low-Pass

$f_c = \frac{1}{2\pi RC}$

$N_{dB} = 20 \log \frac{V_{out}}{V_{in}}$

or

$N_{dB} = 20 \log \frac{X_C}{Z_T}$

RL Low-Pass

$f_c = \frac{R}{2\pi L}$

$N_{dB} = 20 \log \frac{V_{out}}{V_{in}}$

or

$N_{dB} = 20 \log \frac{R}{Z_T}$

RC High-Pass

$f_c = \frac{1}{2\pi RC}$

$N_{dB} = 20 \log \frac{V_{out}}{V_{in}}$

or

$N_{dB} = 20 \log \frac{R}{Z_T}$

RL High-Pass

$f_c = \frac{R}{2\pi L}$

$N_{dB} = 20 \log \frac{V_{out}}{V_{in}}$

or

$N_{dB} = 20 \log \frac{X_L}{Z_T}$

Log-log or semilog graph paper is specified by the number of decades it contains. Each decade is a *graph cycle.* For example, 2-cycle by 4-cycle log-log paper has two decades on one axis and four on the other. The number of cycles must be adequate for the range of data plotted. For example, if the frequency response extends from 25 Hz to 40 kHz, 4 cycles are necessary to plot the frequencies corresponding to the decades 10 Hz to 100 Hz, 100 Hz to 1 kHz, 1 kHz to 10 kHz, and 10 kHz to 100 kHz. A typical sheet of log-log graph paper is shown in Fig. 26–22. Because there are three decades on the horizontal axis and five decades on the vertical axis, this graph paper is called 3-cycle by 5-cycle log-log paper. Notice that each octave corresponds to a 2-to-1 range in values and each decade corresponds to a 10-to-1 range in values. For clarity, several octaves and decades are shown in Fig. 26–22.

When semilog graph paper is used to plot a frequency response, the observed or calculated values of gain (or loss) must first be converted to decibels before plotting. On the other hand, since decibel voltage gain is a logarithmic function, the gain or loss values can be plotted on log-log paper without first converting to decibels.

RC Low–Pass Frequency Response Curve

Figure 26–23a shows an *RC* low-pass filter whose cutoff frequency f_c is 1.592 kHz as determined by Formula 26–3. Figure 26–23b shows its frequency response curve plotted on semilog graph paper. Notice there are 6 cycles on the horizontal axis, which spans a frequency range from 1 Hz to 1 MHz. Notice that the vertical axis specifies the N_{dB} loss, which is the amount of attenuation offered by the filter in decibels. Notice that $N_{dB} = -3$ dB at the cutoff

Figure 26–22 Log-log graph paper. Notice that each octave corresponds to a 2-to-1 range of values and each decade corresponds to a 10-to-1 range of values.

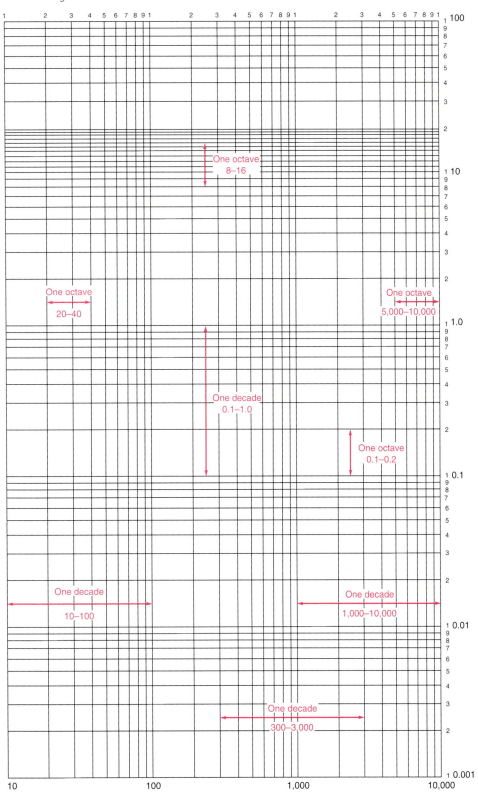

Figure 26–23 *RC* low-pass filter frequency response curve. (*a*) Circuit. (*b*) Frequency response curve.

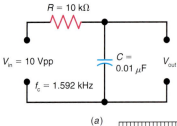

$R = 10\ k\Omega$

$V_{in} = 10\ Vpp$

$C = 0.01\ \mu F$

V_{out}

$f_c = 1.592\ kHz$

(*a*)

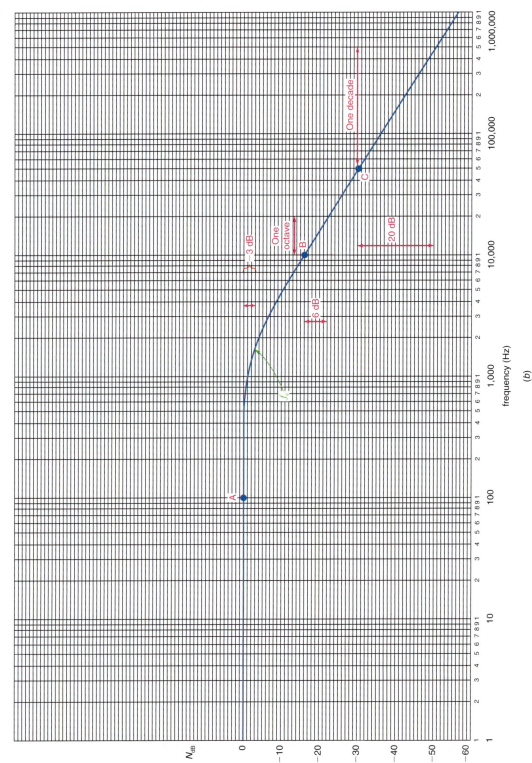

(*b*)

frequency of 1.592 kHz. Above f_c, N_{dB} decreases at the rate of approximately 6 dB/octave, which is equivalent to a rate of 20 dB/decade.

Example 26–9

From the graph in Fig. 26–23b, what is the attenuation in decibels at (a) 100 Hz; (b) 10 kHz; (c) 50 kHz?

ANSWER

a. At $f = 100$ Hz, $N_{dB} = 0$ dB, as indicated by point A on the graph.
b. At $f = 10$ kHz, $N_{dB} = -16$ dB, as indicated by point B on the graph.
c. At $f = 50$ kHz, $N_{dB} = -30$ dB, as indicated by point C.

For filters such as the inverted L, T, or π type, the response curve rolloff is much steeper beyond the cutoff frequency f_c. For example, a low-pass filter with a series inductor and a shunt capacitor has a rolloff rate of 12 dB/octave or 40 dB/decade above the cutoff frequency f_c. To increase the rate of rolloff, more inductors and capacitors must be used in the filter design. Filters are available whose rolloff rates exceed 36 dB/octave.

■ *26–10 Knowledge Check*

> *Answer at end of chapter.*
>
> **What is the rolloff rate for an *RC* low-pass filter above the cutoff frequency?**

■ *26–10 Self-Review*

> *Answers at end of chapter.*
>
> a. At very low frequencies, a low-pass filter provides an attenuation of 0 dB. (True or False)
> b. At the cutoff frequency, a low-pass filter has an N_{dB} loss of -3 dB. (True or False)
> c. On logarithmic graph paper, one cycle is the same as one octave. (True or False)
> d. The advantage of semilog and log-log graph paper is that a larger range of values can be shown in one plot without losing resolution in the smaller values. (True or False)

26–11 Resonant Filters

Tuned circuits provide a convenient method of filtering a band of radio frequencies because relatively small values of L and C are necessary for resonance. A tuned circuit provides filtering action by means of its maximum response at the resonant frequency.

The width of the band of frequencies affected by resonance depends on the Q of the tuned circuit; a higher Q provides a narrower bandwidth. Because resonance is effective for a band of frequencies below and above f_r, resonant filters are called *band-stop* or *band-pass* filters. Series or parallel *LC* circuits can be used for either function, depending on the connections with respect to R_L. In the application of a band-stop filter to suppress certain frequencies, the LC circuit is often called a *wavetrap*.

Series Resonance Filters

A series resonant circuit has maximum current and minimum impedance at the resonant frequency. Connected in series with R_L, as in Fig. 26–24a, the series-tuned LC circuit allows frequencies at and near resonance to produce maximum output across R_L. Therefore, this is band-pass filtering.

When the series LC circuit is connected across R_L as in Fig. 26–24b, however, the resonant circuit provides a low-impedance shunt path that short-circuits R_L. Then there is minimum output. This action corresponds to a shunt bypass capacitor, but the resonant circuit is more selective, short-circuiting R_L just for frequencies at and near resonance. For the bandwidth of the tuned circuit, the series resonant circuit in shunt with R_L provides band-stop filtering.

The series resistor R_S in Fig. 26–24b is used to isolate the low resistance of the LC filter from the input source. At the resonant frequency, practically all of the input voltage is across R_S with little across R_L because the LC tuned circuit then has very low resistance due to series resonance.

Parallel Resonance Filters

A parallel resonant circuit has maximum impedance at the resonant frequency. Connected in series with R_L, as in Fig. 26–25a, the parallel-tuned LC circuit provides maximum impedance in series with R_L at and near the resonant frequency. Then these frequencies produce maximum voltage across the LC circuit but minimum output voltage across R_L. This is a band-stop filter, therefore, for the bandwidth of the tuned circuit.

The parallel LC circuit connected across R_L, however, as in Fig. 26–25b, provides a band-pass filter. At resonance, the high impedance of the parallel LC circuit allows R_L to develop its output voltage. Below resonance, R_L is short-circuited by the low reactance of L; above resonance, R_L is short-circuited by the low reactance of C. For frequencies at or near resonance, though, R_L is shunted by high impedance, resulting in maximum output voltage.

The series resistor R_S in Fig. 26–25b is used to improve the filtering effect. Note that the parallel LC combination and R_S divide the input voltage. At the resonant frequency, though, the LC circuit has very high resistance for parallel resonance. Then most of the input voltage is across the LC circuit and R_L with little across R_S.

Figure 26–24 The filtering action of a series resonant circuit. (*a*) Band-pass filter when L and C are in series with R_L. (*b*) Band-stop filter when LC circuit is in shunt with R_L.

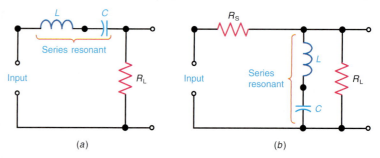

(a)   (b)

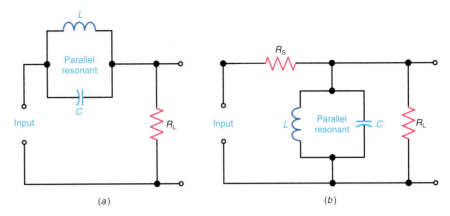

Figure 26-25 The filtering action of a parallel resonant circuit. (*a*) Band-stop filter when *LC* tank is in series with R_L. (*b*) Band-pass filter when *LC* tank is in shunt with R_L.

L-Type Resonant Filter

Series and parallel resonant circuits can be combined in L, T, or π sections for sharper discrimination of the frequencies to be filtered. Examples of an L-type filter are shown in Fig. 26–26.

The circuit in Fig. 26–26*a* is a band-stop filter. The reason is that the parallel resonant L_1C_1 circuit is in series with the load, whereas the series-resonant L_2C_2 circuit is in shunt with R_L. There is a dual effect as a voltage divider across the input source voltage. The high resistance of L_1C_1 reduces voltage output to the load. Also, the low resistance of L_2C_2 reduces the output voltage.

For the opposite effect, the circuit in Fig. 26–26*b* is a band-pass filter. Now the series-resonant L_3C_3 circuit is in series with the load. Here the low resistance of L_3C_3 allows more output for R_L at resonance. Also, the high resistance of L_4C_4 allows maximum output voltage.

Crystal Filters

A thin slice of quartz provides a resonance effect by mechanical vibrations at a particular frequency, like an *LC* circuit. The quartz crystal can be made to vibrate by a voltage input or produce voltage output when it is compressed, expanded, or twisted. This characteristic of some crystals is known as the *piezoelectric*

Figure 26–26 Inverted L filter with resonant circuits. (*a*) Band-stop filtering action. (*b*) Band-pass filtering action.

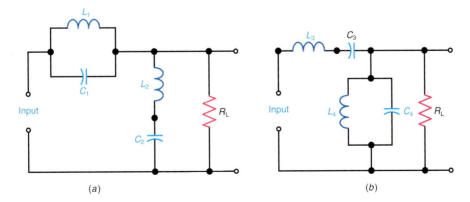

Figure 26–27 Quartz crystal. Size is ½ in. wide.

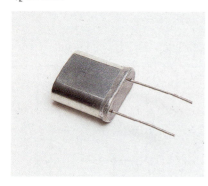

GOOD TO KNOW

In some cases *RF* interference cannot be eliminated by a power line filter because it is radiated from the source. In this case the *RF* energy induces voltage and currents into the circuit in addition to traveling through the 60 Hz ac power line.

Figure 26–28 Power-line filter unit. (*a*) Circuit of balanced L-type low-pass filter. (*b*) Filter unit.

(*a*)

(*b*)

effect. As a result, crystals are often used in place of resonant circuits. In fact, the Q of a resonant crystal is much higher than that of *LC* circuits. However, the crystal has a specific frequency that cannot be varied because of its stability. Crystals are used for radio frequencies in the range of about 0.5 to 30 MHz. Figure 26–27 shows a crystal for the frequency of 3.579545 MHz for use in the color oscillator circuit of a television receiver. Note the exact frequency.

Special ceramic materials, such as lead titanate, can also be used as crystal filters. They have a piezoelectric effect like quartz crystals. Ceramic crystals are smaller and cost less, but they have a lower Q than quartz crystals.

■ *26–11 Knowledge Check*

> *Answer at end of chapter.*
>
> **What is another name for a resonant band-stop filter?**

■ *26–11 Self-Review*

> *Answers at end of chapter.*
>
> a. **A parallel-resonant *LC* circuit in series with the load is a band-stop filter. (True or False)**
> b. **A series resonant *LC* circuit in series with the load is a band-pass filter. (True or False)**
> c. **Quartz crystals can be used as resonant filters. (True or False)**

26–12 Interference Filters

Voltage or current not at the desired frequency represents interference. Usually, such interference can be eliminated by a filter. Some typical applications are (1) low-pass filter to eliminate rf interference from the 60-Hz power-line input to a receiver, (2) high-pass filter to eliminate rf interference from the signal picked up by a television receiving antenna, and (3) resonant filter to eliminate an interfering radio frequency from the desired rf signal. As noted earlier, the resonant band-stop filter is called a *wavetrap.*

Power-Line Filter

Although the power line is a source of 60-Hz voltage, it is also a conductor of interfering rf currents produced by motors, fluorescent lighting circuits, and rf equipment. When a receiver is connected to the power line, the rf interference can produce noise and whistles in the receiver output. The filter shown in Fig. 26–28 can be used to minimize this interference. The filter is plugged into the wall outlet for 60-Hz power, and the receiver is plugged into the filter. An rf bypass capacitor across the line with two series rf chokes forms a low-pass balanced L-type filter. Using a choke in each side of the line makes the circuit balanced to ground.

The chokes provide high impedance for interfering rf current but not for 60 Hz, isolating the receiver input connections from rf interference in the power line. Also, the bypass capacitor short-circuits the receiver input for radio frequencies but not for 60 Hz. The unit then is a low-pass filter for 60-Hz power applied to the receiver while rejecting higher frequencies.

Figure 26–29 A television antenna filter to pass TV channel frequencies above 54 MHz but attenuate lower frequencies that can cause interference

Television Antenna Filter

When a television receiver has interference in the picture resulting from radio frequencies below the television broadcast band that are picked up by the receiving antenna, this rf interference can be reduced by the high-pass filter shown in Fig. 26–29. The filter attenuates frequencies below 54 MHz, which is the lowest frequency for Channel 2.

At frequencies lower than 54 MHz, series capacitances provide increasing reactance with a larger voltage drop, whereas the shunt inductances have less reactance and short-circuit the load. Higher frequencies are passed to the load as the series capacitive reactance decreases and the shunt inductive reactance increases.

Connections to the filter unit are made at the receiver end of the line from the antenna. Either end of the filter is connected to the antenna terminals on the receiver with the opposite end connected to the antenna line.

■ 26–12 Knowledge Check

Answer at end of chapter.

For the power-line filter in Fig. 26–28, does the 60-Hz ac power-line frequency pass through the filter?

■ 26–12 Self-Review

Answers at end of chapter.

a. A wavetrap is a band-stop filter. (True or False)
b. The TV antenna filter in Fig. 26–29 is a high-pass filter with series capacitors. (True or False)

Summary

- A filter can separate high and low frequencies. With input of different frequencies, the high-pass filter allows the higher frequencies to produce output voltage across the load; a low-pass filter provides output voltage at lower frequencies.

- Pulsating or fluctuating direct current varies in amplitude but does not reverse its direction. Similarly, a pulsating or fluctuating dc voltage varies in amplitude but maintains one polarity, either positive or negative.

- Pulsating direct current or voltage consists of a steady dc level, equal to the average value, and an ac component that reverses in polarity with respect to the average level.

The dc and ac can be separated by filters.

- An *RC* coupling circuit is a high-pass filter for pulsating direct current. Capacitance C_C blocks the steady dc voltage but passes the ac component.

- A transformer with an isolated secondary winding also is a high-pass filter. With pulsating direct current in the primary, only the ac component produces output voltage in the secondary.

- A bypass capacitor in parallel with *R* provides a low-pass filter.

- Combinations of *L*, *C*, and *R* can be arranged as L, T, or π filters for more selective filtering. All three arrangements can be used for either

low-pass or high-pass action. See Figs. 26–9 and 26–11.

- In high-pass filters, the capacitance must be in series with the load as a coupling capacitor with shunt *R* or *L* across the line.

- For low-pass filters, the capacitance is across the line as a bypass capacitor, and *R* or *L* then must be in series with the load.

- The cutoff frequency f_c of a filter is the frequency at which the output voltage is reduced to 70.7% of its maximum value.

- For an *RC* low-pass or high-pass filter, $X_C = R$ at the cutoff frequency. Similarly, for an *RL* low-pass or high-pass filter, $X_L = R$ at the cutoff frequency. To calculate f_c

for an RC low-pass or high-pass filter, use the formula $f_c = 1/2\pi RC$. To calculate f_c for an RL low-pass or high-pass filter, use the formula $f_c = R/2\pi L$.

■ For an RC or RL filter, either low-pass or high-pass, the phase angle θ between V_{in} and V_{out} is approximately 0° in the passband. In the stopband, $\theta = \pm 90°$. The sign of θ depends on the type of filter.

■ RC low-pass filters can be combined with RC high-pass filters when it is desired to either pass or block only a certain band of frequencies. These types of filters are called band-pass and band-stop filters, respectively.

■ The decibel (dB) unit of measurement is used to compare two power levels. A passive filter has an attenuation of −3 dB at the cutoff frequency.

■ Semilog and log-log graph paper are typically used to show the frequency response of a filter. On semilog graph paper, the vertical axis uses conventional linear spacing; the horizontal axis uses logarithmically spaced divisions.

■ The advantage of using semilog or log-log graph paper is that a larger range of values can be shown in one plot without losing resolution in the smaller values.

■ A band-pass or band-stop filter has two cutoff frequencies. The band-pass filter passes to the load those frequencies in the band between the cutoff frequencies and attenuates all other frequencies higher and lower than the pass-band. A band-stop filter does the opposite, attenuating the band between the cutoff frequencies, while passing to the load all other frequencies higher and lower than the stopband.

■ Resonant circuits are generally used for band-pass or band-stop filtering with radio frequencies.

■ For band-pass filtering, the series resonant LC circuit must be in series with the load, for minimum series opposition; the high impedance of parallel resonance is across the load.

■ For band-stop filtering, the circuit is reversed, with the parallel resonant LC circuit in series with the load; the series resonant circuit is in shunt across the load.

■ A wavetrap is an application of the resonant band-stop filter.

Important Terms

■ Attenuation - a reduction in signal amplitude.

■ Band-Pass Filter - a filter designed to pass only a specific band of frequencies from its input to its output.

■ Band-Stop Filter - a filter designed to block or severely attenuate only a specific band of frequencies.

■ Bypass Capacitor - a capacitor that bypasses or shunts the ac component of a pulsating dc voltage around a component such as a resistor. The value of a bypass capacitor should be chosen so that its X_C value is one-tenth or less of the parallel resistance at the lowest frequency intended to be bypassed.

■ Crystal Filter - a filter made of a crystalline material such as quartz.

Crystal filters are often used in place of conventional LC circuits because their Q is so much higher.

■ Cutoff Frequency - the frequency at which the attenuation of a filter reduces the output amplitude to 70.7% of its value in the passband.

■ Decade - a 10 to 1 range in frequencies.

■ Decibel (dB) - a logarithmic expression that compares two power levels.

■ Fluctuating DC - a dc voltage or current that varies in magnitude but does not reverse in polarity or direction. Another name for fluctuating dc is pulsating dc.

■ High-Pass Filter - a filter that allows the higher frequency components of the applied voltage to develop appreciable output voltage while at the same time attenuating or eliminating the lower frequency components.

■ Low-Pass Filter - a filter that allows the lower frequency components of the applied voltage to develop appreciable output voltage while at the same time attenuating or eliminating the higher frequency components.

■ Octave - a 2 to 1 range in frequencies.

■ Pulsating DC - a dc voltage or current that varies in magnitude but does not reverse in polarity or direction. Another name for pulsating dc is fluctuating dc.

Related Formulas

RC Low-Pass Filters

$$V_{out} = \frac{X_C}{Z_T} \times V_{in}$$

$$\theta = \arctan(-R/X_C)$$

$$f_c = \frac{1}{2\pi RC}$$

RL Low-Pass Filters

$$V_{out} = \frac{R}{Z_T} \times V_{in}$$

$$\theta = \arctan(-X_L/R)$$

$$f_c = R/2\pi L$$

RC High-Pass Filters

$$V_{out} = \frac{R}{Z_T} \times V_{in}$$

$$\theta = \arctan(X_C/R)$$

$$f_c = 1/2\pi RC$$

RL High-Pass Filters

$$V_{out} = \frac{X_L}{Z_T} \times V_{in}$$

$$\theta = \arctan(R/X_L)$$

$$f_c = R/2\pi L$$

Notch Filter

$$f_N = 1/4\pi R_1 C_1$$

Decibels

$$N_{dB} = 10 \log \frac{P_{out}}{P_{in}}$$

$$N_{dB} = 20 \log \frac{V_{out}}{V_{in}}$$

Self-Test

Answers at back of book.

1. A voltage that varies in magnitude but does not reverse in polarity is called a(n)

   a. alternating voltage.

   b. steady dc voltage.

   c. pulsating dc voltage.

   d. none of the above.

2. The capacitor in an *RC* coupling circuit

   a. blocks the steady dc component of the input voltage.

   b. blocks the ac component of the input voltage.

   c. appears like a short to a steady dc voltage.

   d. will appear like an open to the ac component of the input voltage.

3. The value of a bypass capacitor should be chosen so that its X_C value is

   a. 10 or more times the parallel resistance at the highest frequency to be bypassed.

   b. one-tenth or less the parallel resistance at the lowest frequency to be bypassed.

   c. one-tenth or less the parallel resistance at the highest frequency to be bypassed.

   d. equal to the parallel resistance at the lowest frequency to be bypassed.

4. In an *RC* low-pass filter, the output is taken across the

   a. resistor.

   b. inductor.

   c. capacitor.

   d. none of the above.

5. On logarithmic graph paper, a 10 to 1 range of frequencies is called a(n):

   a. octave.

   b. decibel (dB).

   c. harmonic.

   d. decade.

6. The cutoff frequency, f_c, of a filter is the frequency at which the output voltage is

   a. reduced to 50% of its maximum.

   b. reduced to 70.7% of its maximum.

   c. practically zero.

   d. exactly equal to the input voltage.

7. The decibel attenuation of a passive filter at the cutoff frequency is

   a. −3 dB.

   b. 0 dB.

   c. −20 dB.

   d. −6 dB.

8. To increase the cutoff frequency of an *RL* high-pass filter, you can

   a. decrease the value of *R*.

   b. decrease the value of *L*.

   c. increase the value of *R*.

   d. both b and c.

9. An *RC* low-pass filter uses a 2.2-kΩ *R* and a 0.01-μF *C*. What is its cutoff frequency?

   a. 3.5 MHz.

   b. 72.3 Hz.

   c. 7.23 kHz.

   d. 1.59 kHz.

10. For either an *RC* low-pass or high-pass filter,

    a. $X_c = 0\ \Omega$ at the cutoff frequency.

    b. $X_c = R$ at the cutoff frequency.

    c. X_c is infinite at the cutoff frequency.

    d. none of the above.

11. When a pulsating dc voltage is applied as an input to the primary of a transformer, the output from the secondary contains

    a. only the steady dc component of the input signal.

    b. a stepped up or down version of the pulsating dc voltage.

    c. only the ac component of the input signal.

    d. none of the above.

12. A power-line filter used to reduce rf interference is an example of a

    a. low-pass filter.

    b. high-pass filter.

    c. notch filter.

    d. band-pass filter.

13. On logarithmic graph paper, a 2 to 1 range of frequencies is called a(n)

a. decade.

b. decibel (dB).

c. harmonic.

d. octave.

14. What is the decibel (dB) attenuation of a filter with a 100-mV input and a 1-mV output at a given frequency?

a. −40 dB.

b. −20 dB.

c. −3 dB.

d. 0 dB.

15. In an *RL* high-pass filter, the output is taken across the

a. resistor.

b. inductor.

c. capacitor.

d. none of the above.

16. An *RL* high-pass filter uses a 60-mH *L* and a 1-kΩ *R*. What is its cutoff frequency?

a. 2.65 kHz.

b. 256 kHz.

c. 600 kHz.

d. 32 kHz.

17. A T-type low-pass filter consists of

a. series capacitors and a parallel inductor.

b. series inductors and a bypass capacitor.

c. series capacitors and a parallel resistor.

d. none of the above.

18. A π-type high-pass filter consists of

a. series inductors and parallel capacitors.

b. series inductors and a parallel resistor.

c. a series capacitor and parallel inductors.

d. none of the above.

19. When examining the frequency response curve of an *RC* low-pass filter, it can be seen that the rate of roll-off well above the cutoff frequency is

a. 6 dB/octave.

b. 6 dB/decade.

c. 20 dB/decade.

d. both a and c.

20. For signal frequencies in the passband, an *RC* high-pass filter has a phase angle of approximately

a. 45°.

b. 0°.

c. +90°.

d. −90°.

Questions

1. What is the function of an electrical filter?

2. Give two examples where the voltage has different frequency components.

3. (a) What is meant by *pulsating* direct current or voltage? (b) What are the two components of a pulsating dc voltage? (c) How can you measure the value of each of the two components?

4. Define the function of the following filters in terms of output voltage across the load resistance: (a) High-pass filter. Why is an $R_C C_C$ coupling circuit an example? (b) Low-pass filter. Why is an $R_b C_b$ bypass circuit an example? (c) Band-pass filter. How does it differ from a coupling circuit? (d) Band-stop filter. How does it differ from a band-pass filter?

5. Draw circuit diagrams for the following filter types. No values are necessary. (a) T-type high-pass and T-type low-pass; (b) π-type low-pass, balanced with a filter reactance in both sides of the line.

6. Draw the circuit diagrams for L-type band-pass and L-type band-stop filters. How do these two circuits differ from each other?

7. Draw the response curve for each of the following filters: (a) low-pass cutting off at 20,000 Hz; (b) high-pass cutting off at 20 Hz; (c) band-pass for 20 to 20,000 Hz; (d) band-pass for 450 to 460 kHz.

8. Give one similarity and one difference in comparing a coupling capacitor and a bypass capacitor.

9. Give two differences between a low-pass filter and a high-pass filter.

10. Explain briefly why the power-line filter in Fig. 26–28 passes 60-Hz alternating current but not 1-MHz rf current.

11. Explain the advantage of using semilog and log-log graph paper for plotting a frequency response curve.

12. Explain why an *RC* band-stop filter cannot be designed by interchanging the low-pass and high-pass filters in Fig. 26–18a.

Problems

SECTION 26–1 EXAMPLES OF FILTERING

26–1 Explain the basic function of a

    a. low-pass filter.

    b. high-pass filter.

SECTION 26–2 DIRECT CURRENT COMBINED WITH ALTERNATING CURRENT

26–2 For the values shown in Fig. 26–30,

    a. draw the waveform of voltage which is present across the load, R_L. Indicate the average and peak values on the waveform.

    b. draw the waveform of current which exists in the load, R_L. Indicate the average and peak values on the waveform.

Figure 26–30

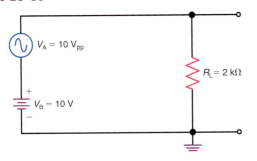

26–3 In Fig. 26–30, how much is the

    a. average dc voltage across the load, R_L?

    b. average dc current through the load, R_L?

26–4 In terms of the ac component in Fig. 26–30, how much is the

    a. peak voltage?

    b. peak-to-peak voltage?

    c. rms voltage?

26–5 In Fig. 26–30, redraw the waveform of voltage present across the load, R_L, if the polarity of V_B is reversed. Indicate the average and peak values on the waveform.

SECTION 26–3 TRANSFORMER COUPLING

26–6 Figure 26–31 shows the application of transformer coupling. Notice that the transformer has a turns ratio of 1:1. How much is the

    a. steady dc voltage in the primary?

    b. steady dc voltage in the secondary?

    c. peak-to-peak ac voltage in the primary?

    d. peak-to-peak ac voltage in the secondary?

26–7 In Fig. 26–31, indicate the peak voltage values for the ac output in the secondary.

26–8 In Fig. 26–31, compare the average or dc value of the primary and secondary voltage waveforms. Explain any difference.

Figure 26–31

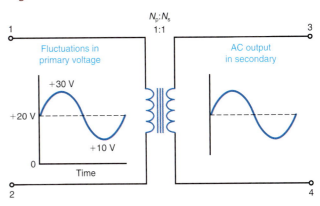

SECTION 26–4 CAPACITIVE COUPLING

26–9 In Fig. 26–32, determine the following:

    a. the X_C value for C_C at $f = 10$ kHz.

    b. the dc voltage across input terminals 1 and 2.

    c. the dc voltage across the coupling capacitor, C_C.

    d. the dc voltage across the resistor, R.

    e. the peak-to-peak ac voltage across terminals 1 and 2.

    f. the approximate peak-to-peak ac voltage across the coupling capacitor, C_C.

    g. the peak-to-peak ac voltage across the resistor, R.

    h. the rms voltage across the resistor, R.

Figure 26–32

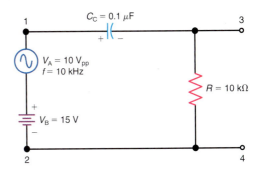

26–10 In Fig. 26–32, draw the voltage waveforms (including average and peak values) that exist across terminals

    a. 1 and 2.

    b. 3 and 4.

26–11 In Fig. 26–32, does C_C charge or discharge during the

    a. positive alternation of V_A?

    b. negative alternation of V_A?

26–12 In Fig. 26–32, is the positive or negative half-cycle of output voltage developed across R when C_C is

    a. charging?

    b. discharging?

26–13 In Fig. 26–32, what is the lowest frequency of V_A that will produce an X_C/R ratio of 1/10?

26–14 An RC coupling circuit is to be designed to couple frequencies above 500 Hz. If $R = 4.7$ kΩ, what is the minimum value for C_C?

SECTION 26–5 BYPASS CAPACITORS

26–15 In Fig. 26–33, determine the following:

a. the X_C value of C_1 at $f = 1$ MHz.

b. the dc voltage across input terminals 1 and 2.

c. the dc voltage across R_1.

d. the dc voltage across R_2.

e. the dc voltage across C_1.

f. the peak-to-peak ac voltage across terminals 1 and 2.

g. the approximate peak-to-peak ac voltage across terminals 3 and 4.

h. the approximate peak-to-peak ac voltage across R_1.

Figure 26–33

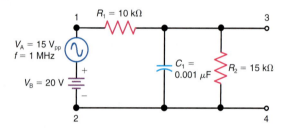

26–16 In Fig. 26–33, will the bypass capacitor, C_1, bypass R_2 if the frequency of V_A is 1 kHz?

26–17 What minimum value of capacitance will bypass a 1-kΩ resistor if the lowest frequency to be bypassed is 250 Hz?

SECTION 26–6 FILTER CIRCUITS

26–18 Classify each of the following as either a low-pass or high-pass filter:

a. transformer coupling (see Fig. 26–31)

b. RC coupling circuit (see Fig. 26–32)

c. bypass capacitor (see Fig. 26–33)

26–19 What type of filter, low-pass or high-pass, uses

a. series inductance and parallel capacitance?

b. series capacitance and parallel inductance?

26–20 Suppose that a low-pass filter has a cutoff frequency of 1 kHz. If the input voltage for a signal at this frequency is 30 mV, how much is the output voltage?

SECTION 26–7 LOW-PASS FILTERS

26–21 For a low-pass filter, define what is meant by the terms

a. passband.

b. stopband.

26–22 Assume that both the RC low-pass filter in Fig. 26–9a and the π-type filter in Fig. 26–9e have the same cutoff frequency, f_c. How do the filtering characteristics of these two filters differ?

SECTION 26–8 HIGH–PASS FILTERS

26–23 Do the terms passband and stopband apply to high-pass filters?

26–24 In Fig. 26–11, does the T-type filter provide sharper filtering than the RC filter? If so, why?

SECTION 26–9 ANALYZING FILTER CIRCUITS

26–25 Identify the filters in each of the following figures as either low-pass or high-pass:

a. Fig. 26–34

b. Fig. 26–35

c. Fig. 26–36

d. Fig. 26–37

Figure 26–34

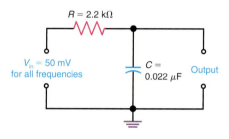

Figure 26–35

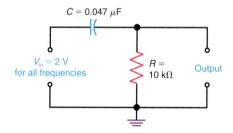

Figure 26–36

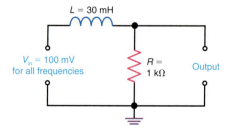

Figure 26-37

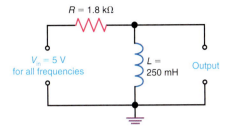

$R = 1.8\ k\Omega$

$V_{in} = 5\ V$
for all frequencies

$L = 250\ mH$

Output

26-26 Calculate the cutoff frequency, f_c, for the filters in each of the following figures:

a. Fig. 26-34

b. Fig. 26-35

c. Fig. 26-36

d. Fig. 26-37

26-27 In Fig. 26-34, calculate the output voltage, V_{OUT}, and phase angle, θ, at the following frequencies:

a. 50 Hz

b. 200 Hz

c. 1 kHz

d. f_c

e. 10 kHz

f. 20 kHz

g. 100 kHz

26-28 In Fig. 26-35, calculate the output voltage, V_{OUT}, and phase angle, θ, at the following frequencies:

a. 10 Hz

b. 50 Hz

c. 100 Hz

d. f_c

e. 1 kHz

f. 20 kHz

g. 500 kHz

26-29 In Fig. 26-36, calculate the output voltage, V_{OUT}, and phase angle, θ, at the following frequencies:

a. 100 Hz

b. 500 Hz

c. 2 kHz

d. f_c

e. 15 kHz

f. 30 kHz

g. 100 kHz

26-30 In Fig. 26-37, calculate the output voltage, V_{OUT}, and phase angle, θ, at the following frequencies:

a. 50 Hz

b. 100 Hz

c. 500 Hz

d. f_c

e. 3 kHz

f. 10 kHz

g. 25 kHz

26-31 For the filters in Figs. 26-34 through 26-37, what is the ratio of V_{OUT}/V_{IN} at the cutoff frequency?

26-32 Without regard to sign, what is the phase angle, θ, at the cutoff frequency for each of the filters in Figs. 26-34 through 26-37?

26-33 For a low-pass filter, what is the approximate phase angle, θ, for frequencies

a. well below the cutoff frequency?

b. well above the cutoff frequency?

26-34 Repeat Prob. 26-33 for a high-pass filter.

26-35 What type of filter is shown in Fig. 26-38?

Figure 26-38

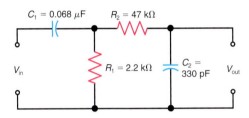

$C_1 = 0.068\ \mu F$ $R_2 = 47\ k\Omega$

V_{in}

$R_1 = 2.2\ k\Omega$ $C_2 = 330\ pF$ V_{out}

26-36 In Fig. 26-38, which components make-up the

a. high-pass filter?

b. low-pass filter?

26-37 In Fig. 26-38, calculate

a. the cutoff frequency, f_{C_1}.

b. the cutoff frequency, f_{C_2}.

c. the bandwidth, $f_{C_2} - f_{C_1}$.

26-38 In Fig. 26-38, why is it important to make R_2 at least 10 times larger than R_1?

26-39 Calculate the notch frequency, f_N, in Fig. 26-39.

Figure 26-39

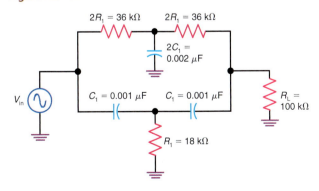

$2R_1 = 36\ k\Omega$ $2R_1 = 36\ k\Omega$

$2C_1 = 0.002\ \mu F$

V_{in}

$C_1 = 0.001\ \mu F$ $C_1 = 0.001\ \mu F$

$R_L = 100\ k\Omega$

$R_1 = 18\ k\Omega$

SECTION 26-10 DECIBELS AND FREQUENCY RESPONSE CURVES

26-40 Calculate the decibel (dB) power gain of an amplifier for the following values of P_{IN} and P_{OUT}:

a. $P_{in} = 1$ W, $P_{out} = 2$ W

b. $P_{in} = 1$ W, $P_{out} = 10$ W

c. $P_{in} = 50$ W, $P_{out} = 1$ kW

d. $P_{in} = 10$ W, $P_{out} = 400$ W

26-41 Calculate the decibel (dB) attenuation of a filter for the following values of P_{in} and P_{out}:

a. $P_{in} = 1$ W, $P_{out} = 500$ mW

b. $P_{in} = 100$ mW, $P_{out} = 10$ mW

c. $P_{in} = 5$ W, $P_{out} = 5\ \mu$W

d. $P_{in} = 10$ W, $P_{out} = 100$ mW

26-42 In Prob. 26-27, you calculated the output voltage for the RC filter in Fig. 26-34 at several different frequencies. For each frequency listed in Prob. 26-27, determine the decibel (dB) attenuation offered by the filter.

26-43 In Prob. 26-28, you calculated the output voltage for the RC filter in Fig. 26-35 at several different frequencies. For each frequency listed in Prob. 26-28, determine the decibel (dB) attenuation offered by the filter.

26-44 What is the rolloff rate of an RC low-pass filter for signal frequencies well beyond the cutoff frequency? Do the values calculated for the decibel (dB) attenuation in Prob. 26-42 verify this rolloff rate?

SECTION 26-11 RESONANT FILTERS

26-45 What determines the width of the band of frequencies that are allowed to pass through a resonant band-pass filter?

26-46 Identify the following configurations as either band-pass or band-stop filters:

a. series LC circuit in series with R_L

b. parallel LC circuit in series with R_L

c. parallel LC circuit in parallel with R_L

d. series LC circuit in parallel with R_L

SECTION 26-12 INTERFERENCE FILTERS

26-47 To prevent the radiation of harmonic frequencies from a radio transmitter, a filter is placed between the transmitter and antenna. The filter must pass all frequencies below 30 MHz and severely attenuate all frequencies above 30 MHz. What type of filter must be used?

26-48 What type of filter should be used to eliminate an interfering signal having a very narrow range of frequencies?

Critical Thinking

26-49 In Fig. 26-40 calculate (a) the cutoff frequency f_c; (b) the output voltage at the cutoff frequency f_c; (c) the output voltage at 50 kHz.

Figure 26-40 Circuit for Critical Thinking Prob. 26-49.

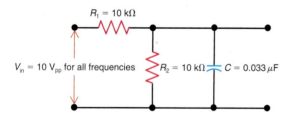

26-50 In Fig. 26-41, calculate the values of L and C required to provide an f_r of 1 MHz and a bandwidth Δf of 40 kHz.

Figure 26-41 Circuit for Critical Thinking Prob. 26-50.

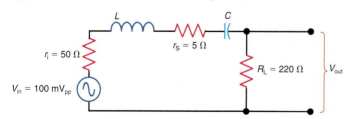

26-51 In Fig. 26-42 calculate the values of L and C required to provide an f_r of 1 MHz and a bandwidth Δf of 20 kHz.

Figure 26-42 Circuit for Critical Thinking Prob. 26-51.

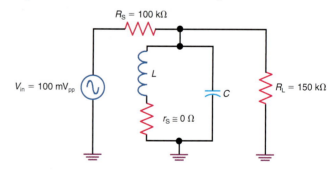

Answers to Knowledge Check Problems

26–1 a low-pass filter

26–2 10 V and 7.07 V

26–3 it blocks dc

26–4 long

26–5 lowest

26–6 7.07 V

26–7 the π-type

26–8 passband

26–9 0°

26–10 6 dB/octave which is the same as 20 dB/decade

26–11 wavetrap

26–12 yes

Answers to Self-Reviews

26–1 a. 500 kHz
    b. 60 Hz

26–2 a. 6 V
    b. 10 and 2 V
    c. 8 V
    d. 4 and 2.8 V

26–3 a. high-pass
    b. 0 V

26–4 a. 0 V
    b. 5 μF

26–5 a. RF
    b. 5 μF

26–6 a. high-pass
    b. low-pass

26–7 a. e and f
    b. low-pass

26–8 a. d
    b. high-pass

26–9 a. F
    b. T
    c. T
    d. T

26–10 a. T
    b. T
    c. F
    d. T

26–11 a. T
    b. T
    c. T

26–12 a. T
    b. T

Cumulative Review Summary (Chapters 25–26)

- Resonance results when reactances X_L and X_C are equal. In series, the net reactance is zero. In parallel, the net reactive branch current is zero. The specific frequency that makes $X_L = X_C$ is the resonant frequency $f_r = 1/(2\pi\sqrt{LC})$.

- Larger values of L and C mean lower resonant frequencies, as f_r is inversely proportional to the square root of L and C. If the value of L or C is quadrupled, for instance, f_r will decrease by one-half.

- For a series resonant LC circuit, the current is maximum. The voltage drop across the reactances is equal and opposite; the phase angle is zero. The reactive voltage at resonance is Q times greater than the applied voltage.

- For a parallel resonant LC circuit, the impedance is maximum with minimum line current, since the reactive branch currents cancel. The impedance at resonance is Q times the X_L value, but it is resistive with a phase angle of zero.

- The Q of the resonant circuit equals X_L/r_S for resistance in series with X_L or R_P/X_L for resistance in parallel with X_L.

- The bandwidth between half-power points is f_r/Q.

- A filter uses inductance and capacitance to separate high or low frequencies. A low-pass filter allows low frequencies to develop output voltage across the load; a high-pass filter does the same for high frequencies. Series inductance or shunt capacitance provides low-pass filtering; series capacitance or shunt inductance provides high-pass filtering.

- A fluctuating or pulsating dc is equivalent to an ac component varying in opposite directions around the average-value axis.

- An RC coupling circuit is effectively a high-pass filter for pulsating dc voltage, passing the ac component but blocking the dc component.

- A transformer with an isolated secondary is a high-pass filter for pulsating direct current, allowing alternating current in the secondary but no dc output level.

- A bypass capacitor in parallel with R is a low-pass filter, since its low reactance reduces the voltage across R at high frequencies.

- The main types of filter circuits are π, L, and T types. These can be high-pass or low-pass, depending on how L and C are connected.

- Resonant circuits can be used as band-pass or band-stop filters. For band-pass filtering, series resonant circuits are in series with the load or parallel resonant circuits are across the load. For band-stop filtering, parallel resonant circuits are in series with the load or series resonant circuits are across the load.

- A wavetrap is an application of a resonant band-stop filter.

- The cutoff frequency of a filter is the frequency at which the output voltage is reduced to 70.7% of its maximum value.

- The cutoff frequency of an RC low-pass or high-pass filter can be calculated from $f_c = 1/2\pi RC$. Similarly, the cutoff frequency of an RL low-pass or high-pass filter can be calculated from $f_c = R/2\pi L$.

- The decibel (dB) is a logarithmic expression that compares two power

levels. In the passband, a passive filter provides 0 dB of attenuation. At the cutoff frequency, a passive filter provides attenuation of −3 dB.

- Semilog and log-log graph paper are typically used to show the frequency response of a filter. The advantage of using logarithmic graph paper is that a wide range of frequencies can be shown in one plot without losing resolution in the smaller values.

Cumulative Review Self–Test

Answers at back of book.

Fill in the numerical answer.

1. An L of 10 H and C of 40 μF has f_r of _____ Hz.

2. An L of 100 μH and C of 400 pF has f_r of _____ MHz.

3. In question 2, if $C = 400$ pF and L is increased to 400 μH, the f_r decreases to _____ MHz.

4. In a series resonant circuit with 10 mV applied across a 1-Ω R, a 1000-Ω X_L and a 1000-Ω X_C at resonance, the current is _____ mA.

5. Imagine a parallel resonant circuit. It has a 1-Ω r_S in series with a 1000-Ω X_L in one branch and a 1000-Ω X_C in the other branch. With 10 mV applied, the voltage across X_C equals _____ mV.

6. In question 5, the Z of the parallel resonant circuit equals _____ MΩ.

7. An LC circuit resonant at 500 kHz has a Q of 100. Its total bandwidth between half-power points equals _____ kHz.

8. A coupling capacitor for 40 to 15,000 Hz in series with a 0.5-MΩ resistor has a capacitance of _____ μF.

9. A bypass capacitor for 40 to 15,000 Hz in shunt with a 1000-Ω R has a capacitance of _____ μF.

10. A pulsating dc voltage varying in a symmetrical sine wave between 100 and 200 V has an average value of _____ V.

11. An RC low-pass filter has the following values: $R = 1$ kΩ, $C = 0.005$ μF. The cutoff frequency f_c is _____.

12. The input voltage to a filter is 10 V$_{pp}$ and the output voltage is 100 μV$_{pp}$. The amount of attenuation is _____ dB.

13. On logarithmic graph paper, a 2-to-1 range of values is called a(n) _____, and a 10-to-1 range of values is called a(n) _____.

14. At the cutoff frequency, the output voltage is reduced to _____ % of its maximum.

Answer True or False.

15. A series resonant circuit has low I and high Z.

16. A steady direct current in the primary of a transformer cannot produce any ac output voltage in the secondary.

17. A π-type filter with shunt capacitances is a low-pass filter.

18. An L-type filter with a parallel resonant LC circuit in series with the load is a band-stop filter.

19. A resonant circuit can be used as a band-stop filter.

20. In the passband, an RC low-pass filter provides approximately 0 dB of attenuation.

21. The frequency response of a filter is never shown on logarithmic graph paper.

Appendix A

Electrical Symbols and Abbreviations

Table A–1 summarizes the letter symbols used as abbreviations for electrical quantities and their basic units. All the metric prefixes for multiple and submultiple values are listed in Table A–2. In addition, Table A–3 shows electronic symbols from the Greek alphabet.

Table A–1	Electrical Quantities	
Quantity	**Symbol***	**Basic Unit**
Current	I or i	ampere (A)
Charge	Q or q	coulomb (C)
Power	P	watt (W)
Voltage	V or v	volt (V)
Resistance	R	ohm (Ω)
Reactance	X	ohm (Ω)
Impedance	Z	ohm (Ω)
Conductance	G	siemens (S)
Admittance	Y	siemens (S)
Susceptance	B	siemens (S)
Capacitance	C	farad (F)
Inductance	L	henry (H)
Frequency	f	hertz (Hz)
Period	T	second (s)

* Capital letters for I, Q, and V are generally used for peak, rms, or dc values, whereas small letters are used for instantaneous values. Small r and g are also used for internal values, such as r_i for the internal resistance of a battery and g_m for the transconductance of a JFET or MOSFET.

Table A–2	Multiples and Submultiples of Units*			
Value		**Prefix**	**Symbol**	**Example**
$1\ 000\ 000\ 000\ 000 = 10^{12}$		tera	T	$THz = 10^{12}\ Hz$
$1\ 000\ 000\ 000 = 10^{9}$		giga	G	$GHz = 10^{9}\ Hz$
$1\ 000\ 000 = 10^{6}$		mega	M	$MHz = 10^{6}\ Hz$
$1\ 000 = 10^{3}$		kilo	k	$kV = 10^{3}\ V$
$100 = 10^{2}$		hecto	h	$hm = 10^{2}\ m$
$10 = 10$		deka	da	$dam = 10\ m$
$0.1 = 10^{-1}$		deci	d	$dm = 10^{-1}\ m$
$0.01 = 10^{-2}$		centi	c	$cm = 10^{-2}\ m$
$0.001 = 10^{-3}$		milli	m	$mA = 10^{-3}\ A$
$0.000\ 001 = 10^{-6}$		micro	μ	$\mu V = 10^{-6}\ V$
$0.000\ 000\ 001 = 10^{-9}$		nano	n	$ns = 10^{-9}\ s$
$0.000\ 000\ 000\ 001 = 10^{-12}$		pico	p	$pF = 10^{-12}\ F$

* Additional prefixes are exa $= 10^{18}$, peta $= 10^{15}$, femto $= 10^{-15}$, and atto $= 10^{-18}$.

Table A–3	Greek Letter Symbols*		
	LETTER		
Name	**Capital**	**Small**	**Uses**
Alpha	A	α	α for angles, transistor characteristic
Beta	B	β	β for angles, transistor characteristic
Gamma	Γ	γ	transistor characteristic
Delta	Δ	δ	Small change in value
Epsilon	E	ϵ	ϵ for permittivity; also base of natural logarithms
Zeta	Z	ζ	
Eta	H	η	η for intrinsic standoff ratio of a unijunction transistor (UJT)

Table A–3	Greek Letter Symbols* (Continued)		
	LETTER		
Name	Capital	Small	Uses
Theta	Θ	θ	Phase angle
Iota	I	ι	
Kappa	K	κ	
Lambda	Λ	λ	λ for wavelength
Mu	M	μ	μ for prefix micro-, permeability, amplification factor
Nu	N	ν	
Xi	Ξ	ξ	
Omicron	O	o	
Pi	Π	π	π is 3.1416 for ratio of circumference to diameter of a circle
Rho	P	ρ	ρ for resistivity
Sigma	Σ	σ	Summation
Tau	T	τ	Time constant
Upsilon	Υ	υ	
Phi	Φ	ϕ	Magnetic flux, angles
Chi	X	χ	
Psi	Ψ	ψ	Electric flux
Omega	Ω	ω	Ω for ohms; ω for angular velocity

* This table includes the complete Greek alphabet, although some letters are not used for electronic symbols.

Appendix B

Solder and the Soldering Process*

From Simple Task to Fine Art

Soldering is the process of joining two metals together by the use of a low-temperature melting alloy. Soldering is one of the oldest known joining techniques, first developed by the Egyptians in making weapons such as spears and swords. Since then, it has evolved into what is now used in the manufacturing of electronic assemblies. Soldering is far from the simple task it once was; it is now a fine art, one that requires care, experience, and a thorough knowledge of the fundamentals.

The importance of having high standards of workmanship cannot be overemphasized. Faulty solder joints remain a cause of equipment failure, and because of that, soldering has become a *critical skill.*

The material contained in this appendix is designed to provide the student with both the fundamental knowledge and the practical skills needed to perform many of the high-reliability soldering operations encountered in today's electronics.

Covered here are the fundamentals of the soldering process, the proper selection, and the use of the soldering station.

The key concept in this appendix is *high-reliability soldering.* Much of our present technology is vitally dependent on the reliability of countless, individual soldered connections. High-reliability soldering was developed in response to early failures with space equipment. Since then the concept and practice have spread into military and medical equipment. We have now come to expect it in everyday electronics as well.

The Advantage of Soldering

Soldering is the process of connecting two pieces of metal together to form a reliable electrical path. Why solder them in the first place? The two pieces of metal could be put together with nuts and bolts, or some other kind of mechanical fastening. The disadvantages of these methods are twofold. First, the reliability of the connection cannot be assured because of vibration and shock. Second, because oxidation and corrosion are continually occurring on the metal surfaces, electrical conductivity between the two surfaces would progressively decrease.

A soldered connection does away with both of these problems. There is no movement in the joint and no interfacing surfaces to oxidize. A continuous conductive path is formed, made possible by the characteristics of the solder itself.

---

* This material is provided courtesy of PACE, Inc., Laurel, Maryland.

The Nature of Solder

Solder used in electronics is a low-temperature melting alloy made by combining various metals in different proportions. The most common types of solder are made from tin and lead. When the proportions are equal, it is known as 50/50 solder—50 percent tin and 50 percent lead. Similarly, 60/40 solder consists of 60 percent tin and 40 percent lead. The percentages are usually marked on the various types of solder available; sometimes only the tin percentage is shown. The chemical symbol for tin is Sn; thus Sn 63 indicates a solder which contains 63 percent tin.

Pure lead (Pb) has a melting point of 327°C (621°F); pure tin, a melting point of 232°C (450°F). But when they are combined into a 60/40 solder, the melting point drops to 190°C (374°F)—lower than either of the two metals alone.

Melting generally does not take place all at once. As illustrated in Fig. B–1, 60/40 solder begins to melt at 183°C (361°F), but it has not fully melted until the temperature reaches 190°C (374°F). Between these two temperatures, the solder exists in a plastic (semiliquid) state—some, but not all, of the solder has melted.

The plastic range of solder will vary, depending on the ratio of tin to lead, as shown in Fig. B–2. Various ratios of tin to lead are shown across the top of this figure. With most ratios, melting begins at 183°C (361°F), but the full melting temperatures vary dramatically. There is one ratio of tin to lead that has no plastic state. It is known as *eutectic solder*. This ratio is 63/37 (Sn 63), and it fully melts and solidifies at 183°C (361°F).

The solder most commonly used for hand soldering in electronics is the 60/40 type, but because of its plastic range, care must be taken not to move any elements of the joint during the cool-down period. Movement may cause a disturbed joint. Characteristically, this type of joint has a rough, irregular appearance and looks dull instead of bright and shiny. It is unreliable and therefore one of the types of joints that is unacceptable in high-reliability soldering.

In some situations, it is difficult to maintain a stable joint during cooling, for example, when wave soldering is used with a moving conveyor line of circuit boards during the manufacturing process. In other cases, it may be necessary to use minimal heat to avoid damage to heat-sensitive components. In both of these situations, eutectic solder is the preferred choice, since it changes from a liquid to a solid during cooling with no plastic range.

Figure B–1 Plastic range of 60/40 solder. Melt begins at 183°C (361°F) and is complete at 190°C (374°F).

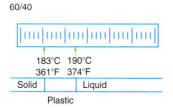

Figure B–2 Fusion characteristics of tin/lead solders.

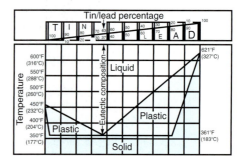

The Wetting Action

To someone watching the soldering process for the first time, it looks as though the solder simply sticks the metals together like a hot-melt glue, but what actually happens is far different.

A chemical reaction takes place when the hot solder comes into contact with the copper surface. The solder dissolves and penetrates the surface. The molecules of solder and copper blend together to form a new metal alloy, one that is part copper and part solder and that has characteristics all its own. This reaction is called *wetting* and forms the intermetallic bond between the solder and copper (Fig. B–3).

Proper wetting can occur only if the surface of the copper is free of contamination and from oxide films that form when the metal is exposed to air. Also, the solder and copper surfaces need to have reached the proper temperature.

Even though the surface may look clean before soldering, there may still be a thin film of oxide covering it. When solder is applied, it acts like a drop of water on an oily surface because the oxide coating prevents the solder from coming into contact with the copper. No reaction takes place, and the solder can be easily scraped off. For a good solder bond, surface oxides must be removed during the soldering process.

The Role of Flux

Reliable solder connections can be accomplished only on clean surfaces. Some sort of cleaning process is essential in achieving successful soldered connections, but in most cases it is insufficient. This is due to the extremely rapid rate at which oxides form on the surfaces of heated metals, thus creating oxide films which prevent proper soldering. To overcome these oxide films, it is necessary to utilize materials, called *fluxes,* which consist of natural or synthetic rosins and sometimes additives called *activators.*

It is the function of flux to remove surface oxides and keep them removed during the soldering operation. This is accomplished because the flux action is very corrosive at or near solder melt temperatures and accounts for the flux's ability to rapidly remove metal oxides. It is the fluxing action of removing oxides and carrying them away, as well as preventing the formation of new oxides, that allows the solder to form the desired intermetallic bond.

Flux must activate at a temperature lower than solder so that it can do its job prior to the solder flowing. It volatilizes very rapidly; thus it is mandatory that the flux be activated to flow onto the work surface and not simply be volatilized by the hot iron tip if it is to provide the full benefit of the fluxing action.

There are varieties of fluxes available for many applications. For example, in soldering sheet metal, acid fluxes are used; silver brazing (which requires a much higher temperature for melting than that required by tin/lead alloys) uses a borax paste. Each of these fluxes removes oxides and, in many cases, serves additional purposes. The fluxes used in electronic hand soldering are the pure rosins, rosins combined with mild activators to accelerate the rosin's fluxing capability, low-residue/no-clean fluxes, or water-soluble fluxes. Acid fluxes or highly activated fluxes should never be used in electronic work. Various types of flux-cored solder are now in common use. They provide a convenient way to apply and control the amount of flux used at the joint (Fig. B–4).

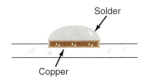

Figure B–3 The wetting action. Molten solder dissolves and penetrates a clean copper surface, forming an intermetallic bond.

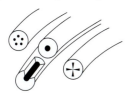

Figure B–4 Types of cored solder, with varying solder-flux percentages.

Soldering Irons

In any kind of soldering, the primary requirement, beyond the solder itself, is heat. Heat can be applied in a number of ways—conductive (e.g., soldering iron, wave, vapor phase), convective (hot air), or radiant (IR). We are mainly concerned with the conductive method, which uses a soldering iron.

Soldering stations come in a variety of sizes and shapes, but consist basically of three main elements: a resistance heating unit; a heater block, which acts as a heat reservoir; and the tip, or bit, for transferring heat to the work. The standard production station is a variable-temperature, closed-loop system with interchangeable tips and is made with ESD-safe plastics.

Controlling Heat at the Joint

Controlling tip temperature is not the real challenge in soldering; the real challenge is to control the *heat cycle* of the work—how fast the work gets hot, how hot it gets, and how long it stays that way. This is affected by so many factors that, in reality, tip temperature is not that critical.

The first factor that needs to be considered is the *relative thermal mass* of the area to be soldered. This mass may vary over a wide range.

Consider a single land on a single-sided circuit board. There is relatively little mass, so the land heats up quickly. But on a double-sided board with plated-through holes, the mass is more than doubled. Multilayered boards may have an even greater mass, and that's before the mass of the component lead is taken into consideration. Lead mass may vary greatly, since some leads are much larger than others.

Further, there may be terminals (e.g., turret or bifurcated) mounted on the board. Again, the thermal mass is increased, and will further increase as connecting wires are added.

Each connection, then, has its particular thermal mass. How this combined mass compares with the mass of the iron tip, the "relative" thermal mass, determines the time and temperature rise of the work.

With a large work mass and a small iron tip, the temperature rise will be slow. With the situation reversed, using a large iron tip on a small work mass, the temperature rise of the work will be much more rapid—even though the *temperature of the tip is the same.*

Now consider the capacity of the iron itself and its ability to sustain a given flow of heat. Essentially, irons are instruments for generating and storing heat, and the reservoir is made up of both the heater block and the tip. The tip comes in various sizes and shapes; it's the *pipeline* for heat flowing into the work. For small work, a conical (pointed) tip is used, so that only a small flow of heat occurs. For large work, a large chisel tip is used, providing greater flow.

The reservoir is replenished by the heating element, but when an iron with a large tip is used to heat massive work, the reservoir may lose heat faster than it can be replenished. Thus the *size* of the reservoir becomes important: a large heating block can sustain a larger outflow longer than a small one.

An iron's capacity can be increased by using a larger heating element, thereby increasing the wattage of the iron. These two factors, block size and wattage, are what determine the iron's recovery rate.

If a great deal of heat is needed at a particular connection, the correct temperature with the right size tip is required, as is an iron with a large enough capacity and an ability to recover fast enough. *Relative thermal mass,* then, is a major consideration for controlling the heat cycle of the work.

A second factor of importance is the *surface condition* of the area to be soldered. If there are any oxides or other contaminants covering the lands or leads, there will be a barrier to the flow of heat. Then, even though the iron tip is the right size and has the correct temperature, it may not supply enough heat to the connection to melt the solder. In soldering, a cardinal rule is that a good solder connection cannot be created on a dirty surface. Before you attempt to solder, the work should always be cleaned with an approved solvent to remove any grease or oil film from the surface. In some cases pretinning may be required to enhance solderability and remove heavy oxidation of the surfaces prior to soldering.

A third factor to consider is *thermal linkage*—the area of contact between the iron tip and the work.

Figure B–5 shows a cross-sectional view of an iron tip touching a round lead. The contact occurs only at the point indicated by the "X," so the linkage area is very small, not much more than a straight line along the lead.

The contact area can be greatly increased by applying a small amount of solder to the point of contact between the tip and workpiece. This solder heat bridge provides the thermal linkage and assures rapid heat transfer into the work.

From the aforementioned, it should now be apparent that there are many more factors than just the temperature of the iron tip that affect how quickly any particular connection is going to heat up. In reality, soldering is a very complex control problem, with a number of variables to it, each influencing the other. And what makes it so critical is *time*. The general rule for high-reliability soldering on printed circuit boards is to apply heat for no more than 2 s from the time solder starts to melt (wetting). Applying heat for longer than 2 s after wetting may cause damage to the component or board.

With all these factors to consider, the soldering process would appear to be too complex to accurately control in so short a time, but there is a simple solution—the *workpiece indicator* (WPI). This is defined as the reaction of the workpiece to the work being performed on it—a reaction that is discernible to the human senses of sight, touch, smell, sound, and taste.

Put simply, workpiece indicators are the way the work talks back to you—the way it tells you what effect you are having and how to control it so that you accomplish what you want.

In any kind of work, you become part of a closed-loop system. It begins when you take some action on the workpiece; then the workpiece reacts to what you did; you sense the change, and then modify your action to accomplish the result. It is in the sensing of the change, by sight, sound, smell, taste, or touch, that the workpiece indicators come in (Fig. B–6).

For soldering and desoldering, a primary workpiece indicator is *heat rate recognition*—observing how fast heat flows into the connection. In practice,

Figure B–5 Cross-sectional view (left) of iron tip on a round lead. The "X" shows point of contact. Use of a solder bridge (right) increases the linkage area and speeds the transfer of heat.

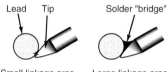

Lead   Tip   Solder "bridge"

Small linkage area   Large linkage area

Figure B–6 Work can be viewed as a closed-loop system (left). Feedback comes from the reaction of the workpiece and is used to modify the action. Workpiece indicators (right)—changes discernible to the human senses—are the way the "work talks back to you."

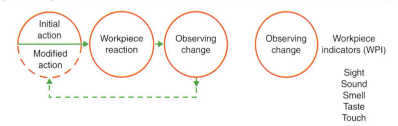

this means observing the rate at which the solder melts, which should be within 1 to 2 s.

This indicator encompasses all the variables involved in making a satisfactory solder connection with minimum heating effects, including the capacity of the iron and its tip temperature, the surface conditions, the thermal linkage between tip and workpiece, and the relative thermal masses involved.

If the iron tip is too large for the work, the heating rate may be too fast to be controlled. If the tip is too small, it may produce a "mush" kind of melt; the heating rate will be too slow, even though the temperature at the tip is the same.

A general rule for preventing overheating is "Get in and get out as fast as you can." That means using a heated iron you can react to—one giving a 1- to 2-s dwell time on the particular connection being soldered.

Selecting the Soldering Iron and Tip

A good all-around soldering station for electronic soldering is a variable-temperature, ESD-safe station with a pencil-type iron and tips that are easily interchangeable, even when hot (Fig. B–7).

The soldering iron tip should always be fully inserted into the heating element and tightened. This will allow for maximum heat transfer from the heater to the tip.

The tip should be removed daily to prevent an oxidation scale from accumulating between the heating element and the tip. A bright, thin tinned surface must be maintained on the tip's working surface to ensure proper heat transfer and to avoid contaminating the solder connection.

Figure B–7 Pencil-type iron with changeable tips.

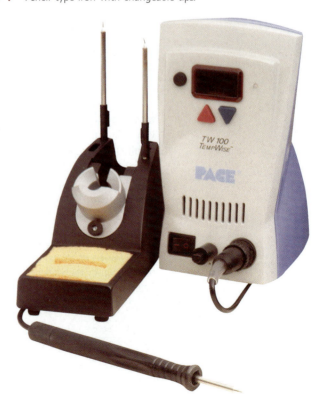

The plated tip is initially prepared by holding a piece of flux-cored solder to the face so that it will tin the surface when it reaches the lowest temperature at which solder will melt. Once the tip is up to operating temperature, it will usually be too hot for good tinning, because of the rapidity of oxidation at elevated temperatures. The hot tinned tip is maintained by wiping it lightly on a damp sponge to shock off the oxides. When the iron is not being used, the tip should be coated with a layer of solder.

Making the Solder Connection

The soldering iron tip should be applied to the area of maximum thermal mass of the connection being made. This will permit the rapid thermal elevation of the parts being soldered. Molten solder always flows toward the heat of a properly prepared connection.

When the solder connection is heated, a small amount of solder is applied to the tip to increase the thermal linkage to the area being heated. The solder is then applied to the opposite side of the connection so that the work surfaces, not the iron, melt the solder. Never melt the solder against the iron tip and allow it to flow onto a surface cooler than the solder melting temperature.

Solder, with flux, applied to a cleaned and properly heated surface will melt and flow without direct contact with the heat source and provide a smooth, even surface, feathering out to a thin edge (Fig. B–8). Improper soldering will exhibit a built-up, irregular appearance and poor filleting. The parts being soldered must be held rigidly in place until the temperature decreases to solidify the solder. This will prevent a disturbed or fractured solder joint.

Selecting cored solder of the proper diameter will aid in controlling the amount of solder being applied to the connection (e.g., a small-gauge solder for a small connection; a large-gauge solder for a large connection).

Removal of Flux

Cleaning may be required to remove certain types of fluxes after soldering. If cleaning is required, the flux residue should be removed as soon as possible, preferably within 1 hour after soldering.

Figure B–8 Cross-sectional view of a round lead on a flat surface.

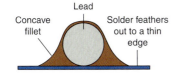

Lead
Concave fillet
Solder feathers out to a thin edge

Appendix C

Components: Values and Specifications

Table C–1	Preferred Resistance Values for Tolerances of ±5%, ±10% and ±20% (Multiple and submultiple values apply to those values shown.)	
+20%	+10%	+5%
10	10	10
		11
	12	12
		13
15	15	15
		16
	18	18
		20
22	22	22
		24
	27	27
		30
33	33	33
		36
	39	39
		43

Table C-1	Preferred Resistance Values for Tolerances of ±5%, ±10% and ±20% (Multiple and submultiple values apply to those values shown.) (*Continued*)	
+20%	**+10%**	**+5%**
47	47	47
		51
	56	56
		62
68	68	68
		75
	82	82
		91
100	100	100

Standard Resistor Values

1.0	1.8	3.3	5.6	10	18	33	56	100	180	330	560	1.0K	1.8K	3.3K	5.6K	10K	18K	33K	56K	100K	180K	330K	560K	1.0M	1.8M	3.3M	5.6M	10M
1.1	2.0	3.6	6.2	11	20	36	62	110	200	360	620	1.1K	2.0K	3.6K	6.2K	11K	20K	36K	62K	110K	200K	360K	620K	1.1M	2.0M	3.6M	6.2M	
1.2	2.2	3.9	6.8	12	22	39	68	120	220	390	680	1.2K	2.2K	3.9K	6.8K	12K	22K	39K	68K	120K	220K	390K	680K	1.2M	2.2M	3.9M	6.8M	
1.3	2.4	4.3	7.5	13	24	43	75	130	240	430	750	1.3K	2.4K	4.3K	7.5K	13K	24K	43K	75K	130K	240K	430K	750K	1.3M	2.4M	4.3M	7.5M	
1.5	2.7	4.7	8.2	15	27	47	82	150	270	470	820	1.5K	2.7K	4.7K	8.2K	15K	27K	47K	82K	150K	270K	470K	820K	1.5M	2.7M	4.7M	8.2M	
1.6	3.0	5.1	9.1	16	30	51	91	160	300	510	910	1.6K	3.0K	5.1K	9.1K	16K	30K	51K	91K	160K	300K	510K	910K	1.6M	3.0M	5.1M	9.1M	

CFR Series

Type (Power Rating)	Dimensions (mm) L	Dimensions (mm) D	Dimensions (mm) d	Max. Working Voltage	Max. Overload Voltage	Dielectric Withstanding Voltage	Rated Ambient Temperature	Temperature Cycling	Resistance Range	Resistance Tolerance
1/8 Watt	3.3 ± 0.4	1.8 ± 0.3	0.5 ± 0.05	150 V	300 V	300 V	70°C	−55°C to 155°C	1.0 Ω to 10 MΩ	±5%
1/4 Watt	6.3 ± 0.5	2.3 ± 0.3	0.6 ± 0.05	250 V	500 V	500 V	70°C	−55°C to 155°C	1.0 Ω to 10 MΩ	±5%
1/2 Watt	9.0 ± 0.5	3.2 ± 0.5	0.6 ± 0.05	350 V	700 V	500 V	70°C	−55°C to 155°C	1.0 Ω to 10 MΩ	±5%

Derating Curve

Features:
- Industry's lowest cost!
- Available in 1/8, 1/4, 1/2 watt; packaged in bulk or tape/reel (except 1/2 watt)
- Exceeds carbon comp MIL-R-11 performance
- Standard tolerance: ±5%
- Exceptional long-term stability

Characteristics: • Operating Temperature Range: −55°C to 155°C • Solderability: 235°C for 5±0.5 seconds
• Short Time Overload: 2.5 Times RCW V for 5 seconds • Resistance To Solvent: Trichroethane for 1 min. with
ultrasonic • Pulse Overload: 4 times RCW V 10000 cycles (1 sec. on, 25 sec. off)

Panasonic E-Series Metallized Polyester Capacitors

GENERAL SPECIFICATIONS:
ELECTRICAL: Rated Voltage 100 V.D.C., 250 V.D.C.,
400 V.D.C., 630 V.D.C. **Capacitance Tolerance** ±10% (K)
Insulation Resistance: Less than 0.33 μF: 9,000 MΩ; More than
0.33 μF: 3,000 MΩ, **Dissipation Factor** = 1.0% at 1 kHz

Withstanding Voltage Rated Voltage × 1.5 (1 min.)
Operating Temperature −40°C ~ +85°C
PANASONIC ECQ-E CROSS REFERENCES:
  Sprague: DF/450P/451P/453P;
  Westlake: 160/168/185/167/184/186;

Mepco: 713A1/719A1/C280; **TRW:** X601PE/X601CPE;
  Packtron: MF/RB; **Secor:** MMKS/MMKR/MMDR;
RIFA: PHE353/PHE354; **Roederstein:** MKT1818/MKT1822;
  WIMA: MKS; **CSF Thompson:** MD/MC;
Siemens: B325X-XX; **Rubycon:** MC/MN; **Nichicon:** DDXM.

Table C–3							
Working Voltage	**Cap. μF**	**Dimensions (mm)**					
		A	**B**	**C**	**F**	**G**	**Dia.**
	.1	12.0	4.0	7.4	10.0	1.0	.6
	.12	12.0	4.5	7.4	10.0	1.0	.6
	.15	12.0	4.0	9.5	10.0	1.0	.6
	.18	12.0	4.5	10.0	10.0	1.0	.6
	.22	12.0	5.0	10.5	10.0	1.0	.6
	.27	17.0	4.5	8.0	15.0	1.0	.6
	.33	17.0	4.5	10.0	15.0	1.0	.6
	.39	17.0	5.0	10.0	15.0	1.0	.6
	.47	17.0	5.5	11.0	15.0	1.0	.6
	.56	17.0	6.5	11.5	15.0	1.0	.6
	.68	17.0	7.0	12.0	15.0	1.0	.8
	.82	17.0	7.5	13.0	15.0	1.0	.8
100	1.0	17.0	8.0	14.0	15.0	1.0	.8
	1.2	17.0	8.5	16.0	15.0	1.0	.8
	1.5	25.5	7.5	14.5	22.5	1.0	.8
	1.8	25.5	8.5	16.0	22.5	1.0	.8
	2.2	25.5	9.5	16.5	22.5	1.0	.8
	2.7	25.5	10.0	18.5	22.5	1.0	.8
	3.3	25.5	10.0	19.0	22.5	1.0	.8
	3.9	30.5	11.0	18.0	27.5	1.5	.8
	4.7	30.5	11.0	20.5	27.5	1.5	.8
	5.6	30.5	12.5	21.0	27.5	1.5	.8

Table C-3 (*Continued*)

Working Voltage	Cap. μF	Dimensions (mm)					
		A	B	C	F	G	Dia.
	6.8	30.5	13.0	24.0	27.5	1.5	.8
100	8.2	30.5	15.0	25.0	27.5	1.5	.8
	10.0	30.5	16.0	27.0	27.5	1.5	.8
	.01	12.5	5.0	8.5	10.0	1.0	.6
	.012	12.5	5.0	8.5	10.0	1.0	.6
	.015	12.5	5.0	8.5	10.0	1.0	.6
	.018	12.5	5.0	8.5	10.0	1.0	.6
	.022	12.5	5.0	8.5	10.0	1.0	.6
	.027	12.5	5.0	8.5	10.0	1.0	.6
	.033	12.5	5.0	8.5	10.0	1.0	.6
	.039	12.5	5.0	8.5	10.0	1.0	.6
	.047	12.5	5.0	8.0	10.0	1.0	.6
	.056	12.5	5.5	8.5	10.0	1.0	.6
	.068	12.5	5.5	8.5	10.0	1.0	.6
	.082	12.5	5.5	10.0	10.0	1.0	.6
	.10	12.5	5.5	10.5	10.0	1.0	.6
	.12	17.5	5.5	10.0	15.0	1.0	.8
250	.15	17.5	6.0	10.5	15.0	1.0	.8
	.18	17.5	6.0	10.5	15.0	1.0	.8
	.22	17.5	6.0	11.0	15.0	1.0	.8
	.27	23.0	6.0	10.5	20.0	1.0	.8
	.33	23.0	6.0	11.0	20.0	1.0	.8
	.39	23.0	6.5	11.5	20.0	1.0	.8
	.47	23.0	7.0	12.0	20.0	1.0	.8
	.56	23.0	7.5	13.0	20.0	1.0	.8

Table C–3 *(Continued)*

Working Voltage	Cap. μF	Dimensions (mm)					
		A	B	C	F	G	Dia.
250	.68	23.0	8.5	14.0	20.0	1.0	.8
	.82	31.0	7.5	13.0	27.5	1.0	.8
	1.0	31.0	8.5	14.0	27.5	1.0	.8
	1.2	31.0	8.5	16.0	27.5	1.0	.8
	1.5	31.0	9.5	17.0	27.5	1.0	.8
	1.8	31.0	9.5	18.5	27.5	1.0	.8
	2.2	31.0	10.5	19.5	27.5	1.0	.8
400	.01	14.0	5.0	8.5	10.5	1.0	.6
	.012	14.0	5.5	8.5	10.5	1.0	.6
	.015	14.0	5.0	8.5	10.5	1.0	.6
	.018	14.0	5.5	8.5	10.5	1.0	.6
	.022	14.0	5.5	10.0	10.5	1.0	.6
	.027	14.0	5.5	10.5	10.5	1.0	.6
	.033	14.0	6.0	11.0	10.5	1.0	.6
	.039	14.0	6.5	11.5	10.5	1.0	.6
	.047	14.0	7.0	12.0	10.5	1.0	.6
	.056	17.0	6.5	11.0	13.5	1.0	.6
	.068	17.0	7.0	12.0	13.5	1.0	.6
	.082	17.0	7.5	12.5	13.5	1.0	.6
	.10	17.0	7.5	14.5	13.5	1.0	.6
	.12	17.0	8.5	15.5	13.5	1.0	.6
	.15	17.0	9.0	16.5	13.5	1.0	.8
	.18	17.0	10.0	17.0	13.5	1.0	.8
	.22	24.5	8.5	15.5	20.5	1.0	.8
	.27	24.5	9.0	16.5	20.5	1.0	.8

Table C–3	(Continued)						
Working Voltage	Cap. μF	Dimensions (mm)					
		A	B	C	F	G	Dia.
400	.33	24.5	9.5	17.0	20.5	1.0	.8
	.39	29.5	9.5	16.5	25.5	1.5	.8
	.47	29.5	10.0	17.5	25.5	1.5	.8
	.56	29.5	11.0	18.0	25.5	1.5	.8
	.68	29.5	12.0	19.0	25.5	1.5	.8
	.82	29.5	13.0	21.0	25.5	1.5	.8
	1.0	29.5	15.0	22.0	25.5	1.5	.8
	1.2	29.5	15.0	25.0	25.5	1.5	.8
	1.5	43.5	13.0	22.0	38.5	1.5	.8
	1.8	43.5	14.5	24.0	38.5	1.5	.8
	2.2	43.5	16.0	25.5	38.5	1.5	.8
630	.01	14.0	5.5	8.5	10.5	1.0	.6
	.015	14.0	6.0	10.5	10.5	1.0	.6
	.018	14.0	6.0	11.0	10.5	1.0	.6
	.022	14.0	6.5	12.0	10.5	1.0	.6
	.027	17.0	5.5	12.5	13.5	1.0	.6
	.033	17.0	6.0	12.0	13.5	1.0	.6
	.039	17.0	6.5	12.0	13.5	1.0	.6
	.047	17.0	7.0	13.5	13.5	1.0	.6
	.056	17.0	7.5	14.0	13.5	1.0	.6
	.068	17.0	8.5	15.0	13.5	1.0	.6
	.082	17.0	9.0	16.0	13.5	1.0	.8
	.10	24.5	7.0	14.0	20.5	1.5	.8
	.12	24.5	8.5	15.5	20.5	1.5	.8
	.15	24.5	9.0	16.5	20.5	1.5	.8

Table C–3	(Continued)						
Working Voltage	**Cap. μF**	**A**	**B**	**C**	**F**	**G**	**Dia.**
				Dimensions (mm)			
630	.18	24.5	10.0	17.0	20.5	1.5	.8
	.22	29.5	9.5	17.0	25.5	1.5	.8
	.27	29.5	10.5	17.5	25.5	1.5	.8
	.33	29.5	11.5	18.5	25.5	1.5	.8
	.39	29.5	12.0	19.5	25.5	1.5	.8
	.47	29.5	13.0	20.5	25.5	1.5	.8
	.56	29.5	15.0	21.5	25.5	1.5	.8
	.68	43.5	12.5	19.5	38.5	1.5	.8
	.82	43.5	14.0	21.0	38.5	1.5	.8
	1.0	43.5	14.5	24.0	38.5	1.5	.8
	1.2	43.5	16.0	25.5	38.5	1.5	.8
	1.5	43.5	18.5	27.0	38.5	1.5	.8

J.W.Miller
M A G N E T I C S

High Quality Epoxy Conformal Coated Inductors

0.51   7.11   25.40   2.79

Operating Temperature: −20°C to +105°C
Working Voltage: 250 VDC max.
Material: Coating-Epoxy Leads-Tinned Copper
Core Material: Ferrite

Table C–4			
L ± 20% μH	**Q Min.**	**Test Freq. MHz**	**I, dc Max. mA**
.10	40	25.0	500
.12	40	25.0	500

Table C–4	(Continued)		
L± 20% μH	Q Min.	Test Freq. MHz	I, dc Max. mA
.15	40	25.0	500
.18	40	25.0	450
.22*	40	25.0	1025
.27*	40	25.0	950
.33*	40	25.0	815
.39*	40	25.0	700
.47*	40	25.0	650
.56*	40	25.0	545
.68*	40	25.0	495
.82*	40	25.0	415
1.0*	40	25.0	385
1.2*	40	7.9	590
1.5*	45	7.9	535
1.8*	50	7.9	455
2.2*	50	7.9	395
2.7*	50	7.9	355
3.3†	50	7.9	270
3.9†	45	7.9	250
4.7†	45	7.9	230
5.6†	45	7.9	185
6.8†	40	7.9	175
8.2†	40	7.9	155
10†	40	2.5	130
12†	45	2.5	155
15†	50	2.5	150

Table C-4	(Continued)		
L± 20% μH	Q Min.	Test Freq. MHz	I, dc Max. mA
18†	50	2.5	145
22†	50	2.5	140
27†	50	2.5	135
33†	50	2.5	193
39†	50	2.5	185
47†	60	2.5	167
56†	60	2.5	150
68†	60	2.5	137
82†	60	2.5	132
100†	60	2.5	125
120†	60	0.79	100
150†	60	0.79	90
180†	60	0.79	84
220†	60	0.79	76
270†	60	0.79	70
330†	60	0.79	65
390†	60	0.79	60
470†	60	0.79	53
560†	60	0.79	51
680†	60	0.79	45
820†	60	0.79	43
1000†	60	0.79	41

* Rated at L ± 10% μH.
† Rated at L ± 5% μH.

Appendix D

Component Schematic Symbols

Voltage and Current Sources

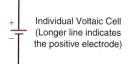 Individual Voltaic Cell
(Longer line indicates
the positive electrode)

 AC Voltage Source

 Standard symbol for a
dc Voltage Source

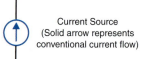 Current Source
(Solid arrow represents
conventional current flow)

 Variable dc Voltage Source

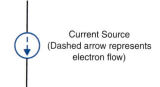 Current Source
(Dashed arrow represents
electron flow)

Ground Symbols

 Earth ground
 Chassis ground
 Common ground

Connections

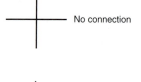 No connection

Dot indicates connection

Connection

No connection

Resistors

Fixed Resistor

Variable Resistor
(Generic symbol)
(2-terminals)

Potentiometer
(3-terminals)

Rheostat
(2 terminals)
R decreases as
wiper is moved up

(No connection)

Rheostat
(2 terminals)
R decreases as
wiper is moved up

(No connection)

Rheostat
(2 terminals)
R increases as
wiper is moved up

Rheostat
(2 terminals)
R increases as
wiper is moved up

Thermistor

Capacitors

Fixed Capacitor

Ganged Capacitors

Variable Capacitor

Electrolytic Capacitor

Variable Capacitor

Electrolytic Capacitor

Inductors (coils)

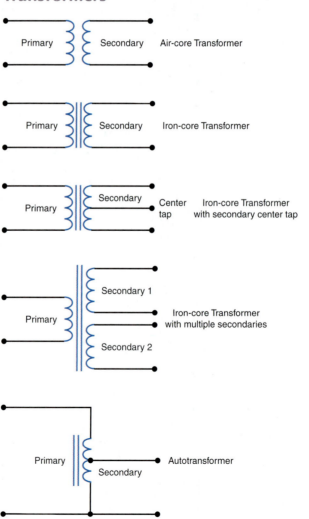

Fixed Inductor (Air-core)

Variable Inductor (Adjustable slug in core)

Variable Inductor (Generic symbol)

Fixed Inductor (iron-core)

Tapped Inductor

Transformers

Primary Secondary Air-core Transformer

Primary Secondary Iron-core Transformer

Primary Secondary Center tap Iron-core Transformer with secondary center tap

Primary Secondary 1 Secondary 2 Iron-core Transformer with multiple secondaries

Primary Secondary Autotransformer

Switches

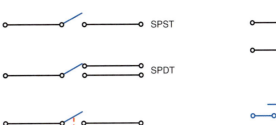

SPST

SPDT

DPST

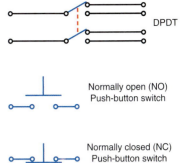

DPDT

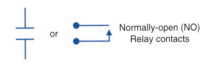

Normally open (NO)
Push-button switch

Normally closed (NC)
Push-button switch

Protective Devices

Fuse

Circuit Breaker

Relays

or

Normally-open (NO)
Relay contacts

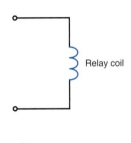

Relay coil

or

Normally-closed (NC)
Relay contacts

Relay Coil

Lamp

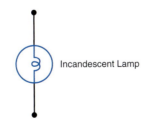

Incandescent Lamp

Test Instruments

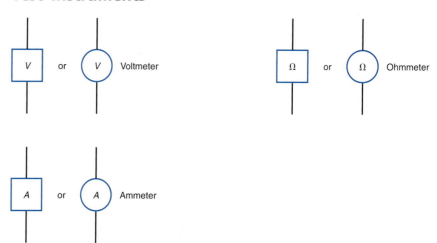

Diodes

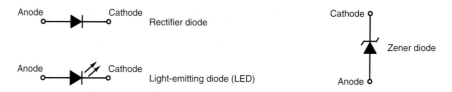

Transistors

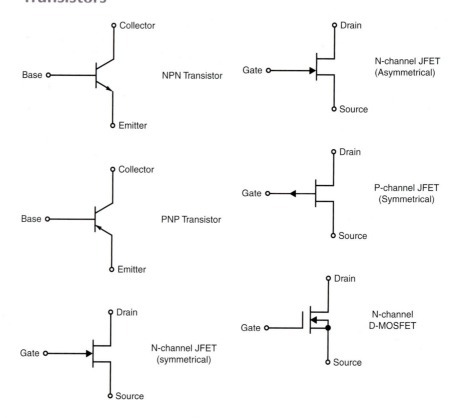

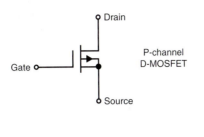

P-channel
D-MOSFET

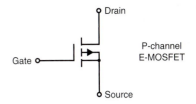

P-channel
E-MOSFET

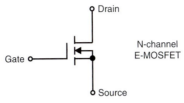

N-channel
E-MOSFET

Thyristors

 Diac

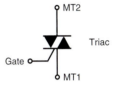 Triac

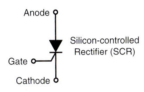 Silicon-controlled
Rectifier (SCR)

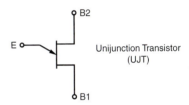 Unijunction Transistor
(UJT)

Operational Amplifier (Op Amp)

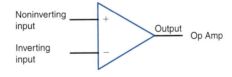

Op Amp

Appendix E

Using the Oscilloscope

Basic Information

The cathode-ray oscilloscope or "scope," as it is commonly known, is one of the most versatile test instruments in electronics. Oscilloscopes are used in a wide variety of applications including consumer electronics repair, digital systems troubleshooting, control system design, and physics laboratories. Oscilloscopes have the ability to measure the time, frequency, and voltage level of a signal, view rapidly changing waveforms, and determine if an output signal is distorted. The technician must therefore be able to operate this instrument and understand how and where it is used.

Oscilloscopes can be classified as either analog or digital. Both types are shown in Fig. E–1. Analog oscilloscopes directly apply the voltage being measured to an electron beam moving across the oscilloscope screen. This voltage deflects the beam up, down, and across, thus tracing the waveform on the screen. Digital oscilloscopes sample the input waveform and then use an analog-to-digital converter (ADC) to change the voltage being measured into digital information. The digital information is then used to reconstruct the waveform to be displayed on the screen.

A digital or analog oscilloscope may be used for many of the same applications. Each type of oscilloscope possesses unique characteristics and capabilities. The analog oscilloscope can display high frequency varying signals in "real time," whereas a digital oscilloscope allows you to capture and store information which can be accessed at a later time or be interfaced to a computer.

WHAT AN OSCILLOSCOPE DOES

An analog oscilloscope displays the instantaneous amplitude of an ac voltage waveform versus time on the screen of a cathode-ray tube (CRT). Basically, the oscilloscope is a graph-displaying device. It has the ability to show how signals change over time. As shown in Fig. E–2, the vertical axis (Y) represents voltage and the horizontal axis (X) represents time. The (Z) axis or intensity is sometimes used in special measurement applications. Inside the cathode-ray tube is an electron gun assembly, vertical and horizontal deflection plates, and a phosphorous screen. The electron gun emits a high-velocity, low-inertia beam of electrons that strike the chemical coating on the inside face of the CRT, causing it to emit light. The brightness (called intensity) can be varied by a control located on the oscilloscope front panel. The motion of the beam over the CRT screen is controlled by the deflection voltages generated in the oscilloscope's circuits outside of the CRT and the deflection plates inside the CRT to which the deflection voltages are applied.

Figure E–3 is an elementary block diagram of an analog oscilloscope. The block diagram is composed of a CRT and four system blocks. These blocks include the display system, vertical system, horizontal system, and trigger system. The CRT provides the screen on which waveforms of electrical signals are viewed. These signals are applied to the vertical input system. Depending on

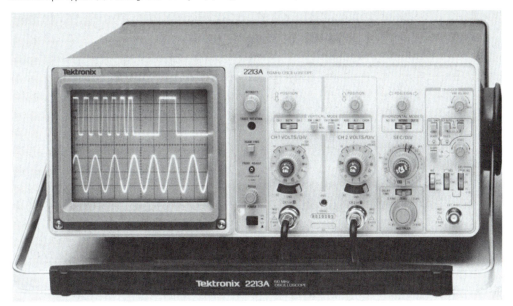

(*a*)

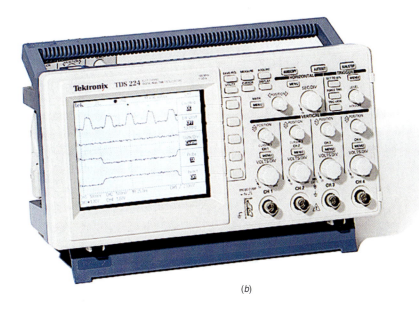

(*b*)

how the volts/div control is set, the vertical attenuator—a variable voltage divider—reduces the input signal voltage to the desired signal level for the vertical amplifier. This is necessary because the oscilloscope must handle a wide range of signal-voltage amplitudes. The vertical amplifier then processes the input signal to produce the required voltage levels for the vertical deflection plates. The signal voltage applied to the vertical deflection plates causes the electron beam of the CRT to be deflected vertically. The resulting up-and-down movement of the beam on the screen, called the trace, is significant in that *the extent of vertical deflection is directly proportional to the amplitude of the signal voltage applied to the vertical, or V, input.* A portion of the input signal, from the vertical amplifier, travels to the trigger system to start or trigger a horizontal sweep. The trigger system determines *when* and *if* the sweep generator will be activated. With the proper LEVEL and SLOPE control adjustment, the

Figure E–2 X, Y, and Z components of a displayed waveform.

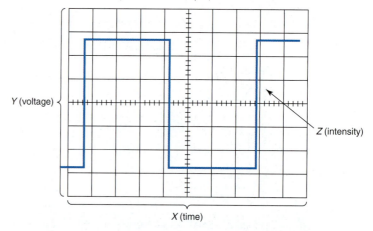

Y (voltage)

Z (intensity)

X (time)

Figure E–3 Analog oscilloscope block diagram.

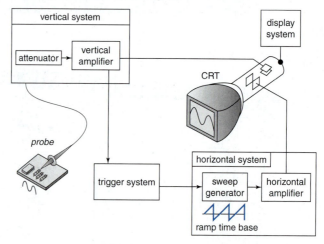

sweep will begin at the same trigger point each time. This will produce a stable display as shown in Fig. E–4. The sweep generator produces a linear time-based deflection voltage. The resulting time-based signal is amplified by the horizontal amplifier and applied to the CRT's horizontal deflection plates. This makes it possible for the oscilloscope to graph a time-varying voltage. The sweep generator may be triggered from sources other than the vertical amplifier. External trigger input signals or internal 60-Hz (line) sources may be selected.

The display system includes the controls and circuits necessary to view the CRT signal with optimum clarity and position. Typical controls include intensity, focus, and trace rotation along with positioning controls.

DUAL–TRACE OSCILLOSCOPES

Most oscilloscopes have the ability to measure two input signals at the same time. These dual-trace oscilloscopes have two separate vertical amplifiers and an electronic switching circuit. It is then possible to observe two time-related waveforms simultaneously at different points in an electric circuit.

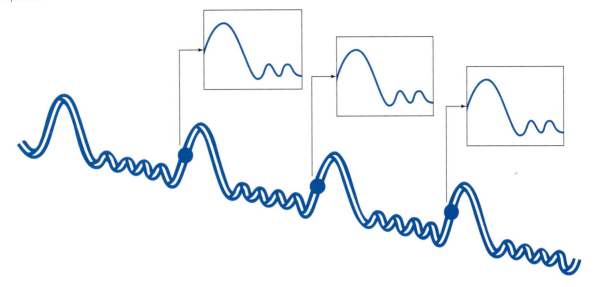

OPERATING CONTROLS OF A TRIGGERED OSCILLOSCOPE

The type, location, and function of the front panel controls of an analog oscilloscope differ from manufacturer to manufacturer and from model to model. The descriptions that follow apply to the broadest range of general-use analog scope models.

INTENSITY. This control sets the level of brightness or intensity of the light trace on the CRT. Rotation in a clockwise (CW) direction increases the brightness. Too high an intensity can damage the phosphorous coating on the inside of the CRT screen.

FOCUS. This control is adjusted in conjunction with the intensity control to give the sharpest trace on the screen. There is interaction between these two controls, so adjustment of one may require readjustment of the other.

ASTIGMATISM. This is another beam-focusing control found on older oscilloscopes that operates in conjunction with the focus control for the sharpest trace. The astigmatism control is sometimes a screwdriver adjustment rather than a manual control.

HORIZONTAL AND VERTICAL POSITIONING OR CENTERING. These are trace-positioning controls. They are adjusted so that the trace is positioned or centered both vertically and horizontally on the screen. In front of the CRT screen is a faceplate called the *graticule*, on which is etched a grid of horizontal and vertical lines. Calibration markings are sometimes placed on the center vertical and horizontal lines on this faceplate. This is shown in Fig. E–5.

VOLTS/DIV. This control attenuates the vertical input signal waveform that is to be viewed on the screen. This is frequently a click-stop control that provides step adjustment of vertical sensitivity. A separate Volts/Div. control is available for each channel of a dual-trace scope. Some scopes mark this control Volts/cm.

Figure E–5 An oscilloscope graticule.

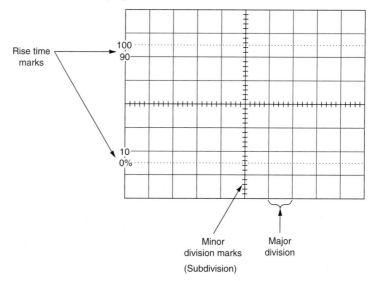

Rise time marks

100
90

10
0%

Minor
division marks
(Subdivision)

Major
division

VARIABLE. In some scopes this is a concentric control in the center of the Volts/Div. control. In other scopes this is a separately located control. In either case, the functions are similar. The variable control works with the Volts/Div. control to provide a more sensitive control of the vertical height of the waveform on the screen. The variable control also has a calibrated position (CAL) either at the extreme counterclockwise or clockwise position. In the CAL position the Volts/Div. control is calibrated at some set value—for example, 5 mV/div., 10 mV/div., or 2 V/div. This allows the scope to be used for peak-to-peak voltage measurements of the vertical input signal. Dual-trace scopes have a separate variable control for each channel.

INPUT COUPLING AC-GND-DC SWITCHES. This three-position switch selects the method of coupling the input signal into the vertical system.

AC—The input signal is capacitively coupled to the vertical amplifier. The dc component of the input signal is blocked.

GND—The vertical amplifier's input is grounded to provide a zero volt (ground) reference point. It does not ground the input signal.

DC—This direct-coupled input position allows all signals (ac, dc, or ac-dc combinations) to be applied directly to the vertical system's input.

VERTICAL MODE SWITCHES. These switches select the mode of operation for the vertical amplifier system.

CH1—Selects only the Channel 1 input signal for display.

CH2—Selects only the Channel 2 input signal for display.

Both—Selects both Channel 1 and Channel 2 input signals for display. When in this position, ALT, CHOP, or ADD operations are enabled.

ALT—Alternately displays Channel 1 and Channel 2 input signals. Each input is completely traced before the next input is traced. Effectively used at sweep speeds of 0.2 ms per division or faster.

CHOP—During the sweep the display switches between Channel 1 and Channel 2 input signals. The switching rate is approximately at 500 kHz. This is useful for viewing two waveforms at slow sweep speeds of 0.5 ms per division or slower.

ADD—This mode algebraically sums the Channel 1 and Channel 2 input signals.

INVERT—This switch inverts Channel 2 (or Channel 1 on some scopes) to enable a differential measurement when in the ADD mode.

TIME/DIV. This is usually two concentric controls that affect the timing of the horizontal sweep or time-base generator. The outer control is a click-stop switch that provides step selection of the sweep rate. The center control provides a more sensitive adjustment of the sweep rate on a continuous basis. In its extreme clockwise position, usually marked CAL, the sweep rate is calibrated. Each step of the outer control is therefore equal to an exact time unit per scale division. Thus, the time it takes the trace to move horizontally across one division of the screen graticule is known. Dual-trace scopes generally have one Time/Div. control. Some scopes mark this control Time/cm.

X-Y SWITCH. When this switch is engaged, one channel of the dual-trace scope becomes the horizontal, or X, input, while the other channel becomes the vertical, or Y, input. In this condition the trigger source is disabled. On some scopes, this setting occurs when the Time/Div. control is fully counterclockwise.

TRIGGERING CONTROLS. The typical dual-trace scope has a number of controls associated with the selection of the triggering source, the method by which it is coupled, the level at which the sweep is triggered, and the selection of the slope at which triggering takes place:

1. *Level Control.* This is a rotary control which determines the point on the triggering waveform where the sweep is triggered. When no triggering signal is present, no trace will appear on the screen. Associated with the level control is an Auto switch, which is often an integral part of the level rotary control or may be a separate push button. In the Auto position the rotary control is disengaged and automatic triggering takes place. In this case a sweep is always generated and therefore a trace will appear on the screen even in the absence of a triggering signal. When a triggering signal is present, the normal triggering process takes over.

2. *Coupling.* This control is used to select the manner in which the triggering is coupled to the signal. The types of coupling and the way they are labeled vary from one manufacturer and model to another. For example, ac coupling usually indicates the use of capacitive coupling that blocks dc; line coupling indicates the 50- or 60-Hz line voltage is the trigger. If the oscilloscope was designed for television testing, the coupling control might be marked for triggering by the horizontal or vertical sync pulses.

3. *Source.* The trigger signal may be external or internal. As already noted, the line voltage may also be used as the triggering signal.

4. *Slope.* This control determines whether triggering of the sweep occurs at the positive going or negative going portion of the triggering signal. The switch itself is usually labeled positive or negative, or simply + or −.

Oscilloscope Probes

Oscilloscope probes are the test leads used for connecting the vertical input signal to the oscilloscope. There are three types: a direct lead that is just a shielded cable, the low-capacitance probe (LCP) with a series-isolating resistor, and a demodulator probe. Figure E–6 shows a circuit for an LCP for an oscilloscope. The LCP usually has a switch to short out the isolating resistor so that the same

Figure E–6 Circuit for low-capacitance probe (LCP) for an oscilloscope.

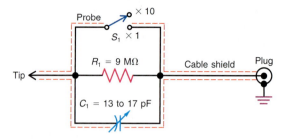

probe can be used either as a direct lead or with low capacitance. (See S_1 in Fig. E–6.)

DIRECT PROBE

The direct probe is just a shielded wire without any isolating resistor. A shielded cable is necessary to prevent any pickup of interfering signals, especially with the high resistance at the vertical input terminals of the oscilloscope. The higher the resistance, the more voltage that can be developed by induction. Any interfering signals in the test lead produce distortion of the trace pattern. The main sources of interference are 60-Hz magnetic fields from the power line and stray RF signals.

The direct probe as a shielded lead has relatively high capacitance. A typical value is 90 pF for 3 ft (0.9 m) of 50-Ω coaxial cable. Also, the vertical input terminals of the oscilloscope have a shunt capacitance of about 40 pF. The total C then is $90 + 40 = 130$ pF. This much capacitance can have a big effect on the circuit being tested. For example, it could detune a resonant circuit. Also, nonsinusoidal waveshapes are distorted. Therefore, the direct probe can be used only when the added C has little or no effect. These applications include voltages for the 60-Hz power line or sine-wave audio signals in a circuit with a relatively low resistance of several kilohms or less. The advantage of the direct probe is that it does not divide down the amount of input signal, since there is no series isolating resistance.

LOW–CAPACITANCE PROBE (LCP)

Refer to the diagram in Fig. E–6. The 9-MΩ resistor in the probe isolates the capacitance of the cable and the oscilloscope from the circuit connected to the probe tip. With an LCP, the input capacitance of the probe is only about 10 pF. The LCP must be used for oscilloscope measurements when

1. The signal frequency is above audio frequencies.
2. The circuit being tested has R higher than about 50 kΩ.
3. The waveshape is nonsinusoidal, especially with square waves and sharp pulses.

Without the LCP, the observed waveform can be distorted. The reason is that too much capacitance changes the circuit while it is being tested.

THE 1:10 VOLTAGE DIVISION OF THE LCP

Refer to the voltage divider circuit in Fig. E–7. The 9-MΩ of R_P is a series resistor in the probe. Also, R_S of 1 MΩ is a typical value for the shunt resistance at the vertical terminals of the oscilloscope. Then $R_T = 9 + 1 = 10$ MΩ. The voltage across R_S for the scope equals R_S/R_T or $1/10$ of the input voltage. For the example in Fig. E–7 with 10 V at the tip of the LCP, 1 V is applied to the oscilloscope.

Figure E-7 Voltage division of 1:10 with a low-capacitance probe.

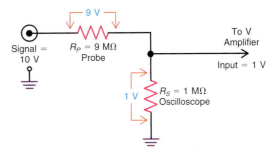

Remember, when using the LCP, multiply by 10 for the actual signal amplitude. As an example, for a trace pattern on the screen that measures 2.4 V, the actual signal input at the probe is 24 V. For this reason, the LCP is generally called the "× 10" probe. Check to see whether or not the switch on the probe is in the direct or LCP position. Even though the scope trace is reduced by the factor of 1/10, it is preferable to use the LCP for almost all oscilloscope measurements to minimize distortion of the waveshapes.

TRIMMER CAPACITOR OF THE LCP

Referring back to Fig. E–6, note that the LCP has an internal variable capacitor C_1 across the isolating resistor R_1. The purpose of C_1 is to compensate the LCP for high frequencies. Its time constant with R_1 should equal the RC time constant of the circuit at the vertical input terminals of the oscilloscope. When necessary, C_1 is adjusted for minimum tilt on a square-wave signal.

CURRENT MEASUREMENTS WITH OSCILLOSCOPE

Although it serves as an ac voltmeter, the oscilloscope can also be used for measuring current values indirectly. The technique is to insert a low R in series where the current is to be checked. Use the oscilloscope to measure the voltage across R. Then the current is $I = V/R$. Keep the value of the inserted R much lower than the resistance of the circuit being tested to prevent any appreciable change in the actual I. Besides measuring the current this way, the waveform of V on the screen is the same as I because R does not affect the waveshape.

Voltage and Time Measurements

In general, an oscilloscope is normally used to make two basic measurements; amplitude and time. After making these two measurements, other values can be determined. Figure E–8 shows the screen of a typical oscilloscope.

As mentioned earlier, the vertical or Y axis represents values of voltage amplitude whereas the horizontal or X axis represents values of time. The volts/division control on the oscilloscope determines the amount of voltage needed at the scope input to deflect the electron beam one division vertically on the Y axis. The seconds/division control on the oscilloscope determines the time it takes for the scanning electron beam to scan one horizontal division. In Fig. E–8 note that there are eight vertical divisions and ten horizontal divisions.

Refer to the sine wave being displayed on the oscilloscope graticule in Fig. E–9. To calculate the peak-to-peak value of the waveform simply count the number of vertical divisions occupied by the waveform and then multiply this number by the volts/division setting. Expressed as a formula,

$$V_{PP} = \text{\# vertical divisions} \times \frac{\text{volts}}{\text{division}} \text{ setting}$$

Figure E–8 Oscilloscope screen (graticule).

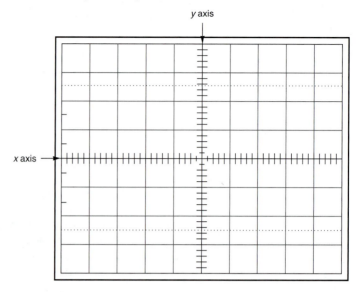

In Fig. E–9, the sine wave occupies 6 vertical divisions. Since the volts/div. setting equals 2 V/division, the peak-to-peak calculations are as follows.

$$V_{PP} = 6 \text{ vertical divisions} \times \frac{2\text{ V}}{\text{division}} = 12\ V_{PP}$$

To calculate the period, T, of the waveform, all you do is count the number of horizontal divisions occupied by one cycle. Then, simply multiply the number of horizontal divisions by the sec./division setting. Expressed as a formula,

$$T = \# \text{ horizontal divisions} \times \frac{\text{sec.}}{\text{division}} \text{ setting}$$

Figure E–9 Determining V_{PP}, T, and f from the sine wave displayed on the scope graticule.

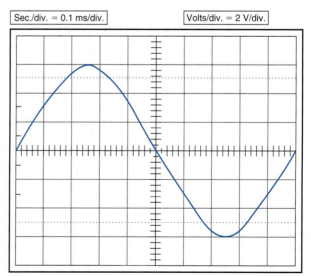

In Fig. E–9, one cycle of the sine wave occupies exactly 10 horizontal divisions. Since the sec./division setting is set to 0.1 ms/div. the calculations for T are as follows:

$$T = 10 \text{ horizontal divisions} \times \frac{0.1 \text{ ms}}{\text{div.}} = 1 \text{ ms}$$

With the period, T, known, the frequency, f, can be found as follows:

$$f = \frac{1}{T}$$

$$= \frac{1}{1 \text{ ms}}$$

$$= 1 \text{ kHz}$$

EXAMPLE 1. In Fig. E–10 determine the peak-to-peak voltage, the period, T, and the frequency, f, of the displayed waveform.

ANSWER. Careful study of the scopes graticule reveals that the height of the waveform occupies 3.4 vertical divisions. With the volts/div. setting at 0.5 V/div. the peak-to-peak voltage is calculated as follows:

$$V_{PP} = 3.4 \text{ vertical divisions} \times \frac{0.5 \text{ V}}{\text{div.}} = 1.7 \text{ } V_{PP}$$

To find the period, T, of the displayed waveform, count the number of horizontal divisions occupied by just one cycle. By viewing the scopes graticule we see that one cycle occupies 5 horizontal divisions. Since the sec./div. control is set to 0.2 ms/div., the period, T, is calculated as:

$$T = 5 \text{ horizontal divisions} \times \frac{0.2 \text{ ms}}{\text{div.}} = 1 \text{ ms}$$

To calculate the frequency, f take the reciprocal of the period, T.

$$f = \frac{1}{T} = \frac{1}{1 \text{ ms}} = 1 \text{ kHz}$$

Figure E–10   Determining V_{PP}, T, and f from the sine wave displayed on the scope graticule.

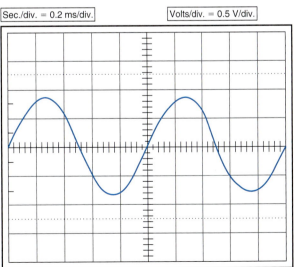

Figure E–11 Determining V_{pk}, tp, prt, prf, and % duty cycle from the rectangular wave displayed on the scope graticule.

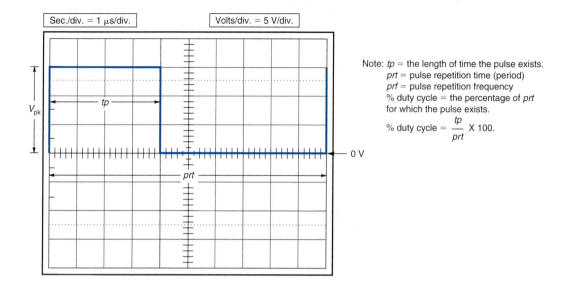

Note: tp = the length of time the pulse exists.
prt = pulse repetition time (period)
prf = pulse repetition frequency
% duty cycle = the percentage of prt for which the pulse exists.

$$\% \text{ duty cycle} = \frac{tp}{prt} \times 100.$$

EXAMPLE 2. In Fig. E–11, determine the pulse time, tp, pulse repetition time, prt and the peak value, V_{pk}, of the displayed waveform. Also, calculate the waveforms % duty cycle and the pulse repetition frequency, prf.

ANSWER. To find the pulse time, tp, count the number of horizontal divisions occupied by just the pulse. In Fig. E–11, the pulse occupies exactly 4 horizontal divisions. With the sec./div. control set to 1 μs/div. the pulse time, tp, is calculated as:

$$tp = 4 \text{ horizontal divisions} \times \frac{1 \ \mu s}{\text{div.}} = 4 \ \mu s$$

The pulse repetition time, prt, is found by counting the number of horizontal divisions occupied by one cycle of the waveform. Since one cycle occupies 10 horizontal divisions, the pulse repetition time, prt, is calculated as follows:

$$prt = 10 \text{ horizontal divisions} \times \frac{1 \ \mu s}{\text{div.}} = 10 \ \mu s$$

With tp and prt known, the % duty cycle is calculated as follows:

$$\% \text{ duty cycle} = \frac{tp}{prt} \times 100$$

$$= \frac{4 \ \mu s}{10 \ \mu s} \times 100$$

$$= 40\%$$

The pulse repetition frequency, prf, is calculated by taking the reciprocal of prt.

$$prf = \frac{1}{prt}$$

$$= \frac{1}{10 \ \mu s}$$

$$= 100 \text{ kHz}$$

The peak value of the waveform is based on the fact that the baseline value of the waveform is 0 V as shown. The positive peak of the waveform is shown to be three vertical divisions above zero. Since the volts/div. setting of the scope is 5 V/div. the peak value of the waveform is:

$$V_{pk} = 3 \text{ vertical divisions} \times \frac{5 \text{ V}}{\text{div.}} = 15 \text{ V}$$

Notice that the waveform shown in Fig. E–11 is entirely positive because the wave-forms pulse makes a positive excursion from the zero volt reference.

PHASE MEASUREMENT

Phase measurements can be made with a dual trace oscilloscope when the signals are of the same frequency. To make this measurement, the following procedure can be used:

1. Preset the scope's controls and obtain a baseline trace (the same for both channels). Set the Trigger Source to whichever input is chosen to be the reference input. Channel 1 is often used as the reference, but Channel 2 as well as External Trigger or Line could be used.
2. Set both Vertical Input Coupling switches to the same position, depending on the type of input.
3. Set the Vertical MODE to Both; then select either ALT or CHOP, depending on the input frequency.
4. Although not necessary, set both Volts/Div. and both Variable controls so that both traces are approximately the same height.
5. Adjust the TRIGGER LEVEL to obtain a stable display. Typically set so that the beginning of the reference trace begins at approximately zero volts.
6. Set the Time/Div. switch to display about one full cycle of the reference waveform.
7. Use the Position controls, Time/Div. switch, and Variable time control so that the reference signal occupies exactly 8 horizontal divisions. The entire cycle of this waveform represents 360° and each division of the graticule now represents 45° of the cycle.
8. Measure the horizontal difference between corresponding points of each waveform on the horizontal graticule line as shown in Fig. E–12.
9. Calculate the phase shift by using the formula:

   Phase Shift = (no. of horizontal difference divisions) × (no. of degrees per division)

As an example, Fig. E–12 displays a difference of 0.6 divisions at 45° per division. The phase shift = (0.6 div.) × (45°/div.) = 27°.

DIGITAL OSCILLOSCOPES

Digital oscilloscopes have replaced analog oscilloscopes in most electronic industries and educational facilities. In addition to being able to make the traditional voltage, time, and phase measurements, digital scopes can also store a measured waveform for later viewing. Digital scopes are also much smaller and weigh less than their analog counterparts. These two advantages alone have prompted many schools and industries to make the switch from analog to digital scopes.

Like any piece of test equipment there is a learning curve involved before you will be totally comfortable operating a digital oscilloscope. The biggest challenge facing you will be familiarizing yourself with the vast number of menus and submenus in which a digital scope uses to access its features and functions. But it's not too bad once you sit down and start with some simple and straightforward measurements. It's always best if you can obtain the operating manual and educational materials for the digital scope you are learning

Figure E-12   Oscilloscope phase shift measurement.

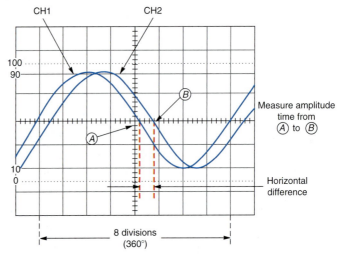

Figure E-13   Four-channel digital oscilloscope.

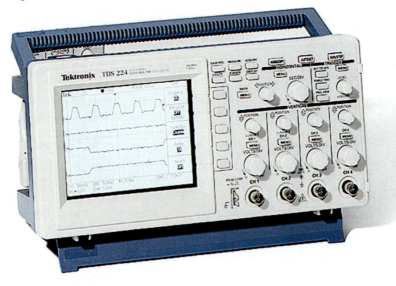

to use. Keep these materials nearby so you can refer to them when you need help in making a measurement. This is not an uncommon practice, even for very experienced users of digital oscilloscopes.

Figure E–13 shows a Tektronix TDS-224 (4-channel) digital oscilloscope. This scope is similar to the one used in *MultiSim*. What follows is a brief explanation of the scope's vertical, horizontal, trigger, and menu and control buttons.

Vertical Controls (See Figure E–14)

CH. 1, 2, 3, 4 AND CURSOR 1 AND 2 POSITION

Positions the waveform vertically. When cursors are turned on and the cursor menu is displayed, these knobs position the cursors. (NOTE: Cursors are horizontal or vertical lines that can be moved up and down or left and right to make either voltage or time measurements.)

Figure E-14

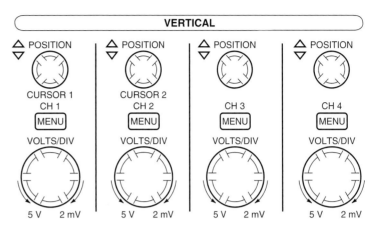

TDS 224

CH. 1, CH. 2, CH. 3, AND CH. 4 MENU

Displays the channel input menu selections and toggles the channel display on and off.

VOLTS/DIV. (CH. 1, CH. 2, CH. 3, AND CH. 4)

Selects calibrated scale factors also referred to as volts/div. settings.

MATH MENU

Displays waveform math operations menu and can also be used to toggle the math waveform on and off.

Horizontal Controls (See Figure E-15)

POSITION

Adjusts the horizontal position of all channels and math waveforms. The resolution of this control varies with the time base.

HORIZONTAL MENU

Displays the horizontal menu.

SEC/DIV.

Selects the horizontal time/div. setting (scale factor).

Trigger Controls (See Figure E-16)

LEVEL AND HOLDOFF

This control has a dual purpose. As an edge trigger level control, it sets the amplitude level the signal must cross to cause an acquisition. As a holdoff control, it sets the amount of time before another trigger event can be accepted. (**NOTE:** The term "acquisition" refers to the process of sampling signals from input channels, digitizing the samples, processing the results into data points, and assembling the data points into a waveform record. The waveform record is stored in memory.)

TRIGGER MENU

Displays the trigger menu.

Figure E-15

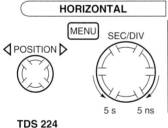

TDS 224

Figure E-16

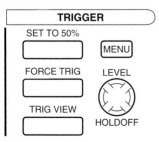

TDS 224

SET LEVEL TO 50%

The trigger level is set to the vertical midpoint between the peaks of the trigger signal.

FORCE TRIGGER

Starts an acquisition regardless of an adequate trigger signal. This button has no effect if the acquisition is already stopped.

TRIGGER VIEW

Displays the trigger waveform in place of the channel waveform while the "TRIGGER VIEW" button is held down. You can use this to see how the trigger settings affect the trigger signal, such as trigger coupling.

Menu and Control Buttons (See Figure E–17)

SAVE/RECALL

Displays the save/recall menu for setups and waveforms.

MEASURE

Displays the automated measurements menu.

ACQUIRE

Displays the acquisition menu.

DISPLAY

Displays the display menu.

CURSOR

Displays the cursor menu. Vertical position controls adjust cursor position while displaying the cursor menu and the cursors are turned on. Cursors remain displayed (unless turned off) after leaving the cursor menu but are not adjustable.

UTILITY

Displays the utility menus.

AUTOSET

Automatically sets the scopes controls to produce a usable display of the input signal.

HARDCOPY

Starts print operations.

RUN/STOP

Starts and stops waveform acquisition.

Since the complexity of the internal operation of a digital oscilloscope is based on many advanced topics which you have not yet covered, we will provide no further explanation of digital scopes in this appendix.

Figure E–17

TDS 224

Appendix F

Introduction to *MultiSim*

In an effort to help the reader understand the concepts presented in this textbook, key examples and problems will be presented through the use of computer simulation using *MultiSim*. *MultiSim* is an interactive circuit simulation package that allows the student to view their circuit in schematic form while measuring the different parameters of the circuit. The ability to create a schematic quickly and then analyze the circuit through simulation makes *MultiSim* a wonderful tool to help students understand the concepts covered in the study of electronics.

This appendix will introduce the reader to the features of *MultiSim* that directly relate to the study of DC, AC, and semiconductor electronics. The topics covered are:

- Work Area
- Opening a File
- Running a Simulation
- Saving a File
- Components
- Sources
- Measurement Equipment
- Circuit Examples

Work Area

The power of this software lies in its simplicity. With just a few steps, a circuit can be either retrieved from disk or drawn from scratch and simulated. The main screen, as shown in Fig. F–1, is divided into three areas: The drop down menu, the tool bars, and the work area.

Figure F–1 Main screen.

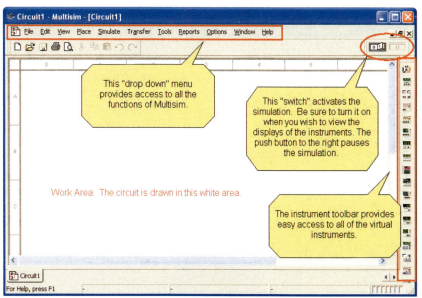

Figure F–2 File drop down menu.

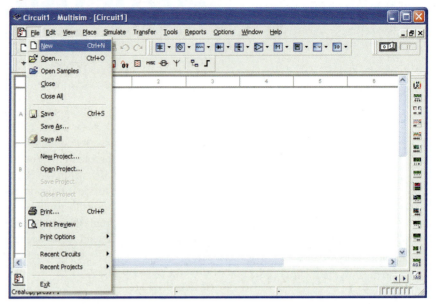

The drop down menu gives the user access to all the functions of the program. Initially, the user will only utilize a few of the different menu selections. Each of the drop down menu main topics can be accessed by either a mouse click or by pressing the <ALT> key and the underlined letter. For example, to access the File menu simply press <ALT><F> at the same time. The File menu will drop down as shown in Fig. F–2.

Initially, there are only two selections from the drop down menu that need to be mastered: Opening a file and saving a file. The rest of the menu options can be explored as time permits.

The tool bars beneath the drop down menu provide access to all of the menu selections. Typically, a user will access them through the tool bars instead of the drop down menus. The most important icon in the assorted tool bars is the on-off switch. The on-off switch starts and stops the simulation. The push button next to the on-off switch will pause the simulation. Pressing the Pause button while the simulation is running allows the viewing of a waveform or meter reading without the display changing.

Opening a File

The circuits referenced in this textbook are included on a compact disk located in the back of this textbook. The files are divided into folders, one for each chapter. The name of the file provides a wealth of information to the user.

Example: A typical file name would be "Ch 4 Problems 4–11". The first part of the file name tells the user that the file is located in the folder labeled "Chapter 4". The second part of the file name tells the user that it is question 11 out of the Problems section at the end of chapter 4.

To open a file, either mouse click on the word "File" located on the drop down menu bar and then mouse click on the Open command or mouse click on the open folder icon located on the tool bar as shown in Fig. F–3. Both methods will cause the Open File dialog box shown in Fig. F–4 to pop open. Navigate to the appropriate chapter folder and retrieve the file needed.

Figure F–3 Opening a File.

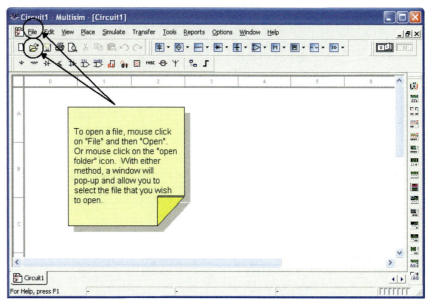

Figure F–4 Open file Dialog Box.

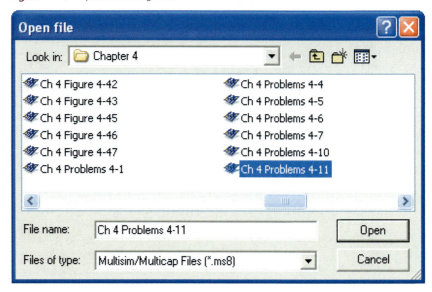

Running a Simulation

The *MultiSim* files developed for this textbook present the circuit in a standard format. The instrumentation is typically connected to the appropriate places within the circuit. If the instrument's display is not visible, double mouse click on the instrument icon and the display will pop up. There is a help screen referred to as the "Description Box" with helpful information relating to the circuit and the instruments within the circuit. This Description Box is opened by pressing <Control><D> while in *MultiSim*.

     The simulation can be started three ways:

- Selecting "Simulate" from the drop down menu and then selecting "Run"
- Pressing the <F5> key
- Pressing the toggle switch with a mouse click

All three of these ways are illustrated in Fig. F–5.

Figure F–5 Starting the simulation.

Saving a File

If the file has been modified, it needs to be saved under a new file name. As shown in Fig. F–6, select "File" with a mouse click located on the drop down menu bar. Then select "Save As" from the drop down menu. This will cause the "Save As" dialog box to open. Give the file a new name and "press" the save button with a mouse click. The process is demonstrated in Fig. F–7.

Components

There are two kinds of component models used in *MultiSim*: Those modeled after actual components and those modeled after "ideal" components. Those modeled after ideal components are referred to as "virtual" components. There is a broad selection of virtual components available, as shown in Fig. F–8.

The difference between the two types of components resides in their rated values. The virtual components can have any of their parameters varied, whereas those modeled after actual components are limited to real world values. For example, a virtual resistor can have any value resistance and percent tolerance as shown in Fig. F–9.

The models of the actual resistors are limited to standard values with either a 1% or 5% tolerance. The same is true for all other components modeled after real components, as shown in Fig. F–10. This is especially important when the semiconductor devices are used in a simulation. Each of the models of actual semiconductors will function in accordance with their data sheets. These components will be listed by their actual device number as identified by the manufacturers. For example, a common diode is the 1N4001. This diode, along with many others, can be found in the semiconductor library of actual components. The parameters of the actual components libraries can also be modified,

Figure F–6 "Save As . . ." screen.

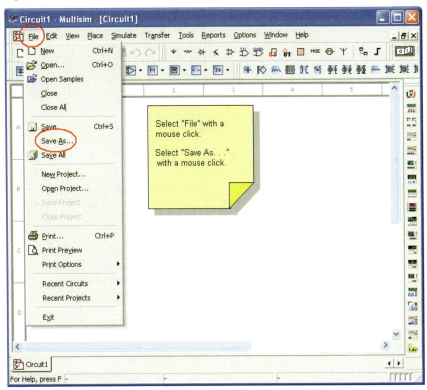

Figure F–7 "Save As . . ." dialog box.

but that requires an extensive understanding of component modeling and is beyond the scope of this appendix.

If actual components are selected for a circuit to be simulated, the measured value may differ slightly from the calculated values as the software will utilize the tolerances to vary the results. If precise results are required, the virtual components can be set to specific values with a zero percent tolerance.

Figure F–8 Virtual component list.

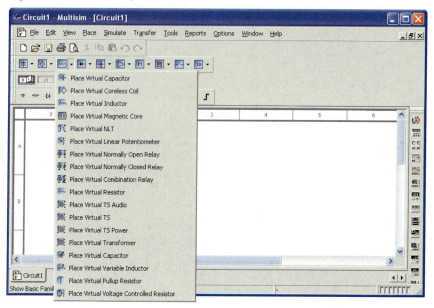

Figure F–9 Configuration screen for a virtual resistor.

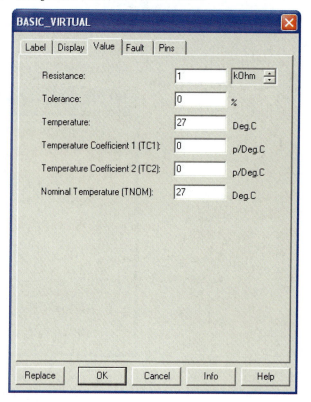

Figure F–10 Component listing for resistors.

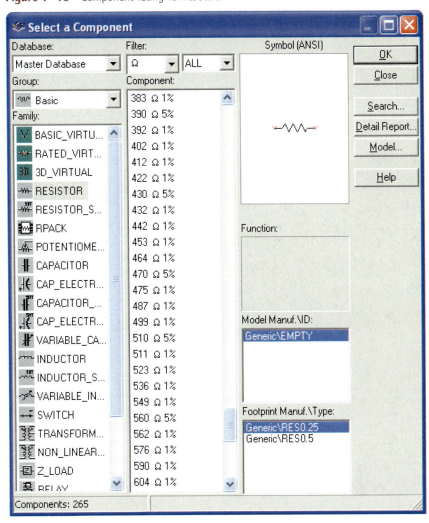

Figure F–11 Switches.

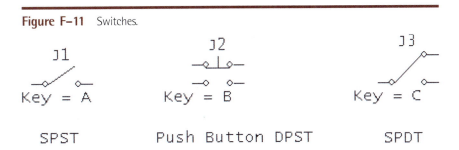

Two of the components require interaction with the user. The switch is probably the most commonly used of these two devices. The movement of the switch is triggered by pressing the key associated with each switch as shown in Fig. F–11. The key is selected while in the switch configuration screen, Fig. F–12. If two switches are assigned the same key, they both will move when the key is pressed.

The second component that requires interaction with the user is the potentiometer. The potentiometer will vary its resistance in predetermined steps with each key press. The pressing of the associated letter on the keyboard will increase the resistance and the pressing of the <Shift> key and the letter will

Figure F–12 Switch configuration screen.

Figure F–13 Potentiometer.

R1

1KΩ_LIN 50%
Key = A

decrease the resistance. As shown in Fig. F–13, the percent of the total resistance is displayed next to the potentiometer. The incremental increase or decrease of resistance is set by the user in the configuration screen. The associated key is also set in the configuration screen as shown in Fig. F–14.

Sources

In the study of DC and AC electronics, the majority of the circuits include either a voltage or current source. There are two main types of voltage sources: DC and AC sources. The DC source can be represented two ways: As a battery in Fig. F–15 and as a voltage supply.

The voltage rating is fully adjustable. The default value is 12 VDC. If the component is double clicked, the configuration screen shown in Fig. F–16 will pop up and the voltage value can be changed.

The voltage supplies are used in semiconductor circuits to represent either a positive or negative voltage supply. Figure F–17 contains the V_{CC} voltage supply used in transistor circuits. FET circuits will utilize the V_{DD} voltage supply as illustrated in Fig. F–18.

Figure F–19 depicts the $+V_{CC}$ and the $-V_{EE}$ voltage sources. These sources are found in operational amplifier circuits. Operational amplifiers typically have two voltage supplies: a negative (V_{EE}) and a positive (V_{CC}) voltage supply, Fig. F–20.

The voltage rating is fully adjustable for all three voltage sources. The default value is $+5$ VDC for V_{CC} and V_{DD}. The default value for V_{EE} is -5 VDC.

Figure F–14 Potentiometer configuration screen.

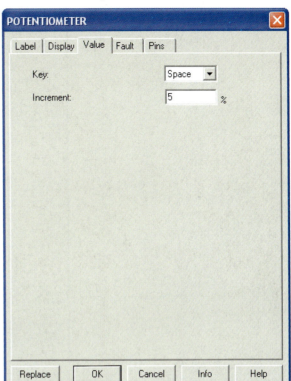

Figure F–15 DC source as a battery.

Figure F–16 Configuration screen for the DC source.

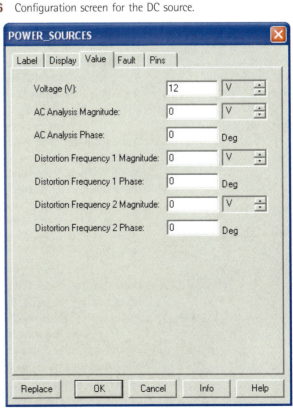

Figure F–17 V_{CC} voltage source.

VCC
⊥ 5V

Figure F–18 V_{DD} voltage source.

VDD
⊥ 5V

Figure F–19 V_{CC} and V_{EE} voltage sources.

VCC   VEE
⊥ 5V   ⊥ −5V

Figure F–20 V_{CC} and V_{EE} Op Amp example.

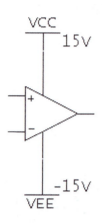

Figure F–21 V_{CC} configuration screen.

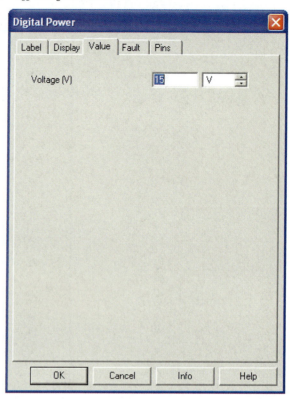

Figure F–22 AC source.

If the component is double clicked, the configuration screen shown in Fig. F–21 will pop up and the voltage value can be changed.

The AC source can be represented as either a schematic symbol or it can take the form of a function generator. The schematic symbol as shown in Fig. F–22 will include information about the AC source. This information will include the device reference number, V_{RMS} value, frequency, and phase shift. These values are fully adjustable. The default values are shown in Fig. F–23. If the component is double clicked, the configuration screen will pop up and the value can be changed.

MultiSim provides two function generators: The generic model and the Agilent model. The Agilent model 33120A has the same functionality as the actual Agilent function generator.

The generic function generator icon is shown in Fig. F–24, along with the configuration screen. The configuration screen is displayed when the function generator icon is double clicked. The generic function generator can produce three types of waveforms: Sinusoidal wave, triangular wave, and a square wave. The frequency, duty cycle, amplitude, and DC offset are all fully adjustable.

The Agilent function generator is controlled via the front panel as shown in Fig. F–25. The buttons are "pushed" by a mouse click. The dial can be turned by dragging the mouse over it or by placing the cursor over it and spinning the wheel on the mouse. The latter is by far the preferred method.

There are two types of current sources: DC and AC sources. The DC current source is represented as a circle with a downward pointing arrow in it. The arrow in Fig. F–26 represents the direction of current flow. The arrow can be pointed downward for conventional current flow. Electron current flow can be simulated by rotating the symbol 180° and the arrow will point upwards.

Figure F–23 Configuration screen for an AC source.

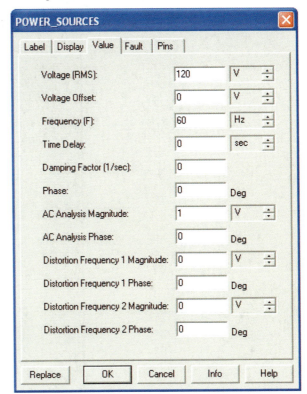

Figure F–24 Generic function generator and configuration screen.

XFG1

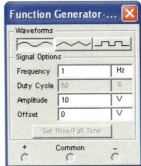

Figure F–25 Agilent function generator.

XFG1

Figure F–26 DC current source.

Figure F–27   DC current source configuration screen.

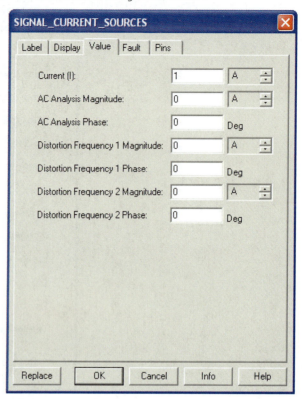

Figure F–28   AC current source.

The current rating is fully adjustable. The default value is 1 A. If the component is double clicked, the configuration screen in Fig. F–27 will pop up and the current value can be changed.

The AC current source is represented as a circle with a downward pointing arrow. There is a sine wave across the arrow. The schematic symbol in Fig. F–28 will include information about the AC current source. This information will include the device reference number, I_{Peak} value, frequency, and phase shift. These values are fully adjustable. The default values are shown in Fig. F–29. If the component is double clicked, the configuration screen will pop up and the values can be changed.

MultiSim requires a ground to be present in the circuit in order for the simulation to function properly. The circuit must contain a ground and all instrumentation must have a ground connection. The schematic symbol for ground is shown in Fig. F–30.

Measurement Equipment

MultiSim provides a wide assortment of measurement equipment. In the study of DC, AC and semiconductor electronics, the three main pieces of measurement equipment are the digital multimeter, the oscilloscope, and the Bode Plotter. The first two pieces of equipment are found in test labs across the world. The Bode Plotter is a fictitious device that automates the task of plotting output voltage versus frequency. This is usually done by taking many measurements and plotting the results in a spreadsheet. The Bode Plotter performs this task for you.

MULTIMETERS

There are two multimeters to choose from: The generic multimeter and the Agilent multimeter. The generic multimeter will measure current, voltage, resistance, and decibels. The meter can be used for both DC and AC measurements.

Figure F–29 AC current source configuration screen.

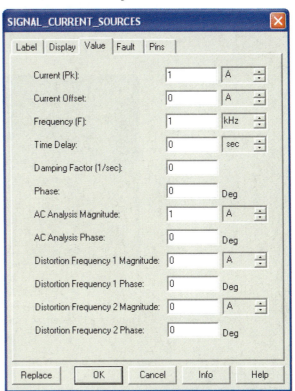

Figure F–30 Ground.

Figure F–31 Generic multimeter icon and configuration screen.

The different functions of the meter are selected by mouse clicking on the icon to the left in Fig. F–31. The mouse click will cause the multimeter display to pop up. The different functions on the display can be selected by "pushing" the different buttons via a mouse click.

The Agilent multimeter icon and meter display are shown in Fig. F–32. The display is brought up by mouse clicking on the Agilent multimeter icon. This multimeter has the same functionality as the actual Agilent multimeter. The different functions are accessed by "pushing" the buttons. This is accomplished by mouse clicking on the button. The input jacks on the right side of the meter display correspond to the five inputs on the icon. If something is connected to the icon, the associated jacks on the display will have a white "X" in them to show a connection.

OSCILLOSCOPES

There are three oscilloscopes to choose from: The generic oscilloscope, the Agilent oscilloscope, and the Tektronix oscilloscope. The generic oscilloscope

Figure F–32 Agilent multimeter icon and meter display.

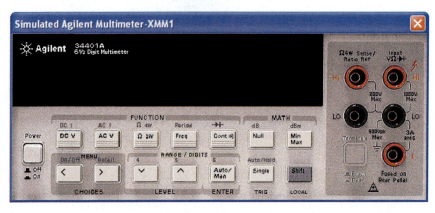

Figure F–33 Generic oscilloscope icon and oscilloscope display.

Figure F–34 Agilent oscilloscope icon.

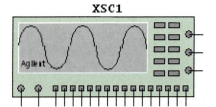

shown in Fig. F–33 is a dual channel oscilloscope. The oscilloscope display is brought up by mouse clicking on the oscilloscope icon. The settings can be changed by clicking in each box and bringing up the scroll arrows.

The Agilent oscilloscope icon in Fig. F–34 has all of the functionality of the model 54622D two channel oscilloscope. The Agilent oscilloscope is controlled via the front panel as shown in Fig. F–35. The buttons are "pushed" by a mouse click. The dials can be turned by dragging the mouse over it or by placing the cursor over it and spinning the wheel on the mouse. The latter is by far the preferred method.

The Tektronix oscilloscope icon shown in Fig. F–36 has all of the functionality of the model TDS2024 four channel digital storage oscilloscope. The color of the four channels is the same as the channel selection buttons on the display: Yellow, blue, purple, and green for channels one through four respectively. The Tektronix oscilloscope is controlled via the front panel as seen in

Figure F–35 Agilent oscilloscope display.

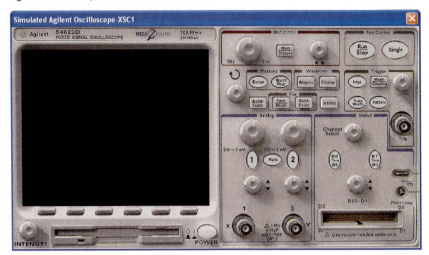

Figure F–36 Tektronix oscilloscope icon.

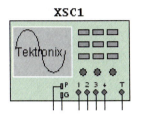

Figure F–37 Tektronix oscilloscope display.

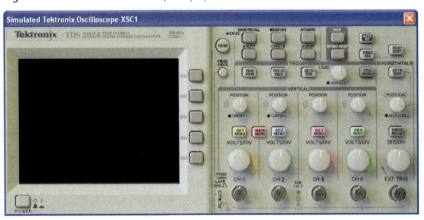

Figure F–38 Voltage and current meters.

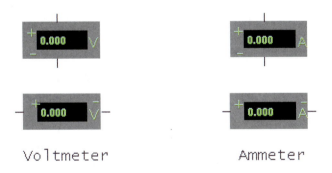

Voltmeter   Ammeter

Fig. F–37. The buttons are "pushed" by a mouse click. The dials can be turned by dragging the mouse over it or by placing the cursor over it and spinning the wheel on the mouse. The latter is by far the preferred method.

VOLTAGE AND CURRENT METERS

When voltage or current need to be measured, *MultiSim* provides very simple voltmeters and ammeters as shown in Fig. F–38. These meters can be placed throughout the circuit. The meters can be rotated to match the polarity needs of the circuit. The default is "DC." If the meters are to be used for AC measurement, then the configuration screen shown in Fig. F–39 must be opened and that parameter changed to reflect AC measurement. To open the configuration screen, double click on the meter.

BODE PLOTTER

The Bode Plotter is used to view the frequency response of a circuit. In the actual lab setting, the circuit would be operated at a base frequency and the output of the circuit measured. The frequency would be incremented by a fixed amount and the measurement repeated. After operating the circuit at a sufficient number of incremental frequencies, the data would be graphed, with independent variable "frequency" on the X axis and dependent variable "amplitude" on the Y axis. This process can be very time consuming. *MultiSim* provides a simpler method of determining the frequency response of a circuit through the use of the virtual Bode Plotter.

In Fig. F–40, the positive terminal of the input is connected to the applied signal source. The positive terminal of the output is connected to the output voltage of the circuit. The other two terminals are connected to ground. The value of the AC source does not matter; the AC source just needs to be in the circuit. The Bode Plotter will provide the input signal. See Fig. F–41 for the display of Bode Plotter as it relates to the circuit connections in F–40.

Circuit Examples

EXAMPLE 1: VOLTAGE MEASUREMENT USING A VOLTMETER IN A SERIES DC CIRCUIT.

A voltmeter in Fig. F–42 is placed in parallel with the resistor to measure the voltage across it. The default is set for "DC" measurement. If AC is required, double

Figure F–40 Bode Plotter.

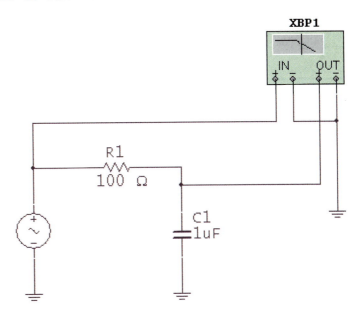

Figure F–41 Bode Plotter display.

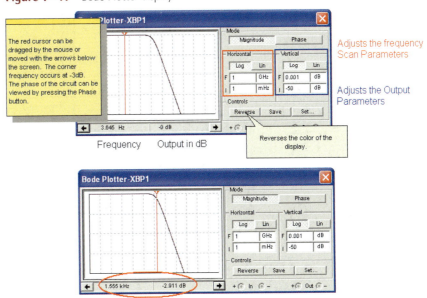

The corner frequency is approximately 1.555 kHz

click on the meter to bring up the configuration screen. All circuits must have a ground. Figure F–43 contains a *Quick Hint* on the use of the voltmeter.

EXAMPLE 2: VOLTAGE MEASUREMENT USING A GENERIC MULTIMETER IN A SERIES DC CIRCUIT.

A generic multimeter is placed in parallel with the resistor to measure the voltage across it. Be sure to double click the generic multimeter icon to bring up the meter display as shown in Fig. F–44. Press the appropriate buttons for "voltage" and then "DC" or "AC" measurement. All circuits must have a ground. Figure F–45 contains a *Quick Hint* on the use of the generic multimeter.

Figure F–42 DC voltage measurement with a voltmeter.

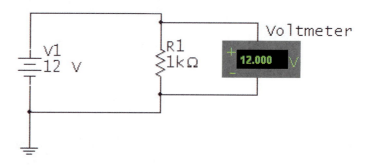

Figure F–43 Voltmeter quick hint.

Voltmeter
(Horizontal)

Voltmeter
(Vertical)

The voltmeter will measure the amount of voltage across a device. It is placed in parallel with the device being measured.

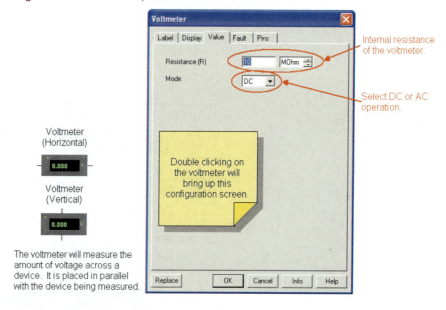

Figure F–44 DC voltage measurement with a generic multimeter.

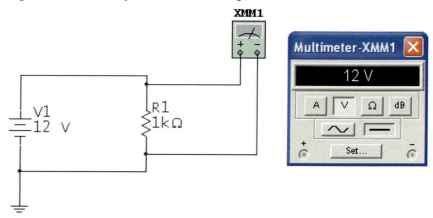

Figure F–45 Generic multimeter quick hint.

Generic Multimeter Icon

Double clicking on the multimeter brings up this display.

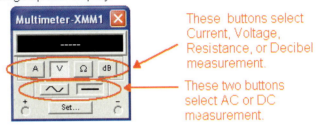

These buttons select Current, Voltage, Resistance, or Decibel measurement.

These two buttons select AC or DC measurement.

Simply mouse click on the appropriate buttons to set up the multimeter for the type of measurement you wish to make.

Figure F–46 DC voltage measurement with an Agilent multimeter.

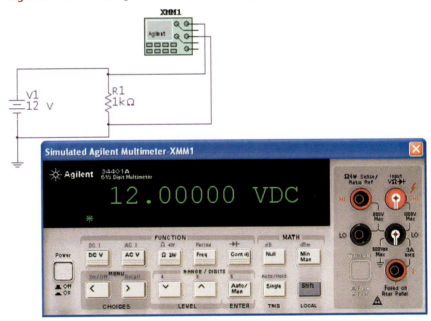

EXAMPLE 3: VOLTAGE MEASUREMENT USING AN AGILENT MULTIMETER IN A SERIES DC CIRCUIT.

In Fig. F–46, an Agilent multimeter is placed in parallel with the resistor to measure the voltage across it. Be sure to double click the Agilent multimeter icon to bring up the meter display. Press the appropriate buttons for "voltage" and then "DC" or "AC" measurement. All circuits must have a ground. Note the two white circles and black x's on the right side of the display to indicate a connection to the meter. Note: This instrument requires that its power button be pressed to "turn on" the meter. Figure F–47 contains a *Quick Hint* on the use of the Agilent multimeter.

Figure F–47 Agilent multimeter quick hint.

Agilent Multimeter

XMM1

These 5 connections represent the connections on the meter face below. The top two of the right hand column are used for voltage measurement.

Double click on the Agilent Multimeter to bring up the meter face.

Simulated Agilent Multimeter-XMM1

Agilent 34401A 6½ Digit Multimeter

Selects DC Voltage

Be sure to turn on the power to the meter, by pressing this button.

The buttons select the type of measurement. Mouse click on a button to "depress" it.

Figure F–48 DC current measurement with an ammeter.

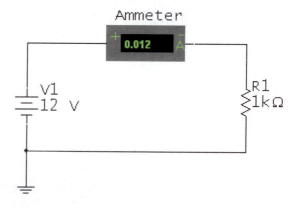

EXAMPLE 4: CURRENT MEASUREMENT USING AN AMMETER IN A SERIES DC CIRCUIT.

In Fig. F–48, an ammeter is placed in series with the resistor and DC source to measure the current flowing through the circuit. The default is set for "DC" measurement. If AC is required, double click on the meter to bring up the configuration screen. All circuits must have a ground. Figure F–49 contains a *Quick Hint* on the use of the ammeter.

EXAMPLE 5: CURRENT MEASUREMENT USING A GENERIC MULTIMETER IN A SERIES DC CIRCUIT.

In Fig. F–50, a generic multimeter is placed in series with the resistor and DC source to measure the current flowing through the circuit. Be sure to double click the generic multimeter icon to bring up the meter display. The current function is selected by clicking on the "A" on the meter display. Since the source is DC, the DC function of the meter is also selected, as indicated by the depressed button. All circuits must have a ground. Figure F–51 contains a *Quick Hint* on the use of the generic multimeter.

Figure F–49 Ammeter quick hint.

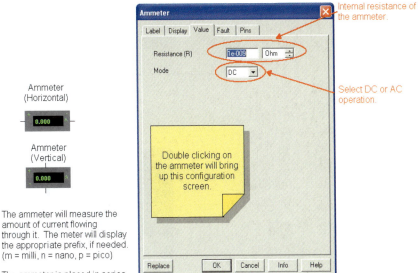

Ammeter
(Horizontal)

Ammeter
(Vertical)

The ammeter will measure the amount of current flowing through it. The meter will display the appropriate prefix, if needed. (m = milli, n = nano, p = pico)

The ammeter is placed in series.

Figure F–50 DC current measurement with a generic multimeter.

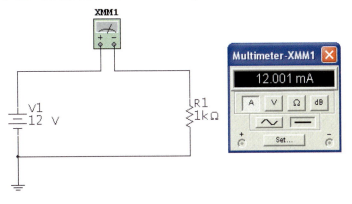

Figure F–51 Generic multimeter quick hint.

Generic Multimeter Icon

Double clicking on the multimeter brings up this display.

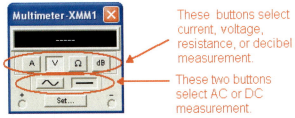

Simply mouse click on the appropriate buttons to set up the multimeter for the type of measurement you wish to make.

Figure F–52 DC current measurement with an Agilent multimeter.

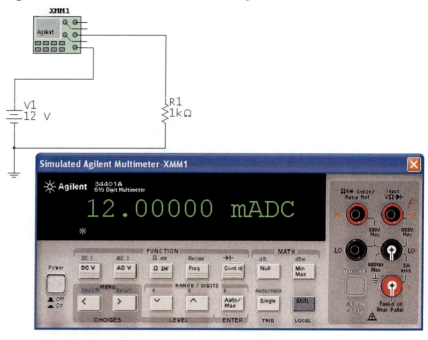

Figure F–53 Agilent multimeter quick hint for current measurement.

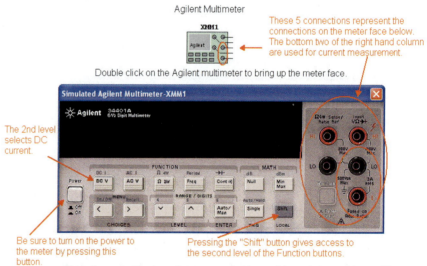

EXAMPLE 6: CURRENT MEASUREMENT USING AN AGILENT MULTIMETER IN A SERIES DC CIRCUIT.

In Fig. F–52, an Agilent multimeter is placed in series with the resistor and source to measure the current flowing through the circuit. Be sure to double click the Agilent multimeter icon to bring up the meter display. Selection of DC current measurement is the second function of the DC voltage measurement button. Be sure to press the "shift" button to access the second function of the voltage button. Note the two white circles and black x's on the right side of the display to indicate a connection to the meter. All circuits must have a ground.

This instrument requires that its power button be pressed to "turn on" the meter. Figure F–53 contains a *Quick Hint* on the use of the Agilent multimeter for current measurement.

Figure F–54 Voltage measurement with a generic oscilloscope.

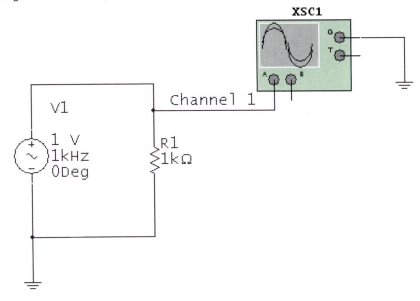

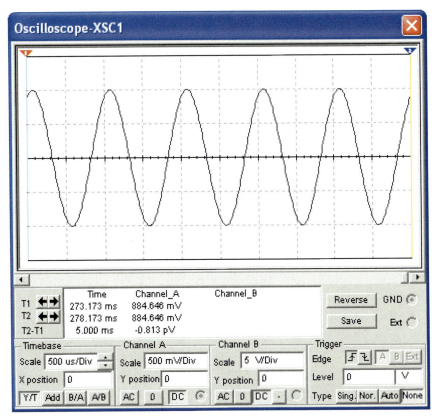

EXAMPLE 7: VOLTAGE MEASUREMENT USING A GENERIC OSCILLOSCOPE IN A SERIES AC CIRCUIT.

In Fig. F–54, channel 1 of the generic oscilloscope is connected to the positive side of the resistor. The ground connection of the scope and the circuit must be grounded. The oscilloscope will use this ground as a reference point. This generic oscilloscope's operation follows that of an actual oscilloscope. The oscilloscope display is brought up by mouse clicking on the oscilloscope icon.

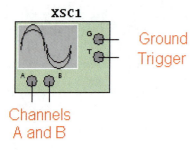

The settings can be changed by clicking in each box and bringing up the scroll arrows. Adjust the volts per division on the channel under measurement until the amplitude of the waveform fills the majority of the screen. Adjust the Timebase so that a complete cycle or two are displayed. Figures F–55, F–56, and F–57 contain *Quick Hints* on the use of the generic oscilloscope.

Figure F–56 Generic Oscilloscope Quick Hint.

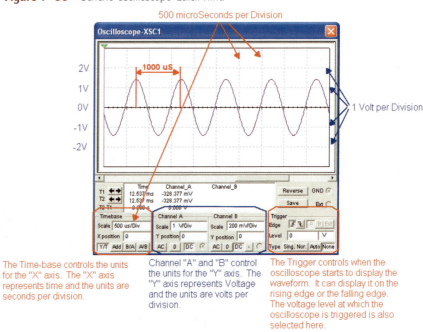

The Time-base controls the units for the "X" axis. The "X" axis represents time and the units are seconds per division.

Channel "A" and "B" control the units for the "Y" axis. The "Y" axis represents Voltage and the units are volts per division.

The Trigger controls when the oscilloscope starts to display the waveform. It can display it on the rising edge or the falling edge. The voltage level at which the oscilloscope is triggered is also selected here.

Figure F–57 Generic oscilloscope quick hint.

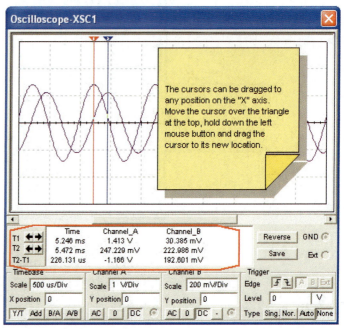

This box and data are from the cursors (1 and 2). The voltage level and time are measured at the point where the waveform passes through the cursor.

EXAMPLE 8: VOLTAGE MEASUREMENT USING THE AGILENT OSCILLOSCOPE IN A SERIES AC CIRCUIT.

In Fig. F–58, channel 1 of the Agilent oscilloscope is connected to the positive side of the resistor. The ground connection of the scope and the circuit must be grounded. The oscilloscope will use this ground as a reference point. This Agilent oscilloscope's operation follows that of a 2-channel, +16 logic channel, 100-MHz bandwidth Agilent model 54622D oscilloscope. The oscilloscope display as shown in Fig. F–59 is brought up by mouse clicking on the oscilloscope icon. The settings can be changed by placing the mouse over the dials and spinning the mouse wheel or by "pressing" the buttons with a mouse click. Adjust the volts per division on the channel under measurement until the amplitude of the waveform fills the majority of the screen. Adjust the time-base in the horizontal section so that a complete cycle or two are displayed. This instrument requires that its power button be pressed to "turn on" the oscilloscope. Figures F–60, F–61 and F–62 contain *Quick Hints* on the use of the Agilent oscilloscope.

Figure F–58   Voltage measurement with an Agilent oscilloscope.

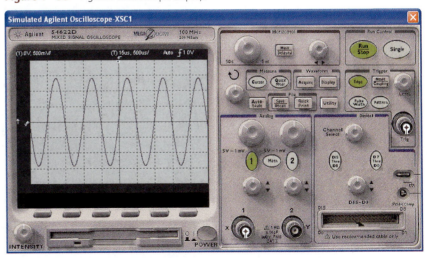

Figure F–59   Agilent oscilloscope display.

Figure F-60 Agilent oscilloscope icon quick hint.

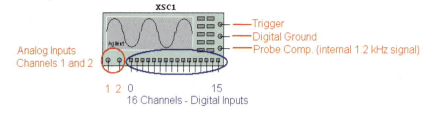

Figure F-61 Agilent oscilloscope quick hint.

The Horizontal section controls the units for the "X" axis. The "X" axis represents time and the units are seconds per division.

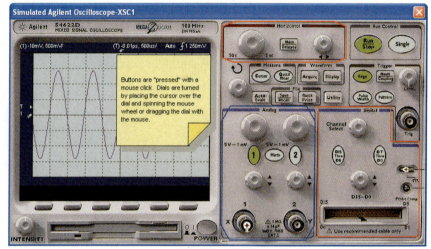

Channel "1" and "2" control the units for the "Y" axis. The "Y" axis represents Voltage and the units are volts per division.

16 Channel Digital Logic Input

Figure F-62 Agilent oscilloscope quick hint.

The Measure section includes cursor operation. The Waveform section allows for waveform storage. The File section saves, recalls, and prints.

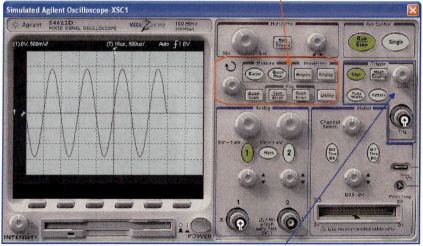

The Trigger controls when the oscilloscope starts to display the waveform. It can display it on the rising edge or the falling edge. The voltage level at which the oscilloscope is triggered is also selected here.

Figure F–63 Voltage and frequency measurement with a Tektronix oscilloscope.

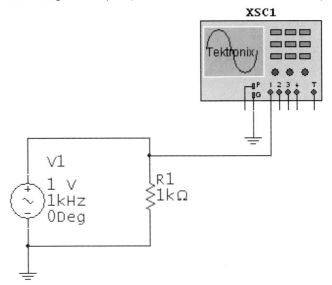

EXAMPLE 9: FREQUENCY AND VOLTAGE MEASUREMENT USING THE TEKTRONIX OSCILLOSCOPE IN A SERIES AC CIRCUIT.

In Fig. F–63, channel 1 of the Tektronix oscilloscope is connected to the positive side of the resistor. The ground connection of the scope and the circuit must be grounded. The oscilloscope will use this ground as a reference point. The Tektronix oscilloscope has all of the functionality of the model TDS2024 four channel digital storage oscilloscope. The oscilloscope display is brought up by mouse clicking on the oscilloscope icon. The settings can be changed by placing the mouse over the dials and spinning the mouse wheel or by "pressing" the buttons with a mouse click. Adjust the volts per division on the channel under measurement until the amplitude of the waveform fills the majority of the screen. Adjust the time-base in the horizontal section so that a complete cycle or two of the waveform is displayed. This instrument requires that its power button be pressed to "turn on" the oscilloscope. The voltage and frequency can be measured by the user or by using the "measure" function of the oscilloscope.

Using volts per division and the seconds per division settings, the amplitude and frequency of the waveform in Fig. F–64 can be determined. The amplitude of the waveform is two divisions above zero volts. (The yellow arrow points to the zero reference point.) The volts per division setting are set to 500 mV per division.

$$V_P = 2 \text{ divisions } \times \frac{500 \text{ mV}}{\text{division}}$$

$$= 1 \text{ V}$$

$$V_{PP} = 4 \text{ divisions } \times \frac{500 \text{ mV}}{\text{division}}$$

$$= 2 \text{ V}$$

Figure F–64 Measurement of the period of the waveform.

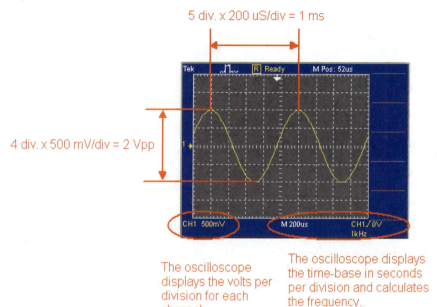

5 div. x 200 uS/div = 1 ms

4 div. x 500 mV/div = 2 Vpp

The oscilloscope displays the volts per division for each channel.

The oscilloscope displays the time-base in seconds per division and calculates the frequency.

The period of the waveform is measured to be 1 ms. Since frequency is the reciprocal of the period, the frequency can be calculated.

$$F = \frac{1}{T}$$

$$= \frac{1}{1 \text{ ms}}$$

$$= 1 \text{ kHz}$$

The Tektronix oscilloscope can also perform the voltage and frequency measurements automatically through the use of the "measure" function. The four steps and the resulting display are shown in Figs. F–65 and F–66. To set-up the oscilloscope to measure these values automatically:

1. Press the Measure button.
2. Select the channel to be measured.
3. Select what is to be measured: V_{pp}, Frequency, etc.
4. Return to the Main Screen.
5. Repeat for other channels and or values.

Figures F–67, F–68, and F–69 contain *Quick Hints* on the use of the Tektronix oscilloscope.

Figure F–65 Tektronix oscilloscope measurement function set-up.

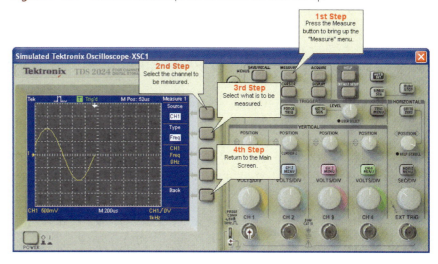

Figure F–66 Tektronix oscilloscope measurement display.

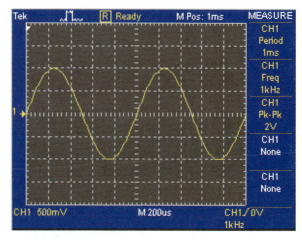

Figure F–67 Tektronix oscilloscope icon quick hint.

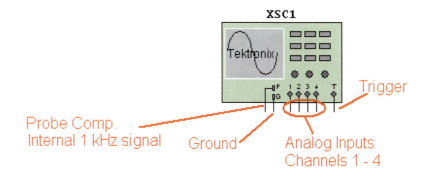

Figure F–68 Tektronix oscilloscope quick hint.

Hint: The buttons are "pressed" by a mouse click. The dials can be turned by dragging them with the cursor or by placing the cursor over the dials and spinning the mouse wheel.

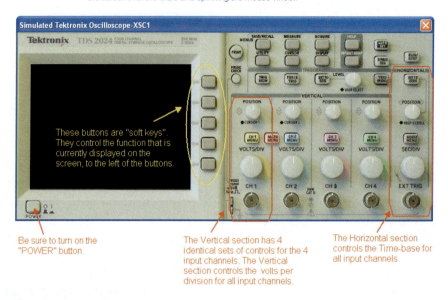

These buttons are "soft keys". They control the function that is currently displayed on the screen, to the left of the buttons.

Be sure to turn on the "POWER" button.

The Vertical section has 4 identical sets of controls for the 4 input channels. The Vertical section controls the volts per division for all input channels.

The Horizontal section controls the Time-base for all input channels.

Figure F–69 Tektronix oscilloscope quick hint.

These buttons bring up the different menus.

This button automatically sets up the oscilloscope.

Be sure to turn on the "POWER" button.

These dials move the traces up and down on the screen.

This dial moves the traces left and right on the screen.

Conclusion

The ability to create a schematic quickly and then analyze the circuit through simulation makes *MultiSim* a wonderful tool to help students understand the concepts covered in the study of DC, AC, and semiconductor electronics.

This appendix introduced the reader to the features of *MultiSim* that directly relate to the topics covered in this textbook. The features covered in this appendix will help the student to utilize the compact disk included with this textbook to its fullest extent.

Glossary

A

ac *See* alternating current.

AC effective resistance, R_e The resistance of a coil for higher frequency alternating current. The value of R_e is more than the dc resistance of the coil because it includes the losses associated with high frequency alternating current in a coil. These losses include skin effect, eddy currents and hysteresis losses.

active component One that can control voltage or current. Examples are transistors and diodes.

acute angle Less than 90°.

A/D converter A device that converts analog input signals to digital output.

admittance (Y) Reciprocal of impedance Z in ac circuits. $Y = 1/Z$.

air gap Air space between poles of a magnet.

alkaline cell or battery One that uses alkaline electrolyte.

alternating current (ac) Current that reverses direction at a regular rate as alternating voltage reverses in polarity. The rate of reversals is the frequency.

alternation One-half cycle of revolution of a conductor loop rotating through a magnetic field. This corresponds to one-half cycle of alternating voltage or current.

alternator AC generator.

amp-clamp probe A meter that can measure ac currents, generally from the 60 Hz ac power line, without breaking open the circuit. The probe of the meter is actually a clamp which fits around the current-carrying conductor.

ampere The basic unit of current. $1\ A = \dfrac{1\ C}{1\ s}$ or $1\ A = \dfrac{1\ V}{1\ \Omega}$.

ampere-hour (A-H) rating A common rating for batteries which indicates how much load current a battery can supply over a specified discharge time. For example, a battery with a 100 A-H rating can deliver 1 A for 100 h or 2 A for 50 h, 4 A for 25 h, etc.

ampere-turn (A-t) Unit of magnetizing force equal to 1 A × 1 turn.

ampere turns/meter $\left(\dfrac{A \cdot t}{m}\right)$ The SI unit of field intensity, H.

analog multimeter A test instrument which is used to make voltage, current and resistance measurements. An analog multimeter uses a moving pointer and a printed scale to display the value of the measured quantity.

antiresonance Term sometimes used for parallel resonance.

apparent power The product of voltage and current VA when V and I are out of phase.

arctangent (arctan) An inverse trigonometric function that specifies the angle, θ, corresponding to a given tangent (tan) value.

armature The part of a generator in which the voltage is produced. In a motor, it is commonly the rotating member. Also, the movable part of a relay.

atom The smallest particle of an element that still has the same characteristics as the element.

atomic number The number of protons, balanced by an equal number of electrons, in an atom.

attenuation A term that refers to a reduction in signal amplitude.

audio frequency (af) Within the range of hearing, approximately 16 to 16,000 Hz.

autotransformer A single, tapped winding used to step up or step down voltage.

average value In sine-wave ac voltage or current, 0.637 of peak value.

B

B-H magnetization curve A graph of field intensity H versus flux density B.

back-off ohmmeter scale Ohmmeter readings from right to left.

balanced bridge A circuit consisting of two series strings in parallel. The balanced condition occurs when the voltage ratio in each series string is identical. The output from the bridge is taken between the centers of each series string. When the voltage ratios in each series string are identical, the output voltage is zero and the bridge circuit is said to be balanced.

bandpass filter Filter that allows coupling a band of frequencies to the load.

bandstop filter Filter that prevents a band of frequencies from being coupled to the load.

bandwidth A range of frequencies that has a resonant effect in LC circuits.

bank Components connected in parallel.

battery Group of cells connected in series or parallel.

bilateral components Electronic components that have the same current for opposite polarities of applied voltage.

bleeder current Steady current from source used to stabilize output voltage with changes in load current.

branch Part of a parallel circuit.

brushes In a motor or generator, devices that provide stationary connections to the rotor.

bypass capacitor One that has very low reactance in a parallel path.

C

C Abbreviation for *coulomb*, the unit of electric charge.

C Symbol for capacitance.

calorie Amount of heat energy needed to raise the temperature of one gram of water by 1°C.

capacitance The ability to store electric charge.

capacitive reactance, X_C A measure of a capacitor's opposition to the flow of alternating current. X_C is measured in ohms. $X_C = \dfrac{1}{2\pi f C}$ or $X_C = \dfrac{V_C}{I_C}$. X_C applies only to sine wave ac circuits.

capacitive voltage divider A voltage divider that consists of series-connected capacitors. The amount of voltage across each capacitor is inversely proportional to its capacitance value.

capacitor Device used to store electric charge.

carbon-composition resistors Resistors made of finely divided carbon or graphite mixed with a powdered insulating material.

carbon-film resistors Resistors made by depositing a thin layer of carbon on an insulated substrate. The carbon film is cut in the form of a spiral.

Celsius scale (°C) Temperature scale that uses 0° for the freezing point of water and 100° for the boiling point. Formerly called *centigrade*.

ceramic Insulator with a high dielectric constant.

cgs Centimeter-gram-second system of units.

charging The effect of increasing the amount of charge stored in a capacitor. The accumulation of stored charge results in a buildup of voltage across the capacitor.

charging The process of reversing the current, and thus the chemical action, in a cell or battery to re-form the electrodes.

chassis ground Common return for all electronic circuits mounted on one metal chassis or PC board. Usually connects to one side of dc supply voltage.

chip capacitor A surface-mounted capacitor.

choke Inductance with high X_L compared with the R of the circuit.

circuit A path for current flow.

circuit breaker A protective device that opens when excessive current flows in circuit. Can be reset.

circular mil (Cmil) Cross-sectional area of round wire with diameter of 1 mil or 0.001 in.

coaxial cable An inner conductor surrounded by an outer conductor that serves as a shield.

coding of capacitors The method used to indicate the value of a capacitor.

coefficient of coupling, k The fraction of total flux from one coil linking another coil nearby.

coil Turns of wire conductor to concentrate a magnetic field.

color code System in which colors are used to indicate values in resistors.

commutator Converts reversing polarities to one polarity.

complex number Has real and j terms; uses form $A + jB$.

compound A combination of two or more elements.

condenser Another (older) name for a capacitor.

conductance (G) Ability to conduct current. It is the reciprocal of resistance, $G = 1/R$. The unit is the siemens (S).

conductor Any material that allows the free movement of electric charges, such as electrons, to provide an electric current.

constant-current source A generator whose internal resistance is very high compared with the load resistance. Because its internal resistance is so high, it can supply a constant current to a load whose resistance value varies over a wide range.

constant-voltage source A generator whose internal resistance is very low compared with the load resistance. Because its internal resistance is so low, it can supply a constant voltage to a load whose resistance value varies over a wide range.

continuity Continuous path for current. Reading of zero ohms with an ohmmeter.

conventional current The direction of current flow associated with positive charges in motion. The current flow direction is from a positive to a negative potential, which is in the opposite direction of electron flow.

corona Effect of ionization of air around a point at high potential.

cosine A trigonometric function of an angle, equal to the ratio of the adjacent side to the hypotenuse in a right triangle.

cosine wave One whose amplitudes vary as the cosine function of an angle. It is 90° out of phase with the sine wave.

coulomb (C) Unit of electric charge. 1 C = 6.25×10^{18} electrons.

counter emf (cemf) A term used to describe the effect of an induced voltage in opposing a change in current.

coupling capacitor Has very low X_C in series path.

cps Cycles per second. Formerly used as unit of frequency. Replaced by hertz (Hz) unit, where 1 Hz = 1 cps.

CRT Cathode-ray tube. A device that converts electric signals to a visual display on a fluorescent screen.

crystal filter A filter that is made up of a crystalline material such as quartz. Crystal filters are often used in place of conventional LC circuits because there Q is so much higher.

curie temperature The temperature at which a magnetic material loses its ferromagnetic properties.

current A movement of electric charges around a closed path or circuit.

current divider A parallel circuit to provide branch I less than the main-line current.

current source Supplies $I = V/r_i$ to load with r_i in parallel.

cutoff frequency The frequency at which the attenuation of a filter reduces the output amplitude to 70.7% of its value in the passband.

cycle One complete set of values for a repetitive waveform.

D

damping Reducing the Q of a resonant circuit to increase the bandwidth.

D'Arsonval meter A dc analog meter movement commonly used in ammeters and voltmeters.

dB Abbreviation for *decibel*. Equals 10 times the logarithm of the ratio of two power levels.

dc *See* direct current.

decade A 10:1 range of values.

decade resistance box A unit for providing any resistance within a wide range of values.

decibel *See* dB.

decimal notation Numbers that are written in standard form without using powers of 10 notation.

degaussing Demagnetizing by applying an ac field and gradually reducing it to zero.

delta (Δ) network Three components connected in series in a closed loop. Same as pi (π) network.

derating curve A graph showing how the power rating of a resistor decreases as its operating temperature increases.

diamagnetic Material that can be weakly magnetized in the direction opposite from the magnetizing field.

dielectric absorption The inability of a capacitor to completely discharge to zero. Dielectric absorption is sometimes called battery action or capacitor memory.

dielectric constant Ability to concentrate the electric field in a dielectric.

dielectric material Insulating material. It cannot conduct current but does store charge.

dielectric strength The ability of a dielectric to withstand a potential difference without internal arcing.

differentiator An RC circuit with a short time constant for pulses across R.

direct current (dc) Current that flows in only one direction. DC voltage has a steady polarity that does not reverse.

discharging The action of neutralizing the charge stored in a capacitor by connecting a conducting path across the capacitor leads.

discharging The process of neutralizing the separated charges on the electrodes of a cell or battery as a result of supplying current to a load resistance.

DMM Digital multimeter. A piece of test equipment used to measure voltage, current, and resistance in an electronic circuit.

double subscript notation A notational system that identifies the points in the circuit where a voltage measurement is to be taken, i.e.; V_{AG}. The first letter in the subscript indicates the point in the circuit where the measurement is to be taken whereas the second letter indicates the point of reference.

DPDT Double-pole double-throw switch or relay contacts.

DPST Double-pole single-throw switch or relay contacts.

dynamometer Type of ac meter, generally for 60 Hz.

E

earth ground A direct connection to the earth usually made by driving copper rods into the earth and then connecting the ground wire of an electrical system to this point. The earth ground connection can serve as a common return path for the current in a circuit.

eddy current Circulating current induced in the iron core of an inductor by ac variations of magnetic flux.

effective value For sine-wave ac waveform, 0.707 of peak value. Corresponds to heating effect of same dc value. Also called *rms value*.

efficiency Ratio of power output to power input × 100%.

EIA Electronic Industries Alliance.

electric field The invisible lines of force between opposite electric charges.

electricity Dynamic electricity is the effect of voltage in producing current in conductors. Static electricity is accumulation of charge.

electrolyte Solution that forms ion charges.

electrolytic capacitor Type with very high C because electrolyte is used to form very thin dielectric. Must be connected with correct polarity in a circuit.

electromagnet Magnet whose magnetic field is associated with electric current in a coil.

electron Basic particle of negative charge in orbital rings around the nucleus in an atom.

electron flow Current of negative charges in motion. Direction is from the negative terminal of the voltage source, through the external circuit, and returning to the positive side of the source. Opposite to the direction of conventional current.

electron valence The number of electrons in an incomplete outermost shell of an atom.

electron volt Unit of energy equal to the work done in moving a charge of 1 electron through a potential difference of 1 V.

element A substance that cannot be decomposed any further by chemical action.

emf Electromotive force; voltage that produces current in a circuit.

engineering notation A form of powers of 10 notation in which a number is expressed as a number between 1 and 1000 times a power of 10 that is a multiple of 3.

equivalent resistance, R_{EQ} In a parallel circuit this refers to a single resistance that would draw the same amount of current as all of the parallel connected branches.

equivalent series resistance (ESR) A resistance in series with an ideal capacitor which collectively represents all of the losses in a capacitor. Ideally, the ESR of a capacitor should be zero.

F

F connector Solderless plug for coaxial cable.

Fahrenheit scale (°F) Temperature scale that uses 32° for the freezing point of water and 212° for the boiling point.

farad (F) Unit of capacitance. Value of one farad stores one coulomb of charge with one volt applied.

Faraday's law For magnetic induction, the generated voltage is proportional to the flux and its rate of change.

ferrite Magnetic material that is not a metal conductor.

ferrite core A type of core which has a high value of flux density, like iron, but is an insulator. A ferrite core used in a coil has minimum eddy current losses due to its high resistance.

ferromagnetic Magnetic properties of iron and other metals that can be strongly magnetized in the same direction as the magnetizing field.

field Group of lines of force; magnetic or electric field.

field intensity (H) The mmf per unit of length.

field winding The part of a motor or generator that supplies the magnetic field cut by the armature.

film capacitor A capacitor that uses a plastic film for its dielectric.

filter Circuit that separates different frequencies.

float charging A method of charging in which the charger and the battery are always connected to each other for supplying current to the load. With this method, the charger provides the current for the load and the current necessary to keep the battery fully charged.

fluctuating direct current Varying voltage and current but no change in polarity.

flux (ϕ) Magnetic lines of force.

flux density (β) Amount of flux per unit area.

flywheel effect Ability of an LC circuit to continue oscillating after the energy source has been removed.

form factor The ratio of the rms to average values. For a sine wave, $\dfrac{rms}{avg} = 1.11$.

free electron Electron that can move freely from one atom to the next.

frequency (f) Number of cycles per second for a waveform with periodic variations. The unit is hertz (Hz).

fuel cell An electrochemical device that converts the chemicals hydrogen and oxygen into water, and in the process produces electricity. A fuel cell provides a steady dc output voltage that can power motors, lights, or other appliances. Unlike a regular battery, however, a fuel cell constantly has chemicals flowing into it so it never goes dead.

fuse Metal link that melts from excessive current and opens a circuit.

G

galvanic cell Electrochemical type of voltage source.

galvanometer Measures electric charge or current.

ganged capacitors Two or three capacitor sections on one common shaft which can be rotated.

gauss (G) Unit of flux density in cgs system equal to one magnetic line of force per square centimeter.

generator A device that produces voltage output. Is a source for either dc or ac V and I.

germanium (Ge) Semiconductor element used for transistors and diodes.

giga (G) Metric prefix for 10^9.

gilbert (Gb) Unit of magnetomotive force in cgs system. One gilbert equals 0.794 ampere-turn.

graph cycle A 10:1 range of values on logarithmic graph paper.

ground Common return to earth for ac power lines. Chassis ground in electronic equipment is the common return to one side of the internal power supply.

H

half-power points Bandwidth defined with 70.7% response for resonant LC circuit.

Hall effect Small voltage generated by a conductor with current in an external magnetic field.

harmonic Exact multiple of fundamental frequency.

henry (H) Unit of inductance. Current change of one ampere per second induces one volt across an inductance of one henry.

hertz (Hz) Unit of frequency. One hertz equals one cycle per second.

high-pass filter A filter that allows the higher frequency components of the applied voltage to develop appreciable output voltage while at the same time attenuating or eliminating the lower frequency components.

holding current The minimum amount of current required to keep a relay energized.

hole Positive charge that exists only in doped semiconductors because of covalent bonds between atoms. Amount of hole charge is the same as that of a proton and an electron.

hole current Motion of hole charges. Direction is the same as that of conventional current, opposite from electron flow.

horsepower (hp) A unit of mechanical power corresponding to 550 ft. · lbs/s. In terms of electric power, 1 hp = 746 W.

hot resistance The R of a component with its normal load current. Determined by V/I.

hot-wire meter Type of ac meter.

hydrometer A device used to check the state of charge of a cell within a lead-acid battery.

hypotenuse Side of a right triangle opposite the 90° angle.

hysteresis In electromagnets, the effect of magnetic induction lagging in time behind the applied magnetizing force.

Hz *See* hertz.

I

imaginary number Value at 90°, indicated by j operator, as in the form jA.

impedance matching Occurs when a transformer is used for its impedance transformation properties. With impedance matching, maximum power is delivered to the load, R_L.

impedance (Z) The total opposition to the flow of current in a sine wave ac circuit. In an RC circuit, the impedance, Z, takes into account the 90° phase relation between X_C and R. Impedance, Z, is measured in ohms.

inductance (L) Ability to produce induced voltage when cut by magnetic flux. Unit of inductance is the henry (H).

induction Ability to generate V or I without physical contact. Electromagnetic induction by magnetic field; electrostatic induction by electric field.

inductive reactance, X_L A measure of an inductor's opposition to the flow of alternating current. X_L is measured in ohms and is calculated as $X_L = 2\pi f L$ or $X_L = \dfrac{V_L}{I_L}$.

inductor Coil of wire with inductance.

insulator A material that does not allow current to flow when voltage is applied, because of its high resistance.

integrator An *RC* circuit with a long time constant. Voltage output across *C*.

internal resistance r_i Limits the current supplied by the voltage source to $I = V/r_i$.

inverse relation Same as reciprocal relation. As one variable increases, the other decreases.

inversely proportional The same as a reciprocal relation; as the value in the denominator increases the resultant quotient decreases. In the formula $X_C = \dfrac{1}{2\pi fC}$, X_C is inversely proportional to both f and C. This means that as f and C increase, X_C decreases.

ion Atom or group of atoms with net charge. Can be produced in liquids, gases, and doped semiconductors.

ionization current A current that results from the movement of ion charges in a liquid or gas.

IR drop Voltage across a resistor.

iron-vane meter Type of ac meter, generally for 60 Hz.

J

j operator Indicates 90° phase angle, as in $j8\ \Omega$ for X_L. Also, $-j8\ \Omega$ is at $-90°$ for X_C.

joule (J) Practical unit of work or energy. One joule equals one watt-second of work.

K

k Coefficient of coupling between coils.

keeper Magnetic material placed across the poles of a magnet to form a complete magnetic circuit. Used to maintain strength of magnetic field.

Kelvin (K) scale Absolute temperature scale, 273° below values on Celsius scale.

kilo (k) Metric prefix for 10^3.

kilowatt-hour A large unit of electrical energy corresponding to 1 kW · 1 h.

Kirchhoff's current law (KCL) The algebraic sum of all currents into and out of any branch point in a circuit must equal zero.

Kirchhoff's voltage law (KVL) The algebraic sum of all voltages around any closed path must equal zero.

L

laminations Thin sheets of steel insulated from one another to reduce eddy-current losses in inductors, motors, etc.

leakage current The current that flows through the dielectric of a capacitor when voltage is applied across the capacitor plates.

leakage flux Any magnetic field lines that do not link two coils which are in close proximity to each other.

leakage resistance A resistance in parallel with a capacitor that represents all of the

leakage paths through which a capacitor can discharge.

Leclanché cell Carbon-zinc primary cell.

left hand rule If the coil is grasped with the fingers of the left hand curled around the coil in the direction of electron flow, the thumb points to the North pole of the coil.

Lenz's law Induced current has magnetic field that opposes the change causing the induction.

linear component An electronic component whose current is proportional to the applied voltage.

linear proportion Straight-line graph between two variables. As one increases, the other increases in direct proportion.

linear resistance A resistance with a constant value of ohms.

load Takes current from the voltage source, resulting in load current.

load currents The currents drawn by the electronic devices and/or components connected as loads in a loaded voltage divider.

loaded voltage The voltage at a point in a series voltage divider where a parallel load has been connected.

loading effect Source voltage is decreased as amount of load current increases.

long-time constant A long-time constant can arbitrarily be defined as one that is five or more times longer than the pulse width of the applied voltage.

loop Any closed path in a circuit.

loop equation An equation that specifies the voltages around a loop.

low-pass filter A filter that allows the lower frequency components of the applied voltage to develop appreciable output voltage while at the same time attenuating or eliminating the higher frequency components.

M

magnetic flux (ϕ) Another name used to describe magnetic field lines.

magnetic pole Concentrated point of magnetic flux.

magnetism Effects of attraction and repulsion by iron and similar materials without the need for an external force. Electromagnetism includes the effects of a magnetic field associated with an electric current.

magnetomotive force (mmf) Ability to produce magnetic lines of force. Measured in units of ampere-turns.

magnitude Value of a quantity regardless of phase angle.

main line The pair of leads connecting all of the individual branches in a parallel circuit to the terminals of the applied voltage, V_A. The main line carries the total current, I_T, flowing to and from the terminals of the voltage source.

make and break Occurs when contacts close and open.

maximum working voltage rating The maximum allowable voltage a resistor can safely withstand without internal arcing.

maxwell (Mx) Unit of magnetic flux equal to one line of force in the magnetic field.

mega (M) Metric prefix for 10^6.

mesh current Assumed current in a closed path, without any current division, for application of Kirchhoff's current law.

mesh The simplest possible closed path within a circuit.

metal-film resistors Resistors made by spraying a thin film of metal onto a ceramic substrate. The metal film is cut in the form of a spiral.

metric prefixes Letter symbols used to replace the powers of 10 that are multiples of 3.

micro (μ) Metric prefix for 10^{-6}.

microfarad A small unit of capacitance equal to 1×10^{-6} F.

milli (m) Metric prefix for 10^{-3}.

Millman's theorem A theorem that provides a short-cut for finding the common voltage across any number of parallel branches with different voltage sources.

mks Meter-kilogram-second system of units.

molecule The smallest unit of a compound with the same chemical characteristics.

motor A device that produces mechanical motion from electric energy.

motor action A motion that results from the net force of two magnetic fields that can aid or cancel each other. The direction of the resultant force is always from a stronger field to a weaker field.

multiplier resistor Resistor in series with a meter movement for voltage ranges.

mutual induction (L_M) Ability of one coil to induce voltage in another coil.

N

nano (n) Metric prefix for 10^{-9}.

nanofarad A small unit of capacitance equal to 1×10^{-9} F.

NC Normally closed relay contacts, or no connection for pinout diagrams.

negative temperature coefficient (NTC) A characteristic of a thermistor indicating that its resistance decreases with an increase in operating temperature.

neutron Particle without electric charge in the nucleus of an atom.

NO Normally open relay contacts.

node A common connection for two or more branch currents.

nonlinear resistance A resistance whose value changes as a result of current producing power dissipation and heat in the resistance.

nonsinusoidal waveform Any waveform that is not a sine or a cosine wave.

Norton's theorem Method of reducing a complicated network to one current source with shunt resistance.

nucleus The massive, stable part of the atom that contains both protons and neutrons.

O

obtuse angle More than 90°.

octave A 2 : 1 range of values.

oersted (Oe) Unit of magnetic field intensity; 1 Oe = 1 Gb/cm.

ohm (Ω) Unit of resistance. Value of one ohm allows current of one ampere with potential difference of one volt.

Ohm's law In electric circuits, $I = V/R$.

ohms per volt rating Sensitivity rating for a voltmeter. High rating means less meter loading.

open circuit One that has infinitely high resistance, resulting in zero current.

open-circuit voltage The voltage present across the output terminals of a voltage source when no load is present.

oscilloscope A piece of test equipment used to view and measure a variety of different ac waveforms.

P

parallel circuit One that has two or more branches for separate currents from one voltage source.

paramagnetic Material that can be weakly magnetized in the same direction from the magnetizing force.

passive component Components such as resistors, capacitors, and inductors. They do not generate voltage or control current.

PC board A device that has printed circuits.

peak-to-peak value (p-p) Amplitude between opposite peaks.

peak value Maximum amplitude, in either polarity; 1.414 times rms value for sine-wave V or I.

permanent magnet (PM) It has magnetic poles produced by internal atomic structure. No external current needed.

permeability Ability to concentrate magnetic lines of force.

phase angle Angle between two phasors; denotes time shift.

phasing dots Used on transformer windings to identify those leads having the same instantaneous polarity.

phasor A line representing magnitude and direction of a quantity, such as voltage or current, with respect to time.

phasor triangle A right triangle which represents the phasor sum of two quantities 90° out of phase with each other.

pickup current The minimum amount of current required to energize a relay.

pico (p) Metric prefix for 10^{-12}.

picofarad A small unit of capacitance equal to 1×10^{-12}F.

polar form Form of complex numbers that gives magnitude and phase angle in the form $A \angle \theta°$.

polarity Property of electric charge and voltage. *Negative* polarity is excess of electrons. *Positive* polarity means deficiency of electrons.

pole The number of completely isolated circuits that can be controlled by a switch.

positive temperature coefficient (PTC) A characteristic of a thermistor indicating that its resistance increases with an increase in operating temperature.

potential difference Ability of electric charge to do work in moving another charge. Measured in volt units.

potentiometer Variable resistor with three terminals connected as a voltage divider.

power (P) Rate of doing work. The unit of electric power is the watt.

power factor Cosine of the phase angle for a sine-wave ac circuit. Value is between 1 and 0.

power supply A piece of test equipment used to supply dc voltage and current to electronic circuits under test.

powers of 10 A numerical representation consisting of a base of 10 and an exponent; the base 10 raised to a power.

preferred values Common values of resistors and capacitors generally available.

primary cell or battery Type that cannot be recharged.

primary winding Transformer coil connected to the source voltage.

principal node A common connection for three or more components in a circuit where currents can combine or divide.

printed wiring Conducting paths printed on plastic board.

proportional A mathematical term used to describe the relationship between two quantities. For example, in the formula $X_L = 2\pi f L$, X_L is said to be directly proportional to both the frequency, f, and the inductance, L. The term proportional means that if either f or L is doubled X_L will double. Similarly, if either f or L is reduced by one-half, X_L will be reduced by one-half. In other words, X_L will increase or decrease in direct proportion to either f or L.

proton Particle with positive charge in the nucleus of an atom.

pulsating dc A dc voltage or current that varies in magnitude but does not reverse in polarity or direction. Another name for pulsating dc is fluctuating dc.

pulse A sharp rise and decay of voltage or current of a specific peak value for a brief period of time.

Q

Q Figure of quality or merit, in terms of reactance compared with resistance. The Q of a coil is X_L/r_i. For an *LC* circuit, Q indicates sharpness of resonance. Also used as the symbol for charge: $Q = CV$.

Q of a coil The quality or figure of merit for a coil. More specifically, the Q of a coil can be defined as the ratio of reactive power in

the inductance to the real power dissipated in the coil's resistance. $Q = \dfrac{X_L}{r_i}$.

quadrature phase A 90° phase angle.

R

R Symbol for resistance.

radian (rad) Angle of 57.3°. Complete circle includes 2π rad.

radio frequency (rf) A frequency high enough to be radiated efficiently as electromagnetic waves, generally above 30 kHz. Usually much higher.

ramp Sawtooth waveform with linear change in V or I.

ratio arm Accurate, stable resistors in one leg of a Wheatstone bridge or bridge circuit in general. The ratio arm fraction, $\dfrac{R_1}{R_2}$, can be varied in most cases, typically in multiples of 10. The ratio arm fraction in a Wheatstone bridge determines two things: the placement accuracy of the measurement of an unknown resistor, R_x, and the maximum unknown resistance, $R_{x(max)}$, that can be measured.

RC phase-shifter An application of a series RC circuit in which the output across either R or C provides a desired phase shift with respect to the input voltage. RC phase-shifter circuits are commonly used to control the conduction time of semiconductors in power control circuits.

reactance Property of L and C to oppose flow of I that is varying. Symbol is X_C or X_L. Unit is the ohm.

real number Any positive or negative number not containing j. $(A + jB)$ is a complex number but A and B by themselves are real numbers.

real power The net power consumed by resistance. Measured in watts.

reciprocal relation Same as inverse relation. As one variable increases, the other decreases.

reciprocal resistance formula A formula that states that the equivalent resistance, R_{EQ}, of a parallel circuit equals the reciprocal of the sum of the reciprocals of the individual branch resistances.

rectangular form Representation of a complex number in the form $A + jB$.

reflected impedance The value of impedance reflected back into the primary from the secondary.

relative permeability (μ_r) The ability of a material to concentrate magnetic flux. Mathematically, relative permeability, designated μ_r, is a ratio of the flux density (B) in a material such as iron to the flux density, B, in air. There are no units for μ_r because it is comparison of two flux densities and the units cancel.

relative permittivity (ε_r) A factor that indicates the ability of an insulator to concentrate electric flux, also known as the dielectric constant, K_e.

relay Automatic switch operated by current in a coil.

relay chatter The vibrating of relay contacts.

resistance (R) Opposition to current. Unit is the ohm (Ω).

resistance wire A conductor having a high resistance value.

resonance Condition of $X_L = X_C$ in an LC circuit to favor the resonant frequency for a maximum in V, I, or Z.

resonant frequency The frequency at which the inductive reactance, X_L and the capacitive reactance, X_C of an LC circuit are equal.

rheostat Variable resistor with two terminals to vary I.

ringing Ability of an LC circuit to oscillate after a sharp change in V or I.

root-mean-square (rms) value For sine-wave ac waveform, 0.707 of peak value. Also called effective value.

rotor Rotating part of generator or motor.

S

sawtooth wave One in which amplitude values have a slow linear rise or fall and a sharp change back to the starting value. Same as a *linear ramp*.

scientific notation A form of powers of 10 notation in which a number is expressed as a number between 1 and 10 times a power of 10.

secondary cell or battery Type that can be recharged.

secondary winding Transformer coil connected to the load.

self-inductance (L) Inductance produced in a coil by current in the coil itself.

semiconductor A material which is neither a good conductor nor a good insulator.

series-aiding A connection of coils in which the coil current produces the same direction of magnetic field for both coils.

series-aiding voltages Voltage sources that are connected so that the polarities of the individual sources aid each other in producing current in the same direction in the circuit.

series circuit One that has only one path for current.

series components Components that are connected in the same current path.

series opposing A connection of coils in which the coil current produces opposing magnetic fields for each coil.

series-opposing voltages Voltage sources that are connected so that the polarities of the individual sources will oppose each other in producing current flow in the circuit.

series string A combination of series resistances.

shield Metal enclosure preventing interference from radio waves.

short circuit Has zero resistance, resulting in excessive current.

short-time constant A short-time constant can arbitrarily be defined as one that is one-fifth or less the time of the pulse width of the applied voltage.

shunt resistor A parallel connection. A device to increase the range of an ammeter.

SI Abbreviation for *Système International*, a system of practical units based on the meter, kilogram, second, ampere, kelvin, mole, and candela.

siemen (S) Unit of conductance. Reciprocal of ohms unit.

silicon (Si) Semiconductor element used for transistors, diodes, and integrated circuits.

sine Trigonometric function of an angle, equal to the ratio of the opposite side to the hypotenuse in a right triangle.

sine wave One in which amplitudes vary in proportion to the sine function of an angle.

skin effect A term that is used to describe current flowing on the outer surface of a conductor at very high frequencies. Skin effect causes the effective resistance of a coil to increase at higher frequencies since the effect is the same as reducing the cmil area of the wire.

slip rings In an ac generator, devices that provide connections to the rotor.

slow-blow fuse A type of fuse that can handle a temporary surge current which exceeds the current rating of the fuse. This type of fuse has an element with a coiled construction and is designed to open only on a continued overload such as short-circuit.

solder Alloy of tin and lead used for fusing wire connections.

solenoid Coil used for electromagnetic devices.

spade lug A type of wire connector.

SPDT Single-pole double-throw switch or relay contacts.

specific gravity Ratio of weight of a substance with that of an equal volume of water.

specific resistance The R for a unit length, area, or volume.

SPST Single-pole single-throw switch or relay contacts.

square wave An almost instantaneous rise and decay of voltage or current in a periodic pattern with time and with a constant peak value. The V or I is on and off for equal times and at constant values.

standard resistor A variable resistor in one leg of a Wheatstone bridge which is varied to provide equal voltage ratios in both series strings of the bridge. With equal voltage ratios in each series string the bridge is said to be balanced.

static electricity Electric charges not in motion.

stator Stationary part of a generator or motor.

steady-state value The V or I produced by a source without any sudden changes. Can be dc or ac value. Final value of V or I after transient.

storage cell or battery Type that can be recharged.

stray capacitance A very small capacitance that exists between any two conductors separated by an insulator. The capacitance can be between two wires in a wiring harness or between a single wire and a metal chassis as examples.

stray inductance The small inductance associated with any length of conductor or component lead. The effects of both stray inductance and stray capacitance are most noticeable with very high frequencies.

string Components connected in series.

superconductivity Very low R at extremely low temperatures.

superposition theorem Method of analyzing a network with multiple sources by using one at a time and combining their effects.

supersonic Frequency above the range of hearing, generally above 16,000 Hz.

surface-mount resistor Resistor made by depositing a thick carbon film on a ceramic base. Electrical connection to the resistive element is made by two leadless solder, end electrodes that are C-shaped.

surface-mount technology Components soldered directly to the copper traces of a printed circuit board. No holes need to be drilled for surface-mounted components.

susceptance (B) Reciprocal of reactance in sine-wave ac circuits; $B = 1/X$.

switch Device used to open or close connections of a voltage source to a load circuit.

switching contacts The contacts that open and close when a relay is energized.

T

tangent (tan) Trigonometric function of an angle, equal to the ratio of the opposite side to the adjacent side in a right triangle.

tank circuit An LC tuned circuit. Stores energy in L and C.

tantalum Chemical element used for electrolytic capacitors.

taper How R of a variable resistor changes with the angle of shaft rotation.

tapered control The manner in which the resistance of a potentiometer varies with shaft rotation. For a linear taper, one-half shaft rotation corresponds to a resistance change of one-half its maximum value. For a nonlinear taper, the resistance change is more gradual at one end, with larger changes at the other end.

taut-band meter Type of construction for meter movement often used in VOM.

temperature coefficient For resistance, how R varies with a change in temperature.

tesla (T) Unit of flux density, equal to 10^8 lines of force per square meter.

thermistor A resistor whose resistance value changes with changes in its operating temperatures.

Thevenin's theorem Method of reducing a complicated network to one voltage source with series resistance.

three-phase power AC voltage generated with three components differing in phase by 120°.

throw The number of closed contact positions that exist per pole on a switch.

time constant Time required to change by 63% after a sudden rise or fall in V and I. Results from the ability of L and C to store energy. Equals RC or L/R.

tolerance The maximum allowable percent difference between the measured and coded values of resistance.

toroid Electromagnet with its core in the form of a closed magnetic ring.

transformer A device that has two or more coil windings used to step up or step down ac voltage.

transient response Temporary value of V or I in capacitive or inductive circuits caused by abrupt change.

trigonometry Analysis of angles and triangles.

troubleshooting A term that refers to the diagnosing or analyzing of a faulty electronic circuit.

tuning Varying the resonant frequency of an LC circuit.

turns ratio Comparison of turns in primary and secondary of a transformer.

twin lead Transmission line with two conductors in plastic insulator.

U

UHF Ultra high frequencies in band of 30 to 300 MHz.

universal time-constant graph A graph which shows the percent change in voltage or current in an RC or RL circuit with respect to the number of time-constants that have elapsed.

V

VAR Unit for volt-amperes of reactive power, 90° out of phase with real power.

Variac Transformer with variable turns ratio to provide different amounts of secondary voltage.

vector A line representing magnitude and direction in space.

VHF Very high frequencies in band of 30 to 300 MHz.

volt (V) Practical unit of potential difference. One volt produces one ampere of current in a resistance of one ohm. $1\ V = \dfrac{1\ J}{1\ C}$

voltage divider A series circuit to provide V less than the source voltage.

voltage drop Voltage across each component in a series circuit. The proportional part of total applied V.

voltage polarity A term that is used to describe the positive and negative ends of a potential difference across a component such as a resistor.

voltage source Supplies potential difference across two terminals. Has internal series r_i.

voltage taps The points in a series voltage divider which provide different voltages with respect to ground.

voltaic cell A device that converts chemical energy into electric energy. The output voltage of a voltaic cell depends on the type of elements used for the electrodes.

volt-ampere (VA) Unit of apparent power, equal to $V \times I$.

volt-ampere characteristic Graph to show how I varies with V.

voltampere reactive (VAR) The voltamperes at the angle of 90°.

voltmeter loading The amount of current taken by the voltmeter acting as a load. As a result, the measured voltage is less than the actual value.

VOM Volt-ohm-milliammeter.

W

watt (W) Unit of real power. Equal to I^2R or $VI \cos \theta$.

watt-hour Unit of electric energy, as power $\times$ time.

wattmeter Measures real power as instantaneous value of $V \times I$.

wavelength (λ) Distance in space between two points with the same magnitude and direction in a propagated wave.

wavetrap An LC circuit tuned to reject the resonant frequency.

weber (Wb) Unit of magnetic flux, equal to 10^8 lines of force.

Wheatstone bridge Balanced circuit used for precise measurements of resistance.

wire gage A system of wire sizes based on the diameter of the wire. Also, the tool used to measure wire size.

wire-wound resistors Resistors made with wire known as *resistance wire* which is wrapped around an insulating core.

work Corresponds to energy. Equal to power $\times$ time, as in kilowatt-hour unit. Basic unit is one joule, equal to one volt-coulomb, or one watt-second.

wye network Three components connected with one end in a common connection and the other ends to three lines. Same as T network.

X

X_C Capacitive reactance equal to $1/(2\pi fC)$.

X_L Inductive reactance equal to $2\pi fL$.

Y

Y Symbol for admittance in an ac circuit. Reciprocal of impedance Z; $Y = 1/Z$.

Y network Another way of denoting a wye network.

Z

Z Symbol for ac impedance. Includes resistance with capacitive and inductive reactance.

zero power resistance The resistance of a thermistor with zero power dissipation, designated R_0.

zero-ohm resistor A resistor whose value is practically $0\ \Omega$. The $0\text{-}\Omega$ value is denoted by a single black band around the center of the resistor body.

zero-ohms adjustment Used with ohmmeter of a VOM to set the correct reading at zero ohms.

Answers

Self-Tests

Introduction to Powers of 10

1. d
2. a
3. b
4. c
5. c
6. a
7. a
8. d
9. b
10. c
11. c
12. b
13. a
14. d
15. c
16. b
17. c
18. a
19. b
20. d

CHAPTER ONE

1. b
2. a
3. c
4. a
5. a
6. d
7. b
8. a
9. c
10. d
11. b
12. a
13. b
14. d
15. c
16. a
17. c
18. d
19. b
20. d
21. c
22. b
23. a
24. d
25. a

CHAPTER TWO

1. b
2. d
3. a
4. c
5. c
6. a
7. a
8. d
9. c
10. c
11. c
12. b
13. d
14. a
15. b

CHAPTER THREE

1. c
2. d
3. a
4. b
5. d
6. b
7. d
8. c
9. a
10. c
11. b
12. c
13. d
14. a
15. d
16. c
17. a
18. b
19. c
20. a

CHAPTER FOUR

1. c
2. b
3. a
4. a
5. c
6. c
7. a
8. b
9. a
10. c
11. b
12. c
13. d
14. b
15. a
16. c
17. b
18. d
19. b
20. c

CHAPTER FIVE

1. a
2. c
3. b
4. d
5. b
6. d
7. c
8. a
9. d
10. a
11. d
12. b
13. c
14. b
15. c
16. b
17. a
18. c
19. d
20. b

CHAPTER SIX

1. c
2. b
3. a
4. c
5. d
6. c
7. a
8. a
9. b
10. d
11. b
12. d
13. a
14. c
15. d
16. b
17. d
18. c
19. a
20. b

REVIEW: CHAPTERS ONE–SIX

1. a
2. c
3. b
4. c
5. c
6. c
7. b
8. c
9. b
10. d
11. b
12. a
13. c
14. b
15. a
16. a
17. a
18. a
19. b
20. a
21. b

CHAPTER SEVEN

1. b
2. c
3. a
4. d
5. b
6. a
7. c

8. a
9. d
10. c

CHAPTER EIGHT

1. a
2. c
3. b
4. c
5. d
6. a
7. b
8. a
9. b
10. d
11. c
12. b
13. c
14. a
15. d
16. b
17. c
18. a
19. b
20. d

REVIEW: CHAPTERS SEVEN AND EIGHT

1. T
2. T
3. T
4. T
5. F
6. F
7. T
8. T
9. T
10. F
11. F
12. T

CHAPTER NINE

1. b
2. a
3. c
4. d
5. c
6. a
7. c
8. b
9. c
10. d
11. d
12. a
13. c
14. a
15. b

CHAPTER TEN

1. d
2. b
3. a
4. b
5. c
6. a
7. b
8. c
9. d
10. d

REVIEW: CHAPTERS NINE AND TEN

1. T
2. T
3. T
4. T
5. T
6. F
7. F
8. T
9. T
10. T
11. T
12. T
13. T
14. T
15. T

CHAPTER ELEVEN

1. b
2. a
3. c
4. b
5. d
6. b
7. a
8. c
9. c
10. d
11. b
12. c
13. d
14. a
15. d

CHAPTER TWELVE

1. d
2. b
3. d
4. a
5. c
6. b
7. a
8. c
9. b

10. c
11. d
12. a
13. a
14. c
15. d

REVIEW: CHAPTERS ELEVEN AND TWELVE

1. d
2. c
3. a
4. c
5. b
6. d
7. b
8. b
9. a
10. d

CHAPTER THIRTEEN

1. c
2. b
3. a
4. d
5. a
6. b
7. d
8. c
9. c
10. b
11. a
12. d
13. c
14. b
15. d
16. a
17. c
18. d
19. d
20. c
21. a
22. c
23. b
24. b
25. d

CHAPTER FOURTEEN

1. c
2. d
3. a
4. b
5. a
6. c
7. b
8. c
9. d

10. d
11. b
12. a
13. d
14. d
15. c
16. b
17. a
18. b
19. b
20. c

CHAPTER FIFTEEN

1. d
2. c
3. a
4. d
5. b
6. c
7. a
8. b
9. c
10. a
11. a
12. c
13. b
14. c
15. b
16. d
17. a
18. b
19. a
20. b
21. a
22. d
23. c
24. d
25. c

REVIEW: CHAPTERS THIRTEEN–FIFTEEN

1. b
2. a
3. c
4. d
5. b
6. d
7. a
8. d
9. c
10. a

CHAPTER SIXTEEN

1. b
2. a
3. d
4. c

5. c
6. d
7. a
8. b
9. c
10. b
11. c
12. a
13. d
14. c
15. a
16. d
17. a
18. c
19. a
20. b
21. a
22. d
23. b
24. c
25. d

CHAPTER SEVENTEEN

1. a
2. c
3. b
4. d
5. b
6. c
7. a
8. d
9. b
10. c
11. b
12. d
13. a
14. b
15. a

CHAPTER EIGHTEEN

1. d
2. c
3. b
4. a
5. c
6. b
7. d
8. a
9. c
10. a
11. c
12. a
13. d
14. c
15. b

REVIEW: CHAPTERS SIXTEEN–EIGHTEEN

1. T
2. T
3. T
4. T
5. T
6. T
7. T
8. T
9. F
10. T
11. T
12. F
13. F
14. T
15. T
16. T
17. F
18. T
19. T
20. T
21. T
22. T
23. T
24. T
25. T
26. T
27. F
28. F
29. T
30. T
31. T
32. T
33. T
34. T
35. T
36. F
37. F
38. T
39. T
40. F

CHAPTER NINETEEN

1. a
2. d
3. b
4. c
5. b
6. a
7. b
8. d
9. c
10. a

11. b
12. d
13. a
14. c
15. d
16. c
17. b
18. a
19. b
20. d

CHAPTER TWENTY

1. b
2. c
3. a
4. d
5. a
6. c
7. b
8. a
9. d
10. b

CHAPTER TWENTY-ONE

1. b
2. a
3. d
4. c
5. c
6. a
7. b
8. d
9. c
10. a
11. c
12. d
13. a
14. d
15. b

CHAPTER TWENTY-TWO

1. c
2. d
3. b
4. a
5. a
6. c
7. b
8. d
9. c
10. b
11. d
12. a
13. d
14. a

15. c
16. b
17. b
18. d
19. c
20. d
21. c
22. d
23. a
24. b
25. a

REVIEW: CHAPTERS NINETEEN–TWENTY-TWO

1. c
2. b
3. d
4. d
5. d
6. d
7. c
8. a
9. b
10. c
11. c
12. a
13. c
14. d
15. b
16. a

CHAPTER TWENTY-THREE

1. d
2. a
3. c
4. b
5. a
6. a
7. b
8. d
9. c
10. b
11. d
12. a
13. c
14. a
15. c

CHAPTER TWENTY-FOUR

1. c
2. a
3. b

4. d
5. b
6. a
7. c
8. b
9. a
10. d
11. d
12. b
13. c
14. c
15. b

REVIEW: CHAPTERS TWENTY–THREE AND TWENTY–FOUR

1. 300
2. 300
3. 300
4. 250
5. 250
6. 200
7. 200
8. 14.1
9. 14.1
10. 1
11. 45°
12. −45°

13. 1
14. 1.41
15. 7.07
16. 600
17. 5.66 $\underline{/45°}$
18. 4 $\underline{/10°}$
19. T
20. T
21. T
22. F

CHAPTER TWENTY–FIVE

1. b
2. c
3. a
4. d
5. b
6. a
7. c
8. c
9. b
10. c
11. a
12. a
13. d
14. b
15. d
16. c

17. a
18. d
19. b
20. a

CHAPTER TWENTY–SIX

1. c
2. a
3. b
4. c
5. d
6. b
7. a
8. d
9. c
10. b
11. c
12. a
13. d
14. a
15. b
16. a
17. b
18. c
19. d
20. b

REVIEW: CHAPTERS TWENTY–FIVE AND TWENTY–SIX

1. 8
2. 0.8
3. 0.4
4. 10
5. 10
6. 1
7. 5
8. 0.08
9. 40
10. 150
11. f_c = 31.83 kHz
12. −100 dB
13. octave, decade
14. 70.7
15. F
16. T
17. T
18. T
19. T
20. T
21. F

Answers

Odd-Numbered Problems and Critical Thinking Problems

Introduction to Powers of 10

SECTION I-1 SCIENTIFIC NOTATION

1. 3.5×10^6
3. 1.6×10^8
5. 1.5×10^{-1}
7. 2.27×10^3
9. 3.3×10^{-2}
11. 7.77×10^7
13. 8.7×10^1
15. 9.5×10^{-8}
17. 6.4×10^5
19. 1.75×10^{-9}
21. 0.000165
23. 863
25. 0.0000000017
27. 1660
29. 0.0000000000033

SECTION I-2 ENGINEERING NOTATION AND METRIC PREFIXES

31. 5.5×10^3
33. 6.2×10^6
35. 99×10^3
37. 750×10^{-6}
39. 10×10^6
41. 68×10^{-6}
43. 270×10^3
45. 450×10^{-9}
47. 2.57×10^{12}
49. 70×10^{-6}
51. 1 kW
53. 35 mV
55. $1 \mu\text{F}$
57. $2.2 \text{ M}\Omega$
59. 1.25 GHz
61. $250 \mu\text{A}$
63. 500 mW
65. $180 \text{ k}\Omega$
67. 4.7Ω
69. $50 \mu\text{W}$

SECTION I-3 CONVERTING BETWEEN METRIC PREFIXES

71. 55 mA
73. $0.0068 \mu\text{F}$
75. $22 \mu\text{F}$
77. $1500 \text{ k}\Omega$
79. $39 \text{ k}\Omega$
81. 7.5 mA
83. $100,000 \text{ W}$
85. 4.7 nF

87. 1.296 GHz
89. $7,500,000 \text{ pF}$

SECTION I-4 ADDITION AND SUBTRACTION INVOLVING POWERS OF 10 NOTATION

91. 7.5×10^4
93. 5.9×10^{-10}
95. 2.15×10^{-3}
97. 5.0×10^7
99. 1.45×10^{-2}
101. 2.6×10^4

SECTION I-5 MULTIPLICATION AND DIVISION INVOLVING POWERS OF 10 NOTATION

103. 1.8×10^6
105. 3.0×10^9
107. 1.0×10^{-5}
109. 2.5×10^4
111. 1.25×10^2
113. 5.0×10^7

SECTION I-6 RECIPROCALS WITH POWERS OF 10

115. 10^{-4}
117. 10^{-1}
119. 10^7
121. 10^{-15}

SECTION I-7 SQUARING NUMBERS EXPRESSED IN POWERS OF 10 NOTATION

123. 2.5×10^7
125. 8.1×10^{11}
127. 1.44×10^{-16}

SECTION I-8 SQUARE ROOTS OF NUMBERS EXPRESSED IN POWERS OF 10 NOTATION

129. 2.0×10^{-2}
131. 6.0×10^{-6}
133. 3.87×10^{-2}

SECTION I-9 THE SCIENTIFIC CALCULATOR

135. Set the display notation to engineering (ENG) by pressing the (2ndF) and ^{FSE}⊙ keys in succession until ENG appears on the LCD display. Next enter the problem using the following keying sequence:
   ① ⑤ (EXP) (±) ③ (×) ① ⊙ ② (EXP) ③ ⊜.
   The calculator will display the answer as 18.000×10^{00}.
137. Set the display notation to engineering (ENG) by pressing the (2ndF) and ^{FSE}⊙

keys in succession until ENG appears on the LCD display. Next enter the problem using the following keying sequence:
   ① ② (÷) ① ⓪ (EXP) ③ ⊜.
   The calculator will display the answer as 1.200×10^{-03}.
139. Set the display notation to engineering (ENG) by pressing the (2ndF) and ^{FSE}⊙ keys in succession until ENG appears on the LCD display. Next enter the problem using the following keying sequence:
   ⑥ ⊙ ⑤ (EXP) ④ ⊕ ② ⑤ (EXP) ③ ⊜.
   The calculator will display the answer as 90.000×10^{03}.

Chapter One

SECTION 1-4 THE COULOMB UNIT OF ELECTRIC CHARGE

1. $+Q = 5 \text{ C}$
3. $+Q = 2 \text{ C}$
5. $-Q = 6 \text{ C}$

SECTION 1-5 THE VOLT UNIT OF POTENTIAL DIFFERENCE

7. $V = 6 \text{ V}$
9. $V = 1.25 \text{ V}$

SECTION 1-6 CHARGE IN MOTION IS CURRENT

11. $I = 4 \text{ A}$
13. $I = 500 \text{ mA}$
15. $I = 10 \text{ A}$
17. $Q = 1 \text{ C}$

SECTION 1-7 RESISTANCE IS OPPOSITION TO CURRENT

19. a. $R = 1 \text{ k}\Omega$
   b. $R = 100 \Omega$
   c. $R = 10 \Omega$
   d. $R = 1 \Omega$
21. a. $G = 5 \text{ mS}$
   b. $G = 10 \text{ mS}$
   c. $G = 20 \text{ mS}$
   d. $G = 40 \text{ mS}$

ANSWERS TO CRITICAL THINKING PROBLEMS

23. $Q = 1.6 \times 10^{-16} \text{ C}$
25. $I = 100 \mu\text{A}$

Chapter Two

SECTION 2-2 RESISTOR COLOR CODING

1. a. $1.5 \text{ k}\Omega, \pm 10\%$
    b. $27 \, \Omega, \pm 5\%$
    c. $470 \text{ k}\Omega, \pm 5\%$
    d. $6.2 \, \Omega, \pm 5\%$
    e. $91 \text{ k}\Omega, \pm 5\%$
    f. $10 \, \Omega, \pm 5\%$
    g. $1.8 \text{ M}\Omega, \pm 10\%$
    h. $1.5 \text{ k}\Omega, \pm 20\%$
    i. $330 \, \Omega, \pm 10\%$
    j. $560 \text{ k}\Omega, \pm 5\%$
    k. $2.2 \text{ k}\Omega, \pm 5\%$
    l. $8.2 \, \Omega, \pm 5\%$
    m. $51 \text{ k}\Omega, \pm 5\%$
    n. $680 \, \Omega, \pm 5\%$
    o. $0.12 \, \Omega, \pm 5\%$
    p. $1 \text{ k}\Omega, \pm 5\%$
    q. $10 \text{ k}\Omega, \pm 10\%$
    r. $4.7 \text{ k}\Omega, \pm 5\%$
3. a. $470 \text{ k}\Omega$
    b. $1.2 \text{ k}\Omega$
    c. $330 \, \Omega$
    d. $10 \text{ k}\Omega$
5. Reading from left to right the colors are:
    a. brown, black, orange, and gold
    b. red, violet, gold, and gold
    c. green, blue, red, and silver
    d. brown, green, green, and gold
    e. red, red, silver, and gold

SECTION 2-3 VARIABLE RESISTORS

7. a. $680,225 \, \Omega$
    b. $8250 \, \Omega$
    c. $18,503 \, \Omega$
    d. $275,060 \, \Omega$
    e. $62,984 \, \Omega$

ANSWERS TO CRITICAL THINKING PROBLEMS

9. Above $250 \text{ k}\Omega$

Chapter Three

SECTION 3-1 THE CURRENT $I = \dfrac{V}{R}$

1. a. $I = 2 \text{ A}$
    b. $I = 3 \text{ A}$
    c. $I = 8 \text{ A}$
    d. $I = 4 \text{ A}$
3. a. $I = 0.005 \text{ A}$
    b. $I = 0.02 \text{ A}$
    c. $I = 0.003 \text{ A}$
    d. $I = 0.015 \text{ A}$
5. Yes, because the current, I, is only 15 A.

SECTION 3-2 THE VOLTAGE $V = IR$

7. a. $V = 50 \text{ V}$
    b. $V = 30 \text{ V}$
    c. $V = 10 \text{ V}$
    d. $V = 7.5 \text{ V}$
9. $V = 10 \text{ V}$

SECTION 3-3 THE RESISTANCE $R = \dfrac{V}{I}$

11. a. $R = 7 \, \Omega$
     b. $R = 5 \, \Omega$

c. $R = 4 \, \Omega$
d. $R = 6 \, \Omega$
13. a. $R = 6000 \, \Omega$
     b. $R = 200 \, \Omega$
     c. $R = 2500 \, \Omega$
     d. $R = 5000 \, \Omega$
15. $R = 8.5 \, \Omega$

SECTION 3-5 MULTIPLE AND SUBMULTIPLE UNITS

17. a. $I = 80 \text{ mA}$
     b. $V = 19.5 \text{ V}$
     c. $V = 3 \text{ V}$
     d. $R = 33 \text{ k}\Omega$
19. $I = 50 \, \mu\text{A}$

SECTION 3-6 THE LINEAR PROPORTION BETWEEN V AND I

21. See Instructor's Manual

SECTION 3-7 ELECTRIC POWER

23. a. $P = 1.5 \text{ kW}$
     b. $P = 75 \text{ W}$
     c. $I = 10 \text{ A}$
     d. $V = 12 \text{ V}$
25. a. $I = 31.63 \text{ mA}$
     b. $I = 2 \text{ mA}$
     c. $P = 150 \, \mu\text{W}$
     d. $V = 200 \text{ V}$
27. $V = 15 \text{ V}$
29. Cost = \$5.04
31. Cost = \$64.80

SECTION 3-8 POWER DISSIPATION IN RESISTANCE

33. a. $P = 1.98 \text{ mW}$
     b. $P = 675 \text{ mW}$
     c. $P = 24.5 \text{ mW}$
     d. $P = 1.28 \text{ W}$
35. $P = 500 \text{ mW}$
37. $P = 2.16 \text{ W}$

SECTION 3-9 POWER FORMULAS

39. a. $I = 5 \text{ mA}$
     b. $R = 144 \, \Omega$
     c. $R = 312.5 \, \Omega$
     d. $V = 223.6 \text{ V}$
41. a. $V = 44.72 \text{ V}$
     b. $V = 63.25 \text{ V}$
     c. $I = 100 \, \mu\text{A}$
     d. $I = 400 \, \mu\text{A}$
43. $I = 2.38 \text{ mA}$
45. $V = 100 \text{ V}$
47. $R = 12 \, \Omega$
49. $V = 50 \text{ V}$
51. $R = 7.2 \, \Omega$

SECTION 3-10 CHOOSING A RESISTOR FOR A CIRCUIT

53. $R = 1.2 \text{ k}\Omega$. Best choice for power rating is 1/4 W.
55. $R = 2 \text{ k}\Omega$. Best choice for power rating is 1 W.
57. $R = 150 \, \Omega$. Best choice for power rating is 1/8 W.
59. $R = 2.2 \text{ M}\Omega$. Best choice for power rating is 1/2 W because it has a 350 V maximum working voltage rating.

ANSWERS TO CRITICAL THINKING PROBLEMS

61. $I = 21.59 \text{ A}$
63. Cost = \$7.52
65. I_{max} at 120°C = 13.69 mA

Chapter Four

SECTION 4-1 WHY I IS THE SAME IN ALL PARTS OF A SERIES CIRCUIT

1. a. $I = 100 \text{ mA}$
    b. $I = 100 \text{ mA}$
    c. $I = 100 \text{ mA}$
    d. $I = 100 \text{ mA}$
    e. $I = 100 \text{ mA}$
    f. $I = 100 \text{ mA}$
3. $I = 100 \text{ mA}$

SECTION 4-2 TOTAL R EQUALS THE SUM OF ALL SERIES RESISTANCES

5. $R_T = 900 \, \Omega$
    $I = 10 \text{ mA}$
7. $R_T = 6 \text{ k}\Omega$
    $I = 4 \text{ mA}$
9. $R_T = 300 \text{ k}\Omega$
    $I = 800 \, \mu\text{A}$

SECTION 4-3 SERIES IR VOLTAGE DROPS

11. $V_1 = 6 \text{ V}$
     $V_2 = 7.2 \text{ V}$
     $V_3 = 10.8 \text{ V}$
13. $R_T = 2 \text{ k}\Omega$
     $I = 10 \text{ mA}$
     $V_1 = 3.3 \text{ V}$
     $V_2 = 4.7 \text{ V}$
     $V_3 = 12 \text{ V}$
15. $R_T = 16 \text{ k}\Omega$
     $I = 1.5 \text{ mA}$
     $V_1 = 2.7 \text{ V}$
     $V_2 = 4.05 \text{ V}$
     $V_3 = 12.3 \text{ V}$
     $V_4 = 4.95 \text{ V}$

SECTION 4-4 KIRCHHOFF'S VOLTAGE LAW (KVL)

17. $V_T = 15 \text{ V}$
19. $V_1 = 7.2 \text{ V}$
     $V_2 = 8.8 \text{ V}$
     $V_3 = 4 \text{ V}$
     $V_4 = 60 \text{ V}$
     $V_5 = 40 \text{ V}$
     $V_T = 120 \text{ V}$

SECTION 4-5 POLARITY OF IR VOLTAGE DROPS

21. a. $R_T = 100 \, \Omega$, $I = 500 \text{ mA}$,
         $V_1 = 5 \text{ V}$, $V_2 = 19.5 \text{ V}$,
         $V_3 = 25.5 \text{ V}$
     b. See Instructor's Manual
     c. See Instructor's Manual
     d. See Instructor's Manual
23. The polarity of the individual resistor voltage drops is opposite to that in Prob. 21. The reason is that the polarity of a resistor's voltage drop depends on the direction of current flow and reversing the polarity of V_T reverses the direction of current.

SECTION 4-6 TOTAL POWER IN A SERIES CIRCUIT

25. $P_1 = 36$ mW
$P_2 = 43.2$ mW
$P_3 = 64.8$ mW
$P_T = 144$ mW

27. $P_1 = 33$ mW
$P_2 = 47$ mW
$P_3 = 120$ mW
$P_T = 200$ mW

SECTION 4-7 SERIES-AIDING AND SERIES-OPPOSING VOLTAGES

29. **a.** $V_T = 27$ V
b. $I = 10$ mA
c. Electrons flow up through R_1.

31. **a.** $V_T = 6$ V
b. $I = 6$ mA
c. Electrons flow up through R_1.

33. **a.** $V_T = 12$ V
b. $I = 400$ mA
c. Electrons flow down through R_1 and R_2.
d. $V_1 = 4.8$ V and $V_2 = 7.2$ V

SECTION 4-8 ANALYZING SERIES CIRCUITS WITH RANDOM UNKNOWNS

35. $I = 20$ mA
$V_1 = 2.4$ V
$V_2 = 2$ V
$V_3 = 13.6$ V
$V_T = 18$ V
$R_3 = 680$ Ω
$P_T = 360$ mW
$P_2 = 40$ mW
$P_3 = 272$ mW

37. $I = 20$ mA
$R_T = 6$ kΩ
$V_T = 120$ V
$V_2 = 36$ V
$V_3 = 24$ V
$V_4 = 40$ V
$R_4 = 2$ kΩ
$P_1 = 400$ mW
$P_2 = 720$ mW
$P_3 = 480$ mW
$P_4 = 800$ mW

39. $R_3 = 800$ Ω
$I = 50$ mA
$V_T = 100$ V
$V_1 = 10$ V
$V_2 = 20$ V
$V_3 = 40$ V
$V_4 = 30$ V
$P_1 = 500$ mW
$P_3 = 2$ W
$P_4 = 1.5$ W
$P_T = 5$ W

41. $R = 1$ kΩ

43. $V_T = 25$ V

SECTION 4-9 GROUND CONNECTIONS IN ELECTRICAL AND ELECTRONIC SYSTEMS

45. $V_{AG} = 18$ V
$V_{BG} = 7.2$ V
$V_{CG} = 1.2$ V

47. $V_{AG} = 20$ V

$V_{BG} = 16.4$ V
$V_{CG} = -9.4$ V
$V_{DG} = -16$ V

SECTION 4-10 TROUBLESHOOTING: OPENS AND SHORTS IN SERIES CIRCUITS

49. $R_T = 6$ kΩ
$I = 4$ mA
$V_1 = 4$ V
$V_2 = 8$ V
$V_3 = 12$ V

51. **a.** $R_T = 3$ kΩ
b. $I = 8$ mA
c. $V_1 = 8$ V, $V_2 = 16$ V, and $V_3 = 0$ V

ANSWERS TO CRITICAL THINKING PROBLEMS

53. $R_1 = 300$ Ω, $R_2 = 600$ Ω, and $R_3 = 1.8$ kΩ

55. $I_{max} = 35.36$ mA

57. $R_1 = 250$ Ω and $V_T = 1.25$ V

ANSWERS TO TROUBLESHOOTING CHALLENGE, TABLE 4-1

Trouble 1: R_2 Open
Trouble 3: R_4 Shorted
Trouble 5: R_1 Shorted
Trouble 7: R_2 Shorted
Trouble 9: R_5 Open
Trouble 11: R_3 Decreased in Value
Trouble 13: R_1 Increased in Value

Chapter Five

SECTION 5-1 THE APPLIED VOLTAGE V_A IS THE SAME ACROSS PARALLEL BRANCHES

1. **a.** 12 V
b. 12 V
c. 12 V
d. 12 V

3. 12 V

SECTION 5-2 EACH BRANCH I EQUALS $\frac{V_A}{R}$

5. I_2 is double I_1 because R_2 is one-half the value of R_1.

7. $I_1 = 600$ mA
$I_2 = 900$ mA
$I_3 = 300$ mA

9. $I_1 = 200$ mA
$I_2 = 15$ mA
$I_3 = 85$ mA
$I_4 = 20$ mA

SECTION 5-3 KIRCHHOFF'S LAW (KCL)

11. $I_T = 300$ mA
13. $I_T = 1.8$ A
15. $I_T = 320$ mA
17. $I_1 = 24$ mA
$I_2 = 20$ mA
$I_3 = 16$ mA
$I_T = 60$ mA

19. **a.** 300 mA
b. 276 mA

c. 256 mA
d. 256 mA
e. 276 mA
f. 300 mA

21. **a.** 2.5 A
b. 500 mA
c. 300 mA
d. 300 mA
e. 500 mA
f. 2.5 A
g. 2 A
h. 2 A

23. $I_2 = 90$ mA

SECTION 5-4 RESISTANCES IN PARALLEL

25. $R_{EQ} = 8$ Ω
27. $R_{EQ} = 20$ Ω
29. $R_{EQ} = 318.75$ Ω
31. $R_{EQ} = 26.4$ Ω
33. $R_{EQ} = 20$ Ω
35. $R_{EQ} = 200$ Ω
37. $R_2 = 1.5$ kΩ
39. $R_{EQ} = 96$ Ω
41. $R_{EQ} = 112.5$ Ω
43. **a.** $R_{EQ} = 269.9$ Ω (Ohmmeter will read 270 Ω approximately.)
b. $R_{EQ} = 256.6$ kΩ (Ohmmeter will read 257 kΩ approximately.)
c. $R_{EQ} = 559.3$ kΩ (Ohmmeter will read 559 kΩ approximately.)
d. $R_{EQ} = 1.497$ kΩ (Ohmmeter will read 1.5 kΩ approximately.)
e. $R_{EQ} = 9.868$ kΩ (Ohmmeter will read 9.87 kΩ approximately.)

SECTION 5-5 CONDUCTANCES IN PARALLEL

45. $G_1 = 2$ mS
$G_2 = 500$ μS
$G_3 = 833.3$ μS
$G_4 = 10$ mS
$G_T = 13.33$ mS
$R_{EQ} = 75$ Ω

47. $G_T = 500$ mS
$R_{EQ} = 2$ Ω

SECTION 5-6 TOTAL POWER IN PARALLEL CIRCUITS

49. $P_1 = 20.4$ W
$P_2 = 1.53$ W
$P_3 = 8.67$ W
$P_4 = 2.04$ W
$P_T = 32.64$ W

51. $P_1 = 13.2$ W
$P_2 = 19.8$ W
$P_3 = 132$ W
$P_T = 165$ W

SECTION 5-7 ANALYZING PARALLEL CIRCUITS WITH RANDOM UNKNOWNS

53. $V_A = 18$ V
$R_1 = 360$ Ω
$I_2 = 150$ mA
$R_{EQ} = 90$ Ω
$P_1 = 900$ mW
$P_2 = 2.7$ W
$P_T = 3.6$ W

55. $R_3 = 500\ \Omega$
$V_A = 75\ \text{V}$
$I_1 = 150\ \text{mA}$
$I_2 = 300\ \text{mA}$
$I_T = 600\ \text{mA}$
$P_1 = 11.25\ \text{W}$
$P_2 = 22.5\ \text{W}$
$P_3 = 11.25\ \text{W}$
$P_T = 45\ \text{W}$

57. $I_T = 100\ \text{mA}$
$I_1 = 30\ \text{mA}$
$I_2 = 20\ \text{mA}$
$I_4 = 35\ \text{mA}$
$R_3 = 2.4\ \text{k}\Omega$
$R_4 = 1.029\ \text{k}\Omega$
$P_1 = 1.08\ \text{W}$
$P_2 = 720\ \text{mW}$
$P_3 = 540\ \text{mW}$
$P_4 = 1.26\ \text{W}$
$P_T = 3.6\ \text{W}$

59. $V_A = 24\ \text{V}$
$I_1 = 20\ \text{mA}$
$I_2 = 30\ \text{mA}$
$I_4 = 24\ \text{mA}$
$R_1 = 1.2\ \text{k}\Omega$
$R_3 = 4\ \text{k}\Omega$
$R_{EQ} = 300\ \Omega$

ANSWERS TO CRITICAL THINKING PROBLEMS

61. $I_{T(max)} = 44.55\ \text{mA}$
63. $R_1 = 2\ \text{k}\Omega, R_2 = 6\ \text{k}\Omega,$ and $R_3 = 3\ \text{k}\Omega$
65. $R_1 = 15\ \text{k}\Omega, R_2 = 7.5\ \text{k}\Omega,$
$R_3 = 3.75\ \text{k}\Omega,$ and $R_4 = 1.875\ \text{k}\Omega$

ANSWERS TO TROUBLESHOOTING CHALLENGE

67. The current meter, M_3, is open; or the wire between points G and H is open. The fault could be isolated by measuring the voltage across points C and D and points G and H. The voltage will measure 36 V across the open points.

69. **a.** M_1 and M_3 will both read 0 A.
b. M_2 will read 0 V.
c. 36 V
d. The blown fuse was probably caused by a short in one of the four parallel branches.
e. With the blown fuse, F_1 still in place; open S_1. (This is an additional precaution.) Connect an ohmmeter across points B and I. The ohmmeter will probably read $0\ \Omega$. Next, remove one branch at a time while observing the ohmmeter. When the shorted branch is removed, the ohmic value indicated by the ohmmeter will increase to a value that is normal for the circuit. Be sure the ohmmeter is set to its lowest range when following this procedure. The reason is that the equivalent resistance, R_{EQ} of this circuit, is normally quite low anyway. Setting the ohmmeter on too high

of a range could result in a reading of $0\ \Omega$ even after the shorted branch has been removed.

71. $0\ \Omega$. One way to find the shorted branch would be to disconnect all but one of the branches along the top at points B, C, D, or E. (When doing this, make certain S_1 is open if the fuse has been replaced.) Next, with F_1 replaced, close S_1. If the only remaining branch blows the fuse, then you know that's the shorted branch. If the fuse F_1 did not blow, open S_1 and reconnect the next branch. Repeat this procedure until the fuse F_1 blows. The branch that blows the fuse is the shorted branch.

73. **a.** 0 V
b. 0 V
75. **a.** M_1 will read 1.5 A and M_3 will read 0 A.
b. 36 V
c. 0 V

Chapter Six

SECTION 6-1 FINDING R_T FOR SERIES-PARALLEL RESISTANCES

1. Resistors R_1 and R_2 are in series and resistors R_3 and R_4 are in parallel. It should also be noted that the applied voltage, V_T, is in series with R_1 and R_2 because they all have the same current.

3. $I_1 = 10\ \text{mA}$
$I_2 = 10\ \text{mA}$
$V_1 = 2.2\ \text{V}$
$V_2 = 6.8\ \text{V}$
$V_3 = 6\ \text{V}$
$V_4 = 6\ \text{V}$
$I_3 = 6\ \text{mA}$
$I_4 = 4\ \text{mA}$

5. **a.** $80\ \Omega$
b. $200\ \Omega$
c. 60 mA
d. 60 mA

7. $P_1 = 432\ \text{mW}$
$P_2 = 230.4\ \text{mW}$
$P_3 = 57.6\ \text{mW}$
$P_T = 720\ \text{mW}$

9. **a.** $250\ \Omega$
b. $200\ \Omega$
c. $450\ \Omega$
d. 40 mA
e. 40 mA

SECTION 6-2 RESISTANCE STRINGS IN PARALLEL

11. **a.** $800\ \Omega$
b. $1.2\ \text{k}\Omega$
c. $I_1 = 30\ \text{mA}$ and $I_2 = 20\ \text{mA}$
d. $I_T = 50\ \text{mA}$
e. $R_T = 480\ \Omega$
f. $V_1 = 9.9\ \text{V}, V_2 = 14.1\ \text{V},$ and $V_3 = 24\ \text{V}$

13. **a.** $300\ \Omega$
b. $900\ \Omega$

c. $I_1 = 120\ \text{mA}$ and $I_2 = 40\ \text{mA}$
d. $I_T = 160\ \text{mA}$
e. $R_T = 225\ \Omega$
f. $V_1 = 12\ \text{V}, V_2 = 24\ \text{V}, V_3 = 27.2\ \text{V},$ and $V_4 = 8.8\ \text{V}$

15. **a.** $I_1 = 8\ \text{mA}, I_2 = 24\ \text{mA}, I_3 = 16$ mA, $I_T = 48\ \text{mA}$
b. $R_T = 500\ \Omega$
c. $V_1 = 8\ \text{V}, V_2 = 16\ \text{V}, V_3 = 24\ \text{V}, V_4 = 8\ \text{V},$ and $V_5 = 16\ \text{V}$

SECTION 6-3 RESISTANCE BANKS IN SERIES

17. **a.** $150\ \Omega$
b. $R_T = 250\ \Omega$
c. $I_T = 100\ \text{mA}$
d. $V_{AB} = 15\ \text{V}$
e. $V_1 = 10\ \text{V}$
f. $I_2 = 25\ \text{mA}$ and $I_3 = 75\ \text{mA}$
g. 100 mA

19. **a.** $R_T = 1.8\ \text{k}\Omega$
b. $I_T = 30\ \text{mA}$
c. $V_1 = 36\ \text{V}, V_2 = 18\ \text{V},$ and $V_3 = 18\ \text{V}$
d. $I_2 = 12\ \text{mA}$ and $I_3 = 18\ \text{mA}$

SECTION 6-4 RESISTANCE BANKS AND STRINGS IN SERIES-PARALLEL

21. $R_T = 500\ \Omega$
$I_T = 70\ \text{mA}$
$V_1 = 8.4\ \text{V}$
$V_2 = 14\ \text{V}$
$V_3 = 5.6\ \text{V}$
$V_4 = 8.4\ \text{V}$
$V_5 = 12.6\ \text{V}$
$I_1 = 70\ \text{mA}$
$I_2 = 14\ \text{mA}$
$I_3 = 56\ \text{mA}$
$I_4 = 56\ \text{mA}$
$I_5 = 70\ \text{mA}$

23. $R_T = 4\ \text{k}\Omega$
$I_T = 30\ \text{mA}$
$V_1 = 30\ \text{V}$
$V_2 = 10\ \text{V}$
$V_3 = 20\ \text{V}$
$V_4 = 30\ \text{V}$
$V_5 = 60\ \text{V}$
$I_1 = 30\ \text{mA}$
$I_2 = 10\ \text{mA}$
$I_3 = 10\ \text{mA}$
$I_4 = 20\ \text{mA}$
$I_5 = 30\ \text{mA}$

25. $R_T = 6\ \text{k}\Omega$
$I_T = 6\ \text{mA}$
$V_1 = 36\ \text{V}$
$V_2 = 5.4\ \text{V}$
$V_3 = 27\ \text{V}$
$V_4 = 9\ \text{V}$
$V_5 = 13.5\ \text{V}$
$V_6 = 4.5\ \text{V}$
$V_7 = 3.6\ \text{V}$
$I_1 = 2.4\ \text{mA}$
$I_2 = 3.6\ \text{mA}$
$I_3 = 2.7\ \text{mA}$

$I_4 = 900 \ \mu A$
$I_5 = 900 \ \mu A$
$I_6 = 900 \ \mu A$
$I_7 = 3.6 \ mA$
27. $R_T = 200 \ \Omega$
$I_T = 120 \ mA$
$V_1 = 6 \ V$
$V_2 = 6 \ V$
$V_3 = 12 \ V$
$V_4 = 12 \ V$
$V_5 = 18 \ V$
$V_6 = 24 \ V$
$I_1 = 40 \ mA$
$I_2 = 30 \ mA$
$I_3 = 10 \ mA$
$I_4 = 20 \ mA$
$I_5 = 10 \ mA$
$I_6 = 80 \ mA$

SECTION 6-5 ANALYZING SERIES-PARALLEL CIRCUITS WITH RANDOM UNKNOWNS

29. $R_T = 300 \ \Omega$
$I_T = 70 \ mA$
$V_T = 21 \ V$
$V_1 = 7 \ V$
$V_2 = 9.24 \ V$
$V_4 = 4.76 \ V$
$I_2 = 42 \ mA$
$I_3 = 28 \ mA$
31. $R_T = 800 \ \Omega$
$I_T = 30 \ mA$
$V_T = 24 \ V$
$V_1 = 5.4 \ V$
$V_2 = 12 \ V$
$V_3 = 3 \ V$
$V_4 = 6.6 \ V$
$V_5 = 2.4 \ V$
$V_6 = 6.6 \ V$
$I_2 = 10 \ mA$
$I_3 = 20 \ mA$
$I_4 = 20 \ mA$
$I_5 = 20 \ mA$
33. $I_T = 30 \ mA$
$R_T = 1 \ k\Omega$
$V_2 = 10.8 \ V$
$V_3 = 10.8 \ V$
$V_4 = 10.8 \ V$
$V_5 = 21.6 \ V$
$R_2 = 600 \ \Omega$
$V_6 = 3 \ V$
$I_2 = 18 \ mA$
$I_3 = 10.8 \ mA$
$I_4 = 7.2 \ mA$
$I_5 = 12 \ mA$
$I_6 = 30 \ mA$

SECTION 6-6 THE WHEATSTONE BRIDGE

35. **a.** $R_X = 6816 \ \Omega$
b. $V_{CB} = V_{DB} = 8.33 \ V$
c. $I_T = 1.91 \ mA$
37. **a.** $R_{X(max)} = 99.999 \ \Omega$
b. $R_{X(max)} = 999.99 \ \Omega$
c. $R_{X(max)} = 9999.9 \ \Omega$
d. $R_{X(max)} = 99,999 \ \Omega$
e. $R_{X(max)} = 999,990 \ \Omega$
f. $R_{X(max)} = 9,999,900 \ \Omega$

39. R_3 must be adjusted to 1 kΩ.
41. **a.** The thermistor resistance is 4250 Ω.
b. T_A has increased above 25°C.

ANSWERS TO CRITICAL THINKING PROBLEMS

43. **a.** $R_1 = 657 \ \Omega$
b. Recommended wattage rating is 25 W approximately
c. $R_T = 857 \ \Omega$
45. With V_T reversed in polarity, the circuit will not operate properly. For example, if the ambient temperature increases: the voltage across points C and D becomes positive. This causes the output voltage from the amplifier to go negative which turns on the heater. This will increase the temperature even more. Unfortunately, the heater will continue to stay on. If the temperature would have decreased initially the voltage across points C and D would have gone negative. This would make the output of the amplifier go positive, thus turning on the air conditioner, making the temperature decrease even further.

ANSWERS TO TROUBLESHOOTING CHALLENGE, TABLE 6-1

Trouble 1: R_6 Open
Trouble 3: R_4 Open
Trouble 5: R_6 Shorted
Trouble 7: R_1 Open
Trouble 9: R_1 Shorted
Trouble 11: R_3 Shorted

Chapter Seven

SECTION 7-1 SERIES VOLTAGE DIVIDERS

1. $V_1 = 3 \ V$
$V_2 = 6 \ V$
$V_3 = 9 \ V$
3. $V_1 = 4 \ V$
$V_2 = 6 \ V$
$V_3 = 8 \ V$
5. $V_1 = 2.5 \ V$
$V_2 = 7.5 \ V$
$V_3 = 15 \ V$
7. **a.** $V_1 = 9 \ V; V_2 = 900 \ mV; V_3 = 100 \ mV$
b. $V_{AG} = 10 \ V; V_{BG} = 1 \ V; V_{CG} = 100 \ mV$
9. $V_1 = 16 \ V; V_2 = 8 \ V; V_3 = 16 \ V, V_4 = 8 \ V$
$V_{AG} = 48 \ V; V_{BG} = 32 \ V; V_{CG} = 24 \ V; V_{DG} = 8 \ V$

SECTION 7-2 CURRENT DIVIDER WITH TWO PARALLEL RESISTANCES

11. $I_1 = 16 \ mA$
$I_2 = 8 \ mA$
13. $I_1 = 64 \ mA$
$I_2 = 16 \ mA$
15. $I_1 = 48 \ mA$
$I_2 = 72 \ mA$

SECTION 7-3 CURRENT DIVISION BY PARALLEL CONDUCTANCES

17. $I_1 = 3.6 \ A$
$I_2 = 2.4 \ A$
$I_3 = 3 \ A$
19. $I_1 = 45 \ mA$
$I_2 = 4.5 \ mA$
$I_3 = 16.5 \ mA$
21. $I_1 = 25 \ \mu A$
$I_2 = 37.5 \ \mu A$
$I_3 = 12.5 \ \mu A$
$I_4 = 75 \ \mu A$

SECTION 7-4 SERIES VOLTAGE DIVIDER WITH PARALLEL LOAD CURRENT

23. The voltage, V_{BG}, decreases when S_1 is closed because R_L in parallel with R_2 reduces the resistance from points B to G. This lowering of resistance changes the voltage division in the circuit. With S_1 closed, the resistance from B to G is a smaller fraction of the total resistance, which in turn means the voltage, V_{BG} must also be less.
25. Resistor R_2

SECTION 7-5 DESIGN OF A LOADED VOLTAGE DIVIDER

27. **a.** $I_1 = 56 \ mA$
$I_2 = 11 \ mA$
$I_3 = 6 \ mA$
$I_T = 66 \ mA$
b. $V_1 = 10 \ V$
$V_2 = 9 \ V$
$V_3 = 6 \ V$
c. $R_1 = 178.6 \ \Omega$
$R_2 = 818.2 \ \Omega$
$R_3 = 1 \ k\Omega$
d. $P_1 = 560 \ mW$
$P_2 = 99 \ mW$
$P_3 = 36 \ mW$
29. **a.** $I_1 = 38 \ mA$
$I_2 = 18 \ mA$
$I_3 = 6 \ mA$
$I_T = 68 \ mA$
b. $V_1 = 9 \ V$
$V_2 = 6 \ V$
$V_3 = 9 \ V$
c. $R_1 = 236.8 \ \Omega$
$R_2 = 333.3 \ \Omega$
$R_3 = 1.5 \ k\Omega$
d. $P_1 = 342 \ mW$
$P_2 = 108 \ mW$
$P_3 = 54 \ mW$

ANSWERS TO CRITICAL THINKING PROBLEMS

31. $R_1 = 1 \ k\Omega$ and $R_3 = 667 \ \Omega$
33. See Instructor's Manual

ANSWERS TO TROUBLESHOOTING CHALLENGE

TABLE 7-2

Trouble 1: R_2 Open
Trouble 3: R_3 Shorted
Trouble 5: R_2 Shorted
Trouble 7: R_3 Open

TABLE 7–3

Trouble 1: R_2 Open
Trouble 3: R_3 Open
Trouble 5: R_1 Open
Trouble 7: R_4 or Load C Shorted

Chapter Eight

SECTION 8-2 METER SHUNTS

1. **a.** $R_S = 50\ \Omega$
   b. $R_S = 5.56\ \Omega$
   c. $R_S = 2.08\ \Omega$
   d. $R_S = 0.505\ \Omega$
3. **a.** $R_S = 1\ k\Omega$
   b. $R_S = 52.63\ \Omega$
   c. $R_S = 10.1\ \Omega$
   d. $R_S = 5.03\ \Omega$
   e. $R_S = 1\ \Omega$
   f. $R_S = 0.5\ \Omega$
5. **a.** $R_{S_1} = 111.1\ \Omega$
   $R_{S_2} = 20.41\ \Omega$
   $R_{S_3} = 4.02\ \Omega$
   b. 1 mA Range; $R_M = 100\ \Omega$
   5 mA Range; $R_M = 20\ \Omega$
   25 mA Range; $R_M = 4\ \Omega$
7. So that the current in the circuit is approximately the same with or without the meter present. If the current meter's resistance is too high, the measured value of current could be significantly less than the current without the meter present.

SECTION 8-3 VOLTMETERS

9. $\dfrac{\Omega}{V}$ Rating $= \dfrac{1\ k\Omega}{V}$

11. $\dfrac{\Omega}{V}$ Rating $= \dfrac{50\ k\Omega}{V}$

13. **a.** $R_1 = 58\ k\Omega$
   $R_2 = 140\ k\Omega$
   $R_3 = 400\ k\Omega$
   $R_4 = 1.4\ M\Omega$
   $R_5 = 4\ M\Omega$
   $R_6 = 14\ M\Omega$
   b. 3 V Range; $R_V = 60\ k\Omega$
   10 V Range; $R_V = 200\ k\Omega$
   30 V Range; $R_V = 600\ k\Omega$
   100 V Range; $R_V = 2\ M\Omega$
   300 V Range; $R_V = 6\ M\Omega$
   1,000 V Range; $R_V = 20\ M\Omega$
   c. $\dfrac{\Omega}{V}$ Rating $= \dfrac{20\ k\Omega}{V}$
15. **a.** $\dfrac{\Omega}{V}$ Rating $= \dfrac{1\ k\Omega}{V}$
   b. $\dfrac{\Omega}{V}$ Rating $= \dfrac{10\ k\Omega}{V}$
   c. $\dfrac{\Omega}{V}$ Rating $= \dfrac{20\ k\Omega}{V}$
   d. $\dfrac{\Omega}{V}$ Rating $= \dfrac{100\ k\Omega}{V}$

SECTION 8-4 LOADING EFFECT OF A VOLTMETER

17. **a.** $V = 7.2\ V$
   b. $V = 7.16\ V$
   c. $V = 7.2\ V$
   Notice that there is little or no voltmeter loading with either meter since R_V is so much larger than the value of R_2.
19. The analog voltmeter with an R_V of 1 MΩ produced a greater loading effect. The reason is that its resistance is less than that of the DMM whose R_V is 10 MΩ.

SECTION 8-5 OHMMETERS

21. **a.** $R_X = 0\ \Omega$
   b. $R_X = 250\ \Omega$
   c. $R_X = 750\ \Omega$
   d. $R_X = 2.25\ k\Omega$
   e. $R_X = \infty\ \Omega$
23. The scale would be nonlinear with values being more spread out on the right-hand side and more crowded on the left-hand side. The ohmmeter scale is nonlinear because equal increases in measured resistance do not produce equal decreases in current.
25. Because the ohms values increase from right to left as the current in the meter backs off from full-scale deflection.
27. On any range the zero-ohms adjustment control is adjusted for zero ohms with the ohmmeter leads shorted. The zero ohms control is adjusted to compensate for the slight changes in battery voltage, V_b, when changing ohmmeter ranges. Without a zero ohms adjustment control, the scale of the ohmmeter would not be properly calibrated.

SECTION 8-8 METER APPLICATIONS

29. The ohmmeter could be damaged or the meter will read an incorrect value of resistance. When measuring resistance, power must be off in the circuit being tested!
31. A current meter is connected in series to measure the current at some point in a circuit. Connecting a current meter in parallel could possibly ruin the meter due to excessive current. Remember, a current meter has a very low resistance and connecting it in parallel can effectively short-out a component.
33. **a.** $0\ \Omega$
   b. infinite $(\infty)\ \Omega$

ANSWERS TO CRITICAL THINKING PROBLEMS

35. $R_1 = 40\ \Omega$, $R_2 = 8\ \Omega$, and $R_3 = 2\ \Omega$
37. $10\ k\Omega/V$

Chapter Nine

SECTION 9-1 KIRCHHOFF'S CURRENT LAW (KCL)

1. $I_3 = 15\ A$
3. $I_3 = 6\ mA$
5. Point X: $6\ A + 11\ A + 8\ A - 25\ A = 0$
   Point Y: $25\ A - 2\ A - 16\ A - 7\ A = 0$
   Point Z: $16\ A - 5\ A - 11\ A = 0$

SECTION 9-2 KIRCHHOFF'S VOLTAGE LAW (KVL)

7. **a.** $4.5\ V + 5.4\ V + 8.1\ V = 18\ V$. This voltage is the same as the voltage V_{R_3}.
   b. $6\ V + 18\ V + 12\ V = 36\ V$. This voltage equals the applied voltage, V_T.
   c. $6\ V + 4.5\ V + 5.4\ V + 8.1\ V + 12\ V = 36\ V$. This voltage equals the applied voltage, V_T.
   d. $18\ V - 8.1\ V - 5.4\ V = 4.5\ V$. This voltage is the same as the voltage, V_{R_4}.
9. $V_{AG} = +8\ V$
   $V_{BG} = 0\ V$
11. $V_{AG} = -50\ V$
   $V_{BG} = -42.5\ V$
   $V_{CG} = -33.5\ V$
   $V_{DG} = -20\ V$
   $V_{AD} = -30\ V$
13. $20\ V - 2.5\ V - 17.5\ V = 0$
15. $17.5\ V + 12.5\ V - 10\ V - 20\ V = 0$

SECTION 9-3 METHOD OF BRANCH CURRENTS

17. **a.** $I_1 + I_3 - I_2 = 0$
   b. $I_3 = I_2 - I_1$
   c. $-V_1 - V_{R_3} + V_{R_1} = 0$ or $-24\ V - V_{R_3} + V_{R_1} = 0$
   d. $-V_2 + V_{R_2} + V_{R_3} = 0$ or $-12\ V + V_{R_2} + V_{R_3} = 0$
   e. $V_{R_1} = I_1 R_1 = I_1 12\ \Omega$ or $12I_1$
   $V_{R_2} = I_2 R_2 = I_2 24\ \Omega$ or $24I_2$
   $V_{R_3} = I_3 R_3 = (I_2 - I_1)12\ \Omega$ or $12(I_2 - I_1)$
   f. Loop 1:
   $-24\ V - 12(I_2 - I_1) + 12I_1 = 0$
   g. Loop 2:
   $-12\ V + 24I_2 + 12(I_2 - I_1) = 0$
   h. Loop 1:
   $24I_1 - 12I_2 = 24\ V$ which can be reduced further to: $2I_1 - I_2 = 2\ V$
   Loop 2:
   $-12I_1 + 36I_2 = 12\ V$ which can be reduced further to:
   $-I_1 + 3I_2 = 1\ V$
   i. $I_1 = 1.4\ A$
   $I_2 = 800\ mA$
   $I_3 = -600\ mA$
   j. No. The assumed direction for I_3 was incorrect as indicated by its negative value. I_3 actually flows downward through R_3.
   k. $V_{R_1} = I_1 R_1 = 1.4\ A \times 12\ \Omega = 16.8\ V$
   $V_{R_2} = I_2 R_2 = 800\ mA \times 24\ \Omega = 19.2\ V$
   $V_{R_3} = I_3 R_3 = 600\ mA \times 12\ \Omega = 7.2\ V$.

l. Loop 1:
$-24\text{ V} + 7.2\text{ V} + 16.8\text{ V} = 0$
Loop 2:
$12\text{ V} + 7.2\text{ V} - 19.2\text{ V} = 0$

m. $I_1 - I_2 - I_3 = 0$ or
$1.4\text{ A} - 800\text{ mA} - 600\text{ mA} = 0$

SECTION 9-4 NODE-VOLTAGE ANALYSIS

19. **a.** $I_2 - I_1 - I_3 = 0$ or
$I_2 = I_1 + I_3$

b. $\dfrac{V_{R_2}}{10\ \Omega} - \dfrac{V_{R_1}}{10\ \Omega} - \dfrac{V_N}{5\ \Omega} = 0$
or
$\dfrac{V_{R_2}}{10\ \Omega} = \dfrac{V_{R_1}}{10\ \Omega} + \dfrac{V_N}{5\ \Omega}$

c. $V_1 + V_{R_3} - V_{R_1} = 0$ or
$10\text{ V} + V_{R_3} - V_{R_1} = 0$

d. $-V_2 + V_{R_3} + V_{R_2} = 0$ or
$-15\text{ V} + V_{R_3} + V_{R_2} = 0$

e. $V_{R_1} = V_1 + V_N$ or $V_{R_1} = 10\text{ V} + V_N$
$V_{R_2} = V_2 - V_N$ or $V_{R_2} = 15\text{ V} - V_N$

f. $\dfrac{15\text{ V} - V_N}{10\ \Omega} = \dfrac{10\text{ V} + V_N}{10\ \Omega} + \dfrac{V_N}{5\ \Omega}$

g. $V_N = 1.25\text{ V}$

h. $V_{R_1} = 10\text{ V} + 1.25\text{ V} = 11.25\text{ V}$
$V_{R_2} = 15\text{ V} - 1.25\text{ V} = 13.75\text{ V}$

i. Yes. Because the solutions for V_{R_1} and V_{R_2} were both positive.

j. $I_1 = 1.125\text{ A}$
$I_2 = 1.375\text{ A}$
$I_3 = 250\text{ mA}$

k. For the loop with V_1 we have
$10\text{ V} + 1.25\text{ V} - 11.25\text{ V} = 0$
going CCW from the positive $(+)$ terminal of V_1.
For the loop with V_2 we have
$-15\text{ V} + 1.25\text{ V} + 13.75\text{ V} = 0$ going CW from the negative $(-)$ terminal of V_2.

l. $I_2 - I_1 - I_3 = 0$ or
$1.375\text{ A} - 1.125\text{ A} - 250\text{ mA} = 0$

SECTION 9-5 METHOD OF MESH CURRENTS

21. **a.** V_1, R_1 and R_3

b. V_2, R_2 and R_3

c. R_3

d. $20I_A - 10I_B = -40\text{ V}$

e. $-10I_A + 25I_B = -20\text{ V}$

f. $I_A = -3\text{ A}$
$I_B = -2\text{ A}$

g. $I_1 = I_A = -3\text{ A}$
$I_2 = I_B = -2\text{ A}$
$I_3 = 1\text{ A}$

h. No. Because the answers for the mesh currents I_A and I_B were negative.

i. I_3 flows in the same direction as I_A or up through R_3.

j. $V_{R_1} = I_1R_1 = 3\text{ A} \times 10\ \Omega = 30\text{ V}$
$V_{R_2} = I_2R_2 = 2\text{ A} \times 15\ \Omega = 30\text{ V}$
$V_{R_3} = I_3R_3 = 1\text{ A} \times 10\ \Omega = 10\text{ V}$

k. $30\text{ V} + 10\text{ V} - 40\text{ V} = 0$

l. $-20\text{ V} - 10\text{ V} + 30\text{ V} = 0$

m. $I_2 + I_3 - I_1 = 0$ or
$2\text{ A} + 1\text{ A} - 3\text{ A} = 0$

ANSWERS TO CRITICAL THINKING PROBLEMS

23. $I_A = 413.3\text{ mA}$, $I_B = 40\text{ mA}$, and
$I_C = 253.3\text{ mA}$
$I_1 = 413.3\text{ mA}$
$I_2 = 373.3\text{ mA}$
$I_3 = 413.3\text{ mA}$
$I_4 = 40\text{ mA}$
$I_5 = 293.3\text{ mA}$
$I_6 = 40\text{ mA}$
$I_7 = 253.3\text{ mA}$
$I_8 = 253.3\text{ mA}$

Chapter Ten

SECTION 10-1 SUPERPOSITION THEOREM

1. $V_P = -6\text{ V}$
3. $V_P = 10\text{ V}$
5. $V_{AB} = 0.2\text{ V}$

SECTION 10-2 THEVENIN'S THEOREM

7. When $R_L = 3\ \Omega$, $I_L = 2.5\text{ A}$ and
$V_L = 7.5\text{ V}$
When $R_L = 6\ \Omega$, $I_L = 1.67\text{ A}$ and
$V_L = 10\text{ V}$
When $R_L = 12\ \Omega$, $I_L = 1\text{ A}$ and
$V_L = 12\text{ V}$

9. When $R_L = 100\ \Omega$, $I_L = 48\text{ mA}$ and
$V_L = 4.8\text{ V}$
When $R_L = 1\text{ k}\Omega$, $I_L = 30\text{ mA}$ and
$V_L = 30\text{ V}$
When $R_L = 5.6\text{ k}\Omega$, $I_L = 10.29\text{ mA}$
and $V_L = 57.6\text{ V}$

11. When $R_L = 200\ \Omega$, $I_L = 45\text{ mA}$ and
$V_L = 9\text{ V}$
When $R_L = 1.2\text{ k}\Omega$, $I_L = 20\text{ mA}$ and
$V_L = 24\text{ V}$
When $R_L = 1.8\text{ k}\Omega$, $I_L = 15\text{ mA}$ and
$V_L = 27\text{ V}$

13. $I_L = 20\text{ mA}$ and $V_L = 24\text{ V}$

SECTION 10-3 THEVENIZING A CIRCUIT WITH TWO VOLTAGE SOURCES

15. $I_3 = 208.3\text{ mA}$ and $V_{R_3} = 3.75\text{ V}$
17. $I_3 = 37.5\text{ mA}$ and $V_{R_3} = 2.1\text{ V}$

SECTION 10-4 THEVENIZING A BRIDGE CIRCUIT

19. $V_{TH} = 10\text{ V}$ and $R_{TH} = 150\ \Omega$
21. $V_{TH} = 9\text{ V}$ and $R_{TH} = 200\ \Omega$

SECTION 10-5 NORTON'S THEOREM

23. $I_N = 5\text{ A}$ and $R_N = 6\ \Omega$
25. $I_N = 1.5\text{ A}$ and $R_N = 10\ \Omega$
27. $I_N = 500\text{ mA}$ and $R_N = 30\ \Omega$
$I_L = 333.3\text{ mA}$
$V_L = 5\text{ V}$

SECTION 10-6 THEVENIN-NORTON CONVERSIONS

29. $V_{TH} = 24\text{ V}$
$R_{TH} = 1.2\text{ k}\Omega$
31. $I_N = 30\text{ mA}$
$R_{TH} = 1.2\text{ k}\Omega$

SECTION 10-7 CONVERSION OF VOLTAGE AND CURRENT SOURCES

33. **a.** See Instructor's Manual
b. $V_T = 30\text{ V}$ and $R = 24\ \Omega$
c. $I_3 = 1\text{ A}$
$V_{R_3} = 6\text{ V}$

SECTION 10-8 MILLMAN'S THEOREM

35. $V_{XY} = -28\text{ V}$
37. $V_{XY} = 0\text{ V}$

SECTION 10-9 T OR Y AND π OR Δ CONNECTIONS

39. $R_A = 17.44\ \Omega$
$R_B = 19.63\ \Omega$
$R_C = 31.4\ \Omega$
41. $R_T = 7\ \Omega$
$I_T = 3\text{ A}$

ANSWERS TO CRITICAL THINKING PROBLEMS

43. $I_L = 0\text{ A}$ and $V_L = 0\text{ V}$
45. $V_{TH} = 21.6\text{ V}$
$R_{TH} = 120\ \Omega$
$I_2 = 67.5\text{ mA}$
$V_2 = 13.5\text{ V}$

Chapter Eleven

SECTION 11-1 FUNCTION OF THE CONDUCTOR

1. **a.** 100 ft.
b. $R_T = 8.16\ \Omega$
c. $I = 14.71\text{ A}$
d. 1.18 V
e. 117.7 V
f. 17.31 W
g. 1.731 kW
h. 1.765 kW
i. 98.1%

SECTION 11-2 STANDARD WIRE GAGE SIZES

3. **a.** 25 cmils
b. 441 cmils
c. 1024 cmils
d. 2500 cmils
e. 10,000 cmils
f. 40,000 cmils

5. **a.** $R = 1.018\ \Omega$
b. $R = 2.042\ \Omega$
c. $R = 4.094\ \Omega$
d. $R = 26.17\ \Omega$

7. A 1000 ft. length of No. 23 gage copper wire.

SECTION 11-3 TYPES OF WIRE CONDUCTORS

9. No. 10 Gage
11. No. 19 Gage

SECTION 11–6 SWITCHES

13. a. 6.3 V
    b. 0 V
    c. No
    d. 0 A
15. a. See Instructor's Manual
    b. See Instructor's Manual

SECTION 11–8 WIRE RESISTANCE

17. a. $R = 2.54\ \Omega$
    b. $R = 4.16\ \Omega$
    c. $R = 24.46\ \Omega$
19. a. $R = 0.32\ \Omega$
    b. $R = 0.64\ \Omega$
21. $R = 0.127\ \Omega$ (approx.)
23. No. 16 gage

SECTION 11–9 TEMPERATURE COEFFICIENT OF RESISTANCE

25. $R = 12.4\ \Omega$
27. $R = 140\ \Omega$
29. $R = 8.5\ \Omega$

ANSWERS TO CRITICAL THINKING PROBLEMS

31. See Instructor's Manual

Chapter Twelve

SECTION 12–6 SERIES AND PARALLEL CONNECTED CELLS

1. $V_L = 3$ V, $I_L = 30$ mA, the current in each cell equals 30 mA.
3. $V_L = 1.25$ V, $I_L = 50$ mA, the current in each cell equals 25 mA.
5. $V_L = 3$ V, $I_L = 300$ mA, the current in each cell equals 100 mA.

SECTION 12–8 INTERNAL RESISTANCE OF A GENERATOR

7. $r_i = 2\ \Omega$
9. $r_i = 6\ \Omega$
11. $r_i = 15\ \Omega$

SECTION 12–9 CONSTANT VOLTAGE AND CONSTANT CURRENT SOURCES

13. a. $I_L = 1\ \mu$A; $V_L = 0$ V
    b. $I_L \approx 1\ \mu$A; $V_L \approx 100\ \mu$V
    c. $I_L \approx 1\ \mu$A; $V_L \approx 1$ mV
    d. $I_L = 0.99\ \mu$A; $V_L = 99$ mV

SECTION 12–10 MATCHING A LOAD RESISTANCE TO THE GENERATOR, r_i

15. a. $I_L = 1.67$ A
       $V_L = 16.67$ V
       $P_L = 27.79$ W
       $P_T = 167$ W
       % Efficiency = 16.64%
    b. $I_L = 1.33$ A
       $V_L = 33.3$ V
       $P_L = 44.44$ W
       $P_T = 133$ W
       % Efficiency = 33.4%
    c. $I_L = 1$ A
       $V_L = 50$ V
       $P_L = 50$ W

$P_T = 100$ W
       % Efficiency = 50%
    d. $I_L = 800$ mA
       $V_L = 60$ V
       $P_L = 48$ W
       $P_T = 80$ W
       % Efficiency = 60%
    e. $I_L = 667$ mA
       $V_L = 66.67$ V
       $P_L = 44.44$ W
       $P_T = 66.7$ W
       % Efficiency = 66.67%
17. a. $R_L = 8\ \Omega$
    b. $P_L = 78.125$ W
    c. % Efficiency = 50%

ANSWERS TO CRITICAL THINKING PROBLEMS

19. a. $R_L = 30\ \Omega$
    b. $P_{L(max)} = 2.7$ W

Chapter Thirteen

SECTION 13–2 MAGNETIC FLUX, ϕ

1. a. 1 Mx = 1 magnetic field line
   b. 1 Wb = 1×10^8 Mx or 1×10^8 magnetic field lines
3. a. 1×10^{-5} Wb
   b. 1×10^{-4} Wb
   c. 1×10^{-8} Wb
   d. 1×10^{-6} Wb or 1 μWb
5. a. 4000 Mx
   b. 2.25×10^{-6} Wb
   c. 8×10^{-4} Wb
   d. 6.5×10^4 Wb
7. 9×10^4 magnetic field lines

SECTION 13–3 FLUX DENSITY, B

9. a. $1\text{G} = \dfrac{1\ \text{Mx}}{\text{cm}^2}$

   b. $1\text{T} = \dfrac{1\ \text{Wb}}{\text{m}^2}$

11. a. 0.4 T
    b. 80 T
    c. 0.06 T
    d. 1 T
13. a. 905 G
    b. 1×10^6 G
    c. 7.5 T
    d. 175 T
15. 0.05 T
17. 4000 G or 4 kG
19. 0.133 T
21. 240,000 Mx
23. 0.02 Wb
25. 1×10^6 magnetic field lines

ANSWERS TO CRITICAL THINKING PROBLEMS

27. a. 40,300 Mx
    b. 403 μWb

Chapter Fourteen

SECTION 14–1 AMPERE-TURN OF MAGNETOMOTIVE FORCE (MMF)

1. a. Gilbert (Gb)
   b. Ampere-Turn (A · t)
3. a. 20 A · t
   b. 2 A · t
   c. 10 A · t
   d. 180 A · t
5. a. $N = 1000$ turns
   b. $N = 4000$ turns
   c. $N = 2500$ turns
   d. $N = 50$ turns
7. a. 100 A · t
   b. 30 A · t
   c. 500 A · t

SECTION 14–2 FIELD INTENSITY (H)

9. a. $500\ \dfrac{A \cdot t}{m}$

   b. $100\ \dfrac{A \cdot t}{m}$

   c. $50\ \dfrac{A \cdot t}{m}$

   d. $200\ \dfrac{A \cdot t}{m}$

11. a. 0.63 Oersteds
    b. 1.89 Oersteds
13. a. 12.6×10^{-6}
    b. 63×10^{-6}
    c. 126×10^{-6}
    d. 630×10^{-6}
    e. 1.26×10^{-3}
15. $\mu_r = 133.3$

SECTION 14–3 B–H MAGNETIZATION CURVE

17. a. 126×10^{-6}
    b. 88.2×10^{-6}

SECTION 14–9 GENERATING AN INDUCED VOLTAGE

19. $v_{ind} = 8$ V
21. $v_{ind} = 2$ kV

ANSWERS TO CRITICAL THINKING PROBLEMS

23. See Instructor's Manual
25. a. $R_W = 1.593\ \Omega$
    b. $R_T = 17.593\ \Omega$
    c. $V_L = 218.3$ V
    d. I^2R power loss = 296.4 W
    e. $P_L = 2.98$ kW
    f. $P_T = 3.27$ kW
    g. % efficiency = 91.1%
27. With a relay, the 1000-ft. length of wire does not carry the load current, I_L, and thus the circuit losses are reduced significantly.

Chapter Fifteen

SECTION 15–2 ALTERNATING VOLTAGE GENERATOR

1. a. 90°
   b. 180°
   c. 270°
   d. 360°

3. **a.** at 90°
   b. at 270°
   c. 0°, 180°, and 360°

SECTION 15-3 THE SINE WAVE

5. **a.** $v = 10$ V
   b. $v = 14.14$ V
   c. $v = 17.32$ V
   d. $v = 19.32$ V
   e. $v = 17.32$ V
   f. $v = -10$ V
   g. $v = -17.32$ V
7. $V_{pk} = 51.96$ V

SECTION 15-4 ALTERNATING CURRENT

9. **a.** counterclockwise
   b. clockwise

SECTION 15-5 VOLTAGE AND CURRENT VALUES FOR A SINE WAVE

11. **a.** 100 V peak-to-peak
    b. 35.35 V rms
    c. 31.85 V average
13. **a.** 56.56 V peak
    b. 113.12 V peak-to-peak
    c. 36 V average
15. **a.** 47.13 mA rms
    b. 66.7 mA peak
    c. 133.3 mA peak-to-peak
    d. 42.47 mA average
17. **a.** 16.97 V peak
    b. 113 V peak
    c. 25 V peak
    d. 1.06 V peak

SECTION 15-6 FREQUENCY

19. **a.** 2000 cps
    b. 15,000,000 cps
    c. 10,000 cps
    d. 5,000,000,000 cps

SECTION 15-7 PERIOD

21. **a.** $T = 500$ μs
    b. $T = 250$ μs
    c. $T = 5$ μs
    d. $T = 0.5$ μs
23. **a.** $f = 200$ Hz
    b. $f = 100$ kHz
    c. $f = 2$ MHz
    d. $f = 30$ kHz

SECTION 15-8 WAVELENGTH

25. **a.** 186,000 mi/s
    b. 3×10^{10} cm/s
    c. 3×10^{8} m/s
27. **a.** 8000 cm
    b. 4000 cm
    c. 2000 cm
    d. 1500 cm
29. 2 m
31. **a.** $f = 1.875$ MHz
    b. $f = 30$ MHz
    c. $f = 17.65$ MHz
    d. $f = 27.3$ MHz

SECTION 15-9 PHASE ANGLE

33. A sine wave has its maximum values at 90° and 270° whereas a cosine wave has its maximum values at 0° and 180°.

SECTION 15-10 THE TIME FACTOR IN FREQUENCY AND PHASE

35. **a.** $t = 83.3$ μs
    b. $t = 125$ μs
    c. $t = 166.7$ μs
    d. $t = 250$ μs

SECTION 15-11 ALTERNATING CURRENT CIRCUITS WITH RESISTANCE

37. $R_T = 250$ Ω
    $I = 40$ mA
    $V_1 = 4$ V
    $V_2 = 6$ V
    $P_1 = 160$ mW
    $P_2 = 240$ mW
    $P_T = 400$ mW
39. $R_T = 900$ Ω
    $I_T = 40$ mA
    $V_1 = 7.2$ V
    $V_2 = 28.8$ V
    $V_3 = 28.8$ V
    $I_2 = 24$ mA
    $I_3 = 16$ mA
    $P_1 = 288$ mW
    $P_2 = 691.2$ mW
    $P_3 = 460.8$ mW
    $P_T = 1.44$ W

SECTION 15-12 NONSINUSOIDAL AC WAVEFORMS

41. **a.** $V = 100$ V peak-to-peak
       $f = 20$ kHz
    b. $V = 30$ V peak-to-peak
       $f = 500$ Hz
    c. $V = 100$ V peak-to-peak
       $f = 2.5$ kHz

SECTION 15-13 HARMONIC FREQUENCIES

43. 1 kHz   First Odd Harmonic
    2 kHz   First Even Harmonic
    3 kHz   Second Odd Harmonic
    4 kHz   Second Even Harmonic
    5 kHz   Third Odd Harmonic
    6 kHz   Third Even Harmonic
    7 kHz   Fourth Odd Harmonic
45. 750 Hz
47. 100 kHz

SECTION 15-14 THE 60 Hz AC POWER-LINE

49. Transformer

ANSWERS TO CRITICAL THINKING PROBLEMS

51. **a.** 68.3 ft.
    b. 2083 cm
53. $f = 4.1$ MHz

Chapter Sixteen

SECTION 16-3 THE FARAD UNIT OF CAPACITANCE

1. **a.** $Q = 50$ μC
   b. $Q = 25$ μC
   c. $Q = 1.5$ μC
   d. $Q = 11$ μC
   e. $Q = 136$ nC
   f. $Q = 141$ nC
3. **a.** $V = 2.5$ V
   b. $V = 6.25$ V
   c. $V = 12.5$ V
   d. $V = 50$ V
   e. $V = 75$ V
5. **a.** $C = 15$ μF
   b. $C = 0.5$ μF or 500 nF
   c. $C = 4$ μF
   d. $C = 0.22$ μF or 220 nF
   e. $C = 0.001$ μF
   f. $C = 0.04$ μF or 40 nF
7. **a.** $C = 1.77$ pF
   b. $C = 2.213$ nF
   c. $C = 44.25$ nF
   d. $C = 106.2$ nF

SECTION 16-6 CAPACITOR CODING

9. Figure 16–32
   a. $C = 0.0033$ μF; $+80\%$, -20%
   b. $C = 0.022$ μF; $+100\%$, -0%
   c. $C = 1800$ pF; $\pm10\%$
   d. $C = 0.0047$ μF; $+80\%$, -20%
   e. $C = 100,000$ pF; $\pm5\%$
   f. $C = 0.15$ μF: $\pm20\%$
11. Figure 16–34
    a. $C = 3900$ pF
    b. $C = 27,000$ pF
    c. $C = 680$ pF
    d. $C = 33,000$ pF
13. Figure 16–36
    a. $C = 2200$ pF
    b. $C = 560$ pF
    c. $C = 27$ pF
    d. $C = 39,000$ pF
15. **a.** 3760 pF to 5640 pF
    b. 25,650 pF to 28,350 pF
    c. 4.6 pF to 6.6 pF
    d. 1620 pF to 1980 pF
    e. 3760 pF to 8460 pF

SECTION 16-7 PARALLEL CAPACITANCES

17. $C_T = 20$ μF
19. $C_T = 0.148$ μF

SECTION 16-8 SERIES CAPACITANCES

21. $C_{EQ} = 0.08$ μF
23. $C_{EQ} = 8561$ pF
25. **a.** $C_{EQ} = 0.02$ μF
    b. $Q_1 = Q_2 = Q_3 = 2.4$ μC
    c. $V_{C_1} = 60$ V, $V_{C_2} = 20$ V, $V_{C_3} = 40$ V
    d. $Q_T = 2.4$ μC

SECTION 16-9 ENERGY STORED IN ELECTROSTATIC FIELD OF CAPACITANCE

27. **a.** $\varepsilon = 1.25$ mJ
    b. $\varepsilon = 5$ mJ
    c. $\varepsilon = 125$ mJ
29. Energy stored by $C_1 = 1.62$ mJ
    Energy stored by $C_2 = 540$ μJ
    Energy stored by $C_3 = 1.08$ mJ

SECTION 16-10 MEASURING AND TESTING CAPACITORS

31. **a.** 15,000 pF
 b. 1000 pF
 c. 680,000 pF
 d. 33 nF
 e. 1000 nF
 f. 560 nF

33. No

SECTION 16-11 TROUBLES IN CAPACITORS

35. **a.** Zero ohms.
 b. Infinite ohms, with the ohmmeter showing no initial charging action.
 c. The ohmmeter will show charging action initially but the final resistance will be much lower than normal.

ANSWERS TO CRITICAL THINKING PROBLEMS

37. $C_1 = 8.08$ nF
 $C_2 = 2.02$ nF
 $C_3 = 161.6$ nF
39. **a.** $\varepsilon = 500$ mJ
 b. $\varepsilon = 250$ mJ
 c. Yes. 250 mJ of energy was lost as heat energy (I^2R) in the wire conductors when the second 100 μF capacitor was connected in part (b).

Chapter Seventeen

SECTION 17-1 ALTERNATING CURRENT IN A CAPACITIVE CIRCUIT

1. **a.** $I = 0$ A
 b. $V_{lamp} = 0$ V
 c. $V_C = 12$ V
3. **a.** $I = 400$ mA
 b. $I = 400$ mA
 c. $I = 400$ mA
 d. $I = 400$ mA
 e. $I = 0$ A
5. The amplitude of the applied voltage, the frequency of the applied voltage, and the amount of capacitance.

SECTION 17-2 THE AMOUNT OF X_C

EQUALS $\dfrac{1}{2\pi fC}$

7. **a.** $X_C = 265.26$ Ω
 b. $X_C = 132.63$ Ω
 c. $X_C = 31.83$ Ω
 d. $X_C = 15.92$ Ω
9. **a.** $f = 33.86$ Hz
 b. $f = 677.26$ Hz
 c. $f = 2.26$ kHz
 d. $f = 67.73$ kHz
11. $f = 776.37$ kHz
13. **a.** $X_C = 20$ kΩ
 b. $X_C = 5$ kΩ
 c. $X_C = 2.5$ kΩ
 d. $X_C = 1$ kΩ
15. **a.** $C = 0.05$ μF
 b. $C = 0.0125$ μF
 c. $C = 0.004$ μF
 d. $C = 1.592$ nF

SECTION 17-3 SERIES OR PARALLEL CAPACITIVE REACTANCES

17. **a.** $X_{CT} = 5$ kΩ
 b. $X_{CT} = 3$ kΩ
 c. $X_{CT} = 150$ kΩ
 d. $X_{CT} = 3$ kΩ

SECTION 17-4 OHM'S LAW APPLIED TO X_C

19. $I = 50$ mA
21. **a.** $X_{CT} = 2.4$ kΩ
 b. $I = 15$ mA
 c. $V_{C_1} = 6$ V, $V_{C_2} = 12$ V, and $V_{C_3} = 18$ V
23. $C_1 = 1.25$ μF, $C_2 = 0.625$ μF, $C_3 = 0.417$ μF, $C_{EQ} = 0.208$ μF
25. **a.** $X_{C_1} = 400$ Ω, $X_{C_2} = 320$ Ω, and $X_{C_3} = 80$ Ω
 b. $I_{C_1} = 60$ mA, $I_{C_2} = 75$ mA, and $I_{C_3} = 300$ mA
 c. $I_T = 435$ mA
 d. $X_{CEQ} = 55.17$ Ω
 e. $C_T = 1.45$ μF

SECTION 17-5 APPLICATIONS OF CAPACITIVE REACTANCE

27. $C = 3.183$ μF
 $C = 159$ nF
 $C = 6.37$ nF
 $C = 31.83$ pF

SECTION 17-6 SINE WAVE CHARGE AND DISCHARGE CURRENT

29. **a.** $i_C = 1$ μA
 b. $i_C = 1$mA
 c. $i_C = 500$ mA
31. For any capacitor, i_C and V_C *are* 90° out of phase with each other, with i_C reaching its maximum value 90° ahead of V_C. The reason that i_C leads V_C by 90° is that the value of i_C depends on the rate of voltage change across the capacitor plates rather than on the actual value of voltage itself.
33. $\dfrac{dv}{dt} = 2.5$ MV/s

ANSWERS TO CRITICAL THINKING PROBLEMS

35. $X_{C_T} = 625$ Ω
 $X_{C_1} = 500$ Ω
 $X_{C_2} = 500$ Ω
 $C_1 = 0.01$ μF
 $C_3 = 0.03$ μF
 $V_{C_1} = 20$ V
 $V_{C_2} = V_{C_3} = 5$ V
 $I_2 = 10$ mA
 $I_3 = 30$ mA

Chapter Eighteen

SECTION 18-1 SINE WAVE V_C LAGS I_C BY 90°

1. **a.** 10 V
 b. 10 mA
 c. 10 kHz
 d. 90° (i_C leads V_C by 90°)
3. **a.** See Instructor's Manual
 b. See Instructor's Manual

SECTION 18-2 X_C AND R IN SERIES

5. **a.** I and V_R are in phase
 b. V_C lags I by 90°
 c. V_C lags V_R by 90°
7. See Instructor's Manual
9. **a.** $V_R = 7.07$ V
 b. $V_C = 7.07$ V
 c. $V_T = 10$ V

SECTION 18-3 IMPEDANCE Z TRIANGLE

11. $Z_T = 25$ Ω
 $I = 4$ A
 $V_C = 80$ V
 $V_R = 60$ V
 $\theta_Z = -53.13°$
13. $Z_T = 21.63$ Ω
 $I = 2.31$ A
 $V_C = 41.58$ V
 $V_R = 27.72$ V
 $\theta_Z = -56.31°$
15. $Z_T = 10.44$ kΩ
 $I = 2.3$ mA
 $V_C = 6.9$ V
 $V_R = 23$ V
 $\theta_Z = -16.7°$
17. $X_C = 5$ kΩ
 $Z_T = 6.34$ kΩ
 $I = 5.68$ mA
 $V_R = 22.15$ V
 $V_C = 28.4$ V
 $\theta_Z = -52°$
19. **a.** X_C increases
 b. Z_T increases
 c. I decreases
 d. V_C increases
 e. V_R decreases
 f. θ_Z increases (becomes more negative)

SECTION 18-4 RC PHASE-SHIFTER CIRCUIT

21. **a.** V_R leads V_T by 28°
 b. V_C lags V_T by 62°
23. **a.** $Z_T = 26.55$ kΩ
 $I = 4.52$ mA
 $V_C = 119.9$ V
 $V_R = 4.52$ V
 $\theta_Z = -87.84°$
 b. V_R leads V_T by 87.84°
 c. V_C lags V_T by 2.16°

SECTION 18-5 X_C AND R IN PARALLEL

25. **a.** 120 V
 b. 120 V
27. $I_R = 3$ A
 $I_C = 4$ A
 $I_T = 5$ A
 $Z_{EQ} = 24$ Ω
 $\theta_1 = 53.13°$
29. $I_R = 2$ A
 $I_C = 4$ A
 $I_T = 4.47$ A
 $Z_{EQ} = 22.37$ Ω
 $\theta_1 = 63.4°$
31. $I_R = 200$ mA
 $I_C = 200$ mA
 $I_T = 282.8$ mA

$Z_{EQ} = 63.65 \ \Omega$
$\theta_I = 45°$

33. a. $I_R = 1 \text{ A}$
$I_C = 1 \text{ A}$
$I_T = 1.414 \text{ A}$
$Z_{EQ} = 35.36 \ \Omega$
$\theta_I = 45°$

b. $I_R = 2 \text{ A}$
$I_C = 200 \text{ mA}$
$I_T = 2.01 \text{ A}$
$Z_{EQ} = 9.95 \ \Omega$
$\theta_I = 5.7°$

c. $I_R = 200 \text{ mA}$
$I_C = 2 \text{ A}$
$I_T = 2.01 \text{ A}$
$Z_{EQ} = 9.95 \ \Omega$
$\theta_I = 84.3°$

35. $X_C = 500 \ \Omega$
$I_R = 20 \text{ mA}$
$I_C = 48 \text{ mA}$
$I_T = 52 \text{ mA}$
$Z_{EQ} = 461.54 \ \Omega$
$\theta_I = 67.38°$

37. a. I_R stays the same
b. I_C decreases
c. I_T decreases
d. Z_{EQ} increases
e. θ_I decreases

SECTION 18-6 RF AND AF COUPLING CAPACITORS

39. $f = 33.86 \text{ kHz}$
$\theta_Z = -5.71°$

SECTION 18-7 CAPACITIVE VOLTAGE DIVIDERS

41. $V_{C_1} = 50 \text{ V}$
$V_{C_2} = 20 \text{ V}$
$V_{C_3} = 10 \text{ V}$

SECTION 18-8 THE GENERAL CASE OF CAPACITIVE CURRENT, i_C

43. See Instructor's Manual

ANSWERS TO CRITICAL THINKING PROBLEMS

45. $I_c = 400 \text{ mA}$
$I_R = 300 \text{ mA}$
$V_A = 36 \text{ V}$
$X_c = 90 \ \Omega$
$C = 5.56 \ \mu\text{F}$
$Z_{EQ} = 72 \ \Omega$

Chapter Nineteen

SECTION 19-1 INDUCTION BY ALTERNATING CURRENT

1. A small current change of 1 to 2 mA.
3. A high frequency alternating current.

SECTION 19-2 SELF-INDUCTANCE L

5. a. $L = 10 \text{ H}$
b. $L = 1.5 \text{ mH}$
c. $L = 1.5 \text{ H}$
d. $L = 6 \text{ mH}$
e. $L = 3 \text{ mH}$

f. $L = 375 \ \mu\text{H}$
g. $L = 15 \text{ H}$
7. $L = 50 \text{ mH}$
9. $L = 2.53 \ \mu\text{H}$

SECTION 19-3 SELF-INDUCED VOLTAGE, V_L

11. $V_L = 500 \text{ V}$
13. a. $V_L = 10 \text{ V}$
b. $V_L = 20 \text{ V}$
c. $V_L = 5 \text{ V}$
d. $V_L = 100 \text{ V}$

SECTION 19-5 MUTUAL INDUCTANCE L_M

15. $k = 0.75$
17. $L_M = 61.24 \text{ mH}$

SECTION 19-6 TRANSFORMERS

19. a. $V_S = 24 \text{ Vac}$
b. $I_S = 2 \text{ A}$
c. $P_{sec} = 48 \text{ W}$
d. $P_{pri} = 48 \text{ W}$
e. $I_P = 400 \text{ mA}$
21. a. $V_{S_1} = 120 \text{ Vac}$
b. $V_{S_2} = 24 \text{ Vac}$
c. $I_{S_1} = 50 \text{ mA}$
d. $I_{S_2} = 1 \text{ A}$
e. $P_{sec1} = 6 \text{ W}$
f. $P_{sec2} = 24 \text{ W}$
g. $P_{pri} = 30 \text{ W}$
h. $I_P = 250 \text{ mA}$
23. a. $\dfrac{N_P}{N_S} = \dfrac{3}{1}$
b. $I_S = 2.5 \text{ A}$
c. $I_P = 833.3 \text{ mA}$
25. % Efficiency = 80%

SECTION 19-7 TRANSFORMER RATINGS

27. The power rating of a transformer is specified in volt-amperes (VA) which is the unit of apparent power.
29. To identify those transformer leads with the same instantaneous polarity.
31. a. $V_{sec1} = 32 \text{ Vac}$
b. $V_{sec2} = 60 \text{ Vac}$
c. $I_{S1(max)} = 1.875 \text{ A}$
d. $I_{S2(max)} = 1.67 \text{ A}$
e. $I_{P(max)} = 1.33 \text{ A}$
33. $I_P = 210 \text{ mA}$

SECTION 19-8 IMPEDANCE TRANSFORMATION

35. a. $Z_P = 200 \ \Omega$
b. $Z_P = 12.5 \ \Omega$
c. $Z_P = 6.25 \text{ k}\Omega$
d. $Z_P = 5 \text{ k}\Omega$
e. $Z_P = 5 \ \Omega$
37. $\dfrac{N_P}{N_S} = 11.18{:}1$

SECTION 19-12 INDUCTANCES IN SERIES OR PARALLEL

39. a. $L_T = 20 \text{ mH}$
b. $L_T = 18 \text{ mH}$
c. $L_T = 1 \text{ mH}$
d. $L_T = 10 \text{ mH}$
41. $L_T = 660 \text{ mH}$

43. a. $L_T = 82.63 \text{ mH}$
b. $L_T = 37.37 \text{ mH}$

SECTION 19-13 ENERGY IN MAGNETIC FIELD OF INDUCTANCE

45. Energy = $243 \ \mu\text{J}$
47. Energy = 675 mJ

ANSWERS TO CRITICAL THINKING PROBLEMS

49. $Z_P = 36.36 \ \Omega$
51. $I_P = 312.6 \text{ mA}$

Chapter Twenty

SECTION 20-1 HOW X_L REDUCES THE AMOUNT OF I

1. $X_L = 0 \ \Omega$ at dc
3. $I_{dc} = 2.5 \text{ A}$
5. a. Because with S_1 in position 2 the inductor has an inductive reactance, X_L, in addition to the dc resistance, r_i, to limit the circuit's current flow. With S_1 in position 1 only the dc resistance of the coil limits current flow since there is no X_L for direct current.
b. $X_L = 4 \text{ k}\Omega$

SECTION 20-2 $X_L = 2\pi fL$

7. a. $X_L = 37.7 \ \Omega$
b. $X_L = 75.4 \ \Omega$
c. $X_L = 1 \text{ k}\Omega$
d. $X_L = 6.28 \text{ k}\Omega$
9. a. $L = 500 \text{ mH}$
b. $L = 100 \text{ mH}$
c. $L = 31.83 \text{ mH}$
d. $L = 25 \text{ mH}$
11. $L = 254.65 \text{ mH}$
13. a. $I = 1 \text{ mA}$
b. $I = 4 \text{ mA}$
c. $I = 1 \text{ mA}$
d. $I = 4 \text{ mA}$
15. a. $X_L = 2.64 \text{ k}\Omega$
b. $X_L = 1.1 \text{ k}\Omega$
c. $X_L = 1 \text{ k}\Omega$
d. $X_L = 1 \text{ k}\Omega$
17. a. $f = 1.99 \text{ kHz}$
b. $f = 530.52 \text{ kHz}$
c. $f = 3.183 \text{ kHz}$
d. $f = 7.96 \text{ kHz}$

SECTION 20-3 SERIES OR PARALLEL INDUCTIVE REACTANCES

19. a. $X_{LEQ} = 720 \ \Omega$
b. $X_{LEQ} = 600 \ \Omega$
c. $X_{LEQ} = 150 \ \Omega$
d. $X_{LEQ} = 133.3 \ \Omega$

SECTION 20-4 OHM'S LAW APPLIED TO X_L

21. a. I increases
b. I decreases
23. $L_1 = 10 \text{ mH}$, $L_2 = 12 \text{ mH}$, $L_3 = 18 \text{ mH}$, and $L_T = 40 \text{ mH}$
25. a. $I_{L_1} = 600 \text{ mA}$
$I_{L_2} = 200 \text{ mA}$
$I_{L_3} = 800 \text{ mA}$

b. $I_T = 1.6$ A
c. $X_{LEQ} = 75\ \Omega$

27. a. $X_{L_1} = 1.6$ kΩ, $X_{L_2} = 6.4$ kΩ, and $X_{L_3} = 1.28$ kΩ
b. $I_{L_1} = 20$ mA, $I_{L_2} = 5$ mA, $I_{L_3} = 25$ mA
c. $I_T = 50$ mA
d. $X_{L_{EQ}} = 640\ \Omega$
e. $L_{EQ} = 16$ mH

SECTION 20-6 WAVESHAPE OF V_L INDUCED BY SINE WAVE CURRENT

29. V_L leads i_L by a phase angle of 90°. This 90° phase relationship exists because V_L depends on the rate of current change rather than on the actual value of current itself.

ANSWERS TO CRITICAL THINKING PROBLEMS

31. $L_1 = 60$ mH
$L_2 = 40$ mH
$L_3 = 120$ mH
$L_T = 90$ mH
$X_{L_1} = 1.2$ kΩ
$X_{L_2} = 800\ \Omega$
$X_{L_T} = 1.8$ kΩ
$V_{L_1} = 24$ V
$V_{L_3} = 12$ V
$I_{L_2} = 15$ mA
$I_{L_3} = 5$ mA

33. $L_1 = 10$ mH
$L_2 = 120$ mH
$L_3 = 40$ mH

Chapter Twenty-One

SECTION 21-1 SINE WAVE i_L LAGS V_L by 90°

1. a. 10 V
b. 10 mA
c. 10 kHz
d. 90°
3. a. See Instructor's Manual
b. See Instructor's Manual

SECTION 21-2 X_L AND R IN SERIES

5. a. 0°
b. 90°
c. 90°
7. See Instructor's Manual
9. a. $V_R = 7.07$ V
b. $V_L = 7.07$ V
c. $V_T = 10$ V

SECTION 21-3 IMPEDANCE Z TRIANGLE

11. $Z_T = 125\ \Omega$
$I = 288$ mA
$V_L = 21.6$ V
$V_R = 28.8$ V
$\theta_Z = 36.87°$
13. $Z_T = 11.18$ kΩ
$I = 10.73$ mA
$V_L = 107.3$ V
$V_R = 53.67$ V
$\theta_Z = 63.44°$

15. $Z_T = 42.43\ \Omega$
$I = 1.18$ A
$V_L = 35.35$ V
$V_R = 35.35$ V
$\theta_Z = 45°$
17. $X_L = 1.8$ kΩ
$Z_T = 3.25$ kΩ
$I = 30.77$ mA
$V_R = 83.1$ V
$V_L = 55.4$ V
$\theta_Z = 33.7°$
19. a. X_L decreases
b. Z_T decreases
c. I increases
d. V_R increases
e. V_L decreases
f. θ_Z decreases

SECTION 21-4 X_L AND R IN PARALLEL

21. a. 0°
b. I_L lags V_A by 90°
c. I_L lags I_R by 90°
23. See Instructor's Manual
25. $I_R = 3$ A
$I_L = 2$ A
$I_T = 3.61$ A
$Z_{EQ} = 33.24\ \Omega$
$\theta_I = -33.7°$
27. $I_R = 4.8$ mA
$I_L = 2$ mA
$I_T = 5.2$ mA
$Z_{EQ} = 4.62$ kΩ
$\theta_I = -22.62°$
29. $Z_{EQ} = 192\ \Omega$
31. a. I_R stays the same
b. I_L decreases
c. I_T decreases
d. Z_{EQ} increases
e. θ_I becomes less negative

SECTION 21-5 Q OF A COIL

33. a. $Q = 3.14$
b. $Q = 6.28$
c. $Q = 10$
d. $Q = 62.83$
35. $R_e = 94.25\ \Omega$

SECTION 21-6 AF AND RF CHOKES

37. a. $L = 4.78$ H
b. $L = 954.9$ mH
c. $L = 11.94$ mH
d. $L = 2.39$ mH
39. a. $V_{out} = 9.95\ V_{PP}$
b. $V_{out} = 7.07\ V_{PP}$
c. $V_{out} = 995\ V_{PP}$

SECTION 21-7 THE GENERAL CASE OF INDUCTIVE VOLTAGE

41. See Instructor's Manual

ANSWERS TO CRITICAL THINKING PROBLEMS

43. $I_T = 6$ mA
$I_R = 3$ mA
$I_L = 5.2$ mA
$X_L = 2.31$ kΩ
$R = 4$ kΩ
$L = 36.77$ mH

Chapter Twenty-Two

SECTION 22-1 RESPONSE OF RESISTANCE ALONE

1. The current, I, reaches its steady-state value immediately because a resistor does not provide any reaction to a change in either voltage or current.
3. The resistor provides 2 Ω of resistance to oppose current from the 12 V source but it does not provide any reaction to the closing or opening of the switch, S_1.

SECTION 22-2 $\frac{L}{R}$ TIME CONSTANT

5. a. $T = 200\ \mu s$
b. 240 mA
c. 0 mA
d. approximately 151.7 mA
e. 1 ms
7. a. either increase L or decrease R
b. either decrease L or increase R

SECTION 22-3 HIGH VOLTAGE PRODUCED BY OPENING AN RL CIRCUIT

9. Without a resistor across S_1 there is no way to determine the time constant of the circuit with S_1 open. This is because there is no way of knowing what the resistance of the open switch is. We do know, however, that the time constant will be very short with S_1 open. This short time constant will result in a very large $\frac{di}{dt}$ value which in turn will produce a very large induced voltage across the open contacts of the switch. This will most likely produce internal arcing across the open switch contacts.

SECTION 22-4 RC TIME CONSTANT

11. a. $V_C = 31.6$ V
b. $V_C = 50$ V
c. $V_C = 50$ V
13. a. $T = 1$ s
b. $T = 1.5\ \mu s$
c. $T = 89.1\ \mu s$
d. $T = 200$ ms
15. a. 25 V
b. $V_C = 40.8$ V
c. $V_C = 50$ V

SECTION 22-5 RC CHARGE AND DISCHARGE CURVES

17. a. 500 μA
b. zero
c. $V_R = 18.4$ V
d. 184 μA

SECTION 22-6 HIGH CURRENT PRODUCED BY SHORT-CIRCUITING RC CIRCUIT

19. a. $T = 100$ ms
b. $T = 250\ \mu s$
21. a. $V_C = 0$ V
b. $V_R = 3$ V
c. $I = 30$ mA
23. $\varepsilon = 4.5$ mJ

SECTION 22-8 LONG AND SHORT TIME CONSTANTS

25. **a.** Long
    b. Short
27. **a.** the output is taken across the capacitor
    b. long

SECTION 22-10 LONG TIME CONSTANT FOR RC COUPLING CIRCUIT

29. **a.** $T = 1$ ms
    b. $\dfrac{tp}{RC} = \dfrac{1}{10}$
    c. See Instructor's Manual

SECTION 22-11 ADVANCED TIME CONSTANT ANALYSIS

31. **a.** $V_C = 0$ V
    b. $V_C = 151$ V
    c. $V_C = 189.6$ V
    d. $V_C = 233.1$ V
    e. $V_C = 259.4$ V
    f. $V_C = 275.4$ V
    g. $V_C = 290.9$ V
33. **a.** $t = 356.7$ ms
    b. $t = 693.1$ ms
    c. $t = 1.1$ s
    d. $t = 1.61$ s
    e. $t = 2.3$ s
35. $T = 7.5$ ms
37. **a.** $V_R = 24$ V
    b. $V_R = 13.17$ V
    c. $V_R = 6.33$ V
    d. $V_R = 3.25$ V
    e. $V_R = 856.5$ mV

SECTION 22-12 COMPARISON OF REACTANCE AND TIME CONSTANT

39. Reactance
41. Long

ANSWERS TO CRITICAL THINKING PROBLEMS

43. **a.** 3 ms
    b. $V_C = 24.35$ V
    c. $V_C = 15$ V
    d. $V_C = 27.54$ V

Chapter Twenty-Three

SECTION 23-1 AC CIRCUITS WITH RESISTANCE BUT NO REACTANCE

1. $R_T = 30\ \Omega$
   $I = 500$ mA
   $V_1 = 6$ V
   $V_2 = 9$ V
3. $I_1 = 3$ A
   $I_2 = 2$ A
   $I_T = 5$ A
   $R_{EQ} = 7.2\ \Omega$

SECTION 23-2 CIRCUITS WITH X_L ALONE

5. $X_{LT} = 250\ \Omega$
   $I = 480$ mA
   $V_1 = 48$ V
   $V_2 = 72$ V
7. $I_1 = 1.2$ A

$I_2 = 300$ mA
$I_T = 1.5$ A
$X_{LEQ} = 80\ \Omega$

SECTION 23-3 CIRCUITS WITH X_c ALONE

9. $X_{CT} = 900\ \Omega$
   $I = 20$ mA
   $V_1 = 4.4$ V
   $V_2 = 13.6$ V
11. $I_1 = 100$ mA
    $I_2 = 400$ mA
    $I_T = 500$ mA
    $X_{CEQ} = 20\ \Omega$

SECTION 23-4 OPPOSITE REACTANCES CANCEL

13. **a.** net $X = X_L = 60\ \Omega$
    b. $I = 400$ mA
    c. $V_L = 72$ V
    d. $V_C = 48$ V
15. **a.** $I_L = 300$ mA
    b. $I_C = 200$ mA
    c. $I_T = I_L = 100$ mA
    d. $X = X_L = 180\ \Omega$

SECTION 23-5 SERIES REACTANCE AND RESISTANCE

17. **a.** $X = X_C = 75\ \Omega$
    b. $Z_T = 125\ \Omega$
    c. $I = 1$ A
    d. $V_R = 100$ V
    e. $V_L = 50$ V
    f. $V_C = 125$ V
    g. $\theta_Z = -36.87°$
19. **a.** $X = X_L = 600\ \Omega$
    b. $Z_T = 750\ \Omega$
    c. $I = 20$ mA
    d. $V_R = 9$ V
    e. $V_L = 36$ V
    f. $V_C = 24$ V
    g. $\theta_Z = 53.13°$

SECTION 23-6 PARALLEL REACTANCE AND RESISTANCE

21. **a.** $I_R = 150$ mA
    b. $I_C = 600$ mA
    c. $I_L = 400$ mA
    d. $I_X = I_C = 200$ mA
    e. $I_T = 250$ mA
    f. $Z_{EQ} = 144\ \Omega$
    g. $\theta_I = 53.13°$
23. **a.** $I_R = 120$ mA
    b. $I_C = 80$ mA
    c. $I_L = 120$ mA
    d. $I_X = I_L = 40$ mA
    e. $I_T = 126.5$ mA
    f. $Z_{EQ} = 94.86\ \Omega$
    g. $\theta_I = -18.44°$

SECTION 23-7 SERIES–PARALLEL REACTANCE AND RESISTANCE

25. **a.** $Z_T = 50\ \Omega$
    b. $I_T = 2$ A
    c. $V_{R_1} = 60$ V
    d. $V_{C_1} = 24$ V, $V_{C_2} = 240$ V, and $V_{C_3} = 96$ V
    e. $V_{L_1} = 144$ V, and $V_{L_2} = 40$ V
    f. $\theta_Z = -53.13°$

SECTION 23-8 REAL POWER

27. **a.** Real Power $= 100$ W
    Apparent Power $= 125$ VA
    PF $= 0.8$
    b. Real Power $= 180$ mW
    Apparent Power $= 300$ mVA
    PF $= 0.6$
    c. Real Power $= 5.4$ W
    Apparent Power $= 9$ VA
    PF $= 0.6$
    d. Real Power $= 1.44$ W
    Apparent Power $= 1.52$ VA
    PF $= 0.947$

ANSWERS TO CRITICAL THINKING PROBLEMS

29. $C = 6.63\ \mu$F or $46.4\ \mu$F

Chapter Twenty-Four

SECTION 24-1 POSITIVE AND NEGATIVE NUMBERS

1. **a.** $0°$
   b. $180°$

SECTION 24-2 THE j OPERATOR

3. The j axis
5. **a.** Real Numbers
   b. Imaginary Numbers
7. **a.** 25 units with a leading phase angle of $+90°$
   b. 36 units with a lagging phase angle of $-90°$

SECTION 24-3 DEFINITION OF A COMPLEX NUMBER

9. Rectangular form
11. **a.** The phase angle is greater than $45°$.
    b. The phase angle is less than $45°$.
    c. The phase angle is $-45°$.
    d. The phase angle is more negative than $-45°$.
    e. The phase angle is less than $45°$.

SECTION 24-4 HOW COMPLEX NUMBERS ARE APPLIED TO AC CIRCUITS

13. $0°$

SECTION 24-5 IMPEDANCE IN COMPLEX FORM

15. **a.** $10\ \Omega + j20\ \Omega$
    b. $15\ \Omega + j10\ \Omega$
    c. $0\ \Omega - j1$ kΩ
    d. 1.5 k$\Omega - j2$ kΩ
    e. $150\ \Omega \pm j0\ \Omega$
    f. $75\ \Omega - j75\ \Omega$

SECTION 24-6 OPERATIONS WITH COMPLEX NUMBERS

17. **a.** $15 + j15$
    b. $40 - j20$
    c. $200 + j150$
    d. $90 - j50$
    e. $36 - j48$

19.
  a. -72
  b. 60
  c. -28
  d. -24
  e. 2
  f. -12.5
  g. 25
  h. -25

21.
  a. $1.19 - j0.776$
  b. $0.188 + j0.188$
  c. $0.461 + j0.194$
  d. $1 - j0.5$

SECTION 24-8 POLAR FORM OF COMPLEX NUMBERS

23.
  a. $14.14\angle 45°$
  b. $12.81\angle -51.34°$
  c. $21.63\angle 56.3°$
  d. $150.4\angle -21.45°$

25.
  a. $3\angle -125°$
  b. $5\angle 150°$
  c. $4\angle 0°$
  d. $4.67\angle 66°$
  e. $30\angle 75°$
  f. $25\angle 80°$
  g. $12.5\angle 50°$

SECTION 24-10 COMPLEX NUMBERS IN SERIES AC CIRCUITS

27.
  a. $Z_T = 30\ \Omega + j40\ \Omega$
  b. $Z_T = 50\angle 53.13°\ \Omega$
  c. $I = 2\angle -53.13°\ A$
  d. $V_R = 60\angle -53.13°\ V$
  e. $V_L = 140\angle 36.87°\ V$
  f. $V_C = 60\angle -143.13°\ V$

SECTION 24-11 COMPLEX NUMBERS IN PARALLEL AC CIRCUITS

29. $Z_T = 33.3\angle 33.69°\ \Omega$ (Polar Form)
  $Z_T = 27.7\Omega + j18.47\ \Omega$ (Rectangular Form)

31. $Y_T = 20\ mS + j6.67\ mS$ (Rectangular Form)
  $Y_T = 21.08\angle 18.44°\ mS$ (Polar Form)
  $Z_T = 47.44\angle -18.44°\ \Omega$ (Polar Form)

33.
  a. See Instructor's Manual
  b. See Instructor's Manual

SECTION 24-12 COMBINING TWO COMPLEX BRANCH IMPEDANCES

35.
  a. $Z_1 = 30\ \Omega - j40\ \Omega = 50\angle -53.13°\ \Omega$
  b. $Z_2 = 20\ \Omega + j15\ \Omega = 25\angle 36.87°\ \Omega$
  c. $Z_T = 22\ \Omega + j4\ \Omega = 22.4\angle 10.3°\ \Omega$

SECTION 24-13 COMBINING COMPLEX BRANCH CURRENTS

37.
  a. $I_1 = 1\angle 53.13°\ A = 600\ mA + j800\ mA$
  b. $I_2 = 2\angle -36.87°\ A = 1.6\ A - j1.2\ A$
  c. $I_T = 2.236\angle -10.3°\ A = 2.2\ A - j400\ mA$

Chapter Twenty-Five

SECTION 25-1 THE RESONANCE EFFECT

1. The condition of equal and opposite reactances in an LC circuit. Resonance occurs at only one particular frequency, known as the resonant frequency.

3. $X_L = X_C = 1\ k\Omega$

SECTION 25-2 SERIES RESONANCE

5.
  a. $X = 0\ \Omega$
  b. $Z_T = 40\ \Omega$
  c. $I = 25\ \mu A$
  d. $\theta = 0°$
  e. $V_L = 50\ mV$
  f. $V_C = 50\ mV$
  g. $V_{rs} = 1\ mV$

7. Because at f_r the total impedance, Z_T is purely resistive.

SECTION 25-3 PARALLEL RESONANCE

9.
  a. Z_{EQ} is maximum
  b. I_T is minimum
  c. $\theta = 0°$

11. The resistance, r_s.

SECTION 25-4 RESONANT FREQUENCY

$$f_r = \frac{1}{2\pi\sqrt{LC}}$$

13.
  a. $f_r = 2.5\ MHz$
  b. $f_r = 400\ kHz$
  c. $f_r = 3\ MHz$
  d. $f_r = 5\ MHz$

15. $C = 70.17\ pF$

17.
  a. $f_r = 5\ MHz$
  b. $X_L = X_C = 628.3\ \Omega$
  c. $Z_T = r_s = 12.56\ \Omega$
  d. $I = 796.2\ \mu A$
  e. $V_L = V_C = 500\ mV$
  f. $\theta_Z = 0°$

19. With C set to 360 pF, $f_r = 1.875$ MHz. To double f_r C must be reduced to 90 pF.

SECTION 25-5 Q MAGNIFICATION FACTOR OF RESONANT CIRCUIT

21.
  a. $f_r = 1\ MHz$
  b. $Q = 100$
  c. $V_L = V_C = 1\ V$

23. $Q = 300$

25.
  a. $f_r = 1.25\ MHz$
  b. $X_L = X_C = 785.4\ \Omega$
  c. $I_L = I_C = 12.73\ mA$
  d. $Q = 100$
  e. $Z_{EQ} = 78.54\ k\Omega$
  f. $I_T = 127.3\ \mu A$

27. $Q = 191$

SECTION 25-6 BANDWIDTH OF RESONANT CIRCUIT

29.
  a. $\Delta f = 12.5\ kHz$

  b. $f_1 = 1.24375$ MHz (exactly) and $f_2 = 1.25625$ MHz (exactly)
  c. $Z_{EQ} = 78.54\ k\Omega$ at f_r, $Z_{EQat}\ f_1$ and $f_2 = 55.53\ k\Omega$

31.
  a. $f_r = 3\ MHz$
  b. $X_L = X_C = 942.5\ \Omega$
  c. $Z_T = 18.85\ \Omega$
  d. $I = 2.65\ \mu A$
  e. $Q = 50$
  f. $V_L = V_C = 2.5\ mV$
  g. $\theta = 0°$
  h. $\Delta f = 60\ kHz,\ f_1 = 2.97\ MHz,$ and $f_2 = 3.03\ MHz$
  i. $I = 1.87\ \mu A$

33. At f_1 I is approximately 70.7% of I at f_r. This is because at f_1, Z_T is approximately 1.41 times the value of Z_T at f_r.

35. At f_1 and f_2 $Z_{EQ} = 138.8\ k\Omega$ and I_T is 14.41 μA.

SECTION 25-7 TUNING

37. No, because as C is varied to provide different resonant frequencies the Q of the circuit varies. Recall that $V_C = Q \times V_{in}$ at f_r. (This assumes that V_{in} remains the same for all frequencies.)

SECTION 25-8 MISTUNING

39.
  a. The circuit appears inductive with a lagging phase angle because $I_L > I_C$.
  b. The circuit appears capacitive with a leading phase angle because $I_C > I_L$.

SECTION 25-9 ANALYSIS OF PARALLEL RESONANT CIRCUITS

41. At f_r, $Q = 125$ which is considered a high Q.

43. $I_L = 1.24$ mA and $I_C = 1.27$ mA. I_L is less than I_C at f_r because the impedance of the inductive branch is greater than X_C or X_L alone.

45. Z_{EQ} is maximum below f_r because this will cause X_C to increase and the impedance of the inductive branch to decrease. At some frequency below f_r the impedance of the inductive branch will equal X_C and Z_{EQ} will be maximum.

SECTION 25-10 DAMPING OF PARALLEL RESONANT CIRCUITS

47.
  a. $Q = 114$
  b. $\Delta f = 8.77\ kHz$

49. $R_P = 196.4\ k\Omega$

$$\frac{X_L^2}{r_s^2} \times r_s^2 = L/C$$

$$X_L^2 = L/C$$

$$X_L = \sqrt{L/C}$$

Chapter Twenty-Six

SECTION 26-1 EXAMPLES OF FILTERING

1. a. A low-pass filter allows the lower frequency signals to pass from its input to its output with little or no attenuation while at the same time severely attenuating or eliminating the higher frequency signals.

b. A high-pass filter does just the opposite of a low-pass filter.

SECTION 26-2 DIRECT CURRENT COMBINED WITH ALTERNATING CURRENT

3. a. 10 V dc
b. 5 mA

5. See Instructor's Manual

SECTION 26-3 TRANSFORMER COUPLING

7. See Instructor's Manual

SECTION 26-4 CAPACITIVE COUPLING

9. a. 159.2 Ω
b. 15 V
c. 15 V
d. 0 V
e. 10 V_{pp}
f. 0 V_{pp}
g. 10 V_{pp}
h. 3.53 V

11. a. C_C charges
b. C_C discharges

13. $f = 1.59$ kHz

SECTION 26-5 BYPASS CAPACITORS

15. a. $X_{C_1} = 159.2$ Ω
b. 20 V
c. 8 V
d. 12 V
e. 12 V
f. 15 V_{pp}
g. 0 V_{pp}
h. 15 V_{pp}

17. $C = 6.37$ μF

SECTION 26-6 FILTER CIRCUITS

19. a. Low-Pass
b. High-Pass

SECTION 26-7 LOW-PASS FILTERS

21. a. The term passband refers to those frequencies below the cutoff frequency of a low-pass filter. Signal frequencies in the passband are allowed to pass from the input to the output of the filter with little or no attenuation.

b. The term stopband refers to those frequencies above the cutoff frequency of a low-pass filter. Signal frequencies in the stopband are severely attenuated as they pass through the filter from input to output.

SECTION 26-8 HIGH-PASS FILTERS

23. Yes, except that for a high-pass filter the passband is above the cutoff frequency and the stopband is below the cutoff frequency.

SECTION 26-9 ANALYZING FILTER CIRCUITS

25. a. Low-Pass
b. High-Pass
c. Low-Pass
d. High-Pass

27. a. $V_{OUT} = 49.99$ mV and $\theta = -0.87°$
b. $V_{OUT} = 49.9$ mV and $\theta = -3.48°$
c. $V_{OUT} = 47.84$ mV and $\theta = -16.91°$
d. $V_{OUT} = 35.35$ mV and $\theta = -45°$
e. $V_{OUT} = 15.63$ mV and $\theta = -71.79°$
f. $V_{OUT} = 8.12$ mV and $\theta = -80.66°$
g. $V_{OUT} = 1.64$ mV and $\theta = -88.12°$

29. a. $V_{OUT} = 99.98$ mV and $\theta = -1.08°$
b. $V_{OUT} = 99.56$ mV and $\theta = -5.38°$

c. $V_{OUT} = 93.57$ mV and $\theta = -20.66°$
d. $V_{OUT} = 70.71$ mV and $\theta = -45°$
e. $V_{OUT} = 33.34$ mV and $\theta = -70.5°$
f. $V_{OUT} = 17.41$ mV and $\theta = -80°$
g. $V_{OUT} = 5.3$ mV and $\theta = -87°$

31. 0.707

33. a. 0°
b. $-90°$

35. Bandpass Filter

37. a. $f_{C_1} = 1.06$ kHz
b. $f_{C_2} = 10.26$ kHz
c. BW = 9.2 kHz

39. $f_N = 4.42$ kHz

SECTION 26-10 DECIBELS AND FREQUENCY RESPONSE CURVES

41. a. $N_{dB} = -3$ dB
b. $N_{dB} = -10$ dB
c. $N_{dB} = -60$ dB
d. $N_{dB} = -20$ dB

43. a. $N_{dB} = -30.6$ dB
b. $N_{dB} = -16.72$ dB
c. $N_{dB} = -10.97$ dB
d. $N_{dB} = -3$ dB
e. $N_{dB} = -0.491$ dB
f. $N_{dB} = 0$ dB
g. $N_{dB} = 0$ dB

SECTION 26-11 RESONANT FILTERS

45. The Circuit Q

SECTION 26-12 INTERFERENCE FILTERS

47. A low-pass filter with a cutoff frequency around 30 MHz.

ANSWERS TO CRITICAL THINKING PROBLEMS

49. a. $f_c = 965$ Hz
b. $V_{out} = 3.535$ V_{pp}
c. $V_{out} = 68.2$ mV_{pp}

51. $L = 191$ μH
$C = 132.63$ pF

Credits

Introduction
Page 17: Courtesy of Sharp Electronics Corporation.

Chapter Openers
© Vol. DT06/Getty Images; © Vol. BS36/Getty Images.

Chapter One
Figure 1.1: © The McGraw-Hill Companies, Inc./Cindy Schroeder, photographer; p.35, 39, 41: © Bettmann/Corbis; 1.16a,b, 1.17a-c: Mitchel Schultz.

Chapter Two
Figure 2.1b: © Corbis; 2.2: Mark Steinmetz; 2.3: © The McGraw-Hill Companies, Inc./Cindy Schroeder, photographer; 2.6: Mark Steinmetz; 2.7b: © The McGraw-Hill Companies, Inc./Cindy Schroeder, photographer; 2.15, 2.16, 2.19: Mark Steinmetz; p. 73: © Vol. DT06/PhotoDisc/Getty Images; 2.25: Mark Steinmetz.

Chapter Three
Page 90, 92: © Bettmann/Corbis; p. 103, 110: © PhotoDisc/Getty Images.

Chapter Four
Figure 4.2c: © The McGraw-Hill Companies, Inc./Cindy Schroeder, photographer.

Chapter Five
Figure 5.3b: © The McGraw-Hill Companies, Inc./Cindy Schroeder, photographer.

Chapter Six
Figure 6.1e: © The McGraw-Hill Companies, Inc./Cindy Schroeder, photographer.

Chapter Eight
Figure 8.1a: Courtesy of MCM Electronics; 8.1b: Courtesy of Fluke Corporation; 8.13: Courtesy of Simpson Electric Company; 8.14: Reproduced by permission of Tektronix, Inc.; 8.15, 8.16: Courtesy of Fluke Corporation.

Chapter Nine
Page 274: © Bettmann/Corbis.

Chapter Eleven
Figure 11.2, 11.4a-e: © The McGraw-Hill Companies, Inc./Cindy Schroeder, photographer; 11.5: Mark Steinmetz; 11.6a-i: © The McGraw-Hill Companies, Inc./Cindy Schroeder, photographer; 11.7a,b, 11.11, 11.13, 11.14, 11.15: Mark Steinmetz; 11.16a-c: © The McGraw-Hill Companies, Inc./Cindy Schroeder, photographer.

Chapter Twelve
Figure 12.1, 12.2, 12.7, 12.8: Mark Steinmetz; 12.9: © The McGraw-Hill Companies, Inc./Cindy Schroeder, photographer; 12.11: Doug Martin; 12.13: Mark Steinmetz.

Chapter Thirteen
Figure 13.4a,b: © Richard Megne/Fundamental Photographs; p. 392: Brown Brothers; p. 394: © Bettmann/Corbis; p. 397: Brown Brothers; 13.12: Mark Steinmetz; p. 403: Culver Pictures; 13.15 : Courtesy of F.W. Bell.

Chapter Fourteen
Page 428: © Novosti/Sovfoto; 14.21: Mark Steinmetz.

Chapter Fifteen
Page 455: © Bettmann/Corbis; 15.22a,b, 15.23: Mark Steinmetz.

Chapter Sixteen
Figure 16.1b: © The McGraw-Hill Companies, Inc./Cindy Schroeder, photographer; p. 491: © Bettmann/Corbis; 16.4b, 16.5b, 16.6, 16.8, 16.9c, 16.10: Mark Steinmetz; 16.25: Courtesy of MCM Electronics; 16.27: Courtesy of Sencore, Inc.

Chapter Nineteen
Page 579: Smithsonian National Museum of American History; 19.4a,b, 19.11, 19.14a: Mark Steinmetz; 19.25: Courtesy of Sencore, Inc.; 19.31: Courtesy of BK Precision.

Chapter Twenty-One
Figure 21.8, 21.10b: Mark Steinmetz.

Chapter Twenty-Six
Figure 26.27, 26.28b, 26.29: Mark Steinmetz.

Appendix B
Appendix B-7: Courtesy of PACE, Inc.

Appendix E
Appendix E-1 a, b, appendix E-13: Reprinted with permission of from Tektronix, Inc.; p. 1147(top): Courtesy of BK Precision.

Index

series, 629–630, 714
    sine wave and, 632–636
    as voltage/current ratio, 625
Inert gases, 29–30
Insulators, 27–28, 326–327
    discharge current and, 347–348
    ferrites and, 402
    relays and, 434
Integrated circuits (ICs), 334
Integrated output, 687
Interference filters, 840–841
Inverse square law, 33
Ions, 47
    alkaline batteries and, 361–362
    current and, 345–347
IR drop, 118–120
    conductor function and, 328–329
    multimeters and, 258
    transformers and, 596–597
    voltage dividers and, 215
    voltmeters and, 244–245
Iron, 28

J

J operator, 738–740
Joule, James Prescott, 92
Joule (J) unit, 36, 92

K

Keeper, 399
Kilowatt-hour (kWh), 93
Kirchhoff, Gustav, 274
Kirchhoff's Laws, 270–271
    algebraic signs and, 272, 274
    current law, 150–153, 272–274, 277–280,
        283–286
    loop equations and, 274–275, 278
    method of branch currents and, 277–280
    method of mesh currents and, 283–286
    network theorems and, 294
    node-voltage analysis and, 281–283
    voltage law, 120–121, 274–277, 281–283
K shell, 29–30

L

Laminated cores, 603
Laptops, 356
Leakage flux, 399
Leclanché cell, 360
Lenz, Heinrich Friedrich Emil, 428
Lenz's law, 428, 584, 662
Light current, 50
Linear current, 297
Linear graphs, 89
Linear resistance, 89
Lithium batteries, 364
Load current, 222–224. *See also* Current
    generators and, 375–376
Loading effect, 247–249
Load resistance, 45
    batteries and, 373–374, 379–380
    generators and, 379–380
Log-log graph paper, 833
Loop equations, 274–275, 278
Loose coupling, 586
Low-pass filters, 819, 822–826
L shell, 29–30

M

Magnetic potential, 412
Magnetism, 389. *See also* Electromagnetism

air gap and, 398–399
B-H curve and, 417–418
Curie temperature and, 400
demagnetization and, 420
diamagnetic materials and, 401
dipole magnet and, 401
discovery of, 388
domains and, 401
electromagnets and, 390, 400
ferromagnetic materials and, 401–402
flux and, 392–399, 402–403, 417–418
gauss (G) unit and, 394–396
Hall effect and, 403
hysteresis and, 418–420
induction and, 397–398, 427–432, 608–609
    (*see also* Induction)
keeper and, 399
loop rotation and, 426
magnetic field and, 49, 390–392
material classification and, 401
maxwell (Mx) unit and, 392–393
north/south poles and, 390–392
paramagnetic materials and, 401
permanent magnets and, 390, 400
permeability and, 398, 402–403, 415
shielding and, 402–403
tesla (T) unit and, 395–396
types of magnets, 400–401
weber (Wb) unit and, 393
Magnetizing force, 412
Magnetomotive force (mmf), 412–414
Main line, 151
Manganin, 342
Mass number, 31
Mathematics
    complex numbers, 738–758
    engineering notation, 7–16
    inverse square law, 33
    metric prefixes, 7–11
    negative powers, 14–15
    reciprocals, 14
    scientific calculator and, 16–17
    scientific notation, 4–6
    sine wave, 448–454
    square roots, 15–16
    squaring, 15
Maxwell, James Clerk, 392
Maxwell (Mx) unit, 392–393
Mercury cells, 362–363
Mesh currents, 283–286, 498–499
Metal-film resistors, 61–62
Meter shunts, 240–243
Metric prefixes, 7–11
Mho unit, 42
Millman's theorem, 311–313
Mistuning, 793
Molecules, 29, 400
Motors
    AC induction, 464
    armature and, 472–473
    brushes and, 473
    commutator and, 473–474
    field winding and, 473
    slip rings and, 473
    universal, 464
Moving-coil meter, 238–240
MP3 players, 356
M shell, 29–30
Multimeters, 236–237, 563
    amp-clamp probe and, 253
    analog, 252–253

applications of, 257–259
capacitors and, 512–513
checking fuses and, 258–259
connectors and, 333–334
continuity checking and, 260–261
damaging, 257
D'Arsonval movement and, 238
decibel scale and, 253
digital, 50–51, 254–257
diodes and, 255
high-voltage probe and, 253
infinity and, 251, 260
low-power ohms and, 252–253
meter shunts and, 240–243
moving-coil meter, 238–240
multiplier resistance and, 244–245
ohmmeters and, 249–252, 260–261
range overload and, 256
resolution and, 255–256
troubleshooting and, 130
voltmeters and, 243–249
Multiplier resistance, 244–245
Mylar, 498

N

National Electrical Code, 329, 471
National Fire Protection Association, 329, 471
Negative ground, 128–129
Negative numbers, 738
Negative polarity, 26, 32
Negative powers, 14–15
Negative temperature coefficient (NTC)
    thermistors, 62, 191–192
Network theorems, 295
    balance and, 317
    delta connections and, 313–317
    Kirchhoff's Laws and, 294
    Millman's, 311–313
    Norton's, 304–307
    pi connections and, 313–317
    superposition, 296–297, 301–302
    T connections and, 313–317
    Thevenin-Norton conversions, 307–309
    Thevenin's, 297–304
    voltage-current source conversion, 309–311
    Y connections and, 313–317
Nichrome, 342
Nickel-cadmium batteries, 368–369
Nickel-metal-hydride (NiMH) batteries, 369
Node equations, 281–282
Noisy controls, 72
No-load voltage, 375
Nonlinear resistance, 90
Nonsinusoidal AC waveforms, 467–469, 672
Normally closed (NC) switch, 337, 433–435
Normally open (NO) switch, 337, 433–435
North-seeking pole, 390–392
Norton, E. L., 304
Norton's theorem, 304–309
N shell, 29–30
Nucleus, 26, 31

O

Oersted (Oe) unit, 414
Ohm, Georg Simon, 41, 80
Ohm-centimeters unit, 342
Ohmmeters, 72–73, 252–253
    back-off scale and, 250–251
    capacitors and, 516–517
    checking continuity with, 260–261
    multiple ranges and, 251

Tesla, Nikola, 395, 397
Tesla (T) unit, 395–396
Thermistors, 62–63, 67, 191–192
Thevenin, M. L., 297
Thevenin's theorem, 297
    bridge circuits and, 302–304
    circuit applications and, 298–304
    Norton conversions and, 307–309
    two voltage sources and, 300–302
Time constants
    advanced analysis and, 691–694
    chokes and, 696
    decay curve equation, 692
    differentiated output and, 687
    induction and, 695–696
    integrated output and, 687
    long, 686, 689–690
    rate of charge/discharge and, 679
    RC circuits and, 678–681,
        686–696
    reactance comparison and, 694–696
    RL circuits and, 674–676
    short, 686–687
Transformers, 470
    autotransformers and, 593
    core losses and, 602–603
    core types and, 603–604
    coupling and, 811–812
    current ratings and, 596–597
    current ratio and, 591
    eddy currents and, 602
    efficiency of, 594
    ferrite, 604
    frequency ratings and, 598
    generators and, 588–589
    hysteresis and, 603
    impedance and, 598–602
    induction and, 588–602
    isolation of secondary, 593–594
    laminated core, 603
    Ohm's law and, 590
    open primary winding and, 612
    open secondary winding and, 612
    powdered-iron, 603
    power ratings and, 597
    primary power and, 592–593
    RF shielding and, 602–603
    secondary current and, 590–591
    secondary power and, 591–593
    short circuits and, 613
    turns ratio and, 589
    voltage ratings and, 594–596
    voltage ratio and, 589–590
Transient response, 674
Transmission lines, 95, 332
Triboelectric effect, 32
Troubleshooting, 102
    capacitors, 516–518
    coils, 612
    connectors, 348–349
    multimeters, 130
    opens in series circuit, 130–131
    parallel circuits, 162–167
    relays, 435
    resistors, 72–73
    series circuits, 130–134
    series-parallel circuits, 193–198
    short circuits, 132–133
    wire, 348–349
Tungsten, 343
Tuning, 790–793, 837–840

U

Ultrasonic frequencies, 455–456
Universal motor, 474

V

Valence, 31
Varactors, 792
Variable resistors, 67–68
Variac, 604–605
Varistors, 62
Vector quantity, 461
Volta, Alessandro, 35–36
Voltage, 27
    applied, 120–121, 148–149
    batteries and, 377–379
    bucking, 584
    capacitors and, 500, 510–511
    current and, 37–49, 131–132, 309–311
    differentiated output and, 687
    electromagnetism and, 429–432
    Faraday's law and, 430
    filters and, 808–811, 822–823 (*see also* Filters)
    ground connections and, 127–130, 471–472
    Hall effect and, 403
    high, 82, 676–678
    induction and, 429–432, 582–585, 646–647
        (*see also* Induction)
    integrated output and, 687
    joule unit and, 92
    Kirchhoff's voltage law and, 120–121,
        274–277, 281–283
    linear proportion with current and, 88–90
    load resistance and, 45
    low, with high current, 83
    maximum working rating for, 99
    multimeters and, 50
    multiple/submultiple units for, 87–88
    network theorems and, 296–317
    no-load, 375
    Ohm's law and, 84–85, 88–90
    open-circuit, 375
    parallel circuits and, 148–149
    phase angle and, 460–463
    phasor triangle and, 553, 648–649
    photoelectricity and, 50
    potential difference and, 35–40
    power dissipation and, 94–96
    RL circuit opening and, 676–678
    series circuits and, 118–125
    series-parallel circuits and, 182, 189–192
    sources in series, 126–127
    three-phase AC power and, 474–476
    transformers and, 589–590, 594–596 (*see also* Transformers)
    Wheatstone bridge and, 189–192
Voltage dividers, 128, 212–213
    capacitive circuits and, 562–564
    design of, 222–224
    load current and, 222–224
    parallel load current and, 220–222
    series, 214–217
    superposition theorem and, 296–297
Voltage drop, 118–119, 752
    generators and, 375–376
    Kirchhoff's voltage law and, 120–121
    polarity and, 121–123
    resistors for, 126
Voltage taps, 215–217
Voltaic cells
    current inside, 258–259
    current outside, 358

electromotive series, 359–360
    internal resistance and, 359
Volt-ampere characteristic, 89
Volt-ampere reactive (VAR) power, 723
Voltmeters
    description of, 243–247
    loading effect of, 247–249
    multiplier resistance and, 244–245
    ohms-per-volt rating and, 246–247
    sensitivity of, 246–247
    typical circuit of, 245
Volt-ohm-milliammeter (VOM), 238, 245,
    252–253, 333–334
Volt (V) unit, 36–37
von Siemens, Ernst, 42

W

Water, 347
Watt, James, 90
Wattmeters, 724–725
Watt (W) unit, 90, 92–93. *See also* Power
Waveforms, 695. *See also* Sine waves
    angular measure and, 447
    capacitive circuits and, 550–551
    capacitive reactance and, 536–540
    cycle and, 446–447
    generators and, 445–448
    inductive reactance and, 632–636
    nonsinusoidal, 467–469
    phase angle and, 460–463
    radian measure and, 447
    RC circuits and, 684–686
    sawtooth, 468
    sharp pulses and, 687–689
    square, 468, 684, 687, 689
Wavelength, 457–460
Weber, Wilhelm Eduard, 393–394
Weber (Wb) unit, 393
Wet cells, 364–368
Wheatstone, Charles, 189
Wheatstone bridge, 189–192
Wire
    cable and, 332
    conductor types and, 331–333
    printed, 334–335
    residential, 470–471
    resistance and, 340–345
    standard gage sizes and, 329–331
    temperature coefficient and, 343–345
    transmission lines and, 332
    troubleshooting of, 348–349
Wire-wound resistors, 60–61
Work
    electron volt (eV) and, 92–93
    kilowatt-hour (kWh) and, 93
    potential difference and, 35–37
    power and, 91–92
    practical units for, 92

X

X axis, 89

Y

Y axis, 89
Y connections, 313–317

Z

Zero current, 131–132
Zero-ohm resistors, 66
Zero-ohms adjustment, 251–252